Whence the Mountains?
Inquiries into the Evolution of Orogenic Systems:
A Volume in Honor of Raymond A. Price

edited by

James W. Sears
Department of Geosciences
The University of Montana
32 Campus Drive #1296
Missoula, Montana 59812-1296
USA

Tekla A. Harms
Department of Geology
Amherst College
Amherst, Massachusetts 01002-5000
USA

Carol A. Evenchick
Natural Resources Canada
625 Robson Street, 14th Floor
Vancouver, British Columbia V6B 5J3
Canada

THE
GEOLOGICAL
SOCIETY
OF AMERICA®

Special Paper 433

3300 Penrose Place, P.O. Box 9140 ▪ Boulder, Colorado 80301-9140 USA

2007

Copyright © 2007, The Geological Society of America (GSA). All rights reserved. GSA grants permission to individual scientists to make unlimited photocopies of one or more items from this volume for noncommercial purposes advancing science or education, including classroom use. For permission to make photocopies of any item in this volume for other noncommercial, nonprofit purposes, contact the Geological Society of America. Written permission is required from GSA for all other forms of capture or reproduction of any item in the volume including, but not limited to, all types of electronic or digital scanning or other digital or manual transformation of articles or any portion thereof, such as abstracts, into computer-readable and/or transmittable form for personal or corporate use, either noncommercial or commercial, for-profit or otherwise. Send permission requests to GSA Copyright Permissions, 3300 Penrose Place, P.O. Box 9140, Boulder, Colorado 80301-9140, USA.

Copyright is not claimed on any material prepared wholly by government employees within the scope of their employment.

Published by The Geological Society of America, Inc.
3300 Penrose Place, P.O. Box 9140, Boulder, Colorado 80301-9140, USA
www.geosociety.org

Printed in U.S.A.

GSA Books Science Editor: Marion E. Bickford

Library of Congress Cataloging-in-Publication Data

Whence the mountains? : inquiries into the evolution of orogenic systems : a volume in
  honor of Raymond A. Price / edited by James W. Sears, Tekla A. Harms, Carol A.
  Evenchick.
      p. cm. — (Special papers (Geological Society of America) ; 433)
   Includes bibliographical references and index.
   ISBN-13: 978-0-8137-2433-1 (pbk.)
     1. Orogeny. 2. Geology, Structural. 3. Plate tectonics. I. Price, Raymond A. II. Sears,
  James W. III. Harms, Tekla A. IV. Evenchick, C. A.
 QE621.W58 2007
 551.8′2—c22   2007032816

Cover: Ray Price leading a Canadian Rockies field trip. Photo courtesy of Robert D. Hatcher Jr. See Figure 5, "Whence the mountains? The contributions of Raymond A. Price" (this volume), p. xiv, for annotated sketch of this view of Cascade Mountain from Price (1972).

Contents

Introduction v
    J.W. Sears, T.A. Harms, and C.A. Evenchick

Whence the mountains?—The contributions of Raymond A. Price xi
    J.W. Sears, T.A. Harms, and C.A. Evenchick

Tectonic Processes

1. Driving mechanism and 3-D circulation of plate tectonics 1
    W.B. Hamilton

2. Petrotectonics of ultrahigh-pressure crustal and upper-mantle rocks—
Implications for Phanerozoic collisional orogens 27
    W.G. Ernst, B.R. Hacker, and J.G. Liou

3. How much strain can continental crust accommodate without developing
obvious through-going faults? 51
    B.C. Burchfiel, C. Studnicki-Gizbert, J.W. Geissman, R.W. King, Z. Chen,
    L. Chen, and E. Wang

4. Mechanics of thin-skinned fold-and-thrust belts: Insights from numerical models 63
    G.S. Stockmal, C. Beaumont, M. Nguyen, and B. Lee

Canada and U.S. Cordillera

5. Lithospheric-scale structures across the Alaskan and Canadian Cordillera:
Comparisons and tectonic implications 99
    P.T.C. Hammer and R.M. Clowes

6. A synthesis of the Jurassic–Cretaceous tectonic evolution of the central and
southeastern Canadian Cordillera: Exploring links across the orogen 117
    C.A. Evenchick, M.E. McMechan, V.J. McNicoll, and S.D. Carr

7. Belt-Purcell Basin: Keystone of the Rocky Mountain fold-and-thrust belt,
United States and Canada 147
    J.W. Sears

8. Thermochronometric reconstruction of the prethrust paleogeothermal gradient and initial
thickness of the Lewis thrust sheet, southeastern Canadian Cordillera foreland belt 167
    S. Feinstein, B. Kohn, K. Osadetz, and R.A. Price

9. *Reconstructing the Snake River–Hoback River Canyon section of the Wyoming thrust belt through direct dating of clay-rich fault rocks* 183
   J.G. Solum and B.A. van der Pluijm

10. *Reinterpretation of fractures at Swift Reservoir, Rocky Mountain thrust front, Montana: Passage of a Jurassic forebulge?* 197
    E.M.G. Ward and J.W. Sears

11. *Structural, metamorphic, and geochronologic constraints on the origin of the Clearwater core complex, northern Idaho* 211
    P.T. Doughty, K.R. Chamberlain, D.A. Foster, and G.S. Sha

Appalachian Regions

12. *Character of rigid boundaries and internal deformation of the southern Appalachian foreland fold-thrust belt* 243
    R.D. Hatcher Jr., P.J. Lemiszki, and J.B. Whisner

13. *Balancing tectonic shortening in contrasting deformation styles through a mechanically heterogeneous stratigraphic succession* 277
    W.A. Thomas

14. *Links among Carolinia, Avalonia, and Ganderia in the Appalachian peri-Gondwanan realm* 291
    J.P. Hibbard, C.R. van Staal, and B.V. Miller

15. *Cat Square basin, Catskill clastic wedge: Silurian-Devonian orogenic events in the central Appalachians and the crystalline southern Appalachians* 313
    A.J. Dennis

Asian and Pacific Regions

16. *A stratigraphic unit converted to fault rocks in the Northland Allochthon of New Zealand: Response of a siliceous claystone to obduction* 331
    K.B. Spörli

17. *Strain rate in Paleozoic thrust sheets, the western Lachlan Orogen, Australia: Strain analysis and fabric geochronology* 349
    D.A. Foster and D.R. Gray

18. *Cenozoic tectonic evolution of Qaidam basin and its surrounding regions (part 2): Wedge tectonics in southern Qaidam basin and the eastern Kunlun Range* 369
    A. Yin, Y. Dang, M. Zhang, M.W. McRivette, W.P. Burgess, and X. Chen

19. *Defining the eastern boundary of the North Asian craton from structural and subsidence history studies of the Verkhoyansk fold-and-thrust belt* 391
    A.K. Khudoley and A.V. Prokopiev

Index 411

Introduction

J.W. Sears
Department of Geosciences, University of Montana, Missoula, Montana 59812, USA
Tekla A. Harms
Department of Geology, Amherst College, Amherst, Massachusetts 01002-5000, USA
Carol A. Evenchick
Natural Resources Canada, 625 Robson Street, 14th Floor, Vancouver, British Columbia V6B 5J3, Canada

Ray Price

The present volume collects a group of 19 original and provocative papers on the tectonic evolution of mountain systems to honor Raymond A. Price on the occasion of the fiftieth anniversary of his first publications on the geological structure of the Canadian Cordillera. These papers demonstrate the wide range of Ray's influence in the fields of structural geology and tectonics, from plate to outcrop scale, from geometry based on mapping and cross-section construction to theoretical processes based on finite-element models, from metamorphic processes to denudation and syntectonic deposition. The authors employ detrital-zircon, argon/argon, and fission-track geochronometry, global positioning system (GPS), deep seismic-reflection and other geophysical methods, strain measurement, kinematic analysis, and classical regional mapping and stratigraphy. Contributors range from distinguished professors to zealous young scientists and graduate students. Study regions range from the North American Cordillera and Appalachians to Siberia, New Zealand, Australia, and China, from Paleozoic to active systems. The common thread of the papers addresses the question, "Whence the mountains?" which Ray posed in an essay during his formative years as a student and geological field assistant in the Canadian Cordillera.

We divide this collection of papers into four parts. The first four papers concern large-scale tectonic processes. The next seven papers examine the structural and tectonic evolution of the North American Cordillera. The following four analyze Appalachian tectonics, and the final four investigate mountains of the southwest Pacific and Asia.

TECTONIC PROCESSES

Chapter 1. W.B. Hamilton sets the stage with the provocative essay, "*Driving mechanism and 3-D circulation of plate tectonics*," in which he proposes that the fundamental controls of plate tectonics are exerted by self-organizing lithospheric dynamics rather than deep-mantle convection. He posits that the ultimate driver of plate tectonics is subduction-hinge rollback resulting from the density inversion of cooling oceanic lithosphere. He makes the case that subducted slabs settle at the mantle transition zone at a depth of 660 km so that plate-tectonic circulation is limited to the upper mantle. Accretionary wedges advance toward trenches as forearc basins extend, and Cordilleran-type margins develop where continents are drawn over retreating trenches. Thus, rollback of the subduction hinges along the eastern Pacific margins initiated the Cordilleran orogenesis and opened the Atlantic Ocean. Hamilton's concepts provide an interesting new context for many of the orogenic processes studied by Ray and discussed in other papers in this volume, from the exhumation of metamorphic rocks to critical wedge tectonics, terrane accretion, and emplacement of disrupted ophiolite sheets.

Chapter 2. W.G. Ernst, B.R. Hacker, and J.G. Liou present an exhaustive global synthesis of ultrahigh-pressure metamorphism, "*Petrotectonics of ultrahigh-pressure crustal and upper-mantle rocks—Implications for Phanerozoic collisional orogens*," including new phase diagrams, that addresses the central questions of how rocks reached depths of 90–140 km in Phanerozoic collisional orogenic belts and then returned to the surface. They present a synthesis that addresses evidence for ultrahigh pressures preserved in inclusions in tough refractory host minerals such as garnet. The authors show that round-trip pressure-temperature paths could have been completed in 5–10 m.y., with approximately equal ascent and descent rates. Rapid ascent preserves the metamorphic assemblages and is driven by buoyancy of low-density crustal material in subduction channels. Exhumation is enabled in extensional environments generated by subduction-hinge rollback, such as those discussed by Hamilton (Ch. 1). Greatest depths are achieved by coherent continental crust that may not decouple as readily as subduction mélange. The concepts in this paper have interesting implications for the observations of Hammer and Clowes (Ch. 5) from deep seismic-reflection studies that tectonic or metamorphic processes produce a relatively constant depth to the Moho across much of the Canadian Cordillera, despite variations of tectonic environments in the overlying crust.

Chapter 3. B.C. Burchfiel, C. Studnicki-Gizbert, J.W. Geissman, R.W. King, Z. Chen, L. Chen, and E. Wang ask "*How much strain can continental crust accommodate without developing obvious through-going faults?*" They present a dilemma in reconciling GPS measurements of actual surface displacement gradients with paleomagnetic and geologic data in the tectonically active eastern syntaxis of the Himalaya of southwest China. This region exhibits a geodetically determined velocity gradient of ~10 mm/yr, which should have accumulated ~100 km of sinistral shear over an ~100-km-wide zone during the past 8–11 m.y. However, the region lacks obvious through-going faults. The displacement appears to manifest through distributed brittle deformation and rotation of the upper crust. Understanding of this process in an active system has fundamental significance for ancient mountain belts given that virtually all orogenic systems include significant transpressional shear zones, such as those that were recognized in the Canadian Cordilleran hinterland by Ray and are examined in the chapters in this book by Evenchick et al. (Ch. 6), Sears (Ch. 7), and Dennis (Ch. 15).

Chapter 4. G.S. Stockmal, C. Beaumont, M. Nguyen, and B. Lee, in "*Mechanics of thin-skinned fold-and-thrust belts: Insights from numerical models*," present refined finite-element models for thrust wedges that build upon Ray's classic cross sections of the southern Canadian Rockies. They develop several models by sequentially incorporating flexural isostatic subsidence, strain softening, multiple detachments, and erosion and sedimentation. Their results beautifully mimic the essential structures of fold-and-thrust belts, including large thrust sheets, duplexes, back thrusts, fault-bend folds, and piggyback basins. Comparisons of their diagrams with cross sections presented in other papers in this volume provide new insight into dynamic controls of horizontal shortening in different settings. For example, the structural style and timing of the Lachlan orogen with its regularly spaced kink-folds presented by Foster and Gray (Ch. 17), and Khudoley and Prokopiev's (Ch. 19) cross section of the Verkhoyansk belt in Siberia match Stockmal et al.'s models that incorporate two detachments. Contrasts with cross sections such as those presented by Sears (Ch. 7) may speak to the influence of original basin geometry, which is not incorporated in these models.

CANADA AND U.S. CORDILLERA

Chapter 5. The second set of papers builds directly upon Ray's work in the Canadian Cordillera. P.T.C. Hammer and R.M. Clowes, in "*Lithospheric-scale structures across the Alaskan and Canadian Cordillera: Comparisons and tectonic implications*," provide a regional perspective along 2500 km of Cordilleran strike length by integrating three lithospheric-scale cross sections. The sections combine deep seismic-reflection and refraction experiments with other geophysical and geochemical investigations, surface mapping, and stratigraphic and geochronological studies. They show basic similarities of the tectonic structures of the southern Canadian Cordillera, northern Canadian Cordillera, and Alaska, including outward-verging crustal-scale décollements and wedges, full-crustal transcurrent faults, and a remarkably flat and shallow Moho, especially under the Canadian sectors. The southern section passes through Ray's original field areas in southeastern British Columbia and southwestern Alberta, and extends his original balanced cross sections to the base of the lithosphere. A comparison with crustal-scale cross sections of central Tibet by Yin et al. (Ch. 18) and the southern Appalachians by Hatcher et al. (Ch. 12) reveals similar patterns of crustal wedging and detachment.

Chapter 6. C.A. Evenchick, M.E. McMechan, V.J. McNicoll, and S.D. Carr present "*A synthesis of the Jurassic–Cretaceous tectonic evolution of the central and southeastern Canadian Cordillera: Exploring links across the orogen,*" in which they provide a succinct crustal-scale interpretation of the southern Canadian Cordillera. They build upon a restoration of 490 km of dextral strike-slip displacement along the Northern Rocky Mountain Trench and Tintina fault system that places the Bowser hinterland basin in a position opposite the Alberta foreland basin. The combination of extensive data sets from the two basins clarifies the tectonic evolution of each and inspires a new tectonic model with two crustal detachments. Their paper extends Ray's original work and methods in the Alberta basin and Rocky Mountain fold-and-thrust belt to the other side of the Omineca crystalline belt, which formed a long-lived tectonic divide in the Cordilleran system. The paper complements the lithospheric-scale view of Hammer and Clowes (Ch. 5) and the detailed examination by Ward and Sears (Ch. 10) of a single site in the foreland basin that recorded passage of the Jurassic forebulge.

Chapter 7. J.W. Sears carries Ray's influence south of the border in "*Belt-Purcell Basin: Keystone of the Rocky Mountain fold-and-thrust belt, United States and Canada.*" He shows that the Mesoproterozoic Belt-Purcell Basin, one of Ray's original Ph.D. research areas, had a controlling influence on the structural evolution of the Montana and southern Canadian Rocky Mountain fold-and-thrust belt. The original geometry of the Belt-Purcell Basin, with two grabens and a horst, defined a mosaic of crustal plates that rotated and sheared against one another along original syndepositional faults. Consideration of thermochronometric data, such as those discussed by Feinstein et al. (Ch. 8), shows that the strong and dense northern plate wedged into the foreland basin. A palinspastic map restores the rifted western edge of the Belt-Purcell Basin and provides a template for testing the details of a paleocontinental restoration of the Belt-Purcell Basin against the northeastern Siberian craton originally advocated by Sears and Price in 1978. The map helps to explain the distribution of Neoproterozoic Windermere rift facies and gives a starting configuration for the kinematics of thrust rotation. The wide shear zones between the rotating crustal blocks may provide ancient examples of the tectonic processes in the active east Himalayan syntaxis discussed by Burchfiel et al. (Ch. 3).

Chapter 8. S. Feinstein, B. Kohn, K. Osadetz, and R.A. Price team up to discuss "*Thermochronometric reconstruction of the prethrust paleogeothermal gradient and initial thickness of the Lewis thrust sheet, southeastern Canadian Cordillera foreland belt.*" They take advantage of deep exposures provided by exhumation of the Lewis thrust sheet in the footwall of the Flathead normal fault to analyze the thermal structure across a 12-km-thick section of the foreland fold-and-thrust belt as it existed prior to displacement on the Flathead fault. They employ low-temperature thermochronometry with fission-track analysis of detrital zircon from the lower part of the Lewis sheet, and vitrinite reflectance of coal from Jurassic-Cretaceous rocks from the upper part and its footwall to interpret the burial and denudation history of the 12–13.5-km-thick Lewis thrust sheet. They uncover surprising results for the thermal history of the thrust sheet, including the discovery that the sheet experienced its thermal maximum during the 15 m.y. before the initiation of thrust movement on the Lewis fault. This supports Ray's hypothesis that circulation of meteoric water driven by hydrologic head in this orogenic system refrigerated and embrittled uplifted thrust sheets, which is an important factor in dynamic models such as those presented by Stockmal et al. (Ch. 4), and raises questions about authigenic illite age dating of fault gouge, such as that presented by Solum and van der Pluijm (Ch. 9).

Chapter 9. J.G. Solum and B.A. van der Pluijm, in "*Reconstructing the Snake River–Hoback River Canyon section of the Wyoming thrust belt through direct dating of clay-rich fault rocks,*" report on provocative new argon-argon dating of authigenic illite in thrust fault gouge of the classic Wyoming fold-and-thrust belt. They sampled the Absaroka, Darby, Bear, and Prospect thrusts near the intersection of the thrust belt with the Gros Ventre Mountains foreland uplift. They interpret their results to suggest nearly simultaneous early to middle Eocene authigenic illite crystallization in thrust faults that were classically thought, based on stratigraphic ages of footwall strata, to have begun to move in Late Cretaceous and Paleocene time in a forward-propagating fashion. Their interpretation implies that internal deformation may have been required to maintain critical taper in this part of the fold-and-thrust belt. Their results are interesting to compare with the finite-element predictions of Stockmal et al. (Ch. 4), which include erosion and sedimentation and also show reactivation of internal thrusts within the wedge.

Chapter 10. In "*Reinterpretation of fractures at Swift Reservoir, Rocky Mountain thrust front, Montana: Passage of a Jurassic forebulge?*" E.M.G. Ward and J.W. Sears show that a systematic set of strike-parallel fractures in the upper Mississippian Madison limestone was widened by karst that penetrated to a few meters below the sub-Jurassic unconformity. Clams bored into the unconformity, as well as into the fracture faces. The fractures and borings were filled by basal Sawtooth sand and a lag conglomerate during the Jurassic transgression into the subsiding foreland basin east of the Cordilleran orogenic belt. The authors propose that the unconformity and fractures may record the passage of the Jurassic forebulge. The significance of this unconformity is discussed in this volume by Evenchick et al. (Ch. 6). The sand-filled, solution-widened fractures (grikes) created secondary porosity that provides a productive hydrocarbon reservoir in the upper Madison Group, sealed by overlying Jurassic shale. Quartz cements in the grikes contain inclusions of hydrocarbon derived from Late Devonian source rocks in the thrust belt.

Chapter 11. P.T. Doughty, K.R. Chamberlain, D.A. Foster, and G.S. Sha present "*Structural, metamorphic, and geochronologic constraints on the origin of the Clearwater core complex, northern Idaho.*" They provide significant new data on the timing of extensional denudation, the geometry of the detachment faults and linking strike-slip faults, and the depth of denudation for an Eocene core complex. This complex, located at an extensional relay zone in the Lewis and Clark line, occurs between the better-known Bitterroot core complex in western Montana and the Priest River complex in northern Washington and Idaho, which P.T. Doughty and T.A. Harms had studied previously for their thesis research with Ray. They use the U-Pb age of Late Cretaceous zircon overgrowths in the external zone of the complex to date the peak metamorphism, and the U-Pb age of late Paleocene zircon overgrowths in the deeper internal zone to date the initiation of extensional denudation that continued through early Eocene, as shown by argon/argon dates. This timing is compatible with findings in the Canadian Cordillera and in the northern Rocky Mountains of Montana that show that the transition from thrusting to extension began at 59 Ma (Evenchick et al., Ch. 6; Sears, Ch. 7; Feinstein et al., Ch. 8), although Solum and van der Pluijm (Ch. 9) interpret argon/argon ages of authigenic illite in fault gouge as showing that thrusting continued in Wyoming into the middle Eocene. The Clearwater core complex is additionally noteworthy because it provides a rare exposure of Proterozoic basement rock within the western Belt-Purcell Basin.

APPALACHIAN REGIONS

Chapter 12. R.D. Hatcher Jr., P.J. Lemiszki, and J.B. Whisner discuss the "*Character of rigid boundaries and internal deformation of the southern Appalachian foreland fold-and-thrust belt.*" They provide a comprehensive analysis of the structural geology and tectonic evolution of the southern Appalachian fold-and-thrust belt, using surface geologic data and a large number of seismic-reflection sections. They contour the basement depths beneath the thrust mass, and the base of the Blue Ridge–Piedmont megathrust sheet, tracing thrust ramps in excess of 10 km high. In 1991, Hatcher and Price discussed striking similarities between the southern Appalachian and southern Canadian Cordilleran fold-and-thrust belts. These are further evident by comparison of this paper with Hammer and Clowes (Ch. 5), Evenchick et al. (Ch. 6), and Sears (Ch. 7). For example, the Blue Ridge–Piedmont megathrust sheet formed a rigid indenter that overrode the thin-skinned southern Appalachian foreland fold-and-thrust belt, much like the Belt-Purcell megathrust sheet overrode the Rocky Mountain fold-and-thrust belt. Both rose over ramps more than 10 km high. A significant difference seems to be that in the Appalachians, a number of extensional faults had offset the crystalline basement during deposition of the Paleozoic shelf section; larger ones later deflected the décollement surface, whereas smaller ones did not, so that some rift basins were beheaded—this has also been recognized farther south by Thomas (Ch. 13). Hatcher et al. provide a preliminary palinspastic map together with kinematic data that raise interesting questions about how megathrust sheets move in orogenic belts.

Chapter 13. W.A. Thomas develops concepts for "*Balancing tectonic shortening in contrasting deformation styles through a mechanically heterogeneous stratigraphic succession,*" through examination of the contrasting structural responses of brittle and plastic strata to horizontal shortening in the Alabama fold-and-thrust belt. The region exhibits two layers of brittle duplexes in stiff lithologies, and ductile duplexes in thick, weak shale. Thomas calls the ductile duplexes "mushwads" because of extreme internal deformation, disharmonic holding, and lateral flowage. The area is further complicated by the presence of the Cambrian Rome graben trough, which offset the basement-cover contact in the foreland, so that the décollement takes on an unusual configuration, as shown farther north by Hatcher et al. (Ch. 12). The behavior of the mushwads calls to mind the disharmonic folding of the Chancellor basin into the Porcupine Creek anticlinorium that Ray demonstrated in his cross sections of the southern Canadian Cordillera, where area balancing must be used instead of line-length balancing in cross-section construction.

Chapter 14. J.P. Hibbard, C.R. van Staal, and B.V. Miller establish "*Links among Carolinia, Avalonia, and Ganderia in the Appalachian peri-Gondwanan realm,*" by providing detailed correlation of the tectonic evolution of major terranes in the Appalachian hinterland. The area is difficult to unravel because of metamorphism and strain, but new isotopic and paleomagnetic data provide additional controls for correlation of isolated segments of terranes. Hibbard et al. conclude that the complex geologic history and composition of the peri-Gondwanan megaterrane Carolinia are more akin to Ganderia than to Avalonia. Their interpretation leads to a more simplified tectonic model, where Carolinia and Ganderia are accreted to Laurentia in the Late Ordovician–Silurian. The terranes and their relationship to the Appalachian foreland bear more than superficial resemblances to the accreted terranes of the Cordillera discussed by Hammer and Clowes (Ch. 5) and Evenchick et al. (Ch. 6), and to the Lachlan belt described by Foster and Gray (Ch. 17). Consideration of all three regions could lead to new understanding of tectonic processes, especially in the wake of Hamilton's (Ch. 1) new interpretation of subduction-hinge rollback.

Chapter 15. In "*Cat Square basin, Catskill clastic wedge: Silurian-Devonian orogenic events in the central Appalachians and the crystalline southern Appalachians,*" A.J. Dennis gives a provocative interpretation based on detrital-zircon geochronology that suggests that a high-grade paragneiss terrane in the southern Appalachian Piedmont may represent a Silurian successor basin that first formed near the Catskill delta in New York and then shifted southwest on the Brevard fault zone, before being driven to the northwest by Alleghanian thrusting. The concept is rather like the well-documented accretion of the Stikine terrane with the Canadian Cordillera, deposition of the Bowser basin, and translation of the terrane and basin along the strike of the orogenic system for hundreds of kilometers discussed by Evenchick et al. (Ch. 6), except, in the Appalachian model presented by Dennis, the basin achieved high-grade metamorphism during terminal continental collision of Gondwana with Laurentia.

ASIAN AND PACIFIC REGIONS

Chapter 16. K.B. Spörli discusses "*A stratigraphic unit converted to fault rocks in the Northland Allochthon of New Zealand: Response of a siliceous claystone to obduction.*" He provides an unusually detailed description of the brittle deformation of an Upper Cretaceous claystone embedded in a young, obducted ophiolite in northern New Zealand. The brittle fabric consists of a fractal array of crisscrossing fractures and faults. He painstakingly unravels the crosscutting relationships among 20 phases of crosscutting fractures and relates them to the processes of growth of an accretionary prism and emplacement of the allochthon. The claystone formed an aquiclude that permitted the buildup of high fluid pressures in the underlying rocks, which facilitated emplacement of the allochthon. The Northland allochthon was obducted onto New Zealand in the Oligocene and Miocene within the complex system of trenches and backarc basins in the boundary zone between the Indian and Pacific plates. The observations are of special interest in view of Hamilton's paper (Ch. 1), which stresses the importance of subduction-hinge rollback and extensional tectonism in convergent margins.

Chapter 17. D.A. Foster and D.R. Gray analyze "*Strain rate in Paleozoic thrust sheets, the western Lachlan Orogen, Australia: Strain analysis and fabric geochronology.*" This orogen-wide strain study integrates argon/argon geochronology, structural fabric analysis, and deep seismic reflection within a kinematic model for large-scale kink folding of the Lachlan orogen above a basement detachment. The authors took detailed measurements of fabrics across the fold belt, sampled syntectonic white mica from known fabric elements, and determined argon crystallization ages for the micas. Since white mica grows below its argon blocking temperature, the ages define the growth of the fabric. They conclude that 67% of shortening occurred over ~16 m.y., for a rate of between 19 and 50 mm/yr. These rates are similar to modern convergent systems, and they are comparable to rates used in the finite-element models of Stockmal et al. (Ch. 4), which produce similar geometry.

Chapter 18. A. Yin, Y. Dang, M. Zhang, M.W. McRivette, W.P. Burgess, and X. Chen present "*Cenozoic tectonic evolution of Qaidam basin and its surrounding regions (part 2): Wedge tectonics in southern Qaidam basin and the eastern Kunlun Range.*" They outline provocative new observations of the near-surface structural geometry of northern Tibet, which show that, contrary to conventional wisdom, the dominant structural vergence of this region is to the south, toward the Tibetan Plateau. Their paper provides previously unavailable ground truth that constrains tectonic modeling of this globally significant region. Yin et al. propose a crustal-scale tectonic wedge model to reconcile their new data with existing tectonic models for the Tibetan Plateau. They found inspiration for their model in Ray's publications on tectonic wedging in the Canadian Cordillera. Their cross section bears a resemblance to wedges shown in the sections of Hammer and Clowes (Ch. 5) and Evenchick et al. (Ch. 6).

Chapter 19. A.K. Khudoley and A.V. Prokopiev, in "*Defining the eastern boundary of the North Asian craton from structural and subsidence history studies of the Verkhoyansk fold-and-thrust belt,*" provide the best summary in western literature on the geology of the Verkhoyansk fold-and-thrust belt, the Siberian analog of the Rocky Mountain fold-and-thrust belt. These belts may have developed on conjugate rift margins. The authors use a variety of geological, geophysical, and theoretical rift subsidence and thrust wedge models to show that the buried eastern rift margin of the North Asian craton follows the boundary between the inner and outer parts of the Verkhoyansk orogenic belt. This improved placement of the rift margin will be important in evaluating paleocontinental restorations of Rodinia. The authors propose that the broad shelf was boudinaged into thin strips during Devonian rifting, and transitional crust evolved in necks of the crustal boudins. Mesozoic crustal shortening then closed the boudins while the sedimentary cover detached and folded.

ACKNOWLEDGMENTS

The editors of this volume would like to formally thank the many reviewers who took their valuable time and surgical skills to evaluate and offer improvements to these manuscripts. Thank you again, Suzanne L. Baldwin, Julia Baldwin, Rebecca Bendick, Steven E. Boyer, Michael Broecker, Jean Pierre Burg, Michael Cecile, Maurice Colpron, Declan G. De Paor, Christopher L. Fergusson, Peter Fermor, Raymond Fletcher, Gary S. Fuis, George Gehrels, Jane A. Gilotti, Laurent Godin, James Hibbard, Dale R. Issler, Thomas K. Kalakay, Paul Kapp, Paul Karabinos, Jeffrey Karson, J. Duncan Keppie, Paul K. Link, Michelle J. Markley, Patrick A. Meere, Elizabeth Miller, Gautam Mitra, Peter Molnar, James Monger, Mike Murphy, R. Damian Nance, Andrew V. Okulitch, Geoff Rait, Edward R. Sobel, Glen S. Stockmal, Robert Thompson, Marian Warren, Tim White, David Wiltschko, Donald U. Wise, and Adolph Yonkee. We offer special thanks to John MacLean, who tirelessly helped Sears with the mechanics of processing the manuscripts, to GSA staff members who were always available to help us, and to GSA Books Editors Abhijit Basu and Pat Bickford for their encouragement through all phases of the project.

Whence the mountains?
The contributions of Raymond A. Price

J.W. Sears
Department of Geosciences, University of Montana, Missoula, Montana 59812, USA
Tekla A. Harms
Department of Geology, Amherst College, Amherst, Massachusetts 01002-5000, USA
Carol A. Evenchick
Natural Resources Canada, 625 Robson Street, 14th Floor, Vancouver, British Columbia V6B 5J3, Canada

This volume is dedicated to Ray Price's long and distinguished quest to understand the origin of mountains. It is the outgrowth of a three-day topical session to celebrate the fiftieth year of Ray's career, which was held at the 2004 annual meeting of The Geological Society of America. At that session, geologists from 16 countries presented 56 papers on diverse aspects of structural geology and tectonics that have been influenced by Ray's career contributions. The session covered 23 orogenic belts of all ages and from five continents. The 122 authors at the session included geologists in every stage of their careers, from young graduate students to mature researchers and retired dignitaries. The selection of papers in this volume is an outgrowth of papers presented at that session and includes structural and tectonic research on global systems as well as individual orogenic belts in Australia, Asia, the Appalachians, and of course, the Canadian Cordillera, where Ray established his career. This essay attempts to outline some of Ray's important contributions to our understanding of mountain systems.

Ray first posed his question, "Whence the mountains?" in 1954, when, as an undergraduate student in the Department of Geology at the University of Manitoba, he wrote a short article with that title for the Faculty of Science news magazine, *The Question Mark*.

Born and raised in Winnipeg, within the flattest part of the Canadian prairie—the floor of former glacial Lake Agassiz—Ray never had an opportunity to see a mountain "close up" until the summer of 1952, when he had the good fortune to obtain a summer job as a junior student assistant with the Geological Survey of Canada field party of Geoff Leech, who was engaged in 1 inch = 1 mile geological mapping of the St. Mary Lake area, in the Purcell Mountains of southeastern British Columbia. Ray's geological training at the time was limited to one introductory course that he had selected to complement his primary interest in physics and chemistry. That outstanding course, presented by a legendary University of Manitoba teacher, Professor Ed Leith, initiated Ray's interest in geology. Subsequently, Geoff Leech nurtured that interest by introducing Ray to field geology and to the history of exploration of the mountains of southeastern British Columbia, and by giving him stimulating and challenging fieldwork assignments. These assignments included mapping, measuring, and describing the unusual Moyie gabbro sills of the lower Purcell Supergroup. The tops of the sills have granophyric veins and pepperite, which subsequently have been shown to be the result of interaction of the magma with wet sediment in the Aldridge Formation. Almost fifty years later, Anderson and Davis U-Pb dated zircons from the granophyre at 1469 Ma. That date finally established the age of deposition of the oldest beds in the Cordillera, which had been one of the great question marks of Ray's early days.

Ray returned to southeastern British Columbia in the summer of 1953 as a student field assistant to Geoff Leech, this time to map Paleozoic rocks in the western Canadian Rockies in the Canal Flats area. That field season stimulated his interest in Late Proterozoic–early Paleozoic stratigraphy and tectonics and led to his first scientific paper, "The base of the Cambrian system in the southeastern Cordillera of Canada," which grew out of an undergraduate term paper and was published in 1956.

Those two summers in the field with Geoff Leech inspired Ray's lifelong interest in the Canadian Cordillera. We are grateful that he decided to major in geology rather than physics and chemistry. Subsequently, Ray built his understanding of the Cordillera outward from its foundations in southeastern British Columbia until it embraced the entire mountain system and its equivalents. Throughout his career he returned to the southeastern Canadian Cordillera, where he helped legions of graduate students learn to map, interpret, and explain the complex structures and difficult stratigraphy of the region.

In 1955, Ray received his B.Sc. degree with honors from the University of Manitoba, along with the Gold Medal in Sciences. He proceeded to Princeton University (Fig. 1) where he became a Procter Fellow and studied under John Maxwell, one of the leading structural geologists of his time. Ray's Ph.D. thesis on the structure, stratigraphy, and tectonic evolution of the Flathead map-area of southwestern Alberta and southeastern British Columbia was a Geological Survey of Canada mapping project supervised by his principal mentor, R.J.W. (Bob) Douglas. Some 40 years later, Ray was to receive the distinguished R.J.W. Douglas Medal from the Canadian Society of Petroleum Geologists for his contributions to our understanding of the Canadian Cordillera. Ray completed the requirements for his Ph.D. in May 1958, and shortly thereafter began geological mapping of the Fernie, east-half, one-degree quadrangle, which included the region surrounding his Ph.D. thesis area. From 1958 to 1968, he worked for the Geological Survey of Canada as research geologist in the southern Canadian Rocky Mountains and in the northern Yukon Territory (Fig. 2). His map of the Flathead area was published in 1959, the accompanying memoir was published in 1965, and the map of the Fernie east-half area and accompanying report was published in 1962.

His early studies were balanced between structural geology and stratigraphy; he published on the lithostratigraphy of the Devonian and Middle Cambrian systems, and on the Purcell Supergroup. He documented the correlation of the Proterozoic Belt-Purcell rocks along the international boundary from the eastern edge of the thrust belt in Waterton and Glacier National Parks, across the Rocky Mountain Trench into the Purcell Mountains. He suggested that the enormous volume of fine-grained sediment in the Belt-Purcell Supergroup represented the immense delta of a major river that had drained a continent-sized basin. That conclusion, published in 1964, was of fundamental importance in later explaining how the Purcell basin captured sediment that was shed from a western continent that has since rifted away. Proof of that concept awaited precise detrital-zircon geochronology and Nd isotope studies that were not possible until decades after Ray's original fieldwork.

By 1967, Ray had also published papers on the structure of the Lewis thrust and on flexural-slip folds and mesoscopic structures of the Crowsnest Pass area. Ray showed that the provenance of the Tertiary deposits in the Kishehnen basin recorded the progressive exhumation of the Lewis thrust sheet in the footwall of the Flathead normal fault, and thus provided important timing constraints for thrusting and the ensuing extensional denudation of this part of the Cordilleran thrust system. More recently, following a collaborative project involving apatite and zircon fission-track analysis of the thermal history of the Lewis thrust sheet and associated rocks, Ray suggested that rapid cooling of the Lewis thrust sheet during thrusting could be attributed to refrigeration by deeply penetrating meteoric water that was driven by increased topographic gradients and was facilitated by enhanced permeability associated with brittle deformation during thrusting and thrust-related folding.

The Crowsnest Pass cross-strike discontinuity and the Crowsnest deflection, across which the trends of the Rocky Mountain folds and thrusts change from northwesterly to northerly, are features of this region that have been of longstanding interest to Ray. In his early work, he showed that the swing-back in the trace of the Lewis thrust at North Kootenay Pass coincided with a lateral ramp in the Lewis thrust, where the fault stepped up to the north from a detachment deep in the Purcell Supergroup into a detachment in the Paleozoic section. Later, he demonstrated that this feature was associated with the Crowsnest Pass cross-strike discontinuity, a through-going transverse structure that includes the Moyie and St. Mary transverse fault zones farther west. The Crowsnest Pass cross-strike discontinuity was a growth fault zone during deposition of the Purcell Supergroup and exhalation of the Sullivan ore body. It was formed by reactivation of the southwestern extension of the Vulcan structure, a Paleoproterozoic suture along the northwest margin of the Archean Medicine Hat block that is marked by a conspicuous magnetic anomaly in the basement of southern Alberta. Ray showed that this transverse structure marks an offset in the margin of the Cordilleran miogeocline that extends 230 km southwest from the Fernie area of southeastern British Columbia to the vicinity of the Columbia River at Chewelah, Washington. That precise jog was later shown to have profound significance in paleocontinental restorations because it exactly fits a conjugate jog in the rifted northern margin of the Siberian craton.

The Geological Survey of Canada (GSC), recognizing Ray's potential, assigned him responsibility for Operation Bow-Athabasca, a two-year, helicopter-supported regional mapping project covering the Canadian Rockies between the Bow and Athabasca Rivers, an area of 31,000 km^2, roughly the size of Switzerland. The project team included six GSC stratigraphers, and five geologists, who were responsible mainly for mapping the geological structure: Ray, his principal collaborator, Eric Mountjoy (Fig. 3), and three Ph.D. candidates. The preliminary mapping was done at 1:50,000 by Ray and Eric, based largely on vertical air photogeologic interpretations of the ice-sculpted mountains: this was supplemented by helicopter and ground observations. Typical helicopter traverses involved filling-in and verifying the air photo interpretations by making short visits to critical sites, or else dropping off a geologist and a student field assistant at the top of a ridge by helicopter and picking them up in the valley at the end of the day. The helicopters were unstable little machines by today's standards. One of them made a forced landing because of an engine failure during the project. The mountains were inhabited by grizzly bears, were largely untracked, and were riven with avalanche chutes, active scree slopes, roaring cascades, and thick forests. Wildlife also included mosquitoes and horseflies.

Ray and Eric established the Canadian Rocky Mountains as one of the world's premier examples of a fold-and-thrust belt

Figure 1. Ray Price as a graduate student, Princeton University, 1958.

Figure 2. Ray Price in an early field season with the Geological Survey of Canada, 1964. Courtesy of Mina Price.

Figure 3. Ray Price and Eric Mountjoy at work on Operation Bow-Athabasca, 1965. Courtesy of Mina Price.

when they integrated the Operation Bow-Athabasca stratigraphic and structural studies with those of numerous other field geologists into a coherent regional synthesis. Their innocuously titled paper: "Geologic structure of the Canadian Rocky Mountains between Bow and Athabasca Rivers—A progress report," published in 1970, has become a classic. In the years following the progress report, Ray, Eric, and co-workers steadily published some three dozen geologic maps from this region at 1:50,000. They are still working on these: the most recent was published in 2005. Balanced structure sections accompany each of the maps.

The Bow-Athabasca paper included the elegant regional balanced cross section of the Canadian fold-and-thrust belt that propelled Ray and the Canadian Rockies onto the world stage (Fig. 4). This was the first cross section that passed from the undisturbed foreland basin, through the highest and widest part of the thrust belt, into the metamorphic hinterland. It showed the Rocky Mountain thrusts flattening downward into a regional décollement above the basement, as had been demonstrated to the south near the U.S. border with seismic-reflection data published by Bally and others in 1966. This downward flattening implied that the displaced imbricate thrust sheets of sedimentary strata had been detached from their basement, and, therefore, that the basement must have extended far to the southwest under the metamorphic hinterland. Ray's cross section indicated as much as 250 km of horizontal shortening of the cover rocks above the uninvolved crystalline basement. He used the better-constrained structure of the foreland to reconstruct the former westward extent of the basement beneath the hinterland.

The Bow-Athabasca paper elegantly integrated the geologic history of the foreland basin, the thrust belt, and the metamorphic hinterland. Given the geochronological data available at the time and the reigning geosynclinal paradigm, the integration was remarkable. It showed that uplifted thrust sheets in the hinterland were undergoing erosion, while sandstones derived from them were being deposited in the foreland basin. The stretching lineations in the metamorphic hinterland were normal to the fold axes in the foreland, suggesting to Ray that diapiric upwelling and gravitational spreading of a mobile infrastructure drove thrust shortening in the brittle suprastructure. This interpretation, which did not require equivalent shortening of the basement beneath the metamorphic hinterland of the southeastern Canadian Cordillera, was consistent with the available data—the geochronologic data were limited to a few K-Ar dates, and geologic history was interpreted from stratigraphy and crosscutting relationships. In 1973, by demonstrating that the crystalline basement had subsided isostatically under the load of the spreading fold-and-thrust belt, and by comparing the gravitational spreading of the fold-and-thrust belt to the flow of a continental ice sheet, Ray showed that the foreland basin had formed by isostatic flexural subsidence of the litho-

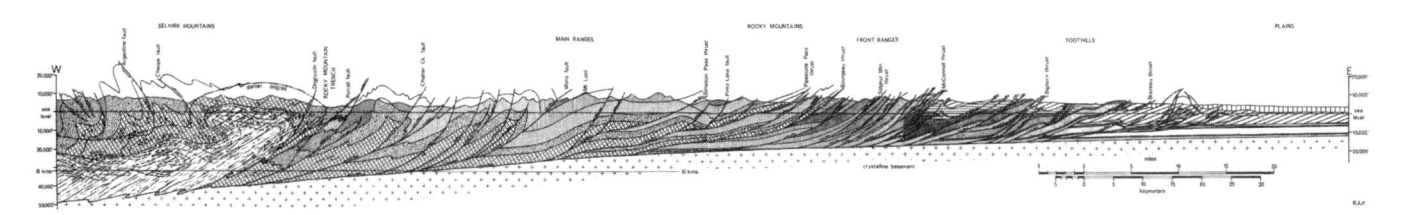

Figure 4. Geologic structure section of the Canadian Rockies from Price and Mountjoy (1970).

sphere that accompanied the advance of the spreading fold-and-thrust belt. He concluded that the migrating foredeep basin was an isostatically induced moat that trapped detrital outwash from the growing fold-and-thrust belt, and, therefore, that episodes of subsidence and sedimentation in the foreland basin should be correlated with orogenic episodes that involved uplift in the hinterland and gravitational spreading of the foreland fold-and-thrust belt. This conclusion was contrary to the then-prevailing view that orogenic episodes were marked by uplift, erosion, and unconformities, not by subsidence and sedimentation. Ray's conceptual model for foreland-basin evolution has developed into an entire field of study of tectonically driven parasequences, with considerable significance for the oil industry because of large reserves in Alberta and other foreland basins worldwide. Ray became internationally famous—his maps and sections were incorporated into the *Encyclopedia Brittanica* and used in structural geology textbooks around the world.

With the advent of plate tectonics and the terrane accretion concept, Ray replaced the diapiric upwelling model in which lateral gravitational spreading of the upwelling of plutonic rocks served as the driver for the gravitational spreading of the fold-and-thrust belt, with an actualistic plate-tectonics model in which gravitational spreading in the fold-and-thrust belt was driven by convergence between Laurentia and the Intermontane terrane. In this model, the fold-and-thrust belt is a critical-taper, accretionary wedge, made up of supracrustal rocks that were scraped off the underriding Laurentian basement by the overriding Intermontane terrane, and the metamorphic complexes were exhumed along extensional detachments during an ensuing episode of ductile crustal extension.

Ray followed up the Bow-Athabasca publication with an excellently illustrated guidebook for field excursions that took place before and after the 24th International Geological Congress, which met in Montreal in 1972. Annotated sketches prepared by an artist from Ray's photographs and instructions illustrate the splendid exposures of fold-and-thrust structures that occur along the main highways throughout the southern Canadian Rockies (Fig. 5). For years, structural geologists have flocked to these famously well-exposed, accessible, and thoroughly inspiring mountains with the little red guidebook in hand to see what thrust structures actually look like in the field. Later, Ray teamed up with Jim Monger to lead several grand field trips along the Trans-Canada Highway across the entire Cordillera. In 2000, they published a large guidebook with maps and data contributed by generations of geologists. The Calgary-to-Vancouver transect is now a global geological destination on par with the Grand Canyon.

In the International Geological Congress guidebook, Ray included a schematic summary stratigraphic section from the foreland to the hinterland, hung on a Late Jurassic sea-level datum at the top of the Fernie Group (Fig. 6). This section showed two sedimentary wedges, both tapering to the east. The lower one was the miogeocline, and showed a tenfold increase in thickness from <2 km in the east to >15 km in the west. It included the Mesoproterozoic (Purcell), Neoproterozoic (Windermere), and Lower Cambrian (Hamill-Gog) clastics to the west, and it showed thick

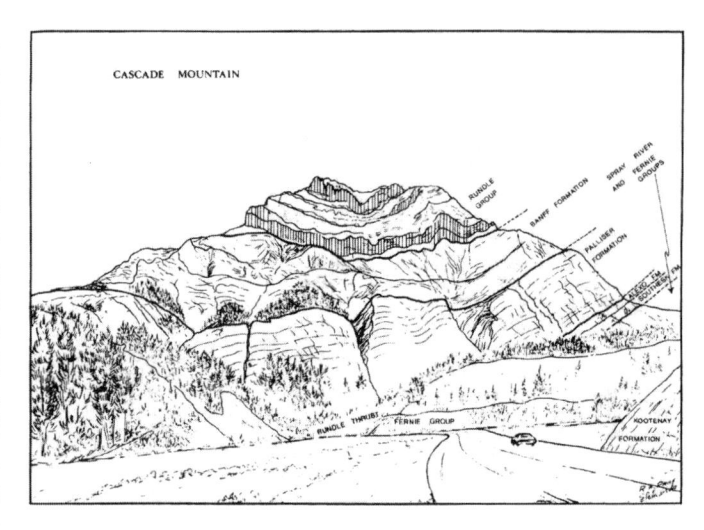

Figure 5. Annotated geological sketch of Cascade Mountain from the Trans-Canada Highway west of Calgary, from Price et al. (1972).

Paleozoic carbonates shaling out to the west. It showed thin unconformity-riddled stratigraphic sections to the east passing into thick, continuous sections in the west. The upper wedge showed that the Late Jurassic–Early Cretaceous foreland basin was located west of the Late Cretaceous–Paleocene foreland basin. The upper sedimentary wedge consisted of the clastic deposits derived from erosion of the rising Cordillera.

Ray moved from the Geological Survey of Canada to the Department of Geological Sciences at Queen's University in Kingston, Ontario, in 1968 and was made Professor and Head of the Department of Geological Sciences in 1972. In his 24 years at Queen's, he supervised 30 graduate students in structural geology and tectonics. With his students, he pursued a mapping campaign across the Cordillera into the Kootenay arc and the metamorphic core-complex country, thesis by thesis. Deep down-plunge projection, polyphase fabric analysis, geobarometry, geothermometry, and geochronology became new tools in the kit. Ray recognized that the reversal in tectonic vergence along the Kootenay arc reflected a fundamental process of thrust-wedging of strong material into weak, seen at scales from outcrops to mountains. He led the way in recognizing the kinematic linkage of Eocene metamorphic core complexes and strike-slip faults, and their relationship to changing plate-tectonic regimes along the western margin of North America.

He was at the forefront of tectonic research in Canada in the early 1970s. In 1972, he and his long-time mentor Bob Douglas edited the essential *Variations in Tectonic Styles in Canada*, which included a chapter on each major tectonic province in the country. In 1979, Jim Monger and Ray recast the Cordillera into the context of plate tectonics. His old stratigraphic section became the Paleozoic continental margin, with the westward transition from shelf to slope to rise. The interior plateaus of British Columbia became oceanic terranes that amalgamated, accreted to the continental shelf, and then drove the shelf eastward into the foreland basin. In 1981, Ray updated his detailed

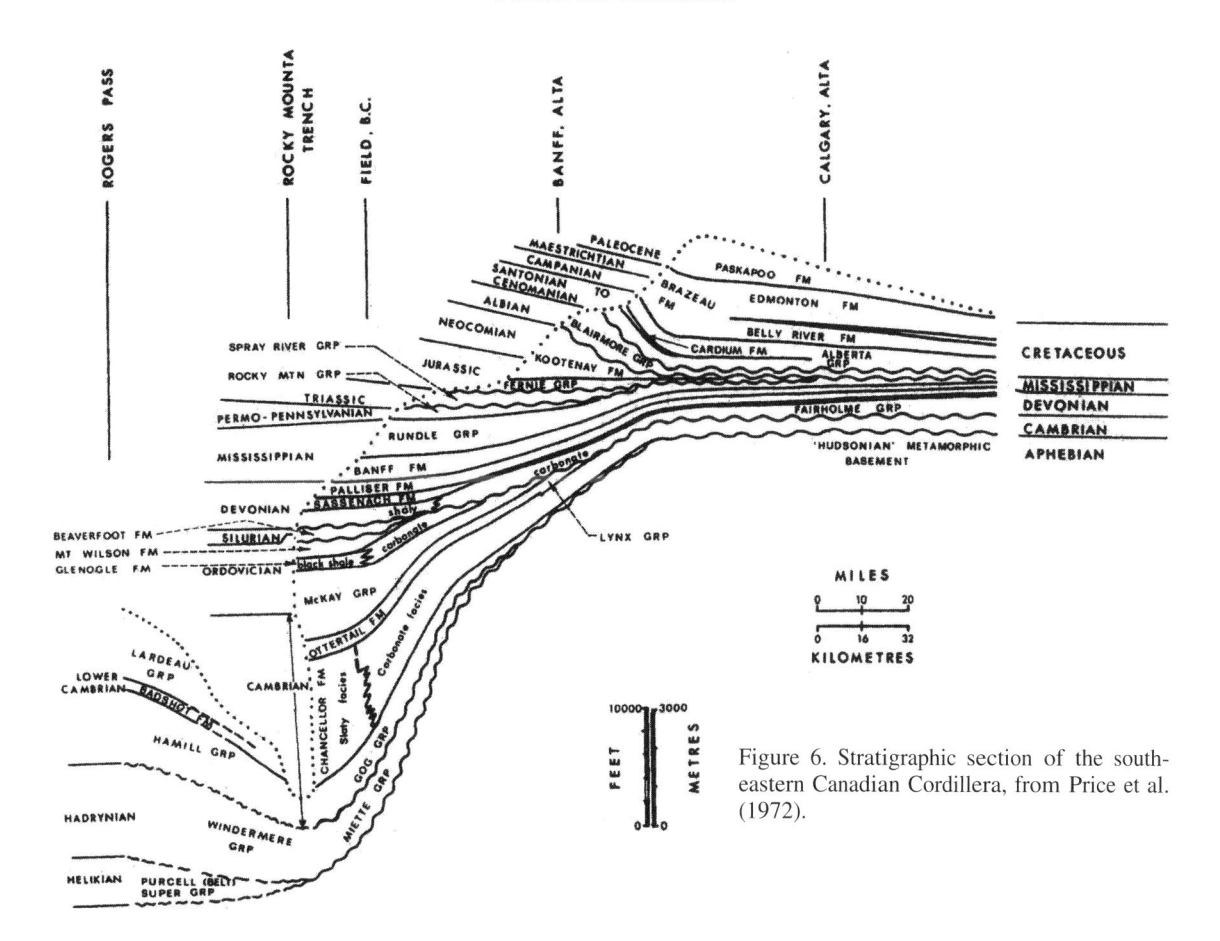

Figure 6. Stratigraphic section of the southeastern Canadian Cordillera, from Price et al. (1972).

structural interpretation of the Cordillera with two new regional balanced cross sections and advanced the idea that Cordilleran anticlinoria represent structurally inverted basins. He now recognized that the Purcell basin of his first days was inverted into the Purcell anticlinorium, and that the Windermere rift basin was inverted into the Selkirk anticlinorium. The Kootenay arc marked a crustal-scale thrust ramp.

Ray's interpretations of the southeastern Canadian Cordillera are readily transferable into the adjacent parts of Montana, Idaho, and Washington, and have greatly illuminated the tectonic evolution of those regions. Integration of structural data across the international boundary and construction of a regional palinspastic map indicate that the thrust mass rotated clockwise during its emplacement.

With his graduate students, Ray dug into the evidence linking the breakup of a Proterozoic supercontinent to the initiation of the Cordilleran margin, with studies ranging from continental-scale comparisons of the Siberian connection to western North America, to detailed maps of Proterozoic grabens and breakup unconformities in the Selkirk Mountains.

Ray returned to the survey in 1981 in order to work full-time on the Canadian contribution to the Geological Society of America Centennial Project, The Decade of North American Geology (Fig. 7). However, when he arrived in Ottawa, there was a major reorganization of the management of the Geological Survey of Canada. Director-General Bill Hutchison was appointed assistant

Figure 7. Ray Price, as Assistant Deputy Minister (earth sciences) of Energy Mines and Resources Canada, with Geoff Leech at the 1988 Geological Survey of Canada Forum, at the Congress Centre in Ottawa. Courtesy of the Geological Survey of Canada.

deputy minister for the Earth Sciences Sector of Energy Mines and Resources Canada. Ray became his replacement as director-general in 1982. And in 1987, when Bill Hutchison suddenly became ill and died, Ray succeeded him as assistant deputy minister for the Earth Sciences Sector. As director-general of the GSC, Ray was an active participant in the establishment of the highly successful Lithoprobe program, Canada's coordinated and integrated multidisciplinary national earth science research project to investigate the three-dimensional structure and evolution of Canada's landmass and continental margins. Spearheaded by seismic and electromagnetic imaging techniques, new magnetic and gravity mapping, and new technology for rapid and precise age determinations, Lithoprobe transects have investigated the major tectonic belts of Canada, from the Cordillera and Pacific margin and the Appalachians and Atlantic margin to the Precambrian belts of the Canadian Shield. Virtually every line has resulted in revolutionary new insight into the tectonics of its region, and often new insights into fundamental crustal processes. The studies have spurred new research that has integrated new mapping, sampling, dating, and geophysical investigations along the transects and promoted interaction among diverse groups of geoscientists.

In addition to his own creative scientific research and his mentoring of students, Ray has done exemplary service for science and society. While chairing the Department of Geological Sciences at Queen's University, Ray initiated the Cordilleran Workshop, which showcases student research into the tectonics and structure of the Canadian Cordillera. Now in its thirtieth year, students come from all over Canada and the United States to participate. His many other contributions include chairmanship or membership on numerous boards to promote advanced geological and other scientific research. A sampling includes the Advisory Committee for the Canadian Institute for Advanced Research Earth System Evolution Program, the Board of Trustees for the Sudbury Neutrino Institute, the United States National Research Council/National Academy of Sciences Board on Earth Sciences and Resources, various committees of the National Science and Engineering Research Council of Canada, the American Geological Institute, the Royal Society of Canada, the Atomic Energy Control Board of Canada, the U.S. Continental Scientific Drilling Program, the Canadian Council for Ocean Drilling Program, Lithoprobe, U.S. National Science Foundation, and the U.S. National Research Council/National Academy of Sciences. He was chairman of the scientific panel to review the proposed deep geological disposal of Canada's nuclear fuel wastes. He served as the president of the Geological Society of America in 1989–1990, as president of the Inter-Union Commission on the Lithosphere (International Council of Scientific Unions) in 1980–1985, and as chairman of the Scientific Committee of the International Geological Correlation Programme in 1992–1994. Ray has greatly contributed to the publication of sound geological interpretations by serving as associate editor for the *Geological Society of America Bulletin*, *Bulletin of Canadian Petroleum Geology*, *Canadian Journal of Earth Sciences*, *Geoscience Canada*, and as editor-in-chief of *Tectonics*.

Ray is a fellow of the Royal Society of Canada, the American Association for the Advancement of Science, The Geological Society of America, and the Geological Association of Canada. He is a foreign associate of the U.S. National Academy of Sciences, an honorary member of the Canadian Society of Petroleum Geologists, and an honorary foreign fellow of the European Union of Geosciences. He has been appointed an Officer of the Order of Canada, and Officier de l'Ordre des Palmes Académiques of France. He has been awarded the Leopold von Buch Medal of the Deutsche Geologische Gesellschaft, the Major Edward D'Ewes Fitzgerald Coke Medal of the Geological Society of London, the Michael T. Halbouty Human Needs Award of the American Association of Petroleum Geologists, the Sir William Logan Medal of the Geological Association of Canada, and the R.J.W. Douglas Medal of the Canadian Society of Petroleum Geologists (Fig. 7). He has received honorary doctor of science degrees from Carleton University and the Memorial University of Newfoundland.

In 1990, Ray retired from the Geological Survey of Canada and returned to teaching at Queen's University. In 1998, he became Professor Emeritus of Geological Sciences and Geological Engineering. He continues to work on his geological maps and research papers and supervise graduate research in the Canadian Cordillera, and he remains active on several advisory boards and committees.

The editors of this volume were graduate students of Ray's. We wish to express our thanks for his guidance in our studies in school and through our careers. His clear mind, steady hand, and enthusiasm for all things tectonic have inspired us as they have the international community of geoscientists. We also acknowledge Ray's wife Mina, his former classmate in geology at the University of Manitoba. After a brief career as a petroleum geologist, Mina has served as his valuable research assistant and a critical reviewer of his maps, sections, and papers, and a warm and generous friend of his graduate students and professional associates.

SELECTED REFERENCES OF R.A. PRICE IN CHRONOLOGICAL ORDER

Price, R.A., 1956, The base of the Cambrian system in the southeastern Cordillera of Canada: Canadian Mining and Metallurgical Bulletin, v. 49, p. 765–771.

Price, R.A., 1959, Flathead, British Columbia and Alberta: Geological Survey of Canada Map 1–1959 (scale: 1 inch = 1 mile).

Price, R.A., 1962, Fernie Map-Area, East-Half, Alberta and British Columbia: Geological Survey of Canada Paper 61–24, 65 p.; accompanied by Geological Survey of Canada Map 35–1961 and structure sections (scale: 1 inch = 2 miles).

Price, R.A., 1962, Geologic structure of the central part of the Rocky Mountains in the vicinity of Crowsnest Pass: Alberta Society of Petroleum Geologists Journal, v. 10, p. 341–351.

Price, R.A., 1964, The Precambrian Purcell System in the Rocky Mountains of southern Alberta and British Columbia: Bulletin of Canadian Petroleum Geology, v. 12, p. 399–426.

Price, R.A., 1964, The Devonian Fairholme-Sassenach succession and evolution of reef-front geometry in the Flathead–Crowsnest Pass area. Alberta and British Columbia: Bulletin of Canadian Petroleum Geology, v. 12, p. 427–451.

Price, R.A., 1965, Flathead Map-Area, British Columbia and Alberta: Geological Survey of Canada Memoir 336, 221 p., including Geological Survey of Canada Map 1154A and 3 structure sections (scale: 1 inch = 1 mile).

Norris, D.K., and Price, R.A., 1966, Middle Cambrian lithostratigraphy of the southeastern Canadian Cordillera: Bulletin of Canadian Petroleum Geology, v. 14, p. 385–404.

Price, R.A., 1967, Tectonic significance of mesoscopic subfabrics in the southern Rocky Mountains of Alberta and British Columbia: Canadian Journal of Earth Sciences, v. 4, p. 39–70.

Price, R.A., and Mountjoy, E.W., 1970, The geological structure of the southern Canadian Rockies between Bow and Athabasca Rivers—A progress report, *in* Wheeler, J.O., ed., A Structural Cross-Section of the Southern Canadian Cordillera: Geological Association of Canada Special Paper 6, p. 7–25.

Price, R.A., 1971, A section through the eastern Cordillera at the latitude of Kicking Horse Pass, in Halliday, I.A.R., and Mathewson, D.H., eds., A Geological Guide of the Eastern Cordillera along the Trans-Canada Highway between Calgary and Revelstoke: Calgary, Alberta, Alberta Society of Petroleum Geologists, p. 17–23.

Douglas, R.J.W., and Price, R.A., 1972, The nature and significance of variations in tectonic style in Canada, in Price, R.A., and Douglas, R.J.W., eds., Variations in Tectonic Styles in Canada: Geological Association of Canada Special Paper 11, p. 625–688.

Price, R.A., Balkwill, H.R., Charlesworth, H.A.K., Cook, D.G., and Simony, P.S., 1972, The Canadian Rockies and Tectonic Evolution of the Southeastern Canadian Cordillera: Montreal, Canada, 24th International Geological Congress Guidebook, Excursions A-15 and C-15, 129 p.

Price, R.A., and Douglas, R.J.W., eds., 1972, Variations in Tectonic Styles in Canada: Geological Association of Canada Special Paper 11, 688 p.

Price, R.A., 1973, Large-scale gravitational flow of supracrustal rocks, southern Canadian Rockies, in de Jong, K.A., and Scholten, R., eds., Gravity and Tectonics: New York, Wiley-Interscience, p. 491–502.

von Engelhardt, W., Goguel, J., Hubbert, M.K., Prentice, J.E., Price, R.A., and Trümpy, R., 1976, Earth resources, time and man—A geoscience perspective: Environmental Geology, v. 1, p. 193–206, doi: 10.1007/BF02407506.

Okulitch, A.V., Price, R.A., and Richards, T.A., 1977, Geology of the Southern Canadian Cordillera—Calgary to Vancouver: Geological Association of Canada, Guidebook, Field Trip 8, 1977 Annual Meeting: Vancouver, Geological Association of Canada, 135 p.

Sears, J.W., and Price, R.A., 1978, The Siberian connection—A case for Precambrian separation of the North American and Siberian cratons: Geology, v. 6, p. 267–270, doi: 10.1130/0091-7613(1978)6<267:TSCACF>2.0.CO;2.

Monger, J.W.H., and Price, R.A., 1979, Geodynamic evolution of the Canadian Cordillera—Progress and problems: Canadian Journal of Earth Sciences, v. 16, p. 770–791.

Price, R.A., 1979, Intracontinental ductile crustal spreading linking the Fraser River and North Rocky Mountain trench transform fault zones, south-central British Columbia and northeast Washington [abs.]: Geological Society of America Abstracts with Programs (1979 Annual Meeting), v. 11, no. 7, p. 499.

Benvenuto, G.L., and Price, R.A., 1980, Evolution of the Hosmer thrust sheet, southeastern British Columbia: Bulletin of Canadian Petroleum Geology, v. 27, no. 3, p. 360–394.

McMechan, R.D., and Price, R.A., 1980, Reappraisal of a reported unconformity in the Paleogene (Oligocene) Kishenehn Formation: Implications for Cenozoic tectonics in the Flathead Valley graben, southeastern British Columbia: Bulletin of Canadian Petroleum Geology, v. 28, p. 37–45.

Price, R.A., 1981, The Cordilleran foreland thrust and fold belt in the southern Canadian Rocky Mountains, in McLay, K.J., and Price, N.J., eds., Thrust and Nappe Tectonics: Geological Society [London] Special Publication 9, p. 427–448.

Price, R.A., Monger, J.W.H., and Muller, J., 1981, Cordilleran cross-section, Calgary to Victoria, in Thompson, R.I., and Cook, D.G., eds., Field Guides to Geology and Mineral Deposits, Calgary '81 Annual Meeting: Calgary, Geological Association of Canada, p. 261–334.

Monger, J.W.H., Price, R.A., and Tempelman-Kluit, D.J., 1982, Tectonic accretion and the origin of the two major metamorphic and plutonic welts in the Canadian Cordillera: Geology, v. 10, p. 70–75, doi: 10.1130/0091-7613(1982)10<70:TAATOO>2.0.CO;2.

McMechan, M.E., and Price, R., 1982, Superimposed low-grade metamorphism in the Mount Fisher area, southeastern British Columbia—Implications for the east Kootenay orogeny: Canadian Journal of Earth Sciences, v. 19, no. 3, p. 476–489.

Porter, J.W., Price, R.A., and McCrossan, R.G., 1982, The Western Canada sedimentary basin: Philosophical Transactions of the Royal Society of London, ser. A, v. 305, p. 169–192.

Price, R.A., and Hatcher, R.D., Jr., 1983, Tectonic significance of similarities in the evolution of the Alabama-Pennsylvania and the Alberta–British Columbia Canadian Cordillera, in Hatcher, R.D., Jr., Williams, H., and Zietz, I., eds., Contributions to the Tectonics and Geophysics of Mountain Chains: Geological Society of America Memoir 158, p. 149–160.

Fermor, P.R., and Price, R.A., 1983, Stratigraphy of the lower part of the Belt-Purcell Supergroup (Middle Proterozoic) in the Lewis thrust sheet of Alberta and British Columbia: Bulletin of Canadian Petroleum Geology, v. 31, no. 3, p. 169–194.

Archibald, D.A., Glover, J.K., Price, R.A., Farrar, E., and Carmichael, D.M., 1983, Geochronology and tectonic implications of magmatism and metamorphism; southern Kootenay arc and neighbouring regions, southeastern British Columbia: Part I. Jurassic to Mid-Cretaceous: Canadian Journal of Earth Sciences, v. 20, no. 12, p. 1891–1913.

Zolnai, A., Price, R.A., and Helmstaedt, H., 1984, Regional cross-section of the Southern Province adjacent to Lake Huron, Ontario: Implications for the tectonic significance of the Murray fault zone: Canadian Journal of Earth Sciences, v. 21, no. 4, p. 447–456.

Price, R.A., 1984, The lithosphere: Laboratory and library, in Malone, T.F., and Roederer, J.G., eds., Global Change: Proceedings of a Symposium sponsored by the International Council of Scientific Unions (ICSU Press Symposium Series No. 5): Cambridge, Cambridge University Press, p. 187–194.

Price, R.A., and Fermor, P.F., 1984, Structure Section of the Cordilleran Thrust and Fold Belt West of Calgary, Alberta: Geological Survey of Canada Paper 84–14, a 108 cm × 145 cm sheet with a 1:250,000 crustal-scale structure section and its palinspastically restored equivalent, plus a 1:50,000 supracrustal-scale version of the eastern half of the structure section.

Price, R.A., Monger, J.W.H., and Roddick, J.A., 1985, Cordilleran cross-section: Calgary to Vancouver, in Tempelman-Kluit, D., ed., Field Guides to Geology and Mineral Deposits in the Southern Canadian Cordillera: Vancouver, B.C., Geological Society of America, Cordilleran Section (Annual Meeting, May '85), p. 3-1–3-85.

Price, R.A., 1986, The southeastern Canada Cordillera: Thrust faulting, tectonic wedging and delamination of the lithosphere: Journal of Structural Geology, v. 8, no. 3/4, p. 239–256, doi: 10.1016/0191-8141(86)90046-5.

Price, R.A., and Carmichael, D.M., 1986, Geometric test for Late Cretaceous–Paleogene intracontinental transform faulting in the Canadian Cordillera: Geology, v. 14, p. 468–471, doi: 10.1130/0091-7613(1986)14<468:GTFLCI>2.0.CO;2.

Price, R.A., 1986, Keynote Address: "Geoscience information—The framework for formulating and implementing policies on resource development," in Shelley, E.P., ed., Proceedings of the Third International Conference on Geoscience Information, Adelaide, Australia, 1–6 June 1986: Adelaide, Australian Mineral Foundation, v. 2, p. 1–14.

Fermor, P.R., and Price, R.A., 1987, Multiduplex structure along the base of the Lewis thrust sheet in the southern Canadian Rockies: Bulletin of Canadian Petroleum Geology, v. 35, no. 2, p. 159–185.

Cook, F.A., Simony, P.S., Coflin, K.C., Green, A.G., Milkereit, B., Price, R.A., Parrish, R., Patenaude, C., Gordy, P.L., and Brown, R.L., 1987, Lithoprobe southern Canadian Cordillera transect; Rocky Mountain thrust belt to Valhalla gneiss complex: Geophysical Journal of the Royal Astronomical Society, v. 89, p. 91–98.

Price, R.A., 1987, Canada's 'new' geological survey: Geotimes, v. 32, no. 12, p. 4.

Price, R.A., Parrish, R.R., Monger, J.W.H., and Roddick, J.A., 1987, Cordilleran cross-section: Calgary to Vancouver: Vancouver, Canada, International Union of Geodesy and Geophysics, Guidebook for Scientific Excursion A1 (XIX General Assembly), 96 p.

Cook, F.A., Green, A.G., Simony, P.S., Price, R.A., Parrish, R., Milkereit, B., Gordy, P.L., Brown, R.L., Coflin, K.C., and Patenaude, C., 1988, Lithoprobe seismic reflection structure of the southeastern Canadian Cordillera: Initial results: Tectonics, v. 7, no. 2, p. 157–180.

Price, R.A., 1988, The mechanical paradox of large overthrusts: Geological Society of America Bulletin, v. 100, no. 12, p. 1898–1908, doi: 10.1130/0016-7606(1988)100<1898:TMPOLO>2.3.CO;2.

Price, R.A., ed., 1989, Origin and Evolution of Sedimentary Basins and Their Energy and Mineral Resources: American Geophysical Union Geophysical Monograph 48 (IUGG Volume 3), 202 p.

Price, R.A., Duke, J.M., and Findlay, D.C., 1989, Government geoscience and exploration, in Garland, G.D., ed., Exploration '87 (Proceedings of an international symposium held in Toronto, September 1987): Toronto, Ontario Geological Survey, p. 761–770.

Price, R.A., 1990, The mechanical paradox of large overthrusts: Alternative interpretation—Reply: Geological Society of America Bulletin, v. 102, p. 531–532, doi: 10.1130/0016-7606(1990)102<0529:TMPOLO>2.3.CO;2.

Apsimon, H., Thornton, I., Fyfe, W., Hong, Yetang, Leggett, J., Nriagu, J.O., Pacyna, J.M., Page, A.L., Price, R., Skinner, B., Steinnes, E., and Yim, W., 1990, Anthropogenically induced global change—Report of Working Group 3, IUGS (International Union of Geological Sciences) Workshop on Global Change Past and Present: Palaeogeography, Palaeoclimatology, Palaeoecology (Global and Planetary Change Section), v. 82, p. 97–111.

Parrish, R.R., Haugerud, R.A., and Price, R.A., 1990, Penrose Conference Report—Eocene tectonic transition, California to Alaska: Geological Society of America (GSA) News and Information, v. 12, p. 124–125.

Cook, F.A., Varsek, J.L., Clowes, R.M., Kanesevich, E.R., Spencer, C.S., Parrish, R.R., Brown, R.L., Johnson, B.J., and Price, R.A., 1992, Lithoprobe crustal reflection cross section of the southern Canadian Cordillera I: Foreland thrust and fold belt to Fraser River fault: Tectonics, v. 11, no. 1, p. 12–35.

Harms, T.A., and Price, R.A., 1992, The Newport fault: Eocene listric normal faulting, mylonitization, and crustal extension in northeast Washington and northwest Idaho: Geological Society of America Bulletin, v. 104, p. 745–761, doi: 10.1130/0016-7606(1992)104<0745:TNFELN>2.3.CO;2.

Harms, T.A., and Price, R.A., 1993, Reply to "The Newport fault: Eocene listric normal faulting, mylonitization, and crustal extension in northeast Washington and northwest Idaho: Discussion," by R.E. Anderson: Geological Society of America Bulletin, v. 105, p. 1512–1514, doi: 10.1130/0016-7606(1993)105<1511:TNFELN>2.3.CO;2.

Price, R.A., 1994, Cordilleran tectonics and the evolution of the Western Canada sedimentary basin, in Mossop, G.D., and Shetsen, I., eds., Geological Atlas of Western Canada: Calgary, Canadian Society of Petroleum Geologists/Alberta Research Council, p. 13–24.

Monger, J.W.H., Clowes, R.M., Cowan, D.S., Potter, C.J., Price, R.A., and Yorath, C.J., 1994, Continent-ocean transition in western North American between latitudes 45 and 56 degrees: Transects B1, B2, B3, in Speed, R.C., ed., Phanerozoic Evolution of North American Continent-Ocean Transition: Boulder, Colorado, Geological Society of America, Decade of North American Geology, Continent-Ocean Transect Volume, p. 357–397.

Colpron, M., and Price, R.A., 1995, Tectonic significance of Kootenay terrane, southeastern Canadian Cordillera: An alternative model: Geology, v. 23, p. 25–28, doi: 10.1130/0091-7613(1995)023<0025:TSOTKT>2.3.CO;2.

Castonguay, S., and Price, R.A., 1995, The southern termination of the Misty thrust sheet, southern Canadian Rocky Mountains: Tectonic heredity and tectonic wedging along an oblique hanging wall ramp: Geological Society of America Bulletin, v. 107, p. 1304–1316, doi: 10.1130/0016-7606(1995)107<1304:THATWA>2.3.CO;2.

Price, R.A. (chairman and editor), Archibald, J.F., Cullimore, D.R., Duquette, D.J., Frind, E.O., Kanasewich, E.R., Kerrich, R., Lind, N.C., Lo, K.Y., Neuman, S.P., Roots, E.F., Seshadri, R., Swanson, S.M., Thérien, N., and Wiles, D.R., 1995, An Evaluation of the Environmental Impact Statement on Atomic Energy of Canada Limited's Concept for the Disposal of Canada's Nuclear Fuel Waste, Report of the Scientific Review Group Advisory to the Nuclear Fuel Waste Management and Disposal Concept Environmental Assessment Panel, Canadian Environmental Assessment Agency, Ottawa, October 6, 1995, 278 p.

Leith, J.A., Price, R.A., and Spencer, J.H., eds., 1995, Planet Earth: Problems and Prospects: Montreal and Kingston, McGill-Queen's University Press, 196 p.

Höy, T., Price, R.A., Legun, A., Grant, B., and Brown, D., 1996, Purcell Supergroup, Southeastern British Columbia, Geological Compilation Map, NTS 82G; 82F/E; 82J/SW; 82K/SE: British Columbia Ministry of Energy Mines and Petroleum Resources, Geological Survey Branch Geoscience Map 1995-1, scale 1:250,000.

Colpron, M., Price, R.A., Archibald, D.A., and Carmichael, D.M., 1996, Middle Jurassic denudation and uplift along the western flank of the Selkirk fan structure: Thermobarometric and thermochronometric constraints from the Illecillewaet synclinorium, southeastern British Columbia: Geological Society of America Bulletin, v. 108, no. 11, p. 1372–1392, doi: 10.1130/0016-7606(1996)108<1372:MJEATW>2.3.CO;2.

Doughty, P.T., Price, R.A., and Parrish, R.R., 1998, Geology and U-Pb geochronology of Archean basement and Proterozoic cover in the Priest River complex, northwestern United States and its implications for Cordilleran structure and Precambrian continent reconstructions: Canadian Journal of Earth Sciences, v. 35, no. 1, p. 39–54, doi: 10.1139/cjes-35-1-39.

Colpron, M., Warren, M., and Price, R.A., 1998, The Selkirk fan structure of the southeastern Canadian Cordillera: Tectonic thickening in response to inherited basement structure: Geological Society of America Bulletin, v. 110, no. 8, p. 1060–1074, doi: 10.1130/0016-7606(1998)110<1060:SFSSCC>2.3.CO;2.

Doughty, P.T., and Price, R.A., 1999, Tectonic evolution of the Priest River complex, northern Idaho and Washington: A reappraisal of the Newport fault with new insights on metamorphic core complex formation: Tectonics, v. 18, no. 3, p. 375–393, doi: 10.1029/1998TC900029.

Colpron, M., Price, R.A., and Archibald, D.A., 1999, ^{40}Ar/^{39}Ar thermochronometric constraints on the tectonic evolution of the southern Clachnacudainn complex, southeastern British Columbia: Canadian Journal of Earth Sciences, v. 36, p. 1989–2006, doi: 10.1139/cjes-36-12-1989.

Sears, J.W., and Price, R.A., 2000, New look at the Siberian connection: No SWEAT: Geology, v. 28, no. 5, p. 423–426, doi: 10.1130/0091-7613(2000)28<423:NLATSC>2.0.CO;2.

Price, R.A., and Monger, J.W.H., 2000, A Transect of the Southern Canadian Cordillera from Calgary to Vancouver: Vancouver, Geological Association of Canada, Cordilleran Section, 164 p.

Monger, J.W.H., and Price, R.A., 2000, A Transect of the Southern Canadian Cordillera from Vancouver to Calgary: Geological Survey of Canada Open-File 3902, 168 p.

Price, R.A., 2000, The southern Canadian Rockies: Evolution of a foreland thrust and fold belt, in GeoCanada 2000: Field Trip Guidebook No. 13: Calgary, GeoCanada2000, 246 p.

Doughty, P.T., and Price, R.A., 2000, Geology of the Purcell Trench rift valley and Sandpoint Conglomerate: Eocene en echelon normal faulting and synrift sedimentation along the eastern flank of the Priest River metamorphic complex, northern Idaho: Geological Society of America Bulletin, v. 112, no. 9, p. 1356–1374, doi: 10.1130/0016-7606(2000)112<1356:GOTPTR>2.0.CO;2.

Price, R.A., 2001, Transverse faults, displacement transfer, and co-evolution of several major thrust faults in the Banff area, Rocky Mountain Front Ranges, Alberta: Journal of Structural Geology, v. 23, p. 1079–1088, doi: 10.1016/S0191-8141(00)00177-2.

Price, R.A., and Sears, J.W., 2001, A preliminary palinspastic map of the Mesoproterozoic Belt/Purcell Supergroup, Canada and U.S.A.: Implications for the tectonic setting and structural evolution of the Purcell anticlinorium and the Sullivan deposit, in Lydon, J.W., Höy, T., Slack, J.F., and Knapp, M., eds., The Geological Environment of the Sullivan Deposit, British Columbia: Geological Association of Canada, Mineral Deposits Division (MDD) Special Volume No. 1, p. 43–64.

Price, R.A., Osadetz, K.G., Kohn, B.P., and Feinstein, S., 2001, Deep refrigeration of an evolving thrust and fold belt by enhanced penetration of meteoric water: The Lewis thrust sheet, southern Canadian Rocky Mountains [abs.]: Geological Society of America, Abstracts with Programs, v. 33, no. 7 (2001 GSA Annual Meeting), p. A-52.

Price, R.A., and Monger, J.W.H., 2003, A Transect of the Southern Canadian Cordillera from Calgary to Vancouver, May 2003 (revised edition): Vancouver, Geological Association of Canada, Cordilleran Section, 165 p.

Sears, J.W., and Price, R.A., 2003, Tightening the Siberian connection to Laurentia: Geological Society of America Bulletin, v. 115, no. 8, p. 943–953, doi: 10.1130/B25229.1.

Sears, J.W., Price, R.A., and Khudoley, A.K., 2004, Linking the Mesoproterozoic Belt-Purcell and Udzha basins across the west Laurentia–Siberia connection: Precambrian Geology, v. 129, p. 291–308, doi: 10.1016/j.precamres.2003.10.005.

Osadetz, K.G., Kohn, B.P., Feinstein, S., and Price, R.A., 2004, Foreland belt thermal history using apatite fission-track thermochronology: Implications for Lewis thrust and Flathead fault in the southern Canadian Cordilleran petroleum province, in Swennen, R., Roure, F., and Granath, J.W., eds., Deformation, Fluid Flow, and Reservoir Appraisal in Foreland Fold and Thrust Belts: Tulsa, American Association of Petroleum Geologists, AAPG Hedberg Series, no. 1, p. 21–48.

Cooley, M.A., Price, R.A., Dixon, J., and Kyser, T.K., 2004, Fault-propagation folding in the Livingstone Range, southern Alberta foothills: Faulting, folding and fluids in a hanging-wall ramp anticline [abs.]: Geological Society of America Abstracts with Programs, v. 36, no. 5, p. 209.

Larson, K.P., and Price, R.A., 2006, The southern termination of the Western Main Ranges of the Canadian Rockies, near Fort Steele, British Columbia: Stratigraphy, structure, and tectonic implications: Bulletin of Canadian Petroleum Geology, v. 54, no. 1, p. 37–61, doi: 10.2113/54.1.37.

Larson, K.P., Price, R.A., and Archibald, D.A., 2006, Tectonic implications of ^{40}Ar/^{39}Ar muscovite dates from the Mt. Haley stock and Lussier River stock, near Fort Steele, British Columbia: Canadian Journal of Earth Sciences, v. 43, no. 11, p. 1673–1684, doi: 10.1139/E06-048

Driving mechanism and 3-D circulation of plate tectonics

Warren B. Hamilton*
Department of Geophysics, Colorado School of Mines, Golden, Colorado 80401, USA

ABSTRACT

A conceptual shift is overdue in geodynamics. Popular models that present plate tectonics as being driven by bottom-heated whole-mantle convection, with or without plumes, are based on obsolete assumptions, are contradicted by much evidence, and fail to account for observed plate interactions. Subduction-hinge rollback is the key to viable mechanisms. The Pacific spreads rapidly yet shrinks by rollback, whereas the subduction-free Atlantic widens by slow mid-ocean spreading. These and other first-order features of global tectonics cannot be explained by conventional models. The behavior of arcs and the common presence of forearc basins on the uncrumpled thin leading edges of advancing arcs and continents are among features indicating that subduction provides the primary drive for both upper and lower plates. Subduction rights the density inversion that is produced when asthenosphere is cooled to oceanic lithosphere: plate tectonics is driven by top-down cooling but is enabled by heat. Slabs sink more steeply than they dip and, if old and dense, are plated down on the 660 km discontinuity. Broadside-sinking slabs push all sublithosphere oceanic upper mantle inward, forcing rapid spreading in shrinking oceans. Down-plated slabs are overpassed by advancing arcs and plates, and thus transferred to enlarging oceans and backarc basins. Plate motions make sense in terms of this subduction drive in a global framework in which the ridge-bounded Antarctic plate is fixed: most subduction hinges roll back in that frame, plates move toward subduction zones, and ridges migrate to tap fresh asthenosphere. This self-organizing kinematic system is driven from the top. Slabs probably do not subduct into, nor do plumes rise to the upper mantle from, the sluggish deep mantle.

Keywords: forearc basins, geodynamics, global tectonics, plate tectonics, rollback, seafloor spreading, subduction.

INTRODUCTION

The reality of plate tectonics, as a description of relative motions of those parts of Earth's outer shell that are internally semirigid, has been proven. Analogy with the systematic relationships of many geologic and magmatic features to modern plate settings demonstrates that plate tectonics has operated in something like its present style for, at most, the past billion years (Stern, 2005, 2007). A key indicator of low-temperature conditions in subduction systems, lawsonite eclogite, has formed only within the past half-billion years (Tsujimori et al., 2006). In still-older assemblages, relationships are progressively less like those now developing, and, in my view (Hamilton, 2003, 2007), neither rigid plates nor subduction—i.e., nothing resembling even a pre-

*whamilto@mines.edu

cursor of plate tectonics—operated before about two billion years ago. Secular cooling of the upper mantle by 75 °C or 100 °C per b.y. is likely (Anderson, 2007; Hamilton, 2007).

Plates commonly are visualized as being driven by whole-mantle convection in a mostly unfractionated mantle: bottom-heated hot mantle rises beneath ridges, diverges, carries lithosphere passively along as it flows laterally while cooling, and sinks at trenches (e.g., Fig. 1). These concepts derive from hypothetical models, not from data, and appear in many variants. Most geodynamic numerical modeling, and all fish-tank modeling, incorporates properties and parameters that enable favored results but that otherwise are incompatible with much information, including the great effects of pressure on physical properties. Observed plate interactions are also among the factors that are largely missing from these popular models. Anderson (2002a, 2002b, 2007), Hamilton (2002, 2003), Hofmeister (2005), and Hofmeister and Criss (2005a) are among the minority of geoscientists arguing instead for an irreversibly fractionated mantle, a top-down drive, and plate circulation limited to the upper mantle.

This paper emphasizes the actual geologic and geophysical products of plate interactions. These features are incompatible with bottom-up mechanisms, whereby plates are driven by bottom-heated convection cells with or without plumes, and instead they indicate that plates are self-organized and are driven by subduction due to top-down cooling. Subduction represents the righting of the density inversion caused by cooling from the top of suboceanic asthenosphere. Plate tectonics is enabled by Earth's internal heat, but convection associated with plate motions is a product, not a cause, of plate motions. Anderson (e.g., 2002a, 2007) has long argued, on geophysical, mineral-physics, and thermodynamic grounds, for a top-down drive, a few other geoscientists now do so also (e.g., Schellart and Lister, 2005), and some geodynamicists add a top-down complication to their dominantly bottom-up explanations. Anderson and I agree that plate-tectonic circulation is largely or entirely limited to the upper mantle—for him, shallower than ~1000 km; for me, shallower than the discontinuity near 660 km. This essay continues the analysis developed in prior papers (e.g., Hamilton, 1979, 1988, 1995, 2002, 2003).

The conventional assumption that plate-tectonic circulation involves the entire mantle and is driven by bottom heating has been repeated and embellished in hundreds of textbooks at all levels and in thousands of scholarly papers, and it has been widely taught as dogma for 30 yr. Contrarians face great inertia in ingrained biases when arguing that none of the conjectural variants of this conventional wisdom is soundly based. I recall my pre–plate-tectonic years as a continental drifter, when most of the American geoscience community was impervious to the powerful evidence for drift because several fine mathematicians had proved (starting from false assumptions) that crustal mobility was impossible.

GLOBAL SPREADING PATTERN

Bottom-up drives are incompatible with the characteristics of plate boundaries and the ages of oceanic crust (Fig. 2)—the broadest features of plate tectonics. The Atlantic spreads slowly and has no subduction about its margins, except for the small Caribbean and Scotia arcs, and the subduction-free Arctic Ocean spreads slower yet. The Pacific spreads rapidly, is mostly rimmed by subduction systems except in the south, and yet is getting smaller by a large fraction of the amount by

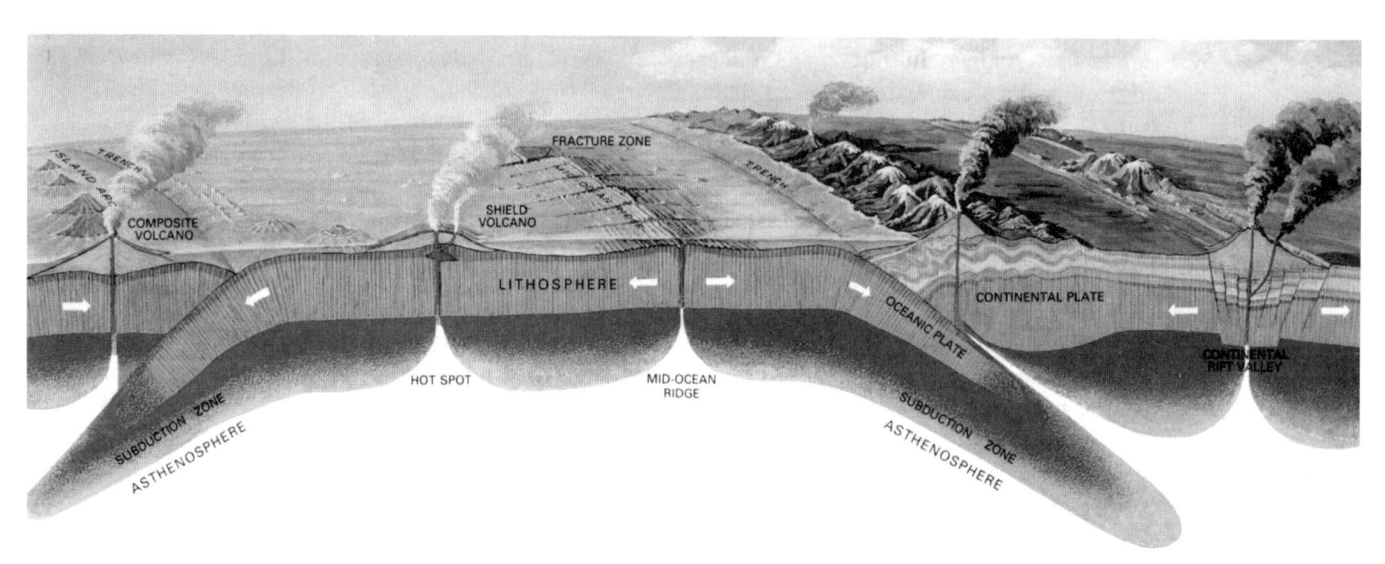

Figure 1. Schematic diagram (from Simkin et al., 1989) that incorporates many popular misconceptions about plate tectonics, such as oceanic lithosphere that spreads away from a fixed ridge, tips sharply downward at a fixed trench, and slides down a slot in the mantle, where an overriding plate is crumpled against the hinge; and the ridge and trench are parallel parts of a cylindrical system. These features do not occur in Earth. (The current version of this figure, in Tilling et al. [2006], retains similar never-happens geometry but does not show the crumpling required by that hypothetical geometry; some accompanying materials are now realistic.)

which the Atlantic gets larger because bounding subduction hinges roll back oceanward. The Indian Ocean spreads slowly in the west, where there is little peripheral subduction, but rapidly in the east, where subduction systems bound it. Antarctica is ringed by spreading ridges that lengthen and change shape as new increments of oceanic lithosphere are added: the ridges migrate relatively away from the continent with varying velocities and initiation ages. The African plate is rimmed on west, south, and east by spreading ridges that similarly lengthen and change shape as they migrate relatively away from the continent as required to stay midway between diverging continents. Ridges enter trenches in a number of places along the east side of the Pacific and abruptly cease spreading as plate boundaries change to accommodate motions between plates separated previously by subducting oceanic plates. For example, the Gulf of California–San Andreas boundary has developed progressively between migrating plate intersections. In the western Pacific, by contrast, some arcs migrate away from continents, opening backarc basins in their wakes, even though the total area of the Pacific plus backarc basins is shrinking.

Such behavior cannot be explained by conventional models (e.g., Fig. 1), and it shows instead that upwelling beneath ridges is a consequence, not a cause, of plate motions. Ridges are not fixed, and cannot mimic shapes of deep-mantle upwellings. (Plume conjecture is anchored to the assumption that much of the Mid-Atlantic Ridge is fixed in whole-Earth space.) Ridges continue to spread primarily at the places where young lithosphere has almost no strength, between plates diverging as a consequence of subduction, and halt when the subduction drive halts. Ridge-ridge transform faults minimize heat loss consequent on spreading. The lack of compressive deformation in young, thin oceanic lithosphere precludes the shortening that would occur were there a ridge-push force. Ridges form where oceanic plates slide apart, and subduction provides the drive. Ridges are in the middles only of nonsubducting oceans, and "mid-ocean ridge" is a misnomer in subduction-bounded oceans.

HINGE ROLLBACK—KEY TO SUBDUCTION

Trenches mark dihedral angles between the tops of subducting oceanic plates and of thin accretionary wedges riding in front of overriding plates (Fig. 3). Seismicity and seismic reflection and refraction show that subducting lithosphere dips gently beneath forearcs and steepens gradually, through broadly

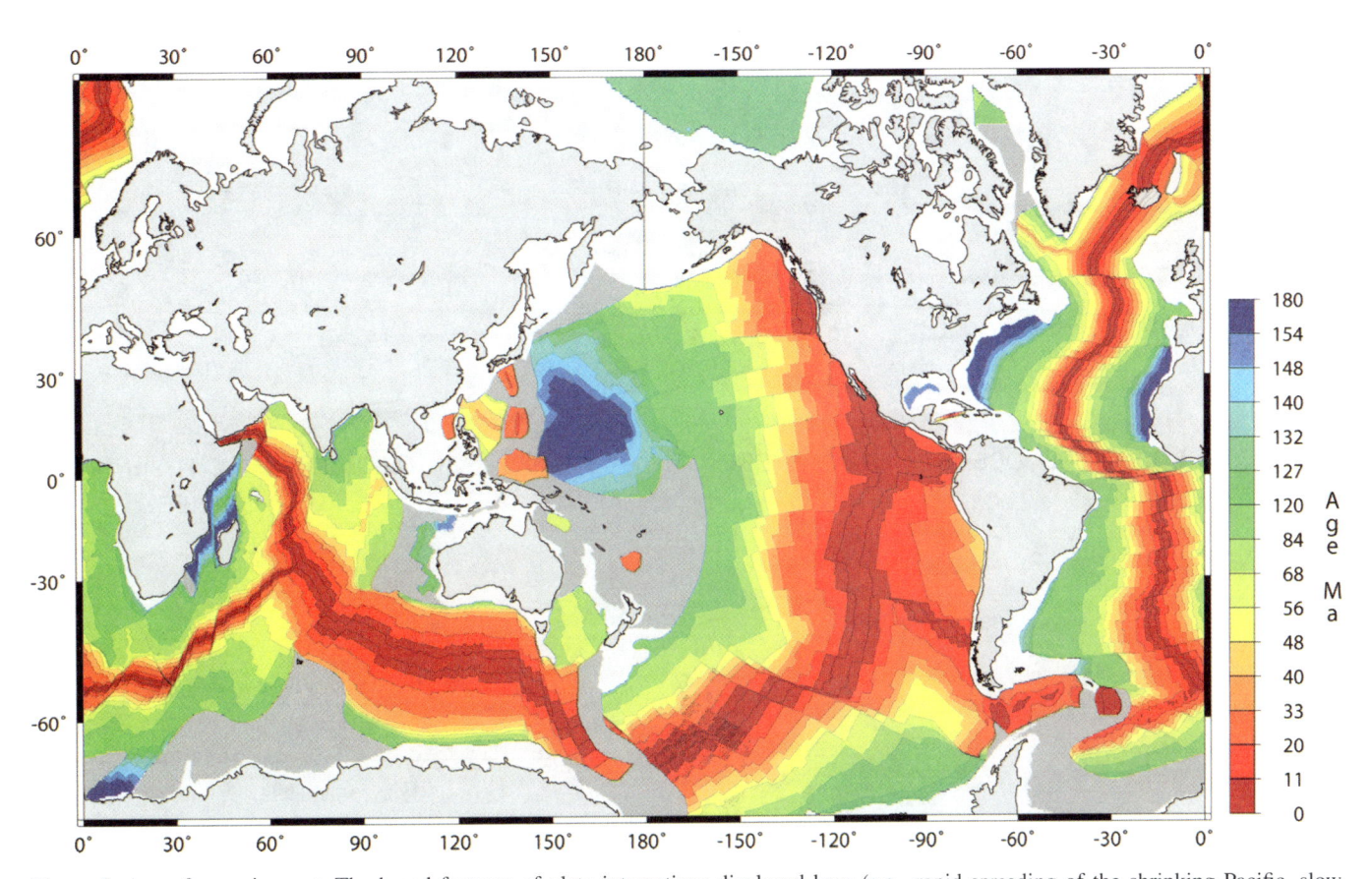

Figure 2. Age of oceanic crust. The broad features of plate interactions displayed here (e.g., rapid spreading of the shrinking Pacific, slow spreading of the enlarging Atlantic, intersections of ridges and trenches) cannot be explained by current models of geodynamics as driven by bottom-heated convection. Isochrons correspond to identified magnetic anomalies. This map was prepared by David Sandwell from data of Müller et al. (1997).

curved hinges, beneath arcs and back arcs. Magmatic arcs typically form ~100 km above subducting slabs.

Subduction is the falling away of oceanic lithosphere, for hinges roll back into subducting plates. Subduction hinges are fundamental structures, whereas trenches are very gentle-sided surficial features, so I use the terms "hinge rollback" and "hinge retreat" in preference to "trench retreat" (e.g., Rizzetto et al., 2004) and "trench migration" (e.g., Faccenna et al., 2001); and "slab rollback" (e.g., Lucente et al., 2006) for me conveys an ambiguous geometry. The shrinking of the Pacific requires hinge retreat on a very large scale. Retreat at smaller scales is shown by many specific interactions: collisions of facing arcs (e.g., Halmahera-Sangihe: Hamilton, 1979), collisions of arcs with passive-margin continents (e.g., Banda with Australia and New Guinea: Hamilton, 1979; Maghrebides with Africa: Lucente et al., 2006; Carpathian with eastern Europe; Manila with Taiwan), advance of small arcs into oceanic plates (e.g., the several Mediterranean arcs: Faccenna et al., 2001; Lucente et al., 2006; Rizzetto et al., 2004; Caribbean; Scotia), migration of oppositely facing arcs apart from each other (e.g., New Hebrides and Tonga), arc reversals (e.g., Sangihe-Mindanao-Sulawesi, and part of Banda: Hamilton, 1979), and arc rotations (e.g., Fiji and New Hebrides: Taylor et al., 2000). The subsurface behavior of sinking lithosphere requires the same conclusion; for example, the sunken slab of the tight-horseshoe Banda arc defines an inward-shallowing basin continuous from south to north rims (Widiyantoro and van der Hilst, 1997).

Hinge retreat requires that subducting slabs sink more steeply than they dip, and that their seismic zones mark transient positions of slabs, not trajectories. Slabs are not injected down inclined slots, although this is the way they commonly are visualized and modeled, but instead sink broadside.

Forearcs

Magmatic arcs commonly are depicted incorrectly (e.g., Fig. 1) as rising directly and steeply from trenches, whereas, in fact, broad systematic structural belts intervene and have characteristics that constrain the process of subduction and hence the driving mechanism of plate tectonics.

The Sumatra-Java-Banda arc of Indonesia (Fig. 4) has a well-defined and continuous forearc system. The trench, the outermost part of the system, is continuous for 4500 km within the area of the figure, and it extends northward in the west another 2500 km, the north half of which is onshore in Bengal and Burma. The trench has typical deep-sea character along Sumatra and south-central Indonesia, but it shallows in the east where the arc has ramped onto continental crust of Australia and New Guinea.

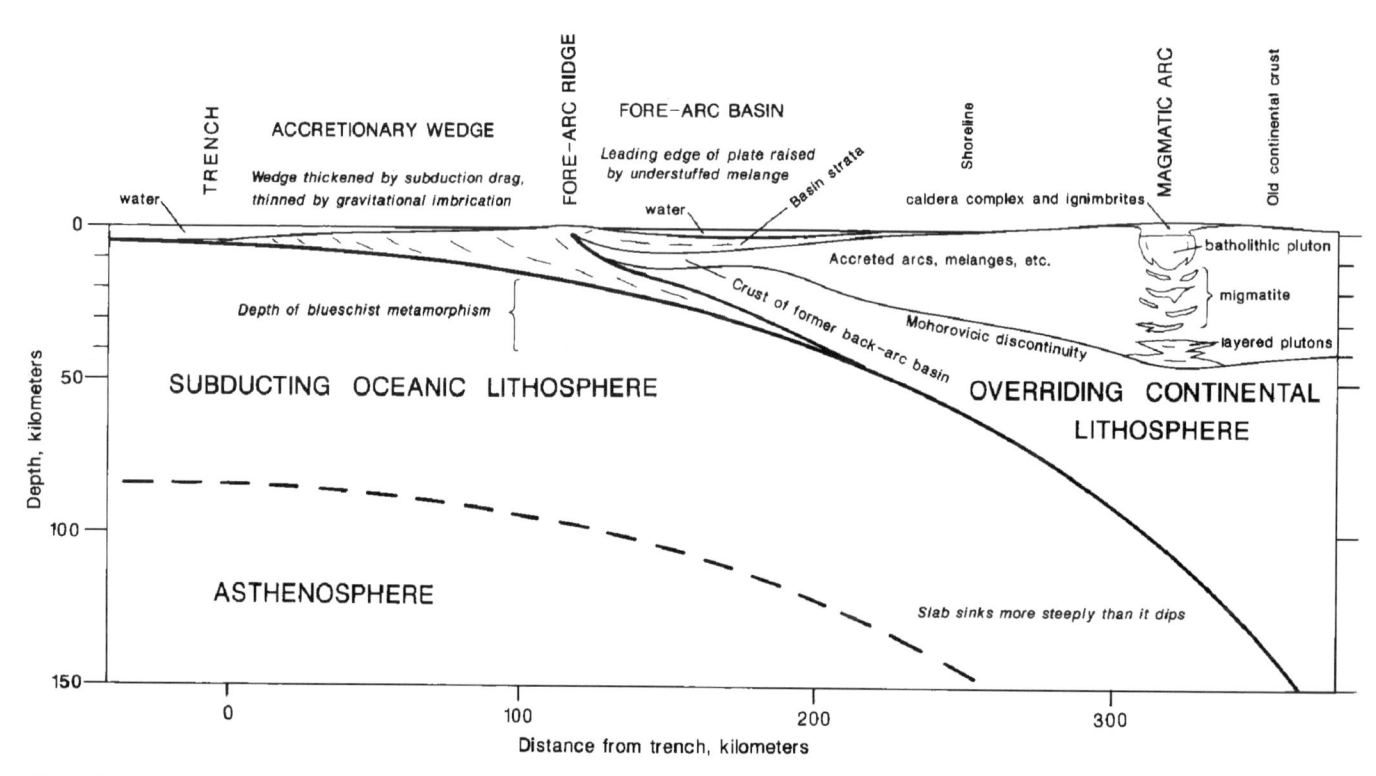

Figure 3. Cross section of convergent margin between continental and oceanic plates. The trench is the broad dihedral angle between the top of the gently inclined subducting oceanic plate and the top of the accretionary wedge of, mostly, scraped-off trench sediments. The presence of a forearc basin indicates that the thin leading edge of the overriding plate was not crumpled during the period recorded by basin sedimentation. This section is scaled to fit modern Sumatra and Cretaceous California, which are dimensionally similar except that the former is currently active and the latter is variably eroded to expose deeper features. After Hamilton (1995); there is no vertical exaggeration.

Within the area of Figure 4, a continuous forearc ridge (mostly submarine) and, behind it, a continuous forearc basin separate trench and magmatic arc, which typically are 300 km apart. The forearc ridge marks the crest of the accretionary wedge: the partly emergent Mentawi ridge off Sumatra in the northwest, the submerged ridge along central Indonesia, and Timor and other islands around the Banda Arc in the east. Many reflection profiles across this arc system were presented by Hamilton (1979, 1995). One of these lines, which has great vertical exaggeration but extends from the outer swell seaward of the trench across accretionary wedge, forearc ridge, and forearc basin, is shown here as Figure 5. (See Figures 3, 6, and 9 for unexaggerated views of such features.) Beaudry and Moore (1985), Kopp et al. (2001, 2002), and Matson and Moore (1992), among others, have published reflection profiles, some of which are accompanied by seismic-refraction and gravity surveys, across this system.

The major shallow earthquakes of subduction systems are caused by ruptures that nucleate at the tops of subducting slabs back under the fronts of overriding plates, follow the tops of the slabs forward, and break up through the accretionary wedges as thrust faults (Wells et al., 2003). The rupture that caused the great Sumatra-Andaman earthquake of December 2004, moment magnitude 9.2, was of this type, and it broke along 1500 km of the length and 150 km of the width of the top of the gently inclined subducting Indian Ocean plate (Chlieh et al., 2007; Lay et al., 2005). The rupture initiated under the Sumatra forearc basin and broke up further forward, through the accretionary wedge, as thrust faults. The large coseismic slip on the December break, and on the rupture of the moment-magnitude 8.7 earthquake that continued the break 300 km southward three months later, was directed toward the trench, oblique to the plate-convergence direction. Lay et al. (2005) ascribed this obliquity to a strike-slip torque, but because such obliquity characterizes coseismic deformation of accretionary wedges where trenches are not perpendicular to convergence directions (e.g., Estabrook and Jacob, 1991; McCaffrey, 1991; Subarya et al., 2006), I take it to indicate instead that strain was released by coseismic slip toward the trench—the gravitationally driven direction—after having accumulated by thickening of the wedge in the plate-convergence direction.

Accretionary Wedges

Thinly tapered accretionary wedges display extreme imbrication (Fig. 6) and internal disruption of trench and pelagic sediments and other surficial, island, and crustal components scraped from subducting plates. Wedges are snowplowed in front of overriding plates and typically have upper surfaces with slopes that steepen toward trenches (Fig. 5), presumably indicative of dynamic equilibrium between thickening in plate-convergence directions and thinning in gravitational, trenchward directions. "Uphill" thrusting within the wedge (as within foreland thrust belts) is driven by the lithostatic head of increasing surface altitude arcward. Not apparent in Figure 6 is the abundant broken formation and polymict melange that typify wedges where drilled at sea or exposed on land. See Hamilton (1978, 1979) for

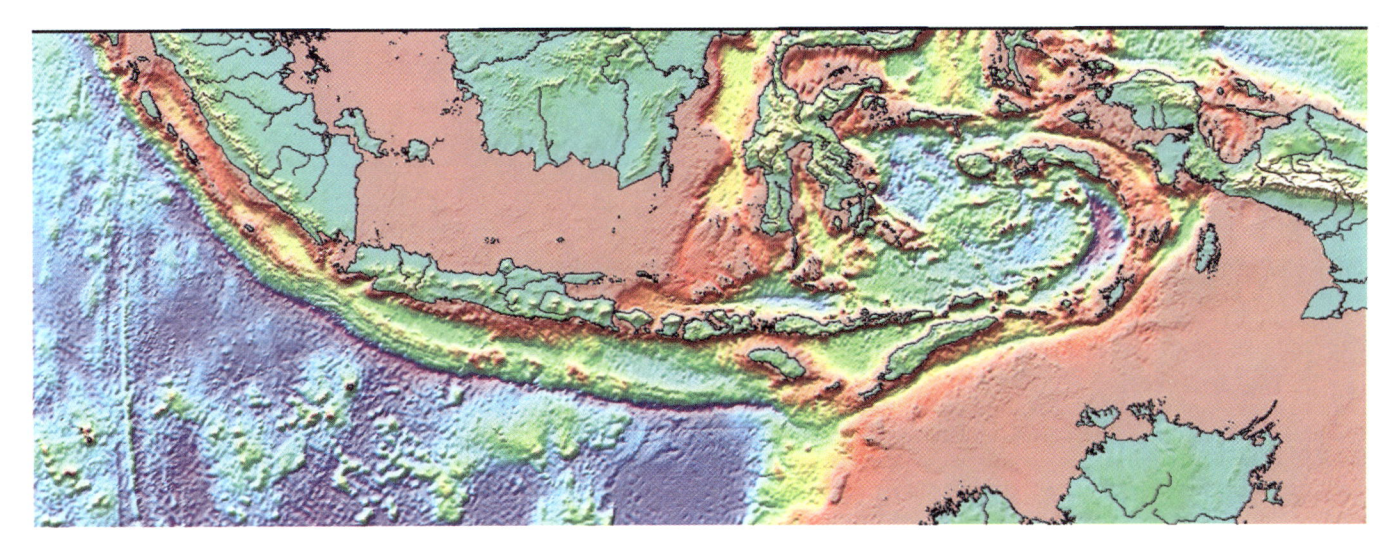

Figure 4. Physiographic map of the active Sumatra-Java-Banda arc system. The trench is the outermost conspicuous arcuate feature, and the magmatic arc (highlands on Sumatra and Java, and islands around the Banda arc) is the innermost. The magmatic arc of Sumatra and Java is superimposed on a geanticline produced by magmatic underplating and injection of the crust. A continuous forearc ridge and a forearc basin lie concentrically between the trench and magmatic arc. The slope from ridge to trench is the surface of the accretionary wedge, which is in front of the overriding plate; the basin is on the thin leading edge of the overriding plate. The arc expanded eastward from Java during Neogene time to form the eastern U-shaped Banda part of the system, the east end was inaugurated long after the west part, and the trench is above continental crust in the east, on the continental shelf of Australia and New Guinea. See Hamilton (1979, 1988) for tectonic synthesis. Area extends from 95°E to 135°E, and from 0° to 15°S. Short-wavelength bathymetry is derived from gravity determined from satellite altimetry, long-wavelength bathymetry is from ship-sounding tracks, and land topography is from EROS Data Center (Earth Resources Observation and Science, U.S. Geological Survey). Map was compiled and provided by David Sandwell.

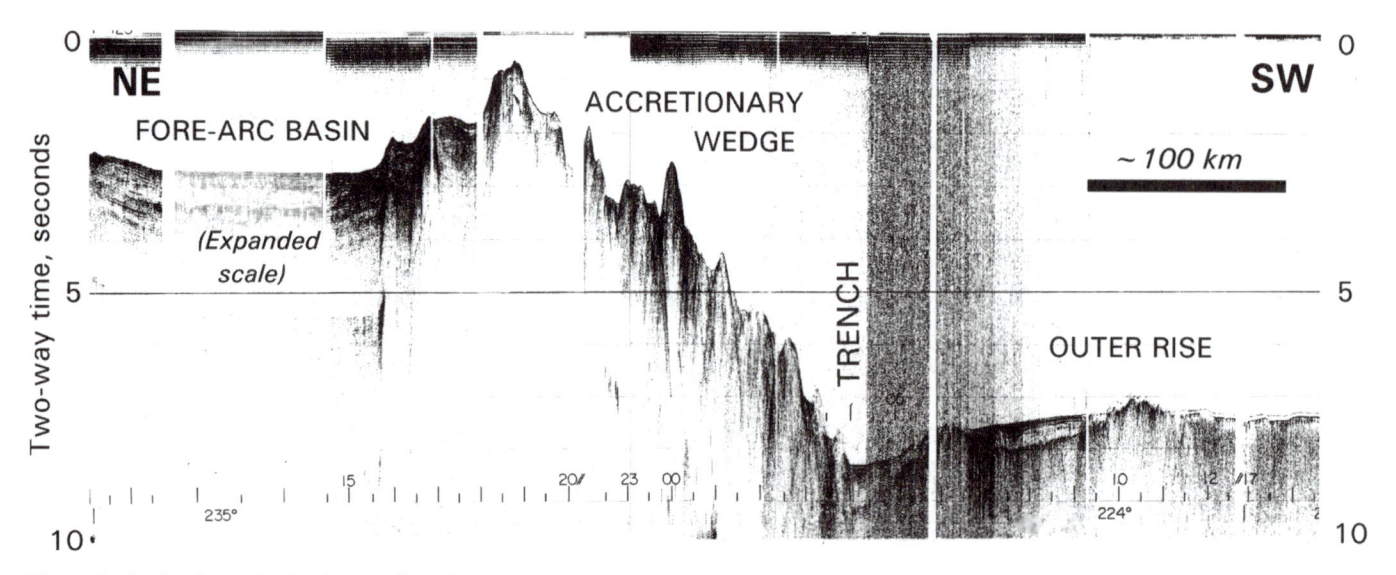

Figure 5. Single-channel reflection profile, with great vertical exaggeration, across the Sumatra forearc and trench. Northeast (left) end of line is 50 km from the coast of south Sumatra at 5°S. The trench is 6 km deep and has very thin turbidites (layered) above oceanic pelagic sediments (transparent). Oceanic crust continues with gentle dip beneath the accretionary wedge but is not resolved on this image; compare with Figure 6. Outer rise is due to waveform depression of elastic crust by weight of the accretionary wedge. Vertical exaggeration is 25 or 30:1, and the actual slope of the top of the accretionary wedge is a few degrees. Profile was provided by Maurice Ewing, Lamont-Doherty Earth Observatory of Columbia University.

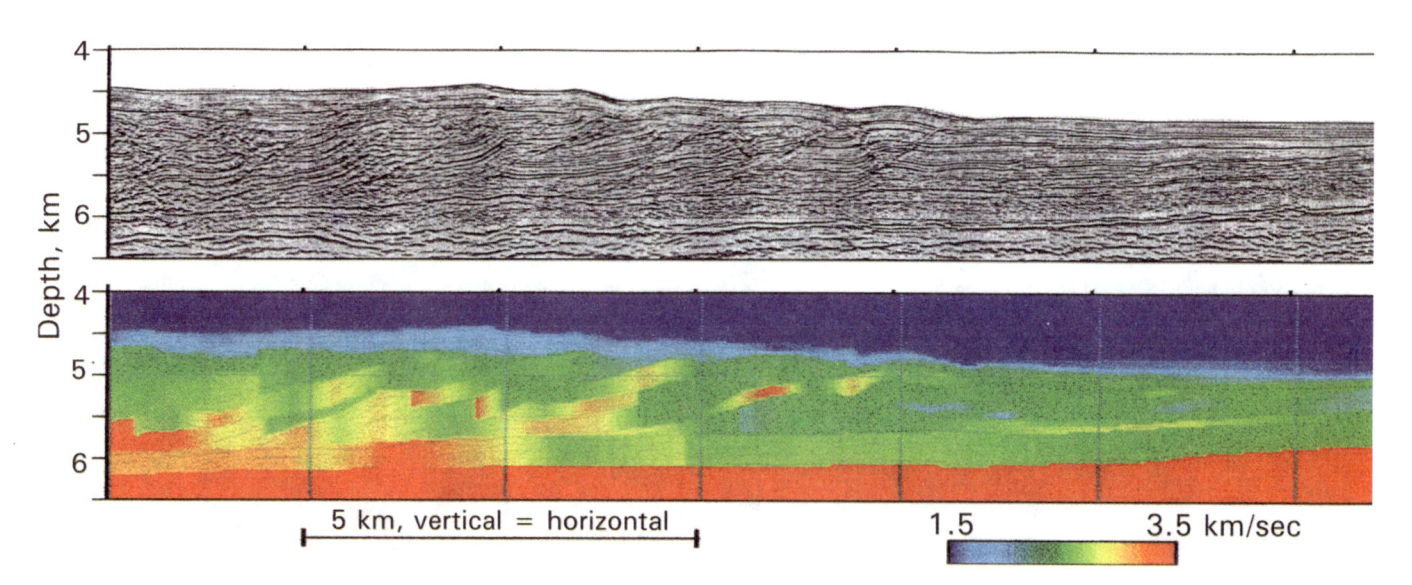

Figure 6. Depth-converted seismic-reflection and velocity profiles across Nankai Trench and toe of accretionary wedge, offshore Shikoku, Japan. Top of oceanic crust (at ~6.3 km depth on left, 5.8 km on right) dips gently beneath the wedge of sheared-off sediments that presses it down. Imbricated sheets of higher-velocity basal-wedge materials illustrate thrusting as wedge is thickened by plate drag but thinned by gravitational outflow. Profiles were provided by Gregory F. Moore; processing and velocity mapping are described by Costa Pisani et al. (2005).

representative field photographs. Many thousands of kilometers of subduction may be recorded by shearing within a wedge.

The shearing and imbrication of the cool, wet, forward parts of accretionary wedges are mostly aseismic. The many thrust faults that deform them typically have trench-parallel strikes (Fig. 7), which confirms that slip within wedges is primarily in the gravitational direction, not the plate-convergence direction. Incoming oceanic plates are flexed as they ride through the standing waves of outer rises and outer trench slopes and are broken by many small down-to-the-trench, trench-parallel, normal faults (Fig. 7; Ranero et al., 2003).

Many accretionary wedges contain blocks or imbricated sheets of exotic high-pressure, low-temperature metamorphic rocks of blueschist and eclogite facies that require depression to depths of 20, 50, or more kilometers, followed by return back up the subduction channel to the surface. Metamorphic ages

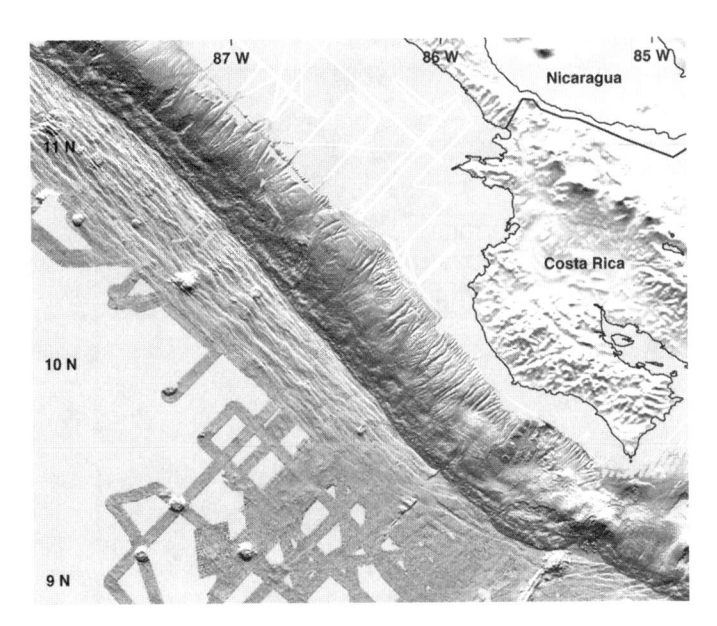

Figure 7. Physiographic map of part of the Middle American Trench. Subaerial digital topography and submarine sidescan sonar are shown in exaggerated shaded relief, illuminated from the north; actual slope of accretionary wedge is very gentle (cf. Fig. 6). The toe of the wedge is imbricated by trench-parallel thrust faults, and the subducting oceanic plate is broken by trench-parallel normal faults as it rolls through the gentle outer rise and into the trench. Area is 330 km wide. Discontinuous forearc basins underlie the continental shelf. This figure was provided by César Ranero (cf. Ranero et al., 2003, their Fig. 1).

of such rocks cluster in narrow ranges within a sector, so the upward return flow is episodic, not continuous. This sporadic return flow may be enabled by relatively rapid periods of hinge retreat and correlative falling away of the subducting slab in front of the advancing plate.

Forearc Basins

Most active arc systems display continuous (Fig. 4) or discontinuous sedimentary basins atop the thin fronts of their overriding plates. Among summaries of basin characteristics are those by Dickinson and Seely (1979) and Dickinson (1995). The basins may be bathymetric as well as sedimentary basins (Figs. 4, 5, 8, 11), or they may be filled to (Figs. 7 and 9), or above, sea level. They may resemble continental shelves or coastal plains in topography, but their basinal character is seen in reflection profiles or outcrop. The characteristics of these basins have profound, but commonly overlooked, implications for the mechanism of subduction. The presence of such a basin demonstrates that the front of the plate has not been crumpled (cf. Fig. 1) during the period recorded by deposition of basin strata. In the case of the Cretaceous and Tertiary forearc basin of central California, this duration was ~100 m.y. (Constenius et al., 2000; Dickinson and Seely, 1979; Hamilton, 1978).

A sampling of other forearc basins, and their equivalent "deep-sea terrace" basins in some arcs around the Pacific, includes many western Melanesian, Indonesian, and southern Philippine examples (Hamilton, 1979), and a basin >700 km long, paired to the Manila Trench of the west Philippines (Hayes and Lewis, 1984; Lewis and Hayes, 1984). That Philippine basin continues north to Taiwan, where it disappears in the collision of the arc with Asia. A similar basin continues north from Taiwan, in the oppositely facing Ryukyu arc (Shyu et al., 2005). The segmented basins of southwest Japan were depicted by Wells et al. (2003; Fig. 8 herein). Suyehiro and Nishizawa (1994) illustrated the forearc basin of Honshu. For forearc basins of the Kuril Islands, Kamchatka, and Aleutian Islands, see Wells et al. (2003); for southwest Alaska, see Dickinson (1995) and Von Huene et al. (1979); for Washington and Oregon, see Wells et al. (2003; Fig. 9 herein); and for Central America, see Hinz et al. (1996), Shor (1974), and Wells et al. (2003). The beaded forearc basins of South America are illustrated in Collot et al. (2002) for Colombia and Ecuador; Krabbenhöft et al. (2004) for Peru; Moberly et al. (1982) for Peru, Ecuador, and Chile; and Mordojovich-K. (1974) for Chile. The reflection profiles of the latter paper were calibrated by drilling: a broad basin underlying the continental shelf has a clastic section thicker than 3 km that extends from Late Cretaceous through Holocene.

The two small active Atlantic arc systems display forearc basins. The forearc basin of the Scotia (South Sandwich) arc is shown by Vanneste et al. (2002). The Tobago Basin is the forearc basin of the southwest part of the Caribbean arc.

The marked profile of Figure 10 extends from the Peru-Chile Trench to a forearc basin of northern Peru, which is known from drilling to contain strata at least as old as Paleocene. Figure 11 is a stratigraphic analysis of the Arica forearc basin of northernmost Chile, which shows landward displacement of depocenters with time. Both illustrations are from Moberly et al. (1982). Similar depocenter migration in other South American forearc basins has been demonstrated by Coulbourn and Moberly (1977). Such depocenter progressions are recorded by reflection profiles across a number of other forearc basins and by stratigraphic studies of onshore forearc basins including that of the Cretaceous and Paleogene in California. Depocenter migration is conspicuous, for example, in the thick basin section south of western Java (Kopp et al., 2002), and in the northwest Luzon basin, for which Lewis and Hayes (1984) presented a stratigraphic analysis similar to that of Figure 11.

As Coulbourn and Moberly (1977), Hamilton (1978, 1979), and Moberly et al. (1982) emphasized, these landward-grading depocenters show that forearc basins typically evolve in response to progressive jacking up of leading edges of overriding plates by accretionary-wedge material stuffed beneath them. Hamilton (1979) explained the great water depths of the forearc basin around the tightly curved Banda arc (east part of Fig. 4) as an elastic response to the much greater frontal uplifts where the leading edge of the overriding plate ramped up onto continental crust of Australia and New Guinea.

Forearc basin strata frequently are seen in reflection profiles to lap onto basement rising to the forearc ridge, but contacts, both top and bottom, of that basement with

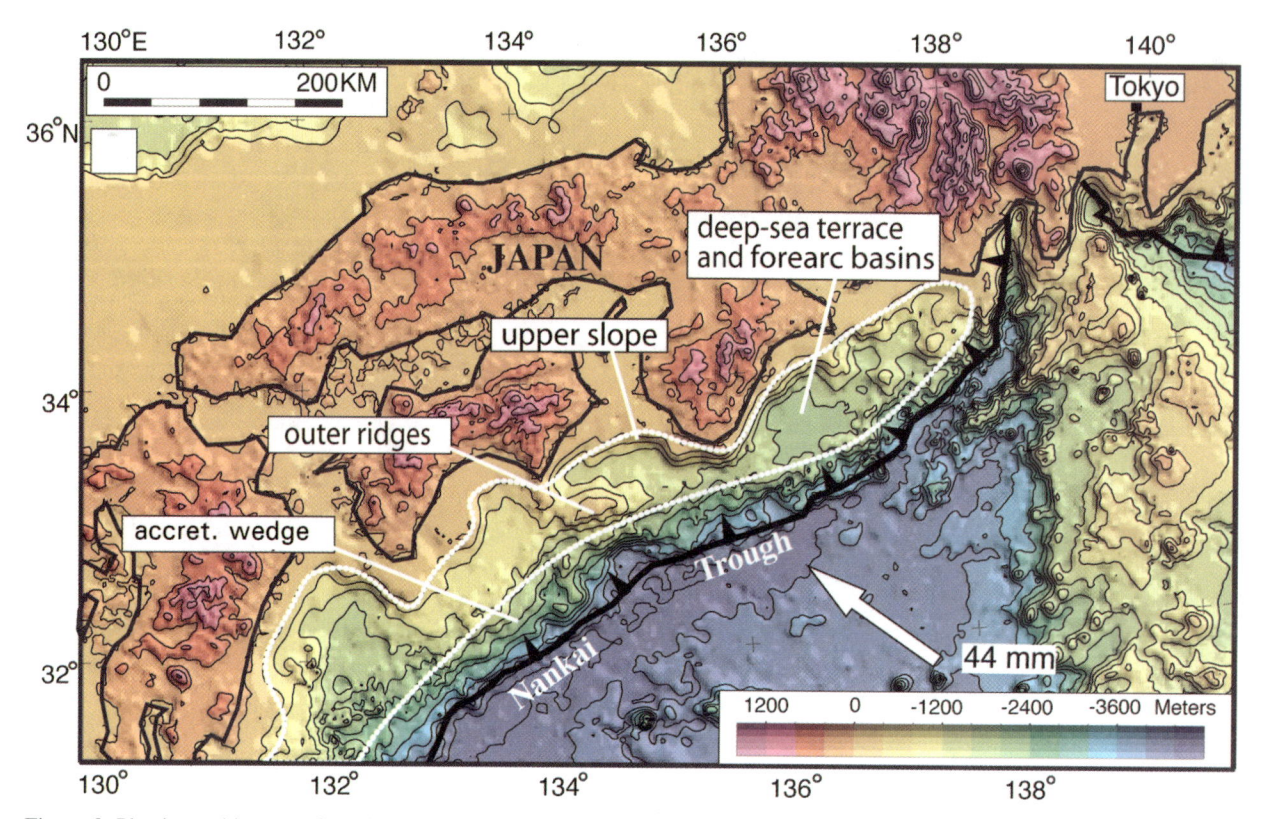

Figure 8. Physiographic map of southwest Japan. Discontinuous forearc basins are separated by ridges. Arrow shows direction of relative convergence, 44 mm/yr, of continental and oceanic plates and does not imply absolute motion. This figure was slightly modified from Wells et al. (2003, their Fig. 2C); figure was provided by Ray Wells.

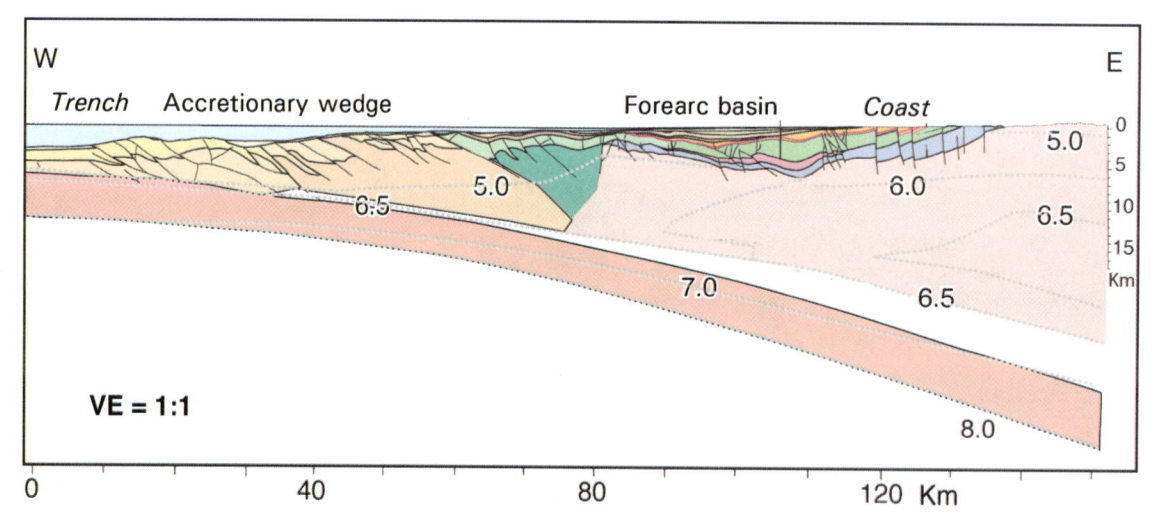

Figure 9. Cross section of convergent margin across central Oregon. Forearc basin is filled by little-deformed Eocene through Quaternary strata. Dotted contours show P-wave velocities, in km/s. Figure is after Wells et al. (2003, their Fig. 19B); figure provided by Ray Wells. No vertical exaggeration (VE).

accretionary-wedge materials are not obvious in the profiles alone. I infer both from submarine topography and from on-land exposures that bedrock fronts commonly are thinly tapered and override wedges (Fig. 3). This agrees also with the progressive uplift of the frontal parts of basins by under-plated mélange. Trenchward-sloping backstops for accretionary wedges are often postulated (e.g., Fig. 9) but do not accord with this constraint. Basements beneath the deep-water portions of the few forearc basins well characterized in on-land exposures are of oceanic basalts (e.g., Cretaceous California,

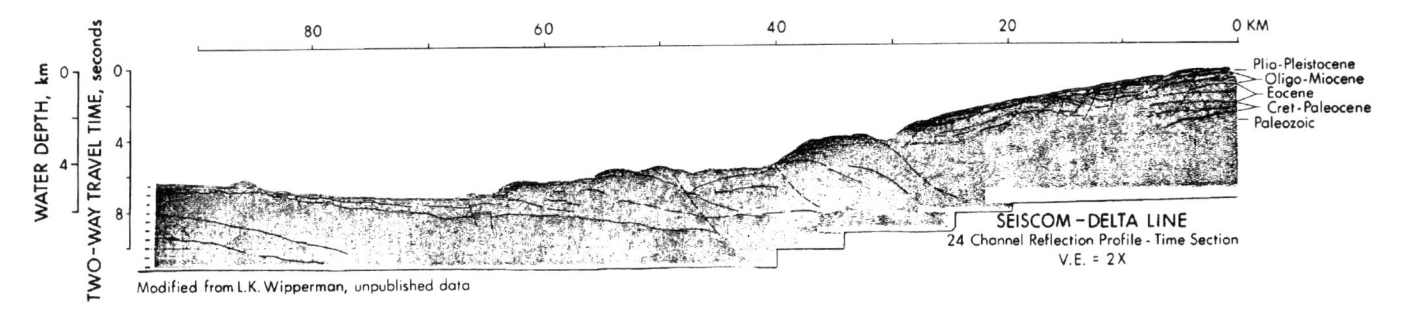

Figure 10. Marked seismic-reflection west-east profile across the northern Peru Trench to a forearc basin on the leading edge of the overriding South American plate. Lines added on left show top of oceanic crust, which can be traced under the small accretionary wedge; an intracrustal reflector; and, at ~9 s on the left edge, probably the Mohorovičić discontinuity. Figure is reprinted from Moberly et al. (1982, their Fig. 6), and reproduced by permission of Ralph Moberly and the Geological Society of London.

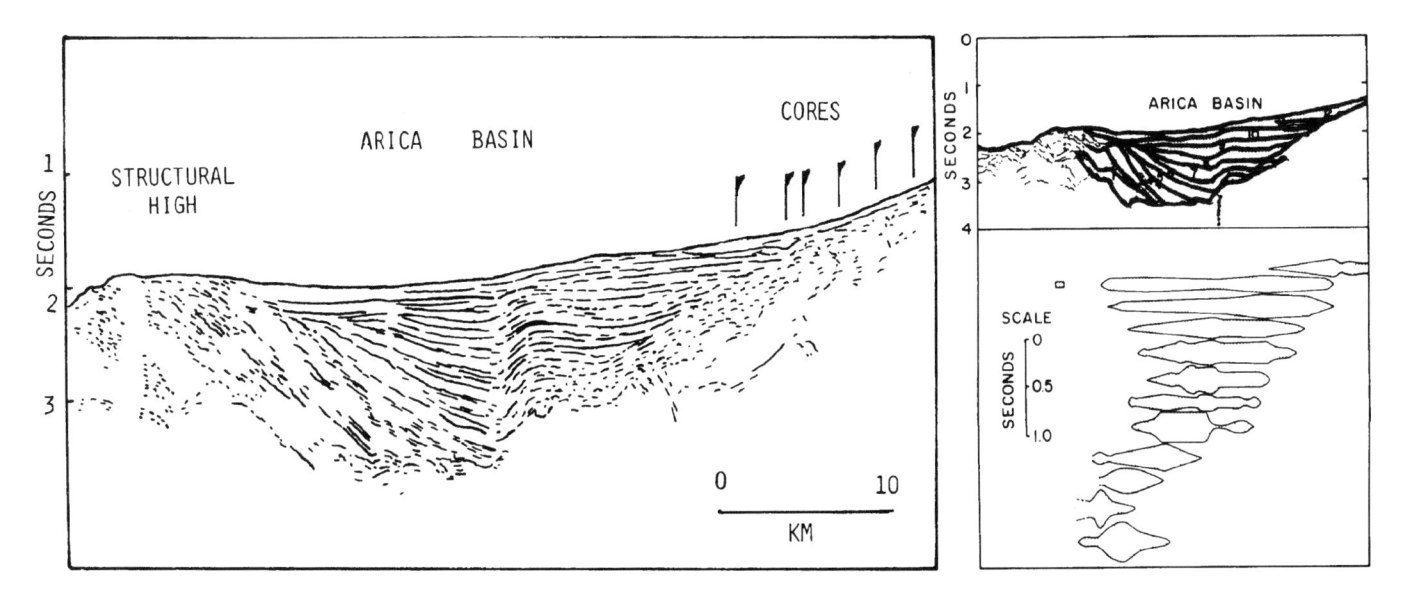

Figure 11. West-east reflection profile and stratigraphic analysis of the forearc ridge and basin of northernmost Chile. Reprinted from Moberly et al. (1982, their Fig. 9), who recognized that the landward migration of depocenters indicates progressive elevation of the leading edge of the overriding South American plate as accretionary-wedge material was stuffed beneath it, and yet that the thin leading edge itself is not crumpled. Figure reproduced by permission of Ralph Moberly and the Geological Society of London.

and Eocene Oregon), although the origins (spreading ridge or backarc) are disputed. The landward parts of basins overlie varied continental and accretionary assemblages.

Deformation of Forearc Basins

The fronts of forearc basins frequently show modest deformation (e.g., Fig. 5). I infer from the many reflection profiles I have seen that this deformation represents variously slight crumpling of the far front of the system, overflow of accretionary-wedge materials into the oceanward part of the basin, diapiric rise of basin shales, extension, and responses to changed plate-motion regimes (such as the Neogene California change from subduction to strike slip). Subduction-erosion from beneath can remove the entire forearc and much of the overriding lithosphere, as in latest Cretaceous and Paleocene time in southern California and adjacent regions.

Shortening across arc systems, discussed in a subsequent section, commonly has little effect on forearc basins and apparently is accomplished by drag on the bases of overriding plates. Forearc basins, whether continuous or discontinuous, prove the general lack of crumpling of the thin leading edges of overriding plates (cf. Fig. 1).

Extensional Arc Systems, Backarc Basins, and Complex Flow

As marine geologists are aware but many landlocked geoscientists are not, arc systems commonly undergo extension. Plate convergence need not produce upper-plate shortening. Karig (1971, and many subsequent papers) recognized early on that island arcs extend internally—even splitting into parts that are left behind and parts that advance—and migrate and increase their trenchward convexity in response to hinge rollback. Oceanic backarc basins open behind the migrating arcs. Hundreds of papers by others since (including many of my own; e.g., Hamilton, 1979) have added documentation and further examples. Intra-arc extension, both oceanic and continental, can involve hundreds of small normal faults (Suter et al., 2001; Taylor et al., 1991). Geodetic data confirm that the Mariana arc is lengthening and increasing its curvature as the Mariana Trough opens behind it (Kato et al., 2003). The highest reported velocities are for the Tonga system: satellite geodesy shows that the north end of the Tonga arc is converging with the subducting Pacific plate at 24 cm/yr, and the northern Lau Basin is opening at 16 cm/yr in the lee of this end of the arc (Bevis et al., 1995). Lengthening and shape change of the New Hebrides arc, as expected from arc-migration analysis, also are shown by geodesy (Calmant et al., 2003). Schellart and Lister (2005) and Schellart et al. (2006) explained various Pacific-margin plate interactions in terms of rollback controls.

The internally deforming Aegean-Turkey plate, with the Aegean arc system in its southwest part, is advancing with curved trajectory (Fig. 12) over the African plate as the subduction hinge in front of it rolls back toward the African continent (Chaumillon and Mascle, 1997; Dilek, 2006; Reilinger et al., 2006; Reston et al., 2002; ten Veen and Kleinspehn, 2003). The overriding plate is bounded by strike-slip faults and diffuse high-strain zones and is deformed internally by distributed and miniplate-bounding extension and strike-slip (Nyst and Thatcher, 2004).

Internal extension increases toward the subduction hinge: the plate is being pulled toward the retreating hinge, not pushed into it. The trench, in front of the hinge from which Mediterranean lithosphere of the African plate subducts relatively northward, is advancing rapidly southward across the Nile abyssal fan in the east and has reached the base of the Libyan continental slope in the west. The accretionary wedge broadens rapidly as Nile sediments are added to it, so the trench is migrating faster than the hinge. Curvature—southward convexity—of arc components increases as they migrate. Africa and the subducting plate are moving much more slowly northward, relative to stable northwest Eurasia, than the Aegean arc system and hinge are migrating southward. (The misnamed Hellenic Trench is a series of extensional features north of the forearc ridge, and it is far behind the subduction-front trench of the system, the also-misnamed Hellenic Trough: Reston et al., 2002.)

The Tyrrhenian-Apennine arc system has steadily lengthened, increased its curvature, rotated counterclockwise, and left continental fragments and newly opened small ocean basins in its wake as it has advanced toward a subduction hinge that is rolling back into Adriatic and Mediterranean lithosphere (Faccenna et al., 1996, 2001; Lucente et al., 2006). Again, migration of the overriding plate toward a hinge that is retreating oceanward is indicated.

The most complex interactions occur in East and Southeast Asia and the Indonesian region. As India crowds northward into

Figure 12. Global Positioning System (satellite geodesy) velocity field of Aegean and Turkish region relative to internally stable northwest Eurasia. Velocities in the overriding plate increase along curving trajectories toward a south-facing subduction system. The trench is not shown but trends near SW and SE corners of map area, is convex southward, and is ~250 km south of Crete, the long island at bottom center. The overriding plate is being extended toward the retreating hinge (not shortened and crumpled against a fixed hinge), the free edge, as it is extruded between the converging Arabian and European plates. High-strain zones of strike slip and extension, not marked here, outline miniplates of lesser internal determination (Nyst and Thatcher, 2004). Other unmarked arc features: the forearc ridge (the crest of the active accretionary wedge) is ~100 km offshore from Peloponnisos (large peninsula of southern Greece, left center), 150 km south of Crete, and 100 km south of Turkish coast at far right; the magmatic arc is convex southward, and is 150 km north of Crete at closest. This image was slightly modified from a figure provided by Wayne Thatcher; cf. Nyst and Thatcher (2004, their Fig. 2).

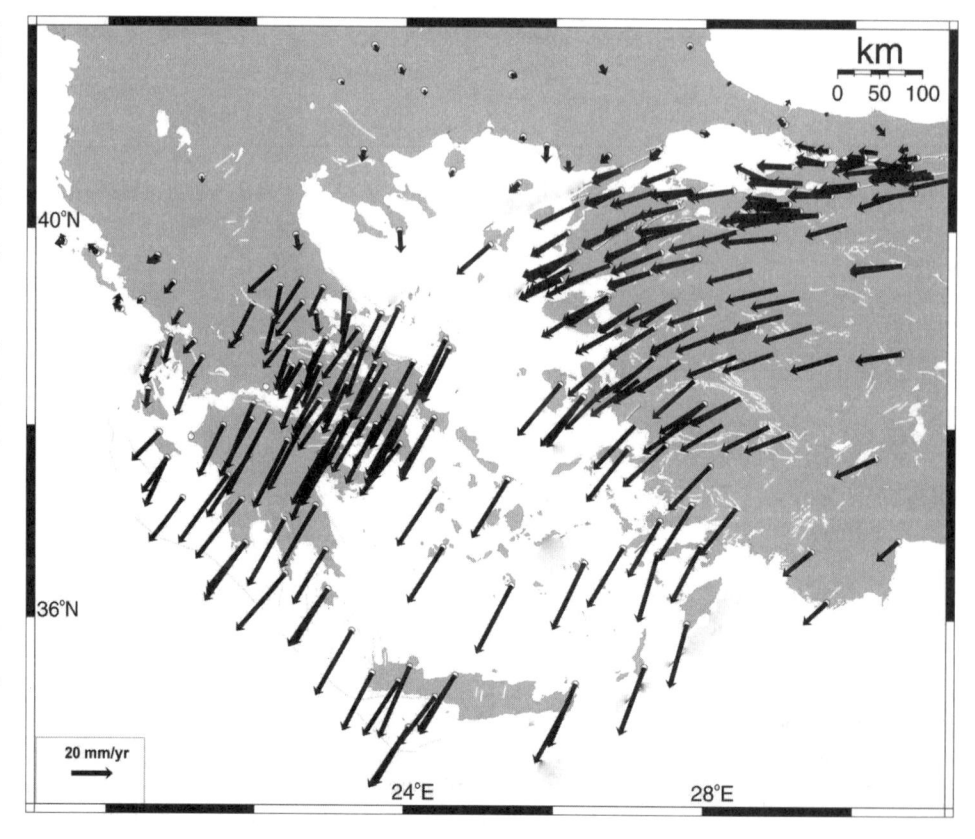

central Asia, extruded crust flows primarily toward free sides and subduction systems—eastward toward the Pacific, relative to stable Eurasia, and also in a giant glacierlike flow southeastward and southward into Southeast Asia, thence westward toward the retreating Andaman-Sumatra hinge (cf. Chen et al., 2000, 2005). Hamilton (1979) and Schellart and Lister (2005) emphasized that this flow is toward retreating extensional subduction systems, which thus provide part of the drive. Socquet et al. (2006) confirmed with satellite geodesy my (Hamilton, 1979) analysis, from observed onshore arcuate strike-slip faults and offshore trenches, of rotating Sulawesi microplates driven by hinge rollback.

Interseismic Locking

Although arcs and plates advancing over subducting oceanic plates are not crumpled from the front, many are shortened by drag at the bottom. The most complete data come from Japan's dense network of permanent Global Positioning System (GPS) stations (e.g., Mazzotti et al., 2000). The northwest (Japan Sea) margin of central Japan, like nearby mainland Asia to the west, is moving slowly eastward relative to stable northwest Eurasia, whereas the southeast part of Japan is converging with Eurasia with 1990s velocities of up to 2 cm/yr. That the net migration of the arc system nevertheless is eastward relative to stable Eurasia is shown by the downplating in the mantle transition zone of subducted Pacific lithosphere beneath China, as discussed subsequently.

GPS data from a number of other continental and composite arc systems also show interseismic shortening. Sumatra and Java may be similar, though less well documented (Michel et al., 2001). Complex New Zealand is rotating clockwise, propeller-fashion about a hub, as strain is transferred from an east-facing arc system, wherein the forearc is moving rapidly eastward away from the rest of the system, in the north, to a west-facing system in the south; the intervening region is subjected to strike-slip faulting and shortening (Beavan and Haines, 2001).

In such cases, only the bottom-frontal part of the overriding plate can be coupled to the slab, for tomography (e.g., Wang and Zhao, 2005; also see subsequent discussion and Fig. 14) shows a low-velocity mantle wedge that intervenes between the plate and slab farther back under the plate. Sumatra modeling by Simoes et al. (2004) indicates that coupling may extend only to a depth of 40 or 50 km.

CAUSE OF SUBDUCTION

Oceanic lithosphere is generated by cooling of asthenosphere from the top. This conversion produces a density inversion, for dense, strong lithosphere overlies light, weak asthenosphere. This inversion is righted by subduction. The temperature of the top of the lithosphere is regulated by the ocean to be a little above 0 °C, whereas the bottom of the lithosphere is defined approximately by the solidus temperature of basalt, which increases with pressure. Oceanic lithosphere does not randomly founder because its strength and mass-supported compression hold it together, and the mass of a plate instead drives it toward a subduction hinge that provides an escape from the surface. The lithostatic head of seafloor relief and of the trenchward inclination of the base of the dense oceanic lithosphere atop the light asthenosphere, products of the thickening of asthenosphere with time as a result of top-down cooling, provides an additional gravitational body force, ridge slide. Oceanic lithosphere is strong in compression but weak in tension, so "slab pull," although often invoked, can be only a minor complication. "Ridge push" is another popular misconception. Body forces do the job.

An overriding plate is carried and driven forward to maintain contact with the retreating hinge and falling slab. This concept is expanded on later in this chapter.

MANTLE CIRCULATION

Both seismicity and high-amplitude tomography show that subducting slabs widely reach the 660 km discontinuity. Only ambiguous evidence, which I question, supports the common view that slabs penetrate this discontinuity, sink deep into the lower mantle, and are balanced by upward flow of hot deep-mantle material as currents, including plumes and megaplumes, and thus that all or most of the mantle circulates together. Like Anderson (2002a, 2002b, 2007), Hamilton (2002, 2003), and Hofmeister and Criss (2005a, 2005b), I argue that, instead, plate circulation is closed within the upper mantle, and heat, but not material, rises from the deeper mantle. I see the deep limit of circulation as the discontinuity at ~660 km, whereas Anderson prefers a limit of near 1000 km on the basis of geoidal wavelength arguments and of his acceptance of tomographic modeling of some subducted material to that depth.

Many arguments that are given for rise of plumes from the deep mantle to the lithosphere are rationalizations of assumptions, and many testable predictions in the conjectures have been disproved. See, for example, papers by various authors in Foulger et al. (2005a) and Foulger and Jurdy (2007), discussions appended to papers in the latter volume, and the extensive discussions and links to published papers, pro and con, at www.mantleplumes.org. Downward flow by subduction into the deep mantle, has, however, been illustrated with cross sections from seismic tomography, which have convinced many geoscientists. If such deep subduction indeed occurs, there of course must also be upward flow, and thus circulation, through all or most of the mantle. Despite its widespread acceptance, tomographic evidence for deep-mantle subduction is, at best, ambiguous and has often been presented in ways that mislead casual observers.

Slabs in the Upper Mantle

The upper-mantle positions of subducting slabs are clearly shown by inclined zones of earthquakes that reach a maximum depth of 690 km where the subducting lithosphere is older than ca. 60 Ma (Fig. 13). Oceanic lithosphere younger than this when it enters a subduction system typically is seismogenic only to a depth of 200 or 300 km (compare Figs. 2 and 13). Young slabs, which are thinner than older ones, lose negative buoyancy and apparently are mixed into the middle upper mantle as they are overridden by advancing arcs and upper plates.

Oceanic lithosphere older than ca. 60 Ma can be tracked in both seismicity and high-amplitude tomographic anomalies into the transition zone, which is bounded by seismic discontinuities at depths of about 410 and 660 km, the lower of which is commonly taken to be the boundary between upper and lower mantle. Many slabs are clearly defined by high-amplitude tomography as subhorizontal on the 660 km discontinuity. This down-plating is a manifestation of the migration of hinges at both tops and bottoms of inclined slabs and is not a result of horizontal injection, although it is often so represented. Fukao et al. (1992) showed that subducted lithosphere of most East and Southeast Asian systems is laid down on or near the 660 km discontinuity for lateral distances of 1000–2000 km beyond the hinges at the bases of inclined slabs, which in turn are 500–1000 km laterally distant from the upper hinges that are rolling back into the Pacific. Huang and Zhao (2006) confirmed this pattern and added still more to the areal extent of laid-down slabs. Deal et al. (1999) found cross-strike down-plating of ~1000 km behind the Tonga subduction system, in addition to the 600–800 km of horizontal distance between lower hinge and trench. Piromallo and Morelli (2003) showed that subducted lithosphere is stranded atop the 660 km discontinuity across a broad Alpine–northern Mediterranean–Balkan–Turkish region. The entire sunken slab of the migrating Apennine-Maghrebide arc is still recognizable in the transition zone (Lucente et al., 2006).

Body-wave tomography is relatively insensitive to horizontal anomalies, and this is critical with regard to whether or not the 660 km discontinuity is penetrated by sinking slabs. To minimize this problem, Zhao (2004) solved for velocities independently within the transition zone, thereby much sharpening the depiction of the settled slabs. Three figures from his global P-wave model are reproduced here, modified slightly in labeling. Note that the colors on Figures 15 and 16 are saturated at calculated P-wave velocity anomalies of less than 0.5% or 1%, although upper-mantle anomalies reach 5% (Fig. 14). Zhao presented both layer maps and vertical sections through his models and regarded them as showing that, although most slabs indeed are plated down on the 660 discontinuity, some penetrate far into the lower mantle,

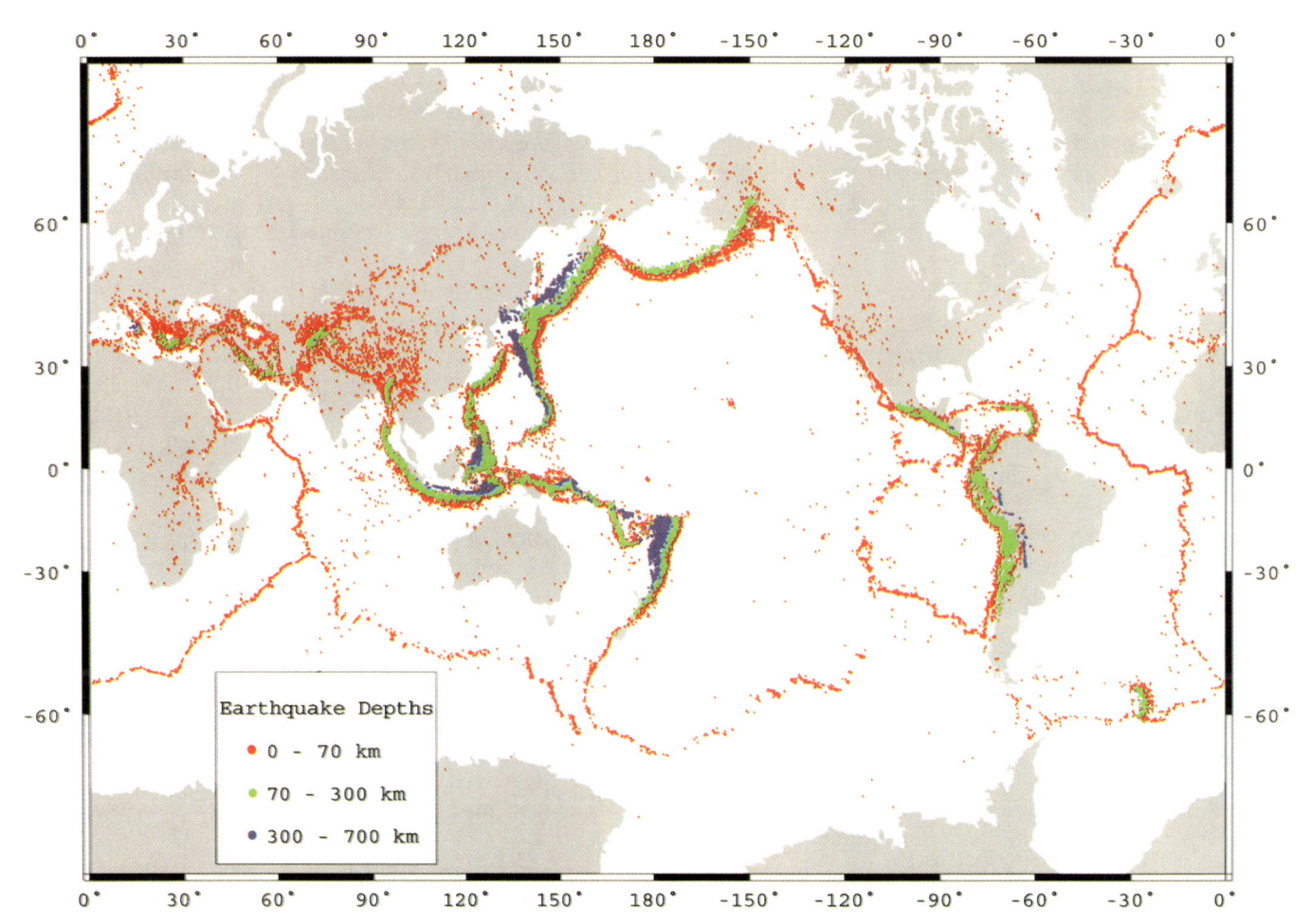

Figure 13. Map of global seismicity. Earthquakes, magnitude >5.1, were plotted from the database of Engdahl et al. (1998) by David Sandwell, who provided the figure.

and that broad upwellings (not the narrow stovepipes of some plumologists) rise from deep in the mantle and are responsible for some volcanic "hotspots"; however, I question these inferences that voluminous material crosses the 660 km discontinuity.

Figure 14, Zhao's upper-mantle profile across central Honshu, shows the position and seismicity of Pacific lithosphere sinking beneath the Japanese arc system. The slab sinks more steeply than it dips from a retreating hinge. The low-velocity region midway between slab and magmatic-arc crust may be hot mantle sucked up above the sinking slab, and it may be the heat source for arc magmatism (Hamilton, 1995; Zhao, 2001). The hot, low-velocity mantle wedge precludes interseismic locking of the slab to the overlying lithosphere plate below shallow depth (see also Wang and Zhao, 2005).

Figure 15 crosses the entire mantle of a larger region that includes Japan. Pacific lithosphere subducted beneath Japan is imaged as plated down on the 660 km discontinuity for 1500 km beneath China, and, with the 1000 km horizontal length of the inclined part of the subducting slab, it records 2500 km of hinge retreat into the advancing Pacific. (The common assumption that flat slabs within the transition zone represent lateral shoving of a beam overlooks hinge rollback.) Active Changbai volcano and other young basaltic tracts of northeast China obviously owe their magmatism to crustal or upper-mantle processes, not to hot material rising beneath them from the lower mantle (Zhao et al., 2004).

Zhao's velocity variations within the transition zone are shown in map view by Figure 16. The broad high-velocity patches presumably include much of the relatively young subducted lithosphere plated down on the 660 km discontinuity in Tethys, Australasia, East and Southeast Asia, and southwest North America. Ritsema's shear-wave tomography (body plus surface waves, in spherical harmonic rendition: Ritsema, 2005, his Fig. 6C) yields similar images for the East Asian–Australasian region, with lesser horizontal resolution, and shows a continuous slab in the transition zone beneath western South America but no similar major anomaly beneath North America.

No Proven Subduction into Lower Mantle

A widely accepted argument for whole-mantle circulation comes from seismic tomography that purportedly shows a great slab inclined downward through most of the lower mantle. This would require that compensatory rising material (plumes are currently favored) come back up through the 660 km discontinuity elsewhere. There is only one visually impressive example of this possible deep subduction: blobs depicted elsewhere are nondescript. Many tomographers believe that the low-amplitude positive-velocity anomalies calculated in the lower mantle that are near regions of surface subduction record deeply subducted slabs, and they conclude that although indeed most old slabs are parked in the transition zone, much other subducted material descends through the 660 km discontinuity and goes varying distances into the lower mantle. Penetration of the 660 km discontinuity is visualized as direct in some cases, the upper-mantle slab

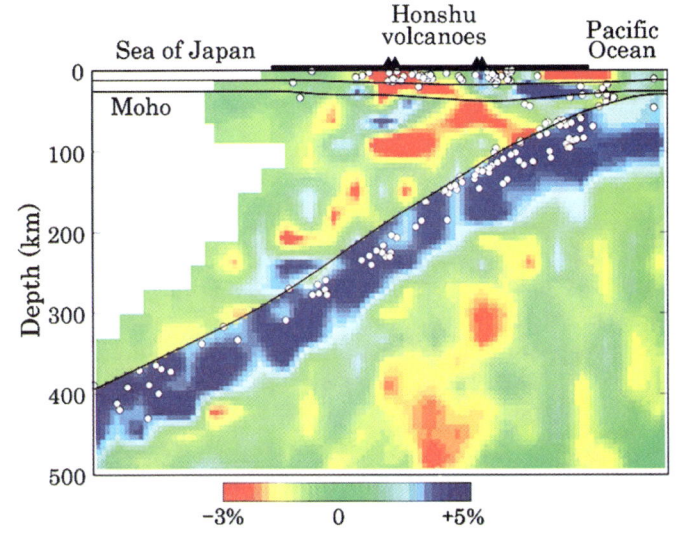

Figure 14. WNW-ESE tomographic section across the subduction system of central Honshu, showing P-wave velocity anomalies of crust and upper mantle. The black bar along the surface represents Honshu, and the trench is just beyond the right end of the profile. The white circles are earthquakes within 40 km of profile. The computer program solved independently for anomalies bounded by marked positions of the Mohorovičić discontinuity, a hypothetical mid-crustal discontinuity, and the top of the subducting slab. This figure is after Zhao (2004, his Fig. 21), who provided the figure.

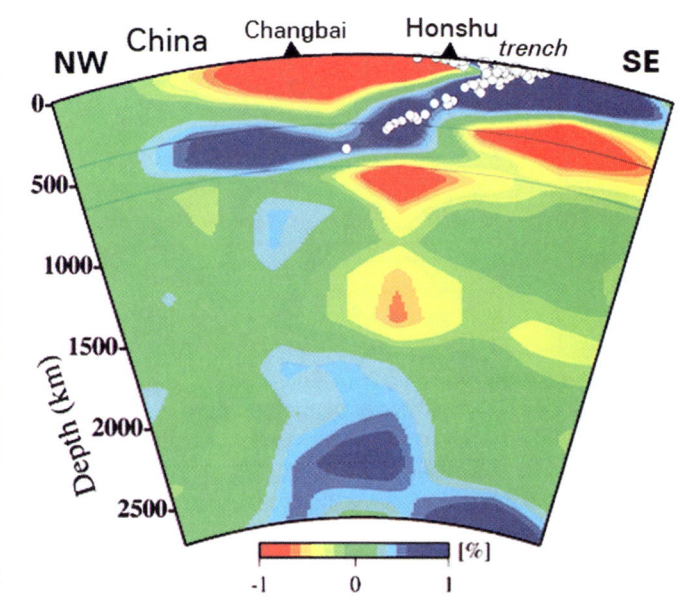

Figure 15. Tomographic profile through northeast China and central Japan of P-wave velocity anomalies. The triangles at the top mark the intraplate Changbai volcano in China and the magmatic arc of Japan. The white circles are earthquakes recorded within 100 km of profile. The 410 km and 660 km discontinuities are also marked. This figure is after Zhao (2004, his Fig. 18), who provided the figure.

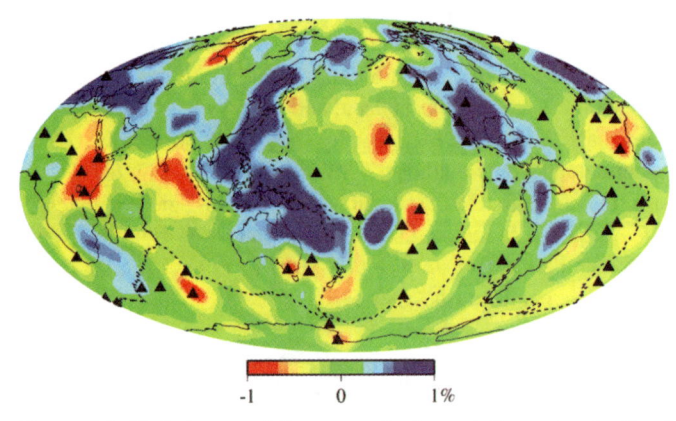

Figure 16. Global map of P-wave velocity variations at a depth of 550 km, in the mantle transition zone. Broad tracts of high velocities, marking oceanic lithosphere plated down on the 660 km discontinuity, apparently are delineated in East Asia, western Pacific, Australasia, and western North America. Triangles mark some hypothetical "hotspots." This figure was provided by Dapeng Zhao; this model is slightly different from that published by Zhao (2004, his Fig. 5).

continuing into the lower mantle at a semiconstant angle, but, more commonly, it is thought to require accumulation of subducted lithosphere in the transition zone until large masses break away downward. Thermal, mechanical, and petrologic rationales given for such subduction are strained.

Tomography

The tomographic anomalies attributed to deep slabs can be interpreted alternatively as artifacts of methodology and sampling deficiencies and as illusions of selective presentation. No deep subduction need be indicated. The illustrations that convince many readers are tomographic profiles that are too often placed only where desired results are shown and too often are truncated downward and laterally to omit anomalies that misfit the message. "Cross sections are for tourists," Bradley Myers, expert geologic-map reader and my longtime colleague, often said with regard to structural geology, and the same applies to much tomography. Colors or patterns on tomographic maps and profiles commonly are saturated for very small-velocity differentials, which make ambiguous low-amplitude, lower-mantle anomalies look similar to upper-mantle ones with amplitudes 10 or 15 times greater. Many ambiguities are inherent in the tomography itself. Most body-wave tomography divides the solid Earth into compartments or grid nodes, calculates raypaths from earthquakes to recorders, determines departures of actual traveltimes from those expected from a standard-Earth radial-velocity model, and inverts to solve for the contribution of velocity within each compartment, or in the vicinity of each grid node. Various smoothing, damping, and sharpening algorithms commonly are applied, and the final result accounts for some fraction, usually unspecified but commonly small, of the observed traveltime variations. The methodology is viable mathematically when each compartment or nodal vicinity is crossed by many rays in many directions, but this requirement commonly is not met. Earthquakes (Fig. 13) and recorders are very irregularly distributed. Most of the mantle beneath the oceans, and the deep mantle almost everywhere south of about latitude 20°S, is poorly sampled. Much other mantle is sampled primarily by subparallel rays from which tomographic methodology generates artifacts. Calculated anomalies are smeared out. No-data regions are populated with interpolated values, or with near-zero default values, or with artifacts required by symmetry in spherical-harmonic renditions. Illusory local anomalies can be generated from actual regional anomalies where coverage is meager.

Profiles showing an irregular low-amplitude anomaly that is 500 km or so in typical thickness, the top of which is inclined eastward deep into the lower mantle beneath the southernmost United States (e.g., Grand et al., 1997), are widely accepted as imaging the subducted Farallon plate, and thus, as proving whole-mantle circulation. Gu et al. (2001) and Dziewonski (2005), however, presented longer profiles, through the same tomographic models and other similar ones, which show the "Farallon anomaly" to be merely part of a highly irregular regional positive anomaly that extends vertically through the entire lower mantle and that extends horizontally 2000 km or so west of North America, where no relation to subduction is plausible. Further, an inclined "Farallon anomaly" can even be arbitrarily designated only within a narrow latitudinal band because to both the north and south, where it should be present if correctly interpreted in terms of long-continued deep eastward subduction, no tidy anomaly is depicted. The near randomness of other such deduced anomalies below the 660 km discontinuity, and the lack of coherent anomalies suggestive of subducting slabs, is obvious on the small-scale global tomographic maps at various mantle depths in several different tomographic models presented by Montelli et al. (2006). (For an elaborate rationalization of highly irregular low-amplitude positive velocity anomalies, deduced in the lower mantle beneath the American continents and the Caribbean Sea, as representing the Farallon and other slabs, see Ren et al., 2007; they did not show their model beneath the adjacent eastern Pacific Ocean, nor did they address the sampling bias discussed next.)

Artifacts due to inadequate sampling may also account for the "Farallon anomaly." About 80% of the earthquakes used in tomography occur in subduction systems (Fig. 13), primarily near the upper surfaces of subducting slabs and in accretionary wedges (Fig. 14). Teleseismic raypaths from these earthquakes mostly exit obliquely downward through the slabs and thereby gain traveltime advances that vary with the thicknesses, orientations, and shapes of the slabs. Most of the waves from which the purported Farallon anomaly is calculated originated near the top of the subducting Andean slab and were recorded at North American stations. The raypaths are clustered, so their shallow near-origin time advances are not canceled out by the simultaneous-inversion process. The Andean slab varies greatly in strike, inclination, and configuration (Brudzinski and Chen, 2005), so within-slab travel distances of diving rays vary complexly on route to different recorders. Also not factored into tomography is the strong velocity anisotropy of slabs. The purported Farallon slab anomaly likely is a methodological artifact due to the scarcity of the crossing rays required for viable body-wave tomography.

Van der Hilst (1995) presented three tomographic profiles, but no maps, across the Tonga and Kermadec arcs. He inferred that slabs subduct deeply westward, beneath lithosphere laid down in the transition zone. As in the "Farallon" example, selective viewing is needed to postulate that slabs are present within irregular regional positive anomalies, which in this case gain volume and complexity where they are truncated at the ends and bottoms of the short and shallow profiles.

Other purported tomographic examples of deep subduction are primarily of low-amplitude positive anomalies in discontinuous tracts of the upper third or half of the lower mantle along the great Tethyan orogenic systems from southwest Asia to Indonesia (e.g., Piromallo and Morelli, 2003; Replumaz et al., 2004; Van der Voo et al., 1999; Widiyantoro and van der Hilst, 1997; Zhao, 2004). These non-slab-like anomalies are mostly irregular blobs that are depicted beneath, or across strike from, only parts of the subduction-generated orogenic terrains, and that have tops often hundreds of kilometers beneath the 660 km discontinuity. The common explanation is a byproduct of erroneous visualization of slabs as pushed laterally along, rather than plated down on, the 660 km discontinuity: the discontinuity resists penetration but is subject to sporadic breakthroughs by piled up slabs. (Similar blobs elsewhere that have no conceivable relationship to young subduction, as beneath East Antarctica and the central Pacific, are not given subduction interpretations.) The earthquakes from which these deep, discontinuous Tethyan anomalies are calculated occur mostly in active subduction systems and in continental orogenic systems that contain or overlie upper-mantle slabs, but, in contrast to the simple Andean-Farallon case, geometries are too complex for casual analysis of raypaths and sampling artifacts. Tomographers could resolve this problem by analyzing raypath directions through their mantle boxes.

Heat

Conventional models of geodynamics require assumptions of temperatures at depth and of easy vertical rise of hot material that are biased in the high directions.

Global Heat Flow

The heat loss from the solid Earth indicated by heat-flow measurements, integrated for continental and oceanic crustal age, is ~31 TW. This is about the maximum amount that can be accounted for as generated concurrently by cosmologically reasonable contents of radioactive elements (Hofmeister and Criss, 2005a). The voluminous numerical-modeling literature nevertheless incorporates a speculative value of ~44 TW, which represents the age-integrated measured heat flow from the continents plus a conjectural amount for the oceans (derived with a half-space cooling model) that is approximately twice the age-integrated measured oceanic heat flow. This gross discrepancy between measurements and model is popularly rationalized as due to loss of half of the oceanic heat flow to circulating seawater, and it is enormously important for evaluating the composition, properties, and behavior of the deep interior. Hofmeister and Criss (2005a, 2005b, 2006) showed that the half-space cooling model grossly overestimates heat loss because of its invalid physical assumptions. One false assumption is that three-dimensional thermal volume expansivity can be used in evaluating the age-dependent subsidence of oceanic lithosphere, whereas the actual problem is of essentially one-dimensional vertical expansivity. This consistent error, repeated in scores of papers over the past 30 yr, exaggerates by a factor of three the effect of lithosphere temperature on seafloor depths. Another false assumption in conventional papers is that thermal conductivity of oceanic lithosphere is constant at its near-surface value, whereas in fact conductivity decreases with temperature and is only about one-third as large at the base of oceanic lithosphere as it is at the top. Both of these errors greatly increase heat flow calculated with the half-space cooling model. Von Herzen et al., 2005, reasserted the assumptions of the half-space model and showed that several of the minor arguments of Hofmeister and Criss (2005b) can be regarded as ambiguous, but they did not address, and hence in effect conceded, the important issues, as Hofmeister and Criss (2005c) made clear. Wei and Sandwell (2006a) deduced, from a variant of the half-space model, a global heat loss of ~42 or 44 TW, but Hofmeister and Criss (2006) demolished this rationale for its mathematical errors, its misuse of three-dimensional (3-D) expansivity, and its implicit assumption of constant thermal conductivity. In their response, Wei and Sandwell (2006b) reasserted their assumptions, made further errors, and did not address the Hofmeister and Criss points and, thus, in my view, conceded the debate. Many conventional modelers of other aspects of lithosphere behavior (e.g., Sleep, 2006) also base their calculations on the false assumptions of constant thermal conductivity and 3-D expansivity.

Global heat loss probably is near the measured 31 TW, not the hypothetical 44 TW, and far less of the total can come from the core than is assumed in popular bottom-heating conjectures.

Magma Temperatures

Another required, but invalid, assumption, commonly misstated as fact, of conventional geodynamics and plumology is that rising columns of hot material produce temperatures at the base of the lithosphere that are several hundred degrees Celsius hotter than ambient asthenospheric mantle (Campbell and Davies, 2006; Sleep, 2006). This assumption commonly incorporates the additional assumption, also misstated as fact, that "hotspot" basalts (e.g., ocean-island basalts [OIB]) erupt at much higher temperatures than "normal" mid-ocean-ridge basalts (N-MORB). In fact, the temperatures of the chemically contrasted end-member N-MORB and OIB magmas are essentially the same (Falloon et al., 2007), as is to be expected from the limited range of seismic properties of oceanic asthenosphere globally.

Geochemistry

The misleading term "normal," which is applied to only about half (N-MORB) of all ridge basalts, reflects model-driven speculation that proper MORB should be derived from "depleted [upper] mantle." The other half of MORBs are geographically interspersed and trend continuously in composition from N-MORB to "ocean-island" (e.g., plots by Debaille et al., 2006, and Fitton, 2007). Rationales in Debaille et al. (2006) typify much basalt geochemistry by invoking mixing of melts from upper and lower mantle sources to explain all local variants, no matter how closely intercalated in space and time. Fitton, by contrast (and despite his conviction that plumes from the deep mantle do rise to the crust), demonstrated that *most* basalts of OIB type, which conventionally are assigned to plume sources, "occur in [many diverse continental and oceanic] situations where mantle plumes cannot provide a plausible explanation."

Two families of explanations provide obvious alternatives to explanations that require lower-mantle sources, and hence whole-mantle circulation, to account for enriched igneous rocks. One family, which is fast gaining in popularity, although it is still a minority view, is that both subduction and crustal delamination recycle fractionated material downward into the upper mantle, where it is selectively melted and is disproportionately represented in small-batch melts (e.g., Anderson, 2007; Herzberg, 2006; McKenzie et al., 2004; Natland, 2007). The second family, which is not exclusive of the first, is that much of the contrast between enriched and depleted basalts is due to different pressure-temperature-composition histories of fractional melting, fractional crystallization, assimilation, and zone refining in contrasting lithospheric settings (e.g., Hamilton, 2002, 2003).

Other Arguments against Whole-Mantle Circulation

There are many reasons to think that through-the-mantle circulation does not operate and that the lower mantle circulates only very sluggishly. Irreversible fractionation of a largely molten planet very early in its history is likely, radioactive heat sources may be almost entirely in the crust and upper mantle, temperature of the lower mantle may be much lower than commonly assumed, and deep-mantle thermal conductivity plus diffusivity may be very high, minimizing transfer of heat in material (Anderson, 2002b, 2007; Hofmeister, 2005; Hofmeister and Criss, 2005a, 2005b). The deep mantle has very low thermal expansivity, so buoyancy effects of lateral temperature variations are minimal. Viscosity increases greatly with pressure, and if the temperature of the lower mantle is lower than the high values commonly assumed to enable rapid convection and plumes, its viscosity may be in part a thousand or more times higher than that of upper mantle, precluding all but the most sluggish circulation. That the low-seismic-velocity regions of the basal mantle, close to the core, owe their velocity retardation to high iron content and high density, not to high temperature, is made likely by their relatively low Vp/Vs ratios and by data from experimental petrology and mineral physics (Ishii and Tromp, 2004; Jacobsen et al., 2004; Trampert et al., 2004), and, if so, they cannot provide the rising hot material attributed to them by plumologists. The conventional assumption of great bottom heating of the mantle by the core may be invalid. The discontinuity near 660 km separates domains of profoundly different seismic patterns that are inconsistent with easy passage of material through it (Gu et al., 2001). The 660 km discontinuity is a phase-change boundary with negative pressure-temperature slope: descending material must be heated to penetrate it, and ascending material must be cooled, thus canceling out thermal buoyancy—although whether this effect is enough to block penetration, if the boundary has not also become compositional as a result of layered circulation, is disputed (e.g., Chudinovskikh and Boehler, 2004). Earth's enormous present heat content must mostly be retained from its early history, although current heat loss may be little more than current radioactive heat generation. Top-concentrated and internal radioactive heat sources, and cooling by subducting slabs, may produce far-from-adiabatic mantle temperatures (Anderson, 2007). Motions of surficial plates and subducted lithosphere have simple possible explanations in terms of flow restricted to the upper mantle, as I elaborate in following sections.

Most current geodynamic (e.g., Jellinek and Manga, 2004) and geochemical models (e.g., Debaille et al., 2006) incorporate the contrary view that there are no significant barriers to whole-mantle circulation and that the evolution and properties of the lower mantle must be such as to enable such circulation. This view is anchored to the dubious assumptions that the lower mantle is mostly unfractionated "primitive mantle" (despite very early and hot separation of the core), whereas the upper mantle has become "depleted mantle" by separation of crust. This notion originated with chemists in the 1950s and early 1960s, when little was known about isotopes or the solar system, and was adopted by dynamicists in the late 1960s to explain plate motions with whole-mantle convection. These assumptions have since become dogma by repetition and self-citation. The common geochemical assignment of "enriched" oceanic volcanic rocks to sources that have risen from hypothetical "primitive [lower] mantle" is one of many circular rationalizations of these assumptions.

Dual Circulation?

Whether or not my skepticism regarding deep-mantle subduction is warranted, the 660 km discontinuity clearly impedes circulation between upper and lower mantle, the properties of the lower mantle require its circulation to be much more sluggish than that of the upper mantle, and simplistic cartoons of whole-mantle flow loops, and of deep-mantle subduction geometrically and temporally similar to that in the upper mantle, cannot be valid. There might, however, be both flow that is mostly restricted to the upper mantle and flow that involves lower and upper regions together. The fact that the shallow-flow–only option permits an elegantly simple representation of plate motions makes it my strong preference.

MECHANISM OF PLATE TECTONICS

The broad patterns of seafloor spreading, rolling-back hinges, and down-plating subducted slabs, and the characteristics of convergent margins all accord with a simple explanation of plate tectonics as top-driven circulation closed within the upper mantle (Fig. 17; Hamilton, 2002, 2003). Oceanic lithosphere is formed by top-down cooling of asthenosphere, and the resulting density inversion is righted by subduction. Oceanic plates are propelled by their mass toward their sinking sides, their only exits from the surface, aided by the lithostatic head represented by seafloor bathymetry and by the common trenchward slope of the base of oceanic lithosphere. Subduction hinges roll back as slabs sink, more steeply than they dip, into the mid-upper mantle if they are young and thin, but into the transition zone if they are old and thick. Overriding arcs and plates are pulled toward the retreating slabs and pass over subducted lithosphere plated down in the transition zone, or mixed into the middle upper mantle.

This process transfers lithosphere from shrinking to expanding oceans. Because slabs sink more steeply than they dip, the entire upper mantle beneath incoming oceanic lithosphere—asthenosphere, transition-zone material, and whatever is between them, an aggregate thickness of 550 km or so—is pushed back under that lithosphere, forcing rapid spreading within shrinking oceans. Advancing arcs and upper plates, by contrast, pass over only subducted lithosphere, perhaps 100 km thick where not doubled up, and a modest amount of entrained material, and only that proportion of mantle material, is thus transferred to expanding oceans and backarc basins, which accordingly widen only slowly. The striking contrast (Fig. 2) between the rapidly spreading, yet more rapidly shrinking Pacific Ocean and the slowly spreading, yet expanding Atlantic has a simple explanation.

In these terms, hinges advance relative to a sluggish lower mantle, and, in the framework that accords with such advance, plate motions should make quantitative sense in terms of subduction as the primary drive. These conditions cannot be satisfied with conventional frameworks, as discussed next, but they are satisfied with an alternative framework.

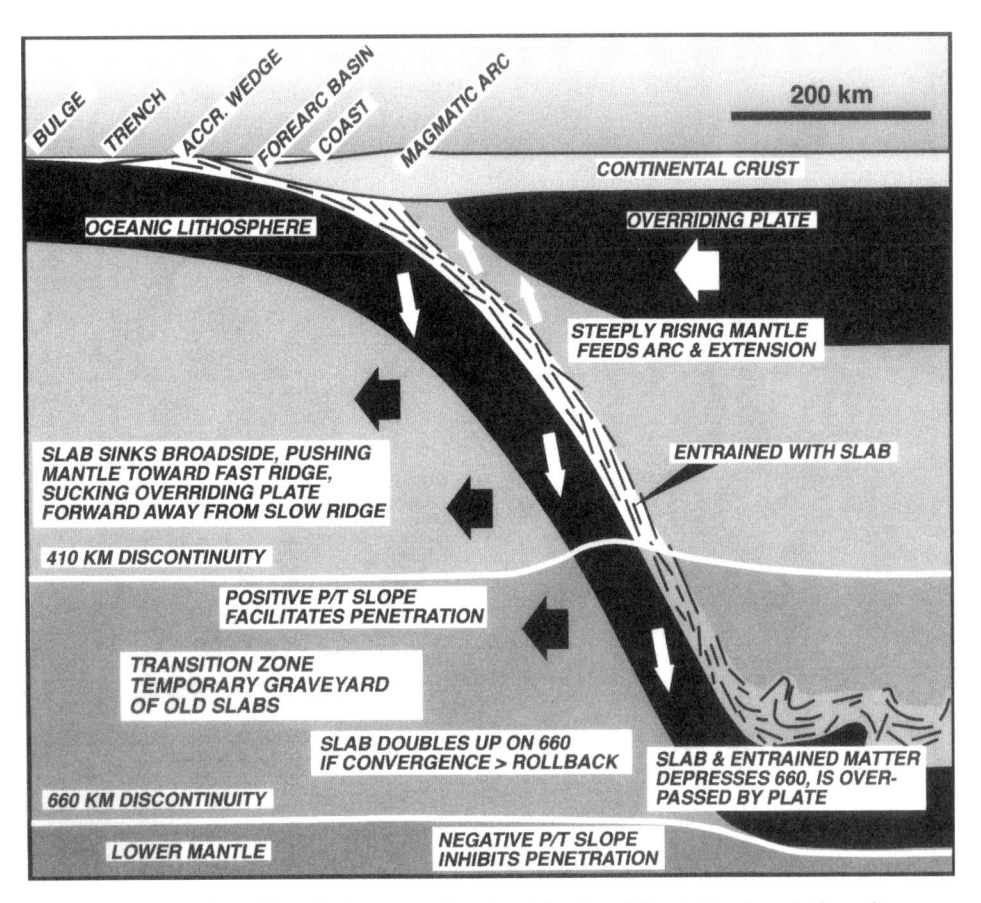

Figure 17. Subduction drive of plate tectonics. A subducting slab, sinking broadside as its upper hinge rolls back, pushes all sublithosphere upper mantle back under the incoming plate, and forces rapid seafloor spreading. The sunken slab is plated down, behind an advancing lower hinge, on the 660 km discontinuity and is over-passed as the overriding plate is sucked forward by the retreating slab. The sunken slab is thus transferred to a slow-spreading ocean behind the continent. Circulation is confined to the upper mantle. Figure is after Hamilton (2003).

FRAMEWORK OF PLATE TECTONICS

If plate tectonics indeed is driven by subduction due to cooling from the top, and its circulation indeed is confined to the upper mantle, then there should be a framework, relative to decoupled and sluggish lower mantle and hence approximately an "absolute" reference frame, of plate motions within which plate interactions make sense. Hinges should roll back, plate motions should relate coherently to subducting margins, and the passive ridges spreading between plates should migrate to tap fresh asthenosphere. These conditions are not remotely met by the popular hotspot and no-net-rotation frameworks, but they are mostly satisfied by an Antarctica-fixed framework.

The geometry of seafloor-spreading anomalies younger than a few million years constrains much of the relative-motion pattern, although it poorly defines the behavior of internally deforming parts of plates and of small-plate interactions. The rapid development of satellite geodesy, and in particular the global positioning system (GPS), in the past 20 yr has done much to close this gap. Despite their short time line, GPS vectors generally accord with the several million years of motion incorporated in seafloor-spreading deductions, showing that most motion is remarkably smooth and continuous (e.g., Sella et al., 2002). Major irregularities are introduced by preseismic and postseismic creep, and interseismic locking.

Published GPS studies relate motions to local frameworks, or to internally stable parts of chosen major plates, or to the international terrestrial reference frame (ITRF), a no-net-rotation frame that incorporates seafloor-spreading data. The result is a confusing array of data sets in different frames. Help is available in the interactive UNAVCO Web site http://jules.unavco.org/VoyagerJr/Earth, maintained by Lou Estey. Either individual GPS velocity vectors or vectors of generalized models of plate velocity, or both, can be displayed on a zoomable world map in a no-net-rotation framework or in a framework wherein any major plate as fixed. The Antarctica-fixed map, Figure 18, is from this source.

No Fixed Hotspots

Evaluation of the driving mechanism of plate tectonics has been long retarded, in my view, by widespread but uncritical acceptance of speculation that plumes of hot, buoyant material rise from fixed positions in the deep mantle and produce surface magmatism and other thermal effects on overpassing lithosphere plates, and thus that a reference frame for motions of plates relative to most of the mantle can be established from surface imprints of those effects. Despite disproof of many aspects of this speculation, the basic conjecture has hardened into dogma. (See historical reviews by Anderson and Natland [2005] and Glen [2005].) Visualization of plumes and their products varies widely and incompatibly between groups and subgroups of investigators—plate dynamicists, geodynamic modelers, fluid dynamicists, seismologists, geochemists, structural geologists—because there are few constraints save imagination. Much conjecture represents ad hoc

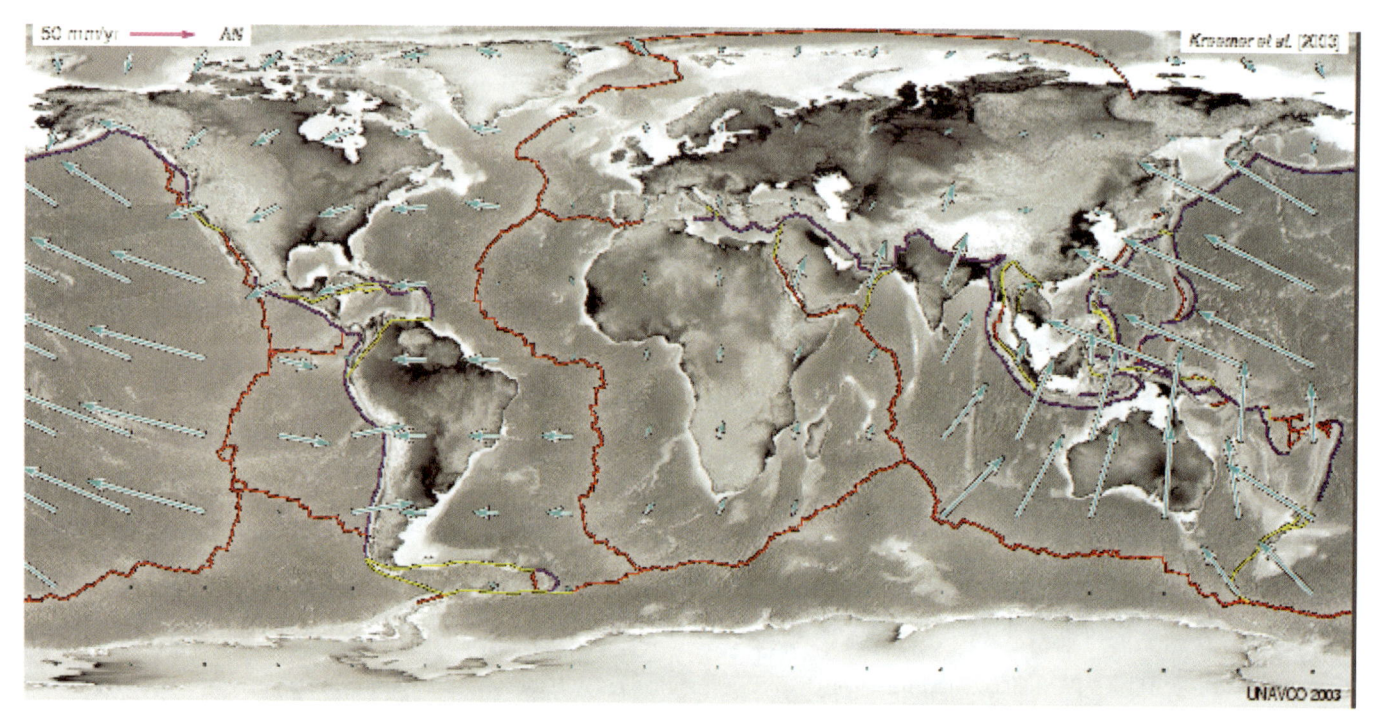

Figure 18. Plate motions relative to a fixed Antarctic plate. Motions of plates and their boundaries in this framework generally accord with subduction, enabled by cooling from the top of oceanic asthenosphere, as the primary drive of plate motions. All ridges migrate to tap fresh asthenosphere. Neither internal deformation of plates nor backarc spreading is incorporated in these vectors. Illustration was provided by L.H. Estey via http://jules.unavco.org/VoyagerJr/Earth.

evasion of disproved predictions: whatever is observed where a plume is postulated to operate is a product of the plume even if unique to that locality, and so the speculation cannot be tested. Some groups (e.g., McKenzie et al., 2004) define all buoyancy-driven upwellings and downwellings as plumes (hence dikes, mid-ocean ridges, salt domes, and subducted slabs), which divorces the term from its common geoscience application to hypothetical narrow upwellings from deep thermal boundary layers. In the ultimate non sequitur, Bourdon et al. (2006) claimed that their evidence for a shallow example of this all-inclusive definition was evidence for the concept of deep-mantle thermal plumes.

Much recent ballyhoo has accompanied the depiction of through-the-mantle plumes with finite frequency ("banana-doughnut") seismic tomography by Montelli and associates (e.g., 2004). For evaluations of the defects in the methodology and in its basis and inconsistent results, and for a statistical demonstration showing that the purported plumes are artifacts of sampling deficiencies, see Julian (2005) and the chronological review, and linked published papers, by four other groups of seismologists at www.mantleplumes.org. (Montelli et al. [e.g., 2006] retain confidence in their method.)

The hotspot reference frame of Gripp and Gordon (2002) is tied to three purportedly fixed hotspots, Hawaii, Yellowstone, and Iceland, to which are added, because they approximately fit the same frame, two more hotspots, regarded as approximately fixed for >5 m.y., and six more that are regarded as fixed but inaugurated only within the past several million years. This short list minimally overlaps conflicting lists of hotspots designated by various other groups, for most other postulated hotspots are unstable within the framework of the accepted few. I see this selection of a small array because they approximately fit the concept as mere wish fulfillment.

Plate kinematics make no sense in this hotspot reference frame. As many subduction hinges migrate forward as retreat in this frame, and some change along strike from advancing to stationary to retreating. Each mode requires different propulsion mechanisms and mantle flow in fixed-hotspot terms, which nevertheless have been thus rationalized by Conrad and Lithgow-Bertelloni (2004), Heuret and Lallemand (2005), and Schellart et al. (2007). The complex mechanisms postulated for such interactions do not account for the observed major and minor features of plate motions and interactions discussed earlier in this essay.

Powerful evidence contradicts the notion of fixed hotspots. The purported spots most important for the hotspot reference frame are Hawaii, Yellowstone, and Iceland. The relative positions of these can be rationalized as fixed relative to one another by selection from ambiguous parameters along the plate circuits between them—but none is viable as the top of a plume fixed relative to the lower mantle.

Hawaii

Geophysics of the Hawaiian region misfits plume predictions (Anderson, 2005). Pacific spreading patterns (Atwater, 1989), paleomagnetism of Emperor seamounts (Tarduno et al., 2003), and paleomagnetic latitudes of cores from the floor of the Pacific plate (Sager, 2007) show independently that the Pacific plate did not change direction by 60° above a fixed hotspot at the time of the Emperor elbow, 50 Ma, as required by fixed Hawaiian plume speculation. Other island and seamount chains once conjectured in the absence of data to fit a Hawaiian trajectory in fact misfit it badly in chronology, trends, and geometry (e.g., Clouard and Bonneville, 2005; Natland and Winterer, 2005). Hawaii and the other chains are properly explained as responses to within-plate stresses (Natland and Winterer, 2005; Norton, 2007; Stuart et al., 2007).

Yellowstone

The east-northeastward progression of late Neogene volcanic centers in the eastern Snake River Plain and Yellowstone region is an anchor for advocates of fixed plumes. Nevertheless, the thermal anomalies, as constrained by high-resolution tomography, are confined to the upper mantle (Humphreys et al., 2000). Plume proponents Waite et al. (2006) made a detailed tomographic study with local seismic arrays and also found no evidence for low velocities deeper than 400 km. A series of magmatic centers that progress west-northwestward into Oregon from the same origin during the same period, and that display a more regular time-distance progression, commonly is ignored by plume proponents because it does not "fit," although plume-advocate Jordan (2005) speculated that the aberrant trend formed by long-distance squirting from the fixed Yellowstone plume.

Iceland

The Iceland thermal anomaly on the northern Mid-Atlantic Ridge also is confined to the upper mantle (Foulger et al., 2001, 2005b). The contrary depiction by Bijwaard and Spakman (1999) of a broad, highly irregular, through-the-mantle low-velocity plume required placement of a cross section in the one orientation through their tomographic model that afforded continuity. Bijwaard and Spakman depicted the hypothetical plume with velocity differentials of only ~0.4% (which are below the actual resolution of their model), and they truncated the profile at both ends where continuation would have shown other low-velocity anomalies in their model where none are plausible in plumological terms (Foulger et al., 2005b). There is no time-progressive hotspot track, and conjectures positing initiations of seafloor spreading by a hotspot are contradicted by geophysical characteristics of the relevant tracts (e.g., Foulger et al., 2005b).

No-Net-Rotation Frame

A framework in which plate motions sum to zero makes a convenient reference, absent other information, although there is no compelling reason to presume such a zero-sum frame to depict "absolute" motions, relative to a sluggishly moving deep mantle, of surficial materials on a spinning Earth. DeMets et al. (1990) depicted plate motions in the framework wherein ridge-spreading rotations less than 3 m.y. old summed to zero, and Sella et al. (2002) depicted in a similar frame the generally similar relative motions defined by space geodesy. Best-station motions in parts of six large plates not undergoing internal deformation, as defined within a slightly modified geodetic zero-sum framework, mostly can be fitted to the ridge-anomaly frame within ~2 mm/yr (Altamimi et al., 2002). Internal deformation of plates, features associated with arc migration, and microplate motions are not generally incorporated in the ridge-anomaly framework, but they are displayed in many regions of the geodetic framework.

A no-net-rotation framework necessarily minimizes relative motions of plate boundaries, so if subduction provides the major drive, any systematic relationships will be obscured by this framework. A subduction drive of plate motions cannot be rationalized from the hodge-podge of boundary motions and stabilities in the no-net-rotation framework.

Antarctica Fixed Frame

If subduction does indeed control motions of both overriding and subducting plates, then there should be a bulk-Earth framework in which hinges mostly roll back and in which rates and orientations of plate rotations make general sense in relation to those hinges. Plates should move toward subduction systems, and plates minimally bounded by such systems should be relatively fixed. An Antarctica-fixed framework approximately fits these criteria because, alone among major plates, the Antarctic plate is bounded almost entirely by spreading ridges that are migrating relatively away from the continent at its center (Fig. 2). The ridges vary much in orientations, spreading rates, and distances from the continent. Both hotspot and no-net-rotation frameworks assign substantial motion to the Antarctic plate, and little support for a subduction drive can be found within either of those frames.

Most global plate motions make sense, in terms of a subduction drive, in an Antarctica-fixed framework (Fig. 18). Most subduction hinges roll back in this framework. Plates move toward hinges, ridges form between plates moving toward different hinges, and, as expected of top-driven motion, ridges migrate to tap fresh asthenosphere. The westward migration of the East Pacific Rise in this framework accords with the asymmetry shown by abundant geophysical data from its south-tropical sector (analysis and references in Hamilton, 2002). Carbotte et al. (2004) and Katz et al. (2004) recognized systematic variations in ridge morphology and magmatism, including petrology, with migration directions of fast-spreading ridges but thought the correlations were poor because they assumed a fixed-hotspot frame, whereas their correlations are good in the Antarctica-fixed frame.

Senses and amounts of rotations of most plates in the Antarctica-fixed frame accord with subduction as their primary drive. Africa rotates slowly counterclockwise toward subduction systems in the northeast, and Arabia rotates faster, pulling away from Africa, into the Makran system, whereas in hotspot and no-net-rotation frameworks, Africa moves rapidly eastward with no likely means of propulsion. Internally stable and subduction-free northern Eurasia also is almost stationary in the Antarctic reference frame, not moving rapidly eastward as in hotspot and no-net-rotation frameworks. East Asian regions, squeezed out in front of advancing India, migrate eastward toward the free edges of Pacific subduction systems. Velocity of the India-Australian plate increases eastward as subduction increases to the north. The Americas move westward, at twice the rate of migration of the Mid-Atlantic Ridge from Europe and Africa, toward bounding subduction systems and, for the nonsubducting margin of part of North America, toward the obliquely retreating Pacific plate. The west-coast arcs of South America, of North America from Oregon into British Columbia, and the Aleutian arc are advancing as hinges roll back in front of them. In the southwest Pacific, the New Hebrides and Tonga hinges are migrating apart, and both are rolling back in the Antarctica-fixed frame, even though Tonga and the Pacific are converging at uniquely high velocity. In Indonesia and western Melanesia, the hinges that bound small plates—Wetar, North Sulawesi, Halmahera, Sangihe, New Guinea, and Manus—appear to be rolling back in the Antarctic frame insofar as data are available (Kreemer et al., 2000; Walpersdorf et al., 1998).

Plate motions have a net westward drift in the Antarctic reference frame. That Earth's rotation is a factor in plate motions is likely because Euler poles (which are independent of framework) of relative motions between large plates are mostly at high latitudes, and relative plate motions are faster at low latitudes than high. Paleomagnetic data are of low resolution compared to kinematic constraints from seafloor spreading and satellite geodesy and provide no limit on net eastward or westward motion, but they are adequate to preclude major true polar wander whereby the entire lithosphere has large net motion through the pole of rotation.

Complications

Although motions of large plates and many small ones accord with subduction drive in the Antarctic frame, a number of arc systems have short-term GPS motions, mostly slow, that are retrograde in that frame. Some of these motions may record transient locking of slabs and overriding plates, and net long-term motions, which include large coseismic slip and associated creep (cf. Melbourne et al., 2002), may commonly be forward in the Antarctic frame. Slow shortening thickens the crust and produces a lithostatic head that results in abrupt coseismic thrusting opposite to the shortening vector. In the Antarctic frame, parts of the Pacific side of Japan have westward vectors of interseismic short-term GPS motion, but, as noted previously, Pacific lithosphere is plated down in the transition zone far westward from Japan, and the Japan Sea side indeed is moving eastward, as required by the relationship of the islands to nearby east-moving mainland Asia and the nonsubducting Sea of Japan. Another candidate for such a seismic-locking explanation may be south-central Alaska, which has at least short-term northward motion absorbed in interior shortening, although the oceanic Aleutian island arc is moving southward toward a rolling-back hinge. Short-term vectors of south-facing Java are mostly northeastward at 1–2 cm/yr (Kreemer et al., 2000), although here also long-term southward and southwestward migration seems likely when large coseismic slips are factored in (cf. Subarya et al., 2006).

Other examples indicate that if the general concepts developed here are valid, lithosphere and subjacent mantle motions much less tidy than the two-dimensional ones of Figure 17 are required in complex arc systems. Explanations likely include mantle flow around the ends of subducting slabs and beneath slabs that are limited to the upper few hundred kilometers of the mantle. West-facing Middle America has a slow eastward GPS motion relative to fixed Antarctica; subduction of young Pacific lithosphere there extends only to shallow depth, and mantle underflow may be indicated. The components of the east-facing Caribbean arc have a slow westward motion in the Antarctic frame, and mantle underflow and end flow both might be involved.

Relationships are complex in the arc systems from Japan to Indonesia. The Philippine Sea plate, which is outlined by arcs on both sides, is rotating clockwise (independent of framework) relative to Asia and the Pacific about an Euler pole near central Japan. This rotation transfers major convergence between the Pacific and Asia from the east side of the Philippine Sea complex in the north to the west side in the south. Along northern Japan, Pacific-Asia convergence is at the Japan trench. Offshore from central Japan, this trench swings south (Bonin Trench), then bows eastward into the Pacific (Mariana Trench), then dribbles out southward in small festoons (inactive Yap and Palau Trenches) and scissors into a slowly spreading ridge. On the west side of the Philippine Sea thus outlined, compensatory convergence increases southward, from near zero along Honshu, thence along the Nankai and Ryukyu Trenches and, beyond them, to a maximum in the southern Philippines and northern Indonesia. On the west side of the Philippine Sea plate, GPS vectors in the Antarctic frame are eastward in the east-facing Ryukyu arc and westward in west-facing Luzon, but the southern Philippines, which face east over the Philippine Trench, also have at least short-term westward vectors (e.g., Walpersdorf et al., 1998). (The Philippines tectonically resemble New Zealand in propeller rotation, although motion is counterclockwise rather than clockwise, with the transition from westward to eastward subduction similarly marked by longitudinal strike-slip faulting.) The east-facing Bonin and Mariana arcs, bounding the Philippine Sea plate on the east, have northwestward GPS velocity vectors to ~6 cm/yr (e.g., Kato et al., 2003), opposite to rollback sense, in the Antarctic frame. South of the Philippines are the miniplate complexities of northern Indonesia.

So many major features fit the general scheme of subduction control in an Antarctica-fixed framework that I assume that the complications just noted can also be fit into it. My specific rationalizations are not substantive enough to warrant further exposition here.

AFTERWORD

A top-down subduction drive, an approximate Antarctic reference frame of "absolute" motion, and closure of plate circulation above the 660 km discontinuity provide a simple explanation for plate motions and interactions. This drive, frame, and closure also are compatible with much additional information, only a little of which was noted here. Extrapolation of the theme predicts that continental collisions and continuing convergence, including that of India with Asia, also are driven by subduction, but reliable and detailed tomographic data with which to test this notion and evaluate geometries are not yet available. Most subducted slabs for which good tomography is available are plated down on the 660 km discontinuity, and so, even if some slabs do sink through that discontinuity, the mechanism postulated here should be generally applicable.

My explanations can be broadly correct only if assumptions now widely accepted in geodynamics, geophysics, and geochemistry are false. Those assumptions derive from conjectures, originated in the 1950s and 1960s, regarding the detailed composition, slow evolution, and bottom-up behavior of Earth. These assumptions predate a huge body of contrary information and yet are now dogma in much of the geoscience community.

The conventional view of Earth's secular evolution is of progressive fractionation. Elsewhere (Hamilton, 2007), I develop the opposite concept: Earth was highly fractionated very early in its history, and its subsequent evolution, consequent on changes enabled by cooling, has involved progressive enrichment of the upper mantle by downward recycling of crustal materials back into it. Plate tectonics provides the current major mode of that recycling, but time-varying processes of delamination may have been dominated by most of Earth's evolution.

ACKNOWLEDGMENTS

It is a pleasure to present this paper in honor of Raymond A. Price, who has done much to make the Canadian Rocky Mountains Earth's best-understood foreland thrust belt, and who has added greatly to comprehension of the rest of the Canadian Cordillera and of other tectonic topics. The tourist parts I saw of the Canadian Rockies in 1942, traveling alone in the period between high school and the U.S. Navy, provided impetus to become a geologist, and my first real field work, in 1947, was in the Wapiti River front range of that mountain system. I have been back to the Canadian Rockies many times since as a geologic field-tripper. I have repeatedly studied Ray's elegantly written work and incorporated his concepts in my own research (as, here in my discussion of accretionary wedges), and Ray and his wife Mina are now my good friends.

Discussions with Don Anderson, William R. Dickinson, Gillian Foulger, Anne Hofmeister, James Natland, and E.L. Winterer have been particularly helpful in developing concepts outlined here. Many of the figures in this report were provided by others: Lou Estey, Maurice Ewing (long ago), Ralph Moberly, Gregory Moore, César Ranero, David Sandwell, Thomas Simkin, Wayne Thatcher, Ray Wells, and Dapeng Zhao. Figure 17 was drawn to my specifications by Dietrich Roeder. Manuscript reviews by Julie Baldwin, Patrick Moore, Edwin Robinson, and Claudio Vita-Finzi resulted in many improvements.

REFERENCES CITED

Altamimi, Z., Sillard, P., and Boucher, C., 2002, ITRF2000—A new release of the international terrestrial reference frame for earth science applications: Journal of Geophysical Research, v. 107, no. B10, paper ETG 2, 19 p., doi: 10.1029/2001JB000561.

Anderson, D.L., 2002a, Plate tectonics as a far-from-equilibrium self-organized system, in Stein, S., and Freymueller, J.T. eds., Plate boundary zones: American Geophysical Union Geodynamics Series, v. 30, p. 411–425.

Anderson, D.L., 2002b, The case for irreversible chemical stratification of the mantle: International Geology Review, v. 44, p. 97–116.

Anderson, D.L., 2005, Scoring hotspots—The plume and plate paradigms, in Foulger, G.R., Natland, J.H., Presnall, D.C., and Anderson, D.L., eds., Plates, Plumes, and Paradigms: Geological Society of America Special Paper 388, p. 31–54, doi: 10.1130/2005.2388(04).

Anderson, D.L., 2007, New Theory of the Earth: New York, Cambridge University Press, 384 p.

Anderson, D.L., and Natland, J.H., 2005, A brief history of the plume hypothesis and its competitors—Concept and controversy, in Foulger, G.R., Natland, J.H., Presnall, D.C., and Anderson, D.L., eds., Plates, Plumes, and Paradigms: Geological Society of America Special Paper 388, p. 119–145, doi: 10.1130/2005.2388(08).

Atwater, T., 1989, Plate tectonic history of the northeast Pacific and western North America, in Winterer, E.L., Hussong, D.M., and Decker, R.W., eds., The Eastern Pacific Ocean and Hawaii: Geological Society of America, The Geology of North America, v. N, p. 21–72.

Beaudry, D., and Moore, G.F., 1985, Seismic stratigraphy and Cenozoic evolution of West Sumatra forearc basin: American Association of Petroleum Geologists Bulletin, v. 69, p. 742–759.

Beavan, J., and Haines, J., 2001, Contemporary horizontal velocity and strain rate fields of the Pacific-Australian plate boundary through New Zealand: Journal of Geophysical Research, v. 106, p. 741–770, doi: 10.1029/2000JB900302.

Bevis, M., Taylor, F.W., Schultz, B.E., Recy, J., Isacks, B.L., Helu, S., Singh, R., Kendrick, E., Stowell, J., Taylor, B., and Calmant, S., 1995, Geodetic observations of very rapid convergence and back-arc extension at the Tonga arc: Nature, v. 374, p. 249–251, doi: 10.1038/374249a0.

Bijwaard, H., and Spakman, W., 1999, Tomographic evidence for a narrow whole-mantle plume below Iceland: Earth and Planetary Science Letters, v. 166, p. 121–126, doi: 10.1016/S0012-821X(99)00004-7.

Bourdon, B., Ribe, N.M., Stracke, A., Saal, A.E., and Turner, S.P., 2006, Insights into the dynamics of mantle plumes from uranium-series geochemistry: Nature, v. 444, p. 713–717, doi: 10.1038/nature05341.

Brudzinski, M.R., and Chen, W.-P., 2005, Earthquakes and strain in subhorizontal slabs: Journal of Geophysical Research, v. 110, no. 8, paper 303, 11 p., doi: 10.1029/2004/B003470.

Calmant, S., Pelletier, B., Lebellegard, P., Bevis, M., Taylor, F.W., and Phillips, D.A., 2003, New insights on the tectonics along the New Hebrides subduction zone based on GPS results: Journal of Geophysical Research, v. 108, no. B6, paper ETG 17, 22 p., doi: 10.1029/2001JB000644.

Campbell, I.H., and Davies, G.F., 2006, Do mantle plumes exist?: Episodes, v. 3, p. 162–168.

Carbotte, S.M., Small, C., and Donnelly, K., 2004, The influence of ridge migration on the magmatic segmentation of mid-ocean ridges: Nature, v. 429, p. 743–746, doi: 10.1038/nature02652.

Chaumillon, E., and Mascle, J., 1997, From foreland to forearc domains—New multichannel seismic reflection survey of the Mediterranean ridge accretionary complex (Eastern Mediterranean): Marine Geology, v. 138, p. 237–259, doi: 10.1016/S0025-3227(97)00002-9.

Chen, Z., Burchfiel, B.C., Liu, Y., King, R.W., Royden, L.H., Tang, W., Wang, E., Zhao, J., and Zhang, X., 2000, Global Positioning System measurements from eastern Tibet and their implications for India/Eurasia intercontinental deformation: Journal of Geophysical Research, v. 105, p. 10,215–10,227.

Chen, Z.-K., Lü, J., Wang, M., and Bürgmann, R., 2005, Contemporary crustal deformation around the southeast borderland of the Tibetan Plateau: Journal of Geophysical Research, v. 110, no. B11, paper 409, 17 p., doi: 10.1029/2004JB003421.

Chlieh, M., Avouac, J.-P., Hjorleifsdottir, V., Song, T.-R.A., Ji, C., Sieh, K., Staden, A., Hebert, H., Prawirodirdjo, L., Bock, Y., and Galetzka, J., 2007, Coseismic slip and afterslip of the great M_w 9.15 Sumatra-Andaman earthquake of 2004: Bulletin of the Seismological Society of America, v. 97, p. S152–S173, doi: 10.1785/0120050631.

Chudinovskikh, L., and Boehler, R., 2004, $MgSiO_3$ phase boundaries measured in the laser-heated diamond cell: Earth and Planetary Science Letters, v. 219, p. 285–296, doi: 10.1016/S0012-821X(04)00005-6.

Clouard, V., and Bonneville, A., 2005, Ages of seamounts, islands, and plateaus on the Pacific plate, in Foulger, G.R., Natland, J.H., Presnall, D.C., and Anderson, D.L., eds., Plates, Plumes, and Paradigms: Geological Society of America Special Paper 388, p. 71–90, doi: 10.1130/2005.2388(06).

Collot, J.-Y., Charvis, P., Gutscher, M.-A., and Operto, S., 2002, Exploring the Ecuador-Colombia active margin and interplate seismic zone: Eos (Transactions, American Geophysical Union), v. 83, p. 185, 189–190, doi: 10.1029/2002EO000120.

Conrad, C.P., and Lithgow-Bertelloni, C., 2004, The temporal evolution of plate driving forces—Importance of "slab suction" versus "slab pull" during the Cenozoic: Journal of Geophysical Research, v. 109, no. B10, paper 407, 14 p., doi: 10.1029/2004JB002991.

Constenius, K.N., Johnson, R.A., Dickinson, W.R., and Williams, T.A., 2000, Tectonic evolution of the Jurassic-Cretaceous Great Valley forearc, California—Implications for the Franciscan thrust-wedge hypothesis: Geological Society of America Bulletin, v. 112, p. 1703–1723, doi: 10.1130/0016-7606(2000)112<1703:TEOTJC>2.0.CO;2.

Costa Pisani, P., Reshef, M., and Moore, G., 2005, Targeted 3-D prestack depth imaging at Legs 190–196 ODP drill sites (Nankai Trough, Japan): Geophysical Research Letters, v. 32, L20309, 4 p., doi: 10.1029/2005GL024191.

Coulbourn, W.T., and Moberly, R., 1977, Structural evidence of the evolution of fore-arc basins off South America: Canadian Journal of Earth Sciences, v. 14, p. 102–116.

Deal, M.M., Nolet, G., and van der Hilst, R.D., 1999, Slab temperature and thickness from seismic tomography: 1. Method and application to Tonga: Journal of Geophysical Research, v. 104, p. 26,789–26,802.

Debaille, V., Blichert-Toft, J., Agranier, A., Doucelance, R., Schiano, P., and Albarede, F., 2006, Geochemical component relationships in MORB from the Mid-Atlantic Ridge, 22–35°N: Earth and Planetary Science Letters, v. 241, p. 844–862, doi: 10.1016/j.epsl.2005.11.004.

DeMets, C., Gordon, R.G., Argus, D.F., and Stein, S., 1990, Current plate motions: Geophysical Journal International, v. 101, p. 425–578.

Dickinson, W.R., 1995, Forearc basins, in Busby, C.J., and Ingersoll, R.V., eds., Tectonics of Sedimentary Basins: Cambridge, Massachusetts, Blackwell Science, p. 221–261.

Dickinson, W.R., and Seely, D.R., 1979, Structure and stratigraphy of forearc regions: American Association of Petroleum Geologists Bulletin, v. 63, p. 2–31.

Dilek, Y., 2006, Collision tectonics of the Mediterranean region—Causes and consequences, in Dilek, Y., and Pavlides, S., eds., Postcollisional Tectonics and Magmatism in the Mediterranean Region and Asia: Geological Society of America Special Paper 409, p. 1–13.

Dziewonski, A.M., 2005, The robust aspects of global seismic tomography, in Foulger, G.R., Natland, J.H., Presnall, D.C., and Anderson, D.L., eds., Plates, Plumes, and Paradigms: Geological Society of America Special Paper 388, p. 147–154, doi: 10.1130/2005.2388(09).

Engdahl, E.R., van der Hilst, R., and Buland, R., 1998, Global teleseismic earthquake relocation with improved travel times and procedures for depth determination: Bulletin of the Seismological Society of America, v. 88, p. 722–743.

Estabrook, C.H., and Jacob, K.H., 1991, Stress indicators in Alaska, in Slemmons, D.B., Engdahl, E.R., Zoback, M.D., and Blackwell, D.D., eds., Neotectonics of North America: Geological Society of America, Decade Map, v. 1, p. 387–399.

Faccenna, C., Davy, P., Brun, J.-P., Funicello, R., Giardini, D., Mattei, M., and Nalpas, T., 1996, The dynamics of back-arc extension—An experimental approach to the opening of the Tyrrhenian Sea: Geophysical Journal International, v. 126, p. 781–795.

Faccenna, C., Funiciello, F., Giardini, D., and Lucente, P., 2001, Episodic back-arc extension during restricted mantle convection in the central Mediterranean: Earth and Planetary Science Letters, v. 187, p. 105–116, doi: 10.1016/S0012-821X(01)00280-1.

Falloon, T.J., Green, D.H., and Danyushevsky, L.V., 2007, Crystallization temperatures of tholeiite parental liquids: Implications for the existence of thermally driven mantle plumes, in Foulger, G.R., and Jurdy, D.M., eds., Plates, Plumes, and Planetary Processes: Geological Society of America Special Paper 430, p. 235–260, doi: 10.1130/2007.2430(12).

Fitton, J.G., 2007, The OIB paradox, *in* Foulger, G.R., and Jurdy, D.M., eds., Plates, Plumes, and Planetary Processes: Geological Society of America Special Paper 430, p. 387–412, doi: 10.1130/2007.2430(20).

Foulger, G.R., and Jurdy, D.M., eds., 2007, Plates, Plumes, and Planetary Processes: Geological Society of America Special Paper 430, doi: 10.1130/2007.2430.

Foulger, G.R., Pritchard, M.J., Julian, B.R., Evans, J.R., Allen, R.M., Nolet, G., Morgan, W.J., Bergeson, B.H., Erlendsson, P., Jakobsdottir, S., Ragnarsson, S., Stefansson, R., and Vogfjord, K., 2001, Seismic tomography shows that upwelling beneath Iceland is confined to the upper mantle: Geophysical Journal International, v. 146, p. 504–530, doi: 10.1046/j.0956-540x.2001.01470.x.

Foulger, G.R., Natland, J.H., Presnall, D.C., and Anderson, D.L., eds., 2005a, Plates, Plumes, and Paradigms: Geological Society of America Special Paper 388, 881 p., doi: 10.1130/2005.2388.

Foulger, G.R., Natland, J.H., and Anderson, D.L., 2005b, Genesis of the Iceland melt anomaly by plate tectonic processes, *in* Foulger, G.R., Natland, J.H., Presnall, D.C., and Anderson, D.L., eds., Plates, Plumes, and Paradigms: Geological Society of America Special Paper 388, p. 595–625, doi: 10.1130/2005.2388(35).

Fukao, Y., Obayashi, M., Inoue, H., and Nenbai, M., 1992, Subducting slabs stagnant in the mantle transition zone: Journal of Geophysical Research, v. 97, p. 4809–4822.

Glen, W., 2005, The origin and early trajectory of the mantle plume quasi-paradigm, *in* Foulger, G.R., Natland, J.H., Presnall, D.C., and Anderson, D.L., eds., Plates, Plumes, and Paradigms: Geological Society of America Special Paper 388, p. 91–117, doi: 10.1130/2005.2388(07).

Grand, S., van der Hilst, R., and Widiyantoro, S., 1997, Global seismic tomography—A snapshot of convection in the Earth: GSA Today, v. 7, no. 4, p. 1–7.

Gripp, A.E., and Gordon, R.G., 2002, Young tracks of hotspots and current plate velocities: Geophysical Journal International, v. 150, p. 321–361, doi: 10.1046/j.1365-246X.2002.01627.x.

Gu, Y.J., Dziewonski, A.M., Su, W.-J., and Ekström, G., 2001, Models of the mantle shear velocity and discontinuities in the patterns of lateral heterogeneities: Journal of Geophysical Research, v. 106, p. 11,169–11,199, doi: 10.1029/2001JB000340.

Hamilton, W.B., 1978, Mesozoic tectonics of the western United States, *in* Howell, D.C., and McDougall, K.A., eds., Mesozoic Paleogeography of the Western United States: Sacramento, Pacific Section, Society of Economic Paleontologists and Mineralogists, Pacific Coast Paleogeography Symposium 2, p. 33–70.

Hamilton, W.B., 1979, Tectonics of the Indonesian region: U.S. Geological Survey Professional Paper 1078, 345 p., and 1:5,000,000 tectonic map.

Hamilton, W.B., 1988, Plate tectonics and island arcs: Geological Society of America Bulletin, v. 100, p. 1503–1527, doi: 10.1130/0016-7606(1988) 100<1503:PTAIA>2.3.CO;2.

Hamilton, W.B., 1995, Subduction systems and magmatism, *in* Smellie, J.L., ed., Volcanism associated with extension at consuming plate margins: Geological Society of London Special Publication 81, p. 3–28.

Hamilton, W.B., 2002, The closed upper-mantle circulation of plate tectonics, *in* Stein, S., and Freymueller, J.T., eds., Plate boundary zones: American Geophysical Union Geodynamics Series, v. 30, p. 359–410.

Hamilton, W.B., 2003, An alternative Earth: GSA Today, v. 13, no. 11, p. 4–12, doi: 10.1130/1052-5173(2003)013<0004:AAE>2.0.CO;2.

Hamilton, W.B., 2007, Earth's first two billion years—The era of internally mobile crust, *in* Hatcher, R.D., Jr., Carlson, M.P., McBride, J.H., and Martínez Catalán, J.R., eds., 4-D Framework of Continental Crust: Geological Society of America Memoir 200, p. 233–296, doi: 10.1130/2007.1200(13).

Hayes, D.E., and Lewis, S.D., 1984, A geophysical study of the Manila Trench, Luzon, Philippines: 1. Crustal structure, gravity, and regional tectonic evolution: Journal of Geophysical Research, v. 89, p. 9171–9195.

Herzberg, C., 2006, Petrology and thermal structure of the Hawaiian plume from Mauna Kea volcano: Nature, v. 444, p. 605–609, doi: 10.1038/nature05254.

Heuret, A., and Lallemand, S., 2005, Plate motions, slab dynamics and back-arc deformation: Physics of the Earth and Planetary Interiors, v. 149, p. 31–51, doi: 10.1016/j.pepi.2004.08.022.

Hinz, K., von Huene, R., and Ranero, C.R., 1996, Tectonic structure of the convergent Pacific margin offshore Costa Rica from multichannel seismic reflection data: Tectonics, v. 15, p. 54–66, doi: 10.1029/95TC02355.

Hofmeister, A.M., 2005, Dependence of diffusive radiative transfer on grain-size, temperature, and Fe-content—Implications for mantle processes: Journal of Geodynamics, v. 40, p. 51–72, doi: 10.1016/j.jog.2005.06.001.

Hofmeister, A.M., and Criss, R.E., 2005a, Heatflow and mantle convection in the triaxial Earth, *in* Foulger, G.R., Natland, J.H., Presnall, D.C., and Anderson, D.L., eds., Plates, Plumes, and Paradigms: Geological Society of America Special Paper 388, p. 289–302, doi: 10.1130/2005.2388(18).

Hofmeister, A.M., and Criss, R.E., 2005b, Earth's heat flux revised and linked to chemistry: Tectonophysics, v. 395, p. 159–177, doi: 10.1016/j.tecto.2004.09.006.

Hofmeister, A.M., and Criss, R.E., 2005c, Reply to comments by R. Von Herzen, E.E. Davis, A.T. Fisher, C.A. Stein, and H.N. Pollack on "Earth's heat flux revised and linked to chemistry": Tectonophysics, v. 409, p. 199–203, doi: 10.1016/j.tecto.2005.08.004.

Hofmeister, A.M., and Criss, R.E., 2006, Comment on "Estimates of heat flow from Cenozoic seafloor using global depth and age data" by M. Wei and D: Sandwell: Tectonophysics, v. 428, p. 95–100, doi: 10.1016/j.tecto.2006.08.010.

Huang, J., and Zhao, D., 2006, High-resolution mantle tomography of China and surrounding regions: Journal of Geophysical Research, v. 111, no. B9, paper 305, 21 p., doi: 10.1920/2005JB004066.

Humphreys, E.D., Dueker, K.G., Schutt, D.L., and Smith, R.B., 2000, Beneath Yellowstone—Evaluating plume and nonplume models teleseismic images of the upper mantle: GSA Today, v. 10, no. 12, p. 1–7.

Ishii, M., and Tromp, J., 2004, Constraining large-scale mantle heterogeneity using mantle and inner-core sensitive normal modes: Physics of the Earth and Planetary Interiors, v. 146, p. 113–124, doi: 10.1016/j.pepi.2003.06.012.

Jacobsen, S.D., Spetzler, H., Reichmann, H.J., and Smyth, J.R., 2004, Shear waves in the diamond-anvil cell reveal pressure-induced instability in (Mg,Fe)O: Proceedings of the National Academy of Sciences of the United States of America, v. 101, p. 5867–5871, doi: 10.1073/pnas.0401564101.

Jellinek, A.M., and Manga, M., 2004, Links between long-lived hot spots, mantle plumes, D", and plate tectonics: Reviews of Geophysics, v. 42, paper RG3002, 35 p.

Jordan, B.T., 2005, Age-progressive volcanism of the Oregon High Lava Plains—Overview and evaluation of tectonic models, *in* Foulger, G.R., Natland, J.H., Presnall, D.C., and Anderson, D.L., eds., Plates, Plumes, and Paradigms: Geological Society of America Special Paper 388, p. 503–515, doi: 10.1130/2005.2388(30).

Julian, B.R., 2005, What can seismology say about hotspots?, *in* Foulger, G.R., Natland, J.H., Presnall, D.C., and Anderson, D.L., eds., Plates, Plumes, and Paradigms: Geological Society of America Special Paper 388, p. 155–169, doi: 10.1130/2005.2388(10).

Karig, D.E., 1971, Origin and development of marginal basins in the western Pacific: Journal of Geophysical Research, v. 76, p. 2542–2561.

Kato, T., Beavan, J., Matsushima, T., Kotake, Y., Camacho, J.T., and Nakao, S., 2003, Geodetic evidence of back-arc spreading in the Mariana Trough: Geophysical Research Letters, v. 30, no. 12, paper 27, 4 p., doi: 10.1029/2002GL016757.

Katz, R.F., Spiegelman, M., and Carbotte, S.M., 2004, Ridge migration, asthenospheric flow and the origin of magmatic segmentation in the global mid-ocean ridge system: Geophysical Research Letters, v. 31, paper L15605, 4 p., doi: 10.1029/2004GL020388.

Kopp, H., Flueh, E.R., Klaeschen, D., Bialas, J., and Reichert, C., 2001, Crustal structure of the central Sunda margin at the onset of oblique subduction: Geophysical Journal International, v. 147, p. 449–474, doi: 10.1046/j.0956-540x.2001.01547.x.

Kopp, H., Klaeschen, D., Flueh, E.R., and Bialas, J., 2002, Crustal structure of the Java margin from seismic wide-angle and multichannel reflection data: Journal of Geophysical Research, v. 107, no. B2, paper ETG-1, 24 p., doi: 10.1029/2000BJ000095.

Krabbenhöft, A., Bialas, J., Kopp, H., Kukowski, N., and Hübscher, C., 2004, Crustal structure of the Peruvian continental margin from wide-angle seismic studies: Geophysical Journal International, v. 159, p. 749–764, doi: 10.1111/j.1365-246X.2004.02425.x.

Kreemer, C., Holt, W.E., Goes, S., and Govers, R., 2000, Active deformation in eastern Indonesia and the Philippines from GPS and seismicity data: Journal of Geophysical Research, v. 105, p. 663–680, doi: 10.1029/1999JB900356.

Kreemer, C., Holt, W.E., and Haines, A.J., 2003, An integrated global model of present-day plate motions and plate boundary deformation: Geophysical Journal International, v. 154, p. 8–34.

Lay, T., Kanamori, H., Ammon, C.J., Nettles, M., Ward, S.N., Aster, R.C., Beck, S.L., Bilek, S.L., Brudzinski, M.R., Butler, R., DeShon, H.R., Ekström, G., Satake, K., and Sipkin, S., 2005, The great Sumatra-Andaman earthquake of 26 December 2004: Science, v. 308, p. 1127–1133, doi: 10.1126/science.1112250.

Lewis, S.D., and Hayes, D.E., 1984, A geophysical study of the Manila Trench, Luzon, Philippines: 2. Forearc basin structural and stratigraphic evolution: Journal of Geophysical Research, v. 89, p. 9196–9214.

Lucente, F.P., Margheriti, L., Piromallo, C., and Barruol, G., 2006, Seismic anisotropy reveals the long route of the slab through the western-central Mediterranean mantle: Earth and Planetary Science Letters, v. 241, p. 517–529, doi: 10.1016/j.epsl.2005.10.041.

Matson, R.G., and Moore, G.F., 1992, Structural influences on Neogene subsidence in the central Sumatra fore-arc basin, in Watkins, J.S., Montadert, L., and Dickerson, P.W., eds., Geology and geophysics of continental margins: American Association of Petroleum Geologists Memoir 53, p. 157–181.

Mazzotti, S., Le Pichon, X., Henry, P., and Miyazaki, S.-I., 2000, Full interseismic locking of the Nankai and Japan-west Kurile subduction zones—An analysis of uniform elastic strain accumulation in Japan constrained by permanent GPS: Journal of Geophysical Research, v. 105, p. 13,159–13,177, doi: 10.1029/2000JB900060.

McCaffrey, R., 1991, Slip vectors and stretching of the Sumatran fore arc: Geology, v. 19, p. 881–884, doi: 10.1130/0091-7613(1991)019<0881:SVASOT>2.3.CO;2.

McKenzie, D., Stracke, A., Blichert-Toft, J., Albarède, F., Grönvold, K., and O'Nions, R.K., 2004, Source enrichment processes responsible for isotopic anomalies in oceanic island basalts: Geochimica et Cosmochimica Acta, v. 68, p. 2699–2724, doi: 10.1016/j.gca.2003.10.029.

Melbourne, T.I., Webb, F.H., Stock, J.M., and Reigber, C., 2002, Rapid postseismic transients in subduction zones from continuous GPS: Journal of Geophysical Research, v. 107, no. B10, paper ETG-10, 10 p., doi: 10.1029/2001JB000555.

Michel, G.W., Becker, M., Reigber, C., Tibi, R., Yu, Y.Q., and Zhu, S.Y., 2001, Regional GPS data confirm high strain accumulation prior to the 2000 June 4 M_w = 7.8 earthquake at southeast Sumatra: Geophysical Journal International, v. 155, p. 221–240.

Moberly, R., Shepherd, G.L., and Coulbourn, W.T., 1982, Forearc and other basins, continental margin of northern and southern Peru and adjacent Ecuador and Chile, in Leggett, J.K., ed., Trench-Forearc Geology—Sedimentation and Tectonics in Modern and Ancient Active Plate Margins: Geological Society of London Special Publication 10, p. 171–189.

Montelli, R., Nolet, G., Dahlen, F.A., Masters, G., Engdahl, E.R., and Hung, S.-H., 2004, Finite frequency tomography reveals a variety of plumes in the mantle: Science, v. 303, p. 338–343, doi: 10.1126/science.1092485.

Montelli, R., Nolet, G., Dahlen, F.A., and Masters, G., 2006, A catalogue of deep mantle plumes—New results from finite-frequency tomography: Geochemistry, Geophysics, Geosystems, v. 7, no. 11, 69 p.

Mordojovich-K., C., 1974, Geology of a part of the Pacific margin of Chile, in Burk, C.A., and Drake, C.L., eds., The Geology of Continental Margins: New York, Springer-Verlag, p. 491–598.

Müller, R.D., Roest, W.R., Royer, J.Y., Gahagan, L.M., and Sclater, J.G., 1997, Digital isochrons of the world's ocean floor: Journal of Geophysical Research, v. 102, p. 3211–3214, doi: 10.1029/96JB01781.

Natland, J.H., 2007, ΔNb and the role of magma mixing at the East Pacific Rise and Iceland, in Foulger, G.R., and Jurdy, D.M., eds., Plates, Plumes, and Planetary Processes: Geological Society of America Special Paper 430, p. 413–449, doi: 10.1130/2007.2430(21).

Natland, J.H., and Winterer, E.L., 2005, Fissure control on volcanic action in the Pacific, in Foulger, G.R., Natland, J.H., Presnall, D.C., and Anderson, D.L., eds., Plates, Plumes, and Paradigms: Geological Society of America Special Paper 388, p. 687–710, doi: 10.1130/2005.2388(39).

Norton, I.O., 2007, Speculations on Cretaceous tectonic history of the Northwest Pacific and a tectonic origin for the Hawaii hotspot, in Foulger, G.R., and Jurdy, D.M., eds., Plates, Plumes, and Planetary Processes: Geological Society of America Special Paper 430, p. 451–470, doi: 10.1130/2007.2430(22).

Nyst, M., and Thatcher, W., 2004, New constraints on the active tectonic deformation of the Aegean: Journal of Geophysical Research, v. 109, no. B11, paper 406, 23 p., doi: 10.1029/2003JB002830.

Piromallo, C., and Morelli, A., 2003, P wave tomography of the mantle under the Alpine-Mediterranean area: Journal of Geophysical Research, v. 108, no. B2, paper ESE 1, 23 p.

Ranero, C.R., Morgan, J.P., and Reichert, C., 2003, Bending-related faulting and mantle serpentinization at the Middle America Trench: Nature, v. 425, p. 367–373, doi: 10.1038/nature01961.

Reilinger, R., and 24 others, 2006, GPS constraints on continental deformation in the Africa-Arabia-Eurasia continental collision zone and implications for the dynamics of plate interactions: Journal of Geophysical Research, v. 111, no. B5, paper 411, 26 p., doi: 10.1029/2005JB004051.

Ren, Y., Stutzmann, E., van der Hilst, R.D.., and Besse, J., 2007, Understanding seismic heterogeneities in the lower mantle beneath the Americas from seismic tomography and plate tectonic history: Journal of Geophysical Research, v. 112, no. B1, 15 p., doi: 10.1029/2005JB004154.

Replumaz, A., Kárason, H., van der Hilst, R.D., Besse, J., and Tapponnier, P., 2004, 4-D evolution of SE Asia's mantle from geological reconstructions and seismic tomography: Earth and Planetary Science Letters, v. 221, p. 103–115, doi: 10.1016/S0012-821X(04)00070-6.

Reston, T.J., Fruehn, J., von Huene, R., and IMERSE Working Group, 2002, The structure and evolution of the western Mediterranean Ridge: Marine Geology, v. 186, p. 83–110, doi: 10.1016/S0025-3227(02)00174-3.

Ritsema, J., 2005, Global seismic structure maps, in Foulger, G.R., Natland, J.H., Presnall, D.C., and Anderson, D.L., eds., Plates, Plumes, and Paradigms: Geological Society of America Special Paper 388, p. 11–18, doi: 10.1130/2005.2388(02).

Rizzetto, C., Marotta, A.M., and Sabadini, R., 2004, The role of trench retreat on the geometry and stress regime in the subduction complexes of the Mediterranean: Geophysical Research Letters, v. 31, L11604, doi: 10.1029/2004GL019889, 4 p.

Sager, W.W., 2007, Divergence between paleomagnetic and hotspot model predicted polar wander for the Pacific plate with implications for hotspot fixity, in Foulger, G.R., and Jurdy, D.M., eds., Plates, Plumes, and Planetary Processes: Geological Society of America Special Paper 430, doi: 10.1130/2007.2430(17).

Schellart, W.P., and Lister, G.S., 2005, The role of the East Asian active margin in widespread extensional and strike-slip deformation in East Asia: Geological Society of London Journal, v. 162, p. 959–972, doi: 10.1144/0016-764904-112.

Schellart, W.P., Lister, G.S., and Toy, V.G., 2006, A Late Cretaceous and Cenozoic reconstruction of the Southwest Pacific region—Tectonics controlled by subduction and slab rollback processes: Earth-Science Reviews, v. 76, p. 191–233, doi: 10.1016/j.earscirev.2006.01.002.

Schellart, W.P., Freeman, J., Stegman, D.R., Moresi, L., and May, D., 2007, Evolution and diversity of subduction zones controlled by slab width: Nature, v. 446, p. 308–311.

Sella, G.F., Dixon, T.H., and Mao, A., 2002, REVEL—A model for Recent plate velocities from space geodesy: Journal of Geophysical Research, v. 107, no. B4, paper ETG-11, 32 p., doi: 10.1029/2000JB000033.

Shor, G.G., Jr., 1974, Continental margin of Middle America, in Burk, C.A., and Drake, C.L., eds., The Geology of Continental Margins: New York, Springer-Verlag, p. 599–602.

Shyu, J.B.H., Sieh, K., Chen, Y.-G., and Liu, C.-S., 2005, Neotectonic architecture of Taiwan and its implications for future large earthquakes: Journal of Geophysical Research, v. 110, no. B8, paper 402, 33 p., doi: 10.1029/2004JB003251.

Simkin, T., Tilling, R.J., Taggart, J.N., Jones, W.J., and Spall, H., 1989, This Dynamic Planet—World Map of Volcanoes, Earthquakes, and Plate Tectonics: U.S. Geological Survey. Map I-2800, scale 1:30,000,000.

Simoes, M., Avouac, J.P., Cattin, R., and Henry, P., 2004, The Sumatra subduction zone—A case for a locked fault zone extending into the mantle: Journal of Geophysical Research, v. 109, doi: 10.1029/2003JB002958.

Sleep, N.H., 2006, Mantle plumes from top to bottom: Earth-Science Reviews, v. 77, p. 231–271, doi: 10.1016/j.earscirev.2006.03.007.

Socquet, A., Simons, W., Vigny, C., McCaffrey, R., Subarya, C., Sarsito, D., Ambrosius, B., and Spakman, W., 2006, Microblock rotations and fault coupling in SE Asia triple junction (Sulawesi, Indonesia) from GPS and earthquake slip vector data: Journal of Geophysical Research, v. 111, no. B8, paper 409, 15 p., doi: 10.1029/2005JB003963.

Stern, R.J., 2005, Evidence from ophiolites, blueschists, and ultrahigh-pressure metamorphic terranes that the modern episode of subduction tectonics began in Neoproterozoic time: Geology, v. 33, p. 557–560, doi: 10.1130/G21365.1.

Stern, R.J., 2007, When and how did plate tectonics begin? Theoretical and empirical considerations: Chinese Science Bulletin, v. 52, p. 578–591.

Stuart, W.D., Foulger, G.R., and Barall, M., 2007, Propagation of the Hawaiian-Emperor volcano chain by Pacific plate cooling stress, *in* Foulger, G.R., and Jurdy, D.M., eds., Plates, Plumes, and Planetary Processes: Geological Society of America Special Paper 430, p. 497–506, doi: 10.1130/2007.2430(24).

Subarya, C., Chlieh, M., Prawirodirdjo, L., Avouac, J.-P., Bock, Y., Sieh, K., Meltzner, A.J., Natawidjaja, D.H., and McCaffrey, R., 2006, Plate-boundary deformation associated with the great Sumatra-Andaman earthquake: Nature, v. 440, p. 46–51, doi: 10.1038/nature04522.

Suter, M., López Martínez, M., Quintero Legorreta, O., and Carillo Martínez, M., 2001, Quaternary intra-arc extension in the central Trans-Mexican volcanic belt: Geological Society of America Bulletin, v. 113, p. 693–703, doi: 10.1130/0016-7606(2001)113<0693:QIAEIT>2.0.CO;2.

Suyehiro, K., and Nishizawa, A., 1994, Crustal structure and seismicity beneath the forearc off northeastern Japan: Journal of Geophysical Research, v. 99, p. 22,331–22,347, doi: 10.1029/94JB01337.

Tarduno, J.A., Duncan, R.A., Scholl, D.W., Cottrell, R.D., Steinberger, B., Thordarson, T., Kerr, B.C., Neal, C.R., Frey, F.A., Torii, M., and Carvallo, C., 2003, The Emperor Seamounts—Southward motion of the Hawaiian hotspot plume in Earth's mantle: Science, v. 301, p. 1064–1069, doi: 10.1126/science.1086442.

Taylor, B., Klaus, A., Brown, G.R., Moore, G.F., Okamura, Y., and Murakami, F., 1991, Structural development of Sumisu Rift, Izu-Bonin arc: Journal of Geophysical Research, v. 96, p. 16,113–16,129.

Taylor, G.K., Gascoyne, J., and Colley, H., 2000, Rapid rotation of Fiji—Paleomagnetic evidence and tectonic implications: Journal of Geophysical Research, v. 105, p. 5771–5782, doi: 10.1029/1999JB900305.

ten Veen, J., and Kleinspehn, K.I., 2003, Incipient continental collision and plate-boundary curvature—Late Pliocene–Holocene transtensional Hellenic forearc, Crete, Greece: Geological Society of London Journal, v. 160, p. 161–181.

Tilling, R.I., Vogt, P.R., Kirby, S.E., Kimberly, P., and Stewart, R.B., 2006, This Dynamic Planet: U.S. Geological Survey Geologic Investigations Series Map I-2800, scale 1:30,000,000.

Trampert, J., Deschamps, F., Resovsky, J., and Yuen, D., 2004, Probabilistic tomography maps chemical heterogeneities throughout the lower mantle: Science, v. 306, p. 853–856, doi: 10.1126/science.1101996.

Tsujimori, T., Sisson, V.B., Liou, J.G., Harlow, G.E., and Sorenson, S.S., 2006, Very-low-temperature record of the subduction process—A review of worldwide lawsonite eclogites: Lithos, v. 92, p. 609–624, doi: 10.1016/j.lithos.2006.03.054.

van der Hilst, R., 1995, Complex morphology of subducted lithosphere in the mantle beneath the Tonga Trench: Nature, v. 374, p. 154–157, doi: 10.1038/374154a0.

Van der Voo, R., Bijwaard, R.H., and Spakman, W., 1999, Tethyan subducted slabs under India: Earth and Planetary Science Letters, v. 171, p. 7–20, doi: 10.1016/S0012-821X(99)00131-4.

Vanneste, L.E., Larter, R.D., and Smythe, D.K., 2002, Slice of intraoceanic arc—Insights from the first multichannel seismic reflection profile across the South Sandwich island arc: Geology, v. 30, p. 819–822, doi: 10.1130/0091-7613(2002)030<0819:SOIAIF>2.0.CO;2.

Von Herzen, R., Davis, K.E., Fisher, A.T., Stein, C.A., and Pollack, H.N., 2005, Earth's heat flux revised and linked to chemistry; Discussion: Tectonophysics, v. 409, p. 193–198.

Von Huene, R., Moore, G.W., Moore, J.C., and Stephens, C.D., 1979, Cross section, Alaska Peninsula–Kodiak Island–Aleutian Trench—Summary: Geological Society of America Bulletin, v. 90, pt. 1, p. 427–430, doi: 10.1130/0016-7606(1979)90<427:CSAPIT>2.0.CO;2.

Waite, G.P., Smith, R.B., and Allen, R.M., 2006, V_P and V_S structure of the Yellowstone hot spot from teleseismic tomography—Evidence for an upper mantle plume: Journal of Geophysical Research, v. 111, no. B4, paper 303, 21 p., doi: 10.1029/2005JB003867.

Walpersdorf, A., Vigny, C., Manuring, P., Sabarya, C., and Sutisna, S., 1998, Determining the Sula block kinematics in the triple junction area in Indonesia by GPS: Geophysical Journal International, v. 135, p. 351–361, doi: 10.1046/j.1365-246X.1998.00641.x.

Wang, Z., and Zhao, D., 2005, Seismic imaging of the entire arc of Tohoku and Hokkaido in Japan using P-wave, S-wave, and sP depth-phase data: Physics of the Earth and Planetary Interiors, v. 152, p. 144–162, doi: 10.1016/j.pepi.2005.06.010.

Wei, M., and Sandwell, D., 2006a, Estimates of heat flow from Cenozoic seafloor using global depth and age data: Tectonophysics, v. 417, p. 325–335, doi: 10.1016/j.tecto.2006.02.004.

Wei, M., and Sandwell, D., 2006b, Reply to comment on "Estimates of heat flow from Cenozoic seafloor using global depth and age data": Tectonophysics, v. 428, p. 101–103, doi: 10.1016/j.tecto.2006.08.007.

Wells, R.E., Blakely, R.J., Sugiyama, Y., Scholl, D.W., and Dinterman, P.A., 2003, Basin-centered asperities in great subduction zone earthquakes—A link between slip, subsidence, and subduction erosion?: Journal of Geophysical Research, v. 108, no. B10, paper ESE 16, 30 p., 2507, doi: 10.1029/2002JB002072.

Widiyantoro, S., and van der Hilst, R., 1997, Mantle structure beneath Indonesian inferred from high-resolution tomographic imaging: Geophysical Journal International, v. 130, p. 167–182.

Zhao, D., 2001, Seismological structure of subduction zones and its implications for arc magmatism and dynamics: Physics of the Earth and Planetary Interiors, v. 127, p. 197–214, doi: 10.1016/S0031-9201(01)00228-X.

Zhao, D., 2004, Global tomographic images of mantle plumes and subducting slabs—Insight into deep Earth dynamics: Physics of the Earth and Planetary Interiors, v. 146, p. 3–34, doi: 10.1016/j.pepi.2003.07.032.

Zhao, D., Lei, J., and Tang, R., 2004, Origin of the Changbai intraplate volcanism in northeast China—Evidence from seismic tomography: Chinese Science Bulletin, v. 49, p. 1401–1408, doi: 10.1360/04wd0125.

MANUSCRIPT ACCEPTED BY THE SOCIETY 22 MARCH 2007

Petrotectonics of ultrahigh-pressure crustal and upper-mantle rocks—Implications for Phanerozoic collisional orogens

W.G. Ernst
Department of Geological and Environmental Sciences, Stanford University, Stanford, California 94305-2115, USA

B.R. Hacker
Department of Geological Sciences, University of California, Santa Barbara, California 93106-9630, USA

J.G. Liou
Department of Geological and Environmental Sciences, Stanford University, Stanford, California 94305-2115, USA

ABSTRACT

Ultrahigh-pressure (UHP) metamorphic terranes in contractional orogens reflect descent of continental crust bonded to a dense, dominantly oceanic plate to depths of 90–140 km. All recognized well-documented UHP complexes formed during Phanerozoic time. Rocks are intensely retrogressed to low-pressure assemblages, with rare relict UHP phases retained in tough, refractory host minerals. Resurrected UHP slabs consist chiefly of quartzofeldspathic rocks and serpentinites; dense mafic + ultramafic lithologies comprise <10% of exhumed masses. Associated garnet-bearing ultramafic lenses are of four general origins: type A peridotite + eclogite pods reflect premetamorphic residence in the mantle wedge; type B masses were mantle-derived ultramafic-mafic magmas that rose into the crust prior to subduction; type C tectonic lenses were present in the oceanic lithosphere prior to underflow; and type D garnet peridotites achieved their deep-seated mantle mineralogy long before—and independent of—the subduction event that produced the UHP-phase assemblages in garnet peridotite types A, B, and C. Geochronology constrains the timing of protolith, peak, and retrograde recrystallization of gneissic, ultramafic, and eclogitic rocks. Round-trip pressure-temperature (P-T) paths were completed in <5–10 m.y., where ascent rates approximated subduction velocities. Exhumation from profound depth involves near-adiabatic decompression through P-T fields of much lower-pressure metamorphic facies. Many complexes consist of thin, allochthonous sheets, but those in eastern China and western Norway are about 10 km thick. Ductilely deformed nappes generated in subduction zones allow heat to be conducted away as sheet-like UHP complexes rise, cooling across both upper and lower surfaces. Thicker UHP massifs also must be quenched. Ascent along the subduction channel is driven mainly by buoyancy of low-density crustal material relative to the surrounding mantle. Rapid exhumation prevents establishment of a more normal geothermal regime in the subduction zone. Lack of H_2O impedes back reaction, whereas its presence accelerates transformation to low-P phase assemblages. Late-stage domal uplifts characterize some collisional terranes; erosion, combined with underplating, contraction, tectonic aneurysms, and/or lithospheric plate shallowing, may further elevate mid-crustal UHP terranes toward the surface.

Keywords: ultrahigh-pressure metamorphism, subduction-zone metamorphism, continental collision, exhumation of UHP rocks.

INTRODUCTION

Most compressional mountain belts form at or near the active edges of continents and/or fringing island arcs. Virtually all result from the underflow of oceanic lithosphere and the consequent transport and descent of spreading centers, oceanic plateaus, island arcs, far-traveled microcontinental terranes, and/or continental crustal salients beneath the continental lithosphere. The downgoing slab is subjected to relatively high-pressure (HP), low-temperature subduction-zone metamorphism, which produces lawsonite and jadeitic pyroxene-bearing assemblages, and mafic blueschists. Long-continued subduction results in the construction of a massive calc-alkaline volcanic-plutonic arc on the crust of the stable, hanging-wall plate, but consumption of a small intervening ocean basin prior to collision does not generate a substantial arc. Most mountain chains are a reflection of their specific geography and unique plate-tectonic history; each orogen tends to exhibit major structural and petrologic contrasts along its length. Some sialic collisional belts contain mineralogic relics reflecting ultrahigh-pressure (UHP) stages of prograde recrystallization. Ultrahigh-pressure conditions are defined as those in which the high-pressure polymorphs of silica and carbon (i.e., coesite and diamond) are stable. Other dense phases and mineral assemblages, including Si- and K-bearing pyroxene, Mg-rich garnet, and eclogite-facies rocks are stable under such remarkable pressure-temperature (P-T) conditions.

Two main end members have been defined, but it is clear that all gradations exist between continent collisional (Alpine-type) and circum-Pacific (Pacific-type) compressional mountain belts. Similar to Pacific HP metamorphic belts, UHP Alpine orogens mark convergent plate junctions (e.g., Hacker et al., 2003a; Ernst, 2005). The former are characterized by the underflow of thousands of kilometers of oceanic lithosphere, whereas the latter involve the consumption of an intervening ocean basin followed by the suturing of an outboard island arc, microcontinent, or promontory of sialic crust against the nonsubducted continental margin. During collision, crustal sections may reach depths approaching 90–140 km, as indicated by the metamorphic crystallization of UHP indicator minerals, phases, and assemblages that are only stable at pressures exceeding ~2.5 GPa. On resurrection, many collisional UHP terranes consist of an imbricate stack of tabular sheets (Ernst et al., 1997). The Dabie-Sulu belt of east-central China, the Western Alps, the Kokchetav Massif of northern Kazakhstan, the western Himalayan syntaxis of northern Pakistan, and the Western Gneiss Region of Norway constitute the best-documented examples of exhumed UHP rocks. In all these complexes, scattered UHP phases are partially preserved in strong, tough, refractory zircon, pyroxene, and garnet—minerals characterized by great tensile strength and low rates of intracrystalline diffusion. Armoring of the UHP inclusions subjects them to high confining pressure, provides spatial separation from the recrystallizing matrix minerals and rate-enhancing intergranular fluids, and thus protects them from back reaction during decompression.

This review tries to assess the nature of the orogenic process from a general petrotectonic viewpoint, concentrating on the architectures and rock assemblages of Phanerozoic UHP complexes. Although Precambrian analogues may have resulted from the operation of comparable lithospheric plate motions, where systematic lithotectonic contrasts were related to the higher geothermal gradients that attended a younger, hotter Earth, the ancient rock record is less clear; for this reason, unambiguously ancient UHP complexes have not yet been well documented; accordingly, we concentrate on Phanerozoic collisional mountain belts in this synthesis.

Exhumation of deeply subducted UHP complexes involves near-adiabatic decompression through the P-T fields of much lower-pressure metamorphic facies. Thus, back reaction, especially where kinetically enhanced by the presence of an aqueous fluid, causes recrystallization and obliteration of the earlier UHP phases. Although volumetrically dominant in exhumed UHP complexes, quartzofeldspathic and pelitic rocks generally retain very few relics of the maximum physical conditions, whereas eclogites and some anhydrous peridotites, because they are relatively impervious to the diffusion of H_2O, have more fully preserved effects of the deep-seated processes (Ernst et al., 1998; Liou et al., 1998). The index minerals coesite and diamond are largely lacking in mafic and ultramafic rock types; hence we attempted to quantify the P-T conditions of putative UHP rocks by employing thermobarometric computations as well as phase-equilibrium experiments on rocks and minerals.

PRESSURE-TEMPERATURE CONDITIONS OF ULTRAHIGH-PRESSURE METAMORPHIC COMPLEXES

HP and UHP terranes are typified by the presence of mafic (and/or ultramafic) eclogite-facies rocks. However, P-T determinations on eclogites are inherently difficult because most contain only two silicate phases, garnet and clinopyroxene. Measuring the Fe-Mg exchange between these two minerals enables calculation of temperature, but additional phases such as phengite or kyanite are required for barometry. Even for simple Fe-Mg exchange reactions, two problems render temperature calculation via this method tenuous: (1) diffusional reequilibration during retrogression ensures that recovery of the peak temperature is unlikely—especially at the highest temperatures; and (2) the P-T position of an Fe-Mg exchange reaction cannot be calculated accurately unless the Fe^{3+}/Fe^{2+} ratios of the iron-bearing phases, particularly clinopyroxene, are known. The former problem is well known (Pattison et al., 2003), but the magnitude of the latter problem perhaps is not widely appreciated. Krogh Ravna and Paquin (2003) summarized the results of half a dozen studies that compared ferrous/ferric ratios calculated by charge balance with ratios measured by Mössbauer, micro-XANES (X-ray absorbtion near edge structure), or titration. They found that Fe-Mg garnet-clinopyroxene temperatures calculated without knowledge of mineral Fc^{3+}/Fe^{2+} typically had uncertainties of ±100 °C. Proyer et al. (2004) used the Mössbauer milliprobe to demonstrate that the problem can be even worse, with apparent temperatures as much as 300 °C too high. Unfortunately, only a handful of Fe^{3+}/Fe^{2+} measurements on UHP eclogites have been made, so this method has not found general application.

A better solution to both of these difficulties with eclogite thermobarometry is to use net-transfer reactions rather than exchange equilibria, although garnet activities for Ca-rich solid solutions also can be problematic. The retrograde diffusional reequilibration problem is solved or at least reduced because the increase in diffusive length scale from grain scale in net-transfer reactions to grain-boundary scale in exchange reactions vastly increases the ability to capture peak temperature, and the problem with ferrous/ferric ratios is solved by using equilibria that involve Mg rather than Fe. In eclogites, the two principal equilibria of choice are (Nakamura and Banno, 1997; Ravna and Terry, 2004):

$$Mg_3Al_2Si_3O_{12} + Ca_3Al_2Si_3O_{12} + KMgAlSi_4O_{10}(OH)_2$$
$$= CaMgSi_2O_6 + KAl_2AlSi_3O_{10}(OH)_2 \quad (1)$$
(pyrope + grossular + celadonite
= diopside + muscovite);

and

$$Mg_3Al_2Si_3O_{12} + Ca_3Al_2Si_3O_{12} + SiO_2$$
$$= Al_2SiO_5 + CaMgSi_2O_6 \quad (2)$$
(pyrope + grossular + coe/qtz
= kyanite + diopside).

Unfortunately, kyanite-phengite eclogites make up only a small portion of the total eclogite population, gravely restricting the applicability of this method. This limitation is offset, however, by the great advantage of the robust pressures and temperatures determined by this method.

We applied this method to calculate accurate eclogite P-T conditions from microprobe mineral analyses presented in the literature. The positions of net-transfer equilibria were calculated using two approaches: (1) using THERMOCALC v. 3.1 (Powell et al., 1998) with the May 2001 updated database of Powell and Holland (1988); and (2) employing the spreadsheet of Ravna and Terry (2004), which depends on the same data set, but involves the Ganguly et al. (1996) garnet activity model rather than the Newton and Haselton (1981) model used by THERMOCALC. Only data from the latter model (Hacker, 2006) are shown in Figure 1; P-T data for the former are similar but are slightly more dispersed. In samples for which a range of mineral compositions was reported, we calculated P-T conditions using the most jadeite-rich omphacite, the most Si-rich white mica, and garnet with the highest $a_{prp}a_{gr}^2$ (prp = pyrope; gr= grossular), following the logic outlined by Carswell et al. (2000). Where possible, we supplemented these data with other robust temperature determinations (e.g., oxygen isotope temperature measurements from Dora Maira by Sharp et al., 1993).

Several important conclusions can be obtained from this diagram. As determined by this technique, the temperature range of UHP kyanite-phengite eclogites is 550–1000 °C, although most values are 600–750 °C; the maximum pressure is slightly in excess of 4 GPa for collisional terranes. This P-T field is smaller than that determined for kyanite- or phengite-free eclogites for which Fe^{3+}/Fe^{2+} had to be assumed. None falls on the high-pressure side of the "forbidden zone," defined as the array of geotherms less than 5 °C/km (Liou et al., 2000; but see Schmid et al., 2003). Within uncertainty, most of the determinations fall along the granite, tonalite, and metasediment solidi. This may indicate that: (1) UHP rocks that experienced hypersolidus temperatures recrystallized continuously in the presence of melt, and then froze in mineral compositions during cooling; (2) UHP rocks that have been subjected to hypersolidus temperatures are rarely exposed at Earth's surface; or (3) few UHP rocks are produced at hypersolidus temperatures. In contrast, as described farther on, some garnet peridotites have computed conditions of crystallization that fall within the high-pressure realm of the "forbidden zone."

GENERATION AND EXHUMATION OF UHP METAMORPHIC COMPLEXES

Ductilely deformed nappes and thrust sheets formed in subduction channels (e.g., Koons et al., 2003; Hacker et al., 2004; Terry and Robinson, 2004) make up the architecture of most recovered HP-UHP complexes; others may represent coherent, non-nappe sections of continental lithosphere (Young et al., 2007). Ascent to shallow crustal levels reflects one or more of several processes: tectonic extrusion (Maruyama et al., 1994, 1996; Searle et al., 2003; Mihalynuk et al., 2004); corner flow blocked by a hanging-wall backstop (Cowan and Silling, 1978; Cloos and Shreve, 1988a, 1988b; Cloos, 1993); underplating combined with extensional or erosional collapse (Platt, 1986, 1987, 1993; Ring and Brandon, 1994, 1999); and/or buoyant ascent (Ernst, 1970, 1988; England and Holland, 1979; Hacker, 1996; Hacker et al., 2000, 2004). Old, thermally relaxed, sinking oceanic lithosphere appears to roll back oceanward more

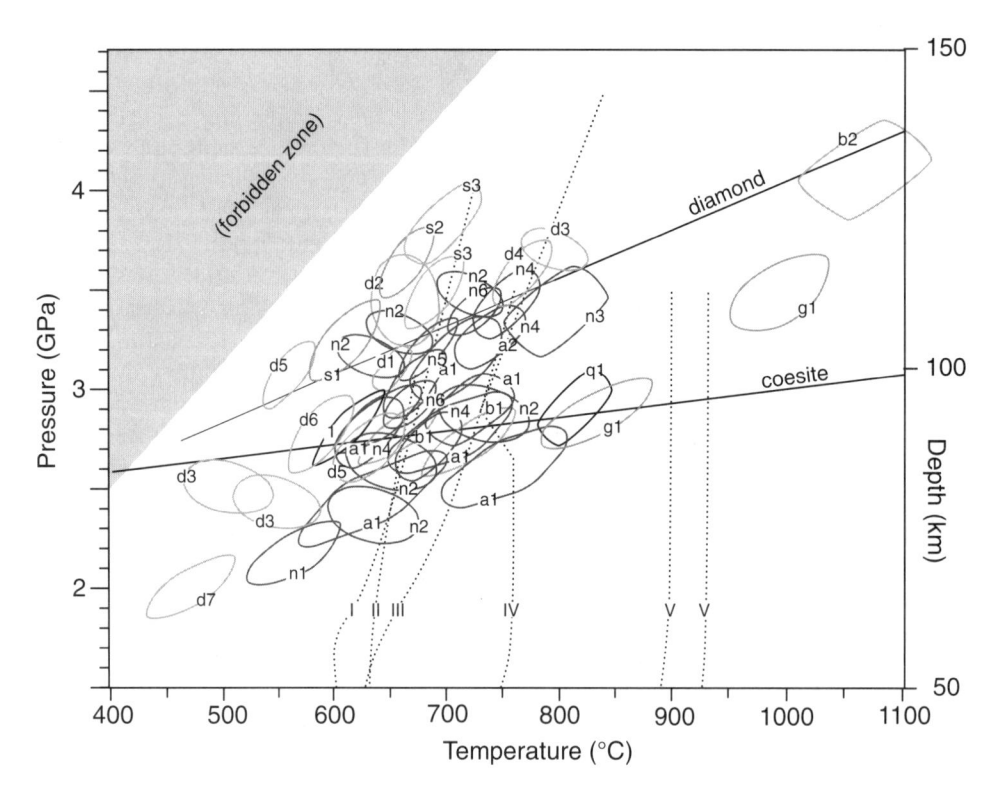

Figure 1. Robust pressure and temperature (*P-T*) conditions of kyanite–phengite eclogites in high-pressure–ultrahigh-pressure terranes (Hacker, 2006) determined using the intersection between garnet–clinopyroxene–muscovite–kyanite–quartz/coesite net-transfer equilibria and the solution models of Krogh Ravna and Terry (2004). The solution models of THERMOCALC result in a similar, but slightly more dispersed set of pressures and temperatures. *P-T* data derived from mineral compositions of: a1—Nowlan (1998), Dora Maira; a2—Kienast et al. (1991), Dora Maira; a3—Coggon and Holland (2002), Dora Maira; b1—Massonne and O'Brien (2003), Münchberg; b2—Massonne and O'Brien (2003), Saidenbach; d1—Okay (1993), Dabie; d2—Proyer et al. (2004), Dabie; d3—Krogh Ravna and Terry (2004), Dabie; d4—Zhang et al. (1995b), Dabie; d5—Okay (1995), Dabie; d6—Zhang and Liou (1994), Hong'an; d7—Eide and Liou (2000), Hong'an; g1—Gilotti and Krogh Ravna (2002), Greenland; n1—Engvik et al. (2000), Norway; n2—Krogh Ravna and Terry (2004), Norway; n3—Terry et al. (2000a), Norway; n4—Wain (1998), Norway; n5—Wain (1998), combined with an Fe^{3+} measurement by C. McCammon of a D. Root sample from Norway; n6—Young et al. (2007), Norway; q1—Song et al. (2003), Qaidam; s1—Zhang et al. (1995a), Sulu; s2—Mattinson et al. (2004), Sulu; s3—Hirajima and Nakamura (2003), Sulu; m1—Caby (1994), Mali. Solidi: I—tonalite (Stern et al., 1975), II—sediment (Nichols et al., 1994), III—granite (Stern et al., 1975), IV—gabbro dehydration (Vielzeuf and Schmidt, 2001), V—mica dehydration (Patiño Douce and McCarthy, 1998).

rapidly than the nonsubducted plate moves forward (Molnar and Atwater, 1978; Seno, 1985; Busby-Spera et al., 1990; Hamilton, 1995), so compression and extrusion of subducted sialic slabs in such convergent plate junctions cannot be responsible for the exhumation unless the oceanic lithosphere tears away. Constriction by a backstop requires buoyancy or tectonic contraction to produce the return flow of subducted sections. Extension and erosion help to unroof HP-UHP terranes once they reach crustal levels, but these processes do not produce the major pressure discontinuities (up to >2 GPa) that mark the major fault boundaries between deeply subducted and nonsubducted crust (Ernst, 1970; Ernst et al., 1970; Suppe, 1972).

Buoyancy coupled with erosional decapitation provides a plausible mechanism for the exhumation of low-density crustal slices propelled upward from great depth by body forces. Geologic relationships, laboratory scale models (Chemenda et al., 1995, 1996, 2000), and numerical simulations (Beaumont et al., 1996, 1999; Pysklywec et al., 2002), illustrated schematically in Figure 2, document this process (see also volumes edited by: Parkinson et al., 2002; Carswell and Compagnoni, 2003; and Malpas et al., 2004). The strengths and integrity of the subducted lithospheric materials, extents of deep-seated devolatilization, and rates of recrystallization strongly influence the characteristics of the resultant UHP metamorphic belts (Ernst et al., 1998). The petrotectonic features of Phanerozoic UHP complexes thus reflect their plate-tectonic settings and *P-T* histories (Table 1).

Attending circum-Pacific subduction of a largely sedimentary mélange, devolatilization and increased ductility cause

(Isacks et al., 1968), may result in rupture and accelerated sinking of the dense oceanic lithosphere. Slab breakoff (Sacks and Secor, 1990; von Blanckenburg and Davies, 1995) increases the net buoyancy of the updip, relatively low-density sialic UHP complex and allows sheets to disengage from the oceanic plate and move back up the subduction channel (van den Beukel, 1992; Davies and von Blanckenburg, 1998). During collision, decoupling and exhumation also may be enhanced as the continental crust warms in the upper mantle and passes through the brittle-ductile transition (Stöckhert and Renner, 1998).

The two-way migration of terranes along subduction channels is well known (Ernst, 1970; Suppe, 1972; Willett et al., 1993). Similar to the subduction of circum-Pacific metaclastic mélanges, low-density sialic crustal sections descend at plate-tectonic rates, and at great depth, generate the distinctive HP-UHP prograde mineralogy of Alpine continental collisional complexes (Peacock, 1995; Ernst and Peacock, 1996). Return of these decoupled sections up the subduction channel during exhumation obviates the need to remove 50–100 km of the overlying hanging wall (the mantle wedge acts as a stress guide) by erosion, extensional collapse, or tectonism.

Densities (g/cm^3) of unaltered oceanic crust, 3.0, continental material, 2.7, and anhydrous mantle, 3.2, increase with elevated pressure, reflecting the transformation of open framework silicates to more compact layer-, chain-, and orthosilicates. Stable UHP mineralogic assemblages and computed rock densities appropriate for burial depths of ~100 km and 700 °C are roughly as follows: metabasaltic eclogite, 3.55; eclogitic granitic gneiss, 3.05; and garnet peridotite, 3.35 (Ernst et al., 1997; Hacker et al., 2003a). Even when transformed completely to a UHP assemblage, K-feldspar + jadeite + coesite-bearing granitic gneiss remains ~0.30 g/cm^3 less dense than garnet lherzolite, whereas metabasaltic eclogite is ~0.20 g/cm^3 denser than upper mantle lithologies. Evidently, subducted packets of UHP metamorphosed sialic crust are buoyant enough to overcome the traction of the oceanic plate carrying them downward because quartzofeldspathic nappes are now exposed at the Earth's surface.

Continental crustal rocks contain muscovite and biotite, minerals stable to 800–1100 °C at subduction depths >140 km (Stern et al., 1975; Nichols et al., 1994; Patiño Douce and McCarthy, 1998), as the main hydrous phases; therefore, such rocks do not devolatilize completely during normal subduction (Ernst et al., 1998). In the absence of a rate-enhancing aqueous fluid, such lithologies are unlikely to transform rapidly, or totally, to UHP mineral assemblages (Hacker, 1996; Austrheim, 1998). In contrast, the main H$_2$O-bearing phase in mafic rocks is hornblende, which is a pressure-limited mineral that devolatilizes at moderate temperatures where depths exceed ~70–80 km. In the presence of this evolving aqueous fluid, metabasaltic eclogites are far more likely to recrystallize to the stable prograde HP-UHP assemblage than are sialic units. Consequently, at upper-mantle depths, continental crust converted completely or incipiently to UHP-phase assemblages remains buoyant relative to the surrounding mantle and should rise to mid-crustal

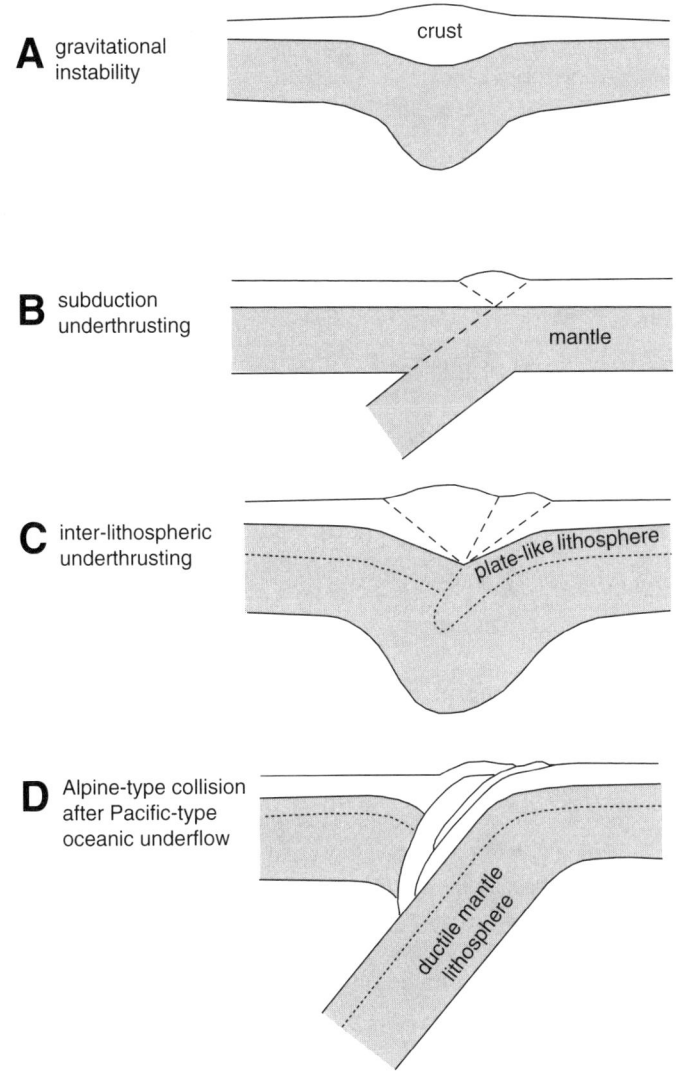

Figure 2. Simplified structural evolution of contractional orogens chiefly based on scale-model experiments (Chemenda et al., 1995, 1996, 2000) and numerical modeling (Beaumont et al., 1996, 1999; Pysklywec et al., 2002). Crust is white; mantle lithosphere is gray. Delamination of mantle lithosphere due to (A) gravitational instability and (B) subduction underthrusting. (C) Modeled deformation of South Island, New Zealand, involving upper-mantle detachment. (D) Himalayan-type nappe imbrication resulting from Pacific-type lithospheric underflow and continental collision; individual décollements are much thinner than that illustrated.

decoupling of subducted HP materials from the downgoing oceanic plate at ~20–50 km, followed by piecemeal ascent. In contrast, for a continental salient well bonded to the lithosphere, disengagement of a coherent crustal slice from the descending oceanic plate may be delayed to a depth of 90–140 km. The insertion of increasing amounts of low-density material into the subduction zone gradually reduces the overall negative buoyancy of the lithosphere. Attainment of neutral buoyancy at moderate upper mantle depths, where the plate is in extension

TABLE 1. SUMMARY DATA FOR ULTRAHIGH-PRESSURE (UHP) METAMORPHIC COMPLEXES[†]

Terrane characteristic	Dabie-Sulu belt, coesite-eclogite unit	Kokchetav Massif, UHP unit	Dora Maira Massif, L. Venasca nappe	Western Gneiss region	Western Himalayan syntaxis, Kaghan V.
Protolith formation age	Chiefly 800–650 Ma	2.3–2.2 Ga	Ca. 300 Ma	1.8–0.4 Ga	>170 Ma
Temperature of metamorphism	650–750 °C	900 ± 75 °C	725 ± 50 °C	600–800 °C	750–780 °C
Depth of metamorphism	90–125 km	~140 km	90–110 km	90–130 km	~100 km
Time of metamorphism	236–226 Ma	535 ± 3 Ma	35 Ma	410–405 Ma	44 Ma
Crustal annealing	230–195 Ma	529 Ma	32 Ma	Ca. 402 Ma	40–42 Ma
Rise time to mid-crust	6 m.y.	6 m.y.	3 m.y.	3–8 m.y.	2–4 m.y.
Exhumation rate[§]	≥10 mm/yr	15–30 mm/yr	~20 mm/yr	8–20 mm/yr	>15 mm/yr
Coesite inclusions	Relatively abundant	Rare, locally abundant	Relatively abundant	Rare	Rare
Diamond inclusions	Very rare	Relatively abundant	Absent	2 localities	Absent
Areal extent	>400 × 50 km	~120 × 10 km	35 km²	165 × 50 km	30 × 70? km
Thickness of individual UHP units	5–15 km	1–3 km	1–2 km	>10 km?	1 km

[†]After Coleman and Wang (1995), Harley and Carswell (1995), Ernst and Peacock (1996), Amato et al. (1999), Hacker et al. (2000, 2003b, 2006), Maruyama and Parkinson (2000), Terry et al. (2000a, 2000b), Hermann et al. (2001), Katayama et al. (2001), Rubatto and Hermann (2001), Massone and O'Brien (2003), Parrish et al. (2003), Rubatto et al. (2003), Treloar et al. (2003), Baldwin et al. (2004), Root et al. (2004, 2005), and Leech et al. (2005).
[§]Average exhumation rates were estimated by dividing depth of UHP metamorphism by time of ascent to 10–15 km crustal depth.

levels; in contrast, eclogitized oceanic crust becomes negatively buoyant compared to both near-surface oceanic basalt and garnet lherzolite and continues to sink. This relationship explains why exhumed HP-UHP terranes worldwide consist of ~90% low-density felsic material and contain only small proportions of dense mafic and anhydrous ultramafic rock types.

Times of UHP recrystallization in well-studied complexes ranges from about 535 Ma in northern Kazakhstan (Sobolev and Shatsky, 1990; Hermann et al., 2001; Katayama et al., 2001; Hacker et al., 2003b) to ~44 Ma in the western Himalayas (Kaneko et al., 2003; Treloar et al., 2003; Schlup et al., 2003), and 35 Ma in the Western Alps (Tilton et al., 1991; Gebauer et al., 1997; Rubatto and Hermann, 2001). Late Proterozoic UHP complexes eventually may be discovered, but Earth's ancient geothermal gradient may have been too high to allow the generation of UHP mineral parageneses during Archean and Early Proterozoic time.

RATE OF ASCENT OF UHP CONTINENTAL COMPLEXES

Considerable effort has been expended to measure the exhumation rates of UHP terranes by radiometric investigations and, to a lesser extent, by diffusion modeling. In general, the most comprehensive studies infer relatively rapid exhumation, approaching plate-tectonic rates. This poses a challenge for geochronologists for several reasons: (1) uncertainties in the decay constants for some radiometric clocks (i.e., ^{40}K and ^{176}Lu) increase the difficulty of obtaining sufficiently accurate ages for pre-Cenozoic rocks; (2) accurate Lu/Hf and Sm/Nd mineral-isochron ages require unzoned, unaltered phases that formed at a single, known P-T stage; and (3) U/Pb ages must have high temporal precision and come from discrete crystal volumes formed at a specific pressure. Advances are being made along all of these fronts, but none of these problems has yet been solved; accurate decay constants (e.g., Begemann et al., 2001) and the ability to analyze subcrystal volumes that can be tied to specific pressures are required. However, the best-documented cases show that exhumation to mid-crustal levels is rapid, with minimum average exhumation rates of tens of millimeters per year (Table 1).

UHP complexes with relatively few geochronological data paint a fairly simple picture. The exhumation rate of the Dora Maira Massif is constrained by U/Pb (chiefly sensitive high-resolution ion microprobe [SHRIMP]), Lu/Hf, and fission-track ages to ~20 mm/yr (see review by Rubatto et al., 2003). The Kokchetav Massif, investigated by Sm/Nd, U/Pb SHRIMP, and ^{40}Ar/^{39}Ar techniques, rose at 15–30 mm/yr (Hermann et al., 2001; Katayama et al., 2001; Hacker et al., 2003b). Sm/Nd and Rb/Sr ages indicate that the Lago di Cignana eclogites of the Lepontine Alps were exhumed at 26 mm/yr (Amato et al., 1999). Pliocene U/Pb ages for UHP rocks in Papua New Guinea indicate exhumation rates of 10–20 mm/yr (Baldwin et al., 2004). The Tso Morari complex of the NW Indian Himalaya was exhumed at 10–15 mm/yr (Massonne and O'Brien, 2003; Leech et al., 2005). The Kaghan Valley eclogites in Pakistan were exhumed within 2–4 m.y. (Treloar et al., 2003), evidently at an average rate approaching 20 mm/yr. Geochronological data from the giant UHP terranes in China and Norway are vastly more abundant, and, as a result, more complex, but exhumation rates in the Dabie-Sulu UHP terrane of China certainly exceeded 10 mm/yr (Hacker et al., 2000, 2006), as did those in the Western Gneiss Region of Norway (Carswell et al., 2003a, 2003b; Root et al., 2004, 2005). Such speedy unloading exceeds present-day regional exhumation and erosion rates (Blythe, 1998), implying that modern erosion rates are mischaracterized or that erosion alone did not expose the known UHP complexes.

CONDUCTIVE COOLING OF UHP CONTINENTAL COMPLEXES

Diffusion modeling studies demonstrate that Himalayan UHP rocks were subjected to temperatures >600 °C for only short times during decompression (O'Brien and Sachan, 2001; Massonne and O'Brien, 2003). These complexes evidently were quenched during exhumation. Poor thermal conductivities of rocks account for high-pressure prograde conditions attending underflow, but this property of Earth materials also dictates that deeply buried units retain heat on rapid exhumation. During decompression, UHP complexes exhibit pervasive mineralogic overprinting and assemblages characteristic of heating (typically granulite-facies), maintenance of constant temperature (amphibolite facies), or only modest cooling (greenschist facies). As an example, Figure 3 illustrates prograde and nearly isothermal retrograde P-T-time (t) trajectories calculated for the Paleogene subduction complex of the western Himalayan syntaxis. On decompression, the presence of a rate-enhancing aqueous fluid would have resulted in virtually complete obliteration of all pre-existing UHP-phase assemblages. Lack of catalytic, grain-boundary H_2O in a complex subjected to rapid ascent substantially decreases the rate of retrogression (Rubie, 1986, 1990; Ernst et al., 1998; Mosenfelder et al., 2005), but even so, heat must be effectively withdrawn from the rocks at some early stage during exhumation while the complex is relatively hot, or mineralogic evidence of former HP-UHP conditions would be lost. The preservation of UHP relics in a rising subduction complex is favored by juxtaposition against cooler rocks, such as by extensional faulting against a colder hanging wall and/or by thrusting against a colder footwall (Hacker and Peacock, 1995). Nevertheless, only in optimally favorable kinetic circumstances are any relict UHP phases and/or mineral assemblages preserved.

To first order, the thermal history of a UHP body during decompression is determined by its minimum dimension (i.e., thickness), its rate of ascent, and the temperature of the medium through which it ascends (e.g., Root et al., 2005). Relatively thin ascending slices will exchange heat more effectively than will thicker units. If a thin UHP body ascends slowly through a typical (cool) subduction thermal gradient, the P-T path during ascent can simply be the reverse of that during compression. However, if a thin UHP body ascends slowly through a zone of much hotter rocks—say, through interior portions of the mantle wedge—it may become hot enough that the evidence of UHP metamorphism is obliterated. Most well-characterized UHP terranes show neither of these types of behavior but, instead, near-isothermal decompression down to ~1 GPa. If a UHP body is thick, the heating or cooling of the body interior will be reduced proportional to the square of its thickness. If the rate of ascent is more rapid, the heating or cooling of the body interior will be reduced, following the square root of the ascent rate. In general terms, for a UHP complex to ascend without significant heating or cooling, its minimum dimension (radius or half-thickness) must exceed the characteristic diffusion distance

$$u = \sqrt{[\kappa \Delta z/(dz/dt)]},$$

where κ is thermal diffusivity, Δz is the vertical ascent distance, and dz/dt is the vertical ascent rate. For example, a UHP body with a minimum dimension of 15 km must ascend at >10 km/m.y., and a UHP body with a minimum dimension of 2 km must ascend at >500 km/m.y. The rapid ascent rates required mean that thin UHP sheets cannot have ascended near-isothermally from 100 km depth in their present shape, but must have cooled, approximating in reverse the subduction-zone prograde P-T trajectory (Chopin, 1984; Rubie, 1984; Ernst, 1988; Ernst and Peacock, 1996). The manner in which thick, decompressing slabs are quenched remains problematic. Of course, for UHP phases to be preserved in even fragmentary form, the ascending complex—thick or thin—must be quenched prior to complete back reaction. Examples of thin and thick UHP sheets are presented in Figures 4–6.

Well-studied exposures in the western syntaxis of the Himalayas (O'Brien et al., 2001; Parrish et al., 2003; Kaneko et al.,

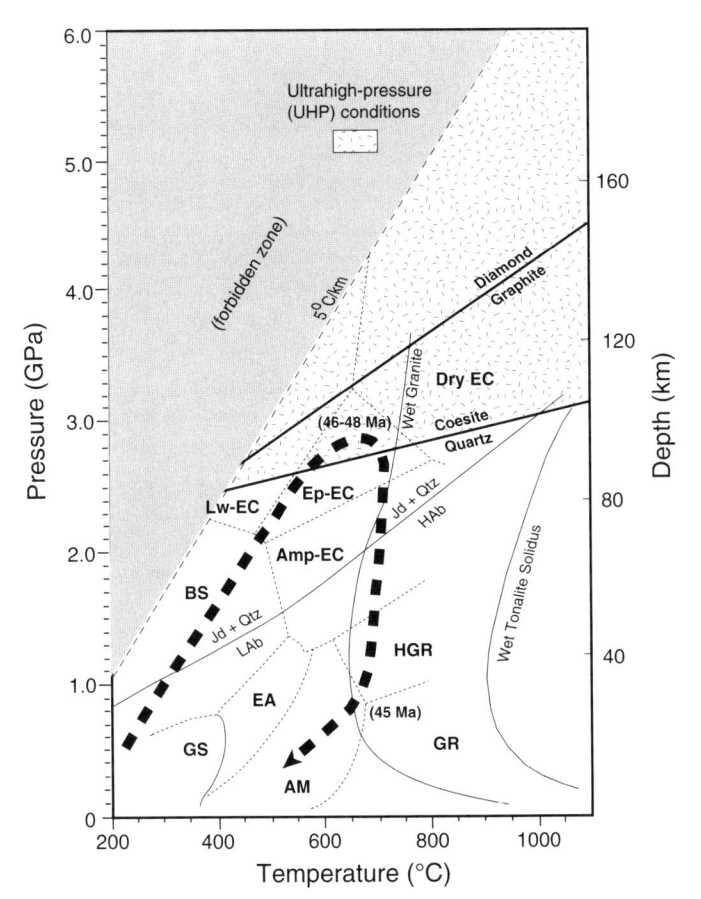

Figure 3. Pressure-temperature history of subduction and nearly isothermal exhumation to mid-crustal levels of ultrahigh-pressure imbricate thrust sheets cropping out in the Kaghan Valley, western Himalayan syntaxis, after O'Brien et al. (2001), Parrish et al. (2003), and Kaneko et al. (2003). For location, geologic map, and cross section, see Figure 4.

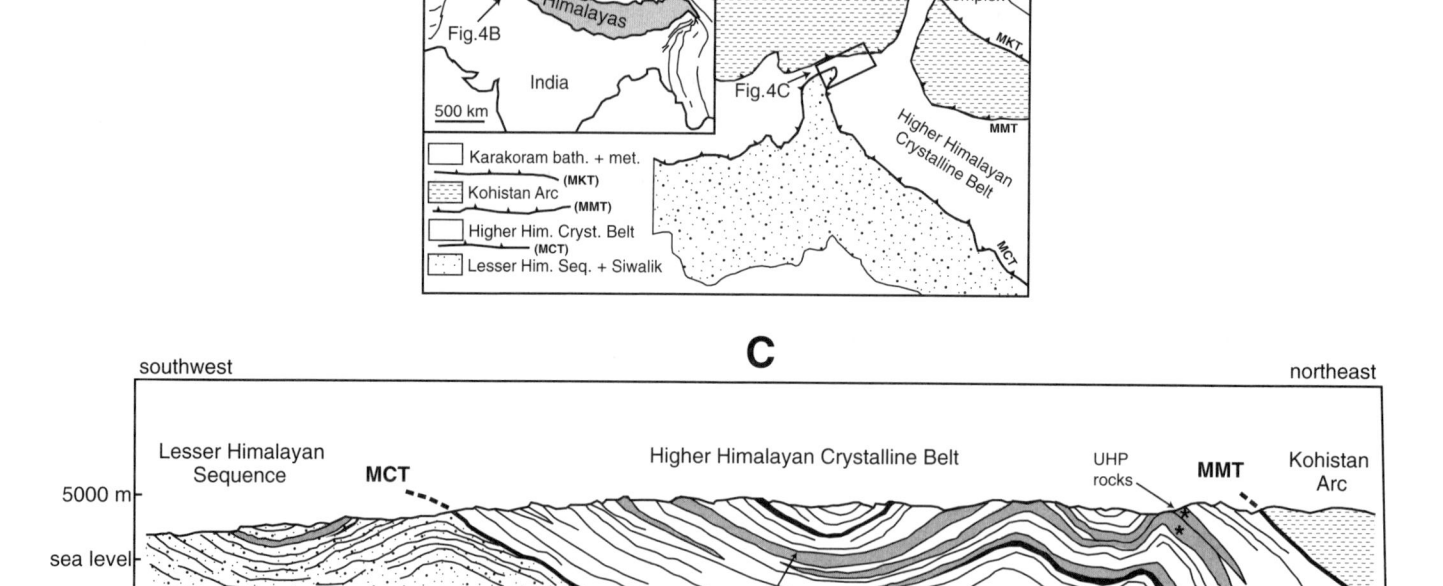

Figure 4. General geologic map and cross section through the Kaghan Valley, western Himalayan syntaxis, Pakistan, from Kaneko et al. (2003). Index maps are shown in A and B. A geologic map and cross section are presented in C. Note in C that the coesite-bearing UHP thrust sheets, indicated by stars, individually are less than about a kilometer thick. MKT—Main Karakorum thrust; MMT—main mantle thrust; MCT—main central thrust.

2003) and the Central and Western Alps (Henry, 1990; Michard et al., 1995) include nappes and imbricate slices of UHP continental crust less than 1–2 km thick. Similar aspect-ratio coesite-bearing thrust sheets have also been documented from the northern Western Gneiss Region of coastal Norway (Terry et al., 2000a, 2000b; Terry and Robinson, 2004), and the Kokchetav Massif of northern Kazakhstan (Kaneko et al., 2000; Maruyama and Parkinson, 2000; see also Dobretsov et al., 2006). In contrast, UHP sections at least 5 km thick have been proven by drilling in the Sulu belt of east-central China (Liu et al., 2004, 2007; Z. Zhang et al., 2006). Moreover, other intensively mapped portions of UHP terranes of comparable thickness include the Hong'an-Dabie terrane of eastern China (Hacker et al., 2000, 2004), and major tracts of the southern Western Gneiss Region (Root et al., 2005). Geologic maps and cross sections of Figures 4–6 illustrate the imbricate nature common to all these UHP complexes; the most striking contrast involves the differing thicknesses of the various UHP nappes.

Schematic relations shown in Figure 7 apply to the underflow and later exhumation of HP-UHP sheets. Descent of the low-density crust occurs only if shear forces caused by underflow (F_s) exceed the combined effects of buoyancy (F_b) and frictional resistance along the hanging wall of the subduction channel (F_r). Here, $F_s > F_b \cos \theta + F_r$. Decoupling and ascent of a slice of the low-density crust take place where buoyancy is greater than the combined effects of shearing along its footwall and resistance to movement along its upper, hanging-wall surface. In this case, $F_b \cos \theta > F_s + F_r$. The mantle wedge guides exhumation, and the rising nappe is emplaced oceanward (outboard) from the site of metamorphism. Where the angle of subduction decreases, the effect of buoyancy lessens during both underflow and exhumation. For HP-UHP complexes to be returned to shallow depths and partly preserved, the rising slab must overcome frictional resistance to sliding, so it must be thick enough for buoyancy-driven ascent, yet thin enough that heat is efficiently removed by conduction across the bounding faults—upper normal and lower reverse. Such kinematic structural relationships have been mapped in many resurrected, relatively thin-aspect-ratio subduction terranes, i.e., the Himalayas (Burchfiel et al., 1989; Searle, 1996; Searle et al., 2001; Kaneko et al., 2003); the Franciscan Complex (Ernst, 1970; Suppe, 1972; Platt, 1986; Jayko et al., 1987); the Western Alps (Henry, 1990; Compagnoni et al., 1995; Michard et al., 1995); the Sanbagawa belt (Kawachi, 1968; Ernst et al., 1970; Banno and Sakai, 1989); and the Kokchetav Massif (Kaneko et al., 2000; Ishikawa et al., 2000; Ota et al., 2000; Maruyama and Parkinson, 2000). Nappes have also been described from the Western Gneiss Region of Norway (Harley and Carswell, 1995; Krogh and Carswell, 1995; Terry et al., 2000a, 2000b); and the Dabie-Sulu belt (Liou et al., 1996; Hacker et al., 1995, 1996, 2000; Webb et al., 1999).

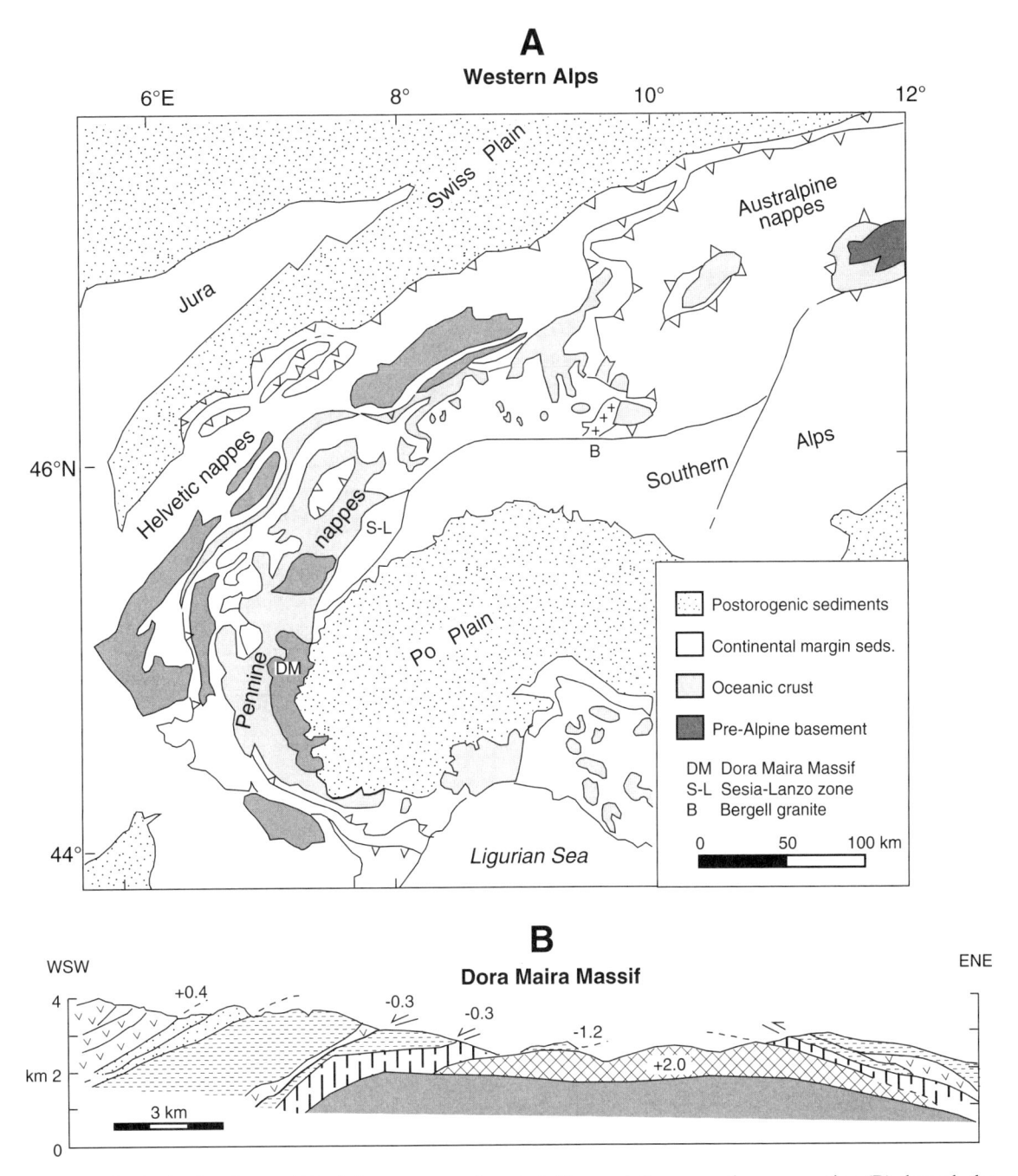

Figure 5. Geologic sketch map (A) of the western and central Alps, and diagrammatic cross section (B) through the southern Dora Maira Massif (DM; after Henry, 1990; Michard et al., 1995). In B, the transect across the Dora Maira Massif, numbers indicate the upward change in recorded pressure in GPa relative to the adjacent underlying unit. The lower Venasca ultrahigh-pressure (UHP) nappe is shown in the gridiron pattern.

Recrystallized, retrogressed UHP complexes, although less dense than anhydrous mantle, become neutrally buoyant at approximately middle levels of the sialic crust (Walsh and Hacker, 2004). In some cases, further exhumation of such slabs may be the product of contractional tectonism (Maruyama et al., 1994, 1996) or low-density crustal underplating—in either case combined with isostatically compensated regional exhumation and erosional decapitation (Platt, 1986, 1987, 1993). In addition, a drop in overall density of the subducting lithosphere after plate breakoff results in a shallowing of the downgoing, increasingly buoyant slab, and may be partly responsible for the late doming recognized in many exhumed convergent plate junction regimes (Ernst et al., 1997; O'Brien, 2001; O'Brien et al., 2001). Yet another unloading mechanism involves the antithetic faulting typical of some con-

Figure 6. High-pressure–ultrahigh-pressure (HP-UHP) domains in the (A) Hong'an area of China and (B) Western Gneiss Region of Norway, after Hacker et al. (2000) and Root et al. (2005), respectively. Gray shades are used to distinguish different units. Cross sections provide a measure of the relatively great thickness of these HP-UHP complexes.

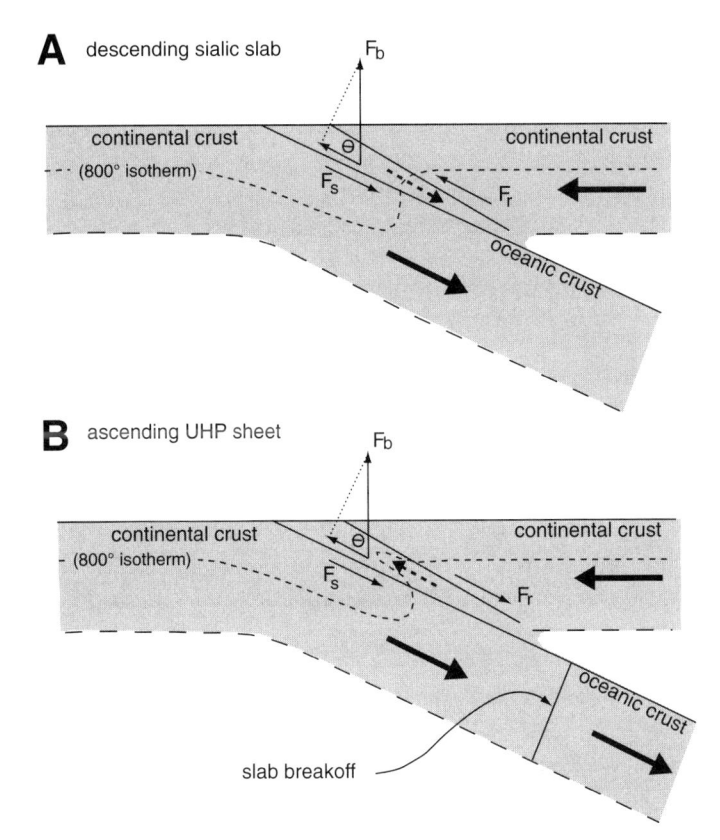

Figure 7. Schematic convergent lithospheric plate-boundary diagram for active subduction, after Ernst and Peacock (1996). (A) Deep burial and thermal structure of a subducted sheet of continental crust. (B) Later decompression cooling of a rising slice of the sialic material. Relative motions of plates and slices are indicated by arrows (the subducting plate actually is sinking and rolling backward; Hamilton, 1995). During ascent of the HP-UHP terrane (thickness exaggerated for clarity), cooling of the upper margin of the sheet takes place where it is juxtaposed against the lower-temperature hanging wall (the mantle wedge); cooling along the lower margin of the sheet takes place where it is juxtaposed against the lower-temperature, subduction-refrigerated lithosphere. Exhumation of low-density slices requires erosive denudation and/or gravitational collapse and a sialic root at depth. The resolutions of forces acting on the sialic slab in stages A and B are discussed in the text. Lithosphere is shaded (crust-mantle boundary not indicated); asthenosphere is unshaded. Degrees in Celsius.

tractional orogens, in which double vergence is produced during terminal stages of the ascent of low-density crust (e.g., Dal Piaz et al., 1972; Ring and Brandon, 1994, 1999).

Exhumation of domal or diapiric bodies of granitic crust appears to be occurring along convergent plate junctions where curvilinear arcs intersect at large angles. At such lithospheric boundary cusps, overthickened continental crust gradually warms and loses strength. Basal portions may partially melt, but in any case, the crust softens, becomes even more buoyant, and rises more-or-less like a salt dome. Such uplifts, shown diagrammatically in Figure 8, have been termed tectonic aneurysms (Zeitler et al., 2001; Koons et al., 2002; Chamberlain et al., 2002). Some appear to be the sites of exhumed UHP terranes (Ernst, 2006).

TECTONIC SIGNIFICANCE OF GARNET PERIDOTITES IN UHP CONTINENTAL COMPLEXES

Studies of volumetrically minor mafic eclogite boudins and layers in subducted continental crust have provided important quantitative constraints regarding the UHP conditions that attended metamorphism of the enclosing, largely quartzofeldspathic complex. The occurrences of spatially associated garnet-bearing peridotite bodies are less well understood. Such ultramafic rocks occur as tectonic massifs, pods, and lenses in many ancient collisional mountain belts. HP-UHP examples include the Caledonian, Variscan, and Alpine orogens of Europe, the Kokchetav Massif of Kazakhstan, and the Triassic Dabie-Sulu terrane in east-central China (for reviews, see Medaris, 1999; Brueckner and Medaris, 2000; O'Brien, 2000). These garnet peridotites are of contrasting origins. Some have been interpreted as mantle-derived bodies tectonically emplaced into sialic crustal sequences (Ernst, 1978; Carswell and Gibb, 1980), whereas, others are regarded as products of prograde HP metamorphism of spinel peridotite, or their serpentinized equivalents, previously emplaced in the crust (e.g., Evans and Trommsdorff, 1978; England and Holland, 1979). Medaris (1999) subdivided garnet peridotites from Eurasian HP-UHP terranes into four general types: (1) serpentinites or ultramafic igneous complexes emplaced in the crust prior to subduction, followed by underflow and UHP metamorphism; (2) mantle wedge spinel and/or garnet peridotites inserted into a downgoing lithospheric plate; (3) low-pressure, high-temperature spinel peridotites that may reflect the upwelling of asthenospheric material; and (4) HP garnet peridotites tectonically extracted from the deepest portions of the continental crust–capped lithosphere.

Quantitative compositional and structural data for the subcontinental lithospheric mantle provide crucial information for the erection of realistic large-scale models describing Earth's geochemical and tectonic evolution (Griffin et al., 1999). Our knowledge of mantle compositions and heterogeneities has been obtained mainly through the study of xenoliths and xenocrysts from kimberlites and volcanic rocks of deep origin. However, detailed, integrated petrochemical, mineralogic, and geochronologic studies of orogenic garnet peridotites provide important constraints on mantle processes, and the chemical-mineralogic compositions and evolution of the mantle wedge overlying a subduction zone. The discovery of phases of very deep origin, such as majoritic garnet, HP clinoenstatite, and olivine containing elevated concentrations of $FeTiO_3$ rods in garnet peridotites from several UHP terranes (e.g., Dobrzhinetskaya et al., 1996; Bozhilov et al., 1999; van Roermund et al., 2000, 2001; Massonne and Bautsch, 2002) has provides important information about mantle dynamics. How these deep-seated (>200 km) mantle rocks were transported to shallow depths, and by what means they were incorporated in subducted continental slabs of contractional mountain belts remain unclear. Some of these garnet peridotites and the enclosing continental crust have been postulated to have undergone subduction-zone UHP metamor-

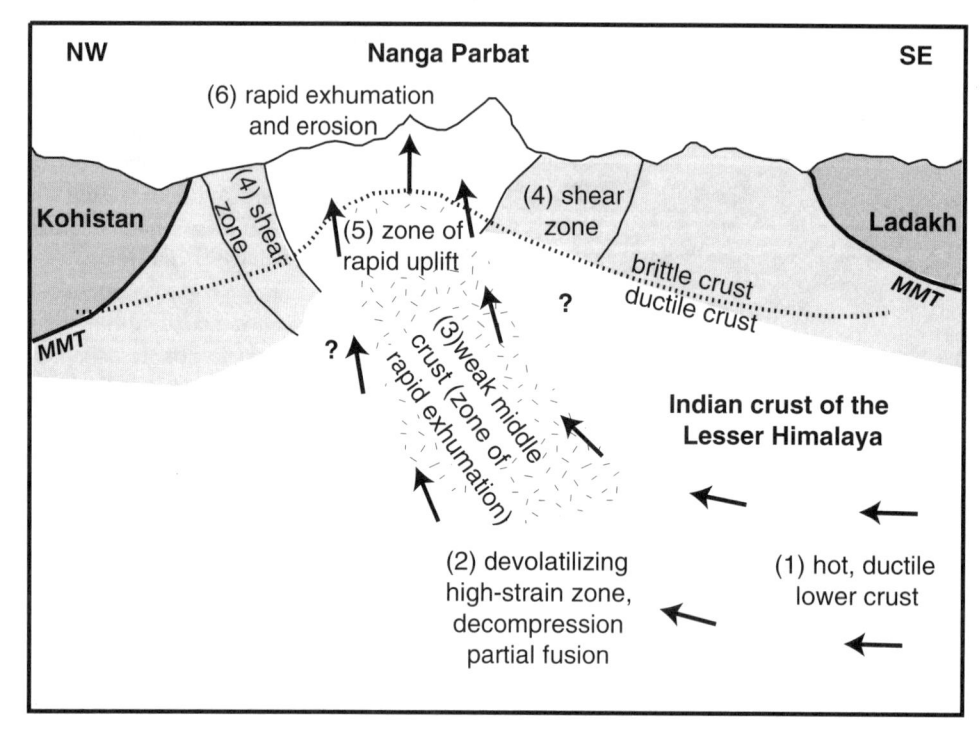

Figure 8. Diagrammatic cross section of the Neogene tectonic aneurysm at the western Himalayan syntaxis, simplified from Zeitler et al. (2001). Erosion-induced rapid unloading of high mountains overlying deep-seated, thickened crust causes upward flow of thermally softened, buoyant crust. Numbered features are as follows: (1) hot, ductile, devolatilizing metamorphosed crust enters flow regime, and (2) passes through high-strain zone, incipiently melting and degassing further. (3) Crust enters region of rapid exhumation as unloading and further melting take place, with granitoids (4) possibly inserted into massif along NW and SE shear zones. (5) Strain focusing leads to accelerated upward advective transport of the lower, ductile crust, carrying along its thermal structure. (6) High topography surmounting the weak diapiric zone is partly removed by vigorous erosion, exposing back-reacted, decompressed migmatites. Also involved laterally is a strong meteoric circulation system (not illustrated). MMT—main mantle thrust.

phism characterized by extremely low thermal gradients, on the order of ≤5 °C/km (e.g., Liou et al., 2000; Zhang et al., 2004). High-pressure experiments reveal that numerous hydrous phases may be stable in such HP environments, the so-called forbidden zone (Liou et al., 1998). Thus, unusually cold subduction zones might well represent the sites of major recycling of H_2O back into the mantle. These findings have advanced our quantitative understanding of the thermal structure of subduction zones and of the return of volatiles to the mantle.

Most Eurasian HP-UHP garnet peridotites are rich in Mg and Cr and represent depleted-upper-mantle materials, but several are more Fe-rich and originated as igneous mafic-ultramafic complexes (Medaris, 1999). These peridotites are polymetamorphic, with UHP garnet-bearing assemblages extensively replaced by a succession of retrograde mineral assemblages generated during exhumation and cooling. Some peridotites also contain evidence for a pre-UHP stage, evidenced by spinel and/or Ti-clinohumite inclusions in garnet. Equilibration conditions of peak-UHP stages have been calculated from garnet-bearing peridotites by employing the olivine-garnet Fe-Mg exchange thermometer and the Al-in-orthopyroxene barometer (but see section dealing with P-T conditions of UHP metamorphism). Garnet peridotites occur as meter- to kilometer-sized blocks and lenses in gneisses; the quartzofeldspathic host rocks also have been subjected to UHP metamorphism and exhibit massive, granoblastic or porphyroblastic textures. Most garnet peridotites are deformed and are partially to almost fully serpentinized. Relict garnet-bearing assemblages on the surface are more completely preserved in the central parts of such ultramafic boudins; some occupy up to 30 vol% of the entire body. Garnet peridotite samples from drill holes, however, tend to be relatively less intensely serpentinized. Exsolution microstructures in olivine, garnet, and diopside, and clinoenstatite polymorphs of orthopyroxene are common (Dobrzhinetskaya et al., 1996; Zhang and Liou, 1999, 2003; Zhang et al., 1999, 2003; Spengler et al., 2006). Geochronologic data for garnet peridotites are poorly constrained, but associated mafic eclogites have been dated by various methods, such as Sm-Nd mineral isochrons and SHRIMP zircon U-Pb dating (Katayama et al., 2003; Zhang et al., 2005a, 2005b; Z. Zhang et al., 2006; Zhao et al., 2007).

Numerous tectonic origins for garnet peridotites in UHP terranes have been proposed (e.g., Brueckner, 1998; Medaris, 1999; Zhang and Liou, 1999). Similar to, but slightly different from the classification of Medaris, we infer contrasting origins for HP-UHP garnet peridotites based on their modes of occurrence, petrochemical characteristics, and tectonic histories; the range of properties for some of these bodies is summarized in Table 2. The four general types of ultramafic rock, now recrystallized to garnet peridotite, are as follows: type-A, hanging-wall (mantle wedge) fragments; type-B, crustal mafic-ultramafic igneous complexes; type-C, tectonic blocks from the footwall mantle lithosphere; and type-D, ancient mantle complexes tectonically emplaced in the crust prior to subduction. Inasmuch as many of the Eurasian garnet peridotites listed in Table 2 are incompletely characterized tectonically, geochemically, and/or are undated (particularly by Re-Os isotopic systematics), our assignment of tectonic type must be considered tentative. Moreover, several types of garnet peridotite may occur in certain HP-UHP terranes. For example, in the Western Gneiss of Norway, type-B garnet peridotites with Caledonian HP assemblages occur in addition to type-D Proterozoic UHP assemblages (Jamtveit, 1987). Global locations of some of these garnet peridotites in UHP metamorphic belts are indicated in Figure 9.

Type-A Garnet Peridotites

These ultramafic rocks originated in the mantle wedge above a subduction zone. Type-A uppermost mantle peridotites are either residual mantle fragments, or they are peridotite and pyroxenite bodies differentiated from mantle-sourced magma; they possess isotopic and geochemical signatures of the hanging-wall mantle. For example, garnet peridotites from eastern China are in fault contact with enclosing country-rock granitic gneisses, they are massive and relatively homogeneous without layering, they exhibit either near-equigranular or porphyroblastic textures, and they contain lenses of bimineralic coesite-bearing eclogite. Most such garnet peridotites belong to the Mg-Cr type of Medaris and Carswell (1990) and contain more MgO and Cr_2O_3 and less fertile elements such as TiO_2, Al_2O_3, CaO, and FeO than primitive mantle as defined by Ringwood (1975). Type-A garnet lherzolites and pyroxenites preserve mantle $\delta^{18}O$ value ranges for garnet, olivine, and clinopyroxene of +4.8‰–+5.6‰, +4.7‰, and +4.5‰–5.6‰, respectively (Zhang et al., 1998, 2000, 2004). They tend to have low $^{87}Sr/^{86}Sr$ (0.7038–0.7044) and $^{143}Nd/^{144}Nd$ (0.5123–0.5124) values. However, some exhibit unusually high isotopic ratios and plot outside the range of mantle values; these anomalous isotopic compositions may be due to later metasomatism and/or contamination by crustal materials. It should be noted that low-pressure, high-temperature garnet peridotites reported by Medaris (1999), such as those from the Bohemia Massif, are included here (see also Carswell and O'Brien, 1993; O'Brien and Rötzler, 2003). Some of these high-temperature bodies were evolved from spinel peridotites, contain abundant inclusions of spinel in garnet, and equilibrated at ~1000–1300 °C.

Type-B Garnet Peridotites

Such bodies were derived from ultramafic portions of pre-subduction crustal mafic-ultramafic complexes. The protoliths were produced by differentiation from mafic magma prior to UHP metamorphism (e.g., Z. Zhang et al., 2006); the continental crustal section was then subjected to underflow and HP-UHP metamorphism. Typically, garnet peridotites are interlayered with eclogites of various compositions. Garnet peridotites from the Dabieshan are characterized by: (1) well-developed compositional banding and/or layering; (2) the occurrence of low-pressure mineral inclusions in garnets (e.g., Okay, 1994); (3) the preservation of relatively light isotopic bulk-rock oxygen compositions ($\delta^{18}O < 5‰$); and (4) an old, presubduction age of intrusion (~500–300 Ma) into the sialic crust, as well as a Triassic (~230–220 Ma) UHP metamorphic age (Chavagnac and Jahn, 1996; Jahn et al., 2003). Type-B mafic-ultramafic igneous complexes exhibit a large range in major-element concentrations, and most contain lower MgO and higher SiO_2, CaO, TiO_2, Al_2O_3, and FeO values than type-A peridotites. Some type-B garnet peridotites contain well-preserved prograde low-pressure mineral assemblages as inclusions in UHP phases. For example, inclusions of sapphirine, corundum, clinochlore, and amphibole occur in garnet porphyroblasts from the Maowu area of the Dabieshan (Okay, 1993). In the Lotru garnet peridotite from the South Carpathians, garnet and orthopyroxene formed as reaction products at the boundary of partly serpentinized olivine and pseudomorphs after plagioclase, now consisting of amphibole, zoisite, and chlorite (Medaris et al., 2003).

Type-C Garnet Peridotites

These tectonic entities were derived from the underlying mantle of subducted oceanic or continental lithosphere. Protoliths of the footwall mantle of a sinking slab in some cases were serpentinized prior to HP-UHP metamorphism. The ultramafic rocks may represent part of an ophiolitic sequence that was emplaced in the downgoing plate prior to deep underflow. Thus, some Alpine garnet peridotites of the Western Alps are associated with eclogites that recrystallized from rodingitized gabbros, and retain geochemical evidence of earlier seawater alteration. Petrochemically, such garnet peridotites are difficult to distinguish from type-A hanging-wall mantle-derived analogues. Accordingly, only a few garnet peridotites from the Lepontine Alps (e.g., Cima de Gagnone and Monte Duria bodies) have been assigned to this group (Evans and Trommsdorff, 1978).

Type-D Garnet Peridotites

These deep-seated mantle fragments were emplaced tectonically at crustal levels prior to subduction. Relict high-pressure, high-temperature peridotitic lenses are present in the Western Gneiss region of coastal Norway; the ultramafic lithologies represent fragments of ancient depleted mantle, the

TABLE 2. CHARACTERISTICS OF GARNET PERIDOTITES IN UHP METAMORPHIC BELTS WORLDWIDE

Terrane	Type	Modes of occurrence	Rock types*	Mineral assemblage	Peak-stage T-P (°C, GPa)	Metamorphic age (Ma)	References
Sulu, Eastern China							
Rongcheng	A	Blocks in gneiss	LZ, DN	Grt+Ol+Opx+Cpx	820–920; 4–6	Triassic	Zhang et al. (1994); Hiramatsu et al. (1995); Jahn (1998); Hacker et al. (1997)
Yangkou		Layers and blocks in gneiss	PD, CP	Ol+Cpx+Opx+Grt+Amp	750 ± 50; >4	Triassic	Zhang et al. (2005a, 2005b)
Rizhao	A	Layers and blocks in gneiss	HZ, CP	Grt+Cpx+Ilm+Chl±Ol	>820; >4	Triassic	Zhang et al. (1994, 2000); Zhang and Liou (2003)
Donghai	A	Blocks in gneiss	LZ, HZ, WR, WB	Grt+Ol+Opx+Cpx±Phl±Mgs	780–980; 4–6.7	SHRIMP U-Pb: 230 ± 5	Zhang et al. (1994, 1995a, 1998, 2000, 2005b)
Dabie, Eastern China							
Bixiling	B	Layered mafic-ultramafic	LZ, WR, WB	Grt+Ol±Opx+Cpx±Chu±Mgs	820–950; 4.7–6.7	Triassic	Zhang et al. (1995b, 2000); Chavanac and Jahn (1996)
Maowu	B	Layered ultramafic	HZ, CP, OP	Ol+Opx+Grt+Cpx+Rt± Mz Mgs	750 ± 50; 4–6	Ca. 220–230	Okay (1994); Liou and Zhang (1998); Zhang et al. (1998, 1999); Jahn et al. (2003)
Raobazhai		Fault block	DN, HZ	Ol+Opx+Cpx+Spn±Grt	>1100; 1.8–2.2	Triassic	Tsai et al. (2000)
Western China							
Altyn	A	Blocks in gneiss	LZ, WR, CP	Grt±Ol+Opx+Cpx±Mgs	890–970; 3.8–5.1	Caledonian	Liu et al. (2002); Zhang et al. (2004)
N Qaidam	A	Block, lens, layers in gneiss	LZ, DN, CP	Grt+Ol+Opx+Cpx	780–850; >2.7–4.5	Caledonian	Yang et al. (2000, 2002); Song et al. (2004, 2005); Zhang et al. (2004)
Indonesia							
Sulawesi	C	Fault slice and xenolith in granite	LZ	Ol+Grt+Opx+Cpx	1025–1200; 2.6–4.8	Cretaceous	Kardarusman and Parkinson (2000)
SW Japan							
Sanbagawa	A	Lens, boudins, or layers	DN, CP, WR, WB	Grt+Ol+Cpx+Cr-Sp±Opx	700–810; 2.9–2.38	Cretaceous	Enami et al. (2004); Mizukami et al. (2004)
Northern Kazakhstan							
Kokchetav	A (?)	Block in gneiss	PD	Ol+Grt+Ti-Chu Mg-Ilm±Cpx±Phl	790–880; 4–6	U-Pb SHRIMP 554–494 (528)	Muko et al. (2002); Katayama et al. (2003)
Western Gneiss Region							
Kalskaret	D	Large body in gneiss	PD	Grt+Ol+Opx+Cpx	890–950; 4.2–4.4	Proterozoic-Archean	Medaris (1984, 1999); Jamtveit et al. (1991); Beyer et al. (2004)
Lien	D	Large body in gneiss	PD	Opx+Grt+Ol+Cpx	850–900; 3.6–3.9	Proterozoic-Archean	Medaris (1980, 1984, 1999)
Rodhaugen	D	Large body in gneiss	PD	Opx+Grt+Ol+Cpx	740–859; 2.3–3.6	Proterozoic-Archean	Medaris (1980, 1999); Carswell (1981)
Sandvika	B, D	Large body in gneiss	PD	Opx+Grt+Ol+Cpx	930–950; 4.4–5.0	Proterozoic-Archean	Medaris (1984, 1999); Jamtveit (1984, 1987)
Raudhaugene, Otroy, Flemsey, Fjortoft	D	Large body in gneiss		Opx+Grt+Ol+Cpx	740–890; 2.3–4.3	Proterozoic-Archean	Carswell (1986); van Roermund et al. (2000, 2001, 2002); Beyer et al. (2004); Spengler et al. (2006)

(*continued*)

TABLE 2. CHARACTERISTICS OF GARNET PERIDOTITES IN UHP METAMORPHIC BELTS WORLDWIDE (continued)

Terrane	Type	Modes of occurrence	Rock types*	Mineral assemblage	Peak-stage T-P (°C, GPa)	Metamorphic age (Ma)	References
Bohemian Massif							
Moldanubian	A, B, C	Lenses in gneiss and granulite	PD, CP	Ol+Grt+Cpx+Opx	815–1330; 2.4–5.6	Variscan (ca. 370–330)	Medaris (1999); Medaris et al. (1990, 2005); Brueckner et al. (1996); Altherr and Kalt (1996); O'Brien and Rötzler (2003)
Erzgebirge	A	Elongate body in gneiss and granulite	PD	Ol+Grt+Opx+Cpx	800–900; 2.9–3.2	Variscan	Schmadicke and Evans (1997)
S. Carpathians	A, B	Lenses in gneiss	PD	Opx+Grt+Ol+Cpx	1150–1300; 2.5–3.2	Variscan (310–360)	Medaris (1999); Medaris et al. (2003)
Ronda	A	Large massif	LZ	Grt+Ol+Opx+Cpx+Dia	1080–1240; 2.1–2.8	Alpine (21–25)	Reisberg et al. (1989); Medaris (1999)
Western Alps							
Lepontine Alps	C	Macroboudins	PD	Grt+Ol+Opx+Cpx	775–820; 3.4–3.7	Alpine	Evans and Trommsdorff (1978)
Alpe Arami	A	Large lense	LZ	Grt+Ol+Opx+Cpx+Chu	840–1130; 3.4–5.2	Alpine	Evans and Trommsdorff (1978); Ernst (1978); Dobrzhinetskaya et al. (1996); Brenker and Brey (1997)
Dominican Republic	C	Boulders		Ol+Grt+Cpx+Spl±Crn	1570–1540; >3.4	Late Cretaceous	Abbott et al. (2005)
British Columbia	C	Detrital Grt, Cpx and Ol in conglomerate	PD		800–950; 3–6	>192–183	MacKenzie et al. (2005)

Note: Mineral abbreviations are after Kretz (1983).
*Rock type abbreviations: CP—clinopyroxenite, DN—dunite, HZ—harzburgite, LZ—lherzolite, OP—orthopyroxenite, PD—peridotite, WB—websterite, WR—wehrlite.

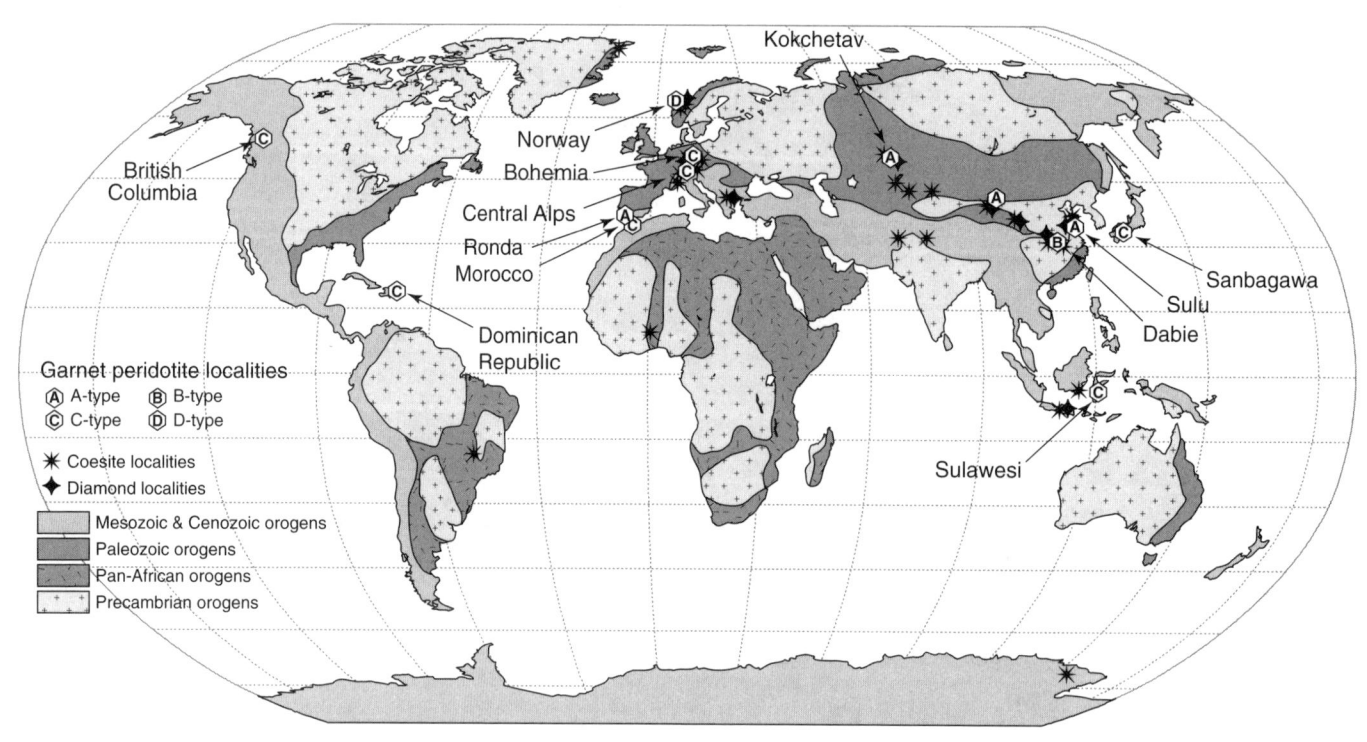

Figure 9. Sketch map by Tsujimori et al. (2006) showing the global distribution of largely continental-crustal ultrahigh-pressure (UHP) metamorphic belts that contain lenses and blocks chiefly of type-A, type-B, type-C, and type-D garnet peridotites (see Table 2 and the text for classification and descriptions). A few type-B ultramafic bodies occur in the Western Gneiss Region of Norway, and type-A and type-B ultramafic bodies are present in the Bohemian Massif.

origins of which are unrelated to the later Caledonian UHP metamorphism (Carswell and van Roermund, 2005). Several garnet peridotite bodies are especially noteworthy because they exhibit evidence of the former stability of megacrystic mineral assemblages that include now-exsolved high-pressure enstatite and majoritic garnet (van Roermund et al., 2000, 2001, 2002; Spengler et al., 2006). Early assemblages in these peridotites possess Sm-Nd Proterozoic ages (Brueckner and Medaris, 1998), close to the igneous crystallization ages of the host granitic gneisses. However, recent in situ Re-Os analysis of sulfides in the garnet peridotites yield a range of Proterozoic and Archean model ages. A Late Archean (3.1–2.7 Ga) protolith age also is supported by whole-rock Re-Os data for dunites from several such bodies (Beyer et al., 2004). The Archean ages bear testament to a process of partial fusion in the mantle that predated formation of the Proterozoic upper crust in the Western Gneiss Region. Apparently, some mantle blocks previously identified as Proterozoic subcontinental lithospheric mantle may represent metasomatized and refertilized Archean mantle.

Contrasting Physical Conditions of Crystallization of Mafic and Ultramafic UHP Rocks?

A comparison of the eclogite thermobarometry described in the text and presented in Figure 1 with computed garnet peridotite conditions of equilibration suggests that although some of the latter rock types (type-D and perhaps some type-A ultramafic bodies) evidently recrystallized under subduction-zone geothermal gradients less than 5 °C/km, none of the eclogites and associated sialic crustal entities can be proven to have formed at so-called forbidden-zone P-T conditions. The reasons for this disparity remain unclear. Possible explanations include: (1) systematic errors were made in the thermobarometric evaluations of peridotites or eclogites, or both lithologies; (2) eclogites and enclosing crustal units re-equilibrated during exhumation-decompression, yielding post-maximum pressures, whereas anhydrous peridotites did not; (3) peridotites characterized by "forbidden-zone" P-T conditions of formation are exotic and formed in a mantle environment unrelated to that of the eclogites. Additional geochemical, geochronologic, and phase-equilibrium investigations are needed to address this problem; what is clear is that at least some zones of continental collision produced subducted HP-UHP assemblages that have been recovered from upper-mantle depths characterized by low prograde geothermal gradients.

CONCLUSIONS FOR PHANEROZOIC CONTRACTIONAL OROGENS

Continental collision involves the essential and substantial consumption of oceanic lithosphere and the transport of a salient of sialic crust, island arc, or microcontinental fragment to the convergent plate junction. Insertion of continental crust and underflow to great depths result in the incipient-to-complete transformation of pre-existing low-pressure quartzofeldspathic and mafic-ultramafic lithologies to UHP-phase assemblages. During prograde metamorphism, evolution of H_2O due to the breakdown of hornblende and serpentine kinetically favors the conversion of mafic and ultramafic rock types to stable eclogitic-garnet peridotitic assemblages, whereas, reflecting the higher-pressure stabilities of biotite and muscovite, micaceous granitic gneisses may persist due to the lack of a free aqueous fluid. The spatial association of volumetrically minor amounts of garnet peridotite and mafic eclogite reflects active participation of both mantle and oceanic crust in the UHP subduction-zone deformation and recrystallization. However, worldwide, chiefly continental materials are regurgitated in exhumed UHP complexes, mirroring the low densities of sialic crust relative to mafic and ultramafic lithologies. Such quartzofeldspathic crustal assemblages are propelled upward by body forces, i.e., buoyancy. Characteristic decompression rates exceeding 10 mm/yr in general are comparable to rates of subduction. Many exposed ultrahigh-pressure complexes consist of ~1–2-km-thick allochthonous sheets, but the largest, in east-central China and western Norway, are ~10 km thick. For thin, ductilely deformed nappes, heat is efficiently conducted away as the UHP complexes rise, cooling the sheets across upper and lower fault-bounded surfaces. For such geometries, the rate of ascent need not be especially rapid. In contrast, the manner in which enormous, much thicker, rapidly decompressing UHP complexes like the Western Gneiss Region and the Dabie-Sulu belt are quenched, preserving relict UHP phases, remains enigmatic. In either case, however, surviving UHP bodies must be relatively dry during the ascent; the absence of a separate aqueous fluid accompanying exhumation would retard back reaction in the complex, allowing the scattered retention of early stage, UHP phases.

The significance of tectonic aneurysms is speculative, but it deserves consideration with regard to the mechanism of final exhumation of UHP terranes. Most recognized UHP collisional complexes bear extensive evidence of prior nappe emplacement, so exposure of the deep-seated terranes may reflect the operation in varying degrees of subduction-zone slab imbrication and/or buoyant massif ascent followed by late domal uplift aided by locally vigorous erosion. Due to relatively rapid decompression at moderately high temperatures, the critical requirement for preservation of UHP relict assemblages in at least fragmentary form is effective heat removal; this, in turn, requires that less rapidly decompressing complexes be characterized by large surface/volume ratios. Massif-type buoyant bodies must rise from great depths at near-adiabatic P-T conditions, i.e., extremely rapidly. For recognizable UHP terranes, transport to mid-crustal levels either in décollement-type structures or as giant slabs must occur first, allowing substantial cooling (quenching) of the UHP mineral assemblages. This event is followed by further exhumation combined with erosional collapse; possible late-stage processes include structural contraction, crustal underplating, shallowing of the dip of the subducting lithosphere, crustal back-folding or faulting, or domal ascent as tectonic aneurysms.

ACKNOWLEDGMENTS

This study was support by Stanford University and National Science Foundation grants EAR-9814889, 0003355, 0003568. and 0510325. Paddy O'Brien and Gordon Medaris provided constructive reviews of a first-draft manuscript. Appreciation is expressed to Hans-Peter Schertl for providing a copy of Elke Nowlan's dissertation. We thank these researchers and institutions for helpful feedback and support.

REFERENCES CITED

Abbott, R.N., Draper, G., and Keshav, S., 2005, UHP magma parageneses, garnet peridotite, and garnet clinopyroxenite: An example from the Dominican Republic: International Geology Review, v. 47, p. 233–247.

Altherr, R., and Kalt, A., 1996, Metamorphic evolution of ultrahigh-pressure garnet peridotites from the Variscan Vosges Mts., France: Chemical Geology, v. 134, p. 27–47, doi: 10.1016/S0009-2541(96)00088-5.

Amato, J.M., Johnson, C., Baumgartner, L., and Beard, B., 1999, Sm-Nd geochronology indicates rapid exhumation of Alpine eclogites: Earth and Planetary Science Letters, v. 171, p. 425–438, doi: 10.1016/S0012-821X(99)00161-2.

Austrheim, H., 1998, Influence of fluid and deformation on metamorphism of the deep crust and consequences for the geodynamics of collision zones, *in* Hacker, B.R., and Liou, J.G., eds., When Continents Collide: Geodynamics and Geochemistry of Ultrahigh-Pressure Rocks: Dordrecht, Kluwer Academic Publishers, p. 297–323.

Baldwin, S.L., Monteleone, B.D., Webb, L.E., Fitzgerald, P.G., Grove, M., and Hill, E.J., 2004, Pliocene eclogite exhumation at plate tectonic rates in eastern Papua New Guinea: Nature, v. 431, p. 263–267, doi: 10.1038/nature02846.

Banno, S., and Sakai, C., 1989, Geology and metamorphic evolution of the Sanbagawa belt, *in* Daly, J.S., Cliff, R.A., and Yardley, B.W.D., eds., Evolution of Metamorphic Belts; Proceedings of the 1987 Joint Meeting of the Metamorphic Studies Group and IGCP Project 235: Geological Society [London] Special Publication 43, p. 519–535.

Beaumont, C., Ellis, S., Hamilton, J., and Fullsack, P., 1996, Mechanical model for subduction-collision tectonics of Alpine-type compressional orogens: Geology, v. 24, p. 675–678, doi: 10.1130/0091-7613-(1996)024<0675:MMFSCT>2.3.CO;2.

Beaumont, C., Ellis, S., and Pfiffner, A., 1999, Dynamics of sediment subduction-accretion at convergent margins: Short-term modes, long-term deformation, and tectonic implications: Journal of Geophysical Research, v. 104, p. 17,573–17,602, doi: 10.1029/1999JB900136.

Begemann, F., Ludwig, K.R., Lugmair, G.W., Min, K., Nyquist, L.E., Patchett, P.J., Renne, P.R., Shih, C.-Y., Villa, I.M., and Walker, R.J., 2001, Call for an improved set of decay constants for geochronological use: Geochimica et Cosmochimica Acta, v. 65, p. 111–121, doi: 10.1016/S0016-7037(00)00512-3.

Beyer, E.B., Brueckner, H.K., Griffin, W.L., O'Reilly, S.Y., and Graham, S., 2004, Archean mantle fragments in Proterozoic crust, Western Gneiss Region, Norway: Geology, v. 32, p. 609–612, doi: 10.1130/G20366.1.

Blythe, A.E., 1998, Active tectonics and ultrahigh-pressure rocks, *in* Hacker, B.R., and Liou, J.G., eds., When Continents Collide: Geodynamics and Geochemistry of Ultrahigh-Pressure Rocks: Dordrecht, Kluwer Academic Publishers, p. 141–160.

Bozhilov, K.N., Green, H.W., and Dobrzhinetskaya, L., 1999, Clinoenstatite in the Alpe Arami peridotite: Additional evidence of very high pressure: Science, v. 284, p. 128–132, doi: 10.1126/science.284.5411.128.

Brenker, F.E., and Brey, G.P., 1997, Reconstruction of exhumation path of the Alpe Arami garnet-peridotite body from depths exceeding 160 km: Journal of Metamorphic Geology, v. 15, p. 581–592, doi: 10.1111/j.1525-1314.1997.00034.x.

Brueckner, H.K., 1998, Sinking intrusion model for the emplacement of garnet-bearing peridotites into continent collision orogens: Geology, v. 26, p. 631–634, doi: 10.1130/0091-7613(1998)026<0631:SIMFTE>2.3.CO;2.

Brueckner, H.K., and Medaris, L.G., Jr., 1998, A tale of two orogens: The contrasting *T-P-t* history and geochemical evolution of mantle in high- and ultrahigh-pressure metamorphic terranes of the Norwegian Caledonides and the Czech Variscides: Schweizerische Mineralogische und Petrographisches Mitteilungen, v. 78, p. 293–307.

Brueckner, H.K., and Medaris, L.G., Jr., 2000, A general model for the intrusion and evolution of "mantle" peridotites in high-pressure and ultrahigh-pressure metamorphic terranes: Journal of Metamorphic Geology, v. 18, p. 123–133, doi: 10.1046/j.1525-1314.2000.00250.x.

Brueckner, H.K., Blusztajn, J., and Bakun-Czubarow, N., 1996, Trace element and Sm-Nd "age" zoning in garnets from peridotites of the Caledonian and Variscan Mountains and tectonic implications: Journal of Metamorphic Geology, v. 14, p. 61–73, doi: 10.1111/j.1525-1314.1996.00061.x.

Burchfiel, B.C., Deng, Q., Molnar, P., Royden, L., Qang, Y., and Zhang, W., 1989, Intracrustal detachment within zones of continental deformation: Geology, v. 17, p. 748–752, doi: 10.1130/0091-7613(1989)017 <0448:IDWZOC>2.3.CO;2.

Busby-Spera, C.J., Mattinson, J.M., Riggs, N.R., and Schermer, E.R., 1990, The Triassic-Jurassic magmatic arc in the Mojave-Sonoran Deserts and the Sierran-Klamath region, in Harwood, D.S., and Miller, M.M., eds., Paleozoic and Early Mesozoic Paleogeographic Relations; Sierra Nevada, Klamath Mountains, and Related Terranes: Geological Society of America Special Paper 255, p. 93–114.

Caby, R., 1994, Precambrian coesite from northern Mali; first record and implications for plate tectonics in the trans-Saharan segment of the Pan-African belt: European Journal of Mineralogy, v. 6, p. 235–244.

Carswell, D.A., 1981, Clarification of the petrology and occurrence of garnet lherzolites, garnet websterites and eclogite in the vicinity of Rodhaugen, Almklovdalen, West Norway: Norsk Geologisk Tidsskrift, v. 61, p. 249–260.

Carswell, D.A., 1986, The metamorphic evolution of Mg-Cr type Norwegian garnet peridotites: Lithos, v. 19, p. 279–297, doi: 10.1016/0024-4937(86)90028-9.

Carswell, D.A., and Compagnoni, R., eds., 2003, Ultrahigh Pressure Metamorphism: Notes in Mineralogy, Volume 5: Budapest, European Union, 508 p.

Carswell, D.A., and Gibb, F.G.F., 1980, The equilibration conditions and petrogenesis of European crustal garnet lherzolites: Lithos, v. 13, p. 19–29, doi: 10.1016/0024-4937(80)90058-4.

Carswell, D.A., and O'Brien, P.J., 1993, Thermobarometry and geotectonic significance of high pressure granulites: Examples from the Moldanubian zone of the Bohemian Massif in lower Austria: Journal of Petrology, v. 34, p. 427–459.

Carswell, D.A., and van Roermund, H.L.M., 2005, On multi-phase mineral inclusions associated with microdiamond formation in mantle-derived peridotite lens at Bardane on Fjørtoft, west Norway: European Journal of Mineralogy, v. 17, p. 31–42, doi: 10.1127/0935-1221/2005/ 0017-0031.

Carswell, D.A., Wilson, R.N., and Zhai, M., 2000, Metamorphic evolution, mineral chemistry and thermobarometry of schists and orthogneisses hosting ultra-high pressure eclogites in the Dabieshan of central China: Lithos, v. 52, p. 121–155, doi: 10.1016/S0024-4937(99)00088-2.

Carswell, D.A., Tucker, R.D., O'Brien, P.J., and Krogh, T.E., 2003a, Coesite micro-inclusions and the U-Pb age of zircons from the Hareidland eclogite in the Western Gneiss Region of Norway: Lithos, v. 67, p. 181–190, doi: 10.1016/S0024-4937(03)00014-8.

Carswell, D.A., Brueckner, H.K., Cuthbert, S.J., Mehta, K., and O'Brien, P.J., 2003b, The timing of stabilisation and the exhumation rate for ultrahigh pressure rocks in the Western Gneiss Region of Norway: Journal of Metamorphic Geology, v. 21, p. 601–612, doi: 10.1046/j.1525-1314. 2003.00467.x.

Chamberlain, C.P., Koons, P.O., Meltzer, A.S., Park, S.K., Craw, D., Zeitler, P.K., and Poage, M.A., 2002, Overview of hydrothermal activity associated with active orogenesis and metamorphism: Nanga Parbat, Pakistan Himalaya: American Journal of Science, v. 302, p. 726–748, doi: 10.2475/ ajs.302.8.726.

Chavagnac, V., and Jahn, B.-M., 1996, Coesite-bearing eclogites from the Bixiling complex, Dabie Mountains, China: Sm-Nd ages, geochemical characteristics and tectonic implications: Chemical Geology, v. 133, p. 29–51, doi: 10.1016/S0009-2541(96)00068-X.

Chemenda, A.I., Mattauer, M., Malavieille, J., and Bokun, A.N., 1995, A mechanism for syn-collisional rock exhumation and associated normal faulting: Results from physical modeling: Earth and Planetary Science Letters, v. 132, p. 225–232, doi: 10.1016/0012-821X(95)00042-B.

Chemenda, A.I., Mattauer, M., and Bokun, A.N., 1996, Continental subduction and a new mechanism for exhumation of high-pressure metamorphic rocks: New modeling and field data from Oman: Earth and Planetary Science Letters, v. 143, p. 173–182, doi: 10.1016/0012-821X(96)00123-9.

Chemenda, A.I., Burg, J.P., and Mattauer, M., 2000, Evolutionary model of the Himalaya-Tibet system: Geopoem based on new modeling, geological and geophysical data: Earth and Planetary Science Letters, v. 174, p. 397–409, doi: 10.1016/S0012-821X(99)00277-0.

Chopin, C., 1984, Coesite and pure pyrope in high-grade blueschists of the western Alps: A first record and some consequences: Contributions to Mineralogy and Petrology, v. 86, p. 107–118, doi: 10.1007/BF00381838.

Cloos, M., 1993, Lithospheric buoyancy and collisional orogenesis: Subduction of oceanic plateaus, continental margins, island arcs, spreading ridges, and seamounts: Geological Society of America Bulletin, v. 105, p. 715–737, doi: 10.1130/0016-7606(1993)105<0715:LBACOS>2.3.CO;2.

Cloos, M., and Shreve, R.L., 1988a, Subduction-channel model of prism accretion, mélange formation and subduction erosion at convergent plate margins: 1. Background and description, in Ruff, L., and Kanamori, H., eds., Subduction Zones: Pure and Applied Geophysics, v. 128, p. 455–500.

Cloos, M., and Shreve, R.L., 1988b, Subduction-channel model of prism accretion, mélange formation and subduction erosion at convergent plate margins: 2. Implications and discussion, in Ruff, L., and Kanamori, H., eds., Subduction Zones: Pure and Applied Geophysics, v. 128, p. 501–545.

Coggon, R., and Holland, T.J.B., 2002, Mixing properties of phengitic micas and revised garnet-phengite thermobarometers: Journal of Metamorphic Geology, v. 20, p. 683–696, doi: 10.1046/j.1525-1314.2002.00395.x.

Coleman, R.G., and Wang, X., 1995, Overview of the geology and tectonics of UHPM, in Coleman, R.G., and Wang, X., eds., Ultrahigh Pressure Metamorphism: Cambridge, Cambridge University Press, p. 1–32.

Compagnoni, R., Hirajima, T., and Chopin, C., 1995, UHPM metamorphic rocks in the western Alps, in Coleman, R.G., and Wang, X., eds., Ultrahigh Pressure Metamorphism: Cambridge, Cambridge University Press, p. 206–243.

Cowan, D.S., and Silling, R.M., 1978, A dynamic, scaled model of accretion at trenches and its implications for the tectonic evolution of subduction complexes: Journal of Geophysical Research, v. 83, p. 5389–5396.

Dal Piaz, G.V., Hunziker, J.C., and Martinotti, G., 1972, La zona Sesia-Lanzo e l'evoluzione tettonico-metamorfica delle Alpi nordoccidentali interne: Memoir Societa Geologia Italia, v. 11, p. 433–460.

Davies, J.H., and von Blanckenburg, F., 1998, Thermal controls on slab breakoff and the rise of high-pressure rocks during continental collisions, in Hacker, B.R., and Liou, J.G., eds., When Continents Collide: Geodynamics and Geochemistry of Ultrahigh-Pressure Rocks: Dordrecht, Kluwer Academic Publishers, p. 97–115.

Dobretsov, N.L., Buslov, M.M., Zhimulev, F.I., Travin, A.V., and Zayachkovsky, A.A., 2006, Vendian–Early Ordovician geodynamic evolution and model for exhumation of ultrahigh-pressure and high-pressure rocks from the Kokchetav subduction-collision zone: Russian Geology and Geophysics, v. 47, p. 424–440.

Dobrzhinetskaya, L., Green, H.W., and Wang, S., 1996, Alpe Arami: A peridotite massif from depths of more than 300 kilometers: Science, v. 271, p. 1841–1846, doi: 10.1126/science.271.5257.1841.

Eide, E., and Liou, J.G., 2000, High-pressure blueschists and eclogites in Hong'an: A framework: Lithos, v. 52, p. 1–22, doi: 10.1016/S0024-4937 (99)00081-X.

Enami, M., Mizukami, T., and Yokoyama, K., 2004, Metamorphic evolution of garnet-bearing ultramafic rocks from the Congen area, Sanbagawa belt, Japan: Journal of Metamorphic Geology, v. 22, p. 1–15, doi: 10.1111/ j.1525-1314.2003.00492.x.

England, P.L., and Holland, T.J.B., 1979, Archimedes and the Tauern eclogites: The role of buoyancy in the preservation of exotic eclogitic blocks: Earth and Planetary Science Letters, v. 44, p. 287–294, doi: 10.1016/0012-821X (79)90177-8.

Engvik, A.K., Austrheim, H., and Andersen, T.B., 2000, Structural, mineralogical and petrophysical effects on deep crustal rocks of fluid-limited polymetamorphism, Western Gneiss Region, Norway: Geological Society [London] Journal, v. 157, p. 121–134.

Ernst, W.G., 1970, Tectonic contact between the Franciscan mélange and the Great Valley sequence, crustal expression of a late Mesozoic Benioff zone: Journal of Geophysical Research, v. 75, p. 886–901.

Ernst, W.G., 1978, Petrochemical study of lherzolitic rocks from the western Alps: Journal of Petrology, v. 28, p. 341–392.

Ernst, W.G., 1988, Tectonic histories of subduction zones inferred from retrograde blueschist *P-T* paths: Geology, v. 16, p. 1081–1084, doi: 10.1130/0091-7613(1988)016<1081:THOSZI>2.3.CO;2.

Ernst, W.G., 2005, Alpine and Pacific styles of Phanerozoic mountain building: Subduction-zone petrogenesis of continental crust: Terra Nova, v. 17, p. 165–188, doi: 10.1111/j.1365-3121.2005.00604.x.

Ernst, W.G., 2006, Preservation/exhumation of ultrahigh-pressure subduction complexes: Lithos, v. 92, p. 321–335, doi: 10.1016/j.lithos.2006.03.049.

Ernst, W.G., and Peacock, S., 1996, A thermotectonic model for preservation of ultrahigh-pressure mineralogic relics in metamorphosed continental crust, *in* Bebout, G.E., Scholl, D.W., Kirby, S.H., and Platt, J.P., eds., Subduction Top to Bottom: American Geophysical Union Geophysical Monograph 96, p. 171–178.

Ernst, W.G., Seki, Y., Onuki, H., and Gilbert, M.C., 1970, Comparative Study of Low-Grade Metamorphism in the California Coast Ranges and the Outer Metamorphic Belt of Japan: Geological Society of America Memoir 124, 276 p.

Ernst, W.G., Maruyama, S., and Wallis, S., 1997, Buoyancy-driven, rapid exhumation of ultrahigh-pressure metamorphosed continental crust: Proceedings of the National Academy of Sciences of the United States of America, v. 94, p. 9532–9537.

Ernst, W.G., Mosenfelder, J.L., Leech, M.L., and Liu, J., 1998, H_2O recycling during continental collision: Phase-equilibrium and kinetic considerations, *in* Hacker, B.R., and Liou, J.G., eds., When Continents Collide: Geodynamics and Geochemistry of Ultrahigh-Pressure Rocks: Dordrecht, Kluwer Academic Publishers, p. 275–295.

Evans, B.W., and Trommsdorff, V., 1978, Petrogenesis of garnet lherzolite, Cima di Gagnone, Lepontine Alps: Earth and Planetary Science Letters, v. 40, p. 333–348, doi: 10.1016/0012-821X(78)90158-9.

Ganguly, J., Cheng, W., and Tirone, M., 1996, Thermodynamics of aluminosilicate garnet solid solution: New experimental data, and optimized model, and thermometric applications: Contributions to Mineralogy and Petrology, v. 126, p. 137–151, doi: 10.1007/s004100050240.

Gebauer, D., Schertl, H.P., Briz, M., and Schreyer, W., 1997, 35 Ma old ultrahigh-pressure metamorphism and evidence for very rapid exhumation in the Dora Maira Massif, western Alps: Lithos, v. 41, p. 5–24, doi: 10.1016/S0024-4937(97)82002-6.

Gilotti, J.A., and Krogh Ravna, E., 2002, First evidence of ultrahigh-pressure metamorphism in the North-East Greenland Caledonides: Geology, v. 30, p. 551–554, doi: 10.1130/0091-7613(2002)030<0551:FEFUPM>2.0.CO;2.

Griffin, W.L., Fisher, N.I., Friedman, J., Ryan, C.G., and O'Reilly, S.Y., 1999, Cr-pyrope garnets in lithospheric mantle: I. Compositional systematics and relations to tectonic setting: Journal of Petrology, v. 40, p. 679–705, doi: 10.1093/petrology/40.5.679.

Hacker, B.R., 1996, Eclogite formation and the rheology, buoyancy, seismicity, and H_2O content of oceanic crust, *in* Bebout, G.E., Scholl, D.W., Kirby, S.H., and Platt, J.P., eds., Subduction Top to Bottom: American Geophysical Union Geophysical Monograph 96, p. 171–178.

Hacker, B.R., 2006, Pressures and temperatures of ultrahigh-pressure metamorphism: Implications for UHP tectonics and H_2O in subducting slabs: International Geology Review, v. 48, p. 1053–1066.

Hacker, B.R., and Peacock, S.M., 1995, Creation, preservation and exhumation of UHPM rocks, *in* Coleman, R.G., and Wang, X., eds., Ultrahigh Pressure Metamorphism: Cambridge, Cambridge University Press, p. 159–181.

Hacker, B.R., Ratschbacher, L., Webb, L., and Dong, S., 1995, What brought them up? Exhuming the Dabie Shan ultrahigh-pressure rocks: Geology, v. 23, p. 743–746, doi: 10.1130/0091-7613(1995)023<0743:WBTUEO>2.3.CO;2.

Hacker, B.R., Wang, X., Eide, E.A., and Ratschbacher, L., 1996, The Qinling-Dabie ultrahigh-pressure collisional orogen, *in* Harrison, T.M., and Yin, A., eds., The Tectonic Development of Asia: Cambridge, Cambridge University Press, p. 345–370.

Hacker, B.R., Sharp, T., Zhang, R.Y., Liou, J., and Hervig, R.L., 1997, Determining the origin of ultrahigh-pressure lherzolite?: Science, v. 278, p. 702–704, doi: 10.1126/science.278.5338.702.

Hacker, B.R., Ratschbacher, L., Webb, L., McWilliams, M.O., Ireland, T., Calvert, A., Dong, S., Wenk, H.-R., and Chateigner, D., 2000, Exhumation of ultrahigh-pressure continental crust in east central China: Late Triassic–Early Jurassic tectonic unroofing: Journal of Geophysical Research, v. 105, p. 13,339–13,364, doi: 10.1029/2000JB900039.

Hacker, B.R., Abers, G.A., and Peacock, S.M., 2003a, Subduction factory: 1. Theoretical mineralogy, density, seismic wave speeds, and H_2O content: Journal of Geophysical Research, v. 108, no. B1, doi: 10.1029/2001JB001127.

Hacker, B.R., Calvert, A.T., Zhang, R.Y., Ernst, W.G., and Liou, J.G., 2003b, Ultra-rapid exhumation of ultrahigh pressure diamond-bearing metasedimentary and meta-igneous rocks of the Kokchetav Massif: Lithos, v. 70, p. 61–75, doi: 10.1016/S0024-4937(03)00092-6.

Hacker, B.R., Ratschbacher, L., and Liou, J.G., 2004, Subduction, collision and exhumation in the ultrahigh-pressure Qinling-Dabie orogen, *in* Malpas, J., Fletcher, C.J.N., Ali, J.R., and Aitchison, J.C., eds., Aspects of the Tectonic Evolution of China: Geological Society [London] Special Publication 226, p. 157–175.

Hacker, B.R., Wallis, S.R., Grove, M., and Gehrels, G., 2006, High-temperature geochronology constraints on the tectonic history and architecture of the ultrahigh-pressure Dabie-Sulu orogen: Tectonics, v. 25, TC5006, doi: 10.1029/2005TC001937.

Hamilton, W., 1995, Subduction systems and magmatism, *in* Smellie, J.L., ed., Volcanism Associated with Extension at Consuming Plate Margins: Geological Society [London] Special Publication 81, p. 3–28.

Harley, S.L., and Carswell, D.A., 1995, Ultradeep crustal metamorphism: A prospective review: Journal of Geophysical Research, v. 100, p. 8367–8380, doi: 10.1029/94JB02421.

Henry, C., 1990, L'unité à coesite du massif Dora-Maira dans son cadre petrologique et structural (Alpes occidentales, Italie): Paris, Université de Paris VI, 453 p.

Hermann, J., Rubatto, D., Korsakov, A., and Shatsky, V.S., 2001, Multiple zircon growth during fast exhumation of diamondiferous, deeply subducted continental crust (Kokchetav Massif, Kazakhstan): Contributions to Mineralogy and Petrology, v. 141, p. 66–82.

Hirajima, T., and Nakamura, D., 2003, The Dabie Shan–Sulu orogen: EMU Notes in Mineralogy, v. 5, p. 105–144.

Hiramatsu, N., Banno, S., Hirajima, T., and Cong, B., 1995, Ultrahigh-pressure garnet lherzolite from Chijiadian, Rongcheng County, in the Su-Lu region of eastern China: The Island Arc, v. 4, p. 324–333, doi: 10.1111/j.1440-1738.1995.tb00153.x.

Isacks, B., Oliver, J., and Sykes, L.R., 1968, Seismology and the new global tectonics: Journal of Geophysical Research, v. 73, p. 5855–5899.

Ishikawa, M., Kaneko, Y., and Yamamoto, H., 2000, Subhorizontal boundary between ultrahigh-pressure and low-pressure metamorphic units in the Sulu-Tjube area of the Kokchetav Massif, Kazakhstan: The Island Arc, v. 9, p. 317–328, doi: 10.1046/j.1440-1738.2000.00281.x.

Jahn, B.-M., 1998, Geochemical and isotopic characteristics of UHP eclogites of the Dabie orogen: Implications for continental subduction and collisional tectonics, *in* Hacker, B.R., and Liou, J.G., eds., When Continents Collide: Geodynamics and Geochemistry of Ultrahigh-Pressure Rocks: Dordrecht, Kluwer Academic Publishers, p. 203–239.

Jahn, B.-M., Fan, Q., Yang, J.-J., and Henin, O., 2003, Petrogenesis of the Maowu pyroxenite-eclogite body from the UHP metamorphic terrane of Dabieshan: Chemical and isotopic constraints: Lithos, v. 70, p. 243–267, doi: 10.1016/S0024-4937(03)00101-4.

Jamtveit, B., 1984, High-*P* metamorphism and deformation of the Gurskebotn garnet peridotite, Sunnmore, western Norway: Norsk Geologisk Tidsskrift, v. 64, p. 97–110.

Jamtveit, B., 1987, Metamorphic evolution of the Eiksunddal eclogite complex, western Norway, and some tectonic implications: Contributions to Mineralogy and Petrology, v. 95, p. 82–99, doi: 10.1007/BF00518032.

Jamtveit, B., Caswell, D.A., and Mearns, E.W., 1991, Chronology of the high-pressure metamorphism of Norwegian garnet peridotites/pyroxenes: Journal of Metamorphic Geology, v. 9, p. 125–139.

Jayko, A.S., Blake, M.C., and Harms, T., 1987, Attenuation of the Coast Range ophiolite by extensional faulting, and nature of the Coast Range "thrust," California: Tectonics, v. 6, p. 475–488.

Kadarusman, A., and Parkinson, C.D., 2000, Petrology and *P-T* evolution of garnet peridotites from central Sulawesi, Indonesia: Journal of Metamorphic Geology, v. 18, p. 193–210, doi: 10.1046/j.1525-1314.2000.00238.x.

Kaneko, Y., Maruyama, S., Terabayashi, M., Yamamoto, H., Ishikawa, M., Anma, R., Parkinson, C.D., Ota, T., Nakajima, Y., Katayama, I., Yamamoto, J., and Yamauchi, K., 2000, Geology of the Kokchetav ultrahigh-pressure–high-pressure metamorphic belt, north-eastern Kazakhstan: The Island Arc, v. 9, p. 264–283, doi: 10.1046/j.1440-1738.2000.00278.x.

Kaneko, Y., Katayama, I., Yamamoto, H., Misawa, K., Ishikawa, M., Rehman, H.U., Kausar, A.B., and Shirashi, K., 2003, Timing of Himalayan ultrahigh-pressure metamorphism: Sinking rate and subduction angle of the

Indian continental crust beneath Asia: Journal of Metamorphic Geology, v. 21, p. 589–599, doi: 10.1046/j.1525-1314.2003.00466.x.

Katayama, I., Maruyama, S., Parkinson, C.D., Terada, K., and Sano, Y., 2001, Ion micro-probe U-Pb zircon geochronology of peak and retrograde stages of ultrahigh-pressure metamorphic rocks from the Kokchetav Massif, northern Kazakhstan: Earth and Planetary Science Letters, v. 188, p. 185–198, doi: 10.1016/S0012-821X(01)00319-3.

Katayama, I., Mukou, A., Iizuka, T., Maruyama, S., Terada, K., Tsutsumi, T., Sano, S., Zhang, R.Y., and Liou, J.G., 2003, Dating of zircon from Ti-clinohumite–bearing garnet peridotite: Implication for timing of mantle metasomatism: Geology, v. 31, p. 713–716, doi: 10.1130/G19525.1.

Kawachi, Y., 1968, Large-scale overturned structure in the Sanbagawa metamorphic zone in central Shikoku, Japan: Geological Society of Japan Journal, v. 74, p. 607–616.

Kienast, J.R., Lombardo, B., Biino, G., and Pinardon, J.L., 1991, Petrology of very-high-pressure eclogitic rocks from the Brossasco-Isasca Complex, Dora-Maira Massif, Italian western Alps: Journal of Metamorphic Geology, v. 9, p. 19–34.

Koons, P.O., Zeitler, P.K., Chamberlain, C.P., Craw, D., and Meltzer, A.S., 2002, Mechanical links between erosion and metamorphism in Nanga Parbat, Pakistan Himalaya: American Journal of Science, v. 302, p. 749–773, doi: 10.2475/ajs.302.9.749.

Koons, P.O., Upton, P., and Terry, M.P., 2003, Three-dimensional mechanics of UHPM terrains and resultant P–T–t paths, in Carswell, D.A., and Compagnoni, R., eds., Ultrahigh Pressure Metamorphism: European Union Notes in Mineralogy, Volume 5: Budapest, European Union, p. 415–441.

Kretz, R., 1983, Symbols for rock-forming minerals: The American Mineralogist, v. 68, p. 277–279.

Krogh, E.J., and Carswell, D.A., 1995, HP and UHP eclogites and garnet peridotites in the Scandinavian Caledonides, in Coleman, R.G., and Wang, X., eds., Ultrahigh Pressure Metamorphism: Cambridge, Cambridge University Press, p. 244–298.

Krogh Ravna, E., and Paquin, J., 2003, Thermobarometric methodologies applicable to eclogites and garnet ultrabasites, in Carswell, D.A., and Compagnoni, R., eds., Ultrahigh Pressure Metamorphism: European Union Notes in Mineralogy, Volume 5: Budapest, European Union, p. 229–259.

Krogh Ravna, E.J., and Terry, M.P., 2004, Geothermobarometry of phengite-kyanite-quartz/coesite eclogites: Journal of Metamorphic Geology, v. 22, p. 579–592, doi: 10.1111/j.1525-1314.2004.00534.x.

Leech, M.L., Singh, S., Jain, A.K., Klemperer, S.L., and Manickavasagam, R.M., 2005, The onset of India-Asia continental collision: Early, steep subduction required by the timing of UHP metamorphism in the western Himalaya: Earth and Planetary Science Letters, v. 234, p. 83–97, doi: 10.1016/j.epsl.2005.02.038.

Liou, J.G., and Zhang, R.Y., 1998, Petrogenesis of ultrahigh-P garnet-bearing ultramafic body from Maowu, the Dabie Mountains, central China: The Island Arc, v. 7, p. 115–134, doi: 10.1046/j.1440-1738.1998.00188.x.

Liou, J.G., Zhang, R.Y., Wang, X., Eide, E.A., Ernst, W.G., and Maruyama, S., 1996, Metamorphism and tectonics of high-pressure and ultrahigh-pressure belts in the Dabie–Sulu region, China, in Yin, A., and Harrison, T.M., eds., The Tectonic Evolution of Asia: Cambridge, Cambridge University Press, p. 300–344.

Liou, J.G., Zhang, R.Y., Ernst, W.G., Rumble, D., III, and Maruyama, S., 1998, High pressure minerals from deeply subducted metamorphic rocks: Reviews in Mineralogy, v. 37, p. 33–96.

Liou, J.G., Hacker, B.R., and Zhang, R.Y., 2000, Into the forbidden zone: Science, v. 287, p. 1215–1216, doi: 10.1126/science.287.5456.1215.

Liu, F.L., Xu, Z.Q., Liou, J.G., and Song, B., 2004, SHRIMP U-Pb ages of ultrahigh-pressure and retrograde metamorphism of gneiss, south-western Sulu terrane, eastern China: Journal of Metamorphic Geology, v. 22, p. 315–326, doi: 10.1111/j.1525-1314.2004.00516.x.

Liu, F.L., Xu, Z.Q., Liou, J.G., Dong, H.L., and Xue, H.M., 2007, Ultrahigh-pressure mineral assemblages in zircons from the surface to 5158 m depth in cores of the main drill hole, Chinese Continental Scientific Drilling Project, SW Sulu belt, China: International Geology Review, v. 49, p. 454–478.

Liu, L., Sun, Y., Xiao, P., Che, Z., Luo, J., Chen, D., Wang, Y., Zhang, A., Chen, L., and Wang, Y., 2002, Discovery of ultrahigh-pressure magnesite-bearing garnet lherzolite (>3.8 GPa) in the Altyn Tagh, northwest China: Chinese Science Bulletin, v. 47, p. 881–886, doi: 10.1360/02tb9197.

MacKenzie, J.M., Canil, D., Johnston, S.T., English, J., Mihalynuk, M.G., and Grant, B., 2005, First evidence for ultrahigh-pressure garnet peridotite in the North American Cordillera: Geology, v. 33, p. 105–108, doi: 10.1130/G20958.1.

Malpas, J., Fletcher, C.J.N., Ali, J.R., and Aitchison, J.C., eds., 2004, Aspects of the Tectonic Evolution of China: Geological Society [London] Special Publication 226, 362 p.

Maruyama, S., and Parkinson, C.D., 2000, Overview of the geology, petrology and tectonic framework of the HP-UHP metamorphic belt of the Kokchetav Massif, Kazakhstan: The Island Arc, v. 9, p. 439–455, doi: 10.1046/j.1440-1738.2000.00288.x.

Maruyama, S., Liou, J.G., and Zhang, R., 1994, Tectonic evolution of the ultrahigh-pressure (UHP) and high-pressure (HP) metamorphic belts from central China: The Island Arc, v. 3, p. 112–121, doi: 10.1111/j.1440-1738.1994.tb00099.x.

Maruyama, S., Liou, J.G., and Terabayashi, M., 1996, Blueschists and eclogites of the world and their exhumation: International Geology Review, v. 38, p. 485–594.

Massonne, H.J., and Bautsch, H.J., 2002, An unusual garnet pyroxenite from the Granulitgebirge, Germany: Origin in the transition zone (>400 km depths) or in a shallower upper mantle region?: International Geology Review, v. 14, p. 779–796.

Massonne, H.J., and O'Brien, P.J., 2003, The Bohemian Massif and the NW Himalaya, in Carswell, D.A., and Compagnoni, R., eds., Ultrahigh Pressure Metamorphism: European Union Notes in Mineralogy, Volume 5: Budapest, European Union, p. 145–187.

Mattinson, C.G., Zhang, R.Y., Tsujimori, T., and Liou, J.G., 2004, Epidote-rich talc-kyanite-phengite eclogites, Sulu terrane, eastern China: The American Mineralogist, v. 89, p. 1772–1783.

Medaris, L.G., Jr., 1980, Petrogenesis of the Lien peridotite and associated eclogites, Almklovdalen western Norway: Lithos, v. 13, p. 339–353, doi: 10.1016/0024-4937(80)90053-5.

Medaris, L.G., Jr., 1984, A geothermobarometric investigation of garnet peridotites in the Western Gneiss Region of Norway: Contributions to Mineralogy and Petrology, v. 87, p. 72–86, doi: 10.1007/BF00371404.

Medaris, L.G., Jr., 1999, Garnet peridotite in Eurasian HP and UHP terranes: A diversity of origins and thermal histories: International Geology Review, v. 41, p. 799–815.

Medaris, L.G., and Carswell, D.A., 1990, Petrogenesis of Mg-Cr garnet-peridotites in European metamorphic belts, in Carswell, D.A., ed., Eclogite Facies Rocks: Glasgow, Blackie and Sons, p. 260–290.

Medaris, L.G., Wang, H.F., Misar, Z., and Jelinek, E., 1990, Thermobarometry, diffusion modeling and cooling rates of crustal garnet peridotites: Two examples from the Moldanubian zone of the Bohemian Massif: Lithos, v. 25, p. 189–202, doi: 10.1016/0024-4937(90)90014-R.

Medaris, L.G., Ducea, M., Ghent, E., and Iancu, V., 2003, Conditions and timing of high-pressure Variscan metamorphism in the South Carpathians, Romania: Lithos, v. 70, p. 141–161, doi: 10.1016/S0024-4937(03)00096-3.

Medaris, L.G., Wang, H., Jelinek, E., Mihaljevic, M., and Jakes, P., 2005, Characteristics and origins of diverse Variscan peridotites in the Gföhl Nappe, Bohemian Massif, Czech Republic: Lithos, v. 82, p. 1–23, doi: 10.1016/j.lithos.2004.12.004.

Michard, A., Henry, C., and Chopin, C., 1995, Structures in UHPM rocks: A case study from the Alps, in Coleman, R.G., and Wang, X., eds., Ultrahigh Pressure Metamorphism: Cambridge, Cambridge University Press, p. 132–158.

Mihalynuk, M.G., Erdmer, P., Ghent, E.D., Cordey, F., Archibald, D.A., Friedman, R.M., and Johannson, G.G., 2004, Coherent French Range blueschist: Subduction to exhumation in <2.5 m.y.?: Geological Society of America Bulletin, v. 116, p. 910–922, doi: 10.1130/B25393.1.

Mizukami, T., Wallis, S.R., and Yamamoto, J., 2004, Natural examples of olivine lattice preferred orientation patterns with a flow-normal a-axis maximum: Nature, v. 427, p. 432–436, doi: 10.1038/nature02179.

Molnar, P., and Atwater, T., 1978, Interarc spreading and Cordilleran tectonics as alternates related to the age of subducted oceanic lithosphere: Earth and Planetary Science Letters, v. 41, p. 330–340, doi: 10.1016/0012-821X(78)90187-5.

Mosenfelder, J.L., Schertl, H.P., Smyth, J.R., and Liou, J.G., 2005, Factors in the preservation of coesite: The importance of fluid infiltration: The American Mineralogist, v. 90, p. 779–789, doi: 10.2138/am.2005.1687.

Muko, A., Okamoto, K., Yoshioka, N., Zhang, R.Y., Parkinson, C.D., Ogasawara, Y., and Liou, J.G., 2002, Petrogenesis of Ti-clinohumite–bearing garnetiferous ultramafic rocks from Kumdy-kol, in Parkinson, C.D., Katayama, I., Liou, J.G., and Maruyama, S., eds., 2002, The Diamond-Bearing

Kokchetav Massif, Kazakhstan: Tokyo, Universal Academy Press, Frontiers Science Series, no. 38, p. 343–360.

Nakamura, D., and Banno, S., 1997, Thermodynamic modeling of sodic pyroxene solid-solution and its application in a garnet–omphacite–kyanite–coesite geothermobarometer for UHP metamorphic rocks: Contributions to Mineralogy and Petrology, v. 130, p. 93–102, doi: 10.1007/s004100050352.

Newton, R.C., and Haselton, H.T., 1981, Thermodynamics of the garnet-plagioclase-Al$_2$SiO$_5$-quartz geobarometer, in Newton, R.C., ed., Thermodynamics of Minerals and Melts: New York, Springer-Verlag, p. 131–147.

Nichols, G.T., Wyllie, P.J., and Stern, C.R., 1994, Subduction zone melting of pelagic sediments constrained by melting experiments: Nature, v. 371, p. 785–788, doi: 10.1038/371785a0.

Nowlan, E.U., 1998, Druck-Temperatur-Entwicklung und Geochemie von Eklogiten des Dora-Maira-Massivs, Westalpen [Ph.D. thesis]: Bochum, Bochum University, 359 p.

O'Brien, P.J., 2000, The fundamental Variscan problem: High-temperature metamorphism at different depths and high-pressure metamorphism at different temperatures, in Franke, W., Haak, V., Onken, O., and Tanner, D., eds., Orogenic Processes: Quantification and Modelling in the Variscan Belt: Geological Society [London] Special Publication 179, p. 369–386.

O'Brien, P.J., 2001, Subduction followed by collision: Alpine and Himalayan examples, in Rubie, D.C., and van der Hilst, R., eds., Processes and Consequences of Deep Subduction: Physics of the Earth and Planetary Interiors, v. 127, p. 277–291.

O'Brien, P.J., and Rötzler, J., 2003, High-pressure granulites: Formation, recovery of peak conditions, and implications for tectonics: Journal of Metamorphic Geology, v. 21, p. 3–20, doi: 10.1046/j.1525-1314.2003.00420.x.

O'Brien, P.J., and Sachan, H.K., 2000, Diffusion modelling in garnet from Tso Morari eclogite and implications for exhumation models: Earth Science Frontiers, v. 7, p. 25–27, China University of Geosciences, Beijing.

O'Brien, P.J., Zotov, N., Law, R., Khan, M.A., and Jan, M.Q., 2001, Coesite in Himalayan eclogite and implications for models of India-Asia collision: Geology, v. 29, p. 435–438, doi: 10.1130/0091-7613 (2001)029<0435:CIHEAI>2.0.CO;2.

Okay, A.I., 1993, Petrology of a diamond and coesite-bearing metamorphic terrain: Dabie Shan, China: European Journal of Mineralogy, v. 5, p. 659–675.

Okay, A.I., 1994, Sapphirine and Ti-clinohumite in ultra-high-pressure garnet-pyroxenite and eclogite from Dabie Shan, China: Contributions to Mineralogy and Petrology, v. 116, p. 145–155, doi: 10.1007/BF00310696.

Okay, A.I., 1995, Paragonite eclogites from Dabie Shan, China: Re-equilibration during exhumation?: Journal of Metamorphic Geology, v. 13, p. 449–460.

Ota, T., Terabayashi, M., Parkinson, C.D., and Masago, H., 2000, Thermobarometric structure of the Kokchetav ultrahigh-pressure–high-pressure massif deduced from a north-south transect in the Kulet and Saldat-kol regions, northern Kazakhstan: The Island Arc, v. 9, p. 328–357, doi: 10.1046/j.1440-1738.2000.00282.x.

Parkinson, C.D., Katayama, I., Liou, J.G., and Maruyama, S., eds., 2002, The Diamond-Bearing Kokchetav Massif, Kazakhstan: Tokyo, Universal Academy Press, Frontiers Science Series, no. 38, 527 p.

Parrish, R.R., Gough, S., Searle, M., and Waters, D., 2003, Exceptionally rapid exhumation of the Kaghan UHP terrane, Pakistan from U-Th-Pb measurements on accessory minerals: Geological Society of America Abstracts with Programs, v. 34, no. 7, p. 556–557.

Patiño Douce, A.E., and McCarthy, T.C., 1998, Melting of crustal rocks during continental collision and subduction, in Hacker, B.R., and Liou, J.G., eds., When Continents Collide: Geodynamics and Geochemistry of Ultrahigh-Pressure Rocks: Dordrecht, Kluwer Academic Publishers, p. 27–55.

Pattison, D.R.M., Chacko, T., Farquhar, J., and McFarlane, C.R.M., 2003, Temperatures of granulite-facies metamorphism; constraints from experimental phase equilibria and thermobarometry corrected for retrograde exchange: Journal of Petrology, v. 44, p. 867–900, doi: 10.1093/petrology/44.5.867.

Peacock, S.M., 1995, Ultrahigh-pressure metamorphic rocks and the thermal evolution of continent collision belts: The Island Arc, v. 4, p. 376–383, doi: 10.1111/j.1440-1738.1995.tb00157.x.

Platt, J.P., 1986, Dynamics of orogenic wedges and the uplift of high-pressure metamorphic rocks: Geological Society of America Bulletin, v. 97, p. 1037–1053, doi: 10.1130/0016-7606(1986)97<1037:DOOWAT>2.0.CO;2.

Platt, J.P., 1987, The uplift of high-pressure, low-temperature metamorphic rocks: Philosophical Transactions of the Royal Society of London, v. 321, p. 87–103.

Platt, J.P., 1993, Exhumation of high-pressure rocks: A review of concepts and processes: Terra Nova, v. 5, p. 119–133.

Powell, R., and Holland, T.J.B., 1988, An internally consistent dataset with uncertainties and correlations: 3. Applications to geobarometry, worked examples and a computer program: Journal of Metamorphic Geology, v. 6, p. 173–204.

Powell, R., Holland, T., and Worley, B., 1998, Calculating phase diagrams involving solid solutions via non-linear equations, with examples using THERMOCALC: Journal of Metamorphic Geology, v. 16, p. 577–588, doi: 10.1111/j.1525-1314.1998.00157.x.

Proyer, A., Dachs, E., and McCammon, C., 2004, Pitfalls in geothermobarometry of eclogites: Fe^{3+} and changes in the mineral chemistry of omphacite at ultrahigh pressures: Contributions to Mineralogy and Petrology, v. 147, p. 305–318, doi: 10.1007/s00410-004-0554-6.

Pysklywec, R.N., Beaumont, C., and Fullsack, P., 2002, Lithospheric deformation during the early stages of continental collision: Numerical experiments and comparison with South Island, New Zealand: Journal of Geophysical Research, v. 107, no. B7, doi: 10.1029/2001JB000252

Reisberg, L., Zindler, A., and Jagoutz, E., 1989, Further Sr and Nd isotopic results from peridotites of the Ronda ultramafic complex: Earth and Planetary Science Letters, v. 96, p. 161–180, doi: 10.1016/0012-821X(89)90130-1.

Ring, U., and Brandon, M.T., 1994, Kinematic data for the Coast Range fault and implications for the exhumation of the Franciscan subduction complex: Geology, v. 22, p. 735–738, doi: 10.1130/0091-7613(1994)022 <0735:KDFTCR>2.3.CO;2.

Ring, U., and Brandon, M.T., 1999, Ductile deformation and mass loss in the Franciscan subduction complex—Implications for exhumation processes in accretionary wedges, in Ring, U., Brandon, M.T., Lister, G.S., and Willett, S.D., eds., Exhumation Processes; Normal Faulting, Ductile Flow and Erosion: Geological Society [London] Special Publication 154, p. 180–203.

Ringwood, A.E., 1975, Composition and Petrology of the Earth's Mantle: New York, McGraw-Hill, 618 p.

Root, D.B., Hacker, B.R., Mattinson, J.M., and Wooden, J.L., 2004, Young age and rapid exhumation of Norwegian ultrahigh-pressure rocks: An ion microprobe and chemical abrasion study: Earth and Planetary Science Letters, v. 228, p. 325–341, doi: 10.1016/j.epsl.2004.10.019.

Root, D.B., Hacker, B.R., Gans, P., Eide, E., Ducea, M., and Mosenfelder, J., 2005, High-pressure allochthons overlie the ultrahigh-pressure Western Gneiss Region, Norway: Journal of Metamorphic Geology, v. 23, p. 45–61, doi: 10.1111/j.1525-1314.2005.00561.x.

Rubatto, D., and Hermann, J., 2001, Exhumation as fast as subduction?: Geology, v. 29, p. 3–6, doi: 10.1130/0091-7613(2001)029<0003:EAFAS>2.0.CO;2.

Rubatto, D., Liati, A., and Gebauer, D., 2003, Dating UHP metamorphism, in Carswell, D.A., and Compagnoni, R., eds., Ultrahigh Pressure Metamorphism: European Union Notes in Mineralogy, Volume 5: Budapest, European Union, p. 341–363.

Rubie, D.C., 1984, A thermal-tectonic model for high-pressure metamorphism in the Sesia zone, western Alps: The Journal of Geology, v. 92, p. 21–36.

Rubie, D.C., 1986, The catalysis of mineral reactions by water and restrictions on the presence of aqueous fluid during metamorphism: Mineralogical Magazine, v. 50, p. 399–415, doi: 10.1180/minmag.1986.050.357.05.

Rubie, D.C., 1990, Role of kinetics in the formation and preservation of eclogites, in Carswell, D.A., ed., Eclogite Facies Rocks: New York, Chapman and Hall, p. 111–140.

Sacks, P.E., and Secor, D.T., Jr., 1990, Delaminations in collisional orogens: Geology, v. 18, p. 999–1002, doi: 10.1130/0091-7613(1990)018 <0999:DICO>2.3.CO;2.

Schlup, M., Carter, A., Cosca, M., and Steck, A., 2003, Exhumation history of eastern Ladakh revealed by ^{40}Ar/^{39}Ar and fission-track ages: The Indus River–Tso Morari transect, NW Himalaya: Geological Society [London] Journal, v. 160, p. 385–399.

Schmädicke, E., and Evans, B.W., 1997, Garnet-bearing ultramafic rocks from the Erzgebirge, and their relation to other settings in the Bohemian Massif: Contributions to Mineralogy and Petrology, v. 127, p. 57–74, doi: 10.1007/s004100050265.

Schmid, R., Wilke, M., Oberhänsli, R., Janssens, K., Falkenberg, G., Franz, L., and Gaab, A., 2003, Micro-Xanes determination of ferric iron and its application in thermobarometry: Lithos, v. 70, p. 381–392, doi: 10.1016/S0024-4937(03)00107-5.

Searle, M.P., 1996, Cooling history, erosion, exhumation, and kinetics of the Himalaya-Karakorum-Tibet orogenic belt, in Yin, A., and Harrison, T.M., eds., The Tectonic Evolution of Asia: Cambridge, Cambridge University Press, p. 110–137.

Searle, M.P., Hacker, B.R., and Bilham, R., 2001, The Hindu Kush seismic zone as a paradigm for the creation of ultrahigh-pressure diamond- and coesite-bearing continental rocks: The Journal of Geology, v. 109, p. 143–153, doi: 10.1086/319244.

Searle, M.P., Simpson, R.L., Law, R.D., Parrish, R.R., and Waters, D.J., 2003, The structural geometry, metamorphic and magmatic evolution of the Everest Massif, High Himalaya of Nepal–South Tibet: Geological Society [London] Journal, v. 160, p. 345–366.

Seno, T., 1985, Age of subducting lithosphere and back-arc basin formation in the western Pacific since the middle Tertiary, in Nasu, N., Kobayashi, K., Uyeda, S., Kushiro, I., and Kagami, H., eds., Formation of Active Ocean Margins: Tokyo, Terrapub, p. 469–481.

Sharp, Z.D., Essene, E.J., and Hunziker, J.C., 1993, Stable isotope geochemistry and phase equilibria of coesite-bearing whiteschists, Dora Maira Massif, western Alps: Contributions to Mineralogy and Petrology, v. 114, p. 1–12, doi: 10.1007/BF00307861.

Sobolev, N.V., and Shatsky, V.S., 1990, Diamond inclusions in garnets from metamorphic rocks: Nature, v. 343, p. 742–746, doi: 10.1038/343742a0.

Song, S., Yang, J., Liou, J.G., Wu, C., Shi, R., and Xua, Z., 2003, Petrology, geochemistry and isotopic ages of eclogites from the Dulan UHPM terrane, the north Qaidam: NW China: Lithos, v. 70, p. 195–211, doi: 10.1016/S0024-4937(03)00099-9.

Song, S.G., Zhang, L.F., and Niu, Y., 2004, Ultra-deep origin of garnet peridotite from the north Qaidam ultrahigh-pressure belt, northern Tibetan Plateau: NW China: American Mineralogist, v. 89, p. 1330–1336.

Song, S.G., Zhang, L.F., Niu, Y., Jian, P., and Liu, D., 2005, Geochronology of diamond-bearing zircons from garnet peridotite in the north Qaidam UHPM belt, northern Tibetan Plateau: A record of complex histories from oceanic lithosphere subduction to continental collision: Earth and Planetary Science Letters, v. 234, p. 99–118, doi: 10.1016/j.epsl.2005.02.036.

Spengler, D., van Roermund, H.L.M., Drury, M.R., Ottolini, L., Mason, P.R.D., and Davies, G.R., 2006, Deep orogen and hot melting of an Archaean orogenic peridotite massif in Norway: Nature, v. 440, p. 913–917, doi: 10.1038/nature04644.

Stern, C.R., Huang, W.L., and Wyllie, P.J., 1975, Basalt-andesite-rhyolite-H$_2$O: Crystallization intervals with excess H$_2$O and H$_2$O-undersaturated liquidus surfaces to 35 kilobars, with implications for magma genesis: Earth and Planetary Science Letters, v. 28, p. 189–196, doi: 10.1016/0012-821X(75)90226-5.

Stöckhert, B., and Renner, J., 1998, Rheology of crustal rocks at ultrahigh pressure, in Hacker, B.R., and Liou, J.G., eds., When Continents Collide: Geodynamics and Geochemistry of Ultrahigh-Pressure Rocks: Dordrecht, Kluwer Academic Publishers, p. 57–95.

Suppe, J., 1972, Interrelationships of high-pressure metamorphism, deformation, and sedimentation in Franciscan tectonics, U.S.A., in 24th International Geological Congress, Reports, Section 3: Montreal, International Geological Congress, p. 552–559.

Terry, M.P., and Robinson, P., 2004, Geometry of eclogite-facies structural features: Implications for production and exhumation of UHP and HP rocks, Western Gneiss Region, Norway: Tectonics, v. 23, doi: 10.1029/2002TC001401.

Terry, M.P., Robinson, P., and Krogh Ravna, E.J., 2000a, Kyanite eclogite thermobarometry and evidence of thrusting of UHP over HP metamorphic rocks, Nordøyane, Western Gneiss Region, Norway: The American Mineralogist, v. 85, p. 1637–1650.

Terry, M.P., Robinson, P., Hamilton, M.A., and Jercinovic, M.J., 2000b, Monazite geochronology of UHP and HP metamorphism, deformation, and exhumation, Nordøyane, Western Gneiss Region, Norway: The American Mineralogist, v. 85, p. 1651–1664.

Tilton, G.R., Schreyer, W., and Schertl, H.P., 1991, Pb-Sr-Nd isotopic behavior of deeply subducted crustal rocks from the Dora-Maira Massif, western Alps: II. What is the age of the ultrahigh-pressure metamorphism?: Contributions to Mineralogy and Petrology, v. 108, p. 22–33, doi: 10.1007/BF00307323.

Treloar, P.J., O'Brien, P.J., Parrish, R.R., and Khan, M.A., 2003, Exhumation of early Tertiary, coesite-bearing eclogites from the Pakistan Himalaya: Geological Society [London] Journal, v. 160, p. 367–376.

Tsai, C.H., Liou, J.G., and Ernst, W.G., 2000, Petrological characterization and tectonic significance of the Raobazhai retrogressed garnet peridotites from the North Dabie Complex, central-eastern China: Journal of Metamorphic Geology, v. 18, p. 181–192, doi: 10.1046/j.1525-1314.2000.00237.x.

Tsujimori, T., Sisson, V.B., Liou, J.G., Harlow, G.E., and Sorensen, S.S., 2006, Very low-temperature record in subduction process: A review of global lawsonite-eclogites: Lithos, v. 92, p. 609–624, doi: 10.1016/j.lithos.2006.03.054.

van den Beukel, J., 1992, Some thermomechanical aspects of the subduction of continental lithosphere: Tectonics, v. 11, p. 316–329.

van Roermund, H.L.M., Drury, M.R., Barnhoorn, A., and De Ronde, A., 2000, Super-silicic garnet microstructures from an orogenic garnet peridotite, evidence for an ultra-deep (>6 GPa) origin: Journal of Metamorphic Geology, v. 18, p. 135–147, doi: 10.1046/j.1525-1314.2000.00251.x.

van Roermund, H.L.M., Drury, M.R., Barnhoorn, A., and De Ronde, A., 2001, Relict majoritic garnet microstructures from ultra-deep orogenic garnet peridotites in western Norway: Journal of Petrology, v. 42, p. 117–130, doi: 10.1093/petrology/42.1.117.

van Roermund, H.L.M., Carswell, D.A., Drury, M.R., and Heyboer, T.C., 2002, Microdiamonds in a megacrystic garnet websterite pod from Bardane on the island of Fjortoft, western Norway: Evidence for diamond formation in mantle rocks during deep continental subduction: Geology, v. 30, p. 959–962, doi: 10.1130/0091-7613(2002)030<0959:MIAMGW>2.0.CO;2.

Vielzeuf, D., and Schmidt, M.W., 2001, Melting relations in hydrous systems revisited: Application to metapelites, metagraywackes and metabasalts: Contributions to Mineralogy and Petrology, v. 141, p. 251–267.

von Blanckenburg, F., and Davies, J.H., 1995, Slab breakoff: A model for syncollision magmatism and tectonics in the Alps: Tectonics, v. 14, p. 120–131, doi: 10.1029/94TC02051.

Wain, A.L., 1998, Ultrahigh-Pressure Metamorphism in the Western Gneiss Region of Norway: [Ph.D. thesis]: Oxford, Oxford University.

Walsh, E.O., and Hacker, B.R., 2004, The fate of subducted continental margins: Two-stage exhumation of the high-pressure to ultrahigh-pressure Western Gneiss complex, Norway: Journal of Metamorphic Geology, v. 22, p. 671–689, doi: 10.1111/j.1525-1314.2004.00541.x.

Webb, L.E., Hacker, B.R., Ratschbacher, L., McWilliams, M.O., and Dong, S., 1999, ^{40}Ar-^{39}Ar thermochronologic constraints on deformation and cooling history of high and ultrahigh-pressure rocks in the Qinling-Dabie orogen: Tectonics, v. 18, p. 621–638, doi: 10.1029/1999TC900012.

Willett, S., Beaumont, C., and Fullsack, P., 1993, Mechanical model for the tectonics of doubly vergent compressional orogens: Geology, v. 21, p. 371–374, doi: 10.1130/0091-7613(1993)021<0371:MMFTTO>2.3.CO;2.

Yang, J.S., Xu, Z.Q., Li, H.B., Wu, C.L., Zhang, J.X., and Shi, R.D., 2000, A Caledonian convergent border along the southern margin of the Qilian terrane, NW China: Evidence from eclogite, garnet-peridotite, ophiolite, and S-type granite: Journal of Geological Society of China, v. 42, p. 142–160.

Yang, J.S., Xu, Z.Q., Zhang, J.X., Song, S.G., Wu, C.L., and Shi, R.D., 2002, Early Palaeozoic north Qaidam UHP metamorphic belt on the northeastern Tibetan Plateau and a paired subduction model: Terra Nova, v. 14, p. 397–404, doi: 10.1046/j.1365-3121.2002.00438.x.

Young, D.J., Hacker, B.R., Andersen, T.B., Corfu, F., Gehrels, G.E., and Grove, M., 2007, Prograde amphibolite facies to ultrahigh-pressure transition along Nordfjord, western Norway: Implications for exhumation tectonics: Tectonics, v. 26, TC1007, doi: 10.1029/2004TC001781.

Zeitler, P.K., Koons, P.O., Bishop, M.P., Chamberlain, C.P., Craw, D., Edwards, M.A., Hamidullah, S., Jan, M.Q., Khan, M.A., Khattak, U.K., Kidd, W.S.F., Mackie, R.L., Meltzer, A.S., Park, S.K., Pecher, A., Poage, M.A., Sarker, G., Schneider, D.A., Seeber, L., and Shroder, J.F., 2001, Crustal reworking at Nanga Parbat, Pakistan: Metamorphic consequences of thermal-mechanical coupling facilitated by erosion: Tectonics, v. 20, p. 712–728, doi: 10.1029/2000TC001243.

Zhang, R.Y., and Liou, J.G., 1994, Coesite-bearing eclogite in Henan Province, central China: Detailed petrology, glaucophane stability and PT path: European Journal of Mineralogy, v. 6, p. 217–233.

Zhang, R.Y., and Liou, J.G., 1999, Exsolution lamellae in minerals from ultrahigh-P rocks: International Geology Review, v. 41, p. 981–993.

Zhang, R.Y., and Liou, J.G., 2003, Clinopyroxenite from the Sulu ultrahigh-pressure terrane, eastern China: Origin and evolution of garnet exsolution in clinopyroxene: The American Mineralogist, v. 88, p. 1591–1600.

Zhang, R.Y., Liou, J.G., and Cong, B., 1994, Petrogenesis of garnet-bearing ultramafic rocks and associated eclogites in the Su-Lu ultrahigh-pressure

metamorphic terrane, China: Journal of Metamorphic Geology, v. 12, p. 169–186.
Zhang, R.Y., Hirajima, T., Banno, S., Cong, B., and Liou, J.G., 1995a, Petrology of ultrahigh-pressure rocks from the southern Su-Lu region, eastern China: Journal of Metamorphic Geology, v. 13, p. 659–675.
Zhang, R.Y., Liou, J.G., and Cong, B., 1995b, Talc-, magnesite- and Ti-clinohumite–bearing ultrahigh-pressure meta-mafic and ultramafic complex in the Dabie Mountains: Journal of Petrology, v. 36, p. 1011–1037.
Zhang, R.Y., Rumble, D., Liou, J.G., and Wang, Q.C., 1998, Low $\delta^{18}O$, ultrahigh-P garnet-bearing mafic and ultramafic rocks from Dabie Shan, China: Chemical Geology, v. 150, p. 161–170, doi: 10.1016/S0009-2541(98)00051-5.
Zhang, R.Y., Shu, J.F., Mao, H.K., and Liou, J.G., 1999, Magnetite lamellae in olivine and clinohumite from Dabie UHP ultramafic rocks, central China: The American Mineralogist, v. 84, p. 564–569.
Zhang, R.Y., Liou, J.G., Yang, J.S., and Yui, T.F., 2000, Petrochemical constraints for dual origin of garnet peridotites from the Dabie-Sulu UHP terrane, eastern-central China: Journal of Metamorphic Geology, v. 18, p. 149–166, doi: 10.1046/j.1525-1314.2000.00248.x.
Zhang, R.Y., Liou, J.G., Yang, J.S., and Ye, K., 2003, Ultrahigh-pressure metamorphism in the forbidden zone: The Xugou garnet peridotite, Sulu terrane, eastern China: Journal of Metamorphic Geology, v. 21, p. 539–550, doi: 10.1046/j.1525-1314.2003.00462.x.
Zhang, R.Y., Liou, J.G., Yang, J.S., Li, L., and Jahn, B.-M., 2004, Garnet peridotites in UHP mountain belts of China: International Geology Review, v. 46, p. 981–1004.
Zhang, R.Y., Liou, J.G., Zheng, J.P., Griffin, W.L., Yui, T.-F., and O'Reilly, S.Y., 2005a, Petrogenesis of the Yangkou layered garnet peridotite complex, Sulu UHP terrane, China: The American Mineralogist, v. 90, p. 801–813, doi: 10.2138/am.2005.1706.
Zhang, R.Y., Yang, J.S., Wooden, J.L., Liou, J.G., and Li, T.F., 2005b, U-Pb SHRIMP geochronology of zircon in garnet peridotite from the Sulu UHP terrane, China: Implication for mantle metasomatism and subduction-zone UHP metamorphism: Earth and Planetary Science Letters, v. 237, p. 729–734, doi: 10.1016/j.epsl.2005.07.003.
Zhang, Z.M., Liou, J.G., Zhao, S., and Shi, Z., 2006, Petrogenesis of Maobei Fe-Ti rich eclogites from the southern Sulu UHP metamorphic belt, east-central China: Journal of Metamorphic Geology, v. 24, p. 727–741, doi: 10.1111/j.1525-1314.2006.00665.x.
Zhao, R., Zhang, R.Y., Liou, J.G., Booth, A.L., Pope, E.C., and Chamberlain, C.P., 2007, Petrochemistry, oxygen isotopes and U-Pb SHRIMP geochronology of mafic-ultramafic bodies from the Sulu UHP terrane, China: Journal of Metamorphic Geology, v. 25, p. 207–224.

MANUSCRIPT ACCEPTED BY THE SOCIETY 22 MARCH 2007

How much strain can continental crust accommodate without developing obvious through-going faults?

B.C. Burchfiel
C. Studnicki-Gizbert
Department of Earth, Atmospheric and Planetary Sciences, Massachusetts Institute of Technology, Cambridge, Massachusetts 02139, USA

J.W. Geissman
Department of Earth and Planetary Sciences, University of New Mexico, Albuquerque, New Mexico 87131, USA

R.W. King
Department of Earth, Atmospheric and Planetary Sciences, Massachusetts Institute of Technology, Cambridge, Massachusetts 02139, USA

Z. Chen
Chengdu Institute of Geology and Mineral Resources, Chengdu, Sichuan, 610082, People's Republic of China

L. Chen
Yunnan Institute of Geological Sciences, No. 87 Dongfong Lane, Kunming, Yunnan 650051, People's Republic of China

E. Wang
Institute of Geology and Geophysics, Chinese Academy of Sciences, Beijing, 100029, People's Republic of China

ABSTRACT

Geologic data combined with global positioning system (GPS) and paleomagnetic data from SW China indicate that continental crust can absorb tens to perhaps at least hundreds of kilometers of horizontal shear without developing either through-going faults or obvious structures capable of accommodating shear strain. The arcuate, left-lateral Xianshuihe-Xiaojiang and Dali fault systems bound crustal fragments that have rotated clockwise around the eastern Himalayan syntaxis. The two fault systems terminate to the south, but faults reappear farther south, and these continue the GPS velocity gradient. The shear must be transmitted across the Lanping-Simao fold belt without forming through-going faults. West of the Longmen Shan, a geodetically determined velocity gradient of ~10 mm/yr at N60°E lies in an area not marked by through-going faults. If this deformation has been active for the past 8–11 m.y., it should have accumulated ~100 km of shear across a belt ~100 km wide. In both regions, there are no obvious structures that are capable of accommodating the shear. Paleomagnetic data from the southern Lanping-Simao belt are interpreted to indicate an unexpected zone of left-lateral shear present (Burchfiel and Wang, 2007) where rotation of crustal material is locally more than 90° across a zone unmarked by any mapped through-going faults. In these examples, the mechanism of deformation is not obvious, but we suggest it is distributed brittle deformation at a range of scales, from closely spaced faults to cataclastic deformation. In older terranes, recognition of such zones potentially adds an unknown uncertainty to field study and tectonic analyses.

Keywords: China, late Cenozoic tectonics, Eastern Himalayan syntaxis, strike-slip faults.

INTRODUCTION

Global positioning system (GPS) studies conducted over the past decade or so in several actively deforming regions have begun to reveal crustal deformation that is not obviously expressed in the geology on the ground (Burchfiel, 2004), and they have revolutionized our approach to understanding the mechanisms and time scales of crustal deformation. These short-term data can often be extrapolated back in time to show that deformation occurred over longer time intervals and produced tens to perhaps hundreds of kilometers of displacement across zones 10–100 km wide. One example of GPS velocity gradients revealing horizontal shear in the absence of through-going structures is the Eastern Mojave shear zone in California (Miller et al., 2001; McClusky et al., 2001, personal commun. [2004] in Burchfiel, 2004). Along this shear zone, crust south of the Garlock fault, as well as the fault itself, are both being rotated in a clockwise fashion, whereas rocks both to the north and south show prominent faults related to right-lateral shear. Potentially, where the appropriate rocks are exposed in the appropriate locations, paleomagnetic studies should be able to demonstrate the rotation of the crust between the shear zones to the north and south of the Garlock fault to identify the through-going nature of the shear (Schermer et al., 1996). However, such data would still not resolve the outstanding problem: how this deformation occurred in apparently unfaulted, yet rotated crust. We submit that the Eastern Mojave shear zone is not an isolated case and additional examples are likely to come to light as geodetic techniques continue to be applied to the study of active crustal deformation.

Recent studies on deformation in the area of the eastern syntaxis of the Himalaya in southwest China have led to the recognition of several variations on the problem of differential shear passing through crustal rocks without obvious through-going structures. This observation presents a major problem for field geologic studies in this and, we suggest, numerous other areas. In several places, geodetic and paleomagnetic data have highlighted broad regions where geologic mapping has failed to identify and/or recognize considerably large magnitudes of regionally important strain.

UNRECOGNIZED SHEAR DEFORMATION IN SOUTHERN YUNNAN, CHINA

Wang et al. (1998) showed that the active left-lateral Xianshuihe-Xiaojiang fault system in the SE part of the Tibetan Plateau begins as a narrow fault zone in western Sichuan (Figs. 1 and 2). It continues to the southeast as a convex-east, arcuate fault zone, and its displacement becomes partitioned progressively onto several faults, until, ~100 km north of the Red River, the zone is more than 100 km wide and consists of numerous subparallel faults (Fig. 2). The faults of the Xianshuihe-Xiaojiang fault system cross numerous older structures and geologic contacts within the Yangtze platform at high angles, which permit accurate measurement of their displacements. Wang et al. (1998) showed that the total displacement across the Xianshuihe-Xiaojiang fault system remained essentially constant at ~60 km whether it was measured across a single, localized fault or summed across numerous faults where the deforming zone is broad. As the Xianshuihe-Xiaojiang fault system near the Red River, the number of mapped faults is fewer, but the displacement on individual faults also becomes less. Only three of the faults reach the Red River valley, at which point each has less than 1 km of displacement and none of them offsets the well-defined contact between the metamorphic rocks of the Ailao Shan and Cenozoic strata to the north (Schoenbohm et al., 2006). Where the fault system intersects the Ailao Shan, the Ailao Shan has a WNW strike compared to the overall NW strike of the entire Ailao Shan. The change in strike was interpreted by Wang et al. (1998) as the result of accommodation of the 60 km of left-lateral shear on the Xianshuihe-Xiaojiang fault system by bending of the Ailao Shan. They further noted that, although the displacement on individual faults within the Xianshuihe-Xiaojiang fault system becomes smaller nearer to the Ailao Shan, the total shear remains approximately constant.

South of the Ailao Shan, in the Lanping-Simao fold belt, several structures, early Cenozoic fold traces and reverse (?) faults, appear to be bent, but no through-going faults parallel to the Xianshuihe-Xiaojiang fault system are present. South of the Lanping-Simao fold belt, several prominent approximately NE-striking left-lateral faults appear and can be traced to the SW for hundreds of kilometers into Indochina (Figs. 1 and 2). Many of these faults show a curved trace convex to the SE. Wang et al. (1998) interpreted the southeastern section of most of these faults (the Dien Bien Phu fault) to be the continuation of the shear manifested in the Xianshuihe-Xiaojiang fault system. The faults north (Xianshuihe-Xiaojiang fault system) and south (Dien Bien Phu fault and others) of the Lanping-Simao were thus interpreted as part of a boundary to a broad crustal area that has been rotating around the eastern Himalayan syntaxis for at least the past ~2–4 m.y., and most probably longer.

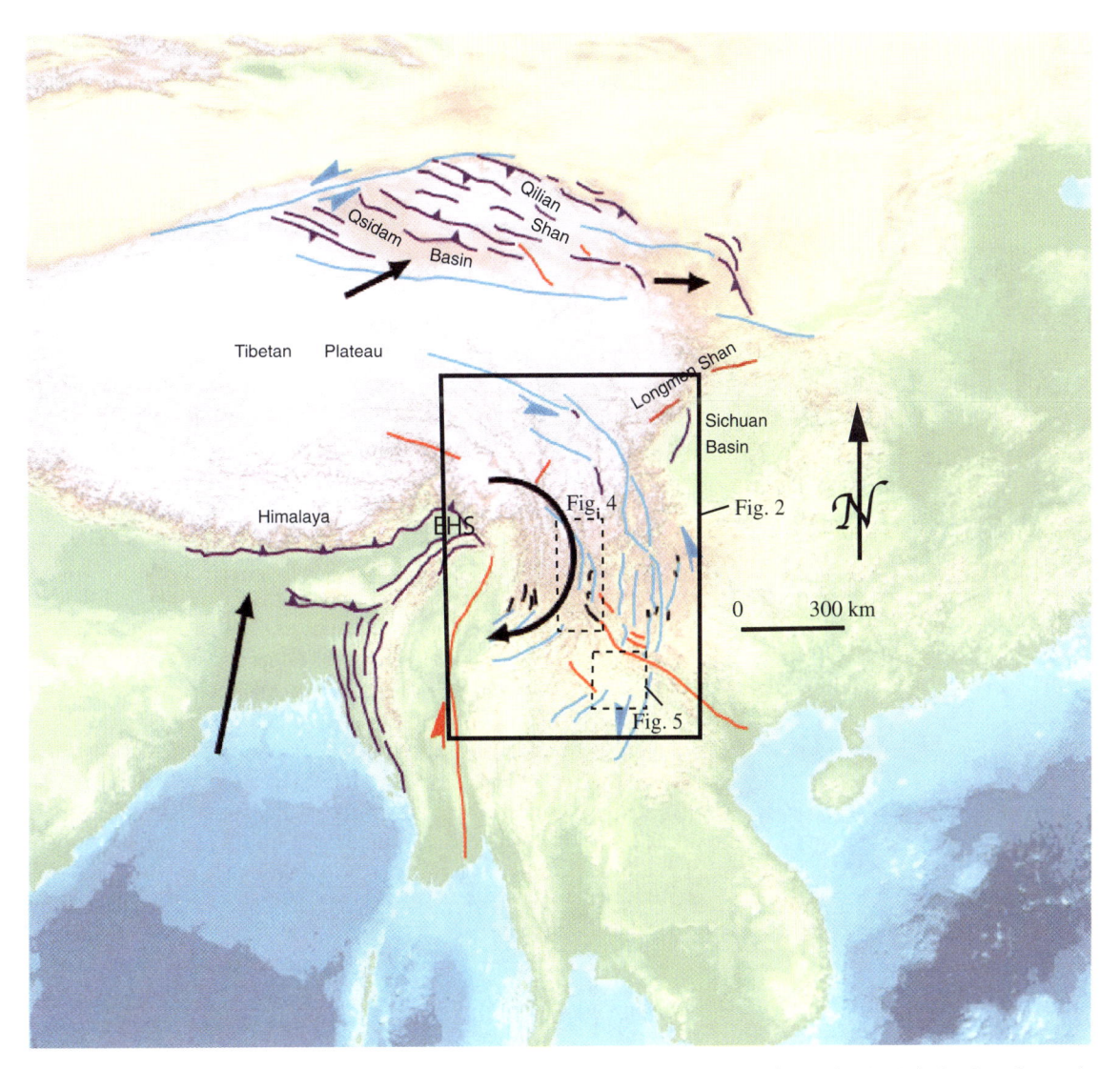

Figure 1. Generalized tectonic map of the eastern part of the Tibetan Plateau and adjacent foreland during late Cenozoic to Holocene time. Black arrows show movement of crustal fragments relative to Eurasia for India and the northeast part of the plateau, and relative to South China in the southeastern part of the plateau. Left-lateral strike-slip faults are shown in blue, right-lateral strike-slip faults are shown in red, shortening structures are shown in purple, and extensional structures are shown by short black lines. EHS—eastern Himalayan syntaxis. Locations of Figures 2, 4, and 5 are shown.

Modern GPS velocities for this region calculated relative to a fixed South China show that the region west of the Xianshuihe-Xiaojiang fault system and its southern continuation is being displaced in a manner that is consistent with clockwise rotation around the eastern Himalayan syntaxis (Fig. 3; King et al., 1997; Chen et al., 2000; Zhang et al., 2004). The GPS velocities parallel the major faults, and elastic dislocation modeling indicates that the geodetically determined velocities can be explained by locking of the few, major faults at a depth of 12–17 km (Studnicki-Gizbert et al., 2004; Meade, 2007). Clearly, the Xianshuihe-Xiaojiang fault system represents the system of structures that accommodates the velocity gradient between the rotating crustal fragment to the west of the fault system and South China. What the data also show is that the velocities and the velocity gradient are continuous across the Lanping-Simao fold belt, which is consistent with the analysis from the geology as well as new block model analysis of the present-day kinematics of the area (Meade, 2007). However, there are no prominent, through-going faults in this region that can be straightforwardly related to the geodetically determined velocities.

Examination of GPS velocities also shows that the translation paths defining crustal rotation are divergent, and there is another velocity gradient with increasing westward component of velocity associated with transtension and left-lateral slip on the active Dali fault system closer to the syntaxis (Fig. 3). The Dali fault system (Fig. 4) consists of several left-lateral faults (the Zhongdian, Jianchuan, Lijiang, and Chenghai faults, among others) that generally parallel the GPS velocities (in a South China frame, see Zhang et al., 2004) and have a convex-east arcuate pattern. The faults are prominent and easily recognizable structures that can be traced in the field and in remote-sensing imagery to the south, where they intersect the eastern boundary of the Lanping-Simao fold belt, and then apparently end. They only displace the boundary locally along the easternmost fault of the Dali fault system (the Chenghai fault; Schoenbohm et al., 2006). On the south side of the Lanping-Simao fold belt, several active, convex-east, left-lateral faults are present that can be regarded as the SW continuation of the left-lateral shear gradient. Like the situation with the Xianshuihe-Xiaojiang fault system to the southeast, there are no faults that mark the shear that must cross the Lanping-Simao fold belt. Thus, like the Xianshuihe-Xiaojiang fault system, the shear gradient along the Dali fault system continues through rocks without any recognizable through-going structures that might accommodate the shear.

The total displacement across the Dali fault system remains poorly determined. Only the Zhongdian fault at the western end of Dali fault system shows a measurable offset of 15–20 km; however, much of this offset is probably early Cenozoic (Burchfiel and Wang, 2007). This fault is associated with 3 km of offset on the Jinsha River. Schoenbohm et al. (2006) has shown an offset of ~7–9 km on the southern end of the Chenghai fault where it offsets the Red River fault. Farther to the north, the same fault offsets the Jinsha River by ~3.5 km, but this represents a minimum value. The Lijiang, Jianchuan, and Heqing faults mostly transfer extension between actively opening basins, the subsidence histories and fill geometries of which are poorly known. The Daju fault (the active continuation of the Zhongdian fault) is associated with 2–4 km of left-slip, constrained by the offset of the pre-incision outlet to the Daju basin. An estimate of >10 km total offset on the Dali fault system is not unreasonable. The total offset across both the Xianshuihe-Xiaojiang fault system (60 km) and the Dali fault system (>10 km) is on the order of ~70 km. Lacassin et al. (1998) have given displacements for the faults south of the Lanping-Simao belt that total 67–75+ km. Based on these observations, we suggest that the short-term GPS velocity gradient is transmitted through the Lanping-Simao belt. In addition, the long-term, total displacement of the Zhongdian-Dali continental fragment and the crust west of the Xianshuihe-Xiaojiang fault system, which rotates about the eastern Himalayan syntaxis, also appears to pass through this belt but is not marked by any through-going structure.

The nature of termination of the faults of the Xianshuihe-Xiaojiang fault system and Dali fault system on either side of the Lanping-Simao belt is different. At their north and northeast boundaries, they stop abruptly at the northern boundary of the Lanping-Simao fold belt (Fig. 2). This boundary is marked by active faults, many of which are normal faults, but there is also right-lateral displacement along part of the Red River fault (Replumaz et al., 2001; Schoenbohm et al., 2006). Those faults of the Xianshuihe-Xiaojiang fault system that reach the Red River area offset active strands of the Red River fault, indicating that both fault sets are active before they reach the northern boundary of the Ailao Shan metamorphic rocks (Schoenbohm et al., 2006). The faults of the Dali fault system end either at normal faults along the Lanping-Simao boundary or terminate within related extensional basins.

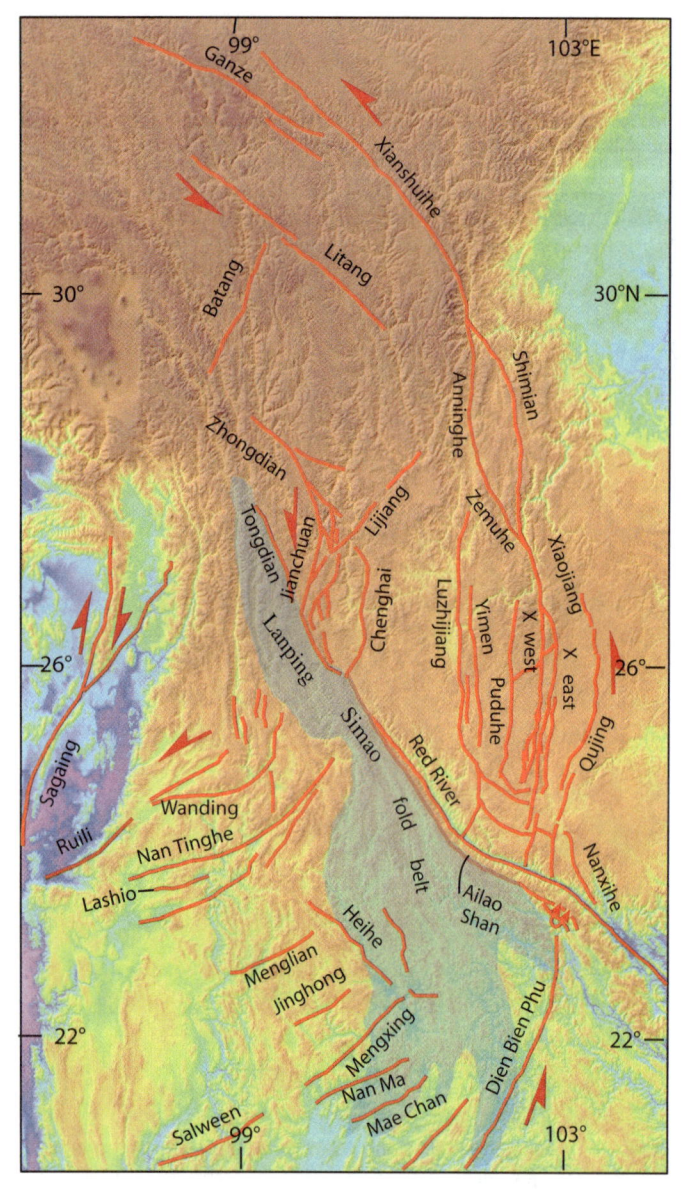

Figure 2. Active faults east of the eastern Himalayan syntaxis. Location of figure is given in Figure 1. The left-lateral Xianshuihe-Xiaojiang fault system consists of a narrow fault zone in the north (Ganze and Xianshuihe faults) and numerous faults between the Luzhijiang and Qujing faults to the south. The left-lateral Dali fault system consists of faults between the Jianchuan and Chenghai faults. The continuation of these fault systems south of the Lanping-Simao fold belt (shaded area) is represented by numerous faults such as the Dien Bien Phu and Menglian faults and the Nantinghe and Wanding faults, respectively, although direct continuations between the two fault systems do not exist. X west and X east are the west and east branches of the Xiaojiang fault, respectively. WS—Wanling Shan.

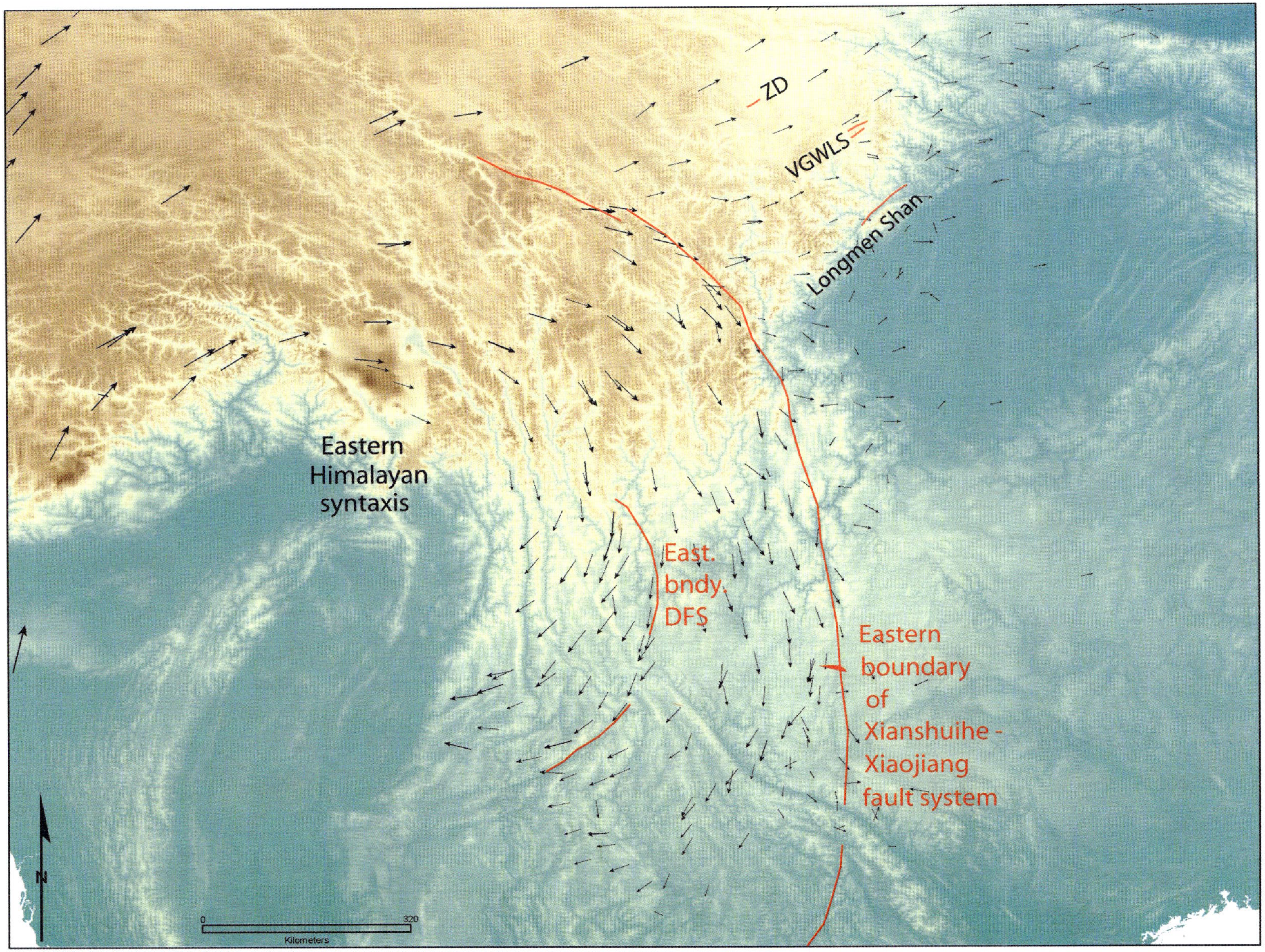

Figure 3. Global positioning system (GPS) velocities around the eastern Himalayan syntaxis and eastern Tibetan Plateau west of the Longmen Shan. Data are from the Massachusetts Institute of Technology (Chen et al., 2000) and Zhang et al. (2004). The eastern boundary faults of the Xianshuihe and Dali fault systems (DFS) are shown. VGWLS—velocity gradient west of Longmen Shan; ZD—Zhe Da fault. Wenchuan-Maowen fault zone = red line just above Longmen Shan. Red lines west of the Longmen Shan are NE-striking faults shown on Chinese maps. Two other short, but discontinuous, faults in this area are shown, but their age and sense of offset are unknown.

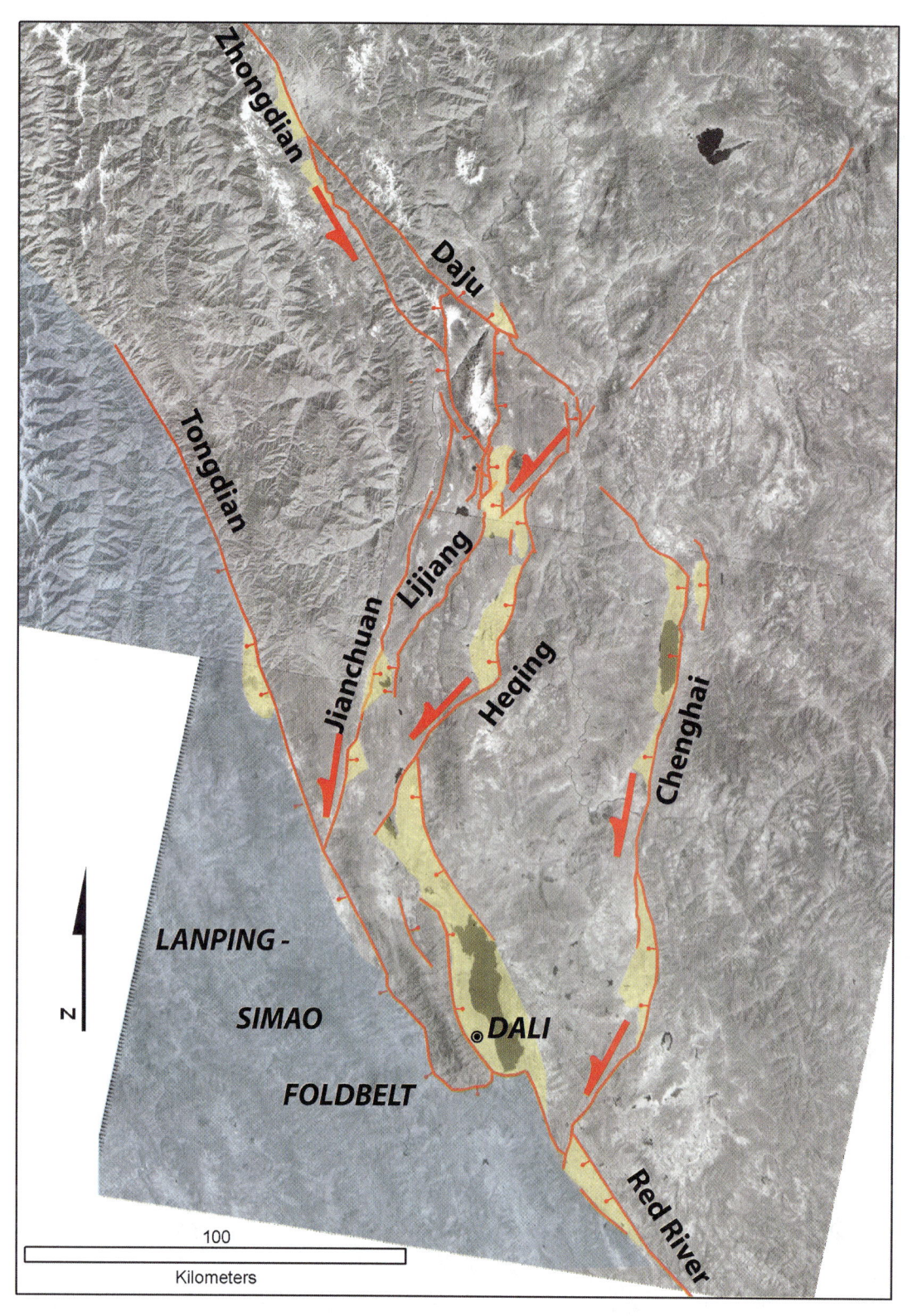

Figure 4. Active faults that make up the Dali fault system and accommodate extension and left-lateral slip. Quaternary basins bounded by transtensional faults are shaded. Note that none of the Dali fault system faults can be traced into the Lanping-Simao fold belt. The boundary between the Lanping-Simao fold belt and South China (Yangtze platform) rocks is roughly coincident with young and perhaps active normal faults. The easternmost fault (the Chenghai fault) is approximately coincident with the location of a prominent divergence of the geodetically determined velocity field. Location of this figure is shown in Figure 1.

On the southwest side of the Lanping-Simao fold belt, the relationship is quite different. The southern boundary of the Lanping-Simao fold belt is marked by a continuous and sinuous (inactive) fault zone. Several of the left-lateral faults strike NE into the boundary but do not cross it, such as the Wanding and Mengliang faults (Fig. 2). Others, such as the Mengxing and NW-striking Heihe faults, cross the boundary a short way but end within the Lanping-Simao rocks. The Nantinghe fault strikes into a prominent bend in the boundary fault at the Wanling Shan (WS in Fig. 2). The Wanling Shan is cored by an anticline of Paleozoic rocks overlain by the Mesozoic strata of the Lanping-Simao fold belt. This fold has a thrust on its north side, and it refolds structures within the western part of the Lanping-Simao belt. We interpret these relations to indicate that the active Nantinghe fault dies out into a vertical-axis fold that folds the boundary fault and some structures within the Lanping-Simao fold belt. The active NW-striking Heihe fault to the south penetrates farther into the Lanping-Simao belt than any of the other faults before dying out along a belt of several faults that cuts through left-laterally sigmoidally bent folds.

The active, left-lateral Dien Bien Phu fault cuts across the entire width of the apparent SE continuation of the Lanping-Simao belt where it enters Vietnam. However, Geissman et al. (2007, personal commun.) have speculated that there may be a substantial difference between the early Cenozoic tectonic development of the SE part of the Lanping-Simao belt and that of the rocks in north Vietnam. Thus, there may be important changes in crustal structure at the location of the Dien Bien Phu fault. Nevertheless, the Dien Bien Phu fault ends before reaching the Ailao Shan metamorphic rocks.

A principal feature of the Lanping-Simao fold belt is the lack of any through-going faults that are associated with the geodetically determined velocity gradient. Seismicity indicates that this region is indeed deforming, but earthquake hypocenter locations do not serve to define any through-going fault zones. Recent block modeling by Meade (2007) also has shown that the best fit to data on active faulting and velocity gradients within the area is accommodated by shear that passes through the Lanping-Simao fold belt. All the data indicate that the shear passes through the Lanping-Simao fold belt but is not manifested by obvious through-going faults.

The key question we pose is the following: by what mechanism does shear deformation continue through the rocks of the southern Yangtze platform and Lanping-Simao belt at the southern end of the Xianshuihe-Xiaojiang fault system? The answer at present is, we do not know. Interpretation of the bedrock geology in the Lanping-Simao belt and adjacent areas is somewhat compromised because of typically poor exposure and access. For example, the area at the southern end of the Xianshuihe-Xiaojiang fault system is characterized by a thick soil cover, often terra rossa, above a well-developed karst surface, and within the Lanping-Simao, there is dense vegetation and considerable agricultural development. Nevertheless, outside of the Lanping-Simao fold belt, in areas of similarly poor exposure, Quaternary faults are well expressed geomorphically and are easily mappable both in the field and on digital elevation models (DEMs) and satellite imagery. Chinese maps of the Lanping-Simao belt show numerous small faults, some parallel to the trend of the Xianshuihe-Xiaojiang fault system, that clearly offset fold axes. Some faults have the appropriate map sense offset, but others do not, and none of these faults is continuous for very far along strike. More importantly, it is unknown if any of the mapped faults in the Lanping-Simao are active: none of these is well expressed geomorphically or is unambiguously associated with historic earthquakes. We suspect, but cannot prove, that shear strain is accommodated by brittle deformation on a broad scale by numerous small faults and gouge zones of variable orientation (thus effectively defining a system of small-scale rotating blocks) and mesoscopically ductile-mode deformation accommodated by diffuse and distributed cataclastic deformation. The problem in the context of field-based geologic observations is that most of this type of deformation would not necessarily go unnoticed, but it would not generally be mapped as parts of a large-scale system and recognized for its regional kinematic importance. However, it does seem clear that this manner of crustal deformation can absorb considerable magnitudes of strain (equivalent to tens or even hundreds of kilometers of offset on a single, localized fault) that might largely remain unrecognized. Such deformation would be increasing difficult to recognize in older more extensively deformed belts.

The Lanping-Simao belt also reveals another example of unrecognized strain. We have completed a regional paleomagnetic study of the Upper Mesozoic to Lower Cenozoic red beds within this belt (Fig. 5; Geissman, 2007, personal commun.). The uppermost crust of the Lanping-Simao belt consists largely of a Jurassic to Oligocene sequence of red beds that were folded during early Cenozoic syn- to postcollisional deformation around the eastern Himalayan syntaxis during a phase of deformation that is older than that emphasized herein but that is superposed on early Cenozoic structures. In combination with interpretations of the regional geology of the area, and our understanding of the timing and sense of displacement along major shear zones along the eastern margin of the eastern syntaxis, the paleomagnetic data are interpreted to demonstrate that during early Cenozoic time, southeastward extrusion of most of the Lanping-Simao fold belt was accompanied by some 60° to 120° of clockwise rotation. However, unexpectedly, the southeasternmost part of the Lanping-Simao belt appears to be unrotated (Fig. 5), although this observation is based on a relatively limited data set. There is a fairly well-defined boundary between rotated and unrotated rocks. Such a boundary has not been inferred to be prominent in any previous tectonic analysis of the Lanping-Simao fold belt. The boundary appears to lie parallel to the curved trend of the folds and faults within the belt. In compilations of Chinese maps we have made of this area, there are no obvious structures present that would define or accommodate such a boundary. The limits of the proposed boundary lie within a belt of arcuate structures in the fold belt and do not follow any mapped through-going fault zone. Perhaps in areas of excellent outcrop, such a boundary

Figure 5. Summary of paleomagnetic results from the southern part of the Lanping-Simao fold belt. Each sampling locality involves multiple (typically ten or more) discrete sites (individual beds) from which minimally five independent samples were obtained from each site. Locality mean paleomagnetic declinations have been transformed into estimates of clockwise rotation (based on a comparison between the observed and expected paleomagnetic declinations for the appropriate age—true north is thus not the expected declination of the magnetization for these rocks), which are indicated by the magnitude of clockwise deflection of the arrow from true north. The approximate location of the sampling locality is the center of the arrow shaft. Although this is based on limited data, the eastern part of the Lanping-Simao belt appears to have not experienced appreciable rotation. The localities that have experienced the greatest magnitude of clockwise rotation lie west of the western shaded line. The available data crudely define a zone between the two shaded lines where rotations are intermediate in magnitude or suggest a progressive increase in magnitude from west to east. Location of this figure is shown in Figure 1.

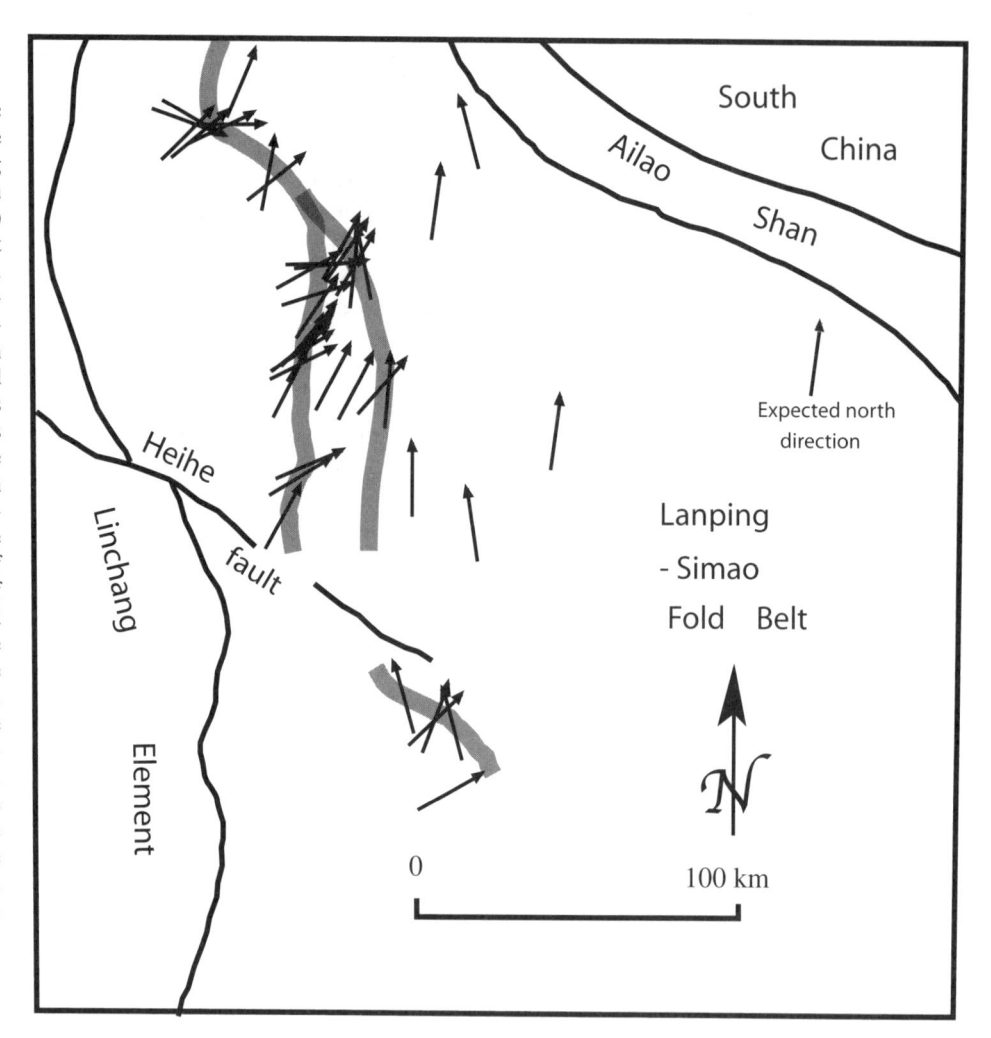

might be readily recognized, but in this vegetated, and relatively highly weathered area, there are no obvious mapped structures that appear to accommodate the differentially rotated rocks.

NORTHEAST TIBET

Before completion of our early GPS investigations within western Sichuan and eastern Qinghai provinces on the Tibetan Plateau, we developed a synthesis of the late Cenozoic tectonic framework of that region (Fig. 1; Burchfiel et al., 1995; Burchfiel, 2004). West of the Longmen Shan, the plateau is underlain by a thick sequence of largely Middle and Upper Triassic flysch of the Songpan Ganze basin intruded by Mesozoic plutons. The flysch is weakly metamorphosed but contains local areas of amphibolite-grade metamorphism (e.g., see Dirks et al., 1994). It is isoclinally folded and locally refolded. Folds trend NW in most of the flysch, and when traced eastward, the folds sweep into the western Longmen Shan in a sense that has been interpreted to have been caused by left-lateral shear of Mesozoic age along the eastern margin of the Songpan Ganze basin (Dirks et al., 1994; Burchfiel et al., 1995).

Our first GPS results from eastern Tibet (King et al., 1997) showed an unexpected velocity gradient west of the Longmen Shan that indicated right-lateral shear between the central Tibetan Plateau and the Longmen Shan (the eastern edge of the plateau). Newer data (VGWLS in Fig. 3; Chen et al., 2000; Zhang et al., 2004) support the existence of the NE-trending velocity gradient; however, this part of the Tibetan Plateau is not covered by an abundance of GPS stations, so the location of the velocity gradient is not well determined. The velocity gradient is ~10 mm/yr across an ~100-km-wide zone (not unlike the dimensions of the Xianshuihe-Xiaojiang fault system discussed previously). There are no mapped through-going structures in that area that can be related to the velocity gradient. Folds and faults in the Triassic rocks have no mapped offsets, and satellite imagery shows no obvious through-going lineaments (Burchfiel, 2004). Our field work in this area recognized an active fault zone at Zhe Da, ~10 km west of the town of Aba (32.9°N, 101.7°E). The fault strikes N60°E, parallel with the velocity gradient. Along this fault, streams are consistently offset right-laterally a few tens of meters (Fig. 6). The mapped fault has not been followed more than 10 km to the NE, and it does not show up as a prominent

Figure 6. Field view looking northeast at the active right-lateral fault at Zhe Da. The fault strikes parallel to the velocity gradient and may be an expression of the shear within this N60°E-striking zone. There are six stream offsets in a row, two of which are shown in the photo, with the trace of the fault shown in white. However, the location of this fault may be too far west to lie within the zone of velocity gradient west of Longmen Shan. The location of this figure is shown on Figure 3 at ZD. Offset of ridge in center of image is ~50 meters.

linear feature on Landsat images. The location of this fault may lie too far to the west to be within the zone of the high GPS velocity gradient. Active right-lateral displacement has also been identified on the Wenchuan-Maowen fault zone within the western part of the Longmen Shan (Burchfiel et al., 1995); however, the trace of this fault appears to be too far east to be related to the GPS velocity gradient. Thus, from all the data available at present, there are no obvious through-going faults that can be unambiguously associated with the geodetically determined velocity gradient. Unfortunately, we do not know how long this velocity gradient has been in existence, but our regional analysis suggests that it may have been active since ca. 8–11 Ma (Burchfiel, 2004; Kirby et al., 2002). If the present geodetic rates can be extrapolated for this entire interval, the deformation would have resulted in ~100 km of right-lateral shear across an ~100-km-wide zone, similar to that of the Xianshuihe-Xiaojiang fault system.

In the region of NE Tibet, outcrop is again rather poor, but we contend that localized, through-going structures that accommodate the shear in this area are simply not present (in part, because faults elsewhere in this region are easily recognizable). These structures do not show up on either geological maps or remote-sensing images. A few short faults with NE strikes are shown in Chinese maps, but none has been investigated except the two mentioned previously, so it is not known if they are active or what their displacement is. We can only speculate that the shear is accommodated by deformation on a broad zone of spaced faults or smaller-scale brittle structures that remain unrecognized. Regardless of the actual mechanism, it is sobering to recognize that the existing geologic information would never have led us to look for such structures or infer such strains; it was only when the GPS results became available that we began to attempt to identify accommodating structures.

DISCUSSION

In the examples given here, there does not appear to be any relation between either rock type or crustal structure through which the velocity gradients pass in the absence of through-going structures. The southern part of the Xianshuihe-Xiaojiang fault system north of the Red River is within the Yangtze platform, which consists of a Neoproterozoic basement of generally low-grade metamorphic and igneous rocks with thin Paleozoic and Mesozoic cover. These rocks were deformed in both Mesozoic and Cenozoic time, and the Xianshuihe-Xiaojiang fault system cuts at a high-angle across older structures. The Lanping-Simao belt consists of at least 6 km of unmetamorphosed Mesozoic and lower Cenozoic red beds that were deformed into a fold-and-thrust belt in early Cenozoic time. The Lanping-Simao red beds were deposited in a Jurassic-Cretaceous to possibly early Cenozoic extensional basin (Yano et al., 1994). Unlike the Yangtze platform, the Lanping-Simao fold belt probably overlies a shallow, regionally extensive décollement and probably does not involve deeper (pre-Paleozoic) basement rocks. The GPS velocity gradients of the Xianshuihe-Xiaojiang and Dali fault systems pass through the structures at a high-angle. The Songpan Ganzi belt in NE Tibet consists of low-grade flysch that is tightly folded and probably also is detached within middle-crustal rocks.

In the three examples we have used here, the velocity gradient does not parallel pre-existing structures but crosses them at a high angle. The only place where older anisotropy is present is in the area of the faults along the southwest margin of the Lanping-Simao belt, and then only locally. Some of the left-lateral faults, such as the Mengxing and the Wanding faults, have early Cenozoic right-lateral displacement that was later reversed during late Cenozoic time (Lacassin et al., 1998). However, other left-lateral faults, such as the Nantinghe fault, appear to be new and were initiated in late Cenozoic time. The faults in this area cut through tectonic assemblages that consist of rocks distinctly different from the Lanping-Simao belt. These assemblages include the Linchang (Permian-Triassic batholith and associated metamorphic rocks), Chengling-Mengliang (blueschist metasedimentary rocks), and Baoshan (thick section of Neoproterozoic to Jurassic sedimentary rocks) units within China.

At least in these three examples, whether or not the GPS velocity gradient is marked by faults, the shear gradients of faults do not appear to be controlled by rock types or preexisting structure. The only thing they have in common is that the GPS gradient passes through a crust at a high-angle to pre-existing structures. Of course, this is also true where the well-expressed Xianshuihe-Xiaojiang fault system cuts across rocks of the Yangtze platform. Alternatively, it might be tempting to suggest that the nature of the basement rocks of the Lanping-Simao somehow controls the localization of shear strain at the surface. This explanation fails, however, to account for the right-lateral velocity gradient west of the eastern plateau margin in the Songpan Ganzi terrane, because where the western part of the Xianshuihe fault cuts through the Songpan Ganzi, it is strongly localized. The mechanism that controls the presence or absence of localized faults expressed at the surface remains unknown.

One of the more sobering consequences of the examples described herein relates to geologic field work in older, inactive settings. Without recourse to geodetic techniques, it is quite possible that regionally significant deformation, if accommodated in a diffuse manner without the development of discrete or mappable structures, may be easily overlooked. Even where bent or apparently folded geologic markers are mapped, it is not always possible to definitively conclude that the mapped geometry of some marker in fact represents deformation imposed on an originally straight marker. A problem of this nature is well represented by field relations in the area north and east of the Himalayan syntaxis. In this region, both the apparent deflection and attenuation of geologic trends and the apparent deflection and lengthening of rivers have led workers (Dewey et al., 1989; Wang and Burchfiel, 1997; Hallet and Molnar, 2001) to speculate that this area represents a broad crustal shear zone, accommodating at least part of the northward movement of the Himalayan syntaxis relative to South China. A considerable component of shear in the area may not be marked by faults. Whether or not this interpretation is correct, unambiguous geologic field evidence of some mechanism or structure that could accommodate this deformation has been difficult to find.

CONCLUSIONS

Short-term GPS data in the eastern part of the Tibetan Plateau and adjacent regions in Sichuan and Yunnan provinces, China, have shown that velocity gradients cross continental crust, often with pronounced regional deformation fabrics at high angles to the gradients, without the development of through-going faults. Near the Xianshuihe-Xiaojiang and Dali fault systems, the velocity gradients are marked by active structures that are several million years old. The Xianshuihe-Xiaojiang fault system has accumulated at least ~60 km of displacement. Strain associated with this fault system and the GPS velocity gradient passes through the Lanping-Simao tectonic unit without any obvious evidence as to how the shear is accommodated. A similar relationship exists in northeast Tibet west of the Longmen Shan; however, the total amount of shear strain accommodated there is uncertain. Paleomagnetic data from the Lanping-Simao fold belt also indicate previously unrecognized crustal shear. These observations are not dissimilar to those from other areas, where contemporary deformation cannot be directly related to extant, through-going structures, such as along the Eastern California shear zone. The results indicate that continental crust can accommodate tens, and perhaps more than a hundred kilometers of shear without developing through-going faults. From one of the examples from China given in this paper (Xianshuihe-Xiaojiang fault system), the shear was recognized during regional geologic analysis and later confirmed by GPS studies. However, it remains unclear how the shear was accommodated. In a second example from NE Tibet, the shear was not recognized until after short-term GPS measurements were made. It appears from these areas that continental crust can accommodate a large amount of shear deformation without developing obvious through-going structures, although the exact conditions under which the crust will deform in such a way remain the object of speculation. We contend that such unrecognized diffuse deformation may be common in many areas of active deformation as well as older, inactive geologic terranes. This adds an unknown but potentially large uncertainty to field study and tectonic analyses of continental deformation. The potential link between this deformation and the middle and lower crust, if not the mantle part of the Indochina lithosphere, requires, if possible, further, geophysical investigation.

ACKNOWLEDGMENTS

This work is the result of cooperative project between scientists at the Chengdu Institute of Geology and Mineral Resources and the Department of Earth, Atmospheric and Planetary Sciences at the Massachusetts Institute of Technology. The data on which the results presented in this paper were based were obtained with the support of National Science Foundation (NSF) grants EAR-0003571 and EAR-8904096 and National Aeronautics and Space Administration (NASA) grant NAGW-2155 awarded to MIT and EAR-9706300 awarded to the University of New Mexico. Support was also provided by the Chengdu Institute of Geology and Mineral Resources.

REFERENCES CITED

Burchfiel, B.C., 2004, New technology, new challenges: Geological Society of America: GSA Today, v. 14, no. 2, p. 4–9, doi: 10.1130/1052-5173-(2004)014<4:PANNGC>2.0.CO;2.

Burchfiel, B.C., and Wang, E., 2002, Northwest-trending, middle Cenozoic, left-lateral faults in southern Yunnan, China, and their tectonic significance: Journal of Structural Geology, v. 25, no. 5, p. 781–792.

Burchfiel, B.C., Chen, Z., Liu, Y., and Royden, L.H., 1995, Tectonics of the Longmen Shan and adjacent regions: International Geological Review, v. 37, no. 8, p. 661–736.

Chen, Z., Burchfiel, B.C., Liu, Y., King, R.W., Royden, L.H., Tang, W., Wang, E., Zhao, J., and Zhang, X., 2000, GPS measurements from eastern Tibet and their implications for India/Eurasia intracontinental deformation: Journal of Geophysical Research, v. 105, p. 16,215–16,227, doi: 10.1029/2000JB900092.

Dewey, J.F., Cande, S., and Pitman, W.C., 1989, Tectonic evolution of the India/Eurasia collision zone: Eclogae Geologicae Helvetiae, v. 82, p. 717–734.

Dirks, P., Wilson, C.J.L., Chen, S., Lou, Z.L., and Liu, S., 1994, Tectonic evolution of the NE margin of the Tibetan Plateau; evidence from the central Longmen Mountains, Sichuan Province, China: Journal of Southeast Asian Earth Sciences, v. 9, p. 181–192, doi: 10.1016/0743-9547(94)90074-4.

Hallet, B., and Molnar, P., 2001, Distorted drainage basins as markers of crustal strain east of the Himalaya: Journal of Geophysical Research, v. 106, doi: 10.1029/2000JB900335

King, R.W., Shen, F., Burchfiel, B.C., Chen, Z., Li, Y., Liu, Y., Royden, L.H., Wang, E., Zhang, X., and Zhao, J., 1997, Geodetic measurement of crustal motion in southwest China: Geology, v. 25, no. 2, p. 179–182, doi: 10.1130/0091-7613(1997)025<0179:GMOCMI>2.3.CO;2.

Kirby, E., Reiners, P.W., Krol, M.A., Whipple, K.X., Hodges, K.V., Farley, K.A., Tang, W., and Chen, Z., 2002, Late Cenozoic evolution of the eastern margin of the Tibetan Plateau: Inferences from $^{40}Ar/^{39}Ar$ and (U/Th) He thermochronology: Tectonics, v. 21, doi: 10.1029/2000TC001246

Lacassin, R., Replumaz, A., and Leloup, P.H., 1998, Hairpin river loops and slip sense inversion on southeast Asian strike-slip faults: Geology, v. 26, p. 703–706, doi: 10.1130/0091-7613(1998)026<0703:HRLASS>2.3.CO;2.

McClusky, S.C., Bjornstad, S.C., Hager, B.H., King, R.W., Meade, G.J., Miller, M.M., Monastero, R.C., and Souter, B.J., 2001, Present day kinematics of the Eastern California shear zone from a geodetically constrained block model: Geophysical Research Letters, v. 28, p. 3369–3372, doi: 10.1029/2001GL013091.

Meade, B.J., 2007, Present-day kinematics at the India-Asia collision zone: Geology, v. 35, no. 1, p. 81–84, doi: 10.1130/G22924A.1.

Miller, M.M., Johnson, D.F., Dixon, T.H., and Dokka, R.K., 2001, Refined kinematics of the Eastern California shear zone from GPS observations 1993–1998: Journal of Geophysical Research, v. 106, p. 2245–2263, doi: 10.1029/2000JB900328.

Replumaz, Z., Lacassin, R., Tapponnier, P., and Leloup, P.H., 2001, Large river offsets and Plio-Quaternary dextral strike-slip rate on the Red River fault (Yunnan, China): Journal of Geophysical Research, v. 106, p. 819–836, doi: 10.1029/2000JB900135.

Schermer, E.R., Luyendyk, B.P., and Cisowski, S., 1996, Late Cenozoic structure and tectonics of the northern Mojave Desert: Tectonics, v. 15, p. 905–932, doi: 10.1029/96TC00131.

Schoenbohm, L.M., Burchfiel, B.C., Chen, L., and Yin, J., 2006, Miocene to present activity along the Red River fault, China, in the context of continental extrusion, upper crustal rotation and lower crustal flow: Geological Society of America Bulletin, v. 118, p. 672–688, doi: 10.1130/b25816.1.

Studnicki-Gizbert, C., Eich, L., King, R., Burchfiel, B., Chen, Z., and Chen, L., 2004, Active transtensional tectonics due to differentially rotating upper crustal blocks east of the eastern Himalayan syntaxis, Yunnan Province: Eos (Transactions, American Geophysical Union), v. 85, no. 47, Fall Meeting supplement, abstract GP42A-03.

Wang, E., and Burchfiel, B.C., 1997, Interpretation of Cenozoic tectonics in the right-lateral accommodation zone between the Ailao Shan shear zone and the eastern Himalayan syntaxis: International Geology Review, v. 39, p. 191–219.

Wang, E., Burchfiel, B.C., Royden, L.H., Chen, L., Chen, J., and Li, W., 1998, Late Cenozoic Xianshuihe-Xiaojiang and Red River Fault Systems of Southwestern Sichuan and Central Yunnan, China: Geological Society of America Special Paper 327, 108 p.

Yano, T., Genyao, W., Mingqing, T., and Shaoli, S., 1994, Tectono-sedimentary development of backarc continental basin in Yunnan, southern China: Journal of Southeast Asian Earth Sciences, v. 9, p. 153–166.

Zhang, P.-Z., Shen, Z., Wang, M., Gan, W., Burgman, R., Molnar, P., and Wang, Q., 2004, Continuous deformation of the Tibetan Plateau from global positioning system data: Geology, v. 32, p. 809–812, doi: 10.1130/G20554.1.

MANUSCRIPT ACCEPTED BY THE SOCIETY 22 MARCH 2007

The Geological Society of America
Special Paper 433
2007

Mechanics of thin-skinned fold-and-thrust belts: Insights from numerical models

Glen S. Stockmal
Geological Survey of Canada, Natural Resources Canada, 3303 33rd Street NW, Calgary, Alberta T2L 2A7, Canada

Christopher Beaumont
Mai Nguyen
Bonny Lee
Dalhousie Geodynamics Group, Department of Oceanography, Dalhousie University, Halifax, Nova Scotia B3H 4J1, Canada

ABSTRACT

In order to investigate the development of structures at scales smaller than that of an entire belt, we examined aspects of the mechanics of thin-skinned fold-and-thrust belts in cross section using an arbitrary Lagrangian-Eulerian frictional-plastic finite-element model. A series of models, beginning with the deformation of a thick uniform layer above a thin weak layer on a fixed base, sequentially illustrates the effects of including flexural isostatic subsidence, strain-softening, multiple layers of strong and very weak materials, and finally erosion and sedimentation. These continuum models develop thin shear zones containing highly sheared material that approximate fault zones. The corresponding structures are similar to those in fold-and-thrust belts and include: far-traveled thrust sheets, irregular-roof and smooth-roof duplexes, back thrusts, pop-ups, detachment folds, fault-bend folds, break thrusts, and piggyback basins. These structures can develop in-sequence or out-of-sequence, remain active for extended periods, or be reactivated.

At the largest scale, the scale of the wedge, the finite-element model results agree with critical wedge solutions, but geometries differ at the sub-wedge scale because the models contain internal structures not predicted by the critical wedge stress analysis. These structures are a consequence of: (1) the complete solution of the governing equations (as opposed to a solution assuming a stress state that is everywhere at yield), (2) the initial finite-thickness layers, (3) the spatial and temporal variations of internal and basal strength, and (4) the coupling between surface processes and deformation of the wedge. The structural styles produced in models involving feedback with surface processes (erosion and sedimentation) are very similar to those mapped in the foothills of the southern Canadian Rockies and elsewhere. Although syndeformational sediments have been removed by postorogenic erosion across the foothills belt, evidence of the interaction between surface processes and deformation is preserved in the structural style.

Keywords: fold-and-thrust belts, thin-skinned, mechanics, surface processes, numerical modeling.

Stockmal, G.S., Beaumont, C., Nguyen, M., and Lee, B., 2007, Mechanics of thin-skinned fold-and-thrust belts: Insights from numerical models, *in* Sears, J.W., Harms, T.A., and Evenchick, C.A., eds., Whence the Mountains? Inquiries into the Evolution of Orogenic Systems: A Volume in Honor of Raymond A. Price: Geological Society of America Special Paper 433, p. 63–98, doi: 10.1130/2007.2433(04). For permission to copy, contact editing@geosociety.org. ©2007 The Geological Society of America. All rights reserved.

INTRODUCTION

The geometry and state of stress of thin-skinned fold-and-thrust belts are described at whole-wedge scales by critical wedge models (e.g., Chapple, 1978; Davis et al., 1983; Dahlen, 1984). These are closed-form solutions where the static equilibrium equations are solved assuming material within the wedge is everywhere at yield. They give the surface and basal dips of wedges that satisfy the yield condition as a function of internal and basal frictional-plastic material properties. Such stress solutions give the geometries that wedges must achieve to exist as critical wedges, but they do not solve the problem of how critical wedges develop from preexisting noncritical geometries, for example, when fold-and-thrust belts grow from preexisting strata. The development and application of critical wedge solutions (e.g., Dahlen et al., 1984; Dahlen and Suppe, 1988; Dahlen, 1988, 1990; among others) to fold-and-thrust belts and accretionary wedges have, however, been among the most significant recent advances in understanding their mechanics at the scale of the wedge. The critical wedge analysis is limited because it is a partial solution for the geometry for a particular stress state and does not include the velocity/strain-rate solution. In addition, the requirement for simultaneous yield conditions throughout the wedge and restrictions on the variability of material properties within the wedge mean that critical wedge solutions are of limited use for understanding the development of structures within tectonic wedges.

To investigate how wedges deform internally to achieve and maintain the large-scale critical geometry, we have developed dynamic numerical models of thin-skinned wedges using a plane-strain, finite-element, continuum mechanics approach (see Fullsack, 1995; Beaumont et al., 2004). These models allow us to examine the sensitivity of wedge growth to its mechanical properties and the feedback from surface processes. In particular, the importance of surface processes (erosion and sedimentation), known to influence the large-scale or whole-wedge evolution of critical wedges (e.g., Beaumont et al., 1992; Willett, 1999; Whipple and Meade, 2004), can be assessed at scales smaller than the entire wedge. Our approach was to design a series of numerical model experiments in which model parameters can be systematically varied to investigate their influence on structural style.

Insights and Constraints from Natural Examples

As emphasized and listed explicitly by Chapple (1978), there are four fundamental characteristics of "folded mountain belts":

1. The belts are thin-skinned, where the basal décollement often, but not always, lies at or just above the crystalline basement.
2. The basal décollement generally occupies relatively weak rock.
3. The fold-and-thrust belt after deformation, as well as the sedimentary prism before deformation, is wedge-shaped, thinning toward the foreland.
4. Motion of the hinterland is permitted by shortening and thickening within the wedge.

Bearing these characteristics in mind, the southern Canadian Rocky Mountains is an archetypal thin-skinned belt (Fig. 1), and we used it for comparison with the model results. The surface of the underlying Archean to Early Proterozoic crystalline basement exhibits little to no relief at the scale of the belt. The basement dips gently toward the Cordilleran orogen, reflecting the regional isostatic response of the lithosphere to supracrustal loading by thrust sheets superimposed on the original basal dip of the platformal and miogeoclinal strata (Price, 1973, 1981; Beaumont, 1981; Price and Fermor, 1985). The basement is overlain by a Paleozoic to middle Mesozoic platform to shelf-margin succession of carbonates and subordinate siliciclastic rocks that increases in stratigraphic thickness from ~2 km at the deformation front to ~3 km at the western edge of the Front Ranges (Fig. 1), ~175 km distant in palinspastic restoration (Price and Fermor, 1985). This dominantly carbonate

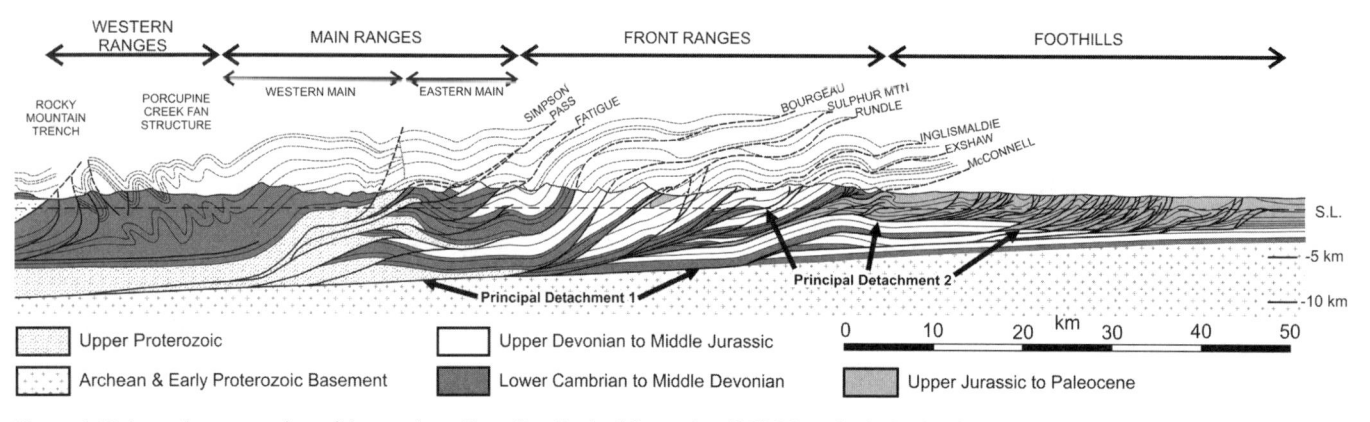

Figure 1. Balanced cross section of the southern Canadian Rocky Mountains (SCRM) at the latitude of Calgary, Alberta (after Price and Fermor, 1985). The southern Canadian Rocky Mountains are considered an archetypal thin-skinned fold-and-thrust belt, and they are used here as a general guide for scaling and interpretation of the numerical models. The positions of two principal detachments are indicated. Postdeformational erosion has removed ~2–3 km of foreland basin strata at the deformation front, and correspondingly more of the deformed wedge farther to the west (left). S.L.—sea level.

succession is overlain by middle Mesozoic to early Cenozoic foreland basin siliciclastic rocks. At the deformation front, the preserved thickness of the foreland succession is ~3 km (Fig. 1), and an additional 2–3 km has been removed by postdeformational erosion (Hacquebard, 1977; Nurkowski, 1984).

These strata vary in strength and form major and minor décollements along weak layers. In the southern Canadian Rocky Mountains, there are two principal detachment horizons (Fig. 1): (1) at the crystalline basement–sedimentary cover contact, where Early Cambrian clastic rocks (Gog Group) are inferred to overlie crystalline rocks, and (2) within the Upper Jurassic Fernie Formation, which is a package of dominantly marine shales that includes the basal units of the flexural foreland basin. Secondary detachment zones, particularly in the foothills (Fig. 1) where foreland basin strata are exposed, are commonly within marine shale units. In this paper, we use the terms "detachment" or "detachment zone" to describe the layer (whether strained or not) within which layer-parallel décollements might propagate, and we reserve the term "décollement" for zones of accumulated high shear strain that are analogous to thrust faults.

MODEL DESIGN

Governing Equations and Model Materials

An arbitrary Lagrangian-Eulerian (ALE) finite-element approach was used to solve the plane-strain deformation of viscous-plastic materials (Fullsack, 1995; Willett, 1999; Huismans and Beaumont, 2003). This creeping-flow deformation is governed by the quasi-static force balance and conservation of mass equations, assuming incompressibility and no inertial forces:

$$\frac{\partial P}{\partial x_j} + \frac{\partial}{\partial x_i}\left(\eta_{\text{eff}}\frac{\partial v_j}{\partial x_i}\right) + \rho g = 0, \ j = 1,2 \qquad (1)$$

$$\frac{\partial v_j}{\partial x_j} = 0, \qquad (2)$$

where P is the pressure, x_j indicate the spatial coordinates, η_{eff} is the effective viscosity, v_j indicates the components of velocity, ρ is the density, g is the (vertical) gravitational acceleration, and repeated indices imply summation. The associated stress tensor is:

$$\sigma_{ij} = P\delta_{ij} + 2\eta_{\text{eff}}\dot{\varepsilon}_{ij}, \qquad (3)$$

where the first term is the mean stress or pressure ($\sigma_{ii}/3$ for incompressible flow), the second term is the deviatoric stress, and δ is the Kronecker delta. The strain rate tensor is:

$$\dot{\varepsilon}_{ij} = \frac{1}{2}\left(\frac{\partial v_i}{\partial x_j} + \frac{\partial v_j}{\partial x_i}\right). \qquad (4)$$

We use the terms "brittle" and "ductile" only to distinguish frictional-plastic flow, which in our models is independent of temperature, from thermally activated creep with a viscous flow law. We recognize that some materials, for example, evaporites like halite, are intrinsically viscous and that in the deeper parts of fold-and-thrust belts, a transition to ductile flow occurs.

We use a Drucker-Prager yield criterion, which is equivalent to the Coulomb yield criterion in plane strain (Fullsack, 1995), for frictional-plastic (brittle) deformation. The plastic yield stress is:

$$(J'_2)^{1/2} = C_0 \cos\varphi_{\text{eff}} + P\sin\varphi_{\text{eff}}, \qquad (5)$$

where $(J'_2)^{1/2} = (\tfrac{1}{2}\sigma'_{ij}\sigma'_{ij})^{1/2}$ is the second invariant of the deviatoric stress, C_0 is the cohesion, where P is the mean stress (following the common convention in earth science that compressive pressure is positive), and φ_{eff} is the effective angle of internal friction.

This angle, φ_{eff}, can include the effects of pore-fluid pressure, P_f, which acts to reduce the mean stress:

$$P\sin\varphi_{\text{eff}} = (P - P_f)\sin\varphi = P(1-\lambda)\sin\varphi, \qquad (6)$$

where φ is the intrinsic angle of internal friction, and $\lambda = P_f/P$ is the pore-fluid pressure ratio. Note that this definition of λ differs from that of Hubbert and Rubey (1959); see Appendix.

As a first-order approximation of material weakening or an increase in effective pore pressure with increasing strain, some model materials discussed here strain soften (φ_{eff} decreases). Postyield softening occurs with increasing strain. We use a parametric model in which φ_{eff} decreases linearly with increasing strain in the range $0.5 < \varepsilon < 1.0$, where ε represents the square root of the second deviatoric strain invariant, $(I'_2)^{1/2}$.

Flow of frictional-plastic material can be modeled as an iterative incompressible viscous flow problem by defining an effective viscosity for those regions that are at yield, while specifying a very large viscosity for "rigid" nonyielding regions (Fullsack, 1995; Willett, 1999). The effective viscosity for plastic flow, η^P_{eff}, is:

$$\eta^P_{\text{eff}} = \frac{1}{2}(J'_2)^{1/2}/(\dot{I}'_2)^{1/2}. \qquad (7)$$

Setting the viscosity to η^P_{eff} for regions that are at yield satisfies the yield condition in Equation 5 and allows the velocity field for viscous creeping flow to be determined iteratively through a finite-element approach. For regions that are not on the failure envelope (i.e., "rigid"), the effective viscosity is many orders of magnitude greater (10^{30} Pa s), which is sufficiently high that negligible deformation occurs on the model time scale.

Numerical Implementation

The ALE formulation allows computation of large-deformation, plane-strain flows in a vertical cross section subject to kinematic velocity boundary conditions (Fig. 2). Velocity and deformation are calculated on an Eulerian finite-element grid

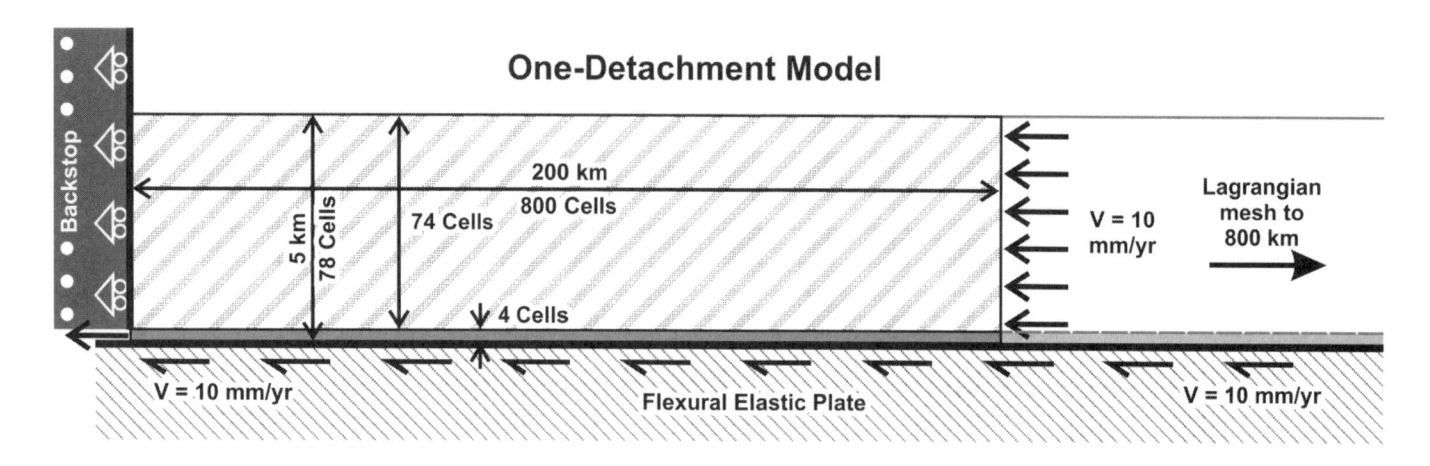

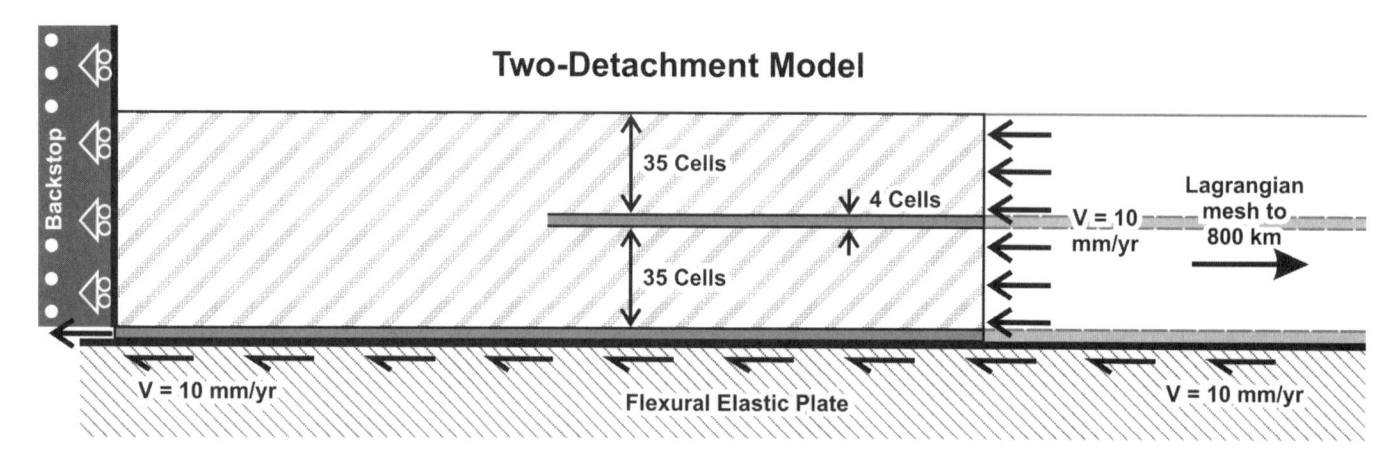

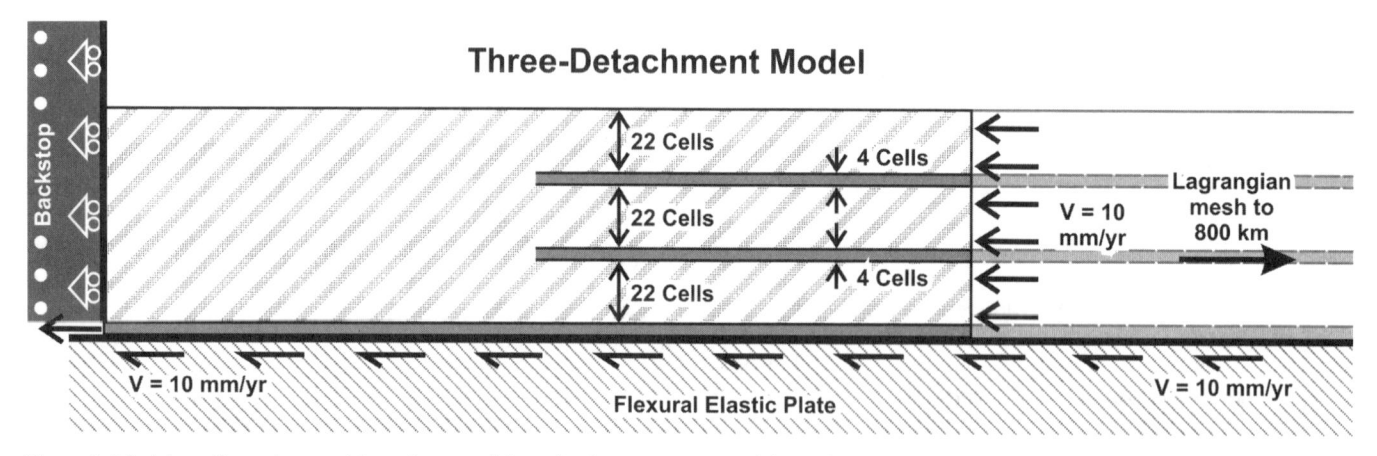

Figure 2. Model configurations and boundary conditions for the one-, two-, and three-detachment models. Vertical exaggeration = 10×. See text for discussion.

that is fixed horizontally with respect to an external reference frame but moves and stretches vertically as the material deforms (Fullsack, 1995). The positions of different materials, their boundaries, and their properties are tracked and updated using a Lagrangian grid (or mesh) that is moved (advected) according to the velocity field calculated on the Eulerian grid. The accuracy of the Lagrangian tracking is improved by inserting passive tracking (tracer) particles into the Lagrangian mesh that are also advected according to the velocity field. The properties of the Eulerian finite elements are updated each model time step by interpolating information from the current Lagrangian mesh and tracer particles. This dual mesh Lagrangian-Eulerian approach allows large displacements and deformation to occur (as calculated on the Lagrangian mesh) without associated distortion of the computational finite elements (the Eulerian grid), which would make it unsuitable for the calculation.

The initially uniform rectangular Lagrangian mesh conveniently displays strain variation within the deforming wedge. In Figures 3–11, a simplified representation (showing every third mesh line, horizontally and vertically) of the Lagrangian mesh is shown as hairlines overlying the Eulerian grid, which is colored according to material type. For visual reference, every twentieth vertical line in the Lagrangian mesh, spaced initially 15 km apart, is bold and labeled with a sequential number. The numbers vertically overlie the upper ends of these lines above the active surface of the model, regardless of whether portions of the Lagrangian mesh are removed by erosion or overlain by later sediment. This can result in the number labels becoming transposed where significant overthrusting or folding occurs.

The initial Eulerian grid for all models presented here has horizontal and vertical dimensions of 200 km × 5 km and encompasses 800 × 78 elements. Neither the number nor the horizontal positions of Eulerian elements changes during a model run, and therefore the initially elongate elements (each one is 250 m × ~64.1 m at the start of each model run) are subsequently stretched vertically to subequal dimensions. At the start of each model run, the Lagrangian mesh overlies the Eulerian grid identically. However, the Lagrangian mesh is four times the width of the Eulerian grid in horizontal extent (800 km) and element number (3200) to allow for material to be carried into the Eulerian grid during model convergence (Fig. 2; Table 1).

In our finite-element approach, the material within the wedge is a continuum. Therefore, there are no discrete faults, although thin zones of very high shear strain develop that emulate faults, resulting in fold-and-thrust belt–like features. For descriptive purposes, it is easiest to use typical fold-and-thrust belt terminology, including calling the thin shear zones "thrusts" that bound "thrust sheets." However, this is merely for convenience, and the distinction should not be forgotten.

Boundary Conditions

Basal Surface: Model materials lie initially on a horizontal surface that is either rigid or responds as an elastic plate representing regional flexural isostasy. This underlying surface, to which the weak basal layer is attached, is moved parallel to the flexed base at a constant rate (Table 1) toward a vertical backstop at the left-hand side of the model (Fig. 2).

Upper Surface: The upper surface is a free surface. In some models, it is subjected to erosion and/or sedimentation (see following).

Right-Hand Boundary: Material enters the Eulerian grid on the right-hand side of the models and is carried at constant velocity on the model base. The convergence rate for all models presented here is 10 mm/yr. Each model time step is 1250 yr, equivalent to 12.5 m of convergence (5% of the Eulerian element width). One million years of model evolution is therefore represented by 800 time steps.

Left-Hand Boundary: The vertical backstop does not support shear stresses. Near its base, a thin slot equal in height to the initial thickness of the weak basal layer allows material to exit the models, which prevents the accumulation of weak material in the corner. In practice, this slot has very little effect on the overall style of model evolution. In natural fold-and-thrust belt systems, the "backstop" will generally be more complex, varying both in time and space depending upon convergence rate, thermal structure, hinterland geology, plate-boundary processes, etc. In many cases no long-term rigid region may exist. We effectively sidestep this difficulty by adopting the approach commonly taken in analogue models.

Model Geometry

The minimum requirements for any realistic model of thin-skinned deformation are one or more material layers above a weaker basal layer (or surface), which in turn overlies a non-involved "basement" (characteristics 1 and 2 of Chapple, 1978). This minimum configuration has been applied by Chapple (1978), Davis et al. (1983), and Stockmal (1983), where in all cases, the weak base has zero thickness. In our continuum mechanical finite-element models, the weak basal layer has a finite thickness.

We designed prototype models to represent layered sedimentary rocks of constant thickness. Although this does not satisfy part of characteristic 3 of Chapple (1978), adoption of the simplest geometry allows us to focus on accretion of a uniformly thick layer at the wedge toe. The thickness of this layer determines a fundamental length scale for the internal deformation of the wedge, and we wanted to investigate systems where this scale is constant before incorporating variable thicknesses. In the case of the southern Canadian Rocky Mountains, the preexisting taper of the platformal succession involved in foothills and front range structures is only 0.33°, using values from Price and Fermor (1985), noted previously. As the models evolve, they develop overall tapered shapes through internal thickening and shortening, satisfying the other part of characteristic 3 as well as characteristic 4. In those models that incorporate syndeformational sedimentation, characteristic 3 develops naturally as the system evolves above a regionally isostatically compensated (flexural) plate. The influences of thickness variations owing to the initial dip and taper of the sedimentary succession are investigated elsewhere.

The initial configurations for basic models with one, two, or three detachment layers are shown in Figure 2. The one-detachment model corresponds to the simplest case of a relatively strong layer overlying a weaker base, as modeled by the closed-form solutions noted already. The strong layers in the two- and three-detachment models are of equal initial thickness. The weak detachment layers, whether on the base or internal, are of equal absolute initial thickness in all models. Each model is initially 5 km thick, similar to the thickness of platformal and foreland strata in the foothills of the southern Canadian Rocky Mountains.

Layering within the two- and three-detachment models does not extend all the way to the backstop in the initial con-

TABLE 1. PROPERTIES, PARAMETERS, AND OPTIONS COMMON TO ALL NUMERICAL MODELS

Materials

	φ_{eff} (°)	C_0 (MPa)	ε range	ρ (kg/m³)
Strong, no strain softening	38	2	n/a	2300
Strong, with strain softening	38→18	2	0.5→1.0	2300
Very weak, basal detachment layer	3.5	2	n/a	2300
Very weak, internal detachment layer	1	2	n/a	2300
Sediment	38→18	2	0.5→1.0	2300

Material strength comparisons

φ_{eff} (°)	$\sin(\varphi_{eff})$	$\sin(\varphi_{eff})/\sin(38°)$	$\tan(\varphi_{eff})$	λ	λ_{HR}*
38	0.616	1.0	0.781	0	0
18	0.309	0.50	0.325	0.50	0.72
3.5	0.0610	0.10	0.0617	0.90	0.96
1	0.0175	0.028	0.0175	0.97	0.99

Initial configuration

	Horizontal Cells (#)	Horizontal Dimension (km)	Vertical Cells (#)	Vertical Dimension (km)
Eulerian mesh	800	200	78	5
Lagrangian mesh	3200	800	78	5

Boundary conditions

Left side	Above 256.41 m from base: zero horizontal velocity, zero shear stress Below 256.41 m from base: material exits to left at rate of 10 mm/yr
Right side	Lagrangian mesh enters from right at rate of 10 mm/yr
Bottom	Bottoms of basal elements move to left at rate of 10 mm/yr
Top	Stress free; subject to erosion and/or sedimentation

Flexure options

Rigid base
Elastic base—either constant or variable flexural rigidity

Surface processes

Erosion—linearly dependent on surface slope
- 1 mm/yr on 1:1 slope

Sedimentation—Dependent upon base-level specification
- No sedimentation, except minor infilling of tight folds, related to surface definition during Eulerian re-gridding
- Filling of all accommodation space up to base level

*The Hubbert-Rubey (HR) pore-fluid pressure ratio assumes that the cohesion is negligible.

figuration (Fig. 2). This minimizes the effects of the backstop boundary conditions by creating a first phase of uniform wedge growth without internal layering that acts as a buffer. This uniform wedge also serves to emphasize the influence of the internal detachment horizons, where a marked change in structural style is observed in all the models at the point where these horizons become involved in the deformation.

The internal angle of friction and cohesion are initially uniform in each of the model layers. No lateral discontinuities or other special effects are used to initiate or control the localization of deformation.

Surface Processes

There is a growing body of literature addressing the coupling of tectonics and landscape evolution (e.g., Braun and Sambridge, 1997; Willett et al., 2001), and how it may relate to regional climatic variables such as orographic rainfall distributions (e.g., Beaumont et al., 2000; Roe et al., 2003). Here, however, we modeled surface processes using very simple, first-order approaches. Our intention was to demonstrate the degree to which even the simplest surface process models can influence structural style.

For erosion, we used a simplified fluvial model where the rate of erosion at a given surface location is directly proportional to the surface slope. For simplicity, we used a spatially constant erosion rate equal to 1 mm/yr on a 1:1 (45°) slope. Lower slopes have proportionally lower erosion rates. More complex, spatially and temporally variable models are easily incorporated but are not justified at the present level of study.

For sedimentation, we considered simple end-member states: either no significant sedimentation or sediments that completely fill the available accommodation space up to a specified base level. Here, we chose the base level to be the initial upper free surface of the model layers prior to any shortening. We did not place any restrictions on conservation of eroded and deposited material within our two-dimensional (2-D) models, because in natural systems, material is commonly transported along strike, and the area modeled represents only the external portion of a potentially much larger orogenic system.

Note that because the rheologies in Table 1 are all frictional-plastic, and therefore rate-independent, the erosion rate and the convergence rate can be scaled together. For example, a model result for a convergence rate of 10 mm/yr and an erosion rate of 1 mm/yr (on a 1:1 slope) will be identical to that for a convergence rate of 1 mm/yr and an erosion rate of 0.1 mm/yr, assuming all other parameters are equal. Under these circumstances, only the elapsed time differs among the scaled models, where the time scale increases with decreasing convergence rate and erosion rate.

MODEL RESULTS

We present a series of eight models that progressively increase in complexity (Table 2).

TABLE 2. SUMMARY OF NUMERICAL MODELS

Model no.	Number of detachments	Strain-softening strong layers	Flexural base	Surface processes
1	1	No	No	No erosion or sedimentation
2	1	No	**Yes*** **10^{23} Nm**	No erosion or sedimentation
3	**2**	No	Yes 10^{23} Nm	No erosion or sedimentation
4	2	**Yes**	Yes 10^{23} Nm	No erosion or sedimentation
5	2	Yes	Yes 10^{23} Nm	**Erosion only**
6	2	Yes	Yes 10^{23} Nm	Erosion **and sedimentation** (to base level)
7	**3**	Yes	Yes 10^{23} Nm	**No erosion or sedimentation**
8	3	Yes	Yes 10^{23} Nm	**Erosion and sedimentation** (to base level)

*The parameters that varied from one model to the next are in **bold italics**.

Model 1

Model 1 (Fig. 3) is an example of the simplest case of a one-detachment configuration (Fig. 2), where the wedge base is horizontal and does not respond isostatically. The strong overlying layer has an effective angle of 38° (a typical "dry" value, with no pore-fluid pressure, comparable to an analogue "sandbox" experiment), whereas the weak basal detachment has an effective angle of internal friction of 3.5° (which can be interpreted to correspond to very high pore-fluid pressures; Eq. 6), resulting in an effective strength ratio of 10:1 (Table 1). There is neither erosion nor sedimentation (Table 2).

The mode of deformation within the wedge, characterized by triangular pop-up structures, is significantly different from that proposed by Dahlen and Suppe (1988) and Dahlen and Barr (1989), in which distributed pure shear strain is specified to maintain an exact, uniformly dipping upper-wedge surface. These pop-up structures, bounded by discrete high-strain shear zones ("faults"), are similar to those observed in analogue models where the basal detachment is weak (e.g., Liu et al., 1992). Pop-ups can also occur in nature for the same reason, for example, where the overburden is underlain by salt (halite) (e.g., Davis and Engelder, 1985). In the case of halite, the basal layer is a weak viscous material as opposed to a weak frictional-plastic material, but the effect is the same. Unlike the sand and other mixtures used in analogue experiments, which initially strain harden and then strain soften according to the initial packing of the granular material, in model 1 there is no strain dependence of the material properties. Therefore, localized deformation is not a consequence of strain-dependent material properties.

The shear zones resemble slip-lines in ideal rigid-plastic materials (e.g., Hill, 1950), which are the characteristics of the hyperbolic equilibrium equations. The ALE finite-element formulation produces minimum rate of work solutions, suggesting that although distributed pure shear solutions are admissible, they are not minimum work solutions.

The size of the pop-up structures clearly scales with the thickness of the undeformed layer. As model 1 grows, the nearly regular nature of newly developing pop-ups is occasionally interrupted, as at 4.5 m.y., between marker numbers 7 and 9 (Fig. 3C). This deviation is probably a numerical artifact because models with more stringent convergence requirements exhibit a more regular progression. The regular pop-up pattern across this segment of the wedge is also largely restored by 6.0 m.y. (Fig. 3D). The wedge subsequently begins to organize into larger, discrete thrust sheets, generally consisting of two adjacent pop-up structures and the underlying intervening triangular regions. This pattern and its development are seen at 12 m.y. and earlier times, where the thrust sheets are being carried on "faults" that intersect the surface near markers 5, 9, 12, and 16 (Fig. 3H). This self-organization and localization of strain occurs in the absence of strain softening or strain hardening and may be related to folding instability (cf. Goff et al., 1996). In some cases, the maintenance of localized high shear strain, in the absence of strain softening or hardening, may be due in part to the advection of small amounts of weak material into the shear zones.

For effective friction angles of 38° for the wedge and 3.5° for the thin basal layer, and a horizontal base, the upper surface slope predicted by the cohesionless wedge solution of Dahlen (1984) is 0.835° (the 2 MPa cohesion is negligible in comparison to wedge strength, except very near the surface). The average slope of model 1 depends on the way the slope is measured: it is very close to the predicted value if the line through the minimum heights is used (Figs. 3A–3D), but it is ~1.6° if the tangent to the tops of the pop-ups is used (Fig. 3C). This result is consistent with a critical wedge solution that requires all parts of the wedge to attain the critical state before the wedge can translate without further internal deformation. The pop-ups exceed this requirement, and the tangent to their tops defines a supercritical wedge. The significance of this result is that a uniform taper critical wedge will not be grown by accreting a finite-thickness layer at the wedge toe. Instead, the fundamental wedge-building units, the pop-ups in this case, are assembled in a manner that creates local supercritical regions of the wedge. Small deviations of the minimum heights from the critical slope line near the rear of the wedge (Fig. 3D) probably reflect slight weakening of the bulk wedge strength due to the advection of weak basal material upward into the wedge along the growing shear zones.

The surface of the model, although not "eroded" as in other models (Table 2), is subject to minor smoothing due to the numerical filtering that occurs because the finite elements have a finite width (in this case 250 m), and therefore it cannot represent shorter-scale topographic variations. A consequence is the numerical infolding of very small volumes of material in the cores of synclinal structures where the top of the Lagrangian tracking mesh has been folded beneath the top of the Eulerian mesh at a length scale shorter than the finite-element width. This material, which appears as "sediment" (e.g., Fig. 3G), is assigned properties identical to the underlying layer (Table 1).

Model 2

Model 2 is identical to model 1 except that the base of the model responds isostatically to wedge thickening (Fig. 4; Table 2). The base behaves like a uniform, thin elastic plate with a flexural rigidity of 10^{23} Nm, equivalent to an effective elastic thickness of ~25 km assuming a Young's modulus of 70 GPa and a Poisson's ratio of 0.25.

The pattern of self-organization into pop-up structures that amalgamate to form thrust sheets is again produced. For example, at 18 m.y., there is a semiregular pattern of thrust sheets carried on faults that intersect the surface near marker numbers 8, 10, 13, 16, 18, and 20 (Fig. 4F). By following the development of these features both backward and forward in time, the pattern of self-organization becomes apparent. The thin zones of instantaneous high shear strain are not fixed with respect to the wedge material but may move as the wedge thickens and adjusts isostatically. This process can result in thick zones of the Lagrangian mesh characterized by moderate to high total shear strain, as seen intersecting the surface at 24 m.y. near markers 8, 14, 19, and 23 (Fig. 4H).

In model 2, the overall wedge taper angle is larger than in model 1 owing to isostatic adjustment and the increase in basal slope. Achievement of this larger taper angle requires larger internal strain, which leads to enhanced development of longer thrust sheets formed as in model 1 by the amalgamation of the pop-up structures. The pop-ups link together in groups of four triangular segments to form coherently deformed and stacked thrust sheets. As for model 1, this self-organization may relate to preferred thrust-sheet length-scale selection as described by Goff et al. (1996) (compare Figs. 3F and 4F, which both have similar wedge toe positions).

Model 3

Model 3 is similar to model 2, but it includes the addition of a second, internal, weak layer in which deformation localizes (Figs. 2 and 5; Table 2). This internal detachment is weaker ($\varphi_{eff} = 1°$) than the basal detachment ($\varphi_{eff} = 3.5°$), leading to an effective strength ratio of ~1:4 between the two (0.28:1; Table 1). The basal décollement steps up from the model base to an intermediate level, as in the archetypal southern Canadian Rocky Mountains (Fig. 1). In southern Canadian Rocky Mountains terms, the two-detachment model could correspond to the thick and relatively strong Mesozoic foreland clastics overlying the thin and weak Fernie Formation (principal detachment 2, Fig. 1), above thick and strong Paleozoic carbonates that overlie the weak basal detachment above the crystalline basement (principal detachment 1, Fig. 1). This simplified view of the southern Canadian Rocky Mountains is of course an approximation owing to the secondary detachments that exist within both the carbonate and clastic-dominated sections.

The inclusion of the internal detachment dramatically influences the structural style, which results in longer thrust sheets in the lower strong layer and associated fault-bend folds and duplex structures. These models tend to produce break thrusts and "snake head" structures at the leading edges of these lower-layer thrust sheets (e.g., at 9 m.y. between markers 10 and 11, Fig. 5C), and internally thicken during transport in a series of pop-ups (e.g., at 18 m.y. between markers 16 and 18, Fig. 5F).

The basal décollement adopts two basic trajectories. The first involves stepping up from the basal detachment layer to the intermediate detachment level at a significant distance back from the toe of the wedge (e.g., at 9 m.y. and 12 m.y., Figs. 5C and 5D). Alternatively, the décollement can lie within the basal detachment layer essentially to the toe of the deforming wedge, as at 6 m.y. (Fig. 5B) and at 15 m.y. and later (Figs. 5E–5H). These two forms, which correspond to basal and frontal accretion, alternate in response to changes in material distribution within the wedge and changes in the basal slope across the model. Although the model fold-and-thrust belt has an overall wedge shape, there are significant sub–wedge-scale departures from the uniform taper predicted for critical Coulomb wedges. These departures occur at length scales that are not equal to the thicknesses of the layers that are accreted at the toe of the wedge and therefore cannot be explained solely by the same mechanism as in model 1.

Model 4

Model 4 (Fig. 6) is identical to model 3 except that the two thick strong layers undergo strain softening (Table 2). In an unstrained to modestly strained state they are as strong as the thick layers in models 1 through 3, but as ε increases from 0.5 to 1.0, φ_{eff} decreases linearly from 38° to 18° (a 50% decrease in strength; Table 1), as described already. There is little sensitivity of model results to the upper strain limit of the strain-softening range, because strain softening leads to strong positive feedback and increased focusing of strain. This positive feedback results in strain being concentrated in fewer shear zones crosscutting the strong layers and, therefore, less tendency of the thrust sheets to deform internally. Due to strain softening, these crosscutting shear zones are significantly weaker than the thick layers they traverse, resulting in fewer pop-up structures within thrust sheets and significantly less layer-parallel shortening of the strong layers (compare Figs. 5 and 6). The strain-softened value of $\varphi_{eff} = 18°$ can be viewed either as a consequence of localized, moderate pore-fluid overpressures (Eq. 6), with a λ value of 0.50 (equivalent to a λ_{HR} value of 0.72; see Appendix and Table 1), or as a consequence of material strain softening of major fault zones, which may occur in nature through a variety of mechanisms (e.g., Wojtal and Mitra, 1986; Bos and Spiers, 2002).

The primary difference between model 4 and model 3 is the contrasting deformation styles in the upper and lower strong layers, both of which undergo equal strain softening. The upper layer develops relatively short thrust sheets with less back thrusting and less internal shortening during transport than the equivalent structures in model 3 (Figs. 6C–6H), due to strain softening and localization within shear zones.

In contrast, very long thrust sheets develop in the lower strong layer, as seen at 18 m.y. and 24 m.y. (Figs. 6F and 6H), and these more closely resemble natural structures. These sheets are transported large distances relative to their length with little internal deformation. They are generally later dismembered by localized, out-of-sequence structures. For example, compare Figure 6F with Figure 6G: out-of-sequence faults are localized above a footwall ramp beneath markers 13 and 14 in Figure 6G. The basal décollement tends to cut up through the lower strong layer to the intermediate level detachment at a significant distance behind (toward the backstop) the deformation front in the upper layer, similar to the southern Canadian Rocky Mountains archetype (Fig. 1). Stacking of the lower thrust sheets gives the wedge a large-scale swell-and-dip surface morphology (Figs. 6D and 6F). In nature, the topographic lows may be occupied by piggyback basins, a process that is investigated in subsequent models.

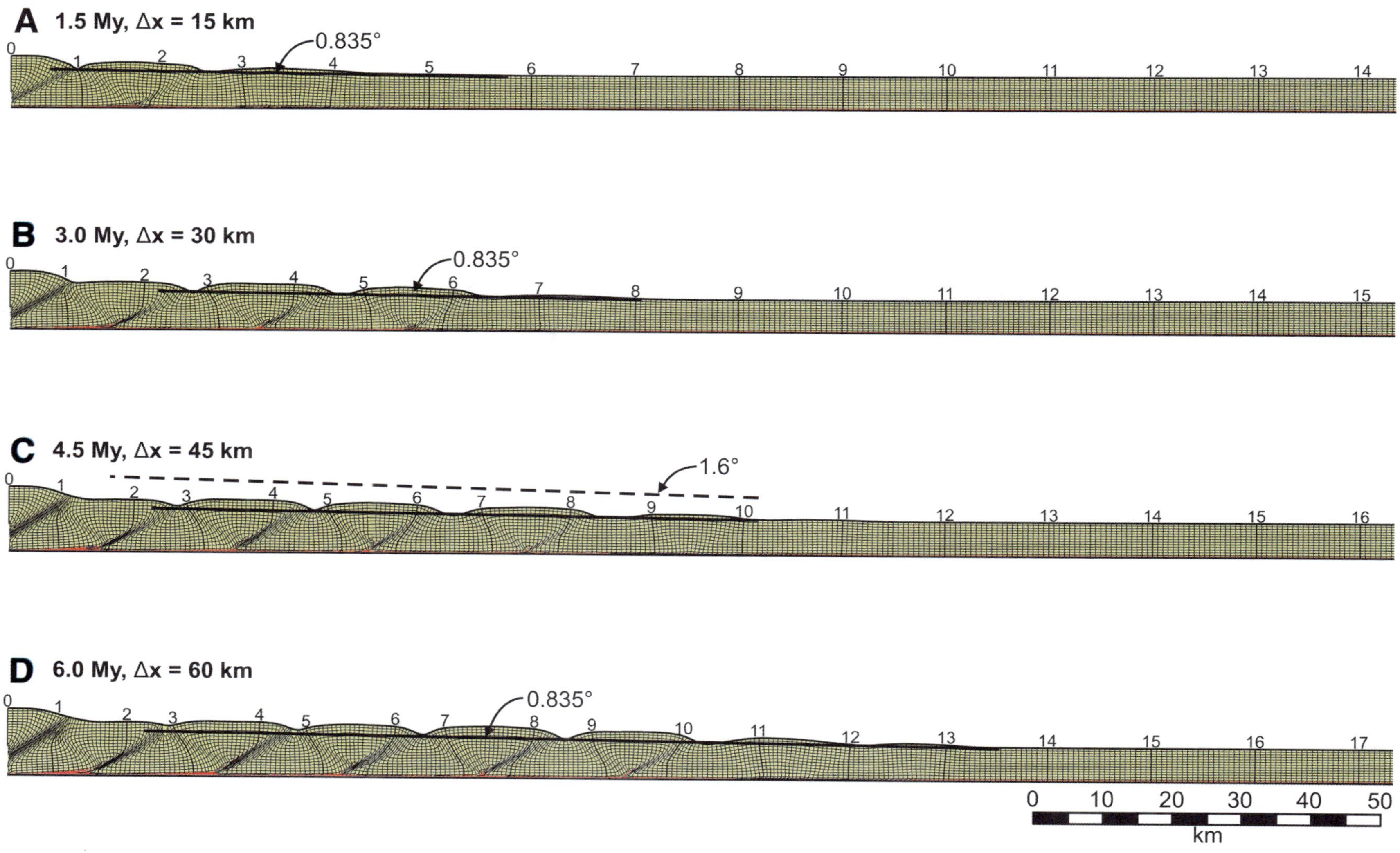

A 1.5 My, Δx = 15 km   0.835°

B 3.0 My, Δx = 30 km   0.835°

C 4.5 My, Δx = 45 km   1.6°

D 6.0 My, Δx = 60 km   0.835°

km 0 10 20 30 40 50

Figure 3 (*on this and following page*). Model 1 results (one-detachment model with rigid, horizontal base); frames are spaced equally in time. Elapsed time since beginning of convergence and magnitude of convergence are shown. Illustrated width is 200 km, encompassing the entire Eulerian grid. Cells within the Eulerian grid are colored according to material type (see Figure 2 and Table 1), and the Lagrangian mesh is shown by black lines (only every third mesh line, horizontally and vertically, is drawn for clarity). Dashed line above wedge in C is a visual fit to the tops of the pop-up structures, with a slope of 1.6°. Solid lines connecting the topographic low points of the upper surfaces in A–D have slopes equal to that predicted by the Dahlen (1984) equations for model 1 materials (0.835°). Arrows in H indicate shear zones with relatively large displacements resulting from self-organization of the wedge.

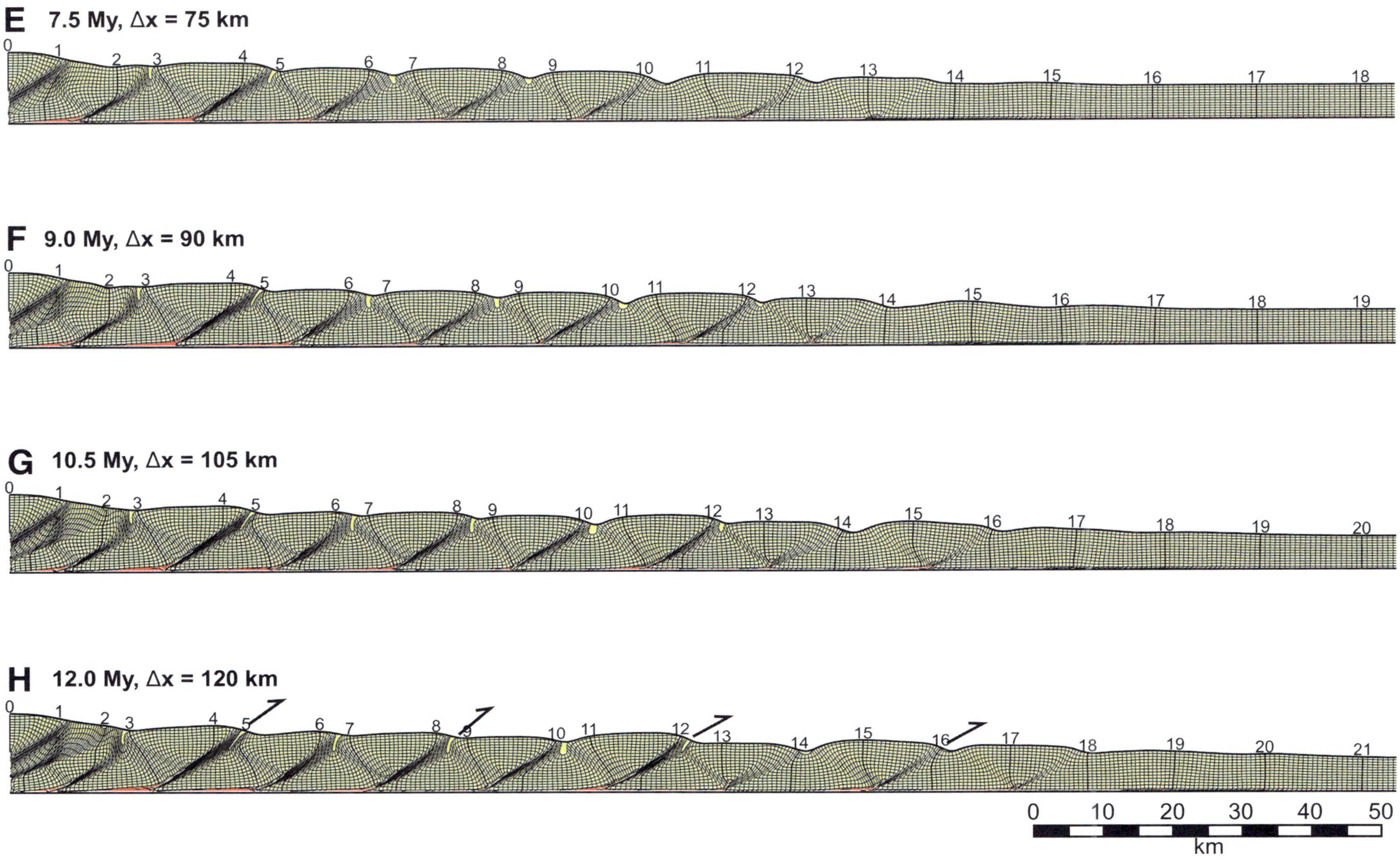

E 7.5 My, Δx = 75 km

F 9.0 My, Δx = 90 km

G 10.5 My, Δx = 105 km

H 12.0 My, Δx = 120 km

Figure 3 (*continued*).

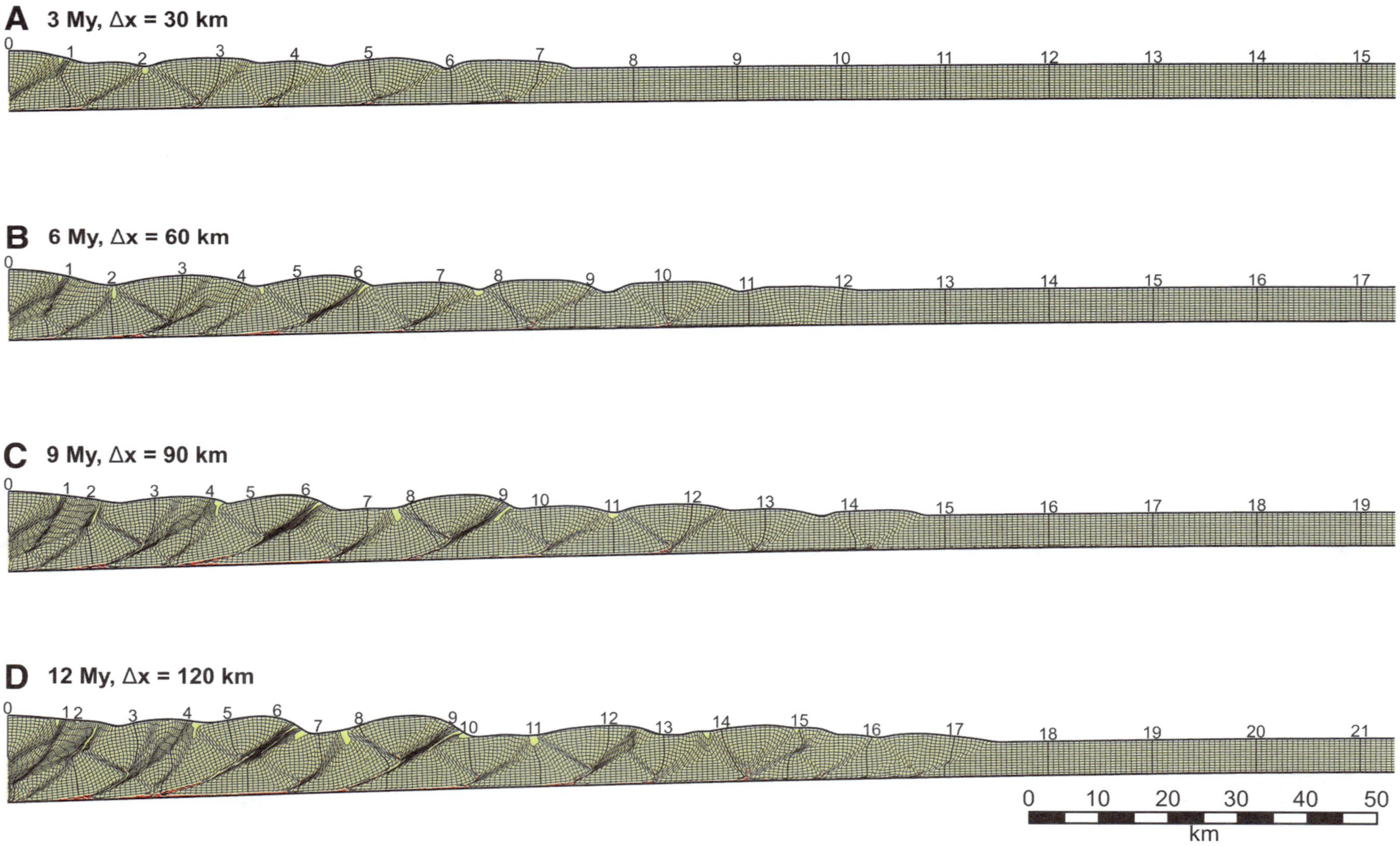

A 3 My, Δx = 30 km

B 6 My, Δx = 60 km

C 9 My, Δx = 90 km

D 12 My, Δx = 120 km

Figure 4 (*on this and following page*). Model 2 results (one-detachment model; addition of flexural isostasy). See Figure 3 caption.

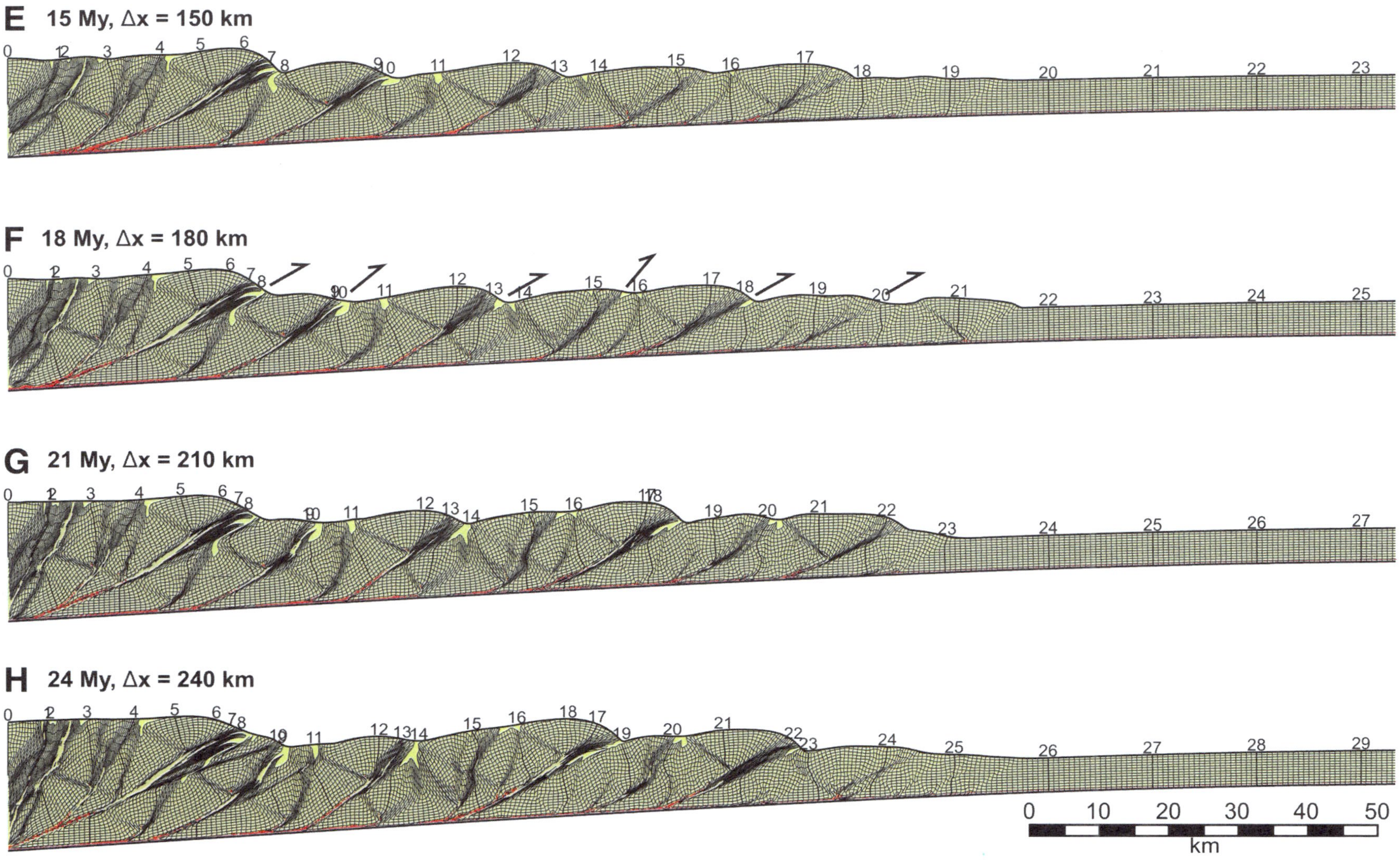

E 15 My, Δx = 150 km

F 18 My, Δx = 180 km

G 21 My, Δx = 210 km

H 24 My, Δx = 240 km

Figure 4 (*continued*).

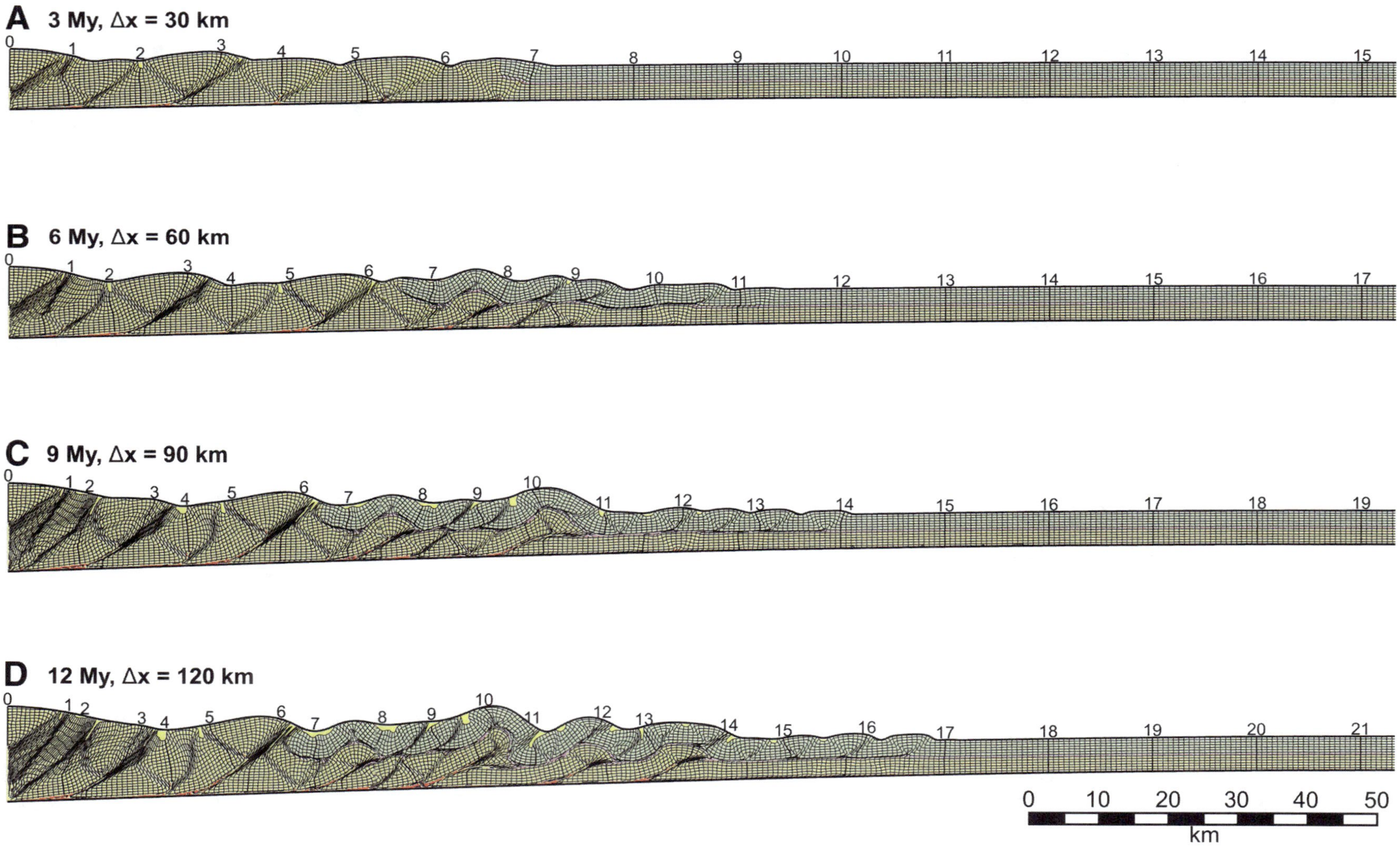

A 3 My, Δx = 30 km

B 6 My, Δx = 60 km

C 9 My, Δx = 90 km

D 12 My, Δx = 120 km

Figure 5 (*on this and following page*). Model 3 results (two-detachment model; addition of an internal detachment layer). See Figure 3 caption.

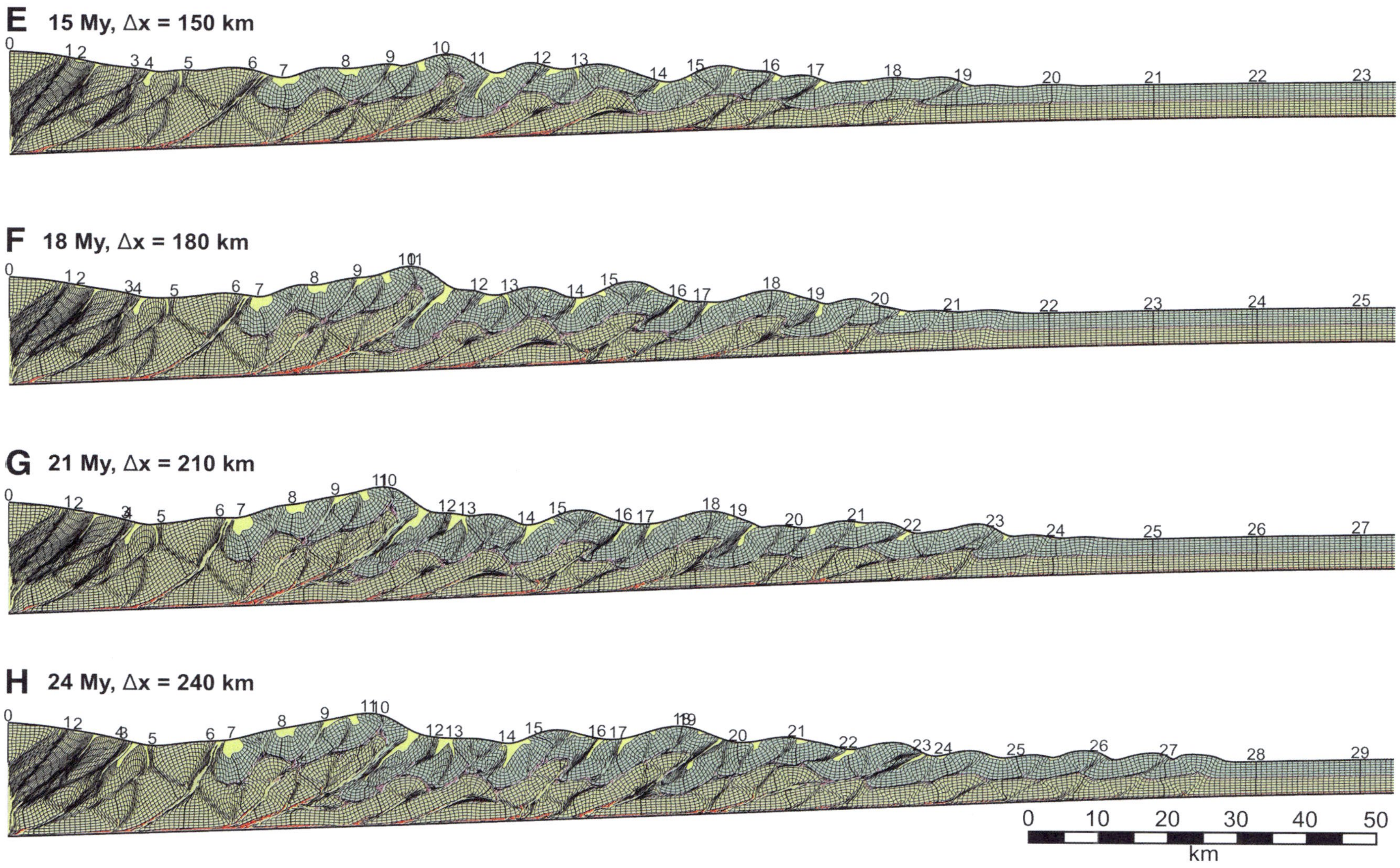

E 15 My, Δx = 150 km

F 18 My, Δx = 180 km

G 21 My, Δx = 210 km

H 24 My, Δx = 240 km

Figure 5 (*continued*).

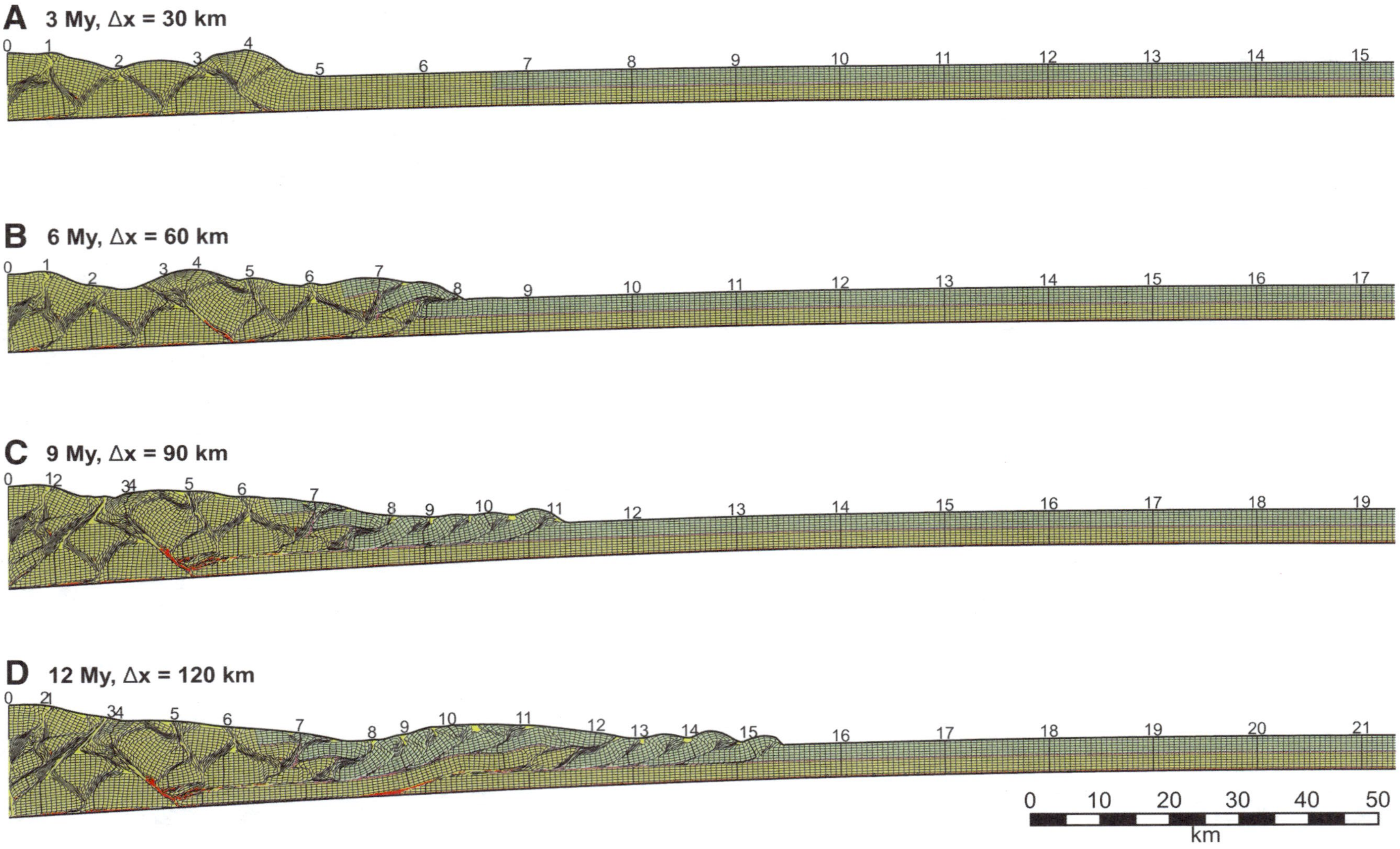

A 3 My, Δx = 30 km

B 6 My, Δx = 60 km

C 9 My, Δx = 90 km

D 12 My, Δx = 120 km

Figure 6 (*on this and following page*). Model 4 results (two-detachment model; addition of strain softening within the thick, strong layers). See Figure 3 caption.

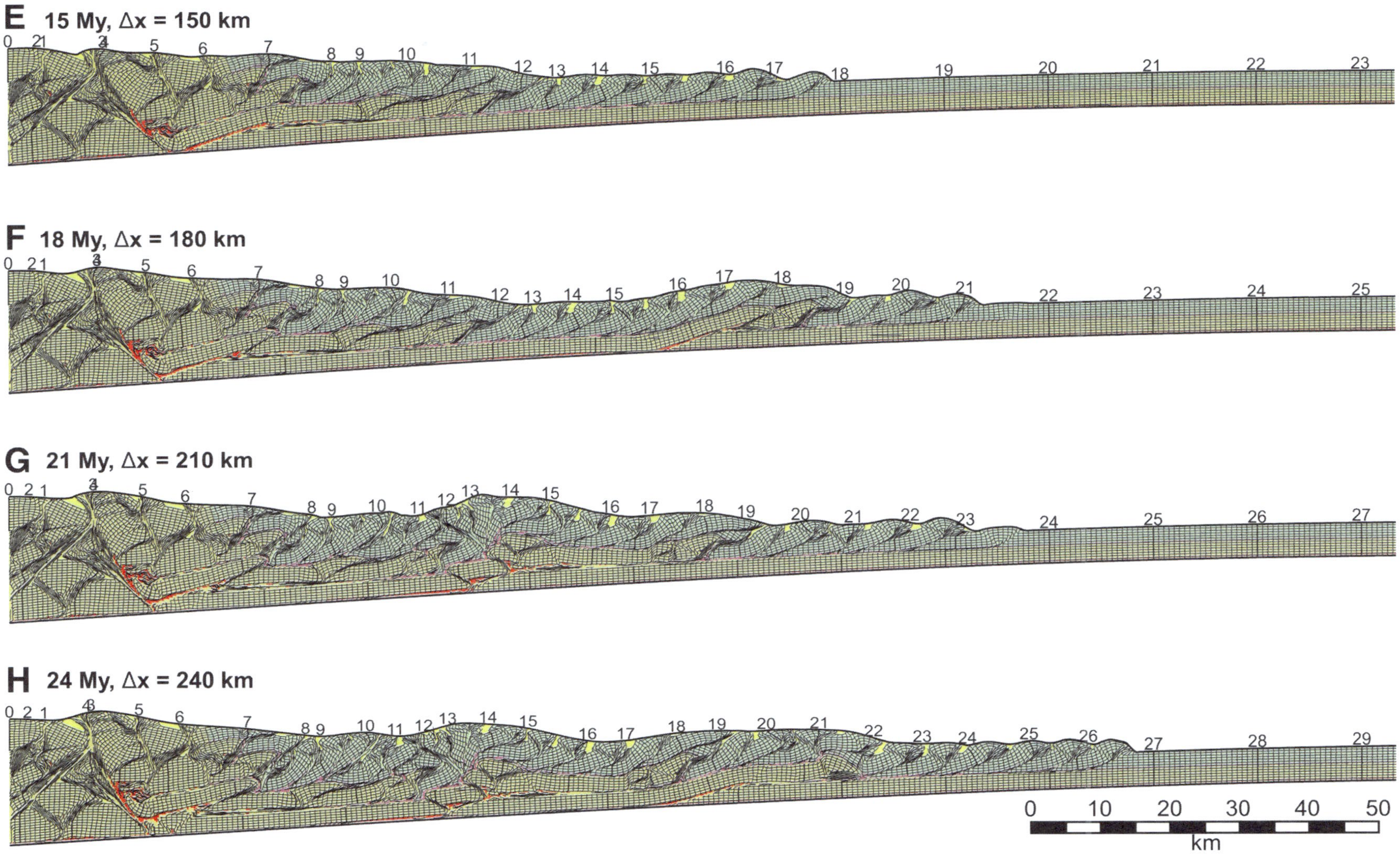

Figure 6 (*continued*).

Model 5

Model 5 (Fig. 7) is identical to model 4 except that the upper surface is subjected to significant erosion (Table 2). A comparison of these models shows that erosion alone reduces the total volume of the wedge at any given time but does not markedly change the overall structural style. A clear though minor influence on structural style is the enhancement of the pop-up structures that develop at the leading edges of short thrust sheets within the upper strong layer (e.g., compare pop-ups in Figs. 7D and 6D). This is an example of positive feedback between structures and surface processes. Wherever there is active shortening within the wedge that results in an increase in surface slope, there is an associated increase in erosion rate that in turn results in a focusing of additional shortening.

Model 6

Model 6 (Fig. 8) is similar to model 5 except that rapid sedimentation results in the filling of all accommodation space below base level at each time step (Table 2) with material that has the same mechanical properties as the upper strong layer (Table 1) and that is fully involved in the deformation of the wedge. Although the syndeformational sediment is not overlain by the initial Lagrangian tracking mesh (Fig. 8), it is fully tracked during deformation for re-gridding onto the Eulerian grid, and the effects of strain softening are computed. For ease of display, the color of sedimentary strata is changed every 2 m.y., and the pattern repeats after five cycles (every 10 m.y.). No attempt was made to balance the volume of sediment against that removed by erosion. In nature, the region we are modeling would be part of a much larger system, with significant potential sediment sources in the hinterland as well as along strike. In model 6, the onset of sedimentation and erosion is delayed until 4.5 m.y. to allow the model to develop the initial buffer wedge as explained previously (in model 5, erosion begins at 0 m.y.).

The influence of syndeformational deposition on model structural style is profound. As the foreland basin deepens and tilts in model 6, the deformation front (wedge toe) steps out into the foreland a distance of several times the thickness of the wedge. This pattern is repeated as the wedge evolves (Fig. 8). As the model base isostatically subsides due to wedge growth, its slope increases. In combination with active sedimentation in front of the deforming wedge (in the foreland basin; Fig. 8B), this results in achievement of overall critical or supercritical taper out to the nascent toe of the thrust wedge without the requirement of internal shortening. This leads to an abrupt forward step in the position of the deformation front and the incorporation of the proximal region of the foreland basin into the deforming wedge. This abrupt step and incorporation of the proximal foreland is seen in Figure 8C (to "wedge toe 1"), where the basin has already been shortened by out-of-sequence structures, and in Figures 8D (to "wedge toe 2") and 8F (to "wedge toe 3"), where the newly incorporated piggyback basins are essentially undeformed.

As convergence continues, shortening across the wedge is concentrated in the more highly strained regions flanking the piggyback basin due to the positive feedback effect of slope-dependent erosion and further deposition of sediments in the piggyback basin, which, respectively, weakens and strengthens these regions. Weakening in these high-strain areas is also enhanced by strain softening of the two thick layers. With further shortening, the cycle that forms piggyback basins repeats, resulting in a structural style where broad, relatively little-deformed synclinal areas are separated by significantly more-deformed anticlinal structures (Figs. 8F, 8G, and 8H).

The coupling of isostatically compensated wedge growth to sedimentation allows the frontal portion of the system to bypass the deformation required to achieve critical taper and move directly to a critical or supercritical state (i.e., translating on the basal décollement, but with effectively no internal strain). For example, the tapers of the frontal portions of the deforming wedges, detached on the weak internal detachment, are very similar in Figure 8D and Figures 7C–7H. However, this taper was achieved in model 5 through internal shortening, whereas in model 6, critical taper developed due to sedimentation in an asymmetrically subsiding foreland basin.

Similar to the structural evolution of models 4 and 5, the basal décollement of model 6 cuts up through the lower strong layer, to the intermediate level detachment, at a significant distance behind the toe of the wedge (Fig. 8). When the basal décollement jumps forward and down-section to incorporate a slice of the strong lower layer into the wedge, the leading edge ramp through this layer is located significantly toward the hinterland of the current toe of the wedge but is more closely aligned with the penultimate wedge toe (see labeled ramps and wedge toes in Figs. 8E and 8F). As contraction continues, a number of out-of-sequence structures crosscut and shorten the piggyback basins (examples labeled in Figs. 8F–8H). Some of these structures are related to the forward jumps in the basal décollement.

Model 7

Model 7 (Fig. 9) is similar to model 4, except that a second internal detachment layer is introduced, resulting in a three-detachment model configuration (Fig. 2; Table 2). As in the previous cases, the two internal detachment layers are weaker than the basal detachment layer, and the thick strong layers strain soften (Table 1). The three-detachment model is similar to areas of the southern Canadian Rocky Mountains that have a regional detachment within the Upper Cretaceous marine shales, in addition to an internal detachment in the Fernie Formation.

Incorporation of a second internal detachment leads to structures similar to those in the simpler two-detachment models, but these are even more reminiscent of natural fold-and-thrust belts. For example, model 7 shows well-developed flat-ramp-flat fault trajectories with associated fault-bend folds, duplexes, and antiformal stacks. Except for the uppermost strong layer, which deforms into a series of short imbricate thrust sheets, the thrust sheets are commonly very long relative to their thickness and are little deformed internally, except near their leading edges. The contrast in deformation style of the two upper strong layers is particularly significant because these layers and their underlying detachments have identical properties. This difference reflects their positions relative to other weak and strong layers (the middle layer is confined below another strong layer, whereas the top layer is not).

The long thrust sheets, composed of slices of the middle strong layer, are accreted to the base of the growing wedge. This basal accretion occurs beneath a shallow-based deforming wedge composed of tight imbricate slices of the upper strong layer (for example, compare Fig. 9E with Fig. 9F). The progressive stepping out of the basal décollement toward the foreland in a flat-ramp-flat fashion leads to a characteristic geometry of a series of anticlinally folded thrust sheets, cored by the thickened leading edges of basal layer thrust sheets, where structural elevation increases steadily toward the back of the wedge (Fig. 9H).

Model 8

Model 8 (Fig. 10) is similar to model 7, except it undergoes both surface erosion and rapid deposition below base level. Model 8 is to model 7 as model 6 is to model 4 (Table 2). As in model 6, sedimentation and erosion begin at 4.5 m.y. to allow the model to develop an initial wedge and form the hinterland edge of the evolving foreland basin.

As seen in model 6, syndeformational deposition has a profound influence on structural style. As the wedge evolves, the wedge toe repeatedly steps out a substantial distance into the foreland, and the intervening segment of the wedge is transported with little or no internal strain. When the upper two strong layers detach together (Figs. 10B and 10E), this distance is larger than when only the uppermost layer detaches (Figs. 10D and 10F). Concurrent erosion and deposition lead to development of out-of-sequence thrusts (Figs. 10D and 10G), as well as to onlap of piggyback basin strata across the leading edges of formerly active and eroded thrust sheets (Figs. 10F–10H). These relationships illustrate the structural complexity that can develop due to feedback among surface processes.

Instantaneous velocities and strain rates for model 8 (Fig. 11) show that deformation patterns vary across the wedge in time as well as space, and large regions are transported with little to no instantaneous strain. Shortening and uplift can be concentrated at or near the toe of the wedge, as in Figures 11A, 11B, 11E, or 11H, or internally, with the toe being nearly passively transported, as in Figures 11C, 11D, 11F, and 11G.

DISCUSSION

Surface Processes and Natural Structures

Many fold-and-thrust belts contain syndeformational piggyback basins, but typical basins of this type are not preserved in the archetypal southern Canadian Rocky Mountains (Fig. 1). Were they ever present? A principal characteristic of models 6 and 8 is broad, relatively little-deformed synforms that underlie the piggyback basins and separate narrower, more heavily shortened antiformal structures. This general structural style of antiforms and synforms is reminiscent of eroded features observed in the foothills of the southern Canadian Rocky Mountains and elsewhere.

Figure 12A is a cross section across the eastern front ranges and the foothills of the southern Canadian Rocky Mountains, 100 km northwest of Calgary, Alberta (after Ollerenshaw, 1978). The Sheep Creek, Burnt Timber, Williams Creek, and Grease Creek synclines comprise a series of broad, relatively little-deformed synforms (the Grease Creek syncline is virtually unfaulted 10 km to the south). Postdeformational erosion is sufficiently large (2–5 km or more; e.g., Osborn et al., 2006) that syndeformational sedimentological evidence has been removed, but a comparison of these synformal structures to those of models 6 and 8 suggests that they may mark the keels of now-eroded piggyback basins. The Alberta syncline, which overlies a blind triangle zone at the edge of the foothills, may also owe its existence to feedback between surface processes and structural development (Stockmal et al., 2001).

Figure 12B is a cross section modified after Mountjoy et al. (2002) across the eastern edge of the front ranges and most of the foothills belt, ~250 km northwest of Calgary, Alberta. These synclines (Bighorn, Black Mountain, and Brazeau synclines, and an unnamed syncline in the hanging wall of the Ancona thrust) are progressively more structurally uplifted to the west. Again, the structural style observed in models 6 and 8 suggests that these broad synforms may mark the axes of now-eroded piggyback basins.

Figure 12C is a cross section from Ramos et al. (2004) across the Sub-Andean belt of northern Argentina, where this structural style is clearly associated with syndeformational piggyback basins. The broad synclines are occupied by synorogenic sediments that document a complex pattern of development. Although there is an initial overall foreland-breaking sequence for the main faults, there is clear sedimentary evidence of out-of-sequence and simultaneous motion on a series of thin-skinned thrusts (Ramos et al., 2004). Similar, less-eroded features have been observed near the leading edges of other orogens, both active and inactive (e.g., Ori and Friend, 1984; Pivnik and Khan, 1996; Leturmy et al., 2000).

The relationship between thin-skinned thrusting and syndeformational sedimentation is well illustrated by Pieri (1989) in seismic lines across the Po Plain, adjacent to the Apennines. Figure 13A is a portion of seismic line B–B′ that shows thrusted Paleogene and Miocene sediments, partly eroded and onlapped by Pliocene and Quaternary deposits, which are in turn slightly deformed. Figures 13B and 13C show portions of model 8 (Figs. 10G and 10H) where similar relationships have evolved. A formerly emergent thrust toe, cut off by an out-of-sequence trailing imbricate, is progressively onlapped and buried by the piggyback basin. These sediments are in turn slightly deformed during later minor motion on the preexisting thrust faults.

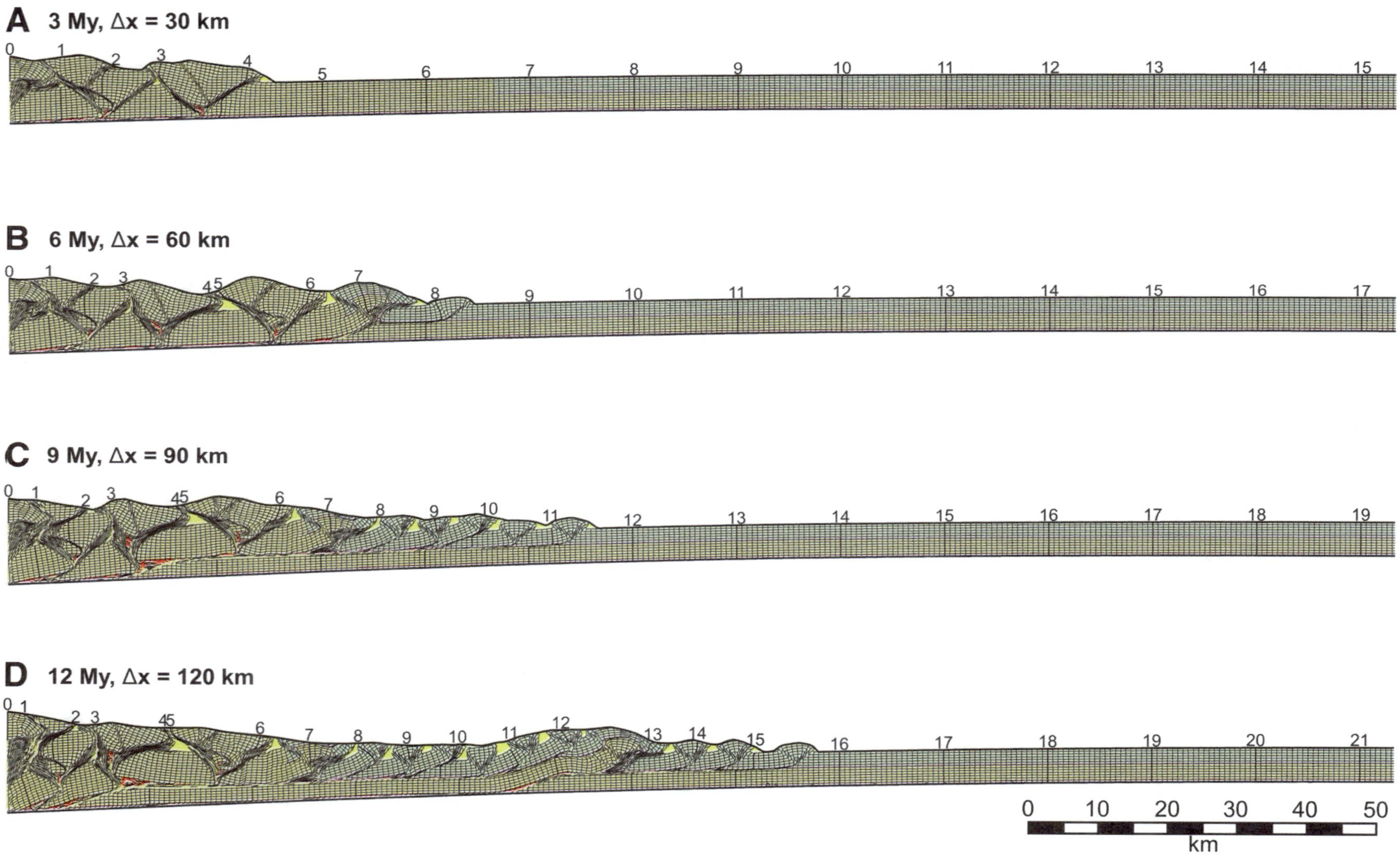

Figure 7 (*on this and following page*). Model 5 results (two-detachment model; addition of significant, slope-dependent surface erosion). See Figure 3 caption.

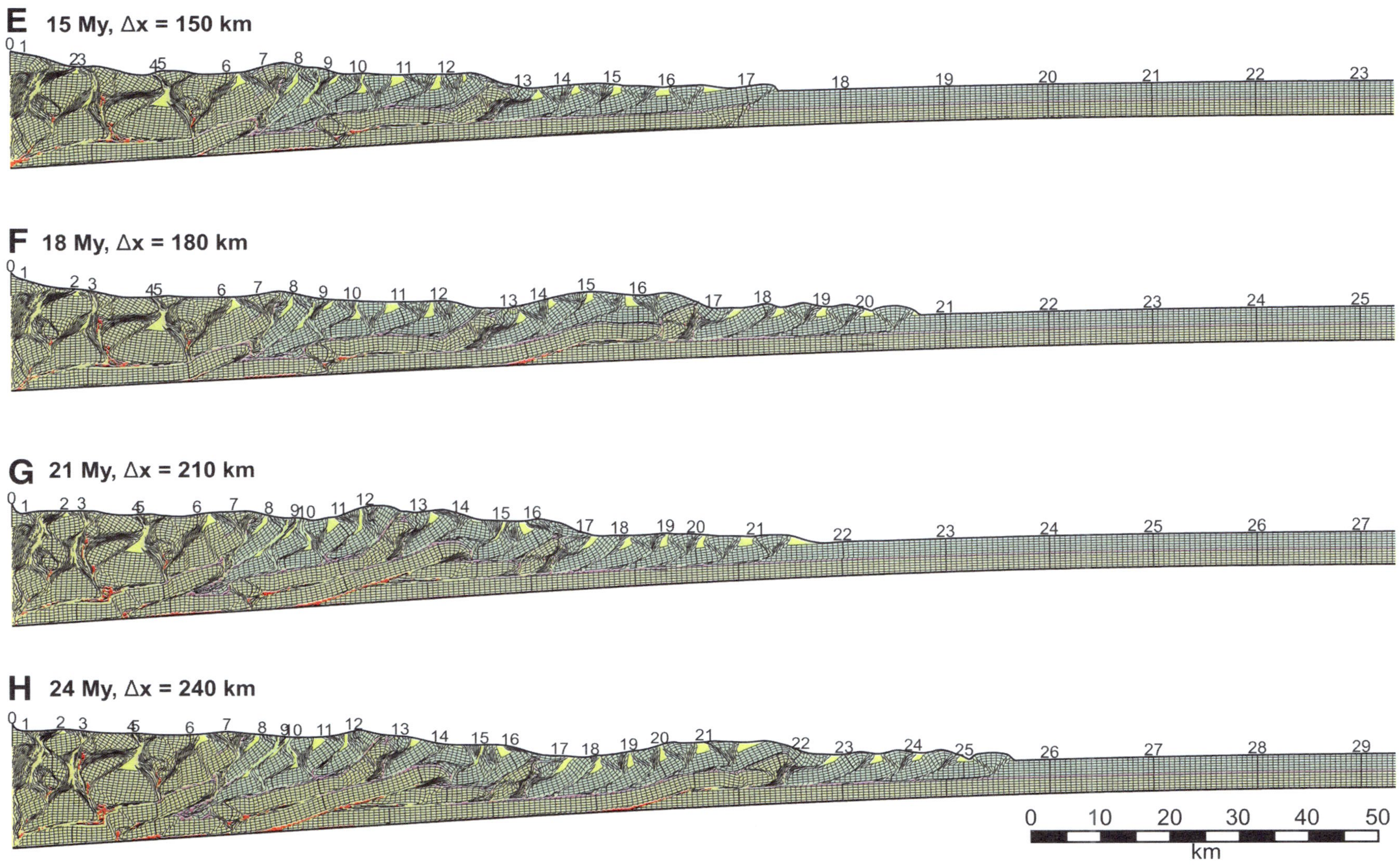

Figure 7 (continued).

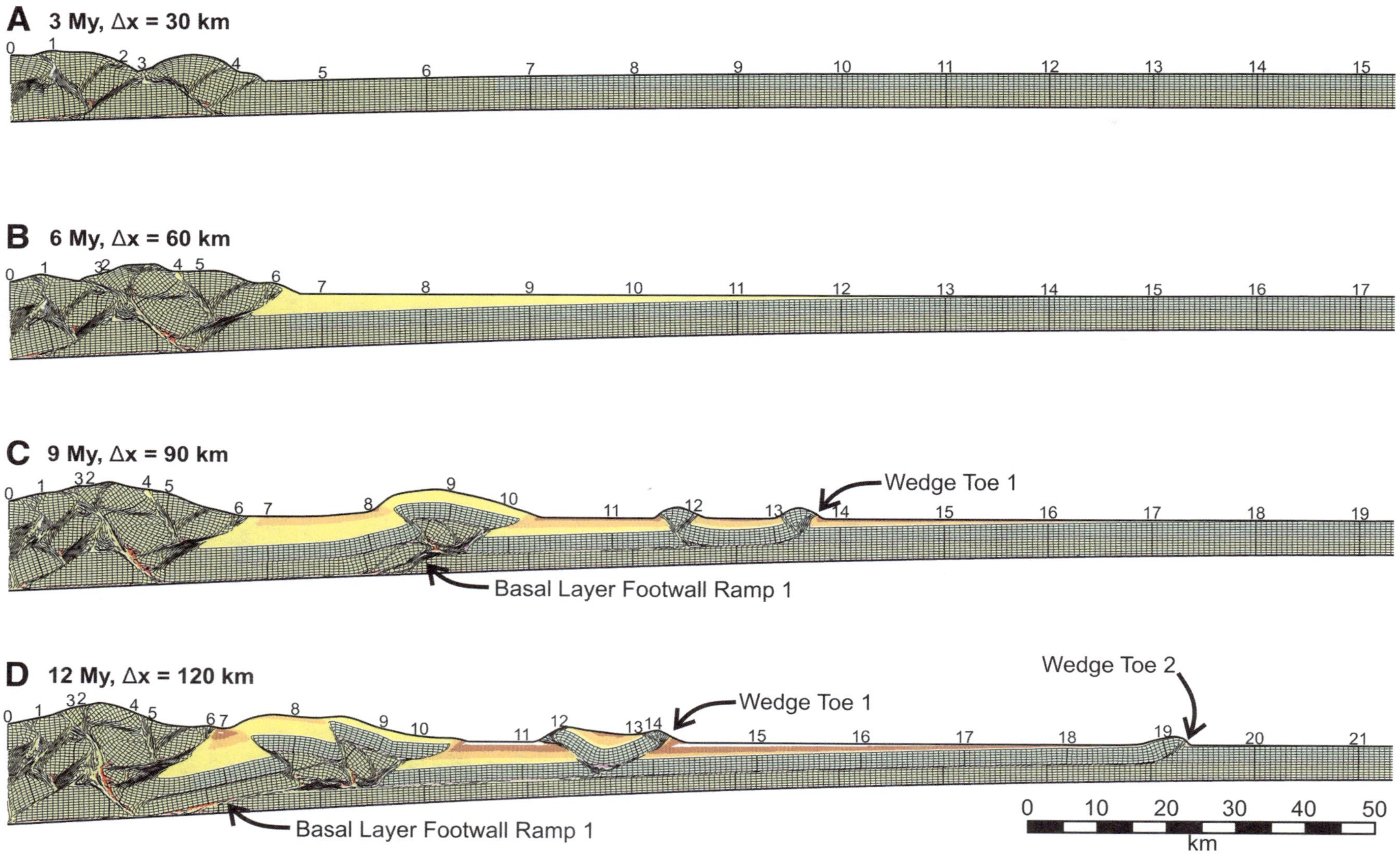

Figure 8 (*on this and following page*). Model 6 results (two-detachment model; addition of sedimentation to specified base level). See Figure 3 caption. OOS—out-of-sequence.

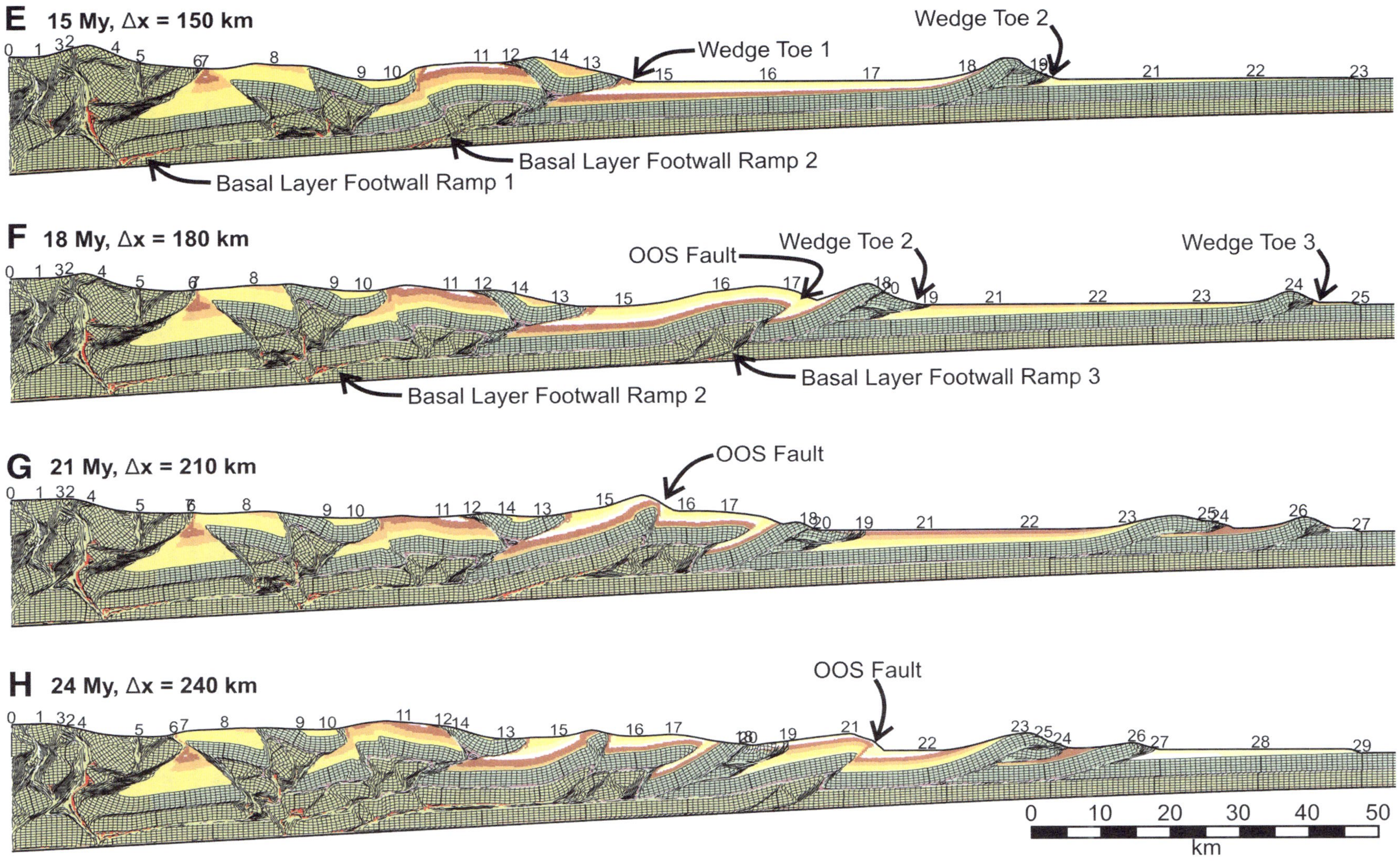

Figure 8 (*continued*).

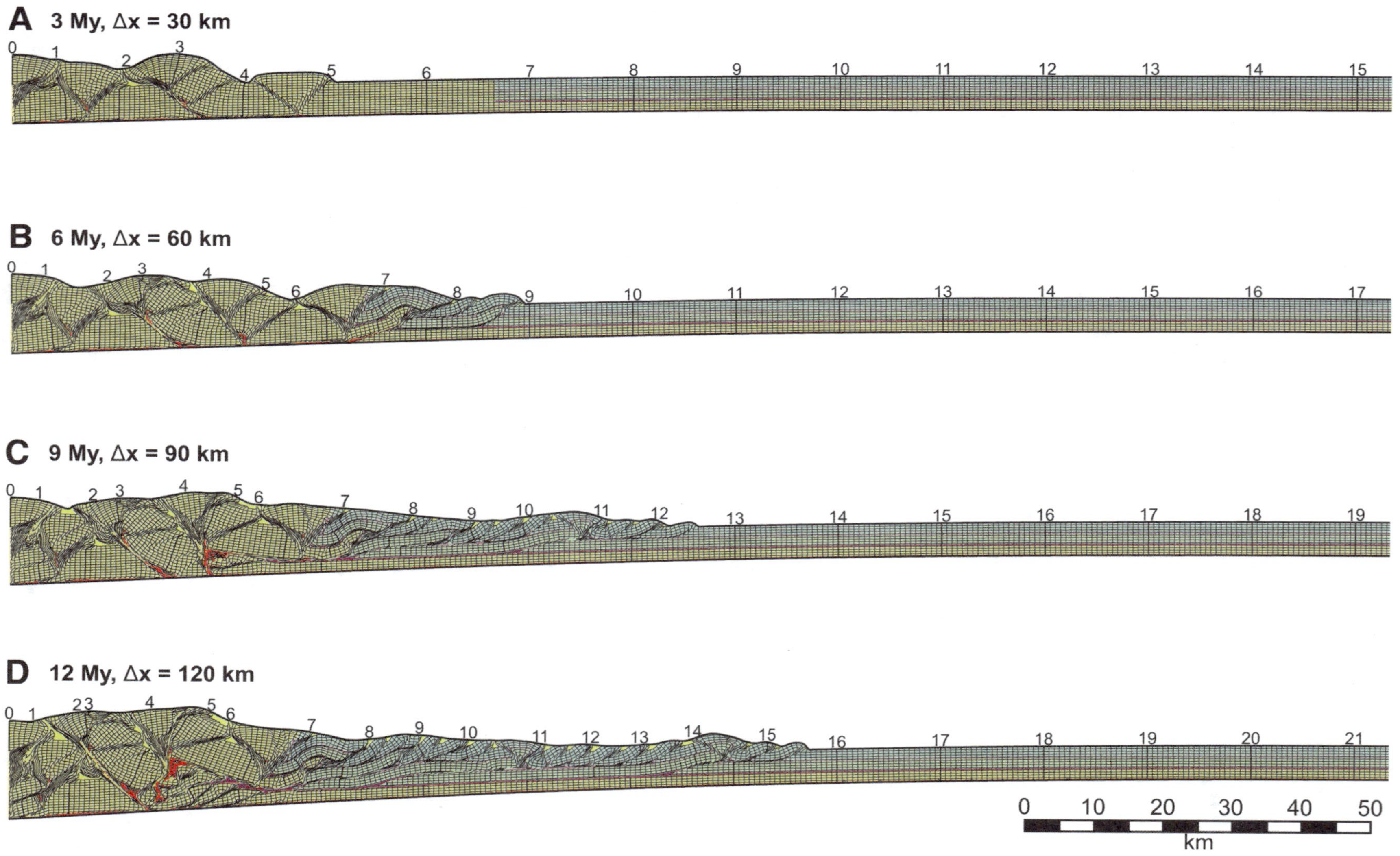

A 3 My, Δx = 30 km

B 6 My, Δx = 60 km

C 9 My, Δx = 90 km

D 12 My, Δx = 120 km

Figure 9 (*on this and following page*). Model 7 results (three-detachment model; erosion only). See Figure 3 caption.

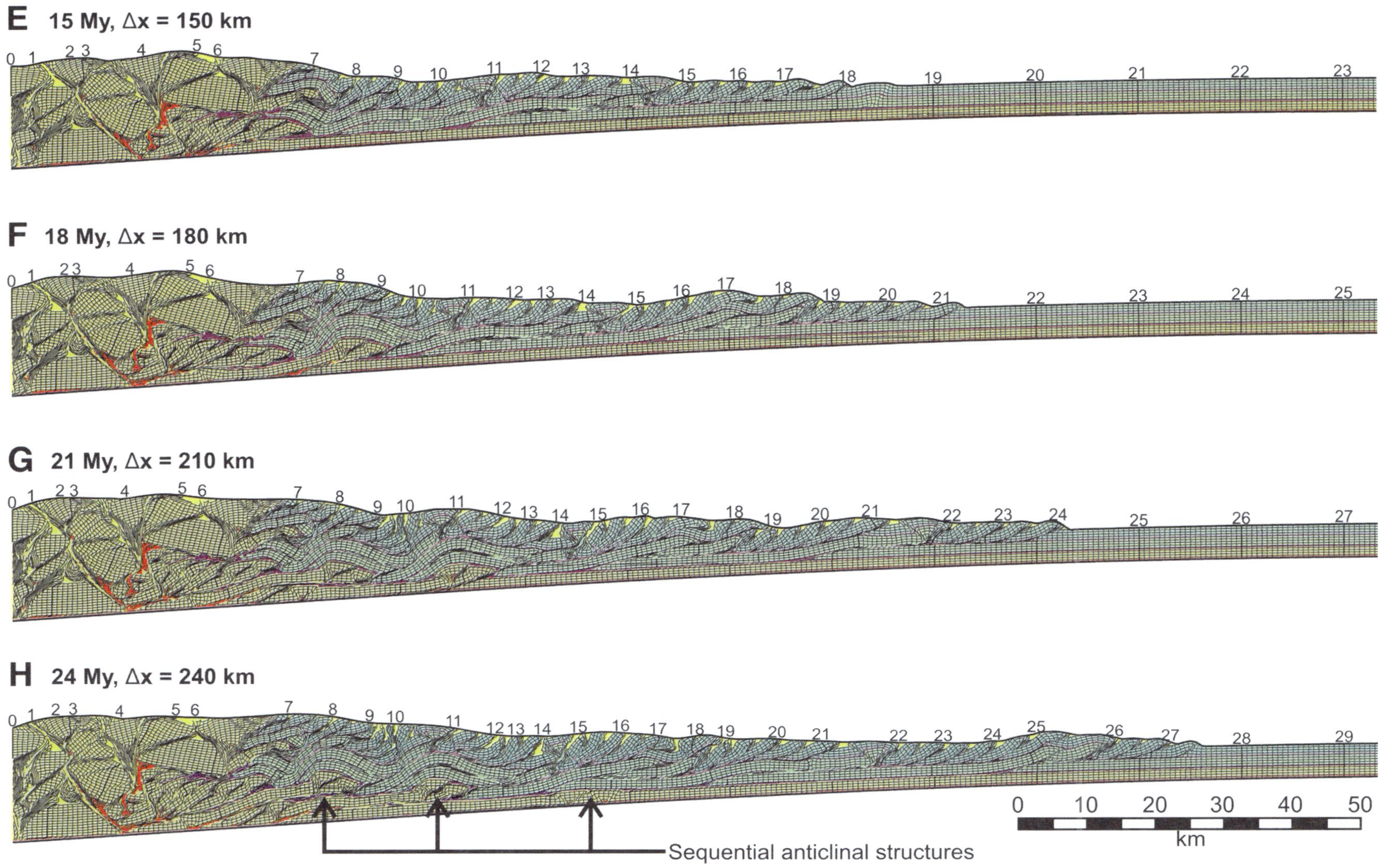

E 15 My, Δx = 150 km

F 18 My, Δx = 180 km

G 21 My, Δx = 210 km

H 24 My, Δx = 240 km

Sequential anticlinal structures

Figure 9 (*continued*).

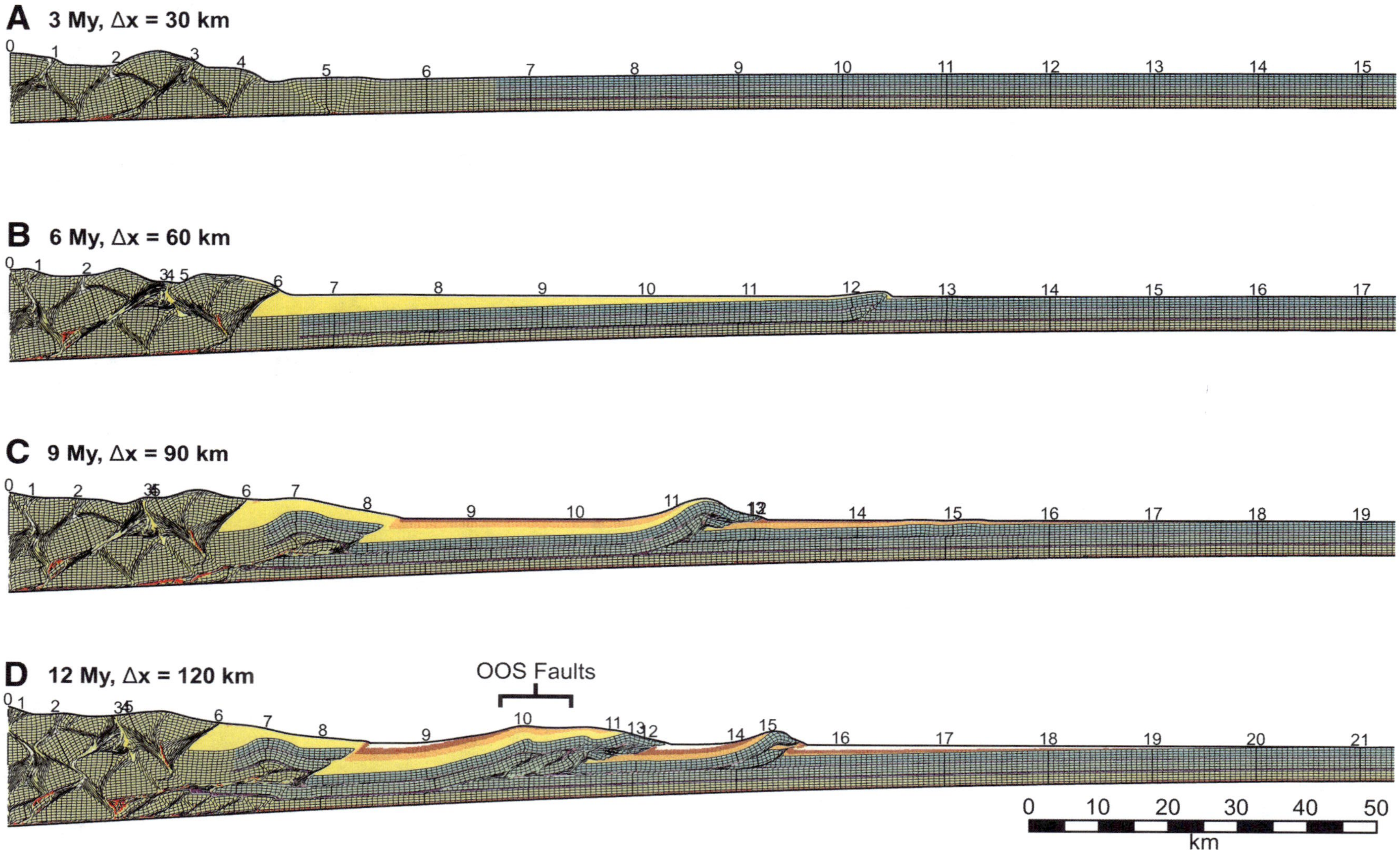

A 3 My, Δx = 30 km

B 6 My, Δx = 60 km

C 9 My, Δx = 90 km

D 12 My, Δx = 120 km

Figure 10 (*on this and following page*). Model 8 results (three-detachment model; erosion and sedimentation). See Figure 3 caption. OOS—out-of-sequence.

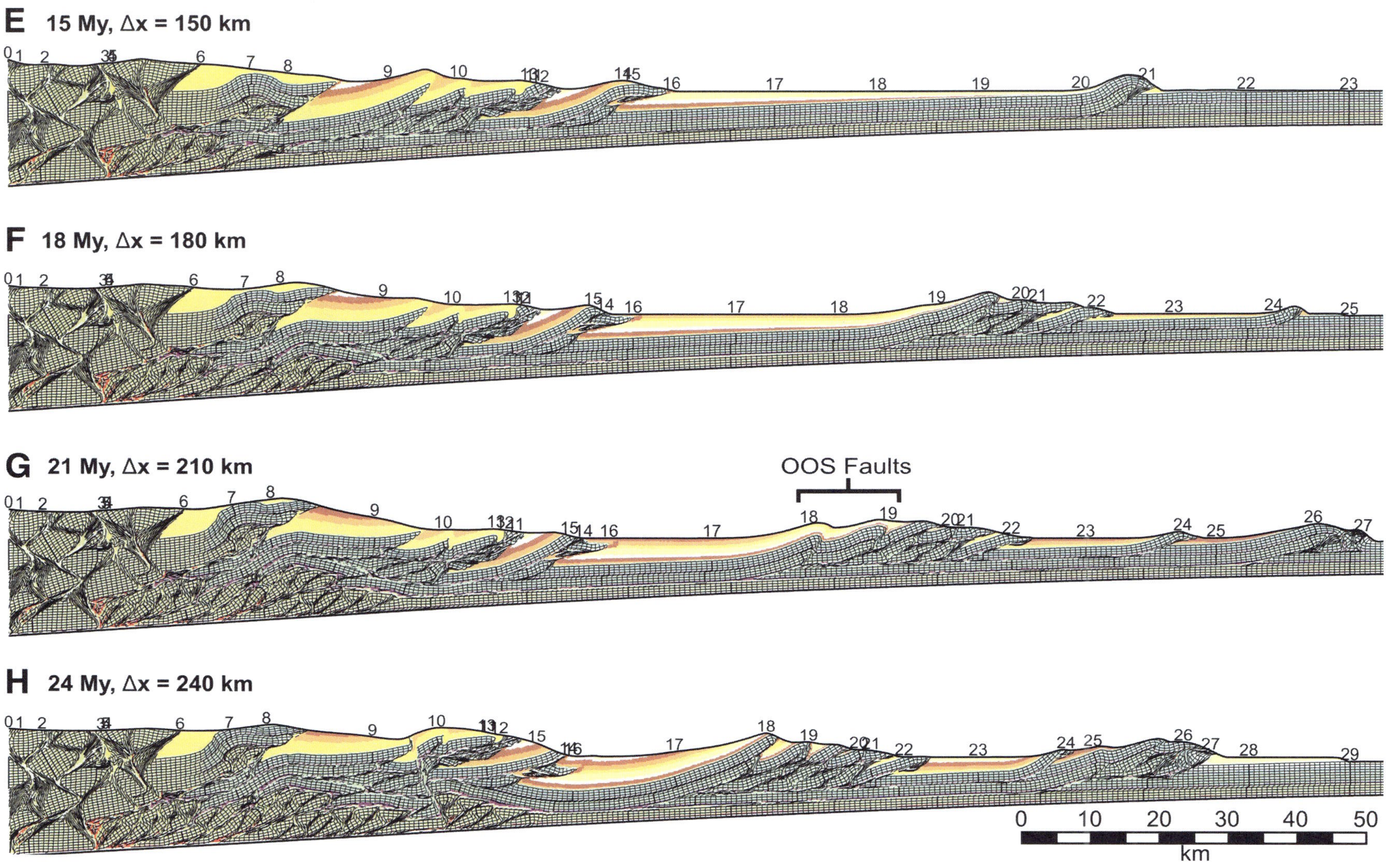

Figure 10 (continued).

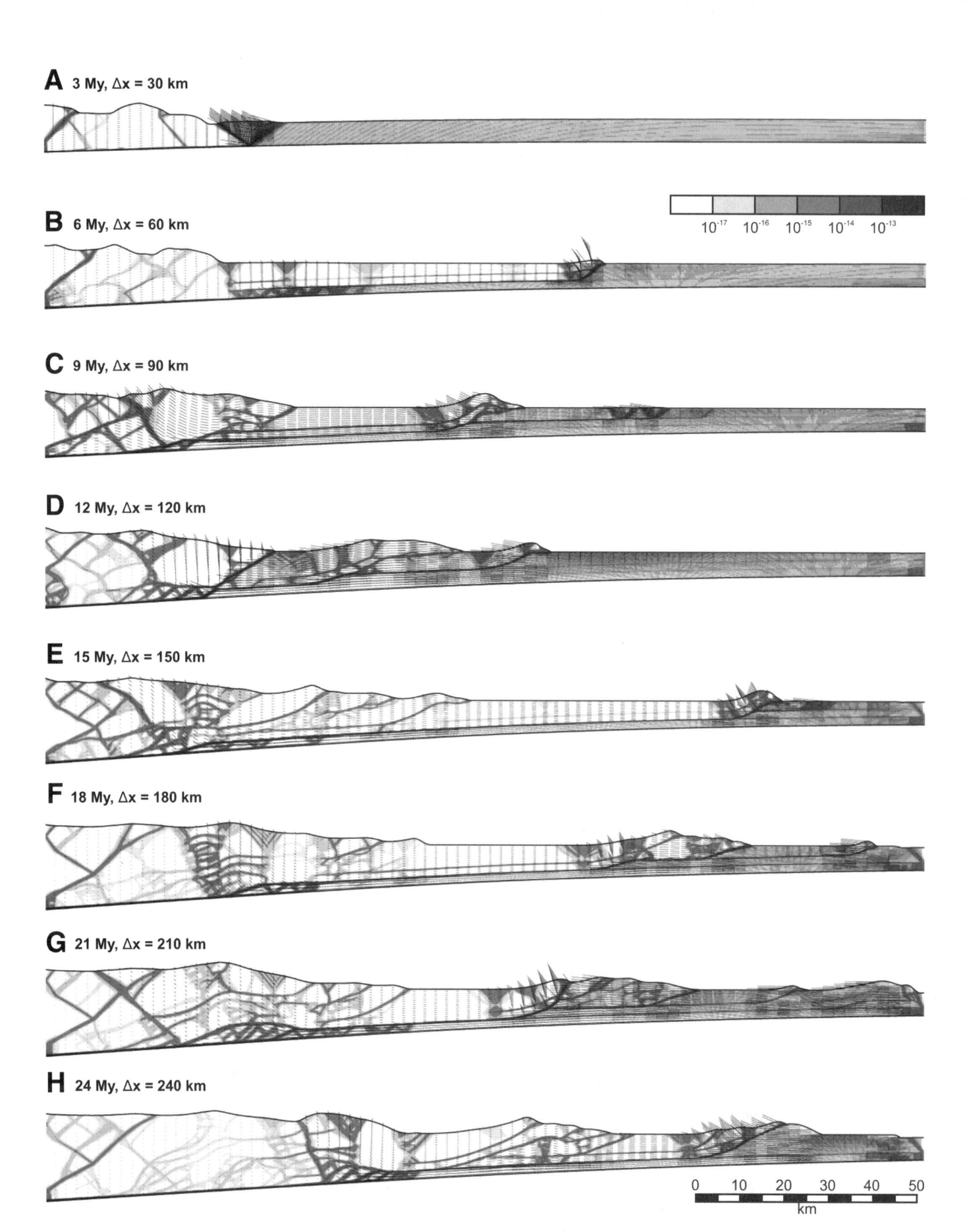

Figure 11. Instantaneous velocities and strain rates for model 8.

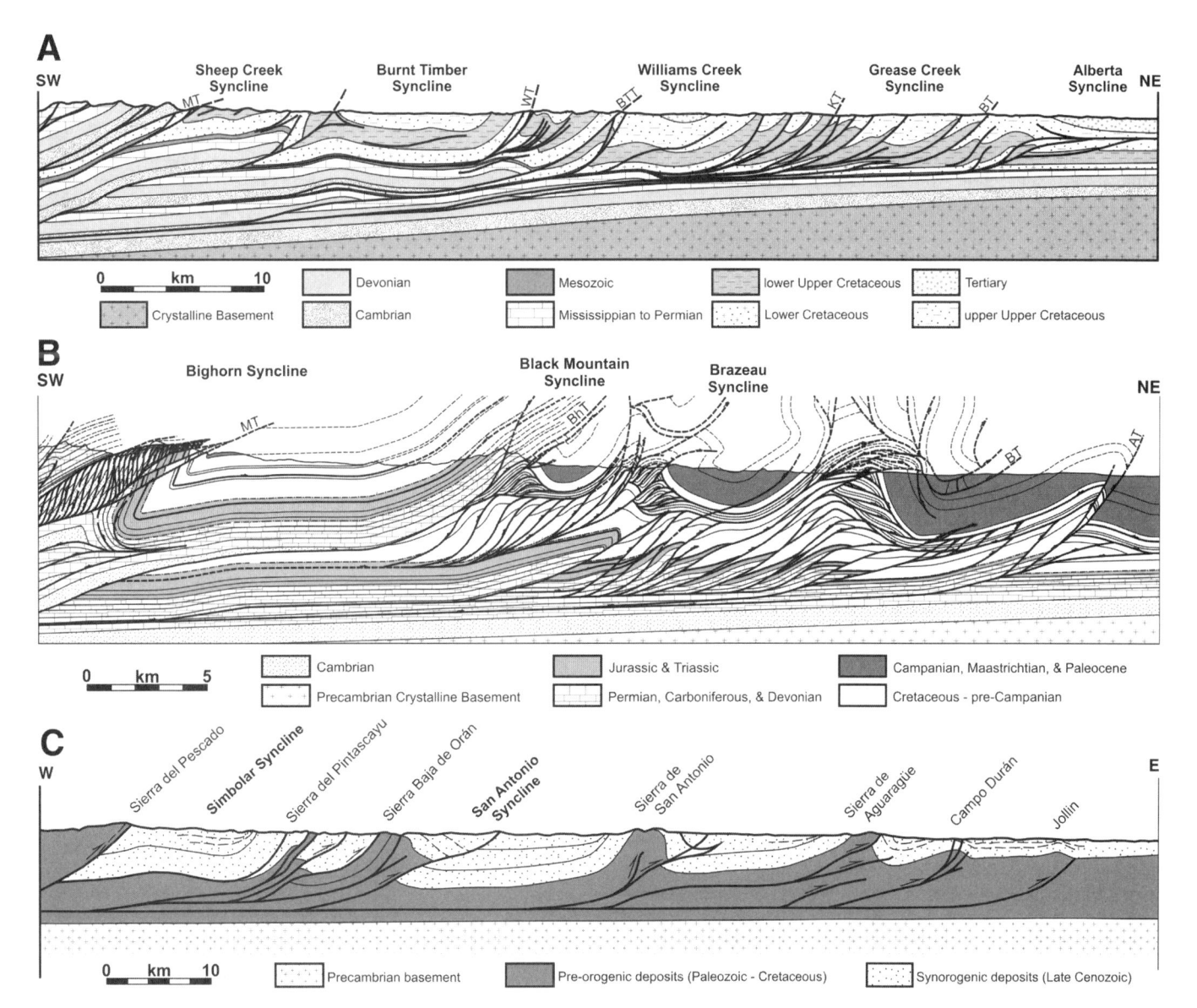

Figure 12. (A) Cross section across the southern Canadian Rocky Mountains foothills, ~100 km NW of Calgary, Alberta (after Ollerenshaw, 1978). (B) Cross section across the southern Canadian Rocky Mountains foothills, ~250 km NW of Calgary, Alberta (after Mountjoy et al., 2002). (C) Cross section across the Sub-Andean belt, northern Argentina, showing broad synclines occupied by synorogenic sediments documenting out-of-sequence and simultaneous motion on a series of thin-skinned thrusts (after Ramos et al., 2004). In all cases, the relatively undeformed synclines, separated by anticlinal culminations characterized by a concentration of contractional structures, are similar to those produced in numerical models incorporating syndeformational sedimentation. AT—Ancoma Thrust, BhT—Bighorn Thrust, BT—Brazeau Thrust, BTT—Burnt Timber Thrust, KT—Keystone Thrust, MT—McConnell Thrust, WT—Waiparous Thrust.

Initial formation of model thrust faults is generally in-sequence, but many are long-lived, reactivated, or out-of-sequence (as illustrated in Fig. 11). Balanced cross sections are commonly constructed assuming not only in-sequence development but with all fault motion essentially complete prior to motion on the next thrust; however, basic critical wedge concepts and the finite-element models confirm that these assumptions are simplistic. Although out-of-sequence or simultaneous motion in the southern Canadian Rocky Mountains is not obvious in Figure 1, Price (2001) has argued that map-scale relationships require simultaneous or temporally overlapping motion on a series of thrust faults in the front ranges, and palinspastic reconstruction suggests out-of-sequence motion on the McConnell thrust through the Mesozoic section (Price and Fermor, 1985). Similar conclusions for the southern Canadian Rocky Mountains as well as the Montana-Idaho-Utah-Wyoming thrust belt have been reached by Boyer (1992, 1995) and DeCelles and Mitra (1995). Models 6 and 8 suggest that some out-of-sequence features may be intimately related to feedback with surface processes.

Comparison to Whole-Wedge Solutions

The models develop overall wedge shapes broadly consistent with the critical wedge solutions of Chapple (1978), Davis et al. (1983), and Dahlen (1984), among others, with significant departures at sub–whole-wedge scales. These closed-form solutions, especially that of Dahlen (1984), have been applied to natural orogens, usually to constrain estimates of one (or more) unknown variable, such as the internal coefficient of friction, basal coefficient of sliding friction, or the internal or basal pore-fluid pressure ratio (e.g., Dahlen, 1990). The calculated values may be valid at the scale of the wedge, but they are unlikely to be representative of structures at smaller scales.

Our models imply that décollements underlying large thrust sheets with little internal deformation are very much weaker than the thrust sheets themselves. When this is not the case, model thrust sheets (unpublished models) shorten internally far more, and are translated far less, than commonly observed in nature in the shallow external portions of fold-and-thrust belts. In contrast, the ratio of bulk internal wedge strength to basal strength derived from whole-wedge solutions (e.g., Davis et al., 1983; Dahlen et al., 1984; Dahlen, 1984, 1990) is typically much smaller than the ratios of strong to weak materials in our numerical models that produce structures that look realistic. This result probably reflects the averaging of material properties at whole-wedge scales: the average strength of the basal décollement will be greater than the local strength within a particular detachment because the décollement ramps up through stronger layers, and the average strength of the wedge will be less than the local strength within an intact thrust sheet owing to weaker detachments and strain-weakened shear zones throughout the wedge.

In model 8, the average whole-wedge surface and basal slopes at 12, 15, and 18 m.y. are 1.86°, 1.47°, and 1.19°, and 5.82°, 3.96°, and 4.09°, respectively (Fig. 14A), where slopes were derived from least-squares linear fits to equally spaced points along the surfaces bounding the deformed wedges. Corresponding values of the internal friction angle of the wedge, φ_{eff}, and basal sliding friction angle, $\varphi_{b,eff}$, using the equations of Dahlen (1984), are shown for these time steps in Figure 14B as solid lines labeled "W-W." In Figure 14B, the range of weak detachment values, 3.5° and 1°, is indicated by the shaded horizontal bar, and the range of possible values of the strong, strain-softening layers, from 38° to 18°, is encompassed by the broad, stippled vertical box. None of the whole-wedge solution curves passes through the overlap area of these two zones.

Local slopes are shown in Figure 14A by dotted lines. The local upper wedge surface is a straight-line fit by eye to topographic low points, similar to the approach illustrated for model 1 (Fig. 3D). The local basal surface is the average décollement dip where it is confined to the gently curving detachment layers.

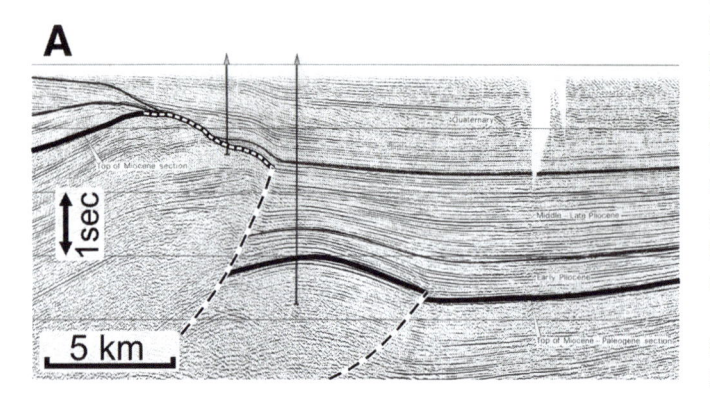

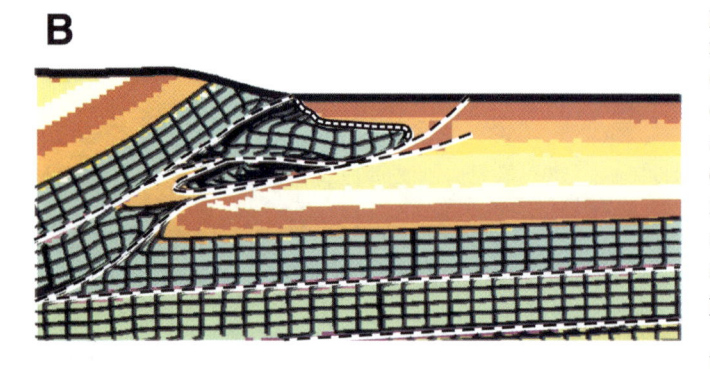

Figure 13. Relationships between contractional structures and syn-deformational sedimentation seen in (A) the Po Plain (Pieri, 1989), in comparison to portions of model 8 at (B) 21 m.y. and (C) 24 m.y.

Figure 14. Comparison between external wedge geometries (thick gray lines) of three time frames from model 8 (12, 15, and 18 m.y.) and critical wedge solutions. (A) Linear regression fits to the upper and basal surfaces of the three time frames (long-dashed lines), with slopes of the upper (α) and basal (β) surfaces indicated, along with slopes determined more locally (dotted lines) by fitting minimum topography on the upper surface and very long detachments on the basal surface, with local slope values in parentheses. The local slopes were determined near the wedge front, and for the 15 and 18 m.y. wedges, also near the wedge middle. (B) Plots of acceptable combinations of the effective angle of internal friction within the wedge, φ_{eff}, and the effective angle of basal sliding friction, $\varphi_{b,eff}$, for each of the three time frames, and for whole-wedge (W-W), wedge-front (W-F), and wedge-middle (W-M) values, using equations of Dahlen (1984).

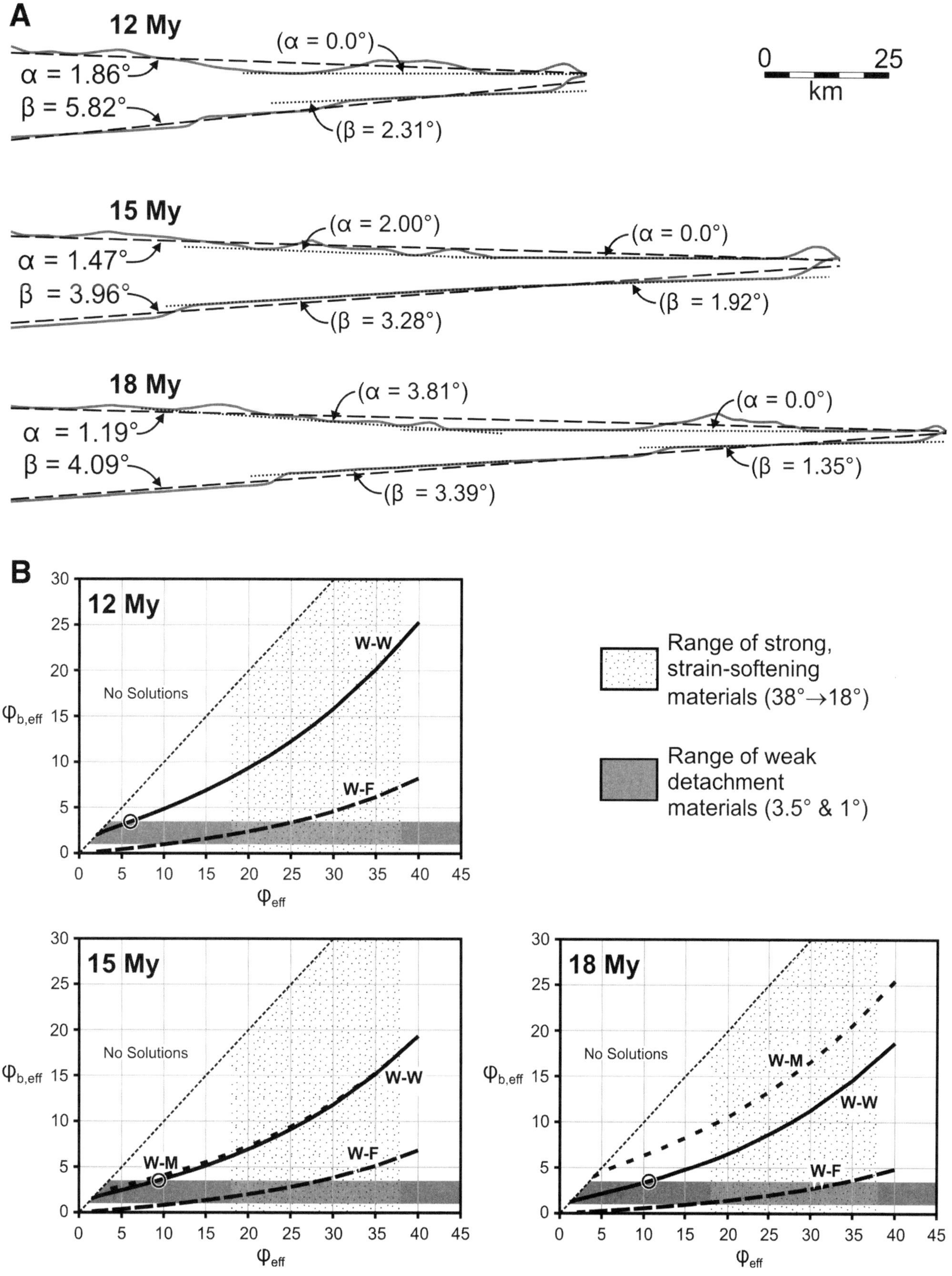

A
12 My
α = 1.86°
β = 5.82°
(α = 0.0°)
(β = 2.31°)
0
25
km
15 My
α = 1.47°
β = 3.96°
(α = 2.00°)
(α = 0.0°)
(β = 3.28°)
(β = 1.92°)
18 My
α = 1.19°
β = 4.09°
(α = 3.81°)
(α = 0.0°)
(β = 3.39°)
(β = 1.35°)
B
12 My
No Solutions
W-W
W-F
φb,eff
φeff
Range of strong, strain-softening materials (38°→18°)
Range of weak detachment materials (3.5° & 1°)
15 My
No Solutions
W-W
W-M
W-F
φb,eff
φeff
18 My
No Solutions
W-M
W-W
W-F
φb,eff
φeff

Local slopes, shown in parentheses in Figure 14A, were determined at the wedge-front (W-F) for all three time frames and at the wedge-middle (W-M) for the 15 and 18 m.y. frames.

In Figure 14B, the curves of $\varphi_{b,eff}$ versus φ_{eff} for all three wedge-front regions intersect the area of overlap between the ranges of model effective friction angles for the strong thick layers and the weak detachments. However, none of these curves has an acceptable solution for $\varphi_{b,eff}$ equal to 1°, which is the appropriate basal effective friction angle at the wedge fronts, where the wedges are detached on one or the other very weak internal detachment layer. This discrepancy between the finite-element model and the Dahlen (1984) solution may be due either to accretion of a finite-thickness layer at the wedge toe, which is not explicitly considered by Dahlen (1984), or to the possible supercritical state of these regions at these times (Fig. 11). The curves of $\varphi_{b,eff}$ versus φ_{eff} for the wedge-middle regions for the 12 and 15 m.y. time frames (Fig. 14B) are either nearly coincident with the whole-wedge curve (12 m.y.) or above it (15 m.y.).

Assuming that the whole-wedge strength of the basal décollement, which lies within detachment horizons with $\varphi_{b,eff}$ of 3.5° and 1°, is at the upper bound of 3.5° (perhaps reflecting the crosscutting of stronger layers), then Figure 14B (circles) shows that the corresponding maximum whole-wedge value of φ_{eff} is ~6–11°. This low value is a measure of the effect of the weak internal detachments and the strain-weakened shear zones on the bulk strength of the entire wedge. This range of values is similar to the harmonic mean value of φ_{eff} for model 8 for all layers above the basal detachment in the undeformed state, which ranges from 7.8° to 6.4° when the value for the thick strong layers is either 38° or 18°, respectively. The harmonic mean value of wedge strength may be useful for comparing layered numerical or analogue models to critical wedge solutions.

Using Equation 6, we can express bulk φ_{eff} in terms of pore-fluid pressure. Assuming φ of 38° for sedimentary rocks, then the pore-fluid pressure ratios ($\lambda = P_f/P$)) required to reduce the whole-wedge φ_{eff} to between 6° and 11° are 0.83–0.69. Using Equation A10 (see Appendix), these values expressed in terms of the familiar Hubbert-Rubey pore-pressure ratio, λ_{HR}, are 0.927–0.853. Similar pore pressure ratios were derived by Davis et al. (1983), Dahlen et al. (1984), and Dahlen (1984, 1990) for some deforming wedges (particularly accretionary wedges). Although excess pore-pressure values are widely observed in active and inactive fold-and-thrust belts, and undoubtedly contribute to weakening in the brittle domain, the pore-pressure ratios derived from critical wedge solutions are likely to overestimate natural values because very weak faults internal to the wedge are not explicitly accounted for in these analyses

Numerical model departures from ideal wedge shapes reflect the distributions of different-strength materials within the wedge and the fact that portions of the wedge are not at yield but are being transported passively. This is most evident in models involving syndeformational deposition (models 6 and 8). Flexural subsidence and foreland deposition can result in a critical or supercritical wedge segment, bypassing the internal deformation commonly required to produce critical taper. If the taper of a portion of a wedge beneath a piggyback basin is supercritical, then the overall wedge geometry will not generally agree with critical whole-wedge solutions. It follows that the evolution of this type of model must be examined as a dynamic system. The mere addition of surface processes to a kinematically constructed critical wedge in such a way that the critical taper is maintained (e.g., Dahlen and Suppe, 1988; Dahlen and Barr, 1989; Whipple and Meade, 2004) misses the fact that the dynamics also create supercritical wedge segments and that the feedback of surface processes to the mechanics may do the same.

Implications of Model Results for Fault Strength

The very weak and thin detachment layers specified in the numerical models host thin zones of very high shear strain that emulate fault zones. As noted previously, our unpublished models, which have significantly less strength contrast between the thin weak and thick strong layers, do not develop these features but rather deform with strain more equally distributed across all layers. Very weak detachments (relative to the thick layers) are essential in the present model formulation to produce results that look like natural fold-and-thrust belts.

The implication is that natural faults must not merely be weak, they must be *extremely* weak, relative to the strength of thrust sheets at the time of deformation. Stronger faults would result in a larger proportion of shortening within the wedge being absorbed *within* thrust sheets rather than being concentrated on relatively few large fault zones. Price (1988) considered the issue of "the mechanical paradox of large overthrusts," and suggested that the explanation may lie in the nature of individual local displacement events (e.g., earthquakes). He concluded that large overthrusts are "not (controlled) by the frictional resistance to sliding integrated over the entire fault surface" (p. 1898) because the entire fault surface does not fail simultaneously. However, if they are considered from the point of view of an energy or mechanical work argument, and results are integrated over *time* as well as space, large overthrusts in the brittle field must have extremely low effective strength.

Although elevated pore-fluid pressures contribute to weakening (Hubbert and Rubey, 1959), it is difficult to understand the extreme difference required in pore-fluid pressures between the fault zones and the thrust sheets that is necessary in order to maintain the high strength contrast. Such large differences imply that high pore-fluid pressures occur locally and are probably related to current deformation (strain rate) associated with fault motion.

Natural faults may evolve to be intrinsically weak by developing materials with low internal angles of friction (e.g., phyllosilicates) or materials dominated by pressure solution, which has a viscous rheology (e.g., Wojtal and Mitra, 1986; Bos and Spiers, 2002). More significant weakening occurs during dynamic slip on faults when displacement rates approach those of seismic events. These conditions result in a profound weakening (e.g., Di Toro et al., 2004), where frictional resistance extrapolates to virtually zero at seismic slip rates of ~1 m/s. Estimations of fault strength that integrate through many earthquake cycles using rate and state friction combined with dynamical weakening processes (such as thermally overpressured fault gouge or flash heating of contact asperities) seem to offer the best prospect for understanding both the dynamic and time-averaged weakness of faults (Rempel and Rice, 2006; Rice 2006).

Model Limitations

1. We are limited to solving the fold-and-thrust belt problem in two dimensions by the computational intensity of the equivalent three-dimensional problem and the lack of suitable software. We understand that fold-and-thrust belts develop fundamental three-dimensional structures and that surface processes are planform, but the best we can anticipate in the immediate future are low-resolution models in three dimensions.

2. The restrictive rigid backstop and basal boundary conditions used here can be eliminated by embedding equivalent models within a coarser-scale finite-element model encompassing the whole orogen.

3. The representation of fault zones by finite-thickness shear zones is a consequence of the continuum mechanics approach. The creeping, viscous-plastic, Stokes flow, finite-element formulation of Fullsack (1995) and Beaumont et al. (2004) does not allow discrete, one-dimensional surfaces. However, the Eulerian finite-element grid is sufficiently fine that the high shear strain zones that emulate faults are very thin (a few elements wide) in comparison to the entire model thickness. Higher-resolution models can improve this situation, but there are practical computer computational limits.

4. The representation of bedding-parallel detachments as very weak layers initially embedded in the models is a compromise, but it parallels approaches often taken in analogue models. Future efforts will involve materials that are initially strong but adopt fault-like (extremely weak) strain- and strain-rate–dependent properties, possibly coupled to dynamical pore-pressure evolution (Morency et al., in press).

5. The very simple surface processes models are sufficient to demonstrate their influence on structural style. Elsewhere, we will examine the sensitivity of structural style to more complex erosional and depositional models, such as those that allow multiple base levels, resulting in perched piggyback basins, and we will offer due consideration of the planform nature of surface processes.

6. Wedge thickening and associated heating through the brittle-ductile transition is not incorporated, but a coupled thermo-mechanical model will be discussed elsewhere. We anticipate that it will reduce boundary-condition effects related to the rigid backstop. See Williams et al. (1994) for an analysis applicable to whole-wedge critical wedges.

7. The initial undeformed model geometry was purposely chosen to be a uniform thickness, unlike the wedge shape of natural systems (characteristic 3 of Chapple [1978] noted already). The effect of tapered initial geometries will be investigated elsewhere.

CONCLUSIONS

Dynamic numerical finite-element models of fold-and-thrust belts illustrate[1] the influences of accretion of a finite-thickness layer, regional flexural isostasy, internal layering and strength contrast, strain softening, and syndeformational erosion and sedimentation. The finite-element approach used here leads to model structures similar to those in natural fold-and-thrust belts, suggesting that it may provide insights into how the natural structures developed. The finite-element models illustrate localization of strain, the associated scales of the much-less-deformed regions (thrust sheets) between shear zones, and the ways in which these features are transported and stacked. The overall results demonstrate the importance of treating thin-skinned fold-and-thrust belts as dynamical coupled systems, where continuous feedback among system components shapes the details of internal structure and surface expression. Our current understanding of these models is at the level of the kinematic organization of fold-and-thrust belts as systems. The results can be used to test the predicted length scales of, for example, thrust sheets based on folding and other instability analyses, which may lead to an improved understanding of the mechanics.

For the simplest models (Figs. 3 and 4), composed of a thick strong layer overlying a thin, very weak basal detachment, the fundamental deformation units (the building blocks of the wedge) are pop-up structures that scale according to the thickness of the strong layer at the accretionary front. The partitioning of deformation into rigid triangular pop-ups bounded by shear zones occurs in the absence of strain softening. By implication, similar structures in analogue sandbox experiments would develop even if the sand did not strain harden or soften.

With flexural isostatic compensation, the simplest model (Fig. 4) has similar pop-up characteristics and a greater tendency to create thrust sheets, most likely because isostatically compensated wedges are thicker, and more shortening is required to achieve critical wedge conditions. Even these simple wedges show significant sub–wedge-scale departures from a uniform taper predicted by critical wedge solutions

Models composed of two equal, thick strong layers, each with a thin, very weak underlying detachment, demonstrate that wedges of this type develop thrust sheets in both the upper and lower layers, with a tendency for longer thrust sheets in the lower layer (Fig. 5). These lower-layer thrust sheets become even longer and have less internal deformation when the thick strong layers can strain soften (Fig. 6).

Significant syndeformational slope-dependent erosion of the surface of the two-layer model with strain softening enhances the development of pop-ups in the upper layer but does not change the overall character of the deformation (Fig. 7). However, the wedge is profoundly modified by the combined effects of slope-dependent erosion and rapid filling of the flexural foreland basin

---

[1]GSA Data Repository Item 2007285, GIF animations of all eight models, is available on request from Documents Secretary, GSA, P.O. Box 9140, Boulder, CO 80301-9140, USA, or editing@geosociety.org, at www.geosociety.org/pubs/ft2007.htm.

with sediment (Fig. 8). Model 6 develops piggyback basins by sequentially accreting the undeformed proximal foreland basin because critical wedge conditions are achieved without the need to thicken the strata.

Models with three thick, strong, strain-softening layers separated by very weak detachments develop structures that most resemble our reference natural example, the southern Canadian Rocky Mountains. In particular, models with slope-dependent erosion and rapidly filled basins develop multiple piggyback basins (Fig. 10), which, when eroded, resemble synclines in the foothills of the southern Canadian Rocky Mountains (Fig. 12).

APPENDIX

For a plane-strain, two-dimensional case, where σ_1 and σ_3 are the maximum and minimum values of the principal stresses, the mean stress, P, is:

$$P = \frac{\sigma_1 + \sigma_3}{2}. \tag{A1}$$

In a Mohr-Coulomb material, parameterized by a cohesion value, C_0, and an angle of internal friction, φ, failure may occur if the effective mean stress is reduced due to an increase in pore-fluid pressure. At failure, in a contractional fold-and-thrust belt setting, the magnitude of the pore-fluid pressure, P_f, is:

$$P_f = \frac{\sigma_1 + \sigma_3}{2} - \frac{\sigma_1 - \sigma_3}{2\sin\varphi} + \frac{C_0}{\tan\varphi}. \tag{A2}$$

In our models, we define the pore-fluid pressure ratio relative to the mean stress:

$$\lambda = \frac{P_f}{P}. \tag{A3}$$

The common Hubbert-Rubey pore pressure ratio is generally defined as:

$$\lambda_{HR} = \frac{P_f}{P_{overburden}}, \tag{A4}$$

where $P_{overburden}$ is the weight of the overburden, equal to $\rho g d$, where ρ, g, and d are the mean density, gravitational acceleration, and depth, respectively. The ratio of λ to λ_{HR} is then:

$$\frac{\lambda}{\lambda_{HR}} = \frac{P_{overburden}}{P}. \tag{A5}$$

The minimum principal stress, σ_3, can be approximated as equal to the overburden weight:

$$\sigma_3 \approx \rho g d = P_{overburden}. \tag{A6}$$

An expression for the maximum principal stress, σ_1, is derived by first substituting Equations A1 and A2 into A3:

$$\lambda\left(\frac{\sigma_1 + \sigma_3}{2}\right) = \frac{\sigma_1 + \sigma_3}{2} - \frac{\sigma_1 - \sigma_3}{2\sin\varphi} + \frac{C_0}{\tan\varphi}, \tag{A7}$$

and then solving for σ_1:

$$\sigma_1 = \frac{\sigma_3(1 - \lambda + \csc\varphi) + 2C_0\cot\varphi}{-1 + \lambda + \csc\varphi}. \tag{A8}$$

Using Equations A6 and A8 in A5, we find:

$$\frac{\lambda}{\lambda_{HR}} = \frac{\rho g d(-1 + \lambda + \csc\varphi)}{\rho g d \csc\varphi + C_0 \cot\varphi}. \tag{A9}$$

Using Equation A9, the Hubbert-Rubey pore-pressure ratio, λ_{HR}, can be expressed in terms of our expression of the ratio, λ, such that:

$$\lambda_{HR} = \lambda\left(\frac{\rho g d + C_0 \cos\varphi}{\rho g d(1 - (1 - \lambda)\sin\varphi)}\right). \tag{A10}$$

Equation A10 can be solved iteratively for λ given a value for λ_{HR}.

Note that when the cohesion is either zero or of negligible magnitude, Equations A9 and A10 are independent of depth. Although the magnitude of λ_{HR} corresponding to saturated (i.e., hydrostatic) conditions is easily determined (ratio of water and rock densities), under overpressured conditions, the correspondence between pore-fluid pressure and mean stress, as in Equation A3, is more intuitive.

For a hydrostatically pressured case, where the average density of the fold-and-thrust belt is 2300 km/m³, the value of λ_{HR} is 1000/2300, or 0.435. When the cohesion is negligible or zero, the corresponding value of λ is 0.278, assuming φ is 30°. In our formulation of effective angle of internal friction (Equation 2, main text), a dry value of 30° would reduce to 21° under these conditions.

ACKNOWLEDGMENTS

We thank Ray Price for many years of inspiration and motivation imparted through his Rocky Mountain maps, articles, lectures, and field trips, and through numerous thoughtful and encouraging discussions. We thank Steve Boyer, Ray Fletcher, and Gautam Mitra for detailed and insightful formal reviews, and Kirk Osadetz and Margot McMechan for internal Geological Survey of Canada reviews. Beaumont acknowledges the support of the Canada Research Chair in Geodynamics, an IBM-Shared University Research Grant, and an NSERC (Natural Science and Engineering Research Council) Discovery Grant. This is Geological Survey of Canada/Earth Sciences Sector contribution 20060607.

REFERENCES CITED

Beaumont, C., 1981, Foreland basins: Geophysical Journal of the Royal Astronomical Society, v. 65, p. 291–329.
Beaumont, C., Fullsack, P., and Hamilton, J., 1992, Erosional control of active compressional orogens, *in* McClay, K.R., ed., Thrust Tectonics: New York, Chapman and Hall, p. 1–18.
Beaumont, C., Kooi, H., and Willett, S., 2000, Coupled tectonic-surface process models with applications to rifted margins and collisional orogens, *in* Summerfield, M.A., ed., Geomorphology and Global Tectonics: Chichester, UK, John Wiley & Sons, p. 29–55.
Beaumont, C., Jamieson, R.A., Nguyen, M.H., and Medvedev, S., 2004, Crustal channel flows: 1. Numerical models with applications to the tectonics of the Himalayan-Tibetan orogen: Journal of Geophysical Research, v. 109, B06406, 29 p.
Bos, B., and Spiers, C.J., 2002, Frictional-viscous flow of phyllosilicate-bearing fault rock: Microphysical model and implications for crustal strength profiles: Journal of Geophysical Research, v. 107, doi: 10.1029/2001JB000301.
Boyer, S.E., 1992, Geometrical evidence for synchronous thrusting in the southern Alberta and northwest Montana thrust belts, *in* McClay, K.R., ed., Thrust Tectonics: New York, Chapman and Hall, p. 377–390.
Boyer, S.E., 1995, Sedimentary basin taper as a factor controlling the geometry and advance of thrust belts: American Journal of Science, v. 295, p. 1220–1254.
Braun, J., and Sambridge, M., 1997, Modelling landscape evolution on geological time scales: A new method based on irregular spatial discretization: Basin Research, v. 9, p. 27–52, doi: 10.1046/j.1365-2117.1997.00030.x.
Chapple, W.M., 1978, Mechanics of thin-skinned fold-and-thrust belts: Geological Society of America Bulletin, v. 89, p. 1189–1198, doi: 10.1130/0016-7606(1978)89<1189:MOTFB>2.0.CO;2.
Dahlen, F.A., 1984, Noncohesive critical Coulomb wedges: An exact solution: Journal of Geophysical Research, v. 89, p. 10,125–10,133.
Dahlen, F.A., 1988, Mechanical energy budget of a fold-and-thrust belt: Nature, v. 331, p. 335–337, doi: 10.1038/331335a0.
Dahlen, F.A., 1990, Critical taper model of fold-and-thrust belts and accretionary wedges: Annual Review of Earth and Planetary Sciences, v. 18, p. 55–99, doi: 10.1146/annurev.ea.18.050190.000415.
Dahlen, F.A., and Barr, T.D., 1989, Brittle frictional mountain building: 1. Deformation and mechanical energy budget: Journal of Geophysical Research, v. 94, p. 3906–3922.
Dahlen, F.A., and Suppe, J., 1988, Mechanics, growth, and erosion of mountain belts, *in* Clark, S.P., Jr., Burchfiel, B.C., and Suppe, J., eds., Processes in Continental Lithospheric Deformation: Geological Society of America Special Paper 218, p. 161–178.
Dahlen, F.A., Suppe, J., and Davis, D., 1984, Mechanics of fold-and-thrust belts and accretionary wedges: Cohesive Coulomb theory: Journal of Geophysical Research, v. 89, p. 10,087–10,101.
Davis, D.M., and Engelder, T., 1985, The role of salt in fold-and-thrust belts: Tectonophysics, v. 119, p. 67–89.
Davis, D., Suppe, J., and Dahlen, F.A., 1983, Mechanics of fold-and-thrust belts and accretionary wedges: Journal of Geophysical Research, v. 88, p. 1153–1172.
DeCelles, P.G., and Mitra, G., 1995, History of the Sevier orogenic wedge in terms of critical taper models, northeast Utah and southwest Wyoming: Geological Society of America Bulletin, v. 107, p. 454–462, doi: 10.1130/0016-7606(1995)107<0454:HOTSOW>2.3.CO;2.
Di Toro, G., Goldsby, D.L., and Tullis, T.E., 2004, Friction falls towards zero in quartz rock as slip velocity approaches seismic rates: Nature, v. 427, p. 436–439, doi: 10.1038/nature02249.
Fullsack, P., 1995, An arbitrary Lagrangian-Eulerian formulation for creeping flows and its application in tectonic models: Geophysical Journal International, v. 120, p. 1–23, doi: 10.1111/j.1365-246X.1995.tb05908.x.
Goff, D.F., Wiltschko, D.V., and Fletcher, R.C., 1996, Décollement folding as a mechanism for thrust-ramp spacing: Journal of Geophysical Research, v. 101, p. 11,341–11,352, doi: 10.1029/96JB00172.
Hacquebard, P.A., 1977, Chapter 3. Rank of coal as an index of organic metamorphism for oil and gas in Alberta, *in* Deroo, G., Powell, T.G., Tissot, B., and McCrossan, R.G., eds., The Origin and Migration of Petroleum in the Western Canadian Sedimentary Basin: Geological Survey of Canada Bulletin, v. 262, p. 11–22.
Hill, R., 1950, The Mathematical Theory of Plasticity: Oxford, Oxford University Press, 355 p.
Hubbert, M.K., and Rubey, W.W., 1959, Role of fluid pressure in mechanics of overthrust faulting: Geological Society of America Bulletin, v. 70, p. 115–166, doi: 10.1130/0016-7606(1959)70[115:ROFPIM]2.0.CO;2.
Huismans, R.S., and Beaumont, C., 2003, Symmetric and asymmetric lithospheric extension: Relative effects of frictional-plastic and viscous strain softening: Journal of Geophysical Research, v. 108, no. B10, p. 2496, doi: 10.1029/2002JB002026
Leturmy, P., Mugnier, J.L., Vinour, P., Baby, P., Colletta, B., and Chabron, E., 2000, Piggyback basin development above a thin-skinned thrust belt with two detachment levels as a function of interactions between tectonic and superficial mass transfer: The case of the Subandean zone (Bolivia): Tectonophysics, v. 320, p. 45–67, doi: 10.1016/S0040-1951(00)00023-8.
Liu, H., McClay, K.R., and Powell, D., 1992, Physical models of thrust wedges, *in* McClay, K.R., ed., Thrust Tectonics: New York, Chapman and Hall, p. 71–81.
Morency, C., Huismans, R.S., Beaumont, C., and Fullsack, P., 2007, A numerical model for coupled fluid flow and matrix deformation with applications to disequilibrium compaction and delta stability: Journal of Geophysical Research (in press).
Mountjoy, E.W., Windh, J., Price, R.A., and Douglas, R.J.W., 2002, Geology, George Creek, Alberta: Geological Survey of Canada Map 1990A, scale 1:50,000, 2 sheets.
Nurkowski, J.R., 1984, Coal quality, coal rank variation and its relation to reconstructed overburden, Upper Cretaceous and Tertiary Plains coals, Alberta: American Association of Petroleum Geologists Bulletin, v. 68, p. 285–295.
Ollerenshaw, N.C., 1978, Geology, Calgary, Alberta–British Columbia: Geological Survey of Canada Map 1457A, scale 1:250,000, 2 sheets.
Ori, G.G., and Friend, P.F., 1984, Sedimentary basins formed and carried piggyback on active thrust sheets: Geology, v. 12, p. 475–478, doi: 10.1130/0091-7613(1984)12<475:SBFACP>2.0.CO;2.
Osborn, G., Stockmal, G., and Haspel, R., 2006, Emergence of the Canadian Rockies and adjacent plains: A comparison of physiography between end-of-Laramide time and the present day: Geomorphology, v. 75, p. 450–477, doi: 10.1016/j.geomorph.2005.07.032.
Pieri, M., 1989, Three seismic profiles through the Po Plain, *in* Bally, A.W., ed., Atlas of Seismic Stratigraphy, Volume 3: American Association of Petroleum Geologists (AAPG) Studies in Geology No. 27, p. 90–110.
Pivnik, D.A., and Khan, M.J., 1996, Transition from foreland- to piggyback-basin deposition, Plio-Pleistocene Upper Siwalik Group, Shinghar Range, NW Pakistan: Sedimentology, v. 43, p. 631–646.
Price, R.A., 1973, Large-scale gravitational flow of supracrustal rocks, southern Canadian Rockies, *in* Jong, K.A., and Scholten, R., eds., Gravity and Tectonics: New York, Wiley, p. 491–502.
Price, R.A., 1981, The Cordilleran foreland thrust and fold belt in the southern Canadian Rocky Mountains, *in* McClay, K.R., and Price, N.J., eds., Thrust and Nappe Tectonics: Geological Society of London Special Publication 9, p. 427–448.
Price, R.A., 1988, The mechanical paradox of large overthrusts: Geological Society of America Bulletin, v. 100, p. 1898–1908, doi: 10.1130/0016-7606(1988)100<1898:TMPOLO>2.3.CO;2.
Price, R.A., 2001, An evaluation of models for the kinematic evolution of thrust and fold belts: Structural analysis of a transverse fault zone in the Front Ranges of the Canadian Rockies north of Banff, Alberta: Journal of Structural Geology, v. 23, p. 1079–1088, doi: 10.1016/S0191-8141(00)00177-2.
Price, R.A., and Fermor, P.R., 1985, Structure Section of the Cordilleran Foreland Thrust and Fold Belt West of Calgary, Alberta: Geological Survey of Canada Paper 84-14, 1 sheet.
Ramos, V.A., Zapata, T., Cristallini, E., and Introcaso, A., 2004, The Andean thrust system—Latitudinal variations in structural style and orogenic shortening, *in* McClay, K.R., ed., Thrust Tectonics and Hydrocarbon Systems: American Association of Petroleum Geologists (AAPG) Memoir 82, p. 30–50.

Rempel, A.W., and Rice, J.R., 2006, Thermal pressurization and onset of melting in fault zones: Journal of Geophysical Research, v. 111, B09314, 18 p., doi: 10.1029/2006JB004314.

Rice, J.R., 2006, Heating and weakening of faults during earthquake slip: Journal of Geophysical Research, v. 111, B05311, 29 p., doi: 10.1029/2005JB004006.

Roe, G.H., Montgomery, D.R., and Hallet, B., 2003, Orographic precipitation and the relief of mountain ranges: Journal of Geophysical Research, v. 108, no. B6, 2315, 11 p., doi: 10.1029/2001JB001521.

Stockmal, G.S., 1983, Modeling of large-scale accretionary wedge deformation: Journal of Geophysical Research, v. 88, p. 8271–8287.

Stockmal, G.S., Lebel, D., McMechan, M.E., and MacKay, P.A., 2001, Structural style and evolution of the triangle zone and external foothills, southwestern Alberta: Implications for thin-skinned thrust-and-fold belt mechanics: Bulletin of Canadian Petroleum Geology, v. 49, p. 472–496, doi: 10.2113/49.4.472.

Whipple, K.X., and Meade, B.J., 2004, Controls on the strength of coupling among climate, erosion, and deformation in two-sided, frictional orogenic wedges at steady state: Journal of Geophysical Research, v. 109, F01011, 24 p., doi: 10.1029/2003JF000019.

Willett, S.D., 1999, Orogeny and orography: The effects of erosion on the structure of mountain belts: Journal of Geophysical Research, v. 104, p. 28,957–28,981.

Willett, S.D., Slingerland, R., and Hovius, N., 2001, Uplift, shortening, and steady state topography in active mountain belts: American Journal of Science, v. 301, p. 455–485.

Williams, C.A., Connors, C., Dahlen, F.A., Price, E.J., and Suppe, J., 1994, Effect of the brittle-ductile transition on the topography of compressive mountain belts on Earth and Venus: Journal of Geophysical Research, v. 99, p. 19,947–19,974, doi: 10.1029/94JB01407.

Wojtal, S., and Mitra, G., 1986, Strain hardening and strain softening in fault zones from foreland thrusts: Geological Society of America Bulletin, v. 97, p. 674–687, doi: 10.1130/0016-7606(1986)97<674:SHASSI>2.0.CO;2.

MANUSCRIPT ACCEPTED BY THE SOCIETY 22 MARCH 2007

Lithospheric-scale structures across the Alaskan and Canadian Cordillera: Comparisons and tectonic implications

P.T.C. Hammer
R.M. Clowes*
Department of Earth and Ocean Sciences, University of British Columbia, Vancouver, British Columbia, Canada V6T 1Z4

ABSTRACT

The North American Cordillera in Canada and Alaska has been investigated through coincident and coordinated geological, geochemical, and geophysical studies along three corridors: (1) the Lithoprobe Southern Cordillera transect, (2) the ACCRETE and Lithoprobe Slave–Northern Cordillera Lithospheric Evolution (SNORCLE) transects, and (3) the Trans-Alaska Crustal Transect (TACT) program. Seismic-reflection and refraction experiments are integral to these studies and contribute to lithospheric-scale models that enable orogen-parallel comparisons to be made. Primary observations include three points: (1) Outward-verging, crustal-scale décollements are characteristic features of the orogen. The three trans-Cordillera transects exhibit decoupling zones that dip away from the Foreland belt to the lowermost crust or Moho. These inboard décollements above an indentor or cratonic backstop extend 500–600 km downdip in the Canadian Cordillera and 250 km downdip in the Alaskan Cordillera. The active subduction megathrusts form opposing décollements and generate structures in the overriding crust that mirror those above the facing intracrustal ramps. (2) Oblique convergence resulting in significant transpressional, transtensional, and orogen-parallel motion has yielded four major transcurrent fault systems that penetrate the entire crust and are associated with tectonic boundaries. (3) Beneath the entire Canadian Cordillera, the Moho remains remarkably flat and shallow despite the variety of ages, terrane compositions, and tectonomagmatic deformations spanned by the seismic corridors. These observations indicate that the Moho is an active, near-solidus, deformation zone that represents a young, re-equilibrated crust-mantle boundary. Beneath Alaska, crustal roots are observed over the subduction zone and at the indentor wedge, but the interior of the orogen also exhibits thin crust.

Keywords: Cordillera, lithosphere, Moho, seismic reflection, seismic refraction.

---

*Also at the Lithoprobe Secretariat, University of British Columbia, Vancouver, Canada.

Hammer, P.T.C., and Clowes, R.M., 2007, Lithospheric-scale structures across the Alaskan and Canadian Cordillera: Comparisons and tectonic implications, *in* Sears, J.W., Harms, T.A., and Evenchick, C.A., eds., Whence the Mountains? Inquiries into the Evolution of Orogenic Systems: A Volume in Honor of Raymond A. Price: Geological Society of America Special Paper 433, p. 99–116, doi: 10.1130/2007.2433(05). For permission to copy, contact editing@geosociety.org. ©2007 The Geological Society of America. All rights reserved.

INTRODUCTION

The Cordillera of Alaska and western Canada has been traversed by three large multidisciplinary transects and smaller adjacent programs (Fig. 1). Carried out over the last two decades, these transects integrate coincident and coordinated geological, geochemical, and geophysical studies. Seismic-reflection and refraction experiments have contributed profiles of crustal and lithospheric mantle structure that provide a unifying framework for integrating the investigations and for advancing models of orogen structure and evolution. The resulting lithospheric-scale models of orogen structure allow along-strike comparisons of large-scale structures, the style of accretion and deformation, and the influence of postorogenic tectonics.

The orogen-crossing transects (Fig. 1) include: (1) the Lithoprobe Southern Cordillera transect, which crosses the Juan de Fuca subduction zone and extends eastward for 1000 km across the orogen at 48°–50°N; (2) the overlapping ACCRETE and Lithoprobe Slave–Northern Cordilleran Lithospheric Evolution (SNORCLE) transects, which cross the strike-slip Queen Charlotte–Fairweather fault and extend northeastward to the Archean core of North America (1800 km; 54°–63°N); and (3) the Trans-Alaska Crustal Transect (TACT) and Aleutian seismic experiment, which cross the Aleutian subduction zone and Yakutat collision zone. TACT extends northward to the Arctic Ocean (1500 km; 145°–150°W). With primary interpretations completed on all three transects, comparisons can be made and extended across the entire orogen. Our comparisons focus on: (1) orogen and margin geometry and structure, (2) crustal thickness and Moho evolution, (3) along-strike deformation and crustal-penetrating faults, (4) upper-mantle heterogeneities and lithospheric thickness, and (5) tectonic development.

GEOTECTONIC OVERVIEW

The North American Cordillera exhibits considerable along-strike variation in structure and tectonic history. Our focus is on the northern portion of the orogen, which extends through Alaska and western Canada. Based on orientation, it can be divided into two sections: a linear, north-northwest–striking component that includes western Canada and the Alaskan panhandle, which we designate the Canadian Cordillera; and the west-striking component made up of much of Alaska, which we designate the Alaskan Cordillera (Fig. 1). An overview of the North American Cordillera is provided by Oldow et al. (1989). Descriptions of the geology and tectonic development of the Canadian Cordillera can be found in Gabrielse and Yorath (1991a, 1991b), Gabrielse et al. (1991), and Monger (1993). The Alaskan Cordillera is discussed in Grantz et al. (1991) and Plafker and Berg (1994a, 1994b). Here, we provide a summary that describes the general tectonic sequence controlling the formation of the orogen and highlights the similarities and differences among the three transect regions.

The northwestern edge of the Laurentian craton developed as a passive margin since the Paleoproterozoic. Through multiple episodes of rifting and compression between 1.74 and 0.38 Ga, a westerly dipping, passive-margin (meta)sedimentary sequence was formed (e.g., Gabrielse et al., 1991; Sears and Price, 2000; Thorkelson et al., 2001, 2005). Since the early Mesozoic, the orogen has developed through a complex, and still incompletely understood, sequence of subduction, oblique terrane accretion, and orogen-parallel deformation. The Canadian and Alaskan Cordillera inspired development of the terrane concept (e.g., Coney et al., 1980; Monger et al., 1982) and is a global archetype for accretionary growth of continents. Subduction, oblique collision, and transcurrent faulting continue along the margins of the Canadian and Alaskan Cordillera.

Canadian Cordillera

The Canadian Cordilleran segment evolved through two major phases of accretion that divide the upper crust of the orogen into three distinct lithotectonic zones: (1) the outboard accreted terranes (Outer terranes and Insular superterrane), (2) the inboard accreted terranes (Intermontane superterrane), and (3) the deformed ancestral North American margin (Foreland belt), which includes Archean (cratonic), Proterozoic, and Paleozoic margin crust (Figs. 1 and 2 [on loose insert accompanying this volume and in the Data Repository[1]]) (Gabrielse et al., 1991). The complex sequence of accretion involving the offshore arcs and accetionary complexes began with the accretion of Quesnellia by 183 Ma (Ghosh, 1995). The other Intermontane terranes (Cache Creek, Stikinia, Dorsey, and Slide Mountain terranes) followed as they were thrust sequentially over the ancestral North American margin (e.g., Monger and Price, 2002). This resulted in the passive-margin sequences being folded, detached from crystalline basement, and thrust onto the edge of ancestral North America to form the foreland fold-and-thrust belt, which occupies the easternmost Cordillera. The Omineca belt, which lies just west of the Foreland belt, represents the metamorphic-plutonic suture zone of this collision and consists of the westernmost section of ancestral North America's miogeocline (Figs. 1 and 2).

Geological evidence indicates that the eastern accreted terranes are thin flakes that occupy only a portion of the crust (e.g., Gabrielse, 1985). This hypothesis is supported throughout the Canadian Cordillera by the seismic profiles (e.g., Clowes et al., 1995, 2005; Cook et al., 1991, 1992, 2004; Evenchick et al., 2005; Hammer and Clowes, 2004; Varsek et al., 1993). The accretion of the Intermontane superterrane began as thin-skinned accretion, where much of the incoming lithosphere was delaminated. The gradual convergence deformed the underlying ancestral North American basement into a wedge that is overlain by tectonically imbricated and deformed accreted and ancestral crust (Cook et al., 2004; Evenchick et al., 2005). Only beneath the western Intermontane superterrane (Stikinia and

---
[1]Figure 2 is available as GSA Data Repository item 2007286, online at www.geosociety.org/pubs/ft2007.htm; on request from editing@geosociety.org or Documents Secretary, GSA, P.O. Box 9140, Boulder, CO 80301-9140, USA.

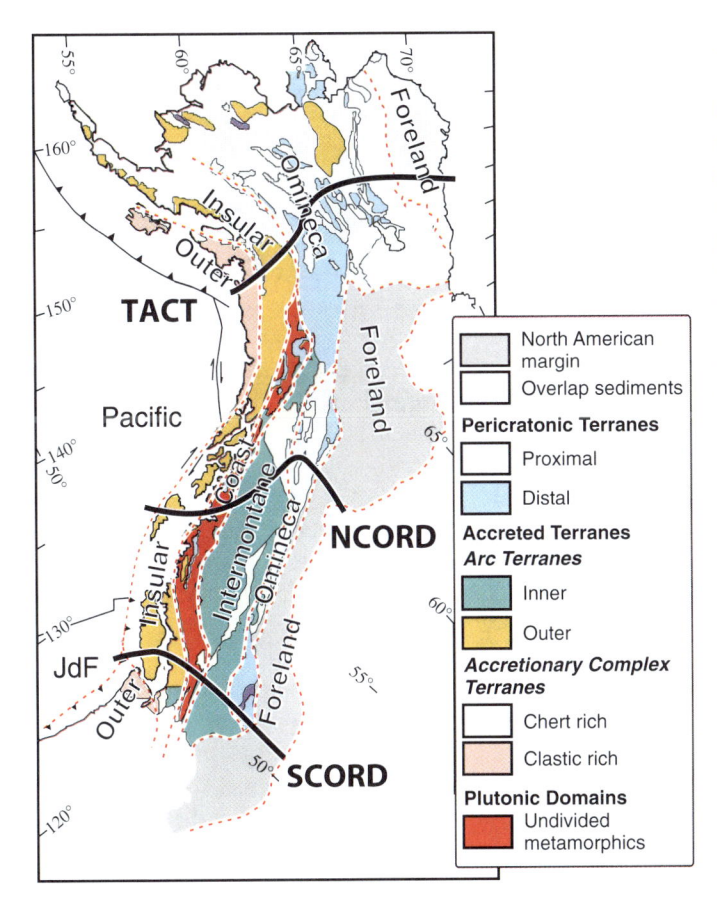

Figure 1. Simplified geological map of northwestern North America locating the corridors traversing the southern Canadian Cordillera (SCORD), northern Canadian Cordillera (NCORD), and Alaskan Cordillera (Trans-Alaska Crustal Transect [TACT]). The geomorphological belts as defined in the Canadian Cordillera are indicated with red dashed lines. For convenience, we apply this nomenclature to the Alaskan Cordillera. For example, we include the North Slope subterrane with the Foreland belt. Omineca is used to represent the North American pericratonic terranes in central Alaska. Further outboard, we include the Peninsular terrane with the Insular belt. The Chugach, Prince William, and Yakutat terranes are classed as Outer terranes. JdF—Juan de Fuca plate.

Cache Creek terranes in the north, Quesnellia in the south) does the accreted Intermontane crust appear to extend to the Moho; to the east, much of the orogen is actually underlain by ancestral North American lithosphere (Fig. 2).

The accretion history of the Outer and Insular terranes varies significantly along-strike; much of the Insular-Intermontane boundary is overprinted by subsequent Coast belt plutonism and orogenesis (Gehrels, 2001; Umhoefer et al., 2002). However, by the Late Cretaceous (90 Ma), the exotic Insular superterrane (Wrangellia and Alexander terranes) had been obliquely accreted to North America, further deforming the interior of the orogen and producing the structural and metamorphic features that characterize the Coast belt suture zone (Figs. 1 and 2). This suture zone evolved for over 50 m.y., concentrating plutonic emplacement, crustal-scale shear zones, and 10–20 km of exhumation (Hollister, 1982). The accretion of the Insular superterrane resulted in the formation of thin-skinned fold-and-thrust belts in the inboard terranes (e.g., Evenchick, 1991). Outboard of the plutonically overprinted Coast belt, the Insular terranes are interpreted to comprise the entire crust. Although both sinistral and dextral motion accompanied the collision, dextral displacements dominate from the mid-Cretaceous to the present; the magnitude of these motions is still under investigation (e.g., Cowan et al., 1997; Enkin et al., 2003; McCausland et al., 2005). By 40 Ma, the accretionary growth of the 52–58°N portion of the orogen was largely complete, and a transform plate boundary had been established.

South of the transcurrent margin, subduction beneath North America continues where the northern remnant of the Farallon plate (the Explorer–Juan de Fuca–Gorda plate) sinks beneath Oregon, Washington, and southwest British Columbia (40°–50°N; Figs. 1 and 2). North of 58°N, the northwest-trending Pacific–North American plate boundary transforms into the subduction zone beneath Alaska and the Aleutian arc. At the complex eastern end of this convergent boundary, the Yakutat terrane (or microplate) subducts or subcretes beneath the North American plate. As it subducts, the Yakutat terrane also overthrusts the subducting Pacific Plate along the Transition fault (Figs. 1 and 2; Eberhart-Phillips et al., 2006). This small terrane, composed of an accretionary prism and thickened oceanic crust, has been indenting the North American margin since the middle Miocene, resulting in large, local crustal uplift and deformation translated laterally across the width of the orogen (Mazzotti and Hyndman, 2002).

Alaskan Cordillera

The current structure of the Alaskan Cordillera exhibits two first-order differences from the relatively linear, northwest-trending Canadian Cordillera: the strike of the orogen rotates from north-northwest in the east to west-southwest in western Alaska, and the orogen lacks a cratonic backstop to the north (Figs. 1 and 2). A number of possible orogenic scenarios have been proposed to account for the oroclinal rotation. Plafker and Berg (1994b) proposed that northern Alaska rotated counterclockwise away from northern Canada between 130 and 115 Ma followed by the counterclockwise rotation of all of Alaska beginning in the Late Cretaceous. Many of these models follow a similar Mesozoic-Cenozoic (180–45 Ma) orogenic history to that currently accepted for the Canadian Cordillera but differ in how the western "rotation" occurred. The underlying differences of these models involve the mechanisms of extension and faulting in the opening of the Canada Basin (Fig. 1) to the north (e.g., Lawver and Scotese, 1990; Plafker and Berg, 1994b; Lane, 1997; Johnston, 2001). Tectonic summaries and detailed references can be found in Oldow et al. (1989), Gabrielse and Yorath (1991a, 1991b), and Moore et al. (1994). In addition to the variety of tectonic hypotheses, terrane and related nomenclature are variable and not entirely consistent with that used in the Canadian Cordillera. To aid comparisons in this paper, we simplify the nomenclature and extend the established Canadian Cordilleran geomorphological belts (e.g., Foreland, Omineca, Insular) through central Alaska (Fig. 1).

During the Late Triassic to Late Jurassic, island arcs formed and terranes amalgamated to the west of North America. During the oblique convergence and accretion of the Intermontane superterrane along the Canadian Cordillera, the Koykuk arc was obducted to the northern North American margin (150–115 Ma), contributing to Foreland belt development through ~500 km of crustal shortening. In contrast with the Canadian Cordillera to the south, intraplate rifting initiated (ca. 130–115 Ma) and detached the telescoped passive margin that underlies the current Brooks Range and much of the northern Foreland belt (Arctic Alaska terrane) from North America. Through a combination of counter-clockwise extensional rotation, transform faulting, and large-scale northward compression resulting in oroclinal folding, the developing passive margin became north-facing (e.g., Lawver and Scotese, 1990; Gabrielse and Yorath, 1991a, 1991b; Moore et al., 1994; Lane, 1997; Plafker and Berg, 1994b; Johnston, 2001). The oroclinal rotation of the Brooks Range and its foreland continued until 113 Ma (Moore et al., 1994). This rotation was followed by extension of the southern Brooks Range and central Alaska (103–96 Ma). In the Late Cretaceous and early Cenozoic, the Insular superterrane (including Wrangellia, Alexander, and Peninsular terranes) was accreted to the southern margin of Alaska. Convergence continued through the Paleocene and Eocene, when the current geometry was established (Figs. 1 and 2). During this period (60–45 Ma), compression and duplexing developed in the Brooks Range above the North Slope (Foreland) block, while continued deformation resulting from northward subduction of the Kula-Pacific plate and accretion of the Outer terranes (including the Chugach, Prince William, and Yakutat terranes) contributed to southward growth and additional deformation and magmatism. Today, northward subduction of the Pacific plate and Yakutat terrane (or microplate) continues. (e.g., Plafker et al., 1994; Mazzotti and Hyndman, 2002; Ferris et al., 2003; Eberhart-Phillips et al., 2006).

Orogen-Parallel Deformation

The oblique convergence of the Intermontane and Insular superterranes with North America resulted in the entire Cordillera experiencing significant along-strike shear. Although both sinistral and dextral motions are documented, right-lateral strike-slip motion dominated and resulted in terrane displacement, deformation, and formation of numerous fault systems. In the Alaskan Cordillera, many faults remain active. In the Canadian Cordillera, the majority of along-strike displacement ceased by 40 Ma. Today, the plate-boundary Queen Charlotte–Fairweather fault remains very active, with continued transform motion on the Denali and other faults.

The amount of dextral and sinistral displacement of the accreted terranes relative to each other and cratonic North America remains unresolved; paleomagnetic and paleontological data require displacements of upward of 3000 km, but geological evidence supports much smaller net dextral motion (e.g., Irving et al., 1996; Cowan et al., 1997; Enkin et al., 2003; McCausland et al., 2005). The ongoing controversy points to fundamental errors in techniques, interpretation of data, and/or the understanding of shear zones.

Of the many important orogen-parallel fault systems, four penetrate the crust and represent important terrane boundaries. The Tintina fault, in the interior of the orogen, separates deformed autochthonous North American rocks from parautochthonous and allochthonous rocks (Figs. 1 and 2 [see footnote 1]). On the basis of petrologic and geophysical contrasts across the Tintina fault, it is interpreted as a lithospheric-scale feature in the northern Canadian Cordillera (Abraham et al., 2001; Lowe et al., 1994; Cook et al., 2004). Recent estimates of displacement across the fault indicate ~425 km of dextral motion since the Paleocene (Murphy and Mortensen, 2003). The Tintina fault system extends through Alaska, where it splays into a number of faults, including the Kaltag fault (Figs. 1 and 2). The Denali fault system, with 350 km of total displacement, also extends through the Alaskan and northern Canadian Cordillera. Beneath the TACT corridor in central Alaska, the Denali fault extends to at least 20 km depth, suggesting that it too may penetrate the crust (Brocher et al., 2004; Fisher et al., 2004) (Fig. 2). A third major shear structure, the Coast shear zone, which is associated with both high-angle and orogen-parallel displacement, is only present in the Canadian Cordillera. This 1200-km-long, lithospheric-scale structure developed between 85–45 Ma during the accretion of the Insular superterrane (Figs. 2A and 2C). Contemporaneous with the Tintina and Denali fault systems, the Coast shear was an important component in strain partitioning across the orogen. The oblique convergence between North America and the Kula plate concentrated exhumation, deformation, and strike-slip motion along the Coast shear zone (Rusmore et al., 2001; Morozov et al., 1998, 2001, 2003; Cowan et al., 1997; Hollister and Andronicos, 1997; Chardon et al., 1999). The fourth major transcurrent fault is the Queen Charlotte–Fairweather fault system, which is the active transform boundary between the North American and Pacific plates that links the Juan de Fuca and Aleutian subduction zones (Fig. 1).

THE SEISMIC SURVEYS

Techniques and acquisition parameters vary for all of the two-dimensional (2-D), controlled-source seismic surveys, but most of the studies include coincident near-vertical incidence (NVI) reflection and refraction/wide-angle reflection (R/WAR) profiles. Details for each experiment are documented in the papers listed in Table 1. In general, air-gun sources were used in the marine environment with NVI data recorded by streamer and R/WAR data recorded onshore. The land NVI reflection surveys were acquired using vibroseis sources, with the exception of the Brooks Range data set, where low-fold NVI data were obtained using explosive sources. In all cases, the long-offset R/WAR surveys incorporated explosive sources. In some areas,

TABLE 1. CONTROLLED-SOURCE SEISMIC EXPERIMENTS IN THE CANADIAN AND ALASKAN CORDILLERA

Transect	NVI reflection	R/WAR	Summary papers
Alaskan Cordillera TACT and other surveys	Ambos et al. (1995); Brocher et al. (1991a, 2004); Fisher et al. (1989, 2004); Wissinger et al. (1997); Levander et al. (1994); Fuis and Plafker (1991); Fuis et al. (1997); Moore et al. (1991)	Beaudoin et al. (1994); Brocher et al. (1989, 1991b, 1994, 2004); Fuis et al. (1991, 1995, 1997); Wolf et al. (1991)	Fuis (1998); Fuis and Clowes (1993)
Northern Canadian Cordillera LITHOPROBE, ACCRETE, and other surveys	Cook et al. (1999, 2004); Evenchick et al. (2005); Snyder et al. (2002)	Fernandez Viejo and Clowes (2003); Fernandez Viejo et al. (1999, 2005); Welford et al. (2001); Creaser and Spence (2005); Hammer and Clowes (2004); Hammer et al. (2000); Morozov et al. (1998, 2001, 2003); Spence and Asudeh (1993); Zelt et al. (2006)	Cook et al. (1999, 2004); Cook and Erdmer (2005); Clowes et al. (2005); Evenchick et al. (2005); Snyder et al. (2002)
Southern Canadian Cordillera LITHOPROBE, SHIPS, and other surveys	Calvert (2002); Calvert and Clowes (1990, 1991); Calvert et al. (2003); Clowes et al. (1983, 1987a, 1987b); Cook (1995); Cook et al. (1992); Cook and van der Velden (1995); Green et al. (1986); Nedimovic et al. (2003); Perz (1993); Varsek et al. (1993); Yorath et al. (1985)	Burianyk and Kanasewich (1995, 1997); Burianyk et al. (1997); Clowes (2002); Drew and Clowes (1990); Ellis et al. (1983); Kanasewich et al. (1994); McMechan and Spence (1983); O'Leary et al. (1993); Ramachandran et al. (2004); Ramachandran et al. (2006); Spence et al. (1985); Spence and McLean (1998); Zelt et al. (1992, 1993, 1995); Zelt and White (1995)	Clowes et al. (1995); Cook et al. (1992); Hyndman et al. (1990); Varsek et al. (1993); Calvert et al. (2003)

NVI—near-vertical incidence; R/WAR—refraction/wide-angle reflection; SHIPS—Seismic Hazards in Puget Sound; TACT—Trans-Alaska Crustal Transect.

particularly over the Aleutian-Wrangell subduction zone and the Cascadia subduction zone, earthquake and teleseismic data provide additional information from which lower resolution, three-dimensional (3-D) models have been developed.

Interpretation of Controlled-Source Seismic Data

Seismic experiments played a central role in the three transects by establishing the present subsurface structure and providing compositional constraints that are essential for extending the geological analyses to depth. Near-vertical incidence (NVI) reflection and refraction/wide-angle reflection (R/WAR) profiles provide complementary information and were therefore run coincidently along most of the transect profiles. NVI reflection data provide detailed geometry and reveal impedance contrasts. In some locations, these can be directly identified through tracing a reflector to the surface where the compositional boundary is mapped. However, in most cases, the source of the reflections is not clearly determined but the reflection fabric characterizes the orientation and scales of structures and heterogeneities. Wide-angle data can also detect large impedance contrasts, but the much sparser coverage, broader scale, and poorer resolution means that fabrics are not routinely imaged. The primary goal of R/WAR experiments is to determine the P- and S-velocity structure of the lithosphere. These velocity structure models not only add structural information but provide constraints for composition and temperature because seismic velocities vary with both parameters. When both P- and S-arrivals are recorded, Vp/Vs or Poisson's ratio models can be generated; these provide the strongest compositional constraints.

One of the key factors in interpreting seismic profiles is consideration of the variable resolution of the images and models. Resolution is defined as the minimum scale at which two structures can be distinguished. As a rule-of-thumb, vertical resolution in reflection profiling is one-quarter the wavelength (the quotient of velocity and frequency). High-frequency energy attenuates more rapidly, and therefore vertical resolution degrades with target depth. Lateral resolution is determined by the Fresnel zone, which depends upon wave speed, frequency, and source distance. For most crustal-scale NVI reflection experiments, rock velocities vary from 3 to 8 km/s, and useful source frequencies range from 10 to 80 Hz. In addition, the technique involves high spatial sampling, with receivers at 25–50 m intervals over distances up to 14 km and source intervals of 50–100 m. Sparser coverage and longer offsets were used in the more rugged Brooks Range segment of the TACT transects (Fuis et al., 1997). Onshore/offshore surveys reverse the geometry by taking advantage of the close shot-spacing provided by an air-gun source (50–100 m) but using more broadly spaced ocean-bottom seismometers and land instruments (5–10 km). Resolution varies with acquisition parameters, but in most of the northern Cordilleran NVI reflection data, structures can be resolved in the upper crust that are on the order of 10 m thick and 500 m apart. In the lowermost crust, vertical resolution is ~200 m, while horizontal resolution is 2.5 km. An additional factor that influences the images and interpretations is coherency filtering, which is often applied to enhance reflector continuity. It can also be used to limit the reflector dips displayed. Most interpretations involve evaluation of stacked, migrated, and coherency-filtered images. However, for clear display of long profiles at publication scales, coherency-filtered images are typically used.

The resolution of wide-angle, crustal-scale seismic data is usually much coarser than that of the NVI data. Typical land-based R/WAR data involve lower-frequency content (2–15 Hz) from the explosive sources, offset ranges varying from zero to 300–800 km, receiver spacings of 1–2 km, and shot spacings of 8–50 km. Traveltimes and amplitudes of individual phases are used to develop velocity structure models using both forward and inverse modeling methods. Forward modeling permits incorporation of known geological structures into the models, thereby yielding sharper and more realistic, but biased, models. Inversion techniques yield the simplest (minimum structure or minimum parameter) models that satisfy the data and provide quantitative estimates of uncertainties in the model. However, these models are much smoother than the real structures and must be interpreted as such. Using either method, structures that are more than 1 km thick and 3–20 km apart typically can be resolved in the upper crust. Lower-crustal velocity structure is often poorly resolved due to limited coverage of refracted arrivals. Moho depth and uppermost mantle velocity structure are generally resolved to better than 2 km vertically and 20–50 km horizontally. Note that in ACCRETE and several of the TACT surveys, better resolution was obtained in the upper crust by using a higher density of sources and/or receivers.

Comparative Lithospheric Cross Sections

We used the coincident R/WAR and NVI seismic profiles, in conjunction with other geological and geophysical data, to develop interpretations of lithospheric structure along the three corridors that cross the Cordilleran orogen. In Figure 2 (see footnote 1), the interpreted velocity-depth structure sections and migrated, coherency-filtered NVI reflection profiles are compiled for each of the three transects. Common distance, depth, and color scales have been incorporated for ease of comparison. The seismic profiles are presented with merged interpretations of lithospheric structure based on the original papers.

DISCUSSION: UNIFYING AND CONTRASTING CHARACTERISTICS

Continental Crustal-Scale Décollements

Crustal-scale décollements within the continental crust are a first-order feature of most, if not all, orogens (Cook and Varsek, 1994). In the Canadian Cordillera, a well-defined, crustal-scale décollement acts as a tectonic accretion surface and separates the accreted terranes from underlying, ancestral North American crust. In Alaska, the indenting wedge of the North Slope (Foreland) block forms a smaller-scale, but analogous crustal-scale ramp above which the already telescoped crust was thrust and duplexed (Fuis et al., 1997). As illustrated in Figures 2 and 3, all three transects exhibit an unconformity or decoupling zone that dips from the Foreland fold-and-thrust belt to the lowermost crust or Moho in the interior of the orogen. Above the décollement, crustal-scale imbrication, antiforms, duplexing, and other deformation were generated by collisional tectonics. In the southern Cordillera, deeper parts of this deformation have been unroofed and exposed as metamorphic core complexes by Eocene extension. Although there is clear evidence for significant shortening in the northern Canadian Cordillera, much of the extensive duplexing and deeper crustal deformation is inferred from the seismic results. Although the décollement formed during the initial major accretionary phase (Intermontane), the continued accretion of the Insular and Outer terranes not only generated thin-skinned deformation across the orogen, but may have linked thrust systems to the main décollement and related thrusts further inboard (e.g., Evenchick, 1991; Evenchick et al., 2005).

These intracrustal décollements exhibit quite different seismic characteristics in each profile (Figs. 2 and 3). In TACT, the duplexed crust above the décollement is highly reflective. The indentor wedge below is defined by a sudden decrease in both near-vertical incidence (NVI) and wide-angle reflectivity (Fuis et al., 1997; Wissinger et al., 1997). P-velocity models do not define a lateral change in lower-crustal composition. In direct contrast with TACT, the crustal décollement beneath the northern Canadian profile is clearly defined by strong NVI reflections, while the less reflective region above the décollement is interpreted as a highly deformed and faulted zone of both North American and accreted crust. Although wide-angle reflections and P-velocities do not require major changes in crustal properties across the décollement, Poisson's ratio models do support a lateral change in lower-crustal composition that coincides with the transition from more mafic Stikinia lower crust to the eastward-thickening wedge of more felsic ancestral North American crust (Fernández-Viejo et al., 2005). In the southern Canadian Cordillera, the crust is generally reflective throughout its depth extent (Fig. 3). Changes in the dips and fabric of this reflectivity, coupled with geological information, help to define the décollement, which is offset by the Slocan Lake normal fault below the Omineca belt (Fig. 2D). Wide-angle reflections define a shallowly dipping surface consistent with the accretionary ramp and below which P-wave velocities are clearly higher than in the overriding material. This region is inferred to represent basement rocks of ancestral North America. These different seismic characteristics of the Cordilleran décollement remain unexplained. Differences in acquisition parameters and conditions may contribute, but composition and tectonic-induced structure are likely the primary factors.

The crustal-scale décollement rising above a smooth, horizontal Moho has implications for estimates of crustal growth. Throughout the Cordillera, the ancestral continental crust extends beneath a

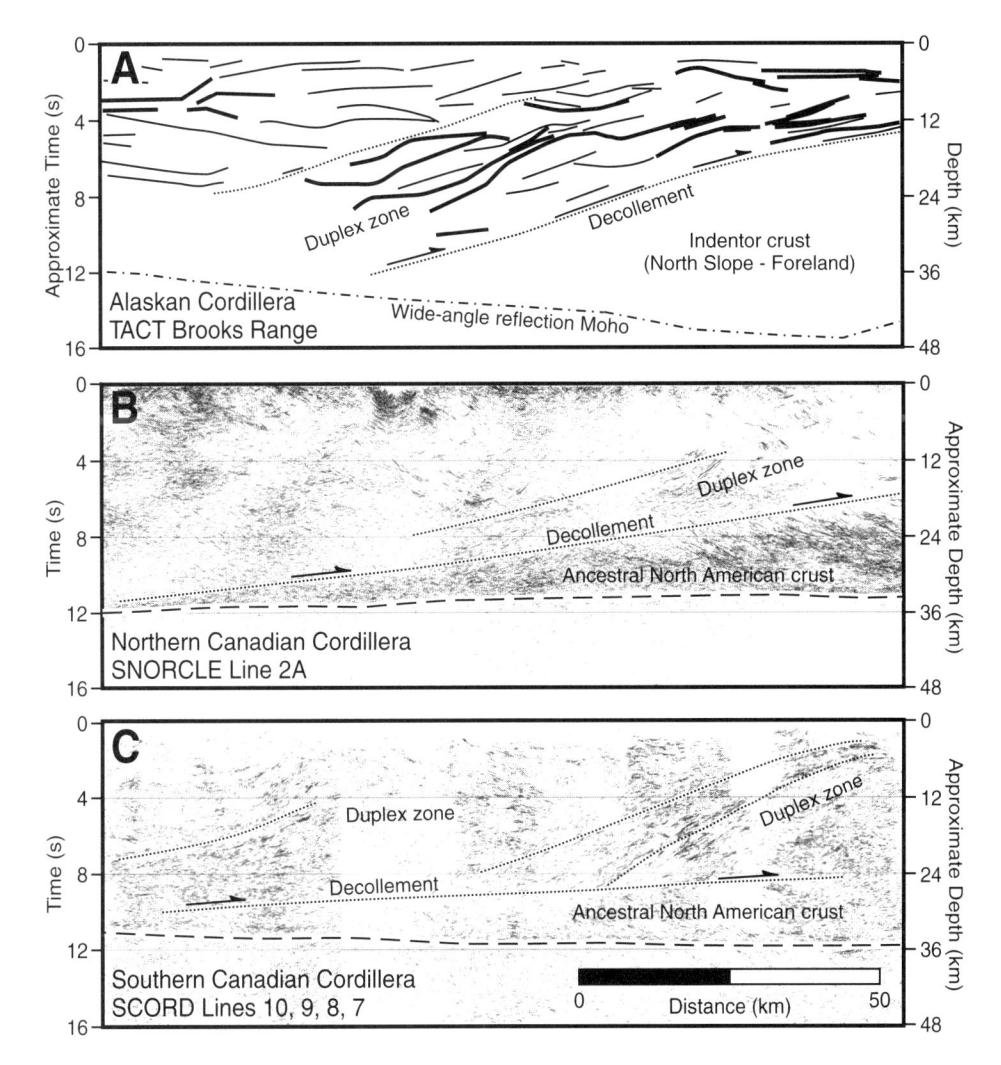

Figure 3. Comparison of seismic near-vertical incidence reflection profiles of crustal-scale décollements. (A) Line drawing of a portion of the Trans-Alaska Crustal Transect (TACT) Brooks Range profile (adapted from Fuis et al., 1997). Bold lines denote stronger reflections. Migrated, coherency-filtered reflection profiles are shown in (B) a portion of line 2A of the SNORCLE transect (Cook et al., 2004), and (C) lines 7, 8, 9, and 10 of the Southern Cordillera Transect (Cook et al., 1992). The near-vertical incidence (NVI) reflection Moho (dashed line) is defined by a decrease in reflectivity and, in the Canadian Cordillera, by subhorizontal reflections at the base of the crust. Dipping reflectivity defines the intracrustal décollement and duplex structures (highlighted by thin dotted lines) as well as structural fabrics. Note that the lower two panels exhibit reflectivity listric into the lower crust and/or Moho.

significant portion of the accreted terranes identified by the surface geology. The seismic interpretations locate the underlying, inboard crustal ramp as extending to: (1) the Fraser fault in the southern Canadian Cordillera, (2) beneath the eastern Stikine terrane in the northern Canadian Cordillera, and (3) the southernmost Brooks Range in the Alaskan Cordillera. Above the ramp and its overlying duplexed and imbricated crust, the accreted terranes in the northern Canadian Cordillera form only a veneer of thin flakes from a few to 10 km thick. In the Alaskan and southern Canadian Cordillera, the "veneer" is 5–25 km thick. In all cases, the volume of accreted or overlying crustal material is much less than would be concluded based on inferences from the surface geology alone.

Although the thickness of the underlying wedge of ancestral North American crust has different interpretations (Snyder et al., 2002; Evenchick et al., 2005), the southern and northern Canadian Cordilleran profiles document that it lies beneath the eastern half of the Cordillera, extending beneath much of the Intermontane superterrane (Fig. 2). The metamorphic grade of the crust accreted above the décollement is generally not high, except in the exposed metamorphic core complexes of the southern Canadian Cordillera. Therefore, the lower crust and lithospheric mantle of the accreted terranes were either detached from their upper parts, subducted and recycled in the mantle, or tectonically underplated beneath the ancient continental margin. These processes would influence magmatism and orogenesis in the overriding crust and likely played an important role in the development of the flat, shallow Moho and high temperatures below the Canadian Cordillera (discussed in the following).

Moore and Wiltschko (2004) discussed how syncollisional delamination in the lower plate will develop at the crust-mantle boundary unless eclogitization of a mafic lower crust creates sufficient density contrast to drive an intracrustal delamination. This model fits well with the Proterozoic Wopmay orogen, where well-defined reflections in the mantle may represent the delaminated, eclogitized lower crust of the Fort Simpson terrane below a crustal wedge of the Hottah terrane (Figs. 2A and 2C; see Cook et al., 1999). In the Cordilleran lithospheric mantle, there are no observations of similar reflectivity (or velocity anomalies) that are clearly linked with the formation of the crustal décollement. The wide-angle seismic profiles only detect discontinuous reflections that are interpreted to be associated with the top of the asthenosphere (Figs. 2C and 2D) (Clowes et al., 1995; Hammer and Clowes, 2004; Harder and Russell, 2006). The absence of evidence for preserved subducted or delaminated lithosphere beneath the Cordilleran décollement is most likely related to current thermal conditions. Another possibility is that detachment at or beneath the Moho may result in substantial translation between coeval crustal and mantle structures (Ellis and Beaumont, 1999). However, studies of mantle xenoliths from below the accreted Stikinia arc terrane in the northern Canadian Cordillera (Figs. 2A and 2C) suggest that the uppermost lithospheric mantle (36–44 km) has an arc chemistry in contrast with the deeper lithospheric mantle (44–70 km) (Harder and Russell, 2006). This is interpreted to indicate that the upper layer represents preserved, accreted Stikinia mantle, which would require little translation at the crust-mantle boundary during accretion of Stikinia and the Insular terranes.

Subduction Megathrust Décollements

Geodynamic modeling (e.g., Ellis and Beaumont, 1999) demonstrates how subduction and accretion generate outward-verging shears in crust within the inboard backstop and above the oceanic plate. This is observed beneath the three profiles where opposing but roughly symmetric deformation patterns are mapped or imaged above the Foreland and the megathrust décollements (Figs. 2B and 2D). The active Cascadia and the Wrangell-Aleutian subduction zones are crossed by the southern Cordilleran and Alaskan profiles, which provide current observations of these oblique convergence zones.

Despite the differences between the Cascadia subduction of the Juan de Fuca–Gorda oceanic plates and the complex subduction zone beneath southern Alaska, which involves the Pacific plate (Aleutian) and the Yakutat terrane (or microplate) (Wrangell), there are also many similarities. General structures of the Cascadia and Alaskan subduction zones are compared in detail by Fuis and Clowes (1993) and Fuis (1998). The seismic and associated geological profiles mapping the Wrangell-Aleutian and Cascadia subduction zones exhibit layered packages of Mesozoic and Cenozoic accretionary complexes that were emplaced by sequential thrusting resulting in subcretion beneath the crustal backstops (e.g., Clowes et al., 1987a, 1987b; Hyndman et al., 1990; Fuis et al., 1991; Ramachandran et al., 2004; Plafker and Berg, 1994b; Eberhart-Phillips et al., 2006). The complex layering of reflectivity and velocity contrasts that dip inboard are interpreted to be layers of metamorphosed sediments and oceanic crust or (serpentinized) upper mantle (Fig. 4). This general outward-verging structural pattern and seismic fabric is a consistent pattern observed over both megathrust décollements (Fig. 2).

In the eastern Aleutian-Wrangell subduction zone, the sharply curved margin and the linked subduction of the Pacific plate and the Yakutat terrane (thickened oceanic crust and a Cretaceous accretionary prism) generate a host of tectonic effects not currently associated with Cascadia. Understanding of the Wrangell-Aleutian subduction zone structure has been improved through earthquake and teleseismic studies (e.g., Ferris et al., 2003). Although the teleseismic data generally yield lower-resolution models of crustal structure than 2-D controlled-source data do, they provide 3-D velocity and Poisson's ratio models with improved resolution in the mantle. A synthesis of the seismic data in the region (Eberhart-Phillips et al., 2006) concludes that the buoyant Yakutat terrane is being shallowly subducted and possibly subcreted beneath the south-verging Outer ter-

Figure 4. Comparison of migrated, coherency-filtered reflection profiles from: (A) the Wrangell subduction zone (Trans-Alaska Crustal Transect [TACT] Chugach line; Fisher et al., 1989), and (B) the Cascadia subduction zone (Vancouver Island line 84-1; Clowes et al., 1987a). The top of the subducting plate as interpreted from controlled-source seismic data is noted by the long dashes (e.g., Hyndman et al., 1990; Fuis et al., 1991; Ramachandran et al., 2006), while the teleseismic interpretation is noted by the dotted line (Nicholson et al., 2005). C and E denote bands of reflectivity. The E layer also corresponds to low-velocity (Ramachandran et al., 2004; Nicholson et al., 2005) and high-conductivity anomalies (Kurtz et al., 1990) and was interpreted by Calvert et al. (2006) as duplexed accreted or oceanic crust.

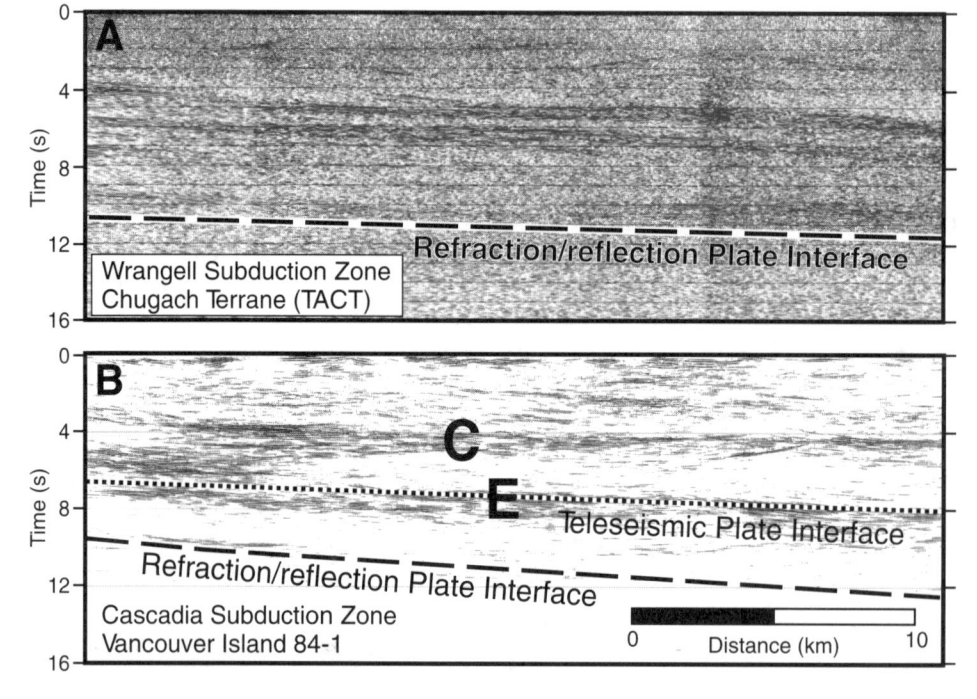

ranes, as well as overthrusting the Pacific plate offshore along the Transition fault (Figs. 2A and 2B). Subduction of the buoyant Yakutat crust has locally thickened the crust to between 50 and 70 km beneath the Alaska and St. Elias–Chugach Ranges (Fuis et al., 1991; Fuis and Plafker, 1991; Eberhart-Phillips et al., 2006), which support topography reaching 6000 m. In addition, the oblique collision with the curved margin results in strain partitioning and dextral extrusion along the Denali and other strike-slip faults (Fig. 2A) and compressional deformation that is influencing central Alaska. Mazzotti and Hyndman (2002) demonstrated that extrusion tectonics propagates the strain produced by the Yakutat collision across the entire orogen to the eastern front of the northern Canadian Cordillera. Similar effects would have occurred throughout the oblique subduction and accretionary development of the Cordillera.

Distinguishing the top of the subducting oceanic plate from the layered, reflective crust above is often difficult and has led to debate, particularly in the southern Canadian Cordillera crossing of the northern Cascadia subduction zone. Conflicting interpretations of the layered bands of reflectivity and velocity changes based on controlled-source (Green et al., 1986; Clowes et al., 1987a, 1987b; Hyndman et al., 1990; Clowes and Hyndman, 2002; Nedimović et al., 2003; Calvert et al., 2003; Ramachandran et al., 2004; Ramachandran et al., 2006; Calvert et al., 2006) and teleseismic (Nicholson et al., 2005) data place the top of the young (<10 Ma) Juan de Fuca plate at depths that differ by as much as 10 km (Fig. 2D). The discrepancies are associated with interpreting the teleseismic and controlled-source responses from the complex structure that results from tectonic underplating and effects of dehydration metamorphism (serpentinization) throughout the crust and mantle wedge. Analysis of the Seismic Hazards in Puget Sound (SHIPS) data set (Ramachandran et al., 2004; Ramachandran et al., 2006) and existing reflection data has led Calvert et al. (2006) to propose a 5–12-km-thick duplexing of the forearc that also results in forearc rock being transported into the forearc mantle (Fig. 2D). This scenario has some similarities with that interpreted for the southern Alaska subduction zone described already, and comparisons may be extended to the duplex structures in the Brooks Range and eastern Canadian Cordillera (Fig. 3). If the duplex interpretation of Calvert et al. (2006) is correct, then the teleseismic anomaly cannot represent the top of the plate. In this case, the anomaly may originate from fluids or a metamorphic front(s) in the overriding crust (e.g., Hyndman et al., 1990). The different interpretations require resolution because they may have implications for understanding megathrust earthquake generation.

Orogen-Parallel Translation, Transpression, and Crustal-Penetrating Faults

Plate-motion reconstructions (e.g., Engebretson et al., 1985) indicate that dextral movement has dominated the oblique convergence since the Middle Cretaceous. Prior to that, sinistral motions may have been important, particularly in southwestern Canada and further south in the Cordillera (e.g., Monger et al., 1994; Umhoefer et al., 2002). The strain partitioning across the Cordillera and the resulting orogen-parallel translation of crustal units are recorded by numerous strike-slip fault systems and shear zones. Determinations of the along-strike displacements within the orogen hinge on reconciling the plate-motion reconstructions and paleomagnetic data with the geological and paleontological observations (Cowan et al., 1997; Enkin et al., 2003; McCausland et al., 2005). The seismic data cannot directly address this question. However, by establishing the subsurface structure in the orogen, including the orientation of major faults at depth, improved models of the orogen that may contribute to the solution can be developed.

Seismic data can be used to image faults in several ways. Dipping faults often generate impedance contrasts related to shear, lithological contrasts, metamorphism, and fluids trapped along the fault plane, and these are often well-defined by near-vertical incidence reflection data. In contrast, vertical faults are difficult to image using NVI reflections. Truncation of reflections and a lack of reflector continuity are often indicators of a near-vertical fault (Fig. 5). Crustal velocity models derived from refraction data can detect dipping or vertical faults only if the rock assemblages offset by the fault have a large enough compositional difference. Often in the upper crust, fault gouge yields extremely low velocities (Fig. 5B). However, in the lower crust, the combination of decreased resolution and distributed ductile deformation often decreases velocity contrasts. The wide-angle refraction data sometimes can indicate faults by offset levels of reflectors within the crust and at the Moho.

Many strike-slip faults slice through the Cordilleran orogen; several have played an important role for extended periods of time. Portions of at least four faults penetrate the crust and possibly the lithosphere. These include: (1) the Queen Charlotte–Fairweather fault, which transforms into the Contact–Chugach–St. Elias–Border Ranges faults in the southern Alaskan Cordillera, (2) the Denali fault, which cuts through the northwestern Canadian Cordillera and the Alaskan Cordillera, (3) the Coast shear zone, which divides the Insular and Intermontane superterranes through much of the Canadian Cordillera, and (4) the Tintina fault, which splays southward into the Rocky Mountain Trench and Fraser fault system in the southern Canadian Cordillera and northward into a number of fault systems through central Alaska.

The Queen Charlotte–Fairweather strike-slip fault, active since 43 Ma, links the Cascadia and Alaskan subduction zones. Current motion is north-northwest at 6.3 cm/yr, primarily dextral transcurrent but with a small component of compression (Engebretson et al., 1985; Hyndman and Ellis, 1981; Yorath and Hyndman, 1983). During its history, small changes in plate motions have resulted in transpression or transtension (Hyndman and Hamilton, 1993). Onshore-offshore R/WAR data yielded a velocity model of the margin that shows rapid thinning of the crust from 27 km beneath the Queen Charlotte Islands to 20 km at the fault zone to 7 km beneath the Pacific oceanic crust (Mackie

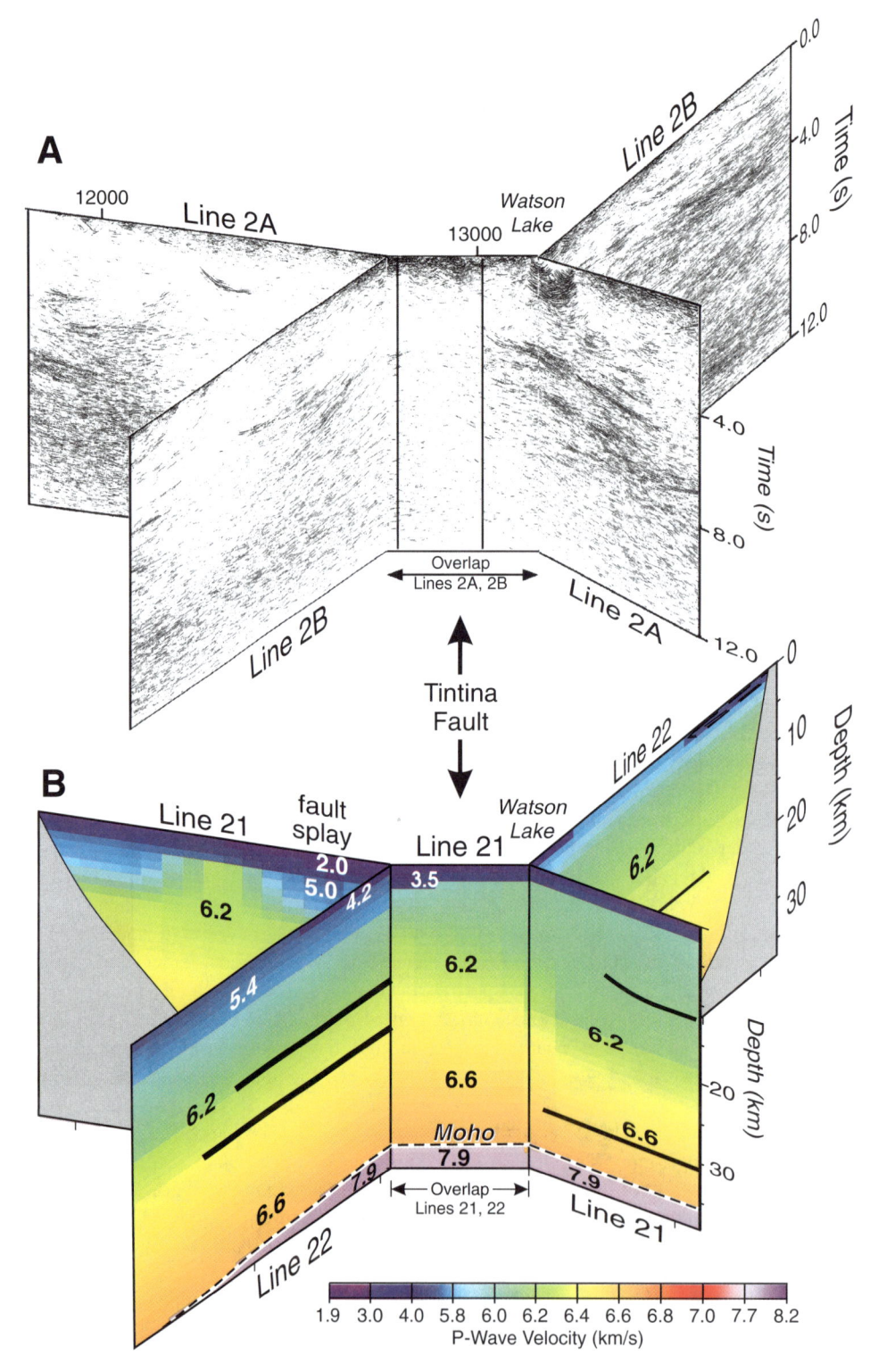

Figure 5. Controlled-source seismic profiles crossing the Tintina fault zone. (A) SNORCLE lines 2A and 2B near-vertical incidence reflection profiles (figure adapted from Cook et al., 2004). The zone of poor reflectivity beneath the fault extends through the entire crust. (B) Velocity structure models derived from SNORCLE lines 21 and 22. Moho (dashed line) and intracrustal (solid lines) wide-angle reflections are shown. Note the extremely slow shallow velocities below both the main fault trace and an adjacent fault splay. Deeper beneath the fault zone, mid- to lower-crustal velocities are elevated, intracrustal wide-angle reflections terminate, and the crust thins.

et al., 1989; Spence and Asudeh, 1993) (Fig. 2C [see footnote 1]). The transition from continental to oceanic crust occurs outboard of the fault. Very low upper-crustal velocities (3.4–4.1 km/s) to depths of 3–4 km below a submarine topographic terrace immediately west of the fault trace indicate that transpression has produced a small accretionary wedge (Dehler and Clowes, 1988).

The Queen Charlotte–Fairweather fault, as discussed previously, takes up the majority of current dextral motion between the North American and Pacific plates offshore western Canada and southeast Alaska. To the north, this fault represents the active margin between North America and the accreting Yakutat terrane. As such, the fault is crustal-penetrating. West of the Alaska-Canada border, the Queen Charlotte–Fairweather fault rotates counterclockwise to a shallower dip and acquires an increasing thrust component, merging with the Contact–Chugach–St. Elias–Border Ranges faults, which are mainly thrust faults. Some of the dextral motion of the Queen Charlotte–Fairweather fault is likely transferred to the Denali-Totshanda fault system. The upper sedimentary crust of the Yakutat terrane is being scraped off against the Chugach–St. Elias thrust, while the basaltic lower crust subducts (e.g., Brocher et al., 1994).

The Contact fault, another important active structure in Alaska, is interpreted to lie along the dipping boundary between the Chugach and Prince William terranes (Figs. 2A and 2B). Seismic models of the Contact fault suggest that it extends to at least 10 km depth (Brocher et al., 1994; Fuis et al., 1991). Analogous to the West Coast fault in the southwestern Canadian Cordillera (Hyndman et al., 1990), the Contact fault may sole into dipping reflective bands that eventually sole into a zone above the subduction thrust interface (Fuis et al., 1991; Wolf et al., 1991).

The eastern border of much of the Insular superterrane in the Canadian Cordillera is defined by the Coast shear zone. This 1200-km-long belt of exhumed high-grade rock records dextral, sinistral, and vertical motions concomitant with latest Cretaceous to Eocene emplacement of the Coast Mountains batholith complex (Klepeis et al., 1998). Neither the Coast shear zone nor the Coast plutonic complex extends into the Alaskan Cordillera. The ACCRETE seismic data (Morozov et al., 1998, 2001, 2003) clearly show that the shear zone penetrates the entire crust. Strong reflections and contrasts in velocity and Poisson's ratio across the zone define a steeply eastward-dipping structure that divides the compositionally distinct Insular superterrane from the Coast belt. The Coast shear zone is listric and merges with the Moho where the crust thickens by 6 km (Fig. 2C). Determining the displacement that occurred across this ductile zone is challenging but may prove to be important for understanding the convergent and orogen-parallel motions in the development of the orogen (e.g., Hollister and Andronicos, 1997). The similarity in geometry of the Coast shear zone with the Contact fault, and perhaps the West Coast fault of the southern Canadian Cordillera (Fig. 2D), suggests a linkage with tectonic processes. Interpretations in all three regions infer that the locus of subduction stepped outboard. Prior to this development, the three faults, and perhaps others, may have been thrust faults associated with accretion and tectonic underplating. Once subduction stepped outboard and the tectonic regime was modified, the faults were reactivated as shear zones that accommodate transpression and transtension.

The Denali fault in the south-central Alaskan Cordillera extends near-vertically to at least 20 km depth as defined by a reflectivity contrast across the fault, low seismic velocities, high conductivities, and low density. The seismic data image a crustal root beneath the Denali fault where the crust reaches 50 km thickness but thins to 32 km to the north and 45 km to the south (Fig. 2B) (Brocher et al., 2004; Fisher et al., 2004).

Further inboard, the Tintina fault generally defines the boundary between the allochthonous and autochthonous terranes (Fig. 2). This dextral fault has accommodated 425 km of motion since the Paleocene (Murphy and Mortensen, 2003). Petrologic studies of basalts and mantle xenoliths indicate that it penetrates the crust (e.g., Abraham et al., 2001). This is confirmed by the seismic profiles, although observations vary considerably along-strike. Where SNORCLE Corridor 2 crosses the Tintina fault (Fig. 2C), the seismic data reveal low upper-crustal velocities, elevated velocities in the lower crust, and a dramatic drop in reflectivity that cuts off prominent reflectivity observed to either side (Fig. 5) (Cook et al., 2004; Hammer and Clowes, 2004; Welford et al., 2001). The Moho remains relatively flat across the Tintina fault; only small variations (1–3 km) are modelled (Hammer and Clowes, 2004; Zelt et al., 2006). Tomographic analyses of the SNORCLE data set, which provides some 3-D coverage, support the interpretation that the Tintina fault is a lithospheric-scale structure since a robust velocity contrast in the upper mantle is required by the data (Zelt et al., 2006). SNORCLE Corridor 3 crosses the Tintina fault 250 km northwest of Corridor 2. This northern crossing of the Tintina shows an even broader fault zone with multiple shallow traces and a broad (30 km) zone of disruption in the lower crust, perhaps associated with anisotropic mylonites (Creaser and Spence, 2005; Snyder et al., 2005).

South of 60°N, offset across the Tintina fault (northern Rocky Mountain Trench) decreases. By 54°N the right-lateral offset is small, and south of 50°N, the southern Rocky Mountain Trench fault becomes an extensional, west-side-down listric fault (e.g., van der Velden and Cook, 1996). The relative motion is distributed among numerous north-south–oriented faults that branch off south of the northern Rocky Mountain Trench. One of the most significant of these is the Fraser–Straight Creek fault, which extends for over 600 km, has experienced ~140 km of offset (Price and Monger, 2000), and penetrates the crust (Figs. 2A and 2D) (Price and Carmichael, 1986; Perz, 1993).

The continuation of the Tintina fault zone northward through the central Alaskan Cordillera also manifests in a splay of faults to the west of 146°W. Several interpretations exist of the tectonic development of this region, each with implications for how the dextral motion of the Tintina has been distributed through Alaska. Beaudoin et al. (1994) found clear evidence that two proposed western splays of the Tintina fault, the Beaver Creek and Victoria Creek (Kaltag) faults, extend to at least mid-crustal depths. The faults do not clearly penetrate the crust, but lateral changes in subhorizontal mid- and lower-crustal reflectors suggest that they extend deeper than 10 km. Other faults proposed to link to the Tintina include the Susulatna, Nixon Fork, and Tozitna fault systems.

Crustal Thickness: Thin Crust and Few Crustal Roots

One of the fundamental observations beneath the Cordilleran transects is that the crust is generally thin and surprisingly uniform in thickness (Fig. 2). In the southern Canadian Cordillera, the Moho lies at depths between 33 and 35 km from the Coast belt to the Omineca belt (a distance of greater than 500 km), and it only increases to depths greater than 40 km below the Rocky Mountains. In the northern Canadian Cordillera, the Moho beneath the eastern two-thirds of the orogen (Intermontane, Omineca, and Foreland belts; ~500 km across strike) is nearly flat, with crustal thicknesses also ranging between 33 and 36 km. Within central Alaska, from the middle of the Yukon-Tanana terrane to the Brooks Range (a distance of ~400 km), the crust also is generally thin, with thickness values of 30–32 km.

Outside of these central areas of the three corridors, crustal thickness is more variable. Thin crust (20–25 km) is observed beneath the outboard terranes along the transform margin in the northern Canadian Cordillera. In contrast, below the Insular superterrane in both the Alaskan and southern Canadian Cordillera, crustal roots are found. In southern Alaska, seismic profiles show that the crust reaches 55 km thickness beneath the Copper River basin (Fig. 2B). As discussed in a previous section, this thickening involves the subduction of the Yakutat terrane; the thickest crust and highest-standing topography are observed near the center of the Yakutat crust. In southern Canada, analyses of combined active source and earthquake data place crustal velocities as deep as 40 km just above the subducting plate. This root may (partially) support the chain of mountains that lies along central Vancouver Island.

Another major crustal root is found below the Brooks Range. The crust in central Alaska thickens by at least 20 km to the north, reaching 50 km beneath the décollement and duplexed crust in northern Alaska (Fuis et al., 1997; Wissinger et al., 1997). The crustal root is associated with the convergence and indentation of the North Slope subterrane. No such prominent roots are observed beneath the Foreland belt in the Canadian Cordillera.

Several smaller changes in crustal thickness are observed in all three transects, and these correlate with major tectonic or geologic boundaries observed on the surface. In the southern Canadian Cordillera, the crust thickens by 8–10 km to the east of the Rocky Mountain Trench, along the Slocan Lake normal fault (SLF, Fig. 2D). This deeper Moho below the eastern Foreland belt extends further east beneath the Western Canada sedimentary basin. At the Fraser River strike-slip fault, the interpretation from refraction data suggests a slight increase in Moho depth from east to west, defining a small root (38 km) below the Coast belt (Zelt et al., 1993). In the northern Canadian Cordillera, significant steps in crustal thickness occur on either side of the Coast belt. The outboard Insular terranes are only 20–25 km thick, stepping down at the Coast shear zone to 30–32 km beneath the Coast belt. Further inboard, the crust thickens to 36 km beneath the eastern margin of the Coast belt and then remains relatively flat for 500 km. The two steps in Moho depth correlate with large lateral changes in P-velocity and Poisson's ratio (Hammer et al., 2000; Morozov et al., 1998, 2001; Hammer and Clowes, 2004). Since the Coast belt represents the locus of oblique collision between the Insular and Intermontane superterranes, the variations in crustal structure across the region reflect differences in the colliding terranes and postcollisonal plutonism and deformation. Changes in crustal thickness are also observed beneath major fault systems. For example, in both the Northern Cordillera and TACT surveys, the depth to Moho decreases by 2–3 km below the Tintina fault system (Fig. 2C) (Hammer and Clowes, 2004; Beaudoin et al., 1994). Further examples and discussion were already noted in the earlier section on margin-parallel faulting.

Flat, Young Moho beneath the Canadian Cordillera

The observation that thin, flat crust makes up a large portion of the Cordilleran orogen has several important implications. First, the lower crust, and the Moho in particular, are weak zones where deformation is concentrated. Most NVI reflection profiles exhibit prominent reflections and reflection fabrics that are listric into the lower crust or Moho (Cook, 1995; Cook et al., 2004), indicating that these regions are important in large-scale deformation (Figs. 2 and 3). This conclusion is consistent with the high temperatures inferred beneath the entire Canadian Cordillera. Discussed in more detail in a later section, both the southern and northern Canadian Cordilleran profiles exhibit slow velocities (Clowes et al., 1995, 2005) and high heat flow (Hyndman and Lewis, 1999a, 1999b; Hyndman et al., 2005) that, when combined with rheological profiles for typical rock types, indicate that the Moho and lower crust are zones of reduced strength. Thus, the flat Moho and horizontal fabrics in the lower crust could be detachments or localized regions of tectonic deformation that partially decouple the crust from the mantle. For example, geodetic and seismicity data indicate that the ongoing Yakutat collision is creating the eastern bulge of the Mackenzie Mountains, 800 km to the northeast (Foreland belt, 63°N, 125°W, Fig. 1; Hyndman et al., 2005). Mazzotti and Hyndman (2002) argued that extrusion strain is taken up not only at the Moho, but by a quasi-rigid displacement of the upper crust over a lower-crustal detachment zone.

Secondly, the thin crust reinforces hypotheses that the Moho of the Canadian Cordillera represents a relatively young transition zone beneath much of the orogen (e.g., Cook, 1995). The complex lower-crustal and upper-mantle structures mapped beneath the Cascadia and Wrangell subduction zones are not observed elsewhere in the orogen. Furthermore, if crustal roots such as that mapped beneath the Brooks Range did exist elsewhere in the collisional orogen, they too have been removed or transformed. The processes that could generate a shallow, flat Moho that extends over 1300 km across Phanerozoic, Proterozoic, and Archean crust (e.g., Clowes et al., 2005) include extension, delamination, and metamorphic re-equilibration.

Extension is often associated with thinning of the crust and generation of a flat, reflective Moho (e.g., the Basin and Range; Jarchow et al., 1993). In Alaska, the extension and related plutonism occurred in the Middle Cretaceous (103–96 Ma) beneath central Alaska and southern Brooks Range (e.g., Beaudoin et al., 1994). Central Alaskan crust remains thin, and the crustal thickening developed beneath the Brooks Range more recently during the Paleogene. The extension and related plutonism in the Yukon-Tanana terrane is interpreted as being related to collapse of the orogen thickened by the accretion of the Intermontane superterrane. Within the northern Canadian Cordillera, no major extensional deformation has occurred, although there have been periods of extension in the last 40 Ma driven by variations in transpression/transtension across the Queen Charlotte–Fairweather fault (Hyndman and Hamilton, 1993; Evenchick et al., 1999; Edwards and Russell, 1999). In contrast, extension has played an important role in the southern Cordillera, with significant extensional deformation, crustal thinning, and exhumation documented in the Eocene (59–40 Ma) (e.g., Gabrielse and Yorath, 1991a, 1991b). Despite the differences in extensional history in the Canadian Cordillera, this entire portion of the orogen has thin crust and a flat Moho. Furthermore, this is observed to continue for 1000 km to the east of the northern Cordillera profile into Proterozoic and Archean crust (Clowes et al., 2005). Therefore, either extension is not the primary cause, or there has been more extension in the north than currently interpreted.

Delamination of the thickened, and perhaps tectonically underplated, crust could yield a shallow, flat Moho. However, delamination would have resulted in large-scale magmatic underplating and lower-crustal modification. The geological observations are not consistent with these processes, and, in general, the velocity models beneath the Canadian Cordillera do not support an extensive high-velocity lower crust that would result from the magmatic underplating (Fig. 2C) (Cook and Erdmer, 2005; Clowes et al., 2005).

Finally, it is possible that the Moho does not represent the petrologic crust-mantle boundary. As discussed by Cook and Vasudevan (2003), eclogitized crust can be seismically indistinguishable from peridotite, and therefore it is possible that in some locations the Moho represents a young metamorphic boundary above an eclogitized crustal root. However, the current thermal regime places the Moho in granulite, not eclogite facies. Therefore, the high velocities observed beneath the Moho would require imbrication of granulite facies crust with ultramafic mantle rock, but this is not supported by the limited mantle xenolith data available (e.g., Abraham et al., 2001, 2005; Harder and Russell, 2006), nor by the seismic-reflection images. Fundamental questions remain. How does the Moho equilibrate? What are the petrologic changes that produce the velocity change that defines the Moho and the typical decrease in reflectivity? Additional V_P/V_S, anisotropy, and xenolith data are required to address these issues.

Temperature and Lithospheric Thickness

Temperature influences the strength and thickness of the lithosphere and thus is an important parameter controlling orogenic deformation and magmatism. Heat-flow measurements and seismic velocities enable estimates to be made of lower-crustal and upper-mantle temperatures. Such observations indicate that the thermal regimes of the Canadian and Alaskan Cordillera are significantly different, which influences deformation.

Most of the Canadian Cordillera is presently hot and weak. Upper-mantle velocities are typically well-constrained and provide a useful proxy for temperature. Beneath the Insular, Coast, Intermontane, and Omineca belts, P-wave velocities are consistently slow, ranging from 7.75 to 7.9 km/s (Figs. 2C and 2D [see footnote 1]) (e.g., Clowes et al., 1995, 2005). Only below the eastern Foreland belt and the transition to ancestral North America do uppermost mantle velocities climb to 8.1–8.2 km/s (Welford et al., 2001; Chandra and Cumming, 1972). Assuming a uniform, mantle composition (peridotite), the sub-7.9 km/s velocities are indicative of high temperatures of ~800–900 °C (Black and Braile, 1982). Such high temperatures act to depress crustal velocities by 0.25 km/s, and this is expressed in the consistently slow mid- and lower-crustal velocities observed throughout most of the Canadian Cordillera (Figs. 2C and 2D). This thermal regime is just approaching partial melt conditions for the upper mantle and granulite facies for the lower crust, an inference consistent with the conclusions made from heat-flow (Hyndman and Lewis, 1999a, 1999b; Lewis et al., 1992, 2003) and petrologic (Harder and Russell, 2006) studies. Therefore, the flat Moho and zones in the lower crust can easily act as detachments or localized regions of tectonic deformation. As discussed earlier, this is consistent with NVI images of reflections listric into the lower crust and Moho (Fig. 3) (e.g., Cook et al., 1992, 2004).

The high temperatures also require a thin lithosphere beneath much of the Canadian Cordillera. Upper-mantle reflections at 50–60 km in the southern Canadian Cordillera (Clowes et al., 1995) and 70 km in the northern Canadian Cordillera (Hammer and Clowes, 2004) correspond closely with magnetotelluric (Jones et al., 2005) and petrologic (Harder and Russell, 2006) depth estimates for the base of the lithosphere. To the east of the Canadian Cordillera, seismic wide-angle (Fernandez-Viejo and Clowes, 2003; Gorman et al., 2006) and magnetotelluric (Boerner et al., 1999; Jones et al., 2005) data are interpreted as showing an eastward thickening and cooling of the lithosphere from beneath the eastern Foreland belt to depths of 150–200 km below the craton (Cook and Erdmer, 2005; Hyndman and Lewis, 1999b). Improved constraints on the transition from Cordilleran to cratonic lithosphere in northwestern Canada should result from teleseismic work in the region (Michael Bostock, 2006, personal commun.).

The mechanisms responsible for the current high heat flow and corresponding weak crust are not clearly understood. In the southern Canadian Cordillera, ongoing subduction could generate asthenospheric back flow (Davis and Lewis, 1984; Gough,

1986). With the addition of the seismic results, Clowes et al. (1995) expanded on this concept and proposed the lithospheric model shown in Figure 2D. Although crustal thickness and high upper-mantle temperatures are very similar in the northern Canadian Cordillera, subduction ended 40 Ma. Therefore, other driving mechanisms must be at play. However, this fails to explain temperatures inferred beneath the rest of the northern Canadian Cordillera, which are higher than those in the southern Cordillera (Lewis et al., 2003). Advective heating resulting from slab-window upwelling (e.g., Thorkelson and Taylor, 1989; Frederiksen et al., 1998; Shi et al., 1998) may provide a possible mechanism.

In contrast, the Alaskan lithosphere is not consistently hot, thin, and weak (Fig. 2B). The region most similar to the Canadian Cordilleran is central Alaska, where Beaudoin et al. (1994) modeled slow velocities in the lower crust (6.4 km/s) and uppermost mantle (7.9 km/s). Tertiary intrusions in this region are a testament to the high upper-mantle temperatures (e.g., Beaudoin et al., 1994). Mantle velocities below the subducting Pacific plate are also slow (7.7–7.8 km/s). A localized complex zone beneath the northern Brooks Range exhibits very slow mantle velocities (7.55–7.9 km/s), but these are associated with the North Slope plate as a wedge indentor and thus are anomalous. The remainder of the TACT profiles all report upper-mantle velocities of 8.2 km/s, indicative of a cooler, more typical continental geotherm.

In summary, the reasons for the significant differences in current lithospheric temperatures and thicknesses are not clearly understood but may relate to subduction slab dip (steep vs. flat), slab-window advection, and/or transpression/transtension effects.

SUMMARY

Three corridors of controlled-source seismic studies traverse the Canadian and Alaskan Cordillera and provide lithospheric-scale profiles of structure, composition, and physical properties. These reflection and refraction data and interpretations are a unifying framework for integration of geological investigations and for advancing models of orogen structure and evolution. An along-strike comparison demonstrates that in Canada up to 50% of the allochthonous terranes overlie ancestral North American crust. The accretion surface over Proterozoic-margin rocks forms a crustal-scale décollement beneath deformed, interleaved, and duplexed crust. In Alaska, no cratonic material is preserved except for a couple of small terranes in southwestern Alaska. Thus, in contrast to the Canadian Cordillera, accretion in Alaska occurred without a cratonic backstop.

Although a crustal root is mapped beneath the Alaskan Cordillera (Brooks Range), the Canadian Cordilleran crust is distinctive because it is uniformly thin, hot, and weak despite significant along-strike variations in terranes, deformation history, elevation, and current tectonics. These observations suggest that the Moho is an active, near-solidus, deformation zone that represents a young, re-equilibrated boundary. Many questions remain regarding how this seismic transition zone evolved, its petrologic significance, and what processes are driving the currently high temperatures beneath much of the orogen. Transcurrent, transpressive, and transtensional deformations have been, and are still, very important in the development of the Cordillera. The seismic profiles provide insight into the current fault geometries at depth and demonstrate that four major fault zones (Tintina, Denali, Coast shear zone, and the Queen Charlotte–Fairweather) penetrate the entire crust. Profiles across the convergent active margin reveal generally similar structure in the Cascadia and Wrangell subduction zones. The use of both controlled- and earthquake-source seismic studies is advancing our understanding of the structure and metamorphic processes in these important regions; resolution of contradictory interpretations between techniques will provide an improved understanding of these regimes and may influence analyses of megathrust earthquake hazard.

ACKNOWLEDGMENTS

This paper draws on over two decades of contributions by hundreds of researchers; we are indebted to them. We are grateful to Gary Fuis and an anonymous referee for their interest and comments, which significantly improved the manuscript. The Natural Sciences and Engineering Research Council of Canada (NSERC) and the Geological Survey of Canada, through their joint support of Lithoprobe, provided principal financial support for the SNORCLE and Southern Cordillera Transects as well as much of the adjacent work. The TACT project was funded primarily by the U.S. Geological Survey, and the ACCRETE project was funded through the National Science Foundation. Funds from an NSERC Discovery grant to Clowes supported preparation of this manuscript.

REFERENCES CITED

Abraham, A.C., Francis, D., and Polve, M., 2001, Recent alkaline basalts as probes of the lithospheric roots of the northern Canadian Cordillera: Chemical Geology, v. 175, p. 361–386, doi: 10.1016/S0009-2541(00)00330-2.

Abraham, A.C., Francis, D., and Poive, M., 2005, Origin of Recent alkaline lavas by lithospheric thinning beneath the northern Canadian Cordillera: Canadian Journal of Earth Sciences, v. 42, p. 1073–1095, doi: 10.1139/e04-092.

Ambos, E.L., Mooney, W.D., and Fuis, G.S., 1995, Seismic refraction measurements within the Peninsular terrane, south central Alaska: Journal of Geophysical Research, v. 100, p. 4079–4095, doi: 10.1029/94JB02621.

Beaudoin, B.C., Fuis, G.S., Lutter, W.J., Mooney, W.D., and Moore, T.E., 1994, Crustal velocity structure of the northern Yukon-Tanana upland, central Alaska: Results from TACT refraction/wide-angle reflection data: Geological Society of America Bulletin, v. 106, p. 981–1001, doi: 10.1130/0016-7606(1994)106<0981:CVSOTN>2.3.CO;2.

Black, P.R., and Braile, L.W., 1982, P_n velocity and cooling of the continental lithosphere: Journal of Geophysical Research, v. 87, p. 10,557–10,568.

Boerner, D.E., Kurtz, R.D., Craven, J.A., Ross, G.M., Jones, F.W., and Davis, W.J., 1999, Electrical conductivity in the Precambrian lithosphere of western Canada: Science, v. 283, p. 668–670, doi: 10.1126/science.283.5402.668.

Brocher, T.M., Fisher, M.A., Geist, E.L., and Christensen, N.I., 1989, A high-resolution reflection/refraction study of the Chugach-Peninsular terrane boundary, southern Alaska: Journal of Geophysical Research, v. 94, p. 4441–4455.

Brocher, T.M., Nokleberg, W.J., Christensen, N.I., Lutter, W.J., Geist, E.L., and Fisher, M.A., 1991a, Seismic reflection/refraction mapping of faulting and regional dips in the eastern Alaska Range: Journal of Geophysical Research, v. 96, p. 10,233–10,249.

Brocher, T.M., Moses, M.J., Fisher, M.A., Stephens, C.D., and Geist, E.L., 1991b, Images of the plate boundary beneath southern Alaska, in Meissner, R., et al., eds., Continental Lithosphere: Deep Seismic Reflections: American Geophysical Union Geodynamics Series 22, p. 241–246.

Brocher, T.M., Fuis, G.S., Fisher, M.A., Plafker, G., Moses, M.J., Taber, J.J., and Christensen, N.I., 1994, Mapping the megathrust beneath the northern Gulf of Alaska using wide-angle seismic data: Journal of Geophysical Research, v. 99, p. 11,663–11,685, doi: 10.1029/94JB00111.

Brocher, T.M., Fuis, G.S., Lutter, W.J., Christensen, N.I., and Ratchkovski, N.A., 2004, Seismic velocity models for the Denali fault zone along the Richardson Highway, Alaska: Bulletin of the Seismological Society of America, v. 94, p. S85–S106, doi: 10.1785/0120040615.

Burianyk, M.J.A., and Kanasewich, E.R., 1995, Crustal velocity structure of the Omineca and Intermontane belts, southeastern Canadian Cordillera: Journal of Geophysical Research, v. 100, p. 15,303–15,316, doi: 10.1029/95JB00719.

Burianyk, M.J.A., and Kanasewich, E.R., 1997, Upper mantle structure in the southeastern Canadian Cordillera: Geophysical Research Letters, v. 24, p. 739–742, doi: 10.1029/97GL00675.

Burianyk, M.J.A., Kanasewich, E.R., and Udey, N., 1997, Broadside, wide-angle seismic studies and three-dimensional structure of the crust in the southeast Canadian Cordillera: Canadian Journal of Earth Sciences, v. 34, p. 1156–1166.

Calvert, A.J., 2002, Seismic reflection imaging of the Cascadia plate boundary offshore Vancouver Island, in Kirby, S., Wang, K., and Dunlop, S,. eds., The Cascadia Subduction Zone and Related Subduction Systems: Geological Survey of Canada Open File 4350 and U.S. Geological Survey Open-File Report 02-328, p. 59–62.

Calvert, A.J., and Clowes, R.M., 1990, Deep, high-amplitude reflections from a major shear zone above the subducting Juan de Fuca plate: Geology, v. 18, p. 1091–1094, doi: 10.1130/0091-7613(1990)018 <1091:DHARFA>2.3.CO;2.

Calvert, A.J., and Clowes, R.M., 1991, Seismic evidence for the migration of fluids within the accretionary complex of western Canada: Canadian Journal of Earth Sciences, v. 28, p. 542–556.

Calvert, A.J., Fisher, M.A., Ramachandran, K., and Trehu, A.M., 2003, Possible emplacement of crustal rocks into the forearc mantle of the Cascadia subduction zone: Geophysical Research Letters, v. 30, doi: 10.1029/2003GL018541.

Calvert, A.J., Ramachandran, K., Kao, H., and Fisher, M.A., 2006, Local thickening of the Cascadia forearc crust and the origin of seismic reflectors in the uppermost mantle: Tectonophysics, v. 420, p. 175–188, doi: 10.1016/j.tecto.2006.01.021.

Chandra, N.N., and Cumming, G.L., 1972, Seismic refraction studies in western Canada: Canadian Journal of Earth Sciences, v. 9, p. 1099–1109.

Chardon, D., Andronicos, C.L., and Hollister, L.S., 1999, Large-scale transpressive shear zone patterns and displacements within magmatic arcs: The Coast plutonic complex, British Columbia: Tectonics, v. 18, p. 278–292, doi: 10.1029/1998TC900035.

Clowes, R.M., 2002, Crustal structure of the northern Juan de Fuca plate and Cascadia subduction zone—New results, old data, in Kirby, S., Wang, K., and Dunlop, S., eds., The Cascadia Subduction Zone and Related Subduction Systems: Geological Survey of Canada Open File 4350 and U.S. Geological Survey Open-File Report 02-328, p. 55–58.

Clowes, R.M., and Hyndman, R.D., 2002, Geophysical studies of the northern Cascadia subduction zone off western Canada and their implications for great earthquake seismotectonics: A review, in Fujinawa, Y., and Yoshida, A., eds., Seismotectonics in Convergent Plate Boundaries: Tokyo, Terra Scientific Publishing, p. 1–23.

Clowes, R.M., Ellis, R.M., and Jones, I.F., 1983, Seismic reflections from the subducting lithosphere?: Nature, v. 303, p. 668–670, doi: 10.1038/303668a0.

Clowes, R.M., Brandon, M.T., Green, A.G., Yorath, C.J., Sutherland-Brown, A., Kanasewich, E.R., and Spencer, C., 1987a, LITHOPROBE—Southern Vancouver Island: Cenozoic subduction complex imaged by deep seismic reflections: Canadian Journal of Earth Sciences, v. 29, p. 1813–1864.

Clowes, R.M., Yorath, C.J., and Hyndman, R.D., 1987b, Reflection mapping across the convergent margin of western Canada: Geophysical Journal of the Royal Astronomical Society, v. 89, p. 79–84.

Clowes, R.M., Zelt, C.A., Amor, J.R., and Ellis, R.M., 1995, Lithospheric structure in the southern Canadian Cordillera from a network of seismic refraction lines: Canadian Journal of Earth Sciences, v. 32, p. 1485–1513.

Clowes, R.M., Hammer, P.T.C., Fernandez-Viejo, G., and Welford, J.K., 2005, Lithospheric structure in northwestern Canada from LITHOPROBE seismic refraction and related studies: A synthesis: Canadian Journal of Earth Sciences, v. 42, p. 1277–1293, doi: 10.1139/e04-069.

Coney, P.J., Jones, D.L., and Monger, J.W.H., 1980, Cordilleran suspect terranes: Nature, v. 288, p. 329–333.

Cook, F.A., 1995, The reflection Moho beneath the southern Canadian Cordillera: Canadian Journal of Earth Sciences, v. 32, p. 1520–1530.

Cook, F.A., and Erdmer, P., 2005, An 1800 km cross section of the lithosphere through the northwestern North American plate: Lessons from 4.0 billion years of Earth's history: Canadian Journal of Earth Sciences, v. 42, p. 1295–1311, doi: 10.1139/e04-106.

Cook, F.A., and van der Velden, A., 1995, Three-dimensional crustal structure of the Purcell anticlinorium in the Cordillera of southwestern Canada: Geological Society of America Bulletin, v. 107, p. 642–664, doi: 10.1130/0016-7606(1995)107<0642:TDCSOT>2.3.CO;2.

Cook, F.A., and Varsek, J.L., 1994, Orogen-scale décollements: Reviews of Geophysics, v. 32, p. 37–60, doi: 10.1029/93RG02515.

Cook, F.A., and Vasudevan, K., 2003, Are there relict crustal fragments beneath the Moho?: Tectonics, v. 22, doi: 10.1029/2001TC001341.

Cook, F.A., Varsek, J.L., and Clowes, R.M., 1991, LITHOPROBE reflection transect of southwestern Canada: Mesozoic thrust and fold belt to mid-ocean ridge, in Meissner, R., Brown, L., Durbaum, H.-J., et al., eds., Continental Lithosphere: Deep Seismic Reflections: American Geophysical Union Geodynamics Series 22, p. 247–255.

Cook, F.A., Varsek, J.L., Clowes, R.M., Kanasewich, E.R., Spencer, C.S., Parrish, R.R., Brown, R.L., Carr, S.D., Johnson, B.J., and Price, R.A., 1992, LITHOPROBE crustal reflection cross section of the southern Canadian Cordillera. I: Foreland thrust and fold belt to Fraser River fault: Tectonics, v. 11, p. 12–35.

Cook, F., van der Velden, A., Hall, K., and Roberts, B., 1999, Frozen subduction in Canada's Northwest Territories: Lithoprobe deep lithospheric reflection profiling of the western Canadian shield: Tectonics, v. 18, p. 1–24, doi: 10.1029/1998TC900016.

Cook, F.A., Clowes, R.M., Snyder, D.B., van der Velden, A.J., Hall, K.W., Erdmer, P., and Evenchick, C.A., 2004, Precambrian crust beneath the Mesozoic northern Canadian Cordillera discovered by Lithoprobe seismic reflection profiling: Tectonics, v. 23, doi: 10.1029/2002TC001412.

Cowan, D.S., Brandon, J.T., and Garver, J.I., 1997, Geologic tests of hypothesis for large coastwise displacements—A critique illustrated by the Baja–British Columbia controversy: American Journal of Science, v. 297, p. 117–171.

Creaser, B., and Spence, G., 2005, Lithospheric structure across the northern Cordillera, Yukon Territory, from seismic wide-angle studies: Omineca belt to Intermontane belt: Canadian Journal of Earth Sciences, v. 42, p. 1187–1203, doi: 10.1139/e04-093.

Davis, E.E., and Lewis, T.J., 1984, Heat flow in a back-arc environment: Intermontane and Omineca crystalline belts, southern Canadian Cordillera: Canadian Journal of Earth Sciences, v. 21, p. 715–726.

Dehler, S.A., and Clowes, R.M., 1988, The Queen Charlotte Islands refraction project. Part I. The Queen Charlotte fault zone: Canadian Journal of Earth Sciences, v. 25, p. 1857–1870.

Drew, J.J., and Clowes, R.M., 1990, A re-interpretation of the seismic structure across the active subduction zone of western Canada, in Green, A.G., ed., Studies of Laterally Heterogeneous Structures Using Seismic Refraction and Reflection Data: Geological Survey of Canada Paper 89-13, p. 115–132.

Eberhart-Phillips, D., Christensen, D.H., Brocher, T.M., Hansen, R., Ruppert, N.A., Haeussler, P.J., and Abers, G.A., 2006, Imaging the transition from Aleutian subduction to Yakutat collision in central Alaska, with local earthquakes and active source data: Journal of Geophysical Research, v. 111, doi: 10.1029/2005JB004240.

Edwards, B.R., and Russell, J.K., 1999, Northern Cordilleran volcanic province: A northern Basin and Range?: Geology, v. 27, p. 243–246, doi: 10.1130/0091-7613(1999)027<0243:NCVPAN>2.3.CO;2.

Ellis, R.M., Spence, G.D., Clowes, R.M., Waldron, D.A., Jones, I.F., Green, A.G., Forsyth, D.A., Mair, J.A., Berry, M.J., Mereu, R.F., Kanasewich, E.R., Cumming, G.L., Hajnal, Z., Hyndman, R.D., McMechan, G.A., and Loncarevic, B.D., 1983, The Vancouver Island seismic project: A COCRUST onshore-offshore study at a convergent margin: Canadian Journal of Earth Sciences, v. 20, p. 719–741.

Ellis, S., and Beaumont, C., 1999, Models of convergent boundary tectonics: Implications for the interpretation of Lithoprobe data: Canadian Journal of Earth Sciences, v. 36, p. 1711–1741, doi: 10.1139/cjes-36-10-1711.

Engebretson, D.C., Cox, A., and Gordon, R.G., 1985, Relative motions between oceanic and continental plates of the Pacific Basin: Geological Society of America Special Paper 206, 59 p.

Enkin, R.J., Mahoney, J.B., Baker, J., Riesterer, J., and Haskin, M.L., 2003, Deciphering shallow paleomagnetic inclinations: 2. Implications from Late Cretaceous strata overlapping the Insular/Intermontane superterrane boundary in the southern Canadian Cordillera: Journal of Geophysical Research, v. 108, doi: 10.1029/2002JB001983.

Evenchick, C.A., 1991, Geometry, evolution and tectonic framework of the Skeena fold belt: North central British Columbia: Tectonics, v. 10, p. 537–546.

Evenchick, C.A., Crawford, J.L., McNicoll, V.J., Currie, L.D., and O'Sullivan, P.B., 1999, Early Miocene or younger normal faults and other Tertiary structures in west Nass River map area, northwest British Columbia, and adjacent parts of Alaska: Current Research: Geological Survey of Canada Paper 1999-1A, p. 1–11.

Evenchick, C.A., Gabrielse, H., and Snyder, D., 2005, Crustal structure and lithology of the northern Canadian Cordillera: Alternative interpretations of SNORCLE seismic reflection lines 2a and 2b: Canadian Journal of Earth Sciences, v. 42, p. 1149–1161, doi: 10.1139/e05-009.

Fernandez-Viejo, G., and Clowes, R.M., 2003, Lithospheric structure beneath the Archaean Slave Province and Proterozoic Wopmay orogen, northwestern Canada, from a LITHOPROBE refraction wide/angle reflection survey: Geophysical Journal International, v. 153, p. 1–19, doi: 10.1046/j.1365-246X.2003.01807.x.

Fernandez-Viejo, G., Clowes, R.M., and Amor, J.R., 1999, Imaging the lithospheric mantle in northwestern Canada with seismic wide-angle reflections: Geophysical Research Letters, v. 26, p. 2809–2812, doi: 10.1029/1999GL005373.

Fernández-Viejo, G., Clowes, R.M., and Welford, K., 2005, Constraints on the composition of the crust and upper mantle in northwestern Canada: Vp, Vs and Poisson's ratio variations along Lithoprobe's SNORCLE transect: Canadian Journal of Earth Sciences, v. 42, p. 1205–1222, doi: 10.1139/e05-028.

Ferris, A., Abers, G.A., Christensen, D.H., and Veenstra, E., 2003, High resolution image of the subducted Pacific (?) plate beneath central Alaska, 50–150 km depth: Earth and Planetary Science Letters, v. 214, p. 575–588, doi: 10.1016/S0012-821X(03)00403-5.

Fisher, M.A., Brocher, T.M., Nokleberg, W.J., Plafker, G., and Smith, G.L., 1989, Seismic reflection images of the crust of the northern part of the Chugach terrane, Alaska: Results of a survey for the Trans-Alaska Crustal Transect (TACT): Journal of Geophysical Research, v. 94, p. 4424–4440.

Fisher, M.A., Nokleberg, W.J., Ratchkovski, N.A., Pellerin, L., Glen, J.M., Brocher, T.M., and Booker, J., 2004, Geophysical investigation of the Denali fault and Alaska Range orogen within the aftershock zone of the October–November 2002, M = 7.9 Denali fault earthquake: Geology, v. 32, p. 269–272, doi: 10.1130/G20127.1.

Frederiksen, A.W., Bostock, M.G., VanDecar, J.C., and Cassidy, J.F., 1998, Seismic structure of the upper mantle beneath the northern Canadian Cordillera from teleseismic travel-time inversion: Tectonophysics, v. 294, p. 43–55, doi: 10.1016/S0040-1951(98)00095-X.

Fuis, G.S., 1998, West margin of North America—A synthesis of recent seismic transects: Tectonophysics, v. 288, p. 265–292, doi: 10.1016/S0040-1951(97)00300-4.

Fuis, G.S., and Clowes, R.M., 1993, Comparison of deep structure along three transects of the western North American continental margin: Tectonics, v. 12, p. 1420–1435.

Fuis, G.S., and Plafker, G., 1991, Evolution of deep structure along the Trans-Alaska Crustal Transect, Chugach Mountains and Copper River Basin, southern Alaska: Journal of Geophysical Research, v. 96, p. 4229–4253.

Fuis, G.S., Ambos, E.L., Mooney, W.D., Christensen, N.I., and Geist, E., 1991, Crustal structure of accreted terranes in southern Alaska, Chugach Mountains and Copper River Basin, from seismic refraction results: Journal of Geophysical Research, v. 96, p. 4187–4227.

Fuis, G.S., Levander, A.R., Lutter, W.J., Wissinger, E.S., Moore, T.E., and Christensen, N.I., 1995, Seismic images of the Brooks Range, Arctic Alaska, reveal crustal-scale duplexing: Geology, v. 23, p. 65–68, doi: 10.1130/0091-7613(1995)023<0065:SIOTBR>2.3.CO;2.

Fuis, G.S., Murphy, J.M., Lutter, W.J., Moore, T.E., Bird, K.J., and Christensen, N.I., 1997, Deep seismic structure and tectonics of northern Alaska: Crustal-scale duplexing with deformation extending into the upper mantle: Journal of Geophysical Research, v. 102, p. 20,873–20,896, doi: 10.1029/96JB03959.

Gabrielse, H., 1985, Major dextral transcurrent displacements along the Northern Rocky Mountain Trench and related lineaments in north-central British Columbia: Geological Society of America Bulletin, v. 96, p. 1–14, doi: 10.1130/0016-7606(1985)96<1:MDTDAT>2.0.CO;2.

Gabrielse, H., and Yorath, C.J., eds., 1991a, Geology of the Cordilleran Orogen in Canada: Geological Survey of Canada, Geology of Canada, No. 4 (also Boulder, Colorado, Geological Society of America, Geology of North America, v. G-2), 844 p.

Gabrielse, H., and Yorath, C.J., 1991b, Tectonic synthesis, in Gabrielse, H., and Yorath, C.J., eds., Geology of the Cordilleran Orogen in Canada: Geological Survey of Canada, Geology of Canada, No. 4 (also Boulder, Colorado, Geological Society of America, Geology of North America, v. G-2), p. 677–705.

Gabrielse, H., Monger, J.W.H., Wheeler, J.O., and Yorath, C.J., 1991, Tectonic framework: Part A. Morphogeologic belts, tectonic assemblages and terranes, in Gabrielse, H., and Yorath, C.J., eds., Geology of the Cordilleran Orogen in Canada: Geological Survey of Canada, Geology of Canada, No. 4 (also Boulder, Colorado, Geological Society of America, Geology of North America, v. G-2), p. 15–28.

Gehrels, G., 2001, Geology of the Chatham Sound region, southeast Alaska and coastal British Columbia: Canadian Journal of Earth Sciences, v. 38, p. 1579–1599, doi: 10.1139/cjes-38-11-1579.

Ghosh, D.K., 1995, Nd-Sr isotopic constraints on the interactions of the Intermontane superterrane with the western edge of North America in southern Canadian Cordillera: Canadian Journal of Earth Sciences, v. 32, p. 1740–1758.

Gorman, A.R., Nemeth, B., Clowes, R.M., and Hajnal, Z., 2006, An investigation of upper mantle heterogeneity beneath the Archaean and Proterozoic crust of western Canada from Lithoprobe controlled-source seismic experiments: Tectonophysics, v. 416, p. 187–207, doi: 10.1016/j.tecto.2005.11.023.

Gough, D.I., 1986, Mantle upflow tectonics in the Canadian Cordillera: Journal of Geophysical Research, v. 91, p. 1909–1920.

Grantz, A., Moore, T.E., and Roeske, S., 1991, Gulf of Alaska to Arctic Ocean: Centennial Continent Ocean Transect 15: Boulder, Colorado, Geological Society of America, 72 p., 3 sheets, scale 1:500,000.

Green, A.G., Clowes, R.M., Yorath, C.J., Spence, C.P., Kanasewich, E.R., Brandon, M.T., and Sutherland-Brown, A., 1986, Seismic reflection imaging of the subducting Juan de Fuca plate: Nature, v. 319, p. 210–213, doi: 10.1038/319210a0.

Hammer, P.T.C., and Clowes, R.M., 2004, Accreted terranes of northwestern British Columbia, Canada: Lithospheric velocity structure and tectonics: Journal of Geophysical Research, v. 109, p. B06305, doi: 10.1029/2003JB002749.

Hammer, P.T.C., Clowes, R.M., and Ellis, R.M., 2000, Crustal structure of N.W. British Columbia and S.E. Alaska from seismic wide-angle studies: Coast plutonic complex to Stikinia: Journal of Geophysical Research, v. 105, p. 7961–7981, doi: 10.1029/1999JB900378.

Harder, M., and Russell, J.K., 2006, Thermal state of the upper mantle beneath the northern Cordilleran volcanic province, British Columbia, Canada: Lithos, v. 87, p. 1–22, doi: 10.1016/j.lithos.2005.05.002.

Hollister, L.S., 1982, Metamorphic evidence for rapid (2 mm/yr) uplift of a portion of the Central gneiss complex, Coast Mountains, BC: Canadian Mineralogy, v. 30, p. 319–332.

Hollister, L.S., and Andronicos, C.L., 1997, A candidate for the Baja–British Columbia fault system in the Coast plutonic complex: GSA Today, v. 7, no. 11, p. 1–7.

Hyndman, R.D., and Ellis, R.M., 1981, Queen Charlotte fault zone: Microearthquakes from a temporary array of land stations and ocean bottom seismographs: Canadian Journal of Earth Sciences, v. 18, p. 776–788.

Hyndman, R.D., and Hamilton, T.S., 1993, Queen Charlotte area Cenozoic tectonics and volcanism and their association with relative plate motions along the northeastern Pacific margin: Journal of Geophysical Research, v. 98, p. 14,257–14,278.

Hyndman, R.D., and Lewis, T.J., 1999a, Review: The thermal regime along the southern Canadian Cordillera Lithoprobe corridor: Canadian Journal of Earth Sciences, v. 32, p. 1611–1617.

Hyndman, R.D., and Lewis, T.J., 1999b, Geophysical consequences of the Cordillera-craton thermal transition in southwestern Canada: Tectonophysics, v. 306, p. 397–422, doi: 10.1016/S0040-1951(99)00068-2.

Hyndman, R.D., Yorath, C.J., Clowes, R.M., and Davis, E.E., 1990, The northern Cascadia subduction zone at Vancouver Island: Seismic structure and tectonic history: Canadian Journal of Earth Sciences, v. 27, p. 313–329.

Hyndman, R.D., Flück, P., Mazzotti, S., Lewis, T.J., Ristau, J., and Leonard, L., 2005, Current tectonics of the northern Canadian Cordillera: Canadian Journal of Earth Sciences, v. 42, p. 1117–1136, doi: 10.1139/e05-023.

Irving, E., Wynne, P.J., Thorkelson, D.J., and Schiarizza, P., 1996, Large (1000 to 4000 km) northward movements of tectonic domains in the northern Cordillera, 83 to 45 Ma: Journal of Geophysical Research, v. 101, p. 17,901–17,916, doi: 10.1029/96JB01181.

Jarchow, C.M., Thompson, G.A., Catchings, R.D., and Mooney, W.D., 1993, Seismic evidence for active magmatic underplating beneath the Basin and Range Province, western United States: Journal of Geophysical Research, v. 98, p. 22,095–22,108.

Johnston, S.T., 2001, The Great Alaskan terrane wreck: Reconciliation of paleomagnetic and geological data in the northern Cordillera: Earth and Planetary Science Letters, v. 193, p. 259–272, doi: 10.1016/S0012-821X (01)00516-7.

Jones, A.G., Ledo, J., Ferguson, I.J., Farquharson, C., Garcia, X., Grant, N., McNeice, G., Roberts, B., Spratt, J., Wennberg, G., Wolynec, L., and Wu, X., 2005, A 1600-km-long magnetotelluric transect from the Archean to the Tertiary: SNORCLE MT overview: Canadian Journal of Earth Sciences, v. 42, p. 1257–1275, doi: 10.1139/e05-080.

Kanasewich, E.R., Burianyk, M.J.A., Ellis, R.M., Clowes, R.M., White, D.J., Côté, T., Forsyth, D.A., Luetgert, J.H., and Spence, G.D., 1994, Crustal velocity structure of the Omineca belt, southeastern Canadian Cordillera: Journal of Geophysical Research, v. 99, p. 2653–2670, doi: 10.1029/93JB03108.

Klepeis, K.A., Crawford, M.L., and Gehrels, G., 1998, Structural history of the crustal-scale Coast shear zone north of Portland Canal, southeast Alaska and British Columbia: Journal of Structural Geology, v. 20, p. 883–904, doi: 10.1016/S0191-8141(98)00020-0.

Kurtz, R.D., DeLaurier, J.M., and Gupta, J.C., 1990, The electrical conductivity distribution beneath Vancouver Island: A region of active plate subduction: Journal of Geophysical Research, v. 95, p. 10,929–10,946.

Lane, L.S., 1997, Canada Basin, Arctic Ocean: Evidence against a rotational origin: Tectonics, v. 16, p. 363–387, doi: 10.1029/97TC00432.

Lawver, L.A., and Scotese, C.R., 1990, A review of tectonic models for the evolution of the Canada Basin, in Grantz, A., Johnson, L., and Sweeney, J.F., eds., The Arctic Ocean Region: Boulder, Colorado, Geological Society of America, Geology of North America, v. L, p. 593–618.

Levander, A., Fuis, G.S., Wissinger, E.S., Lutter, W.J., Oldow, J.S., and Moore, T.E., 1994, Seismic images of the Brooks Range fold and thrust belt, Arctic Alaska, from an integrated seismic reflection/refraction experiment: Tectonophysics, v. 232, p. 13–30, doi: 10.1016/0040-1951(94)90073-6.

Lewis, T.J., Bentkowski, W.H., and Hyndman, R.D., 1992, Crustal temperatures near the southern Canadian Cordillera transect: Canadian Journal of Earth Sciences, v. 29, p. 1197–1214.

Lewis, T.J., Hyndman, R.D., and Flück, P., 2003, Heat flow, heat generation, and crustal temperatures in the northern Canadian Cordillera: Thermal control of tectonics: Journal of Geophysical Research, v. 108, doi: 10.1029/2002JB002090.

Lowe, C., Horner, R.B., Mortensen, J.K., Johnson, S.T., and Roots, C.F., 1994, New geophysical data from the northern Cordillera: Preliminary interpretations and implications for the tectonics and deep geology: Canadian Journal of Earth Sciences, v. 31, p. 891–904.

Mackie, D.J., Clowes, R.M., Dehler, S.A., Ellis, R.M., and Morel-à-l'Huissier, P., 1989, The Queen Charlotte Islands refraction project: Part II. Structural model for transition from Pacific plate to North American plate: Canadian Journal of Earth Sciences, v. 26, p. 1713–1725.

Mazzotti, S., and Hyndman, R.D., 2002, Yakutat collision and strain transfer across the northern Canadian Cordillera: Geology, v. 30, p. 495–498, doi: 10.1130/0091-7613(2002)030<0495:YCASTA>2.0.CO;2.

McCausland, P.J.A., Symons, D.T.A., and Hart, C.J.R., 2005, Rethinking "Yellowstone in Yukon" and Baja–British Columbia: Paleomagnetism of the Late Cretaceous Swede Dome stock, northern Canadian Cordillera: Journal of Geophysical Research, v. 110, doi: 10.1029/2005JB003742.

McMechan, G.A., and Spence, G.D., 1983, P-wave velocity structure of the Earth's crust beneath Vancouver Island: Canadian Journal of Earth Sciences, v. 20, p. 742–752.

Monger, J.W.H., 1993, Canadian Cordilleran tectonics: From geosynclines to crustal collage: Canadian Journal of Earth Sciences, v. 30, p. 209–231.

Monger, J.W.H., and Price, R.A., 2002, The Canadian Cordillera: Geology and tectonic evolution: Canadian Society of Exploration Geophysics Recorder, v. 27, p. 17–36.

Monger, J.W.H., Price, R.A., and Tempelman-Kluit, D.J., 1982, Tectonic accretion and the origin of the two major metamorphic and plutonic welts in the Canadian Cordillera: Geology, v. 10, p. 70–75, doi: 10.1130/0091-7613 (1982)10<70:TAATOO>2.0.CO;2.

Monger, J.W.H., van der Heyden, P., Journeay, J.M., Evenchick, C.A., and Mahoney, J.B., 1994, Jurassic-Cretaceous basins along the Canadian Coast belt; their bearing on pre-Mid-Cretaceous sinistral displacements: Geology, v. 22, p. 175–178, doi: 10.1130/0091-7613(1994)022<0175:JCBATC>2.3.CO;2.

Moore, J.C., Diebold, J., Fisher, M.A., Sample, J., Brocher, T., Talwani, M., Ewing, J., von Huene, R., Rowe, C., Stone, D., Stevens, C., and Sawyer, D., 1991, EDGE deep seismic reflection transect of the eastern Aleutian arc-trench layered lower crust reveals underplating and continental growth: Geology, v. 19, p. 420–424, doi: 10.1130/0091-7613 (1991)019<0420:EDSRTO>2.3.CO;2.

Moore, T.E., Wallace, W.K., Bird, K.J., Karl, S.M., Mull, C.G., and Dillon, J.T., 1994, Geology of northern Alaska, in Plafker, G., and Berg, H.C., eds., The Geology of Alaska: Boulder, Colorado, Geological Society of America, Geology of North America, v. G-1, p. 49–140.

Moore, V.M., and Wiltschko, D.V., 2004, Syncollisional delamination and tectonic wedge development in convergent orogens: Tectonics, v. 23, doi: 10.1029/2002TC001430.

Morozov, I.B., Smithson, S.B., Hollister, L.S., and Diebold, J.B., 1998, Wide-angle seismic imaging across accreted terranes, southeastern Alaska and western British Columbia: Tectonophysics, v. 299, p. 281–296, doi: 10.1016/S0040-1951(98)00208-X.

Morozov, I.B., Smithson, S.B., Chen, J., and Hollister, L.S., 2001, Generation of new continental crust and terrane accretion in southeastern Alaska and western British Columbia from P- and S-wave wide-angle seismic data (ACCRETE): Tectonophysics, v. 341, p. 49–67, doi: 10.1016/S0040-1951 (01)00190-1.

Morozov, I.B., Christensen, N.I., Smithson, S.B., and Hollister, L.S., 2003, Seismic and laboratory constraints on crustal formation in a continental arc (ACCRETE, southeastern Alaska and western British Columbia): Journal of Geophysical Research, v. 108, doi: 10.1029/2001JB001740.

Murphy, D.C., and Mortensen, J.K., 2003, Late Paleozoic and Mesozoic features constrain displacement on Tintina fault and limit large-scale orogen-parallel displacement in the northern Cordillera: Geological Association of Canada–Mineralogical Association of Canada–Society of Economic Geologists Joint Annual Meeting, v. 28 (CD-ROM), no 151.

Nedimović, M.R., Hyndman, R.D., Ramachandran, K., and Spence, G.D., 2003, Reflection signature of seismic and aseismic slip on the northern Cascadia subduction interface: Nature, v. 424, p. 416–420, doi: 10.1038/ nature01840.

Nicholson, T., Bostock, M., and Cassidy, J.F., 2005, New constraints on subduction zone structure in northern Cascadia: Geophysical Journal International, v. 161, p. 849–859, doi: 10.1111/j.1365-246X.2005.02605.x.

Oldow, J.S., Bally, A.W., Avé Lallemant, H.G., and Leeman, W.P., 1989, Phanerozoic evolution of the North American Cordillera; United States and Canada, in Bally, A.W., and Palmer, A.R., eds., The Geology of North America—An Overview: Boulder, Colorado, Geological Society of America, Geology of North America, v. A, p. 139–232.

O'Leary, D.M., Clowes, R.M., and Ellis, R.M., 1993, Crustal velocity structure in the southern Coast belt, British Columbia: Canadian Journal of Earth Sciences, v. 30, p. 2389–2403.

Perz, M.J., 1993, Characterization of the Fraser Fault, Southwestern British Columbia, and Surrounding Geology through Reprocessing of Seismic Reflection Data [M.Sc. thesis]: Vancouver, Canada, University of British Columbia, 191 p.

Plafker, G., and Berg, H.C., eds., 1994a, The Geology of Alaska: Boulder, Colorado, Geological Society of America, Geology of North America, v. G-1, 1021 p.

Plafker, G., and Berg, H.C., 1994b, Overview of the geology and tectonic evolution of Alaska, in Plafker, G., and Berg, H.C., eds., The Geology of Alaska: Boulder, Colorado, Geological Society of America, Geology of North America, v. G-1, p. 989–1021.

Plafker, G., Moore, J.C., and Winkler, G.R., 1994, Geology of the southern Alaska margin, in Plafker, G., and Berg, H.C., eds., The Geology of Alaska: Boulder, Colorado, Geological Society of America, Geology of North America, v. G-1, p. 389–449.

Price, R.A., and Carmichael, D.M., 1986, Geometric test for Late Cretaceous–Paleogene intracontinental transform faulting in the Canadian Cordillera: Geology, v. 14, p. 468–471, doi: 10.1130/0091-7613(1986)14<468:GTFLCI>2.0.CO;2.

Price, R.A., and Monger, J.W.H., 2000, A transect of the southern Canadian Cordillera from Calgary to Vancouver: Geological Association of Canada, Cordilleran Section, Field Trip Guidebook, 164 p.

Ramachandran, K., and Hyndman, R.D., 2007, P- and S-wave velocity structure beneath the northern Cascadia subduction zone: Constraints on serpentinized forearc mantle wedge: Geophysical Journal International (in press).

Ramachandran, K., Dosso, S.E., Zelt, C.A., Spence, G.D., Hyndman, R.D., and Brocher, T.M., 2004, Upper crustal structure of southwestern British Columbia from the 1998 Seismic Hazards Investigation in Puget Sound: Journal of Geophysical Research, v. 109, p. B09303, doi: 10.1029/2003JB002826.

Ramachandran, K., Hyndman, R.D., and Brocher, T.M., 2006, Regional P wave velocity structure of the Northern Cascadia Subduction Zone: Journal of Geophysical Research, v. 111, doi: 10.1029/2005JB004108.

Rusmore, M.E., Gehrels, G., and Woodsworth, G.J., 2001, Southern continuation of the Coast shear zone and Paleocene strain partitioning in British Columbia–southeast Alaska: Geological Society of America Bulletin, v. 113, p. 961–975, doi: 10.1130/0016-7606(2001)113<0961:SCOTCS>2.0.CO;2.

Sears, J.W., and Price, R.A., 2000, New look at the Siberian connection: No SWEAT: Geology, v. 28, p. 423–426, doi: 10.1130/0091-7613(2000)28<423:NLATSC>2.0.CO;2.

Shi, S., Francis, D., Ludden, J., Frederiksen, A., and Bostock, M., 1998, Xenolith evidence for lithospheric melting above anomalously hot mantle beneath the northern Canadian Cordillera: Contributions to Mineralogy and Petrology, v. 131, p. 39–53, doi: 10.1007/s004100050377.

Snyder, D.B., Clowes, R.M., Cook, F.A., Erdmer, P., Evenchick, C.A., van der Velden, A.J., and Hall, K.W., 2002, Proterozoic prism arrests suspect terranes: Insights into the ancient Cordilleran margin from seismic reflection data: GSA Today, v. 12, no. 10, p. 4–10, doi: 10.1130/1052-5173(2002)012<0004:PPASTI>2.0.CO;2.

Snyder, D.B., Roberts, B.J., and Gordey, S.P., 2005, Contrasting seismic characteristics of three major faults in northwestern Canada: Canadian Journal of Earth Sciences, v. 42, p. 1223–1237, doi: 10.1139/e05-027.

Spence, G.D., and Asudeh, I., 1993, Seismic velocity structure of the Queen Charlotte Basin beneath Hecate Strait: Canadian Journal of Earth Sciences, v. 30, p. 787–805.

Spence, G.D., and McLean, N.A., 1998, Crustal seismic velocity and density structure of the Intermontane and Coast belts, southwestern Cordillera: Canadian Journal of Earth Sciences, v. 35, p. 1362–1379, doi: 10.1139/cjes-35-12-1362.

Spence, G.D., Clowes, R.M., and Ellis, R.M., 1985, Seismic structure across the active subduction zone of western Canada: Journal of Geophysical Research, v. 90, p. 6754–6772.

Thorkelson, D.J., and Taylor, R.P., 1989, Cordilleran slab windows: Geology, v. 17, p. 833–836, doi: 10.1130/0091-7613(1989)017<0833:CSW>2.3.CO;2.

Thorkelson, D.J., Mortensen, J.K., Creaser, R.A., Davidson, G.J., and Abbott, J.G., 2001, Early Proterozoic magmatism in Yukon, Canada: Constraints on the evolution of northwestern Laurentia: Canadian Journal of Earth Sciences, v. 38, p. 1479–1494, doi: 10.1139/cjes-38-10-1479.

Thorkelson, D.J., Abbott, J.G., Mortensen, J.K., Creaser, R.A., Villeneuve, M.E., McNicoll, V.J., and Layer, P.W., 2005, Early and Middle Proterozoic evolution of Yukon, Canada: Canadian Journal of Earth Sciences, v. 42, p. 1045–1071, doi: 10.1139/e04-075.

Umhoefer, P.J., Schiarizza, P., and Robinson, M., 2002, Relay Mountain Group, Tyaughton–Methow basin, southwest British Columbia: A major Middle Jurassic to Early Cretaceous terrane overlap assemblage: Canadian Journal of Earth Sciences, v. 39, p. 1143–1167, doi: 10.1139/e02-031.

van der Velden, A.J., and Cook, F.A., 1996, Tectonic development of the southern Rocky Mountain Trench: Tectonics, v. 15, p. 517–544, doi: 10.1029/95TC03288.

Varsek, J.L., Cook, F.A., Clowes, R.M., Journeay, J.M., Monger, J.W.H., Parrish, R.R., Kanasewich, E.R., and Spencer, C.S., 1993, LITHOPROBE crustal reflection structure of the Southern Canadian Cordillera. 2: Coast mountains transect: Tectonics, v. 12, p. 334–360.

Welford, J.K., Clowes, R.M., Ellis, R.M., Spence, G.D., Asudeh, I., and Hajnal, Z., 2001, Lithospheric structure across the craton-Cordilleran transition of northeastern British Columbia: Canadian Journal of Earth Sciences, v. 38, p. 1169–1189, doi: 10.1139/cjes-38-8-1169.

Wissinger, E.S., Levander, A., and Christensen, N.I., 1997, Seismic images of crustal duplexing and continental subduction in the Brooks Range: Journal of Geophysical Research, v. 102, p. 20,847–20,871, doi: 10.1029/96JB03662.

Wolf, L.W., Stone, D.B., and Davies, J.N., 1991, Crustal structure of the active margin, south central Alaska: An interpretation of seismic refraction data from the Trans-Alaska Transect: Journal of Geophysical Research, v. 96, p. 16,455–16,469.

Yorath, C.J., and Hyndman, R.D., 1983, Subsidence and thermal history of Queen Charlotte Basin: Canadian Journal of Earth Sciences, v. 20, p. 135–159.

Yorath, C.J., Green, A.G., Clowes, R.M., Sutherland-Brown, A., Brando, M.T., Kanasewich, E.R., Hyndman, R.D., and Spencer, C.P., 1985, LITHOPROBE, southern Vancouver Island: Seismic reflection sees through Wrangellia to the Juan de Fuca plate: Geology, v. 13, p. 759–763, doi: 10.1130/0091-7613(1985)13<759:LSVISR>2.0.CO;2.

Zelt, B.C., Ellis, R.M., Clowes, R.M., Kanasewich, E.R., Asudeh, I., Luetgert, J.H., Hajnal, Z., Ikami, A., Spence, G.D., and Hyndman, R.D., 1992, Crust and upper mantle velocity structure of the Intermontane belt, southern Canadian Cordillera: Canadian Journal of Earth Sciences, v. 29, p. 1530–1548.

Zelt, B.C., Ellis, R.M., and Clowes, R.M., 1993, Crustal velocity structure in the eastern Insular and southernmost Coast belts, Canadian Cordillera: Canadian Journal of Earth Sciences, v. 30, p. 1014–1027.

Zelt, B.C., Ellis, R.M., Clowes, R.M., and Hole, J.A., 1995, Inversion of three-dimensional wide-angle seismic data from the southwestern Canadian Cordillera: Journal of Geophysical Research, v. 101, p. 8503–8530, doi: 10.1029/95JB02807.

Zelt, C.A., and White, D.J., 1995, Crustal structure and tectonics of the southeastern Canadian Cordillera: Journal of Geophysical Research, v. 100, p. 24,255–24,274, doi: 10.1029/95JB02632.

Zelt, C.A., Ellis, R.M., and Zelt, B.C., 2006, 3-D structure across the Tintina strike-slip fault, northern Canadian Cordillera, from seismic refraction and reflection tomography: Geophysical Journal International, v. 167, p. 1292–1308, doi: 10.1111/j.1365-246X.2006.03090.x.

MANUSCRIPT ACCEPTED BY THE SOCIETY 22 MARCH 2007

Printed in the USA

The Geological Society of America
Special Paper 433
2007

A synthesis of the Jurassic–Cretaceous tectonic evolution of the central and southeastern Canadian Cordillera: Exploring links across the orogen

Carol A. Evenchick*
Geological Survey of Canada, 625 Robson Street, Vancouver, British Columbia V6B 5J3, Canada

Margaret E. McMechan*
Geological Survey of Canada, 3303 33rd Street NW, Calgary, Alberta T2L 2A7, Canada

Vicki J. McNicoll*
Geological Survey of Canada, 601 Booth Street, Ottawa, Ontario K1A 0E8, Canada

Sharon D. Carr*
Ottawa-Carleton Geoscience Centre, Department of Earth Sciences, Carleton University, Ottawa, Ontario K1S 5B6, Canada

ABSTRACT

Restoration of tectonic elements in the central interior of the Canadian Cordillera southward to their paleogeographic position in the Mesozoic permits comparison of data across the active orogen, recognition of the interplay between coeval lithospheric thickening and basin evolution, and new constraints on models of tectonic evolution. The onset of Middle Jurassic clastic sedimentation in the Bowser basin, on the west side of the Jurassic orogen, occurred in response to accretionary events farther inboard. Shortening and thickening of the crust between the Alberta foreland basin on the east side of the Jurassic orogen and Bowser basin on the west side resulted in an Omineca highland between the two basins and lithospheric loading that influenced their Late Jurassic–Cretaceous sedimentation. The provenance of detritus in these basins, and in the Late Cretaceous Sustut basin on the east side of the Bowser basin, reveals migration of drainage divides in the intervening Omineca highland through time. Synchronous and compatible tectonic events within the basins and evolving accretionary orogen, and in rocks of the Stikine terrane and the western margin of North America, suggest that they were kinematically connected above a lower-crust detachment, beginning in the Middle Jurassic. The Coast belt was part of this wide, dynamically linked bivergent orogen from the mid-Cretaceous to earliest Cenozoic,

---

*E-mails: Evenchick: cevenchi@nrcan.gc.ca; McMechan: mmcmecha@nrcan.gc.ca; McNicoll: vmcnicol@nrcan.gc.ca; Carr: scarr@earthsci.carleton.ca.

Evenchick, C.A., McMechan, M.E., McNicoll, V.J., and Carr, S.D., 2007, A synthesis of the Jurassic–Cretaceous tectonic evolution of the central and southeastern Canadian Cordillera: Exploring links across the orogen, *in* Sears, J.W., Harms, T.A., and Evenchick, C.A., eds., Whence the Mountains? Inquiries into the Evolution of Orogenic Systems: A Volume in Honor of Raymond A. Price: Geological Society of America Special Paper 433, p. 117–145, doi: 10.1130/2007.2433(06). For permission to copy, contact editing@geosociety.org. ©2007 The Geological Society of America. All rights reserved.

and the lower-crust detachment rooted near the active plate margin. Nested within the orogen, the east-vergent thin-skinned Skeena fold belt, equivalent in scale to the Rocky Mountain fold-and-thrust belt, was also linked to the detachment system.

Keywords: Canadian Cordillera, tectonic evolution, Bowser basin, Alberta foreland basin, Sustut basin, Omineca belt.

INTRODUCTION

The relationships among basin evolution, fold-and-thrust belt development, and thermotectonic evolution of core zones provide insight into progressive stages of orogenesis. The dynamic link between these realms also provides a basis for geodynamic modeling and feedback for understanding orogen-scale processes. The southeast Canadian Cordillera is an excellent example. The Late Jurassic through earliest Cenozoic inter-relationships among the northeasterly vergent thin-skinned fold-and-thrust belt of the Foreland belt (Fig. 1) on the east side of the Cordilleran orogen, the synorogenic Alberta foreland basin deposits within and east of the Foreland belt, and the internal core zone of the Omineca belt are relatively well understood (Bally et al., 1966; Price and Mountjoy, 1970; Price, 1973, 1981; Brown et al., 1986; McMechan and Thompson, 1989; Fermor and Moffat, 1992; Beaumont et al., 1993). Studies have associated part of their evolution to accretion of terranes in the western Omineca and eastern Intermontane belts during westward underthrusting of the North American plate (e.g., Cant and Stockmal, 1989; Brown et al., 1992a; Price, 1994; Brown and Gibson, 2006; Carr and Simony, 2006). However, analyses commonly extend only as far west as the easternmost of the accreted terranes on the west side of the Omineca belt, and relationships of orogenic processes in the Foreland and Omineca belts to those of the Intermontane and Coast belts farther west have not been explored, nor have their implications for tectonic evolution of the Cordillera as a whole.

New understanding of exhumed mid-crustal rocks exposed in core zones of the Canadian Cordillera (Coast and Omineca belts, Fig. 1) and of depositional and structural histories of the sedimentary basins that flank them, in particular the Bowser, Sustut, and Alberta foreland basins (Fig. 1), is used herein to illustrate links between the major tectonic elements of the orogen. One result is a tectonic reconstruction in which the entire width of the orogen is kinematically linked throughout the Mesozoic–early Cenozoic (Fig. 2). It evolved from a predominantly west-verging "small-cold" (Beaumont et al., 2006) accretionary orogen, ~300 km wide in the Middle Jurassic (Fig. 2A), into an ~1000-km-wide "large-hot" (Beaumont et al., 2006) bivergent orogen in the mid-Cretaceous (Fig. 2B). By the mid-Cretaceous, an unusual geometry developed with two major detachment systems (Figs. 2B and 2C). A lower-crust detachment system extending across the entire orogen rooted near the active plate margin and joined the western magmatic convergent belt (Coast belt) to the eastern front of deformation in the thin-skinned, east-vergent foreland fold-and-thrust belt. A second detachment rooted in the Coast belt rose eastward to relatively high structural levels, forming the basal detachment of the east-vergent Skeena fold-and-thrust belt nested in the interior of the orogen—a fold belt that matches the classic Rocky Mountain fold-and-thrust belt (Foreland belt and eastern Omineca belt) in width and magnitude of horizontal shortening.

In this paper, we include a synthesis of the depositional and structural histories of the two major Mesozoic basins, the Bowser and Sustut basins (Fig. 1), that formed west of the site of initial accretionary orogenesis now preserved in the southern Omineca belt. The Bowser basin is a largely marine basin that formed the Middle Jurassic to Early Cretaceous western continental margin of North America during and following the accretion of its basement, Stikinia (Stikine terrane), to the North American plate (e.g., Evenchick and Thorkelson, 2005, and references therein). The Sustut basin was the site of Late Cretaceous nonmarine, synorogenic clastic sedimentation confined between, and sourced from, the evolving Skeena fold belt on the west and the Omineca highland on the east (Eisbacher, 1974a, 1985; Evenchick, 1991a). Previous work of Ricketts et al. (1992) has shown how early subsidence and sedimentation in the Bowser basin in the latest Early Jurassic or early Middle Jurassic in the central Intermontane belt was related to southwest thrusting of Cache Creek terrane (Fig. 1) over Stikinia, and that these events were contemporaneous with southwest-directed thrust faults in the southern Omineca belt. Eisbacher (1981, 1985) included the depositional and structural history of the basins west of the Omineca belt in a synthesis of depositional patterns across the Cordillera. We build on these works, using data and interpretations from more than 20 yr of research since Eisbacher's analysis, to examine the relationships between Mesozoic depositional and structural events across the orogen. The geological history of the Bowser basin is of fundamental importance to our understanding of the tectonic development of the Cordilleran orogen because it contains the earliest depositional record in response to mid-Mesozoic terrane accretion in the Canadian Cordillera, as well as the record of continued lithospheric response to Jurassic–Cretaceous development of the thickening orogen between the Bowser basin and the Alberta foreland basin; in this regard, the Bowser basin is the western counterpart to the Alberta foreland basin. Also critical to a regional tectonic analysis is the role of the Skeena fold belt, a thin-skinned fold-and-thrust belt that deformed Bowser and Sustut strata at the same time as horizontal shortening occurred at all structural levels in the Coast and Omineca belts and at upper-crustal levels in the Foreland belt (Evenchick, 1991a). The site of the Bowser basin evolved from a region of marine deposition on the western margin of the continent in the Jurassic, to a region of significant horizontal shortening in the interior of the orogen during the Cretaceous, as well as localized Late Cretaceous sedimentation in the Sustut basin. The Bowser and Sustut basins are thus keystones bridging the eastern and western parts of the Cordilleran orogen.

Data used as a basis for the synthesis presented here include: (1) integrated structural and geochronologic studies in the Omineca belt (O1–O10 in Table 1, and references therein), which reveal the diachronous nature of thermotectonic events at different structural levels, the evolution of structural geometry, and the role of metamorphism and plutonism in orogenic processes; (2) similar types of studies of the Coast belt, which have increased our knowledge of its Mesozoic tectonic evolution, and its association with the Intermontane belt (e.g., Crawford et al., 1987; Rubin et al., 1990; Rusmore et al., 2001); (3) a refined stratigraphic and structural framework for the Alberta foreland basin and Foreland belt (e.g., Mossop and Shetsen, 1994) and a new understanding of foreland basin provenance based on detrital zircon geochronology and isotope geochemistry (e.g., Ross et al., 2005); (4) mapping and interpretation of lithofacies assemblages across the Bowser basin, which illustrate the distribution of depositional environments (e.g., Evenchick et al., 2006); (5) integration of the depositional environments with fossil ages, resulting in the first paleontologically constrained depositional history for the Bowser basin (Evenchick et al., 2001; revised herein); (6) revision of the age of Cretaceous Sustut basin strata (A. Sweet, in Evenchick et al., 2001); (7) detrital zircon geochronology studies of the Bowser and Sustut basins, which refine our knowledge of the evolution of source areas (McNicoll et al., 2005); (8) recognition of at least 160 km of horizontal shortening in the Skeena fold belt (Evenchick, 1991a, 1991b, 2001); and (9) refined estimates of the timing and magnitude of Mesozoic–early Cenozoic dextral transcurrent faults east of the Bowser basin (Gabrielse, 1985; Gabrielse et al., 2006).

This paper begins with a review the geology of the morphogeological belts of the Canadian Cordillera (Fig. 1) with emphasis on Middle Jurassic to early Cenozoic evolution (Fig. 3; Table 1 provides sources of information). The focus is on a transect across the Cordillera that predates transcurrent faulting, wherein the Bowser and Sustut basins are restored to their probable site of formation adjacent to Omineca and Foreland belt rocks in the southeast Canadian Cordillera. We summarize the stratigraphy and depositional and structural histories of the Bowser and Sustut basins (Fig. 3), including new provenance data from detrital zircon studies (Fig. 4). The next section contains descriptions of the events occurring across the orogen in a series of successive "time slices," focusing on timing of basin initiation and sedimentation, deformation, magmatism, and metamorphism. These events are illustrated by paleogeographic maps (Fig. 5) and transorogen cross sections (Fig. 6).

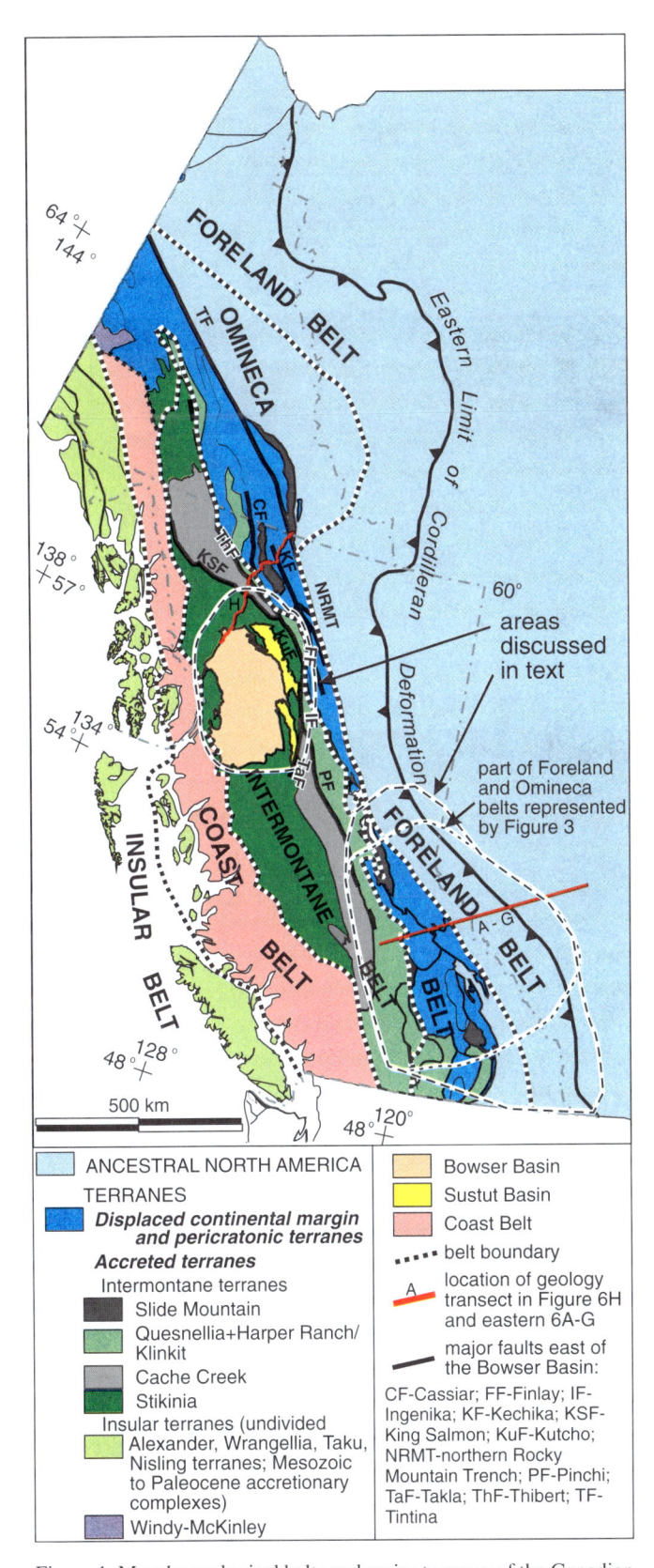

Figure 1. Morphogeological belts and major terranes of the Canadian Cordillera, locations of the Bowser and Sustut basins, major strike-slip faults east of these basins, and outlines of areas discussed in text (modified from Wheeler and McFeely, 1991; Colpron et al., 2006).

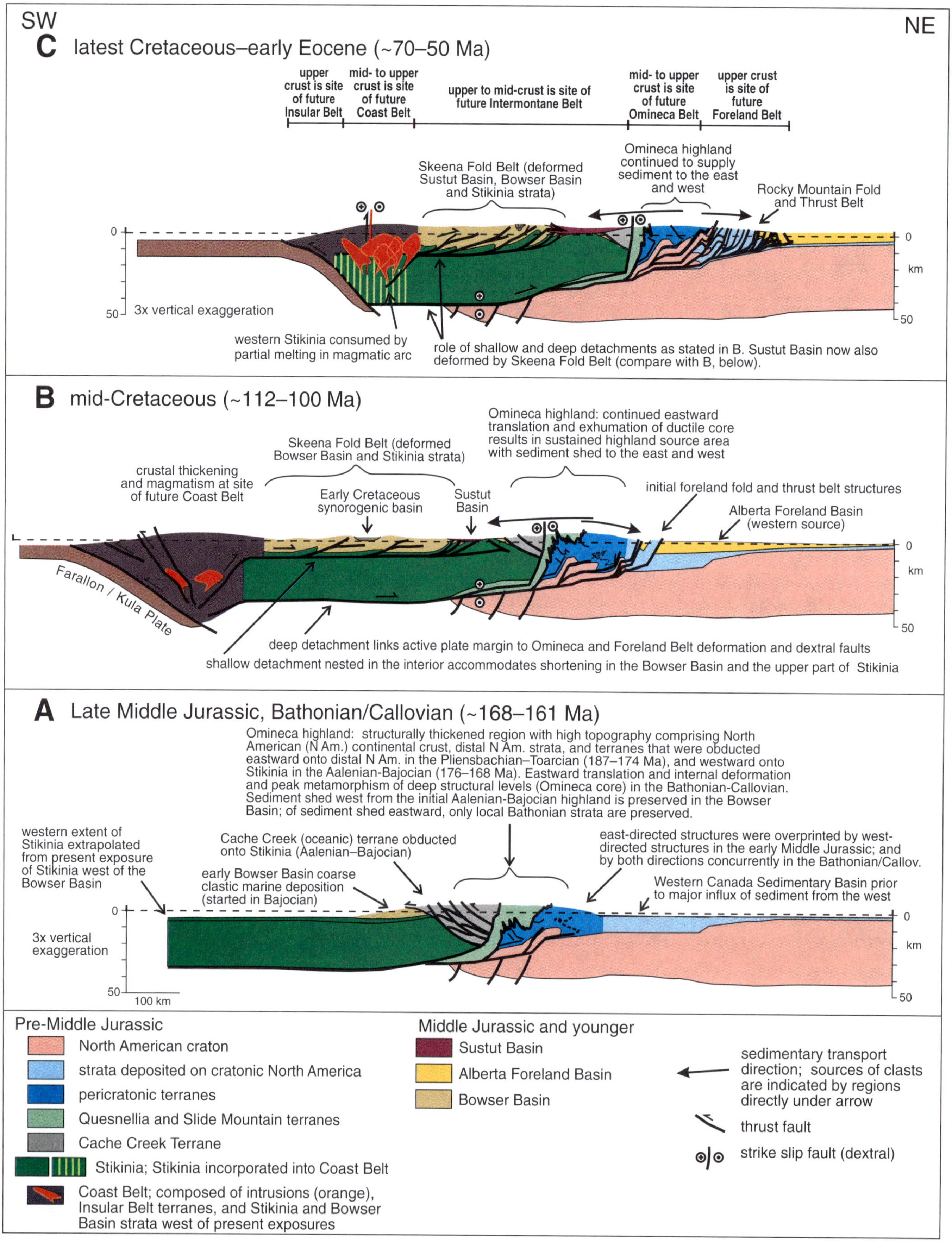
SW
NE
C latest Cretaceous–early Eocene (~70–50 Ma)
upper crust is site of future Insular Belt
mid- to upper crust is site of future Coast Belt
upper to mid-crust is site of future Intermontane Belt
mid- to upper crust is site of future Omineca Belt
upper crust is site of future Foreland Belt
Skeena Fold Belt (deformed Sustut Basin, Bowser Basin and Stikinia strata)
Omineca highland continued to supply sediment to the east and west
Rocky Mountain Fold and Thrust Belt
3x vertical exaggeration
western Stikinia consumed by partial melting in magmatic arc
role of shallow and deep detachments as stated in B. Sustut Basin now also deformed by Skeena Fold Belt (compare with B, below).
B mid-Cretaceous (~112–100 Ma)
Omineca highland: continued eastward translation and exhumation of ductile core results in sustained highland source area with sediment shed to the east and west
Skeena Fold Belt (deformed Bowser Basin and Stikinia strata)
crustal thickening and magmatism at site of future Coast Belt
Early Cretaceous synorogenic basin
Sustut Basin
initial foreland fold and thrust belt structures
Alberta Foreland Basin (western source)
Farallon / Kula Plate
deep detachment links active plate margin to Omineca and Foreland Belt deformation and dextral faults
shallow detachment nested in the interior accommodates shortening in the Bowser Basin and the upper part of Stikinia
A Late Middle Jurassic, Bathonian/Callovian (~168–161 Ma)
Omineca highland: structurally thickened region with high topography comprising North American (N Am.) continental crust, distal N Am. strata, and terranes that were obducted eastward onto distal N Am. in the Pliensbachian–Toarcian (187–174 Ma), and westward onto Stikinia in the Aalenian-Bajocian (176–168 Ma). Eastward translation and internal deformation and peak metamorphism of deep structural levels (Omineca core) in the Bathonian-Callovian. Sediment shed west from the initial Aalenian-Bajocian highland is preserved in the Bowser Basin; of sediment shed eastward, only local Bathonian strata are preserved.
western extent of Stikinia extrapolated from present exposure of Stikinia west of the Bowser Basin
Cache Creek (oceanic) terrane obducted onto Stikinia (Aalenian–Bajocian)
early Bowser Basin coarse clastic marine deposition (started in Bajocian)
east-directed structures were overprinted by west-directed structures in the early Middle Jurassic; and by both directions concurrently in the Bathonian/Callov.
Western Canada Sedimentary Basin prior to major influx of sediment from the west
3x vertical exaggeration
100 km
0
50
km
Pre-Middle Jurassic
North American craton
strata deposited on cratonic North America
pericratonic terranes
Quesnellia and Slide Mountain terranes
Cache Creek Terrane
Stikinia; Stikinia incorporated into Coast Belt
Coast Belt; composed of intrusions (orange), Insular Belt terranes, and Stikinia and Bowser Basin strata west of present exposures
Middle Jurassic and younger
Sustut Basin
Alberta Foreland Basin
Bowser Basin
sedimentary transport direction; sources of clasts are indicated by regions directly under arrow
thrust fault
strike slip fault (dextral)

Figure 2. Generalized cross sections illustrate tectonic elements and key events during the Mesozoic to early Cenozoic evolution of the southern Canadian Cordillera, from a narrow accretionary orogen (A) to a wide orogen with two major detachment systems (B, C). The lower detachment system links structural elements across the entire Cordillera. The upper detachment is the sole fault of the Skeena fold belt nested within the interior of the orogen. The Skeena fold belt is similar in width, magnitude of shortening, and timing of formation to the Rocky Mountain fold-and-thrust belt. Note that zones of penetrative ductile strain are not depicted on the cross sections; please see text and Table 1 for documentation of structural style, particularly in the Omineca and Coast belts.

◀──

Our purposes are to identify coeval and tectonically compatible events that illustrate linkages across the orogen and to use these as a basis for discussing its tectonic evolution.

GEOLOGICAL SETTING

The Canadian Cordillera (Fig. 1) is an amalgam of: (1) predominantly sedimentary deposits formed on and adjacent to the continental margin of ancestral North America, referred to herein as the craton margin; (2) terranes made mostly of arc and oceanic rocks that evolved separately early in their histories and subsequently became part of the Cordilleran "collage" (Coney et al., 1980; Gabrielse and Yorath, 1991); and (3) supracrustal rocks deposited in basins during and after terrane accretion. The five morphogeological belts of the Canadian Cordillera reflect both geological history and physiography (Gabrielse et al., 1991; Fig. 1). The major zones of terrane amalgamation and accretion coincide with the metamorphic-plutonic Omineca and Coast belts (Monger et al., 1982). These are flanked by belts of little metamorphosed or relatively low-grade sedimentary, volcanic, and plutonic rocks that form the Foreland, Intermontane, and Insular belts. The Foreland belt contains strata deposited on the ancient cratonal margin of North America, whereas the Intermontane and Insular belts coincide largely with the accreted terranes. In addition, there are syn- and postaccretion successions that were deformed into fold-and-thrust belts of Cretaceous to early Cenozoic age. The principal period of accretion of terranes of the Intermontane belt to North America was in Early to Middle Jurassic (e.g., Gabrielse and Yorath, 1991). The timing of accretion of Insular belt terranes to those farther inboard was either in the Jurassic or earlier (e.g., van der Heyden, 1992; McClelland et al., 1992), or in the Cretaceous (Monger et al., 1982). During and following amalgamation, there was significant orogen-parallel displacement on transcurrent faults within or marginal to the metamorphic-plutonic belts. Faults were dextral in and bordering the Omineca and Coast belts from mid-Cretaceous to early Cenozoic (e.g., Gabrielse, 1991a; Gabrielse et al., 2006). Prior to that, sinistral displacement of uncertain magnitude is inferred for the Early Cretaceous rocks outboard of the Bowser basin (Evenchick, 2001, and references therein).

Foreland Belt in British Columbia and Alberta

The Foreland belt is composed of Proterozoic to early Mesozoic continental-margin strata of the Western Canada sedimentary basin and of Mesozoic synorogenic strata of the Alberta foreland basin, the part of the Western Canada sedimentary basin that had the uplifting Cordillera as its main source (e.g., Stott and Aitken, 1993; Mossop and Shetsen, 1994). In the Foreland belt south of 58°N, west-derived sediments—a signal of emergence of a western source—first appear in the Oxfordian (F1 in Fig. 3; Table 1), except in the most southwestern part of the basin, where there is evidence for a western source in the Bajocian (Stronach, 1984). A major break in foreland basin sedimentation occurred in the Hauterivian to early Aptian, with a long period of pedimentation (Cadomin conglomerates; White and Leckie, 1999) and development of a low-angle unconformity with a stratigraphic separation that increases to the east (F3 in Fig. 3; Table 1). Clast composition and heavy mineral studies indicate that between 53°N and 57°N, western source areas for most of the coarse clastic units of the uppermost Jurassic to the Turonian include volcanic, plutonic, and medium-grade metamorphic rocks (McMechan and Thompson, 1993, and references therein). In contrast, south of 53°N, clast and heavy mineral studies, detrital zircon studies, and isotope geochemistry indicate that the western source areas were limited to low-grade metasedimentary rocks of the Omineca belt or the Foreland belt (McMechan and Thompson, 1993, and references therein; Ross et al., 2005). Exceptions are in Kimmeridgian strata, where detrital mica indicates a metasedimentary source, and in Albian strata, which have clasts of Intermontane belt volcanic and intrusive rocks (McMechan and Thompson, 1993, and references therein; Ross et al., 2005). Significant sediment accumulation and basin subsidence occurred in the western Alberta foreland basin during the Kimmeridgian to Valanginian (section up to 4 km thick; F2, Fig. 3), and over a broader area during the Campanian through Paleocene (section over 5 km thick; F5 in Fig. 3; Table 1).

Contractional thin-skinned structures of the Foreland belt are kinematically linked to structures in the polydeformed and metamorphosed Omineca belt and are essentially structurally continuous with them south of 53°N (McDonough and Simony, 1988; Kubli and Simony, 1994). Deformation of the foreland fold-and-thrust belt generally progressed from west to east and accommodated up to 200 km of horizontal shortening (Bally et al., 1966; Price and Mountjoy, 1970; Price, 1981; Fermor and Moffat, 1992). Structures near the western margin of the Foreland belt formed prior to ca. 108 Ma near 50°N (Larson et al., 2004) and ca. 100–112 Ma near 53°N (F4 in Fig. 3; McDonough and Simony, 1988). Faults in the east deform the youngest preserved strata (upper Paleocene near 53°N; Demchuk, 1990), and contractional deformation continued into the Eocene (Kalkreuth and McMechan, 1996). Most of the shortening of the southern Foreland belt occurred after the Turonian (younger than 89 Ma), concurrent with strike-slip faulting on the northern Rocky Mountain–Tintina fault system (Fig. 1; Price, 1994; Gabrielse et al., 2006).

TABLE 1. SUMMARY OF TIMING OF EVENTS IN THE INTERMONTANE, OMINECA, AND FORELAND BELTS

Intermontane Belt					Omineca Belt					Foreland Belt			
Fig. 3 code	Event	Timing	Reference		Fig. 3 code	Event	Timing	Reference		Fig. 3 code	Event	Timing	Reference
					O10	Penetrative high strain at the deepest structural levels (e.g., Thor-Odin dome; Frenchman Cap, in part); transition from transpressional to transtensional tectonics in Middle Eocene.	Eocene	Parrish et al. (1988); Gibson et al. (1999); Crowley et al. (2000); Hinchey (2005, and references therein)		F5	Rapid, coarse clastic sedimentation (100 m/m.y.); dominant Foreland belt source; major period of deformation with deformation front migrating eastward across Front Ranges and Foothills.	Late Campanian–Early Eocene (ca. 77–53 Ma)	Price and Mountjoy (1970); Price (1981); Stott and Aitken (1993); McMechan and Thompson (1993, and references therein); Ross et al. (2005)
I6	Initiation of coarse clastic deposition in Sustut basin (Campanian) and continued deformation in Skeena fold belt to affect youngest Sustut rocks (Maastrichtian). Sources same as in I5, but includes deposition of tuff.	Late Campanian to Early Maastrichtian, (ca. 74–68 Ma)	Eisbacher (1974a); Evenchick and Thorkelson (2005); McNicoll (2005, personal commun.); McNicoll et al. (2005, 2006)		O9	South of 52°N: Penetrative polydeformation and metamorphism (e.g., eastern Selkirk fan, Selkirk allochthon north of Monashee Complex; mid-crustal zone south of Thor-Odin), crystalline nappes carried on ductile shear zones (e.g., Gwillim Creek shear zone in Valhalla Complex); zones of high strain (e.g., Monashee décollement in Frenchman Cap dome).	Late Cretaceous–Paleocene	Carr (1992); Scammell (1993); Crowley and Parrish (1999); Gibson et al. (1999); Johnston et al. (2000, and references therein); Gibson (2003); Kuiper (2003); Williams and Jiang (2005); Carr and Simony (2006); Brown and Gibson (2006, and references therein)					
I5	First deposition of easterly derived metamorphic clasts, clasts of cratonic North American lithologies, and detrital zircons of Proterozoic age (Sustut basin).	late Early Cretaceous (Albian; ca. 112–110 Ma)	Eisbacher (1974a); Evenchick et al. (2001); Evenchick and Thorkelson (2005); V.J. McNicoll (2005, personal commun.); McNicoll et al. (2005, 2006)		O8	Purcell thrust system (post-Bearfoot and related thrusts and pre–Horsethief Creek batholith).	Mid-Cretaceous pre-ca. 93 Ma	Archibald et al. (1983); P.S. Simony (2005, personal commun.)		F4	Northeast-directed thrust faulting in westernmost Foreland belt near 53°N; cooling of Yellowjacket gneiss ca. 110 Ma, and 100 Ma deformation thrust faults immediately to east.	110–100 Ma (Albian)	McDonough and Simony (1988, and references therein)
					O7	Emplacement of Malton gneiss basement cored nappes via Bearfoot and related ductile thrusts (post–140–126 Ma as isograds [O6] deflected; pre– or syn–ca. 110 Ma cooling of Yellowjacket gneiss and 100 Ma deformation thrust faults in Foreland belt immediately to east).	late Early Cretaceous	McDonough and Simony (1988, and references therein); Digel et al. (1998); Crowley et al. (2000, and references therein)					

(continued)

TABLE 1. SUMMARY OF TIMING OF EVENTS IN THE INTERMONTANE, OMINECA, AND FORELAND BELTS (continued)

Intermontane Belt

Fig. 3 code	Event	Timing	Reference
I4	Development of piggyback basin within Skeena fold belt signals end of solely east-derived sedimentation and is probably a record of early Skeena fold belt deformation.	early Early to middle Early Cretaceous (ca. 145–135 Ma)	Evenchick et al. (2001); Evenchick and Thorkelson (2005); V.J. McNicoll (2006, personal commun.)
I3	Transition to nonmarine conditions across central and southern Bowser basin (no record in southwest).	Jurassic-Cretaceous boundary (ca. 145 Ma)	Evenchick et al. (2001); Evenchick and Thorkelson (2005); Smith and Mustard (2006); V.J. McNicoll (2005, personal commun.)

Omineca Belt

Fig. 3 code	Event	Timing	Reference
O6	Penetrative polydeformation and metamorphism in the Cariboo, Monashee and Selkirk mountains (51.5–52.5°N latitude). Near line represented by Fig. 3: southeastward imbrication of basement slices at depth with SW-verging folds and faults at higher levels (e.g., Hobson Lake area); fold fans (e.g., Ozalenka fan). South of 52°N and at deep structural levels: generally NE-verging poly-deformation and high-strain zones (e.g., Scammell, 1993).	throughout the Late Jurassic, Early Cretaceous, and Late Cretaceous	Parrish (1995); Currie (1988); Scammell (1993); Digel et al. (1998); Crowley et al. (2000); Reid (2003, and references therein); Gibson (2003)
O5	Penetrative polydeformation and metamorphism—Allan Creek area, Cariboo Mountains.	ca. 143–126 deformation; ca. 135 Ma metamorphism	Parrish (1995); Currie (1988)
O4	SW-verging penetrative deformation and metamorphism—eastern Hobson Lake area, Cariboo Mountains.	ca. 147 Ma syntectonic metamorphism	Reid (2003, and references therein)

Foreland Belt

Fig. 3 code	Event	Timing	Reference
F3	Major period of pedimentation; development of low-angle unconformity with stratigraphic separation increasing eastward except in westernmost Foreland basin.	Hauterivian to Early Aptian (ca. 136–110 Ma)	Stott and Aitken (1993); Stott (1998); White and Leckie (1999)
F2	Rapid, west (Omineca highland) derived sedimentation (to 100 m/m.y.); shallowing- and coarsening-up sequence in Upper Jurassic.	Kimmeridgian-Valanginian (ca. 156–136 Ma)	Poulton (1989); Poulton et al. (1993, 1994b); Stott and Aitken (1993); Stott et al. (1993); Stott (1998)

(continued)

TABLE 1. SUMMARY OF TIMING OF EVENTS IN THE INTERMONTANE, OMINECA, AND FORELAND BELTS *(continued)*

Intermontane Belt					Omineca Belt					Foreland Belt			
Fig. 3 code	Event	Timing	Reference		Fig. 3 code	Event	Timing	Reference		Fig. 3 code	Event	Timing	Reference
I2	Rapid westward migration of facies belts in the Bowser basin.	Late Jurassic (Oxfordian/ Kimmeridgian; ca. 157–151 Ma)	Evenchick et al. (2001); Evenchick and Thorkelson (2005)		O3	Onset and/or formation of significant architecture of major structures such as SW-verging folds, fold fans, belts of NE-verging folds & faults. Peak of metamorphism ca. 165–160 Ma (biotite-grade Hobson Lake, Scrip Nappe).	(3a) 173–164 Ma in Kootenay arc & Purcell anticlinorium; >167 Ma Selkirk Fan (3b) 174–162 Ma in Cariboo Mountains	Archibald et al. (1983); Gerasimoff (1988); Struik (1988); Brown et al. (1992b); Parrish (1995); Warren (1997); Colpron et al. (1998); Gibson (2003); Reid (2003, and references therein)		F1	First regional subsidence and thick W-derived (Omineca highland) sediments; overlies thin, craton-derived, basal transgressive sandstone.	Oxfordian (post–Early Oxfordian (ca. 158–156 Ma)	Poulton (1984, 1989); Poulton et al. (1993, 1994b); Stott (1998)
I1	At least 3 km deposition in Bowser basin focused in northeast trough.	Middle Jurassic (Bathonian to early Oxfordian; ca. 168–158 Ma)	Evenchick et al. (2001); Evenchick and Thorkelson (2005)					Brown and Gibson (2006, and references therein)		F0	Time-transgressive erosive disconformity at base of Alberta foreland basin succession.	Callovian-Oxfordian (ca. 164–160 Ma)	Poulton (1984); Poulton et al. (1993, 1994b)
I0	First east-derived sediment from the Cache Creek terrane deposited on Stikinia is the initial coarse clastic deposition in Bowser basin. Rapid exhumation of Cache Creek.	early Middle Jurassic (Aalenian/ Bajocian; ca. 176–168 Ma)	Gabrielse (1991b); Ricketts et al. (1992); Mihalynuk et al. (2004)		O2	Southwest-verging thrusts and isoclinal recumbent folds south of 51°N.	onset by 175 Ma	Parrish and Wheeler (1983); Klepacki (1985); Smith et al. (1992); Warren (1997); Colpron et al. (1996, 1998); Gibson (2003); Reid (2003)					
					O1	Obduction of Slide Mountain and Quesnellia terranes onto North America pericratonic terranes (e.g., Eureka, Pundata, Stubbs, Waneta faults and related structures).	ca. <187 Ma, 187–173 Ma	Tipper (1984); Parrish and Wheeler (1983); Murphy et al. (1995, and references therein); Beatty et al. (2006)					

Southern Omineca Belt

The southern Omineca belt is composed of predominantly Proterozoic to Paleozoic supracrustal rocks formed on or near the cratonal margin of North America as well as late Paleozoic and early Mesozoic rocks of the most inboard accreted terranes. These rocks were polydeformed and metamorphosed during and following their Early to Middle Jurassic (ca. 187–174 Ma) accretion (O1 in Table 1), and some structural levels were reactivated and overprinted through the Jurassic, Cretaceous, Paleocene, and early Eocene (Table 1; Fig. 3). Protoliths are: (1) Paleoproterozoic basement exposed in structural culminations of the Monashee Complex and as imbricated thrust slices in the Malton Complex; (2) Mesoproterozoic to Paleozoic supracrustal rocks deposited on the western cratonal margin of North America; (3) lower Paleozoic sedimentary, volcanic, and igneous rocks of the pericratonic Kootenay terrane that formed on or adjacent to the North American margin; (4) Permian volcanic and ultramafic rocks of the oceanic Slide Mountain terrane; (5) Triassic to Lower Jurassic volcanic and sedimentary arc-related rocks of Quesnellia; (6) ca. 175–159 Ma calc-alkaline plutons of the Kuskanax and Nelson Suites; and (7) ca. 110–90 Ma granites of the Bayonne Suite. South of 51°N, Paleocene–Eocene peraluminous leucogranites occur in high-grade rocks exhumed from the mid-crust. The youngest rocks, middle Eocene intrusions of the syenitic Coryell Suite, coincide with the locus of significant Eocene east-west extension.

Although there is a record of Paleozoic interaction of offshore terranes with the western margin of North America (cf. Klepacki, 1985), and there is debate about the tectonic setting and paleogeography of the margin (cf. Thompson et al., 2006), our focus is tectonism related to the Late Triassic to Middle Jurassic obduction of terranes, and subsequent shortening and thickening of the orogen. Obduction of the Quesnellia and Slide Mountain terranes starting at ca. 187 Ma was accommodated by crustal shortening, as documented by folding and northeast-directed thrust faulting at low-grade metamorphic conditions, and this resulted in over 40 km of overlap of these terranes onto North American rocks (O1 in Table 1; Fig. 3). In the Middle Jurassic, shortening and thickening was accommodated by west-verging faults and fold systems, including regional-scale isoclinal folds with limbs tens of kilometers long (e.g., Scrip Nappe; Raeside and Simony, 1983; O2 in Table 1; Fig. 3). In the central Omineca belt, southwest-verging structures continued to form in the west (Schiarizza and Preto, 1987; Reid, 2003; Fig. 3). Elsewhere, upright folds, northeast-verging fold systems, regional fan structures, and northeast-directed thrust faults were superimposed on the southwest-verging nappes (O3 in Table 1; Fig. 3). Deformation was accompanied by regional greenschist-facies metamorphism, with some amphibolite-facies assemblages indicating that the crustal thickness was ~50–55 km (Table 1; Archibald et al., 1983; Colpron et al., 1996; Warren, 1997), or greater, and that this pulse of metamorphism reached its peak in the Middle Jurassic and was followed by rapid exhumation in the Middle Jurassic (O3 in Table 1). We follow the blind "tectonic wedge" model (Price, 1986; Struik, 1988; Murphy, 1989; Colpron et al., 1998) in which basement slices or ramps acted as wedges resulting in the formation of large-scale southwest-vergent structures in the detached and deforming overlying supracrustal rocks. Alternative models to explain the southwest-vergent Middle Jurassic structures, based on development of a retro-wedge geometry above a subduction zone (Brown et al., 1993; Brown and Gibson, 2006), are incompatible with timing and geometry of structures documented within the wedge, and the westward location of the magmatic arc and subduction zone relative to the position of deforming rocks (cf. Colpron et al., 1998). In the Late Jurassic and Early Cretaceous, progressive deformation and metamorphism continued in the Omineca belt (O4 and O5 in Table 1), and the foreland became progressively more involved, thus expanding the deforming highland and locus of thickening northeastward relative to the craton. In the Early Cretaceous, the lower structural levels of west-verging fold systems and related higher-level fold fans in the southern Omineca belt were reactivated and tightened during greenschist- and amphibolite-facies metamorphism, and sheets of basement and cover rocks of the Malton Complex were imbricated by northeast-vergent structures (O6 in Fig. 3). By the end of the Albian, the rocks at the latitude of discussion were being cooled and exhumed. However, more deeply exhumed rocks, now exposed south of 52°N, record zones of ductile moderate to extreme strain, folding, and transposition, indicating that mid-Cretaceous–Eocene strain partitioning at deep structural levels in the Omineca belt was concomitant with northeastward propagation of deformation and thin-skinned shortening in the Foreland belt (O9, O10 in Fig. 3).

Central Intermontane Belt

Most of the central Intermontane belt is underlain by Stikinia (Fig. 1), a terrane composed of volcano-plutonic arc assemblages of Devonian to Permian, Late Triassic, and Early Jurassic to early Middle Jurassic age (e.g., Monger and Nokleberg, 1996, and references therein). These strata exhibit a range of styles of deformation, and metamorphism from subgreenschist to greenschist facies. The Cache Creek terrane, in the eastern Intermontane belt (Fig. 1), is an accretionary complex of rocks formed in oceanic environments, with lesser volcanic arc strata. Rocks of oceanic affinity are Mississippian to Early Jurassic (Toarcian) in age and contain Permian Tethyan fauna that indicate an origin far from the North American craton margin (e.g., Struik et al., 2001). The structural style of Cache Creek terrane is dominated by southwest-vergent fold-and-thrust fault systems, superimposed in places on chaotic disrupted structures, and locally overprinted by northeast-verging fold systems (e.g., Gabrielse, 1991b; Struik et al., 2001; Mihalynuk et al., 2004). In the northern segment (Fig. 1), the Cache Creek terrane structurally overlies Stikinia, and locally Bowser basin strata, along the southwest-directed King Salmon fault (KSF, Fig. 1; e.g., Gabrielse, 1991b). In the southern segment (Fig. 1), the western accretionary boundary of Cache Creek terrane with Stikinia is obscured. In both segments, the eastern boundaries of Cache Creek terrane with Quesnellia are Cretaceous and

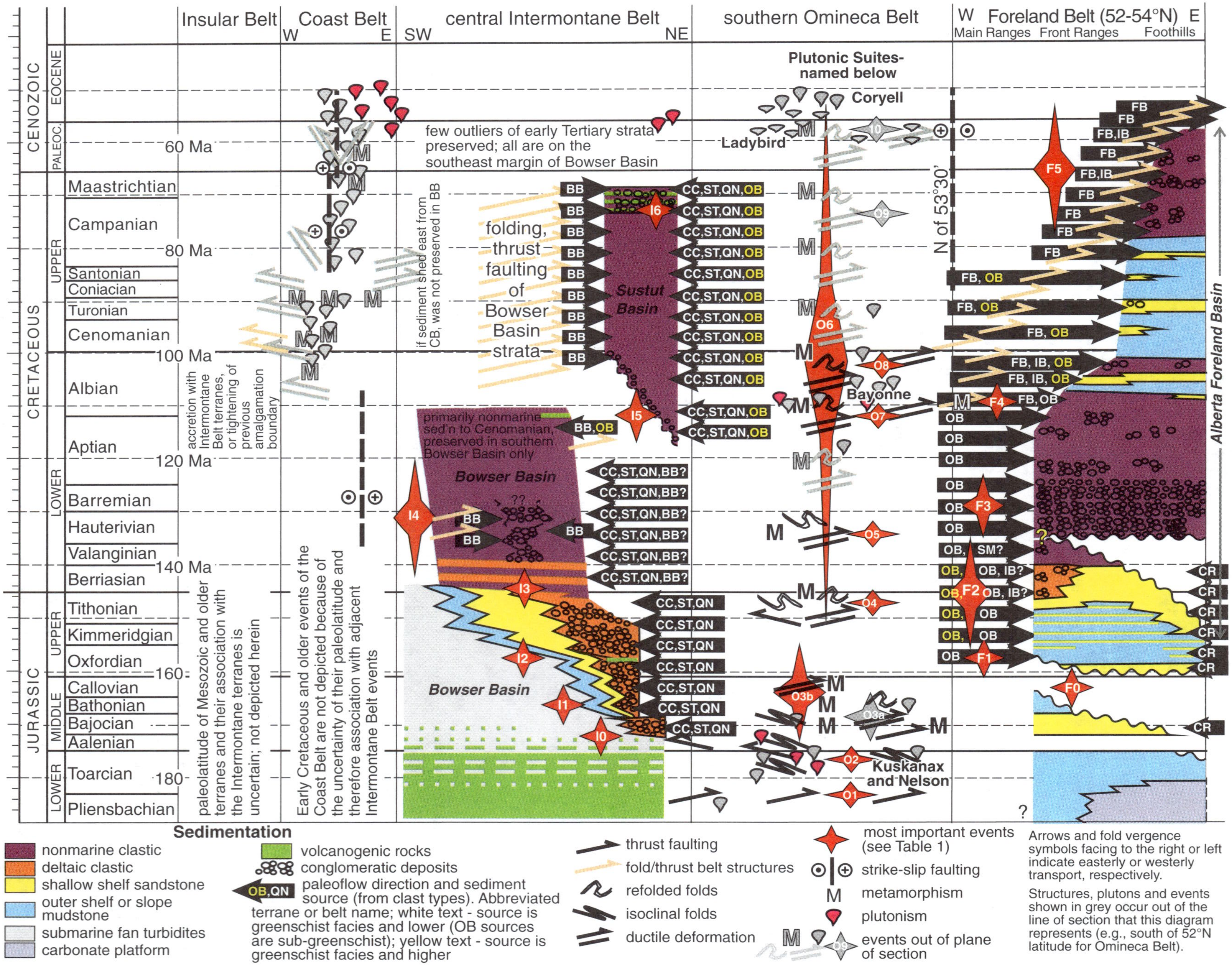

Insular Belt
Coast Belt
W
E
SW
central Intermontane Belt
NE
southern Omineca Belt
W
Foreland Belt (52-54°N)
E
Main Ranges
Front Ranges
Foothills
CENOZOIC
EOCENE
PALEOC.
CRETACEOUS
UPPER
LOWER
JURASSIC
UPPER
MIDDLE
LOWER
Maastrichtian
Campanian
Santonian
Coniacian
Turonian
Cenomanian
Albian
Aptian
Barremian
Hauterivian
Valanginian
Berriasian
Tithonian
Kimmeridgian
Oxfordian
Callovian
Bathonian
Bajocian
Aalenian
Toarcian
Pliensbachian
60 Ma
80 Ma
100 Ma
120 Ma
140 Ma
160
180
Plutonic Suites-named below
Coryell
Ladybird
Bayonne
Kuskanax and Nelson
few outliers of early Tertiary strata preserved; all are on the southeast margin of Bowser Basin
if sediment shed east from CB, was not preserved in BB
folding, thrust faulting of Bowser Basin strata
accretion with Intermontane Belt terranes, or tightening of previous amalgamation boundary
primarily nonmarine sed'n to Cenomanian, preserved in southern Bowser Basin only
Sustut Basin
Bowser Basin
Bowser Basin
paleolatitude of Mesozoic and older terranes and their association with the Intermontane terranes is uncertain; not depicted herein
Early Cretaceous and older events of the Coast Belt are not depicted because of the uncertainty of their paleolatitude and therefore association with adjacent Intermontane Belt events
N of 53°30'
Alberta Foreland Basin
BB
BB,OB
CC,ST,QN,OB
CC,ST,QN,BB?
CC,ST,QN
FB
FB,IB
FB, OB
FB, IB, OB
FB,OB
OB
OB, SM?
OB, IB?
CR
I0
I1
I2
I3
I4
I5
I6
O1
O2
O3a
O3b
O4
O5
O6
O7
O8
O9
O10
F0
F1
F2
F3
F4
F5
M
??
?
Sedimentation
nonmarine clastic
deltaic clastic
shallow shelf sandstone
outer shelf or slope mudstone
submarine fan turbidites
carbonate platform
volcanogenic rocks
conglomeratic deposits
OB,QN
paleoflow direction and sediment source (from clast types). Abbreviated terrane or belt name; white text - source is greenschist facies and lower (OB sources are sub-greenschist); yellow text - source is greenschist facies and higher
thrust faulting
fold/thrust belt structures
refolded folds
isoclinal folds
ductile deformation
most important events (see Table 1)
strike-slip faulting
metamorphism
plutonism
events out of plane of section
Arrows and fold vergence symbols facing to the right or left indicate easterly or westerly transport, respectively.
Structures, plutons and events shown in grey occur out of the line of section that this diagram represents (e.g., south of 52°N latitude for Omineca Belt).

Figure 3. Chart illustrating late Early Jurassic to early Eocene tectonic events in the Canadian Cordillera, with emphasis on the central Intermontane, southern Omineca, and southern Foreland belts. Events labeled with red stars are discussed in the text; a summary of events and sources of information are provided in Table 1 and in the text. CC—Cache Creek; ST—Stikinia; QN—Quesnellia; BB—Bowser basin; OB—Omineca belt; FB—Foreland belt; SM—Slide Mountain; IB—Intermontane belt; CR—craton. Time scale here and elsewhere in text is after Gradstein et al. (2004).

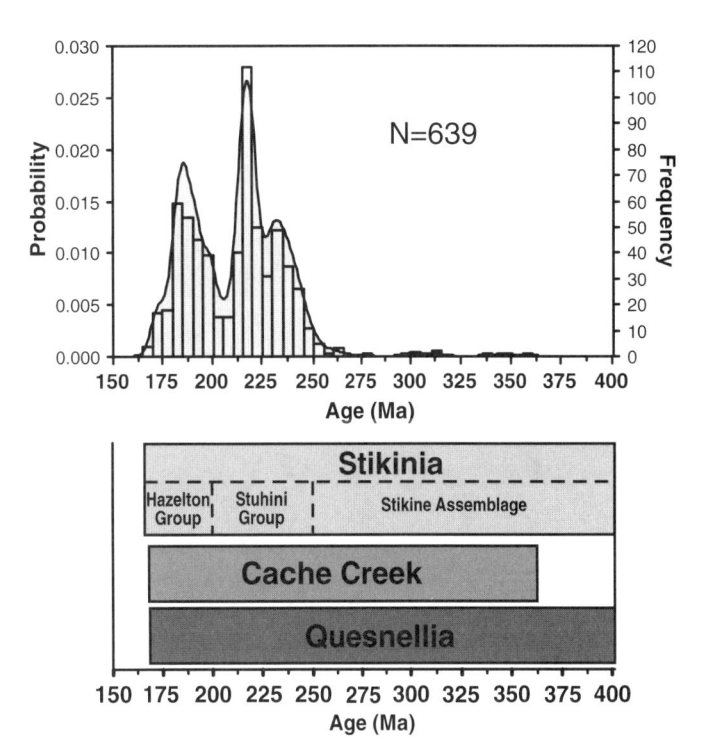

Figure 4. Cumulative probability plot of Bowser basin detrital zircon data for grains older than basin strata (Mississippian to early Middle Jurassic, ca. 360–169 Ma), from all Bowser basin samples ($n = 21$). Ages of rocks in Stikinia, Cache Creek, and Quesnellia are shown below. Of over 1435 single-grain analyses, 639 have ages older than the depositional age of strata in the basin. This diagram shows only the older Bowser basin data to highlight the ages of pre–Bowser basin sources. Potential source rock ages in Stikinia, Cache Creek and Quesnellia are from: Monger et al. (1991) and references therein; Monger and Nokleberg (1996) and references therein; Mihalynuk et al. (2004) and references therein; and Evenchick and Thorkelson (2005) and references therein.

Cenozoic dextral strike-slip faults such as the Thibert and Pinchi faults, in the north and south, respectively (Fig. 1; Gabrielse, 1985; Struik et al., 2001). The latter is inferred to overprint the Triassic–Jurassic Pinchi suture (Struik et al., 2001). Quesnellia is an arc terrane similar to Stikinia in general age and lithology (e.g., Monger and Nokleberg, 1996; Beatty et al., 2006). Significant differences occur in their pre-Triassic stratigraphy (e.g., Monger and Nokleberg, 1996), and Quesnellia has stratigraphic and structural ties with pericratonic Kootenay terrane, Slide Mountain terrane and the North American continental margin, which demonstrate that it probably formed adjacent to the continent (Beatty et al., 2006, and references therein). In addition, although Quesnellia has considerable north-south extent (Fig. 1), its map area is small relative to Stikinia. On the seismic-reflection profile that crosses most of the terranes in northern British Columbia, Quesnellia and Cache Creek are interpreted to form thin sheets (~2.5 km thick and ~7.5 km thick, respectively), whereas Stikinia is interpreted to be relatively thick (~35 km), comprising the entire crust above the Moho (Cook et al., 2004).

Bowser Basin

Much of northern Stikinia is overlain by strata of the Bowser Lake Group (Fig. 1), which were deposited in the Bowser basin and comprise a widespread upper Middle Jurassic to mid-Cretaceous marine and nonmarine clastic succession at least 6000 m thick (Tipper and Richards, 1976; Eisbacher, 1981; Evenchick and Thorkelson, 2005). Skeena Group, Cretaceous in age, contains rocks similar to coeval facies of the Bowser Lake Group, but includes volcanic successions, and many exposures occur south of its northern limit in the southern Bowser basin (Bassett and Kleinspehn, 1997; Smith and Mustard, 2006). Sand and pebble clasts in Bowser basin strata are dominated in many places by radiolarian chert derived from Cache Creek terrane; these clasts demonstrate the stratigraphic link between Cache Creek terrane and Stikinia and record the final stages of closure of the Cache Creek ocean (Gabrielse, 1991b; Evenchick and Thorkelson, 2005; I0 in Fig. 3).

The Bowser Lake Group is a monotonous assemblage of sandstone, siltstone, and conglomerate lacking laterally continuous stratigraphic markers. Lithofacies assemblages interfinger laterally and repeat vertically on a range of scales (I0 to I3 in Fig. 3). Strata were deposited in submarine fan and interfan, slope, shallow-marine, deltaic, fluvial, and lacustrine environments from southeast, east, and northeasterly sources in an overall regressive basin history (Tipper and Richards, 1976; Evenchick and Thorkelson 2005, and references therein).

Integration of timing constraints from index fossils with the distribution of major lithofacies assemblages demonstrates the migration of facies boundaries through time (Evenchick et al., 2001; Fig. 3). From Bathonian through early Oxfordian time, the major depocenter was restricted to the north-northeastern part of the basin (I1 in Table 1; Fig. 3) and only a condensed marine section formed at the western side. Sections in the south are also relatively thin and fine grained compared to those in the northeast (Tipper and Richards, 1976). Between mid-Oxfordian and early Kimmeridgian time, there was rapid south and southwest migration of facies proximal to the source over more distal facies (I2 in Fig. 3). New mapping in the central Bowser basin (Evenchick et al., 2006) indicates that the shelf–slope break migrated ~200 km southwestward during this time. The result was a wide, shallow, marine shelf bounded on the southwest by a region of submarine fan deposition at least 80 km wide that, based on its marine fauna, was probably

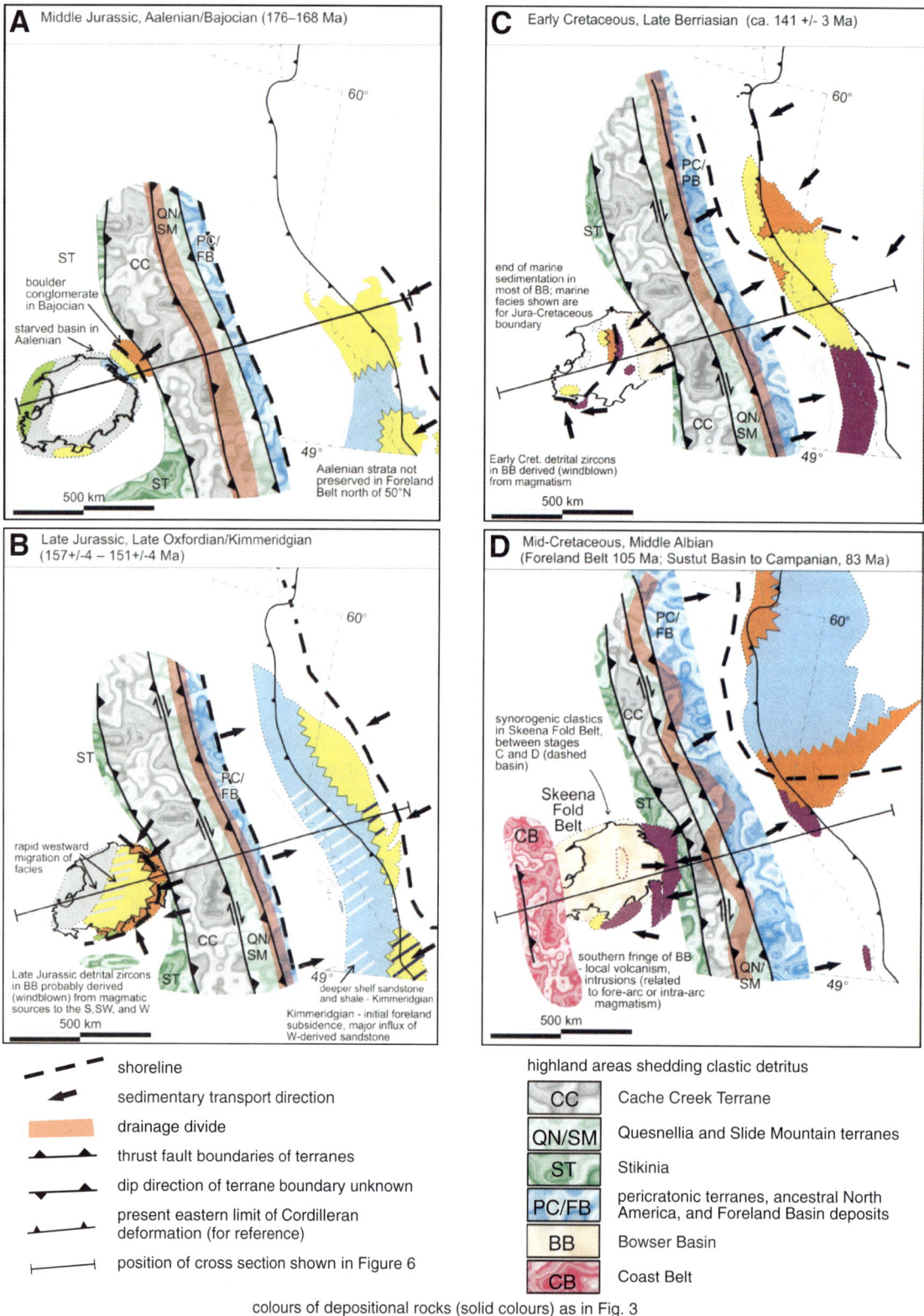

Figure 5 (*on this and following page*). Paleogeographic maps for the early Middle Jurassic through Late Cretaceous showing the inferred position of Bowser basin relative to cratonic North America (as discussed in text; North America is shown in present geographic coordinates), areas and types of sedimentation in the Bowser basin and Foreland belt, inferred source areas and drainage divide, and location of cross sections in Figure 6. The outline of the Bowser basin, with 50% shortening of the Skeena fold belt (SFB) restored, is shown for reference. Sources of information for deposition in the Bowser and Sustut basins are Eisbacher (1974a), Tipper and Richards (1976), Bassett and Kleinspehn (1997), Evenchick et al. (2001; and revisions based on new mapping in Evenchick et al., 2006), and Evenchick and Thorkelson (2005). Sources of information for facies belts in the Alberta foreland basin are Stott (1982, 1998), Hall (1984), Smith (1994), Poulton et al. (1994b), Stott et al. (1993), Leckie and Burden (2001), and McMechan et al. (2006).

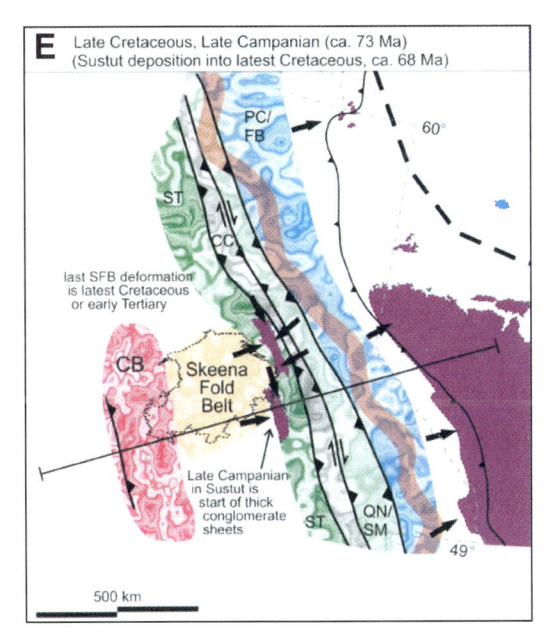

Figure 5 (*continued*).

open to the Pacific Ocean (Poulton et al., 1994a). Facies boundaries in the southern basin also migrated more rapidly than previously (Tipper and Richards, 1976). The shelf–slope break in the central and northern basin remained in about the same position into the latest Jurassic or earliest Cretaceous (I3 in Fig. 3); younger strata are absent in the western basin. In earliest Cretaceous time, deltaic and nonmarine strata were deposited in large parts of the northern basin, and nonmarine strata were deposited in the southern basin (Bassett and Kleinspehn, 1997; Smith and Mustard, 2006; V.J. McNicoll, 2006, personal commun.). By middle Early Cretaceous, these strata included thick conglomeratic braided river and alluvial fan deposits in the north-central Bowser basin (Eisbacher, 1974b; I4 in Fig. 3). Preserved strata of mid-Cretaceous age are restricted to the southern basin and are largely nonmarine clastic, but they include local volcanic centers and minor marine strata (Bassett and Kleinspehn, 1997). The youngest fluvial systems were probably continuous with the oldest Sustut fluvial systems deposited farther northeast. Ages of most Early Cretaceous strata are not narrowly constrained (Bassett and Kleinspehn, 1997; Evenchick and Thorkelson, 2005, and references therein; V.J. McNicoll, 2004, 2006, personal commun.). Deposition of detrital muscovite in the southern basin, likely starting in Aptian or Albian time, signals initiation of metamorphic Omineca belt detritus shed to the west (Bassett and Kleinspehn, 1997), and its initiation was roughly coeval with deposition of muscovite in the Sustut basin farther northeast (Eisbacher, 1974a; Evenchick and Thorkelson, 2005).

New provenance data that refine our understanding of the depositional history of the Bowser basin are provided by detrital zircons analyzed from sandstone samples that have paleontologically well-constrained depositional ages and from diverse ages, map units, and areas of the basin. The results, based on U-Pb sensitive high-resolution ion microprobe (SHRIMP) analyses of over 1435 detrital zircons from 21 samples of Bowser Lake Group, ranging from Bathonian to earliest Cretaceous age, show that the Bowser basin was receiving detritus mainly from sources of Early Triassic age to as young as the depositional age of the rock sampled (Fig. 4; McNicoll et al., 2005). The source regions indicated by paleocurrents, clast types, and facies distribution of Bowser basin strata are Cache Creek terrane, Quesnellia, and Stikinia (e.g., Evenchick and Thorkelson, 2005, and references therein). These are potential sources for the detrital zircons that are older than Bowser basin strata (Fig. 4). However, the age of the youngest zircon population in each rock is indistinguishable from the paleontologically determined depositional age of the rock, and it is interpreted to have originated from wind-blown ash from sources south, southwest, and/or possibly west of the Bowser basin (McNicoll et al., 2005).

Regional relationships indicate that basin subsidence for the Jurassic and earliest Cretaceous Bowser basin was controlled, in part, by flexural subsidence resulting from sediment load and obduction of Cache Creek terrane, and by thermal subsidence resulting from cessation of arc-related magmatism within northern Stikinia (e.g., Eisbacher, 1981; Ricketts et al., 1992; Evenchick and Thorkelson, 2005); the role of dynamic subsidence in Bowser basin evolution is unknown. An outlier of Early Cretaceous braided river and alluvial fan deposits in the northern Bowser basin (I4 in Fig. 3) is interpreted to represent synorogenic deposition within the Skeena fold belt (Evenchick and Thorkelson, 2005).

Sustut Basin

The Sustut Group is composed of more than 2000 m of Late Cretaceous nonmarine clastic strata (Eisbacher, 1974a). Since Eisbacher's (1974a) description and interpretation of the group, new constraints on the ages of units (A. Sweet, in Evenchick and Thorkelson, 2005), and additional stratigraphic and structural relationships, have clarified its tectonic significance (Evenchick and Thorkelson, 2005).

The lower of two formations, the Tango Creek Formation, is up to 1400 m thick and overlies Triassic to Late Jurassic units, including Bowser Lake Group, with angular unconformity (Eisbacher, 1974a). It is dominated by sandstone, siltstone, and mudstone, and in the upper part, by mudstone, calcareous siltstone, and calcareous sandstone. Paleocurrents, which in the lowest part are to the south and southwest, and high quartz clast content of sandstone are both consistent with the interpretation of derivation from a northeastern, Omineca belt source (Eisbacher, 1974a). Quartzite characteristic of Lower Cambrian miogeoclinal strata occurs as pebbles; these were also likely derived from the Omineca belt. In the middle and upper part of the formation, paleoflow to the northeast was accompanied by an increase in chert content in sandstone, interpreted as Cache Creek clasts recycled from Bowser Lake Group strata. The base of the formation is diachronous, ranging from Barremian–early Albian, to Coniacian–Campanian, and the upper age limit is late Campanian (I5 in Fig. 3).

Figure 6 (*on this and following page*). Cross sections illustrating major depositional, structural, metamorphic, and plutonic events in the region discussed in the text. Abbreviations are as in Figure 3. Note that zones of penetrative ductile strain are not depicted on the cross sections; please see text and Table 1 for documentation of structural style, particularly in the Omineca and Coast belts. The set with no vertical exaggeration shows structural relationships. The set with 3× vertical exaggeration is included to show the depositional units. The location of cross sections A, C, D, and E are given in Figure 5.

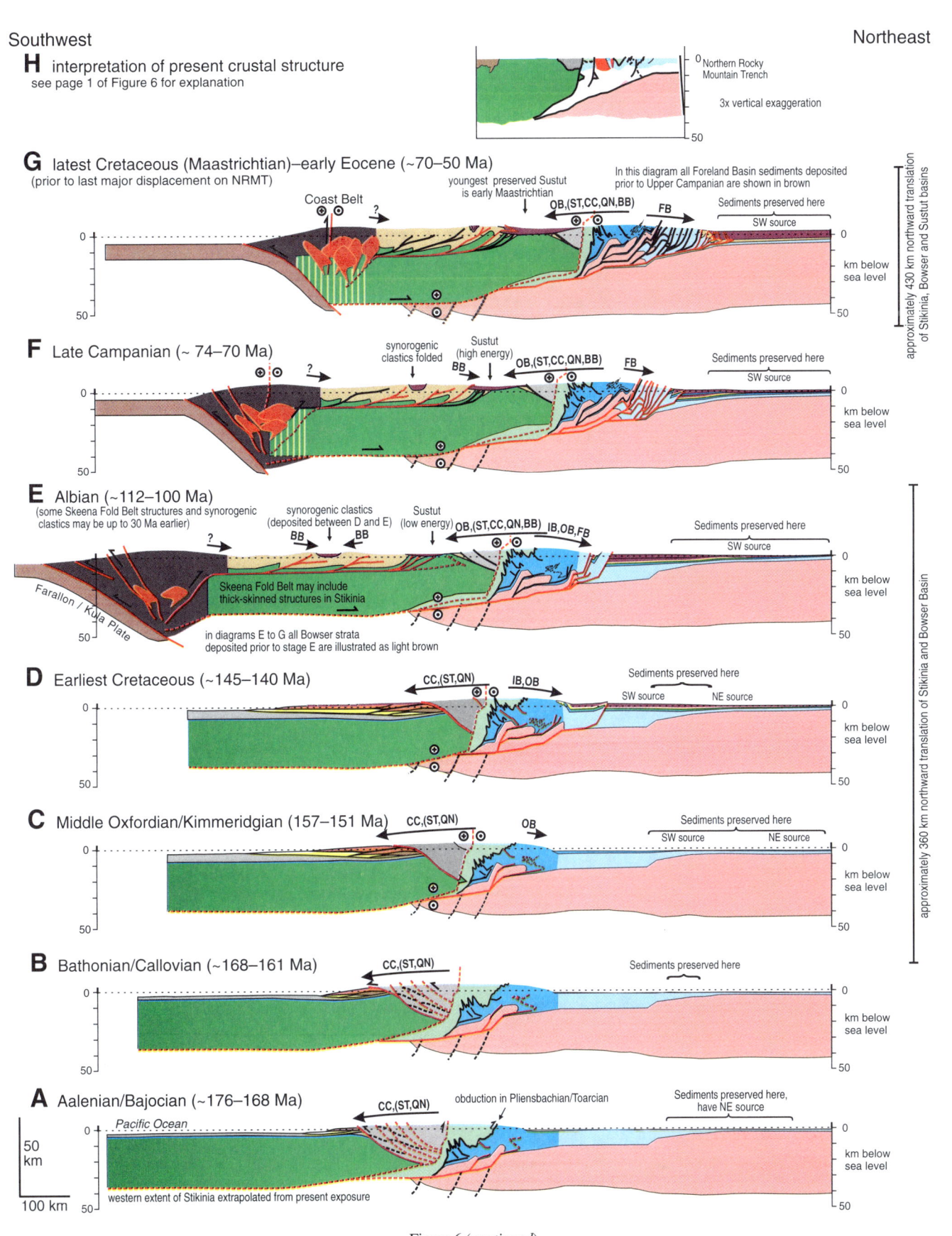

Figure 6 (continued).

The Tango Creek Formation is abruptly and conformably overlain by the Brothers Peak Formation, which is characterized by polymict conglomerate, sandstone, and felsic tuff (Eisbacher, 1974a). A basal conglomeratic succession is commonly more than 50 m thick. Paleocurrents were primarily southeast, longitudinally down the basin, with input from the north, east, and west (Eisbacher, 1974a). Strata are late Campanian to late early Maastrichtian (A. Sweet, in Evenchick et al., 2001). Two felsic tuff layers in the formation have been dated at ca. 75 and ca. 71 Ma (V.J. McNicoll, 2004, personal commun.). The dramatic change upward from the mudstone and siltstone of the upper Tango Creek Formation to the conglomerate-rich base of the Brothers Peak Formation (I6 in Fig. 3) indicates a marked increase in energy of Sustut fluvial systems.

Paleocurrent and clast types show that initial Sustut basin deposits had an eastern Omineca belt source that continued throughout the depositional history; however, early in this history, development of the Skeena fold belt provided an additional, southwest, source of sediment (Eisbacher, 1974a). Detrital zircon data show that the Sustut basin shared similar Triassic to Early Cretaceous sources with the Bowser basin, and/or zircons were recycled from the Bowser basin (McNicoll et al., 2005). In contrast, the Sustut basin also received Archean, Paleoproterozoic, Mesoproterozoic, Paleozoic, and Late Cretaceous zircons that are clearly distinct from Bowser basin sources (McNicoll et al., 2005) and are consistent with provenance studies of the Sustut Group that illustrate an Omineca belt source in the late Early Cretaceous (Eisbacher, 1974a).

Skeena Fold Belt

The Skeena fold belt is a regional fold-and-thrust belt that is best expressed in the thinly bedded clastic rocks of the Bowser and Sustut basins, but also involves Stikinia, as shown by folded contacts, structural culminations of Stikinian rocks within the fold belt, and klippen of early Mesozoic Stikinian rocks on Cretaceous Sustut basin strata. Details of the geometry, magnitude of shortening, and timing of the fold belt have been documented by Evenchick (1991a, 1991b, 2001) and Evenchick and Thorkelson (2005). The fold belt has accommodated a minimum of 44% (160 km) northeasterly shortening, it locally terminates to the northeast in a triangle zone within the Sustut basin, and is inferred to root to the west in the Coast belt.

Folds of a range of scales, from several hundred meters to a kilometer or more in wavelength, are the most obvious structures, and northwest-trending fold trains occupy most of the fold belt. Most verge northeast, and they vary from upright to overturned. Thrust faults in Bowser strata are apparent where they juxtapose the Bowser Lake Group against other map units, but they are difficult to recognize in most of the basin unless hanging-wall or footwall cutoffs are exposed. Thrust faults and/or detachment zones are required by the style of folding and are inferred to be largely bedding-parallel, blind thrusts (Evenchick, 1991b).

The age of the fold belt is constrained by regional stratigraphic relationships. The youngest folded marine Bowser basin strata, deposited at the Jurassic–Cretaceous boundary, lack a western source or other indications that the fold belt had evolved significantly. Sustut basin strata unconformably overlie contractional structures that involve Bowser basin and Stikinia strata, illustrating contractional deformation and erosion of Stikinian and Bowser strata in the northeast prior to the Albian (I5 in Fig. 3). Uppermost Sustut Group strata are the youngest deformed rocks. Western Skeena fold belt structures are overlain by flat-lying Pliocene volcanic rocks (Evenchick and Thorkelson, 2005). These relationships demonstrate that the fold belt was initiated in the Early Cretaceous with at least some deformation prior to the Albian, and that it ended in latest Cretaceous or early Cenozoic time.

Direct constraints on the magnitude of shortening for specific structures and/or time periods are sparse. A ca. 84 Ma post-tectonic pluton near the central Bowser basin intruded Late Jurassic strata and constrains deformation there to between Late Jurassic and Campanian (Evenchick and McNicoll, 1993). Synorogenic clastic rocks include an Early Cretaceous piggyback basin in the north-central part of the fold belt, which records Early Cretaceous (I4 in Fig. 3) subaerial erosion of topographic highs in the central and eastern fold belt. The Sustut Group itself is synorogenic and records a western source (Skeena fold belt) from Albian through early Maastrichtian time. Angular unconformities within the lowest Sustut strata are indications of Albian or Cenomanian tectonism in the northeastern fold belt. The dramatic change within the Sustut Group from low- to high-energy fluvial systems at the base of the Brothers Peak Formation (I6 in Fig. 3) indicates increased relief in the late Campanian.

Coast and Insular Belts

The Coast belt is composed mainly of Middle Jurassic to early Cenozoic plutonic rock and lesser amounts of greenschist- to amphibolite-, and locally granulite-facies metamorphic rock (e.g., Gabrielse et al., 1991). In the northern Coast belt (Fig. 1), Cretaceous continental arc magmatism, interpreted to be associated with accretion of terranes and subduction of Farallon, Kula, and possibly Resurrection plates, migrated eastward, resulting in a western 105–90 Ma arc and an eastern 80–50 Ma arc, separated by the Coast shear zone (e.g., Crawford et al., 2005, and references therein). Contractional deformation of these arcs resulted in significant crustal thickening in the mid-Cretaceous, continuing to ca. 60 Ma; arc igneous activity ended at ca. 50 Ma with the beginning of a period of extension, and the eastern side of the arc was exhumed on bounding shear zones (e.g., Andronicos et al., 2003; Crawford et al., 2005). Farther west, the Insular belt consists of little metamorphosed Late Proterozoic to early Mesozoic volcanic arc terranes (mainly Wrangellia and Alexander) that are stratigraphically distinct from the more inboard terranes (e.g., Monger and Nokleberg, 1996, and references therein). Additional components of the Insular belt are minor late Mesozoic and Cenozoic accretionary complexes (e.g., Gabrielse et al., 1991) and Proterozoic to Paleozoic metamorphosed continental-margin sequences (e.g., Gehrels and Boghossian, 2000, and references therein). The western boundaries of Stikinia and the Bowser basin are within the Coast belt, but their relationships with Insular belt terranes are obscured by the large volume of intrusive rocks, medium- to high-grade metamorphism, and high-strain zones; thus, the timing of amalgamation of terranes of the Intermontane and Insular belts is controversial. It was either in the Jurassic or earlier (e.g., van der Heyden, 1992; McClelland et al., 1992) or the mid-Cretaceous (e.g., Monger et al., 1982). Uncertainty in the accretion history and in magnitude of postaccretion orogen-parallel translation leads to uncertainties about the Middle Jurassic to early Cenozoic paleogeography of these terranes. For these reasons we discuss only Cretaceous and early Cenozoic features of significant strike length, and only in general terms. Examples are: Cretaceous northeast-vergent contractional structures on the east side of the Coast belt that merge with contractional structures in the Intermontane belt (e.g., Evenchick, 1991a, 1991b; Journeay and Friedman, 1993; Rusmore and Woodsworth, 1994); mid-Cretaceous crustal thickening and magmatism (e.g., Crawford et al., 1987, 2005); and west-vergent contractional structures on the west side of the Coast belt (e.g., Rubin et al., 1990; Journeay and Friedman, 1993).

MESOZOIC TECTONIC EVOLUTION

Paleogeographic Reconstructions—Restoration of Jurassic to Eocene Orogen-Parallel Faults

Discussion of the Mesozoic evolution of the Bowser basin and Skeena fold belt in the context of tectonism in the Omineca and Foreland belts requires restoration of displacement on orogen-parallel strike-slip faults within and bordering the Omineca belt (Fig. 1). The discrepancies between geological and paleomagnetic estimates of dextral motion in the Late Cretaceous and Eocene have been reviewed and discussed by Gabrielse et al. (2006). The amount of Late Cretaceous to Eocene dextral displacement on the array of faults including the Tintina, Northern Rocky Mountain Trench fault, and related splays (Fig. 1) is estimated to be 490 km, with 430 km primarily in the Eocene, and ~60 km on the Northern Rocky Mountain Trench fault in early Late Cretaceous (Gabrielse et al., 2006). Additional strike-slip faults of this age are the Kechika-Spinel (80 km; a splay of the Northern Rocky Mountain Trench fault), and the Pinchi fault. The magnitudes and specific ages of older displacement on these faults, and of older faults, are less well constrained. Offset of lower Paleozoic facies boundaries suggest a total of ~700 km displacement (Gabrielse et al., 2006) on the Northern Rocky Mountain Trench fault; if 490 km of this was in the Late Cretaceous to Eocene, earlier displacement must have been ~210 km. Other Late Cretaceous or older fault systems to consider are (see Fig. 1): (1) the pre-Eocene part of the Kechika-Sifton fault, a splay of Northern Rocky Mountain Trench (~90 km); (2) the Kechika-Thudaka-Finlay-Ingenika-Takla system (~110 km); and (3) the Cassiar-Kutcho-Thibert system, which must be younger than the early Middle Jurassic (or younger) contractional structures and Early Jurassic intrusions that they displace. Faults of the latter two fault systems that cut the Cassiar batholith are considered to be synchronous with intrusion of the batholith at ca. 95–110 Ma (Gabrielse et al., 2006).

Figure 5A reconstructs the paleogeography of the early Middle Jurassic to account for the faults listed previously. Their continuation to the south as discrete structures is problematic (see Gabrielse et al., 2006), which makes an accurate paleogeographic reconstruction challenging. Accordingly, we restore Stikinia and the Bowser basin southward to approximately the latitude indicated by strike-slip faults in the north (~800 km), yet outboard enough to allow for shortening in the Foreland and Omineca belts (~300 km), stacking of terranes in the Omineca belt, and Eocene extension. Following Gabrielse (1985), we depict the Cache Creek terrane and Quesnellia in Figure 5A as continuous belts, with Cache Creek in thrust fault contact with Stikinia. In doing so, we assume that most of the later dextral fault displacement occurred within and/or between Cache Creek terrane and Quesnellia and is now obscured by Paleogene and younger strata of the southern Intermontane belt. Stikinia and the Bowser basin are restored with clockwise rotation from the present orientation, which in part reflects the restoration of known faults and allows for rotation about the Euler pole during northward translation (Price and Carmichael, 1986). To depict the minimum original width of Stikinia and the Bowser and Sustut basins, horizontal shortening of the Skeena fold belt, estimated at ~50%, is restored from across the Bowser basin, from more than half of the Sustut basin and most of Stikinia.

This reconstruction puts the Bowser basin adjacent to the site of the southern Canadian Cordillera at the time of final closure of the Cache Creek ocean in the Aalenian-Bajocian, just prior to the onset of major clastic deposition. In the absence of constraints on net displacements on strike-slip faults for specific periods, we assume, from Gabrielse et al. (2006), that dextral displacements were (1) ~300 km between Middle Jurassic and mid-Cretaceous time, with ~100 km on the fault systems west of the Northern Rocky Mountain Trench fault and 200 km on the Northern Rocky Mountain Trench fault, and (2) ~500 km on the Northern Rocky Mountain Trench–Tintina faults and related splays in the Late Cretaceous to late Eocene.

Relationship of Structures in Neighboring Intermontane, Omineca, and Foreland Belts, and Development of Basins and Fold-and-Thrust Belts Flanking the Omineca Belt

In the following sections, the major deformation, metamorphic, magmatic, and depositional events in the north-central and southeastern Canadian Cordillera are summarized for a series of time periods from the Middle Jurassic to Early Cretaceous, with emphasis on the regions that became the Intermontane, Omineca, and Foreland belts. The morphogeologic belt terminology applies to the Cordillera today; however, we use the terms as a convenient way to refer to regions that eventually became the belts. Cross sections (Fig. 6) for each time period show the tectonic evolution and relationships between events in neighboring belts, such as links between highland sources and basins of deposition. Although the cross sections are drawn with the east side fixed relative to eastward translation of terranes and structures, events described are the result of the North American plate moving westward relative to the hot-spot reference frame, with its western convergent plate boundary interacting with, and accumulating, parts of microcontinents, volcanic arcs, and intervening ocean basins of the Pacific Ocean—the future terranes of the Cordillera (e.g., Engebretson et al., 1985; Coney and Evenchick, 1994). Deposition in the Bowser and Alberta Foreland basins, shown in Figure 5, and development of mid-crustal structures in the Omineca belt and those in the upper crust in the Foreland belt, shown in Figure 6, are relatively well constrained because they are now exposed, are confidently inferred by plunge projection of structures, or have been drilled by hydrocarbon exploration wells. The geometry of the boundary between the Cache Creek terrane and Quesnellia in Figure 6, however, is highly uncertain because the only contacts known are the dextral faults that may obscure possible earlier low-angle faults. Other geometries that satisfy the constraint of the Cache Creek terrane forming a major part of the source region for clasts deposited in the Bowser basin are possible. The thickness of Stikinia in Figure 6 is interpreted from seismic sections (Cook et al., 2004), but its thickness through time is poorly constrained. It could have been significantly thinner in the Jurassic and later thickened by a combination of thin- and thick-skinned contractional structures during formation of the Skeena fold belt, with inversion of early Mesozoic extensional structures. The nature and position of the original western boundary of Stikinia are poorly constrained, and thus only the palinspastically restored width of exposed Stikinia strata is represented Figure 6.

Early Jurassic–Pliensbachian/Toarcian (ca. 190–176 Ma)

Prior to the Middle Jurassic, Stikinia was separated from cratonic North America by a region of marginal or pericratonic terranes, Quesnellia, and the last vestiges of the Cache Creek ocean, which was closing probably as a result of southwest subduction beneath Stikinia and northeast subduction beneath Quesnellia (e.g., Mortimer, 1987; Marsden and Thorkelson, 1992). Upper parts of Cache Creek terrane were delaminated to form the accretionary complex that now sits in thrust contact above Stikinia. Deposition in the Cache Creek ocean ended in the latest Early Jurassic (Toarcian; e.g., Struik et al., 2001), and northern Stikinia evolved from a region of widespread subduction-related volcanism in the Pliensbachian and earlier, to a largely marine clastic environment with minor volcanism by the Aalenian-Bajocian (Fig. 3; e.g., Marsden and Thorkelson, 1992; Anderson, 1993). The structural overlap of Quesnellia over North American basement at this time is documented by changes in the geochemistry of volcanic-arc rocks (Ghosh and Lambert, 1995) and the geochemistry and composition of detritus in volcaniclastic sediments of the arc (Petersen et al., 2004). Early Jurassic I-type intrusions of the Kuskanax and Nelson Suites indicate the presence of a subduction zone, and their isotope geochemistry indicates primitive signatures contaminated with that of continental North America (Armstrong, 1988; Ghosh and Lambert, 1995). In northern Quesnellia, there was increasing cratonal influence on granitic rocks of Triassic to mid-Cretaceous age (Gabrielse, 1998). Quesnellia and Slide Mountain terranes were obducted onto the North American pericratonic terranes and imbricated and thrust eastward along the Eureka, Pundata, Stubbs, and related faults between ca. 187 and 173 Ma (O1 in Fig. 3; Murphy et al., 1995, and references therein). Obduction was closely followed by the onset of southwest-directed thrusts and isoclinal recumbent folding, which continued in the Middle Jurassic (O2 in Fig. 3). Dating of east-vergent structures in eastern Quesnellia in the east-central Intermontane belt indicates that the initial stages of obduction there occurred at ca. 186 Ma (Nixon et al., 1993). In the Foreland belt, a Sinemurian to middle Toarcian cherty carbonate platform south of 54°N changed westward and northward into a narrow belt of shale and carbonate sandstone and then into a westward thinning unit of phosphatic mudstone and limestone (Asgar-Deen et al., 2004). All facies in the Foreland belt were overlain by a thin, euxinic, black shale deposited during the worldwide Toarcian (anoxic black shale) "transgressive" event (Poulton et al., 1994b).

Early Middle Jurassic–Aalenian/Bajocian (ca. 176–168 Ma)

An early phase of Bowser basin deposition began with subsidence in the northeast, marked by a starved phase in the Aalenian, and followed in the Bajocian by deposition of subaerially eroded Cache Creek strata in the northeasternmost part of the basin (I0 in Fig. 3; Fig. 6A). These events are interpreted to be a result of southwest thrusting of Cache Creek strata onto Stikinia (Gabrielse, 1991b; Ricketts et al., 1992). Stacking of Quesnellia and pericratonic terranes onto or against Cache Creek probably facilitated crustal thickening in the source area (Fig. 6A). Some of the first Cache Creek chert clasts deposited in the Bowser basin are close in age to the youngest blueschist (173.7 ± 0.8 Ma) resulting from Cache Creek subduction, a relationship interpreted by Mihalynuk et al. (2004) to indicate rapid exhumation of northern Cache Creek strata at ca. 174–171 Ma. In the northern part of the south segment of Cache Creek terrane, westward obduction of Cache Creek onto Stikinia occurred between 190 and 165 Ma, and west-vergent structures were overprinted by east-vergent ones (Struik et al., 2001).

In the Omineca belt, the formation of large southwest-verging, recumbent isoclinal folds and polyphase deformation record progressive crustal thickening and low- to medium-grade regional metamorphism (O1–O3 in Table 1; Fig. 3). By ca. 173–168 Ma, northeast-verging fold and fault systems were superimposed on southwest-verging fold systems, south of 52°N, as indicated by the relationships of syn- and post-tectonic ca. 173–168 Ma plutons of the Kuskanax and Nelson plutonic suites in the Selkirk and Purcell Mountains (Armstrong, 1988; Parrish and Wheeler, 1983; Colpron et al., 1998; Gibson, 2003; O2 in Table 1; Fig. 3).

North of 52°N, polyphase southwest-verging fold systems in the Cariboo Mountains and thrust faults at higher structural levels were coeval with northeastward thrusting on shear zones and penetrative deformation at depth (O3b in Table 1; Fig. 3; Struik, 1988). Following Price (1986), Struik (1988), and Murphy (1989), we use a tectonic wedge model with southwest-directed back thrusting and folding of the cover above northeast-directed detachments and thrusting of the craton to explain the geometry and structural evolution of the internally deforming and thickening crust (Fig. 6A). In the Foreland belt, 30–90 m of shelf mudstone and northeasterly derived sandstone were deposited in the Bajocian and possibly Aalenian (Fig. 3; Hall, 1984).

The western Omineca belt and eastern Intermontane belt collectively defined a growing region of structurally thickened Cache Creek, Quesnellia, and pericratonic terranes, the growth of which was concomitant with the onset of westward deposition of Cache Creek detritus into the Bowser basin. Changes in structural vergence are present in both Intermontane (Cache Creek) and Omineca belt rocks and may indicate the onset of decoupling of supracrustal rocks from the westward underthrusting of North America. Coincident timing of southwest-verging structures in the Omineca belt with southwest thrusting of Cache Creek on Stikinia has been used to suggest a link between the southern Omineca belt and Stikinia at this time (Ricketts et al., 1992). The size and types of clasts that were deposited in the Bowser basin record deposition from the west side of the Omineca highland, in high-energy conditions, during rapid exhumation of Cache Creek strata, and they provide further evidence for linkage between these realms. The structurally thickened crust and emerging highland had little effect on the sedimentary record preserved in the Foreland belt over 350 km to the east (Figs. 3, 5A, 6A), suggesting that the loading occurred on weak lithosphere too far to the west to have elastically depressed the thick craton to the east.

Late Middle Jurassic–Bathonian/Callovian (ca. 168–161 Ma)

A major increase in deposition of chert-rich detritus in the northeastern Bowser basin began in the Bathonian, with up to 3000 m of strata deposited in base-of-slope to deltaic environments, and expansion of the extent of Cache Creek chert clasts ~130 km farther southwest than in the Bajocian (I1 in Table 1; Figs. 3 and 6B). Rapid denudation of northern Stikinia or the Cache Creek terrane is indicated by deposition of ca. 161 Ma dacite boulders in the early Callovian slope assemblage of the Bowser basin (Ricketts and Parrish, 1992). The development of fan deltas in the northern part of the basin demonstrates high sedimentation rates (Ricketts and Evenchick, 1991, 2007). Sections of this age elsewhere in the basin are considerably finer grained and thinner and lack the spectacular submarine channel deposits present in the north. This scenario continued into the early Oxfordian, with ~30 km of south and southwest migration of facies boundaries. Southern Bowser basin facies migrated northerly away from a westerly trending arch of Stikinia that defined the south margin of the basin (Tipper and Richards, 1976).

In the Omineca belt, the crust was 50–55 km thick or greater (O3 in Table 1), and it likely formed a broad highland that was internally deforming as it was being translated toward the craton (Table 1; Fig. 6B). Southwest-verging polyphase folding was ongoing in the Cariboo and Monashee Mountains (Reid, 2003; O3b in Fig. 3), but elsewhere, upright folds and northeastward-verging fold systems and faults dominated after ca. 168–167 Ma (Warren, 1997; Colpron et al., 1998; Gibson, 2003; O3a in Fig. 3). The peak of regional metamorphism occurred at ca. 165–160 Ma, although at higher structural levels, it had started to wane (Gerasimoff, 1988; Warren, 1997; Parrish, 1995). Plutons of ca. 167–159 Ma age, within the Nelson Suite, crosscut large-scale belts of folds (e.g., Scrip nappe, early Selkirk fan, Dogtooth structure, and Kootenay arc), indicating that the architecture of these belts had largely formed by the end of the Middle Jurassic (Warren, 1997, and references therein). In the southwest Foreland belt, an 80-m-thick Bathonian section appears to provide the first record of increased subsidence due to tectonic loading (Poulton et al., 1993), and the local presence of cherty quartz-arenites provides the first evidence of a western source area (Stronach, 1984). Elsewhere in the Foreland belt, the first preserved record of west-derived sediments occurs in the Upper Jurassic (Bally et al., 1966; Poulton, 1984; Poulton et al., 1993; Fig. 3). A period of pre-Oxfordian uplift removed much of the Middle Jurassic record in the eastern Foreland belt.

In summary, links across the orogen may be inferred from evidence for continued growth of the Omineca highland, which loaded the lithosphere and affected the sedimentation patterns of adjacent basins as it was translated inboard from distal transitional crust that was significantly thinned during Proterozoic and early Paleozoic rifting, onto thicker, more rigid transitional crust. This is expressed in the northeastern Bowser basin by substantially thicker and more widespread deposits sourced from the Cache Creek terrane, and in the Foreland belt by the first, but limited, westerly derived foreland basin sediments. A time-transgressive Callovian–Oxfordian erosive disconformity developed across the Foreland belt (F0 in Fig. 3) and adjacent craton, marking the northeastward migration of the forebulge (Poulton, 1984). Detrital zircon analyses from Bowser basin strata indicate that in addition to wind-blown ash, the source areas were Triassic to Middle Jurassic in age (McNicoll et al., 2005). These data, combined with paleocurrents, facies distribution, and clast types in the Bowser basin indicate that the Cache Creek, and Stikine, and/or Quesnel terranes (Fig. 4) formed the upper structural levels of at least the western Omineca highland, and that no cratonic North American detritus was being shed to the west (Fig. 6B).

Late Jurassic–Middle Oxfordian/Kimmeridgian (ca. 158–151 Ma)

Starting in the middle Oxfordian, the Bowser basin changed from a relatively narrow northeastern depocenter of base-of-slope to deltaic deposits, to widespread deposition when the shelf–slope break migrated ~200 km west (I2 in Table 1; Figs. 3, 5B, and 6C). The result was accumulation of up to 2 km of shelf deposits in the central basin, and over 4 km of submarine fan deposits in the western basin. Deltas migrated westward to occupy much of the northeastern basin. Facies boundaries in the southern basin also migrated rapidly, and pebbly deposits became widespread there for the first time, accompanied by local volcanic flows in the Oxfordian (Tipper and Richards, 1976). Regionally, the dominant pebble and sand type remained Cache Creek radiolarian chert (e.g., Evenchick and Thorkelson, 2005, and references therein), but at the basin's southern margin, clasts reflect a local source of Stikinia strata from the arch to the south.

In the Omineca belt between 52°N and 53°N, a zone of medium-grade metamorphism that trends approximately north-northwest across the Cariboo, Monashee, and Selkirk Mountains represents the deepest exposed levels of rocks at this latitude. These rocks preserve evidence of penetrative shortening throughout the Late Jurassic and Early Cretaceous accompanied by metamorphism, folding, shearing, and reactivation of structures (Ferguson, 1994; Currie, 1988; Digel et al., 1998; Crowley et al., 2000; Reid, 2003; Ghent and Simony, 2005, and references therein). Deformation progressed northeastward and carried the accreted terranes farther onto the craton (Fig. 6C). South of 52°N, post-tectonic plutonism and cooling indicate rapid exhumation and quenching at higher structural levels in the Late Jurassic (Warren, 1997, and references therein), providing a source for sediments deposited in the Alberta foreland basin. Pronounced flexural subsidence and development of a two-sided foredeep trough in the Alberta foreland basin began in the Oxfordian. Immense quantities of west-derived silty mud were transported into the trough along its western side at the same time as thin, craton-derived, basal, transgressive sandstone was deposited along its eastern side (Poulton, 1984; Stott, 1998; F1 in Fig. 3). Pronounced subsidence continued through the Kimmeridgian, and sands containing detrital chert, stretched quartz, and mica derived from sedimentary or metasedimentary rocks were deposited into the trough from the west (Hamblin and Walker, 1979). These are the first indications of an Omineca belt source for the Alberta foreland basin; however, isotope geochemistry studies of these strata have not detected juvenile material from the accreted terranes (Ross et al., 2005; F2 in Fig. 3; Fig. 6C).

We conclude that links across the orogen in the Oxfordian can be inferred from events within the Omineca belt and significant changes in regions bordering it. The Omineca highland was likely maintained by continued internal structural thickening during northeastward translation of the Omineca belt core relative to the North American craton. To the west, the Bowser basin experienced a dramatic westward migration of facies boundaries, indicating sustained topography in the source areas, and to the east, the Foreland belt experienced its first significant flexural subsidence and deposition. The latter resulted from the elastic response of the first widespread loading of the North American craton by thickened North American supracrustal rocks and the western thickened lithosphere of pericratonic terranes, Quesnellia, Cache Creek, and possibly Stikinia. Detrital zircons in Bowser basin strata indicate that the sources were still within the accreted terranes (Cache Creek, and Stikinia and/or Quesnellia) that were incorporated into the western Omineca highland, whereas sources for Alberta foreland basin deposition included mica and quartz from exhumed metamorphic rocks of continental origin then exposed at the surface of the eastern Omineca highland. Accordingly, the drainage divide in the Omineca highland is shown on Figures 5B and 6C as lying in the eastern part of the carapace of accreted terranes such that sediments were transported from the exhumed terranes west to the Bowser basin, and Alberta foreland basin drainages had access only to the eastern Omineca highland.

Early Early Cretaceous (ca. 145–135 Ma)

In the early Early Cretaceous, deposition in the Bowser basin changed from widespread marine and marginal marine, to nonmarine, including floodplain deposition marginal to deltas, and low-energy fluvial systems (I4 in Table 1; Figs. 3 and 6D). The only widespread strata preserved of this age are in the north-central part of the basin. Other strata deposited in this stage, or prior to the Albian, include the synorogenic coarse clastic deposits in the north-central Bowser basin and nonmarine strata in the southern basin (Evenchick and Thorkelson, 2005; Bassett and Kleinspehn, 1997; V.J. McNicoll, 2006, personal commun.; Smith and Mustard, 2006).

In the Omineca belt, Early Cretaceous zones of penetrative deformation and metamorphism occurred at mid-crustal levels within the existing edifice (Currie, 1988; Digel et al., 1998; Crowley et al., 2000; Reid, 2003; O4–O6 in Fig. 3), while imbrication of the basement and northeastward translation of the belt occurred on shear zones at depth, near the base of the edifice (Fig. 6D). Middle Jurassic plutons, such as the Hobson Lake and Fang plutons, which crosscut early southwest-verging structures, record Early Cretaceous quenching to low temperatures as they were progressively exhumed to higher structural levels. The interplay between exhumation of older structures and renewed metamorphism and penetrative deformation at depth is consistent with a model of progressive shortening of the Omineca belt during northeastward translation.

In the Foreland belt, west-derived sands and conglomerates flooded into the Alberta foreland basin, locally filling the western foreland trough with over 1.5 km of lower Lower Cretaceous sediments (Stott, 1998; Fig. 3). Sedimentation changed from marine to nonmarine in the late Tithonian south of 52°N, and during the late Berriasian to early Valanginian further north (52°N–55°N). In the Berriasian, a west-sourced delta system with rivers carrying volcanic and metamorphic clasts and abundant radiolarian chert entered the basin near 54°N, indicating that the depositional system had sources in the Slide Mountain terrane and metamorphic parts of the Omineca belt (McMechan et al., 2006). Relatively rapid Late Jurassic to Early Cretaceous sedimentation (Valanginian; up to 100 m/m.y.) was followed by pedimentation and conglomeratic sedimentation in the Hauterivian and Barremian (F3 in Fig. 3). There was substantial erosion of lower Lower Cretaceous and Upper Jurassic sediments in the eastern part of the Alberta foreland basin, and the conglomerates were derived solely from sedimentary sources (Gibson, 1985, and references therein; Ross et al., 2005). The westernmost Alberta foreland basin sediments were being deformed and eroded by the end of the Valanginian (Fig. 6D). Near the line of section in Figure 6D, northeast-directed thrusting likely reached the western Foreland belt in the early Early Cretaceous with the initiation of the Malton Gneiss basement slice (O7 in Table 1).

From the material presented here, we conclude that the early Early Cretaceous was a time of significant change across the Cordillera and that events in adjacent regions were kinematically linked within the developing orogen. The basins flanking the Omineca highland became regions of exclusively nonmarine deposition, with >1.5 km of sediment deposited in the western Alberta foreland basin, and probably >2 km of sediment deposited in the Bowser basin. A major middle Early Cretaceous unconformity in the Alberta foreland basin developed within all but the westernmost areas. This period also records the first significant deformation of western Alberta foreland basin deposits, and possibly the initial deformation in the Skeena fold belt. The age of the latter is not narrowly constrained, but the presence of locally derived early to middle Early Cretaceous coarse synorogenic clastic rocks in the north-central Bowser basin suggests that deformation started in this period, or shortly thereafter. Detrital zircons in Early Cretaceous Bowser basin strata have similar source ages as the older strata, indicating that the drainage divide remained in the western Omineca–eastern Intermontane components of the Omineca highland, and detritus of only Cache Creek, Quesnellia, and Stikinia was carried westward (Fig. 5C and 6D).

Albian (ca. 112–100 Ma)

By the mid-Cretaceous, the Insular belt terranes were accreted to western Stikinia, and a major phase of ductile deformation, metamorphism, and magmatism began in the Coast belt (e.g., Monger et al., 1982; Crawford et al., 1987, 2005; Fig. 6E). In the west-central Coast belt, greenschist metamorphism prior to ca. 98 Ma was followed by burial locally to 8 kbar (800 MPa; 30 km), and regional west-directed thrusting started at ca. 100 Ma (e.g., Crawford et al., 1987; Rubin et al., 1990). Farther east in the central Coast belt, crustal thickening by thrust faulting and emplacement of tabular syntectonic plutons resulted in 6 kbar (600 MPa) metamorphism by ca. 90 Ma (Crawford et al., 1987). Rocks in the central Coast Belt, west of the southern Bowser basin, presently overlie 30 km of crust (Morozov et al., 1998; Hammer et al., 2000), suggesting that in the mid-Cretaceous, part of the central Coast belt crust was up to 60 km thick (e.g., Crawford et al., 1987). Ductile east-directed thrusting involving Stikinia on the east side of the central and southern Coast belt began at ca. 90 Ma, about the same time as the major crustal thickening described previously (e.g., Rusmore and Woodsworth, 1994; Rusmore et al., 2000; Crawford et al., 2005).

Shortening of Bowser basin strata in the Skeena fold belt started before the Albian, and there is no record of Albian or younger Bowser basin deposition except at the southern margin of the basin (Evenchick, 1991a; Evenchick and Thorkelson, 2005; Bassett and Kleinspehn, 1997; I5 in Table 1; Fig. 3). Widespread fluvial deposition in the Sustut basin began in the Albian with an eastern (Omineca belt) source of detrital micas and clasts distinctive of the early Paleozoic Cordilleran margin (Eisbacher, 1981), including recycled detrital zircons of Archean, Paleoproterozoic, Mesoproterozoic, and Paleozoic ages initially derived from the North America craton (McNicoll et al., 2005; I5 in Fig. 3; Fig. 6E). Omineca belt sources may be represented in the southern Bowser basin earlier than Albian (Bassett and Kleinspehn, 1997). The depositional overlap of Skeena fold belt structures by basal Sustut Group strata illustrates the extent of contractional deformation, which reached far to the northeast part of the Bowser basin prior to Albian time (Eisbacher, 1981; Evenchick, 1991a). Clasts of chert recycled from the Bowser Lake Group into the Sustut basin in this period suggest that the Skeena fold belt formed highlands southwest of the Sustut basin (5e in Fig. 3; Fig. 5D; Eisbacher, 1981).

In the Omineca belt, the general structural style established in the Early Cretaceous prevailed, whereby zones of penetrative deformation in the mid-crust and movement on deep-seated shear zones accommodated deformation and translation across the belt (O7 in Table 1; Fig. 6E) and transferred shortening across the orogen from the plate margin to the active foreland. Emplacement and stacking of imbricated basement-cored nappes of the Malton Complex occurred in the Albian, as bracketed by the ca. 140–120 Ma isograds south of the complex, which were deflected during emplacement of the complex, and the ca. 110–100 Ma cooling dates for the complex (O7 in Table 1). The ca. 105–90 Ma Cassiar batholith, now located in the northern Omineca belt, was probably emplaced in the southern Omineca belt and translated northward. Geochemistry indicates derivation mainly from melting of continental crust (Driver et al., 2000), which was associated with, and facilitated by, movement on mid-Cretaceous transcurrent faults within a transpressive environment (Gabrielse et al., 2006).

In the Foreland belt, renewed subsidence of the Alberta foreland basin occurred during the Albian. The greatest subsidence was associated with extensional faulting near 56°N above the ancestral Peace River Arch, following a pattern established in the Aptian (Stott, 1993). Several regional transgressive-regressive cycles in the north caused an alternation of marine, coastal, and nonmarine environments, whereas nonmarine deposition and disconformities developed in the south (Fig. 5D; Stott, 1993; Smith, 1994). Rock fragments and heavy minerals in Albian strata indicate a mixed sedimentary, metamorphic, and volcanic/intrusive source between 53°N and 56°N, and a dominant volcanic and intrusive Quesnellian source south of 53°N (McMechan and Thompson, 1993; Ross et al., 2005). Exhumation and cooling of metasedimentary rocks and basement slices above northeast-directed thrust faults occurred in the Albian near 53°N (F4 in Fig. 3; McDonough and Simony, 1988).

We conclude that the Albian was a milestone in tectonic development across the southern Canadian Cordillera, including the Coast belt. A fundamental change in detrital sources for regions west of the Omineca highlands was marked by a flood of clasts derived from deeper levels of the Omineca belt, including the first clasts of exhumed metamorphic rocks that had originally been deposited on the Paleozoic margin of cratonic North America (Fig. 6E). Either exhumation and erosion of the Omineca belt reduced the carapace of accreted terranes on the west side of the highland, or increased surface uplift of the Omineca belt shifted the drainage divide eastward. After an initial flood of clastics into the Sustut basin from the east (Omineca belt), deposition in the basin was focused in a northwest-trending trough, confined between the Omineca highland and highlands of the Skeena fold belt (Fig. 6E). Widespread sedimentation in the Alberta foreland basin driven by flexural and dynamic processes (Beaumont et al., 1993) produced a westward-thickening wedge, locally up to 1.7 km thick, that extended 1000 km eastward onto the craton. In most of the southern Alberta foreland basin, sediments were derived from the Omineca and Foreland highlands. In contrast, south of 51°N, the Albian was the only time when rivers flowed unimpeded from a region underlain by Quesnellia strata in the western Omineca belt into the foreland basin with little input from central Omineca or Foreland belt strata (Leckie and Krystinik, 1995). The structural architecture formed in the Jurassic to Early Cretaceous within the Omineca and eastern Intermontane belts was carried northeastward by deep-seated zones of deformation, and in the Foreland belt deformation occurred mainly on northeast-directed thrust faults (Fig. 6E). The net result of mid-Cretaceous tectonism in the Coast belt was significant crustal thickening by stacking of thick crustal slabs. Involvement of Stikinia in the east-directed ductile thrust system on the east side of the Coast belt is the basis for the interpretation that the high-level structures of the Skeena fold belt, also involving Stikinia, root in the Coast belt (Fig. 6E; Evenchick, 1991a). We speculate that the crustal thickening in the Coast belt may have provided a western source of sediment deposited over deformed Bowser basin strata of the western Skeena fold belt. Strata of this age are not preserved, but apatite fission-track and vitrinite reflectance data suggest that since the latest Cretaceous–early Cenozoic, 4.4–7 km of section has been eroded from the northwest Bowser basin–Skeena fold belt (O'Sullivan et al., 2005).

Late Campanian/Maastrichtian–Early Eocene (ca. 74–50 Ma)

The contractional ductile deformation of the Coast belt that began in the mid-Cretaceous continued into the earliest Cenozoic with emplacement of large volumes of magma and an eastward migration of magmatism; deformation included the development of large recumbent nappes in the core of the Coast belt (younger than ca. 85 Ma), with the result that thick crust was created by latest Cretaceous time (e.g., Crawford et al., 1987, 2005; Fig. 6G). These events occurred in an environment of dextral transpression that lasted into earliest Cenozoic time (Rusmore et al., 2001; Andronicos et al., 2003). Exhumation began during the Cretaceous during contraction (e.g., Crawford et al., 1987). Extension, pluton emplacement, and rapid (2 mm/yr) exhumation followed in the Paleocene and early Eocene (Hollister, 1982; Andronicos et al., 2003, and references therein).

The last major depositional change in the Sustut basin was in the late Campanian, when the relatively low-energy fluvial systems of the Tango Creek Formation were succeeded by high-energy systems that deposited sheets of conglomerate of the Brothers Peak Formation (Eisbacher, 1981; I6 in Table 1; Figs. 3 and 6F). Deformation of uppermost Sustut strata constrains the youngest Skeena fold belt deformation to post-Maastrichtian. Preliminary apatite fission-track thermochronology results (O'Sullivan et al., 2005) and thermal maturity data (Stasiuk et al., 2005) suggest that a few kilometers of strata have been eroded from above present exposures of Sustut and northeastern Bowser strata, and 4.4–7 km has been eroded from northwestern Bowser strata (O'Sullivan et al., in 2005) since the latest Cretaceous–early Cenozoic; some of this "missing" section may have been mid-Cretaceous age, as described earlier, but some, or all, may have been late Maastrichtian to early Eocene age, in part supplied from the west during the rapid exhumation of the Coast belt.

Continued internal deformation and northeastward translation of the Omineca belt (O9 in Table 1; Fig. 6F) is manifested by out-of-sequence structures, such as the pre–ca. 93 Ma Purcell thrust (Archibald et al., 1983; P.S. Simony, 2005, personal commun.), by crystalline thrust nappes, such as the Gwillim Creek shear zone in Valhalla complex, which carried metamorphic rocks inboard and ramped them onto cold basement (Carr and Simony, 2006), and by belts of penetrative deformation within mid-crustal rocks, such as those exposed in the eastern Selkirk Mountains and in the Monashee Mountains north, west, and south of the Monashee Complex (Johnston et al., 2000; Gibson, 2003; Hinchey, 2005; Williams and Jiang, 2005; Brown and Gibson, 2006). Dextral displacement primarily in the Eocene, on faults within and transecting the western Omineca highlands, amounted to ~430 km (Gabrielse et al., 2006) and resulted in the final northward movement of the Bowser-Sustut region and western Omineca highlands (including Cassiar batholith) relative to cratonic North America.

A major change from dominantly marine shale with pulses of westerly derived deltaic sand to dominantly nonmarine coarse clastics occurred during the Santonian in the western Alberta foreland basin south of 51°N (Stott, 1963; Leahy and Lerbekmo, 1995; Payenberg et al., 2002); a similar change occurred at the base of the late Campanian elsewhere in the western foreland basin (Dawson et al., 1994; 4e in Fig. 3). At the eastern margin of the Foreland belt, rapid, nonmarine sedimentation probably continued into the early Eocene, with up to 4 km of late Campanian to Paleocene strata preserved (Stott and Aitken, 1993; F5 in Fig. 3). An additional 2–2.5 km accumulation of Paleocene and Eocene strata, no longer preserved, is inferred for the eastern edge of the Foreland belt, south of 54°N, from coal reflectance data (Nurkowski, 1984; Kalkreuth and McMechan, 1996). Most Santonian to Paleocene sediments were derived from the Foreland highlands and from volcanic airfall (Ross et al., 2005). Two exceptions occur near 53°N (Fig. 3), where local conglomerates, one late Maastrichtian and the other middle Paleocene, contain a few andesitic pebbles (Jerzykiewicz, 1985), which indicate that the drainage divide locally extended into the Intermontane belt. The leading edge of the thrust system progressed eastward from the central Foreland belt to its eastern edge during the late Campanian to early Eocene (Fig. 6G), doubling its width and resulting in exhumation, erosion, and cannibalization of the Alberta foreland basin wedge (Price and Mountjoy, 1970). Motion on major thrust faults in the Eastern Front Ranges (e.g., Lewis, McConnell) deformed late Campanian strata, and major thrust faults in the Foothills (e.g., Bighorn, Brazeau) deformed Paleocene strata (Price, 1981; McMechan and Thompson, 1993). Alberta foreland basin subsidence ended with the cessation of contractional deformation in the early Eocene (Fig. 6G).

The late Campanian to earliest Eocene was the last period of trans-Cordilleran horizontal shortening and sedimentation that may be attributed to kinematic links across the orogen. It was characterized by northeastward translation of the Omineca belt on mid-crustal structures at the same time as significant nonmarine deposition in basins on the east and west sides of the Omineca highland, thickening and horizontal shortening at high crustal levels in the Skeena fold belt and Foreland belt, and crustal thickening and eastward migration of magmatism in the Coast belt. The start of this period marks one of the fundamental changes in sedimentation in the basins flanking the Omineca highlands. In the Sustut basin, the change was from low- to high-energy fluvial deposition, and in the Alberta foreland basin, it changed from deltaic marine pulses to entirely nonmarine deposition. Confinement of Sustut Group deposition to a linear northwest-trending trough with southeast paleocurrents suggests that the Skeena fold belt and Omineca highland continued to form topographic barriers bounding the Sustut basin. During the late Campanian to early Maastrichtian, ~1.5 km of strata accumulated in the Sustut basin, while up to 2 km accumulated in the Alberta foreland basin. Deposition in the Alberta foreland basin continued into the early Eocene, forming a late Campanian to Eocene westward-thickening wedge over 6 km thick at the eastern margin of the Foreland belt. Deposition of strata, no longer preserved, across the Sustut basin and/or a broader and younger basin and originating in part from exhumation of the Coast belt, may have continued into the latest Cretaceous or earliest Cenozoic.

LINKS ACROSS THE OROGEN AND THEIR SIGNIFICANCE

The previous section demonstrated the kinematic and dynamic links across the orogen through time, illustrated by Figure 3 and tectonic interpretations in Figure 6. To summarize:

(1) Middle Jurassic obduction of Quesnellia and Slide Mountain terranes onto pericratonic terranes and distal North America was followed closely by southwest obduction of Cache Creek onto Stikinia, which resulted in initiation of flexural subsidence in the Bowser basin and deposition of coarse Cache Creek detritus in the Bowser basin (I0, I1, O1, O2 in Table 1);

(2) Ongoing deformation, crustal thickening, and/or exhumation in the Omineca belt from the Late Jurassic to Paleocene occurred while sediment was shed from the Omineca highland eastward into the Alberta foreland basin, and westward into the Bowser and Sustut basins (I2–I6, O3–O9, F1–F5 in Table 1; expanded on herein);

(3) Mid-Cretaceous initiation of thin-skinned shortening in the Foreland belt, major crustal thickening in the Coast belt, thin-skinned shortening in the Skeena fold belt, initiation of the Sustut basin, and continued eastward translation of the exhuming core of the Omineca belt above a basal detachment system (I5, O6, O7, O8, F4 in Table 1; Crawford et al., 1987)—in summary, shortening across the width of the Cordillera—was accommodated at different structural levels;

(4) Approximately synchronous pulses of synorogenic coarse clastics were deposited in the Sustut basin and Alberta foreland basin in the late Campanian–early Maastrichtian, during a period of major structural thickening and denudation (I6, O9, F5 in Table 1); and

(5) Horizontal shortening across all belts lasted into the latest Cretaceous to early Cenozoic (I6, O9, O10, F5 in Table 1; e.g., Crawford et al., 1987).

Such links are permissible, despite uncertainties in timing and magnitude of superimposed transcurrent faults, because fundamental aspects of each belt have considerable strike length. Although the ~300-km-long Bowser basin is the most restricted tectonic element discussed, it was the major center of sedimentation west of the evolving orogen in the Middle and Late Jurassic, and it was probably localized adjacent to the region of maximum crustal thickening in the Omineca belt.

From these examples of synchronous and compatible tectonic events in adjacent belts, we suggest that the Intermontane, Omineca, and Foreland belts were kinematically connected from the Middle Jurassic to early Cenozoic, and that the Insular and Coast belts were included in this kinematic connection from mid-Cretaceous to early Cenozoic. Both the Bowser and Alberta Foreland basins received sediment from a persistent source in the Omineca highland, which was composed of the eastern accreted terranes, pericratonic terranes, and supracrustal rocks of western North American. This source lasted for ~120 m.y., from earliest Bowser basin deposition to the final Alberta foreland basin deposition. Throughout the evolution of the orogen, changes in source rocks for each basin reflect changes in the level of exhumation in the intervening highland, which progressively exposed deeper rocks, and migration of the drainage divide. The relationship between basin evolution and lithospheric loading for the Foreland and Omineca belts has been well established (e.g., Beaumont et al., 1993). This relationship is less direct for the Bowser basin because much of northern Stikinia was already the site of marine deposition, although limited in thickness, prior to the Bajocian/Bathonian. Subsidence was caused, at least in part, by cooling of Stikinia, which probably continued through at least the early phases of coarse Bowser basin deposition. Facies relationships indicate that the northeastern part of the basin was the primary site of deposition during the late Middle and early Late Jurassic, presumably in response to lithospheric loading by the growing accretionary orogen to the east and by accumulating sediment. The large scale of both the Foreland and Bowser basins and widespread contractional structures are incompatible with the alternative explanation of basin formation in a transtensional regime. Therefore, the basins are most likely related, in differing degrees, to subsidence due to flexure and progressive loading of the lithosphere by the thickened crust between them and by deposition of sediments from the same persistent source, the elevation of which was maintained by the processes that resulted in continued exhumation of the Omineca belt.

Shortening of the orogen and inferred linkages across the orogen were likely accommodated by a lower-crustal detachment as illustrated in Figure 6. Support for this interpretation is provided by interpretation of SNORCLE line 2a, which displays a low-angle boundary between Stikinia and North American rocks that rises eastward from the lower crust to middle crust (Fig. 6H; Cook et al., 2004; Evenchick et al., 2005). This boundary may be part of the lower-crustal detachment illustrated in Figure 6. The deformation and structural architecture of mid- and upper-crustal rocks in the western Foreland and eastern Omineca belts were established in the Middle Jurassic to Early Cretaceous (Carr and Simony, 2006, and references therein); therefore, linkage of mid-Cretaceous to Eocene structures in the Foreland and Omineca belts must have occurred via a detachment that passed beneath the western Foreland belt. This interpretation is consistent with geophysical data that are, in part, controlled by outcrop and drill-hole information (Cook et al., 1988, 1992; Cook and van der Velden, 1995).

We suggest that the southern Canadian Cordillera evolved from a relatively narrow, doubly vergent, "small-cold" orogen in the Jurassic with the core centered in the Omineca belt (Figs. 6A–6D) to a much broader doubly vergent, "large-hot" orogen in the mid-Cretaceous (terminology of Koons, 1990; Beaumont et al., 2006). The mid- and Late Cretaceous orogen (Figs. 6E–6G) may be viewed as a wide doubly vergent orogen with the predominantly east-directed structures in the Omineca and Foreland belts on the eastern, retro-wedge side (terminology of Willett et al., 1993), and west-directed structures of the accretionary orogen of the Coast belt on the pro-wedge side, at the active oblique subduction margin, all linked by a basal detachment. At this time, the Jurassic core of the orogen was mainly translated eastward and exhumed as part of the retro-wedge, which included the active thrust front in the eastern Foreland belt. However, within this first-order, large-scale geometry, there was a detachment beneath the east-directed upper-crustal Skeena fold belt, which soled westward into mid- and lower-crustal ductile structures in the eastern Coast belt. Thus, there were two coeval cratonward-verging upper-crust fold-and-thrust belts, equally large in cross section, at the same latitude in the Cordilleran orogen: the Rocky Mountain fold-and-thrust belt and the Skeena fold belt.

The Mesozoic history of Stikinia as an arc and (or) back arc may have facilitated development of the lower-crust detachment. Hyndman et al. (2005) argued that back arcs or recent back arcs are hot as a result of transfer of convective heat below thin lithosphere, and that the high temperature results in weak lower crust, facilitating development of lower-crust detachments, which separate crustal elements above from underlying lithosphere in the manner of "orogenic float" (Oldow et al., 1990). The decay in temperature after the source is removed is slow enough that former back-arc regions may remain weak long after cessation of arc activity (Hyndman et al., 2005). From Hyndman et al.'s (2005) calculations of temperature decay, and Monger and Nokleberg's (1996) review of the evolution of arc development in the Cordillera, Stikinia probably remained relatively hot for the period discussed herein. To illustrate their model, Hyndman et al. (2005) explained the relationship between the modern collision of the Yakutat block in the Gulf of Alaska and shortening in the Mackenzie Mountains at the front of the thrust belt as being facilitated by a detachment in a weak lower crust. We suggest that the connectivity of tectonic elements across the Cordillera in the Cretaceous is an ancient example of this phenomenon.

Recognition of a deep detachment across the orogen may provide a broader context for understanding structural relationships. For example, Rusmore et al. (2001) posed the problem of accommodation in the Coast belt of the large horizontal displacement inferred from plate motions. They concluded that reverse motion on the Paleocene Coast shear zone represents the orthogonal component of oblique convergence, but we question the magnitude of shortening that may be accommodated by this structure, and instead suggest that a lower-crust detachment transferred a component of shortening eastward. This is also an effective way to accommodate the regional transpression inferred by Rusmore et al. (2001) who associated dextral faults in the Coast belt with dextral faults in the western Omineca belt–Northern Rocky Mountain Trench.

A question that arises from consideration of the Cordillera as one kinematically connected orogen is the perplexing thickness of Stikinia, which is interpreted from seismic-reflection data as being ~35 km thick (e.g., Cook et al., 2004). Was this Paleozoic–early Mesozoic arc terrane, possibly built partly on rifted fragments of continental margin, always thick, or was it substantially thinner prior to its accretion to North America and then thickened during Cretaceous Cordilleran-wide contraction?

In this interpretation, the mid-orogen Skeena fold belt was carried piggyback above a lower-crust detachment connected to the Rocky Mountain fold-and-thrust belt. This scenario is unusual in modern or ancient orogens. A factor that may have contributed to the geometry is the mechanical effect of the stratigraphy within the Intermontane belt. The Bowser succession, formed just prior to deformation, is a thinly bedded succession with substantial mechanical heterogeneity, and therefore it was relatively weak. In contrast, Stikinia is composed of units of limited lateral continuity, such as volcanic edifices and surrounding sedimentary units with rapid facies changes, and associated plutons. Pre-Triassic strata underwent at least two phases of deformation and local low-grade metamorphism, and Early Jurassic intrusions pierce all sub-Bowser stratigraphy. Compared to the Bowser succession, Stikinia contains few laterally continuous horizontal weak layers. An exception is the thinly layered early Middle Jurassic clastic succession, immediately below the Bowser Lake Group, which, along with underlying layered Early Jurassic volcanic successions, was intimately involved in thin-skinned deformation. We infer that the relative strength of Stikinia, combined with a relatively weak lower crust, as discussed previously, localized a deep detachment, whereas the mechanical heterogeneity of the Bowser succession and immediately underlying strata of Stikinia facilitated an upper-crustal detachment leading to the Skeena fold belt in uppermost Stikinia, Bowser, and Sustut strata.

CONCLUSIONS

Synthesis of the Jurassic and Cretaceous depositional and tectonic histories of the central Intermontane belt, and the southern Omineca and Foreland belts, when considered in paleogeographic context, reveals sedimentation and structural linkages across the orogen and highlights the tectonic interplay among crustal thickening, basin formation, and topographic evolution. We conclude that coeval and tectonically compatible events in regions considered to have been in close proximity require kinematic linkage across the entire orogen. In our model for the mid- and Late Cretaceous, a detachment in the lower crust extended from the active plate boundary and Coast belt eastward below the Intermontane belt and then rose into the middle crust in the Omineca belt and, ultimately, to the upper crust at the front of the orogen in the Rocky Mountain fold-and-thrust belt. It was also connected to dextral strike-slip faults, facilitating regional transpression partitioned across the orogen in response to oblique plate convergence. In the Intermontane belt, this lower-crust detachment carried Stikinia as well as the Skeena fold belt, an upper-crust and craton-verging fold-and-thrust belt nested in the interior of the orogen. The basal detachment of the Skeena fold belt rooted in ductile structures on the east side of the Coast belt. Development of this nested fold belt was in part a consequence of the mechanical stratigraphy of Stikinia and overlying basins. Consideration of the orogen as a whole, in paleogeographic context, should lead to improved tectonic models of Cordilleran evolution.

ACKNOWLEDGMENTS

Research support was provided to CAE, MEM, and VJM by the Geological Survey of Canada, Natural Resources Canada, and the British Columbia Ministry of Energy and Mines, and to SDC by a NSERC research grant. We thank the members of the Cordilleran community for discussions and for sharing their insights and expertise, in particular Hubert Gabrielse and Philip Simony. We especially thank Ray Price for his considerable scientific contributions in tectonics and Cordilleran geology. We would like to acknowledge the importance of his professional leadership and the significance of his role in mentoring students and colleagues, many of whose work provided the platform for this synthesis. We appreciate the helpful comments on an early version of the manuscript by Hubert Gabrielse, David Ritcey, and Jim Monger, and the thoughtful and constructive reviews of Maurice Colpron and Glen Stockmal, and editor Tekla Harms. This is Geological Survey of Canada contribution 2005540.

REFERENCES CITED

Anderson, R.G., 1993, A Mesozoic stratigraphic and plutonic framework for northwestern Stikinia (Iskut River area), northwestern British Columbia, Canada, in Dunne, G., and McDougall, K., eds., Mesozoic Paleogeography of the Western United States II: Field Trip Guidebook—Pacific Section, Society of Economic Paleontologists and Mineralogists, v. 71, p. 477–494.

Andronicos, C.L., Chardon, D.H., Hollister, L.S., Gehrels, G.E., and Woodsworth, G.J., 2003, Strain partitioning in an obliquely convergent orogen, plutonism, and synorogenic collapse: Coast Mountains Batholith, British Columbia, Canada: Tectonics, v. 22, no. 2, p. 1012, doi: 10.1029/2001TC001312.

Archibald, D.A., Glover, J.K., Price, R.A., Farrar, E., and Carmichael, D.M., 1983, Geochronology and tectonic implications of magmatism and metamorphism, southern Kootenay arc and neighbouring regions, southeastern British Columbia. Part I: Jurassic to Mid-Cretaceous: Canadian Journal of Earth Sciences, v. 20, p. 1891–1913.

Armstrong, R.L., 1988, Mesozoic and early Cenozoic magmatic evolution of the Canadian Cordillera, in Clark, S.D., Jr., Burchfiel, B.C., and Suppe, J., eds., Processes in Continental Lithospheric Deformation: Geological Society of America Special Paper 218, p. 55–91.

Asgar-Deen, M., Riediger, C., and Hall, R., 2004, The Gordondale Member: Designation of a new member in the Fernie Formation to replace the informal "Nordegg Member" nomenclature of the subsurface of west-central Alberta: Bulletin of Canadian Petroleum Geology, v. 52, p. 201–214.

Bally, A.W., Gordy, P.L., and Stewart, G.A., 1966, Structure, seismic data, and orogenic evolution of the southern Canadian Rocky Mountains: Bulletin of Canadian Petroleum Geology, v. 14, p. 337–381.

Bassett, K.N., and Kleinspehn, K.L., 1997, Early to Middle Cretaceous paleogeography of north-central British Columbia; stratigraphy and basin analysis of the Skeena Group: Canadian Journal of Earth Sciences, v. 34, p. 1644–1669.

Beatty, T.W., Orchard, M.J., and Mustard, P.S., 2006, Geology and tectonic history of the Quesnel terrane in the area of Kamloops, British Columbia, in Colpron, M., and Nelson, J.L., eds., Paleozoic Evolution and Metallogeny of Pericratonic Terranes at the Ancient Pacific Margin of North America, Canadian and Alaskan Cordillera: Geological Association of Canada Special Paper 45, p. 483–504.

Beaumont, C., Quinlan, G.M., and Stockmal, G.S., 1993, The evolution of the Western Interior Basin: Causes, consequences and unsolved problems, in Caldwell, W.G.E., and Kauffman, E.G., eds., Evolution of the Western Interior Basin: Geological Association of Canada Special Paper 39, p. 97–117.

Beaumont, C., Nguyen, M.H., Jamieson, R.J., and Ellis, S., 2006, Crustal flow modes in large hot orogens, in Law, R.D., Searle, M., and Godin, L., eds., Channel Flow, Ductile Extrusion and Exhumation in Continental Collision Zones: Geological Society [London] Special Publication 268, p. 91–145.

Brown, R.L., and Gibson, H.D., 2006, An argument for channel flow in the southern Canadian Cordillera and comparison with Himalayan tectonics, in Law, R.D., Searle, M., and Godin, L., eds., Channel Flow, Ductile Extrusion and Exhumation in Continental Collision Zones: Geological Society [London] Special Publication 268, p. 543–559.

Brown, R.L., Journey, J.M., Lane, L.S., Murphy, D.C., and Rees, C.J., 1986, Obduction, backfolding and piggyback thrusting in the metamorphic hinterland of the southeastern Canadian Cordillera: Journal of Structural Geology, v. 8, p. 255–268, doi: 10.1016/0191-8141(86)90047-7.

Brown, R.L., Carr, S.D., Johnson, B.J., Coleman, V.J., Cook, F.A., and Varsek, J.L., 1992a, The Monashee décollement of the southern Canadian Cordillera: A crustal scale shear zone linking the Rocky Mountain Foreland belt to lower crust beneath accreted terranes, in McClay, K.R., ed., Thrust Tectonics: London, UK, Chapman and Hall, p. 357–364.

Brown, R.L., McNicoll, V.J., Parrish, R.R., and Scammell, R.J., 1992b, Middle Jurassic plutonism in the Kootenay terrane, northern Selkirk Mountains, British Columbia, *in* Radiogenic and Isotopic Studies, Report 5: Geological Survey of Canada Paper 91-2, p. 135–141.

Brown, R.L., Beaumont, C., and Willett, S.D., 1993, Comparison of the Selkirk fan structure with mechanical models: Implications for interpretation of the southern Canadian Cordillera: Geology, v. 21, p. 1015–1018, doi: 10.1130/0091-7613(1993)021<1015:COTSFS>2.3.CO;2.

Cant, D.J., and Stockmal, G.S., 1989, The Alberta foreland basin: Relationship between stratigraphy and Cordilleran terrane-accretion events: Canadian Journal of Earth Sciences, v. 26, p. 1964–1975.

Carr, S.D., 1992, Tectonic setting and U-Pb geochronology of the early Tertiary Ladybird leucogranite suite, Thor-Odin-Pinnacles area, southern Omineca belt, British Columbia: Tectonics, v. 11, p. 258–278.

Carr, S.D., and Simony, P.S., 2006, Ductile thrusting versus channel flow in the southeastern Canadian Cordillera: Evolution of a coherent crystalline thrust sheet, *in* Law, R.D., Searle, M., and Godin, L., eds., Channel Flow, Ductile Extrusion and Exhumation in Continental Collision Zones: Geological Society [London] Special Publication 268, p. 561–587.

Colpron, M., Price, R.A., Archibald, D.A., and Carmichael, D.M., 1996, Middle Jurassic exhumation along the western flank of the Selkirk fan structure: Thermobarometric and thermochronometric constraints from the Illecillewaet synclinorium, southeastern British Columbia: Geological Society of America Bulletin, v. 108, p. 1372–1392, doi: 10.1130/0016-7606(1996)108<1372:MJEATW>2.3.CO;2.

Colpron, M., Warren, M.J., and Price, R.A., 1998, Selkirk fan structure, southeastern Canadian Cordillera: Tectonic wedging against an inherited basement ramp: Geological Society of America Bulletin, v. 110, p. 1060–1074, doi: 10.1130/0016-7606(1998)110<1060:SFSSCC>2.3.CO;2.

Colpron, M., Nelson, J.L., and Murphy, D.C., 2006, A tectonostratigraphic framework for the pericratonic terranes of the northern Cordillera, *in* Colpron, M., and Nelson, J.L., eds., Paleozoic Evolution and Metallogeny of Pericratonic Terranes at the Ancient Pacific Margin of North America, Canadian and Alaskan Cordillera: Geological Association of Canada Special Paper 45, p. 1–23.

Coney, P.J., and Evenchick, C.A., 1994, Consolidation of the American Cordilleras: Journal of South American Earth Sciences, v. 7, no. 3/4, p. 241–262, doi: 10.1016/0895-9811(94)90011-6.

Coney, P.J., Jones, D.L., and Monger, J.W.H., 1980, Cordilleran suspect terranes: Nature, v. 288, p. 329–333, doi: 10.1038/288329a0.

Cook, F., and van der Velden, A., 1995, Three-dimensional crustal structure of the Purcell anticlinorium in the Cordillera of southwestern Canada: Geological Society of America Bulletin, v. 107, p. 642–664, doi: 10.1130/0016-7606(1995)107<0642:TDCSOT>2.3.CO;2.

Cook, F., Green, A., Simony, P., Price, R., Parrish, R., Milkereit, B., Brown, R., Coflin, K., and Patenaude, C., 1988, Lithoprobe seismic reflection structure of the southeastern Canadian Cordillera: Initial results: Tectonics, v. 7, p. 157–180.

Cook, F., Varsek, J., Clowes, R., Kanasewich, E., Spencer, C., Parrish, R., Brown, R., Carr, S., Johnson, B., and Price, R., 1992, Lithoprobe crustal reflection cross section of the southern Canadian Cordillera: 1. Foreland thrust and fold belt to Fraser River fault: Tectonics, v. 11, p. 12–35.

Cook, F.A., Clowes, R.M., Snyder, D.B., van der Velden, A.J., Hall, K.W., Erdmer, P., and Evenchick, C.A., 2004, Precambrian crust beneath the Mesozoic northern Canadian Cordillera discovered by Lithoprobe seismic reflection profiling: Tectonics, v. 23, no. 2, TC2010, doi: 10.1029/2002TC001412, 28 p., 2 sheets.

Crawford, M.L., Hollister, L.S., and Woodsworth, G.J., 1987, Crustal deformation and regional metamorphism across a terrane boundary, Coast Plutonic Complex, British Columbia: Tectonics, v. 6, p. 343–361.

Crawford, M.L., Crawford, W.A., and Lindline, J., 2005, 105 million years of igneous activity, Wrangell, Alaska, to Prince Rupert, British Columbia: Canadian Journal of Earth Sciences, v. 42, p. 1097–1116, doi: 10.1139/e05-022.

Crowley, J.L., and Parrish, P.R., 1999, U-Pb isotopic constraints on diachronous metamorphism in the northern Monashee Complex, southern Canadian Cordillera: Journal of Metamorphic Geology, v. 17, p. 483–502, doi: 10.1046/j.1525-1314.1999.00210.x.

Crowley, J.L., Ghent, E.D., Carr, S.D., Simony, P.S., and Hamilton, M.A., 2000, Multiple thermotectonic events in a continuous metamorphic sequence, Mica Creek area, southeastern Canadian Cordillera: Geological Materials Research, v. 2, p. 1–45.

Currie, L.D., 1988, Geology of the Allan Creek Area, British Columbia [M.Sc. thesis]: Calgary, University of Calgary, 152 p.

Dawson, F.M., Kalkreuth, W.D., and Sweet, A.R., 1994, Stratigraphy and coal resource potential of the Upper Cretaceous to Tertiary strata of northwestern Alberta: Geological Survey of Canada Bulletin 466, 64 p.

Demchuk, T.D., 1990, Palynostratigraphic zonation of Paleocene strata in the central and south-central Alberta Plains: Canadian Journal of Earth Sciences, v. 27, p. 1263–1269.

Digel, S.G., Ghent, E.D., Carr, S.D., and Simony, P.S., 1998, Early Cretaceous kyanite-sillimanite metamorphism and Paleocene sillimanite overprint near Mount Cheadle, southeastern British Columbia: Geometry, geochronology, and metamorphic implications: Canadian Journal of Earth Sciences, v. 35, p. 1070–1087, doi: 10.1139/cjes-35-9-1070.

Driver, L.A., Creaser, R.A., Chacko, T., and Erdmer, P., 2000, Petrogenesis of the Cretaceous Cassiar batholith, Yukon–British Columbia, Canada: Implications for magmatism in the North American Cordillera interior: Geological Society of America Bulletin, v. 112, p. 1119–1133, doi: 10.1130/0016-7606(2000)112<1119:POTCCB>2.3.CO;2.

Eisbacher, G.H., 1974a, Sedimentary History and Tectonic Evolution of the Sustut and Sifton Basins, North-Central British Columbia: Geological Survey of Canada Paper 73-31, 57 p.

Eisbacher, G.H., 1974b, Deltaic Sedimentation in the Northeastern Bowser Basin: Geological Survey of Canada Paper 73-33, 13 p.

Eisbacher, G.H., 1981, Late Mesozoic–Paleogene Bowser Basin molasse and Cordilleran tectonics, western Canada, *in* Miall, A.D., ed., Sedimentation and Tectonics in Alluvial Basins: Geological Association of Canada Special Paper 23, p. 125–151.

Eisbacher, G.H., 1985, Pericollisional strike-slip faults and synorogenic basins, Canadian Cordillera, *in* Biddle, K.T., and Christie-Blick, N., eds., Strike-Slip Deformation, Basin Formation, and Sedimentation: Society of Economic Paleontologists and Mineralogists Special Publication 37, p. 265–282.

Engebretson, D.C., Cox, A., and Gordon, R.G., 1985, Relative motions between oceanic and continental plates in the Pacific Basin: Geological Society of America Special Paper 206, 59 p.

Evenchick, C.A., 1991a, Geometry, evolution, and tectonic framework of the Skeena fold belt, north-central British Columbia: Tectonics, v. 10, p. 527–546.

Evenchick, C.A., 1991b, Structural relationships of the Skeena fold belt west of the Bowser Basin, northwest British Columbia: Canadian Journal of Earth Sciences, v. 28, p. 973–983.

Evenchick, C.A., 2001, Northeast-trending folds in the western Skeena fold belt, northern Canadian Cordillera: A record of Early Cretaceous sinistral plate convergence: Journal of Structural Geology, v. 23, p. 1123–1140, doi: 10.1016/S0191-8141(00)00178-4.

Evenchick, C.A., and McNicoll, V.J., 1993, U-Pb ages for Late Cretaceous and early Tertiary plutons in the Skeena fold belt, north-central British Columbia, *in* Radiogenic Age and Isotopic Studies, Report 7: Geological Survey of Canada Paper 93-2, p. 99–106.

Evenchick, C.A., and Thorkelson, D.J., 2005, Geology of the Spatsizi River Map Area, North-Central British Columbia: Geological Survey of Canada Bulletin 577, 276 p.

Evenchick, C.A., Poulton, T.P., Tipper, H.W., and Braidek, I., 2001, Fossils and facies of the northern two-thirds of the Bowser Basin, northern British Columbia: Geological Survey of Canada Open-File 3956, scale 1:250,000, 1 sheet, 103 p. text.

Evenchick, C.A., Gabrielse, H., and Snyder, D., 2005, Crustal structure and lithology of the northern Canadian Cordillera: Alternative interpretations of SNORCLE seismic reflection lines 2a and 2b: Canadian Journal of Earth Sciences, v. 42, p. 1149–1161, doi: 10.1139/e05-009.

Evenchick, C.A., Mustard, P.S., McMechan, M.E., Ferri, F., Ritcey, D.H., and Smith, G.T., 2006, Compilation of geology of Bowser and Sustut Basins draped on shaded relief map, north-central British Columbia: Geological Survey of Canada Open-File 5313, and British Columbia Ministry of Energy, Mines and Petroleum Resources, Petroleum Geology Open-File 2006-1, scale 1:500,000.

Ferguson, C.A., 1994, Structural Geology and Stratigraphy of the Northern Cariboo Mountains between Isaac Lake and Fraser River, British Columbia [Ph.D. thesis]: Calgary, University of Calgary, 331 p.

Fermor, P.R., and Moffat, I.W., 1992, The Cordilleran collage and the foreland fold-and-thrust belt, *in* Macqueen, R.W., and Leckie, D.A., eds., Foreland

Basins and Fold Belts: American Association of Petroleum Geologists Memoir 55, p. 81–105.

Gabrielse, H., 1985, Major dextral transcurrent displacements along the Northern Rocky Mountain Trench and related lineaments in north-central British Columbia: Geological Society of America Bulletin, v. 96, p. 1–14, doi: 10.1130/0016-7606(1985)96<1:MDTDAT>2.0.CO;2.

Gabrielse, H., 1991a, Structural styles, Chapter 17, in Gabrielse, H., and Yorath, C.J., eds., Geology of the Cordilleran Orogen in Canada: Geological Survey of Canada, Geology of Canada, no. 4, p. 571–675 (also Geological Society of America, Geology of North America, v. G-2).

Gabrielse, H., 1991b, Late Paleozoic and Mesozoic terrane interactions in north-central British Columbia: Canadian Journal of Earth Sciences, v. 28, p. 947–957.

Gabrielse, H., 1998, Geology of Dease Lake (104J) and Cry Lake (104I) Map Areas, North-Central British Columbia: Geological Survey of Canada Bulletin 504, 147 p.

Gabrielse, H., and Yorath, C.J., 1991, Tectonic synthesis, Chapter 18, in Gabrielse, H., and Yorath, C.J., eds., Geology of the Cordilleran Orogen in Canada: Geological Survey of Canada, Geology of Canada, no. 4, p. 679–705 (also Geological Society of America, Geology of North America, v. G-2).

Gabrielse, H., Monger, J.W.H., Wheeler, J.O., and Yorath, C.J., 1991, Part A. Morphogeological belts, tectonic assemblages, and terranes: Tectonic framework, Chapter 2, in Gabrielse, H., and Yorath, C.J., eds., Geology of the Cordilleran Orogen in Canada: Geological Survey of Canada, Geology of Canada, no. 4, p. 15–28 (also Geological Society of America, the Geology of North America, v. G-2).

Gabrielse, H., Murphy, D.C., and Mortensen, J.K., 2006, Cretaceous and Cenozoic dextral orogen-parallel displacements, magmatism, and paleogeography, north-central Canadian Cordillera, in Haggart, J.W., Enkin, R.J., and Monger, J.W.H., eds., Paleogeography of the North American Cordillera: Evidence For and Against Large-Scale Displacements: Geological Association of Canada Special Paper 46, p. 255–276.

Gehrels, G.E., and Boghossian, N.D., 2000, Reconnaissance geology and U-Pb geochronology of the west flank of the Coast Mountains between Bella Coola and Prince Rupert, coastal British Columbia, in Stowell, H.H., and McClelland, W.C., eds., Tectonics of the Coast Mountains, Southeastern Alaska and British Columbia: Geological Society of America Special Paper 343, p. 61–75.

Gerasimoff, M.D., 1988, The Hobson Lake Pluton, Cariboo Mountains, and Its Significance to Mesozoic and Early Tertiary Cordilleran Tectonics [M.Sc. thesis]: Kingston, Queen's University, 188 p.

Ghent, E.D., and Simony, P.S., 2005, Geology of isogradic, isothermal, and isobaric surfaces: Interpretation and application: The Canadian Mineralogist, v. 43, p. 295–310.

Ghosh, D.K., and Lambert, R., St.J., 1995, Nd-Sr isotope geochemistry and petrogenesis of Jurassic granitoid intrusives, southeast British Columbia, Canada, in Miller, D.M., and Busby, C., eds., Jurassic Magmatism and Tectonics of the North American Cordillera: Geological Society of America Special Paper 299, p. 141–157.

Gibson, D.W., 1985, Stratigraphy, Sedimentology and Depositional Environments of the Coal-Bearing Jurassic–Cretaceous Kootenay Group, Alberta and British Columbia: Geological Survey of Canada Bulletin, v. 357, 108 p.

Gibson, H.D., 2003, Structural and Thermal Evolution of the Northern Selkirk Mountains, Southeastern Canadian Cordillera: Tectonic Development of a Regional-Scale Composite Structural Fan [Ph.D. thesis]: Ottawa, Carleton University, 298 p.

Gibson, H.D., Brown, R.L., and Parrish, R.R., 1999, Deformation-induced inverted metamorphic field gradients: An example from the southeastern Canadian Cordillera: Journal of Structural Geology, v. 21, p. 751–767, doi: 10.1016/S0191-8141(99)00051-6.

Gradstein, F.M., Ogg, J.G., Smith, A.G., Agterberg, F.P., Bleeker, W., Cooper, R.A., Davydov, V., Gibbard, P., Hinnov, L.A., House, M.R., Lourens, L., Luterbacher, H.P., McArthur, J., Melchin, M.J., Robb, L.J., Shergold, J., Villeneuve, M., Wardlaw, B.R., Ali, J., Brinkhuis, H., Hilgen, F.J., Hooker, J., Howarth, R.J., Knoll, A.H., Laskar, J., Monechi, S., Plumb, K.A., Powell, J., Raffi, I., Röhl, U., Sadler, P., Sanfilippo, A., Schmitz, B., Shackleton, N.J., Shields, G.A., Strauss, H., Van Dam, J., van Kolfschoten, T., Veizer, J., and Wilson, D., 2004, A Geologic Time Scale 2004: Cambridge, Cambridge University Press, 589 p.

Hall, R.L., 1984, Lithostratigraphy and biostratigraphy of the Fernie Formation (Jurassic) in the southern Canadian Rocky Mountains, in Glass, D.J., and Stott, D.F., eds., The Mesozoic of Middle North America: Canadian Society of Petroleum Geologists Memoir 9, p. 233–247.

Hamblin, A.P., and Walker, R.G., 1979, Storm-dominated shallow marine deposits: The Fernie–Kootenay (Jurassic) transition, southern Rocky Mountains: Canadian Journal of Earth Sciences, v. 16, p. 1673–1690.

Hammer, P.T.C., Clowes, R.M., and Ellis, R.M., 2000, Crustal structure of NW British Columbia and SE Alaska from seismic wide-angle studies: Coast Plutonic Complex to Stikinia: Journal of Geophysical Research, v. 105, p. 7961–7981, doi: 10.1029/1999JB900378.

Hinchey, 2005, Thor-Odin Dome: Constraints on Paleocene–Eocene Anatexis and Deformation, Leucogranite Generation and the Tectonic Evolution of the Southern Omineca Belt, Canadian Cordillera [Ph.D. thesis]: Ottawa, Carleton University, 230 p.

Hollister, L.D., 1982, Metamorphic evidence for rapid (2 mm/yr) uplift of a portion of the Central Gneiss Complex, Coast Mountains, B.C.: Canadian Mineralogist, v. 20, p. 319–332.

Hyndman, R.D., Currie, C.A., and Mazzotti, S.P., 2005, Subduction zone backarcs, mobile belts, and orogenic heat: GSA Today, v. 15, no. 2, p. 4–10.

Jerzykiewicz, T., 1985, Tectonically deformed pebbles in the Brazeau and Paskapoo Formations, central Alberta foothills, Canada: Sedimentary Geology, v. 42, p. 159–180, doi: 10.1016/0037-0738(85)90043-0.

Johnston, D.H., Williams, P.F., Brown, R.L., Crowley, J.L., and Carr, S.D., 2000, Northeastward extrusion and extensional exhumation of crystalline rocks of the Monashee complex, southeastern Canadian Cordillera: Journal of Structural Geology, v. 22, p. 603–625, doi: 10.1016/S0191-8141(99)00185-6.

Journeay, J.M., and Friedman, R.M., 1993, The Coast belt thrust system: Evidence of Late Cretaceous shortening in southwest British Columbia: Tectonics, v. 12, p. 756–775.

Kalkreuth, W., and McMechan, M., 1996, Coal rank and burial history of Cretaceous–Tertiary strata in the Grande Cache and Hinton areas, Alberta, Canada: Implications for fossil fuel exploration: Canadian Journal of Earth Sciences, v. 33, p. 938–957.

Klepacki, D.W., 1985, Stratigraphy and Structural Geology of the Goat Range Area, Southeastern British Columbia [Ph.D. thesis]: Boston, Massachusetts Institute of Technology, 268 p.

Koons, P.O., 1990, Two-sided orogen: Collision and erosion from the sandbox to the Southern Alps, New Zealand: Geology, v. 18, p. 679–682, doi: 10.1130/0091-7613(1990)018<0679:TSOCAE>2.3.CO;2.

Kubli, T.E., and Simony, P.S., 1994, The Dogtooth Duplex, a model for the structural development of the northern Purcell Mountains: Canadian Journal of Earth Sciences, v. 31, p. 1672–1686.

Kuiper, Y.D., 2003, Isotopic Constraints on Timing of Deformation and Metamorphism in the Thor-Odin Dome, Monashee Complex, Southeastern British Columbia [Ph.D. thesis]: Ottawa, Carleton University, 321 p.

Larson, K.P., Price, R.A., and Archibald, D.A., 2004, The southern termination of the Western Ranges and Main Ranges of the southern Canadian Rocky Mountains: Tectonic and geochronologic implications of new stratigraphic and structural data and detailed mapping: Geological Society of America Abstracts with Programs, v. 36, no. 5, p. 208.

Leahy, G.D., and Lerbekmo, J.F., 1995, Macrofossil magnetobiostratigraphy from the upper Santonian–lower Campanian interval in the Western Interior of North America: Comparisons with European stage boundaries and planktonic foraminiferal zonal boundaries: Canadian Journal of Earth Sciences, v. 32, p. 247–260.

Leckie, D., and Burden, E.T., 2001, Stratigraphy, sedimentology, and palynology of the Cretaceous (Albian) Beaver Mines, Mill Creek, and Crowsnest Formations (Blairmore Group) of southwestern Alberta: Geological Survey of Canada Bulletin 563, 103 p.

Leckie, D.A., and Krystinik, L.F., 1995, Cretaceous igneous-clast conglomerate in the Blairmore Group, Rocky Mountain foothills and adjacent subsurface (Bow Island Formation), Alberta, Canada: Bulletin of Canadian Petroleum Geology, v. 43, p. 320–342.

Marsden, H., and Thorkelson, D.J., 1992, Geology of the Hazelton volcanic belt in British Columbia: Implications for the Early to Middle Jurassic evolution of Stikinia: Tectonics, v. 11, p. 1266–1287.

McClelland, W.C., Gehrels, G.E., and Saleeby, J.B., 1992, Upper Jurassic–Lower Cretaceous basinal strata along the Cordilleran margin: Implications for the accretionary history of the Alexander–Wrangellia–Peninsular terrane: Tectonics, v. 11, p. 823–835.

McDonough, M.R., and Simony, P.S., 1988, Structural evolution of basement gneisses and Hadrynian cover, Bulldog Creek area, Rocky Mountains, British Columbia: Canadian Journal of Earth Sciences, v. 25, p. 1687–1702.

McMechan, M.E., and Thompson, R.I., 1989, Structural style and history of the Rocky Mountain fold-and-thrust belt, Chapter 4, in Ricketts, B.D., ed., Western Canada Sedimentary Basin: A Case History: Calgary, Alberta, Canadian Society of Petroleum Geologists, p. 47–71.

McMechan, M.E., and Thompson, R.I., 1993, The Canadian Cordilleran fold-and-thrust belt south of 66°N and its influence on the Western Interior Basin, in Caldwell, W.G., and Kauffman, E.G., eds., Evolution of the Western Interior Basin: Geological Association of Canada Special Paper 39, p. 73–90.

McMechan, M., Anderson, B., Creaser, R., and Ferri, F., 2006, Clasts from the past: Latest Jurassic–earliest Cretaceous foreland basin conglomerates, northeast British Columbia and northwest Alberta: Geological Survey of Canada Open-File 5086, 1 sheet.

McNicoll, V.J., Evenchick, C.A., and Mustard, P.S., 2005, Provenance studies on the depositional histories of the Bowser and Sustut Basins and their implications for tectonic evolution of the northern Canadian Cordillera: Geological Association of Canada Annual Meeting Abstracts, v. 30, p. 133.

Mihalynuk, M.G., Erdmer, P., Ghent, E.D., Cordey, F., Archibald, D.A., Friedman, R.M., and Johannson, G.G., 2004, Coherent French Range blueschist; subduction to exhumation in <2.5 m.y.?: Geological Society of America Bulletin, v. 116, p. 910–922, doi: 10.1130/B25393.1.

Monger, J.W.H., and Nokleberg, W.H., 1996, Evolution of the northern North American Cordillera: Generation, fragmentation, displacement and accretion of successive North American plate margin arcs, in Coyner, E.R., and Fahey, P.L., eds., Geology and Ore Deposits of the American Cordillera: Geological Society of Nevada Symposium Proceedings, April 1995: Reno/Sparks, Nevada, Geological Society of Nevada, v. III, p. 1133–1152.

Monger, J.W.H., Price, R.A., and Tempelman-Kluit, D.J., 1982, Tectonic accretion and the origin of two major metamorphic and plutonic welts in the Canadian Cordillera: Geology, v. 10, p. 70–75, doi: 10.1130/0091-7613 (1982)10<70:TAATOO>2.0.CO;2.

Monger, J.W.H., Wheeler, J.O., Tipper, H.W., Gabrielse, H., Harms, T., Struik, L.C., Campbell, R.B., Dodds, C.J., Gehrels, G.E., and O'Brien, J., 1991, Upper Devonian to Middle Jurassic assemblages, Chapter 8: Part B. Cordilleran terranes, in Gabrielse, H., and Yorath, C.J., eds., Geology of the Cordilleran Orogen in Canada: Geological Survey of Canada, Geology of Canada, no. 4, p. 281–327 (also Geological Society of America, Geology of North America, v. G-2).

Morozov, I.B., Smithson, S.B., Hollister, L.S., and Diebold, J.B., 1998, Wide-angle seismic imaging across accreted terranes, southeastern Alaska and western British Columbia: Tectonophysics, v. 299, p. 281–296, doi: 10.1016/S0040-1951(98)00208-X.

Mortimer, N., 1987, The Nicola Group: Late Triassic and Early Jurassic subduction-related volcanism in British Columbia: Canadian Journal of Earth Sciences, v. 24, p. 2521–2536.

Mossop, G.D., and Shetsen, I., compilers, 1994, Geological Atlas of the Western Canada Sedimentary Basin: Calgary, Alberta, Canadian Society of Petroleum Geologists and Alberta Research Council, 510 p.

Murphy, D.C., 1989, Crustal paleorheology of the southeastern Canadian Cordillera and its influence on the kinematics of Jurassic convergence: Journal of Geophysical Research, v. 94, p. 15,723–15,739.

Murphy, D.C., van der Heyden, P., Parrish, R.R., Klepacki, D.W., McMillan, W., Struik, L.C., and Gabites, J., 1995, New geochronological constraints on Jurassic deformation of the western edge of North America, southeastern Canadian Cordillera, in Miller, D.M., and Busby, C., eds., Jurassic Magmatism and Tectonics of the North American Cordillera: Geological Society of America Special Paper 299, p. 159–171.

Nixon, G.T., Archibald, D.A., and Heaman, L.M., 1993, [40]Ar-[39]Ar and U-Pb geochronometry of the Polaris Alaskan-type complex, British Columbia: Precise timing of Quesnellia–North America interaction: Geological Association of Canada Annual Meeting Abstracts, v. 18, p. A–76.

Nurkowski, J.R., 1984, Coal quality, coal rank variation and its relation to reconstructed overburden, Upper Cretaceous and Tertiary plains coals, Alberta, Canada: American Association of Petroleum Geologists Bulletin, v. 68, p. 285–295.

Oldow, J.S., Bally, A.W., and Ave Lallement, H.G., 1990, Transpression, orogenic float and lithospheric balance: Geology, v. 18, p. 991–994, doi: 10.1130/0091-7613(1990)018<0991:TOFALB>2.3.CO;2.

O'Sullivan, P.B., Osadetz, K.G., Evenchick, C.A., Ferri, F., and Donelick, R.A., 2005, Apatite fission track thermochronology (AFTT) constraints on the Cenozoic thermal history of the Bowser and Sustut Basins, British Columbia: Geological Association of Canada Annual Meeting Abstracts, v. 30, p. 148.

Parrish, R.R., 1995, Thermal evolution of the southeastern Canadian Cordillera: Canadian Journal of Earth Sciences, v. 32, p. 1618–1642.

Parrish, R.R., and Wheeler, J.O., 1983, A U-Pb zircon age from the Kuskanax Batholith, southeastern British Columbia: Canadian Journal of Earth Sciences, v. 20, p. 1751–1756.

Parrish, R.R., Carr, S.D., and Parkinson, D.L., 1988, Eocene extensional tectonics and geochronology of the southern Omineca belt, British Columbia and Washington: Tectonics, v. 7, p. 181–212.

Payenberg, T.H.D., Braman, D.R., Davis, D.W., and Miall, A.D., 2002, Litho- and chronostratigraphic relationships of the Santonian–Campanian Milk River Formation in southern Alberta and Eagle Formation in Montana utilising stratigraphy, U-Pb geochronology, and palynology: Canadian Journal of Earth Sciences, v. 39, p. 1553–1577, doi: 10.1139/e02-050.

Petersen, N.T., Smith, P.L., Mortensen, J.K., Creaser, R.A., and Tipper, H.W., 2004, Provenance of Jurassic sedimentary rocks of south-central Quesnellia, British Columbia; implications for paleogeography: Canadian Journal of Earth Sciences, v. 41, p. 103–125, doi: 10.1139/e03-073.

Poulton, T.P., 1984, The Jurassic of the Canadian Western Interior from 49°N latitude to Beaufort Sea, in Glass, D.J., and Stott, D.F., eds., The Mesozoic of Middle North America: Canadian Society of Petroleum Geologists Memoir 9, p. 15–41.

Poulton, T.P., 1989, Upper Absaroka to lower Zuni: The transition to the Foreland basin, in Ricketts, B.D., ed., Western Canada Sedimentary Basin: A Case History: Calgary, Alberta, Canadian Society of Petroleum Geologists, p. 233–247.

Poulton, T.P., Braun, W.K., Brooke, M.M., and Davies, E.H., 1993, Jurassic, subchapter 4H, in Stott, D.F., and Aitken, J.D., eds., Sedimentary Cover of the Craton in Canada: Geological Survey of Canada, Geology of Canada, no. 5, p. 321–357 (also Geological Society of America, Geology of North America, v. D-1).

Poulton, T.P., Hall, R.L., and Callomon, J.H., 1994a, Ammonite and bivalve assemblages in Bathonian through Oxfordian strata of northern Bowser Basin, northwestern British Columbia, Canada: Geobios, v. 27, p. 415–421, doi: 10.1016/S0016-6995(94)80162-2.

Poulton, T.P., Christopher, J.E., Hayes, B.J.R., Losert, J., Tittermore, J., and Gilchrist, R.D., 1994b, Jurassic and lowermost Cretaceous strata of the Western Canada sedimentary basin, in Mossop, G., and Shetsen, I., eds., Geological Atlas of the Western Canada Sedimentary Basin: Calgary, Alberta, Canadian Society of Petroleum Geologists and Alberta Research Council, p. 297–316.

Price, R.A., 1973, Large-scale gravitational flow of supracrustal rocks, southern Canadian Rockies, in de Jong, K.A., and Scholten, R., eds., Gravity and Tectonics: New York, Wiley, p. 491–502.

Price, R.A., 1981, The Cordilleran thrust and fold belt in the southern Canadian Rocky Mountains, in McClay, K.R., and Price, N.J., eds., Thrust and Nappe Tectonics: Geological Society [London] Special Publication 9, p. 427–448.

Price, R.A., 1986, The southeastern Canadian Cordillera: Thrust faulting, tectonic wedging, and delamination of the lithosphere: Journal of Structural Geology, v. 8, p. 239–254, doi: 10.1016/0191-8141(86)90046-5.

Price, R.A., 1994, Cordilleran tectonics and the evolution of the Western Canada sedimentary basin, in Mossop, G., and Shestin, I., eds., Geological Atlas of the Western Canada Sedimentary Basin: Calgary, Alberta, Canadian Society of Petroleum Geologists and Alberta Research Council, p. 13–24.

Price, R.A., and Carmichael, D.M., 1986, Geometric test for Late Cretaceous–Paleogene intracontinental transform faulting in the Canadian Cordillera: Geology, v. 14, p. 468–471, doi: 10.1130/0091-7613(1986) 14<468:GTintina faultLCI>2.0.CO;2.

Price, R.A., and Mountjoy, E.W., 1970, Geologic structure of the Canadian Rocky Mountains between Bow and Athabasca Rivers: A progress report, in Wheeler, J.O., ed., Structure of the Southern Canadian Cordillera: Geological Association of Canada Special Paper 6, p. 7–25.

Raeside, R.P., and Simony, P.S., 1983, Stratigraphy and deformational history of the Scrip Nappe, Monashee Mountains, British Columbia: Canadian Journal of Earth Sciences, v. 20, p. 639–650.

Reid, L.F., 2003, Stratigraphy, Structure, Petrology, Geochronology and Geochemistry of the Hobson Lake Area (Cariboo Mountains, British

Columbia) in Relation to the Tectonic Evolution of the Southern Canadian Cordillera [Ph.D. thesis]: Calgary, University of Calgary, 221 p.

Ricketts, B.D., and Evenchick, C.A., 1991, Analysis of the Middle to Upper Jurassic Bowser Basin, northern British Columbia, in Current Research: Geological Survey of Canada Paper 91-1A, p. 65–73.

Ricketts, B.D., and Evenchick, C.A., 2007, Evidence of different contractional styles along foredeep margins provided by Gilbert deltas; examples from Bowser Basin, British Columbia, Canada: Bulletin of Canadian Petroleum Geology, v. 55 (in press).

Ricketts, B.D., and Parrish, R.R., 1992, Age and provenance of felsic clasts in the Bowser Basin northern British Columbia, in Radiogenic age and isotopic studies, Report 5: Geological Survey of Canada Paper 92-2, p. 141–144.

Ricketts, B.D., Evenchick, C.A., Anderson, R.G., and Murphy, D.C., 1992, Bowser Basin, northern British Columbia: Constraints on the timing of initial subsidence and Stikinia North America terrane interactions: Geology, v. 20, p. 1119–1122, doi: 10.1130/0091-7613(1992)020<1119:BBNBCC>2.3.CO;2.

Ross, G.M., Patchett, P.J., Hamilton, M., Heaman, L., DeCelles, P.G., Rosenberg, E., and Giovanni, M.K., 2005, Evolution of the Cordilleran orogen (southwestern Alberta, Canada) inferred from detrital mineral geochronology, geochemistry, and Nd isotopes in the foreland basin: Geological Society of America Bulletin, v. 117, p. 747–763, doi: 10.1130/B25564.1.

Rubin, C.M., Saleeby, J.B., Cowan, D.S., Brandon, M.T., and McGroder, M.F., 1990, Regionally extensive mid-Cretaceous west-vergent thrust system in the northwestern Cordillera: Implications for continental margin tectonism: Geology, v. 18, p. 276–280, doi: 10.1130/0091-7613(1990)018<0276:REMCWV>2.3.CO;2.

Rusmore, M.E., and Woodsworth, G.J., 1994, Evolution of the eastern Waddington thrust belt and its relation to the mid-Cretaceous Coast Mountains arc, western British Columbia: Tectonics, v. 13, p. 1052–1067, doi: 10.1029/94TC01316.

Rusmore, M.E., Woodsworth, G.J., and Gehrels, G.E., 2000, Late Cretaceous evolution of the eastern Coast Mountains, Bella Coola, British Columbia, in Stowell, H.H., and McClelland, W.C., eds., Tectonics of the Coast Mountains, Southeastern Alaska and British Columbia: Geological Society of America Special Paper 343, p. 89–105.

Rusmore, M.E., Gehrels, G., and Woodsworth, G.J., 2001, Southern continuation of the Coast shear zone and Paleocene strain partitioning in British Columbia–southeast Alaska: Geological Society of America Bulletin, v. 113, p. 961–975, doi: 10.1130/0016-7606(2001)113<0961:SCOTCS>2.0.CO;2.

Scammell, R.J., 1993, Mid-Cretaceous to Tertiary Thermotectonic History of Former Mid-Crustal Rocks, Southern Omineca Belt, Canadian Cordillera [Ph.D. thesis]: Kingston, Queen's University, 576 p.

Schiarizza, P., and Preto, V.A., 1987, Geology of the Adams Plateau–Clearwater–Vavenby Area: British Columbia Ministry of Energy, Mines and Petroleum Resources Paper 1987-2, 88 p.

Smith, D.G., 1994, Paleogeographic evolution of the Western Canada Foreland basin, in Mossop, G., and Shetsen, I., eds., Geological Atlas of the Western Canada Sedimentary Basin: Calgary, Alberta, Canadian Society of Petroleum Geologists and Alberta Research Council, p. 277–296.

Smith, G.T., and Mustard, P.S., 2006, Supporting evidence for a conformable southern contact of the Bowser Lake and Skeena Groups, in Summary of Activities 2006: Resource Development and Geoscience Branch, British Columbia Ministry of Energy and Mines, Victoria, British Columbia, p. 125–134.

Smith, M.T., Gehrels, G.E., and Klepacki, D.W., 1992, 173 Ma U-Pb age of felsite sills (Kaslo River intrusives) west of Kootenay Lake, southeastern British Columbia: Canadian Journal of Earth Sciences, v. 29, p. 531–534.

Stasiuk, L.D., Evenchick, C.A., Osadetz, K.G., Ferri, F., Ritcey, D., Mustard, P.S., and McMechan, M., 2005, Regional thermal maturation and petroleum stage assessment using vitrinite reflectance, Bowser and Sustut Basins, north-central British Columbia: Geological Survey of Canada Open-File 4945, scale 1:500,000, 1 sheet, 13 p. text.

Stott, D.F., 1963, The Cretaceous Alberta Group and equivalent rocks, Rocky Mountain Foothills, Alberta: Geological Survey of Canada Memoir 317, 306 p.

Stott, D.F., 1982, Lower Cretaceous Fort St. John Group and Upper Cretaceous Dunvegan Formation of the Foothills and Plains of Alberta, British Columbia, District of Mackenzie and Yukon Territory: Geological Survey of Canada Bulletin 328, 124 p.

Stott, D.F., 1993, Evolution of Cretaceous foredeeps: A comparative analysis along the length of the Canadian Rocky Mountains, in Caldwell, W.G.E., and Kauffman, E.G., eds., Evolution of the Western Interior Basin: Geological Association of Canada Special Paper 39, p. 131–150.

Stott, D.F., 1998, Fernie Formation and Minnes Group (Jurassic and Lowermost Cretaceous), Northern Rocky Mountain Foothills, Alberta and British Columbia: Geological Survey of Canada Bulletin 516, 516 p.

Stott, D.F., and Aitken, J.D., eds., 1993, Sedimentary Cover of the Craton in Canada: Geological Survey of Canada, Geology of Canada, no. 5, 825 p. (also Geological Society of America, Geology of North America, v. D-1).

Stott, D.F., Caldwell, W.G.E., Cant, D.J., Christopher, J.E., Dixon, J., Koster, E.H., McNeil, D.H., and Simpson, F., 1993, Cretaceous, in Stott, D.F., and Aitken, J.D. eds., Sedimentary Cover of the Craton in Canada: Geological Survey of Canada, Geology of Canada, no. 5, subchapter 4I, p. 358–438 (also Geological Society of America, Geology of North America, v. D-1).

Stronach, N.J., 1984, Depositional environments and cycles in the Jurassic Fernie Formation, southern Canadian Rocky Mountains, in Glass, D.J., and Stott, D.F., eds., The Mesozoic of Middle North America: Canadian Society of Petroleum Geologists Memoir 9, p. 43–67.

Struik, L.C., 1988, Crustal evolution of the eastern Canadian Cordillera: Tectonics, v. 7, p. 727–747.

Struik, L.C., Schiarizza, P., Orchard, M.J., Cordey, F., Sano, H., MacIntyre, D.G., Lapierre, H., and Tardy, M., 2001, Imbricate architecture of the upper Paleozoic to Jurassic oceanic Cache Creek terrane, central British Columbia, in Struik, L.C., and MacIntyre, D.G., eds., The Nechako NATMAP Project of the Central Canadian Cordillera: Canadian Journal of Earth Sciences, v. 38, p. 495–514.

Thompson, R.I., Glombic, P., Erdmer, P., Heaman, L.M., Lemieux, Y., and Daughtry, K.L., 2006, Evolution of the ancestral Pacific margin, southern Canadian Cordillera: Insights from new geologic maps, in Colpron, M., and Nelson, J.L., eds., Paleozoic Evolution and Metallogeny of Pericratonic Terranes at the Ancient Pacific Margin of North America, Canadian and Alaskan Cordillera: Geological Association of Canada Special Paper 45, p. 483–504.

Tipper, H.W., 1984, The age of the Jurassic Rossland Group, in Current Research Part A: Geological Survey of Canada Paper 84-1A, p. 631–632.

Tipper, H.W., and Richards, T.A., 1976, Jurassic Stratigraphy and History of North-Central British Columbia: Geological Survey of Canada Bulletin 270, 73 p.

van der Heyden, P., 1992, A Middle Jurassic to early Tertiary Andean–Sierran arc model for the Coast belt of British Columbia: Tectonics, v. 11, p. 82–97.

Warren, M.J., 1997, Crustal Extension and Subsequent Crustal Thickening Along the Cordilleran Rifted Margin of Ancestral North America, Western Purcell Mountains, Southeastern British Columbia [Ph.D. thesis]: Kingston, Queen's University, 361 p.

Wheeler, J.O., and McFeely, P., 1991, Tectonic assemblage map of the Canadian Cordillera and adjacent parts of the United States of America: Geological Survey of Canada Map 1712A, scale 1:2,000,000, 3 sheets.

White, J.M., and Leckie, D.A., 1999, Palynological age constraints on the Cadomin and Dalhousie Formations in SW Alberta: Bulletin of Canadian Petroleum Geology, v. 47, p. 199–222.

Willett, S., Beaumont, C., and Fullsack, P., 1993, Mechanical model for the tectonics of doubly vergent compressional orogens: Geology, v. 21, p. 371–374, doi: 10.1130/0091-7613(1993)021<0371:MMFTTO>2.3.CO;2.

Williams, P.F., and Jiang, D., 2005, An investigation of lower crustal deformation: Evidence for channel flow and its implications for tectonics and structural studies: Journal of Structural Geology, v. 27, p. 1486–1504, doi: 10.1016/j.jsg.2005.04.002.

MANUSCRIPT ACCEPTED BY THE SOCIETY 22 MARCH 2007

Belt-Purcell Basin: Keystone of the Rocky Mountain fold-and-thrust belt, United States and Canada

J.W. Sears*

University of Montana, Department of Geosciences, Missoula, Montana 59812, USA

ABSTRACT

The Mesoproterozoic Belt-Purcell Basin of the United States–Canadian Rocky Mountains formed in a complex intracontinental-rift system. The basin contained three main fault blocks: a northern half-graben, a central horst, and a southern graben. Each had distinct internal stratigraphy and mineralization that influenced Phanerozoic sedimentation; the northern half-graben and horst formed a platform with a condensed section, whereas the southern graben formed the subsiding Central Montana trough. They formed major crustal blocks that rotated clockwise during Cordilleran thrust displacement, with transpressional shear zones deforming their edges. The northern half-graben was deepest and filled with a structurally strong prism of quartz-rich sedimentary rocks and thick mafic sills that tapered toward the northeast from >15-km-thick near the basin-bounding fault. This strong, dense prism was driven into the foreland basin as a ready-made, critically tapered tectonic wedge and was inverted into the Purcell anticlinorium. Erosion did not breech the Belt-Purcell Supergroup in this prism during thrusting. The southern graben was thinner, weaker, lacked mafic sills, and was engorged with sheets of granite during thrusting. It was internally deformed to achieve critical taper and shed thick deposits of syntectonic Belt-Purcell–clast conglomerate into the foreland basin.

A palinspastic map of the basin combined with a detailed paleocontinental map that juxtaposes the northeastern corner of the Siberian craton against western North America indicates that the basin formed at the complicated junction of three continental-scale rift zones.

Keywords: Belt-Purcell Basin, Rocky Mountains, fold-and-thrust belt, basin inversion, Mesoproterozoic.

INTRODUCTION

The Mesoproterozoic Belt-Purcell Basin (Fig. 1) dominates the geology of the northern Rocky Mountains of Montana and parts of Idaho, Washington, British Columbia, and Alberta (Price, 1964; Harrison, 1972; Winston, 1986a; Lydon, 2000). The basin was one of Professor Price's great research interests, and over the past half-century, he and his students and colleagues have contributed greatly to our understanding of the geometry of this mineral-rich basin and its influence on the tectonic evolution of the Cordilleran miogeocline and Rocky Mountain fold-and-thrust belt.

*james.sears@umontana.edu

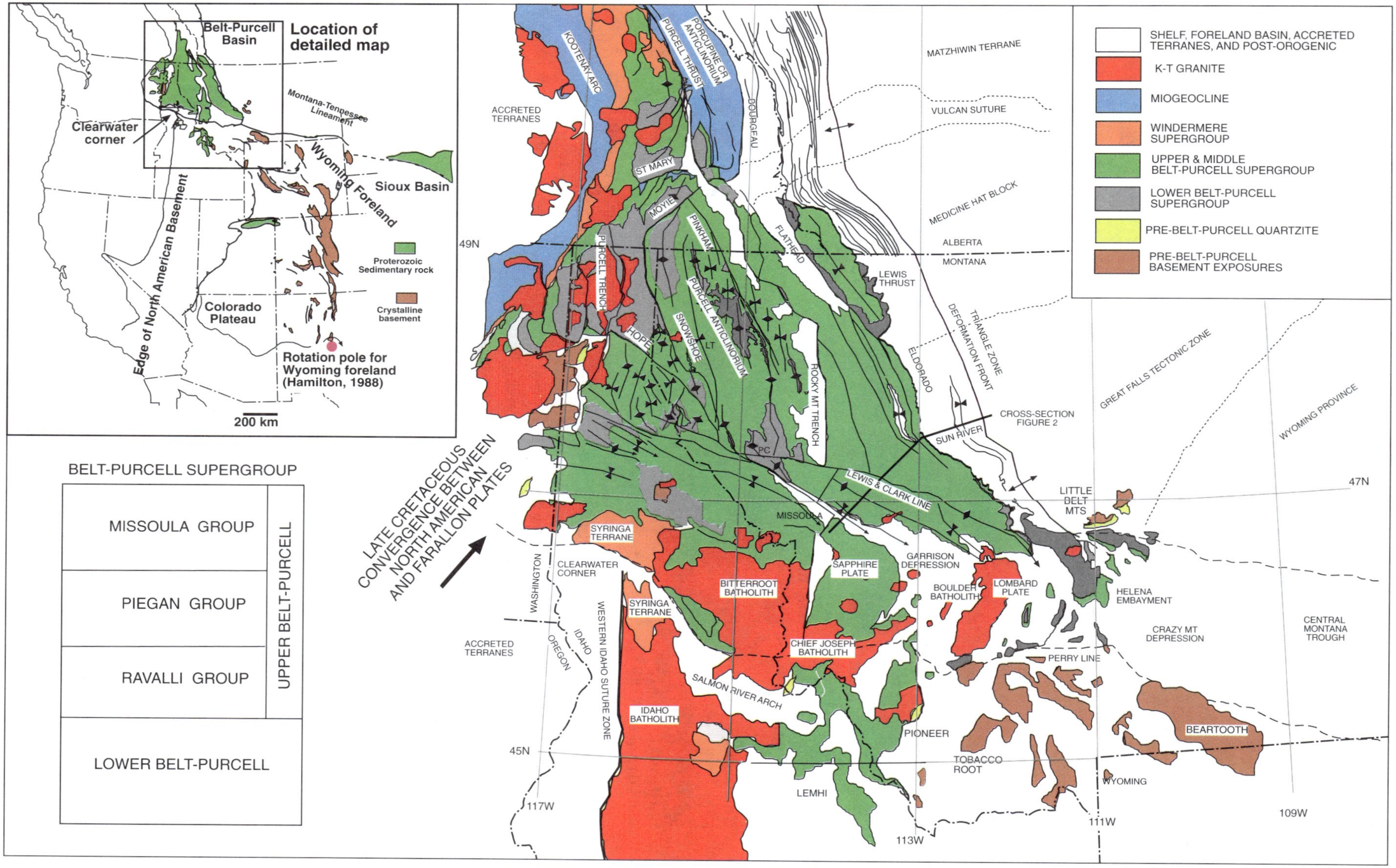

Figure 1. Tectonic map of Mesoproterozoic Belt-Purcell Basin and environs (compiled from U.S. Geological Survey 1° × 2° geologic maps; Price, 1981; Link et al., 1993; Reed et al., 1993). Inset map shows location of basin in northern Rocky Mountains (after Hamilton, 1988). Inset column shows main divisions of Belt-Purcell Supergroup. PC—Perma culmination, LT—Libby trough.

The present paper builds upon Professor Price's work and develops a comprehensive new model for the origin and tectonic evolution of the Belt-Purcell Basin. The model palinspastically restores the basin to its Mesoproterozoic geometry and proposes that it evolved from two large grabens and an intervening horst block. Each of these fault blocks had distinct stratigraphy and unique mineralization; massive sulfides formed in the northern graben, stratabound silver-copper occurs in the central horst, and copper, gold, and silver are associated with Cretaceous granites in the southern graben. The paper develops the origin of the basin by placing it in a paleocontinental reconstruction against the Udzha and Khastakh troughs of the northeastern Siberian craton (Sears and Price, 2003). It concludes that the basin formed at an intracontinental rift triple-junction, where the rifts intersected in a complex triangular pattern because of the interference of pre-existing basement structures. This reconstruction suggests that the mosaic of fault blocks rotated clockwise during Cordilleran thrust displacement, with transpressional shear zones forming along their edges.

TECTONIC SETTING

The Belt-Purcell Supergroup was deposited from 1500 to 1370 Ma (Evans et al., 2000; Lydon, 2000), and it has a thickness >15 km in the northern part of the basin, where the base is not exposed (Harrison, 1972). It includes four stratigraphic divisions: from the base up, the Lower Belt-Purcell, Ravalli Group, Piegan Group (formerly Middle Belt Carbonate; cf. Winston, 2007), and Missoula Group (Link, 1993).

The exposed part of the basin is entirely allochthonous, having been transported northeastward over Phanerozoic strata in the Rocky Mountain fold-and-thrust belt, and in foreland uplifts in central-western Montana. Figure 2A presents a typical structural cross section of the northern part of the basin, and Figure 2B restores the section to its configuration at the conclusion of thrusting, ca. 60 Ma. Cordilleran thrusts and transverse faults reactivated syndepositional extensional faults and divided the basin into a mosaic of structural blocks (Winston, 1986a). Thrusting was driven by accretion of volcanogenic terranes along the Cordilleran margin of the North American plate from Middle Jurassic to late Paleocene time (Monger and Price, 1979; Evenchick et al., this volume; Hammer and Clowes, this volume).

Cretaceous and Paleogene granitic plutons of the Cordilleran magmatic arc engulfed the western and southern parts of the basin (Archibald et al., 1984; Hyndman et al., 1988). Basement-cored ranges of the Wyoming foreland province border the Belt-Purcell rocks on the southeast against the Perry Line, a major basin-bounding syndepositional fault (McMannis, 1963; Winston, 1986a). The Supergroup also occupies an isolated region south of the Chief Joseph batholith in central Idaho and southwestern Montana (Link and Fanning, 2003).

A number of Tertiary listric normal faults reactivated thrust décollements within the Belt-Purcell Basin (Price, 1965; Sears, 2001a; Constenius, 1996). Eocene metamorphic core complexes emerged in the west (Doughty et al., 1998; Coney and Harms, 1984; Harms and Price, 1992; Doughty et al., this volume) and exposed the deeper parts of the Supergroup. Neogene Basin-and-Range faults broke up the Belt-Purcell rocks south of the Lewis and Clark Line and along the Rocky Mountain Trench (Lund et al., 2003; Janecke, 1994; Sears and Ryan, 2003). Some remain active today (Stickney et al., 2000; Sears and Fritz, 1998).

The Belt-Purcell Basin extended across two northeast-trending basement structural zones of western Laurentia (Reed, 1993; Ross, 1991a). The Vulcan suture joined the Archean Matzhiwin terrane and Medicine Hat block, and the Great Falls tectonic zone sheared the Medicine Hat block against the Archean Wyoming province (Hoffman, 1988; Boerner et al., 1998; Mueller et al., 2002). Reactivation of parts of these structures influenced the internal architecture of the Belt-Purcell Basin.

Frost and Winston (1987), Cressman (1989), Ross et al. (1992), and Ross and Villeneuve (2003) showed that much of the Belt-Purcell sediment was derived from a continental landmass to the west of the present edge of the basin. This source region was removed by the continental rifting that initiated the Cordilleran miogeocline (Stewart, 1972). The basin thus provides an important piercing point for reconstruction of the Neoproterozoic supercontinent Rodinia and its Mesoproterozoic predecessor. Sears and Price (2003) suggested that the Udzha-Khastakh Basin of northeastern Siberia represents the conjugate partner of the original Belt-Purcell Basin. That model is further developed in this essay to propose a tectonic mechanism for initiation of the basin.

ARCHITECTURE OF THE RESTORED BELT-PURCELL BASIN

Figure 3 palinspastically restores the Belt-Purcell Basin to its possible original Mesoproterozoic configuration by subtracting the effects of Tertiary extension, Cordilleran thrusting, and Neoproterozoic–early Paleozoic continental rifting. The proposed restoration divides the Belt-Purcell Basin into three large blocks: a northern half-graben, a central horst, and a southern graben (Fig. 4). These three blocks accumulated significantly different thicknesses and sedimentary facies of Belt-Purcell strata and formed distinct structural panels that sheared against one another during Cordilleran thrust displacement.

Northern Half-Graben

The northern part of the basin comprised a deep half-graben, tilted toward the southwest. The Snowshoe fault represents a remnant of the basin-bounding fault zone on its southwest side. The Snowshoe fault was later reactivated as a southwest-verging thrust during Cordilleran orogenesis (Fillipone and Yin, 1994). The sedimentary fill of the half-graben thickened to the southwest to >15 km from a depositional zero-edge along its northeast side. In the center of the graben, the sedimentary fill exhibits an exponentially decaying accumulation rate, consistent with a continental-

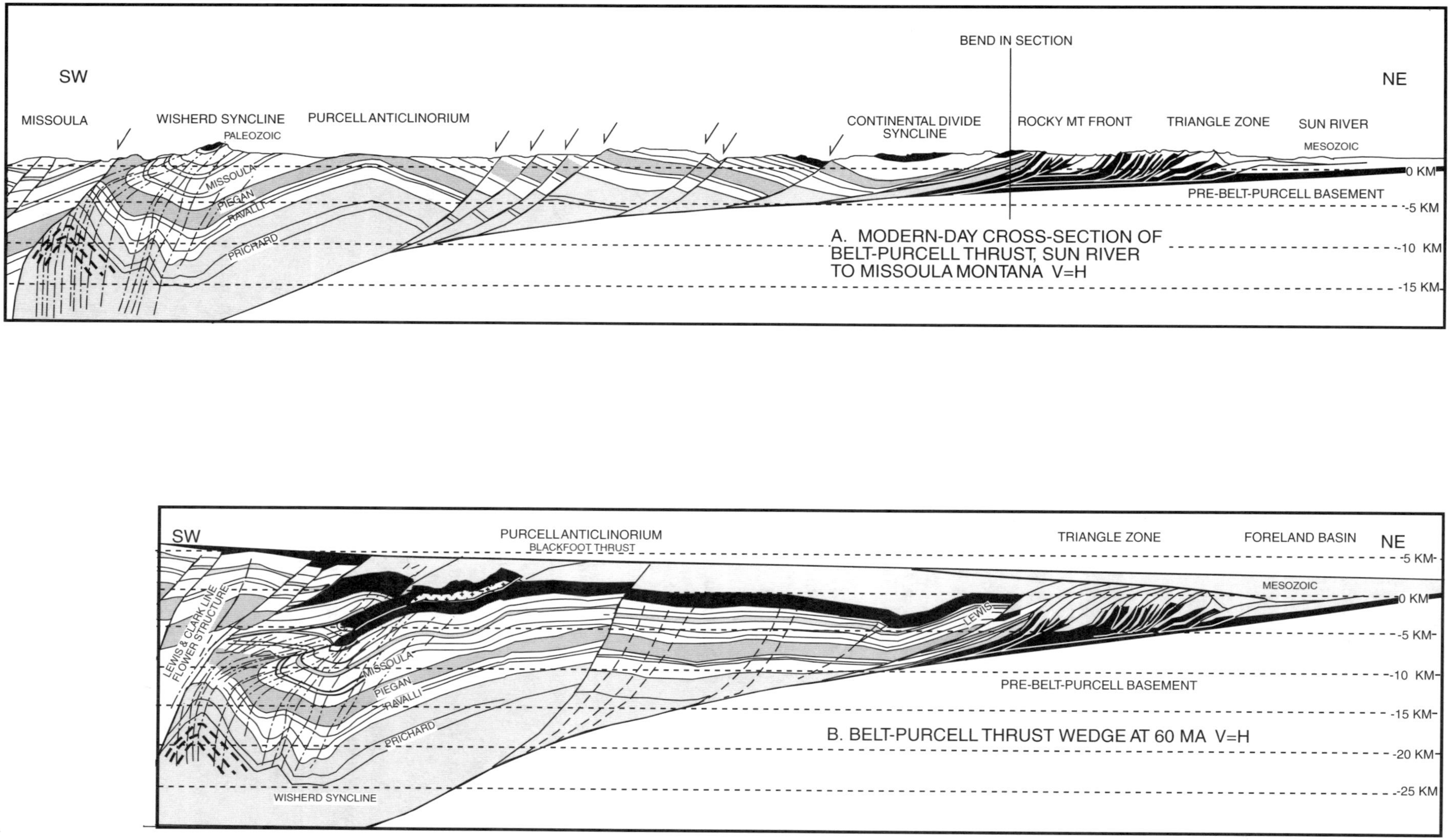

Figure 2. True-scale cross section of northern half-graben inverted into Purcell anticlinorium. See Figure 1 for location. (A) Present-day cross section. (B) Restored to conclusion of thrusting, ca. 60 Ma. Extensional faults are restored, and the plate is rotated downward so that Belt-Purcell rocks are kept underground, as indicated by sedimentary studies in foreland basin. Upper part of the southwestern end of the section was projected upward into the line of the section from deep Garrison depression to southeast. Western end was modified from Sears and Hendrix (2004), center, from Sears (2000), east end, from Sears et al. (2005).

rift origin (Sears, 2007; Lydon, 2000). The rapid early accumulation rate records the active rift phase that initiated the graben. The slower later rate records passive thermal sagging and sediment loading (Chandler, 2000). Mafic pulses were associated with basin-deepening events, suggesting renewed rift phases. During deposition of the lower three divisions of the Belt-Purcell Supergroup, the siliciclastic sediment was largely derived from the west and entered the basin from a point near Spokane, Washington (Cressman, 1989). The sediment source shifted to the south during deposition of the Missoula Group (Winston, 1986b; Farooqui, 1997), but full continental separation did not occur at this time.

The northern graben was largely inverted into the Purcell anticlinorium (Fig. 2) after it was detached from its basement and thrust northeastward over the relatively flat surface of the foreland basin (Price, 1981; Cook and Van der Velden, 1995). The anticlinorium was carried by a series of large thrust faults that emerge along its east edge. These include the Purcell, MacDonald, Lewis, and Eldorado thrusts. The Purcell thrust placed Belt-Purcell rocks against highly strained Lower Paleozoic rocks of the Porcupine Creek anticlinorium (Price, 1981). The Lewis thrust placed lower Belt-Purcell rocks over Upper Cretaceous rocks of the foreland basin (Willis, 1902). The Lewis thrust forms a rectilinear salient in Waterton and Glacier National Parks at the international boundary. The salient is gently folded into a broad syncline by a duplex in underlying Paleozoic and Mesozoic strata (Boyer and Elliott, 1982; Fermor and Moffat, 1992). South of Glacier Park, the Lewis thrust transfers displacement to the Eldorado thrust, which lies to the east, and the Hoadley thrust, which lies to the west (Mudge et al., 1982; Mudge and Earhart, 1983). The Rocky Mountain triangle zone, a tectonic-wedge structure, lies at the east edge of the thrust mass (Price, 1981, 1986; Stockmal et al., 1996; Sears et al., 2005).

The Rocky Mountain Trench and Flathead normal faults cut the east limb of the anticlinorium. They dip west and flatten downward into the thrust décollement (Bally et al., 1966; Price, 1965, 1981; van der Velden and Cook, 1996). The Rocky Mountain Trench merges downward into the basin-margin ramp. The Flathead fault was localized on the west side of a large duplex stack beneath the Lewis thrust plate (Fermor and Moffat, 1992; Feinstein et al., this volume).

In British Columbia, the west limb of the anticlinorium includes the Kootenay arc, a steep, west-facing monocline with 15 km of structural relief (Price, 1981). The monocline marks the western edge of the Belt-Purcell Basin and of Precambrian North America. It includes the westward progradation of the Paleozoic–early Mesozoic continental shelf. In northwestern Montana, the west limb of the anticlinorium faces into the Libby trough, and in central-western Montana, into the Garrison depression. In these regions, it drapes a footwall ramp that defines the undisturbed position of the northeastern margin of the Belt-Purcell Basin (Sears, 2000). In Figure 3, the leading edges of the Lewis and Eldorado thrusts are restored against the footwall ramps. The restoration is consistent with the cumulative horizontal shortening in the Rocky Mountain fold-and-thrust belt to the east (Price and Sears, 2000).

The core of the Purcell anticlinorium reveals the deepest part of the northern half-graben. It is characterized by >6-km-thick basinal turbidites and mafic sills of the Lower Belt-Purcell Supergroup—the Prichard Formation in the United States and the equivalent Aldridge Formation in Canada (Lydon, 2000). Magnetotelluric surveys indicate that attenuated and imbricated basement may occur beneath the Prichard Formation in the anticlinorium (Box et al., 2006). In the Perma culmination near the southern edge of the half-graben, the lower part of the Prichard Formation records shoaling of the basin from deep-water turbidites (Members A–D) to shallow-water, mud-cracked and ripple-marked quartz-arenite (Member E). Cressman (1989) calculated that the basin tectonically collapsed by an estimated 3 km, back into deep water, to deposit Member F turbidites. The collapse was accompanied by widespread injection of mafic sills into wet Prichard and Aldridge sediments at 1469 Ma (Sears et al., 1998; Anderson and Davis, 1995; Höy et al., 2000). In the Perma culmination, adjacent to the graben-bounding normal fault, a major syndepositional fault in the lower Prichard Formation flattened into down-to-the-basin soft-sediment folds and was intruded by fluidized breccia during the intrusion (Buckley and Sears, 1997). It is on trend with the Snowshoe fault. Individual sills in the Prichard and Aldridge Formations are as much as 1 km thick, and exposed sills aggregate 2 km in thickness (Anderson and Parrish, 2000). An additional 6 km of sills are interpreted from seismic-reflection profiles to underlie exposed sill sections in southern British Columbia (Cook and Van der Velden, 1995). The thermal content of these sills and possibly an underlying gabbroic body drove static greenschist metamorphism in the basinal facies of the Lower Belt-Purcell and Ravalli Group (Poage et al., 2000); randomly oriented biotite porphyroblasts occur throughout the basinal facies of the Prichard and Aldridge Formations and lower Ravalli Group in the northern half-graben (Norwick, 1972). Chlorite and 2-M mica occurs throughout the Belt-Purcell Supergroup (Ryan, 1991), indicating that the thermal content remained high throughout the diagenetic history of the northern part of the basin. The sills and metamorphism greatly increased the brittle strength of the sedimentary wedge and enabled it to be displaced en masse in great thrust sheets during Cordilleran orogenesis.

The Purcell anticlinorium has internal thrusts, culminations, and depressions, reflecting original syndepositional faults and inverted sub-basins (Höy et al., 2000). The inverted sub-basins are highlighted on Figure 1 by the distribution of culminations of Lower Belt-Purcell rocks. The internal thrusts include the Mt. Forster, Hall Lake, Libby, Pinkham, Hefty, and Wigwam thrusts. These decrease in displacement southeastward, indicating clockwise rotation of the fault blocks. The Pinkham thrust has a component of sinistral shear, as shown by tracts of en echelon folds in the hanging wall and footwall. The Pinkham thrust may represent an inverted shelf-edge normal fault; shelf facies lie mostly to the east, and basinal facies lie to the west.

The east limb of the anticlinorium corresponds to the inverted northeastern margin of the graben. The Belt-Purcell rocks thin dramatically across the east limb to <2 km thick along the Rocky

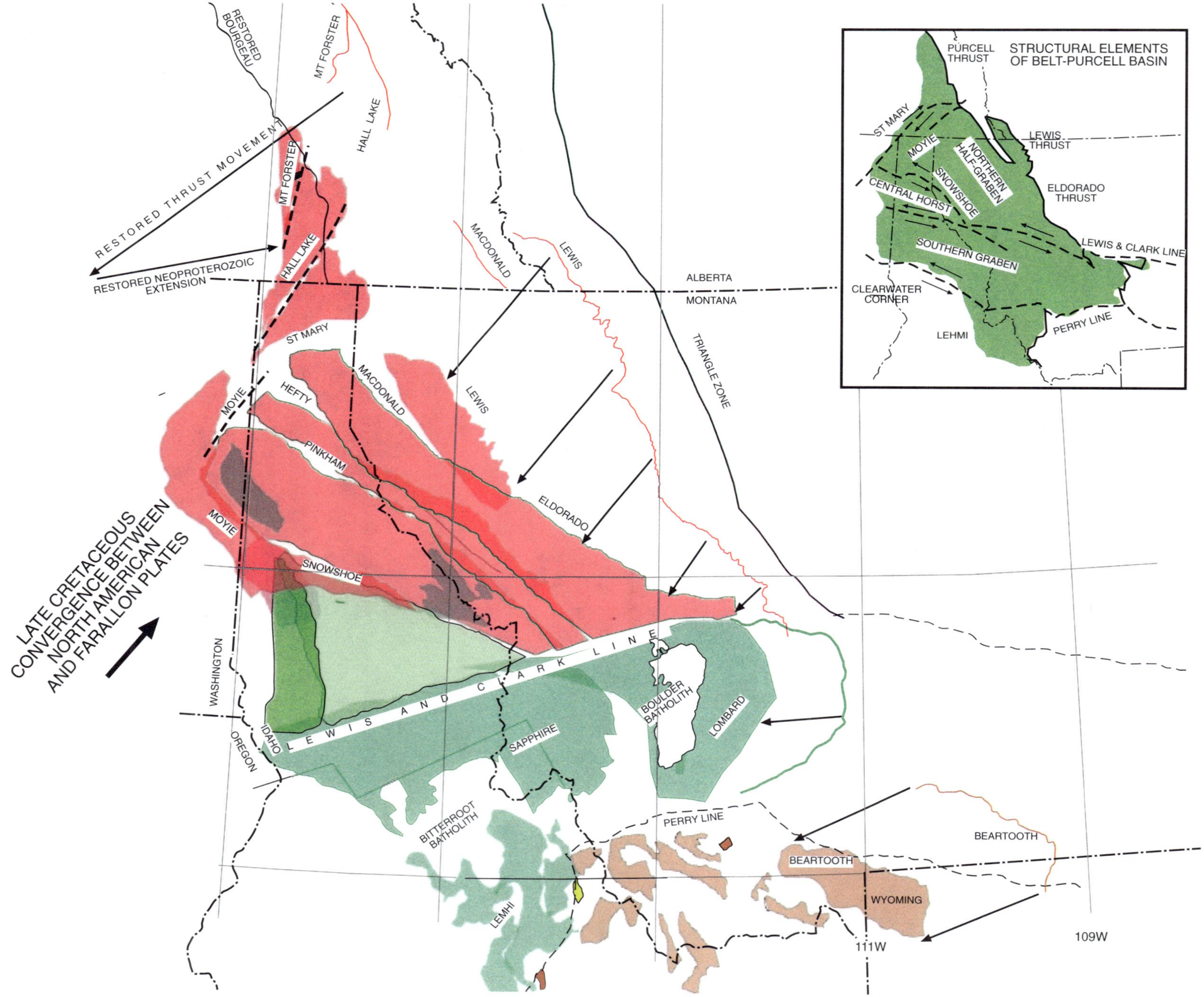
STRUCTURAL ELEMENTS OF BELT-PURCELL BASIN
PURCELL THRUST
LEWIS THRUST
ELDORADO THRUST
LEWIS & CLARK LINE
PERRY LINE
LEHMI
CLEARWATER CORNER
ST MARY
MOYIE
NORTHERN HALF-GRABEN
SNOWSHOE
CENTRAL HORST
SOUTHERN GRABEN
RESTORED BOURGEAU
MT FORSTER
HALL LAKE
RESTORED THRUST MOVEMENT
RESTORED NEOPROTEROZOIC EXTENSION
MACDONALD
LEWIS
ALBERTA
MONTANA
TRIANGLE ZONE
ST MARY
HEFTY
PINKHAM
ELDORADO
SNOWSHOE
LATE CRETACEOUS CONVERGENCE BETWEEN NORTH AMERICAN AND FARALLON PLATES
WASHINGTON
OREGON
IDAHO
LEWIS AND CLARK LINE
BOULDER BATHOLITH
LOMBARD
SAPPHIRE
BITTERROOT BATHOLITH
PERRY LINE
LEMHI
BEARTOOTH
WYOMING
111W
109W

Figure 3. Palinspastic restoration of Belt-Purcell Basin (modified after Price and Sears, 2000). Red lines show traces of major thrusts of northern basin that restore back to southwest as shown by arrows. Northern segment of basin restores to southwest to account for thrusting, then back to northeast to account for Neoproterozoic–early Paleozoic extension. It meets restored trace of Bourgeau thrust. Green line shows trace of Lombard thrust that restores to west as shown by arrow. Brown line shows trace of Beartooth fault that restores as shown by arrows. Large displacement shown for Beartooth includes cumulative displacement of additional foreland faults to east not shown on map. Darker shading—restored overlap on normal faults. Inset shows major structural elements of the basin. Rectilinear line in central Idaho represents restored core complexes.

Mountain front, where Lower Belt-Purcell shelf facies consist of thin shale, stromatolitic carbonate, and fluvial quartzite (Cressman, 1989; Price and Fermor, 1986; Höy et al., 2000).

The Lower Belt-Purcell Supergroup grades upward into the Ravalli Group (Creston Formation in Canada), marking a long-term shoaling of the northern half-graben basin into mud-cracked and ripple-marked siliciclastic facies (Cressman, 1989; Chandler, 2000). Western facies of the Ravalli Group include fine-grained, tabular Revett Formation quartzites that were spread eastward across a broad alluvial apron by massive sheet floods (Winston, 1986c). Eastern facies of the Ravalli Group include coarse, white sandstone lenses in rippled-marked and mud-cracked red argillite. Sedimentology and detrital-zircon geochronology indicate that the white sandstone was derived from the Canadian Shield to the east, possibly reworked from the Neihart Quartzite blanket (Winston, 1991; Ross and Villeneuve, 2003). The two facies of the Ravalli Group mingled along the line of the Purcell anticlinorium.

The Piegan Group includes the Wallace and Helena Formations (Winston, 2007). The Wallace Formation is thicker and more prevalent in the west. It has a large siliciclastic component that was derived from the west; its hallmark sand-carbonate cycles commonly include hummocky cross-stratification, indicating that the floor of the basin, while subaqueous, was within reach of storm waves. By contrast, the Helena Formation facies is dominated by carbonate, with significant stromatolitic horizons, and is more prevalent in the eastern part of the basin. The division between the dominantly siliciclastic and carbonate facies approximates the axial trace of the Purcell anticlinorium (Wallace et al., 1999).

The Missoula Group heralds a return to dominantly siliciclastic facies, and a change in provenance to the south. Units become thicker, coarser grained, more immature and feldspathic to the south (Farooqui, 1997; Winston and Wheeler, 2006). A change in detrital zircon ages indicates that the western source was cut off, and a new southern source was tapped that correlates with the Granite-Rhyolite Province (Van Schmus and Bickford, 1993; Ross and Villeneuve, 2003; Sears, 2007). These changes appear to have been driven by renewed rifting recorded by faulting and eruption of the 1443 Ma Purcell and Nicol Creek basalts in the northern part of the basin (Evans et al., 2000; Höy et al., 2000). Another rifting event is indicated by the large 1370 Ma Shoup gabbro-granophyre complex in central Idaho (Evans et al., 2000; Doughty and Chamberlain, 1996).

The St. Mary, Kimberly, and Moyie transverse faults cross the northern part of the Purcell anticlinorium. These faults followed a major crustal discontinuity, the Vulcan suture zone (Price and Sears, 2000). In the subsurface east of the Rocky Mountains, the Vulcan suture zone appears to be overlain by a basin filled with Proterozoic strata and bounded by steep fault zones (Clowes et al., 1997). Höy et al. (2000) and Turner et al. (2000) showed that the faults were syndepositional transform faults normal to the axis of the northern branch of the Belt-Purcell Basin; they bounded deep sub-basins during deposition of the lower Belt-Purcell Supergroup, the injection of mafic sills, and the exhalation of massive sulfide deposits, including the world-class Sullivan ore body. The faults were reactivated during Neoproterozoic–early Paleozoic rifting, when the northernmost part of the Belt-Purcell basin shifted relatively southwest, opening the Chancellor Basin and Eager trough (Price and Sears, 2000). During the Middle Jurassic–Early Cretaceous Columbian orogeny, the Hall Lake, Mt. Forster, Purcell, and St. Mary faults experienced dextral transpression, where the Belt-Purcell rocks were wedged into the Chancellor Basin to strongly cleave the shales and invert the basin into the Porcupine Creek anticlinorium. The Bourgeau thrust reactivated a west-dipping normal fault that had defined the east edge of the Chancellor rift basin. This same normal fault evidently had originally formed the east edge of this part of the Belt-Purcell rift basin. A crosscutting stock shows that thrust movement ended before 94 Ma (Price and Sears, 2000). Further displacement of the entire system during the Laramide orogeny (75–60 Ma) may have bent the Crowsnest deflection in the Rocky Mountain fold-and-thrust belt (Price and Mountjoy, 1970), although this may represent an inherited feature of the original geometry of the thrust sheets.

The Lewis and Eldorado thrusts carried the wedge-edge of the half-graben over the Paleozoic shelf and Mesozoic foreland basin rocks. Their displacement decreases southward from ~140 km at the international boundary to a few kilometers near Helena, indicating clockwise rotation about an Euler pole near Helena (Sears, 1994). Small circles of rotation were normal to fold hinges and parallel to slickenfibers in the deformed foreland basin sedimentary rocks (Sears et al., 1997; Price, 1967). The rotation was tangent to the Late Cretaceous convergence direction between the North American and Farallon plates (Coney, 1978; Sears, 2006).

The Lewis and Eldorado thrusts began to move at ca. 75 Ma, as shown by the ages of the youngest overridden sediments and andesite sills, the onset of potassium metasomatism in bentonites, and apatite fission-track systematics in the footwall (Osadetz et al., 2000; Sears, 2001a; Feinstein et al., this volume). Displacement continued until ca. 60 Ma, when dikes cut the thrust trace near Rogers Pass, Montana (Schmidt, 1978). Despite its considerable movement over this time span, the thrust plate did not shed Belt-Purcell lithologies into the adjacent foreland basin. Rather, the sediment deposited between 75 and 60 Ma in the fore-

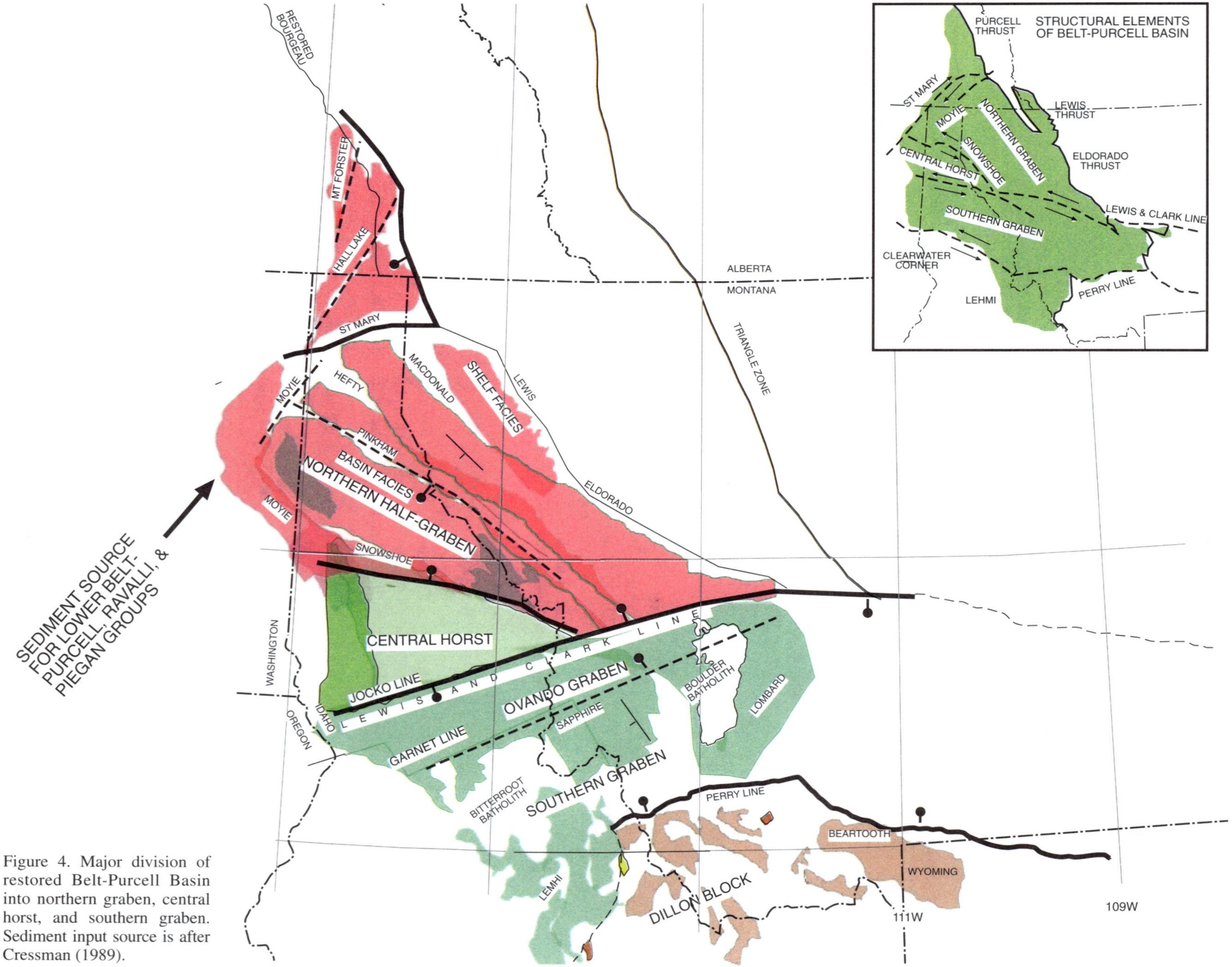

Figure 4. Major division of restored Belt-Purcell Basin into northern graben, central horst, and southern graben. Sediment input source is after Cressman (1989).

land basin was derived from the Paleozoic and Mesozoic cover of the thrust plate (Dyman et al., 1995; Mack and Jerzykiewicz, 1989; Catuneanu and Sweet, 1999; Catuneanu et al., 1997). The thrusts evidently did not surface, but rather were emplaced into the foreland basin subcutaneously, as a strong, dry tectonic wedge (Davis et al., 1983). North of the termination of the Belt-Purcell wedge at ~53°N, local Maastrichtian and middle Paleocene conglomerates contain andesite pebbles derived from the Intermontane terrane farther to the west (Evenchick et al., this volume); evidently, the position of the drainage divide was influenced by the mechanical emplacement of the wedge.

The restored cross section (Fig. 2B) suggests that the present Rocky Mountain front near Sun River was part of the triangle-zone tectonic-wedge during its formation where it climbed into the Upper Cretaceous foreland basin strata. Fluid expulsion from beneath the Belt-Purcell thrust plate emplaced hydrocarbon deposits in the triangle zone in Canada, but it has not been adequately explored in Montana (Stockmal et al., 1996; Oliver, 1986; Osadetz et al., 2000).

The northern Belt-Purcell wedge had inherent brittle strength because the thick sedimentary rocks were quartz-rich and had been dehydrated by static Mesoproterozoic greenschist metamorphism, and the basal half of the section was laced with dense gabbro sills. The metamorphism had not destroyed the layer-anisotropy of the wedge. Most of the wedge was in its brittle temperature field at the time of thrusting, as shown by brittle fabrics and scarcity of cleavage (Sears, 2001a). The bulk density of the wedge had been enhanced by mafic sills, and the basement ramp at the edge of the basin subsided isostatically under the load of the thrust plate as it advanced into the foreland basin (Fig. 2B). The load of the thrust plate contributed to subsidence of the neighboring foreland basin due to the flexural strength of the continent (Price, 1973; Beaumont et al., 1993; Stockmal et al., this volume).

The Belt-Purcell rocks of the northern part of the basin were denuded by deep erosion and extension in the footwalls of normal faults such as the Rocky Mountain Trench fault and faults of the Lewis and Clark Line. Upon cessation of thrusting and the onset of isostatic rebound, Belt-Purcell clasts thereafter swamped the fluvial-alluvial systems flowing outward from the region. The rebound restored the relief on the basin-bounding ramp, raising and extending the east limb of the anticlinorium (Sears, 2001a).

Central Horst

The central horst is a triangular block bounded by transpressional sinistral shear zones on the northeast and south, and by the Purcell trench normal fault on the west (Fig. 4). The sinistral shear fabrics were associated with Cordilleran orogenesis, and were overprinted by brittle Tertiary dextral-transtensional faults. The Snowshoe, Hope, and Moyie faults were on the northeast side of the horst. The Hope fault, a major strike-slip fault that branches from the Lewis and Clark Line in western Montana, trends northwest into northern Idaho near the northeastern edge of the block (Harrison et al., 1974). These faults bound sinistral arrays of folds that involved the 71 Ma Dry Creek stock (Fillipone and Yin, 1994).

On the south, the Lewis and Clark Line was a major sinistral shear zone. Sears and Hendrix (2004) related the shear zone to clockwise thrust-rotation of adjacent parts of the Belt-Purcell Basin. The line extends for 800 km, from near Spokane, Washington, to central Montana (Bennett and Venkatakrishnan, 1982; Billingsley and Locke, 1939). Its Late Cretaceous–late Paleocene fabric is characterized by en echelon flower structures, in which material at depth squeezed plastically upward during horizontal shortening, while material near the surface shortened into box folds (Sears and Hendrix, 2004). It is marked by a 30- to 50-km-wide cleavage band (Yin and Oertel, 1995), which is notable because cleavage is rare in most of the northern part of the basin. Cleavage bends into southeasterly trends and becomes pervasive in the band, in a pattern diagnostic of sinistral shear zones (Ramsay, 1980). Strain is greater west of the intersection of the Hope fault, where Ravalli quartzites are pervasively mylonitized in a narrow band. In the Coeur d'Alene district, the mylonitic foliation dips steeply, has strong downdip stretching lineations, and is kink-folded (T. Raup, 2006, personal commun.; White, 1997). Fluids driven out of the batholiths and metamorphic zones to the south may have risen along the Lewis and Clark Line to facilitate the shearing (Hyndman et al., 1988; Sears and Clements, 2000). In west-central Montana, folds within the band display a consistent southeasterly plunge that resulted from draping of the strained surficial rocks of the Lewis and Clark Line over a major basement ramp, which marks the northeastern depositional margin of the northern graben (Sears, 1988). The eastern end of the Lewis and Clark Line records two distinct periods of sinistral shear. The first period accompanied Campanian deposition in the foreland basin. The second period deformed Maastrichtian rocks but predated Eocene volcanics. An en echelon set of valleys follow post–middle Eocene brittle dextral-normal fault zones along the Lewis and Clark Line (Sears and Hendrix, 2004). The extension was related to postorogenic rebound of the basement ramp (Sears, 2001a).

The internal structure of the central horst block is characterized by a chaotic array of folds (White, 1997). It includes the Noxon arch of White (2006), which appears to have been a high-standing block during Belt-Purcell deposition because distinct units of the Ravalli, Piegan, and Missoula Groups thin across it (Harrison, 1972; Winston, 1991), the Prichard Formation is characterized by shelf facies rather than deep turbidite facies (Finch and Baldwin, 1984), and thick gabbroic sills are absent (Höy et al., 2000). The Jocko Line was a down-to-the-south syn–Belt-Purcell growth fault across which several stratigraphic units thickened dramatically off the horst block; it coincides with the Lewis and Clark Line at the southern edge of the horst (Winston, 1991).The horst contains the world-class Revett copper-silver belt of northern Idaho and northwestern Montana (Harrison, 1972; Hobbs et al., 1965; White, 2006; Boleneus et al., 2006). Fillipone and Yin (1994) determined that the fault experienced kilometers of west-side-down Eocene dip-slip.

Southern Graben

The southern graben forms a deep structural trough that sweeps from central Montana to eastern Washington. Its sedimentary fill was not thrusted out of the basin during Cordilleran orogenesis and remains largely buried in the trough. The graben was bounded on the north by the Lewis and Clark Line, and its antecedent, the Jocko growth fault (Winston, 1991). The Ovando graben insets the north edge of the larger graben, down-faulted between the Jocko and Garnet Lines (Fig. 4). These faults are marked by soft-sediment slump folds and rapid increases in thickness of several units in the Belt-Purcell Supergroup (Winston, 1991).

Structural components of the southern graben include the Central Montana trough, Crazy Mountain basin, Helena embayment, and Garrison depression. The Central Montana trough contains autochthonous Belt-Purcell rocks in the subsurface to the east of Billings (Peterson, 1991; Mallory, 1972; Link, 1993). The Helena embayment forms an east-projecting arm of the Belt-Purcell Basin in the eastern Rocky Mountains along the south flank of the Belt Mountains (Harrison, 1972; Winston, 1986a, 1991). The Helena embayment plunges west into the deep Garrison trough, which crosses the Rocky Mountain thrust belt to western Idaho. The Garrison trough contains the Lombard and Sapphire thrust plates, the Boulder batholith, and the Bitterroot batholith, which rests in a syncline; the trough has at least 15–20 km of structural relief (Sears, 1988; Hyndman et al., 1988). A cross section suggests ~80 km of eastward horizontal movement for the Lombard thrust (Burton et al., 1997).

The Belt-Purcell section of the southern graben is much thinner than that of the northern graben; the Supergroup totals only 7 km thick in the Helena embayment and may be only 3 km thick in the area of the Lombard thrust. The sub-Cambrian unconformity truncated the Ravalli Group east of the Sapphire plate. Farther west, the Piegan and Missoula Groups successively appear beneath the Cambrian, indicating that the entire graben was tilted west prior to Cambrian erosion. The Belt-Purcell section may thicken to 10 km in the far west (Peterson, 1991). The tilt caused the relative sense of displacement to change from north-side-down to south-side-down along the trace of the Lewis and Clark Line (Fig. 4). The tilting may partly reflect block faulting that simultaneously depressed the west edge of the basin, cut off the western sediment-source, and opened a new sediment transport path recorded in the Missoula Group (Sears, 2007). Fluvial-alluvial fans in the Missoula Group spread northward along the western side of the basin from the southern graben, across the central horst, and across the northern graben (Winston and Wheeler, 2006).

The Perry Line (Fig. 4) defines the southern boundary of the trough. The Perry fault lifted the Archean basement of southwest Montana, which shed thick subaqueous debris flows and turbidites into the Lahood Formation of the lower Belt-Purcell Supergroup (McMannis, 1963; Ross and Villeneuve, 2003). The Perry fault had great Precambrian structural relief; to the south, Cambrian layers directly overlie crystalline basement rocks; to the north, Cambrian rocks rest on >6 km of Belt-Purcell rocks (Winston, 1991). The Perry Line was reactivated as the Montana transverse shear zone during Cordilleran orogenesis (Schmidt et al., 1988; O'Neill, 1997; Sears, 2006).

The Perry Line may pass westward across the Chief Joseph and Bitterroot batholiths to the Clearwater corner of western Idaho. The Clearwater corner (Fig. 1) defines a jog in the western Idaho suture zone between North American rocks and accreted terranes (Strayer et al., 1989; Fleck and Criss, 1985; Burchfiel et al., 1992). The Syringa terrane of western Idaho, a Neoproterozoic metamorphic complex, appears to be split and sinistrally offset by ~100 km at the Clearwater corner (Lewis et al., 2004). Facies changes in the Belt-Purcell Supergroup indicate that this region may approximate the original southwest margin of the Belt-Purcell Basin (Winston, 2003, personal commun.; Finch and Baldwin, 1984).

South of the Perry Line, Precambrian basement rocks core several ranges of southwestern Montana and Wyoming (Kulik and Schmidt, 1988). These basement ranges exhibit significant structural relief as measured against the base of the Cambrian (Brown, 1988). For example, the vertical structural relief between the Crazy Mountains depression, one of the deepest in the Rocky Mountains, and the high Beartooth Plateau sums to 10–15 km at the base of the Cambrian (King, 1969; Muehlberger, 1996). The horizontal component of displacement on these basement thrusts is large (Smithson et al., 1979; Blackstone, 1983; Gries, 1983). Slip along the Perry Line accommodated this horizontal shortening (Sears, 2006).

An isolated body of allochthonous Belt-Purcell rocks occurs in southwestern Montana and adjacent Idaho, bounded on the east by thrust faults, and engulfed on the west and north by Late Cretaceous batholiths (Ruppel et al., 1981). Due to Cretaceous metamorphic overprint, these rocks have been difficult to correlate with the Belt-Purcell Supergroup to the north; indeed, they were once considered to have been deposited in a separate basin (Harrison, 1972; Ruppel et al., 1981). Link (1993) showed, however, that the rocks have close stratigraphic affinities with the Belt-Purcell Supergroup. Detrital-zircon correlations support the lithostratigraphic matches (Link and Fanning, 2003).

The southern graben contains a thicker and deeper-water sequence of Paleozoic sedimentary rocks than the northern graben. The Central Montana trough extends miogeoclinal facies eastward from Idaho into southwest Montana and is evident in several isopach maps of Paleozoic units (Mallory, 1972; Peterson, 1991). By contrast, the northern graben underlies "Montania," a high-standing platform with a condensed Paleozoic section (Price and Sears, 2000).

Syntectonic granite sheets engorged the southern graben; they may have been guided far to the east of the Cordilleran subduction zone by west-dipping bedding anisotropy of the Belt-Purcell Supergroup (Hyndman et al., 1988; Foster and Fanning, 1997; Kalakay et al., 2000). The Boulder batholith hosts the world-class Butte porphyry copper deposit (Robinson et al., 1968).

Andesitic magma rose to the east and erupted from the Helena embayment to produce the Elkhorn Mountains, Two Medicine, and Adel Mountains volcanics (Smedes, 1966; Schmidt, 1978; Viele and Harris, 1965). The emplacement of magma altered the thrust dynamics by lowering the strength and density and increasing the gravitational potential of the thrust wedge (Davis et al., 1983). Upper amphibolite–facies metamorphism and fold nappes formed in Belt-Purcell rocks on the margins of the Bitterroot and Chief Joseph batholiths (Hyndman et al., 1988). Several thrusts in southwest Montana coincide with granitic sills, either because the rocks failed on the sills, or because the sills intruded thrust planes. Thrust movement was linked to cleavage development and metamorphism in rocks that were heated by the magma (Sears et al., 1989). Tectonic slides and extreme attenuation of stratigraphy occurred on the flanks of some batholiths (Zen, 1988). Part of the thrust displacement was channelized down the axis of the trough and spilled over the edges along the Montana transverse zone (Schmidt et al., 1988).

In contrast with the northern half-graben, massive syntectonic Beaverhead conglomerates with high percentages of Belt-Purcell clasts occur in the foreland of the southern graben (Nicols et al., 1985; Janecke et al., 2000; Perry et al., 1988; Ryder and Scholten, 1973). The 3-km-thick Late Cretaceous–Paleocene Beaverhead Group and correlative formations contain thick andesite members derived from eruptions of the Boulder batholith (Gwinn and Mutch, 1965; Hamilton and Meyers, 1967). Evidently, the combined effects of thrusting and granitic intrusion led to exposure and erosion of the Belt-Purcell rocks, perhaps because the thermally weakened rocks required a steeper critical taper, which raised the Belt-Purcell rocks to the surface.

Siberian Connection

A paleocontinental restoration (Fig. 5) that places the northeastern corner of the Siberian craton against the Belt-Purcell Basin illuminates aspects of the origin of the basin, the breakup of the continent, and the kinematics of displacement Belt-Purcell thrust plates.

Late Proterozoic to Early Cambrian rifting and seafloor spreading broke up the Precambrian supercontinent Rodinia and removed the continental block that once lay to the west of the Belt-Purcell Basin (Stewart, 1972; Burchfiel et al., 1992; Ross et al., 1992). Although Moores (1991) proposed that Antarctica was the western continent, Sears and Price (1978, 2003) identified the Siberian craton (Kosygin and Parvenof, 1975) as the western continental fragment (Fig. 5, inset). Evidence supporting this proposal includes a tight geographical match of the conjugate margins, ten correlative basement piercing-points, aligned dike swarms (Okrugin et al., 1990; Wooden et al., 1978), matching intracontinental basins (Sears et al., 2006), correlative rifting and thermal subsidence histories of the conjugate margins (Khudoley and Serkina, 2002; Levy and Christie-Blick, 1991), provenance data, Early Cambrian faunal similarities to the genus level, and permissive paleomagnetic data (Sears and Price, 2003). The Siberian connection places the Belt-Purcell Basin against the Udzha-Khastakh Basin of the northeastern Siberian craton (Surkov et al., 1991). It implies that the Belt-Purcell Basin formed at an intracontinental triple-rift junction, similar in scale and internal structure to the Afar triple-rift junction of East Africa (Burke and Dewey, 1973). Final rifting separated the conjugate continents by late Neoproterozoic to Early Cambrian time. Today, the deformed Cordilleran miogeocline and accreted oceanic terranes underlie the regions to the west.

A detailed contour map of the rifted basement surface of the northeastern Siberian craton neatly accommodates the configuration of the restored Belt-Purcell thrust plates (Fig. 5). The continental reconstruction closely aligns the St. Mary and Moyie faults with the Siberian Udzha trough, the Ovando graben (Winston, 1991) with the Siberian Khastakh trough, and the northern Belt-Purcell half-graben with the Taimyr trough. It locates the central horst block as a prolongation of the horst block between the Udzha and Khastakh troughs.

Preexisting basement structures may have influenced the geometry of the triple-rift junction. The Udzha trough closely follows the Hapshan suture, bending from south to southeast, in Siberian coordinates. The west edge of the central horst follows the edge of the Great Falls tectonic zone and abuts a fracture zone in Siberia that follows the west edge of the equivalent Aekit province. A fault followed the eastern edge of the Great Falls tectonic zone and its extension into Siberia, at the southern limit of the Lemhi domain.

The rifts were parts of a continental-scale network (Fig. 5, inset). The northern branch coincides with the Rocky Mountain–Tintina trench, which continues from western Montana to the Yukon. Cook and Van der Velden (1993) showed that the trench coincides with a Proterozoic rift structure for much of its length in Canada. The southern branch is part of the Montana-Tennessee lineament (Hatcher et al., 1987; Paulsen and Marshak, 1994; Marshak et al., 2000), which curves gently across the mid-continent. The western branch parallels a Proterozoic basement structure for 2000 km from the Udzha trough in northeastern Siberia to the Sette-Daban trough in southeastern Siberia (Kosygin and Parvenof, 1975; Rosen et al., 1994). Sears (2001b) proposed that the fractures were part of a larger icosahedral system of fractures that resulted from uniform horizontal extension of the lithosphere of a Mesoproterozoic supercontinent.

The Udzha trough is a narrow rift, 400 km long, interpreted to be filled with >9 km of Mesoproterozoic sediments (Zonenshain et al., 1990), although only a few kilometers are exposed along river banks in the low-relief landscape of the northern Siberian craton. The Khastakh trough also remains mostly in the subsurface, with some small outcrops in the Olenek uplift (Sears et al., 2004).

The reconstruction implies that the line of eventual separation between the Siberian and North American cratons began as a growth fault during Belt-Purcell sedimentation. The Belt-Purcell Basin appears to have occupied the hanging wall of this fault, whereas the Siberian side, with its condensed section, occupied the uplifted footwall. The point source for the sedi-

Figure 5. Restoration of northeast Siberian craton against palinspastically restored Belt-Purcell Basin. Structural contours of basement surface relative to sea level are after Parvenof and Kuzmin (2001). Note tight fit of western rifted edges of Belt-Purcell blocks against eastern rifted edge of northeast Siberia, close alignment of Khastakh trough with Ovando graben, and close alignment of Udzha trough with Moyie and St. Mary faults. Medium lines with arrows show inferred transport paths for Belt-Purcell sediment. Thin dashed lines are basement boundaries that correlate across Siberian connection. Inset shows Siberia-Laurentia connection proposed by Sears and Price (2003), with box showing location of enlarged figure. Heavy lines in inset map show three continental-scale rift systems that intersect at Belt-Purcell Basin.

ment fans in the northern graben coincides with the mouth of the Udzha trough on the restoration. The source for fans in the southern graben may have been the Khastakh trough. The Udzha and Khastakh troughs may have captured sediment flow from an epicontinental region and delivered it into the Belt-Purcell Basin. Tectonic activity within the rift system may then have shunted sediment between the Udzha and Khastakh troughs (Sears et al., 2004; Sears, 2007). The sediment source shifted to the south and a large fluvial-alluvial fan system spread a thick section of Missoula Group sediments northward across the entire basin, indicating that the basin as a whole was down-faulted and tilted to the southwest along a northwest-trending block fault (Winston and Wheeler, 2006). The tilt of the basin toward the fault accounts for the pre-Flathead truncation of the Belt-Purcell Supergroup across the basin, with the young strata preserved toward the west. The fault activity may have been accompanied by eruption of the Purcell–Nicol Creek volcanics at 1443 Ma (Evans et al., 2000), and the permissively correlative Unguokhtakh volcanics in the Udzha trough (Sears et al., 2004).

Continental Breakup

The reconstruction (Fig. 6) predicts that the fault at the western edge of the Belt-Purcell basin was reactivated during breakup of the continent, with depositional onlap of rift and miogeoclinal sediments in a graben system that eventually opened into an ocean basin.

The Neoproterozoic Windermere Supergroup and Hamill-Gog Group overlie the Belt-Purcell Supergroup in British Columbia, Washington, and Idaho (Ross, 1991b). These rocks include continental-rift facies that record several episodes of rifting and magmatism between 780 and 550 Ma (Lund et al., 2003). The 780 Ma Gunbarrel large igneous province includes flows, sills, and dikes that have been mapped along the trend of the eastern Cordillera from northwestern Canada to the Teton Range of Wyoming (Harlan et al., 1997, 2003). The well-known Purcell sill near Logan Pass in Glacier National Park is part of this system. Other sills and dikes of this age occur near Missoula (Burtis et al., 2007). The Windermere Supergroup is commonly several kilometers thick and is characterized by basal diamictite, pillow basalt, quartz- and sulfide-rich turbidite, slate, and minor allodapic carbonate (Ross, 1991b). These units were deposited in deep northwest-trending grabens (Warren, 1997) and in large turbidite fans (Ross, 1991b). Lis and Price (1976) determined a minimum of 13 km of vertical displacement across the St. Mary transverse fault during Windermere rifting.

The Hamill-Gog Group is as much as a few kilometers thick, and it consists of thick, cross-bedded quartz-arenite, shale, carbonate, mafic volcanics, and sills, U-Pb dated to 570 Ma (Devlin, 1989; Devlin and Bond, 1986; Colpron et al., 2002). Locally, it exhibits an angular unconformable relationship with the underlying Windermere rocks, indicating that renewed rifting occurred before and during its deposition in northwest-trending grabens (Kubli and Simony, 1992; Warren, 1997). Skolithus appears in the upper Gog Group, showing that it passes into the Early Cambrian (Lickorish and Simony, 1995).

The Neoproterozoic rifting reactivated some extensional and transverse faults that originated during Belt-Purcell deposition (Price and Sears, 2000). The rifting opened the Chancellor Basin in British Columbia, apparently reactivating a down-to-the-west Belt-Purcell basin-margin fault in that region. The opening of the Chancellor Basin was accommodated by sinistral offset of the northernmost part of the Belt-Purcell Basin along the Moyie fault. This movement also opened small rift basins along the Eager trough between the Moyie and St. Mary faults. This extensional displacement was recovered, and the Belt-Purcell facies were approximately realigned by Middle Jurassic–Early Cretaceous Cordilleran shortening, when the Chancellor Basin was inverted into the Porcupine Creek anticlinorium (Price, 1986).

By late Early Cambrian time, rifting had progressed to full-scale continental separation with accumulation of the Cordilleran miogeocline along the resulting passive margin (Bond and Kominz, 1984; Fritz et al., 1991). Thermal subsidence of the passive margin was highlighted by burial of an Early Cambrian archeocyathan reef beneath thick, black, sulfidic shales of the Lardeau Group (Sears and Price, 2003), and deposition of thick calcareous shale in the Chancellor Basin (Norris and Price, 1966; Gardner et al., 1976). The Belt-Purcell region occupied this passive continental margin for the remainder of Paleozoic and the first half of Mesozoic time. The Windermere Supergroup is onlapped by the Hamill-Gog Group, which is onlapped by the Paleozoic miogeocline in the Kootenay arc.

By Middle Cambrian time, shallow-marine sandstone transgressed eastward across the Belt-Purcell Basin to directly overlie pre–Belt-Purcell basement east of the basin. The base of the Middle Cambrian Flathead Sandstone thus provides an excellent regional horizontal datum for palinspastic restoration of the Belt-Purcell Basin (Price, 1956, 1981). Renewed rifting during Devonian time produced 5–10 km of vertical offset along the Moyie fault, so that Upper Devonian strata locally rest directly on Belt-Purcell rocks to the south, while 7-km-thick Lower Paleozoic sedimentary rocks occur to the north beneath the Upper Devonian strata (Price and Sears, 2000).

Cordilleran Orogenesis and Thrust Rotations

As suggested in Figure 6, the removal of the Siberian craton exposed the western edge of the Belt-Purcell Basin to the Cordilleran margin. There it presented an anomalous set of strong, layered rocks to collisional orogenesis when the Cordilleran margin became tectonically active. Ultimately, the displacement of the Belt-Purcell Basin was driven by easterly convergence of accreted volcanogenic terranes against the Cordilleran miogeocline. The inherited architecture of the Belt-Purcell rift basin, however, strongly influenced the kinematics of thrusting in the northern Rocky Mountain fold-and-thrust belt. Syndepositional faults discussed in the preceding sections had compartmentalized the basin into a mosaic of structural blocks of different thick-

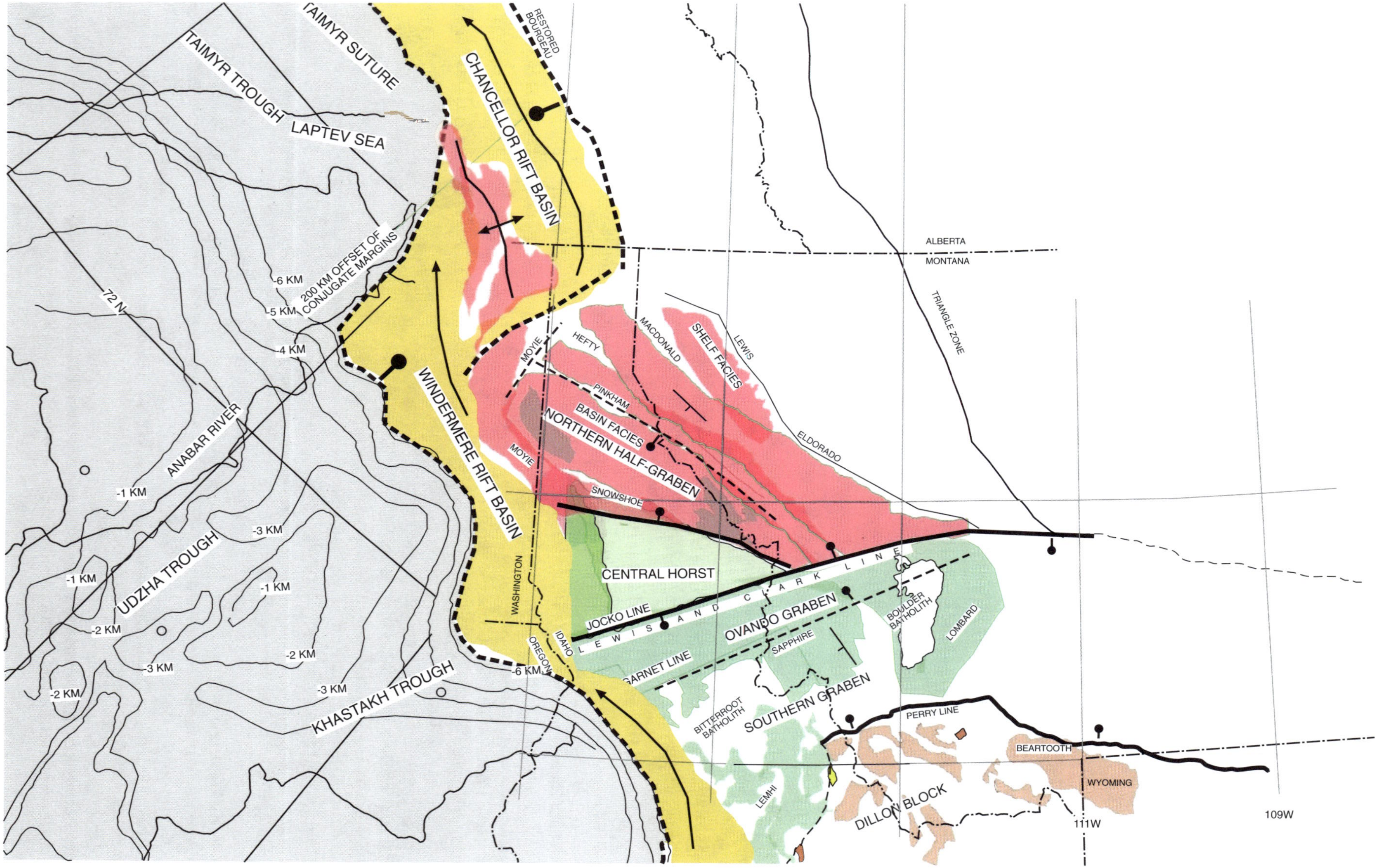

Figure 6. Inferred opening of Windermere rift along western edge of Belt-Purcell Basin. Note that Windermere rift splits around northern part of basin, opening Chancellor graben, and the asymmetry of rift-detachment shifts from west-dipping, north of the St. Mary fault, to east-dipping, south of the St. Mary fault.

nesses, densities, strengths, aspects, and orientations. Reactivation of the syndepositional faults as transpressional shear zones accommodated rotation of the individual structural blocks during thrust displacement, as shown by paleomagnetic studies (Priest, 2000; Elston et al., 2002). The character of the blocks controlled the taper of the critical thrust wedge (Mitra, 1997; Boyer, 1991). It also controlled tectonic loading and fluid expulsion in the foreland basin (Miall, 1997).

The northern half-graben comprised an exceptionally strong and dense, northeast-tapering wedge. It was not heavily intruded by granite, so it remained cool enough to maintain high brittle strength during thrust displacement. It trended at a high angle to tectonic convergence (Fig. 3), and so was thrust out of its basin of deposition and tectonically inverted into the Purcell anticlinorium. By contrast, the southern graben trended subparallel to tectonic convergence. It was engulfed with sheets of syntectonic magma that followed the basin axis and thermally softened the rocks. It was not disgorged from its basin and remains in its original depositional trough.

The entire mosaic of structural blocks experienced clockwise thrust rotation, as exemplified by the palinspastic map (Fig. 3). The cumulative displacement of the fold-and-thrust belt increases northward from a few kilometers near Helena to 250 km at the latitude of Calgary (Price and Sears, 2000). The increased cumulative displacement resulted from greater displacement on each of a number of individual thrusts and from penetrative strain within the strongly cleaved Porcupine Creek anticlinorium. Because stress had to be transmitted from the plate boundary eastward across all of the structural blocks in order for them to move, all fault contacts between the blocks were transpressive. Clockwise rotation of the northern branch of the basin required commensurate clockwise rotation of the remainder of the basin, as well as of the Wyoming foreland south of the Perry Line (Sears, 2006).

The central horst block transferred sinistral transpression between the northern and southern parts of the basin. It was bounded by sinistral shear zones along the Hope and Snowshoe faults on the north, and by the Lewis and Clark Line on the south. The Lewis and Clark Line trended at a low angle to the Late Cretaceous convergence direction between the North American and Farallon plates. This may have induced sinistral transpression along the Lewis and Clark Line as the southern branch of the Belt-Purcell Basin rotated about an Euler pole that was eccentric to the line (Sears and Hendrix, 2004).

The Wyoming foreland experienced clockwise thrust rotation about a pole in central New Mexico, according to a model by Hamilton (1988) (Fig. 1, inset). The uplift of the basement blocks along the Perry Line may have resulted from cumulative rotation of the Wyoming foreland (Mitra and Mount, 1998; Sears, 2006). This would have compressed the southern branch of the Belt-Purcell Basin and contributed to the overall clockwise thrust rotation of the basin.

In summary, the anomalous geometry, strength, and position of the Belt-Purcell Basin on the continental margin created a structural keystone in the Rocky Mountain fold-and-thrust belt.

ACKNOWLEDGMENTS

I sincerely thank Ray Price for his guidance as my major professor at Queens University and for many years of encouragement and collaboration in studies of the tectonics of the Belt-Purcell Basin and beyond. I also thank my colleague and friend Don Winston for sharing his enthusiasm and knowledge on exciting geological field trips to all parts of the basin. This research was partially funded by National Science Foundation (NSF) grant EAR-0107024. Many thanks are due to Peter Fermor and Andy Okulitch for careful reviews and helpful comments.

REFERENCES CITED

Anderson, H.E., and Davis, D.W., 1995, U-Pb geochronology of the Moyie sills, Purcell Supergroup, southeastern British Columbia: Implications for the Mesoproterozoic geologic history of the Purcell (Belt) Basin: Canadian Journal of Earth Sciences, v. 32, p. 1180–1193.

Anderson, H.E., and Parrish, R.R., 2000, U-Pb geochronological evidence for the geological history of the Belt-Purcell Supergroup, southeastern British Columbia, Chapter 7, in Lydon, J.W., Höy, T., Slack, J.F., and Knapp, M.E., eds., The Geological Environment of the Sullivan Deposit, British Columbia: Geological Association of Canada, Mineral Deposits Division (MDD) Special Publication No. 1, p. 61–81.

Archibald, D.A., Krogh, T.E., Armstrong, R.L., and Farrar, E., 1984, Geochronology and tectonic implications of magmatism and metamorphism, southern Kootenay arc and neighbouring regions, southeastern British Columbia, Part II: Mid-Cretaceous to Eocene: Canadian Journal of Earth Sciences, v. 21, p. 567–583.

Bally, A.W., Gordy, P.L., and Stewart, G.A., 1966, Structure, seismic data, and orogenic evolution of southern Canadian Rocky Mountains: Bulletin of Canadian Petroleum Geology, v. 14, p. 337–381.

Beaumont, C., Quinlan, G.M., and Stockmal, G.S., 1993, The evolution of the Western Interior Basin: Causes, consequences and unsolved problems, in Caldwell, W.G.E., and Kauffman, E.G., eds., Evolution of Western Interior Basin: Geological Association of Canada Special Paper 39, p. 97–117.

Bennett, E.H., and Venkatakrishnan, R., 1982, A palinspastic reconstruction of the Coeur d'Alene mining district based on ore deposits and structural data: Economic Geology and the Bulletin of the Society of Economic Geologists, v. 77, p. 1851–1866.

Billingsley, P., and Locke, A., 1939, Structure of ore districts in the continental framework: New York, American Institute of Mineralogical and Metallurgical Engineering, 51 p.

Blackstone, D.L., Jr., 1983, Laramide compressional tectonics, southeastern Wyoming: Contributions to Geology, v. 22, p. 1–38.

Boerner, D.E., Craven, J.A., Kurtz, R.D., Ross, G.M., and Jones, F.W., 1998, The Great Falls tectonic zone: Suture or intracontinental shear zone?: Canadian Journal of Earth Sciences, v. 35, p. 175–183, doi: 10.1139/cjes-35-2-175.

Boleneus, D., Appelgate, L.M., Zientek, M.L., Carlson, M.H., Assmus, K.C., and Chase, D.W., 2006, Stratigraphic control of copper-silver deposits in the Mesoproterozoic Revett Formation, Idaho and Montana: Northwest Geology, v. 35, p. 57–60.

Bond, G.C., and Kominz, M.A., 1984, Construction of tectonic subsidence curves for the early Paleozoic miogeocline, southern Canadian Rocky Mountains: Implications for subsidence mechanisms, age of breakup, and crustal thinning: Geological Society of America Bulletin, v. 95, p. 155–173, doi: 10.1130/0016-7606(1984)95<155:COTSCF>2.0.CO;2.

Box, S.E., Bedrosian, P.A., and Pellerin, L., 2006, Deep crustal structure revealed by magnetotelluric transect between the Selkirk crest, Idaho, and the Whitefish Range, Montana: Northwest Geology, v. 35, p. 103–104.

Boyer, S.E., 1991, Sedimentary basin taper as a factor controlling structural variation in thrust belts: Geological Society of America Abstracts with Programs, v. 23, no. 1, p. 10.

Boyer, S.E., and Elliott, D., 1982, Thrust systems: American Association of Petroleum Geologists Bulletin, v. 66, p. 1196–1230.

Brown, W.G., 1988, Deformational style of Laramide uplifts in the Wyoming foreland, in Schmidt, C.J., and Perry, W.J., Jr., eds., Interaction of the

Rocky Mountain Foreland and the Cordilleran Thrust Belt: Geological Society of America Memoir 171, p. 1–25.

Buckley, S.N., and Sears, J.W., 1997, Emplacement of mafic sills into wet sediments in the Prichard Formation, Middle Proterozoic Belt Supergroup, Perma area, western Montana, *in* Berg, R., ed., Proceedings of Belt Symposium III: Montana Bureau of Mines and Geology Special Paper 112, p. 32–43.

Burchfiel, B.C., Cowan, D.S., and Davis, G.A., 1992, Tectonic overview of the Cordilleran orogen in the western United States, *in* Burchfiel, B.C., Lipman, P.W., and Zoback, M.L., eds., The Cordilleran Orogen: Conterminous U.S.: Boulder, Colorado, Geological Society of America, Geology of North America, v. G-3, p. 407–478.

Burke, K., and Dewey, J.F., 1973, Plume-generated triple junctions: Key indicators in applying plate tectonics to old rocks: The Journal of Geology, v. 81, p. 406–433.

Burtis, E., Sears, J.W., and Chamberlain, K.R., 2007, Age and petrology of Neoproterozoic intrusions in the Northern Rocky Mountains, U.S.A., *in* Link, P.K., and Lewis, R.S., eds., Proterozoic Geology of Western North America and Siberia: SEPM (Society for Sedimentary Geology) Special Publication No. 86, p. 175–192.

Burton, B.R., Ballard, D.W., Lageson, D.R., Perkins, M., Schmidt, C.J., and Warne, J.R., 1997, Deep drilling results and new interpretations of the Lombard thrust, southwest Montana, *in* Berg, R., ed., Proceedings of Belt Symposium III: Montana Bureau of Mines and Geology Special Paper 112, p. 229–243.

Catuneanu, C., and Sweet, A.R., 1999, Maastrichtian-Paleocene foreland-basin stratigraphies, western Canada: A reciprocal sequence architecture: Canadian Journal of Earth Sciences, v. 36, p. 685–703, doi: 10.1139/cjes-36-5-685.

Catuneanu, C., Sweet, A.R., and Miall, A.D., 1997, Reciprocal architecture of Bearpaw T-R sequences, uppermost Cretaceous, Western Canada sedimentary basin: Canadian Petroleum Geology Bulletin, v. 45, p. 75–94.

Chandler, F.W., 2000, The Belt-Purcell Basin as a low-latitude passive rift: Implications for the geological environment of Sullivan-type deposits, *in* Lydon, J.W., Höy, T., Slack, J.F., and Knapp, M.E., eds., The Geological Environment of the Sullivan Deposit, British Columbia: Geological Association of Canada, Mineral Deposits Division (MDD) Special Publication No. 1, p. 82–112.

Clowes, R., Hammer, P., Mandler, H., Ross, G.M., Cook, F., and Eaton, D., 1997, The Vulcan low, Matzhiwin high and related domains revealed by SALT 95 reflection data, *in* Ross, G.M., ed., 1997 Alberta Basement Transects Workshop, Lithoprobe Report 59: Vancouver, University of British Columbia, p. 15–22.

Colpron, M., Logan, J.M., and Mortensen, J.K., 2002, U-Pb zircon age constraint for Late Neoproterozoic rifting and initiation of the lower Paleozoic passive margin of western Laurentia: Canadian Journal of Earth Sciences, v. 39, p. 133–143, doi: 10.1139/e01-069.

Coney, P.J., 1978, Mesozoic-Cenozoic Cordilleran plate tectonics, *in* Smith, R.B., and Eaton, G.P., eds., Cenozoic Tectonics and Regional Geophysics of the Western Cordillera: Geological Society of America Memoir 152, p. 33–50.

Coney, P.J., and Harms, T.A., 1984, Cordilleran metamorphic core complexes; Cenozoic extensional relics of Mesozoic compression: Geology, v. 12, p. 550–554, doi: 10.1130/0091-7613(1984)12<550:CMCCCE>2.0.CO;2.

Constenius, K.N., 1996, Late Paleogene extensional collapse of the Cordilleran foreland fold and thrust belt: Geological Society of America Bulletin, v. 108, p. 20–39, doi: 10.1130/0016-7606(1996)108<0020:LPECOT>2.3.CO;2.

Cook, F.A., and Van der Velden, A.J., 1993, Proterozoic crustal transition beneath the Western Canada sedimentary basin: Geology, v. 21, p. 785–788, doi: 10.1130/0091-7613(1993)021<0785:PCTBTW>2.3.CO;2.

Cook, F.A., and Van der Velden, A.J., 1995, Three dimensional crustal structure of the Purcell anticlinorium in the Cordillera of southwestern Canada: Geological Society of America Bulletin, v. 107, p. 642–664, doi: 10.1130/0016-7606(1995)107<0642:TDCSOT>2.3.CO;2.

Cressman, E.R., 1989, Reconnaissance Stratigraphy of the Prichard Formation (Middle Proterozoic) Near Plains, Sanders County, Montana: U.S. Geological Survey Professional Paper 1490, 80 p.

Davis, D., Suppe, J., and Dahlen, F.A., 1983, Mechanics of fold-and-thrust belts and accretionary wedges: Journal of Geophysical Research, v. 88, p. 1153–1172.

Devlin, W.J., 1989, Stratigraphy and sedimentology of the Hamill Group, in the northern Selkirk Mountains, British Columbia: Evidence for latest Proterozoic–Early Cambrian extensional tectonism: Canadian Journal of Earth Sciences, v. 26, p. 515–533.

Devlin, W.J., and Bond, G.C., 1986, The initiation of the early Paleozoic Cordilleran miogeocline; evidence from the uppermost Proterozoic–Lower Cambrian Hamill Group of southeastern British Columbia: Canadian Journal of Earth Sciences, v. 25, p. 1–19.

Doughty, P.T., and Chamberlain, K.R., 1996, Salmon River arch revisited: New evidence for 1370 rifting near the end of deposition in the Middle Proterozoic Belt basin: Canadian Journal of Earth Sciences, v. 33, p. 1037–1052.

Doughty, P.T., Price, R.A., and Parrish, R.R., 1998, Geology and U-Pb geochronology of Archean basement and Proterozoic cover in the Priest River complex, northwestern United States, and their implications for Cordilleran structure and Precambrian continental restorations: Canadian Journal of Earth Sciences, v. 35, p. 39–54, doi: 10.1139/cjes-35-1-39.

Doughty, P.T., Chamberlain, K.R., Foster, D.A., and Sha, G.S., 2007, this volume, Structural, metamorphic, and geochronologic constraints on the origin of the Clearwater core complex, northern Idaho, *in* Sears, J.W., Harms, T.A., and Evenchick, C.A., eds., Whence the Mountains? Inquiries into the Evolution of Orogenic Systems: A Volume in Honor of Raymond A. Price: Geological Society of America Special Paper 433, doi: 10.1130/2007.2433(11).

Dyman, T.S., Porter, K.W., Tysdal, R.G., Cobban, W.A., Fox, J.E., Hammond, R.H., Nichols, D.J., Perry, W.J., Jr., Rice, D.D., Setterholm, D.R., Shurr, G.W., Haley, J.C., Lane, D.E., Anderson, S.B., and Campen, E.B., 1995, West-east stratigraphic transect of Cretaceous rocks in the northern Rocky Mountains and Great Plains regions, southwestern Montana to southwestern Minnesota: U.S. Geological Survey Miscellaneous Investigations Map I-2474-A.

Elston, D.P., Enkin, R.J., Baker, J., and Kisilevsky, D.K., 2002, Tightening the Belt: Paleomagnetic-stratigraphic constraints on deposition, correlation, and deformation of the Middle Proterozoic (ca. 1.4 Ga) Belt-Purcell Supergroup, United States and Canada: Geological Society of America Bulletin, v. 114, p. 619–638, doi: 10.1130/0016-7606(2002)114<0619:TTBPSC>2.0.CO;2.

Evans, K.V., Aleinikoff, J.N., Obradovich, J.D., and Fanning, C.M., 2000, SHRIMP U-Pb geochronology of volcanic rocks, Belt Supergroup, western Montana: Evidence for rapid deposition of sedimentary strata: Canadian Journal of Earth Sciences, v. 37, p. 1287–1300, doi: 10.1139/cjes-37-9-1287.

Evenchick, C.A., McMechan, M.E., McNicoll, V.J., and Carr, S.D., 2007, this volume, A synthesis of the Jurassic–Cretaceous tectonic evolution of the central and southeastern Canadian Cordillera: Exploring links across the orogen, *in* Sears, J.W., Harms, T.A., and Evenchick, C.A., eds., Whence the Mountains? Inquiries into the Evolution of Orogenic Systems: A Volume in Honor of Raymond A. Price: Geological Society of America Special Paper 433, doi: 10.1130/2007.2433(06).

Farooqui, M.A., 1997, Provenance of the Bonner Formation (Middle Proterozoic Belt Supergroup), Montana, and the Buffalo Hump Formation (Deer Trail Group), Washington: Petrographic evidences for multiple source terrains, *in* Berg, R.B., ed., Belt Symposium III Abstracts: Montana Bureau of Mines and Geology Special Publication 381, p. 17–19.

Feinstein, S., Kohn, B., Osadetz, K., and Price, R.A., 2007, this volume, Thermochronometric reconstruction of the prethrust paleogeothermal gradient and initial thickness of the Lewis thrust sheet, southeastern Canadian Cordillera foreland belt, *in* Sears, J.W., Harms, T.A., and Evenchick, C.A., eds., Whence the Mountains? Inquiries into the Evolution of Orogenic Systems: A Volume in Honor of Raymond A. Price: Geological Society of America Special Paper 433, doi: 10.1130/2007.2433(08).

Fermor, P.R., and Moffat, I.W., 1992, Tectonics and structure of the western Canada foreland basin, *in* Macqueen, R.W., and Leckie, D.A., eds., Foreland Basins and Fold Belts: American Association of Petroleum Geologists Memoir 55, p. 81–105.

Fillipone, J.A., and Yin, A., 1994, Age and regional tectonic implications of Late Cretaceous thrusting and Eocene extension, Cabinet Mountains, northwest Montana and northern Idaho: Geological Society of America Bulletin, v. 106, p. 1017–1032, doi: 10.1130/0016-7606(1994)106<1017:AARTIO>2.3.CO;2.

Finch, J.C., and Baldwin, D.O., 1984, Stratigraphy of the Prichard Formation, Belt Supergroup, *in* Hobbs, S.W., ed., The Belt, Belt Symposium II, Abstracts with Summaries: Montana Bureau of Mines and Geology Special Publication 90, p. 5–7.

Fleck, R.J., and Criss, R.C., 1985, Strontium and oxygen isotope variations in Mesozoic and Tertiary plutons of central Idaho: Contributions to Mineralogy and Petrology, v. 90, p. 291–308, doi: 10.1007/BF00378269.

Foster, D.A., and Fanning, M., 1997, Geochronology of the northern Idaho batholith and the Bitterroot metamorphic core complex: Magmatism preceding and contemporaneous with extension: Geological Society of America Bulletin, v. 109, p. 379–394, doi: 10.1130/0016-7606(1997)109<0379:GOTNIB>2.3.CO;2.

Fritz, W.H., Cecile, M.P., Norford, B.S., Morrow, D., and Geldsetzer, H.H.J., 1991, Cambrian to Middle Devonian assemblage, in Gabrielse, H., and Yorath, C.J., eds., Geology of the Cordilleran Orogen in Canada: Ottawa, Canada, Geological Survey of Canada, Geology of Canada, no. 4, p. 151–218.

Frost, C.D., and Winston, D., 1987, Nd isotope systematics of coarse- and fine-grained sediments: Examples from the Middle Proterozoic Belt-Purcell Supergroup: The Journal of Geology, v. 95, p. 309–327.

Gardner, D.A.C., Price, R.A., and Carmichael, D.M., 1976, The petrology and structural fabric of some Lower Paleozoic calcareous pelites in the Porcupine Creek anticlinorium near Golden, British Columbia: Geological Survey of Canada Paper 76–1A, p. 137–140.

Gries, R.R., 1983, Oil and gas prospecting beneath the Precambrian of foreland thrust plates in the Rocky Mountains: American Association of Petroleum Geologists Bulletin, v. 67, p. 1–26.

Gwinn, V.E., and Mutch, T.A., 1965, Intertongued Upper Cretaceous volcanic and nonvolcanic rocks, central-western Montana: Geological Society of America Bulletin, v. 76, p. 1125–1144, doi: 10.1130/0016-7606(1965)76[1125:IUCVAN]2.0.CO;2.

Hamilton, W.B., 1988, Laramide crustal shortening, in Schmidt, C.J., and Perry, W.J., Jr., eds., Interaction of the Rocky Mountain Foreland and the Cordilleran Thrust Belt: Geological Society of America Memoir 171, p. 1–25.

Hamilton, W.B., and Meyers, W.B., 1967, The nature of batholiths: U.S. Geological Survey Professional Paper 554-C, p. 509–549.

Hammer, P.T.C., and Clowes, R.M., 2007, this volume, Lithospheric-scale structures across the Alaskan and Canadian Cordillera: Comparisons and tectonic implications, in Sears, J.W., Harms, T.A., and Evenchick, C.A., eds., Whence the Mountains? Inquiries into the Evolution of Orogenic Systems: A Volume in Honor of Raymond A. Price: Geological Society of America Special Paper 433, doi: 10.1130/2007.2433(05).

Harlan, S.S., Geissman, J.W., and Snee, L.W., 1997, Paleomagnetic and $^{40}Ar/^{39}Ar$ geochronologic data from late Proterozoic mafic dikes and sills, Montana and Wyoming: U.S. Geological Survey Paper 1580, 16 p.

Harlan, S.S., Heaman, L., LeCheminant, A.N., and Premo, W.R., 2003, Gunbarrel mafic magmatic event: A key 780 Ma time marker for Rodinia plate reconstructions: Geology, v. 31, p. 1053–1056, doi: 10.1130/G19944.1.

Harms, T.A., and Price, R.A., 1992, The Newport fault: Eocene listric normal faulting, mylonitization, and crustal extension in northeast Washington and northwest Idaho: Geological Society of America Bulletin, v. 104, p. 745–761, doi: 10.1130/0016-7606(1992)104<0745:TNFELN>2.3.CO;2.

Harrison, J.E., 1972, Precambrian Belt basin of the northwestern United States, its geometry, sedimentation, and copper occurrences: Geological Society of America Bulletin, v. 83, p. 1215–1240, doi: 10.1130/0016-7606(1972)83[1215:PBBONU]2.0.CO;2.

Harrison, J.E., Griggs, A.B., and Wells, J.D., 1974, Tectonic features of the Precambrian Belt basin and their influence on post-Belt structures: U.S. Geological Survey Professional Paper 866, 15 p.

Hatcher, R.D., Jr., Zietz, I., and Litehiser, J.J., 1987, Crustal subdivisions of the eastern and central United States and a seismic boundary hypothesis for eastern seismicity: Geology, v. 15, p. 528–532, doi: 10.1130/0091-7613(1987)15<528:CSOTEA>2.0.CO;2.

Hobbs, S.W., Griggs, A.B., Wallace, R.E., and Campbell, A.B., 1965, Geology of the Coeur d'Alene District, Shoshone County, Idaho: U.S. Geological Survey Professional Paper 478, 139 p.

Hoffman, P.J., 1988, The united plates of America: Annual Review of Earth and Planetary Sciences, v. 16, p. 543–603.

Höy, T., Anderson, D., Turner, R.J.W., and Leitch, C.H.B., 2000, Tectonic, magmatic, and metallogenic history of the early synrift phase of the Purcell Basin, southeastern British Columbia, Chapter 4, in Lydon, J.W., Höy, T., Slack, J.F., and Knapp, M.E., eds., The Geological Environment of the Sullivan Deposit, British Columbia: Geological Association of Canada, Mineral Deposits Division (MDD) Special Publication No. 1, p. 32–60.

Hyndman, D.W., Alt, D., and Sears, J.W., 1988, Post-Archean metamorphic and tectonic evolution of western Montana and northern Idaho, in Ernst, W.G., ed., Metamorphism and Crustal Evolution of the Western United States: Englewood Cliffs, New Jersey, Prentice-Hall, Rubey Volume VII, p. 332–361.

Janecke, S.U., 1994, Sedimentation and paleogeography of an Eocene to Oligocene rift zone, Idaho and Montana: Geological Society of America Bulletin, v. 106, p. 1083–1095, doi: 10.1130/0016-7606(1994)106<1083:SAPOAE>2.3.CO;2.

Janecke, S.U., VanDenburg, C.J., Blankenau, J.J., and M'Gonigle, J.W., 2000, Long-distance longitudinal transport of gravel across the Cordilleran thrust belt of Montana and Idaho: Geology, v. 28, p. 439–442, doi: 10.1130/0091-7613(2000)28<439:LLTOGA>2.0.CO;2.

Kalakay, T.J., John, B.E., and Foster, D.A., 2000, Granite emplacement during contemporaneous thrust faulting in the Sevier fold and thrust belt of southwest Montana, in Roberts, S., and Winston, D., eds., Geologic Field Trips, Western Montana and Adjacent Areas: Missoula, University of Montana, p. 157–179.

Khudoley, A.K., and Serkina, G.G., 2002, Early Paleozoic rifting of the east margin of Siberian craton: Comparison of geological data and subsidence curves, in Karakin, Yu.V., ed., Tectonics and Geophysics of Lithosphere: Moscow, GEOS, p. 288–291 (in Russian).

King, P.B., 1969, Tectonic Map of North America: U.S. Geological Survey, scale 1:5,000,000.

Kosygin, Y.U., and Parvenof, L.M., 1975, Structural evolution of eastern Siberia and adjacent areas: American Journal of Science, v. 275-A.

Kubli, T.E., and Simony, P.S., 1992, The Dogtooth high, northern Purcell Mountains, British Columbia: Bulletin of Canadian Petroleum Geology, v. 40, p. 36–51.

Kulik, D.M., and Schmidt, C.J., 1988, Region of overlap and styles of interaction of Cordilleran thrust belt and Rocky Mountain foreland, in Schmidt, C.J., and Perry, W.J., Jr., eds., Interaction of the Rocky Mountain Foreland and the Cordilleran Thrust Belt: Geological Society of America Memoir 171, p. 75–98.

Levy, M., and Christie-Blick, N., 1991, Tectonic subsidence of the early Paleozoic passive continental margin in eastern California and southern Nevada: Geological Society of America Bulletin, v. 103, p. 1590–1606, doi: 10.1130/0016-7606(1991)103<1590:TSOTEP>2.3.CO;2.

Lewis, R.S., Vervoort, J.D., McClelland, W.C., and Chang, Z., 2004, Age constraints on metasedimentary rocks northwest of the Idaho batholith based on detrital zircons and intrusive sills: Geological Society of America Abstracts with Programs, v. 36, no. 4, p. 87.

Lickorish, W.H., and Simony, P.S., 1995, Evidence for late rifting of the Cordilleran margin outlined by stratigraphic division of the Lower Cambrian Gog Group, Rocky Mountain Main Ranges, British Columbia and Alberta: Canadian Journal of Earth Sciences, v. 32, p. 860–874.

Link, P.K., ed., 1993, Middle and Late Proterozoic stratified rocks of western US Cordillera, Colorado Plateau, Basin and Range Province, in Reed, J.C., et al., eds., Precambrian: Continental U.S.: Boulder, Colorado, Geological Society of America, Geology of North America, v. C-2, p. 187–208.

Link, P.K., and Fanning, C.M., 2003, Detrital zircon ages from the Yellowjacket, Apple Creek and Gunsight Formations, Blackbird mining district, Salmon River Mountains, central Idaho: Northwest Geology, v. 32, p. 206–207.

Lis, M.G., and Price, R.A., 1976, Large-scale block faulting during deposition of the Windermere Supergroup (Hadrynian) in southeastern British Columbia: Geological Survey of Canada Paper 76–1A, p. 135–136.

Lund, K., Aleinikoff, J.N., Evans, K.V., and Fanning, C.M., 2003, SHRIMP U-Pb geochronology of Neoproterozoic Windermere Supergroup, central Idaho: Implications for rifting of western Laurentia and synchroneity of Sturtian glacial deposits: Geological Society of America Bulletin, v. 115, p. 349–372, doi: 10.1130/0016-7606(2003)115<0349:SUPGON>2.0.CO;2.

Lydon, J.W., 2000, A synopsis of the current understanding of the geological environment of the Sullivan deposit, in Lydon, J.W., Höy, T., Slack, J.F., and Knapp, M.E., eds., The Geological Environment of the Sullivan Deposit, British Columbia: Geological Association of Canada, Mineral Deposits Division (MDD) Special Publication No. 1, p. 12–31.

Mack, G.H., and Jerzykiewicz, T., 1989, Provenance of post-Wapiabi sandstones and its implications for Campanian to Paleocene tectonic history of the southern Canadian Cordillera: Canadian Journal of Earth Sciences, v. 26, p. 665–676.

Mallory, W.W., ed., 1972, Geologic Atlas of the Rocky Mountain Region: Denver, Rocky Mountain Association of Geologists, 331 p.

Marshak, S., Karlstrom, K., and Timmons, J.M., 2000, Inversion of Proterozoic extensional faults: An explanation for the pattern of Laramide and ancestral Rockies intracratonic deformation, United States: Geology, v. 28, p. 735–738, doi: 10.1130/0091-7613(2000)28<735:IOPEFA>2.0.CO;2.

McMannis, W.J., 1963, LaHood Formation: A coarse facies of the Belt series in southwestern Montana: Geological Society of America Bulletin, v. 74, p. 407–436, doi: 10.1130/0016-7606(1963)74[407:LFCFOT]2.0.CO;2.

Miall, A.D., 1997, Reciprocal architecture of Bearpaw T-R sequences, uppermost Cretaceous, Western Canada sedimentary basin: Bulletin of Canadian Petroleum Geology, v. 45, p. 75–94.

Mitra, G., 1997, Evolution of salients in a fold-and-thrust belt: The effects of sedimentary basin geometry, strain distribution, and critical taper, in Sengupta, S., ed., Evolution of Geological Structures in Micro- to Macro-Scales: London, Chapman and Hall, p. 59–97.

Mitra, S., and Mount, V.S., 1998, Foreland basement involved structures: American Association of Petroleum Geologists (AAPG) Bulletin, v. 82, p. 70–109.

Monger, J.W.H., and Price, R.A., 1979, Geodynamic evolution of the Canadian Cordillera—Progress and problems: Canadian Journal of Earth Sciences, v. 16, p. 770–791.

Moores, E.M., 1991, Southwest U.S.–East Antarctic (SWEAT) connection: A hypothesis: Geology, v. 19, p. 425–428, doi: 10.1130/0091-7613(1991)019<0425:SUSEAS>2.3.CO;2.

Mudge, M.R., and Earhart, R.L., 1983, Bedrock Geologic Map of Part of the Northern Disturbed Belt, Lewis and Clark, Teton, Pondera, Glacier, Flathead, Cascade, and Powell Counties, Montana: U.S. Geological Survey Miscellaneous Investigations Map I-1375, scale 1:125,000.

Mudge, M.R., Earhart, R.L., Whipple, J.W., and Harrison, J.E., 1982, Geologic and Structure Map of the Choteau 1×2 Degree Quadrangle, Western Montana: U.S. Geological Survey Miscellaneous Investigations Map I-1300, scale 1:250,000.

Muehlberger, W.R., 1996, Tectonic Map of North America: American Association of Petroleum Geologists, Tulsa, Oklahoma, scale 1:5,000,000.

Mueller, P.A., Heatherington, A.L., Kelly, D.M., Wooden, J.L., and Mogk, D.W., 2002, Paleoproterozoic crust within the Great Falls tectonic zone: Implications for the assembly of southern Laurentia: Geology, v. 30, p. 127–130, doi: 10.1130/0091-7613(2002)030<0127:PCWTGF>2.0.CO;2.

Nichols, D.J., Perry, W.J., Jr., and Haley, J.C., 1985, Reinterpretation of the palynology and age of Laramide syntectonic deposits, southwestern Montana, and revision of the Beaverhead Group: Geology, v. 13, p. 149–153, doi: 10.1130/0091-7613(1985)13<149:ROTPAA>2.0.CO;2.

Norris, D.K., and Price, R.A., 1966, Middle Cambrian lithostratigraphy of southeastern Canadian Cordillera: Bulletin of Canadian Petroleum Geology, v. 14, p. 385–404.

Norwick, S.A., 1972, The Regional Metamorphic Facies of the Prichard Formation of Western Montana and Northern Idaho [Ph.D. thesis]: Missoula, University of Montana, 129 p.

Okrugin, A.V., Oleinikov, B.V., Savvinov, V.T., and Tomshin, M.D., 1990, Late Precambrian dyke swarms of the Anabar massif, Siberian platform, USSR, in Parker, A.J., Rickwood, P.C., and Tucker, D.H., eds., Mafic Dykes and Emplacement Mechanisms: Rotterdam, Balkema, p. 529–533.

Oliver, J., 1986, Fluids expelled tectonically from orogenic belts: Their role in hydrocarbon migration and other geologic phenomena: Geology, v. 14, p. 99–102, doi: 10.1130/0091-7613(1986)14<99:FETFOB>2.0.CO;2.

O'Neill, J.M., 1997, Stratigraphic character and structural setting of the Belt Supergroup in the Highland Mountains, southwestern Montana, in Berg, R.B., ed., Belt Symposium III: Montana Bureau of Mines and Geology Special Publication 112, p. 12–16.

Osadetz, K.G., Stockmal, G.S., and Lebel, D., 2000, Waterton Lakes to Crowsnest Pass via Highway 6 and Highway 3 with foothills analyses, in Schalla, R.A., and Johnson, E.H., eds., Montana/Alberta Thrust Belt and Adjacent Foreland: Billings, Montana Geological Society, 50th Anniversary Symposium, v. II, p. 69–162.

Parvenof, L.M., and Kuzmin, M.I., 2001, Tectonics, Geodynamics, and Metallogeny of Yakutia: Moscow, Nauka/Interperiodica, 114 p. (in Russian).

Paulsen, T., and Marshak, S., 1994, Cratonic weak zone in the U.S. continental interior: The Dakota-Carolina corridor: Geology, v. 22, p. 15–18, doi: 10.1130/0091-7613(1994)022<0015:CWZITU>2.3.CO;2.

Perry, W.J., Jr., Haley, J.C., Nichols, D.J., Hammons, P.M., and Ponton, J.D., 1988, Interactions of the Rocky Mountain foreland and Cordilleran thrust belt in Lima region, southwest Montana, in Schmidt, C.J., and Perry, W.J., Jr., eds., Interactions of the Rocky Mountain Foreland and the Cordilleran Thrust Belt: Geological Society of America Memoir 171, p. 267–291.

Peterson, J.A., 1991, General stratigraphy and regional paleotectonics of the western Montana overthrust belt, in Peterson, J.A., ed., Sedimentation and Tectonics: American Association of Petroleum Geologists Memoir 41, p. 57–86.

Poage, M.A., Hyndman, D.W., and Sears, J.W., 2000, Petrology, geochemistry, and diabase-granophyre relations of a thick basaltic sill emplaced into wet sediments, western Montana: Canadian Journal of Earth Sciences, v. 37, p. 1109–1119, doi: 10.1139/cjes-37-8-1109.

Price, R.A., 1956, The base of the Cambrian system in the southeastern Cordillera of Canada: Canadian Mining and Metallurgical Bulletin, v. 49, p. 765–771.

Price, R.A., 1964, The Precambrian Purcell system in the Rocky Mountains of southern Alberta and British Columbia: Bulletin of Canadian Petroleum Geology, v. 12, Special Issue, p. 399–426.

Price, R.A., 1965, Flathead Map-Area, British Columbia and Alberta: Geological Survey of Canada Memoir 336, 221 p.

Price, R.A., 1967, The tectonic significance of mesoscopic subfabrics in the southern Canadian Rocky Mountains of Alberta and British Columbia: Canadian Journal of Earth Sciences, v. 4, p. 39–70.

Price, R.A., 1973, Large-scale gravitational flow of supracrustal rocks, southern Canadian Rockies, in De Jong, K.A., and Scholten, R., eds., Gravity and Tectonics: New York, Wiley, p. 491–502.

Price, R.A., 1981, The Cordilleran thrust and fold belt in the southern Canadian Rocky Mountains, in McClay, K.R., and Price, N.J., eds., Thrust and Nappe Tectonics: Geological Society [London] Special Publication 9, p. 427–448.

Price, R.A., 1986, The southeastern Canadian Cordillera: Thrust faulting, tectonic wedging, and delamination of the lithosphere: Journal of Structural Geology, v. 8, p. 239–254, doi: 10.1016/0191-8141(86)90046-5.

Price, R.A., and Fermor, P.R., 1986, Structure section of the Canadian foreland thrust and fold belt west of Calgary, Alberta: Geological Survey of Canada, paper 84-14, 1 sheet.

Price, R.A., and Mountjoy, E.W., 1970, Geologic structure of the Canadian Rocky Mountains between Bow and Athabasca Rivers: A progress report, in Wheeler, J.O., ed., Structure of the Southern Canadian Cordillera: Geological Association of Canada Special Publication 6, p. 7–25.

Price, R.A., and Sears, J.W., 2000, A preliminary palinspastic map of the Mesoproterozoic Belt-Purcell Supergroup, Canada and USA: Implications for the tectonic setting and structural evolution of the Purcell anticlinorium and the Sullivan deposit, in Lydon, J.W., Höy, T., Slack, J.F., and Knapp, M.E., eds., The Geological Environment of the Sullivan Deposit, British Columbia: Geological Association of Canada, Mineral Deposits Division (MDD) Special Publication No. 1, p. 61–81.

Priest, B.M., 2000, Structural and paleomagnetic study of thrust rotation of a late Cretaceous sill, Gibson Reservoir, Bob Marshall Wilderness, Montana [M.S. thesis]: Missoula, University of Montana, 120 p.

Ramsay, J.G., 1980, Shear zone geometry, a review: Journal of Structural Geology, v. 2, p. 83–89, doi: 10.1016/0191-8141(80)90038-3.

Reed, J.C., Jr., 1993, Map of the Precambrian rocks of the conterminous United States and some adjacent parts of Canada: Boulder, Colorado, Geological Society of America, Geology of North America, v. C-2, scale 1:2,500,000.

Reynolds, M.W., and Brandt, T.R., 2007, Geologic map of the Canyon Ferry 30×60 quadrangle, west-central Montana: U.S. Geological Survey, Scientific Investigations Map 2860, scale 1:100,000.

Robinson, G.D., Klepper, M.R., and Obradovich, J.D., 1968, Overlapping plutonism, volcanism, and tectonism in the Boulder batholith region, western Montana, in Coats, R.R., Hay, R.L., and Anderson, C.A., eds., Studies in Volcanology: A Memoir in Honor of Howel Williams: Geological Society of America Memoir 116, p. 557–575.

Rosen, O.M., Condie, K.C., Natapov, L.M., and Nozhkin, A.D., 1994, Archean and early Proterozoic evolution of the Siberian craton: A preliminary assessment, in Condie, K.C., ed., Archean Crustal Evolution: Developments in Precambrian Geology, v. 11, p. 411–459.

Ross, G.M., 1991a, Precambrian basement in the Canadian Cordillera: An introduction: Canadian Journal of Earth Sciences, v. 28, p. 1133–1139.

Ross, G.M., 1991b, Tectonic setting of the Windermere Supergroup, revisited: Geology, v. 19, p. 1125–1128, doi: 10.1130/0091-7613(1991)019<1125:TSOTWS>2.3.CO;2.

Ross, G.M., and Villeneuve, M., 2003, Provenance of the Mesoproterozoic (1.45 Ga) Belt basin (western North America): Another piece in the pre-Rodinia paleogeographic puzzle: Geological Society of America Bulletin, v. 115, p. 1191–1217, doi: 10.1130/B25209.1.

Ross, G.M., Parrish, R.R., and Winston, D., 1992, Provenance and U-Pb geochronology of the Mesoproterozoic Belt Supergroup (northwestern United States): Implications for age of deposition and pre-Panthalassa plate reconstructions: Earth and Planetary Science Letters, v. 113, p. 57–76, doi: 10.1016/0012-821X(92)90211-D.

Ruppel, E.T., Wallace, C.A., Schmidt, R.G., and Lopez, D.A., 1981, Preliminary interpretation of the thrust belt in southwest and west-central Montana and east-central Idaho, in Tucker, T.E., ed., Field Conference and Symposium Guidebook to Southwest Montana: Billings, Montana Geological Society, p. 139–159.

Ryan, P., 1991, Structural variations in illite and chlorite in the Belt Supergroup, western Montana and northern Idaho [M.S. thesis]: Missoula, University of Montana, 48 p.

Ryder, R.T., and Scholten, R., 1973, Syntectonic conglomerates in southwestern Montana: Their nature, origin, and tectonic significance: Geological Society of America Bulletin, v. 84, p. 773–796, doi: 10.1130/0016-7606(1973)84<773:SCISMT>2.0.CO;2.

Schmidt, C.J., O'Neill, J.M., and Brandon, W.C., 1988, Influence of Rocky Mountain foreland uplifts on the development of the frontal fold and thrust belt, southwestern Montana, in Schmidt, C.J., and Perry, W.J., eds., Interaction of the Rocky Mountain Foreland and the Cordilleran Thrust Belt: Geological Society of America Memoir 171, p. 171–201.

Schmidt, R.G., 1978, Rocks and Mineral Resources of the Wolf Creek Area, Lewis and Clark and Cascade Counties, Montana: U.S. Geological Survey Bulletin 1441, 91 p.

Sears, J.W., 1988, Two major thrust slabs in the west-central Montana Cordillera, in Schmidt, C., and Perry, W.J., eds., Interactions of the Rocky Mountain Foreland and the Cordilleran Thrust Belt: Geological Society of America Memoir 171, p. 165–170.

Sears, J.W., 1994, Thrust rotation of the Belt basin, Canada and United States: Northwest Geology, v. 23, p. 81–92.

Sears, J.W., 2000, Rotational kinematics of the Rocky Mountain thrust belt of northern Montana, in Schalla, R.A., and Johnson, E.H., eds., Montana/Alberta Thrust Belt and Adjacent Foreland: Billings, Montana, Montana Geological Society, 50th Anniversary Symposium, v. 1, p. 143–150.

Sears, J.W., 2001a, Emplacement and denudation history of the Lewis-Eldorado-Hoadley thrust slab in the northern Montana Cordillera, USA: Implications for steady-state orogenic processes: American Journal of Science, v. 301, p. 359–373, doi: 10.2475/ajs.301.4-5.359.

Sears, J.W., 2001b, Icosahedral fracture tessellation of early Mesoproterozoic Laurentia: Geology, v. 29, no. 4, p. 327–330, doi: 10.1130/0091-7613(2001)029<0327:IFTOEM>2.0.CO;2.

Sears, J.W., 2006, The Montana transform: Rotation along the northern boundary of the Wyoming foreland: Rocky Mountain Geology, v. 41, p. 65–76.

Sears, J.W., 2007, Destabilization of a Proterozoic epi-continental pediment by rifting: A model for the Belt-Purcell Basin, North America, in Link, P.K., and Lewis, R.S., eds., Proterozoic Geology of Western North America and Siberia: SEPM (Society for Sedimentary Geology) Special Publication No. 86, p. 55–64.

Sears, J.W., and Clements, P.S., 2000, Geometry and kinematics of the Blackfoot thrust fault and Lewis and Clark Line, Bonner, Montana, in Roberts, S., and Winston, D., eds., Geological Field Trips, Western Montana and Adjacent Areas for Geological Society of America Rocky Mountain Section Meeting, Missoula: Missoula, University of Montana, p. 123–130.

Sears, J.W., and Fritz, W.J., 1998, Cenozoic tilt-domains in southwest Montana: Interference among three generations of extensional fault systems, in Faulds, J.E., and Stewart, J.H., eds., Accommodation Zones and Transfer Zones: The Regional Segmentation of the Basin and Range Province: Geological Society of America Special Paper 323, p. 241–248.

Sears, J.W., and Hendrix, M., 2004, Lewis and Clark Line and the rotational origin of the Alberta and Helena salients, North American Cordillera, in Sussman, A., and Weil, A., eds. Orogenic Curvature: Geological Society of America Special Paper 383, p. 173–186.

Sears, J.W., and Price, R.A., 1978, The Siberian connection: A case for the Precambrian separation of the North American and Siberian cratons: Geology, v. 6, p. 267–270, doi: 10.1130/0091-7613(1978)6<267:TSCACF>2.0.CO;2.

Sears, J.W., and Price, R.A., 2003, Tightening the Siberian connection to western Laurentia: Geological Society of America Bulletin, v. 115, p. 943–953, doi: 10.1130/B25229.1.

Sears, J.W., and Ryan, P., 2003, Cenozoic evolution of the Montana Cordillera: Evidence from paleovalleys, in Raynolds, R., and Flores, J., eds., Cenozoic Paleogeography of Western US: Denver, Colorado, Rocky Mountain Section, Society of Exploration Paleontologists and Mineralogists, p. 289–301.

Sears, J.W., Schmidt, W.J., Dresser, H.W., and Hendrix, T., 1989, A geologic transect from the Highland Mountains foreland block, through the southwest Montana thrust belt, to the Pioneer batholith: Tobacco Root Geological Society Field Conference, 1989: Northwest Geology, v. 18, p. 1–20.

Sears, J.W., Jahn, B.R., Porder, S.J., Beck, M.A., Goodwin, D.H., Murphy, S.P., Schaffer, C.M., and Waddell, A.M., 1997, Kinematic test of a rotational thrust model for the Lewis-Hoadley-Eldorado plate near Wolf Creek, Montana: Northwest Geology, v. 27, p. 39–46.

Sears, J.W., Chamberlain, K.R., and Buckley, S.N., 1998, Structural and U-Pb geochronologic evidence for 1.47 Ga rifting event in the Belt Basin, western Montana: Canadian Journal of Earth Sciences, v. 35, p. 467–475, doi: 10.1139/cjes-35-4-467.

Sears, J.W., Price, R.A., and Khudoley, A.K., 2004, Linking the Mesoproterozoic Belt-Purcell and Udzha Basins across the west Laurentia-Siberia connection: Precambrian Research, v. 129, p. 291–308, doi: 10.1016/j.precamres.2003.10.005.

Sears, J.W., Braden, J., Edwards, J., Geraghty, E., Janiszewski, F., McInenly, M., McLean, J., Riley, K., and Salmon, E., 2005, Rocky Mountain foothills triangle zone, Sun River, northwest Montana, in Thomas, R., ed., Proceedings, Annual Meeting Tobacco Root Geological Society: Northwest Geology, v. 34, p. 45–70.

Sears, J.W., Khudoley, A.K., Prokopiev, A.V., Chamberlain, K.R., and MacLean, J.S., 2006, Lithostratigraphic comparison of Meso- and Neoproterozoic strata between SE Siberia and SW Laurentia, in Stone, D., ed., Contributions to Tectonics of Northeast Russia: Fairbanks, Alaska, University of Alaska, Geophysical Institute Report (CD-ROM).

Smedes, H.W., 1966, Geology and Igneous Petrology of the Northern Elkhorn Mountains, Jefferson and Broadwater Counties, Montana: U.S. Geological Survey Professional Paper 510, 116 p.

Smithson, S.B., Brewer, J.A., Kaufman, S., Oliver, J.E., and Hurich, C.A., 1979, Structure of the Laramide Wind River uplift, Wyoming, from COCORP deep reflection data and from gravity data: Journal of Geophysical Research, v. 84, p. 5955–5972.

Stewart, J.H., 1972, Initial deposits of the Cordilleran geosyncline: Evidence of late Precambrian (<850 m.y.) continental separation: Geological Society of America Bulletin, v. 83, p. 1345–1360, doi: 10.1130/0016-7606(1972)83[1345:IDITCG]2.0.CO;2.

Stickney, M.C., Haller, K.M., and Machette, M.N., 2000, Quaternary Faults and Seismicity in Western Montana: Montana Bureau of Mines and Geology, Special Publication 114. Scale 1:750,000.

Stockmal, G.S., MacKay, P.A., Lawton, D.C., and Spratt, D.A., 1996, The Oldman River triangle zone: A complicated tectonic wedge delineated by new structural mapping and seismic interpretation: Bulletin of Canadian Petroleum Geology, v. 44, no. 2, p. 202–214.

Stockmal, G.S., Beaumont, C., Nguyen, M., and Lee, B., 2007, this volume, Mechanics of thin-skinned fold-and-thrust belts: Insights from numerical models, in Sears, J.W., Harms, T.A., and Evenchick, C.A., eds., Whence the Mountains? Inquiries into the Evolution of Orogenic Systems: A Volume in Honor of Raymond A. Price: Geological Society of America Special Paper 433, doi: 10.1130/2007.2433(04).

Strayer, L.M., IV, Hyndman, D.W., Sears, J.W., and Myers, P.E., 1989, Direction and shear sense during suturing of the Seven Devils/Wallowa terrane: Western Idaho: Geology, v. 17, p. 1025–1028, doi: 10.1130/0091-7613(1989)017<1025:DASSDS>2.3.CO;2.

Surkov, V.S., Grishin, M.P., Larichev, A.I., Lotyshev, V.I., Melnikov, N.V., Kontorovich, A.E.H., Trofimuk, A.A., and Zolotov, A.N., 1991, The Riphean sedimentary basins of the eastern Siberia province and their petroleum potential: Precambrian Research, v. 54, p. 37–44, doi: 10.1016/0301-9268(91)90067-K.

Turner, R.J.W., Leitch, C.H.B., and Delaney, G., 2000, Syn-rift structural controls on the paleoenvironmental setting and evolution of the Sullivan orebody, Chapter 32, in Lydon, J.W., Höy, T., Slack, J.F., and Knapp, M.E., eds., The Geological Environment of the Sullivan Deposit, British Columbia: Geological Association of Canada, Mineral Deposits Division (MDD) Special Publication No. 1, p. 582–616.

van der Velden, A.J., and Cook, F.A., 1996, Structure and tectonic development of the southern Rocky Mountain Trench: Tectonics, v. 15, p. 517–544, doi: 10.1029/95TC03288.

Van Schmus, W.R., and Bickford, M.E., 1993, Transcontinental Proterozoic provinces, *in* Reed, J.C., et al., eds., Precambrian: Continental U.S.: Boulder, Colorado, Geological Society of America, Geology of North America, v. C-2, p. 171–334.

Viele, G.W., and Harris, F.G., 1965, Montana Group stratigraphy, Lewis and Clark County, Montana: American Association of Petroleum Geologists Bulletin, v. 49, p. 379–417.

Wallace, C.A., Harrison, J.E., and Lidke, C.J., 1999, Lithofacies of the Helena and Wallace Formations (Belt Supergroup, middle Proterozoic) Montana and Idaho, *in* Berg, R.B., ed., Belt Symposium III Abstracts: Montana Bureau of Mines and Geology Special Publication 381, p. 61–64.

Warren, M.J., 1997, Tectonic significance of stratigraphic and structural contrasts between the Purcell anticlinorium and the Kootenay arc, Duncan Lake area, British Columbia [Ph.D. thesis]: Kingston, Ontario, Canada, Queen's University, 316 p.

White, B.G., 1997, Diverse tectonism in the Coeur d'Alene mining district, Idaho, *in* Berg, R., ed., Proceedings of Belt Symposium III: Montana Bureau of Mines and Geology Special Paper 112, p. 254–265.

White, B.G., 2006, Collection and concentration of stratabound metals during progressive reduction of oxidized Revett strata: A 180 degree variation of the redbed source model: Northwest Geology, v. 35, p. 49–56.

Willis, B., 1902, Stratigraphy and structure of the Lewis and Livingston Ranges, Montana: Geological Society of America Bulletin, v. 14, p. 69–84.

Winston, D., 1986a, Sedimentation and tectonics of the Middle Proterozoic Belt Basin and their influence on Phanerozoic compression and extension in western Montana and northern Idaho, *in* Peterson, J.A., ed., Sedimentation and Tectonics: American Association of Petroleum Geologists Memoir 41, p. 87–118.

Winston, D., 1986b, Stratigraphic correlation and nomenclature of the Middle Proterozoic Belt Supergroup, Montana, Idaho, and Washington, *in* Roberts, S.M., ed., Belt Supergroup: Montana Bureau Mines and Geology Special Paper 94, p. 69–84.

Winston, D., 1986c, Sedimentology of the Ravalli Group, Middle Belt Carbonate, and Missoula Group, Middle Proterozoic Belt Supergroup, Montana, Idaho, and Washington, *in* Roberts, S.M., ed., Belt Supergroup: Montana Bureau Mines and Geology Special Paper 94, p. 85–124.

Winston, D., 1991, Evidence for intracratonic, fluvial, and lacustrine settings of Middle to Late Proterozoic basins of western U.S.A., *in* Gower, C.F., Rivers, T., and Ryan, B., eds., Mid-Proterozoic Laurentia-Baltica: Geological Association of Canada Special Paper 38, p. 535–564.

Winston, D., 2007, Revised stratigraphy and depositional history of the Helana and Wallace Formations, Mid-Proterozoic Piegan Group, Belt Supergroup, Montana and Idaho, *in* Link, P.K., and Lewis, R.S., eds., Proterozoic geology of western North America and Siberia: SEPM (Society for Sedimentary Geology) Special Publication No. 86, p. 65–100.

Winston, D., and Wheeler, T., 2006, Shepard, Mount Shields, and Bonner Formations of the Middle Proterozoic Missoula Group in Libby syncline, Kootenai Falls, northwest Montana: Northwest Geology, v. 35, p. 63–76.

Wooden, J.L., Vitaliano, C.J., Koehler, S.W., and Ragland, P.C., 1978, The late Precambrian mafic dikes of the southern Tobacco Root Mountains, Montana: Geochemistry, Rb-r geochronology, and relationship to Belt tectonics: Canadian Journal of Earth Sciences, v. 15, p. 467–479.

Yin, A., and Oertel, G., 1995, Strain analysis of the Ninemile fault zone, western Montana: Insights into multiply deformed regions: Tectonophysics, v. 247, p. 133–143, doi: 10.1016/0040-1951(95)00012-C.

Zen, E-an, 1988, Bedrock Geology of the Vipond Park 15-Minute, Stine Mountain 7 1/2-Minute, and Maurice Mountain 7 1/2-Minute Quadrangles, Pioneer Mountains, Beaverhead County, Montana: U.S. Geological Survey Bulletin 1625, 49 p.

Zonenshain, L.P., Kuzmin, M.I., and Natapov, L.M., 1990, Geology of the USSR: A plate-tectonic synthesis: American Geophysical Union Geodynamic Monograph 21, 242 p.

MANUSCRIPT ACCEPTED BY THE SOCIETY 22 MARCH 2007

The Geological Society of America
Special Paper 433
2007

Thermochronometric reconstruction of the prethrust paleogeothermal gradient and initial thickness of the Lewis thrust sheet, southeastern Canadian Cordillera foreland belt

Shimon Feinstein*
Department of Geological and Environmental Sciences, Ben Gurion University of the Negev, Beer Sheva 84105, Israel

Barry Kohn
School of Earth Sciences, University of Melbourne, Victoria 3010, Australia

Kirk Osadetz
Natural Resources Canada, Earth Sciences Sector, Geological Survey of Canada–Calgary, 3303 33rd St. NW, Calgary, Alberta T2L 2A7, Canada

Raymond A. Price
Department of Geological Sciences and Geological Engineering, Queen's University, Kingston, Ontario K7L 3N6, Canada

ABSTRACT

The Lewis thrust, which is >225 km long and has a maximum displacement of >80 km, is a major Foreland belt structural element in the southeastern Canadian Cordillera. We use low-temperature thermochronometry in the preserved Lewis thrust sheet stratigraphic succession to constrain variations in both paleogeothermal gradient and Lewis thrust sheet thickness immediately prior to motion on the Lewis thrust fault. Fission-track and vitrinite reflectance data combined with stratigraphic data suggest that maximum Phanerozoic burial and heating occurred in the Lewis thrust sheet during a short interval (<15 m.y.) in late Campanian time immediately prior to thrusting (ca. 75 Ma). The data suggest that the late predeformational Lewis thrust sheet paleogeothermal gradient was between ~18 and 22.5 °C/km, which is higher than that inferred for subsequent syn- and postdeformational intervals by other studies. The inferred paleotemperatures and geothermal gradients indicate that the preserved Lewis thrust sheet stratigraphic succession was overlain by ~4–5.5 km of additional Late Cretaceous strata that were subsequently removed by erosional denudation. We estimate that the Lewis thrust sheet was ~12–13.5 km thick when

---

*shimon@bgu.ac.il

Feinstein, S., Kohn, B., Osadetz, K., and Price, R.A., 2007, Thermochronometric reconstruction of the prethrust paleogeothermal gradient and initial thickness of the Lewis thrust sheet, southeastern Canadian Cordillera foreland belt, in Sears, J.W., Harms, T.A., and Evenchick, C.A., eds., Whence the Mountains? Inquiries into the Evolution of Orogenic Systems: A Volume in Honor of Raymond A. Price: Geological Society of America Special Paper 433, p. 167–182, doi: 10.1130/2007.2433(08). For permission to copy, contact editing@geosociety.org. ©2007 The Geological Society of America. All rights reserved.

thrusting commenced. Deposition of the Late Cretaceous succession was terminated by the onset of displacement on the Lewis thrust (ca. 75 ± 5 Ma) and was followed by intervals of erosional denudation that are constrained stratigraphically by both early Oligocene and current erosion surfaces on the Lewis thrust sheet.

Keywords: vitrinite reflectance, zircon fission-track dating, Lewis thrust sheet, Canadian Cordillera, fold-and-thrust belt.

INTRODUCTION

Reconstruction of burial depth and thermal history are important steps for both tectonic modeling and economic prospect assessments, particularly related to petroleum exploration. However, the evidence required for geohistory reconstruction is often severely restricted because of extensive erosional denudation, especially in strongly deformed terrains. Nevertheless, a "memory" of the paleoburial and its associated thermal history may be preserved in the rocks below the missing section, and this "memory" may be retrievable from various paleothermometric and paleobarometric parameters (e.g., Feinstein et al., 1996). In this study, we use paleothermometric constraints provided by both vitrinite reflectance (VR) in the Lower Cretaceous Mist Mountain Formation (Kootenay Group) and zircon fission-track (ZFT) data primarily from the Mesoproterozoic Grinnell and Appekunny Formations (Purcell Supergroup) to reconstruct the paleogeothermal gradient and thickness of the Lewis thrust sheet in the southern Canadian Cordilleran Foreland belt (Figs. 1 and 2) just prior to initiation of displacement on the Lewis thrust fault.

VR is a maximum-temperature paleothermometric tool (Barker, 1989; Feinstein et al., 1992). The time and temperature of maximum coalification can be constrained using the variations in VR with stratigraphic depth and the pattern of VR variations relative to deformed stratigraphic markers (Teichmuller and Teichmuller, 1966; Hacquebard and Donaldson, 1974; Cook and Kantsler, 1980), whereas the maximum temperature can be inferred from time-temperature (t-T) kinetic VR models (Burnham and Sweeney, 1989). VR has been described previously throughout the study region (Hacquebard and Donaldson, 1974; England, 1984; Kalkreuth and McMechan, 1984; Pearson and Grieve, 1985; England and Bustin, 1986; Grieve, 1987; Bustin and England, 1989; Bustin, 1991).

Zircon fission-track (ZFT) dating is applicable in this study because at sufficiently high temperatures and over long enough time spans, all existing fission tracks in zircon will gradually shorten (anneal) and eventually disappear, effectively resetting the fission-track clock (e.g., Gallagher et al., 1998). The wide temperature interval over which most of the gradual annealing or fading of fission tracks occurs for a specific mineral led to the concept of a partial annealing zone or partial stability zone (Wagner, 1972). Unlike lower-temperature apatite fission-track (AFT) annealing, there is no clear consensus regarding the temperature range for the zircon partial annealing zone, partly because no samples have been recovered and analyzed from the base of the present zircon partial annealing zone. Further, it has been shown that the zircon partial annealing zone temperature range (and effective closure temperature) varies with cooling rate and degree of α-radiation damage (e.g., Brandon et al., 1998; Rahn et al., 2004). Tagami et al. (1998) and Tagami (2005) summarized geological constraints on ZFT annealing, including the studies of the effects of both distance from intrusive contacts (Tagami and Shimada, 1996) and variations in deep borehole profiles (Zaun and Wagner, 1985; Green et al., 1996; Tagami et al., 1996), which can be compared to laboratory kinetic annealing models (Yamada et al., 1995). The base of the zircon partial annealing zone is inferred to occur at ~350 °C and 325 °C for heating times on the order of 10^6 and 10^7 yr, respectively. Brix et al. (2002) inferred higher temperatures (e.g., ~350–400 °C for heating of 4 ± 2 m.y.). Brandon et al. (1998) suggested a lower temperature limit of ~180–250 °C for the zircon partial annealing zone for grains with α-radiation damage when heating intervals are ~10^7 yr. However, such grains, with a considerable degree of α-radiation damage, are also the most prone to annealing due to their decreased track thermal stability (Kasuya and Naeser, 1988). Many grains collected and investigated in this study were metamict, and these were unsuitable for fission-track analysis. We therefore restricted our analysis to the grains exhibiting the least α-radiation damage and used temperature limits for the zircon partial annealing zone suitable for the low α-radiation damage model (Tagami, 2005).

GEOLOGICAL SETTING

The Canadian Rocky Mountain foreland fold-and-thrust belt (Gabrielse and Yorath, 1992) is a northeastward-tapering deformational belt. It consists of Mesoproterozoic, Paleozoic, and Mesozoic strata from several different depositional basins that were scraped off the under-riding North American craton and accreted to the over-riding Intermontane terrane during Late Jurassic to Paleocene convergence between them (Monger and Price, 1979; Price, 1981; McMechan and Thompson, 1989; Price, 1994). The geological structure within the deformational belt is dominated by thrust faults, most of which are listric and northeasterly or easterly verging. The thrust faults commonly follow long bedding-parallel detachments that are separated by ramps along which the faults change stratigraphic level rapidly (Douglas, 1950; Bally et al., 1966; Dahlstrom, 1970; Fermor, 1999). The thrusts merge downward with a basal detachment that converges with the contact between the sedimentary rocks

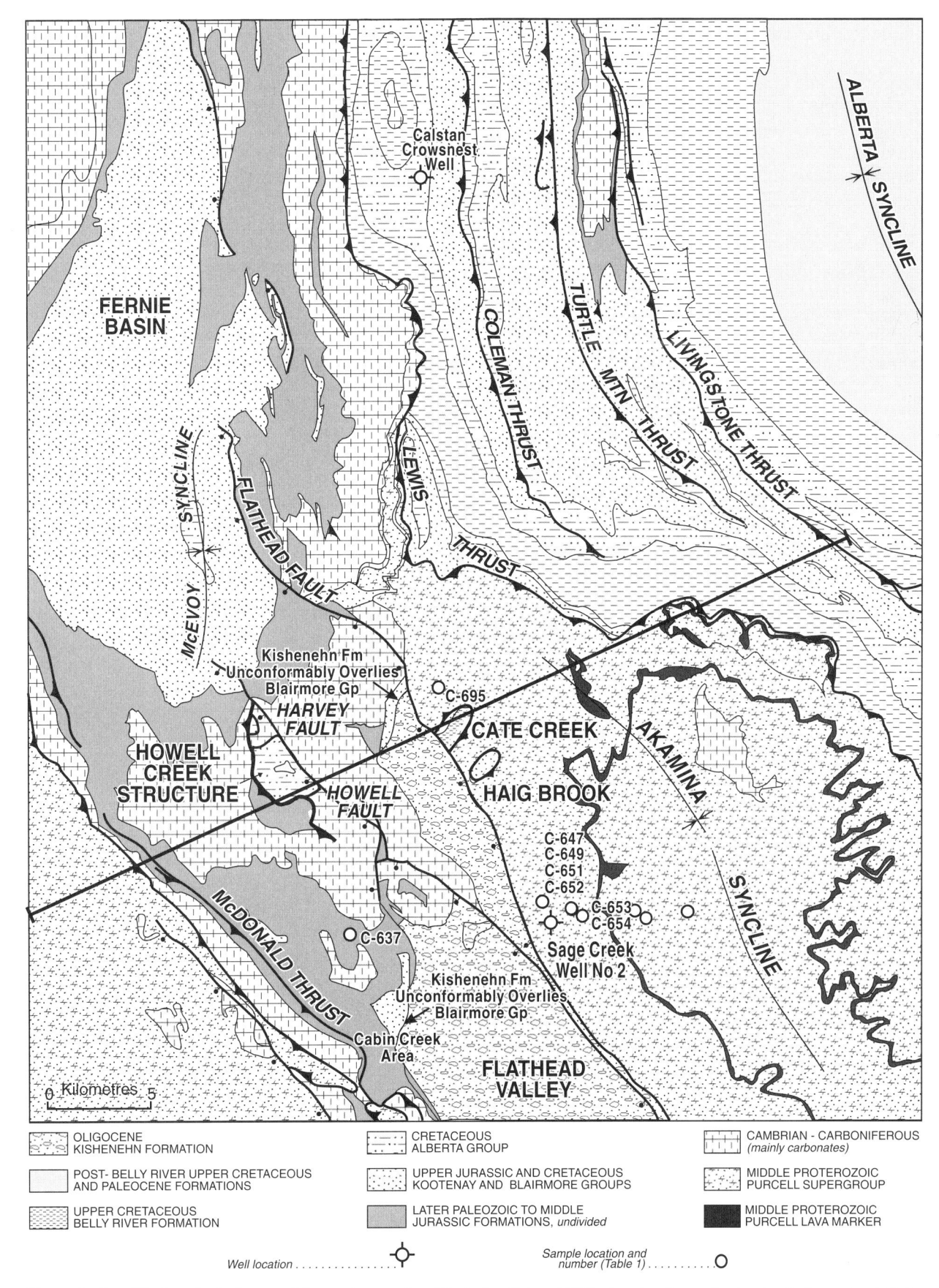

Figure 1. Simplified geological map of the study area (modified after Price, 1962), showing the main structural elements, vitrinite reflectance (VR) and zircon fission-track (ZFT) sample locations, and the location of the cross section in Figure 2.

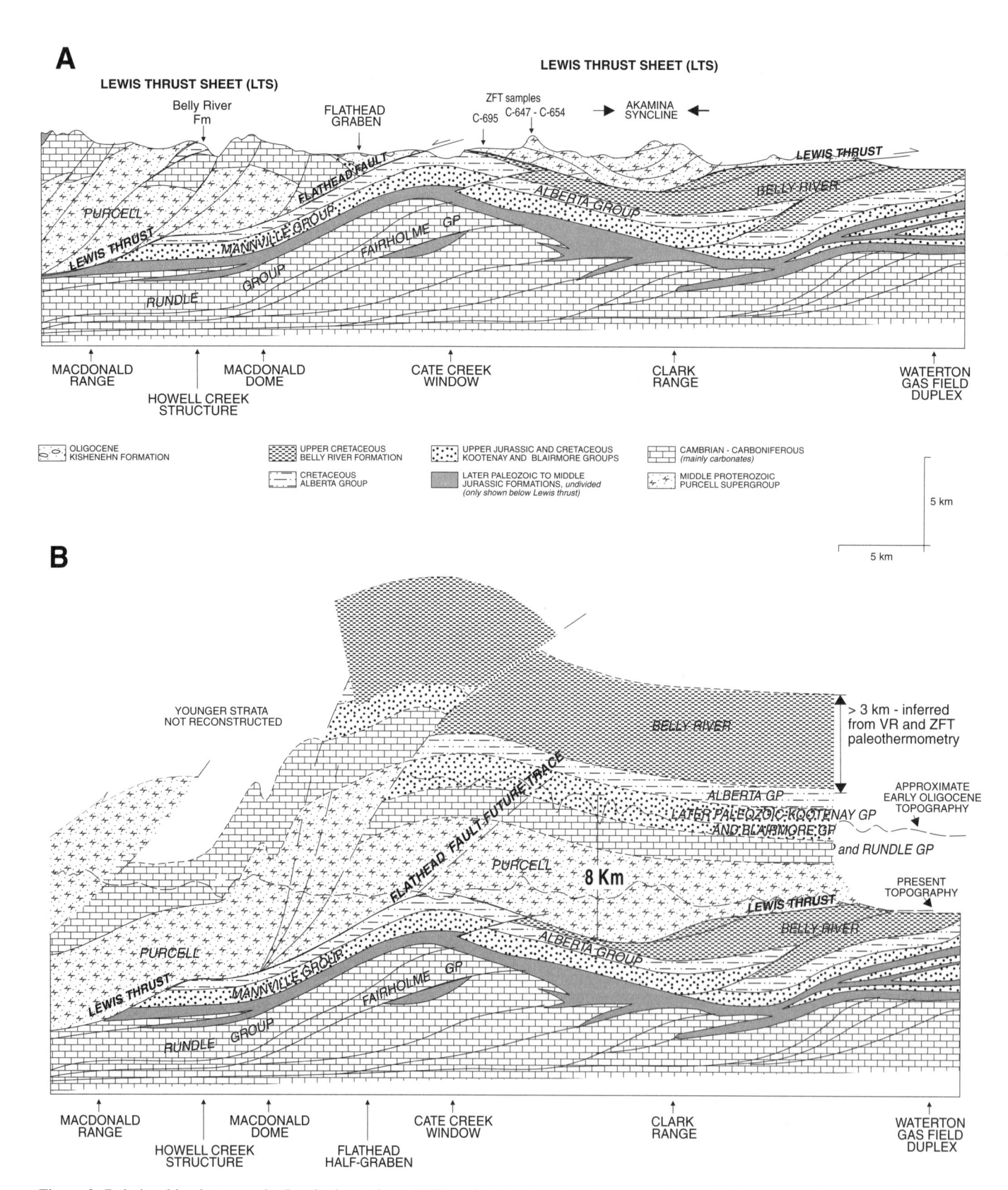

Figure 2. Relationships between the Lewis thrust sheet (LTS) and the Flathead listric normal fault: (A) Present-day balanced E-W structural section (from Osadetz et al., 2004; modified after Fermor and Moffat, 1992; for location, see Fig. 1); and (B) the same section restored to remove the effects of Paleogene displacement on the Flathead normal fault (from Osadetz et al., 2004). Inferred thicknesses of eroded strata and the inferred early Oligocene erosion surface are discussed in the text.

and the underlying crystalline basement. The westward-dipping basal detachment extends into the Cordilleran metamorphic core at mid-crustal levels (e.g., Price and Fermor, 1985). The thrust faults generally cut up through the stratigraphic succession in the direction of displacement of the hanging wall, and they generally juxtapose older strata over younger, causing stratigraphic repetitions. The supracrustal cover has been vertically thickened and horizontally shortened.

The total horizontal shortening across the foreland fold-and-thrust belt is localized mainly on a small number of major thrust faults with large displacement. One of the largest of these faults is the Lewis thrust (Figs. 1 and 2). The Lewis thrust is >225 km long; it has a maximum displacement of >80 km; and it has juxtaposed a >8-km-thick Mesoproterozoic to Cretaceous succession, which is the subject of this study, over Cretaceous and older strata (Price, 1964, 1965; Mudge and Earhart, 1980; Fermor and Moffat, 1992; Sears, 2001). Using apatite fission-track thermochronology, Osadetz et al. (2004) inferred that major motion on the Lewis thrust fault occurred at ca. 75 Ma. Their result was consistent with the characteristics of a synorogenic, but locally predeformational chemical remagnetization fabric in Paleozoic carbonates of the Lewis thrust sheet and other Front Range and Foothills structures (Enkin et al., 2000). Fault-bend, fault-propagation, and detachment folding occurred in conjunction with the thrusting. The amount of detachment folding varies along strike with stratigraphically controlled changes in mechanical properties of the stratigraphic succession. In the southern Foreland belt, folds make a relatively small contribution to the overall horizontal shortening; however, many thrust faults are folded with underlying strata as a result of thrust-related folding. Thrusting and folding increased the thickness within the northeastward-tapering wedge of supracrustal rocks by >10 km along the western flank of the North American craton. The resulting tectonic load produced a lithospheric flexure in which up to 8 km of Upper Jurassic, Cretaceous, and Paleocene foreland basin sediments accumulated concurrently with the thrusting (Price, 1973; Beaumont, 1981; Mack and Jerzykiewicz, 1989; Peper, 1993). Older foreland basin deposits were incorporated into the fold-and-thrust belt as it expanded toward the craton.

Late Cretaceous and Paleocene thrusting in the southern Canadian Rocky Mountains Foothills and eastern Front Ranges was associated with oblique, right-hand convergence between the Intermontane terrane and the North American craton (Price, 1994). In the south-central Canadian Cordillera, the oblique right-hand transpression was replaced by Late Paleocene to Middle Eocene right-hand transtension (Ewing, 1980; Price, 1979, 1994) involving conspicuous east-west crustal extension and the tectonic exhumation of mid-crustal metamorphic rocks, including Archean and Paleoproterozoic basement rocks (Carr et al., 1987; Parrish et al., 1988; Doughty and Price, 1999, 2000). This change from regional horizontal compression to regional horizontal extension has been inferred to mark the end of both thrusting and folding in the Foreland belt and the flexural subsidence of the foreland basin (Price, 1994).

Late Paleocene–middle Eocene and younger crustal extension and the uplift of mid-crustal metamorphic core complexes in the Omineca belt (Carr, 1992; Lorencak et al., 2001; Doughty and Price, 1999, 2000; Vanderhaeghe et al., 2003) were accompanied by westward increasing erosional exhumation of the Foreland fold-and-thrust belt and the adjacent margin of the undeformed Interior Platform. The thickness of strata eroded increases westward from ~2–3 km at the eastern edge of the fold-and-thrust belt (Magara, 1976; Hacquebard, 1977; Nurkowski, 1984; England and Bustin, 1986; Issler et al., 1990; Majorowicz et al., 1990), to >8 km in the uplifted footwall of the Flathead normal fault (Figs. 1 and 2). Farther west, in the metamorphic core complexes of the Omineca crystalline belt, isostatic uplift of mid-crustal rocks is commonly >25 km, and the basal detachment of the Cordilleran Foreland fold-and-thrust belt is exposed locally at >2 km above sea level (Parrish et al., 1988; Cook et al., 1992).

Upper Cretaceous strata occur in the footwall of the Lewis thrust along most of the eastern edge of the thrust sheet in southwestern Alberta and adjacent Montana (Price, 1962; Mudge and Earhart, 1980) and also in windows eroded through the Lewis thrust adjacent to the Flathead fault at Cate Creek and Haig Brook in southeastern British Columbia (Figs. 1 and 2; Price, 1962). Thus, in the area north and east of the Cate Creek and Haig Brook windows, the footwall of the Lewis thrust evidently follows a detachment zone within Campanian-Maastrichtian strata (Fermor and Price, 1987; Fermor and Moffat, 1992; Jerzykiewicz et al., 1996). Northeast of the Flathead fault, the Lewis thrust fault and the strata in its hanging wall have been deformed into a large fault-bend fold (the Akamina syncline) by two antiformally stacked, thrust duplex culminations involving underlying Paleozoic carbonate rocks (Bally et al., 1966; Yin and Kelty, 1991; Boyer, 1992). The duplex structures contain large petroleum accumulations (Fermor and Moffat, 1992). The Akamina syncline, the duplex culminations, and the entrapment of the petroleum within them are inferred to postdate the main displacement on the Lewis thrust. These relationships mean that the thermal history of the Lewis thrust sheet is directly relevant to the analysis of foreland petroleum occurrences; moreover, they provide a relative sequence of events leading to the formation of the petroleum pools.

In the Fernie area (Fig. 1), stratigraphic constraints on the timing of the displacement on the Lewis thrust are broad. Displacement began after the deposition of the marine shales and sandstones of the Cenomanian to early Campanian Alberta Group, which is the youngest stratigraphic unit in the hanging wall of the Lewis thrust (Fig. 1, in McEvoy syncline) as well as in the footwall (Price 1962, 1965); some of the displacement occurred after the deposition of the late Campanian–early Maastrichtian Bearpaw Formation, which is the youngest stratigraphic unit that is cut by the Lewis thrust (Jerzykiewicz et al., 1996); and all of the displacement occurred prior to the displacement on the Flathead normal fault and the resulting deposition of the late Eocene–early Oligocene Kishenehn Formation, which unconformably overlies deformed Lower Cretaceous to Upper Paleozoic strata of the

Lewis thrust sheet in the Flathead graben (Price, 1962, 1965). However, the upper limit on time of displacement on the Lewis thrust is also constrained by the regional change in the southern Canadian Cordillera from the thrusting and folding that was associated with dextral transpression between North America and outboard accreted terranes to the crustal extension and exhumation of metamorphic core complexes that was associated with dextral transtension (Price, 1994). On the basis of U-Pb dating of zircons from various deformed and crosscutting mid-crustal granitic rocks in south-central British Columbia, Carr (1992) concluded that this change occurred at ca. 59 Ma. Moreover, in northern Washington, Idaho, and adjacent Montana, in the area in which the Mesoproterozoic Belt-Purcell rocks of the Lewis thrust sheet of southern Canada were deposited (Price and Sears, 2000), the transition from thrusting and folding to crustal stretching resulted in diachronous, rapid cooling of the Priest River metamorphic core complexes during their progressive extensional exhumation; contours of K-Ar and ^{40}Ar/^{39}Ar cooling dates from biotite show that the cooling occurred between >55 Ma and ca. 40 Ma (Doughty and Price, 2000). These constraints imply that all of the displacement on the Lewis thrust, and also on all the younger thrusts that developed below the Lewis thrust sheet and have deformed it, occurred between ca. 75 Ma and 59 Ma. These limits on time of displacement on the Lewis thrust in southern Canada are consistent with several indications for the age of significant motion on the Lewis thrust, including the limits established near the south end of the Lewis thrust in northwestern Montana, provided by radiometric dating of prethrusting volcanic rocks and a post-thrusting dike (Sears, 2004)—74 Ma and 59 Ma, respectively—and the rapid, profound cooling of Lewis thrust sheet rocks ca. 75 Ma inferred from AFT thermochronology (Osadetz et al., 2004).

Vrolijk and van der Pluijm (1999) and van der Pluijm et al. (2001) obtained a K-Ar age of 51.5 ± 3.5 Ma for the authigenic component in a mixed detrital-authigenic clay in the gouge along the Lewis thrust from a locality ~40 km north of the area of our study. They inferred that the authigenic illite had formed during thrust displacement and, therefore, that it dated displacement on the Lewis thrust. Because this date for displacement on the Lewis thrust is anomalous in that it is substantially younger than dates they obtained from the underlying, more easterly Rundle and McConnell thrusts, they made the ad hoc suggestion that the Lewis thrust is an out-of-sequence thrust (Vrolijk and van der Pluijm, 1999). The fact that the Lewis thrust has been folded as a result of displacement on underlying younger thrusts, as illustrated, for example, in Figure 2, demonstrates that it is not an out-of-sequence thrust. Moreover, the K-Ar "age" may record the time of crystallization of authigenic illite in the fault gouge, but, in the absence of evidence that the illite crystallized during and not subsequent to thrust displacement, it does not date the time of emplacement of the Lewis thrust.

The study area is unique because of the preservation of a Paleogene erosion surface beneath the late Eocene–early Oligocene and younger fanglomerates and tilted lacustrine and fluvial strata of the Kishenehn Formation, which accumulated on the hanging wall during displacement on the Flathead fault (Price, 1962; McMechan and Price, 1980; McMechan, 1981; Constenius, 1996). The Flathead fault is a listric southwest-dipping normal fault with an estimated maximum dip slip of ~15 km (Constenius, 1988). It cuts and offsets the Lewis thrust sheet, but it merges with the underlying Lewis thrust fault without offsetting the strata in its footwall (Bally et al., 1966). The Kishenehn strata form a northeastward-thickening wedge on the flank of a rollover anticline in the hanging wall of the Flathead fault. They were tilted to the northeast during the listric displacement on the Flathead fault (Fig. 2; Price, 1962; McMechan, 1981; Constenius, 1996). Lower Cretaceous strata of the Blairmore Group are preserved beneath the unconformity at the base of the Kishenehn Formation (Fig. 3). This shows that in late Eocene time, when the Kishenehn Formation began to accumulate on the hanging wall of the Flathead fault, the Lewis thrust sheet was still stratigraphically at least 7 km thick in the vicinity of the Akamina syncline. East of the Flathead fault, where petroleum exploration boreholes penetrate stacked thrust sheets of Upper Cretaceous and older Phanerozoic strata below the Lewis thrust, the thrust-faulted Mesozoic and Paleozoic succession is stratigraphically ~4 km thick and is similar to that in the vicinity of the Waterton gas field thrust duplex (Fig. 2). Fault-bounded Upper Cretaceous strata occur at one locality within the Lewis thrust sheet southwest of the Flathead fault, in the vicinity of Howell Creek (Fig. 2). These Upper Cretaceous strata have been interpreted as (1) a part of the footwall of the Lewis thrust that has been displaced by a minor out-of-sequence thrust that cuts the Lewis thrust, and by later offset by normal faults associated with the Flathead fault (Price, 1962, 1965); (2) part of the Lewis thrust sheet that occurs in the footwall of a higher thrust fault and has been down-dropped by a Tertiary normal fault (Lebreque and Shaw, 1973); and (3) large rock slides (Jones, 1977).

By combining the section of the Lewis thrust sheet that is exposed beneath the Kishenehn Formation in the hanging wall of the Flathead normal fault with the section of the Lewis thrust sheet that is exposed above the Cate Creek and Haig Brook windows in the footwall of the Flathead fault, and with the section below the Lewis thrust fault that has been penetrated by petroleum exploration wells, it is possible to sample a section ~12 km thick through the Foreland fold-and-thrust belt as it existed prior to displacement on the Flathead normal fault (Fig. 2).

ANALYTICAL METHODS

Vitrinite Reflectance (VR)

Samples of coal were ground to 850 μm and mounted in epoxy resin. The particulate blocks were polished according to the ICCP (1971) recommendation and examined under a Zeiss MPM II microscope fitted with white (halogen) and fluorescent (HBO 100) light sources. Maximum reflectance measurements (% Ro$_{max}$) were taken on vitrinite using an oil-immersion objective (N.A. 40/0.90; n$_{oil}$ = 1.518 at 546 nm).

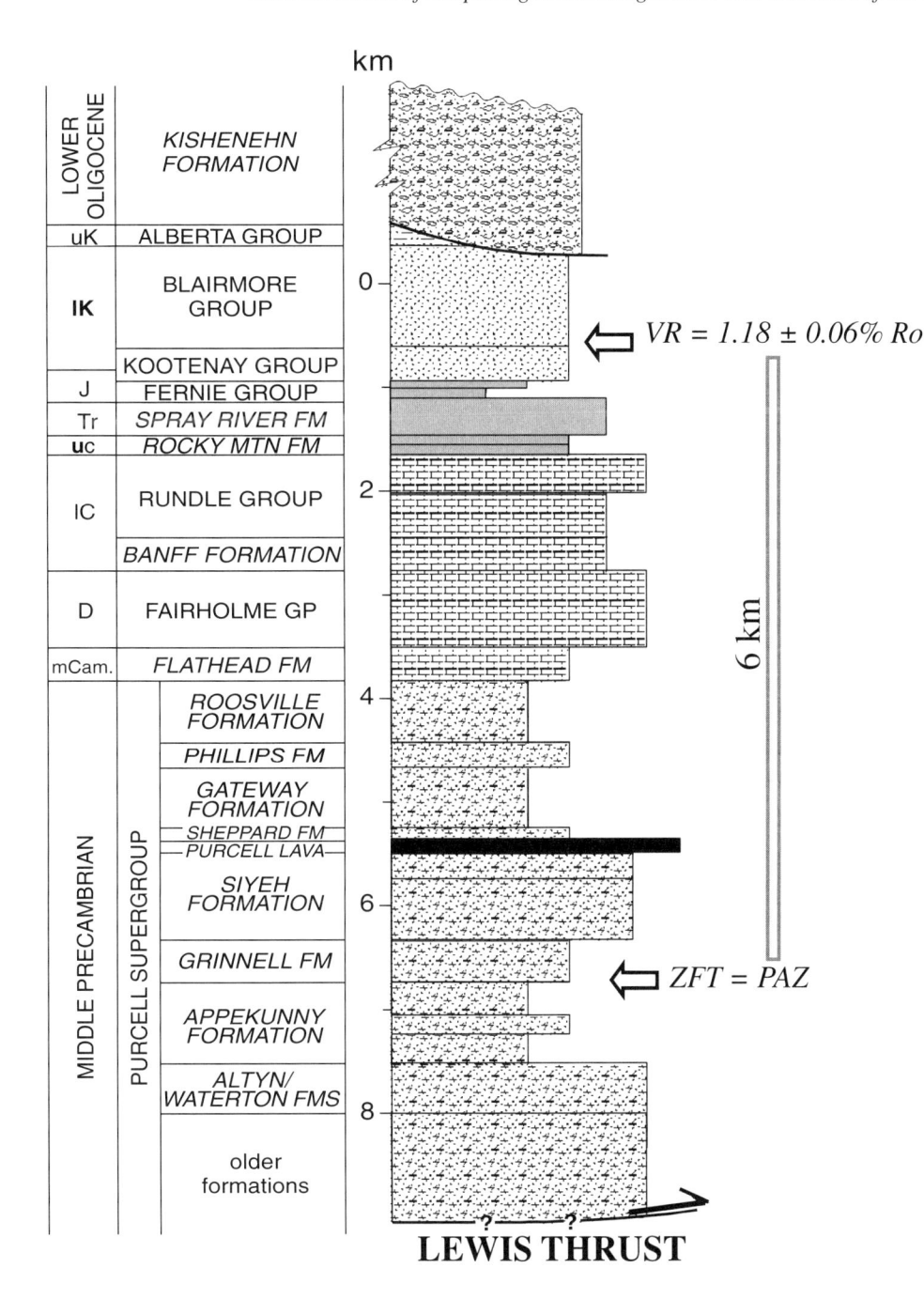

Figure 3. Composite columnar stratigraphic section of the Mesoproterozoic to Upper Cretaceous succession that is preserved in the Lewis thrust sheet (modified after Osadetz et al., 2004). Also shown are the stratigraphic positions of the Mist Mountain Formation (Kootenay Group) and the Grinnell and Appekunny Formations (Purcell Supergroup), from which, respectively, VR and ZFT data were obtained, and the maximum 6 km of vertical stratigraphic separation between them in the reconstructed section. PAZ—partial annealing zone, VR—vitrinite reflectance, ZFT—zircon fission track; uK—upper Cretaceous, lK—lower Cretaceous, J—Jurassic, Tr—Triassic, uC—upper Carboniferous, lC—lower Carboniferous, D—Devonian, mCam—middle Cambrian.

ZFT

Samples were crushed, and heavy minerals were concentrated on a Wilfley Table. Zircon grains in the 63–250 μm size range were separated using standard magnetic and heavy liquid techniques. Zircon concentrates were mounted in PFA Teflon discs, ground, polished to an optical finish, and etched in a eutectic KOH-NaOH melt at 220 °C (Gleadow et al., 1976). In order to obtain a high-quality etch, each sample required a different etching period, which ranged from 5 to 23 h. Neutron irradiations were carried out in the well-thermalized (high thermal/fast neutron flux ratio) RT-4 facility of the National Institute of Standards and Technology (NIST, formerly National Bureau of Standards) reactor at Gaithersburg, Maryland. Fission-track ages were measured using the external detector method, and Brazil Ruby muscovite was used to record induced tracks (Gleadow, 1981). The muscovite detectors were etched for 30 min in 48% HF at room temperature to reveal the induced tracks. Thermal neutron fluences were monitored by measuring the track density in muscovites attached to NIST SRM-962a (formerly NBS-612). Counting was carried out with a transmitted light polarizing microscope using a dry 80× objective at a total magnification of 1250×. Only tracks on grains with sharp polishing scratches were counted. Ages were calculated using the zeta calibration method, following procedures described by Hurford and Green (1983). To correct for the difference between track registration geometries of zircon internal surfaces and the external muscovite detector, a factor of 0.5 was used (Gleadow and Lovering, 1977). Errors were calculated using the "conventional method" of Green (1981) and are expressed as one standard deviation (Table 1).

RESULTS

VR measurements obtained from the Lower Cretaceous Mist Mountain Formation of the Kootenay Group in the Lewis thrust sheet, sampled at Cabin Creek and Sage Creek (Fig. 1) are summarized in Table 2. The VR obtained for the Cabin Creek sample was ~1.18% mean Ro_{max}, with a 0.059 standard deviation (Table 2). This VR is consistent with observations of 1.1%–1.25% Ro_{max} from similar stratigraphic levels in the Kootenay Group in the adjacent Fernie Basin and Elk Valley area (Pearson and Grieve, 1985).

ZFT data obtained from a limited number of grains from seven samples of the Mesoproterozoic Grinnell and Appekunny Formations of the Purcell Supergroup and a single Upper Paleozoic Rocky Mountain Formation sample from different stratigraphic, topographic, and structural positions within the Lewis thrust sheet are summarized in Table 1. The sample from Paleozoic strata was collected from the hanging wall of the Flathead normal fault, while the Mesoproterozoic samples are from the footwall of Flathead fault. The difference in sample structural and stratigraphic setting is not significant for our study, since our analysis pertains only to predeformational time intervals and geothermal environments.

Analytical data for each sample are presented in Table 1, where all grain data for a particular sample are expressed as the central age, which represents essentially a weighted mean of individual single-grain ages (Galbraith and Laslett, 1993). Central ZFT ages obtained from the Purcell Supergroup samples ranged between 184 ± 17 Ma and 486 ± 61 Ma, while the sample from Upper Paleozoic strata was dated at 184 ± 12 Ma (Table 1). Also included in Table 1 are the results of the chi-square test, which detects the presence of uncertainty additional to that allowed by Poissonian variation (because the probability of spontaneous fission occurring is a random process) in track counts (Galbraith, 1981). The test thus provides some measure of the dispersion of count data and evaluates the presence of multiple populations in single-grain ages. For data presented in Table 1, all samples (except for sample 651) pass the chi-square test, i.e., >5%, suggesting that for most samples, the Poissonian distribution is the only source of variation in the overall sample age (Galbraith, 1981).

Sample 637, from the Upper Paleozoic Rocky Mountain Formation, is the only sample that has a significant number of tracks suitable for horizontal confined fission-track length measurement. In that sample, 22 length measurements, with a mean of ~8.23 ± 0.37 µm and a standard deviation (SD) of 1.74, show a bimodal distribution (Table 1). The tracks suitable for length measurement in the other zircon samples from the Purcell Supergroup (Table 1) were too few for robust length analysis, but those observed were of moderate length (~8–9 µm), similar to sample 637. While sample 637 represents the "best" individual data regarding zircon partial annealing, we used the largest stratigraphic interval over which subsequent estimates of paleogeothermal gradient and maximum stratal thickness could be estimated, as provided by the combination of VR data in Lower Cretaceous strata and ZFT data in Mesoproterozoic strata.

In Figure 4, using a radial plot, we graphically display a mixing model for the Mesoproterozoic Grinnell and Appekunny Formation zircons analyzed for this study. Single-grain ZFT ages of the 31 grains from Purcell Supergroup samples ranged between 144 Ma to 641 Ma (Fig. 4). If the measured single-grain fission-track ages from a sedimentary sample, like the ones studied here, are detrital ages, they may be composed of multiple populations derived from different source areas that have contributed to the sedimentary mix. In order to resolve a mixture of detrital grains in a given data set into significant populations, we used a statistical approach called "mixture modeling" (Sambridge and Compston, 1994). This approach is also useful for defining distinct components of fission-track ages in partially annealed samples. In such samples, the degree of annealing of fission tracks in each grain depends on the inherited variation of grain chemistry, provenance history, and post-depositional thermal history, and may result in a wide spread in single-grain ages. The youngest population identified provides an estimate of the maximum time at which cooling occurred from maximum paleotemperatures. For these 31 grains, from 7 samples, the zircons exhibit two distinct populations with mean ages of 183 ± 13 Ma and 341 ± 24 Ma, respectively (Fig. 4).

DISCUSSION

Constraints on the Time of Maximum Paleotemperatures

The time of maximum paleotemperatures, paleogeothermal gradient, and maximum burial depth of the Lewis thrust sheet succession can be constrained by spatial and temporal distribution of AFT, ZFT, and VR data. Time-temperature models based on AFT data from various Mesoproterozoic and Phanerozoic strata indicate cooling of the Lewis thrust sheet at 75 ± 5 Ma from >110 °C. Subsequent cooling history varied between the hanging wall and footwall of the Flathead normal fault (Osadetz et al., 2004). AFT time-temperature (t-T) models show that samples from the hanging wall of the Flathead fault, including clasts in conglomerate of the Upper Eocene–Lower Oligocene Kishenehn Formation, which were eroded from higher levels in the Lewis thrust sheet in the footwall of the Flathead normal fault, record rapid cooling at 75 ± 5 Ma to <60 °C (i.e., through the apatite partial annealing zone). Samples from deeper levels in the footwall of the Flathead fault record cooling at that time only to temperatures within the apatite partial annealing zone

TABLE 1. ZIRCON FISSION-TRACK DATA FROM LEWIS THRUST SHEET

Sample no.	Unit	Stratigraphic age	Elevation (m)	No. of grains	Standard track density (×10⁶ cm⁻²)	Fossil track density (×10⁶ cm⁻²)	Induced track density (×10⁶ cm⁻²)	Uranium content (ppm)	Chi-square probability (%)	Age dispersion (%)	Fission-track age (±1σ) (Ma)*	Mean track length ± std. error (μm)	Std. dev. (μm)
637	Rocky Mtn	Permian-Pennsylvanian	1905	9	0.167 (1910)	8.807 (1815)	1.130 (104)	96	55.0	0.12	184 ± 12[†]	8.23 ± 0.37 (22)	1.74
647	Grinnell	Proterozoic	2445	4	0.164 (1902)	9.798 (967)	1.560 (154)	117	96.2	0.00	184 ± 17	—	—
649	Appekunny	Proterozoic	2184	5	0.135 (1923)	8.976 (1051)	0.658 (77)	60	64.7	0.01	326 ± 39	—	—
651	Appekunny	Proterozoic	2062	7	0.160 (1902)	8.426 (1440)	0.796 (136)	61	0.01	38.15	301 ± 52	—	—
652	Appekunny	Proterozoic	2015	5	0.156 (1902)	10.21 (1254)	0.570 (70)	45	93.9	0.00	486 ± 61	—	—
653	Appekunny	Proterozoic	1399	3	0.128 (1923)	8.336 (426)	0.939 (48)	91	87.8	0.00	202 ± 31	—	—
654	Appekunny	Proterozoic	1397	5	0.170 (1902)	7.475 (760)	0.610 (62)	44	99.5	0.00	368 ± 49	—	—
695	Appekunny	Proterozoic	2040	2	0.162 (1902)	13.13 (497)	1.294 (49)	176	91.3	0.00	291 ± 44	—	—

Note: Number of tracks or lengths counted is shown in parentheses. Standard and induced track densities were measured on mica external detectors (g = 0.5) and fossil track density was measured on internal surfaces.
*All ages are central ages (Galbraith and Laslett, 1993). Zircon ages were calculated by B. Kohn using zeta = 362 for dosimeter glass SRM-612.
[†]Zircon age calculated by R. Brown using zeta = 127 for dosimeter glass CN-2.

TABLE 2. VITRINITE REFLECTANCE DATA MEASURED IN THE KOOTENAY GROUP, LEWIS THRUST SHEET

Locality and samples identification	Formation	No. of observations	Minimum reflectance (%)	Maximum reflectance (%)	Mean reflectance (%)	Standard deviation	Coefficient of variation (%)
Crowsnest Pass (surface; sample 602, Figure 2)	Mist Mountain	50	0.979	1.138	1.081	0.028	2.59
Cabin Creek Adit (surface; near Grieve [1987], p. 362, sample 75-D-02; sample 676, Figure 2)	Mist Mountain	50	1.102	1.311	1.183	0.059	4.99
Sage Creek No. 2 well (Figure 2, 2035 m)	Beaver Mines	50	1.047	1.205	1.120	0.039	3.48
Sage Creek No. 2 well (Figure 2, 2352 m)	Mist Mountain	50	1.038	1.243	1.157	0.063	5.44
Sage Creek No. 2 well (Figure 2, 2400 m)	Mist Mountain	50	1.055	1.208	1.116	0.036	3.23
Footwall of Lewis thrust (surface; sample 616, Figure 2)	Mist Mountain	50	0.709	0.777	0.738	0.016	2.17

176   Feinstein et al.

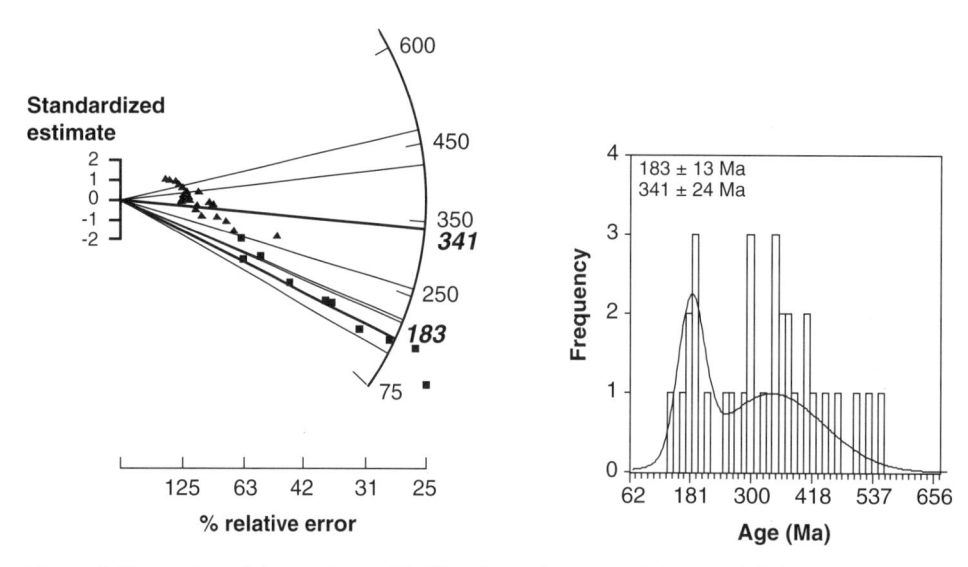

Figure 4. Estimation of the maximum likelihood set of age populations and their mean ages using 31 grain zircon fission-track age measurements in 7 samples from the Proterozoic Grinnell and Appekunny Formations in the Lewis thrust sheet. The mean age of each population was determined assuming a Gaussian error and by applying the mixture modeling approach described by Sambridge and Compston (1994). The zircon fission-track ages show two distinct populations with mean ages of 183 ± 17 Ma and 341 ± 24 Ma. The radial plot (Galbraith, 1990) graphically displays the mean zircon fission-track age (in italics) and associated error of each population. The fission-track age of any datum can be read by extrapolating a straight line from the origin (0) at the left of the plot to intercept the radial scale around the perimeter of the plot. The position of a grain on the x-axis records the uncertainty of an individual age estimate, while each age has the same standard error on the y-axis (shown as ±2σ). The further a point plots to the right of the origin, the more precise the individual grain age measurement. For further explanation of mixture modeling, see text.

range (110 °C > T > 60 °C), but later in Eocene time, they cooled to <60 °C. Osadetz et al. (2004) attributed the cooling at 75 ± 5 Ma to the onset of displacement on the Lewis thrust fault and the later Eocene cooling phase to displacement on the Flathead normal fault and resulting footwall uplift and erosional denudation. It is stressed that the AFT t-T models indicate that postdeformation temperatures in the Lewis thrust sheet never exceeded those that predated the onset of 75 ± 5 Ma thrusting.

The constraint imposed on the estimated time of maximum heating by the ZFT data is relatively limited. ZFT central ages from all samples (486 ± 61 Ma to 184 ± 12 Ma; Table 1) and the range of single-grain ZFT ages from Mesoproterozoic strata (641 Ma to 144 Ma; Fig. 4) are substantially younger than the age of their host strata depositional ages, but they are older than the time indicated by both VR and AFT data for maximum heating. This suggests that the ZFT ages in the succession from Upper Paleozoic to Mesoproterozoic strata record heating to zircon partial annealing zone temperatures at some time later than the youngest age population (i.e., ca. 180 Ma), but earlier than the onset of cooling at 75 ± 5 Ma recorded in the AFT data. This analysis is consistent with the ZFT parameters of the Paleozoic Rocky Mountain Formation sample.

Various studies (e.g., Hacquebard and Donaldson, 1974; Bustin, 1983; Pearson and Grieve, 1985; and new data presented in Table 2) indicate that VR in and below the Lewis thrust sheet in the study region correlates with stratigraphic position rather than postdeformational tectonic burial. Predeformation coalification in the study area is further corroborated by the relationship between tectonically induced fracturing and coalification (Cameron and Kalkreuth, 1982) and vitrinite reflectance optical axis orientation with respect to bedding planes (Ting, 1984). The rapid cooling and low thermal gradient recorded by the AFT data (Osadetz et al., 2004) in the study area represent unfavorable conditions for major postdeformation coalification. Some regional studies show examples where coalification profiles were modified either completely or partially due to tectonic burial (England and Bustin, 1986; Bustin and England, 1989). However, effective postdeformation coalification is indicated only below the Lewis thrust fault and not in the Lewis thrust sheet, which is the focus of this study (Fig. 2).

In order to reach maximum paleotemperatures indicated by VR in Mesozoic strata, without totally resetting ZFT ages in the Mesoproterozoic succession, our analysis indicates that a relatively thick, younger, but no longer preserved, Mesozoic succession was deposited on the Lewis thrust sheet in a moderate geothermal gradient environment prior to the onset of thrusting and folding of the Lewis thrust fault. Strata of the Lower Cretaceous Blairmore Group and the Upper Cretaceous of the Alberta Group preserved in the Fernie Basin (Fig. 1) can be reasonably inferred to have been deposited regionally. However, the cumulative thickness of the preserved Mesozoic succession overlying

Mist Mountain Formation, ~2.5 km, is insufficient to account for the observed Mist Mountain Formation VR without assuming a high paleogeothermal gradient, which would exceed ZFT total annealing temperatures if extrapolated into the Mesoproterozoic Grinnell and Appekunny Formations. However, ZFT data from these formations (Purcell Supergroup) near the bottom of the Lewis thrust sheet suggest only partial rather than complete ZFT annealing at the time of peak paleotemperatures and peak coalification (VR) in Lower Cretaceous strata. Therefore, it is reasonable to assume that peak predeformational temperatures in the Lewis thrust sheet were achieved as the result of burial by a thicker Mesozoic succession, much of which is not preserved, in a lower predeformational geothermal gradient environment. Estimates of the thickness of the strata no longer preserved vary inversely with estimates of the zircon partial annealing zone temperature and the paleogeothermal gradient (Fig. 5). This interpretation is consistent with the results observed in the Rocky Mountain Formation ZFT sample, which indicates that zircon partial annealing zone temperatures extended over a large stratigraphic interval, into the Paleozoic succession.

The youngest strata preserved within the Lewis thrust sheet, marine shales and sandstone (possibly the lower part of the Cardium Formation) of the Alberta Group (Price, 1962), occur in the McEvoy syncline (Fig. 1) and are of late Cenomanian to early Campanian age. This age, the VR measured in Mesozoic strata, and the AFT data and their time-temperature models (Osadetz et al., 2004) constrain the maximum Phanerozoic burial and heating in the Lewis thrust sheet to have occurred in the Late Cretaceous within a time interval of <15 m.y., i.e., between the deposition of the youngest preserved part of the Alberta Group (ca. 90 Ma) and the onset of thrusting and folding at 75 ± 5 Ma (Osadetz et al., 2004). We infer that the Mesoproterozoic rocks deeper in the Lewis thrust sheet succession also reached their maximum zircon partial annealing zone temperatures during the same time interval.

Constraints on Maximum Temperatures

The temperature range implied for the Mist Mountain Formation during coalification can be constrained to lie between the temperature required for the lower VR recorded (1.1% Ro_{max}) to evolve under long duration of heating (e.g., 25 m.y.) and that required for the higher VR measured (1.25% Ro_{max}) to evolve over a short duration of heating (1 m.y.). Using the Burnham and Sweeney (1989) coalification model, the lower and upper thermal limits estimated for the Mist Mountain Formation lie between ~140 °C and ~170 °C, respectively. The actual coalification temperature for the Mist Mountain Formation from the 1.18% mean Ro_{max} value and <15 m.y. residence time at maximum temperature is estimated to be 150 °C.

Further constraints on determination of the degree of heating within the zircon partial annealing zone require the measurement of horizontal confined fission-track lengths. The track length data available (Table 1) are insufficient to estimate a discrete paleo-

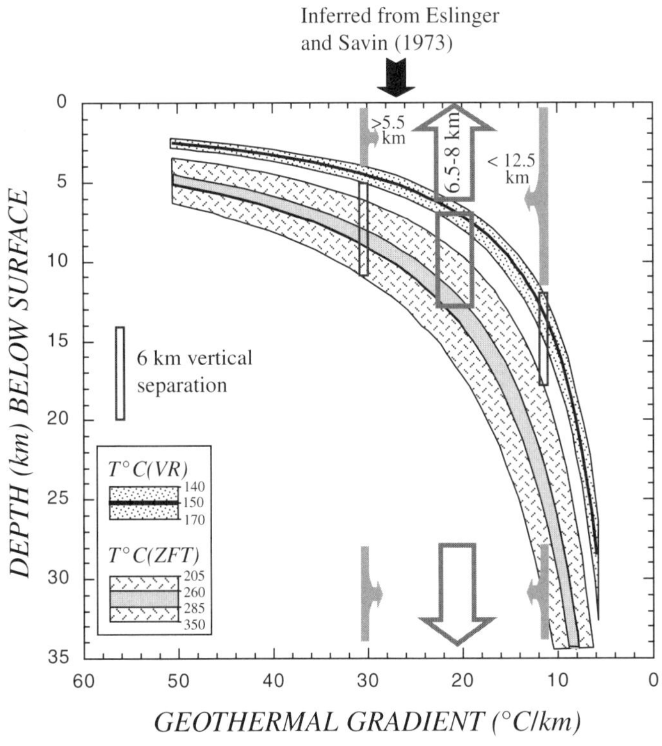

Figure 5. Isothermal curves for temperatures at the time of maximum heating, as inferred from vitrinite reflectance (VR) data from the Upper Jurassic–Lower Cretaceous Mist Mountain Formation (Kootenay Group) and zircon fission-track (ZFT) data from the Mesoproterozoic Grinnell and Appekunny Formations (Purcell Supergroup), plotted in depth versus geothermal gradient space. There are three curves for each stratigraphic sampling level: upper and lower limits of the temperature range estimated assuming short (1 m.y.) and long (25 m.y.) duration of heating, and within this range, the paleotemperature believed to have actually prevailed estimated with <15 m.y. duration of heating. The vertical (stratigraphic) separation between the VR and ZFT sampling levels at the time of maximum heating was 6 km (Fig. 3), as indicated by the rectangular boxes. The solid gray vertical bars with merging horizontal arrows represent constraints. The two bars at the lower part show estimated constraints on the maximum and minimum paleogeothermal gradients (respectively, 30 and 11 °C/km) at time of maximum heating obtained using the maximum and minimum paleotemperatures estimated from the observed VR and ZFT and 6 km of vertical section between them. The equivalent symbols at the upper part show the constraints on minimum and maximum paleoburial (respectively, 5.5 and 12.5 km) at the time of maximum heating estimated for, respectively, the maximum and minimum paleogeothermal gradients shown below. Likewise, the thick down-pointing arrow represents the range of geothermal gradients, 18–22.5 °C/km estimated for the ZFT and VR data (respectively ~260–285 °C and 150 °C), and the up-pointing arrow shows the paleoburial estimate, 6.5–8 km, at that time. The down-pointing solid black arrow at the top indicates, in general, the range of paleogeothermal gradients and burial inferred from the work of Eslinger and Savin (1973), who estimated a peak paleotemperature in the Mesoproterozoic succession of 310 °C.

temperature in the Mesoproterozoic succession, but they do provide a limited basis for estimation of the temperature range implied using the ZFT data. Similar to the calculation made for the coalification temperatures in the Mist Mountain Formation, the paleotemperature range of the Mesoproterozoic Grinnell and Appekunny Formations can be constrained between the temperature required for the top zircon partial annealing zone level (~10 μm track length) under long heating at maximum temperature (25 m.y.) and the base zircon partial annealing zone (~6 μm track length) under short heating at maximum temperature (1 m.y.). These conditions constrain the temperature range for the zircon partial annealing zone to between ~205 °C and ~350 °C (Tagami, 2005). On the basis of the indicated <15 m.y. duration at maximum temperature for the Lewis thrust sheet, from stratigraphic and coalification data, and the number of zircon fission-track length measurements that appear to fall between 8 and 9 μm, we infer that the maximum Phanerozoic temperature of the Mesoproterozoic Grinnell and Appekunny Formations at the time of peak coalification in Mesozoic strata was ~260–285 °C.

The temperature range inferred for the Mesoproterozoic succession for our study is similar to estimates of maximum paleotemperatures of ~310 °C determined for the same Mesoproterozoic formations in adjacent Montana using quartz-illite oxygen isotope geothermometry and illite crystallinity transformations (Eslinger and Savin, 1973). However, Eslinger and Savin (1973) assumed that the incipient metamorphism of Mesoproterozoic strata occurred over a short heating period during Proterozoic time, although their study gives no clear indication of how they constrained the actual time of peak temperatures in Proterozoic strata. Based on the thermochronometric data and analysis discussed here, we suggest that the incipient metamorphism of the Purcell Supergroup observed in their study gives temperatures that are similar to the peak temperatures inferred for ZFT analysis and that the incipient metamorphic reactions might have occurred coincident with peak coalification in Mesozoic strata, at times that are considerably younger than were assumed previously.

Paleogeothermal Gradient and Burial Depth at Maximum Temperatures

Figure 5 presents an estimation of paleogeothermal gradient and depth of burial from the coalification and fission-track thermochronometric data. The plot shows the constrained paleotemperature ranges estimated for the Mist Mountain Formation (upper envelope) and for the Mesoproterozoic Grinnell and Appekunny Formations (lower envelope) based on, respectively, VR (minimum and maximum) data and zircon partial annealing zone boundaries. We prefer to infer the predeformational paleogeothermal gradient over the longest possible stratigraphic interval and use only ZFT data from the Mesoproterozoic Grinnell and Appekunny Formations for this purpose (Fig. 5). The heavy curves within the envelopes represent the actual maximum estimated paleotemperatures, 150 °C in the Mist Mountain Formation (the upper envelope), and 260 °C to 285 °C in the Mesoproterozoic Grinnell and Appekunny Formations (the lower envelope). Figure 5 shows the variation in depth of the isothermal curves estimated as a function of geothermal gradient. We used a linear paleogeothermal gradient for the geothermal profile approximation following previous studies (e.g., Majorowicz et al., 1990; Osadetz et al., 1992), which showed that present and paleo–thermal profiles in the study region are well approximated by a linear and piece-wise linear geothermal profile.

Palinspastic reconstruction of fault displacements (Figs. 2B and 3) shows a stratigraphic separation of ~6 km between the Mesoproterozoic Grinnell and Appekunny Formations from which the ZFT data were obtained and the Lower Cretaceous portion of the Mist Mountain Formation from which the VR measurements were obtained. Dividing the difference between the maximum range of possible paleotemperatures estimated for the zircon partial annealing zone and the VR for 1 m.y. of heating (i.e., respectively, 350 °C and 170 °C) by the 6 km vertical stratigraphic separation gives a maximum paleogeothermal gradient of ~30 °C/km for the Lewis thrust sheet prior to the displacement and cooling at 75 ± 5 Ma (Fig. 5, bottom left gray bar). Likewise, the minimum paleogeothermal gradient, obtained for paleotemperatures calculated for 25 m.y. of heating (i.e., 205 °C and 140 °C), is estimated to be ~11 °C/km (Fig. 5, bottom right gray bar). This appears to be an anomalously low geothermal gradient, particularly for an interval that is inferred to have been the time of maximum Phanerozoic heating. Using paleotemperatures of 260–285 °C for the zircon partial annealing zone as estimated for mean horizontal confined track length (8–9 μm) together with the 150 °C estimated from the VR (1.18% Ro_{max}) obtained for <15 m.y. of heating, the paleogeothermal gradient in the Lewis thrust sheet during the Phanerozoic maximum heating is estimated to have been between 18 °C/km and 22.5 °C/km (Fig. 5, bottom central open arrow).

The depth of burial of the Mist Mountain Formation and underlying formations at the time of maximum heating can also be estimated by linear extrapolation of an estimated thermal gradient to surface temperatures. Assuming an average surface temperature of ~5 °C and, respectively, the maximum and minimum constraints on the geothermal gradients, the depth of burial of the Mist Mountain Formation during peak heating could have been between >5.5 km and <12.5 km (Fig. 5, respectively, the upper left and right gray bars). An ~2.5-km-thick Cretaceous succession is preserved in the Lewis thrust sheet above the Mist Mountain Formation where VR was measured (Figs. 2 and 3). Hence, a younger sedimentary succession between >3 km and <10 km thick, which is no longer preserved, should have overlain the preserved strata of the Lewis thrust sheet after deposition of the Alberta Group and before the displacement on the Lewis thrust. The estimated limit for the maximum paleoburial depth and/or thickness of the eroded succession is probably much greater than the actual depth and thickness because of the anomalously low minimum paleothermal gradient obtained from the paleothermometric models. Assuming that the degree of annealing within the zircon partial annealing zone from the Grinnell and Appekunny Formations is represented by an average horizontal confined track length of ~8–9 μm, which is consistent with all the observed ZFT data (Table 1), and that the Mist Mountain Formation VR is represented by 1.18% Ro$_{max}$, while the duration of heating was <15 m.y., the more constrained paleoburial is estimated to have been between 6.5 and 8 km (Fig. 5, upper central open arrow) and the Upper Cretaceous succession that is not preserved is estimated to have been between ~4 and 5.5 km thick. A comparable thickness of eroded Mesozoic foreland basin deposits has been estimated for the laterally equivalent Lewis-Eldorado-Hoadley thrust sheet in northern Montana (Sears, 2004).

The thickness of the Lewis thrust sheet at the time of onset of displacement on the Lewis thrust fault can be calculated from the cumulative thickness of the succession preserved in the footwall and hanging wall of the Flathead normal fault, (~8 km; Figs. 2B and 3) together with the part of the thrust sheet that has been completely removed by erosion, but which has been reconstructed from the paleothermometric constraints on depth of burial at the time of maximum heating (Fig. 5). Thus, at the onset of displacement on the Lewis thrust, the original thickness of the Lewis thrust sheet was >11 km, likely between 12.0 and 13.5 km. The rapid accumulation of an estimated ~4–5.5-km-thick Upper Cretaceous succession during a time interval not exceeding 15 m.y. contrasts markedly with the accumulation of the ~8-km-thick succession preserved in the Lewis thrust sheet in the time interval from Mesoproterozoic to Late Cretaceous. The very thick Upper Cretaceous succession, a major part of which was later eroded, represents synorogenic foreland basin sedimentation that was the harbinger of the eastward-propagating deformation that formed the Lewis thrust fault, and the underlying structures of the Cordilleran foreland fold-and-thrust belt. The strong correlation between VR and stratigraphic position, despite the extensive deformation and long displacement during subsequent thrusting and folding, suggests that a similar thickness and geothermal gradient prevailed over a wide area. We infer that the Coniacian-Campanian marine shales and sandstones of the Alberta Group that were deposited on the Lewis thrust sheet before displacement on the Lewis thrust began must have been overlain by a substantial thickness of Belly River strata, some of which may have accumulated in a "wedge-top" (piggyback) basin during the initial stages of the displacement of the Lewis thrust sheet, and the emergence and erosion of the toe of the sheet. The upper time limit for the removal of most of this >3 km succession from the Lewis thrust sheet is constrained by the deposition of Lower Oligocene fanglomerates of the Kishenehn Formation, which unconformably overlie Blairmore Group strata in the Flathead Valley graben (Figs. 1 and 3).

The predeformational geothermal gradient in the strata of the Lewis thrust sheet, which is inferred to have been between ~18 °C/km and 22.5 °C/km, is similar to the current geothermal gradient of ~17 °C/km, as estimated from corrected bottom hole logging temperatures (courtesy Shell Canada Resources Ltd., K. Root, 1992, personal commun.). On the other hand, syndeformational geothermal gradients are inferred from AFT thermochronology to have been considerably lower, ~8.6–12.0 °C/km (Osadetz et al., 2004). Thus VR, AFT, and ZFT paleothermometry indicate significant variations in Lewis thrust sheet paleogeothermal gradients during Late Cretaceous deformation compared to the late predeformational peak temperatures and present-day geothermal gradients.

CONCLUSIONS

The integration of coalification data and fission-track thermochronology provides constraints for the investigation of variations in the late predeformational paleogeothermal gradient and thickness in the Lewis thrust sheet in the Foreland belt of the southeastern Canadian Cordillera. Thermochronometric and stratigraphic data constrain the maximum burial and heating in the Lewis thrust sheet to have occurred during the Campanian, over a time interval of <15 m.y., prior to the onset of displacement on the Lewis thrust at 75 ± 5 Ma.

Paleotemperatures derived from vitrinite reflectance in the Upper Jurassic–Lower Cretaceous Mist Mountain Formation (Kootenay Group) and zircon fission-track data in the underlying Mesoproterozoic Grinnell and Appekunny Formations (Purcell Supergroup) constrain the possible predeformational paleogeothermal gradient to a range between <30 to >11 °C/km. The preferred predeformational paleothermal gradient estimate at the time of peak coalification and maximum temperatures is inferred to have been between 18 and 22 °C/km, which is considerably higher than the syndeformational paleothermal gradient estimate obtained from apatite fission-track analysis (Osadetz et al., 2004).

The inferred paleotemperatures and geothermal gradients indicate that the currently preserved Lewis thrust sheet succession was overlain by an additional >3 km (probably between 4 and 5.5 km) of Late Cretaceous strata that were subsequently denuded. Along with the preserved ~8-km-thick succession, this indicates that the Lewis thrust sheet was ~12–13.5 km thick just before thrusting commenced.

The accumulation of the thick Late Cretaceous sedimentary cover took place relatively quickly, in <15 m.y., and this deposition was terminated by the onset of displacement on the Lewis thrust at ca. 75 ± 5 Ma (Osadetz et al., 2004). The subsequent syn- and postdeformation period of extensive erosional denudation is constrained by stratigraphy to have ceased west of the Flathead Fault by early Oligocene time, when the Lewis thrust sheet in the footwall of the Lewis thrust fault became the focus of rapid erosional denudation.

Our analysis indicates that the combination of paleothermometric indicators can provide additional constraints on foreland belt geohistory that improve the description of critical time intervals and successions not preserved in the rock record, but which are essential to the analysis and modeling of processes accompanying foreland belt evolution.

ACKNOWLEDGMENTS

This study was partially funded by the Australian Research Council and the Australian Institute of Nuclear Science and Engineering, the National Science and Engineering Research Council of Canada, and the Department of Natural Resources Canada. We thank T. Gentzis for making the vitrinite reflectance measurements available. S. Willett, P. Vrolik, and B. van der Pluijm provided constructive comments on an earlier version of this work. The paper has greatly benefited from helpful reviews provided by Dale Issler and Ed Sobel. This is Geological Survey of Canada contribution no. 2003099.

REFERENCES CITED

Bally, A.W., Gordy, P.L., and Stewart, G.A., 1966, Structure, seismic data, and orogenic evolution of southern Canadian Rocky Mountains: Bulletin of Canadian Petroleum Geology, v. 14, p. 337–381.

Barker, C.E., 1989, Temperature and time in the thermal maturation of sedimentary organic matter, in Naeser, N.D., and McCullough, T.H., eds., Thermal History of Sedimentary Basins: Methods and Case Histories: New York, Springer-Verlag, p. 73–98.

Beaumont, C., 1981, Foreland basins: Geophysical Journal of the Royal Astronomical Society, v. 65, p. 291–329.

Boyer, S.E., 1992, Geometric evidence for synchronous thrusting in the southern Alberta and northwest Montana thrust belts, in McClay, K.R., ed., Thrust Tectonics: New York, Chapman and Hall, p. 377–390.

Brandon, M.T., Roden-Tice, M.K., and Garver, J.I., 1998, Late Cenozoic exhumation of the Cascadia accretionary wedge in the Olympic Mountains, northwest Washington State: Geological Society of America Bulletin, v. 110, p. 985–1009, doi: 10.1130/0016-7606(1998)110 <0985:LCEOTC>2.3.CO;2.

Brix, M.R., Stokhert, B., Seidel, E., Theye, T., Thomson, S.N., and Kuster, M., 2002, Thermobarometric data from fossil zircon partial annealing zone in high pressure–low temperature rocks of eastern and central Crete, Greece: Tectonophysics, v. 349, p. 309–326, doi: 10.1016/S0040-1951(02)00059-8.

Burnham, A.K., and Sweeney, J.J., 1989, A chemical kinetic model of vitrinite maturation and reflectance: Geochimica et Cosmochimica Acta, v. 53, p. 2649–2657, doi: 10.1016/0016-7037(89)90136-1.

Bustin, R.M., 1983, Heating during thrust faulting in the Rocky Mountains: Friction or fiction: Tectonophysics, v. 95, p. 309–328, doi: 10.1016/0040-1951(83)90075-6.

Bustin, R.M., 1991, Organic maturation of the Western Canada sedimentary basin: International Journal of Coal Geology, v. 19, p. 319–358, doi: 10.1016/0166-5162(91)90026-F.

Bustin, R.M., and England, T.D.J., 1989, Timing of organic maturation (coalification) relative to thrust faulting in the southeastern Canadian Cordillera: International Journal of Coal Geology, v. 13, p. 327–339, doi: 10.1016/0166-5162(89)90098-0.

Cameron, A.R., and Kalkreuth, W.D., 1982, Petrological characteristics of Jurassic-Cretaceous coals in the Foothills and Rocky Mountains of western Canada: Utah Geological and Mineral Survey Bulletin, v. 118, p. 163–167.

Carr, S., 1992, Tectonic setting and U-Pb geochronology of the early Tertiary Ladybird leucogranite suite, Thor-Odin–Pinnacles area, southern Omineca belt, British Columbia: Tectonics, v. 11, p. 258–278.

Carr, S.D., Parrish, R.R., and Brown, R.L., 1987, Eocene structural development of the Valhalla complex, southeastern British Columbia: Tectonics, v. 6, p. 175–196.

Constenius, K.D., 1988, Structural configuration of the Kishenehn basin delineated by geophysical methods, northwestern Montana and southeastern British Columbia: The Mountain Geologist, v. 25, p. 13–28.

Constenius, K.D., 1996, Late Paleogene extensional collapse of the Cordilleran foreland fold and thrust belt: Geological Society of America Bulletin, v. 108, p. 20–39, doi: 10.1130/0016-7606(1996)108 <0020:LPECOT>2.3.CO;2.

Cook, A.C., and Kantsler, A.J., 1980, The maturation history of the epicontinental basins of western Australia: Tech. Bulletin—U.N. Econ. Soc. Comm. Asia Pac., Comm. Co-ord. Jt. Prospect. Miner. Resour. South Pac. Offshore Areas, v. 3, p. 171–195.

Cook, F.A., Varsek, J.L., Clowes, R.M., Kanasevich, E.R., Spencer, C.S., Parrish, R.R., Brown, R.L., Carr, S.D., Johnson, B.J., and Price, R.A., 1992, Lithoprobe crustal reflection cross section of the southern Canadian Cordillera: 1. Foreland thrust and fold belt to Fraser River: Tectonics, v. 11, p. 12–35.

Dahlstrom, C.D.A., 1970, Structural geology in the eastern margin of the Canadian Rocky Mountains: Bulletin of Canadian Petroleum Geology, v. 18, p. 332–406.

Doughty, P.T., and Price, R.A., 1999, Tectonic evolution of the Priest River Complex, Northern Idaho and Washington: A reappraisal of the Newport fault with new insights on metamorphic core complex formation: Tectonics, v. 18, no. 3, p. 375–393, doi: 10.1029/1998TC900029.

Doughty, P.T., and Price, R.A., 2000, Geology of the Purcell Trench rift valley and Sandpoint Conglomerate: Eocene en echelon normal faulting and synrift sedimentation along the eastern flank of the Priest River metamorphic complex, northern Idaho: Geological Society of America Bulletin, v. 112, p. 1356–1374, doi: 10.1130/0016-7606(2000)112<1356:GOTPTR>2.0.CO;2.

Douglas, R.J.W., 1950, Callum Creek, Langford Creek and Gap Map-Areas: Geological Survey of Canada Memoir 255, 124 p.

England, T.D.J., 1984, Thermal maturation of the Western Canadian sedimentary basin in the Rocky Mountains foothills and plains of Alberta south of the Red Deer River [M.Sc. thesis]: Vancouver, University of British Columbia, 171 p.

England, T.D.J., and Bustin, R.M., 1986, Thermal maturation of the Western Canadian sedimentary basin south of the Red Deer River: 1. Alberta Plains: Bulletin of Canadian Petroleum Geology. v. 37, p. 71–90.

Enkin, R.J., Osadetz, K.G., Baker, J., and Kisilevsky, D., 2000, Orogenic remagnetizations in the front ranges and inner foothills of the southern Canadian Cordillera: Chemical harbinger and thermal handmaiden of Cordilleran deformation: Geological Society of America Bulletin, v. 112, p. 929–942.

Eslinger, E.V., and Savin, S.M., 1973, Oxygen isotope geothermometry of the burial metamorphic rocks of the Precambrian Belt Supergroup, Glacier National Park, Montana: Geological Society of America Bulletin, v. 84, p. 2549–2560, doi: 10.1130/0016-7606(1973)84<2549:OIGOTB>2.0.CO;2.

Ewing, T.E., 1980, Paleogene tectonic evolution of the Pacific Northwest: The Journal of Geology, v. 88, p. 619–638.

Feinstein, S., Kisch, H.J., and Shagam, R., 1992, Pre-Cretaceous coalification in southern Israel: Contribution to time-temperature coalification modeling: Journal of Petroleum Geology, v. 15, p. 327–344.

Feinstein, S., Issler, D.R., Snowdon, L.R., and Williams, G.K., 1996, Characterization of major unconformities by paleothermometric and paleobarometric methods: Application to the Mackenzie Plain, Northwest Territories, Canada: Bulletin of Canadian Petroleum Geology, v. 44, p. 55–71.

Fermor, P.R., 1999, Aspects of the three-dimensional structure of the Alberta Foothills and Front Ranges: Geological Society of America Bulletin, v. 111, p. 317–346, doi: 10.1130/0016-7606(1999)111<0317:AOTTDS>2.3.CO;2.

Fermor, P.R., and Moffat, I.W., 1992, Tectonics and structure of the Western Canada foreland basin, in Macqueen, R.W., and Leckie, D.A., eds., Foreland Basins and Fold Belts: American Association of Petroleum Geologists Memoir 55, p. 81–105.

Fermor, P.R., and Price, R.A., 1987, Multiduplex structure along the base of the Lewis thrust sheet in the southern Canadian Rockies: Bulletin of Canadian Petroleum Geology, v. 35, p. 159–185.

Gabrielse, H., and Yorath, C.J., eds., 1992, Geology of the Cordilleran Orogen in Canada: Ottawa, Geological Survey of Canada, Geology of Canada, no. 4, 844 p.

Galbraith, R.F., 1981, On statistical models for fission track counts: Mathematical Geology, v. 13, p. 471–478, doi: 10.1007/BF01034498.

Galbraith, R.F., 1990, The radial plot: Graphical assessment of spread in ages: International Journal of Radiation Measurement and Applied Instrumentation–Part D: Nuclear Tracks and Radiation Measurement, v. 17, p. 207–214, doi: 10.1016/1359-0189(90)90036-W.

Galbraith, R.F., and Laslett, G.M., 1993, Statistical models for mixed fission track ages: Nuclear Tracks, v. 21, p. 459–470.

Gallagher, K., Brown, R.W., and Johnson, C., 1998, Fission track analysis and its applications to geological problems: Annual Review of Earth and Planetary Sciences, v. 26, p. 519–572, doi: 10.1146/annurev.earth.26.1.519.

Gleadow, A.J.W., 1981, Fission track dating methods: What are the real alternatives?: Nuclear Tracks, v. 5, p. 169–174, doi: 10.1016/0191-278X(81)90039-1.

Gleadow, A.J.W., and Lovering, J.F., 1977, Geometry factor for external detectors in fission track dating: Nuclear Track Detection, v. 1, p. 99–106, doi: 10.1016/0145-224X(77)90003-5.

Gleadow, A.J.W., Hurford, A.J., and Quaife, D.R., 1976, Fission track dating of zircon: Improved etching techniques: Earth and Planetary Science Letters, v. 33, p. 273–276, doi: 10.1016/0012-821X(76)90235-1.

Green, P.F., 1981, A new look at statistics in fission track dating: Nuclear Tracks, v. 5, p. 77–86, doi: 10.1016/0191-278X(81)90029-9.

Green, P.F., Hegarty, K.A., Duddy, I.R., Foland, S.S., and Gorbachev, V., 1996, Geological constraints on fission track annealing in zircon, in International Workshop on Fission Track Dating Abstracts: Ghent, p. 44.

Grieve, D.A., 1987, Coal rank distribution, Flathead coal-field, southeastern British Columbia (82G/2, 82G/7), in Geological Fieldwork, 1986: British Columbia Ministry of Energy, Mines and Petroleum Resources Paper 1987-1, p. 361–364.

Hacquebard, P.A., 1977, Rank of coal as an index of organic metamorphism for oil and gas in Alberta, in Deroo, G., Powell, T.G., Tissot, B., and McCrossan, R.G., eds., The Origin and Migration of Petroleum in the Western Canadian Sedimentary Basin, Alberta—A Geochemical and Thermal Maturation Study: Geological Survey of Canada Bulletin 262, p. 11–22.

Hacquebard, P.A., and Donaldson, J.R., 1974, Rank studies of coals in the Rocky Mountains and inner Foothills belt, Canada, in Dutcher, R.R., Hacquebard, P.A., Schopf, J.M., and Simon, J.A., eds., Carbonaceous Materials as Indicators of Metamorphism: Geological Society of America Special Paper 153, p. 75–94.

Hurford, A.J., and Green, P.F., 1983, A guide to fission track dating calibration: Chemical Geology (Isotope Geoscience Section), v. 1, p. 285–317.

International Committee for Coal Petrology (ICCP), 1971, International Handbook of Coal Petrology (suppl. to 2nd edition, 1971): Paris, Centre Nationale de la Recherche Scientifique, unpaginated.

Issler, D.R., Beaumont, C., Willet, S.D., Donelick, R.A., Moore, J., and Grist, A., 1990, Preliminary evidence from apatite fission-track data concerning the thermal history of the Peace River Arch region, Western Canada sedimentary basin: Bulletin of Canadian Petroleum Geology, v. 38A, p. 250–269.

Jerzykiewicz, T., Sweet, A.R., and McNeil, D.H., 1996, Shoreface of the Bearpaw Sea in the footwall of the Lewis thrust, southern Canadian Cordillera, Alberta, in Current Research 1996-A: Ottawa, Geological Survey of Canada, p. 15–163.

Jones, P.B., 1977, The Howell Creek structure—A Paleogene rock slide in the southern Canadian Rocky Mountains: Bulletin of Canadian Petroleum Geology, v. 25, p. 868–881.

Kalkreuth, W., and McMechan, M.E., 1984, Regional pattern of thermal maturation as determined from coal rank studies, Rocky Mountains Foothills and Front Ranges north of Grande Cache, Alberta—Implications for petroleum exploration: Bulletin of Canadian Petroleum Geology, v. 32, p. 249–271.

Kasuya, M., and Naeser, C., 1988, The effect of α-damage on fission track annealing in zircon: Nuclear Tracks and Radiation Measurements, v. 14, p. 477–480, doi: 10.1016/1359-0189(88)90008-8.

Lebreque, J.E., and Shaw, E.W., 1973, Restoration of basin and range faulting across the Howell Creek window and Flathead Valley of southeastern British Columbia: Bulletin of Canadian Petroleum Geology, v. 21, p. 117–122.

Lorencak, M.D., Seward, D., Vanderhaeghe, O., Teyssier, C., and Burg, J.P., 2001, Low temperature cooling history of the Shuswap metamorphic core complex, British Columbia: Constraints from apatite and zircon fission-track ages: Canadian Journal of Earth Sciences, v. 38, p. 1615–1625, doi: 10.1139/cjes-38-11-1615.

Mack, G.H., and Jerzykiewicz, T., 1989, Provenance of post-Wapiabi sandstones and its implications for Campanian to Paleocene tectonic history of the Canadian Cordillera: Canadian Journal of Earth Sciences, v. 26, p. 665–676.

Magara, K., 1976, Thickness of removed sedimentary rocks, paleopore pressure and paleotemperature, southwestern part of Western Canada basin: American Association of Petroleum Geologists Bulletin, v. 60, p. 554–565.

Majorowicz, J.A., Jones, F.W., Ertman, M.E., Osadetz, K.G., and Stasiuk, L.D., 1990, The relationship between thermal maturation gradients, geothermal gradients and estimates of the thickness of the eroded Tertiary section, southern Alberta Plains, Canada: Marine and Petroleum Geology, v. 7, p. 138–152, doi: 10.1016/0264-8172(90)90037-H.

McMechan, R.D., 1981, Stratigraphy, Sedimentology, Structure and Tectonic Implications of the Oligocene Kishenehn Formation, Flathead Valley Graben, Southeastern British Columbia [Ph.D. thesis]: Kingston, Queen's University, 327 p.

McMechan, R.D., and Price, R.A., 1980, Reappraisal of reported unconformity in the Palaeogene (Oligocene) Kishenehn Formation: Implications for Cenozoic tectonics in the Flathead valley graben, southeastern British Columbia: Bulletin of Canadian Petroleum Geology, v. 28, p. 37–45.

McMechan, M.E., and Thompson, R.I., 1989, Chapter 4: Structural style and history of the Rocky Mountain fold and thrust belt, in Ricketts, B.D., ed., Western Canada Sedimentary Basin: A Case History: Canadian Society of Petroleum Geologists Special Paper 30, p. 47–76.

Monger, J.W.H., and Price, R.A., 1979, Geodynamic evolution of the Canadian Cordillera—Progress and problems: Canadian Journal of Earth Sciences, v. 16, p. 771–791.

Mudge, M.R., and Earhart, R.L., 1980, Lewis Thrust Fault and Related Structures in the Disturbed Belt, Northwestern Montana: U.S. Geological Survey Professional Paper 1174, 18 p.

Nurkowski, J.R., 1984, Coal quality, coal rank variation and its relation to reconstructed overburden, Upper Cretaceous and Tertiary plains coals, Alberta, Canada: American Association of Petroleum Geologists Bulletin, v. 68, p. 285–295.

Osadetz, K.G., Jones, F.W., Majorowicz, J.A., Pearson, D.E., and Stasiuk, L.D., 1992, Thermal history of the Cordilleran foreland basin in western Canada: A review, in Macqueen, R.W., and Leckie, D.A., eds., Foreland Basins and Fold Belts: American Association of Petroleum Geologists Memoir 55, p. 259–278.

Osadetz, K.G., Kohn, B., Feinstein, S., and Price, R., 2004, Foreland belt thermal history using apatite fission track thermochronology: Implications for Lewis thrust and Flathead fault in the southern Canadian Cordilleran petroleum province, in Swennen, R., Roure, F., and Granath, W., eds., Deformation, Fluid Flow, and Reservoir Appraisal in Foreland Fold and Thrust Belts: American Association of Petroleum Geologists Hedberg Series, no. 1, p. 21–48.

Parrish, R.R., Carr, S.D., and Parkinson, D.L., 1988, Eocene extensional tectonics and geochronology of the southern Omineca belt, British Columbia and Washington: Tectonics, v. 7, p. 181–212.

Pearson, D.E., and Grieve, D.A., 1985, Rank variation, coalification and coal quality in the Crowsnest coalfield, British Columbia: Canadian Institute of Mining and Metallurgy Bulletin, v. 78, no. 881, p. 39–46.

Peper, T., 1993, Tectonic control on the sedimentary record in foreland basins: Inference from quantitative subsidence analysis and stratigraphic modeling [thesis and supplement]: Den Haag, CIP-DATA Doninklijke Bibliotheek, Vrije Universiteit Amsterdam, 187 p.

Price, R.A., 1962, Fernie Map Area, East Half, Alberta and British Columbia: Geological Survey of Canada Paper 61–24, 65 p.

Price, R.A., 1964, The Precambrian Purcell system in the Rocky Mountains of southern Alberta and British Columbia: Bulletin of Canadian Petroleum Geology, v. 12, p. 399–426.

Price, R.A., 1965, Flathead Map-Area, British Columbia and Alberta: Geological Survey of Canada Memoir 336, 221 p.

Price, R.A., 1973, Large-scale gravitational flow of supracrustal rocks, southern Canadian Rockies, in de Jong, K.A., and Scholten, R., eds., Gravity and Tectonics: New York, Wiley-Interscience, p. 491–502.

Price, R.A., 1979, Intracontinental ductile crustal stretching linking the Fraser River and northern Rocky Mountain Trench transform fault zones: Geological Society of America Abstracts with Programs, v. 11, no. 7, p. 499.

Price, R.A., 1981, The Cordilleran thrust and fold belt in the southern Canadian Rocky Mountains, in McClay, K.R., and Price, N.J., eds., Thrust and Nappe Tectonics: Geological Society [London] Special Publication 9, p. 427–448.

Price, R.A., 1994, Chapter 2, Cordilleran tectonics and the evolution of the Western Canada sedimentary basin, in Mossop, G., and Shetsen, I., compilers, Geological Atlas of the Western Canada Sedimentary Basin: Calgary and Edmonton, Canadian Society of Petroleum Geologists and Alberta Research Council, p. 13–24.

Price, R.A., and Fermor, P.R., 1985, Structure Section of the Cordilleran Foreland Thrust and Fold Belt West of Calgary, Alberta: Geological Survey of Canada Paper 84–14, 1 p.

Price, R.A., and Sears, J.W., 2000, A preliminary palinspastic map of the Mesoproterozoic Belt-Purcell Supergroup, Canada and USA: Implications for the tectonic setting and structural evolution of the Purcell anticlinorium and the Sullivan deposit; Chapter 5, in Lydon, J.W., Høy, T., Slack, J.F., and Knapp M.E., eds., The Geological Environment of the Sullivan Deposit, British Columbia: Geological Association of Canada, Mineral Deposits Division (MDD) Special Publication, No. 1, p. 61–81.

Rahn, M.K., Brandon, M.T., Batt, G.E., and Garver, J.I., 2004, A zero-damage model for fission-track annealing in zircon: The American Mineralogist, v. 89, p. 473–484.

Sambridge, M.S., and Compston, W., 1994, Mixture modelling of multi-component data sets with application to ion-probe zircon ages: Earth and Planetary Science Letters, v. 128, p. 373–390, doi: 10.1016/0012-821X(94)90157-0.

Sears, J.W., 2001, Emplacement and denudation history of Lewis-Eldorado-Hoadley thrust slab in the northern Montana Cordillera, USA: Implications for steady-state orogenic processes: American Journal of Science, v. 301, p. 359–373, doi: 10.2475/ajs.301.4-5.359.

Sears, J.W., 2004, Three-dimensional reconstruction of the Lewis-Eldorado-Hoadley thrust slab at the conclusion of Paleocene thrusting, Montana Cordillera: Geological Society of America Abstract with Programs, v. 36, no. 5, p. 208.

Tagami, T., 2005, Zircon fission-track thermochronology and applications to fault studies, in Reiners, P.W., and Ehlers, T.A., eds., Low-Temperature Thermochronology: Techniques, Interpretations, and Applications: Reviews in Mineralogy and Geochemistry, v. 58, p. 95–122.

Tagami, T., and Shimada, C., 1996, Natural long-term annealing of the zircon fission track system around a granitic pluton: Journal of Geophysical Research, v. 101, p. 8245–8255, doi: 10.1029/95JB02885.

Tagami, T., Carter, A., and Hurford, A.J., 1996, Natural long-term annealing of the zircon fission track system in Vienna Basin deep borehole samples: Constraints upon the partial annealing zone and closure temperature: Chemical Geology, v. 130, p. 147–157, doi: 10.1016/0009-2541(96)00016-2.

Tagami, T., Galbraith, R.F., Yamada, R., and Laslett, G.M., 1998, Revised annealing kinetics of fission tracks in zircon and geological implications, in Van den haute, P., and De Corte, F., eds., Advances in Fission-Track Geochronology: Dordrecht, Netherlands, Kluwer Academic Publishers, p. 99–112.

Teichmuller, M., and Teichmuller, R., 1966, Geological causes of coalification, in Coal Science: Advanced Chemistry Series, 55, Washington, D.C., p. 133–155.

Ting, F.T.C., 1984, Paragenetic relationship of thermal maturation (coalification) and tectonic framework of some Canadian Rocky Mountain coals: Organic Geochemistry, v. 5, p. 279–281, doi: 10.1016/0146-6380(84)90015-9.

Vanderheeghe, O.C., McDougal, T.I., and Dunlap, W.J., 2003, Cooling and exhumation of the Shuswap metamorphic core complex constrained by $^{40}Ar/^{39}Ar$ thermochronology: Geological Society of America Bulletin, v. 115, p. 200–216, doi: 10.1130/0016-7606(2003)115<0200:CAEOTS>2.0.CO;2.

van der Pluijm, B.A., Hall, C.M., Vrolijk, P.J., Peaver, D.R., and Covey, M.C., 2001, The dating of shallow faults in the Earth's crust: Nature, v. 412, p. 172–175, doi: 10.1038/35084053.

Vrolijk, P.J., and van der Pluijm, B.A., 1999, Clay gouge: Journal of Structural Geology, v. 21, p. 1039–1048, doi: 10.1016/S0191-8141(99)00103-0.

Wagner, G.A., 1972, The geological interpretation of fission track ages: Transactions of the American Nuclear Society, v. 15, p. 117.

Yamada, R., Tagami, T., Nishimura, S., and Ito, H., 1995, Annealing kinetics of fission tracks in zircon: An experimental study: Chemical Geology, v. 122, p. 249–258, doi: 10.1016/0009-2541(95)00006-8.

Yin, A., and Kelty, T.K., 1991, Structural evolution of Lewis plate in Glacier National Park, Montana: Implications for regional tectonic development: Geological Society of America Bulletin, v. 103, p. 1073–1089, doi: 10.1130/0016-7606(1991)103<1073:SEOTLP>2.3.CO;2.

Zaun, P.E., and Wagner, G.W., 1985, Fission-track stability in zircon under geological conditions: Nuclear Tracks and Radiation Measurements, v. 10, p. 303–307, doi: 10.1016/0735-245X(85)90119-X.

MANUSCRIPT ACCEPTED BY THE SOCIETY 22 MARCH 2007

The Geological Society of America
Special Paper 433
2007

Reconstructing the Snake River–Hoback River Canyon section of the Wyoming thrust belt through direct dating of clay-rich fault rocks

John G. Solum*
Ben A. van der Pluijm
*Department of Geological Sciences, University of Michigan,
1100 N. University, 2534 C.C. Little Bldg., Ann Arbor, Michigan 48109, USA*

ABSTRACT

Quantification of fault-related illite neomineralization in clay gouge allows periods of fault activity to be directly dated, complementing indirect fault dating techniques such as dating synorogenic sedimentation. Detrital "contamination" of gouge is accounted for through the use of illite age analysis, where gouge samples are separated into at least three size fractions, and the proportions of detrital and authigenic illite are determined using illite polytypism ($1M_d$ = neoformed, $2M_1$ = detrital). Size fractions are dated using the $^{40}Ar/^{39}Ar$ method, representing a significant improvement over earlier methods that relied on K-Ar dating. The percentages of detrital illite are then plotted against the age of individual size fractions, and the age of fault-related neoformed material (i.e., 0% detrital/100% neoformed illite) is extrapolated.

The sampled faults and their ages are the Absaroka thrust (47 ± 9 Ma), the Darby thrust (46 ± 10 Ma), and the Bear thrust (50 ± 12 Ma). Altered host rock along the frontal Prospect thrust gives an age of 85 ± 12 Ma, indicating that the 46–50 Ma ages are not related to a regional fluid-flow event. These ages indicate that the faults in the Snake River–Hoback River Canyon section of the Wyoming thrust belt were active at the same time, indicating that a significant segment of the thrust belt (100 km²+) was active and therefore critically stressed in Eocene time.

Keywords: illite, clay gouge, $^{40}Ar/^{39}Ar$ dating, Sevier orogeny, Wyoming thrust belt.

INTRODUCTION AND GEOLOGIC SETTING

Wyoming Thrust Belt

The Wyoming thrust belt (Royse et al., 1975; Dixon, 1982; Wiltschko and Dorr, 1983) (Figs. 1 and 2) formed during the Late Cretaceous to Eocene Sevier orogeny, which affected western North America from Canada to Mexico (Armstrong 1968; Wiltschko and Dorr, 1983; Price, 1986; Burchfiel et al., 1992; Miller, 2004). The orogeny occurred as the result of the subduction of the Kula and Farallon plates beneath the North American plate (Livaccari and Perry, 1993; Bird, 1998). The deformation style is generally characterized by east-verging, shallowly dipping thrusts with detachments in Mesozoic and Paleozoic shale-rich horizons (DeCelles, 1994).

It has long been proposed that the generally foreland-younging sequence of faults in thrust belts ("in-sequence" faults),

*Present address: Department of Geography and Geology, Sam Houston State University, Campus Box 2148, Huntsville, Texas 77341, USA.

Solum, J.G., and van der Pluijm, B.A., 2007, Reconstructing the Snake River–Hoback River Canyon section of the Wyoming thrust belt through direct dating of clay-rich fault rocks, *in* Sears, J.W., Harms, T.A., and Evenchick, C.A., eds., Whence the Mountains? Inquiries into the Evolution of Orogenic Systems: A Volume in Honor of Raymond A. Price: Geological Society of America Special Paper 433, p. 183–196, doi: 10.1130/2007.2433(09). For permission to copy, contact editing@geosociety.org. ©2007 The Geological Society of America. All rights reserved.

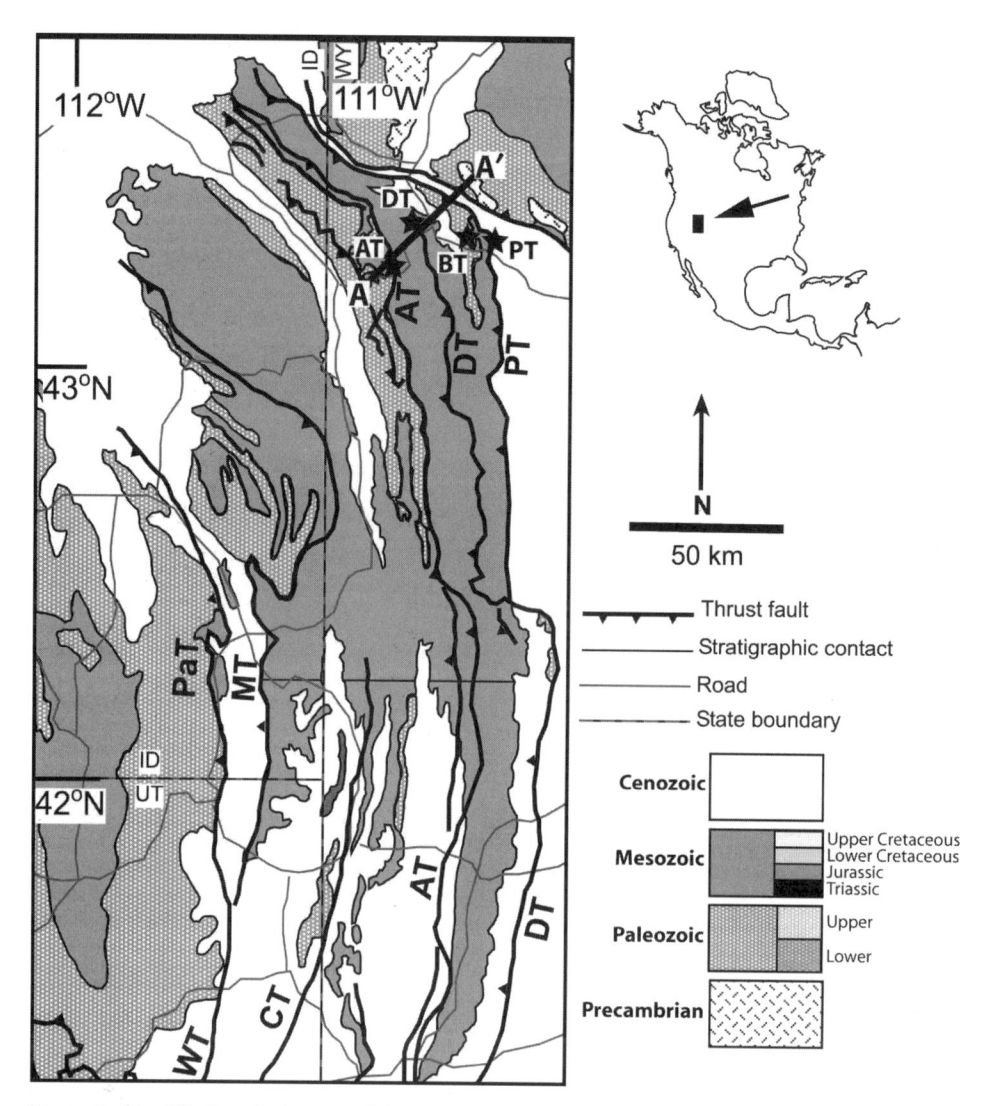

Figure 1. Simplified geologic map of the Idaho-Wyoming thrust belt (after Mitra et al., 1988). Sampling locations are shown with stars. AT—Absaroka thrust; BT—Bear thrust; CT—Crawford thrust; DT—Darby thrust; GCT—Game Creek thrust; HF—Hoback normal fault; MT—Meade thrust; PaT—Paris thrust; PT—Prospect thrust; WT—Willard thrust. Details of sampling locations are described in the text.

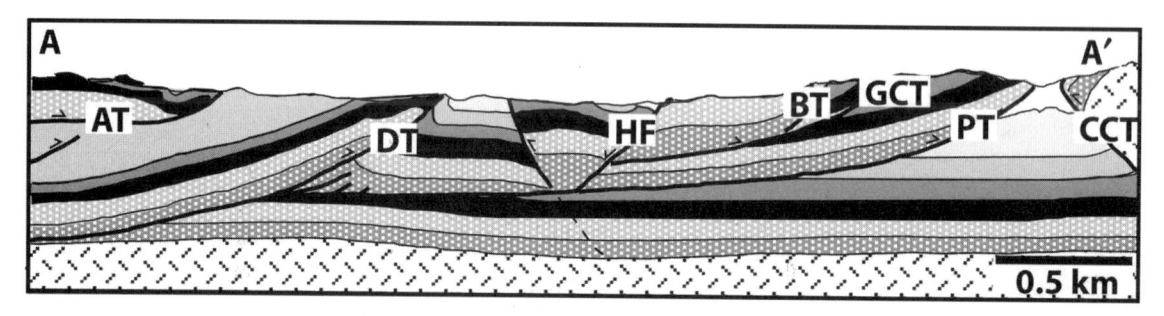

Figure 2. Cross section through the Snake River–Hoback River Canyon section of the Wyoming thrust belt (after Craddock et al., 1988). Reconstructions of this section that are based on timings inferred from synorogenic sedimentation suggest that most of the shortening occurred prior to the Eocene. The results presented in this paper suggest that Eocene shortening was more important than has previously been thought. Fault labels and lithologies are defined in Figure 1.

and faults that violate this trend ("out-of-sequence" faults) can be explained by progressive wedge evolution (Price, 1981; Dahlen, 1984). In order to reconstruct wedge evolution, it is critical to determine the ages of the involved faults. Thick sequences of synorogenic sediments shed from the Sevier thrust belt are preserved in Utah, Idaho, and Wyoming, which have allowed the deformation history of associated faults to be determined (Dorr, 1958; Royse et al., 1975; Wiltschko and Dorr, 1983; DeCelles, 1994; DeCelles and Mitra, 1995). Based on histories inferred from this sedimentation, it is thought that the evolution of the Utah-Idaho-Wyoming thrust belt was controlled by maintenance of critical wedge taper (Wiltschko and Dorr, 1983; DeCelles, 1994; DeCelles and Mitra, 1995). However, in addition to uncertainties in stratigraphic correlation, this type of reconstruction cannot be applied to fault systems where no record of synorogenic sedimentation is preserved. It is therefore desirable to directly date fault rocks whenever possible. The considerations, complications, limitations, and techniques necessary to directly date clay-rich fault rocks are discussed in this contribution using new results from the Wyoming segment of the Sevier thrust, followed by the implications of results from the study area.

Character of Illite and Illite-Smectite Neomineralization

It has long been recognized that the clay fraction of a sample is composed of a mixture of clays with multiple origins (i.e., detrital and authigenic). Velde and Hower (1963) noted that the <1 μm fraction of illite-bearing samples was relatively enriched in $1M_d$ illite. Using the phase relations of Smith and Yoder (1956), which indicate that $2M_1$ illite is stable at higher temperatures than 1M and $1M_d$, they concluded that the $1M_d$ fraction of their samples formed at a lower temperature than the $2M_1$ fraction, likely during diagenesis. They similarly inferred that the $2M_1$ component of their samples was detrital in origin. Hower et al. (1963) noted that the K-Ar ages of size fractions from shale samples decreased with decreasing grain size and that the finer size fractions contained more $1M_d$ illite than the coarser fractions. They concluded that the finer material formed some time after deposition of the shale, but that the age was difficult to interpret because the ages of all size fractions reflected mixtures of clays with different origins. More recent studies (Pevear et al., 1997; Clauer et al., 1997; Pevear, 1999; Srodon, 2002) have shown that finer fractions of shales tend to be more enriched in mixed-layer illite-smectite (I-S) than coarser fractions, and therefore that I-S is authigenic and that discrete illite is detrital. The ability to extract geologically meaningful ages from clay-rich rocks requires additional characterization of the clay populations in a sample. Whereas the fine fraction of a shale or gouge may be dominated by authigenic material, it still generally contains some detrital material, and therefore the age of that fraction will be a mixing age. Quantification of the amounts of detrital and authigenic phases is therefore required to extract the neoformation age.

Originally the amounts of detrital and authigenic illite were quantified by measuring the amounts of discrete illite and interlayered illite-smectite (Pevear, 1994; van der Pluijm et al., 2001), a procedure called illite age analysis. Additionally, illite age analysis can also incorporate changes in illite polytypism to quantify the detrital and authigenic components, based on the observation that the $2M_1$ polytype is detrital, while the $1M_d$ is authigenic (Grathoff and Moore, 1996; Grathoff et al., 1998, 2001; Solum et al., 2005). This study of Wyoming thrust fault rocks principally uses changes in polytypism to quantify the proportions of detrital and neoformed clays in fault rocks.

Direct Dating of Fault Rocks

The age of a shallow-crustal fault is usually constrained by bracketing, relying on the age of features that are cut by the fault, or by dating synorogenic sedimentation. Although minerals suitable for radiometric dating are common in many fault rocks (the K-bearing clay mineral illite), direct dating of fault rocks is complicated by the observation that such rocks are a mixture of detrital and neoformed fault-related phases. The occurrence of multiple illite polytypes in clay-bearing samples, and, as discussed already, variations between polytype composition and age, has long been recognized (e.g., Hower et al., 1963; Velde and Hower, 1963), but the use of detailed polytype quantification and some form of illite age analysis has been used sparingly to date fault rocks (Vrolijk and van der Pluijm, 1999; Ylagan et al., 2002; Solum et al., 2005) and undeformed shales and mudstones (Grathoff et al., 2001).

Multiple gouge (or shale) size fractions coupled with clay characterizations have been used previously for dating (Hower et al., 1963; Lyons and Snellenburg, 1971; Chen et al., 1988; Wang et al., 1990; Parry et al., 2001; Zwingmann et al., 2004; Zwingmann and Mancktelow, 2004). However, these studies did not use the illite age analysis technique to extrapolate an age for neoformed clays (Pevear et al., 1997; van der Pluijm et al., 2001, 2006; Solum et al., 2005). The finest size fraction in shales and clay gouges contains the least amount of detrital illite, so its age is closest to the age of authigenic mineralization. However, even the finest fraction contains some detrital material, so the age of the finest fraction is only the maximum possible age of authigenic mineralization. In some special circumstances, detailed characterizations of the clay fractions may not be necessary. If the age of all three size fractions is the same, that age is the age of faulting, as this phenomenon indicates that all of the illite has been reset (Parry et al., 2001). Such age relations, however, are rarely observed.

Few studies have simultaneously used both quantifications of $2M_1/1M_d$ illite and discrete illite/mixed-layer I-S to quantify detrital and authigenic material, but, based on such studies, it appears that polytype quantification is the more generally applicable of the two approaches. Pevear (1999) conducted a study using both approaches on shales. Ylagan et al. (2002) conducted a study for the Canadian Rockies in Alberta and British Columbia, and Solum et al. (2005) studied the Moab normal fault in east-central Utah. As pointed out by Ylagan et al. (2002), use of polytypism quantification instead of discrete illite/mixed-layer I-S has the advantage of making use of the entire three-dimensional crystal structure, and not just relying on 001 reflections. In addition, it is not possible to use discrete illite/mixed-layer I-S quantification if a sample contains little or no I-S, and the use of discrete illite/mixed-layer I-S dating may also be complicated by fault-related neoformation of illite. For example, gouge along the Lewis thrust (Vrolijk and van der Pluijm, 1999; Yan et al., 2001) and the Moab fault (Solum et al., 2005) is enriched in illite relative to protolith, indicating that when I-S–bearing material is deformed, discrete illite may form. Note that this contradicts the assumption that all discrete illite in a sample is detrital. The formation of illite from smectite during burial is also well-documented (Ahn and Peacor, 1986; Freed and Peacor, 1989, 1992; Lindgreen et al., 1991; Buatier et al., 1992). Moreover, in order to use discrete illite/mixed-layer I-S quantification to extrapolate authigenic ages, the detrital/authigenic ratios must be normalized based on the percentage of illite interlayers in the mixed-layer I-S (Ylagan et al., 2002), which accounts for the general lack of potassium in the smectite in I-S (not including exchangeable cation sites). Therefore, the smectite interlayers should not be included in age-related calculations. If this renormalization is not conducted, the proportion of the detrital phase is underestimated, resulting in invalid extrapolations (Srodon, 1999), although with illite-rich I-S, this renormalization results in only minor adjustments. For these reasons, polytype quantification may be preferred over discrete illite/mixed-layer I-S quantification when dating fault rocks in most settings.

METHODS

Sampling Locations

Fault rock samples were collected along the Snake River and Hoback River sections of the Wyoming segment of the Idaho-Wyoming thrust belts (Dorr et al., 1987; Coogan, 1992) (Figs. 1 and 2). A generalized stratigraphic section of the region is shown in Figure 3. The Absaroka thrust was sampled ~22.5 km (14 mi) west of Hoback Junction, Wyoming, in the Snake River Canyon along the north side of Highway 26. At this location, the Mississippian Madison limestone (intensely fractured with abundant calcite veins within 2–4 m of the fault) is faulted against the Cretaceous Bear River Formation. Small outcrops of limestone in the hanging wall and sandstone in the footwall make the location of the fault easy to determine; however, clay gouge is not well-exposed. We sampled a footwall outcrop of faulted clay-rich gouge, with abundant slickenlines and polished surfaces, ~5 m from the main fault.

The Darby thrust was sampled along Fall Creek Canyon, ~1 km (0.6 mi) along a dirt road that forms off from Fall Creek Road, ~5.1 km (3.2 mi) from the intersection of Fall Creek Road and Highway 26 (~6.8 km [4.2 mi] west of Hoback Junction, Wyoming). The Jurassic Nugget sandstone, brecciated for a distance of 5–7 m from the fault, is in fault contact with a Mesozoic carbonate, which is similarly intensely fractured near the fault. The gouge is up to 0.3 m wide, well-consolidated, and composed of a red/yellow sandy clay and carbonate with anastomosing foliations.

The Bear thrust was sampled at Red Creek, along Highway 191, ~4.8 km (3 mi) east of the mouth of the Hoback River Canyon. Here, the Pennsylvanian Tensleep Formation was emplaced over the Triassic Chugwater Formation. Gouge along the main fault is not exposed, but many small thrusts with displacements from ~1–3 m are exposed in the Chugwater Formation along the banks of the Hoback River, below the main thrust.

No suitable exposures of gouge along the Prospect thrust were found on the west side of Granite Creek, ~1.4 km (0.9 mi) from the junction of Granite Creek with the Hoback River, ~18 km (11.2 mi) east of Hoback Junction in the Hoback River Canyon, where the Jurassic Nugget and Triassic Chugwater Formations are emplaced over the late Paleocene Hoback Formation. For comparison with fault rock samples, a clay-rich rock ~10 m from the fault contact in the hanging wall of the fault within the Chugwater Formation was collected. At this location, detrital muscovite grains were common and visible to the naked eye.

Sample Preparation

Unconsolidated samples were placed in beakers filled with distilled water and allowed to disaggregate for approximately one week. Consolidated samples were crushed using a jaw crusher and then treated in the same manner as unconsolidated samples. After initial soaking, the samples were suspended and allowed to settle multiple (three or more) times. After each settling, the remaining clear liquid was decanted to remove dissolved salts. Samples were treated with a small amount of powdered sodium carbonate to aid deflocculation. Following rinsing, the <4 μm fraction was separated using gravity settling (hereafter referred to as the "clay fraction"). The settling was repeated approximately three times to increase the volume of the fraction. The <2 μm fraction was separated from samples in which the <4 μm fraction contained K-feldspar, as indicated by X-ray diffraction (XRD). The clay fraction was then placed in multiple 50 mL plastic centrifuge aliquots, each of which was treated with an ultrasonic probe for 5 min. Following this treatment, the <0.5 μm fraction was separated using centrifugal settling. A similar procedure was used to extract the <0.05 μm fraction from the sample. In this fashion, a gouge sample was separated into coarse (4–0.5 μm), intermediate (2–0.5 μm), medium (0.5–0.05 μm), and fine (<0.05 μm) fractions.

During extraction, clay samples are commonly treated with weak acids and organic solvents to remove carbonates, iron oxides, and organic matter. Moore and Reynolds (1997) cautioned that possible effects of each of these treatments on mixed-layer clays, such as illite-smectite, have not been quantitatively evaluated, and so we did not apply these treatments to our samples. Moreover, while the presence of these phases can

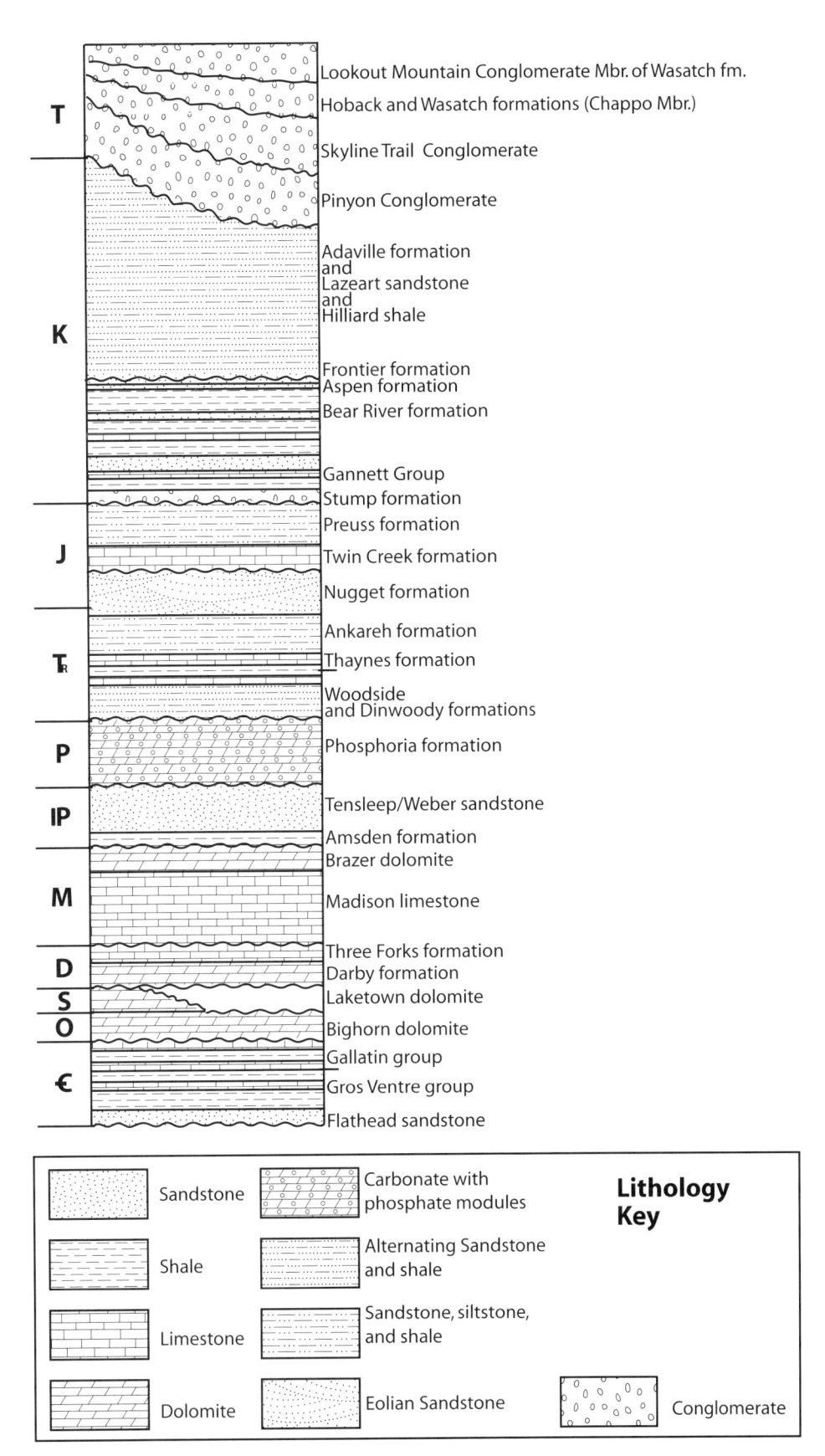

Figure 3. Simplified stratigraphic section for the general area of the Wyoming thrust belt (modified from Dixon, 1982; Wiltschko and Dorr, 1983). T—Tertiary; K—Cretaceous; J—Jurassic; TR—Triassic; P—Permian; IP—Pennsylvanian; M—Mississippian; D—Devonian; S—Silurian; O—Ordovician; €—Cambrian.

impair sample preparation or mineral identification, the small grain-size fractions limited the concentration of these contaminating phases in our medium and fine fractions.

Clay Characterizations

Samples for polytypism analysis required additional sample preparation. The $2M_1$-specific peaks all have non-(001) indices, and so will not be visible on an X-ray diffractogram from a sample with a strong preferred orientation. Consequently, it is necessary to prepare a sample with a random orientation, which can be difficult in the case of clay minerals due to their platy habit. A near-random preparation was achieved by using an end-packer device (Moore and Reynolds, 1997), in which a dried, powdered sample was loosely tamped into a milled plate of aluminum covered with a glass slide. Once tamping was complete, the glass slide was removed, and the aluminum holder was placed in the diffractometer. Care was taken to compact the powder as loosely as possible, as it was found through trial and error that strong tamping increased the preferred orientation in the sample. Grathoff and Moore (1996) noted that it is possible to evaluate the effectiveness of the random sample preparation by comparing the relative intensity of the (020) and (002) peaks on random and oriented samples. If the relative intensity of the (002) peak is reduced, then the powder pattern may be used for polytype quantification. If the relative intensity is not reduced, then the sample must be repacked and rescanned.

Similar to analyses of discrete illite and interlayered I-S (Solum et al., 2005), illite polytypism was quantified by modeling X-ray powder diffraction patterns. The program WILDFIRE (Reynolds, 1993a, 1993b) was used to generate synthetic patterns of various types of $2M_1$ and $1M_d$ illite (detailed in the following). WILDFIRE allows the number of continuous interlayers to be varied, as well as allowing for some degree of preferred orientation in the sample through the use of an orientation factor (the "Dollase factor"). There are several more parameters that can be varied in the case of $1M_d$, such as the proportion of cis- and trans-vacant interlayers, the number of smectite interlayers, and the degree of ordering in the stacking sequence (i.e., it is not assumed that stacking is completely random). WILDFIRE was used to calculate patterns of $1M_d$ with ordering that varied from random to complete (i.e., classic $1M_d$ to classic 1M). Patterns for $2M_1$ illite were generated assuming a minimum number of continuous interlayers of 5, and maximum numbers of 10–30 in increments of 5, using Dollase factors of 1, 0.9, and 0.8 (15 patterns). Patterns for $1M_d$ illite were calculated using a probability of a zero rotation of 0.33–1.0 (0.33 = pure $1M_d$; 1.0 = pure 1M) in increments of 0.1, using Dollase factors of 1, 0.9, and 0.8. The ratio of trans-vacant to cis-vacant interlayers was set at 0, 0.5, and 1, and the fraction of expandable interlayers was set to 0.1, 0.3, 0.5, 0.7, and 0.9, and interlayer rotations of multiples of 60° were used (360 patterns).

Spreadsheets using the WILDFIRE-generated reference libraries were used to quantify the concentrations of the $2M_1$ and $1M_d$ polytypes in a sample. Patterns of each of the possible types of $2M_1$ illite were mixed with each of the types of $1M_d$ illite from 100% to 0% in increments of 10% of $2M_1$. The variance between the synthetic mixtures and the gouge sample and the synthetic mixture was calculated, and mixtures with the lowest variance were further refined by creating patterns from 100% to 0% of $2M_1$ in increments of 1%. The sample with the lowest variance was selected as the best match. Quartz, carbonate, plagioclase, and hematite peaks in the sample were excluded. Representative XRD powder patterns for $2M_1$ and $1M_d$ illites with WILDFIRE-generated overlays are shown in Figure 4, and examples of best matches for gouge size fractions are shown in Figure 5. Extrapolated illite ages were calculated using a York regression (York, 1968), which takes into account uncertainties in both variables.

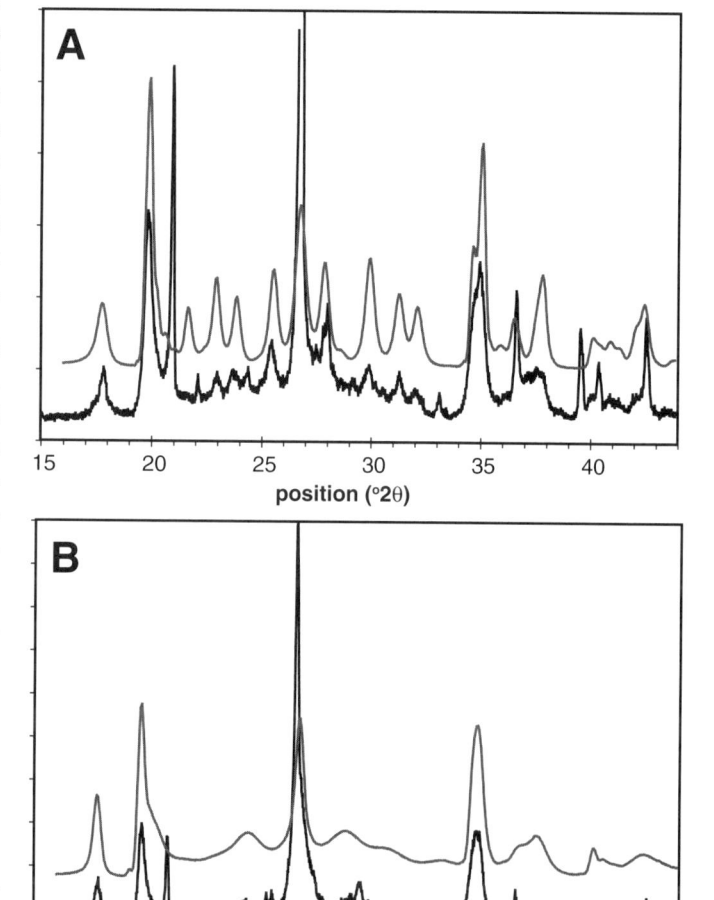

Figure 4. Illite polytype standards overlain with WILDFIRE synthetic patterns. (A) $2M_1$ illite and (B) $1M_d$ illite. Unmatched peaks in X-ray diffraction (XRD) patterns are quartz and feldspars. Note the greater number of peaks for the $2M_1$ illite as a result of coherent stacking of clay interlayers. The matching technique described in this paper matches mixtures of these standards to ~2%.

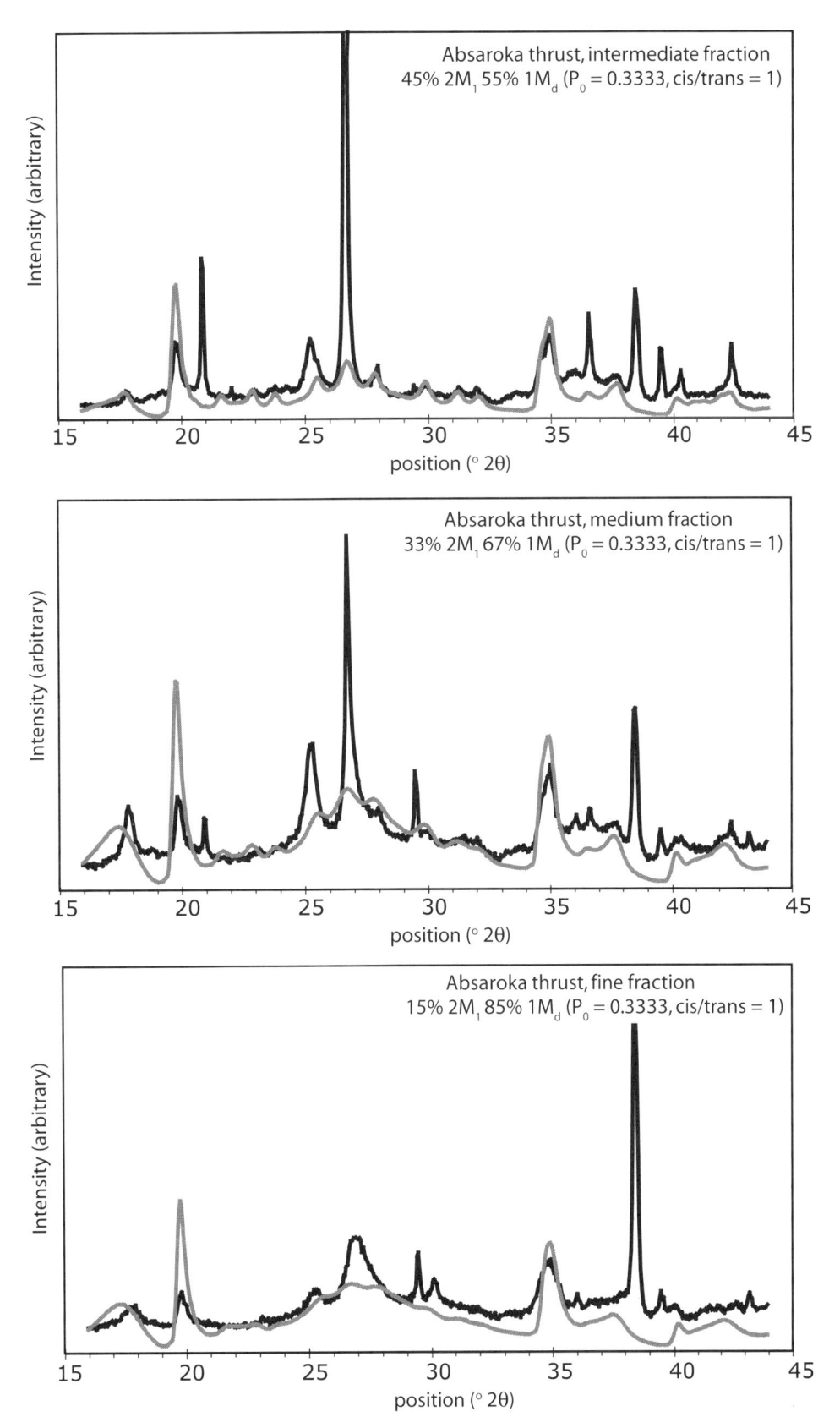

Figure 5. Best-fit WILDFIRE-generated powder patterns for X-ray diffraction (XRD) patterns of size fractions for a representative fault gouge sample from the Absaroka thrust. Synthetic patterns are based on varying proportions and crystal structures of $2M_1$ and $1M_d$ illite. The crystallographic parameters of the $1M_d$ illite are given in parentheses. As discussed in the text, unmatched peaks are quartz, feldspars, and non-illite clays.

It is also possible to determine the concentration of the $2M_1$ polytype by measuring the area of peaks that are unique to the $2M_1$ polytype and finding the ratio of them against the area of a peak at ~0.258 nm (2.58 Å)/35 °2θ (Cu Kα), which is common to both polytypes (Grathoff and Moore, 1996; Grathoff et al., 1998). This technique has been successfully used to generate quantifications in dating fault rocks (Solum et al., 2005) and hydrothermal events (Grathoff et al., 2001); however, the quantification approach used in this paper offers the advantage of matching an entire powder pattern instead of separate, isolated peaks.

In contrast to earlier studies, we do not include classic 1M illite in our analyses (cf. Grathoff and Moore, 1996; Grathoff et al., 1998), because transmission emission microscopy (TEM) investigations of shales, mudstones, and slates inferred to contain 1M illite based on XRD analysis have almost exclusively been unable to find that polytype (Peacor et al., 2002). This is also in agreement with the work of Zoller and Brockamp (1997), who concluded that $2M_1$ and 1M illite differ in composition and therefore are not polytypes in a strict sense. Therefore, instead of modeling gouge as a mixture of $2M_1$, 1M, and $1M_d$ illite, we model gouge as a mixture of $2M_1$ and $1M_d$ illite with varying probabilities of a 0° rotation, as discussed already.

Errors of ±2% absolute were assigned for each of the polytype characterizations. The magnitude of this error was based on the characterization of a 50/50 mixture of $2M_1$ illite from an unknown site in Illinois and $1M_d$ illite (Illite from Silver Hill, Montana, Clay Minerals Society Source Clay IMt-1) using the approach described already. Our quantification of this standard yielded a result of 52% $2M_1$/48% $1M_d$.

RESULTS

The Absaroka, Darby, and Bear thrusts are located along the Snake River and Hoback River Canyons, reflecting a geographical progression from west to east. A sample from wall rock in the most frontal segment of the belt, in the hanging wall of the Prospect fault, is also shown for comparison with fault rock. Age spectra for all samples are shown in Figures 6 and 7. The extrapolated ages and statistical parameters of our analyses are shown in Figure 8 and in Table 1. Individual faults are discussed next here.

Absaroka Thrust

Based on the record of synorogenic sedimentation, the Absaroka thrust was initiated with a period of activity in the Late Cretaceous (the "Early Absaroka" of Royse et al., 1975), it experienced a period of major activity in the late Paleocene, and it underwent a final period of activity in the Eocene (DeCelles and Mitra, 1995). The earlier episodes have been inferred to be major displacements (Wiltschko and Dorr, 1983; DeCelles and Mitra, 1995), whereas the latter two have been inferred to be relatively minor (DeCelles and Mitra, 1995). The illite age for fault rock of the Absaroka thrust in the Snake River Canyon is 47 ± 9 Ma, indicating a significant period of activity at that location in the lowermost middle Eocene.

Darby Thrust

The age of the Darby thrust is poorly defined due to a lack of well-constrained synorogenic sedimentation, but it is considered to be the youngest thrust in the belt. Wiltschko and Dorr (1983) concluded that the Darby thrust was active around the middle Paleocene to the early Eocene, and perhaps involved multiple periods of motion. The polytype-derived authigenic age of fault rock of the Darby thrust along Fall Creek is 46 ± 10 Ma, indicating a period of activity along the fault at that location in the middle Eocene.

Bear Thrust

The Bear thrust outcrops along the Hoback River and is an imbricate splay from the frontal Prospect thrust (e.g., Craddock et al., 1988) and must, therefore, be younger than earliest Eocene on stratigraphic grounds (Wiltschko and Dorr, 1983; Dorr and Steidtmann, 1977). Indeed, the illite age for subsidiary faults in the footwall of the Bear thrust along Red Creek gives an early to middle Eocene age of 50 ± 12 Ma for the fault rocks.

Wall Rock at the Prospect Thrust

The similarity in ages from fault rocks as discussed here may indicate regional resetting of the Ar clock in illite and not fault-related neomineralization, so we also examined a sample of host rock for comparison. If fault rock ages reflect regional resetting, through fluid and/or thermal activity, nearby host rock ages of similar composition should give the same age range as fault rocks. The age of the Prospect thrust (the Cliff Creek thrust of Dorr et al., 1977) was constrained to be early Eocene by Wiltschko and Dorr (1983), as the fault cuts the late Paleocene Skyline trail conglomerate and is overlapped by the early Eocene Lookout Mountain conglomerate. This constraint is supported by our radiometric age of the Bear fault splay (see previous). In contrast, the polytype-derived authigenic age for host rock near the Prospect thrust is 85 ± 12 Ma. This older illite age supports a non–fault related alteration of Late Cretaceous age.

DISCUSSION

The illite ages from the Snake River–Hoback River thrusts show that all of the main faults (Absaroka, Darby, and Bear faults), combined with the stratigraphically constrained Eocene age of the Prospect thrust in the most easterly segment of the

Figure 6. Argon age spectra for individual size fractions from clay gouge samples. Extrapolated ages of fault-related authigenesis are shown in Figure 8. All size fractions are composed of a mixture of clays with multiple origins (detrital and authigenic), and so plateaus are not expected.

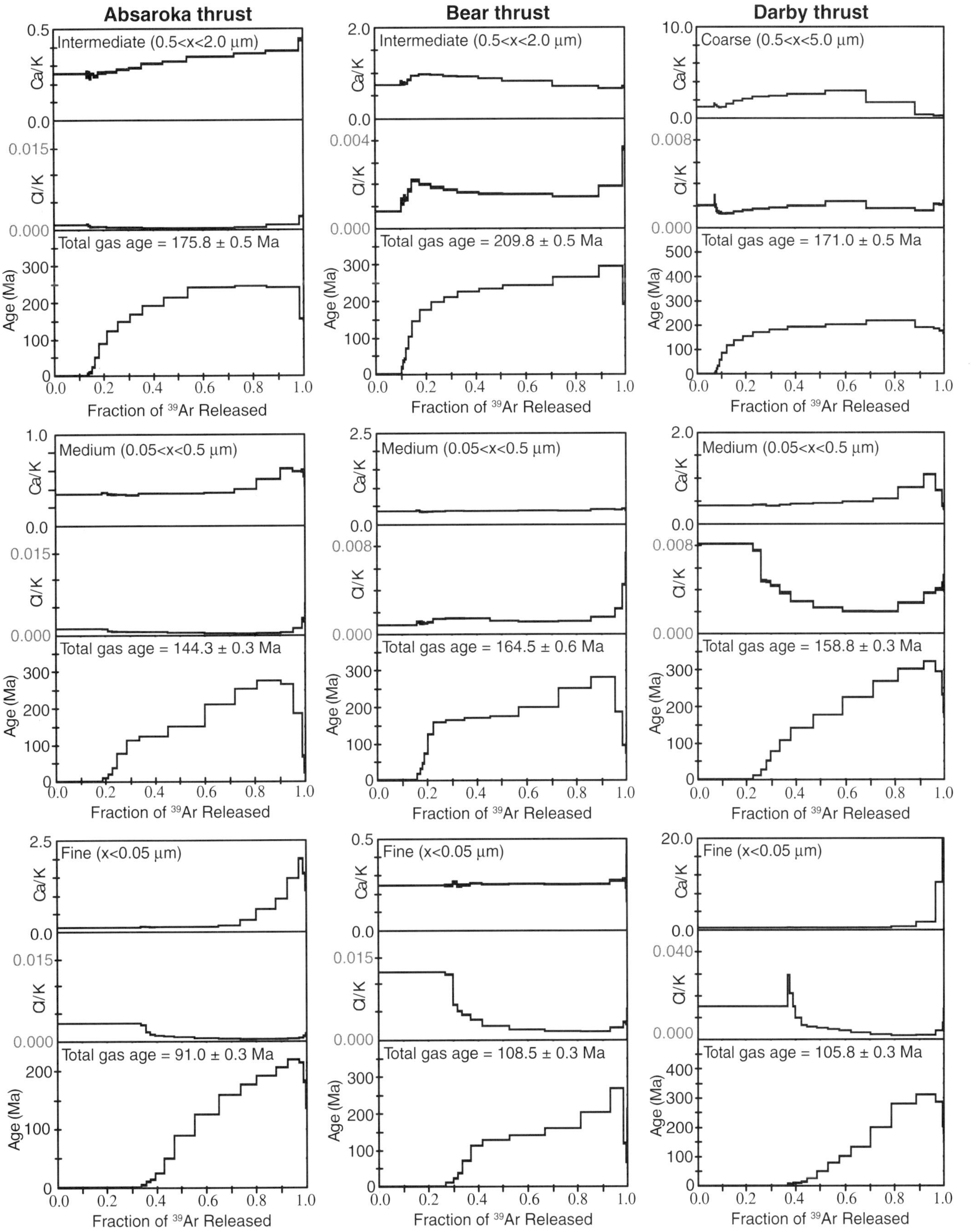
Absaroka thrust
Bear thrust
Darby thrust
Intermediate (0.5<x<2.0 µm)
Coarse (0.5<x<5.0 µm)
Medium (0.05<x<0.5 µm)
Fine (x<0.05 µm)
Ca/K
Cl/K
Age (Ma)
Total gas age = 175.8 ± 0.5 Ma
Total gas age = 209.8 ± 0.5 Ma
Total gas age = 171.0 ± 0.5 Ma
Total gas age = 144.3 ± 0.3 Ma
Total gas age = 164.5 ± 0.6 Ma
Total gas age = 158.8 ± 0.3 Ma
Total gas age = 91.0 ± 0.3 Ma
Total gas age = 108.5 ± 0.3 Ma
Total gas age = 105.8 ± 0.3 Ma
Fraction of 39Ar Released

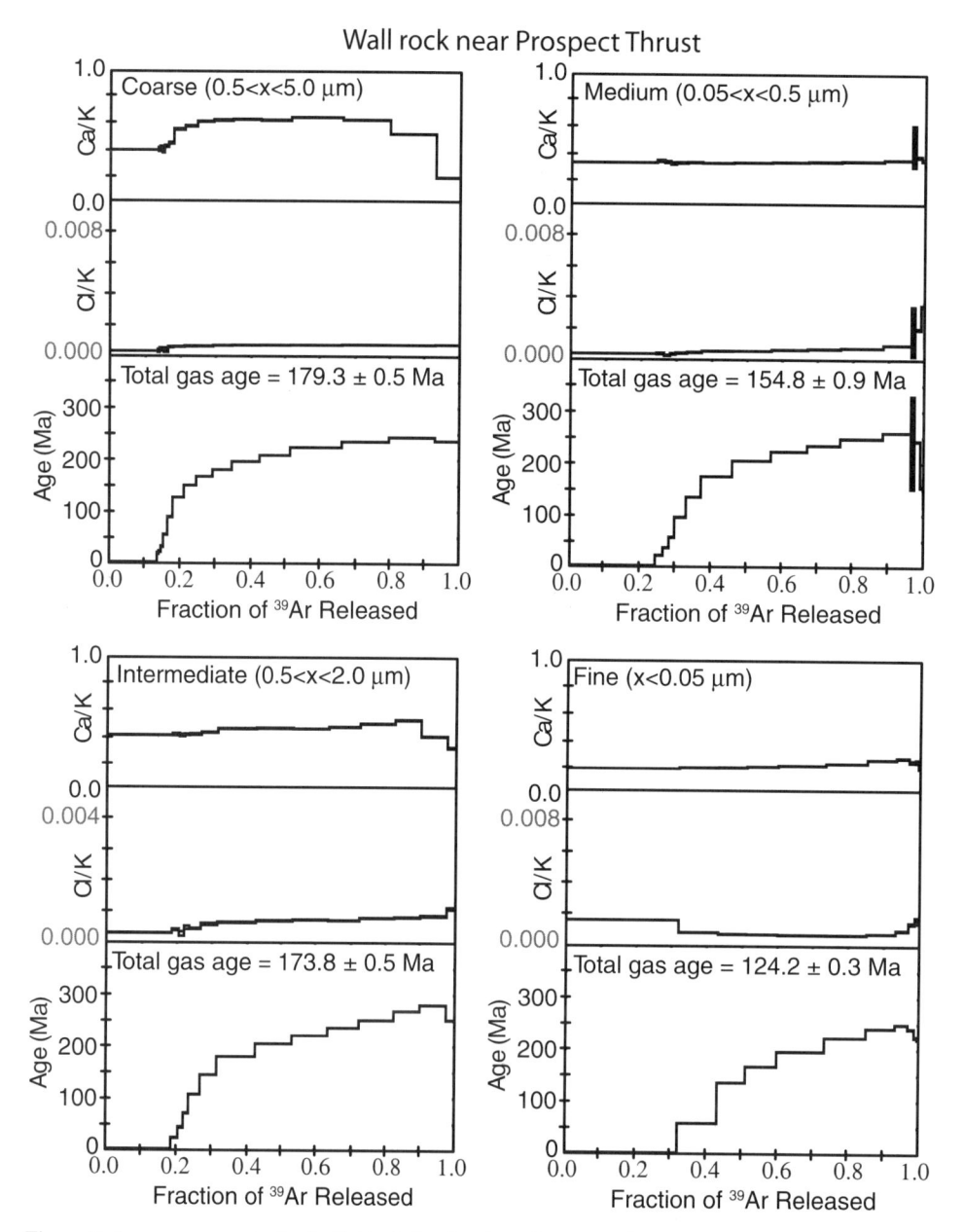

Figure 7. Argon age spectra for individual size fractions from samples of wall rock near the Prospect thrust. Extrapolated ages of fault-related authigenesis are shown in Figure 8. All size fractions are composed of a mixture of clays with multiple origins, so plateaus are not expected.

section, representing a section of 30+ km, were active at the same time in the Eocene. This section was shortened by >15% from the Cretaceous to the Eocene, representing approximately one-third of the total shortening of the thrust belt (Royse et al., 1975, their plate IV). Within our error estimates, this implies that thrusts were not progressively abandoned as the belt propagated into the foreland; rather, this segment of the thrust belt was active during the same period in the early to middle Eocene. Our results suggest that more shortening occurred in the Eocene than has previously been thought.

As noted previously, Cretaceous activity, as well as an Eocene period of activity, are inferred for the Absaroka thrust based on synorogenic sedimentation near the town of Kemmerer (Royse et al., 1975), located ~180 km south of the Snake River Canyon. DeCelles and Mitra (1995) concluded that in southern Wyoming, the Cretaceous periods of activity were major, while the Eocene period was minor. Since no record of Cretaceous activity is shown by the fault rock ages along the Snake River Canyon section of the Absaroka thrust, this indicates that either the fault was active in the Eocene, or that clays that were formed

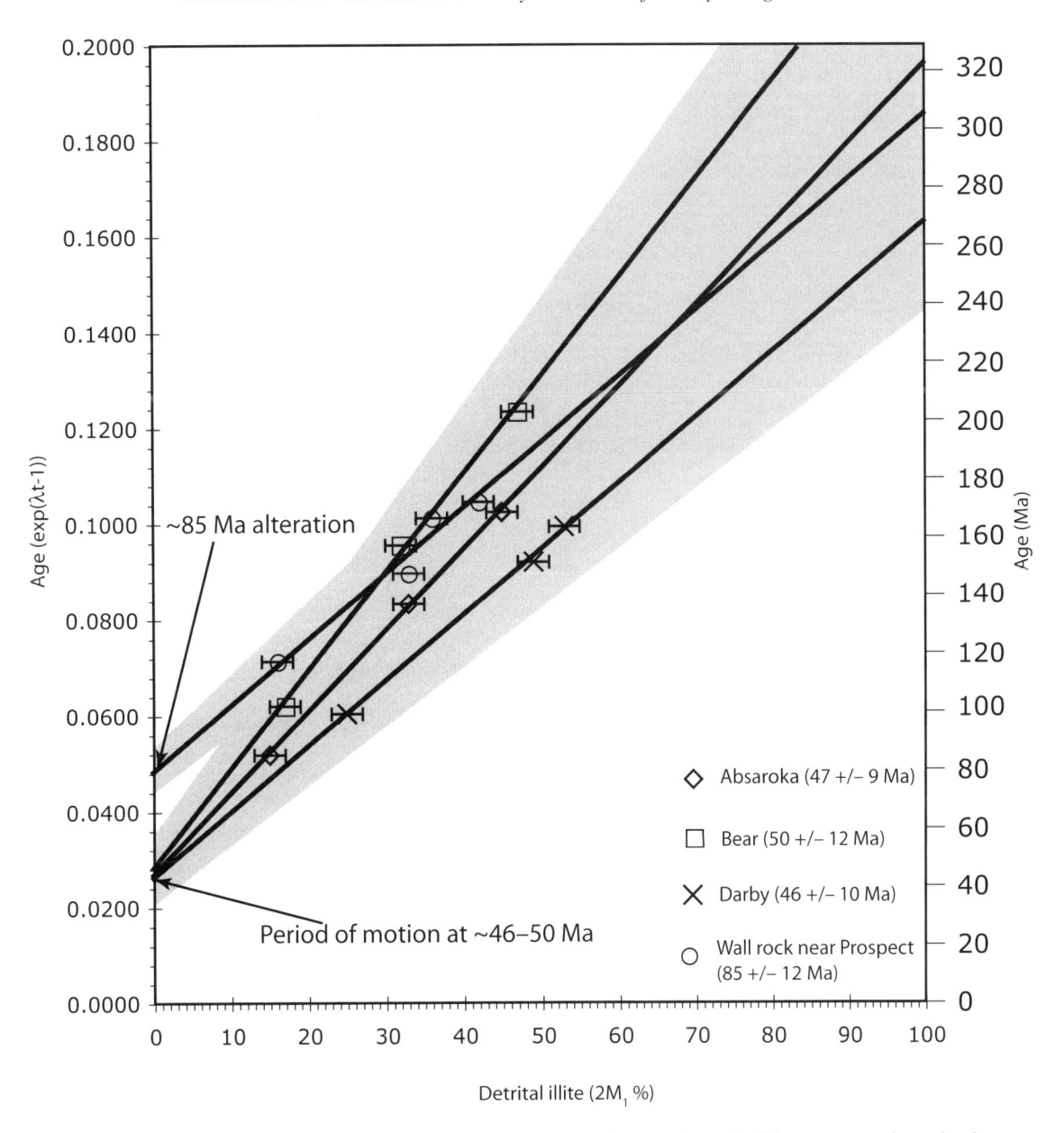

Figure 8. Illite age analysis plot for three fault rock samples and one host rock sample. The errors are shows by the gray bands surrounding best-fit lines to the data. The relationships of these results to ages derived from the record of synorogenic sedimentation are discussed in the text.

during the inferred Cretaceous events were reworked, causing the fault rock ages to be reset. Regardless of resetting, the early to middle Eocene must have been a period of major fault activity in this segment of the thrust belt.

Coeval periods of activity along multiple thrusts have been proposed as a means by which critical taper of a thrust wedge can be maintained (Boyer, 1992; Jordan et al., 1993). This scenario has been inferred for the thrust belt in northeast Utah and southwest Wyoming (DeCelles, 1994), although DeCelles (1994) noted that more internal deformation, including reactivation along existing faults, is required than has previously been recognized. The illite ages reported in this study suggest that internal deformation may be more widespread than previously thought, increasing the applicability of critical taper models to the Sevier thrust belt.

The observation that the 46–50 Ma ages occur in fault rocks from several different faults, and not in adjacent protolith, is a strong indication that those ages reflect a period of regional fault activity. The age of ca. 85 Ma from clay-rich alteration in the Chugwater Formation in the hanging wall of

TABLE 1. MINERALOGIC QUANTIFICATIONS AND ASSOCIATED ILLITE AGE ANALYSES FOR SAMPLES ALONG THE SNAKE RIVER–HOBACK RIVER SECTION OF THE WYOMING SEGMENT OF THE SEVIER THRUST BELT

	Detrital illite ($2M_1$,%)	Error	Total gas age (Ma)	Error	Age [exp(λt) − 1]
Absaroka thrust					
Size fraction					
Intermediate	45.00	2.00	175.80	0.50	0.10
Medium	33.00	2.00	144.30	0.30	0.08
Fine	15.00	2.00	91.00	0.30	0.05
Slope	0.00169	0.00016			
Intercept	0.02662	0.00530			
Authigenic age	47 ± 9				
Bear thrust					
Size fraction					
Intermediate	38.00	2.00	209.80	0.50	0.12
Medium	22.00	2.00	164.50	0.60	0.10
Fine	17.00	2.00	108.50	0.30	0.06
Slope	0.00205	0.00019			
Intercept	0.02798	0.00664			
Authigenic age	50 ±12				
Darby thrust					
Size fraction					
Coarse	53.00	2.00	171.00	0.50	0.10
Medium	49.00	2.00	158.80	0.30	0.09
Fine	25.00	2.00	105.80	0.30	0.06
Slope	0.00137	0.00013			
Intercept	0.02595	0.00566			
Authigenic age	46 ±10				
Wall rock near the Prospect thrust					
Size fraction					
Coarse	42.00	2.00	179.30	0.50	0.10
Intermediate	36.00	2.00	173.80	0.50	0.10
Medium	33.00	2.00	154.80	0.90	0.09
Fine	16.00	2.00	124.20	0.30	0.07
Slope	0.00137	0.00015			
Intercept	0.04805	0.00485			
Authigenic age	85 ±12				

the Prospect thrust is clearly unrelated to the timing of motion along that fault, which is well-constrained to be Eocene in age based on dating of synorogenic sediments. This highlights the importance of obtaining samples directly from gouge zones in order to date fault rocks, since even clay-rich samples close to the gouge zone may not record fault-related mineralization. Most importantly, this age demonstrates that the 46–50 Ma ages from fault rocks are not related to a regional diagenetic event. If this were the case, then protolith as well as fault rocks would exhibit the same authigenic age. It is possible that the ca. 85 Ma age of host rock may represent neomineralization associated with fluids migrating toward the foreland during Cretaceous motion along more inboard thrusts, such as the Paris, Meade, or Crawford thrusts, which produced synorogenic conglomerate of Late Cretaceous age (Gannet Group; e.g., DeCelles and Mitra, 1995).

SUMMARY AND CONCLUSIONS

Illite ages using $2M_1/1M_d$ polytype quantification of clay-rich rocks provide a reliable means of directly dating periods of activity along shallow faults. The illite ages obtained in the area and the record of synorogenic sedimentation are generally compatible; however, the illite ages indicate that motion may occur along a fault without leaving a record of synorogenic sedimentation or that some stratigraphic ages are incompletely constrained. Moreover, the ages suggest that Eocene shortening in this section of the Wyoming thrust belt may have been more substantial than has previously been suspected. The illite ages indicate that the faults in the Snake River–Hoback River Canyon section of the thrust belt were simultaneously active at ca. 48 Ma. The Absaroka, Darby, and Bear/Prospect thrusts are presently separated by 30+ km, and, when restored, were separated by 45 or more kilometers. These ages therefore indicate that a considerable segment of the Wyoming thrust belt was pervasively active in early to middle Eocene times. The coeval period of faulting in this segment of the thrust belt supports experiments and theoretical models (e.g., Dahlen, 1984; Dahlen et al., 1984; Wiltschko and Dorr, 1983; DeCelles and Mitra, 1995) of a critically stressed thrust belt, where frontal deformation of the Sevier thrust belt was controlled by the maintenance of a critical taper (e.g., DeCelles, 1994).

ACKNOWLEDGMENTS

The authors extend their thanks to Adolph Yonkee for assistance and helpful conversations in the field. Chris Hall is thanked for discussion and supervision of Ar analysis in the University of Michigan's Radiogenic Isotope Geochemistry Laboratory; XRD analysis was carried out at the University of Michigan Electron Microbeam Analysis Laboratory; sample preparation was aided by Tracy Kolb. Georg Grathoff and Bob Ylagan are thanked for conversations about clay mineralogy. We thank two anonymous reviewers for their helpful comments. Support for this work was provided by National Science Foundation (NSF) grant EAR-0230055, a Grant-In-Aid from the American Association of Petroleum Geologists, and the Scott M. Turner Fund at the University of Michigan.

REFERENCES CITED

Ahn, J.H., and Peacor, D.R., 1986, Transmission and analytical electron microscopy of the smectite-to-illite transition: Clays and Clay Minerals, v. 34, p. 165–179, doi: 10.1346/CCMN.1986.0340207.

Armstrong, R.L., 1968, Sevier orogenic belt in Nevada and Utah: Geological Society of America Bulletin, v. 79, p. 429–458, doi: 10.1130/0016-7606(1968)79[429:SOBINA]2.0.CO;2.

Bird, P., 1998, Kinematic history of the Laramide orogeny in latitudes 35°–49°N, western United States: Tectonics, v. 17, p. 780–801, doi: 10.1029/98TC02698.

Boyer, S.E., 1992, Geometric evidence for synchronous thrusting in the southern Alberta and northwest Montana thrust belts, in McClay, K.R., ed., Thrust Tectonics: New York, Chapman and Hall, p. 377–390.

Buatier, M.D., Peacor, D.R., and O'Neil, J.R., 1992, Smectite-illite transition in Barbados accretionary wedge sediments: TEM and AEM evidence for dissolution/crystallization at low temperature: Clays and Clay Minerals, v. 40, p. 65–80, doi: 10.1346/CCMN.1992.0400108.

Burchfiel, B.C., Cowan, D.S., and Davis, G.D., 1992, Tectonic overview of the Cordilleran orogen in the western United States, in Burchfiel, B.C., Lipman, P.W., and Zoback, M.L., eds., The Cordilleran Orogen: Conterminous U.S.: Boulder, Colorado, The Geological Society of America, Geology of North America, v. G-3, p. 407–426.

Chen, W.J., Ji, F.J., Li, Q., Li, D.M., Wang, Q.L., and Xin, W., 1988, Geochronological implication of K-Ar, FT and TL systems of fault gouge from Yi-Shu fault zone: Dizhen Dizhi (Seismology and Geology), v. 10, p. 191–198.

Clauer, N., Srodon, J., Francu, J., and Sucha, V., 1997, K-Ar dating of illite fundamental particles separated from illite-smectite: Clay Minerals, v. 32, p. 181–196, doi: 10.1180/claymin.1997.032.2.02.

Coogan, J.C., 1992, Structural evolution of piggyback basins in the Wyoming-Idaho-Utah thrust belt, in Link, P.K., Kentz, M.A., and Platt, L.B., eds., Regional Geology of Eastern Idaho and Western Wyoming: Geological Society of America Memoir 179, p. 55–81.

Craddock, J.P., Kopania, A.A., and Wiltschko, D.V., 1988, Interaction between the northern Idaho-Wyoming thrust belt and bounding basement blocks, central western Wyoming, in Schmidt, C.J., and Perry, W.J., Jr., eds., Interaction of the Rocky Mountain Foreland and the Cordilleran Thrust Belt: Geological Society of America Memoir 171, p. 333–351.

Dahlen, F.A., 1984, Noncohesive critical Coulomb wedges: An exact solution: Journal of Geophysical Research, v. 89, p. 10,125–10,133.

Dahlen, F.A., Suppe, J., and Davis, D., 1984, Mechanics of fold-and-thrust belts and accretionary wedges: Cohesive Coulomb theory: Journal of Geophysical Research, v. 89, p. 10,087–10,101.

DeCelles, P.G., 1994, Late Cretaceous–Paleocene synorogenic sedimentation and kinematic history of the Sevier thrust belt, northeast Utah and southwest Wyoming: Geological Society of America Bulletin, v. 106, p. 32–56, doi: 10.1130/0016-7606(1994)106<0032:LCPSSA>2.3.CO;2.

DeCelles, P.G., and Mitra, G., 1995, History of the Sevier orogenic wedge in terms of critical taper models, northeast Utah and southwest Wyoming: Geological Society of America Bulletin, v. 107, p. 454–462, doi: 10.1130/0016-7606(1995)107<0454:HOTSOW>2.3.CO;2.

Dixon, J., 1982, Regional structural synthesis, Wyoming salient of the Western overthrust belt: American Association of Petroleum Geologists Bulletin, v. 66, p. 1560–1580.

Dorr, J.A., 1958, Early Cenozoic vertebrate paleontology, sedimentation, and orogeny in central western Wyoming: Geological Society of America Bulletin, v. 69, p. 1217–1243, doi: 10.1130/0016-7606(1958)69[1217:ECVPSA]2.0.CO;2.

Dorr, J.A., Jr., and Steidtmann, J.R., 1977, Stratigraphic-tectonic implications of a new, earliest Eocene, mammalian fanule from central western Wyoming, in Heisey, E.L., Lawson, D.E., Norwood, E.R., Wach, P.H., and Hale, L.A., eds., Rocky Mountain Thrust Belt Geology and Resources: Casper, Wyoming, Wyoming Geological Association, 29th Annual Field Conference, p. 327–337.

Dorr, J.A., Jr., Spearing, D.R., and Steidtmann, J.R., 1977, The tectonic and synorogenic depositional history of the Hoback basin and adjacent areas, in Heisey, E.L., Lawson, D.E., Norwood, E.R., Wach, P.H., and Hale, L.A., eds., Rocky Mountain Thrust Belt Geology and Resources: Casper, Wyoming, Wyoming Geological Association, 29th Annual Field Conference, p. 549–562.

Dorr, J.A., Jr., Spearing, D.R., Steidtmann, J.R., Wiltschko, D.V., and Craddock, J.P., 1987, Hoback River Canyon, central western Wyoming, in Beus, S.S., ed., Geological Society of America Centennial Field Guide: Boulder, Colorado, Geological Society of America, Rocky Mountain Section, p. 197–200.

Freed, R.L., and Peacor, D.R., 1989, Variability in temperature of the smectite/illite reaction in Gulf Coast sediments: Clay Minerals, v. 24, p. 171–180, doi: 10.1180/claymin.1989.024.2.05.

Freed, R.L., and Peacor, D.R., 1992, Diagenesis and the formation of authigenic illite-rich I/S crystals in Gulf Coast shales: TEM study of clay separates: Journal of Sedimentary Petrology, v. 62, p. 220–234.

Grathoff, G.H., and Moore, D.M., 1996, Illite polytype quantification using WILDFIRE-calculated patterns: Clays and Clay Minerals, v. 44, p. 835–842, doi: 10.1346/CCMN.1996.0440615.

Grathoff, G.H., Moore, D.M., Lay, R.L., and Wemmer, K., 1998, Illite polytype quantification and K/Ar dating of Paleozoic shales: A technique to quantify diagenetic and detrital illite, in Schieber, J., Zimmerle, W., and Sethi, P., eds., Shales and Mudstones II: Stuttgart, E. Schweizerbart'sche Verlagsbuchhandlung (Nägele u. Obermiller), p. 161–175.

Grathoff, G.H., Moore, D.M., Hay, R.L., and Wemmer, K., 2001, Origin of illite in the lower Paleozoic of the Illinois basin: Evidence for brine migrations: Geological Society of America Bulletin, v. 113, p. 1092–1104, doi: 10.1130/0016-7606(2001)113<1092:OOIITL>2.0.CO;2.

Hower, J., Hurley, P.M., Pinson, W.H., and Fairbairn, H.W., 1963, The dependence of K-Ar age on the mineralogy of various particle size ranges in a shale: Geochimica et Cosmochimica Acta, v. 27, p. 405–410, doi: 10.1016/0016-7037(63)90080-2.

Jordan, T.E., Allmendinger, R.W., Damanti, J.F., and Drake, R.E., 1993, Chronology of motion in a complete thrust belt: The Precordillera, 30–31°S, Andes Mountains: The Journal of Geology, v. 101, p. 135–156.

Lindgreen, H., Jacobsen, H., and Jakobsen, H.J., 1991, Diagenetic structural transformations in North Sea Jurassic illite/smectite: Clays and Clay Minerals, v. 39, p. 54–69, doi: 10.1346/CCMN.1991.0390108.

Livaccari, R.F., and Perry, F.V., 1993, Isotopic evidence for preservation of Cordilleran lithospheric mantle during the Sevier-Laramide orogeny, western United States: Geology, v. 21, p. 719–722, doi: 10.1130/0091-7613(1993)021<0719:IEFPOC>2.3.CO;2.

Lyons, J.B., and Snellenburg, J., 1971, Dating faults: Geological Society of America Bulletin, v. 82, p. 1749–1751, doi: 10.1130/0016-7606(1971)82[1749:DF]2.0.CO;2.

Miller, E.L., 2004, The North American Cordillera, in van der Pluijm, B.A., and Marshak, S., eds., Earth Structure (second edition): New York: W.W. Norton and Company, p. 557–565.

Mitra, G., Hull, J.M., Yonkee, W.A., and Protzman, G.M., 1988, Comparison of mesoscopic and microscopic deformational styles in the Idaho-Wyoming thrust belt and the Rocky Mountain Foreland, in Schmidt, C.J., and Perry, W.J., Jr., eds., Interaction of the Rocky Mountain Foreland and the Cordilleran Thrust Belt: Geological Society of America Memoir 171, p. 119–141.

Moore, D.M., and Reynolds, R.C., Jr., 1997, X-ray Diffraction and the Identification and Analysis of Clay Minerals: New York, Oxford University Press, 378 p.

Parry, W.T., Bunds, M.P., Bruhn, R.L., Hall, C.M., and Murphy, J.M., 2001, Mineralogy, $^{40}Ar/^{39}Ar$ dating and apatite fission track dating of rocks along the Castle Mountain fault, Alaska: Tectonophysics, v. 337, p. 149–172, doi: 10.1016/S0040-1951(01)00117-2.

Peacor, D.R., Bauluz, B., Dong, H., Tillick, D., and Yan, Y., 2002, TEM and AEM evidence for high Mg contents of 1M illite: Absence of 1M polytypism in normal prograde diagenetic sequences: Clays and Clay Minerals, v. 50, p. 757–765, doi: 10.1346/000986002762090281.

Pevear, D.R., 1994, Potassium-Argon Dating of Illite Components in an Earth Sample: Houston, Texas, U.S. Patent Office, patent 5.288.695.

Pevear, D.R., 1999, Illite and hydrocarbon exploration: Proceedings of the National Academy of Sciences of the United States of America, v. 96, p. 3440–3446, doi: 10.1073/pnas.96.7.3440.

Pevear, D.R., Vrolijk, P.J., and Longstaffe, F.J., 1997, Timing of Moab fault displacement and fluid movement integrated with burial history using radiogenic and stable isotopes, in Hendry, J., Carey, P., Parnell, J., Ruffell, A., and Worden, R., eds., Geofluids II '97: Contributions to the Second International Conference on Fluid Evolution, Migration and Interaction in Sedimentary Basins and Orogenic Belts: Belfast, The Queen's University of Belfast, p. 42–45.

Price, R.A., 1981, The foreland thrust and fold belt in relation to Cordilleran tectonics, in Dickinson, W.R., and Payne, W.D., eds., Relations of Tectonics to Ore Deposits in the Southern Cordillera: Arizona Geological Society Digest, v. 14, p. 287.

Price, R.A., 1986, The southeastern Canadian Cordillera; thrust faulting, tectonic wedging, and delamination of the lithosphere: Journal of Structural Geology, v. 8, p. 239–254, doi: 10.1016/0191-8141(86)90046-5.

Reynolds, R.C., Jr., 1993a, Three-dimensional powder X-ray diffraction from disordered illite: Simulation and interpretation of the diffraction patterns, in Reynolds, R. C., Jr., and Walker, J.R., eds., Computer Applications to X-Ray Powder Diffraction Analysis of Clay Minerals: Aurora, Colorado, The Clay Minerals Society, CMS Workshop Lectures, v. 5, p. 43–78.

Reynolds, R.C., Jr., 1993b, WILDFIRE—A computer program for the calculation of three-dimensional powder X-ray diffraction patterns for mica polytypes and their disordered variations: Hanover, New Hampshire, Reynolds (lab manual), 38 p.

Royse, F., Jr., Warner, M.A., and Reese, D.L., 1975, Thrust belt structural geometry and related stratigraphic problems, Wyoming–Idaho–northern Utah, in Bolyard, D.W., ed., Deep Drilling Frontiers in the Central Rocky Mountains: Denver, Colorado, Rocky Mountain Association of Geologists, p. 41–54.

Smith, J.V., and Yoder, H.S., 1956, Experimental and theoretical studies of the mica polymorphs: Mineralogical Magazine, v. 31, p. 209–231, doi: 10.1180/minmag.1956.031.234.03.

Solum, J.G., van der Pluijm, B.A., and Peacor, D.R., 2005, Neocrystallization, fabrics and age of clay minerals from an exposure of the Moab fault, Utah: Journal of Structural Geology, v. 27, p. 1563–1576, doi: 10.1016/j.jsg.2005.05.002.

Srodon, J., 1999, Extracting K-Ar ages from shales: A theoretical test: Clay Minerals, v. 34, p. 375–378, doi: 10.1180/000985599546163.

Srodon, J., 2002, Quantitative mineralogy of sedimentary rocks with emphasis on clays and with applications to K-Ar dating: Mineralogical Magazine, v. 66, p. 677–687, doi: 10.1180/0026461026650055.

van der Pluijm, B.A., Hall, C.M., Vrolijk, P.J., Pevear, D.R., and Covey, M.C., 2001, The dating of shallow faults in the Earth's crust: Nature, v. 412, p. 172–175, doi: 10.1038/35084053.

van der Pluijm, B.A., Hall, C.M., Pevear, D.R., Solum, J.G., and Vrolijk, P.J., 2006, Fault dating in the Canadian Rocky Mountains: Evidence for Late Cretaceous and early Eocene orogenic pulses: Geology, v. 34, p. 837–840, doi: 10.1130/G22610.1.

Velde, B., and Hower, J., 1963, Petrological significance of illite polymorphism in Paleozoic sedimentary rocks: The American Mineralogist, v. 48, p. 1239–1254.

Vrolijk, P., and van der Pluijm, B.A., 1999, Clay gouge: Journal of Structural Geology, v. 21, p. 1039–1048, doi: 10.1016/S0191-8141(99)00103-0.

Wang, Y., Hu, Z., and Zheng, Y., 1990, The characteristics and K-Ar age of clay minerals in the fault gouges in the Yunmeng Mountains, Beijing, China: Physics and Chemistry of the Earth, v. 17, p. 25–32, doi: 10.1016/0079-1946(89)90005-0.

Wiltschko, D.V., and Dorr, J.A., Jr., 1983, Timing of deformation in Overthrust belt and foreland of Idaho, Wyoming, and Utah: American Association of Petroleum Geologists Bulletin, v. 67, p. 1304–1322.

Yan, Y., van der Pluijm, B.A., and Peacor, D.R., 2001, Deformation microfabrics of clay gouge, Lewis thrust, Canada: A case for fault weakening from clay transformation, in Holdsworth, R.E., Strachan, R.A., Magloughlin, J.F., and Knipe, R.J., eds., The Nature and Tectonic Significance of Fault Zone Weakening: Geological Society of London Special Publication 186, p. 103–112.

Ylagan, R.F., Kim, C.S., Pevear, D.R., and Vrolijk, P.J., 2002, Illite polytype quantification of accurate K-Ar age determination: The American Mineralogist, v. 87, p. 1536–1545.

York, D., 1968, Least squares fitting of a straight line with correlated errors: Earth and Planetary Science Letters, v. 5, p. 320–324, doi: 10.1016/S0012-821X(68)80059-7.

Zoller, M., and Brockamp, O., 1997, 1M and $2M_1$ illites: Different minerals and not polytypes: European Journal of Mineralogy, v. 9, p. 821–827.

Zwingmann, H., and Mancktelow, N., 2004, Timing of Alpine fault gouges: Earth and Planetary Science Letters, v. 223, p. 415–425, doi: 10.1016/j.epsl.2004.04.041.

Zwingmann, H., Offler, R., Wilson, T., and Cox, S.F., 2004, K-Ar dating of fault gouge in the northern Sydney Basin, NSW, Australia—Implications for the breakup of Gondwana: Journal of Structural Geology, v. 26, p. 2285–2295, doi: 10.1016/j.jsg.2004.03.007.

MANUSCRIPT ACCEPTED BY THE SOCIETY 22 MARCH 2007

Reinterpretation of fractures at Swift Reservoir, Rocky Mountain thrust front, Montana: Passage of a Jurassic forebulge?

E.M. Geraghty Ward*
J.W. Sears
Department of Geosciences, University of Montana, Missoula, Montana 59812, USA

ABSTRACT

A sub–Middle Jurassic unconformity is exhumed at Swift Reservoir, in the Rocky Mountain fold-and-thrust belt of Montana. The unconformity separates late Mississippian Sun River Dolomite of the Madison Group (ca. 340 Ma) from the transgressive basal sandstone of the Middle Jurassic (Bajocian-Bathonian) Sawtooth Formation (ca. 170 Ma). North-northwest–trending, karst-widened fractures (grikes) filled with cherty and phosphatic sandstone and conglomerate of the basal Sawtooth Formation penetrate the Madison Group for 4 m below the unconformity. The fractures link into sandstone-filled cavities along bedding planes. Clam borings, filled with fine-grained Sawtooth sandstone, pepper the unconformity surface and some of the fracture walls. Sandstone-filled clam borings also perforate rounded clasts of Mississippian limestone that lie on the surface of the unconformity within basal Sawtooth conglomerate. After deposition of the overlying foreland basin clastic wedge, the grikes were stylolitized by layer-parallel shortening and then buckled over fault-propagation anticlinal crests in the Late Cretaceous–Paleocene fold-and-thrust belt. We propose that the grikes record uplift and erosion followed by subsidence as the Rocky Mountain foreland experienced elastic flexure in response to tectonic loading at the plate boundary farther to the west during the Middle Jurassic. The forebulge opened strike-parallel fractures in the Madison Group that were then karstified. The sandstone-filled karst system contributes secondary porosity and permeability to the upper Madison Group, which is a major petroleum reservoir in the region. The recognition of the fractures as pre–Middle Jurassic revises previous models that have related them to Cretaceous or Paleocene fracturing over the crests of fault-propagation folds in the fold-and-thrust belt, substantially changing our understanding of the hydrocarbon system.

Keywords: foreland, thrust, forebulge, deformation, Montana, hydrocarbons.

*e_geraghty@umwestern.edu

INTRODUCTION

Fractures create secondary porosity and permeability in carbonate reservoirs. Modern horizontal-drilling methods exploit such fractures and increase oil and gas productivity. The Waterton and Turner Valley fields of Alberta produce oil and gas from fractures in the uppermost part of the Mississippian carbonate that are sealed by overlying Jurassic shale (Boyer, 1992) (Fig. 1). Petroleum geologists and geophysicists have for decades enrolled in field seminars along the Rocky Mountain front to observe fractured-carbonate reservoirs. One key locality is at Swift Reservoir, southwest of Browning, Montana (Fig. 2), where exposed structures within a fault-propagation anticline include systems of fractures that are subparallel to the fold axis. These structures have been interpreted to be kinematically related to extension over the anticlinal crest (Stearns and Friedman, 1972) (Fig. 3).

As documented in this paper, however, many of the nearly strike-parallel fractures in this locality are stratigraphically determined to be Middle Jurassic (Bajocian) in age and, hence, are not the consequence of folding, but were instead deformed by folding some 100 m.y. later. The prethrust age of these fractures requires reevaluation of the mechanics of development of fractured hydrocarbon reservoirs in Mississippian carbonate rocks in the region.

In this paper, we suggest that the Bajocian fractures exposed at Swift Reservoir comprise part of a regional set that opened as a result of flexure of the Madison Group over a forebulge during early history of the Rocky Mountain foreland. Bradley and Kidd (1991) modeled analogous forebulge extension and karstification in the Ordovician Taconic foreland of the northern Appalachians.

STRUCTURAL SETTING

The Rocky Mountain fold-and-thrust belt of northwestern Montana consists primarily of northwest-trending, southwest-dipping thrust faults and associated folds that span a 130 × 40 km region (Mudge, 1972b). The thrust belt is divided into three physiographic and geologic regions: the Foothills to the east, the Front Ranges in the center, and the Main Ranges to the west. This classification is similar to the fold-and-thrust belt described in Canada by Bally et al. (1966) and Price and Mountjoy (1970). The Montana Foothills province is characterized by large, flat thrust sheets of Paleozoic strata at depth, overlain by thrust-imbricated Mesozoic strata. It is bounded on the east by a wedge-shaped triangle zone (Sears et al., 2005). The Montana Front Range province comprises imbricated thrust sheets of Paleozoic and Mesozoic strata. The Montana Main Ranges are underlain by large thrust sheets of the Mesoproterozoic Belt-Purcell Supergroup (see chapters by Feinstein et al., this volume; Sears, this volume). The thrusted rocks in these provinces are underlain by undisturbed basement. The Sawtooth Range of the Montana Front Ranges province contains the field sites for this study (Fig. 4).

The Sawtooth Range is generally interpreted to be an exhumed duplex; the thrust faults now exposed at the surface are thought to have originally formed beneath the Lewis-Eldorado thrust, which placed the Middle Proterozoic Belt-Purcell Supergroup over Upper Cretaceous rocks (Willis, 1902; Childers, 1963; Ross, 1959, *in* Hoffman et al., 1976; Boyer and Elliott, 1982; Mitra, 1986; Dolberg, 1986; Sears, this volume). Potassium-argon dating of potassium-bentonites indicates that thrust burial of the Upper Cretaceous section occurred from 74 to 59 Ma (Hoffman et al., 1976; corrected for revised decay constant). The dates of thrusting thus coincide with stratigraphic evidence for the Late Cretaceous onset of thrusting in the foreland basin (Mudge, 1972b; Robinson et al., 1968, *in* Hoffman et al., 1976; Catuneanu and Sweet, 1999; Sears, 2001; Feinstein et al., this volume). Osborn et al. (2006) summarized evidence for the emplacement of the Rocky Mountain fold-and-thrust belt and inferred that the end of thrusting occurred at 60–55 Ma (latest Paleocene).

The Swift Reservoir site provides an excellent exposure of one of the thrust sheets within the northern portion of the Sawtooth duplex. Mississippian, Jurassic, and Lower Cretaceous rocks structurally overlie Upper Cretaceous rocks on a west-dipping thrust fault (Fig. 5). A northeast-verging, northwest-plunging hanging-wall anticline represents a fault-propagation fold cut by the thrust fault. The anticline folded the thick-bedded Madison Group carbonate predominantly by flexural slip.

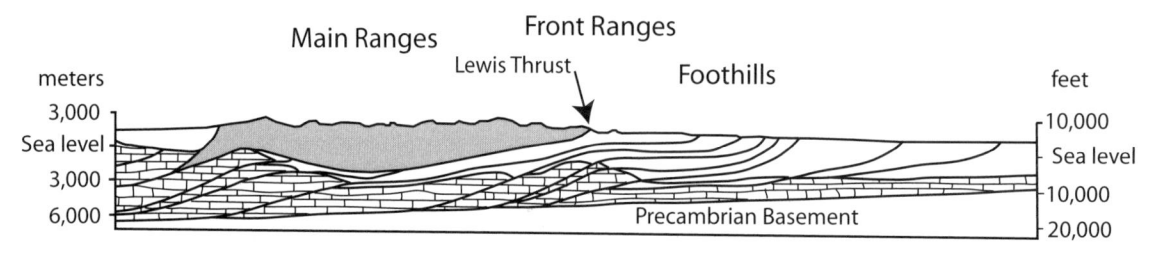

Figure 1. Cross section through the Front Ranges and Foothills of the Rocky Mountain thrust front of Canada. Hydrocarbon-producing Mississippian unit is marked by brick texture. This cross section, located just north of the Montana border, depicts the structural style that can be found in both southern Canada and northwest Montana. The difference between the two areas is that the Mississippian carbonate duplex is beneath the surface (and Lewis thrust) in Canada, and it is exposed at the surface in the thrust belt of northwest Montana (location of primary study area for this paper) (modified from Boyer, 1992).

SUB–MIDDLE JURASSIC (BAJOCIAN) UNCONFORMITY

Figure 6 shows the stratigraphic section and its regional correlations. The unconformity between the Mississippian Sun River Dolomite and the overlying Middle Jurassic rocks is exhumed across ~2 km² on a broad dip-slope on the west limb of the anticline at Swift Reservoir (Fig. 7). The excellent outcrops are washed by the rising and falling reservoir and are best seen at low water. The stratigraphic contact of the Middle Jurassic (Bajocian-Bathonian) Sawtooth Formation on the Sun River Dolomite is exposed at the north and south ends of the exhumed outcrop of the unconformity on the gentle western limb of the anticline, as well as on the steep eastern limb (Fig. 8). The Sawtooth Formation contains a <3-m-thick basal sandstone, overlain by shale and calcareous siltstone. The basal sandstone is fine-grained, medium-light-gray, calcareous cherty quartz-arenite that contains some pyrite, black phosphatic pebbles, and, in areas, is conglomeratic with limestone pebbles and cobbles (Cobban, 1945; Mudge, 1972a). Because of the erosional weakness of the Jurassic shales and sandstones and the resistance of the Sun River Dolomite, the Jurassic has been cleanly swept away from

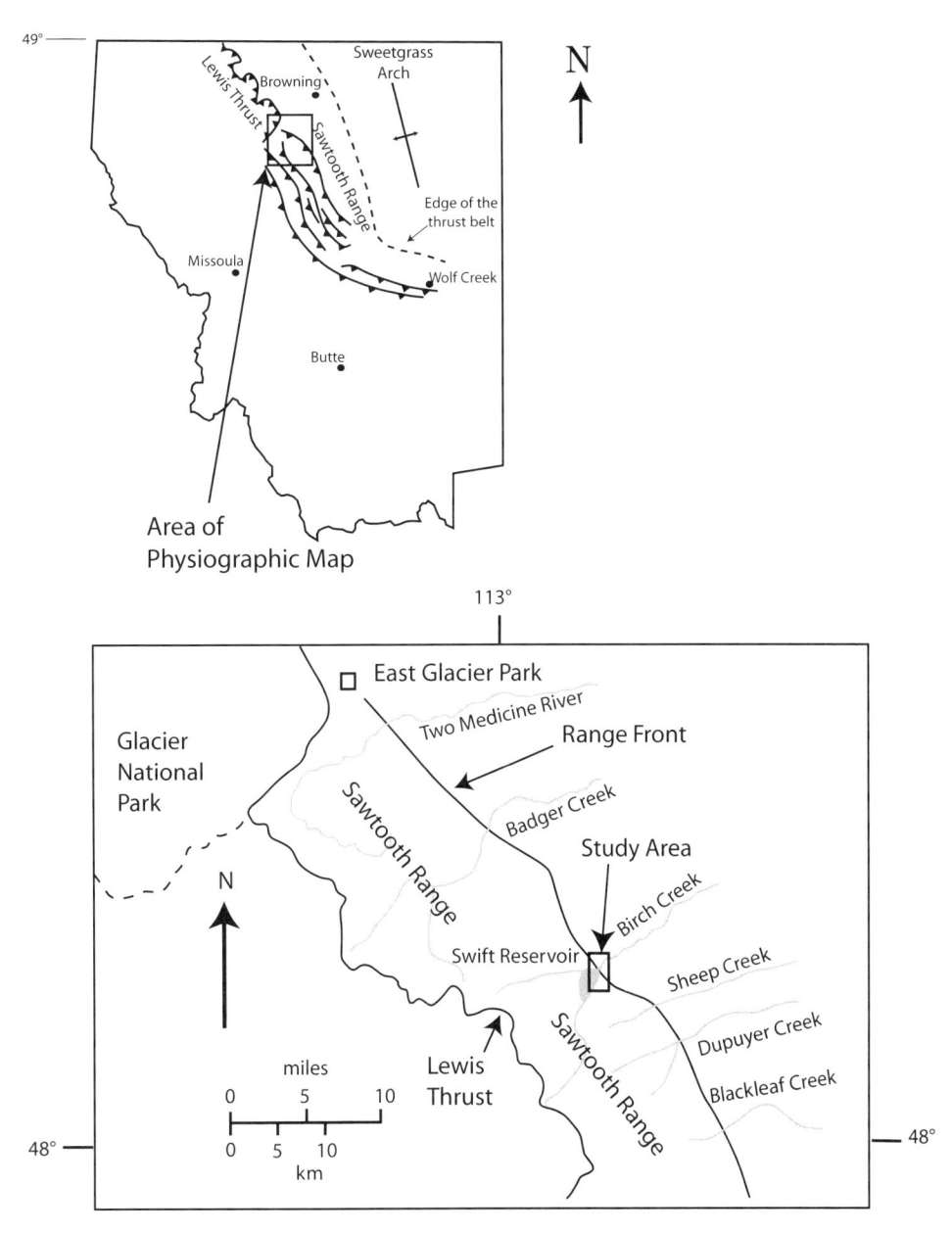

Figure 2. Geologic setting and location of primary study site, Swift Reservoir, in northwest Montana. Site is located at the easternmost extent of thrusting of Mississippian carbonate exposed at the surface in the Sawtooth Range (modified from Hoffman et al., 1976; Singdahlsen, 1986).

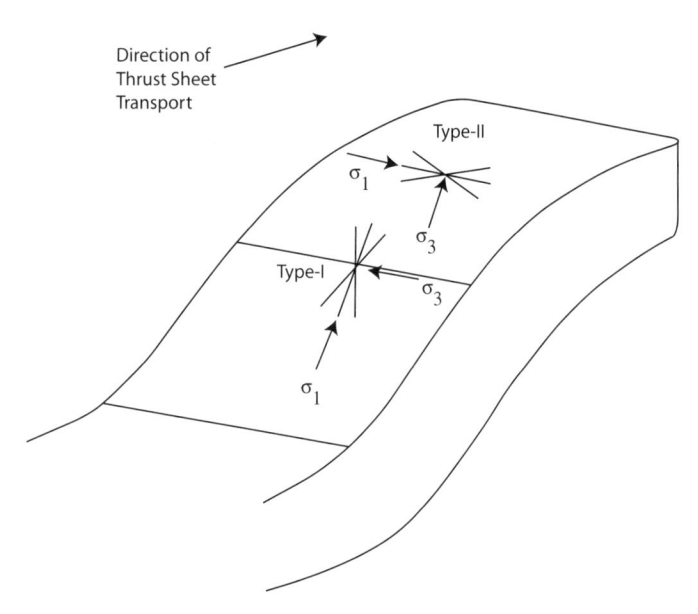

Figure 3. Stearns model depicting "type-I" and "type-II" fractures kinematically linked to folding of strata (from Goldburg, 1984). Solution-widened fractures (e.g., grikes) are oriented parallel to the strike of the folded carbonate strata and look similar to "type-II" fractures depicted in this diagram. Ages differ between grikes and "type-II" fractures related to Cretaceous thrust-related folding because grike walls exhibit Jurassic-age clam borings and have been infilled with Middle Jurassic sand of the Sawtooth Formation.

the dolomite, revealing minute details of the unconformity surface. Downstream from Swift Dam, the canyon of Birch Creek exposes 50 m of rock beneath the unconformity, providing a three-dimensional view (Fig. 9).

Details of the unconformity include densely spaced cylindrical borings up to 0.5 cm in diameter and 1 cm deep that are filled with fine-grained, brown, cherty Sawtooth sandstone (Fig. 10C). These borings occur on bedding planes and on karst-widened fracture faces up to a few decimeters below the flat surface of the unconformity. At Sun River, 70 km south of the study site, the borings also penetrate the surfaces of rounded cobbles of dolomite within the basal Sawtooth conglomerate. The basal conglomerate is interbedded with fossiliferous shale (Mudge, 1972a). The borings are generally normal to the surfaces that they penetrate; on bedding-plane surfaces, they are normal to bedding; on fractures, they are normal to fracture walls; and in clasts, they are normal to the surfaces of the clasts. They occur on all sides of rounded cobbles, including the bases, and therefore predate burial of the clasts in the conglomerate. The borings are interpreted to have been made by carbonate-boring clams that occupied the substrate, open fractures, and lag gravel. As currents tumbled the lag gravel, the borings in the clasts were periodically exposed, as seen in some modern beach cobbles. The unconformity regionally truncated >250 m of the Sun River Dolomite (Mudge, 1972a), but the borings penetrate only a few centimeters below

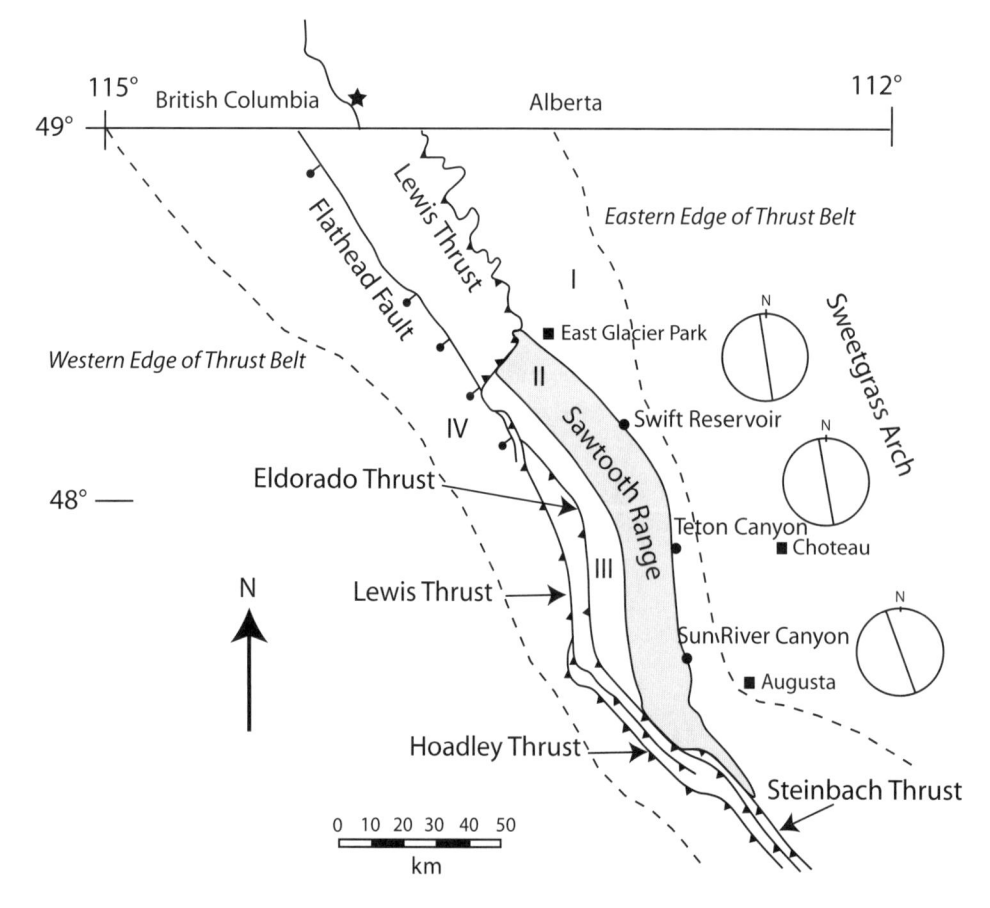

Figure 4. Regional map of northern Montana depicting major structural divisions. Structural "subbelt" subdivisions are labeled I–IV (Mudge, 1972b). Thrust faults labeled on this diagram displace Precambrian Belt-Purcell rocks. Within the Sawtooth Range (location of study area), multiple thrust faults displace younger Paleozoic and Mesozoic rocks but are not delineated in this diagram. Swift Reservoir, Teton Canyon, and Sun River Canyon research locations are labeled with compasses depicting the average grike-stylolitic fracture azimuth from field measurements on the unconformity surface. Star marks the approximate location of the cross section in Figure 1.

the unconformity surface. They clearly invaded the erosional surface and its lag gravel and are thus trace fossils that refer to the age of the surface, not to the depositional age of the dolomite. Since the borings are filled with Sawtooth sandstone, and cobbles with borings are enclosed in the matrix of the basal Sawtooth conglomerate, there is little doubt that the borings were made by clams as the Sawtooth sea transgressed across the region, but before the rock habitat was buried by Sawtooth sand and mud. Numerous pelecypod and ammonite fossils collected from the Sawtooth Formation in the area establish its depositional age as Bajocian-Bathonian (Cobban, 1945; Imlay, 1945, 1952; Peterson, 1957; Mudge, 1972a). Mudge (1972a, p. A41) similarly con-

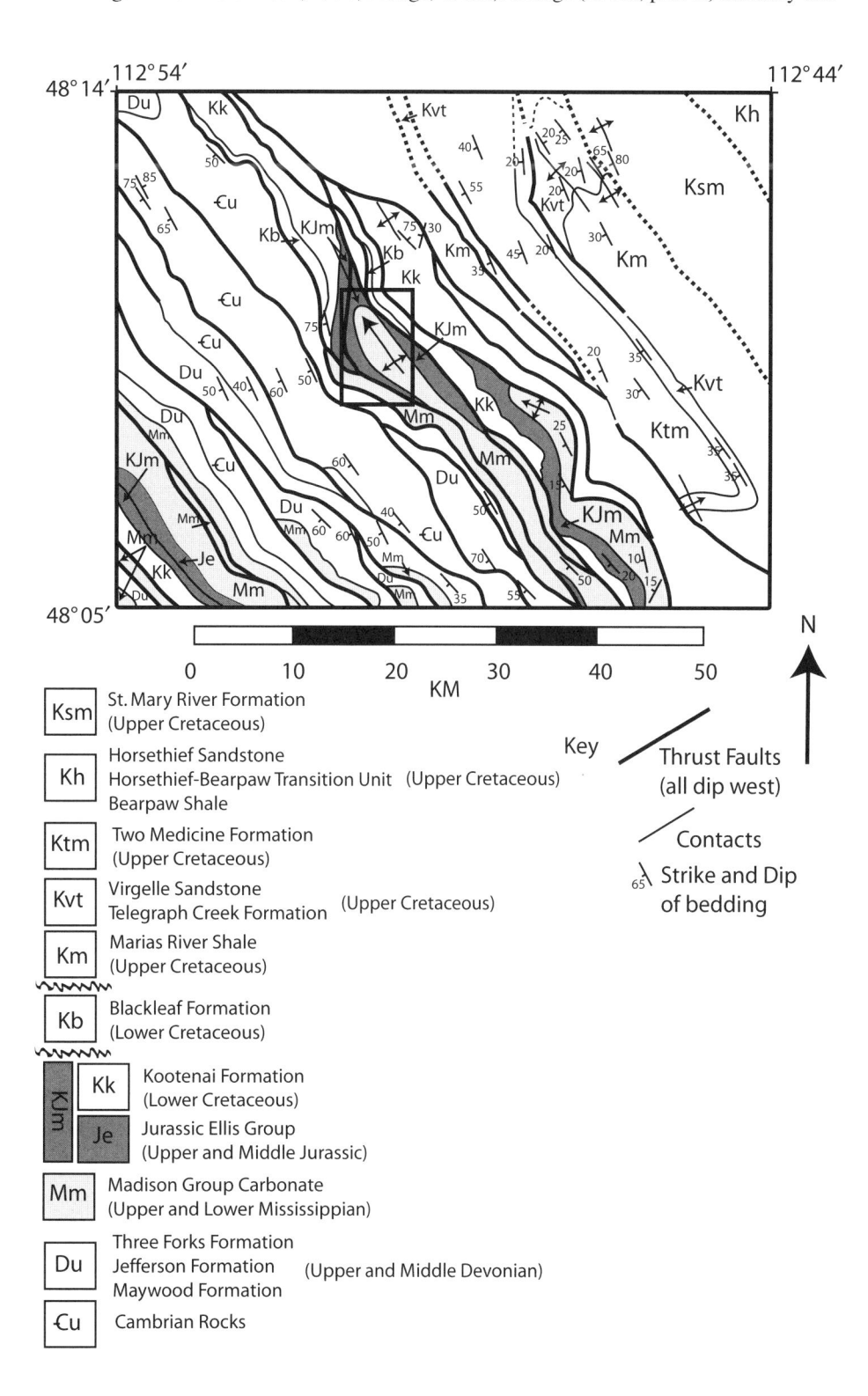

Figure 5. Geologic map of the Swift Reservoir site, northwest Montana (modified from Mudge and Earhart, 1983). Swift Reservoir structure and research area are contained within the rectangle.

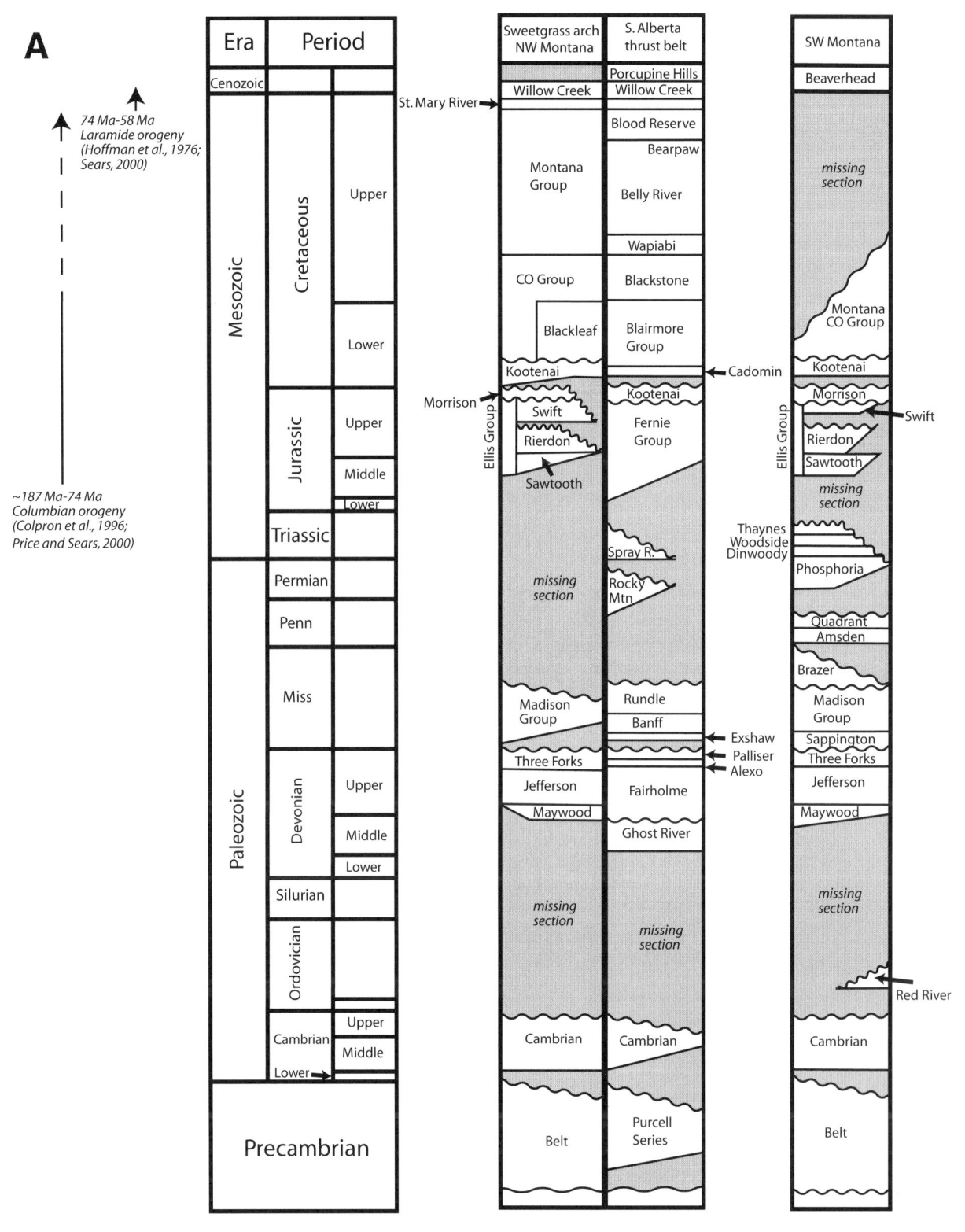

Figure 6 (*on this and following page*). (A) Generalized stratigraphic sections for northwest Montana, Sweetgrass arch, southwest Montana, and southern Alberta (from Hammond and Trapp, 1959; Mudge, 1972a). The sub–Middle Jurassic unconformity is apparent in both northwest Montana and Sweetgrass arch areas. Ellis Group units thin eastward, perhaps atop forebulge limb. The regional unconformity marking the forebulge development in Alberta is at the base of the Fernie Group. Flexure within the crust began as terranes accreted onto North America during Early–Middle Jurassic marking onset of Columbian orogeny in Canada (ca. 187 Ma) (Colpron et al., 1996). Deformation continued through the Mesozoic, with Laramide orogeny marking a later pulse, which began ca. 74 Ma in northern Montana (Hoffman et al., 1976).

Figure 6 (*continued*). (B) Stratigraphy of Sawtooth Range, Montana, discussed in text (Mudge, 1972a). Mississippian–Jurassic unconformity surface represents ~130 m.y. Nowhere along Rocky Mountain Front does basal sandstone of Sawtooth exceed 3 m, and it is locally absent in areas (Cobban, 1945).

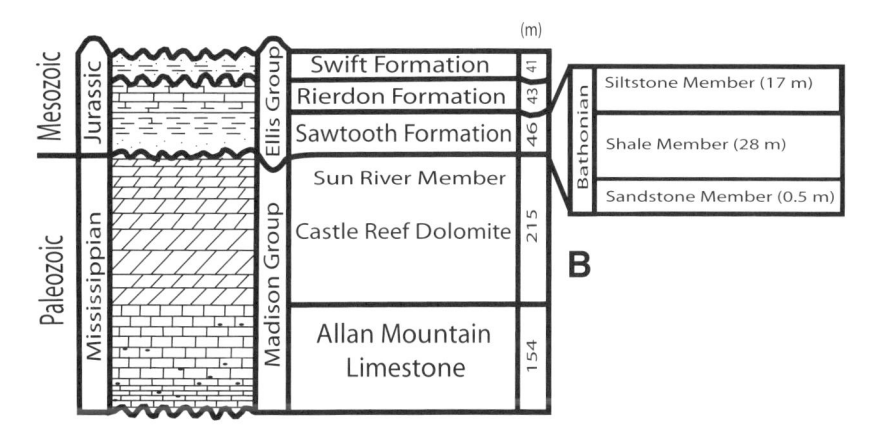

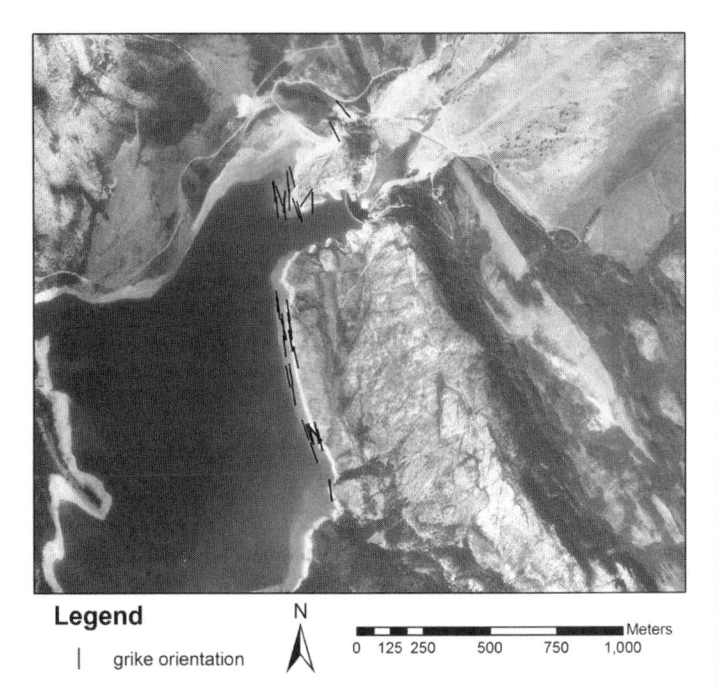

Figure 7. Aerial photograph showing the locations of stations containing grikes at Swift Reservoir field site, northwest Montana. Location of grikes (*n* = 44) are hash marks oriented parallel with the azimuth direction of grikes. Where multiple grikes are present at one station, the average azimuth direction is depicted.

Figure 8. Jurassic (Je)–Mississippian (Mm) unconformity is well exposed on the steeply dipping east limb of the anticline (right). The basal sandstone of Sawtooth Formation is rather thin at this location (<1 m). The dashed line approximates the top of the unit.

cluded: "Mississippian rocks were widened by solution and filled with sand during Middle Jurassic sedimentation to depths of 20 ft or more...the top of the Madison contains abundant small borings that are filled with fine-grained Jurassic sandstone."

BAJOCIAN GRIKES (SANDSTONE-FILLED FRACTURES)

Fractures riddle the exhumed Sun River Dolomite on the gentle west limb of the anticline. A set of these fractures trends north-northwest, nearly parallel with the fold axis and normal to bedding (Fig. 11). These have generally been ascribed to extension over the hinge of the anticline. A number of these fractures, however, contain irregular 1–10-cm-wide seams of brown cherty sandstone or chert, phosphate, and limestone-clast conglomerate (Fig. 10A). Others also contain secondary quartz, dolomite, calcite, and hydrocarbon as indicated from petrographic and X-ray diffraction analyses (Ward, 2007). Clam borings penetrate the surfaces of some of these fractures and are filled with brown cherty sandstone. The irregular, pocked surfaces of the fractures indicate that they were widened by solution before being bored into and

Figure 9. View looking south upon northeast-verging anticline in cross section (fold form highlighted with dashed line). Folding of thick, competent Mississippian-age Madison Group (Mm) carbonate rocks is accommodated by flexural slip along polished and slickensided bedding planes (marked by arrow).

then filled with sand. Such sandstone-filled, solution-widened fractures are termed grikes. The fractures link downward into solution-widened bedding planes that are similarly filled with sand or conglomerate. The fractures and bedding plane cavities form a network that penetrates at least 4 m below the upper surface of the unconformity, as reported by Mudge (1972a) in the Sun River area. The sandstone that fills the grikes is most likely the basal sandstone member of the Sawtooth Formation, based upon stratigraphic relationships at this study site and the similarity of the grike sandstone and the basal Sawtooth sandstone. Sand evidently spilled into a cave system in the Madison carbonate as the Middle Jurassic sea transgressed.

The grikes occur as wide, continuous features ≤50 m long, or as feathery structures that trace into stylolites (Fig. 10C). Stylolitization occurs at grike walls where the grike fill comes in contact with the carbonate (Fig. 10). The fractures pass into traditional stylolites where grike fill is discontinuous along the length of the fractures. (Fig. 10B). Stylolitic fractures are significantly more common than grikes in the area, but because the grikes pass laterally and vertically into stylolitic fractures, they comprise a single genetic set of fractures, only some of which were karst-widened before stylolitization. Commonly, less continuous grikes that grade into stylolites occur in clusters (generally less than five) with spacings of less than one meter. More prominent grikes occur alone or in pairs with spacings of 1–10 m. Secondary ptygmatic folding of the grikes occurs where the sandstone fill has quartz-cement overgrowths (Fig. 10B). On the steep forelimb of the anticline, the fractures appear to have been reactivated as small faults to accommodate flexural-slip folding (Fig. 11).

The stylolitization of the grikes indicates that they predated east-west layer-parallel shortening. Layer-parallel shortening is also exhibited as strike-parallel, bedding-normal stylolitic cleavage in marly carbonate of the Jurassic Rierdon Formation, which stratigraphically overlies the Sawtooth Formation in the region. The stylolitic cleavage is coaxially folded with the bedding in structures of the Rocky Mountain fold-and-thrust belt.

Mean intersection orientations of stylolitic fractures and bedding were determined for the Swift Reservoir (351°, $n = 66$), Teton (350°, $n = 39$), and Sun River (340°, $n = 6$) localities (Figs. 4 and 12). Of all the locations, the Swift Reservoir site provides the clearest exposure of the unconformity surface and the grikes. Although the Teton Canyon exposure of the unconformity surface is extensive, no definitive grikes were observed. Stylolitic fractures with similar orientations were present but lacked sandstone fill. Although Sun River Canyon provided excellent exposures of the unconformity surface along the dip-slopes of the carbonate thrust sheets, only a few grikes were observed. The more apparent features were the limestone-clast conglomerates and randomly distributed, sand-filled fracture networks at the unconformity surface.

FOREBULGE MODEL

We deduce the following sequence of events from our observations of the Sun River Dolomite at Swift Reservoir: (1) fracturing produced vertical joint sets, (2) north-northwest–trending joints and bedding planes were karstified, (3) clams bored the rock surfaces, (4) sand filled the caves and borings, (5) secondary cementation, mineralization, and entrapment of hydrocarbon inclusions occurred, (6) north-northwest–trending joints and grikes were stylolitized, and (7) the unconformity was folded and faulted.

Parallelism of the grikes, stylolitic fractures, folds, and thrusts indicates a genetic relationship among these structures. The grikes exploited a regional north-northwest–trending joint set that was normal to the horizontal extensional axis before the beds were folded. The subsequent stylolitization of the grike joint set indicates that the horizontal extensional axis was converted into the horizontal compressional axis. Coaxial buckling of the

Figure 10. Photos of grike characteristics. (A) A grike contained within Mississippian-age carbonate (Mm) that has been widened by solution and filled by black cherty sand of Sawtooth Formation (Je) (marker for scale). The arrow points to an irregular grike wall that has undergone stylolitization. (B) Two grikes (hammer for scale). The sand-filled grike at the bottom of the photo shows similar characteristics to that of A; however, where a fracture was not solution-widened, it became a stylolite (marked by arrow). The mineralized grike in this photo contains a mix of Sawtooth sand that has been cemented primarily by quartz. The dark tone is in part due to bitumen contained within the mineralized grike. Quartz within the grikes contained hydrocarbon inclusions. Secondary folding (marked by arrow) was characteristic only of mineralized grikes, which may be an effect of the contrast between carbonate and quartz-cemented grike fill. Stylolitization and secondary folding are evidence of grike deformation after sand-filling and mineralization of grikes took place (see thin section photo). (C) Clam-bored grike wall (field book for scale). The solution widening of the fractures allowed more surface area for clam activity and provides relative age constraint for grikes. The unconformity surface also exhibits clam borings that have been subsequently filled with basal sand of the Sawtooth Formation.

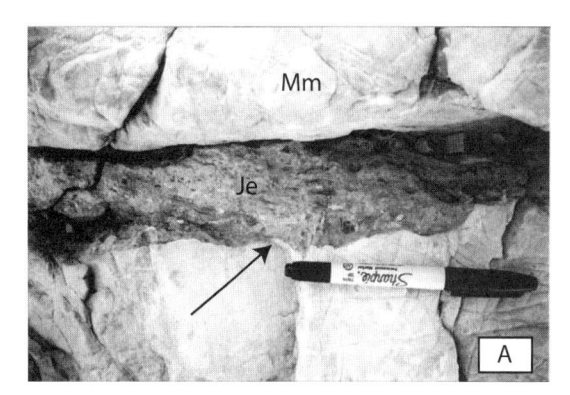

beds during thrusting and folding resulted in strike-parallel, bedding-stylolite intersections. Stylolitized planes are normal to bedding on both limbs of the anticline.

The overall strike of the thrust belt paralleled the western margin of the Cordilleran miogeocline and the collisional belts that deformed it (Monger and Price, 1979). We suggest that the grike system records elastic flexure of the continental shelf across a forebulge in response to tectonic loading on the continental margin to the west (e.g., Flemings and Jordan, 1990). Forebulge regions are areas of erosion and/or nondeposition caused by crustal flexure and intraplate stress, resulting in unconformities (Beaumont, 1981; Dorobek, 1995).

Bradley and Kidd (1991) described an analogous example of foreland basin development associated with the Taconic orogeny of the northern Appalachians and suggested that a forebulge may have initiated a positive topographic feature with extension faulting. The forebulge then sunk beneath shallow-marine waters during advance of the Taconic foredeep. Their model indicates that the forebulge region marked an intertidal-supratidal zone, where extension, erosion, and karst processes occurred (Fig. 13).

Stockmal and Beaumont (1987) developed an elastic flexural model to interpret evolution of the Alberta basin in southern Canada (Fig. 14). Their model predicted that a forebulge migrated in advance of the foreland basin due to the flexural rigidity of the lithosphere. Important conclusions are: (1) prior to terrane accretion in the Cordillera of northwestern North America, sediment accumulated on the passive continental shelf from the late Neoproterozoic to the Early Jurassic, (2) regional sub-Jurassic unconformities mark the passage of the forebulge, (3) synorogenic sediments in the form of east-tapering clastic wedges separated by unconformities mark phases of Jurassic to early Paleogene Cordilleran orogenesis, and (4) episodic uplift and subsidence of the foreland reciprocated tectonic loading and rebound in the orogenic belt.

Figure 11. View west onto steeply dipping limb of anticline. Fractures within Mississippian carbonate rocks (Mm) are well exposed on unconformity surface and are reactivated as small faults on the steep limb of the anticline (see arrow), perhaps to accommodate folding of the carbonate unit. Where offset, exposed fracture walls are polished. (Field of view ~10 m.)

Gillespie and Heller (1995) presented tectonic subsidence curves that document the history of the southern Canada and northern Montana foreland basin (Fig. 15). A westerly derived orogenic clastic wedge appears in the Bathonian in the southwestern Canadian Rockies, as shown by increased tectonic subsidence and the appearance of cherty quartz-arenites (Poulton et al., 1993). A progressive Callovian–early Oxfordian erosive disconformity is attributed to the northeastward migration of the forebulge across the Alberta foreland (Poulton, 1984). A clastic wedge followed the bulge across the Alberta foreland basin in the middle Oxfordian–Kimmeridgian as the basin tectonically collapsed under the load of the advancing orogenic belt (Fermor and Moffat, 1992; Evenchick et al., this volume).The initial flexure was caused by the load of tectonically thickened crust farther to the west in the Kootenay arc and Purcell anticlinorium, where folding began by 175 Ma; that thickening drove Barrovian metamorphism, which peaked at 165–160 Ma (Colpron et al., 1996; Evenchick et al., this volume).

Jordan (1981), DeCelles and Currie (1996), White et al. (2002), DeCelles (2004), and Parcell and Williams (2005) all discussed the Middle and Late Jurassic evolution of the Sevier forebulge-foredeep system to the south, in the thrust belt of Idaho-Wyoming and Utah-Nevada.

The sub-Jurassic unconformity in the northern Montana Rockies represents a missing depositional record that may have included younger Mississippian, Pennsylvanian, and Permian shelf sediments that are preserved in southern, south-central, and western Montana, and Triassic and Lower Jurassic rocks found in eastern and southwestern Montana (Mudge, 1972a). Pre-Triassic erosion may account for some of the missing section on the shelf; a thin Triassic section oversteps Permian and Pennsylvanian strata to rest directly on Mississippian beds in the Williston basin of northeastern Montana (Mallory, 1972). The Cordilleran miogeocline in the western Canadian Rockies preserves the Pennsylvanian and Permian Rocky Mountain Group, Triassic Spray River Group, and Lower-Middle Jurassic basal Fernie Group (Price and Mountjoy, 1970). Some of these rocks also occur in the Rocky Mountain Trench in northern Montana (Harrison et al., 1992).

The Swift Reservoir site underwent uplift, fracturing, and karstification sometime before the Bajocian (ca. 170 Ma), when it subsided and received sediment from the orogenic belt. The subsidence slightly preceded the first evidence of a westerly derived clastic wedge in the southwestern Canadian foreland basin in Bathonian, but it correlates with initiation of crustal thickening in the Kootenay arc and Purcell anticlinorium. A palinspastic map by Price and Sears (2000) places the Swift Reservoir site 300–400 km east of the Kootenay arc, which lay on the west side of the restored Mesoproterozoic Belt-Purcell Basin. The Belt-Purcell Supergroup formed an unusually stiff and dense crustal element (Sears, this volume) that may have been elastically flexed by this load.

The sub-Jurassic unconformity can be traced in the subsurface from the Montana thrust belt eastward to the Sweetgrass arch, which may have represented the Bathonian position of the forebulge (Dolson et al., 1993). At the beginning of the upper Bathonian, the Sweetgrass arch formed a broad, low uplift from the Little Belt Mountains north to Canada (Cobban, 1945). The sea transgressed over the arch to deposit the Rierdon Formation in the Callovian Stage, then retreated in the late Callovian but returned in the Oxfordian to deposit the Swift Formation. The Swift Formation was followed by deposition of the main part of the foreland basin clastic wedge: the nonmarine upper Jurassic (Kimmeridgian; Imlay, 1952) Morrison Formation, nonmarine Lower Cretaceous clastics, marine Upper Cretaceous shales, and mostly nonmarine Late Cretaceous and Paleocene clastics (Price and Mountjoy, 1970; Cobban, 1945). The depositional history of the forebulge region may be a proxy for the pulses of tectonic thickening and rebound within the orogenic belt farther to the west (e.g., Catuneanu and Sweet, 1999).

HYDROCARBON IMPLICATIONS

The sand-filled grikes exposed at Swift Reservoir have implications for the fractured hydrocarbon reservoirs of the region. The grikes enhance the permeability of the uppermost section of the Mississippian carbonate rock, which provides a significant reservoir in this area.

The grikes contain hydrocarbon fluid inclusions in quartz-cement overgrowths as detailed in Ward (2007). The quartz cements were deformed by stylolitization, so the hydrocarbons entered the system before it was horizontally shortened. Microthermometry of two-phase (liquid-vapor) hydrocarbon inclusions records average homogenization temperatures of ~75 °C. Fluorescence data yielded average fluid density at ~32° API (American Petroleum Institute) or 0.8654 g/mL. Oil trapped within the Upper Mississippian carbonate reservoir of the Sweetgrass arch (~90 km east of the

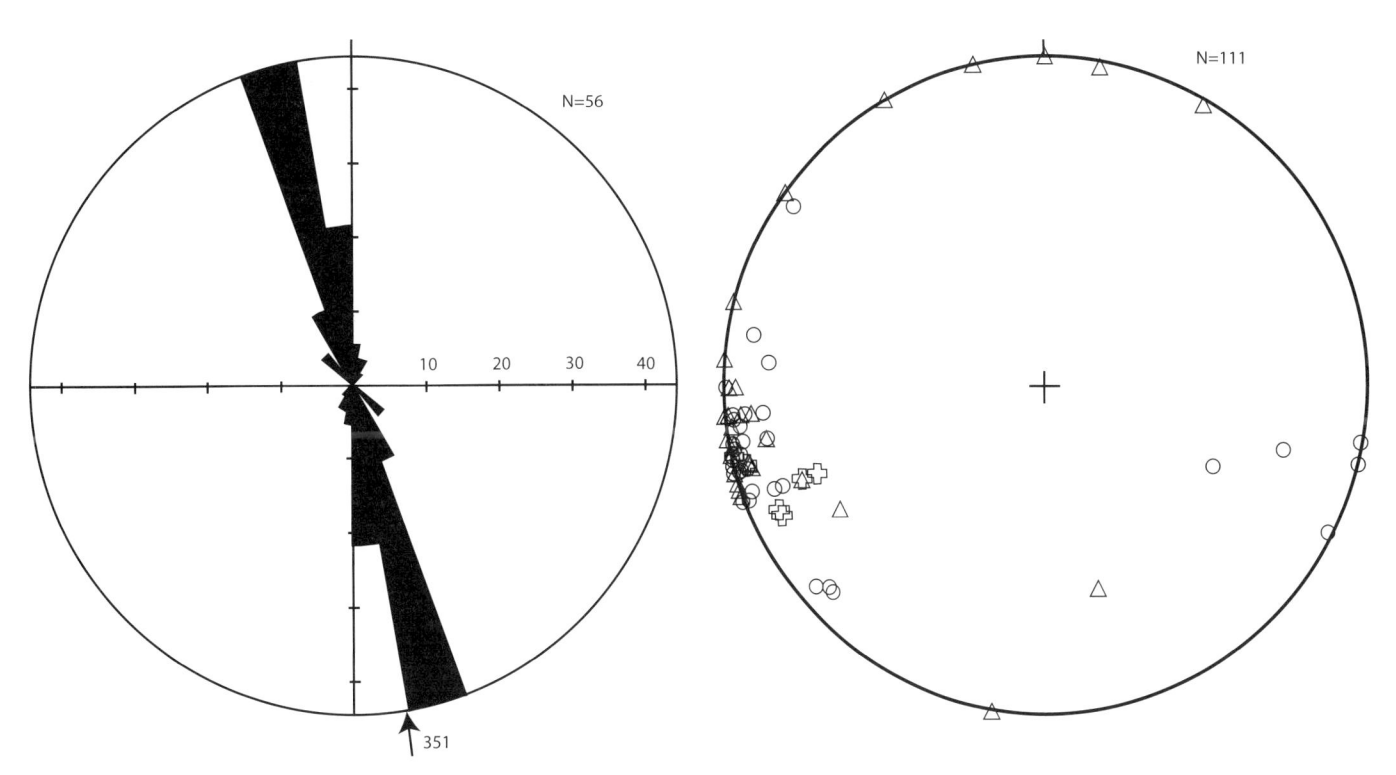

Figure 12. Rose diagram depicting the grike azimuth at Swift Reservoir and stereonet for grikes found within the thrust belt of Montana. The arrow located on the rose diagram shows mean grike azimuth (351°) for 56 measurements from Swift Reservoir. Stereonet represents grike orientation at Swift Reservoir (open circles), Sun River (open crosses), and Teton (open triangles) locales (*n* = 111).

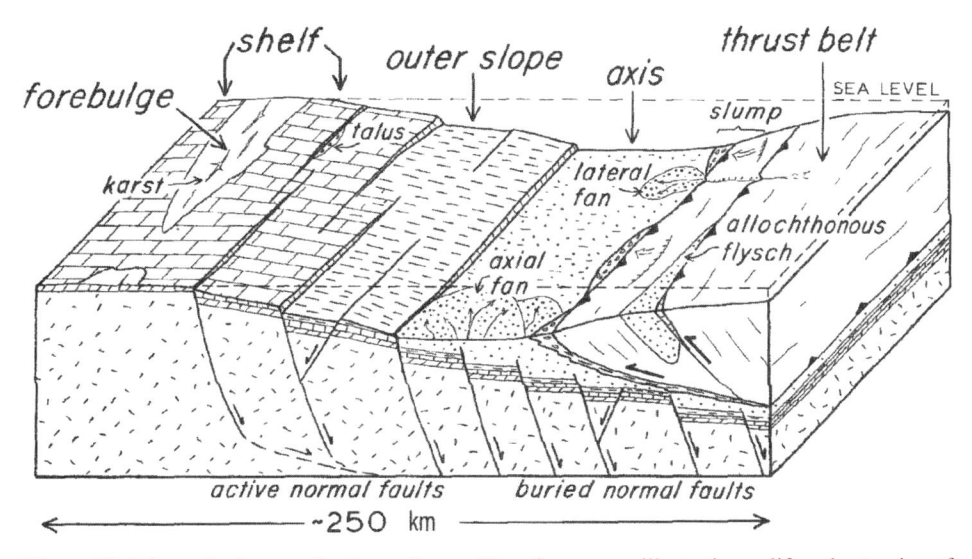

Figure 13. Schematic diagram for the analogous Taconic orogeny illustrating uplift and extension of the forebulge and associated fracture and karst in that region (Bradley and Kidd, 1991). The foreland basin development associated with the Taconic orogeny suggests that the forebulge initiated as a positive topographic feature that then sunk beneath shallow-marine waters during a marine transgression. The model indicates that the forebulge region would mark an intertidal-supratidal zone, where extension, erosion, and karst processes occur. A similar scenario can be explained for the northern Rockies foreland basin system. Uplift of the forebulge in the Swift Reservoir study area may have caused erosion of sediments overlying the Madison Group to create the sub–Middle Jurassic unconformity and created extensional deformation within carbonates at the crest of the forebulge that would match the strike-parallel orientation of the grikes.

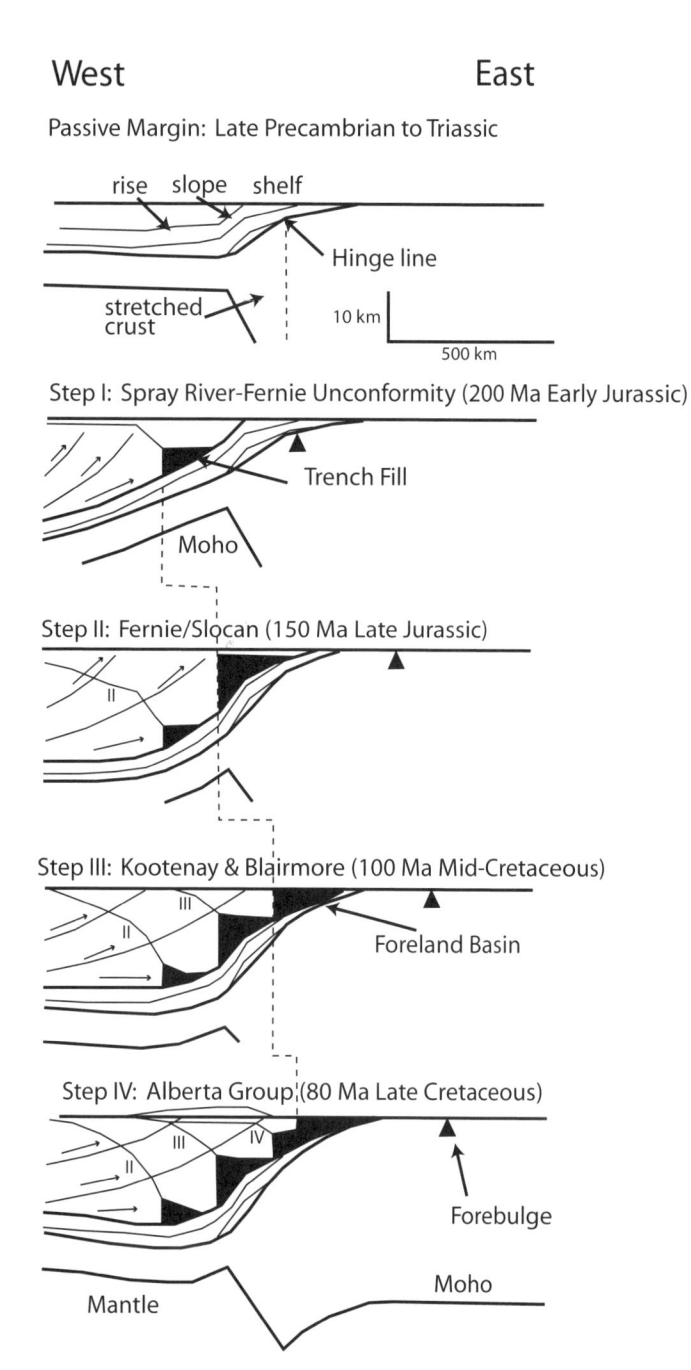

Figure 14. Model for foreland basin development in the Alberta Basin of Canada. The triangle marks the region of the forebulge (Stockmal and Beaumont, 1987). Step II during the Early Jurassic depicts the development of the Fernie unconformity due to tectonic thickening to west. This model could be applied to the Rocky Mountain thrust front of northern Montana. Accretion of island arcs onto the North American continent during Early and Middle Jurassic time loaded crust and caused forebulge flexure. A regional unconformity marking forebulge development in northern Montana occurs at the base of the Middle Jurassic Ellis Group. As accretion and tectonic thickening continued, the foreland basin migrated eastward marked by onlap of Ellis Group units toward the Sweetgrass arch.

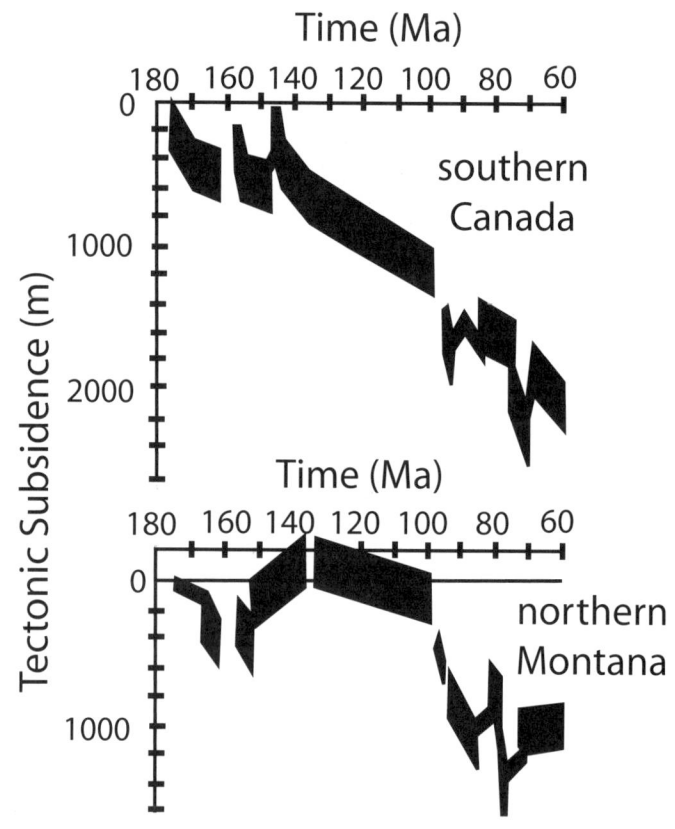

Figure 15. Subsidence curves for Alberta and northern Montana foreland basins (Gillespie and Heller, 1995). Curves support initial subsidence of northern Montana foreland basin by Middle Jurassic time (ca. 170 Ma).

study area) is of similar density (~34° API). Geochemical analyses tie the Sweetgrass arch oil to a Devonian source in the footwall of the thrust belt to the west (Dolson et al., 1993), suggesting that thrust emplacement during the Laramide orogeny provided the conditions necessary for oil maturation and migration.

The study site may provide a snapshot of the eastward migration of hydrocarbon fluids; from footwall sources beneath Laramide-age thrusts, vertically through the fractured carbonate and grike system, and eastward along the sub–Middle Jurassic unconformity until trapped within the Sweetgrass arch. The permeable grike system is sealed by overlying Jurassic shale. This migration model is consistent with other fluid-migration events recorded at the onset of the Laramide orogeny as documented by other fluid-inclusion analyses and paleomagnetic studies (Enkin et al., 2000).

Clay mineral assemblages within the sandstone-filled grikes at the Swift Reservoir site indicate that the rocks experienced moderate temperatures during deformation (Ward, 2007). X-ray diffraction analyses of 17 samples taken from sandstone-filled grikes indicate that the temperature reached at least 180 °C, which is comparable to fluid-inclusion data from the same grike fills (trapping temperatures calculated to be 110–170 °C). Overall, the temperature values are consistent with other published temperature data that indicate that this area of the Rocky Mountain thrust front did not exceed 200 °C (Hoffman and Hower, 1979).

CONCLUSIONS

The existence of a system of karst-widened fractures filled with distinctive cherty sandstone of the Sawtooth Formation calls for a reevaluation of the models for fracture development in the Mississippian carbonate located in the Rocky Mountains of northern Montana. Because the grikes were filled by sand and were subsequently stylolitized in concert with layer-parallel shortening and then folded, our model proposes that grike development predates fracturing over anticlinal crests during Laramide deformation. The grikes may record passage of the forebulge through this region during the initiation of the northern Rockies foreland basin.

We propose the following sequence of events:

(1) pre-Bajocian—uplift on forebulge, extensional fracturing, karstification to depth of 4 m;

(2) Bajocian—subsidence of forebulge, infestation of dolomite surfaces by carbonate-boring clams, infilling of karst by sand and conglomerate of basal Sawtooth Formation;

(3) Bathonian–Late Cretaceous—deposition of foreland basin succession;

(4) Late Cetaceous–late Paleocene—hydrocarbon migration followed by layer-parallel shortening and thrusting and folding of foreland basin.

We conclude that the grikes predated folding over the crests of anticlines in the fold-and-thrust belt by >100 Ma. They are not restricted to the hinge areas of anticlines. We suggest that grike fracture permeability may be better preserved east of the stylolitization front, for example in the Sweetgrass arch, where fractures were not "healed" by pressure solution. The consistent orientation of the grikes may aid in enhanced hydrocarbon recovery from this productive horizon.

REFERENCES CITED

Bally, A.W., Gordy, P.L., and Stewart, G.A., 1966, Structure, seismic data, and orogenic evolution of the southern Canadian Rocky Mountains: Bulletin of Canadian Petroleum Geology, v. 14, no. 3, p. 337–381.

Beaumont, C., 1981, Foreland basins: Geophysical Journal of the Royal Astronomical Society, v. 65, p. 291–329.

Boyer, S.E., 1992, Geometric evidence for synchronous thrusting in the southern Alberta and northwest Montana thrust belts, in McClay, K.R., ed., Thrust Tectonics: New York, Chapman and Hall, p. 377–390.

Boyer, S.E., and Elliott, D., 1982, Thrust systems: American Association of Petroleum Geologists Bulletin, v. 66, p. 1196–1230.

Bradley, D.C., and Kidd, W.S.F., 1991, Flexural extension of the upper continental crust in collisional foredeeps: Geological Society of America Bulletin, v. 103, p. 1416–1438, doi: 10.1130/0016-7606(1991)103<1416:FEOTUC>2.3.CO;2.

Catuneanu, C., and Sweet, A.R., 1999, Maastrichtian-Paleocene foreland-basin stratigraphies, western Canada: A reciprocal sequence architecture: Canadian Journal of Earth Sciences, v. 36, p. 685–703, doi: 10.1139/cjes-36-5-685.

Childers, M.O., 1963, Structure and stratigraphy of the southwest Marias Pass area Flathead County, Montana: Geological Society of America Bulletin, v. 74, p. 141–163, doi: 10.1130/0016-7606(1963)74[141:SASOTS]2.0.CO;2.

Cobban, W.A., 1945, Marine Jurassic Formations of Sweetgrass Arch, Montana: Bulletin of the American Association of Petroleum Geologists, v. 29, no. 9, p. 1262–1303.

Colpron, M., Price, R., Archibald, D.A., and Carmichael, D.M., 1996, Middle Jurassic exhumation along the western flank of the Selkirk fan structure: Thermobarometric and thermochronometric constraints from the Illecillewaet synclinorium, southeastern British Columbia: Geological Society of America Bulletin, v. 108, no. 11, p. 1372–1392, doi: 10.1130/0016-7606(1996)108<1372:MJEATW>2.3.CO;2.

DeCelles, P.G., 2004, Late Jurassic to Eocene evolution of the Cordilleran thrust belt and foreland basin system, western U.S.A.: American Journal of Science, v. 304, p. 105–168, doi: 10.2475/ajs.304.2.105.

DeCelles, P.G., and Currie, B.S., 1996, Long-term sediment accumulation in the Middle Jurassic–early Eocene Cordilleran retroarc foreland-basin system: Geology, v. 24, no. 7, p. 591–594, doi: 10.1130/0091-7613(1996)024<0591:LTSAIT>2.3.CO;2.

Dolberg, D.M., 1986, A duplex beneath a major overthrust plate in the Montana Disturbed Belt; surface and subsurface data [master's thesis]: Missoula, University of Montana, 57 p.

Dolson, J., Piombino, J., Franklin, M., and Harwood, R., 1993, Devonian oil in Mississippian and Mesozoic reservoirs—Unconformity controls on migration and accumulation, Sweetgrass Arch, Montana: The Mountain Geologist, v. 30, no. 4, p. 125–146.

Dorobek, S.L., 1995, Synorogenic carbonate platforms and reefs in foreland basins: Controls on stratigraphic evolution and platform/reef morphology, in Dorobek, S.L., and Ross, G.M., eds., Stratigraphic Evolution of Foreland Basins: Society of Economic Paleontologists and Mineralogists Special Publication 52, p. 127–147.

Enkin, R.J., Osadetz, K.G., Baker, J., and Kisilvesky, D., 2000, Orogenic remagnetizations in the Front Ranges and Inner Foothills of the southern Canadian Cordillera: Chemical harbinger and thermal handmaiden of Cordilleran deformation: Geological Society of America Bulletin, v. 112, no. 6, p. 929–942, doi: 10.1130/0016-7606(2000)112<0929:ORITFR>2.3.CO;2.

Evenchick, C.A., McMechan, M.E., McNicoll, V.J., and Carr, S.D., 2007, this volume, A synthesis of the Jurassic–Cretaceous tectonic evolution of the central and southeastern Canadian Cordillera: Exploring links across the orogen, in Sears, J.W., Harms, T.A., and Evenchick, C.A., eds., Whence the Mountains? Inquiries into the Evolution of Orogenic Systems: A Volume in Honor of Raymond A. Price: Geological Society of America Special Paper 433, doi: 10.1130/2007.2433(06).

Feinstein, S., Kohn, B., Osadetz, K., and Price, R.A., 2007, this volume, Thermochronometric reconstruction of the prethrust paleogeothermal gradient and initial thickness of the Lewis thrust sheet, southeastern Canadian Cordillera foreland belt, in Sears, J.W., Harms, T.A., and Evenchick, C.A., eds., Whence the Mountains? Inquiries into the Evolution of Orogenic Systems: A Volume in Honor of Raymond A. Price: Geological Society of America Special Paper 433, doi: 10.1130/2007.2433(08).

Fermor, P.R., and Moffat, I.W., 1992, Tectonics and structure of the Western Canada foreland Basin, in Macqueen, R.W., and Leckie, D.A., eds., Foreland Basins and Fold Belts: American Association of Petroleum Geologists Memoir 55, p. 81–105.

Flemings, P.B., and Jordan, T.E., 1990, Stratigraphic modeling of foreland basins: Interpreting thrust deformation and lithospheric rheology: Geology, v. 18, p. 430–434, doi: 10.1130/0091-7613(1990)018<0430:SMOFBI>2.3.CO;2.

Gillespie, J.M., and Heller, P.L., 1995, Beginning of foreland subsidence in the Columbian-Sevier belts, southern Canada and northwest Montana: Geology, v. 23, no. 8, p. 723–726, doi: 10.1130/0091-7613(1995)023<0723:BOFSIT>2.3.CO;2.

Goldburg, B.L., 1984, Geometry and Styles of Displacement Transfer, Eastern Sun River Canyon Area, Sawtooth Range, Montana [master's thesis]: College Station, Texas A&M University, 128 p.

Hammond, C.R., and Trapp, H., eds., 1959, Proceedings, Sawtooth-Disturbed Belt Area 10th Anniversary Field Conference: Billings, Billings Geological Society, 17 p.

Harrison, J.E., Cressman, E.R., and Whipple, J.W., 1992, Geologic and Structure Maps of the Kalispell 1 × 2 Quadrangle, Montana, and Alberta and

British Columbia.: U.S. Geological Survey Miscellaneous Investigations Series Map I-2267, scale 1:250,000.

Hoffman, J., and Hower, J., 1979, Clay mineral assemblages as low grade metamorphic geothermometers: Application to the thrust faulted Disturbed Belt of Montana, U.S.A.: Society of Economic Paleontologists and Mineralogists Special Publication 26, p. 55–79.

Hoffman, J., Hower, J., and Aronson, J.L., 1976, Radiometric dating of the time of thrusting in the Disturbed Belt of Montana: Geology, v. 4, p. 16–20, doi: 10.1130/0091-7613(1976)4<16:RDOTOT>2.0.CO;2.

Imlay, R.W., 1945, Occurrence of Middle Jurassic rocks in Western Interior of the United States: Bulletin of the American Association of Petroleum Geologists, v. 29, no. 9, p. 1019–1027.

Imlay, R.W., 1952, Correlation of the Jurassic formations of North America, exclusive of Canada: Geological Society of America Bulletin, v. 63, p. 953–992.

Jordan, T.E., 1981, Thrust loads and foreland basin evolution, Cretaceous, western United States: American Association of Petroleum Geologists Bulletin, v. 65, no. 12, p. 2506–2520.

Mallory, W.W., ed., 1972, Geologic Atlas of the Rocky Mountain Region: Denver, Colorado, Rocky Mountain Association of Geologists, 331 p.

Mitra, S., 1986, Duplex structures and imbricate thrust systems: Geometry, structural position, and hydrocarbon potential: American Association of Petroleum Geologists Bulletin, v. 70, no. 9, p. 1087–1112.

Monger, J.W.H., and Price, R.A., 1979, Geodynamic evolution of the Canadian Cordillera—Progress and problems: Canadian Journal of Earth Sciences, v. 16, p. 770–791.

Mudge, M.R., 1972a, Pre-Quaternary Rocks in the Sun River Canyon Area, Northwestern Montana: U.S. Geological Survey Professional Paper 663-A, 142 p.

Mudge, M.R., 1972b, Structural Geology of the Sun River Canyon and Adjacent Areas, Northwestern Montana: U.S. Geological Survey Professional Paper 663-B, p. B1–B52.

Mudge, M.R., and Earhart, R.L., 1983, Bedrock Geologic Map of Part of the Northern Disturbed Belt, Lewis and Clark, Teton, Pondera, Glacier, Flathead, Cascade, and Powell Counties, Montana: U.S. Geological Survey Miscellaneous Investigations Map I-1375, scale 1:125,000.

Osborn, G., Stockmal, G., and Haspel, R., 2006, Emergence of the Canadian Rockies and adjacent plains: A comparison of physiography between end-of-Laramide time and present day: Geomorphology, v. 75, p. 450–477, doi: 10.1016/j.geomorph.2005.07.032.

Parcell, W.C., and Williams, M.K., 2005, Mixed sediment deposition in a retro-arc foreland basin: Lower Ellis Group (M. Jurassic), Wyoming and Montana, U.S.A.: Sedimentary Geology, v. 177, p. 175–194, doi: 10.1016/j.sedgeo.2005.02.007.

Peterson, J.A., 1957, Marine Jurassic of northern Rocky Mountains and Williston Basin: American Association of Petroleum Geologists Bulletin, v. 41, p. 399–440.

Poulton, T.P., 1984, The Jurassic of the Canadian western interior from 49°N latitude to Beaufort Sea, in Glass, D.J., and Stott, D.F., eds., The Mesozoic of Middle North America: Canadian Society of Petroleum Geologists Memoir 9, p. 15–41.

Poulton, T.P., Braun, W.K., Brooke, M.M., and Davies, E.H., 1993, Jurassic; Subchapter 4H, in Stott, D.F., and Aitken, J.D., eds., Sedimentary Cover of the Craton in Canada: Ottawa, Geological Survey of Canada, Geology of Canada, no. 5, p. 321–557.

Price, R.A., and Mountjoy, E.W., 1970, The geological structure of the Southern Canadian Rockies between Bow and Athabasca Rivers—A progress report, in Wheeler, J.O., ed., A Structural Cross-Section of the Southern Canadian Cordillera, Geological Association of Canada Special Paper 6, p. 7–25.

Price, R.A., and Sears, J.W., 2000, A preliminary palinspastic map of the Mesoproterozoic Belt-Purcell Supergroup, Canada and USA: Implications for the tectonic setting and structural evolution of the Purcell anticlinorium and the Sullivan deposit, Chapter 5, in Lydon, J.W., Höy, T., Slack, J.F., and Knapp, M.E., eds., The Geological Environment of the Sullivan Deposit, British Columbia: Geological Association of Canada, Mineral Deposits Division (MDD) Special Publication, No. 1, p. 61–81.

Sears, J.W., 2000, Rotational Kinematics of the Rocky Mountain Thrust Belt of Northern Montana, in Montana/Alberta Thrust Belt and Adjacent Foreland: Montana Geological Society 50th Anniversary Symposium, v. 1, p. 143–150.

Sears, J.W., 2001, Emplacement and denudation history of the Lewis-Eldorado-Hoadley thrust slab in the northern Montana Cordillera, USA: Implications for steady state orogenic processes: American Journal of Science, v. 301, p. 359–373, doi: 10.2475/ajs.301.4-5.359.

Sears, J.W., 2007, this volume, Belt-Purcell Basin: Keystone of the Rocky Mountain fold-and-thrust belt, United States and Canada, in Sears, J.W., Harms, T.A., and Evenchick, C.A., eds., Whence the Mountains? Inquiries into the Evolution of Orogenic Systems: A Volume in Honor of Raymond A. Price: Geological Society of America Special Paper 433, doi: 10.1130/2007.2433(07).

Sears, J.W., Braden, J., Edwards, J., Geraghty, E., Janiszewski, F., McInenly, M., MacLean, J., Riley, K., and Salmon, E., 2005, Rocky Mountain foothills triangle zone, Sun River, northwest Montana, in Thomas, R., ed., Proceedings of Annual Meeting of the Tobacco Root Geological Society: Northwest Geology, v. 34, p. 45–70.

Singdahlsen, D.S., 1986, Structural Geology of the Swift Reservoir Culmination, Sawtooth Range, Montana [master's thesis]: Missoula, University of Montana, 124 p.

Stearns, D.W., and Friedman, M., 1972, Reservoirs in fractured rock, in Stratigraphic Oil and Gas Fields: Classification, Exploration Methods, and Case Histories: American Association of Petroleum Geologists Memoir 16, p. 82–106.

Stockmal, G.S., and Beaumont, C., 1987, Geodynamic models of convergent margin tectonics: The southern Canadian Cordillera and the Swiss Alps, in Beaumont, C., and Tankard, A.J., eds., Sedimentary Basins and Basin-Forming Mechanisms: Canadian Society of Petroleum Geologists Memoir 12, p. 393–411.

Ward, E.M.G., 2007, Development of the Rocky Mountain Foreland Basin: Combined Structural, Mineralogical, and Geochemical Analysis of Basin Evolution, Rocky Mountain Thrust Front, Northwest Montana [Ph.D. thesis]: Missoula, University of Montana, 171 p.

White, T., Furlong, K., and Arthur, M., 2002, Forebulge migration in the Cretaceous Western Interior basin of the central United States: Basin Research, v. 14, p. 43–54, doi: 10.1046/j.1365-2117.2002.00165.x.

Willis, B., 1902, Stratigraphy and structure of the Lewis and Livingston Ranges, Montana: Geological Society of America Bulletin, v. 14, p. 69–84.

MANUSCRIPT ACCEPTED BY THE SOCIETY 22 MARCH 2007

Printed in the USA

# Structural, metamorphic, and geochronologic constraints on the origin of the Clearwater core complex, northern Idaho

P. Ted Doughty*
Department of Geology, Eastern Washington University, 130 Science Hall, Cheney, Washington 99004, USA

Kevin R. Chamberlain
Department of Geology and Geophysics, University of Wyoming, Dept. 3006, 1000 University Ave., Laramie, Wyoming 82071, USA

David A. Foster
Department of Geological Sciences, University of Florida, P.O. Box 112120, Gainesville, Florida 32611, USA

Grant S. Sha
Halliburton Energy Services, 1125 17th Street, Suite 1900, Denver, Colorado 80202, USA

ABSTRACT

New structural, metamorphic, and geochronologic data from the Clearwater complex, north-central Idaho, define the origin and exhumation history of the complex. The complex is divisible into an external zone bound by normal faults and strike-slip faults of the Lewis and Clark Line, and an internal zone of Paleoproterozoic basement exposed in two shear zone–bounded culminations. U-Pb sensitive high-resolution ion microprobe (SHRIMP) dating of metamorphic zircon overgrowths from the external zone yield zircon growth at ca. 70–72 Ma and 80–82 Ma, during peak metamorphism and before tectonic exhumation of the external zone. U-Pb SHRIMP dating of metamorphic zircon rims from the internal zone record growth at ca. 64 and between 59 and 55 Ma. The older ages record pre-extension metamorphism. The younger rim ages were derived from fractured zircons in the Jug Rock shear zone, and they document the beginning of exhumation of the internal zone along deep-seated shear zones that transported the basement rocks to the west. The $^{40}Ar/^{39}Ar$ ages record quenching of the external zone starting ca. 54 Ma and the internal zone between 53 and 47 Ma by movement along the bounding faults and internal shear zones. After ca. 47 Ma, extension was accommodated via a west-dipping detachment that was active until after ca. 41 Ma. The Clearwater complex is interpreted as an Eocene metamorphic core complex that formed in an extensional relay zone between faults of the Lewis and Clark Line.

Keywords: geochronology, core complex, Idaho, Eocene, Lewis and Clark, tectonics.

*ted.doughty@mail.ewu.edu

Doughty, P.T., Chamberlain, K.R., Foster, D.A., and Sha, G.S., 2007, Structural, metamorphic, and geochronologic constraints on the origin of the Clearwater core complex, northern Idaho, *in* Sears, J.W., Harms, T.A., and Evenchick, C.A., eds., Whence the Mountains? Inquiries into the Evolution of Orogenic Systems: A Volume in Honor of Raymond A. Price: Geological Society of America Special Paper 433, p. 211–241, doi: 10.1130/2007.2433(11). For permission to copy, contact editing@geosociety.org. ©2007 The Geological Society of America. All rights reserved.

INTRODUCTION

The formation of metamorphic core complexes along low-angle normal faults in the Cordillera has been known since their recognition some 30 yr ago (e.g., summaries of Coney [1980] and Armstrong [1982]). While initially viewed as isolated domal structures that formed solely in response to overthickened crust, continued studies have shown that many core complexes are kinematically linked with other fault systems. In some instances, core complexes are associated with relays or jogs in strike-slip faults (i.e., Death Valley, California; Burchfiel and Stewart, 1966). In the U.S. Cordillera, there is abundant evidence of core complexes forming along active strike-slip faults, but there are few examples older than 20 Ma (e.g., Faulds and Stewart, 1998).

The Northern Rockies is an ideal locale to examine the dynamic and kinematic relationships between metamorphic core complexes and coeval large-magnitude strike-slip faults in older rocks and across all levels of the crust. Eocene extension in northern Washington and Idaho has resulted in north-south–trending core complexes like the Kettle Dome and Priest River complex. The locus of crustal extension ceases abruptly at the Columbia Plateau and jumps 100 km to the southeast in western Montana (Fig. 1), where the Bitterroot and Anaconda complexes record large-magnitude east-west extension.

The eastward jump in the position of core complexes coincides with the Lewis and Clark Line, a >300-km-long, strike-slip fault system traceable from western Idaho to east-central Montana (Harrison et al., 1974; Reynolds, 1979; Hyndman et al., 1988). It moved in a right-lateral sense during core complex formation and behaved like a continental-scale transform fault that linked two offset domains of crustal extension during Eocene time (Sheriff et al., 1984; Hyndman et al., 1988; Doughty and Sheriff, 1992; Foster and Fanning, 1997; Sears et al., 2000; Lewis et al., 2002; Foster et al., 2007).

The Clearwater complex, also known as Boehls Butte–Goat Mountain, is a unique area of anomalously high-grade metamorphic rocks and Precambrian basement that lies within the southern part of the Lewis and Clark Line as it passes through the northern border zone of the Idaho Batholith in north-central Idaho (Figs. 1 and 2) (Hietanen, 1984; Grover et al., 1992; Burmester and Lewis, 1999; Doughty and Buddington, 2002). These rocks have been interpreted as having been uplifted in the hanging wall of Mesozoic thrust faults (Harrison et al., 1986; Skipp, 1987), but the proximity to the Lewis and Clark Line led Seyfert (1984), Doughty and Sheriff (1992), and Burmester and Lewis (1999) to propose that the basement gneisses could have been exhumed in a releasing step in the Lewis and Clark fault zone. Petrologic and isotopic data from the high-grade gneisses reveal that the rocks have been subjected to a late episode of nearly isothermal decompression (Grover et al., 1992; Larson and Sharp, 1998; Mora et al., 1999), which is consistent with their exhumation in an extensional rather than contractional setting.

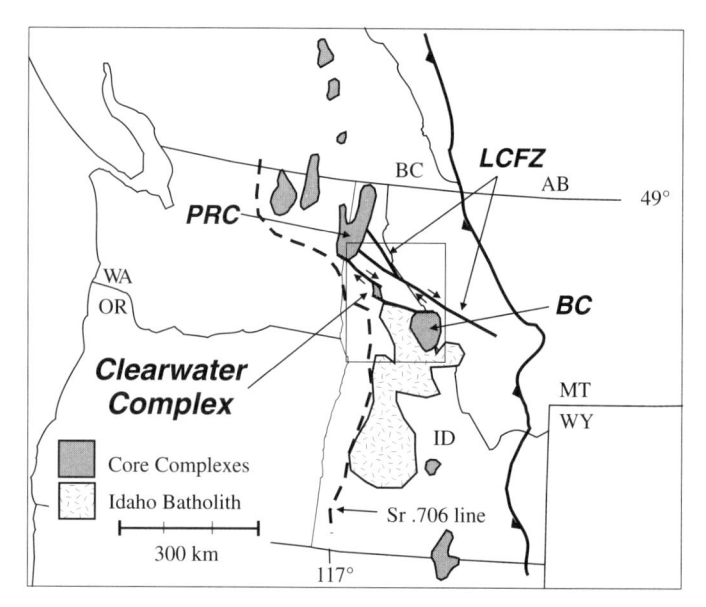

Figure 1. Geologic sketch map of the northern Rockies illustrating the distribution of Middle Eocene core complexes (shaded) and the Lewis and Clark fault zone (modified from Coney, 1980). The box outlines the area shown in Figure 2. BC—Bitterroot complex; PRC—Priest River complex; LCFZ—Lewis and Clark fault zone.

This paper addresses the structure of the Clearwater complex and the style, kinematics, and chronology of shearing within the complex. We also address the timing of deformation, depth of faulting, and cooling history of the complex.

GEOLOGIC SETTING

Rock Types and Metamorphism

The Clearwater complex lies in the St. Joe–Clearwater region of north-central Idaho, along the northern margin of the Idaho Batholith (Fig. 1). This part of the Cordilleran orogen is composed principally of high-grade metamorphic rocks derived from the lower part of the Middle Proterozoic Belt Supergroup and satellites of the Late Cretaceous to early Tertiary Idaho Batholith and epizonal Eocene plutons (Fig. 2) (Hietanen, 1963a, 1963c; Hyndman et al., 1988; Marvin et al., 1984; Burmester et al., 2004). The dominant pattern of metamorphism, classified as M2 by Lang and Rice (1985a), is concentric to the Idaho Batholith and is Barrovian. The peak phase of M2 metamorphism in the Clearwater complex and Bitterroot complexes reached the kyanite-sillimanite-muscovite zone (650–750 °C, 800–1100 MPa) (Grover et al., 1992; House et al., 1997; Foster et al., 2001). East of the Clearwater complex, near Snow Peak, M2 metamorphism reached 600 MPa and 565 °C (Lang and Rice, 1985b). M2 spans between ca. 117 Ma to 82 Ma in the Clearwater region (Grover et al., 1993). Foster et al. (2001) and House et al. (1997) found that peak metamorphism in the northern Bitterroot complex occurred at ≥80–56 Ma. Both estimates

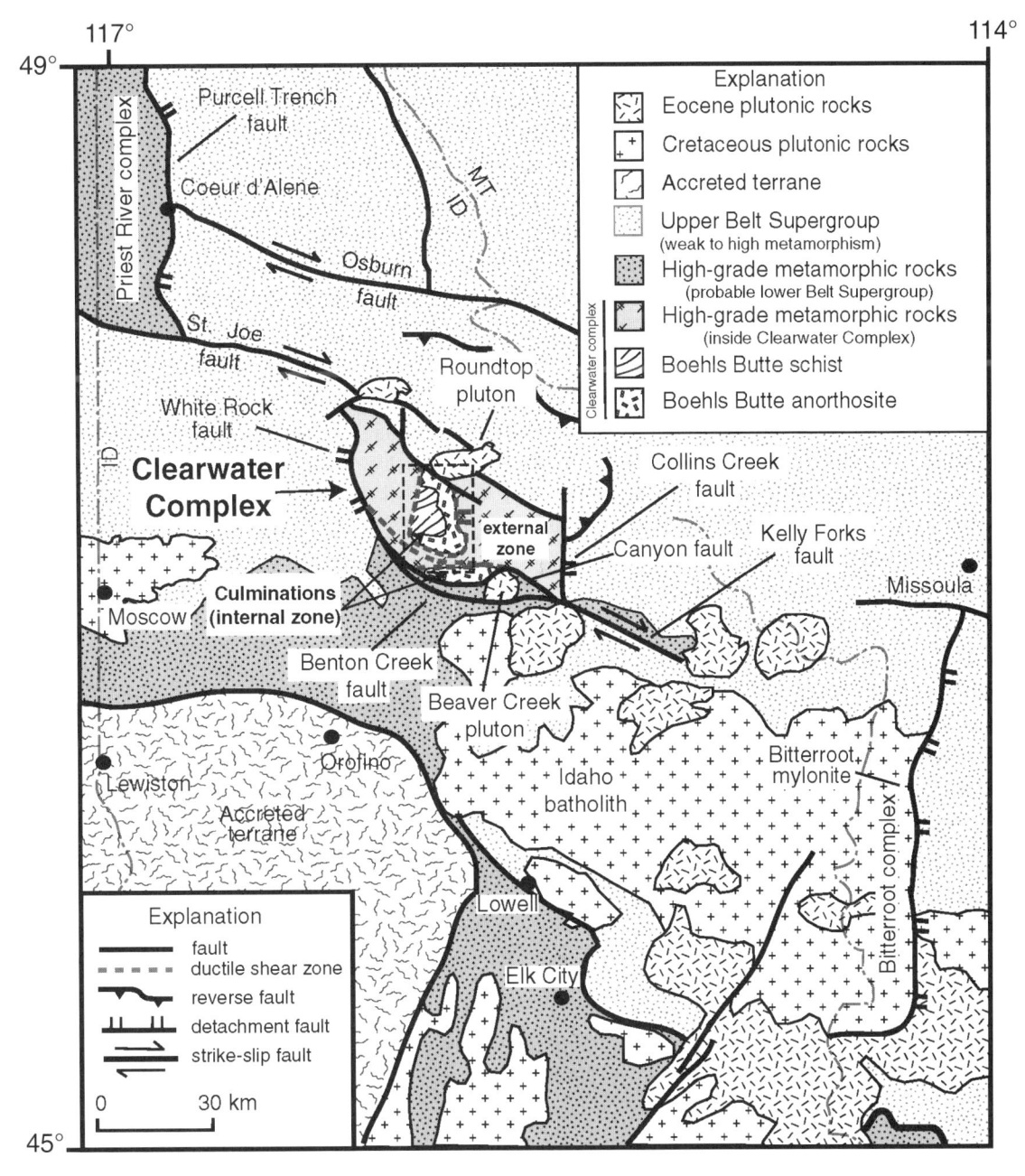

Figure 2. Simplified geologic map of the northern border zone of the Idaho Batholith. Surficial deposits and Oligocene and Miocene volcanic rocks are not shown. The Clearwater complex (outlined in darker shading) occupies a pull-apart structure between the St. Joe fault and the Kelly Forks–Benton Creek fault of the Lewis and Clark Line. This fault system links middle Eocene extension in the Bitterroot complex with the Priest River complex. The external zone of the Clearwater complex is shaded dark gray and patterned. The internal zone of the Clearwater complex, which occurs in two culminations inside the external zone, is white and patterned. Thick dashed gray lines denote ductile shear zones inside the Clearwater complex. The box outlines the extent of Figure 3. This figure was modified from R. Lewis (2004, personal commun.).

largely coincide with intrusion of the main phase of the Idaho Batholith and accretion of exotic terranes to the west.

A third metamorphic event, classified as M3, has been found only within the core of the Clearwater complex. This event, exhibited by complex disequilibrium mineral textures between low-pressure and high-pressure minerals, resulted from isothermal decompression of the Clearwater complex (Lang and Rice, 1985a; Grover et al., 1992) at around 50 Ma (Larson and Sharp, 1998). Similar mineral parageneses are found in the Priest River complex (Rhodes, 1986; Doughty and Price, 1999), and northern Bitterroot complex (Cheney, 1975; Foster et al., 2001).

Geologic Structures

The high-grade metamorphic rocks exposed in the northern border zone of the Idaho Batholith contain east-directed thrust faults, northwest-trending strike-slip faults of the Lewis and Clark Line, and north-trending normal faults (Fig. 2). Many of the thrust faults are ductile synmetamorphic structures that coincide with peak metamorphism and are identified on the basis of juxtaposed rock types. Other thrust faults postdate peak metamorphism and are identified by offset metamorphic isograds.

Most fault zones of the Lewis and Clark Line in the vicinity of the Clearwater complex have steeply dipping slip surfaces bearing subhorizontal mineral lineations and slip-surface striations. Childs (1982) found that brecciation and low-grade alteration locally overprint an older mylonitic fabric, suggesting that the faults were active across a range of crustal depths. The faults have also been intruded by syntectonic hypabyssal dikes (Childs, 1982) that are related to epizonal middle Eocene plutons like the Beaver Creek, Bungalow, and Roundtop plutons. These intrusions yield 52–46 Ma crystallization ages and date the right-lateral motion as middle Eocene in age (Burmester et al., 2004; Lewis et al., 2002; Marvin et al., 1984). Older left-lateral or transpressional movement is reported on parts of the Lewis and Clark Line but is not well documented in the Clearwater region (i.e., Kell and Childs, 1999; Yin and Oertel, 1995; Sears et al., 2000).

Three large west- to northwest-trending strike-slip faults of the Lewis and Clark fault zone bound the Clearwater complex (Fig. 2). The St. Joe fault is the northern fault, and it can be traced from the southern end of the Priest River complex to the southeast along the northern margin of the Clearwater complex. Near the Clearwater complex, the St. Joe fault is expressed as a poorly defined zone of strike-slip faults and shear that has mylonitized the southern border of the ca. 52 Ma Roundtop pluton (Marvin et al., 1984). The southern boundary of the Clearwater complex is formed by the Canyon fault and Benton Creek fault (Hietanen, 1984), which is part of an 8-km-wide zone of east-west–trending strike-slip faults (Lewis et al., 2002, 2007) that extend westward from the Kelly Forks fault identified along the northwestern corner of the Bitterroot complex (Childs, 1982). Other faults, including the Clugs Jumpoff fault, splay off of the Canyon fault near the Clearwater complex. To the west, the fault system bends northward and merges with the White Rock fault on the western side of the Clearwater complex.

A relay connection between the St. Joe and the Canyon fault–Benton Creek fault system is completed by two north-south–trending normal faults, the White Rock fault on the western side and the Collins Creek fault on the eastern side of the Clearwater complex (Fig. 2). Both faults juxtapose the high-grade metamorphic rocks of the Clearwater complex against lower-grade metamorphosed Middle Belt Supergroup strata (Hietanen, 1963b; Lewis et al., 1992). The White Rock fault is a ductile-brittle detachment fault with top-to-the-west mylonites and chloritic brecciation that offset metamorphic isograds, whereas the Collins Creek fault is a top-to-the-east normal fault that has not been studied in any detail (R. Lewis, 2004, personal comm.).

CLEARWATER COMPLEX

In this paper, we define the Clearwater complex as a large tract of amphibolite-facies metamorphic rocks that lies between the east-west–trending St. Joe and Canyon strike-slip faults and the north-trending White Rock and Collins Creek normal faults (Fig. 2). Previous studies by Hietanen (1963a, 1984), Grover et al. (1992), Seyfert (1984), and Doughty and Sheriff (1992) considered the Clearwater complex to be a small structure centered on the high-grade Paleoproterozoic basement rocks that occur in the core of the complex. A largely intact fault block, the Boehls Butte–Goat Mountain fault block, was envisioned to host the basement rocks (Hietanen, 1984; Grover et al., 1992). Slip along the margins of this block was invoked to explain the isothermal decompression of the rocks within, as determined from metamorphic mineral assemblages and oxygen isotopes. The existence and location of these faults was inferred because of poor exposure, difficult access, and rock-type similarity across the area.

New geologic mapping during this study and by Lewis et al. (2007) significantly revises the prior mapping of Hietanen (1963a, 1963b, 1984) and the interpretations based on that mapping. Our mapping, as well as differences in rock type, strain, apparent metamorphic grade, and thermochronology, shows that the Clearwater complex is best described in terms of isolated domes, or culminations, and tectonic slivers of high-grade basement rocks that are surrounded by metasedimentary rocks. Rocks along the margins of the culminations are sheared to varying degrees, with mylonites well preserved along the eastern side of each culmination. We assign the rocks within the culminations to the internal zone of the Clearwater complex, and the rocks between the culminations and the bounding strike-slip and normal faults to the external zone of the Clearwater complex.

Internal Zone

Rocks of the internal zone are exposed in two dome-like culminations and as fault slivers along the Clugs Jumpoff fault (Fig. 3). Internal zone rocks are composed of metamorphosed 1787 Ma Paleoproterozoic anorthosite and aluminum- and magnesium-enriched gneiss (Al-Mg–rich gneiss) that are interleaved with sill-like bodies of 1587 Ma amphibolite (Hietanen, 1956, 1963a, 1969b, 1984; Nord, 1973; Juras, 1974; Doughty and Chamberlain, 2007). Intercalation of the basement rocks with pelitic schists derived from the Belt Supergroup occurs along synmetamorphic thrust faults within the internal zone (Doughty and Chamberlain, 2007) (Fig. 3).

Metastable mineral assemblages and disequilibrium textures in metamorphic rocks of the internal zone record a complex polymetamorphic history. The oldest documented metamorphism is poorly understood and Mesoproterozoic in age, ca. 1.1 Ga (Sha et al., 2004). M1, the first regional metamorphic event, is poorly preserved in the internal zone and is best seen east of the Clearwater complex (Lang and Rice, 1985a). The second event, M2, is the peak metamorphic event, and it is characterized by the growth of kyanite and sillimanite with muscovite in Al-Mg gneisses. Quantitative estimates of the conditions for M2 range from 800 to 1100 MPa at 650 °C to 750 °C (mean of 699 °C) (Grover et al., 1992). Larson and Sharp (1998) reported temperatures for M2, based on oxygen isotopes, of 700–775 °C. So far, evidence of M3 has only been found where the unusual composition of the Al-Mg gneisses facilitated the replacement of strained kyanite with andalusite and other unique mineral assemblages (Carey et al., 1992). Conditions of metamorphism during M3 range from 400 to 600 MPa at the same temperature as M2 (Grover et al., 1992). The similarity in temperatures for M2 and M3 led Grover et al. (1992) to conclude that isothermal decompression occurred between M2 and M3.

External Zone

The part of the Clearwater complex that surrounds the culminations of internal-zone basement rocks and extends out to the bounding White Rock and Collins Creek faults is defined as the external zone (Figs. 2 and 3). Hietanen (1963a, 1984) placed some of the rocks that we assign to the external zone within the uplifted Boehls Butte–Goat Mountain fault block.

Rocks of the external zone consist of rusty weathering mica schists, quartzites, and amphibolites derived from the Prichard Formation of the Middle Proterozoic Belt Supergroup metamorphosed in the garnet through kyanite-sillimanite zones of the amphibolite facies (Hietanen, 1963a, 1968; Doughty and Chamberlain, 2007). There are also several masses of Cretaceous(?) orthogneiss within this zone. These rocks are generally not sufficiently aluminous to contain aluminum silicates, although Hietanen (1963a, 1968, 1984) reported a few occurrences of kyanite or sillimanite.

Rocks along the eastern and southern flanks of the northern culmination are distinct from those found elsewhere in the external zone (Fig. 3). They consist of quartzite and fine-grained semischist that are locally calcareous enough to produce calc-silicate minerals and marbles. These rocks are very fine-grained with a weak foliation that lies at a high angle to compositional layering (relict bedding) (Fig. 4A). In many locations, the rocks exhibit very low amounts of strain, as indicated by randomly oriented porphyroblasts of tremolite or zoisite, and small poikilitic spots of orthoclase (cf. Hietanen, 1963a) (Fig. 4B). Many of the rocks, especially east of Monumental Buttes, are fine-grained, rusty weathering, siltites that look identical to the weakly metamorphosed Prichard Formation seen across much of Montana and Idaho. Small plugs of pyroxene gabbro, containing a border zone of amphibolite, intrude the metasedimentary rocks. These mafic intrusions crystallized ca. 1465 Ma (Doughty and Chamberlain, 2007).

The rocks of the external zone appear to display significant differences in composition and metamorphic history compared to the gneisses in the internal zone. Most rocks in the external zone have simple textures and low-variance mineral assemblages, and only local evidence for polymetamorphism (i.e., Lang and Rice, 1985a).

SHEAR-ZONE CHARACTERISTICS AND KINEMATICS

The contact between rocks of the internal zone and rocks of the external zone is marked by zones of intense noncoaxial shear. There are excellent exposures of these sheared contacts in glaciated cirques and ridges along Monumental Buttes, Goat Mountain, and Crescendo Peak. Elsewhere, the exposure is very poor due to intense surface weathering and thick forest growth. We have examined the contact at six different localities (Figs. 3 and 5): Jug Rock, Cedar Creek Canyon, North Monumental, and Floodwood Creek localities along the borders of the northern culmination; and Little North Fork of the Clearwater River, and Aquarius localities along the borders of the southern culmination. We also examined the Clugs Jumpoff fault as it passes through the Smith Ridge syncline (Fig. 5). Detailed structural analysis of rock fabrics and the kinematics of these shear zones are discussed next and in Sha (2004).

Northern Culmination (Jug Rock Shear Zone)

Exposures of the sheared contact between the internal and external zones are superb along the eastern edge of the northern culmination in the high country along Monumental Buttes and the Little Goat Mountains (Figs. 3 and 4). Here, shear zones are exposed on the north, east, and southern margins of the internal zone. In this paper, we define these interlinked faults and shear zones as the Jug Rock shear zone for the excellent and readily accessible exposures along the northern side of Jug Rock (Fig. 3). Along this stretch, the zone is a shallowly east-dipping ductile shear zone with downdip mineral lineations. To the north, northwest of Monumental Buttes, the Jug Rock shear zone bends sharply to the west and follows the northern edge of the anorthosite. Along this contact, it is a shallow to steep northeast-dipping shear zone with subhorizontal mineral lineations. South of Jug Rock, past Crescendo Peak, the Jug Rock shear zone turns sharply to the west and follows the southern margin of the anorthosite. Here, again, it is a steeply dipping east-west–trending zone with subhorizontal mineral lineations (Fig. 5).

The north-south–trending segment of the Jug Rock shear zone is ~500 m thick, with gradational upper and lower boundaries. The upper boundary is a narrow zone of decreasing strain at the contact between internal zone rocks and the overlying quartzites and semischists of the external zone, which maintain a mineral lineation for a short distance above the contact. The lower boundary is a more diffuse zone of decreasing strain that occurs in the upper part of the anorthosite bodies, which change from being strongly foliated within the shear zone to massive below.

Protomylonites, mylonites, and ultramylonites are recognized within the Jug Rock shear zone. Protomylonites are developed primarily in quartz-poor lithologies, like anorthosite and plagioclase-rich Al-Mg gneiss. Mylonites and ultramylonites are best developed in mica schist and quartz-bearing Al-Mg gneiss. An increase in grain-size reduction in the mylonites correlates well with a decrease in the proportion of plagioclase in the protolith, and there is considerable heterogeneity in the character of the mylonitic fabrics across the shear zone (Sha et al., 2003).

Microstructures of the Jug Rock Shear Zone

The primary foliation in rocks of the Jug Rock shear zone formed at conditions near peak metamorphic conditions. The foliation is defined by coarse-grained reddish biotite, muscovite, fibrolite, and quartz (Fig. 4F). The quartz typically forms coarse-grained equigranular mosaics with interlobate grain boundaries indicative of static recrystallization after shearing. The micas display undulose extinction and fish-like forms. The foliation wraps around large porphyroclasts of strained kyanite, aggregates of andalusite (inverted from kyanite), garnet, and plagioclase (Fig. 4F). The porphyroclasts exhibit asymmetric sigma morphologies with tails of recrystallized quartz and pale-green biotite. Strained kyanite porphyroclasts are boudinaged with muscovite and pale-green biotite growing in the pull-apart zones. Locally, aluminosilicate porphyroclasts are surrounded by coronas of cordierite that separate the aluminosilicate from adjacent biotite. Cordierite also forms equidimensional aggregates, surrounding relict grains of andalusite, parallel to the foliation.

A second foliation overprints the primary foliation at a shallow angle (30°–40°) (Figs. 4G and 4H). Carey (1985) recognized this second fabric as a crenulation cleavage with hartschiefer texture, but these features are better interpreted as a secondary shear-band foliation related to mylonitization. The second foliation is defined by thin bands of fine-grained reddish biotite intergrown with fibrolite and dynamically recrystallized quartz grains (Figs. 4G and 4H). The average grain sizes, 0.08–0.04 mm, contrast sharply with the coarse-grained size of minerals that form the primary foliation. Most of the quartz grains have recrystallized into a fine-grained granoblastic aggregate, but locally, there are preserved quartz ribbons with undulose extinction and small elongate subgrains (0.005 mm).

Bands of ultramylonite, 2–10 cm wide, occur locally in mica schists, Al-Mg gneiss, and anorthosite. In outcrop, they form glassy, black resistant bands. The ultramylonites have a matrix of very fine-grained dark-brown biotite (0.0025 mm) enclosing small (0.5-.08 mm) rounded porphyroclasts of quartz, tourmaline, garnet, kyanite/andalusite, and aggregates of dynamically recrystallized, unannealed quartz (Figs. 4I and 4J). Sigma-type porphyroclasts of andalusite often contain tails of pale-green biotite or chlorite (Fig. 4K). Dynamic recrystallization has produced an S-C fabric with subgrains that display a strong preferred orientation at a high angle to the primary foliation.

Initial formation of the primary foliation, S1, coincides with the growth of peak M2 minerals like fibrolite and the micas, which define the foliation. Kyanite, which contains epitaxial overgrowths of fibrolite and forms strained porphyroclasts, could predate the formation of S1, although Carey (1985) concluded that it grew during the early stages of S1 development. Because S1 contains asymmetric porphyroclasts and mica fish that yield the same sense of shear as that recorded by S2 fabrics, we infer that mylonitization began during uplift of the rocks from the kyanite field into the sillimanite field. Minerals deformed by this foliation are normally annealed, which reflects the high temperatures present during mylonitization. The shear-band foliation, S2, coincides with the replacement of strained kyanite by andalusite, growth of secondary biotite, muscovite, and chlorite, and unannealed mylonitic textures. It also contains mats of fibrolite that appear to have grown along S2. Thus, S2 initially formed in the sillimanite stability field but continued to develop as the rocks were uplifted into the stability field of andalusite and chlorite, at considerably less pressure and temperature than S1. It records the strain associated with the exhumation of the Goat Mountain area to the surface. These relationships indicate that the Jug Rock shear zone formed during progressive unroofing of the rocks in the Clearwater complex.

S-C fabrics, shear-band foliations, and rotated porphyroclasts from all three segments of the Jug Rock shear zone were used to constrain the kinematics of motion. The north-trending segment of the Jug Rock shear zone moved in a consistent down-dip top-to-the-east direction along an azimuth of 100° (Fig. 5). The east-west–trending northern segment of the Jug Rock shear zone, northwest of Monumental Buttes, moved horizontally in a right-lateral sense along an azimuth of 108° (Fig. 5). The east-west–trending southern segment of the Jug Rock shear zone, northwest of Cedar Creek, moved subhorizontally in a left-lateral sense along an azimuth of 273° (Fig. 5). These kinematic indicators document exhumation of the internal zone by east-directed transport on the Jug Rock shear zone.

Figure 3. (A) Simplified geologic map of the northern culmination of the Clearwater complex. The two culminations of polymetamorphosed anorthosite flanked by Al-Mg–rich schist comprise the internal zone of the Clearwater complex. Ductile shear zones, like the Jug Rock shear zone, wrap around these culminations and juxtapose these basement rocks against metasedimentary rocks of the external zone. Unpatterned areas of the external zone are coarse-grained mica schist. Ruled pattern areas of the external zone are fine-grained, seemingly weakly strained schist, quartzite, calc-silicate, and marble. The Jug Rock shear zone and related structures are thicker than portrayed on the map. Sense of shear is shown with arrows and barbs on the hanging wall. Geochronology samples are shown with solid circles, and prominent peaks are shown with open triangles. Bold numbers adjacent to sample localities are ^{40}Ar/^{39}Ar cooling ages in Ma. The fault that juxtaposes Al-Mg schist over quartzite and schist within the northern culmination (dashed line with teeth) is postulated on the basis of detrital zircon geochronology (Doughty and Chamberlain, 2007) and is interpreted to be a synmetamorphic thrust fault. This figure is based on unpublished mapping by the authors, Lewis et al. (2007), Sha (2004), Hietanen (1963a), and Nord (1973). (B) Schematic cross section illustrating subsurface geologic structure.

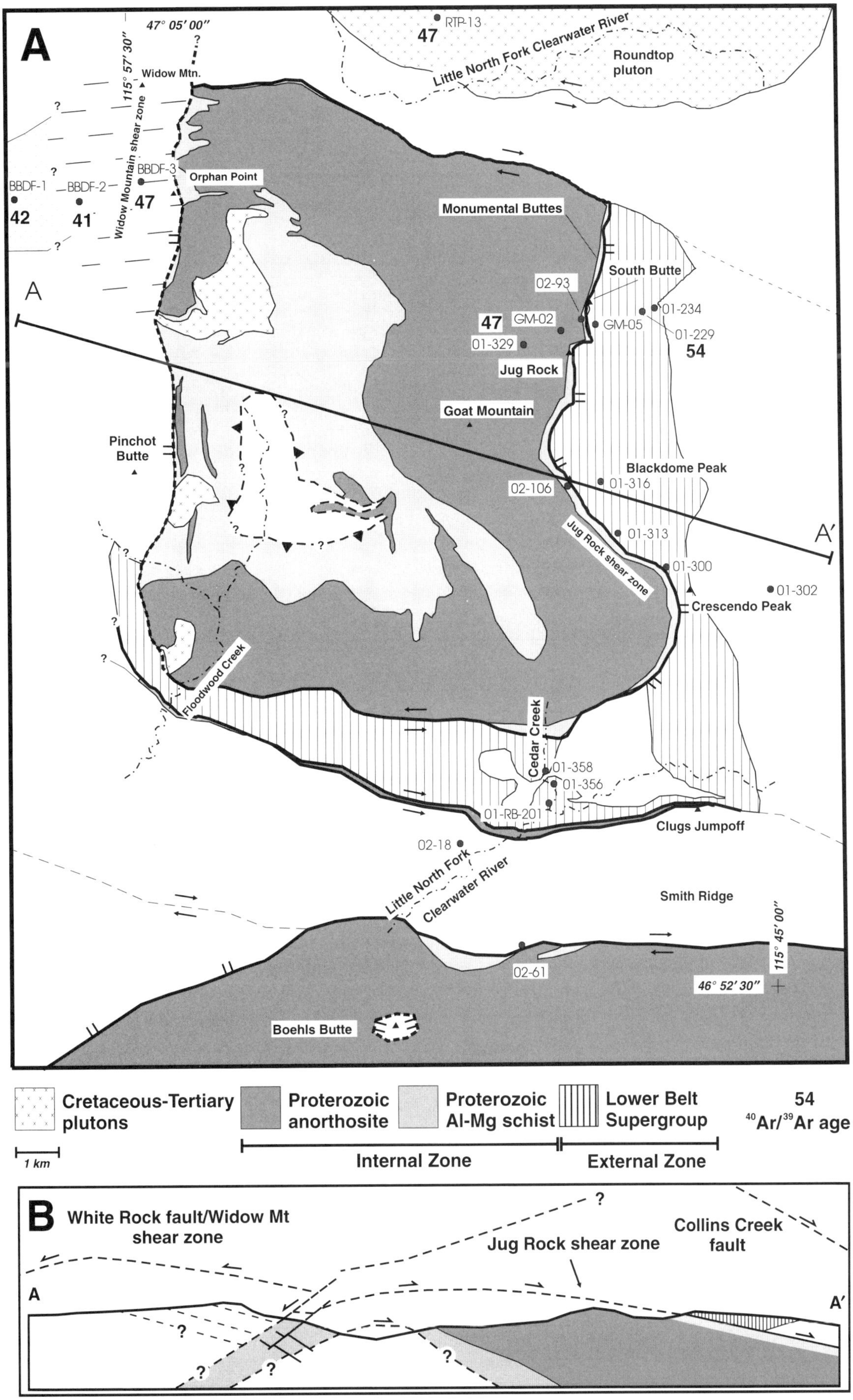
A
47° 05′ 00″
115° 57′ 30″
Widow Mtn.
Widow Mountain shear zone
RTP-13
47
Little North Fork Clearwater River
Roundtop pluton
BBDF-1
42
BBDF-2
41
BBDF-3
47
Orphan Point
Monumental Buttes
South Butte
02-93
GM-02
47
GM-05
01-234
01-229
54
01-329
Jug Rock
Goat Mountain
Pinchot Butte
Blackdome Peak
02-106
01-316
01-313
Jug Rock shear zone
01-300
01-302
Crescendo Peak
Floodwood Creek
Cedar Creek
01-358
01-356
01-RB-201
Clugs Jumpoff
02-18
Little North Fork Clearwater River
Smith Ridge
115° 45′ 00″
02-61
46° 52′ 30″
Boehls Butte
A
A′
Cretaceous-Tertiary plutons
Proterozoic anorthosite
Proterozoic Al-Mg schist
Lower Belt Supergroup
54
$^{40}Ar/^{39}Ar$ age
1 km
Internal Zone
External Zone
B
White Rock fault/Widow Mt shear zone
Jug Rock shear zone
Collins Creek fault
A
A′

Selected Photomicrographs

External zone rocks above the Jug Rock shear zone

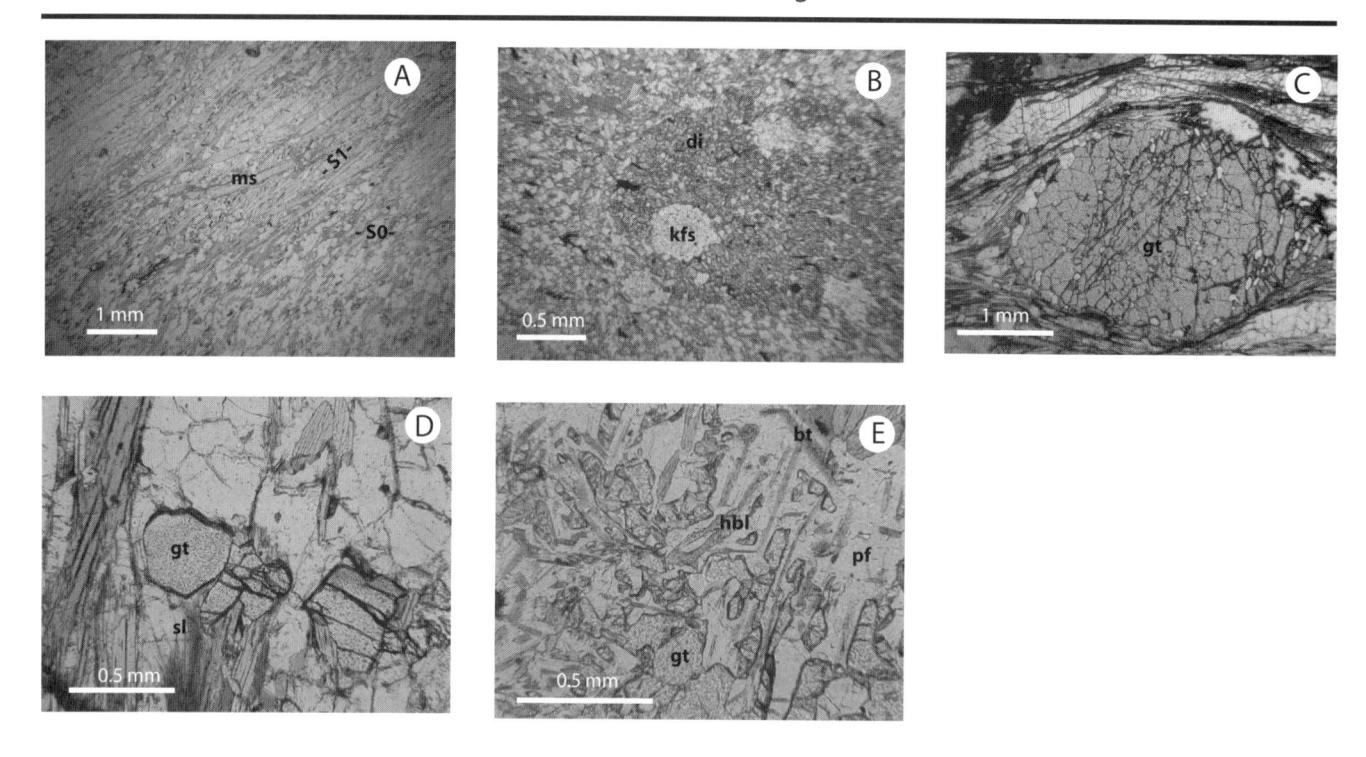

Mylonites: Jug Rock Shear Zone

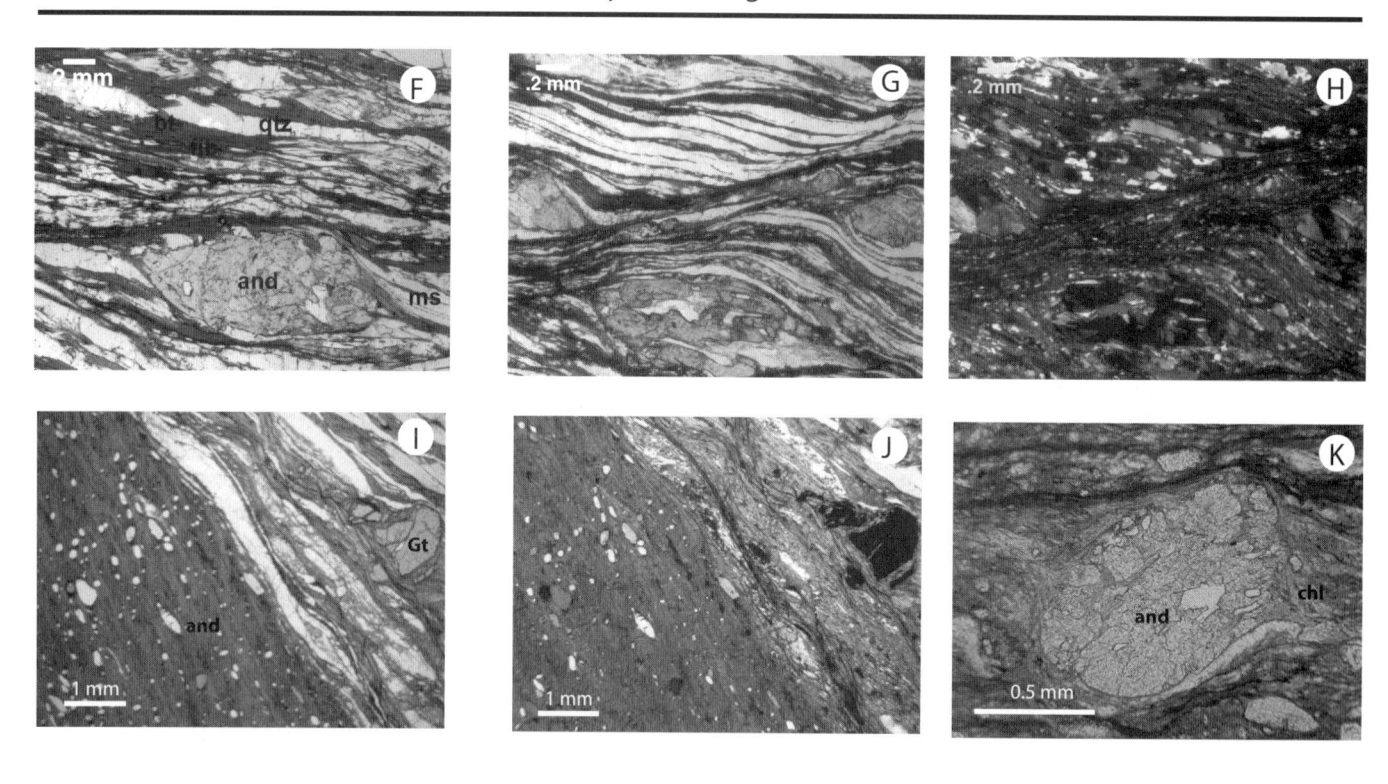

Figure 4. Photomicrographs of rocks from Clearwater complex. (A–E) Samples from the external zone illustrating the fine grain size, apparent low strain, and simple metamorphic textures in this domain. (A) External zone schist (0.5 km above Jug Rock shear zone) showing foliation at a high angle to compositional layering. (B) Randomly oriented poikilitic porphyroblasts of diopside (gray) and orthoclase (clear) in a poorly foliated semischist collected 0.5 km above Jug Rock shear zone. (C) Garnet porphyroblast from sample 300, illustrating the simple metamorphic textures of rocks in the external zone. (D) Foliated fibrolite and fine-grained biotite overgrowing small anhedral garnet porphyroblast(s) from sample 02-18. (E) Symplectite intergrowth of hornblende, biotite, and plagioclase surrounding embayed garnet, sample 01-358. (F–K) Photomicrographs of mylonites from the Jug Rock shear zone. (F) Primary mylonitic foliation, illustrating coarse grain size with foliation-parallel fibrolite, muscovite, and biotite, partially annealed fabrics, and sigmoidal porphyroclasts of kyanite pseudomorphing into andalusite. These textures are consistent with shear at near-peak metamorphic conditions. Top-to-the-east sense of shear. (G–H) Extensional shear bands in coarse-grained mylonite schist. Shear bands are finer grained and less annealed than the primary foliation, consistent with exhumation during mylonitization. Top-to-the-east sense of shear (G—plane light; H—crossed nichols). (I–J) Ultramylonite composed of small porphyroclasts of andalusite (pseudomorphing kyanite), plagioclase, and garnet in a very fine-grained foliated matrix of biotite and quartz (I—plane light; J—crossed nichols). (K) Porphyroclast of kyanite (replaced by andalusite) in unannealed mylonite from Jug Rock shear zone. Tails contain pale-green chlorite. The unannealed texture and presence of chlorite demonstrate that mylonitization occurred during decreasing metamorphic conditions and exhumation. Key to mineral annotations: and (andalusite); chl (chlorite); bt (biotite); di (diopside); fib (fibrolite); gt (garnet); hbl (hornblende); kfs (K-feldspar); ms (muscovite); pf (plagioclase); qtz (quartz).

Floodwood Creek (Western Boundary of Northern Culmination)

The western boundary of the northern culmination was interpreted by Hietanen (1963a, 1984) as a steep north-south–trending normal fault (Orphan Point fault), largely on the basis of a sharp demarcation between anorthosite and external zone rocks to the west. This fault was mapped as extending to the north and south of the northern culmination. During the course of our mapping, we found no obvious field evidence (such as brecciation, alteration) or minor structures supporting the presence of a steep brittle fault along the western edge of the northern culmination. The exposure, however, is poor, and there are no locations where the contact is exposed. Our interpretation of the nature of the contact is based on examination of structures and fabrics in rocks adjacent to the contact.

Rocks west of the contact in the external zone vary from weakly foliated fine-grained semischists in the south (Floodwood Creek) to mylonitic granite and metasedimentary rocks along the rugged alpine ridges of Orphan Point and Widow Mountain in the north (Fig. 3). The mylonites in the north are herein named the Widow Mountain shear zone for the excellent exposures on that mountain. The mylonites dip gently to the east with downdip lineations, but they have kinematic indicators with updip, top-to-the-west sense of shear. The mylonite zone is wholly contained within the external zone and appears to be unrelated to the juxtaposition between internal and external zone rocks to the east (but see following). The fault's updip kinematics suggest that it could be Mesozoic in age, or a younger rotated structure. If the latter, it could be part of the top-to-the-west White Rock detachment fault, which bounds the western margin of the Clearwater complex ~20 km to the west (R. Lewis, 2004, personal commun.).

The dominant fabric present in rocks east of the contact is a gently west-dipping annealed and recrystallized foliation in anorthosite. Xenoliths of foliated anorthosite within sills of foliated hornblende-biotite orthogneiss show that this fabric is older than the orthogneiss; it is also older than stocks of foliated biotite granite along the western side of the northern culmination (Fig. 3). Lewis et al. (2007) inferred that the orthogneisses are Cretaceous in age, but without actual dates, this old fabric could range from Precambrian to Eocene in age.

A younger second, and possibly third, deformation event is evident in these rocks. Spaced centimeter-scale bands of ultramylonite crosscut the foliated anorthosite and sills of orthogneiss, and stocks of granite are penetratively deformed into protomylonites, mylonites, and centimeter-scale zones of ultramylonite. Many of these mylonitic granite stocks are similar to granites dated at 48 Ma along the southern border of the complex (Burmester et al., 2004). We conclude that this fabric is Eocene in age and it is related to exhumation of the Clearwater complex. The unannealed nature of the fabrics and presence of chlorite within mineral pull-aparts and within the mylonitic foliation show that these second-event mylonites formed at significantly lower temperatures and pressures than the primary metamorphic foliation. The second-event fabrics strike north with gentle east or west dips and downdip mineral lineations (Fig. 5). Crosscutting relationships suggest that some of the top-to-the-east mylonites are younger than the top-to-the-west mylonites. Kinematic indicators yield consistent downdip shear in all cases.

Unlike the Jug Rock shear zone, the western boundary of the internal zone along Floodwood Creek lacks a unidirectional sense of shear. There is no systematic difference between the second event west-directed and east-directed fabrics, and all rock types contain mylonite fabrics with both senses of shear. These mylonites could have formed as conjugates of one another in a regime of pure shear, or they could indicate episodes of deformation with opposing shear sense within the core of the Clearwater complex. Our preferred interpretation is that the dominant west-dipping fabric is Eocene in age and part of the Jug Rock shear zone that has been arched and back-rotated during formation of the Clearwater complex. The younger second-event mylonites formed after the initial arching of the internal zone and record overlapping top-to-the-west and top-to-the-east shear in the footwall of the White Rock–Widow Mountain shear zone and Collins Creek fault. Similar back-dipping mylonite zones with a shear antithetic to the main detachment are well known from near the mylonitic front of other core complexes (Reynolds and Lister, 1990; Axen and Bartley, 1997), including the Bitterroot complex (see cross section in Foster et al., 2007).

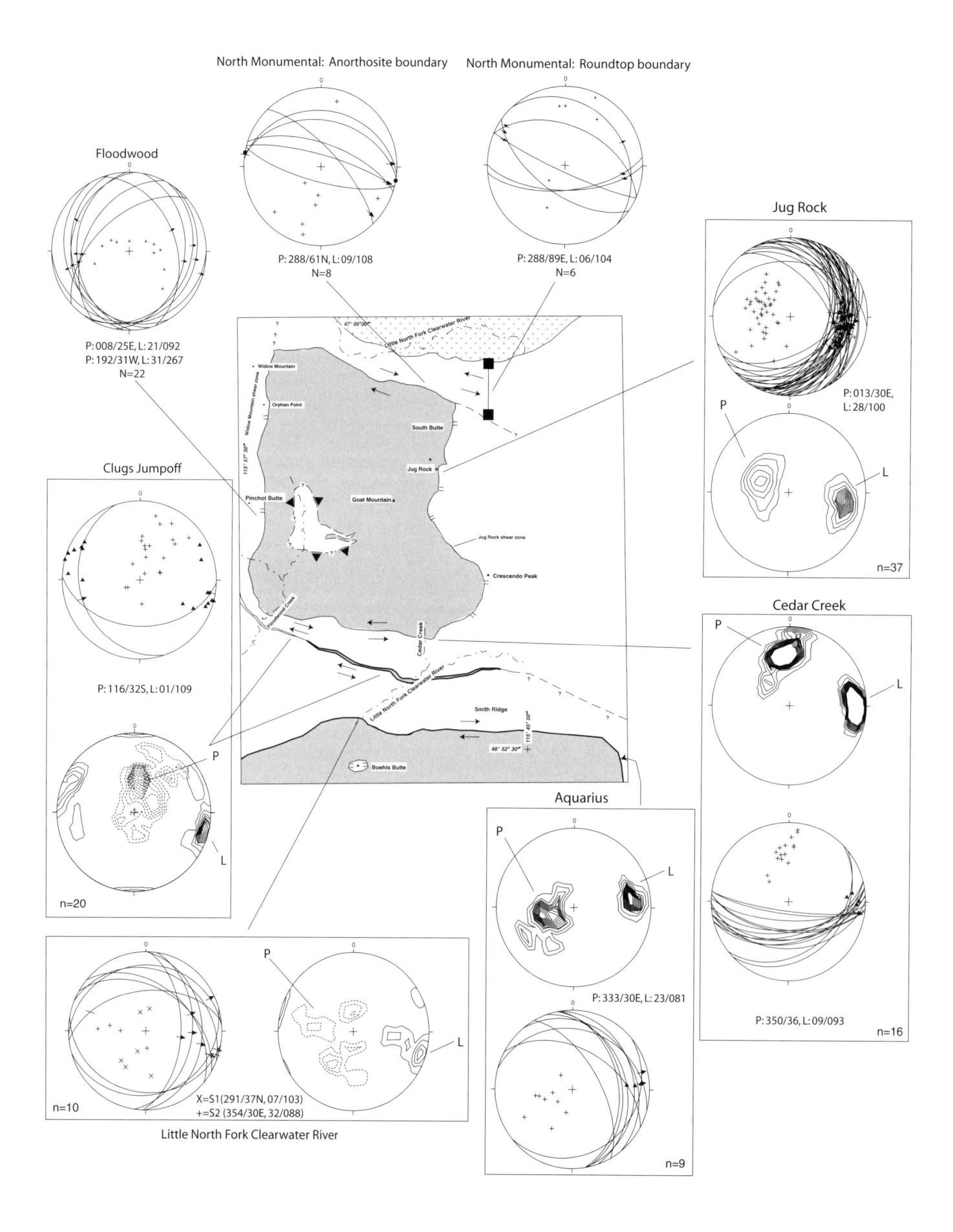

Floodwood
P: 008/25E, L: 21/092
P: 192/31W, L: 31/267
N=22

North Monumental: Anorthosite boundary
P: 288/61N, L: 09/108
N=8

North Monumental: Roundtop boundary
P: 288/89E, L: 06/104
N=6

Jug Rock
P: 013/30E,
L: 28/100
n=37

Clugs Jumpoff
P: 116/32S, L: 01/109
n=20

Cedar Creek
P: 350/36, L: 09/093
n=16

Aquarius
P: 333/30E, L: 23/081
n=9

Little North Fork Clearwater River
X = S1 (291/37N, 07/103)
+ = S2 (354/30E, 32/088)
n=10

Map labels: Widow Mountain, Orphan Point, Widow Mountain shear zone, Pinchot Butte, South Butte, Jug Rock, Goat Mountain, Jug Rock shear zone, Floodwood Creek, Cedar Creek, Crescendo Peak, Little North Fork Clearwater River, Smith Ridge, Boehls Butte, 47° 05′ 00″, 115° 57′ 30″, 115° 45′ 00″, 46° 52′ 30″

Figure 5. Lower-hemisphere projections of mylonitic fabrics from major shear zones in the Clearwater complex. Foliations are denoted by plusses and crosses, and lineations are denoted by solid triangles and arrows. Arrows show the sense of motion of the hanging wall relative to the footwall for data sets with kinematic indicators. Numbers adjacent to each stereonet diagram are the orientations of the contoured maxima for foliations (P) and lineations (L) for each data set. Dark shading denotes internal zone of the Clearwater complex. Most shear zones record a unidirectional sense of shear that is compatible with translation of the rocks of the internal zone to the west relative to the overlying external zone rocks. The western boundary of the northern culmination is marked by conjugate mylonites that formed at the back-rotated mylonitic front of the Jug Rock shear zone or that record overlapping shear zones of different age (see text). The azimuth of shear (southeast) for all shear zones is parallel to movement along the Lewis and Clark zone and compatible with formation of the Clearwater complex as a mid-crustal pull-apart between relaying faults of the Lewis and Clark zone.

Southern Culmination

Due to very poor exposure, there were only two localities where we examined the boundary of the southern culmination in any detail. These were the Little North Fork of the Clearwater River at Dworshak Reservoir and Aquarius along the northern and eastern margins of the southern culmination, respectively (Fig. 5). At both locations, the contact is characterized by decimeters of sheared Al-Mg gneiss and anorthosite. The sheared rocks exhibit gradational upper and lower boundaries, partially annealed microstructures, heterogeneous distribution related to protolith composition, and consistent shear sense among kinematic indicators.

Little North Fork of the Clearwater River

The northern boundary of the southern culmination is best exposed where the Little North Fork of the Clearwater River enters Dworshak Reservoir (Fig. 5). At this locality, mylonitized plagioclase-rich gneiss exhibits a predominately gently northeast-dipping foliation with nearly subhorizontal mineral lineations (Fig. 5). The foliation is defined by coarse-grained sillimanite, quartz, and kinked kyanite; most of the kyanite has been replaced by andalusite, except where wholly surrounded by plagioclase porphyroclasts. A second foliation, which is north-south–oriented with gentle east dips and down-dip lineations, crenulates the first foliation. This fabric is defined by unannealed quartz ribbons, and fibrolite intergrown with fine-grained muscovite and talc. The second foliation formed during mylonitization at lower pressures and temperatures than that preserved in the first foliation, consistent with its development during uplift and exhumation of the anorthosite in the southern culmination.

Kinematic indicators yield a subhorizontal, right-lateral sense of shear along an azimuth of 103° for the older foliation and a downdip, normal (top-to-the-east) sense of shear along an azimuth of 088° for the second foliation. Both of these fabrics are consistent with shearing of the anorthosite during exhumation of the southern culmination by translation westward relative to the flanking rocks.

Aquarius

The eastern boundary of the southern culmination is well exposed along the north side of Dworshak Reservoir, near where the North Fork of the Clearwater River enters the reservoir at Aquarius (Fig. 5). At this locality, anorthosite beneath the contact exhibits a gently east-northeast–dipping foliation with a nearly east-trending mineral lineation. It contains the typical internal zone assemblage (kyanite-andalusite-sillimanite-rutile), textures, and parageneses found at Goat Mountain and in the Jug Rock shear zone. Metasedimentary rocks above the anorthosite display a similar attitude with an east-dipping foliation and east-trending lineation defined by fibrolite, biotite, and recrystallized quartz. The foliation wraps around porphyroclasts of plagioclase, andalusite, and minor muscovite. Rare kyanite grains are preserved within the core of plagioclase porphyroclasts. The andalusite differs from that typically found within the inner core zone. They are commonly embayed remnants of euhedral andalusite porphyroclasts with marked pleochroism. Fibrolite and muscovite surround and overgrow the andalusite.

Asymmetric porphyroclasts and crenulated foliations indicate a downdip, top-to-the-east, sense of shear along an azimuth of 081° for the rocks along this contact, very similar to the fabrics in the north-south segment of the Jug Rock shear system, and exhumation of the internal zone rocks of the southern culmination by translation to the west relative to the overlying external zone rocks (Fig. 5).

Smith Ridge Syncline

Clugs Jumpoff Fault

The area between the two large culminations of anorthosite and Al-Mg gneiss is underlain by an east-west–trending band of metasedimentary and meta-igneous rocks of the external zone, referred to as the Smith Ridge syncline by Lewis et al. (2007). Recent mapping by us and Lewis et al. (2007) has identified a major fault within this band of external zone rocks. This fault, termed the Clugs Jumpoff fault by Lewis et al. (2007) (Figs. 3 and 4), has been recognized by the presence of sparse east-west–trending masses of anorthosite and Al-Mg–rich gneiss along its very poorly exposed trace. The Clugs Jumpoff fault zone runs along the southern edge of the fine-grained quartzites and marbles that flank the northern culmination and subdivides the Smith Ridge Syncline in two (Figs. 3 and 4). The eastern and western extents of the fault are not completely mapped, but it extends past Floodwood Creek to the west and as far east as Clugs Jumpoff. We postulate that it continues its east-west trend and merges with the Canyon fault–Kelly Forks fault system somewhere east of the southern mass of anorthosite, but we do not have any idea where it extends to the west (Fig. 3).

Sheared rocks within the Clugs Jumpoff fault are shallowly southwest dipping with subhorizontal lineations along an azimuth of 109° (Fig. 5). A kinematic indicator on the north side of one anorthosite body yields a left-lateral sense of shear, whereas a kinematic indicator on the south side of another anorthosite body yields a right-lateral sense of shear. In the absence of more complete or compelling data, we conclude that the anorthosite was translated to the west relative to the rocks on either side of the Clugs Jumpoff fault.

METAMORPHIC THERMOBAROMETRY

Samples

Eight metamorphic rocks from the external zone of the Clearwater complex contain mineral assemblages suitable for the application of metamorphic thermobarometers. We employed quantitative thermobarometry in an attempt to better characterize the metamorphic conditions of the external zone. Six samples were from the domain of fine-grained metasediments and pyroxene gabbros adjacent to the Jug Rock shear zone (Fig. 3). Two samples (01-300, 01-316) were fine-grained quartz-rich schists that had a weak foliation at a high angle to relict bedding (Fig. 4A). Garnet porphyroblasts (1.5–3 mm in diameter) occur in one of two forms, subhedral, slightly embayed with inclusions of quartz, or subhedral with a core rich in quartz inclusions surrounded by a wide inclusion-poor rim. The typical mineral assemblage is garnet-biotite-muscovite-plagioclase-quartz-oxide (ilmenite, or magnetite) + pyrite or pyrrhotite. Some of the more calcareous samples contain tremolite, zoisite, or diopside (Fig. 4B). Two samples (01-313, 01-234) were garnet-bearing amphibolite from the border zone of pyroxene gabbro stocks. These amphibolites are very fine grained and contain small, 1 mm garnets in a foliated matrix of hornblende-plagioclase-quartz ± biotite (Fig. 4C). A third sample of amphibolite (01-358) was collected from Cedar Creek Canyon, south of the northern mass of anorthosite. This amphibolite is composed of coarse-grained hornblende-plagioclase-quartz-biotite with large embayed garnets surrounded by a symplectite corona of plagioclase and hornblende (Fig. 4E).

Sample (01-302) was from coarse-grained garnet-biotite-muscovite schist 2 km east of the Jug Rock shear zone. This sample is typical of the coarse-grained metamorphic rocks that characterize much of the external zone of the Clearwater complex. These schists have the same composition and mineral assemblages as the previously described samples.

Sample (02-18) was from coarse-grained mica schist that occurs between the Clugs Jumpoff fault and the southern mass of anorthosite (Fig. 3). Sample 02-18 is a coarse-grained quartzofeldspathic gneiss containing a metamorphic foliation defined by biotite, minor muscovite, fibrolite, and thin (1–2 mm) segregations of quartz and feldspar. Garnets in 02-18 are very small, anhedral, and embayed in part. The foliation wraps around the garnets, and fibrolite needles and small biotite crystals locally grow around the garnet in pressure shadows at a high angle to foliation (Fig. 4D).

Methods

Mineral compositions were analyzed at the Washington State University (WSU) Geoanalytical Laboratory, and results are reported in Table 1, A–E. Sample 02-18 contains fibrolite, and we employed the relatively robust garnet-aluminosilicate-plagioclase-quartz (GASP) barometer, using the calibration of Koziol and Newton (1988) for that sample. No other metasedimentary samples contained aluminosilicate, and we employed the less robust garnet-biotite-muscovite-plagioclase-quartz (GPMQ, GPMB) barometers using the calibrations of Hodges and Crowley (1985). This calibration was chosen because it allows a direct comparison with the results of Carey (1985), who applied the same barometer to rocks in the internal zone. For amphibolites, we employed the garnet-hornblende barometer (GHPQ) of Kohn and Spear (1990). Temperatures were constrained with the garnet-biotite (GARB) exchange thermometer, using the calibration of Ferry and Spear (1978) with the garnet mixing model of Berman (1990), and the garnet-hornblende thermometer (GAHB) of Graham and Powell (1984). We also compared our computed temperatures with the Ganguly and Saxena (1984) and Indares and Martignole (1985) calibrations, which were used by previous workers in the area. Calculations were performed with the program GTB, available from Frank Spear (http://ees2.geo.rpi.edu/MetaPetaRen/GTB_Prog/GTB.html). All iron was assumed to be ferrous, based on very low ferric iron content (~0.0019) in coexisting ilmenite and the report of very little ferric iron in similar rocks near Snow Peak (Lang and Rice, 1985b). Garnets are almandine-rich (70%–78% Fe), with small amounts of spessartine component (3%–19% Mn) and (Ca + Mn)/total cation ratios between 0.55% and 0.08%. The biotites contain 0.1%–0.03% $(Al^6 + Ti)/(Al^6 + Ti + Fe + Mg)$ and Al^6 contents between 0.194 and 0.476. Plagioclase ranges in composition from An_{17} to An_{87}. Mineral compositions are generally within the range of compositions required for the application of these calibrations.

Results

Results of the thermobarometric calculations are reported in Table 2, A–B, and Figure 6. Despite complexities, seven of the eight samples equilibrated within or near the kyanite stability field. Sample 02-18, which contains fibrolite, equilibrated in the sillimanite stability field as expected. The computed pressure and temperature conditions are consistent with the distribution of regional metamorphic isograds that places these rocks above the staurolite breakdown reaction, below the muscovite out reaction, and within or near the stability field of kyanite (Hietanen, 1963a, 1968; Lang and Rice, 1985b; Grover et al., 1992). Transects across garnets from 01 to 300 and 01-302 display a flat major-element profile in the core, with increases in Ca, Fe/Mg ratio, and Mn at the rim. This flat zoning profile is characteristic of garnets from high-grade rocks that have undergone high-temperature homogenization (e.g., Tracy et al., 1976; Frost and Chacko, 1989).

TABLE 1A. GARNET COMPOSITIONS

Pelite	Na$_2$O	MgO	Al$_2$O$_3$	SiO$_2$	FeO	MnO	TiO$_2$	K$_2$O	CaO	Totals
01-316c	0.04	2.39	21.56	36.94	16.37	8.52	0.03	0.00	13.16	99.01
01-316r	0.01	3.41	21.73	38.03	17.93	5.59	0.01	0.00	13.33	100.03
01-302r	0.02	3.06	21.23	36.99	34.92	1.35	0.02	0.00	3.05	100.63
01-302c	0.03	3.59	21.68	36.97	35.05	1.41	0.01	0.00	1.70	100.45
01-300c	0.01	3.91	21.52	35.63	31.49	5.44	0.00	0.00	1.41	99.42
01-300r	0.01	4.45	21.46	36.67	31.74	3.13	0.01	0.01	2.09	99.57
02-18ravg	0.02	2.25	20.90	37.04	31.51	6.29	0.00	0.02	2.21	100.25
02-18cavg	0.00	3.52	21.14	37.62	32.99	1.93	0.00	0.01	3.29	100.51
02-18c3	0.00	3.50	21.00	37.78	33.68	2.00	0.00	0.01	3.00	100.97
02-18r3	0.03	1.97	20.86	36.80	30.93	7.41	0.00	0.03	1.96	100.00
Amphibolite										
01-234r	0.04	3.62	21.82	38.24	22.00	2.69	0.04	0.00	12.53	100.98
01-234c	0.06	3.10	21.50	37.62	20.83	2.47	0.31	0.00	13.92	99.82
01-313r	0.00	3.98	21.74	38.11	21.18	2.19	0.04	0.00	12.50	99.75
01-313c	0.00	4.08	21.77	38.42	21.69	2.47	0.12	0.01	11.75	100.32
01-358c	0.05	3.74	21.41	37.78	24.00	2.00	0.20	0.00	9.85	99.02
01-358r	0.03	3.72	21.40	38.18	23.65	2.12	0.16	0.00	10.81	100.06

Pelite	CNa	CMg	CAl	CSi	CFe	CMn	CTi	CK	CCa	Total
01-316c	0.01	0.28	2.02	2.94	1.09	0.57	0.00	0.00	1.12	8.05
01-316r	0.00	0.40	2.00	2.97	1.17	0.37	0.00	0.00	1.12	8.03
01-302r	0.00	0.37	2.01	2.96	2.34	0.09	0.00	0.00	0.26	8.03
01-302c	0.00	0.43	2.04	2.96	2.34	0.10	0.00	0.00	0.15	8.02
01-300c	0.00	0.47	2.06	2.90	2.14	0.37	0.00	0.00	0.12	8.07
01-300r	0.00	0.53	2.03	2.95	2.13	0.21	0.00	0.00	0.18	8.04
02-18ravg	0.00	0.27	1.99	2.99	2.13	0.43	0.00	0.00	0.19	8.01
02-18cavg	0.00	0.42	1.99	3.00	2.20	0.13	0.00	0.00	0.28	8.01
02-18c3	0.00	0.41	1.97	3.00	2.24	0.13	0.00	0.00	0.26	8.01
02-18r3	0.01	0.24	2.00	2.99	2.10	0.51	0.00	0.00	0.17	8.02
Amphibolite										
01-234r	0.01	0.42	2.00	2.97	1.43	0.18	0.00	0.00	1.04	8.04
01-234c	0.01	0.36	1.99	2.95	1.37	0.16	0.02	0.00	1.17	8.04
01-313r	0.00	0.46	2.00	2.98	1.38	0.15	0.00	0.00	1.05	8.02
01-313c	0.00	0.47	1.99	2.99	1.41	0.16	0.01	0.00	0.98	8.01
01-358c	0.01	0.44	2.00	2.99	1.59	0.13	0.01	0.00	0.83	8.00
01-358r	0.00	0.43	1.98	2.99	1.55	0.14	0.01	0.00	0.91	8.01

Molar proportions of the cations, 12 oxygens
r—rim, c—core, m—matrix phases, rm—rim of matrix phases, cm—core of matrix phases, avg—average. All Fe reported as Fe+2 except for plagioclase compositions for 02-18 and 01-358.

TABLE 1B. BIOTITE COMPOSITIONS

Pelite	Na$_2$O	MgO	Al$_2$O$_3$	SiO$_2$	FeO	MnO	TiO$_2$	K$_2$O	CaO	Totals
01-316	0.08	15.95	15.58	38.06	13.26	0.42	1.82	9.08	0.04	94.29
01-302	0.29	8.48	18.88	35.65	20.86	0.16	2.49	8.22	0.00	95.03
01-302green	0.24	8.00	18.69	34.48	21.26	0.05	2.45	8.16	0.00	93.32
01-300	0.14	10.44	18.96	35.20	17.12	0.32	2.31	8.97	0.01	93.47
01-300r	0.22	10.24	20.24	34.60	18.26	0.27	1.18	9.20	0.01	94.21
02-18r3	0.17	7.86	20.45	34.44	21.24	0.28	0.90	9.64	0.01	94.99
02-18m3	0.32	7.64	19.41	34.15	21.36	0.22	1.65	9.16	0.02	93.93
02-18m5	0.32	8.28	19.37	34.10	18.52	0.23	2.89	8.17	0.17	92.04
02-18cavg	0.27	7.52	19.75	34.22	20.65	0.23	2.19	9.40	0.01	94.25
Amphibolite										
01-313	0.14	12.99	15.74	35.79	16.52	0.24	2.26	8.68	0.05	92.41
01-358corona	0.18	11.31	15.89	34.80	16.24	0.13	1.98	5.24	0.71	86.49
01-358m	0.17	12.65	15.14	36.11	14.27	0.16	3.35	9.10	0.01	90.97

Pelite	CNa	CMg	CAl	CSi	CFe	CMn	CTi	CK	CCa	Total
01-316	0.01	1.77	1.37	2.84	0.83	0.03	0.10	0.86	0.00	7.81
01-302	0.04	0.96	1.70	2.72	1.33	0.01	0.14	0.80	0.00	7.71
01-302green	0.04	0.93	1.72	2.69	1.39	0.00	0.14	0.81	0.00	7.73
01-300	0.02	1.19	1.71	2.70	1.10	0.02	0.13	0.88	0.00	7.76
01-300r	0.03	1.17	1.83	2.65	1.17	0.02	0.07	0.90	0.00	7.83
02-18r3	0.03	0.91	1.86	2.66	1.37	0.02	0.05	0.95	0.00	7.84
02-18m3	0.05	0.89	1.79	2.67	1.40	0.01	0.10	0.91	0.00	7.82
02-18m5	0.05	0.97	1.79	2.67	1.21	0.02	0.17	0.82	0.01	7.70
02-18cavg	0.04	0.87	1.81	2.66	1.34	0.01	0.13	0.93	0.00	7.80
Amphibolite										
01-313	0.02	1.50	1.44	2.77	1.07	0.02	0.13	0.86	0.00	7.81
01-358corona	0.03	1.37	1.52	2.82	1.10	0.01	0.12	0.54	0.06	7.58
01-358m	0.03	1.47	1.39	2.82	0.93	0.01	0.20	0.91	0.00	7.75

Molar proportions of the cations, 12 oxygens
r—rim, c—core, m—matrix phases, rm—rim of matrix phases, cm—core of matrix phases, avg—average. All Fe reported as Fe+2 except for plagioclase compositions for 02-18 and 01-358.

TABLE 1C. MUSCOVITE COMPOSITIONS

Pelite	Na$_2$O	MgO	Al$_2$O$_3$	SiO$_2$	FeO	MnO	TiO$_2$	K$_2$O	CaO	Totals
02-18	0.26	0.97	35.40	47.69	1.22	0.01	1.03	6.46	0.00	93.03
01-300	0.79	1.37	33.80	46.61	1.24	0.03	1.02	10.24	0.01	95.11
01-302r	1.00	1.09	33.82	45.67	1.81	0.00	0.85	9.95	0.00	94.19
01-302c	0.91	1.15	33.92	45.66	1.56	0.00	0.89	10.20	0.00	94.30

Pelite	CNa	CMg	CAl	CSi	CFe	CMn	Cti	CK	CCa	Total
02-18	0.03	0.10	2.76	3.15	0.07	0.00	0.05	0.54	0.00	6.71
01-300	0.10	0.14	2.66	3.11	0.07	0.00	0.05	0.87	0.00	7.00
01-302r	0.13	0.11	2.69	3.08	0.10	0.00	0.04	0.86	0.00	7.02
01-302c	0.12	0.12	2.70	3.08	0.09	0.00	0.05	0.88	0.00	7.02

Molar proportions of the cations, 11 oxygens

TABLE 1D. PLAGIOCLASE COMPOSITIONS

Pelite	Na$_2$O	MgO	Al$_2$O$_3$	SiO$_2$	FeO	MnO	TiO$_2$	K$_2$O	CaO	Totals
01-300c	8.97		23.56	62.29	0.02			0.28	4.82	99.94
01-300r	8.65		23.94	61.92	0.07			0.17	5.27	100.00
01-302r	9.29		23.31	62.29	0.00			0.16	4.36	99.43
01-302c	9.88		22.81	63.25	0.03			0.18	3.79	99.92
02-18(3)	7.50	0.00	26.12	58.60	0.09	0.03	0.00	0.20	6.94	99.49
02-18avg	8.40	0.00	25.15	60.18	0.05	0.02	0.00	0.19	5.76	99.76
Amphibolite										
01-234r	6.99		26.23	58.43	0.00			0.25	7.78	99.68
01-234c	7.98		25.10	60.04	0.00			0.25	6.75	100.13
01-313r	2.72		33.09	48.21	0.19			0.10	15.75	100.07
01-313c	4.76		29.85	52.77	0.07			0.13	11.92	99.51
01-358m	7.91	0.00	25.41	61.44	0.05	0.02	0.01	0.24	6.42	101.52
01-358corona	4.96	0.00	30.08	52.44	0.12	0.00	0.02	0.09	11.16	98.89

Pelite	CNa	CMg	CAl	CSi	CFe	CMn	CTi	CK	CCa	Total
01-300c	0.77		1.23	2.76	0.00			0.02	0.23	5.01
01-300r	0.74		1.25	2.75	0.00			0.01	0.25	5.00
01-302r	0.80		1.22	2.77	0.00			0.01	0.21	5.02
01-302c	0.85		1.19	2.80	0.00			0.01	0.18	5.03
02-18(3)	0.65	0.00	1.38	2.63	0.00	0.00	0.00	0.01	0.33	5.01
02-18avg	0.73	0.00	1.32	2.68	0.00	0.00	0.00	0.01	0.28	5.02
Amphibolite										
01-234r	0.61		1.39	2.62	0.00			0.01	0.37	5.00
01-234c	0.69		1.32	2.67	0.00			0.01	0.32	5.02
01-313r	0.24		1.79	2.21	0.01			0.01	0.77	5.02
01-313c	0.42		1.60	2.40	0.00			0.01	0.58	5.01
01-358m	0.67	0.00	1.31	2.69	0.00	0.00	0.00	0.01	0.30	4.98
01-358corona	0.44	0.00	1.62	2.40	0.00	0.00	0.00	0.01	0.55	5.02

Molar proportions of the cations, 8 oxygens.
r—rim, c—core, m—matrix phases, rm—rim of matrix phases, cm—core of matrix phases, avg—average. All Fe reported as Fe+2 except for plagioclase compositions for 02-18 and 01-358.

TABLE 1E. HORNBLENDE COMPOSITIONS

Amphibolite	Na$_2$O	MgO	Al$_2$O$_3$	SiO$_2$	FeO	MnO	TiO$_2$	K$_2$O	CaO	Totals
01-234m	1.57	11.32	13.96	43.56	12.87	0.20	0.69	0.74	11.82	96.73
01-234r	1.16	8.61	13.56	43.03	16.68	0.53	0.39	0.87	11.90	96.71
01-313m	0.78	13.74	10.91	46.16	10.85	0.13	0.62	0.93	12.33	96.46
01-313r	0.92	12.01	13.65	44.15	11.77	0.33	0.77	1.21	12.34	97.15
01-358rm	1.49	10.05	14.24	44.10	12.63	0.29	0.84	0.76	11.61	96.01
01-358cm	1.60	10.75	13.26	45.58	11.70	0.40	0.93	0.73	11.48	96.41
01-358avg corona	2.48	8.59	11.61	49.06	10.37	0.47	0.41	0.29	11.43	94.73

(continued)

TABLE 1E. HORNBLENDE COMPOSITIONS (continued)

Amphibolite	CNa	CMg	Cal	CSi	CFe	CMn	CTi	CK	CCa	Totals
01-234m	0.45	2.50	2.44	6.45	1.59	0.03	0.08	0.14	1.87	15.55
01-234r	0.34	1.94	2.41	6.50	2.11	0.07	0.04	0.17	1.93	15.50
01-313m	0.22	3.00	1.89	6.77	1.33	0.02	0.07	0.17	1.94	15.41
01-313r	0.26	2.63	2.36	6.49	1.45	0.04	0.09	0.23	1.94	15.49
01-358rm	0.43	2.23	2.49	6.55	1.57	0.04	0.09	0.14	1.85	15.39
01-358cm	0.45	2.36	2.30	6.70	1.44	0.05	0.10	0.14	1.81	15.35
01-358avg corona	0.70	1.90	2.01	7.23	1.29	0.06	0.05	0.06	1.80	15.1

Molar proportions of the cations, 23 oxygens
r—rim, c—core, m—matrix phases, rm—rim of matrix phases, cm—core of matrix phases, avg–average. All Fe reported as Fe+2 except for plagioclase compositions for 02-18 and 01-358.

TABLE 2A. THERMOBAROMETRY RESULTS: PELITES

Sample, UTM coordinates	GARB-GASP P (MPa)	GARB-GASP T (°C)	GARB-GPMQ P (MPa)	GARB-GPMQ T (°C)	GARB-GPMB P (MPa)	GARB-GPMB T (°C)	GARB T (°C)
01-300, 11, 592547E, 5201723N			5.5	595	6.25	600	
01-302, 11, 595081E, 5201177N			8.5	655	8.8	660	
01-316, 11, 590960E, 5203750N							575
02-18 (3), 11, 587511E, 5195225N	500	630	525	595	470	620	
02-18avg, same	>1200	>800	1200	775	1200	775	

Note: GARB (garnet-biotite thermometer); GASP (garnet-aluminosilicate-plagioclase barometer); GPMQ (garnet-plagioclase-muscovite-quartz barometer); GPMB (garnet-plagioclase-muscovite-biotite barometer); see text for calibrations. (3)—circle 3 analyses; avg—average mineral compositions.

TABLE 2B. THERMOBAROMETRY RESULTS: AMPHIBOLITES

Sample, UTM coordinates	GARB-GAPQ P (MPa)	GARB-GAPQ T (°C)	GAHB-GAPQ P (MPa)	GAHB-GAPQ T (°C)
01-234, 11, 592288E, 5207817N			10	700
01-313, 11, 591373E, 5202527N	630	675	600	650
01-358 (mtrx), 11, 589550E, 5196945N	900	575	950	680

Note: GARB (garnet-biotite thermometer); GAPQ (garnet-amphibole-plagioclase-quartz barometer); GAHB (garnet-hornblende thermometer); see text for calibrations. (mtrx)—matrix mineral compositions.

The garnet rim compositions are believed to best represent the conditions of peak metamorphism in these rocks, although some of the variability in our results could be due to partial reequilibration of the garnet rims during cooling.

Metamorphic temperatures calculated with rim compositions range between 595 °C and >800 °C, depending on the calibration and thermometer used (Fig. 6). The garnet-hornblende thermometer, applied to three amphibolites with the calibration of Graham and Powell (1984), gives temperatures between 650 °C and 700 °C, with an average of 677 °C. Temperatures calculated with the garnet-biotite thermometer vary widely depending on the calibration used. The calibration of Ferry and Spear (1978) with Berman's (1990) mixing properties of garnet yields temperatures between 580 °C to 850 °C (average of 683 °C), which is anomalously high for some samples. Grover et al. (1992) and Carey et al. (1992) also struggled with anomalously high calculated temperatures in rocks of the internal zone, which they believed was due to oxidation of biotite during the last metamorphic event (M3). To overcome this problem, Grover et al. (1992) employed the Indares and Martignole (1985) calibration, whereas Carey (1985) employed the Ganguly and Saxena (1984) calibration. We examined these other calibrations in order to potentially obtain a better estimate of the temperatures and to facilitate direct comparisons of peak metamorphic

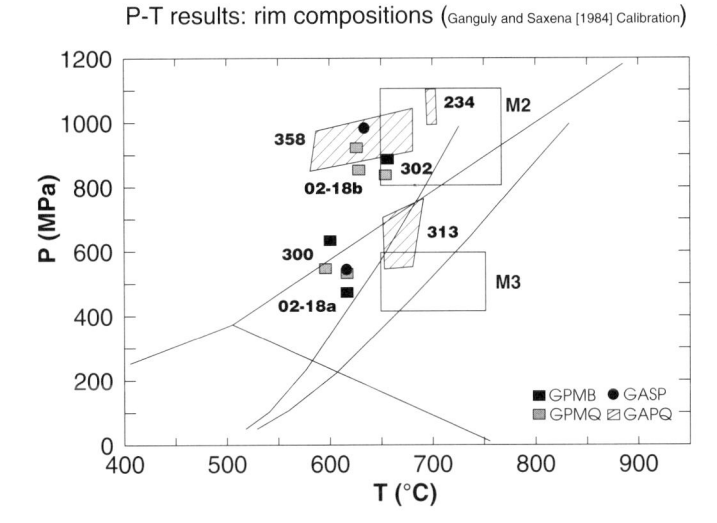

Figure 6. Pressure and temperature estimates of metamorphic conditions from rocks in the external zone of the Clearwater complex based on the intersection of geothermobarometers discussed in the text. Stability fields of the aluminosilicate phases are shown for reference, as well as the staurolite and muscovite breakdown reactions. Open boxes outline the conditions of M2 and M3 metamorphism reported by Grover et al. (1992). Hatched boxes define the area of amphibolite pressure and temperature estimates. Conditions for sample 02-18 were calculated with (A) mineral compositions obtained from a small part of the thin section (circle 3), or (B) pressures derived from averaged mineral compositions in thin section (see text for discussion). Metaphoric barometers: GPMB—garnet-plagioclase-muscovite-biotite; GASP—garnet-aluminosilicate-plagioclase-quartz; GPMQ—garnet-plagioclase-muscovite-quartz; GAPQ—garnet-aluminosilicate-plagioclase-quartz.

pressures between our study and theirs. The Indares and Martignole (1985) calibration yields widely scattered temperatures between 100 °C less to 200 °C higher than the Ferry and Spear (1978) and Berman (1990) calibration. The Ganguly and Saxena (1984) calibration yields a tight cluster of temperatures between 575 °C and 660 °C (average of 621 °C) with only two temperatures above 850 °C. This calibration appears to provide the most robust estimate of temperature in these rocks and overlaps with temperatures derived from the garnet-hornblende thermometer, and it is consistent with the presence of primary muscovite, which constrains temperatures to be below the breakdown of muscovite + quartz at ~700 °C. Our preferred estimate of the temperature, which includes both the biotite-garnet and garnet-hornblende thermometers, is 634 °C.

Calculated pressures fall naturally into two groups that vary between ~900–1000 MPa or 500–600 MPa (Fig. 6). Four samples comprise the high-pressure group. Schist 01-302 and amphibolite 01–234, from east of the Jug Rock shear zone, have simple metamorphic textures that grew during one metamorphic event (Fig. 4C). Sample 01-302 equilibrated at ~870 MPa and 660 °C, based on the GPMQ and GPMB barometers. Amphibolite 01–234 gives a higher pressure and temperature of 1000 MPa at 700 °C with the GAPQ barometer. Ziegler (1991) also obtained high pressures (1050 and 1200 MPa) from two amphibolites east of the study area.

Two of the samples from the high-pressure group (02-18 and 01-358) lie between the two culminations and have more complex textures that suggest more than one metamorphic event. Sample 01-358 is a coarse-grained amphibolite with symplectite intergrowths of hornblende and plagioclase surrounding embayed garnets (Fig. 4E). Matrix phases and core garnet compositions give a pressure and temperature of ~925 MPa and 630 °C using the GAPQ barometer and GARB/GAHB thermometers (Fig. 6). Reliable estimates of the conditions of equilibration during symplectite growth of hornblende and plagioclase were not obtained.

Paragneiss 02-18, which is the only sample to contain an aluminosilicate phase in our suite of samples, gives conflicting results. The average composition of garnet, biotite, and plagioclase from sample 02-18 yields temperatures greater than 800 °C and pressures greater than 1200 MPa with the GASP, GPMQ, and GPMB barometers (Fig. 6). If more reasonable temperatures are used (the average of 634 °C for all samples), the pressure only drops to 950 MPa. One small part of the sample contains fibrolite overgrowing a small anhedral garnet porphyroblast (Fig. 4D). Mineral compositions from this part of the thin section yield a tight intersection of calculated equilibria at ~475 MPa and 600 °C (Fig. 6). These data suggest that final equilibration at ~5 MPa and 600 °C occurred after an earlier high-pressure metamorphic event.

Two samples (01-300, 01-313), from right above the Jug Rock shear zone, comprise the lower-pressure group. Both samples have very simple, fine-grained textures indicative of only one episode of metamorphic growth (Figs. 4A and 4C). Geothermobarometers applicable to semischist 01-300 and amphibolite 01-313 are in good agreement and give pressures and temperatures of ~600 MPa and 619 °C.

Interpretation

Quantitative geothermobarometry, coupled with petrographic observation, suggests that both high-pressure and intermediate-pressure metamorphic events are present within rocks of the external zone. The presence of nonequilibrium textures in two samples and the scattered distribution of sample localities make it unlikely that a geologic structure is solely responsible for all of the different metamorphic pressures observed in the external zone. High-pressure and intermediate-pressure metamorphism occurs in both amphibolites and metasediments, with apparently simple metamorphic histories. In these rocks, either some of the pressure determinations are erroneously high (the lack of aluminosilicate in most metasediments is a concern), or recrystallization of older metamorphic minerals during subsequent metamorphic events was highly variable. Samples of this type with the lowest pressures (01-300 and 01-313) lie close to the top of the Jug Rock shear zone and could have been affected by their proximity to the fault. Perhaps the strongest evidence for more than one metamorphic event comes from the coronas of intergrown hornblende and plagioclase around embayed garnet porphyroblasts in sample 01-358 (Fig. 4E). These coronas are strikingly similar to garnet coronas documented from other core complexes that have undergone nearly isothermal decompression (e.g., Ziegler, 1991; House et al., 1997), and they argue strongly that at least part of the external zone has undergone an episode of exhumation and high-temperature decompression. If true, the process by which some samples could either escape recrystallization or undergo complete recrystallization during this event requires further investigation.

The high-pressure metamorphism recorded in the external zone occurred at conditions of ~900 MPa and 650 °C, which is similar to the conditions of M2 in the internal zone, and this suggests that the two are correlative. Our estimate for the conditions of M2 includes pressures based on the GAPQ barometer from two amphibolites (01-234, 01-358). Ziegler (1991) showed that that the GAPQ barometer overestimated the metamorphic pressures by 100–300 MPa relative to the more robust GRIPS barometer when applied to amphibolites in both the internal and external zones of the Clearwater complex. If this is the case, our estimates for M2 in the external zone could be 100–200 MPa too high. The low-pressure metamorphism in three samples records conditions of around 600 MPa and 623 °C, which could either correlate with the older M1 event (i.e., Lang and Rice, 1985b) in some samples or, more likely, the M3 exhumation event that is so well documented in the internal zone (Carey et al., 1992).

U-Pb SHRIMP GEOCHRONOLOGY

Strategy

In order to place constraints on the timing of metamorphism and deformation in rocks of the Clearwater Complex, we report new U-Pb dating of metamorphic overgrowths on zircons from rocks within the Jug Rock shear zone. These data were collected during a study of detrital and igneous zircon ages for rocks within the Clearwater complex (Doughty and Chamberlain, 2007), and more detailed sample descriptions and localities are described in that publication and in Figure 3.

We collected a total of 29 analyses of metamorphic overgrowths on zircons from four localities. Three samples (GM01-02, 02-93, and 02-106B) were collected from rocks below or within the Jug Rock shear zone. One sample (GM01-05) was collected from above the Jug Rock shear zone.

Analytical Methods

Zircon grains were extracted and concentrated with conventional mineral separation techniques at the University of Wyoming. All samples were analyzed at the Stanford–U.S. Geological Survey facility using the sensitive high-resolution ion microprobe (SHRIMP-RG [Reverse Geometry]) instrument. Data reduction followed Ludwig (1988, 1991, 2003). We report ^{206}Pb/^{238}U dates corrected by the ^{207}Pb method (e.g., Williams, 1998) and 2σ errors, but our interpretations are based on concordia intercepts from total Pb Tera-Wasserburg (1972) plots to eliminate the influence of common-Pb correction choices. Although many of the analyses have high common Pb (4%–30% of ^{206}Pb; Table 3), each age population has at least one concordant to nearly concordant analysis with low common Pb (less than 2%), which strengthens our interpretation that discordance is related to common Pb. Inheritance is unlikely in these data because each spot was positioned in a discrete cathodoluminescent (CL) domain, and the SHRIMP-RG pits are only a few microns deep.

The interpretation of metamorphic origin for these rims relies on the relatively homogeneous, unzoned nature of the domains in CL images (Fig. 7) and distinctive Th/U values (Table 3). The rims are either bright or dark in CL, but they are always distinct from the zoned magmatic and detrital cores of the grains (Fig. 7). Low Th/U values (≤0.02) are often characteristic of metamorphic zircon growth (e.g., Williams, 1998) and can be diagnostic especially when combined with textural evidence. Low Th/U zircons can also grow in some magmatic systems, although values of 0.2–0.7 are more typical. In the samples from Clearwater complex, the combination of low Th/U, CL evidence for overgrowths, and the tectonic interpretation of a Mesoproterozoic origin for these rocks (Doughty and Chamberlain, 2007) leads us to interpret the rims as metamorphic zircon growth.

Results (Internal Zone)

Sample GM01-02 (Amphibolite of Moses Butte)

Sample GM01-02 is a garnet amphibolite collected ~500 m west of the Jug Rock shear zone (Fig. 3). We analyzed metamorphic overgrowths on seven zircons from this sample (Table 3; Fig. 8). The zircons in sample GM01-02 contain small embayed and bleb-like, zoned cores that are dark in CL, Mesoproterozoic in age (Doughty and Chamberlain, 2007), and are surrounded by banded to complexly zoned U-rich, euhedral rims (Fig. 7C) and pale, unzoned rims (Fig. 7D). In many grains, the core composes only 20% of the grain (Fig. 7D), and in some grains, there are no distinct cores at all. Data from eight spots fall into two groups with concordia intercepts of 64 ± 1.0 and 59 ± 1.9 Ma (Fig. 8A). These spots sampled a variety of CL domains and included edges, centers, and homogeneous interiors of grains, although the two youngest spots sampled unzoned rims (Fig. 7D). Zircon growth in this rock appears to be dominantly metamorphic at 65–57 Ma. Analyses 2.1 and 2.2 are from the center and rim of the same grain and yield the same ages within error. Six of the eight analyses are nearly concordant (Fig. 8A), with ^{206}Pb/^{238}U dates that range from 58.9 ± 1.6 Ma to 64.9 ± 2.6 Ma (Table 3). On a Tera-Wasserburg (1972) plot of total Pb, the data are consistent with two periods of zircon growth and varying degrees of common Pb producing linear scatter. Both age groups contain concordant analyses with low common Pb. Coupled with the CL evidence, we interpret the data to indicate that metamorphism occurred in two pulses ca. 64 and 59 Ma.

TABLE 3. U-Pb SHRIMP DATA FROM METAMORPHIC ZIRCON RIMS, CLEARWATER COMPLEX

Grain-spot	Domain	U (ppm)	Th (ppm)	$\frac{^{232}Th}{^{238}U}$	Rad ^{206}Pb (ppm)	Common ^{206}Pb (%)	$\frac{^{206}Pb}{^{204}Pb}$	Total $\frac{^{238}U}{^{206}Pb}$	2σ (%)	Total $\frac{^{207}Pb}{^{206}Pb}$	2σ (%)	^{207}Corr $\frac{^{206}Pb}{^{238}U}$ age (Ma)	2σ (abs. err)
INTERNAL ZONE													
GM-01-2, Amphibolite of Moses Butte, east of Moses Butte (11T, 589972E, 5207309N)													
6.2	br rim	205	1	0.004	1.7	9.59	188	103.36	3.2	0.0925	16.6	58.6	2.4
1.1	pale	331	2	0.005	2.6	1.92	942	107.78	2.6	0.0561	9.5	58.9	1.6
3.2	br rim	405	22	0.056	3.9	16.25	111	89.93	2.6	0.1560	8.4	61.8	3.1
4.1	pale	161	1	0.006	1.4	3.77	479	101.19	3.6	0.0670	14.3	61.9	2.4
8.1	pale	232	1	0.007	2.0	2.69	670	98.61	2.9	0.0746	9.2	62.9	2.0
2.2	zoned rim	391	2	0.006	3.4	1.73	1040	99.50	2.4	0.0599	11.1	63.5	1.6
2.1	mott core	364	3	0.008	3.2	2.11	856	97.38	2.4	0.0645	9.7	64.5	1.6
5.2	pale zoned	141	1	0.005	1.3	6.60	273	96.07	3.7	0.0703	19.0	64.9	2.7
02-93, Schist of South Butte (quartzite in schist), south of Monumental Buttes (11T, 590513E, 5207526N)													
2.1	outer br	52	1	0.019	0.4	30.53	53	101.02	5.6	0.1796	12.5	54.2	4.0
3.1	outer br	100	0	0.003	0.8	3.66	445	109.70	4.1	0.0712	14.8	56.9	2.5
5	inner dr	83	1	0.014	0.7	2.50	652	101.11	4.1	0.0465	18.3	63.5	2.7
3.2	inner dr	132	1	0.009	1.1	1.43	1134	99.97	3.2	0.0472	13.0	64.2	2.1
02-106B, Schist of Blackdome Peak (quartzite in schist), southwest of Blackdome Peak (11T, 590053E, 5203560N)													
28.3	dk rim	2332	4	0.002	21.3	5.37	303	94.04	1.3	0.0644	3.8	66.9	1.0
EXTERNAL ZONE													
GM-01-5, Quartzite of Monumental Buttes, south of Monumental Buttes (11T, 590667E, 5207391N)													
14.1	br rim	2657	39	0.015	30.2	16.67	98	75.60	0.6	0.1934	5.8	71.0	3.3
18.1	dk rim	3314	53	0.016	32.7	2.30	707	87.06	0.6	0.0645	11.5	72.2	0.8
1.2	dk rim	2063	26	0.013	20.6	2.12	769	86.10	0.8	0.0625	6.0	73.2	0.7
1	dk rim	2141	27	0.013	21.8	3.20	509	84.30	1.3	0.0715	2.8	74.0	1.1
17.1	dk rim	2834	42	0.015	30.9	8.49	192	78.68	0.7	0.1211	12.6	74.8	2.1
16.1	dk rim	2483	33	0.014	31.6	16.37	99	67.51	0.9	0.1841	6.0	80.5	3.5
14.2	dk rim	5402	130	0.025	60.8	1.57	1039	76.30	0.4	0.0619	4.5	82.6	0.5

Notes: Domain characteristics: br rim—cathodoluminescent (CL) bright rim, pale—CL pale gray, mott core—CL mottled interior, outer br—CL bright, inner dr—CL dark, dk rim—CL dark rim; errors given at 2σ for both ratios and ages; Rad—radiogenic. Total ratios involving ^{206}Pb include common Pb, ^{207}Pb corr ^{206}Pb/^{238}U ages are corrected for initial Pb by the ^{207}Pb method following Williams (1998) using Stacey and Kramers (1975) model Pb values. Dates were calculated using the ^{238}U and ^{235}U decay constants recommended by Steiger and Jäger (1977).

Sample 02-106b (Schist of Blackdome Peak) and Sample 02-93 (Schist of South Butte)

Samples 02-106b and 02-93 are both quartz-rich schists from the internal zone that are caught up in the Jug Rock shear zone. We analyzed metamorphic overgrowths on one grain from sample 02-106b and four grains from sample 02-93 (Table 3; Fig. 7). Sample 02-106b contains rounded detrital zircon cores overgrown by thin euhedral rims, which are bright in CL. The zircons from sample 02-93 are substantially more complex, with both banded, dark overgrowths and bright homogeneous rims (Figs. 7A, 7B, and 7H). Fractured cores overgrown by bright overgrowths are also present, suggesting that deformation occurred prior to metamorphic zircon growth (Figs. 7A, 7B, and 7H). One grain (3) exhibits two periods of metamorphic rim growth. The inner overgrowth (3.2) surrounds a rounded detrital core and displays a complex pattern of zoning. The outer rim (3.1) is a light-colored banded rim identical to those observed on the other grains (Fig. 7B). One analysis of the outer metamorphic overgrowth from sample 02-106b yielded a ^{206}Pb/^{238}U date of 66.9 ± 1 Ma (Table 3; Fig. 8C). Two concordant analyses of the inner metamorphic overgrowth from two grains in 02-93 have ^{206}Pb/^{238}U ages of 63.5 ± 2.6 and 64.2 ± 2 Ma (Table 3; Fig. 8C). Outer, bright overgrowths from sample 02-93 have ages of 54.2 ± 4 and 56.9 ± 2.4 Ma, with an intercept age of 57.5 ± 2.9 Ma. The textural evidence from 02-93, combined with two distinct rim ages, supports the interpretation from GM01-02 that the internal zone was flushed by at least two pulses of metamorphic fluids ca. 64 and 59–55 Ma.

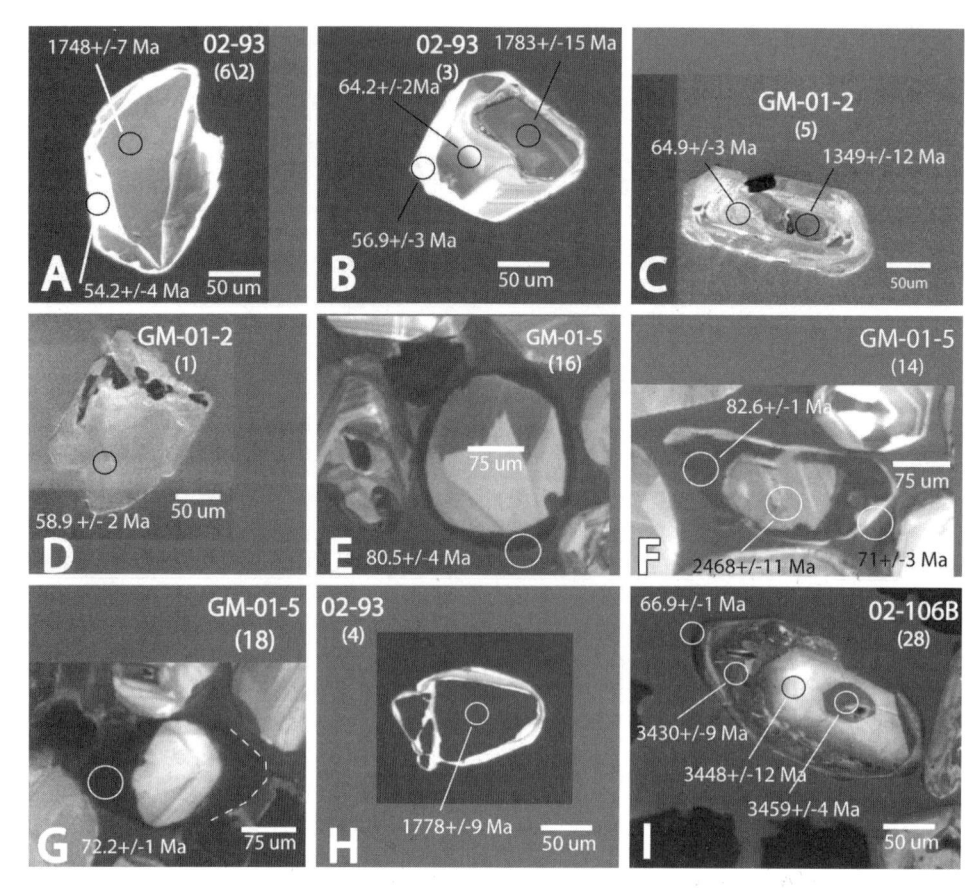

Figure 7. Cathodoluminescence images of representative zircons analyzed in this study. Locations of sensitive high-resolution ion microprobe (SHRIMP) pits are denoted by 30-μm-diameter circles. Bright areas have higher amounts of U than darker areas. Annotated ages are $^{207}Pb/^{206}Pb$ for ages greater than 800 Ma and $^{206}Pb/^{238}U$ ages for those less than 800 Ma. Errors are 2σ. (A–B) Complex banded, dark overgrowths and bright homogeneous rims that have overgrown detrital cores with Mesoproterozoic ages from sample 02-93. Grain 02-93-3 (B) contains two periods of zircon overgrowth. The older inner rim (3.2) shows a complex pattern of zoning around a rounded detrital core. The younger outer rim (3.1) is a light-colored banded rim identical to those observed on the other grains (A and H). (C–D) Embayed magmatic zircon cores overgrown by younger euhedral metamorphic rims from GM01-02. (C) Gray magmatic core, Mesoproterozoic in age, surrounded by banded to complexly zoned, U-rich rim. (D) Dark-gray small embayed and bleb-like core surrounded by complexly zoned U-rich, euhedral rim. (E–G) Examples of two periods of metamorphic rim growth from GM01-05. The inner overgrowth is dark-colored under cathodoluminescence (CL) (due to low U content) and well-faceted to embayed (E–G). The outer rim is very thin, light-colored (due to higher U content), and euhedral (F). (H) Fractured detrital core overgrown by bright overgrowth from sample 02-93, suggesting that deformation occurred prior to metamorphic zircon growth. (I) Rounded detrital zircon core with complex internal zones overgrown by euhedral rim from sample 02-106B.

Figure 8. U-Pb sensitive high-resolution ion microprobe (SHRIMP) results. (A) Tera-Wasserburg concordia plot for SHRIMP analysis of metamorphic zircon overgrowths from sample GM01-02. Six of the eight analyses are nearly concordant, and $^{206}Pb/^{238}U$ dates range from 58.9 ± 1.6 Ma to 64.9 ± 2.6 Ma. The data define two chords with intercepts of 64.0 ± 1.0 Ma and 59.0 ± 1.9 Ma. (B) Weighted mean $^{206}Pb/^{238}U$ date diagram for the older group of nearly concordant analyses from sample GM01-02. These yield a weighted mean $^{206}Pb/^{238}U$ date of 63.6 ± 0.86 Ma (2σ), outside error of the younger analysis. We interpret the data to indicate that metamorphism occurred ca. 64 and 59 Ma. (C) Tera-Wasserburg concordia plot for SHRIMP analysis of metamorphic zircon overgrowths from sample 02-93. Two concordant analyses of the inner metamorphic overgrowth from two grains in 02-93 have $^{206}Pb/^{238}U$ ages of 63.5 ± 2.6 and 64.2 ± 2 Ma (Table 2). Outer, bright overgrowths from sample 02-93 have ages of 54.2 ± 4 and 56.9 ± 2.4 Ma, with an intercept age of 57.5 ± 2.9 Ma. The textural evidence from 02-93, combined with two distinct rim ages, supports the interpretation from GM01-02 that the internal zone of the Clearwater complex was flushed by at least two pulses of metamorphic fluids ca. 64 and 59–56 Ma. (D) Tera-Wasserburg concordia plot for SHRIMP analysis for the two stages of metamorphic zircon overgrowth observed in sample GM01-05. Seven analyses of the inner dark overgrowth from five grains cluster into two ages, 80–82 Ma and 72–74 Ma, with intercepts at 82.8 ± 0.4 Ma and 73.1 ± 1.2 Ma. Analysis of one bright rim on one of these grains (14, not shown) overlaps the age of the younger group. (E) Weighted mean $^{206}Pb/^{238}U$ date plot for analyses from GM01-05. The analyses yield weighted mean ages for the two groups of 82 ± 4.0 Ma and 73.1 ± 1.1 Ma, respectively. These results establish that metamorphic zircon growth occurred in at least two pulses in the external zone and that they were distinctly older than the metamorphic growths in the internal zone. MSWD—mean square of weighted deviations.

Internal Zone

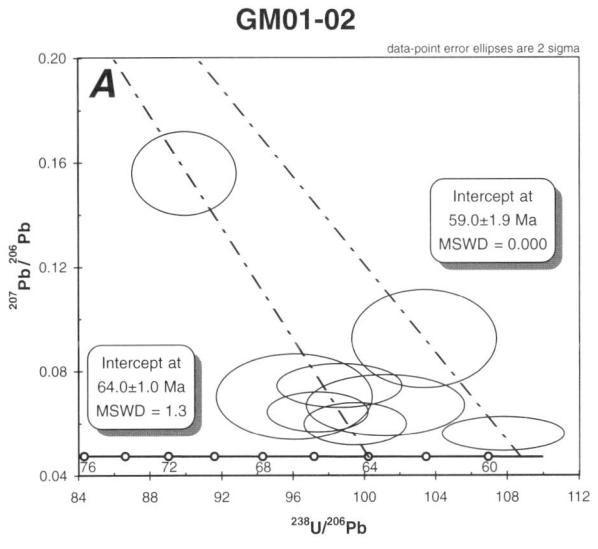

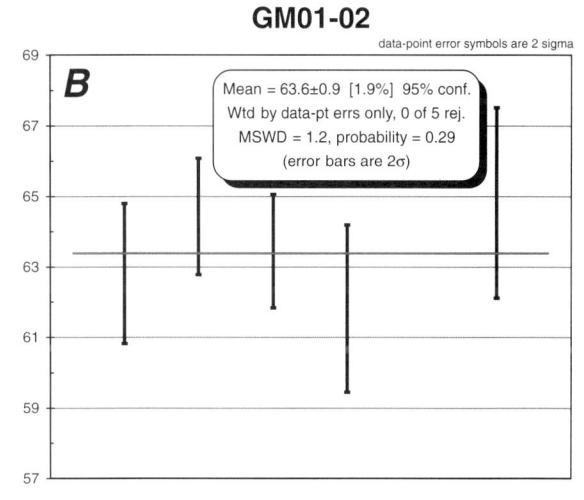

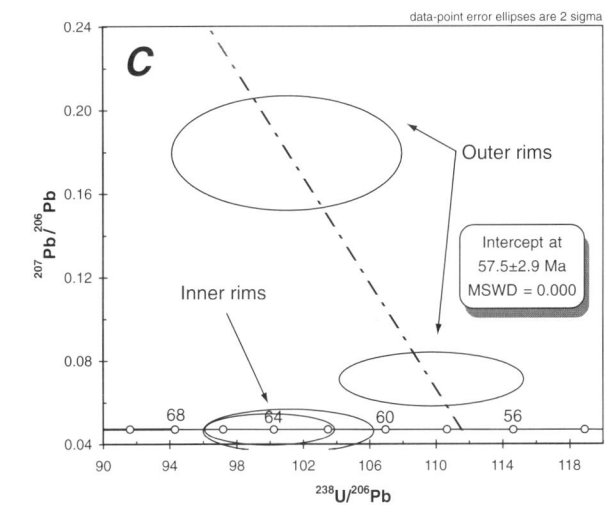

GM01-02

A — Intercept at 64.0±1.0 Ma, MSWD = 1.3; Intercept at 59.0±1.9 Ma, MSWD = 0.000

GM01-02

B — Mean = 63.6±0.9 [1.9%] 95% conf. Wtd by data-pt errs only, 0 of 5 rej. MSWD = 1.2, probability = 0.29 (error bars are 2σ)

C — Outer rims; Inner rims; Intercept at 57.5±2.9 Ma, MSWD = 0.000

External Zone

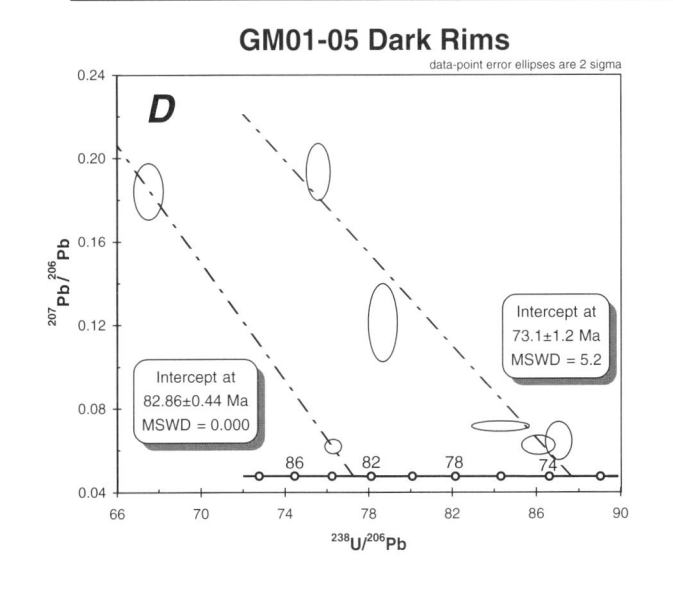

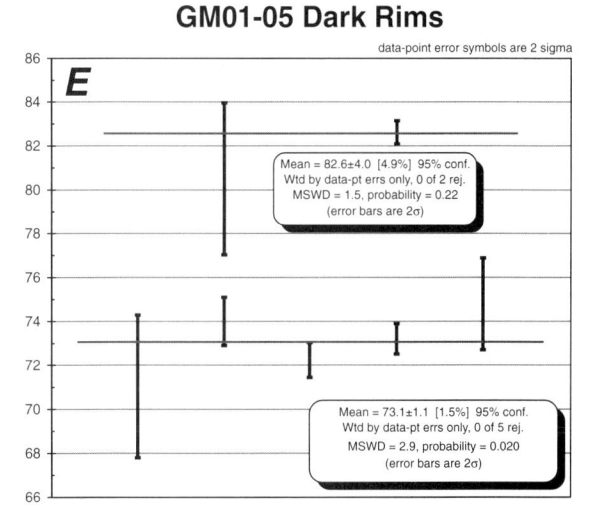

GM01-05 Dark Rims

D — Intercept at 82.86±0.44 Ma, MSWD = 0.000; Intercept at 73.1±1.2 Ma, MSWD = 5.2

GM01-05 Dark Rims

E — Mean = 82.6±4.0 [4.9%] 95% conf. Wtd by data-pt errs only, 0 of 2 rej. MSWD = 1.5, probability = 0.22 (error bars are 2σ)

Mean = 73.1±1.1 [1.5%] 95% conf. Wtd by data-pt errs only, 0 of 5 rej. MSWD = 2.9, probability = 0.020 (error bars are 2σ)

Results (External Zone)

Sample GM01-05 (Quartzite of Monumental Buttes)

Sample GM01-05 is a clean quartzite within the metasedimentary rocks east of the Jug Rock shear zone (Fig. 3). We analyzed seven metamorphic overgrowths on five detrital grains from sample GM01-05 (Table 3; Figs. 8D and 8E). Sample GM01-05 contains rounded detrital cores that have been overgrown by two periods of metamorphic rim growth (Figs. 7E, 7F, and 7G). The inner overgrowth is dark-colored in CL due to low U content and well-faceted to embayed. The outer rim is very thin, light-colored (due to higher U content), and euhedral. We dated the inner dark overgrowth from five grains with six analyses. The results cluster into two ages, 80–82 Ma and 72–74 Ma (Table 3; Figs. 8D and 8E). Analysis of one bright rim on one of these grains (14, Fig. 7F) overlaps the age of the younger group. Although these analyses have variable and fairly high concentrations of common Pb, the two ages are robust, and each group contains nearly concordant analyses with low common Pb. As a whole, these results establish that metamorphic zircon growth occurred in at least two pulses in the external zone and that they are distinctly older than the metamorphic growths in the internal zone.

$^{40}Ar/^{39}Ar$ THERMOCHRONOLOGY

Strategy

In order to place constraints on the timing of cooling and tectonic exhumation of the Clearwater complex, we report new $^{40}Ar/^{39}Ar$ dating of micas from metamorphic and igneous rocks within both the internal and external zones. Seven samples were collected from igneous (RTP-13, BBDF-01, -02, -03) and metasedimentary rocks (01-356, 01-229, 01-224) from the external zone. One sample (01-329) was collected from mica schist in the internal zone.

Methods

Biotite and muscovite were separated from whole-rock samples using conventional crushing, magnetic, and heavy liquid methods followed by hand selecting grains. Samples were wrapped in Al foil and stacked in a fused silica tube with the neutron flux monitor GA1550 biotite (98.5 ± 0.8 Ma: Spell and McDougall, 2003). Samples were irradiated at the Oregon State reactor facility. Correction factors for interfering neutron reactions on K and Ca were determined by analysis of K-glass and optical-grade CaF_2 included in the irradiation, and the following values were used: $(^{40}Ar/^{39}Ar)_K = 2.66 \times 10^{-2}$, $(^{36}Ar/^{37}Ar)_{Ca} = 2.70 \times 10^{-4}$, and $(^{39}Ar/^{37}Ar)_{Ca} = 6.76 \times 10^{-4}$. Following irradiation, mica grains or groups of grains were heated using a CO_2 laser. The laser beam was defocused to ensure roughly uniform heating, and step heating was performed by changing the power output of the laser. Reactive gases were removed by two SAES GP-50 getters prior to expansion into to a MAP215–50 mass spectrometer. Peak intensities were measured using a Balzers electron multiplier. Mass spectrometer discrimination and sensitivity were monitored by analysis of atmospheric argon aliquots from an online pipette system. The sensitivity of the mass spectrometer was ~6×10^{-17} mol mV^{-1}. The data were also corrected for line blanks analyzed at regular intervals between the unknowns. Ages were calculated using a ^{40}K decay constant of 5.543×10^{-10} yr^{-1} and are reported with 2σ errors.

Results (External Zone)

Sample RTP-13 (Roundtop Pluton)

Sample (RTP-13) is a coarse-grained hornblende-biotite granite obtained from the Roundtop pluton, intruded into the external zone ca. 52 Ma (Marvin et al., 1984) (Fig. 3). Biotite separated from sample RTP-13 gave a well-defined plateau age of 47.2 ± 0.8 Ma for ~90% of the ^{39}Ar released (Fig. 9A). The inverse isochron age for these steps is 47.6 ± 1.3 Ma (Fig. 9B).

Sample 01-356 (Schist of Cedar Creek)

Sample 01-356 is a coarse-grained calcareous schist exposed along the Cedar Creek Canyon road. Biotite from sample 01-356 gave a strongly discordant age spectrum with a total fusion age of 180.6 ± 2.2 Ma. The low- and high-temperature apparent ages are >200 Ma and drop down to a minimum age of ca. 102 Ma for the intermediate-temperature step (Table 4). The shape of this age spectra is indicative of significant excess argon contamination, and, therefore, no thermochronological information is given by this sample.

Samples 01-229 and 01-224 (Schist of Monumental Buttes)

Samples 01-229 and 01-224 are from fine-grained mica schist-semischist ~2 km east of the Jug Rock shear zone (Fig. 3). Muscovite separated from sample 01-229 gave a relatively flat age spectrum with a total fusion age of 58.2 ± 3.8 Ma (Fig. 9C). An inverse isochron for the four steps gives an age of 53.7 ± 2.5 Ma and an initial ratio of $^{40}Ar/^{36}Ar$ ratio greater than atmosphere (Fig. 9D). The isochron age is considered to be the best estimate of the cooling age for this muscovite because the trapped component is homogeneous and nonatmospheric in composition.

A total fusion age for two grains of muscovite from sample 01-224 gave an age of 57.8 ± 3.2 Ma (Table 4).

Samples BBDF-01, -02, -03 (Granitoids)

Three samples of foliated granitoids were from the western external zone, west of the Widow Mountain shear zone and east of the White Rock fault. Samples BBDF-01 and BBDF-02 were closest to the White Rock fault, and they give concordant plateau ages of 42.4 ± 0.5 Ma and 41.4 ± 0.7 Ma, respectively (Figs. 10A and 10B). BBDF-03, which came from just west of the Widow Mountain shear zone, gives a plateau age of 46.5 ± 1.0 Ma (Fig. 10C). All three of these plateau ages comprise great than 98% of the gas released from the respective samples and are considered to be robust cooling ages.

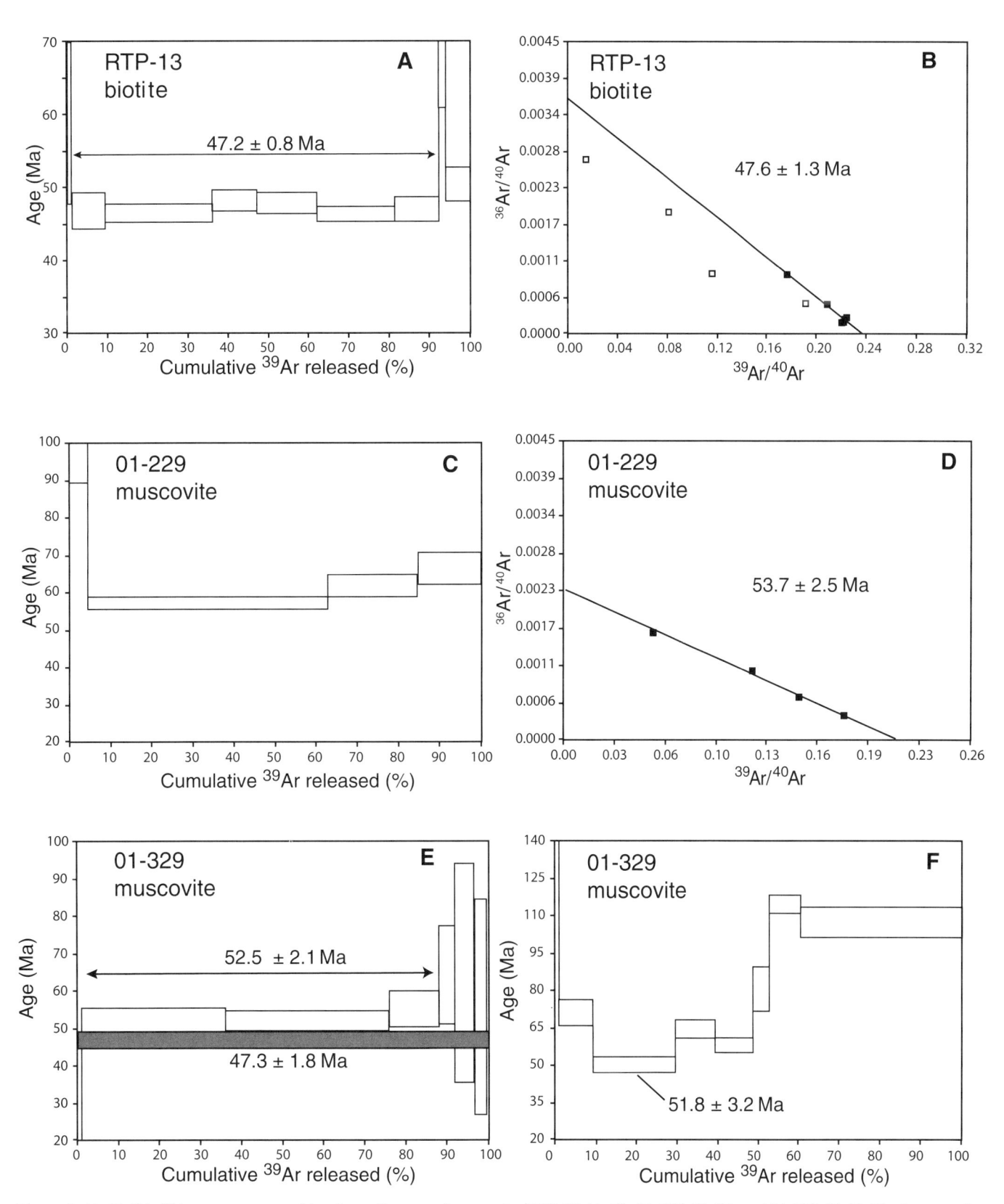

Figure 9. (A–F) ^{40}Ar/^{39}Ar age spectra and isochron diagrams for samples RTP-13 (A–B), 01-229 (C–D), and 01-329 (E–F). The best age for RTP-13 is the plateau age of 47.2 ± 0.8 Ma (2σ errors). This isochron age of 53.7 ± 2.5 Ma for 01-229 (D) is interpreted as the best estimate of the cooling age. 01-329 was analyzed three times and yielded ages between 47.3 ± 1.8 Ma (total fusion age in gray) and 52.5 ± 2.1 Ma (plateau age).

TABLE 4. ^{40}Ar/^{39}Ar DATA

Step no.	^{36}Ar (volts)	^{37}Ar (volts)	^{39}Ar$_{(K)}$ (volts)	^{40}Ar* (volts)	^{40}Ar* (%)	^{40}Ar*/^{39}Ar$_K$	^{39}Ar (%)	Calculated age (Ma)	Error in age (±2 s.d.)
External Zone									
RTP-13 Biotite; J = 0.00647, Round Top pluton (11T, 586845E, 5214566N)									
1	0.0014	0.0000	0.0079	0.1089	20.38	13.85	0.27	154.85	75.73
2	0.0009	0.0000	0.0371	0.2008	44.11	5.41	1.28	62.10	14.11
3	0.0012	0.0000	0.2364	0.9637	72.39	4.08	8.16	46.96	2.42
4	0.0018	0.0010	0.7672	3.1111	85.18	4.05	26.49	46.72	1.27
5	0.0003	0.0000	0.3213	1.3516	93.39	4.21	11.09	48.45	1.44
6	0.0004	0.0000	0.4308	1.7973	93.29	4.17	14.87	48.05	1.43
7	0.0007	0.0000	0.5547	2.2425	91.21	4.04	19.15	46.58	1.03
8	0.0003	0.0097	0.3145	1.2906	92.07	4.10	10.86	47.28	1.68
9	0.0004	0.0100	0.0469	0.2894	71.69	6.17	1.62	70.60	9.55
10	0.0005	0.0645	0.1796	0.7906	84.87	4.40	6.20	50.67	2.32
01-356 Biotite; J = 0.00647, Calc-silicate, Cedar Creek Canyon (11T, 589766E, 5196610N)									
1	0.0019	0.0000	0.0008	0.0000	0.00	0.00	1.04	0.00	0
2	0.0003	0.0000	0.0062	0.1401	60.91	22.71	8.39	247.29	53.82
3	0.0005	0.0000	0.0217	0.1944	55.91	8.95	29.52	101.59	14.49
4	0.0004	0.0000	0.0194	0.2082	62.64	10.75	26.32	121.31	19.44
5	0.0004	0.0000	0.0108	0.1630	56.80	15.13	14.64	168.50	32.02
6	0.0003	0.0000	0.0048	0.1557	60.59	32.45	6.52	343.79	74.03
7	0.0013	0.0000	0.0100	0.3361	46.40	33.67	13.57	355.53	128.94
01-229 Muscovite; J = 0.00647, Semischist, near North Butte (11T, 591972E, 5207742N)									
1	0.0009	0.0063	0.0010	0.0000	0.00	0.00	0.12	0.00	0.00
2	0.0010	0.0000	0.0356	0.3336	51.84	9.37	4.17	106.14	16.87
3	0.0011	0.0000	0.4971	2.4781	88.21	4.99	58.24	57.27	1.62
4	0.0008	0.0000	0.1879	1.0128	80.07	5.39	22.02	61.83	3.09
5	0.0012	0.0000	0.1318	0.7644	68.60	5.80	15.45	66.43	4.26
01-224 Muscovite; J = 0.00647, Semischist near North Butte (11T, 592755E, 5207026N)									
1	0.0017	0.0000	0.9560	4.8071	90.24	5.03	100.00	57.76	3.24
BBDF-01 Biotite; J = 0.000418, Granite along Freezeout Ridge (11T, 571192E, 5207450N)									
1	0.00152	0.00000	0.00303	0.0000	0.00	0.00	1.60	0.00	0.00
2	0.00107	0.00000	0.02526	1.4585	82.11	57.74	13.38	43.02	0.77
3	0.00066	0.00000	0.04347	2.4512	92.58	56.39	23.02	42.03	0.65
4	0.00042	0.00000	0.03985	2.2627	94.85	56.78	21.10	42.32	0.67
5	0.00027	0.00000	0.02992	1.7024	95.53	56.90	15.84	42.41	0.90
6	0.00024	0.00000	0.02652	1.5194	95.52	57.30	14.04	42.70	0.78
7	0.00025	0.00000	0.02078	1.1751	94.13	56.54	11.01	42.14	0.74
BBDF-02 Biotite; J = 0.000418, Granite along Freezeout Ridge (11T, 575105E, 5207091N)									
1	0.00032	0.00000	0.00098	0.0000	0.00	0.00	1.19	0.00	0.00
2	0.00043	0.00000	0.02311	1.2677	90.98	54.86	28.07	40.90	0.72
3	0.00018	0.00000	0.02175	1.2103	95.73	55.65	26.42	41.48	0.64
4	0.00021	0.00000	0.02220	1.2237	95.27	55.11	26.97	41.09	0.91
5	0.00003	0.00000	0.00734	0.4216	98.22	57.46	8.91	42.82	1.16
6	0.00007	0.00000	0.00694	0.3871	94.81	55.80	8.43	41.60	1.31

(*continued*)

TABLE 4. ^{40}Ar/^{39}Ar DATA (continued)

Step no.	^{36}Ar (volts)	^{37}Ar (volts)	^{39}Ar$_{(K)}$ (volts)	^{40}Ar* (volts)	^{40}Ar* (%)	^{40}Ar*/^{39}Ar$_K$	^{39}Ar (%)	Calculated age (Ma)	Error in age (±2 s.d.)
External Zone									
BBDF-03 Biotite; J = 0.000418, Granite near Widow Mountain (11T, 579534E, 5210702N)									
1	0.00031	0.00000	0.00158	0.0000	0.00	0.00	0.54	0.00	0.00
2	0.00041	0.00000	0.01678	1.0731	89.92	63.93	5.75	47.58	1.34
3	0.00047	0.00000	0.04168	2.6729	95.04	64.12	14.29	47.72	1.15
4	0.00047	0.00000	0.04420	2.8323	95.31	64.08	15.15	47.69	0.98
5	0.00020	0.00000	0.04074	2.6240	97.78	64.41	13.97	47.93	0.77
6	0.00015	0.00000	0.05435	3.3742	98.67	62.09	18.63	46.22	0.59
7	0.00012	0.00000	0.03806	2.3056	98.47	60.57	13.05	45.11	0.90
8	0.00013	0.00000	0.05426	3.3057	98.87	60.92	18.60	45.37	0.71
Internal Zone									
01-329 Muscovite; J = 0.00647 (Run 1), Al-Mg schist, north of Goat Mountain (11T, 589075E, 5206901N)									
1	0.001	0.000	0.002	0.658	67.79	352.96	0.08	2144.98	221.55
2	0.002	0.000	0.021	1.624	72.83	76.41	0.86	724.73	31.40
3	0.002	0.000	0.200	1.248	70.60	6.23	8.14	71.28	5.09
4	0.002	0.003	0.500	2.204	80.34	4.41	20.31	50.76	3.22
5	0.001	0.000	0.236	1.333	88.07	5.65	9.59	64.74	3.54
6	0.001	0.000	0.238	1.207	85.38	5.08	9.65	58.35	2.83
7	0.000	0.000	0.101	0.715	83.68	7.05	4.12	80.43	8.73
8	0.000	0.000	0.183	1.850	97.03	10.11	7.43	114.36	3.63
9	0.004	0.000	0.980	9.272	87.29	9.46	39.82	107.18	5.87
01-329 Muscovite; J = 0.00647 (Run 3)									
1	0.0039	0.0000	2.0796	8.4953	87.64	4.09	100.00	47.28	1.77
01-329 Muscovite; J = 0.00647 (Run 2)									
1	0.0002	0.3355	0.0046	0.0000	0.00	0.00	1.15	0.00	0
2	0.0008	0.0000	0.1416	0.6348	72.03	4.48	35.00	51.59	3.94
3	0.0000	0.0000	0.1622	0.7341	97.51	4.53	40.09	52.08	2.61
4	0.0000	0.0000	0.0481	0.2308	99.45	4.80	11.88	55.20	4.84
5	0.0000	0.0000	0.0169	0.0947	99.53	5.61	4.17	64.36	13.22
6	0.0001	0.0000	0.0178	0.1011	84.85	5.67	4.41	64.98	29.47
7	0.0000	0.0000	0.0133	0.0647	99.45	4.85	3.30	55.74	29

Results (Internal Zone)

Sample 01-329 (Schist of Goat Mountain)

Sample 01-329 is from a coarse-grained mica schist exposed along the north flank of Goat Mountain below the Jug Rock shear zone. Analyses of three grains of muscovite from sample 01-329 gave a discordant age spectrum with a total fusion age of 92.9 ± 2.7 Ma (Fig. 9E). A second analysis of this sample gave a plateau age of 52.5 ± 2.1 Ma for nearly 90% of the gas released (Fig. 9E). An addition total fusion analysis of one grain of muscovite from this sample gave a total fusion age of 47.3 ± 1.8 Ma (Table 4). The discordant age spectrum for the first analysis was due to excess argon. The time that this sample cooled below ~400–350 °C was probably either ca. 53 Ma or as young as ca. 47 Ma.

Interpretation

Excess argon was a significant problem for several of the samples, and it provides an explanation for the spread in K-Ar ages obtained by Hietanen (1969a). This required analyzing small single grains, or two to three grains at a time to find some without excess argon. This resulted in most of the age spectra having a limited number of steps. The cooling age of the eastern part of the external zone below muscovite closure was probably ca. 54 Ma, based on sample 01-229. The internal zone was probably cooling at the same time but may not have cooled below ~300 °C until ca. 47 Ma, based on the total fusion age of the small single grain from sample 01-329.

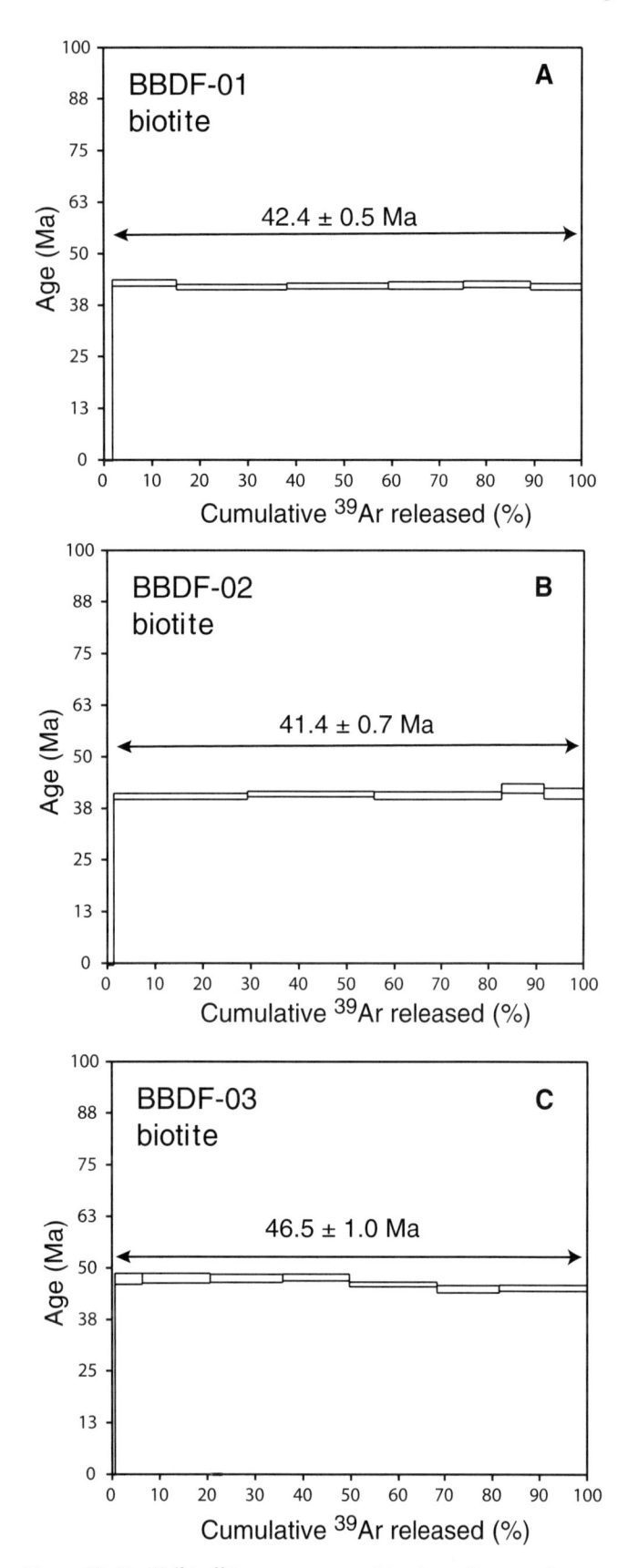

Figure 10. (A–C) ^{40}Ar/^{39}Ar age spectra and isochron diagrams for samples BBDF-01 to BBDF-03; 2σ errors are given for the plateau ages.

The youngest apparent ages are recorded by samples in the footwall of the White Rock fault in the western part of the external zone (BBDF-01, -02, and -03). Samples closest to the White Rock fault did not cool below ~350–300 °C until ca. 41–42 Ma.

SYNTHESIS

The timing and pattern of exhumation of the Clearwater complex are bracketed by ^{40}Ar/^{39}Ar mica cooling ages from the complex that are between ca. 54 and 42 Ma (Fig. 3). Rocks in the eastern half of the external zone cooled first, below the muscovite closure temperature at ca. 54 Ma, based on sample 01-229 (Fig. 11). The internal zone of the complex cooled below muscovite closure slightly later, between 53 and 47 Ma, based on sample 01-329. The western half of the external zone cooled through biotite closure shortly thereafter in a west to east direction between 47 Ma to ca. 41 Ma, based on samples BFDF-01 to BFDF-03 (Figs. 3 and 11). This pattern of westward progressive cooling is consistent with exhumation and cooling of the Clearwater complex primarily by the west-directed White Rock fault and underlying Widow Mountain shear zone. A sharp break in cooling ages across the Jug Rock shear zone suggests that it was active during the early stages of exhumation, but was abandoned when the White Rock fault became the dominant structure sometime between 54 and 47 Ma (Fig. 11). The role of the Collins Creek fault in exhuming the complex is poorly constrained by our sparse data set, but, it was likely active during formation and exhumation of the Clearwater complex as well.

Metamorphic events in the Clearwater complex are bracketed by new U-Pb dating of metamorphic zircon overgrowths. The external zone preserves two pulses of Late Cretaceous metamorphism (ca. 72–74 Ma and 80–82 Ma) in rocks that equilibrated under conditions of ~900 MPa and ~650 °C before exhumation and local reequilibration of some mineral phases at ~600 MPa and 623 °C prior to 54 Ma. Our estimate for the timing of peak M2 metamorphism in the Clearwater region is consistent with that of Grover et al. (1993), and the timing of peak metamorphism in the Shuswap complex and Priest River complex (i.e., Parrish, 1995; Doughty et al., 1998), and the early phase of high-grade metamorphism in the Bitterroot complex between 75 and 80 Ma (Toth and Stacey, 1992; Foster and Fanning, 1997; Foster et al., 2001). In contrast, rocks inside the culminations, which contain M2 minerals strongly overprinted by M3 metamorphism, underwent later metamorphic zircon growth during two episodes of fluid migration around 64 and between 59 and 55 Ma, and there is no evidence of Cretaceous metamorphism. The age of metamorphism inside the culminations documented here overlaps with the age of "main-phase" igneous activity and high-grade metamorphism in the Bitterroot complex between 65 and 53 Ma (House et al., 1997, 2002; Foster et al., 2001) and was probably related to that regional event. The youngest metamorphic zircon growth, however, could be related to an influx of hot meteoric water into the Clearwater complex during the initial phase of exhumation (i.e., Walker, 1993; Larson and Sharp, 1998; Mora et al., 1999).

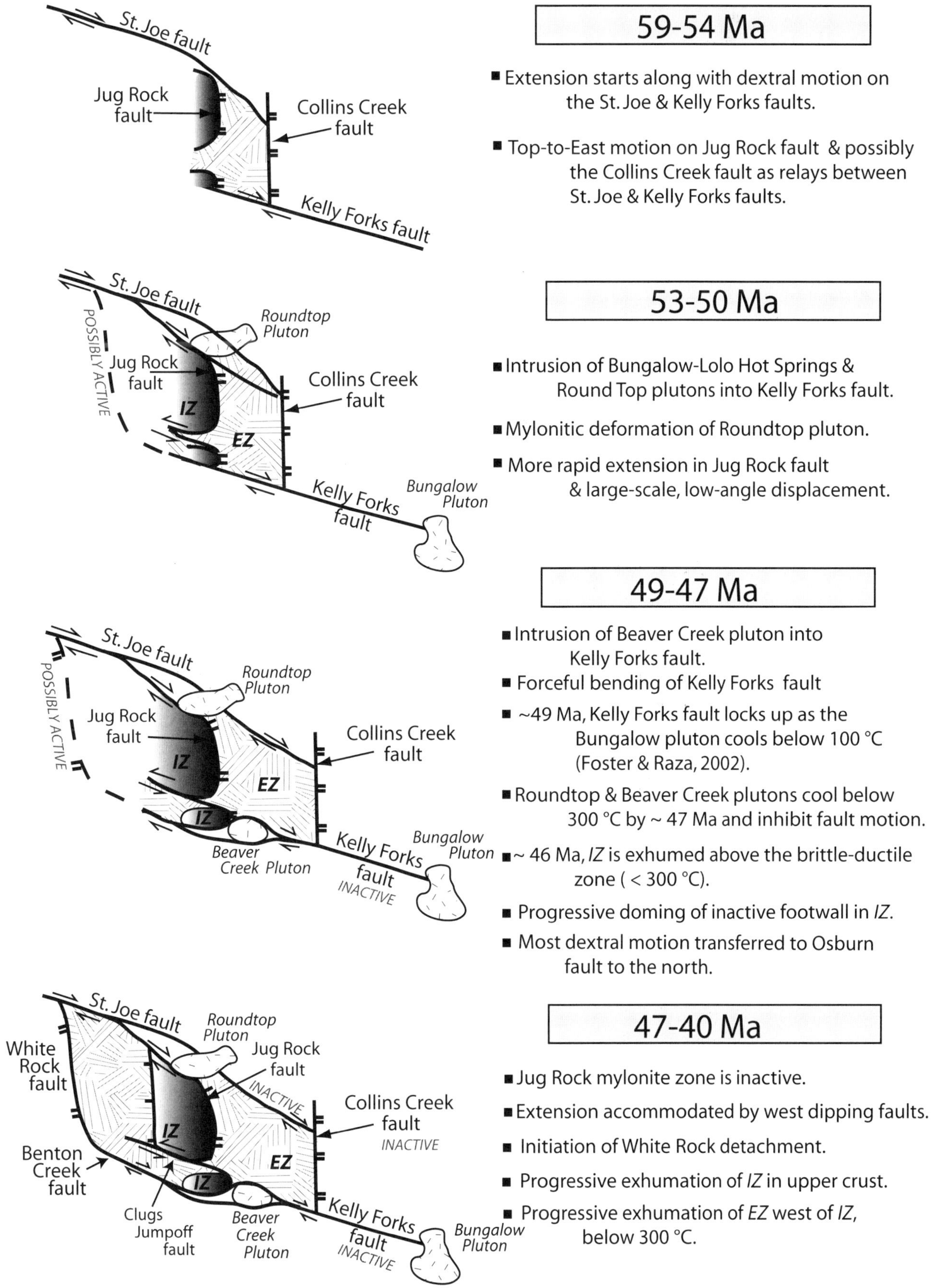

59-54 Ma

- Extension starts along with dextral motion on the St. Joe & Kelly Forks faults.
- Top-to-East motion on Jug Rock fault & possibly the Collins Creek fault as relays between St. Joe & Kelly Forks faults.

53-50 Ma

- Intrusion of Bungalow-Lolo Hot Springs & Round Top plutons into Kelly Forks fault.
- Mylonitic deformation of Roundtop pluton.
- More rapid extension in Jug Rock fault & large-scale, low-angle displacement.

49-47 Ma

- Intrusion of Beaver Creek pluton into Kelly Forks fault.
- Forceful bending of Kelly Forks fault
- ~49 Ma, Kelly Forks fault locks up as the Bungalow pluton cools below 100 °C (Foster & Raza, 2002).
- Roundtop & Beaver Creek plutons cool below 300 °C by ~ 47 Ma and inhibit fault motion.
- ~ 46 Ma, IZ is exhumed above the brittle-ductile zone (< 300 °C).
- Progressive doming of inactive footwall in IZ.
- Most dextral motion transferred to Osburn fault to the north.

47-40 Ma

- Jug Rock mylonite zone is inactive.
- Extension accommodated by west dipping faults.
- Initiation of White Rock detachment.
- Progressive exhumation of IZ in upper crust.
- Progressive exhumation of EZ west of IZ, below 300 °C.

Figure 11. Sketch maps depicting our interpretation of the movement history of the faults bounding the Clearwater metamorphic core complex and exhumation history of rocks in the footwalls of the detachment faults during Eocene time. IZ—internal zone; EZ—external zone.

The internal structure of the Clearwater complex is dominated by two doubly-plunging culminations of basement rocks (internal zone) that have been exhumed relative to the surrounding metasedimentary rocks of the external zone. Each basement culmination is mantled by shear zones that show a top-to-the-east sense of shear on the gently east-dipping eastern boundary, or dextral or sinistral shear on the steeply dipping northern and southern boundaries, respectively (Fig. 5). Thus, the overlying external zone metasedimentary rocks were transported to the east-southeast relative to the underlying Paleoproterozoic basement. The shear zones, as typified by the Jug Rock shear zone, record initial movement in the amphibolite facies with continued deformation into the greenschist facies. A lack of brittle structures shows that the shear zones ceased moving prior to reaching the upper levels of the crust.

The differences in metamorphic ages and cooling between rocks of the external and internal zones argue for significant displacement along the bounding shear zones. However, the exact timing of motion is difficult to quantify. The presence of fractured zircons overgrown by euhedral metamorphic rims within the shear zone (Figs. 7A and 7H) suggests that that metamorphic zircon growth could be related to fluid migration along the Jug Rock shear zone at 64 Ma or between 59 and 55 Ma. Although motion on the Jug Rock shear zone at this time, and earlier as a compressional structure, is allowed by the geochronologic data, we note that the youngest ages for metamorphic growth and possible deformation on the Jug Rock shear zone overlap with extension in the Valhalla complex at 59–54 Ma (Carr et al., 1987) and with the beginning of exhumation in the Bitterroot complex at 55–53 Ma (i.e., Coyner et al., 2001; House et al., 2002; Foster et al., 2007). On this basis, we conclude that the Jug Rock shear zone was active as extensional structure during the earliest phase of exhumation in the Clearwater complex (Fig. 11). Further support for this interpretation includes decreasing metamorphic conditions during shear and a sharp break in cooling ages across the shear zone. Both of these observations are more compatible with movement during extension and exhumation than during compression. Movement on the Jug Rock shear zone may have continued until around 47 Ma, when the internal zone cooled through the closure temperature for argon, and greenschist-facies mylonitic fabrics and ultramylonites formed as the internal zone and external zone were finally juxtaposed. The internal shear zones of the Clearwater complex compare favorably with the Valkyr shear zone in the Valhalla complex (Carr et al., 1987) and the Spokane dome mylonite zone in the Priest River complex (Doughty and Price, 1999), both of which are ductile extensional shear zones exposed in the core of these complexes.

The amount of uplift of the internal zone culminations relative to the outlying rocks of the external zone along these shear zones is not resolvable with metamorphic minerals and geothermobarometry alone. Average high-pressure (M2) conditions for rocks in the external zone are slightly lower than that of the internal zone, but they overlap within the uncertainty typically associated with these thermobarometers (±100 MPa and 50 °C). If the amphibolites yield erroneously high estimates of M2 pressure, as suggested by Ziegler (1991), and are excluded from the calculated estimates, there is about a 200 MPa drop in pressure across the Jug Rock shear zone. If temperatures based on oxygen isotopes are used for the internal zone (i.e., Larson and Sharp, 1998), then a temperature drop of ~100 °C exists across this boundary. The inability of these thermobarometers to accurately define the amount of offset along the Jug Rock shear zone is partly due to the composition of the external zone rocks, which forced us to use the less robust GMBP barometer for most samples, and uncertainty associated with the complex metamorphic history and age of metamorphic events between the internal and external zones.

We envision the following sequence of events that led to creation of the Clearwater complex (Fig. 11): At the end of the Cretaceous, rocks now exposed in the Clearwater complex were undergoing metamorphism and deformation within the thickened crust of the Cordilleran orogen. Exhumation of the Clearwater complex may have begun in the Paleocene (64 Ma), but was definitely happening along the deep-seated shear zones like the Jug Rock shear zone between 59 and 55 Ma and along the bounding detachment faults by 54 Ma (Fig. 11). The St. Joe and Kelly Forks faults of the Lewis and Clark fault zone were active at this time, as was the Bitterroot complex. The eastern side of the external zone was exhumed first and cooled at around 54 Ma. Unroofing of the Clearwater complex continued on both the internal ductile shear zones and outer bounding faults until between 54 and 47 Ma, when the Jug Rock shear zone was abandoned and the west-directed White Rock–Widow Mountain fault became the dominant detachment fault that unroofed the Clearwater complex. Slip continued on the White Rock–Widow Mountain fault until ca. 42–41 Ma as a relay between the St. Joe and Kelly Forks faults. The opposing shear sense of the deep-seated mylonites and overlying ductile-brittle detachment fault is not common within most core complexes, which are typically dominated by a uniform sense of shear at all levels within the complex (i.e., Wernicke, 1985). The Priest River complex is a notable exception and provides a possible explanation for the observed structure within the Clearwater complex. Arching of the metamorphic infrastructure in the Priest River complex caused a large antithetic mylonite zone (Newport fault zone) to form on the backside of the infrastructure culmination, which broke the culmination into discrete crustal blocks. A similar scenario may have occurred in the Clearwater complex, when arching of the older, deep-seated shear zones initiated formation of the west-directed White Rock–Widow Mountain fault, which then dominated exhumation of the complex.

CONCLUSIONS

A combination of geologic mapping, kinematic analysis, and geochronology described in this paper provides a compelling case for interpreting the Clearwater complex as an Eocene metamorphic core complex. The direction of slip along shear zones within the complex parallels the direction of slip on the Lewis and Clark Line and in the Bitterroot complex, and ^{40}Ar/^{39}Ar mica cooling ages show that the complex was exhumed and cooled between 54 and 41 Ma. The cooling of the Clearwater complex coincided with an episode of crustal extension and core complex formation in the Northern Rockies (i.e., Parrish et al., 1988; Foster et al., 2007) and with motion on the Lewis and Clark Line in the vicinity (Burmester et al., 2004; Lewis et al., 2002). We conclude that the Clearwater complex formed as a mid-crustal pull-apart structure associated with a step-over in the Lewis and Clark Line.

ACKNOWLEDGMENTS

We thank Andrew Buddington, Brian Boothe, Rebecca Pitts, Darren Tollstrup, Brandt Halver, Travis Kumm, Jason Burt, and Andrew Wiser for superlative field assistance. Potlatch Corporation and the state of Idaho provided access and some of the only outcrops available. The U.S. Geological Survey–Eastern Washington University Cooperative Program loaned us field vehicles and motorbikes. We thank Joe Wooden and staff at Stanford–U.S. Geological Survey Micro-Isotopic Analytical Center for help with collecting the SHRIMP data; Mike Hartley, Jim Vogl, and Warren Grice for assistance with the ^{40}Ar/^{39}Ar analyses at University of Florida; and Scotty Cornelius at the WSU (Washington State University) Geoanalytical Laboratory for help with microprobe data acquisition. A special thanks to Andrew Buddington (Spokane Community College), whose optimistic attitude and enthusiasm made the field work enjoyable, and John Watkinson for supervising Sha's thesis and providing an open and conducive academic environment. Joe Hull, the sole attendee of the NAGT field conference, was also immensely helpful. This paper benefited substantially from constructive reviews by G. Gehrels and T. Kalakay. Reed Lewis and Russ Burmester also deserve special thanks for discussions and support. They surely don't agree with all of our interpretations. This work was supported by National Science Foundation (NSF) grant EAR01532280 to Doughty and EAR0107088 to Chamberlain.

REFERENCES CITED

Armstrong, R.L., 1982, Cordilleran metamorphic core complexes—From Arizona to southern Canada: Annual Review of Earth and Planetary Sciences, v. 10, p. 129–154, doi: 10.1146/annurev.ea.10.050182.001021.

Axen, G.J., and Bartley, J.M., 1997, Field tests of rolling hinges; existence, mechanical types, and implications for extensional tectonics: Journal of Geophysical Research, ser. B, Solid Earth and Planets, v. 102, p. 20,515–20,537, doi: 10.1029/97JB01355.

Berman, R.G., 1990, Mixing properties of Ca-Mg-Fe-Mn garnets: The American Mineralogist, v. 75, p. 328–344.

Burchfiel, B.C., and Stewart, J.H., 1966, 'Pull-apart' origin of the central segment of Death Valley, California: Geological Society of America Bulletin, v. 77, p. 439–441, doi: 10.1130/0016-7606(1966)77[439:POOTCS]2.0.CO;2.

Burmester, R.F., and Lewis, R.S., 1999, The Boehls Butte core complex and the St. Joe fault, a ridge-transform system of Eocene extension: Geological Society of America Abstracts with Programs, v. 31, no. 4, p. A-6.

Burmester, R.F., McClelland, W.C., and Lewis, R.S., 2004, U-Pb dating of plutons along the transfer zone between the Bitterroot and Priest River metamorphic core complexes: Geological Society of America Abstracts with Programs, v. 36, no. 4, p. 72.

Carey, J.W., 1985, Petrology and Metamorphic History of Metapelites in the Boehls Butte Quadrangle [M.Sc. thesis]: Corvallis, University of Oregon, 169 p.

Carey, J.W., Rice, J.M., and Grover, T.W., 1992, Petrology of aluminous schist in the Boehls Butte region of northern Idaho: Geologic history and aluminum-silicate phase relations: American Journal of Science, v. 292, p. 455–473.

Carr, S.D., Parrish, R.R., and Brown, R.L., 1987, Eocene structural development of the Valhalla complex, southeastern British Columbia: Tectonics, v. 6, p. 175–196.

Cheney, J.T., 1975, Kyanite, sillimanite, phlogopite, cordierite layers in the Bass Creek anorthosites, Bitterroot Range, Montana: Northwest Geology, v. 4, p. 77–82.

Childs, J.F., 1982, Geology of the Precambrian Belt Supergroup and the Northern Margin of the Idaho Batholith, Clearwater County, Idaho [Ph.D. dissertation]: Santa Cruz, University of California, 491 p.

Coney, P.J., 1980, Cordilleran metamorphic core complexes—An overview, in Crittenden, M.D., Jr., Coney, P.J., and Davis, G.H., eds., Cordilleran Metamorphic Core Complexes: Geological Society of America Memoir 153, p. 7–31.

Coyner, S.J., Foster, D.A., and Fanning, C.M., 2001, Emplacement chronology of the Paradise Pluton: Implications for the development of the Bitterroot metamorphic core complex, Montana/Idaho: Northwest Geology, v. 30, p. 10–20.

Doughty, P.T., and Buddington, A.B., 2002, Eocene structural evolution of the Boehls Butte anorthosite and Clearwater core complex, north-central Idaho: A basement involved extensional strike-slip relay: Geological Society of America Abstracts with Programs, v. 34, no. 6, p. 332.

Doughty, P.T., and Chamberlain, K.R., 2007, Age of Paleoproterozoic basement and related rocks in the Clearwater complex, northern Idaho, U.S.A., in Link, P.K., and Lewis, R.L., eds., Proterozoic Geology of Western North America and Siberia: Society of Economic Paleontologists and Mineralogists Special Publication 86, p. 9–35.

Doughty, P.T., and Price, R.A., 1999, Tectonic evolution of the Priest River complex, northern Idaho and Washington: A reappraisal of the Newport fault with new insights on metamorphic core complex formation: Tectonics, v. 18, p. 375–393, doi: 10.1029/1998TC900029.

Doughty, P.T., and Sheriff, S.D., 1992, Paleomagnetic evidence for Eocene en echelon extension and crustal rotations in western Montana and Idaho: Tectonics, v. 11, p. 663–671.

Doughty, P.T., Price, R.A., and Parrish, R.R., 1998, Geology and U-Pb geochronology of Archean basement and Proterozoic cover in the Priest River complex, northwestern United States, and their implications for Cordilleran structure and Precambrian continent reconstructions: Canadian Journal of Earth Sciences, v. 35, p. 39–54, doi: 10.1139/cjes-35-1-39.

Faulds, J.E., and Stewart, J.H., eds., 1998, Accommodation Zones and Transfer Zones: The Regional Segmentation of the Basin and Range Province: Geological Society of America Special Paper 323, 257 p.

Ferry, J.M., and Spear, F.S., 1978, Experimental calibration of the partitioning of Fe and Mg between biotite and garnet: Contributions to Mineralogy and Petrology, v. 66, p. 113–117, doi: 10.1007/BF00372150.

Foster, D.A., and Fanning, C.M., 1997, Geochronology of the northern Idaho Batholith and the Bitterroot metamorphic core complex: Magmatism preceding and contemporaneous with extension: Geological Society of America Bulletin, v. 109, p. 379–394, doi: 10.1130/0016-7606(1997)109<0379:GOTNIB>2.3.CO;2.

Foster, D.A., and Raza, A., 2002, Low-temperature thermochronological record of exhumation of the Bitterroot metamorphic core complex, northern Cordilleran orogen: Tectonophysics, v. 349, p. 23–36, doi: 10.1016/S0040-1951(02)00044-6.

Foster, D.A., Schafer, C., Fanning, M.C., and Hyndman, D.W., 2001, Relationships between crustal partial melting, plutonism, orogeny, and exhumation: Idaho-Bitterroot Batholith: Tectonophysics, v. 342, p. 313–350, doi: 10.1016/S0040-1951(01)00169-X.

Foster, D.A., Doughty, P.T., Kalakay, T.J., Fanning, C.M., Coyner, S., Grice, W.C., and Vogl, J., 2007, Kinematics and timing of exhumation of metamorphic core complexes along the Lewis and Clark fault zone, northern Rocky Mountains, USA, in Till, A.B., Roeske, S.M., Sample, J.C., and Foster, D.A., eds., Exhumation Associated with Continental Strike-Slip Fault Systems: Geological Society of America Special Paper 434, p. 205–229, doi: 10.1130/2007.2434(10).

Frost, R.B., and Chacko, T., 1989, The granulite uncertainty principle: Limitation on thermobarometry in granulites: The Journal of Geology, v. 97, p. 435–450.

Ganguly, J., and Saxena, S.K., 1984, Mixing properties of aluminosilicate garnets: Constraints from natural and experimental data, and applications to geothermo-barometry: The American Mineralogist, v. 69, p. 88–97.

Graham, C.M., and Powell, R., 1984, A garnet-hornblende geothermometer: Calibration, testing, and application to the Pelona Schist, Southern California: Journal of Metamorphic Geology, v. 2, p. 13–21.

Grover, T.W., Rice, J.M., and Carey, J.W., 1992, Petrology of aluminous schist in the Boehls Butte region of northern Idaho: Phase equilibria and P-T evolution: American Journal of Science, v. 292, p. 474–507.

Grover, T.W., Rice, J.W., Snee, L.W., and Unruh, D.M., 1993, Constraints on the tectonometamorphic evolution of the St. Joe–Clearwater region, northern Idaho: Geological Society of America Abstracts with Programs, v. 25, no. 6, p. 173.

Harrison, J.E., Griggs, A.B., and Wells, J.B., 1974, Tectonic Features of the Precambrian Belt Basin and Their Influence on Post-Belt Structures: U.S. Geological Survey Professional Paper 866, 15 p.

Harrison, J.E., Griggs, A.B., and Wells, J.B., 1986, Geologic and Structure Maps of the Wallace 1° × 2° Quadrangle, Montana and Idaho: U.S. Geological Survey Miscellaneous Investigations Series Map I-1509-A, scale 1:250,000.

Hietanen, A., 1956, Kyanite, andalusite, and sillimanite in the schist in Boehls Butte quadrangle, Idaho: The American Mineralogist, v. 41, p. 1–27.

Hietanen, A., 1963a, Anorthosite and Associated Rocks in the Boehls Butte Quadrangle and Vicinity, Idaho: U.S. Geological Survey Professional Paper 344-B, 78 p.

Hietanen, A., 1963b, Metamorphism of the Belt Series in the Elk River–Clarkia Area, Idaho: U.S. Geological Survey Professional Paper 344-C, 49 p.

Hietanen, A., 1963c, Idaho Batholith near Pierce and Bungalow Clearwater County, Idaho: U.S. Geological Survey Professional Paper 344-D, 42 p.

Hietanen, A., 1968, Belt Series in the Region around Snow Peak and Mallard Peak, Idaho: U.S. Geological Survey Professional Paper 344-E, 34 p.

Hietanen, A., 1969a, Distribution of Fe and Mg between garnet, staurolite, and biotite in aluminum-rich schist in various metamorphic zones north of the Idaho Batholith: American Journal of Science, v. 267, p. 422–456.

Hietanen, A., 1969b, Metamorphic environment of anorthosite in the Boehls Butte area, Idaho, in Isachsen, Y.W., ed., Origin of Anorthosite and Related Rocks: New York State Museum and Science Service Memoir 18, p. 371–386.

Hietanen, A., 1984, Geology along the Northwest Border Zone of the Idaho Batholith, Northern Idaho: U.S. Geological Survey Bulletin 1608, 17 p.

Hodges, K.V., and Crowley, P.D., 1985, Error estimation and empirical geothermobarometry for pelitic systems: The American Mineralogist, v. 70, p. 702–709.

House, M.A., Hodges, K.V., and Bowring, S.A., 1997, Petrological and geochronological constraints on regional metamorphism along the northern border of the Bitterroot Batholith: Journal of Metamorphic Geology, v. 15, p. 753–764, doi: 10.1111/j.1525-1314.1997.00052.x.

House, M.A., Bowring, S.A., and Hodges, K.V., 2002, Implications of middle Eocene epizonal plutonism for the unroofing history of the Bitterroot metamorphic core complex, Idaho-Montana: Geological Society of America Bulletin, v. 114, p. 448–461, doi: 10.1130/0016-7606(2002)114<0448:IOMEEP>2.0.CO;2.

Hyndman, D.W., Alt, D., and Sears, J.W., 1988, Post-Archean metamorphic and tectonic evolution of western Montana and northern Idaho, in Ernst, G.W., ed., Metamorphism and Crustal Evolution of the Western United States, Rubey Volume VII: Englewood Cliffs, New Jersey, Prentice Hall, p. 333–361.

Indares, A., and Martignole, J., 1985, Biotite-garnet geothermometry in the granulite facies: The influence of Ti and Al in biotite: The American Mineralogist, v. 70, p. 272–278.

Juras, D.S., 1974, The Petrofabric Analysis and Plagioclase Petrography of the Boehls Butte Anorthosite [Ph.D. dissertation]: Moscow, University of Idaho, 132 p.

Kell, R.E., and Childs, J.F., 1999, Mesozoic-Cenozoic structural events affecting the Belt Basin between the Idaho Batholith and the Lewis and Clark Line; implications on identifying syndepositional belt-age structure, in Berg, R.B., ed., Belt Symposium III Abstracts: Montana Bureau of Mines Open-File Report 1993, p. 32–34.

Kohn, M.J., and Spear, F.S., 1990, Two new barometers for garnet amphibolites with applications to eastern Vermont: The American Mineralogist, v. 75, p. 89–96.

Koziol, A.M., and Newton, R.C., 1988, Redetermination of the anorthite breakdown reaction and improvement of the plagioclase-garnet-Al_2SiO_5-quartz barometer: The American Mineralogist, v. 73, p. 216–223.

Lang, H.M., and Rice, J.M., 1985a, Metamorphism of pelitic rocks in the Snow Peak area, northern Idaho: Sequence of events and regional implications: Geological Society of America Bulletin, v. 96, p. 731–736, doi: 10.1130/0016-7606(1985)96<731:MOPRIT>2.0.CO;2.

Lang, H.M., and Rice, J.M., 1985b, Geothermometery, geobarometry, and T-X(Fe-Mg) relations in metapelites, Snow Peak, northern Idaho: Journal of Petrology, v. 26, p. 889–924.

Larson, P.B., and Sharp, Z.D., 1998, Mineral oxygen isotope ratios for the Boehls Butte–Goat Mountain metamorphic complex, Idaho: Evidence for fast cooling: American Journal of Science, v. 298, p. 572–593.

Lewis, R.S., Burmester, R.F., McFaddan, M.D., Eversmeyer, B.A., Wallace, C.A., and Bennett, E.H., 1992, Geologic Map of the Upper North Fork of the Clearwater River Area, Northern Idaho: Idaho Geological Survey Geologic Map Series, scale 1:100,000.

Lewis, R.S., Burmester, R.F., Frost, T.P., and McClelland, W.C., 2002, Newly mapped Eocene strike-slip faults south of the Boehls Butte anorthosite, northern Idaho: Geological Society of America Abstracts with Programs, v. 34, no. 5, p. A-84.

Lewis, R.S., Burmester, R.F., McFaddan, M.D., Kauffman, J.D., Doughty, P.T., Oakley, W.L., and Frost, T.P., 2007, Geologic Map of the Headquarters 30′ × 60′ Quadrangle, Idaho: Idaho Geological Survey Digital Web Map 92, scale 1:100,000.

Ludwig, K.R., 1988, PBDAT for MS-DOS: A Computer Program for IBM-PC Compatibles for Processing Raw Pb-U-Th Isotope Data, Version 1.24: U.S. Geological Survey Open-File Report 88-542, 32 p.

Ludwig, K.R., 1991, ISOPLOT for MS-DOS: A Plotting and Regression Program for Radiogenic-Isotope Data, for IBM-PC Compatible Computers, Version 2.75: U.S. Geological Survey Open-File Report 91-445, 45 p.

Ludwig, K.R., 2003, ISOPLOT 3.00: A Geochronological Toolkit for Microsoft Excel: Berkeley Geochronology Center Special Publication 4, 70 p.

Marvin, R.F., Zartman, R.E., Obradovich, J.D., and Harrison, J.E., 1984, Geochronometric and Lead Isotope Data on Samples from the Wallace 1° × 2° Quadrangle, Montana and Idaho: U.S. Geological Survey Miscellaneous Field Studies Map MF-1354-G, 1:250,000.

Mora, C.I., Riciputi, L.R., and Cole, D.R., 1999, Short-lived oxygen diffusion during hot, deep seated meteoric alteration of anorthosite: Science, v. 286, p. 2323–2325, doi: 10.1126/science.286.5448.2323.

Nord, G.L., 1973, The Origin of the Boehls Butte Anorthosite and Related Rocks, Shoshone County, Idaho [Ph.D. dissertation]: Berkeley, University of California, 159 p.

Parrish, R.R., 1995, Thermal evolution of the southeastern Canadian Cordillera: Canadian Journal of Earth Sciences, v. 32, p. 1618–1642.

Parrish, R.R., Carr, S.D., and Parkinson, D.L., 1988, Eocene extensional tectonics and geochronology of the southern Omineca belt, British Columbia and Washington: Tectonics, v. 7, p. 181–212.

Reynolds, M.W., 1979, Character and extent of Basin-Range faulting, western Montana and east-central Idaho, in Newman, G.W., and Goode, H.D., eds., Basin and Range Symposium and Great Basin Field Conference: Denver, Rocky Mountain Association of Geologists, p. 185–193.

Reynolds, S.J., and Lister, G.S., 1990, Folding of mylonitic zones in Cordilleran metamorphic core complexes: Evidence from near the mylonitic front: Geology, v. 18, p. 216–219, doi: 10.1130/0091-7613(1990)018<0216:FOMZIC>2.3.CO;2.

Rhodes, B.P., 1986, Metamorphism of the Spokane dome mylonite zone, Priest River complex: Constraints on the tectonic evolution of northeastern Washington and northern Idaho: The Journal of Geology, v. 94, p. 539–556.

Sears, J.W., Hendrix, M., Waddell, A., Webb, B., Nixon, B., King, T., Roberts, E., and Lerman, R., 2000, Structural and stratigraphic evolution of the Rocky Mountain foreland basin in central-western Montana, in Roberts, S., and Winston, D., eds., Geologic Field Trips, Western Montana and Adjacent Areas: Missoula, Rocky Mountain Section of the Geological Society of America, p. 131–155.

Seyfert, C.K., 1984, The Clearwater core complex, a new Cordilleran metamorphic core complex and its relation to a major continental transform fault: Geological Society of America Abstracts with Programs, v. 16, no. 6, p. 651.

Sha, G., 2004, Tectonic Evolution of the Eastern and Southern Boundaries of the Boehls Butte–Clearwater Core Complex, North-Central Idaho [M.Sc. thesis]: Pullman, Washington State University, 143 p.

Sha, G.S., Watkinson, A.J., and Doughty, P.T., 2003, Exhumation of the Boehls Butte–Clearwater metamorphic core complex, north central Idaho: Heterogeneous strain distribution and compositionally controlled mylonitization: Geological Society of America Abstracts with Programs, v. 35, no. 6, p. 180.

Sha, G.S., Vervoort, J., Watkinson, A.J., Doughty, P.T., Prytulak, J., Lee, R.G., and Larson, P.B., 2004, Geochronologic constraints on the tectonic evolution of the Boehls Butte–Clearwater core complex: Evidence from 1.01 Ga garnets: Geological Society of America Abstracts with Programs, v. 36, no. 4, p. 72.

Sheriff, S.D., Sears, J.W., and Moore, J.N., 1984, Montana's Lewis and Clark fault zone; an intracratonic transform fault system: Geological Society of America Abstracts with Programs, v. 16, no. 6, p. 653–654.

Skipp, B., 1987, Basement thrust sheets in the Clearwater orogenic zone, central Idaho and western Montana: Geology, v. 15, p. 220–224, doi: 10.1130/0091-7613(1987)15<220:BTSITC>2.0.CO;2.

Spell, T.L., and McDougall, I., 2003, Characterization and calibration of $^{40}Ar/^{39}Ar$ dating standards: Chemical Geology, v. 198, p. 189–211, doi: 10.1016/S0009-2541(03)00005-6.

Stacey, J.S., and Kramers, J.D., 1975, Approximation of terrestrial lead isotope evolution by a two-stage model: Earth and Planetary Science Letters, v. 26, p. 207–221.

Steiger, R.H., and Jäger, E., 1977, Subcommission on geochronology: Convention on the use of decay constants in geo- and cosmochronology: Earth and Planetary Science Letters, v. 36, p. 359–362.

Tera, F., and Wasserburg, G.J., 1972, U-Th-Pb systematics in three Apollo basalts and the problem of initial Pb in lunar rocks: Earth and Planetary Science Letters, v. 14, p. 281–304.

Toth, M.I., and Stacey, J.S., 1992, Constraints on the Formation of the Bitterroot Lobe of the Idaho Batholith, Idaho and Montana, from U-Pb Zircon Geochronology and Feldspar Pb Isotopic Data: U.S. Geological Survey Bulletin 2008, 14 p.

Tracy, R.J., Robinson, P., and Thompson, A.B., 1976, Garnet composition and zoning in the determination of temperature and pressure of metamorphism, central Massachusetts: The American Mineralogist, v. 61, p. 762–775.

Walker, K.D., 1993, Mineralogy and Oxygen Isotope Composition of the Boehls Butte Anorthosite, Northern Idaho [M.Sc. thesis]: Knoxville, University of Tennessee, 111 p.

Wernicke, B., 1985, Uniform-sense normal simple shear of the continental lithosphere: Canadian Journal of Earth Sciences, v. 22, p. 108–125.

Williams, I.S., 1998, U-Th-Pb geochronology by ion microprobe, *in* McKibben, M.A., Shanks, W.C., III, and Ridley, W.I., eds., Applications of Microanalytical Techniques to Understanding Mineralizing Processes: Reviews in Economic Geology, v. 7, p. 1–35.

Yin, A., and Oertel, G., 1995, Strain analysis of the Ninemile fault zone, western Montana: Insights into multiply deformed regions: Tectonophysics, v. 247, p. 133–143.

Ziegler, R.D., 1991, Petrology of Garnet Amphibolites from the Boehls Butte Quadrangle, Northern Idaho [M.Sc. thesis]: Corvallis, University of Oregon, 116 p.

MANUSCRIPT ACCEPTED BY THE SOCIETY 22 MARCH 2007

Character of rigid boundaries and internal deformation of the southern Appalachian foreland fold-thrust belt

Robert D. Hatcher Jr.*
Department of Earth and Planetary Sciences and Science Alliance Center of Excellence, University of Tennessee, Knoxville, Tennessee 37996-1410, USA

Peter J. Lemiszki
Tennessee Division of Geology, 3711 Middlebrook Pike, Knoxville, Tennessee 37921-6538, USA

Jennifer B. Whisner
Department of Earth and Planetary Sciences and Science Alliance Center of Excellence, University of Tennessee, Knoxville, Tennessee 37996-1410, USA

> A realistic model for overthrust faulting must be compatible with the nature of real overthrust faults, and...with the way...displacements...occur on real faults.
> —Raymond A. Price, 1988, p. 1898

ABSTRACT

The deformed wedge of Paleozoic sedimentary rocks in the southern Appalachian foreland fold-thrust belt is defined by the configurations of the undeformed basement surface below and the base of the Blue Ridge–Piedmont megathrust sheet above, together with the topographic free surface above the thrust belt. The base of the Blue Ridge–Piedmont sheet and undeformed basement surface have been contoured using industry, academic, and U.S. and state geological survey seismic-reflection and surface geologic data. These data reveal that the basement surface dips gently SE in the Tennessee embayment from Virginia to Georgia, and it contains several previously unrecognized normal faults and an increase in dip on the basement surface, which produces a topographic gradient. The basement surface is broken by many normal faults beneath the exposed southern Appalachian foreland fold-thrust belt in western Georgia and Alabama closer to the margin and beneath the Blue Ridge–Piedmont sheet in Georgia and the Carolinas. Our reconstructions indicate that small-displacement normal faults form beheaded basins over which thrust sheets were not deflected, whereas large-displacement normal faults (e.g., Tusquittee fault) localized regional facies changes in the early Paleozoic section and major Alleghanian (Permian) structures. These basement structures correlate with major changes in southern Appalachian foreland fold-thrust belt structural style from Virginia to Alabama.

---

*bobmap@utk.edu

Hatcher, R.D., Jr., Lemiszki, P.J., and Whisner, J.B., 2007, Character of rigid boundaries and internal deformation of the southern Appalachian foreland fold-thrust belt, *in* Sears, J.W., Harms, T.A., and Evenchick, C.A., eds., Whence the Mountains? Inquiries into the Evolution of Orogenic Systems: A Volume in Honor of Raymond A. Price: Geological Society of America Special Paper 433, p. 243–276, doi: 10.1130/2007.2433(12). For permission to copy, contact editing@geosociety.org. ©2007 The Geological Society of America. All rights reserved.

Several previously unrecognized structures along the base of the Blue Ridge–Piedmont sheet have been interpreted from our reconstructions. Large frontal duplexes composed of rifted-margin clastic and platform rocks obliquely overridden along the leading edge of the Blue Ridge–Piedmont sheet are traceable for many kilometers beneath the sheet. Several domes within the Blue Ridge–Piedmont sheet also likely formed by footwall duplexing of platform sedimentary rocks beneath, which then arched the overlying thrust sheet. The thickness and westward limit of the Blue Ridge–Piedmont sheet were estimated from the distribution of low-grade footwall metamorphic rocks, which were observed in reentrants in Georgia and southwestern Virginia, but are not present in simple windows in Tennessee. These indicate that the original extent of the sheet is near its present-day trace, whereas in Georgia, it may have extended some 30 km farther west.

The southern Appalachian foreland fold-thrust belt consists mostly of a stack of westward-vergent, mostly thin-skinned thrusts that propagated westward into progressively younger units as the Blue Ridge–Piedmont sheet advanced westward as a rigid indenter, while a few in northeastern Tennessee and southwestern Virginia involved basement. Additional boundary conditions include temperatures <300 °C and pressures <300 MPa over most of the belt. The southern Appalachian foreland fold-thrust belt thrusts, including the Blue Ridge–Piedmont megathrust sheet, reach >350 km displacement in Tennessee and decrease both in displacement and numbers to the SW and NE.

Much of the Neoproterozoic to Early Cambrian rifted-margin succession was deformed and metamorphosed during the Taconic orogeny, and it is considered part of the rigid indenter. Only the westernmost rocks of the rifted-margin succession exhibit ideal thin-skinned behavior and thus are part of the southern Appalachian foreland fold-thrust belt. Palinspastic reconstructions, unequal thrust displacements, and curved particle trajectories suggest that deformation of the belt did not occur by plane strain in an orogen that curves through a 30° arc from northern Georgia to SW Virginia. Despite the balance of many two-dimensional cross sections, the absence of plane strain diminishes their usefulness in quantifying particle trajectories. Coulomb behavior characterizes most individual faults, but Chapple's perfectly plastic rheology for the entire thrust belt better addresses the particle trajectory problem. Neither, however, addresses problems such as the mechanics of fault localization, out-of-sequence thrusts, duplex formation, three-dimensional transport, and other southern Appalachian foreland fold-thrust belt attributes.

Keywords: Appalachians, foreland thrust-belt, fault mechanics, nonplane strain.

INTRODUCTION

Foreland fold-thrust belts are components of collisional orogens worldwide. Besides being thin-skinned, these belts are characterized by: listric thrusts that rise from a basal weak zone near the basement-cover contact; faults that ramp more steeply across strong units to higher detachments, or to the surface; and an initial undeformed, continentward-thinning sedimentary wedge. Furthermore, mechanical models for thrust-belt growth require that the wedge shape be maintained as it is shortened and thickened into the deformed state (Chapple, 1978; Dahlen, 1990). Because intrinsic and extrinsic space-time variables exist, important individual differences and along-strike variations occur in foreland fold-thrust belts. Structural style is determined by intrinsic variables including stratigraphic parameters, anisotropy, bonding between lithologies, fluids, original shapes of the continental margin and platform, shape of the indenter, and basement irregularities. Thrust spacing initially appears to be related to stratigraphic thickness, but this may be an oversimplification. Extrinsic variables relate to the deformational history of a region and include changes in burial depth, temperature, fluid pressure, and strain rate during thrusting, along with the lithospheric thermal regime at the time of collision. These and other variables give rise to parts of foreland fold-thrust belts that are fold dominated and contain relatively few faults (e.g., Jura and MacKenzie Mountains, Alabama and central Appalachians), while other segments are thrust dominated and contain several master faults and numerous subsidiary faults (e.g., southern Canadian Rockies, Bergamesc Alps, Tennessee southern Appalachians). Our purpose here is to discuss the regional shapes of the bounding surfaces (fixed boundary conditions) that existed during the development of the southern Appalachian foreland fold-thrust belt and their influence on the final shape of this part of the thrust belt. We also intend to explore these influences and to examine the deformed volume from an orogen-scale perspective to better understand the kinematics and mechanics that produced the southern Appalachian foreland fold-thrust belt that we see today.

Thrust-sheet deformation can occur by folding and faulting, with local internal pressure-solution strain (layer-parallel shortening) (e.g., Geiser and Engelder, 1983). The concept of structural-lithic units (SLU) has been applied to the southern part of the southern Appalachian foreland fold-thrust belt at different scales to emphasize the interplay between structural style and deforming unit properties (Willis, 1893; Currie et al., 1962; Woodward, 1988). Specifically, Currie et al. (1962) defined a structural-lithic unit as part of a stratigraphic section that has a characteristic response to deformation. The strain mechanism, however, may also change for a particular structural-lithic unit, because deformational conditions change through time. As a result of interactions between these and other variables (see previous paragraphs), numerous exceptions exist to the predicted deformational style exhibited by a particular rock unit or segment of a foreland fold-thrust belt, and make it difficult to apply the structural-lithic unit concept at the thrust-system scale. Geometric models based on relatively few variables are appealing, but they are ultimately not useful for detailed analysis because they oversimplify the system and yield unrealistic structures (see, for example, Ramsay, 1991).

The boundaries considered to be most fixed in many foreland fold-thrust belts (e.g., southern and central Appalachians, Alps, Scandinavian Caledonides) are the shapes of the bounding surfaces—the base of the rigid indenter and the top of the rigid basement (Fig. 1). The shapes of these surfaces can potentially influence not only structures that form near the indenter front and near the basement surface, but, because they are so fundamental, they also can influence structures that form below the free (topographic) surface within the deforming wedge far from the boundaries. Therefore, many tectonic structures in a thrust belt are the products of superimposed local- and orogen-scale stresses within the deforming wedge that are generated by the rigid bounding surfaces. When reconstructing the structural evolution of a thrust system, the conditions at the boundaries and within the deforming wedge should be identified. Subsurface data were examined to determine the shape of the basement surface beneath the southern Appalachian foreland fold-thrust belt and the present shape of the Blue Ridge–Piedmont megathrust indenter. Pressure and temperature data were used to evaluate the forelandward extent of Blue Ridge–Piedmont sheet loading, to better outline the original indenter shape, and the evolving eastward extent of the free surface of the wedge. These results will influence present and future three dimensional (3-D) reconstructions of the southern Appalachian foreland fold-thrust belt.

The southern Appalachian foreland fold-thrust belt (Fig. 2) consists of a wedge-shaped stack of mostly west-vergent, mostly thin-skinned thrusts located above the undeformed Mesoproterozoic basement (lower bounding surface) that occur in the exposed foreland and beneath the Blue Ridge–Piedmont composite crystalline megathrust sheet (upper bounding surface) from Alabama to Maryland. The Blue Ridge–Piedmont sheet was the rigid indenter that deformed the foreland during the Carboniferous–Permian Alleghanian orogeny, and it consists of rifted-margin and platform sedimentary rocks and some Mesoproterozoic basement. The third boundary is the unconfined topographic free surface (Fig. 1). The classic Appalachian foreland fold-thrust belt at first appears amenable to two-dimensional (2-D) balanced cross-section analysis and fits current models for foreland fold-thrust belt development (e.g., Davis et al., 1983; Dahlen, 1990). The southern Appalachians are a strongly curved, thrust-dominated orogen that describes a convex-west arc of 30° from northern Georgia through Tennessee into southwestern Virginia (Fig. 3); a similar arc exists in Pennsylvania (Geiser, 1988; Geiser *in* Hatcher et al., 1989a; Hatcher et al., 1990; Faill, 1998). Across-strike thrust populations vary from three to five across the belt in NE Alabama and SW Virginia to 12 across the Tennessee segment (Fig. 2A). The segment with the greatest number of thrusts is also the one that contains the thrusts with the largest displacements: >50 km on at least three (Fig. 4), with a restorable displacement of >350 km on the Blue Ridge–Piedmont sheet.

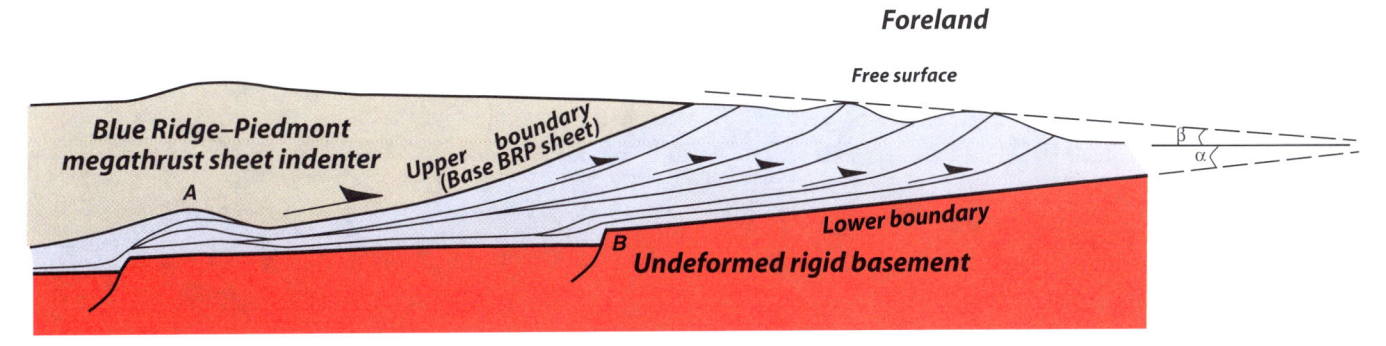

Figure 1. Primary boundaries of foreland fold-thrust belts, using the southern Appalachians as an example. Additional boundary conditions include temperatures <300 °C and pressures <400 MPa. (A) Ancient rift fault beneath the crystalline indenter nucleates an antiformal stack duplex that arches the crystalline sheet. (B) Rift fault beneath the platform nucleates a foreland fold-thrust belt thrust. BRP—Blue Ridge–Piedmont megathrust sheet; α—dip on basement beneath foreland fold-thrust belt; β—topographic slope angle.

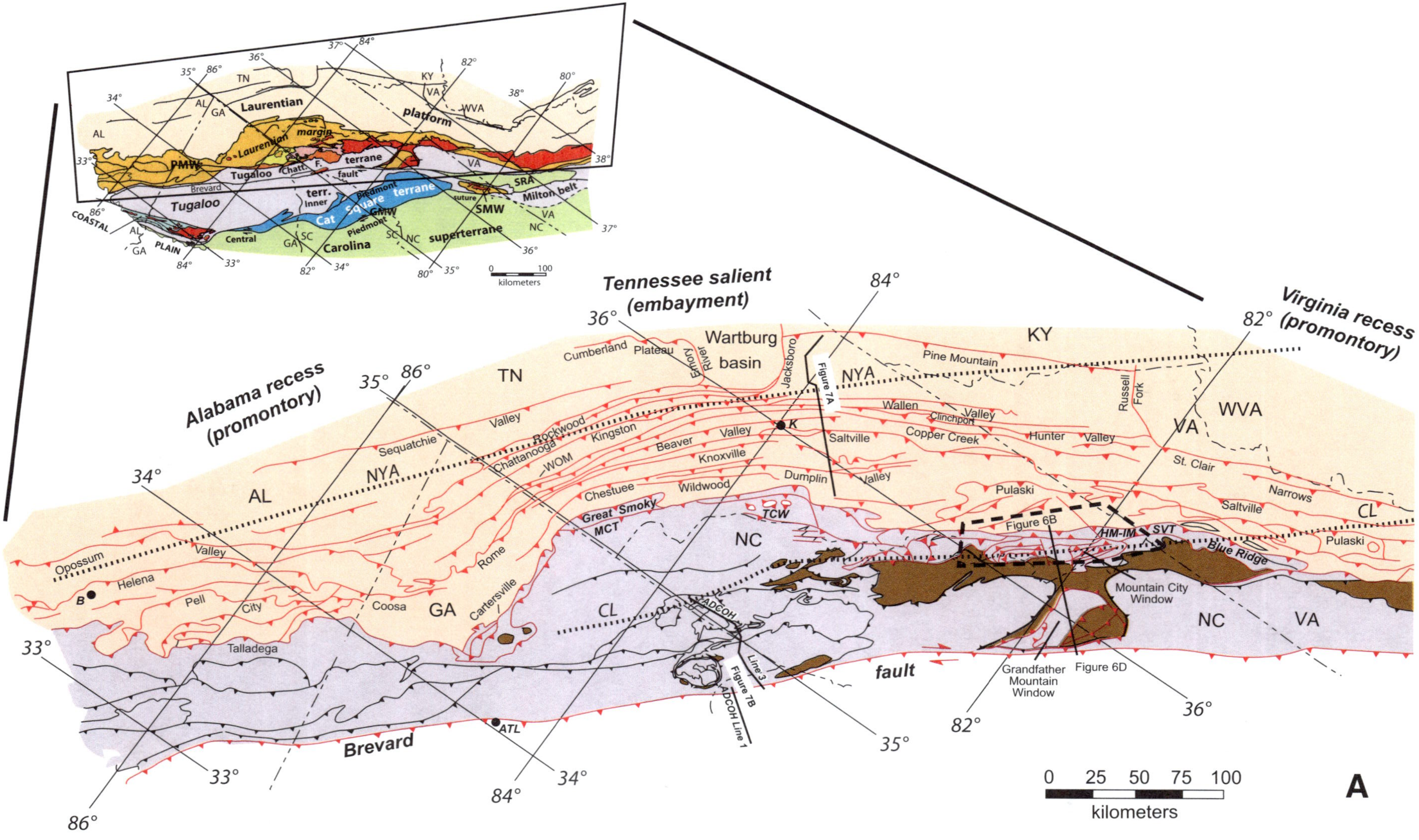

Figure 2 (on this and following page). (A) Alleghanian faults (Pennsylvanian-Permian; red) in the southern Appalachian foreland fold-thrust belt; names of faults discussed in the text are indicated, along with several others. Older faults are shown in black. HM-IM—Holston Mountain–Iron Mountain fault (underlies the Shady Valley thrust sheet; see Fig. 6); MCT—Miller Cove thrust; SVT—Shady Valley thrust sheet; TCW—Tuckaleechee Cove (center), Cades Cove (to west), and Wear Cove (to east) windows; WOM—Whiteoak Mountain fault; NYA—New York-Alabama magnetic lineament of King and Zietz (1978); CL—Clingman magnetic lineament of Nelson and Zietz (1983). Abbreviations and unlabeled colors in index map: PMW—Pine Mountain window; SRA—Smith River allochthon; SMW—Sauratown Mountains window; GMW—Grandfather Mountain window; red—Middle Proterozoic Grenville basement; yellow—Cowrock terrane; pink—Cartoogechaye terrane. Note that all of the Laurentian margin rocks, except the most frontal thrust sheet, were deformed and metamorphosed during earlier events, so they consist of Paleozoic basement and, together with the pre-Paleozoic components, make up the indenter in the Alleghanian thrust system. Lavender—Blue Ridge–Piedmont sheet NW of the Brevard fault. Brown areas in the Blue Ridge and Piedmont indenter consists of Grenville basement plus Paleozoic basement generated during Taconian, Neoacadian, and early Alleghanian deformation and metamorphism (Hatcher, 1999). Tan—southern Appalachian foreland fold-thrust belt. Black toothed lines are earlier thrusts. ATL—Atlanta; B—Birmingham; K—Knoxville. Figure was modified from Hatcher et al. (1990).

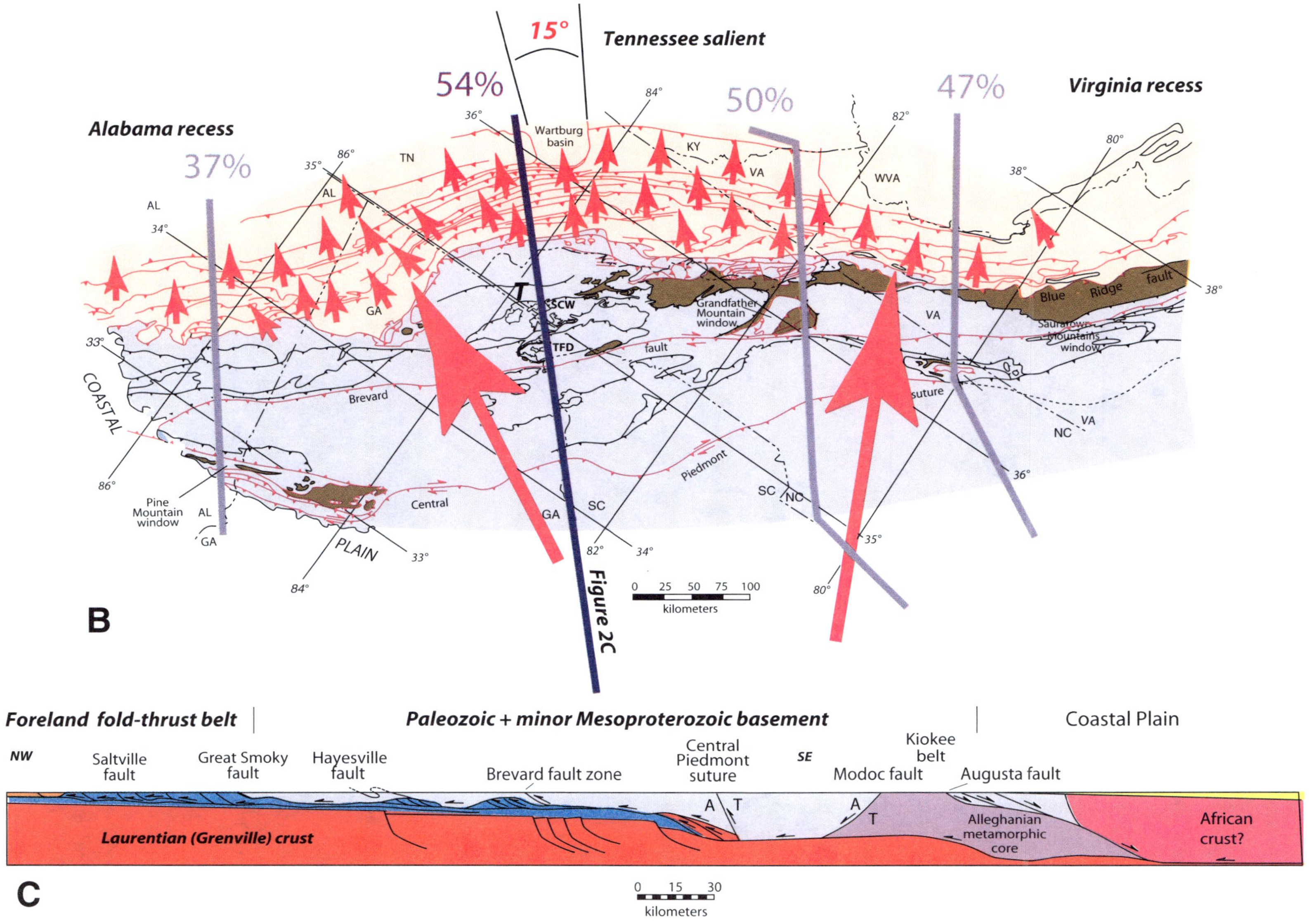

Figure 2 (*continued*). (B) Simplified tectonic map of the southern Appalachians. Shortening (in percent) is shown along four restored cross sections that extend from the Cumberland-Allegheny Plateau to the Coastal Plain (see Hatcher et al., 1989b, 1990, for all sections). Large arrows indicate transport directions, particularly within the Blue Ridge–Piedmont composite crystalline thrust sheet; those along the westernmost Cumberland Plateau and Pine Mountain blocks were determined by Kilsdonk and Wiltschko (1988). Small arrows represent normal to the hinges of narrow macroscopic folds (compiled from published state geologic maps of Alabama [Osborne et al., 1989], Georgia [Pickering, 1976], Tennessee [Hardeman, 1966], and Virginia [Johnson, 1993]). The 15° divergence between the traces of the Jacksboro and Emory River faults may indicate both divergence in transport direction and the orientation of lateral ramps. Colors are the same as in A. T—location of Tusquittee, North Carolina; SCW—Shooting Creek window; TFD—Tallulah Falls dome. (C) Cross section with >380 km shortening illustrating thin-skinned structural style for Alleghanian deformation across the southern Appalachians; section indicated in B with 54% shortening (simplified from section B–B′ in Hatcher et al., 1990). A—away from observer; T—toward observer.

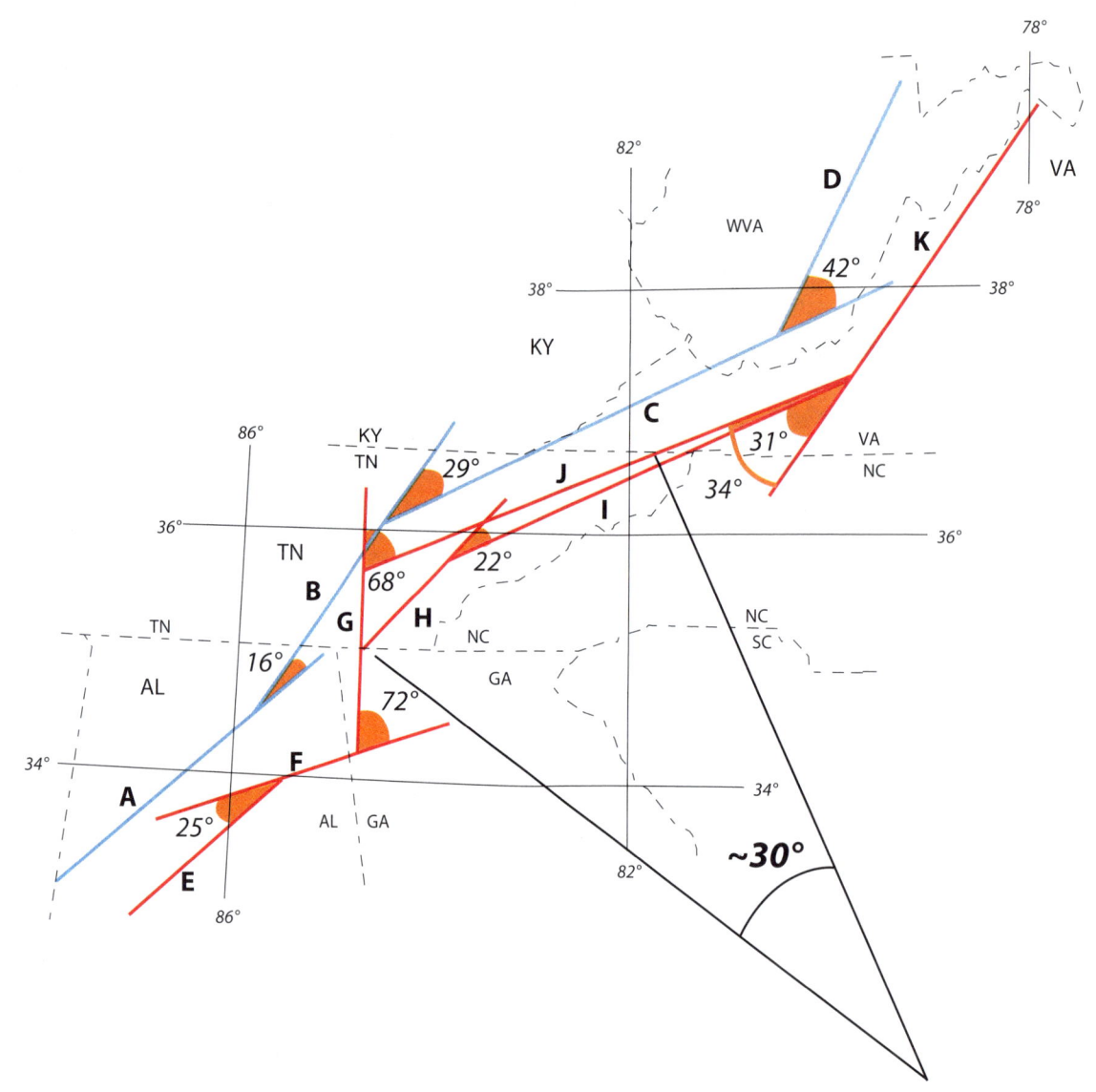

Figure 3. Angles between curved segments within the southern Appalachians. Red lines represent segments of the Blue Ridge front (Talladega, Cartersville, Great Smoky, Miller Cove, Holston–Iron Mountain [Shady Valley sheet], and Blue Ridge faults); blue lines represent segments measured along the eastern edge of the Cumberland (Allegheny) Plateau, also a structural front. Note that, if the updip projection of the Blue Ridge–Piedmont sheet is implemented as suggested in Figure 8, the angle on the front from Georgia to Virginia is 75°–80°, and not ~105° as it is at the present erosion level. The arc of the entire Tennessee salient is ~30°, measured from the inflection point in Georgia to that in Virginia. Also note that the Georgia segment is truly curved, but the Virginia transition consists of overlap between two straight segments (C–D and J–K). A—Alabama Plateau; B—Alabama-Tennessee Plateau; C—Tennessee-Virginia Plateau; D—West Virginia Plateau; E—Alabama Blue Ridge; F—Alabama-Georgia Blue Ridge; G—Georgia-Tennessee Blue Ridge; H—Tennessee-Georgia Blue Ridge; I—Tennessee-Virginia Blue Ridge; J—Tennessee-Virginia Blue Ridge; K—Virginia Blue Ridge.

Properly delimiting the shapes of the rigid bounding surfaces is important when restoring a curved thrust belt. Early attempts to restore the southern Appalachian foreland fold-thrust belt relied on surface geologic data and limited subsurface control to construct relatively unconstrained 2-D cross sections (Rodgers, 1964; Milici, 1970; Harris and Milici, 1977; Roeder et al., 1978; Woodward, 1985). Palinspastic reconstruction of a curved thrust system, however, cannot be accomplished with "balanced" 2-D cross sections, because parts of converging dip sections restore to the same area (Geiser, 1988; Hindle and Burkhard, 1999), yielding a 3-D room problem that requires changes in line length (or out-of-section material volume) from undeformed to the deformed states (Marshak, 1988). Therefore, palinspastic reconstruction of the southern Appalachian foreland fold-thrust belt requires a priori evaluation of all factors influencing thrust-belt curvature.

Figure 4. SE-dipping Saltville thrust exposed in road-cut on Broadway Avenue 100 m north of I-640 in Knoxville, Tennessee. Light-colored carbonate in footwall is upper Knox Group dolomite; multicolored unit in the hanging wall is the Lower Cambrian Rome Formation. Steep dip here is related to later arching of the hanging wall as other thrust sheets propagated NW of the Saltville. (Photo was taken in 1978; Ray Price is in yellow rain jacket in foreground.)

The problem of reconstruction of curved segments of orogens has been addressed in a number of studies, but without a satisfactory solution (e.g., Geiser, 1988; Laubscher, 1988; Marshak, 1988, 2004). A kinematic solution may be achieved by adherence to principles of conservation of volume, but a mechanical solution is not possible without assuming a rheology for body deformation of the rock mass (Chapple, 1978). Some success with kinematics has been achieved by estimating body strain (e.g., with the calcite strain gauge), which has yielded the conclusion that transport-parallel simple shear (line length not constant), rather than curved particle paths, is the dominant mechanism in part of the northern Subalpine Chain in France (Ferrill and Groshong, 1993). The Jura Mountains and Helvetic Alps, however, display a radiating pattern of transport vectors (Platt et al., 1989; Hindle and Burkhard, 1999). Several variables influence thrust-belt curvature: (1) shape of the indenter; (2) basement obstacles; and (3) the number of deformation events. Indenter shape exerts a strong control on the amount of local shortening in the foreland and influences development of the regional stress field, both of which are manifested in the curvature of the thrust belt. For example, the shape of the southern Appalachian foreland fold-thrust belt generally mimics the geometry of the Blue Ridge–Piedmont sheet front (Figs. 2B and 3). The map view shape of the Blue Ridge–Piedmont sheet has been affected by erosion, so additional data, such as burial-depth estimates of foreland rock units, are needed to correct for this. Foreland basement obstacles that existed prior to or developed during thrusting are considered to be primary controls that can influence thrust-belt curvature (Macedo and Marshak, 1999). Evaluation of primary basement controls of southern Appalachian foreland fold-thrust belt curvature was not possible before now for lack of subsurface data. Local stratigraphic anomalies that correlate with basement features may become piercing points for thrust-system reconstructions. Even if individual thrust-sheet displacements are poorly known, the magnitude and direction of shortening for a portion of the thrust system may then be identified and provide an important limit for reconstructions.

Multiple deformation events manifested as changes in the regional shortening direction through time have been documented in the central and southern Appalachians (Roeder, 1975; Geiser and Engelder, 1983; Evans, 1994). Roeder (1975), using local structural analysis, argued for polyphase thrusting in the Tennessee portion of the southern Appalachian foreland fold-thrust belt, but his conclusions remain unsubstantiated. Bartholomew and Schultz (1980) observed coaxial superposed mesoscopic folds in the Pulaski thrust sheet in southwestern Virginia. These studies emphasize the difficulty of separating local overprinting from incremental variations in regional shortening direction.

Deformation of the Appalachian and other foreland fold-thrust belts at outcrop scale is clearly brittle and involves: fold mechanisms, such as buckling and flexural slip; brittle faulting, which produces cataclasis, rigid-body rotation and translation, and contractional and extensional faulting (Norris, 1958; Harris and Milici, 1977); and other fracturing mechanisms. The presence of semiductile deformation mechanisms, such as flexural flow, cleavage, and body strain, suggests that these mechanisms also contribute to overall foreland fold-thrust belt deformation. Chapple (1978) formulated a model for foreland fold-thrust belts based on plastic rheology. Today, based on the papers by Davis et al. (1983), Dahlen (1990), and others, most assume that the deformation in foreland fold-thrust belts occurs by brittle (Coulomb) processes. Elliott (1976) suggested that thrust faults propagate as a "ductile bead" along the tip line of a propagating thrust, and the displacement on a thrust propagates incrementally as a smeared-out (Somigliana) dislocation (Price, 1988), reintroducing the possibility of ductile deformation at meso- and microscales. Wojtal (1992), using southern Appalachian foreland fold-thrust belt examples, constructed a model for continuous flow in mesoscale thrust zones involving both plane and nonplane strain that assumed a viscous rheology on the scale of loose lines in individual thrust sheets. Woodward (1987) also pointed out the difficulties of applying critical-wedge concepts developed in sandbox experiments and accretionary complexes to foreland fold-thrust belts—in this case, models developed for brittle deformation.

TECTONIC HISTORY AND PRESENT-DAY STRUCTURAL FRONTS

The southern Appalachian foreland fold-thrust belt has a sinuous trace that extends from the Virginia recess through the Tennessee salient to the Alabama recess, and these features had to have evolved from promontories and embayments in the pre-Appalachian continental margin (Thomas, 1977, 1991) (Fig. 2B). The western deformation limit of the belt occurs in the Cumberland-Allegheny Plateau (e.g., Hatcher et al., 1990), and the SE border of the exposed belt occurs at the present-day surface along the trace of the Blue Ridge–Piedmont sheet. The belt contains a stratigraphic succession that formed after the late Neoproterozoic to early Paleozoic breakup of Rodinia followed by sediment accumulation along the irregular Laurentian margin. This rifted margin evolved into a carbonate platform (shelf) that, by the late Early Cambrian, faced the open Iapetus ocean. By the end of Early Ordovician time, the carbonate platform in most of eastern North America was exposed by a eustatic sea-level drop, was subaerially eroded, producing as much as 150 m of topographic relief and extensive karst topography, and then was submerged again with renewed carbonate deposition. The Neoproterozoic to Ordovician margin was inverted, and the eastern carbonate platform was buried in the Middle Ordovician by formation of a foredeep basin. This basin was filled by a westward-advancing clastic wedge derived from uplifted parts of the eastern platform as Taconian arcs and deep-water sediments collided with, were obducted onto, and loaded the outer margin. Clastic-wedge deposition persisted into the Silurian (molasse) until the southern Appalachian platform was uplifted. Erosion removed most Middle Silurian to Middle Devonian deposits, but clastic deposition resumed in the Late Devonian and continued into the early Mississippian (Ettensohn, 2004). These units probably represent the Neoacadian clastic wedge in the southern Appalachians, which are capped by middle Mississippian carbonates. The Pennsylvanian-Permian Alleghanian clastic wedge advanced westward across the foreland and into the Appalachian foreland basin as final construction of the orogen began during the north-to-south zippered joining of Africa with Laurentia (Hatcher, 2002). More detailed summaries of the Paleozoic tectonic history of the U.S. Appalachians can be found in Hatcher et al. (1989b).

The Blue Ridge–Piedmont crystalline sheet master décollement formed within the ductile-brittle transition in previously deformed and metamorphosed Paleozoic crust, and propagated upward into bedding thrusts in the unmetamorphosed weak Lower Cambrian clastic rocks (Sandsuck Formation and Chilhowee Group). The master décollement then propagated into the clastic-evaporite succession (Rome Formation), which rests on Grenvillian basement. Middle- and upper-level detachments in the foreland that formed in Cambrian, Ordovician, Devonian-Mississippian, and Pennsylvanian shale and coal are linked by ramps that formed in thick strong units of massive Cambrian-Ordovician carbonate and younger sandstone. The assembled rifted margin, carbonate platform, and eastward-thickening clastic wedges were driven westward in snowplow fashion in front of the advancing Blue Ridge–Piedmont crystalline thrust-sheet indenter (Figs. 1, 2, and 5). Folded faults and map patterns support the contention that thrusts propagated outward from the interior of the orogen, but Milici (1975) used the truncating relationships among thrusts in the Tennessee Valley and Ridge to conclude that thrusts propagated from the foreland toward the hinterland. The presence of a continuous sequence

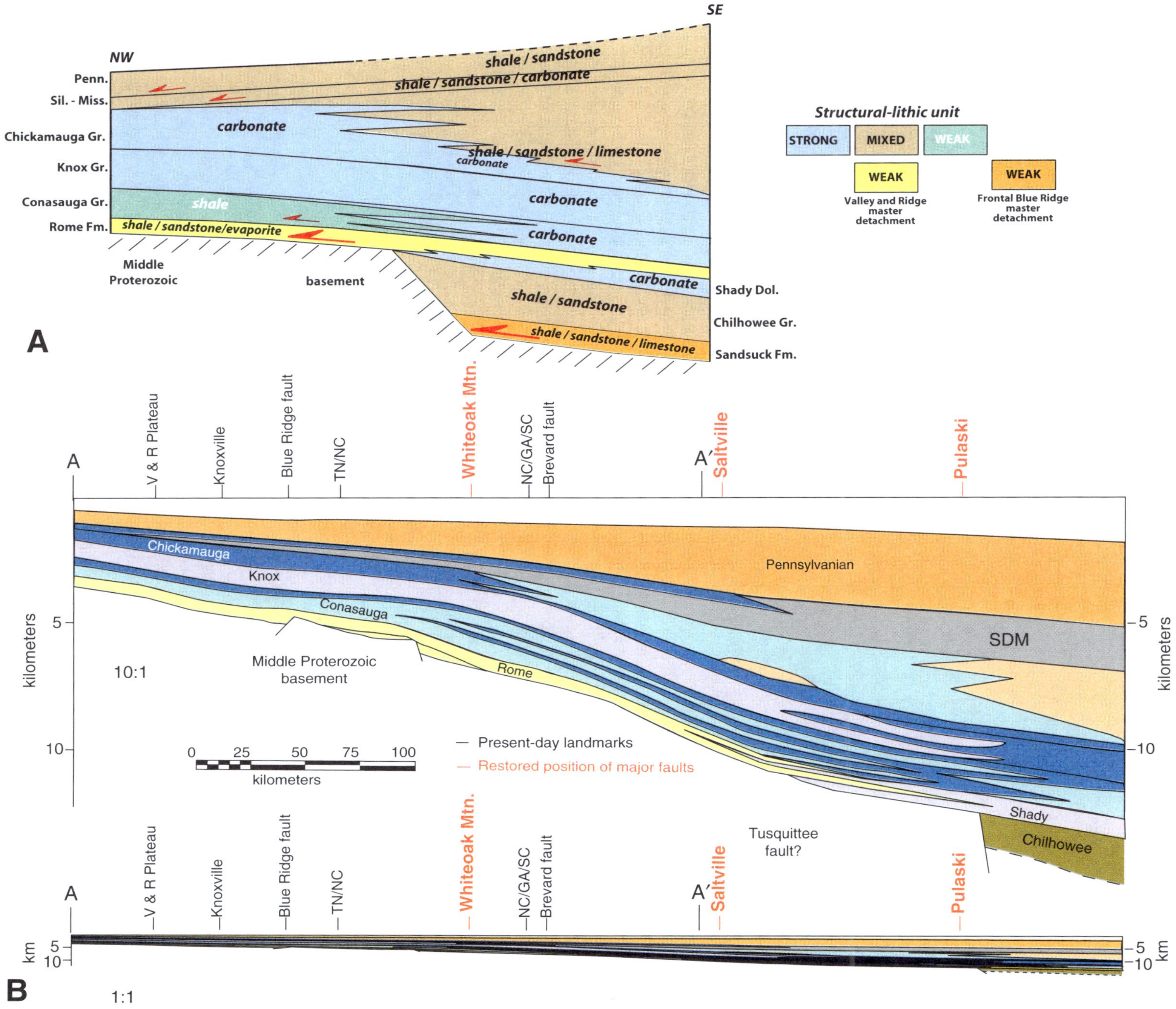

Figure 5. (A) Simplified restored stratigraphic section through the Tennessee southern Appalachian foreland fold-thrust belt. Dip section shows location of detachment zones, and definition of structural-lithic units is based on changes in lithology of the depositional sequence. Red arrows indicate regional detachment units. (B) Palinspastic facies diagram across the southern Appalachian foreland fold-thrust belt is reconstructed approximately at the latitude of Knoxville, Tennessee, quantitatively reconstructed by retrodeforming major faults and facies boundaries. The Pulaski fault is projected southward into the section. Blue—limestone facies; lavender—dolomite facies; green—shale facies; light orange—coarse clastics and turbidites; gray—siltstone facies; light yellow—sandstone, shale, and dolomite intertidal facies. SDM—Silurian, Devonian, and Mississippian rocks.

from crystalline basement into the Ordovician (Knox) carbonate in the Shady Valley syncline in the northeastern Valley and Ridge of Tennessee (Hardeman, 1966) suggests that other processes, such as out-of-sequence thrusting, may have created most of these truncations (Woodward, 1985).

The shape of the Blue Ridge structural front, defined by changes in strike of the (Blue Ridge–Piedmont) frontal thrust, is quite variable from Virginia to Alabama (Fig. 2A). Smaller variations occur along the Cumberland-Allegheny Plateau front, but the changes in trend are subparallel and bound an arcuate fold-thrust belt. From Tennessee northeastward into the Virginia recess, the strike of Valley and Ridge structures rotates through an arc of 30° (Fig. 3). Rodgers (1970) pointed out that the change in strike does not take place smoothly. Instead, central and southern Appalachian structures interfinger and appear to overprint each other in a broad displacement transfer zone in southwestern Virginia (Couzens et al., 1993; Spraggins and Dunne, 2002). These strike changes in the southern Appalachians are smoothly curving arcs with little to no surface expression of structures that developed to accommodate the change in shortening direction. Thomas (1991) suggested that cratonward thrust propagation was greater through a thicker sedimentary wedge deposited in the Tennessee and Pennsylvania embayments compared with the thinner succession deposited on the promontories. This affected the sinuous trace of the southern Appalachian foreland fold-thrust belt, so that salients of the fold-thrust belt coincide with embayments and recesses coincide with promontories in the ancient continental margin. While this concept may explain part of the structural relationships in salients and recesses, most of the rifted-margin succession was penetratively deformed and metamorphosed earlier, so by the Alleghanian, this assemblage was Paleozoic basement, and the master detachment propagated through it along the ductile-brittle transition.

A substantial across-strike change in structural style within the southern Appalachian foreland fold-thrust belt occurs at the boundary between the Cumberland-Allegheny Plateau and the Valley and Ridge. The front is defined either by a fault, foreland syncline, or a frontal monocline. This frontal structure locally has a triangle-zone geometry (Gwinn, 1964; Gordy et al., 1977; Jones, 1987, 1996; Hatcher and Whisner, 1998). The boundary between the Valley and Ridge and the Blue Ridge coincides approximately with the northwestern limit of Mesoproterozoic basement rocks in Virginia and late Neoproterozoic rift-fill sedimentary and volcanic rocks farther south in the hanging wall of the Blue Ridge–Piedmont thrust sheet. The Blue Ridge front consists of a series of faults named (from SW to NE): Talladega, Cartersville, Great Smoky, Miller Cove, Holston–Iron Mountain, and Blue Ridge faults (Fig. 2A).

REGIONAL STRATIGRAPHIC FRAMEWORK

The entire sedimentary sequence in the southern Appalachian foreland fold-thrust belt has an eastward-thickening wedge-shaped geometry below which the autochthonous Mesoproterozoic basement dips SE toward and beneath the Blue Ridge–Piedmont crystalline thrust sheet (Figs. 2C and 5). Throughout most of the southern Appalachian foreland fold-thrust belt, the basal décollement propagated through the Early to Middle Cambrian rifted-margin clastic rocks immediately above the basement (Rome Formation and Conasauga Group) (Rodgers, 1953; Milici, 1975; Thomas, 1988) (Fig. 5). The Rome Formation is composed of shale and siltstone, with lesser amounts of sandstone, dolostone, limestone, and evaporite. This sequence had a westward clastic source, and it becomes more carbonate-rich toward the east. Although the exact thickness of the Rome Formation is difficult to determine because its base is mostly faulted across the exposed southern Appalachian foreland fold-thrust belt, isopach maps indicate that the thickness reaches 600 m (Harris and Milici, 1977; Rankin et al., 1989). The basement was active during the time when the Rome Formation was deposited because sedimentation was affected by normal faults bounding the subsurface Rome trough in Kentucky (Gao et al., 2000) and by basement faulting in Alabama (Thomas, 1986, 1991; Thomas and Bayona, 2005).

Conasauga Group facies also change regionally from clastic-rich to the west to carbonate-rich to the east, with thickness and lithologic variations across the Appalachian basin (Rodgers, 1953) (Fig. 5). Palinspastic maps indicate a general thickness increase toward the SE. A local depocenter has been identified in the middle of the Tennessee salient (Luttrell basin of Hasson and Haase, 1988). These major thickness and lithologic changes in the Early (Rome) to Middle Cambrian (Conasauga) clastic interval may reflect depositional responses to basement highs and lows.

The Late Cambrian to Early Ordovician Knox Group extends across the entire southern Appalachian foreland fold-thrust belt and is the mechanically strong stratigraphic unit in the carbonate sequence. The carbonate sequence thickens at the base (eastward) where older Conasauga Group facies become carbonate rich, and also thickens at the top where the younger Chickamauga Group is dominantly carbonate (Fig. 5). The Knox Group in Tennessee may also contain a local thickness increase related to the Luttrell basin (Rankin et al., 1989). Greater carbonate-sequence thickness, which produced greater mechanical strength, localized ramps between lower and upper detachments and formed the strut for southern Appalachian foreland fold-thrust belt thrust sheets.

A middle detachment exists in mudstone-shale–rich intervals in the Middle Ordovician Chickamauga Group. Near the Blue Ridge front, the detachment occurs in the shaly Lenoir Limestone or the overlying Sevier Shale, but farther NW, the detachment is in the Moccasin Formation (top of Stones River [Black River] Group). An important upper detachment horizon in northeastern Tennessee and Virginia is the interval including the Devonian-Mississippian Chattanooga Shale and Silurian clastic rocks. The Chattanooga Shale thins from ~700 m in southwestern Virginia and 300 m in northeastern Tennessee to 16 m farther west along the Tennessee-Kentucky border (Harris and Milici, 1977). As a result, the Pine Mountain fault, which ramps into the Chattanooga Shale (Rich, 1934), locally steps down stratigraphically toward the SW into Silurian clastic rocks (Harris, 1970; Milici, 1975). Farther SW in the Sequatchie

Valley–Cumberland Plateau overthrust sheet, lower Mississippian Grainger Formation and early Pennsylvanian Gizzard Group shale and coal act as upper detachment surfaces, presumably because the Chattanooga to Silurian clastic interval is too thin (Wilson and Stearns, 1958; Harris, 1970). The unfaulted Wartburg basin separates the Pine Mountain thrust sheet from the Cumberland Plateau overthrust sheet (Fig. 2A).

REGIONAL STRUCTURAL FRAMEWORK

The southern Appalachian foreland fold-thrust belt is a classic thin-skinned foreland fold-thrust belt (Fig. 1). The anisotropy produced by stratigraphic units with contrasting strength and/or thickness was recognized early as a major influence on both fold and fault style (Hayes, 1891; Willis, 1893; Butts, 1927; Rich, 1934; Suppe, 1983). Hayes (1891), working in the southern Appalachian foreland fold-thrust belt in Georgia, was one of the first to attempt palinspastic reconstruction of thrust sheets. Butts (1927), and later Rich (1934), however, used the deformation style of the Pine Mountain thrust sheet to explain bedding-parallel and bedding-oblique thrust faults (ramps) in sedimentary strata. Rich's ramp-flat thrust model has been applied with great success both to the Appalachians and other thrust belts (Bally et al., 1966; Dahlstrom, 1970; Harris, 1970; Price and Mountjoy, 1970; Royse et al., 1975; Boyer and Elliott, 1982). The Blue Ridge–Piedmont indenter thus consists of basement—crust consisting of rocks deformed and metamorphosed during a previous orogeny (Hatcher, 1999). This includes all rocks penetratively deformed and metamorphosed during the Grenvillian, Taconian, Neoacadian, and early Alleghanian orogenies.

The southern Appalachian foreland fold-thrust belt is composed of three to more than ten major thrust sheets (Fig. 2A), where several master faults (e.g., Rome-Saltville, Copper Creek, Whiteoak Mountain–Clinchport–Hunter Valley–Wallen Valley, Pulaski) are traceable along strike for hundreds of kilometers. Thrust faults generally strike NE, dip toward the SE, and are responsible for over 120 km of shortening (Roeder et al., 1978). Several major faults end in the Virginia recess and increase horizontal displacement and stratigraphic separation southwestward from zero to several kilometers. Within the Virginia recess, Carboniferous rocks are preserved in synclines in several major thrust sheets and in windows (Johnson, 1993). In the Tennessee salient, 12 subparallel thrust faults crop out in an across-strike distance of only 50 km, recording more than 300 km of shortening. Southward from the Tennessee salient, the thrust belt changes character as: (1) the number of thrust faults and the displacement on them decrease; (2) Carboniferous rocks are again abundant in footwall synclines; and (3) folds are again an important part of the structural style.

Direct evidence for the influence of basement features on thrust-belt development is sparse (Wiltschko *in* Hatcher et al., 1989a), although the location of thrust ramps may be influenced by pre-existing normal faults that form steps in the basement surface (Wiltschko and Eastman, 1983; Thomas, 1986; Schedl and Wiltschko, 1987). Jacobeen and Kanes (1974) documented a relationship between basement geometry and thrust-ramp position in the central Appalachians. Previous data for the depth and geometry of the basement surface beneath the southern Appalachian foreland fold-thrust belt consisted of a few drill holes; short, isolated seismic-reflection profiles; and depth estimates from magnetic surveys (e.g., Watkins, 1964). In general, the top of the basement was considered to be a smooth, gently SE-dipping surface (Milici, 1970, his Figure 3; Harris, 1979; Roeder et al., 1978). Probably the most conclusive evidence of the influence of basement faults on thrust-belt structure is in Alabama, where Thomas (1982, 1986, 2001) identified a down-to-the-SE basement normal fault that subsequently localized the ramp beneath the Birmingham anticlinorium. Both stratigraphic variations and soft-sediment deformation in several units across the Birmingham anticlinorium indicate activity of the basement fault throughout the Paleozoic. Harris and Milici (1977) suggested that several thrust faults in Tennessee were deflected upward from the basal décollement by warps in the basement surface. Woodward (1988) concluded that the presence of basement in thrust sheets around the Mountain City window did not influence thrust-belt structural style.

The Shady Valley thrust sheet (Holston–Iron Mountain fault) west of the Mountain City window (Fig. 6), however, contains stratigraphy ranging from Mesoproterozoic basement (or the Neoproterozoic Mt. Rogers Formation in NE Tennessee and SW Virginia) to a continuous section from the Lower Cambrian Chilhowee Group into the Lower Ordovician (Knox Group) (King and Ferguson, 1960; Hardeman, 1966). Stratigraphy in this thrust sheet thereby ties the thin-skinned Valley and Ridge structural style from the Valley and Ridge into the western Blue Ridge. We therefore attach much greater significance to this fault system and its relationships to thrust sheets to the NW and SE than did Woodward (1988). The Shady Valley thrust sheet arches above the Mountain City window and joins the Stone Mountain thrust sheet, which transports mostly basement, around the NE end of the Mountain City window (King and Ferguson, 1960, their Plate 1). The Shady Valley–Stone Mountain sheet represents a transition from the strictly thin-skinned sheets to the NW (Pulaski, Saltville) into the Stone Mountain basement sheet to the SE, bringing with them implications not only for the southern Appalachian foreland fold-thrust belt, but for foreland fold-thrust belts worldwide. This illustrates that the traditional thin-skinned thrusting concept should be modified to include thrusts that propagate from a subhorizontal detachment in the ductile-brittle transition into the sedimentary cover (e.g., Hatcher, 2004). The SE-dipping Holston–Iron Mountain thrust propagated from the basement through the upper Unicoi Formation (basal Chilhowee Group) into the Cambrian rift-to-drift sequence, suggesting that this fault system preserves a hanging-wall ramp from the ductile-brittle transition in the basement (Fig. 6; cross sections in King and Ferguson, 1960, their Plate 17). This thrust sheet thus preserves the lower detachment that formed in the basement and undeformed remnants of the overlying rifted-margin succession (Figs. 5 and 6). The Shady Valley–Stone Mountain thrust sheet thus provides an important structural tie from the Blue Ridge–Piedmont sheet into the southern Appalachian foreland fold-thrust belt.

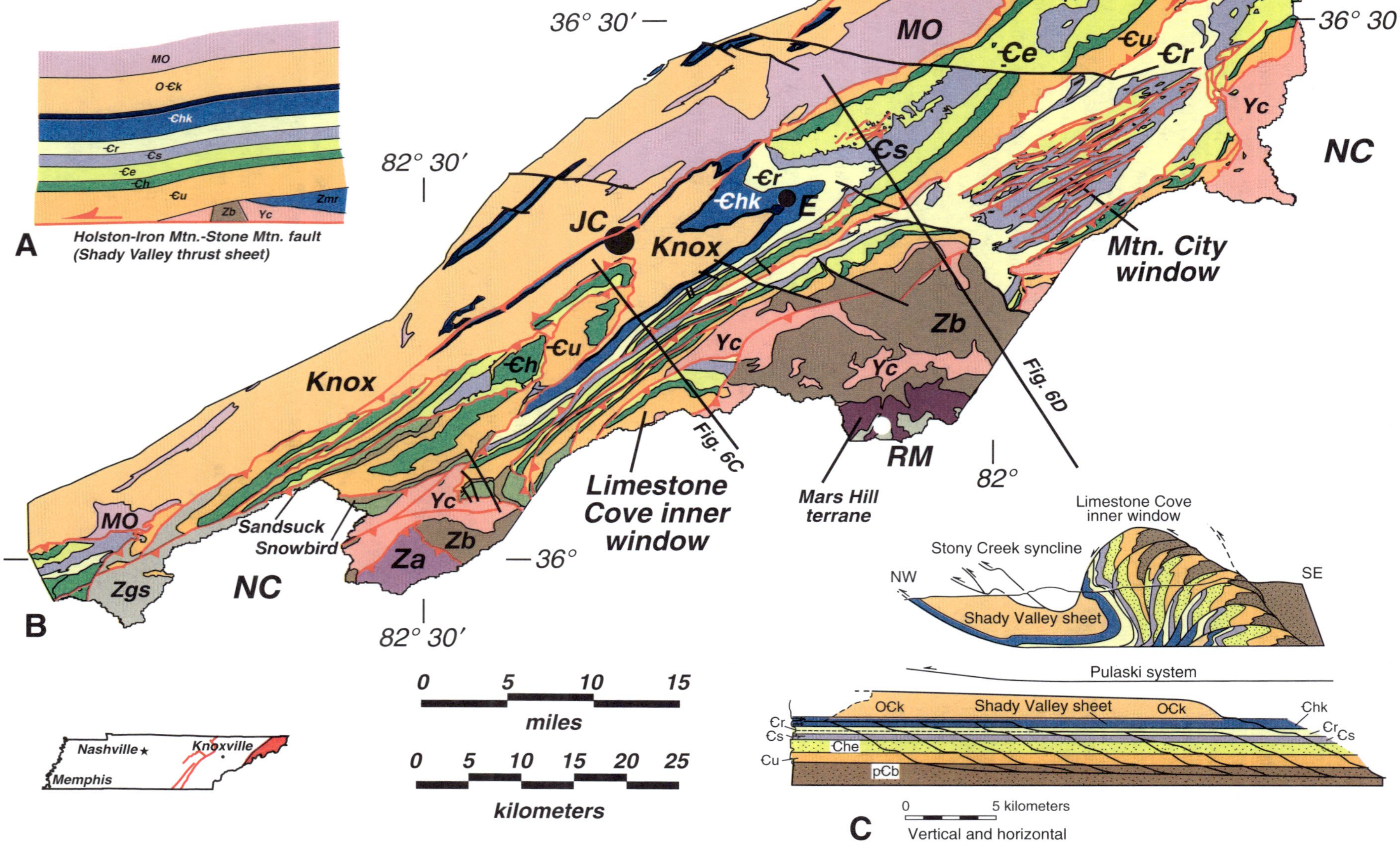

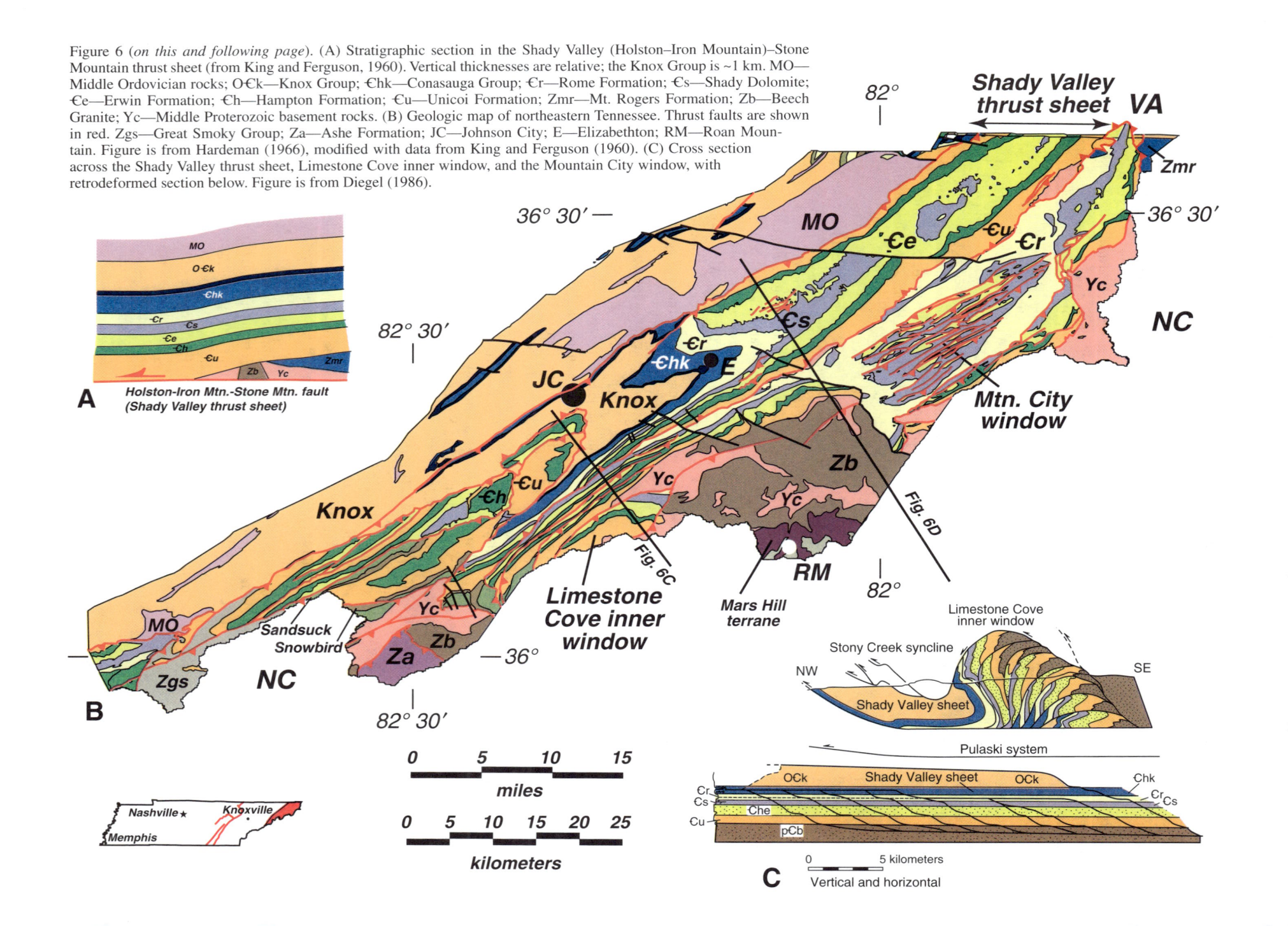

Figure 6 (on this and following page). (A) Stratigraphic section in the Shady Valley (Holston–Iron Mountain)–Stone Mountain thrust sheet (from King and Ferguson, 1960). Vertical thicknesses are relative; the Knox Group is ~1 km. MO—Middle Ordovician rocks; O€k—Knox Group; €hk—Conasauga Group; €r—Rome Formation; €s—Shady Dolomite; €e—Erwin Formation; €h—Hampton Formation; €u—Unicoi Formation; Zmr—Mt. Rogers Formation; Zb—Beech Granite; Yc—Middle Proterozoic basement rocks. (B) Geologic map of northeastern Tennessee. Thrust faults are shown in red. Zgs—Great Smoky Group; Za—Ashe Formation; JC—Johnson City; E—Elizabethton; RM—Roan Mountain. Figure is from Hardeman (1966), modified with data from King and Ferguson (1960). (C) Cross section across the Shady Valley thrust sheet, Limestone Cove inner window, and the Mountain City window, with retrodeformed section below. Figure is from Diegel (1986).

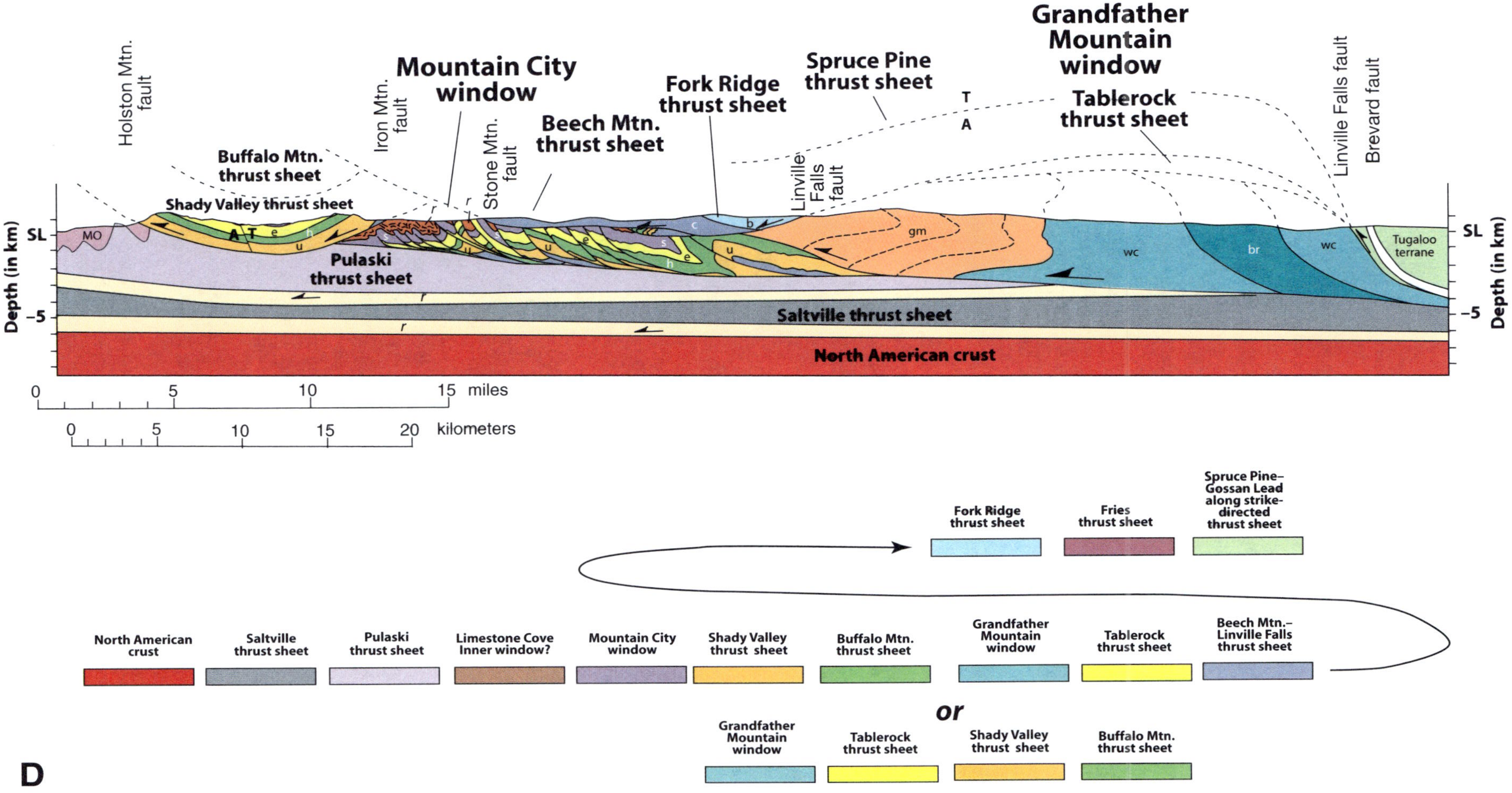

Figure 6 (continued). (D) Cross section and retrodeformed stacked thrust sheets from the Valley and Ridge southeastward across the Brevard fault zone in northeastern Tennessee and western North Carolina. Figure was modified from cross sections by King and Ferguson (1960), Bryant and Reed (1971), Boyer and Elliott (1982), and Diegel (1986). Section location is shown in Figures 2B and 6A. Abbreviations: c—Cranberry and related basement gneisses; wc—Wilson Creek Gneiss; br—Blowing Rock Gneiss; b—Beech Granite; gm—Grandfather Mountain Formation; u—Unicoi Formation; h—Hampton Formation; e—Erwin Formation; s—Shady Dolomite; r—Rome Formation; MO—Middle Ordovician rocks; SL—sea level. Below: Retrodeformed thrust sheets without regard to geometry or magnitude of displacement. Note the two options for the relative positions of the Shady Valley, Buffalo Mountain, Grandfather Mountain window, and Tablerock thrust sheets. Boxes representing relative positions of thrust sheets below do not follow the color scheme of the cross section.

The Great Smoky thrust sheet farther southwest (Fig. 2A) less spectacularly preserves a lower detachment in the Cambrian(?) Sandsuck Formation (Walden Creek Group, Ocoee Supergroup), much of the Sandsuck Formation, all of the Chilhowee Group, Shady Dolomite, and part of the Rome Formation (Hardeman, 1966; King et al., 1968, their Plate 1; Carter, 1994, his Plate 1). Nevertheless, this bedding thrust provides another tie to southern Appalachian foreland fold-thrust belt geology, representing the easternmost thrust sheet in the Valley and Ridge. The Great Smoky thrust sheet is cut off by the Miller Cove thrust sheet, which transports earlier deformed and metamorphosed Ocoee Supergroup rocks, and some Mesoproterozoic basement. The Miller Cove fault (and equivalents) is thus an abrupt western limit of the Blue Ridge–Piedmont sheet from the latitude of Knoxville, Tennessee, southward into Alabama (Fig. 2A), but the Shady Valley–Stone Mountain thrust sheet to the NE provides a less clear transition from the Blue Ridge–Piedmont sheet into the southern Appalachian foreland fold-thrust belt. Other frontal duplexes and remnants of Chilhowee Group rifted-margin sequence rocks occur from northern Georgia to Alabama (Costello, 1984; Sears, 1988), providing additional ties between Blue Ridge and southern Appalachian foreland fold-thrust belt structure.

BOUNDARIES OF SOUTHERN APPALACHIAN FORELAND FOLD-THRUST BELT DEFORMATION

Initial parameters prior to southern Appalachian foreland fold-thrust belt deformation consist of the basement surface, the basal geometry of the future Blue Ridge–Piedmont thrust sheet, and the mechanical properties of the rocks that comprise the deforming mass (Fig. 1). Temperature-pressure conditions, fluid pressure, and strain rate are additional influences on mechanical properties. The erosion surface maintained during deformation was a free surface of the deforming wedge (Davis et al., 1983), whereas the base of the Blue Ridge–Piedmont sheet geometry evolved during emplacement.

We have constructed contour maps of the top of the basement surface beneath the southern Appalachian foreland fold-thrust belt, and the base of the Blue Ridge–Piedmont sheet from industry, academic, and U.S. and state geological survey seismic-reflection profiles (e.g., Fig. 7), surface geologic maps, and well data (Figs. 8 and 9). Crustal seismic lines in the more internal parts of the orogen (Cook et al., 1983; Costain et al., 1989) further delimit basement surface geometry even where thrust displacement on the Blue Ridge–Piedmont megathrust sheet exceeds 300 km. The basement and Blue Ridge–Piedmont maps were contoured by hand because of the nature and distribution of data.

The data employed are of variable quality and consist of unmigrated and migrated seismic-reflection profiles, collected using either vibroseis or dynamite sources (Fig. 7). The base of the sedimentary section (on top of basement) produces a strong reflector package that separates a lower, nearly transparent (reflection-free) basement from the highly reflective sedimentary section (Fig. 7). Although the base of the sedimentary section is relatively easy to see, few basement normal faults are apparent in the data (Figs. 7 and 8B). The depth to the top of the basement was determined by employing a constant velocity of 2.75 km/s (two-way traveltime, TWT) for Valley and Ridge seismic lines and 3.0 km/s TWT for seismic lines in the Blue Ridge and Piedmont. Two basement holes in Tennessee (Fig. 8) provided local calibration of estimated velocity.

Basement Surface Reconstruction

The basement surface beneath the southern Appalachian foreland fold-thrust belt (Fig. 8A) represents the surface formed during the Neoproterozoic as the Grenville mountains were eroded, and is also a post-Neoproterozoic to Early Cambrian rifting erosion surface. This surface has been imaged in 2-D seismic-reflection profiles, but no attempt has been made, before now, to construct a contoured map of this surface from these data. Watkins (1964) attempted a reconstruction using aeromagnetic data, but data limitations prevented him from recognizing many details. The basement surface consists of a gently SE-dipping surface cut by several (previously unrecognized) large- and small-displacement normal faults (Fig. 8A). Most faults dip toward the margin.

The most prominent basement feature beneath the southern Appalachian foreland fold-thrust belt and Blue Ridge–Piedmont sheet is a large normal fault traceable from Alabama to Virginia (Fig. 8A). This fault was first recognized by Thomas (1986) near Birmingham, Alabama, and was called the "Alabama basement fault" by Rankin et al. (1989). We suggest renaming the entire fault the Tusquittee fault for a small town and mountains located in North Carolina about halfway along its mapped extent (Fig. 2B). Although several data gaps exist along the fault trace, the Tusquittee fault has a consistent 2.5 km down-to-the-SE displacement. Shumaker (1996) mapped a normal fault in southern West Virginia that connects northward with a major basement fault system in the central Appalachians; this fault may connect to the south with the northern end of the Tusquittee fault, where Shumaker estimated a displacement of ~300 m. The Tusquittee fault passes beneath the Blue Ridge–Piedmont thrust sheet between northern Georgia and Virginia, reappears in the Valley and Ridge in northeastern Tennessee, then trends back beneath the Blue Ridge–Piedmont sheet just north of the Tennessee-Virginia state line near the margin of the map. Thomas (1986) recognized the importance of the Tusquittee fault in localizing both sedimentation and subsequent deformation in the Alabama foreland fold-thrust belt.

The other dominant feature in the basement reconstruction is the prominent NE-SW–trending gradient located along the entire SE margin of the map (Fig. 8A). This gradient probably outlines part of the ancient Laurentian margin modified by Neoproterozoic-Cambrian rifting. The shallowing trend toward the SW may outline the transition from the Alabama promontory to the faulted "Tennessee" embayment (actually located in the Carolinas and NE Georgia). The gradient is truncated by the Tusquittee fault, particularly on its SE side. Moreover, the change in orientation of the gradient mimics the shape of the Alleghanian Blue Ridge structural front in Georgia, Tennessee,

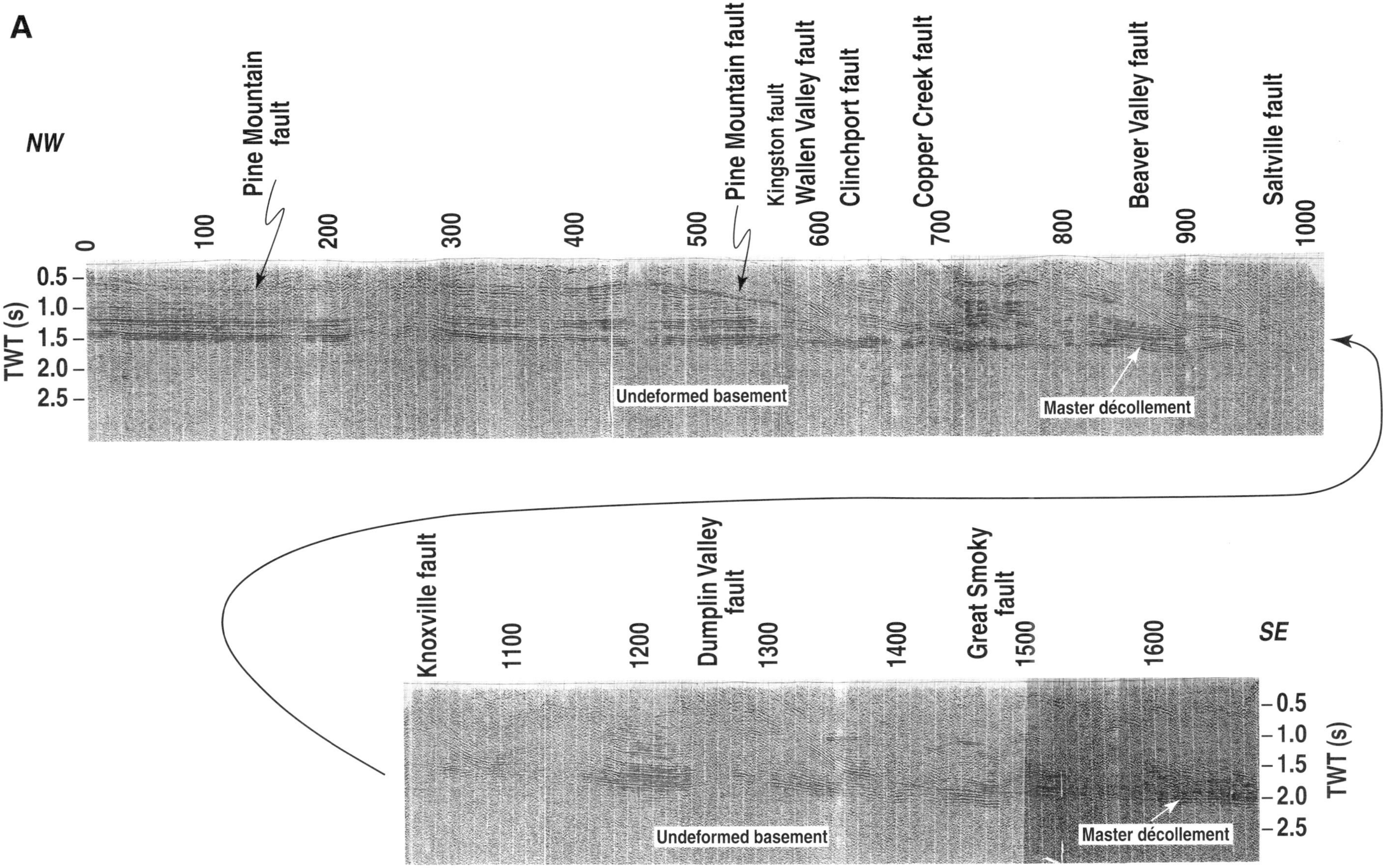

Figure 7 (on this and following page). (A) Industry across-strike seismic-reflection profile from the Pine Mountain block southeastward to the Great Smoky fault SE of Knoxville. Note the steeper dip of faults SE of the Pine Mountain fault. We assume that steepening occurred as the master décollement propagated westward, rotating thrusts that had already formed to the SE to steeper dips. Also, note the gentle SE dip of the basement surface (immediately beneath the series of strong reflectors—Rome Formation shale and sandstone—that rests on basement) from NW to SE. TWT—two-way traveltime.

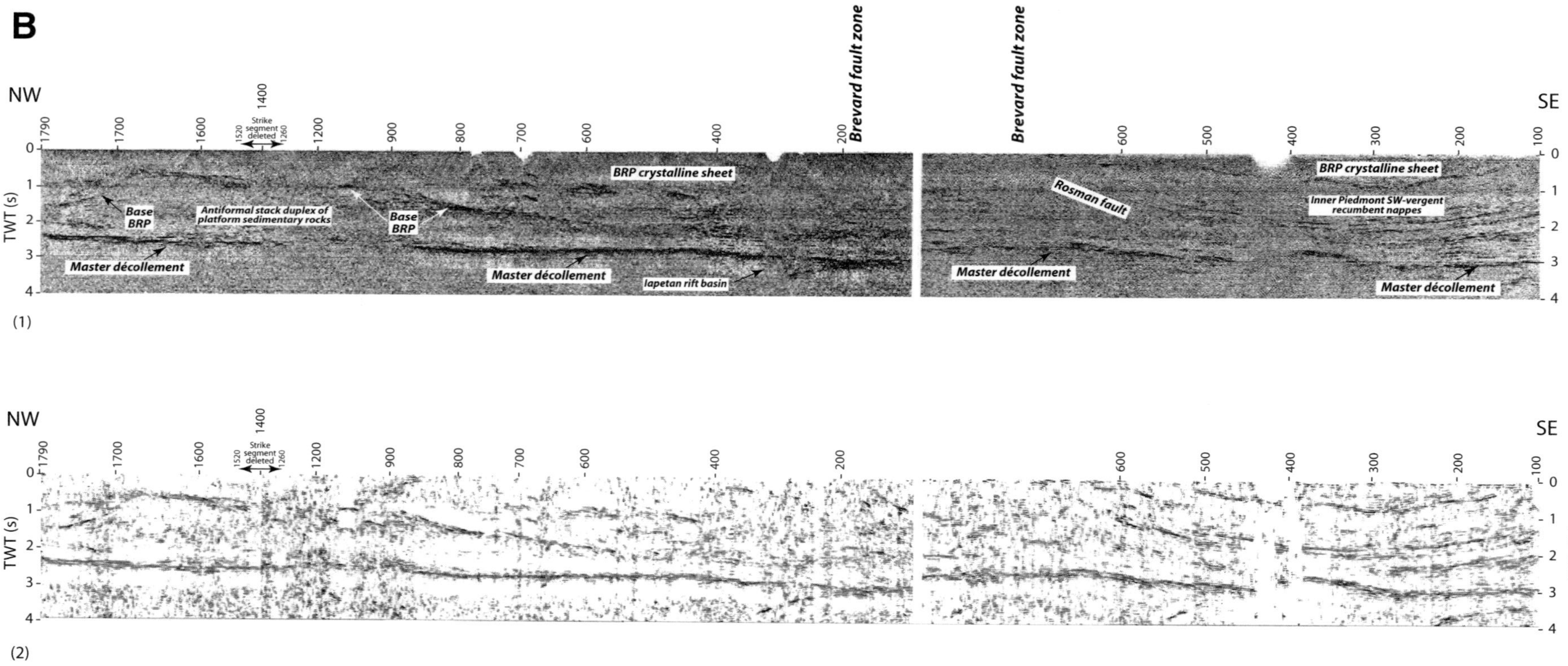

Figure 7 (*continued*). (B) ADCOH (Appalachian Ultradeep Core Hole Project site investigation): (1) seismic-reflection line (west segment, Line 3, processing included automatic gain control and whitening; east segment, Line 1, included all of that in Line 3 and time migration) and (2) automatic line drawing. (Figure is from Costain et al. [1989] and Hatcher et al. [1989b, Plate 8].) Despite the location of B deep in the crystalline Appalachians, the gently SE-dipping set of Rome Formation reflectors resting on basement is still visible here, along with a strongly arched segment of the Blue Ridge–Piedmont (BRP) sheet above an antiformal stack duplex. Locations of both seismic-reflection lines are shown in Figure 2A.

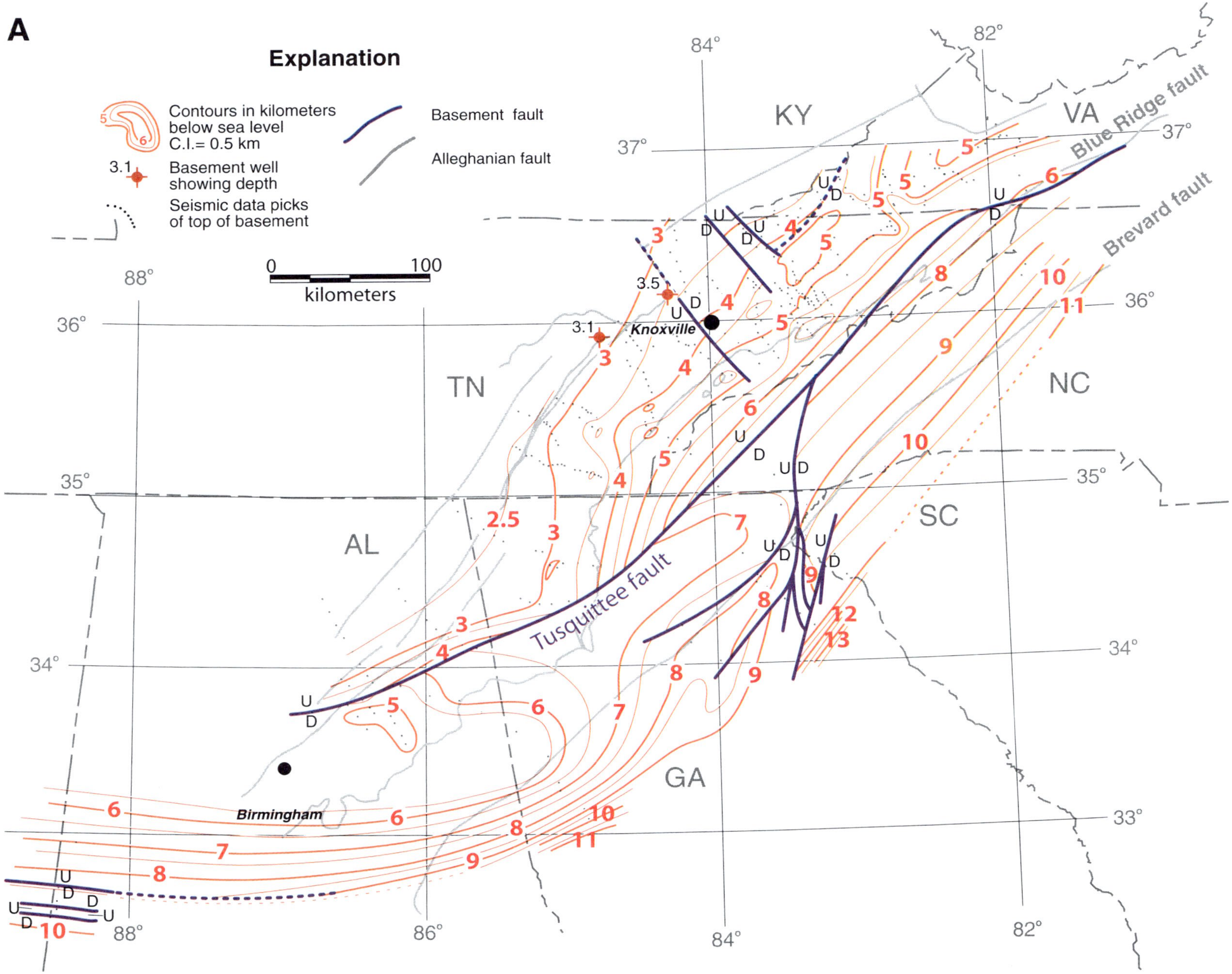

Figure 8 (*on this and following page*). (A) Basement depths in the southern Appalachians estimated from seismic-reflection data, assuming a velocity of 2.75 km/s in the southern Appalachian foreland fold-thrust belt and 3.0 km/s in the Blue Ridge–Piedmont sheet. (Compiled by RDH and PJL 1996–1997; modified through 2007 by JBW.) C.I.—contour interval.

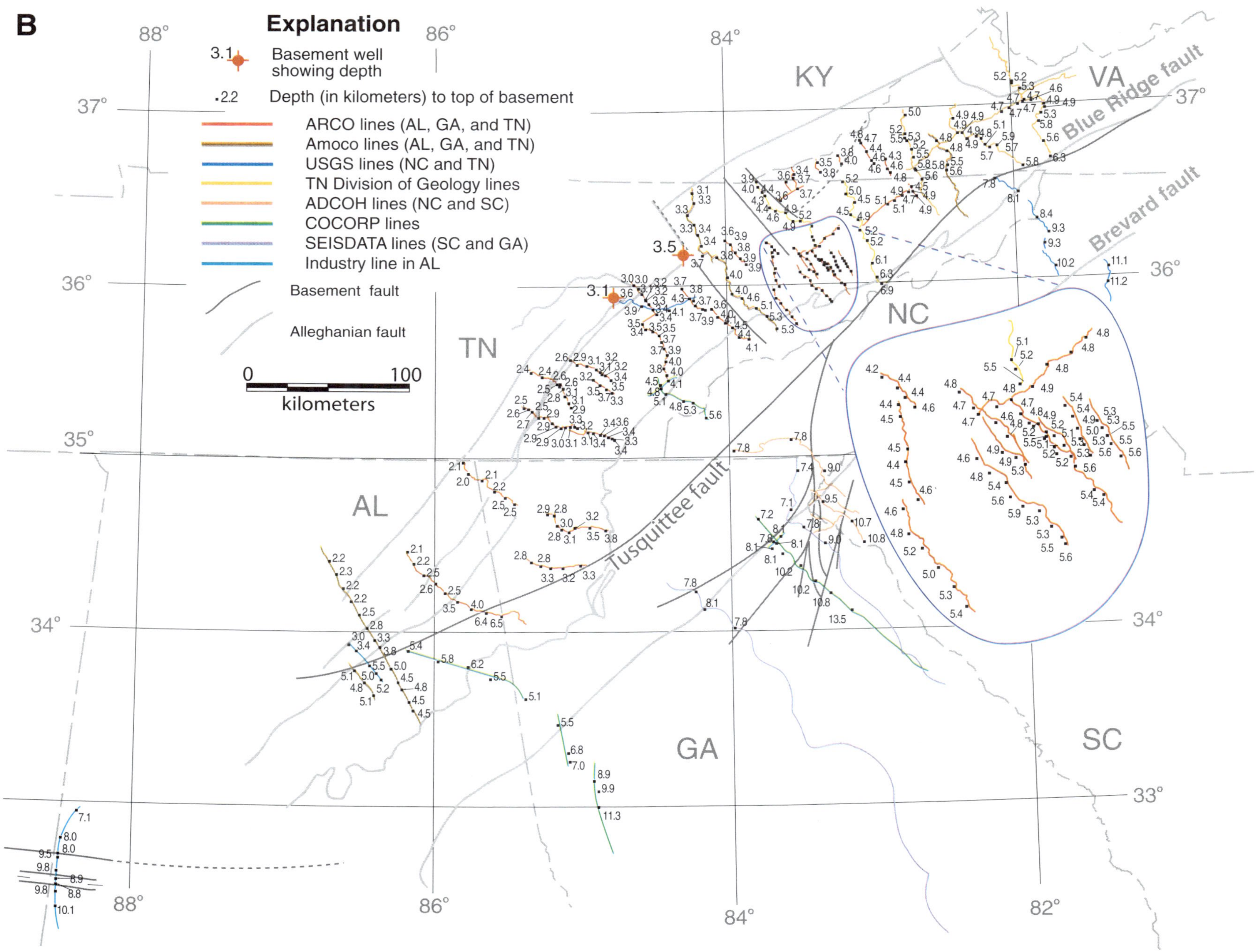

Figure 8 (*continued*). (B) Picks and distribution of data used in construction of the basement surface.

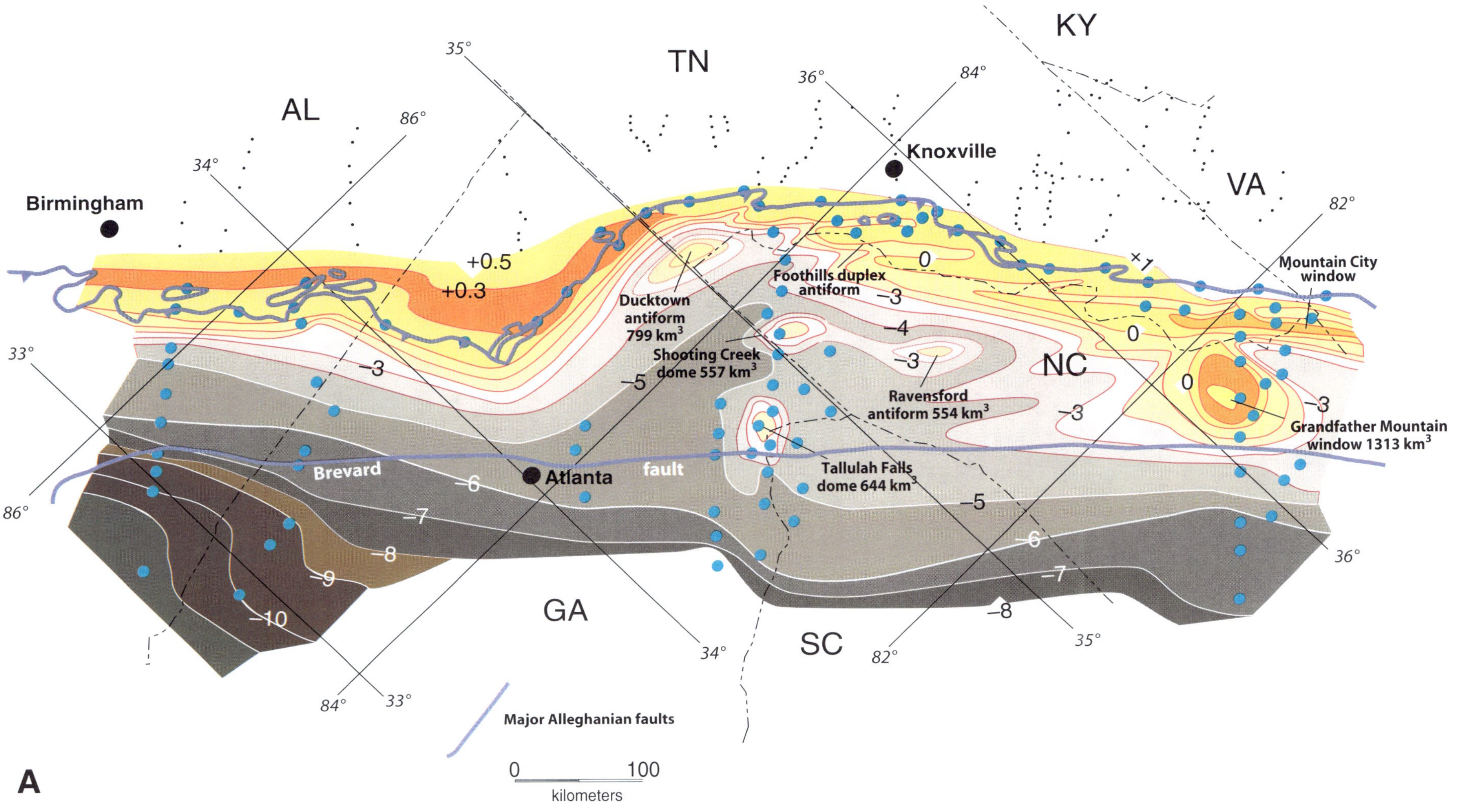

Figure 9 (*on this and following page*). (A) Roof geometry of the southern Appalachian foreland fold-thrust belt (base of Blue Ridge–Piedmont thrust sheet): Preliminary estimate from seismic-reflection and surface data in the southern Appalachians. Indicated volumes of closed structures (subthrust duplexes) were estimated using GeoSec™ 3D.

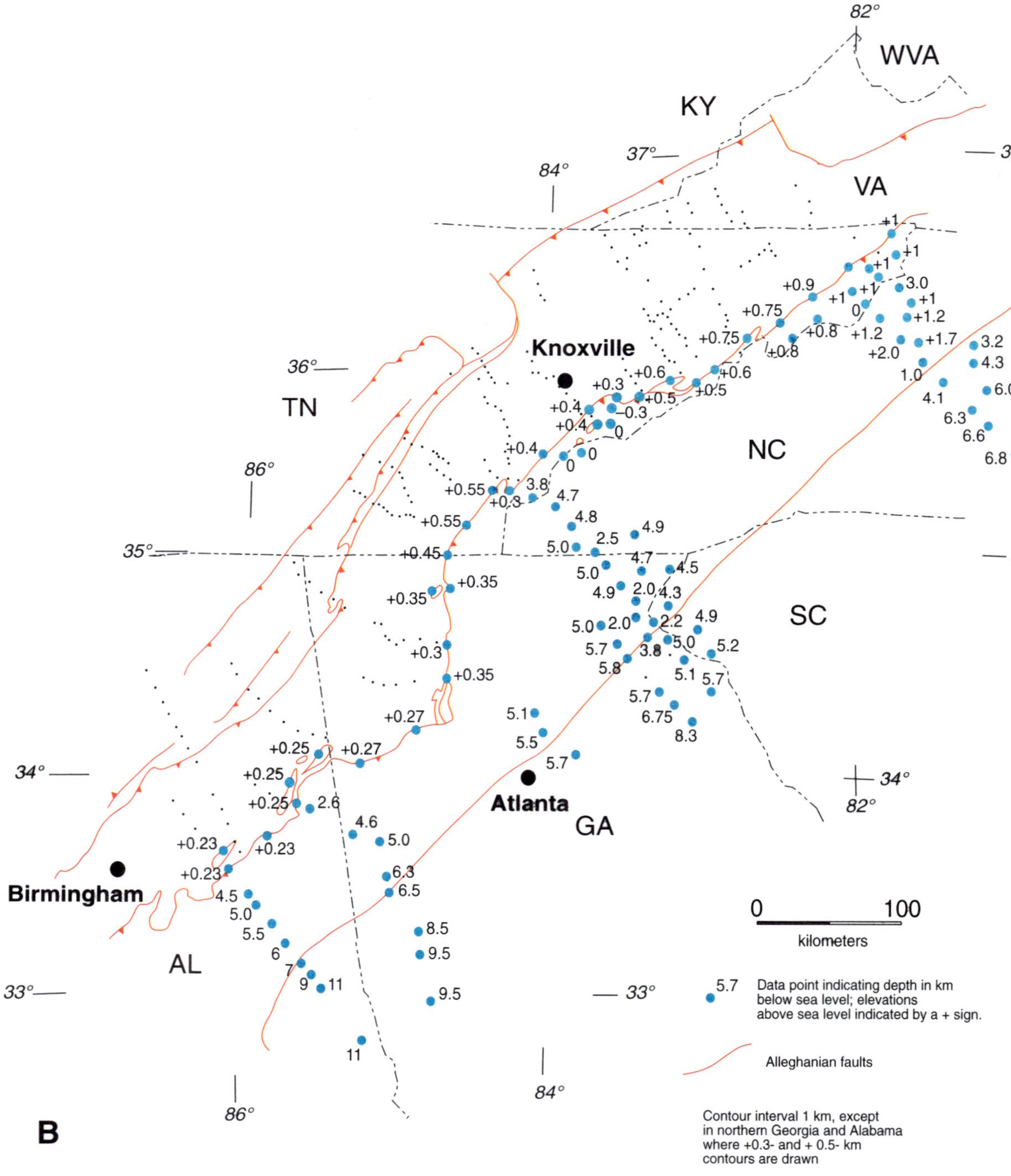

Figure 9 (*continued*). (B) Picks and distribution of data used in construction of the base of the Blue Ridge–Piedmont sheet. Blue dots indicate exact locations of points used for depth determination in seismic-reflection profiles.

and Alabama. The palinspastic location of this feature may have influenced sedimentation patterns and subsequent deformation in the Appalachian basin in Georgia, Tennessee, and southwestern Virginia. The Tusquittee fault and accompanying increased slope of the basement surface do not correlate with the location of the "Appalachian gravity gradient" that Rankin (1975), and Hatcher and Zietz (1980), correlated with the location of the eastern subsurface limit of the platform at the latitude of the southern Appalachian foreland fold-thrust belt. Hatcher and Zietz (1980) recognized that the gravity gradient is a more complex feature because it corresponds to the location of the central Piedmont suture in the hanging wall of the Blue Ridge–Piedmont sheet.

Smaller normal faults break the basement surface beneath the southern Appalachian foreland fold-thrust belt in Georgia, the Carolinas, and Tennessee (Fig. 8A). A concentration of NE-trending basement normal faults with displacements of 2 km or less was clearly imaged in Appalachian Ultradeep Core Hole Project (ADCOH) (Costain et al., 1989) and Consortium for Continental Reflection Profiling (COCORP) (Cook et al., 1983) deep seismic-reflection profiles, and in potential field data (Favret and Williams, 1988), beneath the Blue Ridge–Piedmont sheet in NE Georgia and adjacent South Carolina. One N-S fault in this system may join the Tusquittee fault.

Four small-displacement basement normal faults have been mapped in NE Tennessee and SW Virginia (Mitra, 1988) (Fig. 8A), and they extend from the Pine Mountain thrust sheet to the present Blue Ridge front. These faults strike NE and NW and have 0.25 to 1 km displacements. Available seismic-reflection data, however, poorly delimit their northeastward extent into Virginia, but the data indicate that they may end NE of Knoxville, Tennessee. We interpret one of these faults to lie directly beneath the Jacksboro fault. This fault may extend southeastward across the southern Appalachian foreland fold-thrust belt to immediately SW of the present surface location of the simple windows (Tuckaleechee Cove, Cades Cove, etc., see TCW in Fig. 2A) in the frontal Blue Ridge–Piedmont sheet. This fault steps down to the NE, and, together with two other NW-striking faults that step down to the SW, it defines a low in the basement surface that extends SE from beneath the Pine Mountain thrust sheet. These three faults have a similar orientation to cross-strike structural discontinuities and lineaments documented elsewhere in the Appalachian basin (Wheeler et al., 1979; Thomas, 1990). The fourth fault in this area steps down to the SE and strikes subparallel to the Tusquittee fault. This fault is interpreted to merge with one of the NW-striking faults and defines a triangular-shaped basement high beneath the Pine Mountain thrust sheet. How far this fault extends into Virginia is presently unknown.

Base of the Blue Ridge–Piedmont Sheet Reconstruction

The base of the Blue Ridge–Piedmont sheet was reconstructed from seismic-reflection and surface geologic data. It consists of an undulating surface that maintains a gentle homoclinal SE dip interrupted by numerous antiforms and synforms, some of which have several kilometers of structural relief (Fig. 9A). Several antiforms, including the Grandfather Mountain and Sauratown Mountains windows (in North Carolina), and the Pine Mountain window (in west-central Georgia and east-central Alabama), are now breached by erosion, and others, like the Tallulah Falls dome and Shooting Creek dome (and window) (Fig. 2B), have been crossed by seismic reflection lines (e.g., Costain et al., 1989). They reveal that early Paleozoic rifted-margin and platform rocks beneath the Blue Ridge–Piedmont sheet are deformed into antiformal stack duplexes (Boyer and Elliott, 1982; Hatcher et al., 1989a; Hatcher, 1991) (Figs. 7 and 9A).

These sub–Blue Ridge–Piedmont sheet antiforms are traceable far beyond their initially apparent limited extent (Fig. 9A). Frontal linear duplexes also developed obliquely and en echelon beneath the leading edge of the Blue Ridge–Piedmont sheet and are traceable for considerable distances beneath the sheet.

Additional Boundary Conditions

The deformation conditions for the eastern part of the southern Appalachian foreland fold-thrust belt deformed wedge beneath the Blue Ridge–Piedmont sheet in northern Georgia can be estimated from our reconstruction and existing data, including the threshold of greenschist-facies metamorphism (~300 °C, 400 MPa; Weaver and Broekstra, 1984). Because of its very large displacement, the Blue Ridge–Piedmont sheet doubtlessly was an erosion (emergent) thrust that accompanied surface exposure of much of the southern Appalachian foreland fold-thrust belt in the unconfined part of the deforming wedge. The updip limit of the belt is unknown, but conodont alteration index (CAI) data (Fig. 10A) and mineral assemblages provide important information about burial depth. Low-grade metamorphic rocks with CAI values >4.0 are exposed in a large reentrant in Georgia (Epstein et al., 1977). Rocks in Tuckaleechee Cove and other simple windows farther north or along the frontal edge in Tennessee are not metamorphosed, where CAI values are 3.5 or less (Orndorf et al., 1988). These relationships suggest that the updip westward extent of the Blue Ridge–Piedmont sheet is near the present-day fault trace and that the sheet was never very thick here. In Georgia, however, it may have extended as much as 30 km farther west of the present-day trace, and was at least 3 km thicker than the frontal 30 km in Tennessee, as suggested by the metamorphic grade of footwall rocks. This close correlation of Blue Ridge–Piedmont sheet thickness with CAI data in the southern Appalachian foreland fold-thrust belt is important: much lower CAI values in the more western parts of the belt indicate much lower pressure and temperature conditions during deformation. Higher CAI and illite crystallinity index values in northeastern Georgia, southwestern Virginia, and northeasternmost Tennessee indicate not only higher pressure and temperature conditions during deformation, but also much greater westward extent of the Blue Ridge–Piedmont sheet in these two areas immediately following emplacement. Additional temperature data from burial curves (e.g., Steinhauff and Walker, 1997) indicate that Cambrian

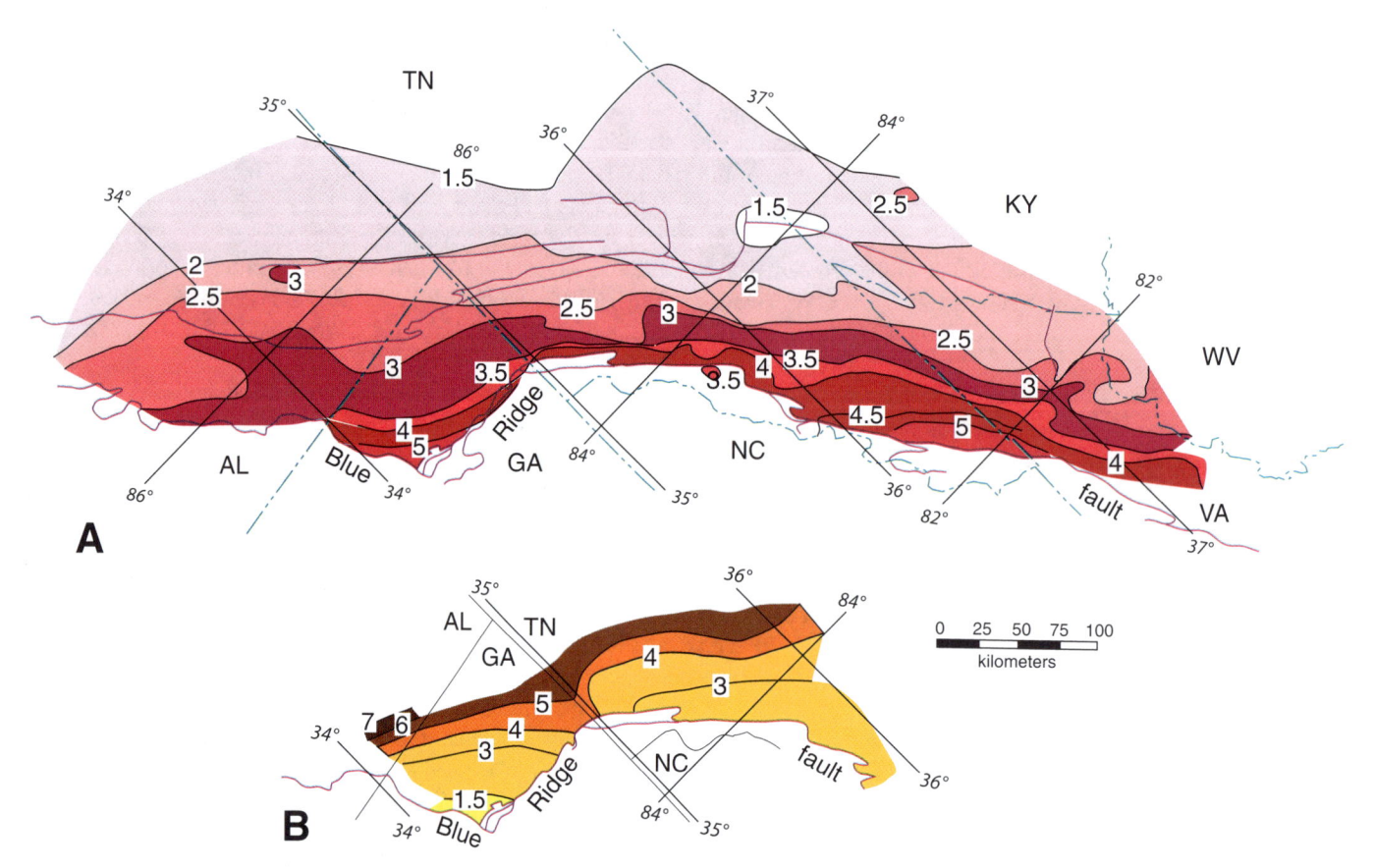

Figure 10. Thermal maturation indicators for the southern Appalachian foreland fold-thrust belt. (A) Conodont alteration index (CAI) map for the southern Appalachian foreland fold-thrust belt (after Epstein et al., 1977; Harris and Milici, 1977; Orndorf et al., 1988). Note that higher CAI values indicate higher grade. (B) Illite crystallinity index values for part of the southern Appalachian foreland fold-thrust belt (after Weaver and Broekstra, 1984). Note that lower illite crystallinity index values indicate higher grade and that values less than 1.5 are in the chlorite zone of metamorphism.

and Ordovician rocks in the western part of the belt in Tennessee were subjected to temperatures of 120–175 °C, whereas the same rocks in the eastern exposed belt were subjected to temperatures of 145–200 °C, again consistent with CAI data. Illite crystallinity index studies by Weaver and Broekstra (1984) from the southern Appalachian foreland fold-thrust belt immediately west of the Blue Ridge–Piedmont sheet in northwestern Georgia yield results that parallel the CAI data: illite crystallinity index values decrease systematically toward the SE and identify an anchizone to epizone metamorphism plateau SE of a steeper west-to-east gradient from lower to higher grade across the Georgia southern Appalachian foreland fold-thrust belt (Fig. 10B).

DISCUSSION

We have described the upper and lower bounding surfaces of the southern Appalachian foreland fold-thrust belt, along with many properties of the deformed belt. These surfaces are not only bounding surfaces, but they are also structures that strongly influenced the internal geometry of the deformed volume of platform sedimentary rocks that lie between. The wedge geometry that existed before deformation was likely deformed in an inside-out fashion, but with some smaller displacement, out-of-sequence thrusts forming in Tennessee where maximum shortening occurs (Dumplin Valley, Chestuee, Wildwood). This suggests subcritical Coulomb wedge behavior (e.g., Woodward, 1987) in the southern Appalachian foreland fold-thrust belt until shortening approached maximum values, then deformation of the SE part of the wedge may have switched to supercritical behavior. The Blue Ridge–Piedmont sheet free surface would have developed with $\alpha = 0$ as the Blue Ridge–Piedmont sheet propagated along the ductile-brittle transition; the surface slope (β) would likewise have developed with an angle of 0°. As the Blue Ridge–Piedmont sheet ramped from the ductile-brittle transition first into unmetamorphosed weak shales of the rifted-margin succession, and then into the Rome Formation resting on Mesoproterozoic basement, dip on the detachment would have increased on the ramp, then decreased once the Blue Ridge–Piedmont thrust propagated into the Rome Formation and arrived on the platform. The topographic free surface of the Blue Ridge–Piedmont thrust probably maintained a near-uniform elevation throughout once it passed the crustal ramp. The brittle Rosman fault component of the Brevard fault zone that lies within the Blue Ridge–Piedmont sheet also is an out-of-sequence thrust (e.g., Hatcher, 2001), but this fault does not extend past the projected region of maximum displacement of the Blue Ridge–Piedmont sheet—between Atlanta and the Grandfather Mountain window (Fig. 2A). It likely propagated into a suitably oriented part of the existing Brevard fault zone as the Blue Ridge–Piedmont sheet approached the crustal ramp (Hatcher, 1971, 2001; Thomas, 1986).

Correlations of stratigraphic patterns and basement structures are needed for accurate reconstruction of a deeply eroded curved orogen. If unique stratigraphic patterns or anomalies are identified, such as a thickness increase related to a local depocenter, thrust-system restoration could be guided by basement features that may have caused thickness changes, such as basement faults (Thomas, 1986).

Timing

All available data suggest that the basement faults in our basement surface reconstruction are old (post–565 Ma; Aleinikoff et al., 1995), rift-related features that formed during development of the Neoproterozoic-Cambrian Laurentian continental margin. There is no evidence of Neoproterozoic or younger reactivation of the New York–Alabama aeromagnetic gravity lineament (King and Zietz, 1978), or the Clingman aeromagnetic lineament (Nelson and Zietz, 1983) (Fig. 2A). Correlations between palinspastically restored Cambrian and Ordovician isopach thicknesses in the southern Appalachian foreland fold-thrust belt (e.g., Rodgers, 1953; Hasson and Haase, 1988) and the Tusquittee fault suggest that thickness trends may be tied to early Paleozoic movement on the fault. None of the basement faults were reactivated during Alleghanian folding and thrusting, but they may have served as important passive features that influenced sedimentation and structural development (e.g., Thomas, 1986). Inversion of basement structures and the development of foreland basement uplifts, like those in the Rockies and Andes, did not occur here during deformation, although crustal loading has been linked to the evolution of cratonic arches and basins, such as the Cincinnati arch and Appalachian basin (Quinlan and Beaumont, 1984). Subduction polarity with respect to the North American craton probably was not the same in the southern and central Appalachians as in these Tertiary orogens, and we suggest that A-type subduction was dominant in the Alleghanian orogen.

Primary and Secondary Basement Controls

Our basement surface reconstruction permits evaluation of the primary and secondary controls of basement structure on the structural development of the southern Appalachian foreland fold-thrust belt, but the data currently define the basement surface transition from the Tennessee salient to the Alabama recess better than the transition into the Virginia recess. For example, the change in orientation of the subsurface basement gradient from Tennessee into Alabama is partly reflected in the curvature of the overlying thrust belt (Figs. 2 and 3). The concave map geometry toward the craton and thinned stratigraphies of recesses (e.g., Thomas, 1977) may be viewed as primary basement obstacles that influenced regional stress/strain fields, particle trajectories, and shortening directions within the deforming southern Appalachian foreland fold-thrust belt. Particular attributes of the basement surface that may have affected the structural development of the thrust belt are: (1) changes in gradient, both parallel and oblique to thrust-belt structural strike; (2) basement obstacles; and (3) basement steps caused by Neoproterozoic to Cambrian rift-related normal faulting.

Boyer (1995) and Mitra (1997) proposed that a direct relationship exists between the dip of the basal detachment and the width of a thrust belt. We define the width of the southern Appalachian foreland fold-thrust belt as the distance between the Blue Ridge–Piedmont sheet and the Pine Mountain or Cumberland Plateau faults, but we recognize that the trailing edge of the foreland has been overridden by the Blue Ridge–Piedmont sheet and that the thrust belt does not extend past the Wartburg basin (Fig. 2A). Furthermore, we assume that the increased slope in the basement surface represents the original dip of the basal detachment, because the Rome Formation is in the exposed hanging walls of all major thrust sheets in the southern and central Appalachians, except the Pine Mountain and Cumberland Plateau thrust sheets (and Plateau sheets farther north), which formed along an upper-level detachment (Gwinn, 1964; Harris and Milici, 1977). There also had to have been negative isostatic effects of overthrusting of the Blue Ridge–Piedmont sheet onto the platform and the load imposed, but these may have been mitigated by rebound as the Appalachians were eroded. The basement surface gradient NW of the Tusquittee fault is a fairly consistent 1.5° except for the area beneath and to the SE of the Pine Mountain thrust sheet, where it is interrupted by several normal faults that strike parallel and perpendicular to the strike of the southern Appalachian foreland fold-thrust belt. Ignoring the narrowing of the belt near the Wartburg basin, the consistent basement surface gradient correlates well with the consistent width of the thrust belt. The gradient flattens slightly in northern Georgia, which corresponds to a slight narrowing of the width of the fold-thrust belt, especially when the forelandward extent of the Blue Ridge–Piedmont sheet is restored. Therefore, there may be a direct but subtle relationship between the width of the fold-thrust belt and the basement surface gradient NW of the Tusquittee fault.

Although data are sparse, the basement surface gradient SE of the Tusquittee fault is quite variable, and it is disrupted by the N-S–striking normal-fault complex imaged in the ADCOH seismic-reflection lines (Fig. 6). In NE Tennessee and northwestern North Carolina, the U.S. Geological Survey seismic-reflection data (Harris et al., 1981) indicate that the basement surface slope dip doubles to ~3° SE. In the area between the Tusquittee fault and the normal-fault complex, there may be a subtle basement high, and the gradient may even slope gently back to the NW toward the Tusquittee fault. The relatively flat basement surface defining the basement high begins to steepen onto the Alabama promontory where the gradient slopes gently to the NE toward the normal-fault complex. The basement gradient on the SE side of the Alabama promontory is similar to that in NE Tennessee. If we again assume that the basement surface gradient parallels the dip of the basal detachment, the taper angle of the advancing thrust wedge will decrease as it advances over the Tusquittee fault (or elsewhere) onto the less steep basement gradient to the NW. This would cause the thrust wedge to thicken internally for it to be able to continue advancing onto the foreland. Such a scenario may be responsible for the late (Permian?) reactivation of the Brevard fault, which occurs within the "rigid" indenter at this point in the development of the southern Appalachian foreland fold-thrust belt. It may also be responsible for the development of the antiformal-stack duplexes in platform rocks beneath the Blue Ridge–Piedmont thrust sheet.

Steps in the basement surface may also influence the development and location of thrust-belt structures. A down-to-the-SE step on a normal fault, such as the Tusquittee fault, will form a notch in the basement surface. A notch can form a perturbation in the regional stress field that initiates development of a thrust fault at the front of an advancing thrust-belt wedge. The Pulaski fault is considered to be one of the major southern Appalachian foreland fold-thrust belt thrust faults that extends from Virginia into Tennessee, where it is overridden by the Blue Ridge–Piedmont thrust sheet. Palinspastic restorations of the thrust belt restore the Pulaski fault to approximately overlie the step in the basement surface produced by the Tusquittee fault. Although speculative, it is possible that nucleation of the Pulaski fault occurred by a perturbation in the regional stress field as the thrust wedge advanced over the Tusquittee fault.

Thomas (1986) similarly concluded that the Tusquittee fault is responsible for the out-of-sequence, brittle, dip-slip Alleghanian reactivation of the Brevard fault zone (Rosman fault) as the Blue Ridge–Piedmont thrust sheet passed up the ramp and temporarily locked at the inflection point at the crest of the ramp. This also accompanied transportation of numerous Knox Group horses (see Hatcher et al., 1989a, their Figure 31; Hatcher, 2001) into the Brevard fault zone.

Secondary basement structural controls on sedimentation patterns also exist in the southern Appalachian platform margin. These effects produced lithologic and thickness variations that changed the mechanical characteristics of the stratigraphy. Isopach maps based on existing palinspastic reconstructions of part of the southern Appalachian foreland fold-thrust belt contain thickness trends from the Rome Formation through the Knox Group that correlate with mapped basement features (Read *in* Rankin et al., 1989). Regional thickness gradients and depocenters, however, do not directly overlie mapped basement features, with the exception of the southwestern extent of the Tusquittee fault near Birmingham, Alabama (Thomas, 1986). Potentially correlative basement features that may be responsible for these anomalies lie just SE of these thickness trends. This would be expected, because the isopach maps employ palinspastic base maps that used poorly constrained data on the amounts and directions of thrust displacement. Although the estimated displacement of each thrust fault may be questioned, correlation of a stratigraphic anomaly with a basement feature may improve estimates of total shortening in the southern Appalachian foreland fold-thrust belt.

Major facies changes and the westward limit of thick rifted-margin sedimentation, along with the locus of the Cambrian rift-to-drift transition, were probably controlled by the shape of the basement surface. Rodgers (1953) pointed out that major facies changes occur in the Conasauga (Middle-Upper Cambrian) and Knox (Upper Cambrian–Lower Ordovician) Groups at the Pulaski fault in Tennessee and SW Virginia. The abrupt westward disappearance of pre–Rome Formation clastics in thrust sheets west of the Blue Ridge frontal faults has long been known, with the exception of those beneath the Pulaski sheet in southwestern Virginia (Johnson, 1993). Palinspastic restoration of the thrust sheets in the Tennessee southern Appalachian foreland fold-thrust belt by Roeder and Witherspoon (1978) and Finney et al. (1996) provided rough estimates of the original locations of major faults that record these changes. The Tusquittee fault appears to be located west of the major facies change in the retrodeformed Cambrian-Ordovician carbonate bank. This basement fault not only influenced facies development but may also have served as a fault that nucleated several smaller-displacement faults in the Birmingham anticlinorium in Alabama (Thomas, 1986).

Internal Deformation of the Southern Appalachian Foreland Fold-Thrust Belt

The broad arcuate trace of the Blue Ridge–Piedmont sheet is probably partly related to the original shape of the continental margin, as suggested by basement surface geometry (Thomas, 1991) (Fig. 8A), because the Blue Ridge–Piedmont sheet is largely composed of the already deformed and metamorphosed rifted-margin succession and older basement, with only the frontal Great Smoky fault system behaving as a thinned-skinned thrust. Additional factors that probably contributed to the curvature of this system are heterogeneous strain that produced nonparallel particle trajectories within the Blue Ridge–Piedmont sheet and southern Appalachian foreland fold-thrust belt.

The shape of the Blue Ridge–Piedmont indenter was partly determined by the shape of the Neoproterozoic-Cambrian continental margin, but its earlier Paleozoic history prior to being transported onto the Laurentian platform also influenced its shape. The rocks comprising proximal parts of the thrust sheet were polydeformed and metamorphosed and were likely accreted to the most distal part of the Laurentian margin during the Taconic orogeny. A group of suspect and exotic terranes was later accreted to the Taconian collage during the Late Devonian–Mississippian Neoacadian orogeny (e.g., Merschat et al., 2005). Both the Taconian and Neoacadian events formed new continental crust—basement—that was cold and rigid by the time the late Alleghanian collision occurred to complete formation of the indenter and transport it onto the Laurentian margin as the Blue Ridge–Piedmont sheet (Hatcher, 1999, 2004).

A curved foreland fold-thrust belt, like the southern Appalachian (Fig. 2), is not likely to have deformed by straight-line particle trajectories and plane strain within thrust sheets, because particle trajectories should curve in an orogen that curves through an arc of 30° from northern Georgia to southwestern Virginia (Figs. 2A and 3). In addition to an originally irregular continental margin, basement faults, stratigraphic thickness and facies changes, and the original shape of the Blue Ridge–Piedmont sheet indenter also influenced emplacement trajectories of both individual thrust sheets and total volume strain of the southern Appalachian foreland fold-thrust belt (Fig. 2B). Intuitively, 3-D heterogeneous deformation should be favored, but Ferrill and Groshong (1993) determined that displacement vectors are parallel in a curved portion of the French Subalpine chain. Displacement vectors in the Jura Mountains, Helvetic Alps, and Penninic Alps, however, display a radiating pattern (Platt et al., 1989; Hindle and Burkhard, 1999). Displacement vectors in the southern Appalachian foreland fold-thrust belt can be approximated using normals to hinge lines of map-scale folds (Fig. 2B), and available mesoscopic data (folds, slickenlines, tectonic stylolites). These data indicate curved thrust-emplacement pathways (Fig. 2B). Despite apparent the balance of many 2-D cross sections, the absence of plane strain diminishes their utility for quantifying particle-displacement trajectories.

Displacement vectors and magnitudes for the Pine Mountain and Cumberland Plateau faults (Fig. 2B) were determined from the orientations of the Russell Fork, Jacksboro, and Emory River tear faults (Fig. 2A). Although these faults may not all be kinematically linked, they probably are representative of the kinematics in the part of the southern Appalachian foreland fold-thrust belt where they occur and when they formed. Displacement vectors are assumed to parallel motion on tear faults, which are generally orthogonal to the strike of the thrust faults. The Emory River and Jacksboro faults distend a 15° arc across the Wartburg basin (Figs. 2A and 2B). Both the Cumberland Plateau and Pine Mountain faults reach maximum displacement on the tear faults at the SW (Jacksboro) and NE (Emory River) ends of each fault block, respectively. Displacement along the Russell Fork fault at the NE end of the Pine Mountain block (Fig. 2A) is ~7 km but increases southwestward to 17 km along the Jacksboro fault (Rich, 1934). The Cumberland Plateau fault system reaches its maximum displacement of 5 km at its NE end along the Emory River tear fault. Linear displacement particle-path trajectories with unequal displacement are probably responsible for creating the 15° arc between the two fault systems, and these faults may represent the last displacement path trajectory of the southern Appalachian foreland fold-thrust belt.

Alternatively, both the curvature in the southern Appalachian foreland fold-thrust belt and the 15° angular discordance may be attributed to rotations about vertical Euler poles, analogous to the analysis of Price and Sears (2000) and Sears and Hendrix (2004) for the northern Rockies. This alternative model would also require formation of the Tennessee segment of the southern Appalachian foreland fold-thrust belt on a promontory, not a reentrant in the continental margin.

Determination of regional slip-line orientations may be possible in a foreland fold-thrust belt using Hansen's (1971) method, which employs fold hinge-line orientations and shear sense of asymmetric mesoscopic folds. We have attempted to apply the method using particle trajectories derived from both mesoscopic (slickenlines, tectonic stylolites, asymmetric folds) and macroscopic structures (hinges of narrow mapped folds) (Fig. 2B). If this is a correct indicator of trajectories within the thrust sheets in which the synclines and other indicators occur, the southern Appalachian foreland fold-thrust belt sheets restore to an area to the SE in the Carolinas and Georgia beneath the present Blue Ridge–Piedmont sheet (Figs. 11 and 12B). A mechanism involving homogeneous strain and outward propagation from the Blue Ridge–Piedmont indenter would predict that the thrusts in this region have smaller displacements toward the outer convex arcs of the southern Appalachian foreland fold-thrust belt and greater displacement in the inner convex arcs of the thrust belt. This is not the case, however, because the southern Appalachian foreland fold-thrust belt in the Tennessee salient is dominated by two or three large thrusts separated by a number of smaller-displacement thrusts. The outermost thrusts, Pine Mountain and Sequatchie Valley–Cumberland Plateau, do have small displacements.

An array of cross sections has been constructed around the Tennessee salient and subsequently retrodeformed (Figs. 11 and 12). Different fault systems can be traced though the cross-section array from Virginia to SE Tennessee, with smaller-displacement thrusts terminating within the array. It is clear from the sections that the master detachment propagated along the Rome Formation and ramped across the stronger carbonate section. Most southern Appalachian foreland fold-thrust belt thrusts were oversteepened at present outcrop following NW propagation of the next more western thrust, and their listric character is evident in both seismic-reflection profiles (e.g., Fig. 7A) and these cross sections. The overlapping retrodeformed sections and map (Figs. 12A and 12B) again point out the difficulty of understanding 3-D deformation in the southern Appalachian foreland fold-thrust belt or any strongly curved segment of a foreland fold-thrust belt. Dominance of certain faults, like the Saltville, occurs from southwestern Virginia to Georgia, but there also is a change in character of the Saltville at the latitude of Knoxville, where much of its displacement from Knoxville southward is transferred to the Beaver Valley fault. The Whiteoak Mountain fault dominates the western Valley and Ridge from Georgia to NW of Knoxville, where it branches and displacement is transferred to the Wallen Valley, Hunter Valley, and Clinchport faults. These faults are then traceable northeastward into Virginia, where they terminate. The relatively small-displacement, moderate along-strike length thrusts (Dumplin Valley, Chestuee, Wildwood; Fig. 2A) probably formed as out-of-sequence thrusts in a domain otherwise dominated by folding (Rodgers, 1953; and see Hardeman [1966] for map relationships) (Fig. 11).

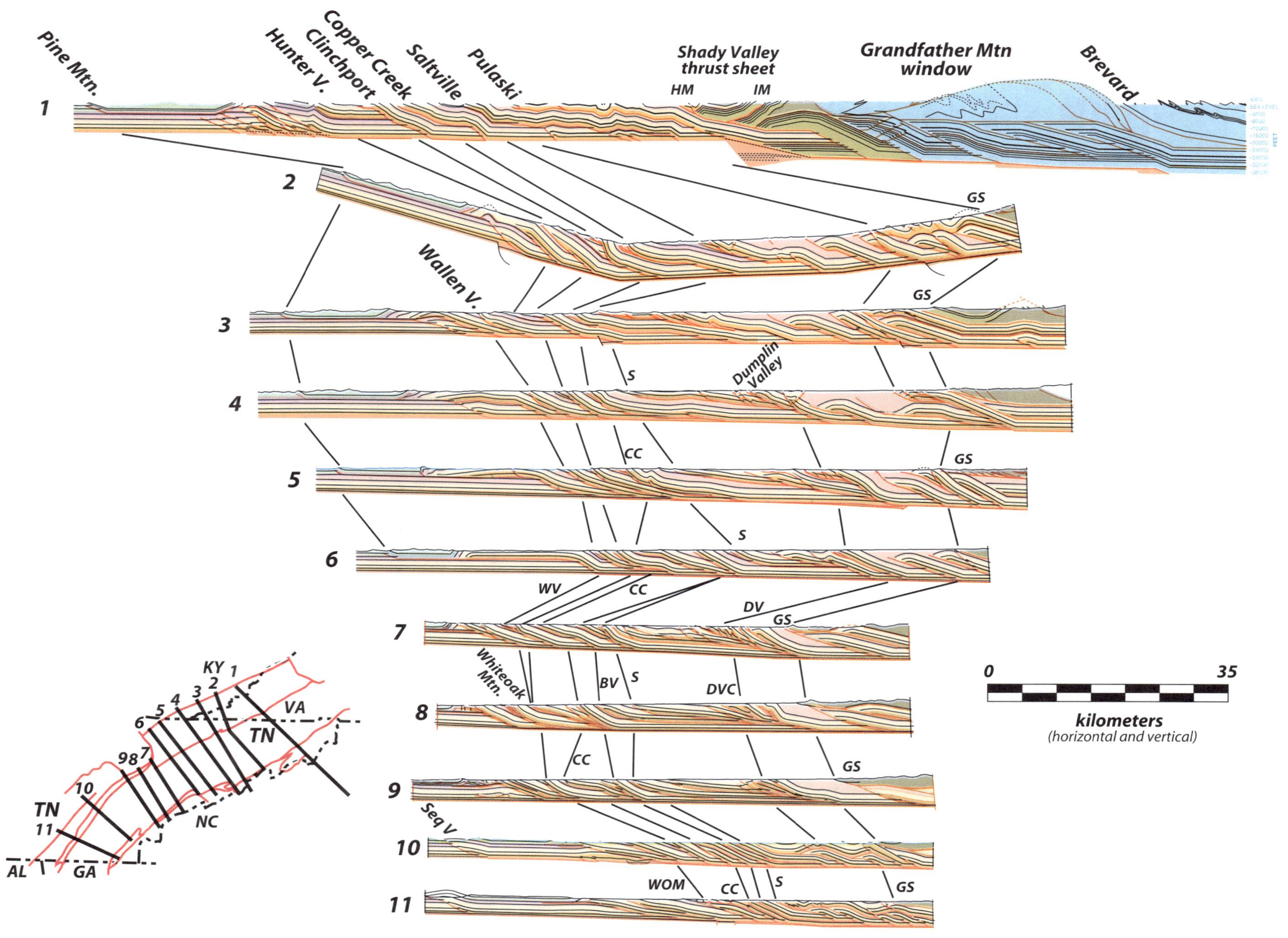

Pine Mtn.
Hunter V.
Clinchport
Copper Creek
Saltville
Pulaski
Shady Valley thrust sheet
HM
IM
Grandfather Mtn window
Brevard
1
2
GS
Wallen V.
3
S
Dumplin Valley
4
CC
5
6
WV
DV
7
Whiteoak Mtn.
BV
DVC
8
9
Seq V
10
WOM
11
0
35
kilometers
(horizontal and vertical)
KY
VA
TN
NC
AL
GA

Figure 11. Cross sections around the Tennessee salient constructed using surface and seismic-reflection data. Note how a few faults occur in all of the sections, while most faults end somewhere in the sections, but motion is transferred to another fault wherever this occurs. Ocoee Supergroup and Chilhowee Group rocks of the Blue Ridge–Piedmont sheet are shades of greenish brown. The light blue color in section 1 consists of basement and Grandfather Mountain Formation of the Grandfather Mountain window and frontal thrust sheets (see Figure 6D for detailed section), Ashe–Tallulah Falls Formation rocks along the SE flank of the Grandfather Mountain window and in the Inner Piedmont SE of the Brevard fault zone, and the imbricated platform and rifted-margin sedimentary rocks in the subsurface beneath the Blue Ridge–Piedmont sheet. The Rome Formation and overlying Conasauga Group are shades of brown. The Knox Group is tan. Middle and Upper Ordovician rocks are pink. Silurian, Devonian, and Mississippian rocks are purple. Mississippian rocks where separated are blue. Pennsylvanian rocks are green. WV—Wallen Valley fault; WOM—Whiteoak Mountain fault; CC—Copper Creek fault; BV—Beaver Valley fault; S—Saltville fault; DV—Dumplin Valley fault; DVC—Dumplin Valley–Chestuee fault system; HM—Holston Mountain fault; IM—Iron Mountain fault; GS—Great Smoky fault; Seq V—Sequatchie Valley.

Retrodeformation of Valley and Ridge thrusts and the Blue Ridge–Piedmont sheet (upper bound) above the basement (lower bound) has permitted us to gain a better understanding of the shortening in the southern Appalachian foreland fold-thrust belt region of maximum total displacement (Figs. 2C and 13). The retrodeformed across-strike displacement of each thrust sheet through which a 2-D section is constructed across the southern Appalachian foreland fold-thrust belt (or any foreland fold-thrust belt) should have a relationship to the indenter, here the Blue Ridge–Piedmont sheet, such that

$$X = \sum_{n=i} x_i < Y_{gs}, \quad (1)$$

where X is total displacement estimated from retrodeformed foreland thrust sheets in a 2-D cross section, x_i is the displacement of individual faults in the section, and Y_{gs} is total displacement on the Blue Ridge–Piedmont sheet (or any indenter in any foreland fold-thrust belt). Equation 1 can be expanded to express the displacements of individual thrust sheets across the southern Appalachian foreland fold-thrust belt as

$$X_f = x_1 + x_2 + x_3 + \ldots + x_n < Y_{gs}. \quad (2)$$

For individual thrust sheets in the Figure 13 cross section (from west to east), x_1 is the Sequatchie Valley thrust sheet, x_2 is the Rockwood thrust, x_3 is the Chattanooga thrust sheet, etc. (Table 1). If X_f is the width of retrodeformed foreland thrust sheets placed end to end, the width of the deformed thrust sheets in the southern Appalachian foreland fold-thrust belt would sum to less than the retrodeformed width. The cold Blue Ridge–Piedmont sheet, however, is shortened internally only by out-of-sequence thrusts (e.g., the Rosman fault, with ~20 km displacement; Hatcher, 2001), so the Blue Ridge–Piedmont indenter did not shorten proportionally, and

$$X'_f \ll Y'_{gs} + Y'_{Ros}, \quad (3)$$

where X'_f is the deformed width of the foreland fold-thrust belt, Y'_{gs} is the displaced emergent Blue Ridge–Piedmont sheet indenter, and Y'_{Ros} is the displacement on the Rosman fault. The internal isolated domes in the Blue Ridge–Piedmont sheet probably also formed by interactive footwall-duplex arching of the advancing Blue Ridge–Piedmont thrust sheet (Hatcher, 1991) as it drove the southern Appalachian foreland fold-thrust belt deformation in front of and beneath it. These antiforms and synforms may ultimately have partially locked the Blue Ridge–Piedmont thrust sheet, transferring motion to the deforming southern Appalachian foreland fold-thrust belt footwall rocks. They also probably contain additional tens to >100 km of shortening not initially accounted for in Equations 1, 2, and 3. Taking this component into account, Equation (3) can be modified to

$$X'_f + X'_d = Y'_{gs} + Y'_{Ros}, \quad (4)$$

where X'_d is the amount of shortening in the sedimentary section beneath the Blue Ridge–Piedmont sheet.

Expansion of this discussion to a detailed 3-D reconstruction of the internal parts of the southern Appalachian foreland fold-thrust belt wedge will permit better understanding of the kinematics and mechanics for development of the thrust stack in this and other curved foreland fold-thrust belts. Our recognition of the role played by a major crystalline thrust sheet and the other boundaries of the thrust system as controlling elements in the formation of the southern Appalachian foreland fold-thrust belt may represent some progress toward understanding the 3-D development of this and other foreland fold-thrust belts. Even so, present-day 3-D balancing techniques are at best "semiquantitative considerations of material balance" (Laubscher, 1988, p. 1313).

CONCLUSIONS

1. Boundaries of the southern Appalachian foreland fold-thrust belt consist of the basement surface below the belt, the base of the Blue Ridge–Piedmont sheet above, and the eroded topographic free surface of both the southern Appalachian foreland fold-thrust belt and Blue Ridge–Piedmont sheet. Additional boundary conditions consist of temperature and pressure maxima.

2. The basement surface observed beneath the southern Appalachian foreland fold-thrust belt is likely the same as it was throughout the Paleozoic, except for the effects of Alleghanian thrust loading. It is interrupted by several down-to-the-SE normal faults with displacements up to 3 km. The extensive Tusquittee fault is traceable from Alabama to Virginia. In addition, a major steepening of slope of the basement surface has been identified farther east on the Laurentian margin and similarly has been traced from Alabama to Virginia.

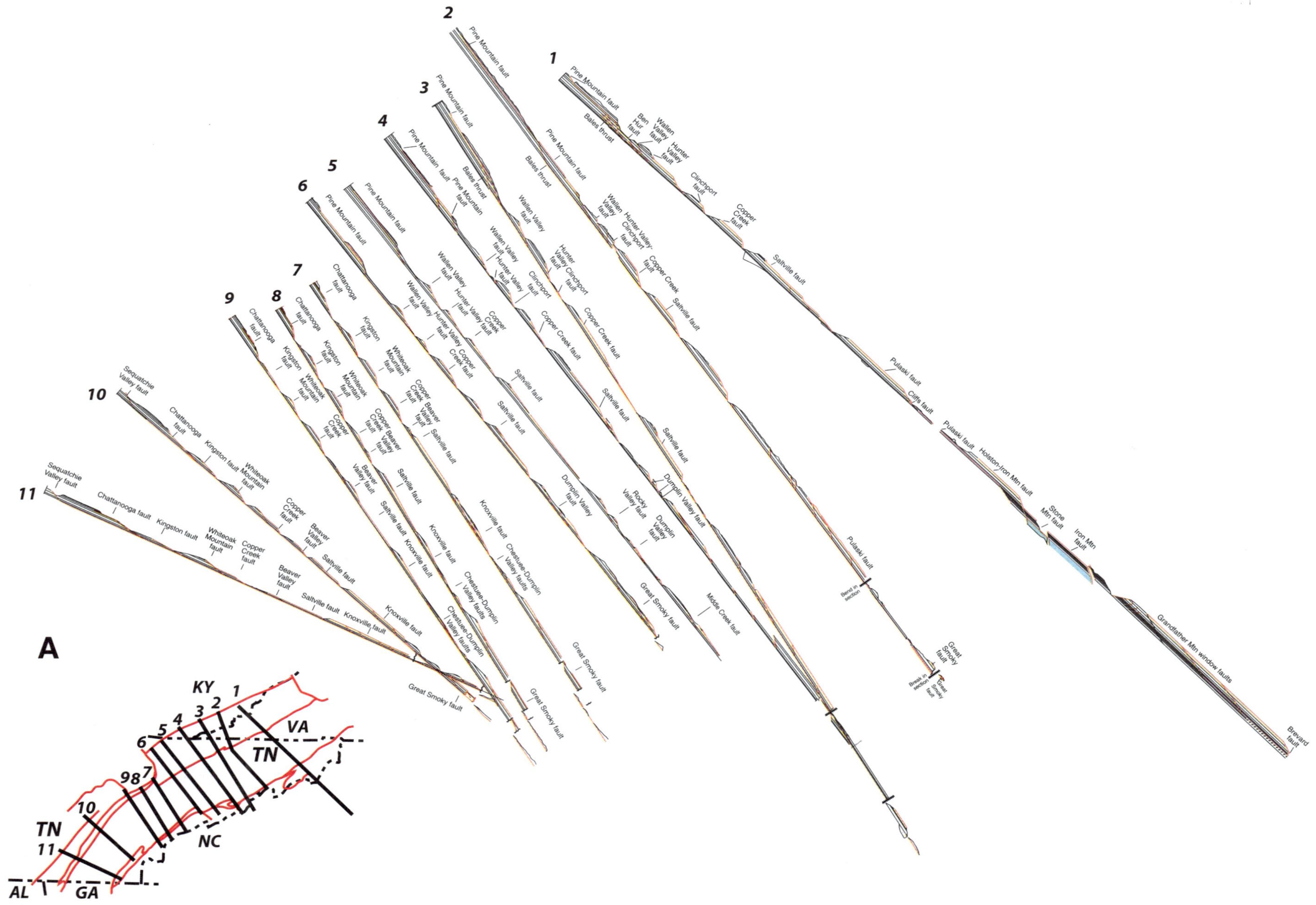

Figure 12 (*on this and following page*). (A) Restored sections shown in Figure 11 arrayed in map view around the Tennessee salient. Note the overlap of some sections.

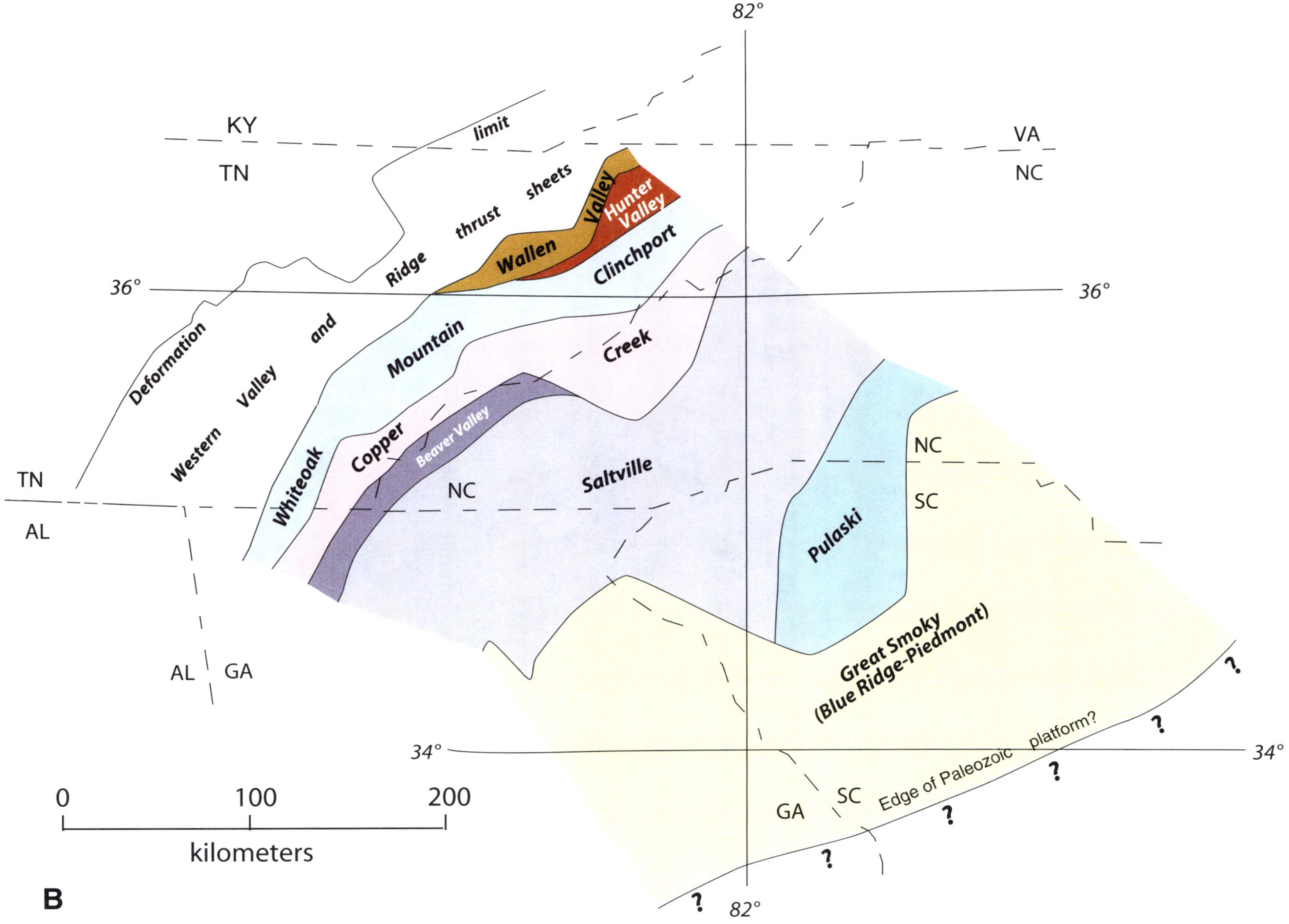

Figure 12 (*continued*). (B) Palinspastic map constructed from sections in A. Note the sinuous geometry of the Great Smoky thrust near the southwestern edge of the map in the area where the retrodeformed sections overlap.

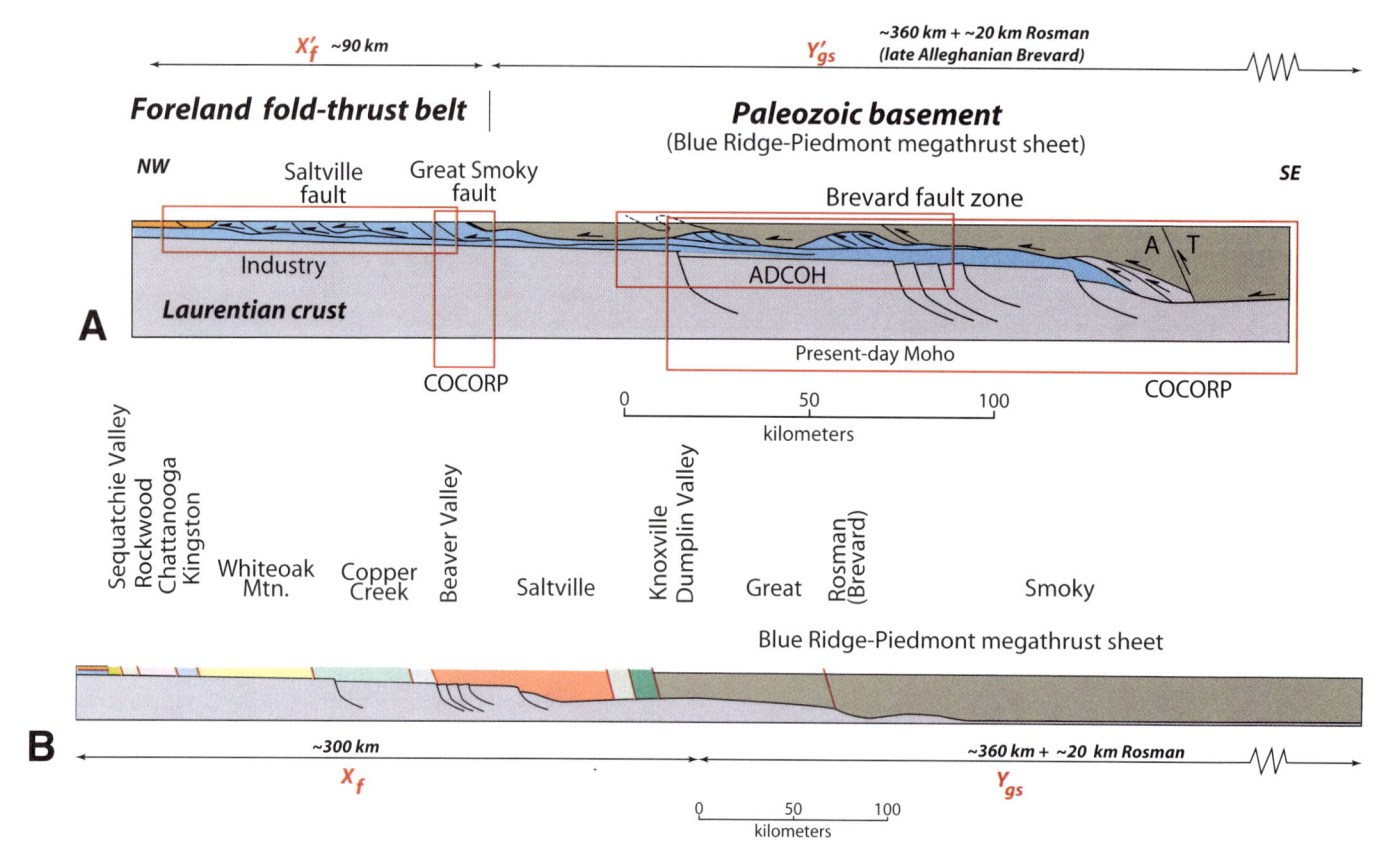

Figure 13. Restoration of the Blue Ridge–Piedmont section in Figure 2B. (A) Unrestored section showing locations of seismic-reflection data (red boxes indicate sources of data) employed for construction of the section. X'_f is the deformed width of the foreland fold-thrust belt; Y'_{gs} is the deformed width of the Great Smoky (Blue Ridge–Piedmont) thrust sheet. (B) Restored sections constructed by pulling apart the foreland thrusts and placing tectonic units end to end according to estimated displacements, but without respect to geometry, with the Blue Ridge–Piedmont thrust sheet shown intact. X_f is the undeformed width of the foreland fold-thrust belt; Y_{gs} is the undeformed width of the Blue Ridge–Piedmont thrust sheet.

TABLE 1. ESTIMATED DISPLACEMENTS FOR SOUTHERN APPALACHIAN THRUST SYSTEM NAMED FAULTS ALONG REGIONAL CROSS SECTION IN FIGURE 13

Fault	Minimum total displacement (heave + throw) (km)	Basis
Sequatchie Valley–Cumberland Plateau overthrust	5	Last ramp toward foreland in system, and throw at ramp takes up most displacement; very small displacement west of ramp
Rockwood	7	Width; short strike length
Chattanooga	20	Width; short strike length
Kingston	10	Width; short strike length
Whiteoak Mountain	80	Mid-Ordovician facies boundary; throw; long strike length
Copper Creek	50	Width; moderate strike length
Beaver Valley	10	Motion transferred from Saltville west of Knoxville; short strike length
Saltville	100	Width; very long strike length (Virginia to Alabama); facies changes in Mid-Ordovician
Knoxville–Rocky Valley	10	Width; short strike length
Dumplin Valley–Chestuee	10	Width; short strike length; fold-related; may be out of sequence
Foreland total	302	
Great Smoky	360	Long strike length; measurable subsurface width based on seismic-reflection data
Rosman	20	Footwall Knox horses along Rosman fault, and geometry from ADCOH seismic-reflection profile
BRP total	380	

BRP—Blue Ridge–Piedmont megathrust; ADCOH—Appalachian Ultradeep Core Hole Project.

3. Palinspastic locations of southern Appalachian foreland fold-thrust belt thrusts suggest that these basement surface features may have profoundly affected original facies distributions and subsequent localization of faults and other structures as the southern Appalachian foreland fold-thrust belt was deformed.

4. The base of the Blue Ridge–Piedmont sheet is a SE-dipping surface interrupted by numerous antiforms that represent duplexes composed of southern Appalachian foreland fold-thrust belt footwall rocks. As these duplexes were constructed, they deformed the Blue Ridge–Piedmont thrust sheet as it moved forward. Many of these antiforms along the front of the sheet are quite linear, have relatively low structural relief (<2 km), and are traceable for many tens of kilometers; others farther into the sheet have more of an elliptical shape and several kilometers of structural relief.

5. The original shape of the basement surface exerted major influence on the deformed shape of the southern Appalachian foreland fold-thrust belt, as well as on the displacement vectors that tracked thrust-sheet emplacement.

6. Particle trajectories must curve in an orogen that traces a 30° arc from the inflection point of maximum curvature in N Georgia to the corresponding inflection point in SW Virginia, where thrusts have unequal displacements, so plane strain may have occurred locally but not regionally.

7. Plastic rheology for the entire thrust belt better addresses the particle trajectory problem than Coulomb behavior, but neither addresses problems of mechanics of localization of fault families, out-of-sequence thrusts, thrust spacing, and other attributes.

8. Displacement vectors within the southern Appalachian foreland fold-thrust belt may represent slip lines that track body deformation in the deforming wedge. Heterogeneous strain is responsible for the 3-D deformation plan of the southern Appalachian foreland fold-thrust belt, and probably other curved foreland fold-thrust belts.

9. Displacements of southern Appalachian foreland fold-thrust belt thrusts sum approximately to the minimum restorable displacement on the Blue Ridge–Piedmont megathrust sheet. Sub–Blue Ridge–Piedmont deformation of southern Appalachian foreland fold-thrust belt rocks (not quantifiable), along with out-of-sequence thrusting within the Blue Ridge–Piedmont sheet, increases the total displacement on the thrust system.

ACKNOWLEDGMENTS

Support for this research was provided by the U.S. Department of Energy National Petroleum Technology Laboratory (contract DE-FC26-02NT15341), the University of Tennessee Science Alliance Center of Excellence, ConocoPhillips, Tengasco, Inc., and Coal Creek Mining and Manufacturing Company. We thank Camilo Montes for calculating the volumes beneath several domes in Figure 9A. The original versions of cross sections one through five in Figure 11 were constructed in 2003 by Arthur J. Merschat, Matthew P. Gatewood, Donald W. Stahr III, John G. Bultman, and Jonathan C. Evenick as part of a graduate course in cross-section construction. These have been modified some and retrodeformed by Whisner. Earlier versions of this paper were improved by reviews from John Rodgers, William A. Thomas, Richard H. Groshong, and Catherine L. Hanks. Constructive reviews of this manuscript by James W. Sears and Robert I. Thompson significantly improved both the science and the writing. The authors, however, remain culpable for all errors of fact or interpretation.

REFERENCES CITED

Aleinikoff, J.N., Zartman, R.E., Rankin, D.W., and Burton, W.C., 1995, U-Pb ages of metarhyolites of the Catoctin and Mt. Rogers Formations, central and southern Appalachians: Evidence of two pulses of Iapetan rifting: American Journal of Science, v. 295, p. 428–454.

Bally, A.W., Gordy, P.L., and Stewart, G.A., 1966, Structure, seismic data, and orogenic evolution of southern Canadian Rocky Mountains: Bulletin of Canadian Petroleum Geology, v. 14, p. 337–381.

Bartholomew, M.J., and Schultz, A.P., 1980, Geologic Structure and Hydrocarbon Potential along the Saltville and Pulaski Thrust Sheets in Southwestern Virginia and Northeastern Tennessee: Part B. Deformation in the Hanging Wall of the Pulaski Thrust Sheet near Ironto, Montgomery County, Virginia: Virginia Division of Mineral Resources Publication 23, Sheets 3 and 4.

Boyer, S.E., 1995, Sedimentary basin taper as a factor controlling the geometry and advance of thrust belts: American Journal of Science, v. 295, p. 1220–1254.

Boyer, S.E., and Elliott, D., 1982, Thrust systems: American Association of Petroleum Geologists Bulletin, v. 66, p. 1196–1230.

Bryant, B., and Reed, J.C., Jr., 1971, Geology of the Grandfather Mountain Window and Vicinity, North Carolina and Tennessee: U.S. Geological Survey Professional Paper 615, 190 p.

Butts, C., 1927, Fensters in the Cumberland Overthrust Block in Southwestern Virginia: Virginia Geological Survey Bulletin 28, 12 p.

Carter, M.W., 1994, Stratigraphy and Structure of Part of the Blue Ridge Foothills, Polk and Monroe Counties, Tennessee [M.S. thesis]: Knoxville, University of Tennessee, 233 p.

Chapple, W.M., 1978, Mechanics of thin-skinned fold-and-thrust belts: Geological Society of America Bulletin, v. 89, p. 1189–1198, doi: 10.1130/0016-7606(1978)89<1189:MOTFB>2.0.CO;2.

Cook, F.A., Brown, L.D., Kaufman, S., and Oliver, J.E., 1983, The COCORP Seismic Reflection Traverse across the Southern Appalachians: American Association of Petroleum Geologists Studies in Geology 14, 61 p.

Costain, J.K., Hatcher, R.D., Jr., and Çoruh, C., 1989, Appalachian Ultradeep Core Hole (ADCOH) Project site investigation: Regional seismic lines and geologic interpretation, in Hatcher, R.D., Jr., Viele, G.W., and Thomas, W.A., eds., The Appalachian–Ouachita Orogen in the United States: Boulder, Colorado, Geological Society of America, Geology of North America, v. F-2, Plate 8.

Costello, J.O., 1984, Relationships between the Cartersville Fault and Great Smoky Fault in the Southern Appalachians: A Reinterpretation [M.S. thesis]: Columbia, University of South Carolina, 75 p.

Couzens, B.A., Dunne, W.M., Onasch, C.M., and Glass, R., 1993, Strain variations and three-dimensional strain at the transition from the southern to the central Appalachians: Journal of Structural Geology, v. 15, p. 451–464, doi: 10.1016/0191-8141(93)90140-6.

Currie, J.B., Patnode, H.W., and Trump, R.P., 1962, Development of folds in sedimentary strata: Geological Society of America Bulletin, v. 73, p. 655–674.

Dahlen, F.A., 1990, Critical taper model of fold-and-thrust belts and accretionary wedges: Annual Review of Earth and Planetary Sciences, v. 18, p. 55–99, doi: 10.1146/annurev.ea.18.050190.000415.

Dahlstrom, C.D.A., 1970, Structural geology in the eastern margin of the Canadian Rocky Mountains: Bulletin of Canadian Petroleum Geology, v. 18, p. 332–406.

Davis, D., Suppe, J., and Dahlen, F.A., 1983, Mechanics of fold-and-thrust belts and accretionary wedges: Journal of Geophysical Research, v. 88, p. 1153–1172.

Diegel, F.A., 1986, Topological constraints on imbricate thrust networks, examples from the Mountain City window, Tennessee, U.S.A.: Journal of Structural Geology, v. 8, p. 269–280, doi: 10.1016/0191-8141(86)90048-9.

Elliott, D., 1976, The energy balance and deformation mechanisms of thrust sheets: Royal Society of London Philosophical Transactions, Series A, v. 283, p. 289–312.

Epstein, A.G., Epstein, J.B., and Harris, L.D., 1977, Conodont Color Alteration—An Index to Organic Matter Metamorphism: U.S. Geological Survey Professional Paper 995, 27 p.

Ettensohn, F.R., 2004, Modeling the nature and development of major Paleozoic clastic wedges in the Appalachian basin, USA: Journal of Geodynamics, v. 37, p. 657–681, doi: 10.1016/j.jog.2004.02.009.

Evans, M.A., 1994, Joints and décollement zones in Middle Devonian shales: Evidence for multiple deformation events in the central Appalachian Plateau: Geological Society of America Bulletin, v. 106, p. 447–460, doi: 10.1130/0016-7606(1994)106<0447:JADCZI>2.3.CO;2.

Faill, R.T., 1998, A geologic history of the north-central Appalachians. Part 3: The Alleghany orogeny: American Journal of Science, v. 298, p. 131–179.

Favret, P.D., and Williams, R.T., 1988, Basement beneath the Blue Ridge and Inner Piedmont in northeastern Georgia and the Carolinas: A preserved Late Proterozoic rifted continental margin: Geological Society of America Bulletin, v. 100, p. 1999–2007, doi: 10.1130/0016-7606 (1988)100<1999:BBTBRA>2.3.CO;2.

Ferrill, D.A., and Groshong, R.H., Jr., 1993, Kinematic model for the curvature of the northern Subalpine Chain, France: Journal of Structural Geology, v. 15, p. 523–541, doi: 10.1016/0191-8141(93)90146-2.

Finney, S.C., Grubb, B.J., and Hatcher, R.D., Jr., 1996, Graphic correlation of Middle Ordovician graptolite shale, southern Appalachians: An approach for examining the subsidence and migration of a Taconic foreland basin: Geological Society of America Bulletin, v. 108, p. 355–371, doi: 10.1130/0016-7606(1996)108<0355:GCOMOG>2.3.CO;2.

Gao, D., Shumaker, R.C., and Wilson, T.O., 2000, Along-axis segmentation and growth history of the Rome trough in the central Appalachian basin: American Association of Petroleum Geologists Bulletin, v. 84, p. 75–99.

Geiser, P., 1988, Kinematics of deformation in the construction and analysis of geological cross sections in deformed terranes, in Mitra, G., and Wojtal, S., eds., Geometries and Mechanisms of Thrusting: Geological Society of America Special Paper 222, p. 47–77.

Geiser, P., and Engelder, T., 1983, The distribution of layer-parallel shortening fabrics in the Appalachian foreland of New York and Pennsylvania: Evidence for two coaxial phases of the Alleghanian orogeny, in Hatcher, R.D., Jr., Williams, H., and Zietz, I., eds., Contributions to the Tectonics and Geophysics of Mountain Chains: Boulder, Colorado, Geological Society of America Memoir 158, p. 161–176.

Gordy, P.L., Frey, F.R., and Norris, D.K., 1977, Geological Guide for the Canadian Society of Petroleum Geologists and 1977 Waterton–Glacier Park Field Conference: Calgary, Canadian Society of Petroleum Geologists, 93 p.

Gwinn, V.E., 1964, Thin-skinned tectonics in the Plateau and northwestern Valley and Ridge Provinces of the central Appalachians: Geological Society of America Bulletin, v. 75, p. 863–900, doi: 10.1130/0016-7606 (1964)75[863:TTITPA]2.0.CO;2.

Hansen, E., 1971, Strain Facies: New York Springer-Verlag, 207 p.

Hardeman, W.D., 1966, Geologic Map of Tennessee: Nashville, Tennessee Division of Geology, scale 1:250,000.

Harris, L.D., 1970, Details of thin-skinned tectonics in parts of Valley and Ridge and Cumberland Plateau Provinces of the southern Appalachians, in Fisher, G.W., Pettijohn, F.J., Reed, J.C., Jr., and Weaver, K.N., eds., Studies of Appalachian Geology: Central and Southern: New York, Wiley-Interscience, p. 161–173.

Harris, L.D., 1979, Similarities between the thick-skinned Blue Ridge anticlinorium and the thin-skinned Powell Valley anticline: Geological Society of America Bulletin, v. 90, p. 525–539, doi: 10.1130/0016-7606 (1979)90<525:SBTTBR>2.0.CO;2.

Harris, L.D., and Milici, R.C., 1977, Characteristics of Thin-Skinned Style of Deformation in the Southern Appalachians, and Potential Hydrocarbon Traps: U.S. Geological Survey Professional Paper 1018, 40 p.

Harris, L.D., Harris, A.G., de Witt, W., Jr., and Bayer, K.C., 1981, Evaluation of the southern overthrust belt beneath Blue Ridge–Piedmont thrust: American Association of Petroleum Geologists Bulletin, v. 65, p. 2497–2505.

Hasson, K.O., and Haase, C.S., 1988, Lithofacies and paleogeography of the Conasauga Group (Middle and Late Cambrian) in the Valley and Ridge province of East Tennessee: Geological Society of America Bulletin, v. 100, p. 234–246, doi: 10.1130/0016-7606(1988)100<0234:LAPOTC> 2.3.CO;2.

Hatcher, R.D., Jr., 1971, Structural, petrologic, and stratigraphic evidence favoring a thrust solution to the Brevard problem: American Journal of Science, v. 270, p. 177–202.

Hatcher, R.D., Jr., 1991, Interactive property of large thrust sheets with footwall rocks—The subthrust interactive duplex hypothesis: A mechanism of dome formation in thrust sheets: Tectonophysics, v. 191, p. 237–242, doi: 10.1016/0040-1951(91)90059-2.

Hatcher, R.D., Jr., 1999, Crust-forming processes, in Sinha, A.K., ed., Basement Tectonics Volume 13: Dordrecht, Netherlands, Kluwer Academic Publishers, p. 99–118.

Hatcher, R.D., Jr., 2001, Rheological partitioning during multiple reactivation of the Palaeozoic Brevard fault zone, southern Appalachians, USA, in Holdsworth, R.E., Strachan, R.A., Macloughlin, J.F., and Knipe, R.J., eds., The Nature and Significance of Fault Zone Weakening: Geological Society [London] Special Publication 186, p. 255–269.

Hatcher, R.D., Jr., 2002, The Alleghanian (Appalachian) orogeny, a product of zipper tectonics: Rotational transpressive continent-continent collision and closing of ancient oceans along irregular margins, in Catalán, J.R.M., Hatcher, R.D., Jr., Arenas, R., and García, F.D., eds., Variscan-Appalachian Dynamics: The Building of the Late Paleozoic Basement: Geological Society of America Special Paper 364, p. 199–208.

Hatcher, R.D., Jr., 2004, Properties of thrusts and the upper bounds for the size of thrust sheets, in McClay, K.R., ed., Thrust Tectonics and Hydrocarbon Systems: American Association of Petroleum Geologists Memoir 82, p. 18–29.

Hatcher, R.D., Jr., and Whisner, J.B., 1998, Deformation in the core of the southern Appalachian triangle zone: Geological Society of America Abstracts with Programs, v. 30, no. 7, p. A-234.

Hatcher, R.D., Jr., and Zietz, I., 1980, Tectonic implications of regional aeromagnetic and gravity data from the southern Appalachians, in Wones, D.R., ed., The Caledonides in the USA: Virginia Polytechnic Institute Department of Geological Sciences Memoir 2, p. 235–244.

Hatcher, R.D., Jr., Thomas, W.A., Geiser, P.A., Snoke, A.W., Mosher, S., and Wiltschko, D.V., 1989a, Alleghanian orogen, Chapter 5, in Hatcher, R.D., Jr., Thomas, W.A., and Viele, G.W., eds., The Appalachian–Ouachita Orogen in the United States: Boulder, Colorado, Geological Society of America, Geology of North America, v. F-2, p. 233–318.

Hatcher, R.D., Jr., Thomas, W.A., and Viele, G.W., eds., 1989b, The Appalachian-Ouachita Orogen in the United States: Boulder, Colorado, Geological Society of America, Geology of North America, v. F-2, 767 p.

Hatcher, R.D., Jr., Osberg, P.H., Robinson, P., and Thomas, W.A., 1990, Tectonic map of the U.S. Appalachians: Boulder, Colorado, Geological Society of America, Geology of North America, v. F-2, Plate 1, scale 1:2,000,000.

Hayes, C.W., 1891, The overthrust faults of the southern Appalachians: Geological Society of America Bulletin, v. 2, p. 141–154.

Hindle, D.A., and Burkhard, M., 1999, Strain, displacement and rotation associated with the formation of curvature in fold belts; the example of the Jura arc: Journal of Structural Geology, v. 21, p. 1089–1101, doi: 10.1016/ S0191-8141(99)00021-8.

Jacobeen, F., Jr., and Kanes, W.H., 1974, Structure of the Broadtop synclinorium and its implications for Appalachian structural style: American Association of Petroleum Geologists Bulletin, v. 58, p. 362–375.

Johnson, S.S., 1993, Geologic Map of Virginia: Charlottesville, Virginia Division of Mineral Resources, scale 1:500,000.

Jones, P.B., 1987, Quantitative Geometry of Thrust and Fold Belt Structures: American Association of Petroleum Geologists, Methods in Exploration Series No. 6, 26 p.

Jones, P.B., 1996, Triangle zone geometry, terminology and kinematics: Bulletin of Canadian Petroleum Geology, v. 44, p. 139–152.

Kilsdonk, M.W., and Wiltschko, D.V., 1988, Deformation mechanisms in the southeast ramp region of the Pine Mountain block, Tennessee: Geological Society of America Bulletin, v. 100, p. 653–664, doi: 10.1130/0016-7606 (1988)100<0653:DMITSR>2.3.CO;2.

King, E.R., and Zietz, I., 1978, The New York–Alabama lineament: Geophysical evidence for a major crustal break in the basement beneath the Appalachian basin: Geology, v. 6, p. 312–318, doi: 10.1130/0091-7613(1978)6 <312:TNYLGE>2.0.CO;2.

King, P.B., and Ferguson, H.W., 1960, Geology of Northeasternmost Tennessee: U.S. Geological Survey Professional Paper 311, 136 p.

King, P.B., Neuman, R.B., and Hadley, J.B., 1968, Geology of the Great Smoky Mountains National Park, Tennessee and North Carolina: U.S. Geological Survey Professional Paper 587, 23 p.

Laubscher, H.P., 1988, Material balance in Alpine orogeny: Geological Society of America Bulletin, v. 100, p. 1313–1328, doi: 10.1130/0016-7606 (1988)100<1313:MBIAO>2.3.CO;2.

Macedo, J., and Marshak, S., 1999, Controls on the geometry of fold-thrust belt salients: Geological Society of America Bulletin, v. 111, p. 1808–1822, doi: 10.1130/0016-7606(1999)111<1808:COTGOF>2.3.CO;2.

Marshak, S., 1988, Kinematics of orocline and arc formation in thin-skinned orogens: Tectonics, v. 7, p. 73–86.

Marshak, S., 2004, Salients, recesses, arcs, oroclines, and syntaxes—A review of ideas concerning the formation of map-view curves in fold-thrust belts, in McClay, K.R., ed., Thrust Tectonics and Hydrocarbon Systems: American Association of Petroleum Geologists Memoir 82, p. 131–156.

Merschat, A.J., Hatcher, R.D., Jr., and Davis, T.L., 2005, 3D deformation, kinematics, and crustal flow in the northern Inner Piedmont, southern Appalachians, USA: Journal of Structural Geology, v. 27, p. 1252–1281, doi: 10.1016/j.jsg.2004.08.005.

Milici, R.C., 1970, The Allegheny structural front in Tennessee and its regional tectonic implications: American Journal of Science, v. 268, p. 127–141.

Milici, R.C., 1975, Structural patterns in the southern Appalachians—Evidence for a gravity glide mechanism for Alleghanian deformation: Geological Society of America Bulletin, v. 86, p. 1316–1320, doi: 10.1130/0016-7606(1975)86<1316:SPITSA>2.0.CO;2.

Mitra, G., 1997, Evolution of salients in a fold-and-thrust belt: The effects of sedimentary basin geometry, strain distribution and critical taper, in Sengupta, S., ed., Evolution of Geological Structures in Micro- to Macro-scales: London, Chapman & Hall, p. 59–90.

Mitra, S., 1988, Three-dimensional geometry and kinematic evolution of the Pine Mountain thrust system, southern Appalachians: Geological Society of America Bulletin, v. 100, p. 72–95, doi: 10.1130/0016-7606(1988)100 <0072:TDGAKE>2.3.CO;2.

Nelson, A.E., and Zietz, I., 1983, The Clingman lineament, other aeromagnetic features, and other major lithotectonic units in part of the southern Appalachian Mountains: Southeastern Geology, v. 24, p. 147–157.

Norris, D.K., 1958, Structural Conditions in Canadian Coal Mines: Geological Survey of Canada Bulletin 44, 54 p.

Orndorf, R.C., Harris, A.G., and Schultz, A.P., 1988, Reevaluation of Conodont Color Alteration Patterns in Ordovician Rocks, East-Central Valley and Ridge and Western Blue Ridge Provinces, Tennessee: U.S. Geological Survey Bulletin 1839, p. D1–D10.

Osborne, W.E., Szabo, M.W., Copeland, C.W., Jr., and Neathery, T.L., 1989, Geologic Map of Alabama: Tuscaloosa, Alabama Geological Survey, scale 1:500,000.

Pickering, S.M., Jr., 1976, Geologic Map of Georgia: Atlanta, Georgia Geologic Survey, scale 1:500,000.

Platt, J.P., Behrmann, J.H., Cunningham, P.C., Dewey, J.F., Helman, M., Parisch, M., Shepley, M.G., Wallis, S., and Weston, P.J., 1989, Kinematics of the Alpine arc and the motion history of Adria: Nature, v. 337, p. 158–161, doi: 10.1038/337158a0.

Price, R.A., 1988, The mechanical paradox of large overthrusts: Geological Society of America Bulletin, v. 100, p. 1898–1908, doi: 10.1130/0016-7606 (1988)100<1898:TMPOLO>2.3.CO;2.

Price, R.A., and Mountjoy, E.W., 1970, Geologic structure of the Canadian Rocky Mountains between Bow and Athabaska Rivers—A progress report: Geological Association of Canada Special Paper 6, p. 7–25.

Price, R.A., and Sears, J.W., 2000, A preliminary palinspastic map of the Mesoproterozoic Belt-Purcell Supergroup, Canada and USA: Implications for the tectonic setting and structural evolution of the Purcell anticlinorium and the Sullivan deposit, Chapter 5, in Lydon, J.W., Höy, T., Slack, J.F., and Knapp, M.E., eds., The Geological Environment of the Sullivan Deposit, British Columbia: Geological Association of Canada, Mineral Deposits Division, MDD Special Publication No. 1, p. 61–81.

Quinlan, G.M., and Beaumont, C., 1984, Appalachian thrusting, lithospheric flexure, and the Paleozoic stratigraphy of the eastern interior of North America: Canadian Journal of Earth Sciences, v. 21, p. 973–996.

Ramsay, J.M., 1991, Some geometric problems of ramp-flat thrust models, in McClay, K.R., ed., Thrust Tectonics: London, Chapman and Hall, p. 191–200.

Rankin, D.W., 1975, The continental margin of eastern North America in the southern Appalachians: The opening and closing of the Proto-Atlantic Ocean: American Journal of Science, v. 275-A, p. 298–336.

Rankin, D.W., Drake, A.A., Jr., Glover, L., III, Goldsmith, R., Hall, L.M., Murray, D.P., Ratcliffe, N.M., Read, J.F., Secor, D.T., Jr., and Stanley, R.S., 1989, Pre-orogenic terranes, in Hatcher, R.D., Jr., Thomas, W.A., and Viele, G.W., eds., The Appalachian-Ouachita Orogen in the United States: Boulder, Colorado, Geological Society of America, Geology of North America, v. F-2, p. 7–100.

Rich, J.L., 1934, Mechanics of low-angle overthrust faulting as illustrated by Cumberland thrust block, Virginia, Kentucky, and Tennessee: American Association of Petroleum Geologists Bulletin, v. 18, p. 1584–1596.

Rodgers, J., 1953, Geologic Map of East Tennessee with Explanatory Text: Nashville, Tennessee Division of Geology Bulletin 58, Part II, 168 p.

Rodgers, J., 1964, Basement and no-basement hypotheses in the Jura and the Appalachian Valley and Ridge, in Lowry, W.D., ed., Tectonics of the Southern Appalachians: Virginia Polytechnic Institute Department of Geological Sciences Memoir 1, p. 71–80.

Rodgers, J., 1970, The Tectonics of the Appalachians: New York, Interscience Publishers, 271 p.

Roeder, D.H., 1975, Polyphase thrusting in the Valley and Ridge: Geological Society of America Abstracts with Programs, v. 7, no. 4, p. 527–528.

Roeder, D.H., and Witherspoon, W.R., 1978, Palinspastic map of East Tennessee: American Journal of Science, v. 278, no. 4, p. 543–550.

Roeder, D.H., Gilbert, O.E., Jr., and Witherspoon, W.R., 1978, Evolution and macroscopic structure of Valley and Ridge Province thrust belt, Tennessee and Virginia: Knoxville, Tennessee, University of Tennessee, Department of Geological Sciences, Studies in Geology no. 2, p. 25.

Royse, F., Jr., Warner, M.A., and Reese, D.L., 1975, Thrust belt of Wyoming, Idaho, and northern Utah; structural geometry and related problems, in Balyard, D.W., ed., Deep Drilling Frontiers in Central Rocky Mountains: Denver, Colorado, Rocky Mountain Geological Association Symposium, p. 41–54.

Schedl, A., and Wiltschko, D.V., 1987, Possible effects of pre-existing basement topography on thrust fault ramping: Journal of Structural Geology, v. 9, p. 1029–1038, doi: 10.1016/0191-8141(87)90011-3.

Sears, J.W., 1988, Origin and palinspastic significance of a fault duplex near Cartersville, Georgia, in Mitra, G., and Wojtal, S., eds., Geometries and Mechanisms of Thrusting, with Special Reference to the Appalachians: Geological Society of America Special Paper 222, p. 179–184.

Sears, J.W., and Hendrix, M., 2004, Lewis and Clark Line and the rotational origin of the Alberta and Helena salients, North American Cordillera, in Sussman, A., and Weil, A., eds., Orogenic Curvature: Geological Society of America Special Paper 383, p. 173–186.

Shumaker, R.C., 1996, Structural history of the Appalachian basin, in Roen, J.B., and Walker, B.J., eds., The Atlas of Major Appalachian Gas Plays: Morgantown, West Virginia, U.S. Department of Energy Morgantown Energy Center, p. 8–25.

Spraggins, S.A., and Dunne, W.M., 2002, Deformation history of the Roanoke recess, Appalachians, USA: Journal of Structural Geology, v. 24, p. 411–433, doi: 10.1016/S0191-8141(01)00077-3.

Steinhauff, D.M., and Walker, K.R., 1997, Relationship of Middle Ordovician late-stage diagenesis to diagenesis of Lower Ordovician Upper Knox Group carbonates in East Tennessee: Lexington, Kentucky, American Association of Petroleum Geologists, Eastern Section Annual Meeting Abstracts, p. 84–86.

Suppe, J., 1983, Geometry and kinematics of fault-bend folding: American Journal of Science, v. 283, p. 648–721.

Thomas, W.A., 1977, Evolution of Ouachita-Appalachian salients and recesses from reentrants and promontories in the continental margin: American Journal of Science, v. 277, p. 1233–1278.

Thomas, W.A., 1982, Stratigraphy and structure of the Appalachian fold and thrust belt in Alabama, in Thomas, W.A., and Neathery, T.L., eds., Appalachian Thrust Belt in Alabama: Tectonic and Sedimentation: Guidebook for Field Trip No. 13, 9th Annual Meeting of the Geological Society of America: Tuscaloosa, Alabama Geological Society, p. 55–66.

Thomas, W.A., 1986, A Paleozoic synsedimentary structure in the Appalachian fold-thrust belt in Alabama, in McDowell, R.C., and Glover, L., III, eds., The Lowry Volume: Studies in Appalachian Geology: Blacksburg,

Virginia, Virginia Tech Department of Geological Sciences Memoir 3, p. 1–12.

Thomas, W.A., 1988, Stratigraphic framework of the geometry of the basal décollement of the Appalachian-Ouachita fold-thrust belt: Geologische Rundshau, v. 77, p. 183–190, doi: 10.1007/BF01848683.

Thomas, W.A., 1990, Controls on locations of transverse zones in thrust belts: Eclogae Geologicae Helvetiae, v. 83, p. 727–744.

Thomas, W.A., 1991, The Appalachian-Ouachita rifted margin of southeastern North America: Geological Society of America Bulletin, v. 103, p. 415–431, doi: 10.1130/0016-7606(1991)103<0415:TAORMO>2.3.CO;2.

Thomas, W.A., 2001, Mushwad: Ductile duplex in the Appalachian thrust belt in Alabama: American Association of Petroleum Geologists Bulletin, v. 85, p. 1847–1969.

Thomas, W.A., and Bayona, G., 2005, The Appalachian Thrust Belt in Alabama and Georgia: Thrust-Belt Structure, Basement Structure, and Palinspastic Reconstruction: Geological Survey of Alabama Monograph 16, 48 p.

Watkins, J.S., 1964, Regional Geologic Implications of the Gravity and Magnetic Fields of a Part of East Tennessee and Southern Kentucky: U.S. Geological Survey Professional Paper 516–A, 17 p.

Weaver, C.E., and Broekstra, B.R., 1984, Illite-mica, Chapter 4, in Weaver, C.E., ed., Shale-slate Metamorphism in Southern Appalachians: New York, Elsevier Science Publishers, p. 67–97.

Wheeler, R.L., Winslow, M., Horner, R., Dean, S., Kulander, B., Drahovzal, J.A., Gold, D.P., Gilbert, O.E., Jr., Werner, E., Sites, R., and Perry, W.J., 1979, Cross-strike structural discontinuities in thrust belts, mostly Appalachian: Southeastern Geology, v. 20, p. 193–203.

Willis, B., 1893, The Mechanics of Appalachian Structure: Washington, D.C., U.S. Geological Survey 13th Annual Report, Part 2, p. 211–281.

Wilson, C.W., Jr., and Stearns, R.G., 1958, Structure of the Cumberland Plateau, Tennessee: Geological Society of America Bulletin, v. 69, p. 1283–1296, doi: 10.1130/0016-7606(1958)69[1283:SOTCPT]2.0.CO;2.

Wiltschko, D.V., and Eastman, D.B., 1983, Role of basement warps and faults in localizing thrust ramps, in Hatcher, R.D., Jr., Williams, H., and Zietz, I., eds., Contributions to the Tectonics and Geophysics of Mountain Chains: Geological Society of America Memoir 158, p. 177–190.

Wojtal, S., 1992, One-dimensional models for plane and non-plane power-law flow in shortening and elongating thrust zones, in McClay, K.R., ed., Thrust Tectonics: London, Chapman and Hall, p. 41–52.

Woodward, N.B., ed., 1985, Balanced Structure Cross Sections in the Appalachians (Pennsylvania to Alabama): Knoxville, University of Tennessee, Department of Geological Sciences, Studies in Geology No. 12, 63 p.

Woodward, N.B., 1987, Geological applicability of critical-wedge thrust-belt models: Geological Society of America Bulletin, v. 99, p. 827–832, doi: 10.1130/0016-7606(1987)99<827:GAOCTM>2.0.CO;2.

Woodward, N.B., 1988, Primary and secondary basement controls on thrust sheet geometries, in Schmidt, C.J., and Perry, W.L., Jr., Interaction of the Rocky Mountain Foreland and the Cordilleran Thrust Belt: Geological Society of America Memoir 171, p. 353–366.

MANUSCRIPT ACCEPTED BY THE SOCIETY 22 MARCH 2007

Balancing tectonic shortening in contrasting deformation styles through a mechanically heterogeneous stratigraphic succession

William A. Thomas*
Department of Geological Sciences, University of Kentucky, Lexington, Kentucky 40506-0053, USA

ABSTRACT

Multiple levels of frontal ramps and detachment flats accommodate tectonic shortening in contrasting deformation styles at different levels in a mechanically heterogeneous stratigraphic succession in a foreland thrust belt. The late Paleozoic Appalachian thrust belt in Alabama exhibits a balance of shortening in contrasting deformation styles at different stratigraphic levels. The regional décollement is in a weak unit (Cambrian shale) near the base of the Paleozoic succession above Precambrian crystalline basement rocks. Basement faults, now beneath the décollement, controlled the sedimentary thickness of the Cambrian shale and the location of high-amplitude frontal ramps of the regional stiff layer (Cambrian-Ordovician massive carbonate); shortening in a mushwad (ductile duplex) from thick Cambrian shale is balanced by translation of the regional stiff layer at a high-amplitude frontal ramp above a basement fault. A trailing, high-amplitude, brittle duplex of the regional stiff layer has a floor on the regional décollement and a roof that is also the floor of an upper-level, lower-amplitude, brittle duplex. The roof of the upper-level brittle duplex is a diffuse ductile detachment below an upper-level mushwad, with which parts of the brittle duplex are imbricated. The basal detachment of the upper-level mushwad changes along strike into a frontal ramp at a location coincident with a sedimentary facies change in the weak shale unit that hosts the mushwad. The roof of the upper-level mushwad is a brittle massive sandstone. Shortening on the regional décollement is balanced successively upward through contrasting tectonic styles in successive mechanically contrasting stratigraphic units.

Keywords: mushwad, duplex, thrust belt, décollement, Appalachians, Alabama.

INTRODUCTION

Foreland thrust belts in sedimentary rocks comprise a three-dimensional system of interconnected fault surfaces, including detachment flats, frontal ramps, and cross-strike links (lateral ramps, transverse faults, displacement transfers, and displacement gradients) (e.g., Rich, 1934; Price, 1981; Boyer and Elliott, 1982; Suppe, 1983; Thomas, 1990; Thomas and Bayona, 2002). Detachment flats parallel bedding in generally thin, weak (ductile) layers; whereas, frontal ramps cut across bedding in stiff (brittle) layers, up section in the direction of tectonic transport. Cross-strike links transfer displacement from one frontal ramp to another across strike and terminate frontal ramps along strike. An idealized system incorporates frontal ramps rising from a detachment flat (regional décollement) in an areally extensive weak layer at the base of a stiff sedimentary cover succession above a

*geowat@uky.edu

planar top of crystalline basement rocks in the footwall. Matching the corresponding hanging-wall and footwall cutoffs in the frontal ramps of imbricated thrust sheets provides for construction of balanced cross sections to measure tectonic shortening in the thrust belt (e.g., Dahlstrom, 1969; Price, 1981).

Structure of the sub-décollement basement rocks and variations in cover stratigraphy impose modifications to an idealized system of thrust-belt fault geometry. An alternation of weak and stiff layers in the stratigraphic succession allows multiple levels of detachment flats, and multilevel detachments provide roof and floor thrusts for a duplex (e.g., Boyer and Elliott, 1982). Lateral thickness variations and pinch-outs of detachment-host weak layers localize frontal ramps or cross-strike links, depending on orientation of the stratigraphic gradient with respect to the direction of tectonic transport. Fault-bounded blocks of sub-décollement crystalline basement rocks localize ramps in the overlying thrust belt (e.g., Wiltschko and Eastman, 1983, 1988; Thomas and Bayona, 2002), and amplitude of the thrust ramp generally corresponds to magnitude of the underlying basement fault. Style of thrust-belt structure (e.g., detachment fold or ramp-related fold) depends on the relative thicknesses of weak and stiff layers (e.g., Suppe, 1983; Suppe and Medwedeff, 1990; Groshong and Epard, 1994). These potential variations yield a complex three-dimensional array of multiple-level detachment flats, detachment folds, frontal ramps, ramp-related folds, duplexes, and cross-strike links. Overall tectonic shortening in the thrust belt must be balanced through differing structural styles at successive levels.

The northeast-striking, northwest-verging structures of the Appalachian thrust belt in Alabama (Figs. 1, 2) provide an opportunity to test the balance of shortening through a range of variations in stratigraphy, basement structure, and thrust-belt structure. The classic geometry of frontal ramps and detachment flats generally characterizes the thrust belt; however, detailed balanced cross sections reveal modifications in geometry imposed by stratigraphic variations and basement structure. The large-scale thrust sheets illustrate a variety of structural styles at different stratigraphic levels, requiring that tectonic shortening be balanced through differing structural styles at different stratigraphic levels, using both line-length and area balancing. The cross sections used in this article are derived from a set of 18 balanced cross sections of the Appalachian thrust belt in Alabama and Georgia, the database for which includes geologic map patterns, bedding attitudes, stratigraphic thicknesses, deep wells, and seismic reflection profiles (Thomas and Bayona, 2005). The along-strike continuity of frontal ramps is partitioned by cross-strike links, most of which are aligned within four regional transverse zones across strike of the thrust belt (Fig. 1B) (Thomas, 1990; Thomas and Bayona, 2002). The transverse zones in the Appalachian thrust belt in Ala-

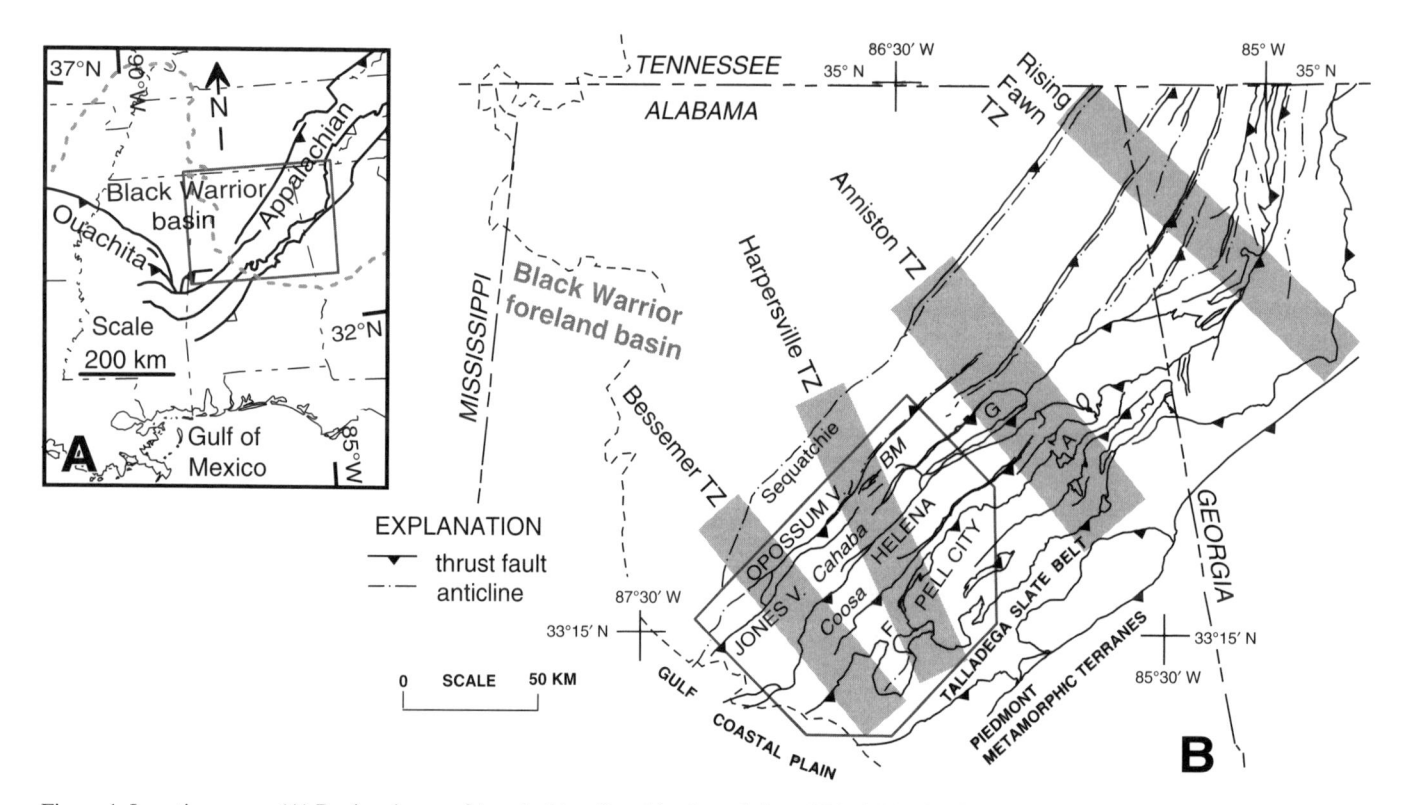

Figure 1. Location maps. (A) Regional map of Appalachian-Ouachita thrust belt and Black Warrior foreland basin. Dashed gray line shows limit of Gulf Coastal Plain. Gray rectangle shows area of Figure 1B. (B) Outline map of structural geology of Appalachian thrust belt in Alabama and Georgia. Abbreviations: A—Angel lateral ramp; BM—Blount Mountain syncline; F—Fourmile Creek lateral ramp anticline; G—Gadsden mushwad; TZ—transverse zone; V.—Valley. Names of regional synclinoria shown in italics. Gray outline shows area of Figure 2.

bama and Georgia are structurally similar to well-documented transverse zones in other thrust belts, such as in the southern Canadian Rocky Mountains (Benvenuto and Price, 1979). This article will focus on the area between the Bessemer and Anniston transverse zones, including the Harpersville transverse zone, in the Appalachian thrust belt in Alabama (Fig. 1B).

STRATIGRAPHIC SUCCESSION, MECHANICAL PROPERTIES, AND STRUCTURAL STYLES

The Appalachian thrust belt in Alabama incorporates a succession of Paleozoic strata ranging in age from Cambrian to Pennsylvanian (e.g., Thomas and Bayona, 2005), and the Pennsylvanian age of the youngest displaced rocks documents that the foreland thrusting was a component of the late Paleozoic Alleghanian orogeny (e.g., Hatcher et al., 1989). The Paleozoic succession encompasses a range of rock types, mechanical properties, and lateral variations, which characterize four distinct lithotectonic/mechanical units[1] (Fig. 3).

The regional décollement is in Cambrian strata (lithotectonic/mechanical unit 1, Fig. 3) near the base of the cover succession above Precambrian crystalline basement rocks. In the footwall

---
[1]The term, "lithotectonic unit," was defined by Wood and Bergin (1970) and expanded by Wiltschko and Geiser (p. 247–250, in Hatcher et al., 1989) to identify subdivisions with distinctive mechanical properties within a stratigraphic succession, and the term is used commonly in that sense in sedimentary thrust belts. The same term, "lithotectonic unit," also has been used for rock assemblages with origins in a specific tectonic environment through a defined time span (e.g., Hibbard et al., 2006). To avoid possible confusion and to recognize common usage, a modified term, "lithotectonic/mechanical unit," is used here.

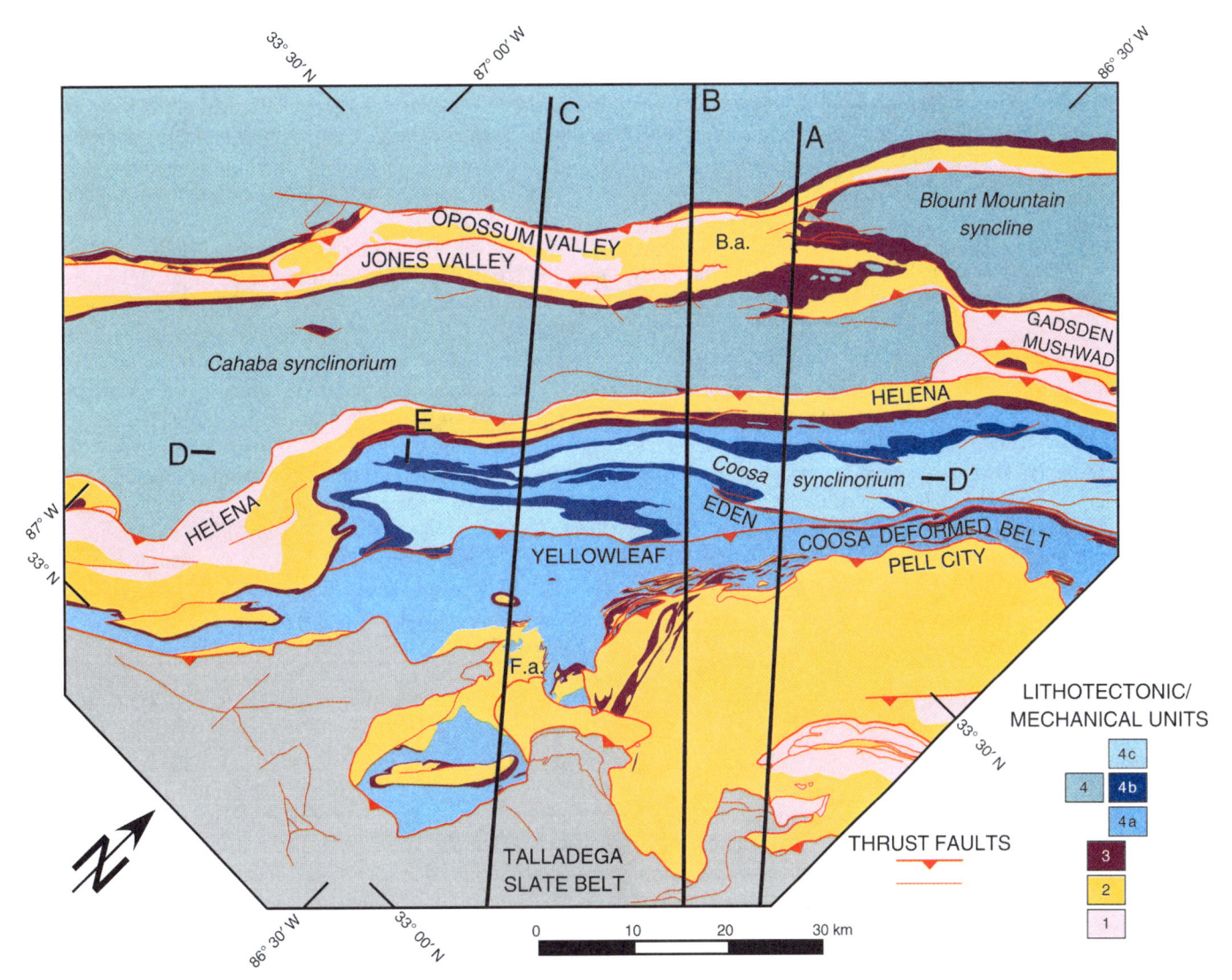

Figure 2. Geologic map of part of Appalachian thrust belt in Alabama (modified from Osborne et al., 1988, and Szabo et al., 1988). Location of map shown in Figure 1B. Straight lines show locations of strike-perpendicular cross sections A, B, and C (Fig. 4) and E (Fig. 7); labels show end points of strike-parallel cross section D–D′. Abbreviations: B.a.—Birmingham anticlinorium; F.a.—Fourmile Creek lateral ramp anticline.

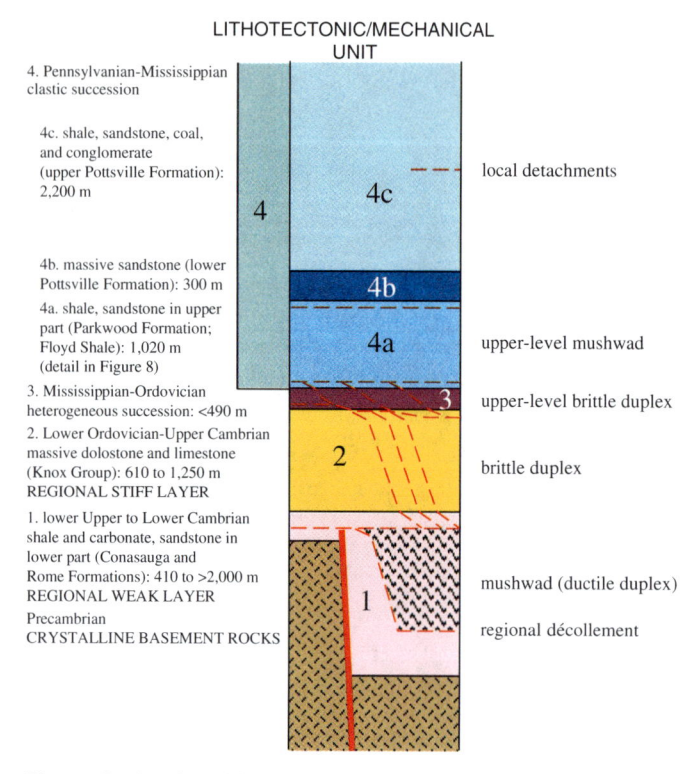

Figure 3. Stratigraphic column, showing lithotectonic/mechanical units used in construction of cross sections of Figures 4, 5, 6, 7, and 9. Dashed red lines show stratigraphic levels of detachments and duplexes. Units are identified by number in the text and by color on the cross sections. Thicknesses are approximate regional averages.

of the regional décollement, steep faults displace the basement rocks, producing a horst-and-graben system beneath the décollement in the cover strata (Fig. 4) (Thomas and Bayona, 2005). At the base of the Paleozoic cover stratigraphy, Lower Cambrian sandstone and overlying carbonate pinch out northwestward toward the foreland, and the overlying upper Lower Cambrian mudstone-sandstone redbeds (Rome Formation) downlap northwestward onto basement rocks (Kidd and Neathery, 1976). A complex facies mosaic of carbonate and shale in the Middle to lower Upper Cambrian Conasauga Formation records an upward transition from dominantly clastic to dominantly carbonate deposition; however, the stratigraphic level of the transition varies laterally (Osborne et al., 2000). Distributions of facies and thickness have a systematic relationship to basement horsts and grabens, indicating synsedimentary extensional fault movement during a late phase of continental rifting and opening of the Iapetus Ocean (Thomas, 1991; Thomas et al., 2000). A very thick succession of shale and thin-bedded limestone fills the Birmingham basement graben, whereas a massive carbonate marks the northwest shoulder of the graben. The regional décollement is in the basal Cambrian sandstone only in the more interior (southeastern) thrust sheets and rises stratigraphically toward the foreland through the Rome Formation redbeds into shale-dominated parts of the Conasauga Formation in the frontal thrust sheets (Thomas and Bayona,

2005). Depending on stratigraphic thickness of the weak layer, the style of structure ranges from a ramp-related fold to a detachment fold to a mushwad, which is a ductile duplex of weak-layer rocks (Thomas, 2001). In contrast to a brittle duplex, a mushwad consists of disharmonically folded and faulted strata with no internally coherent large horses or fault surfaces. Tectonic thickening of the weak layer uplifts and distorts the structural geometry of the stiff layer in the roof of a mushwad. Evolution of a mushwad from the merger of cores of detachment folds depends on availability of a thick succession (large volume) of bulk ductile strata. Floor and roof thrusts bound a mushwad at top and bottom, like the floor and roof faults of a brittle duplex. The concept of mushwad structure originated in the Appalachian thrust belt in Alabama; however, such structures are well documented elsewhere in evaporite weak layers and may be more common than recognized in shale weak layers (Thomas, 2001, and references cited therein).

A thick, massive Upper Cambrian–Lower Ordovician carbonate unit (Knox Group, lithotectonic/mechanical unit 2, Fig. 3) constitutes the dominant regional stiff layer that controls geometry of large frontal ramps (Thomas and Bayona, 2005). Massive dolomitized carbonate rocks in the lower part of the Knox Group grade upward into limestone in the upper part (Raymond, 1993). A large-scale brittle duplex has a floor on the regional décollement and a roof in or above the limestone beds of the upper part of the Knox Group.

A relatively thin succession from Middle Ordovician through Lower Mississippian (lithotectonic/mechanical unit 3, Fig. 3) is highly variable in composition and thickness (summary in Thomas and Bayona, 2005). Four regional unconformities punctuate the succession. The succession includes units of carbonate, chert, shale, and sandstone, which represent a range of depositional settings from passive-margin shelf facies to Taconic and Acadian synorogenic deposits. In part of the thrust belt, the layered, heterogeneous, Ordovician-Mississippian succession hosts local detachment flats at several different stratigraphic levels; and, along with detachment flats in the upper part of the underlying Knox Group, the local detachment flats provide the floor of an upper-level duplex with horses of the heterogeneous Ordovician-Mississippian succession.

Above Lower Mississippian shelf-carbonate rocks, Upper Mississippian and Lower Pennsylvanian synorogenic clastic wedges reflect Ouachita orogenesis on the southwest and Appalachian orogenesis on the southeast (e.g., Mack et al., 1983; Thomas, 1988). In the Appalachian thrust belt and foreland in Alabama, the Upper Mississippian and Lower Pennsylvanian clastic facies thicken southwestward and prograde northeastward (Thomas, 1972, 1995; Mars and Thomas, 1999); the Upper Mississippian clastic facies intertongue with and grade northeastward into a shallow-marine carbonate facies. Linear deposits of carbonate grainstone and quartzose sandstone parallel the northwest-trending facies boundary between Mississippian clastic and carbonate facies (Thomas, 1972). Facies and thickness distributions indicate down-to-southwest subsidence of the Black Warrior basin in the Ouachita orogenic foreland

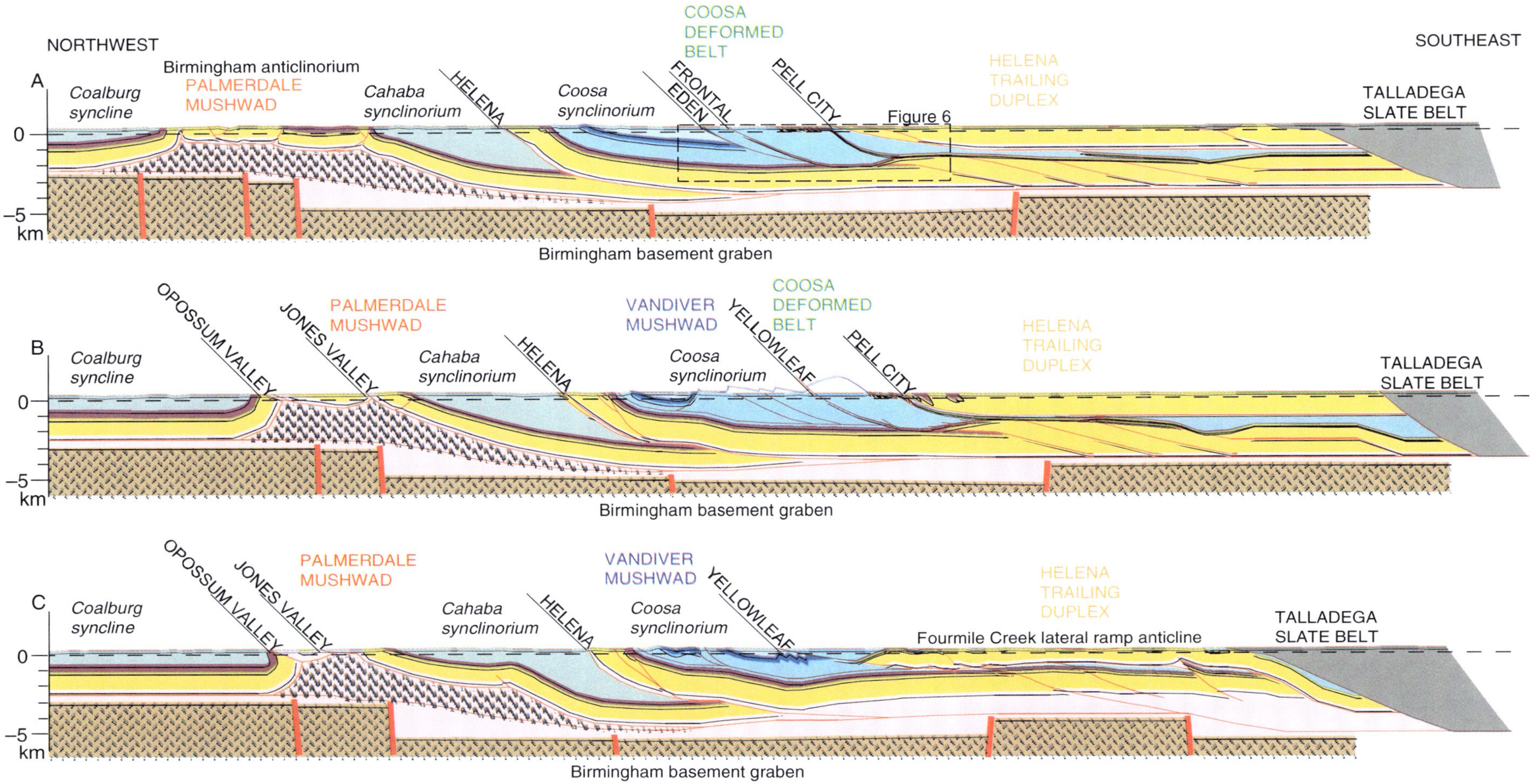

Figure 4. Strike-perpendicular, balanced, restorable structural cross sections A, B, and C, showing the Palmerdale mushwad, Helena thrust sheet and trailing duplex, Coosa deformed belt upper-level duplex, and Vandiver upper-level mushwad. Lines of cross sections are shown in Figure 2. Color symbols for lithotectonic/mechanical units are explained in Figure 3. Blue line above topographic surface in cross section B represents minimum projected elevation of the deformed top of unit 4a. Dashed outline on cross section A shows area of Figure 6.

(Fig. 1A), which included the area of the palinspastic locations of the Appalachian thrust sheets (Thomas, 1995); and the northwest-trending facies boundaries are approximately perpendicular to the subsequent Appalachian thrust-belt strike. In contrast, southeastward thickening and coarsening of the upper part of the Lower Pennsylvanian succession indicate initiation of down-to-southeast subsidence and northwestward progradation of clastic facies in response to initial Appalachian orogenic thrust loading of the foreland on the southeast during Early Pennsylvanian time (Thomas, 1988; Whiting and Thomas, 1994), later than both Ouachita thrust loading on the southwest and Alleghanian foreland thrust loading farther northeast along Appalachian strike (Fig. 1). Within the northeastward prograding clastic wedge, shale in the lower part grades upward to a shale-dominated succession that contains relatively thin and discontinuous sandstone units (lithotectonic/mechanical unit 4a, Fig. 3). In contrast, the overlying Lower Pennsylvanian succession includes relatively thick, massive, quartzose sandstone (lithotectonic/mechanical unit 4b, Fig. 3). Above the massive sandstone, the youngest preserved strata in the Appalachian thrust belt in Alabama include shale, sandstone, coal, and conglomerate (lithotectonic/mechanical unit 4c, Fig. 3), part of which represents the Appalachian provenance on the southeast (Mack et al., 1983; Pashin, 1994). In the frontal structures of the thrust belt, the entire Upper Mississippian–Lower Pennsylvanian succession was transported passively in large-scale thrust sheets and is not differentiated in the structural cross sections (lithotectonic/mechanical unit 4, Fig. 3). In parts of the thrust belt, shale-dominated intervals within the Mississippian-Pennsylvanian clastic wedge include local detachment flats (e.g., Pashin and Groshong, 1998). The thick Mississippian shale-dominated succession (unit 4a) locally hosts an upper-level mushwad. The massive Lower Pennsylvanian sandstone (unit 4b) constitutes a stiff layer, and the stratigraphically higher shale-sandstone-coal-conglomerate succession (unit 4c) hosts locally discontinuous detachments.

STRUCTURAL GEOLOGY

Jones Valley Thrust Sheet and Palmerdale Mushwad

The Jones Valley thrust fault and a leading splay, the Opossum Valley thrust fault, form the most northwesterly (most frontal) large-scale frontal ramp in the Appalachian thrust belt in Alabama, and structures farther northwest have low amplitude (Figs. 2, 4) (Thomas and Bayona, 2005). The high-amplitude frontal ramp rises northwestward in the direction of tectonic transport over the down-to-southeast, northwest-boundary fault of the Birmingham basement graben (Fig. 4). The Jones Valley and Opossum Valley faults end northeastward along strike through displacement gradients at the Harpersville transverse zone (Figs. 1B, 2) (Thomas and Bayona, 2005). Displacement increases southwestward along strike; however, the Opossum Valley fault ends abruptly southwestward at a lateral ramp in the footwall of the Jones Valley fault within the Bessemer transverse zone (Figs. 1B, 2). The large-scale Cahaba synclinorium comprises the trailing part of the Jones Valley thrust sheet. With the exceptions of local detachments in the Pennsylvanian coal-bearing succession (unit 4c) in the Cahaba synclinorium in the trailing part of the thrust sheet (Pashin and Groshong, 1998) and of small-amplitude folds associated with forward and back thrusts in the leading part of the thrust sheet (Thomas and Bayona, 2005), strata above the regional stiff layer (unit 2) were transported passively within the Jones Valley thrust sheet.

The Jones Valley and Opossum Valley thrust sheets comprise the roof of the Palmerdale mushwad (ductile duplex), the floor of which is the regional décollement in the shale-dominated Conasauga Formation (unit 1) (Fig. 4). At the Bessemer transverse zone, where the Opossum Valley thrust sheet ends southwestward, an along-strike change in structure of the leading edge distinguishes the Palmerdale mushwad from the otherwise laterally continuous Bessemer mushwad along strike to the southwest (Thomas and Bayona, 2005). Northeastward at the Harpersville transverse zone, a dextral offset from the leading edge of the Palmerdale mushwad to that of the Gadsden mushwad constitutes a displacement transfer at the mushwad level (Thomas, 2001). The leading part of the Palmerdale mushwad ends northeastward along strike, where the surface expression is the northeast-plunging southwest end of the Blount Mountain syncline (Fig. 2) above the roof thrust of the mushwad (Thomas, 2001). Northeastward along strike, the roof of the mushwad descends to the level of the regional décollement beneath the Blount Mountain syncline in concert with thinning and pinch-out of the mushwad. The trailing part of the Palmerdale mushwad merges northeastward along strike with the leading part of the Gadsden mushwad. The trailing part of the Gadsden mushwad pinches out southwestward beneath the southwest-plunging northeast end of the Cahaba synclinorium (Fig. 2), where the mushwad roof descends southwestward to the regional décollement (Thomas, 2001). Together, the Gadsden, Palmerdale, and Bessemer mushwads are aligned along strike of the Birmingham basement graben, which is beneath the regional décollement (Fig. 4) (Thomas, 2001; Thomas and Bayona, 2005).

The Jones Valley and Opossum Valley thrust faults in the roof of the Palmerdale mushwad have detachments in the Conasauga Formation (unit 1), and the stiff-layer roof includes some upper Conasauga strata attached to the stratigraphically overlying stiff-layer Knox Group carbonate rocks (unit 2) (Fig. 4). Where the thrust faults end northeastward along strike, the roof of the Palmerdale mushwad forms the crest of the Birmingham anticlinorium; however, several laterally discontinuous thrusts and back thrusts break the stiff layer in the mushwad roof (cross section A, Fig. 4). The Palmerdale mushwad is entirely in the subsurface beneath the Jones Valley and Opossum Valley thrust sheets. The Gadsden mushwad (abbreviated label G, Fig. 1B; Fig. 2) farther northeast is partly exposed and has been penetrated by exploratory drilling, providing documentation for the internal structural style of a mushwad (Thomas, 2001). By analogy with the Gadsden mushwad, the Palmerdale and Bessemer mushwads in the subsurface are inferred to consist of ductilely deformed shale-dominated Conasauga Formation (unit 1). All of the mushwads have similar expression in seismic reflection profiles, lacking internal coherent reflectors.

Helena Thrust Sheet

In a large-scale frontal ramp at the trailing cutoff of the Jones Valley thrust sheet, the Helena fault cuts up section from the regional décollement in the Cambrian Rome Formation (unit 1) through the entire Paleozoic stratigraphic succession to the top of the Pennsylvanian coal-bearing succession (unit 4c) (Fig. 4). The Helena fault conforms to the simple geometry of a single detachment flat and large-scale frontal ramp. A large trailing syncline, the Coosa synclinorium, ends southwestward at the Bessemer transverse zone in northeast-plunging beds over a high-amplitude footwall lateral ramp (Fig. 5), which is expressed in an abrupt sinistral bend in the trace of the Helena fault (Fig. 2). Northeast of the lateral ramp at the Bessemer transverse zone, a nonplunging flat extends ~29 km along strike of the Coosa synclinorium to a structurally lower lateral ramp at the Harpersville transverse zone, where northeast-plunging beds descend into the deepest depression of the synclinorium (Thomas, 2001; Thomas and Bayona, 2005).

Internally the Helena thrust sheet includes a trailing, high-amplitude, lower-level, brittle duplex of unit 2 (regional stiff layer); a low-amplitude, upper-level brittle duplex mostly of unit 3 (Ordovician-Mississippian heterogeneous strata); and an upper-level mushwad of unit 4a (Mississippian shale-dominated succession) with associated internal folds in the Coosa synclinorium (Fig. 4). In the footwall of the Helena fault, truncation of southwest-plunging beds at the up-plunge northeast end of the Cahaba synclinorium in the trailing part of the Jones Valley thrust sheet (Fig. 2) indicates a significant component of break-back (out-of-sequence) thrusting (Thomas and Bayona, 2005).

Helena Duplex

The leading part of the Helena thrust sheet (Figs. 2, 4), between the Bessemer transverse zone and the Anniston transverse zone, appears as a rigid beam of the regional stiff layer (unit 2), rising northwestward over a frontal ramp (the trailing cutoff of the Jones Valley thrust sheet on the southwest, and the trailing cutoffs of other thrust sheets northeast of the Harpersville transverse zone). The Helena frontal ramp dips down to the regional décollement beneath the trailing Coosa synclinorium. The trailing part of the Helena thrust sheet in the subsurface southeast of the Coosa synclinorium includes a brittle duplex of most of unit 2 (regional stiff layer) (Fig. 4). The floor of the duplex is the regional décollement in unit 1, and the roof thrust ranges in stratigraphic level from the upper part of unit 2 (regional stiff layer) to various horizons within unit 3 (Middle Ordovician–Lower Mississippian heterogeneous succession). Seismic reflection profiles faintly image the separate horses and more clearly image thickening by imbrication. The duplex style of deformation in the trailing part of the Helena thrust sheet is illustrated in the exposed Western Coosa thrust sheet, which is an up-plunge counterpart of the Helena thrust sheet northeast of the Angel lateral ramp (abbreviated label A, Fig. 1B) at the Anniston transverse zone (Thomas and Bayona, 2005). Southwest of the sinistral bend in trace of the Helena fault at the Bessemer transverse zone (Fig. 1B), the exposed leading part of the Helena thrust sheet (Fig. 2) has a duplex style with a floor in unit 1 (Cambrian shale) beneath the regional stiff layer (unit 2). The balanced cross sections (Fig. 4) use the exposed duplexes in the Western Coosa and southwestern Helena thrust sheets as analogs for the geometry of the duplex in the trailing part of the subsurface Helena thrust sheet (Thomas and Bayona, 2005).

Coosa Deformed Belt

The Coosa deformed belt merges two distinct, but tectonically imbricated, structural styles in mechanically distinct stratigraphic units. The structurally and stratigraphically lower part is a brittle duplex, of which the floor thrust ranges stratigraphically from the upper part of unit 2 through several different horizons within unit 3 (Figs. 3, 6). The floor thrust of the upper-level brittle duplex corresponds to the roof thrust of the lower-level brittle duplex in unit 2 in the trailing part of the Helena thrust sheet (Fig. 4). The roof of the upper-level duplex is above the top of unit 3 near the base of the thick shale-dominated succession in the lower part of the Mississippian-Pennsylvanian clastic wedge (unit 4a); however, the roof apparently is a diffuse zone of disharmonic, bulk-ductile deformation rather than a discrete fault.

The stratigraphically thin horses of unit 3 and uppermost unit 2 in the Coosa deformed belt are arranged in three tiers across strike. One to eight internal imbricate thrust sheets comprise the duplex in each tier (e.g., Fig. 6). Tectonically thickened, disharmonically deformed masses of the shale-dominated lower part of the Mississippian-Pennsylvanian clastic wedge (unit 4a) are imbricated with the three tiers of the upper-level brittle duplex in the Coosa deformed belt (Figs. 4, 6). The most frontal (northwestern) tier is laterally continuous along strike, but the two trailing tiers consist of laterally discontinuous imbricate thrust sheets (the interior tier is restricted to the northeast of cross section A, Figs. 2, 4). The footwall beneath the floor thrust of the frontal tier is the Eden thrust sheet (cross section A, Fig. 4; Fig. 6), which consists of disharmonically deformed beds of the shale-dominated lower part of the Mississippian-Pennsylvanian clastic wedge (unit 4a), stratigraphically and structurally similar to the rocks between the imbricate tiers of the brittle upper-level duplex in the Coosa deformed belt. Southwestward along strike at the Harpersville transverse zone, unit 3 strata in the frontal tier plunge beneath the land surface (near cross section A, Figs. 2, 4), and the floor thrust is mapped as the Yellowleaf fault, which has unit 4 strata in both hanging wall and footwall (Figs. 2, 4). Southwestward across the Harpersville transverse zone, a lateral ramp of the trailing part of the Yellowleaf fault cuts down section from the upper-level

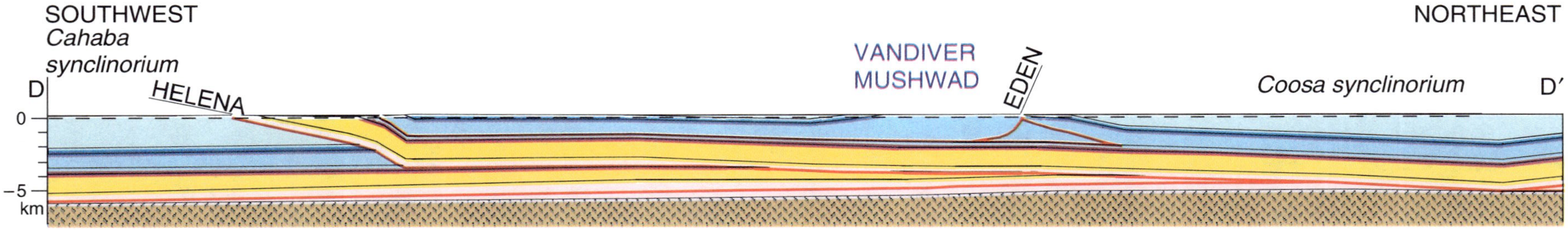

Figure 5. Cross section D–D' parallel to regional strike along the Coosa synclinorium from the footwall lateral ramp on the southwest at a sinistral bend in the Helena fault to the deepest depression of the synclinorium, showing relation of the Vandiver mushwad to along-strike variations in structure of the Coosa synclinorium. The cross section crosses the lateral ramp in the Helena fault and the lateral ramp where the Eden fault bends from a frontal ramp on the northeast to the floor of the Vandiver mushwad on the southwest. End points of cross section are shown in Figure 2. Color symbols for lithotectonic/mechanical units are explained in Figure 3.

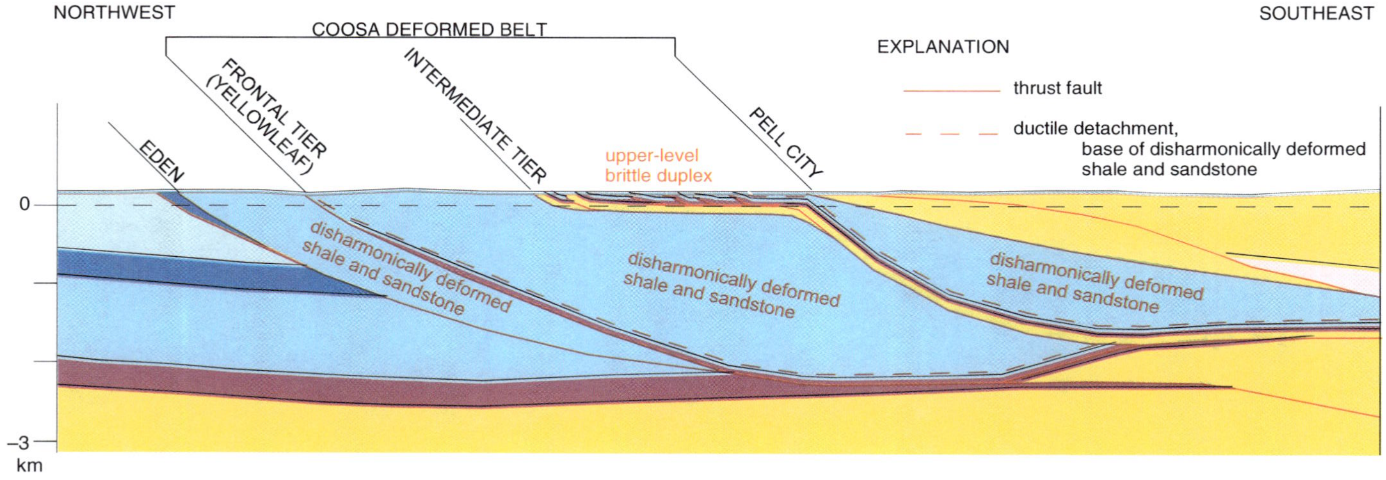

Figure 6. Cross section of Coosa deformed belt upper-level brittle duplex imbricated with disharmonically deformed shale and sandstone of unit 4a, enlarged from cross section A of Figure 4.

detachment near the base of unit 3 to the regional décollement in unit 1 beneath the northeast-plunging Fourmile Creek lateral ramp anticline (abbreviated label F, Fig. 1B; abbreviated label F.a., Fig. 2; cross section C, Fig. 4) (Thomas and Bayona, 2005). Similarly, northeastward at the Anniston transverse zone, an oppositely directed lateral ramp of the floor thrust cuts down section to the regional décollement beneath the southwest-plunging Angel lateral ramp (abbreviated label A, Fig. 1B). Stratigraphic variations both along and across the Coosa deformed belt suggest substantial shortening within the highly variable stratigraphic succession. The trailing parts of the Coosa deformed belt, including most of the interior tier, are truncated by the Pell City fault, which has a break-back sequence of thrusting with respect to the Coosa deformed belt (Fig. 4).

Eden Fault and Upper-Level Vandiver Mushwad

Between the Anniston and Harpersville transverse zones, a footwall frontal ramp of the Eden fault cuts stratigraphically upward from the upper-level detachment in Mississippian shale (unit 4a) toward the foreland to the top of the preserved coal-bearing Pennsylvanian strata (unit 4c) at the trailing cutoff of the Coosa synclinorium in the Helena thrust sheet (cross section A, Fig. 4; Fig. 6). The Eden frontal ramp rises from the stratigraphic level of the roof of the upper-level brittle duplex in the Coosa deformed belt, and the Eden hanging wall consists of disharmonically deformed, shale-dominated rocks of the lower part of the Mississippian-Pennsylvanian clastic wedge (unit 4a), similar to the rocks between the imbricate tiers of the upper-level duplex in the Coosa deformed belt (Fig. 6). The relationship of the Eden fault to beds in the footwall (the Coosa synclinorium), however, changes dramatically southwestward along strike. At the Harpersville transverse zone, the trace of the Eden fault curves westward across regional thrust-belt strike (between cross section A and cross section B, Figs. 2, 4), and the Eden thrust sheet merges southwestward into disharmonically deformed shale and sandstone (unit 4a) in the core of a detachment anticline between synclines in the unit 4b massive sandstone (cross sections B and C, Figs. 2, 4). The Eden fault evidently merges downward and southwestward with a diffuse, ductile detachment flat in the lower part of the Mississippian shale (unit 4a). In effect, the Eden frontal ramp on the northeast is deflected downward at a lateral ramp to the extensive ductile detachment on the southwest. The detachment flat extends the Eden fault northwestward across strike in the direction of tectonic transport at a stratigraphic level similar to that of the roof of the brittle upper-level duplex in the Coosa deformed belt. The detachment of the Eden fault is stratigraphically above unit 3 and is not associated with an underlying brittle duplex of unit 3 like that in the Coosa deformed belt. The Eden detachment in unit 4a extends northwestward across strike beneath the Pennsylvanian massive sandstone (unit 4b) in the Coosa synclinorium (cross sections B and C, Fig. 4).

Southwest of the Harpersville transverse zone, ductilely deformed shale and shale with thin units of sandstone (unit 4a) above the detachment flat (Eden fault) comprise the tectonically thickened Vandiver mushwad, which is exposed in a wide area across the southwestern part of the Coosa synclinorium (locally mapped as the Fungo Hollow deformed zone, Osborne and Ward, 1996). The outcrop areal extent of deformed shale and sandstone is irregular between outcrops of synclinally folded, massive Pennsylvanian sandstone (unit 4b) (Figs. 2, 4) (Osborne and Ward, 1996; Irvin et al., 2002; Rindsberg et al., 2003; Ward et al., 2004), which constitutes the roof of the upper-level Vandiver mushwad (Fig. 7). Outcrop data and seismic reflection profiles define folds of intermediate wavelength and amplitude in the massive Pennsylvanian sandstone. The folds are disharmonic with respect to the long frontal ramp of the Helena fault expressed in the regional stiff layer (unit 2), and the disharmony is accommodated in ductilely deformed, shale-dominated Mississippian strata (unit 4a) (cross sections B and C, Fig. 4; Fig. 7). Although the Eden fault can be projected as the floor of the mushwad, no distinct continuous roof thrust is recognizable. The roof may be a folded detachment at the stratigraphic contact between the ductile shale-dominated succession (unit 4a) and the more brittle, massive Pennsyl-

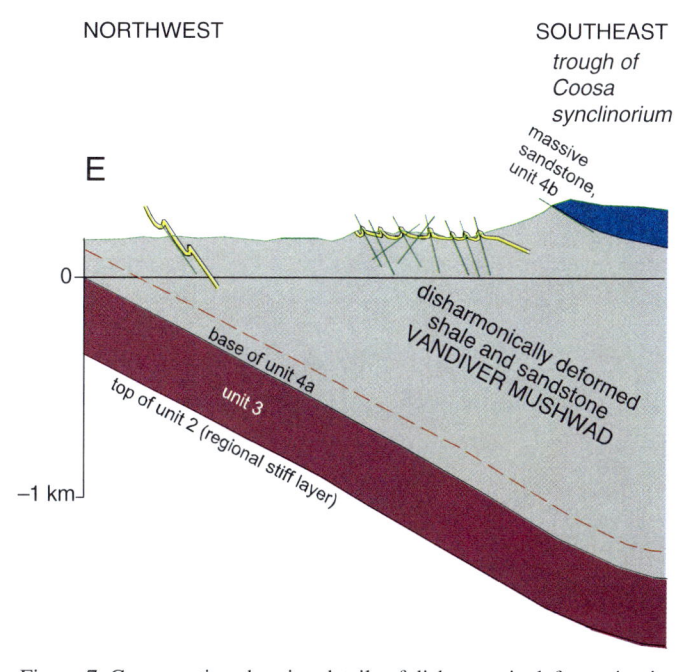

Figure 7. Cross section showing details of disharmonic deformation in shale and sandstone of unit 4a in the Vandiver mushwad on the northwest limb of the Coosa synclinorium. Line of cross section is shown in Figure 2. Color symbols for lithotectonic/mechanical units 3 and 4b are explained in Figure 3. For unit 4a, yellow color shows mappable deformed sandstone interbeds, and gray color shows poorly exposed disharmonically deformed shale and sandstone. Green lines show dip angles. Projections above and below ground surface of the shapes of folds in the sandstone are based on up- and down-plunge views from outcrop pattern in detailed mapping. Dashed red line shows approximate level of ductile detachment at the floor of the Vandiver mushwad; the base of the Pennsylvanian massive sandstone (unit 4b) defines the roof of the mushwad.

vanian sandstone (unit 4b). Across strike, the upper-level mushwad extends from the leading frontal ramp of the Yellowleaf fault (corresponding to the leading frontal ramp of the Coosa deformed belt) to the present eroded limit of the mushwad-host shale. Along strike, the northeast limit of the mushwad (defined by the along-strike change in dip of the Eden fault) coincides with the lower-level lateral ramp in the Coosa synclinorium, and the mushwad extends along strike of the flat between the two northeast-plunging footwall lateral ramps in the Coosa synclinorium (Fig. 5).

The internal structural style of the Vandiver upper-level mushwad is illustrated in local detailed mapping, which shows tight, upright folds in sandstone units (Fig. 7). Some of the folds are tightly appressed, expelling all of the shale from the cores of folds. Some of the exposed appressed anticlines appear as single successions of steeply dipping sandstone beds along ridge crests, and the hinges can be recognized only in canyons through the ridges where the sandstone beds in the opposite limbs diverge at lower elevations.

The along-strike change in style of the Eden fault from the floor of the Vandiver upper-level mushwad northeastward across the Harpersville transverse zone to the frontal ramp at the trailing cutoff of the Coosa synclinorium is localized at a facies change in the mushwad-host stratigraphy. On the southwest, the floor of the mushwad is in a Mississippian shale succession (Fig. 8). Toward the northeast (more distally in the Ouachita foreland, and along strike of subsequent Appalachian thrust faults), the Mississippian shale facies intertongues with limestone (replaced by chert) and quartzose sandstone that pinch out southwestward at the stratigraphic level of the floor of the mushwad farther southwest (Fig. 8). The change from ductile, shale-dominated stratigraphy to brittle carbonate and sandstone coincides spatially with the deflection of the mushwad floor thrust upward through a lateral ramp into the frontal ramp of the Eden fault northeast of the Harpersville transverse zone.

MEASURE OF TECTONIC SHORTENING IN DIFFERENT STRUCTURAL STYLES AT DIFFERENT STRATIGRAPHIC LEVELS

Palinspastically restored cross sections (Fig. 9), constructed by using line-length and area balancing, provide a measure of tectonic shortening (e.g., Price, 1981). The restored cross sections (Fig. 9) are derived from balanced structural cross sections (Fig. 4), which are based on outcrop geology, deep wells, and seismic reflection profiles (Thomas and Bayona, 2005). The seismic reflection profiles constrain the depth and geometry of subsurface structures, including frontal ramps and detachment flats, as well as the top of crystalline basement rocks. Where fault-related folds respond by flexure and interstratal slip, and where penetrative strain is small in comparison to the size of folds and thrust faults, a palinspastic reconstruction will conserve the length of stratigraphic markers from the deformed state to the undeformed reconstruction (Price, 1981). Shortening effects of micro- and mesoscopic structures are accommodated in bulk through area balancing of ductile units but are not separately incorporated in these measures of shortening.

Palinspastic restoration of the Palmerdale mushwad relies on area balancing to restore the ductilely deformed rocks as the fill of the Birmingham basement graben, and restoration by area balancing of the mushwad is balanced by bed-length restoration of the overlying stiff layer (Fig. 9). The shallow-marine depositional setting of the Cambrian-Ordovician carbonate stiff layer (unit 2) requires a horizontal restoration over the graben; however, post–Early Ordovician reactivation/inversion of the basement faults and synorogenic flexural subsidence deformed the stiff layer prior to Alleghanian thin-skinned thrusting (as shown in restored cross sections, Fig. 9). Where displacement on the Jones Valley and Opossum Valley faults decreases northeastward along strike to zero, and where cutoffs of the smaller faults are preserved within the crest of the Birmingham anticlinorium, the entire length of the stiff layer can be restored by bed-length balancing, providing the best possible measure of shortening (cross section A, Figs. 4, 9). Along strike to the southwest, where the leading edges (including the leading cutoffs) of the Jones Valley and Opossum Valley thrust sheets have been eroded, only a minimum estimate of shortening of the stiff layer can be documented by bed-length balancing. Area balancing of the subsurface mushwad, however, provides an independent estimate of shortening on the roof thrust. Successful area balancing of the mushwad in several cross sections along strike (Thomas and Bayona, 2005)

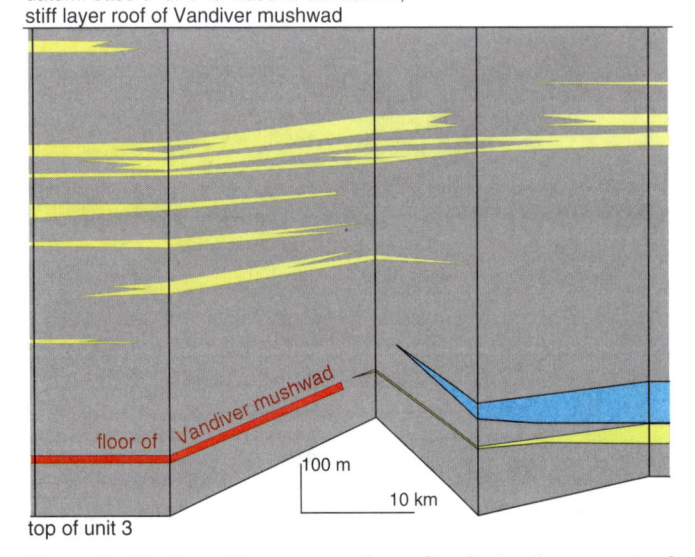

Figure 8. Stratigraphic cross section of unit 4a (lower part of Mississippian-Pennsylvanian synorogenic clastic wedge) along present structural strike of the Coosa synclinorium (from detailed stratigraphic sections in Thomas, 1972). Line of cross section approximates line of cross section D–D′ (Figs. 2, 5) but extends farther to the northeast. Colors: gray—shale, yellow—sandstone, blue—limestone (replaced by chert). Datum is base of unit 4b.

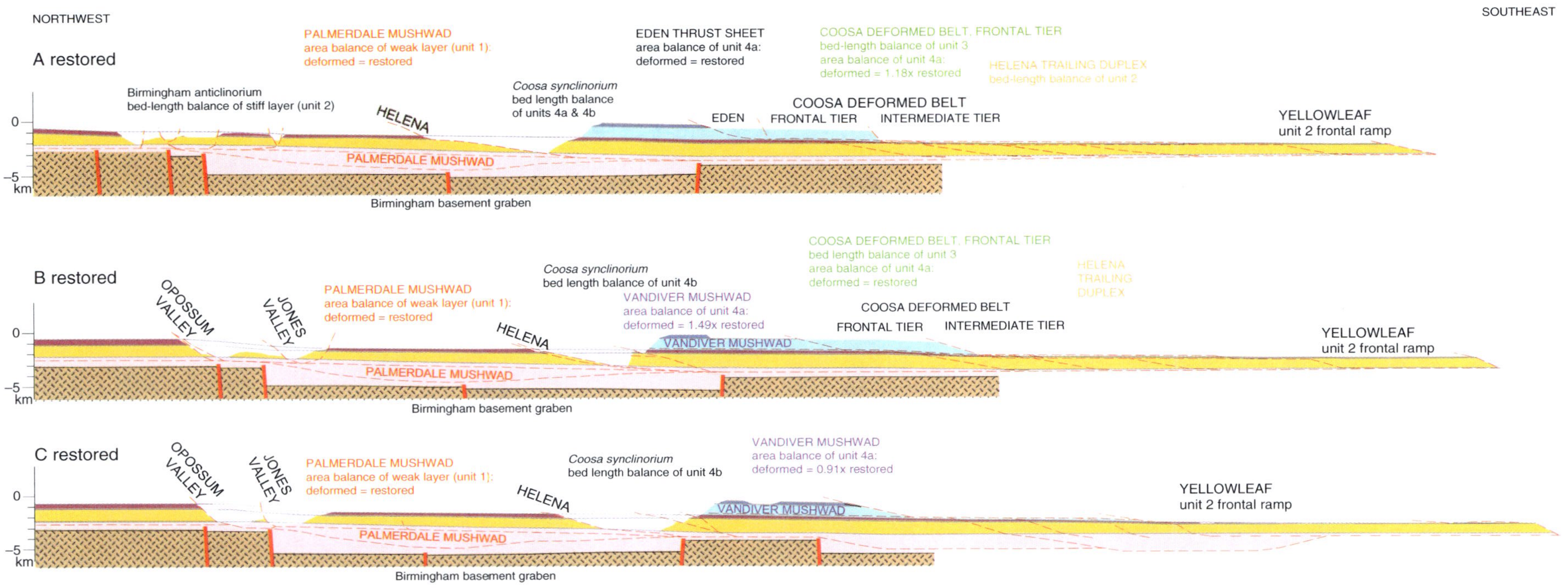

Figure 9. Palinspastic restorations of cross sections A, B, and C of Figure 4. Dashed red lines show projection of thrust faults. The footwall beneath the regional décollement is not shown southeast of the limit of seismic reflection profiles.

indicates that highly likely out-of-plane ductile displacement is compensated in three dimensions along strike.

Most of the tectonic shortening transmitted from the Alleghanian hinterland was accommodated in the high-amplitude Jones Valley (plus Opossum Valley) and Helena frontal ramps and in the Palmerdale mushwad, as well as in the trailing duplex of the Helena thrust sheet (Figs. 4, 9). Only a small component of tectonic shortening was transmitted into the foreland along the regional décollement, and the low-amplitude Sequatchie detachment anticline (Fig. 1B) marks the tip of the blind detachment (Thomas and Bayona, 2005). A buttressing effect (e.g., Wiltschko and Eastman, 1983, 1988) of the down-to-southeast, northwest-boundary fault of the Birmingham basement graben evidently concentrated shortening of the regional stiff layer in the high-amplitude frontal ramps and mushwad, rather than in foreland thrusting. Buttressing by the basement fault is further indicated by the break-back thrusting sequences of the Helena and trailing Pell City faults (Thomas and Bayona, 2005). Tectonic shortening in the Helena trailing duplex must be accommodated in the stratigraphy above the regional stiff layer, either by internal shortening or by foreland translation over frontal ramps that rise ultimately to the present erosion surface (Fig. 9).

The roof of the Helena trailing duplex of the regional stiff layer (unit 2) corresponds to the floor of the upper-level brittle duplex in the Coosa deformed belt (Figs. 4, 6, 9). Tectonic shortening within the Coosa deformed belt includes imbrication of the thin, brittle horses within each tier, as well as imbrication of each tier of the brittle duplex with the ductilely deformed, shale-dominated rocks (unit 4a) above the ductile roof of the upper-level brittle duplex. Bed-length balancing of the brittle horses in the upper-level duplex in the Coosa deformed belt equals the bed-length balance of the larger-scale, brittle horses in the lower-level Helena trailing duplex of the regional stiff layer (unit 2) (Fig. 9). Lack of a coherent cover bed precludes area balancing of the ductilely deformed rocks between the tiers of the thin brittle duplex of the Coosa deformed belt. Nevertheless, a minimal area balance, using an arbitrary unconstrained horizontal top, yields a restored stratigraphic thickness that is consistent with regional thickness distribution of the lower part of the Mississippian-Pennsylvanian clastic wedge (unit 4a), where it is not disharmonically deformed (Thomas, 1972).

Along the Eden fault at the frontal ramp between the Anniston and Harpersville transverse zones, tectonic shortening is transmitted upward to the present erosion surface (cross section A, Figs. 4, 9). In contrast, along strike southwestward across the Harpersville transverse zone, tectonic shortening is transmitted into the Vandiver upper-level mushwad (cross sections B and C, Figs. 4, 9). Tectonic shortening in the Vandiver mushwad is measured by area balancing of the disharmonically deformed, shale-dominated, lower part of the Mississippian-Pennsylvanian clastic wedge (unit 4a) and by the corresponding bed-length balancing of the massive Pennsylvanian sandstone stiff layer (unit 4b). Along-strike comparison of balanced cross sections shows an imbalance of area of mushwad rocks with respect to bed length of the stiff layer. Near the deflection of the Eden fault from the frontal ramp on the northeast to mushwad floor on the southwest, the area of the deformed mushwad is 149% of the restored area (cross section B, compare Fig. 4 and Fig. 9), indicating significant volume gain. In contrast, farther southwest along strike, the area of the deformed mushwad is 91% of the restored area (cross section C, compare Fig. 4 and Fig. 9), indicating volume loss. The along-strike northeastward gradient from volume loss (thinning) to volume gain (thickening) of the mushwad indicates northeastward along-strike ductile displacement of the rocks in the Vandiver mushwad. The deflection of the Eden fault structurally upward to the northeast along strike at a lateral ramp into the frontal-ramp geometry defines the northeast end of the Vandiver mushwad, and the deflected fault surface at the lateral ramp may have provided a buttress against which the ductile core of the mushwad was tectonically thickened.

CONCLUSIONS

An alternation of weak and stiff layers in the cover stratigraphy, as well as fault blocks in sub-décollement basement rocks, imposes variations in structural style of a thin-skinned thrust belt. Where a basement fault constitutes a buttress, thrust shortening is concentrated, and break-back sequences nucleate. Tectonic shortening is transmitted through the cover stratigraphy above the regional décollement, either by frontal ramps that cut up section to the top of the stratigraphic succession or by internal shortening in the stratigraphy. The Appalachian thrust belt in Alabama illustrates shortening on a regional décollement beneath a dominant regional stiff layer; however, variations in mechanical properties within the stratigraphy are reflected in upper-level detachments at which structural style depends on the stratigraphy.

Tectonic shortening, buttressed at a basement fault, is accommodated in a lower-level mushwad (ductile duplex) at the regional décollement and in high-amplitude frontal ramps of the regional stiff layer (Fig. 9). Tectonic shortening on the lower detachment and mushwad is transferred from the trailing floor detachment through the mushwad to the leading roof detachment, and shortening is balanced by translation of the stiff layer at the frontal ramps. A high-amplitude, brittle, trailing duplex of the regional stiff layer has a roof thrust near the top of the stiff layer, and displacement is transferred to the rocks above the roof (Fig. 9). The roof of the lower-level duplex forms the floor of an upper-level, brittle duplex involving a thinner stratigraphic succession; and the upper-level brittle duplex is imbricated with ductilely deformed rocks above a ductile roof (Figs. 4, 9). The roof thrust of the upper-level duplex ramps up to the surface along part of the leading edge; but along strike, the frontal ramp is deflected downward at a lateral ramp and extends as the floor of an upper-level mushwad (Figs. 4, 9). The roof of the upper-level mushwad is a disharmonically folded massive sandstone. Shortening on the regional décollement is balanced successively upward through the stratigraphic succession and through contrasting tectonic styles at successive stratigraphic levels. Along-strike changes in structural styles are related to along-strike changes in stratigraphy.

ACKNOWLEDGMENTS

This article was prepared while I was a Visiting Scientist at the Geological Survey of Alabama on sabbatical leave from the University of Kentucky, and I acknowledge both sources of support. Acknowledgment is made to the donors of the Petroleum Research Fund (38965), administered by the American Chemical Society, for partial support of this research. The cross sections, on which this article is based, are derived from a set of 18 balanced structural cross sections of the Appalachian thrust belt in Alabama and Georgia published as Monograph 16 of the Geological Survey of Alabama (Thomas and Bayona, 2005). My work in the Appalachian thrust belt in Alabama has benefited greatly from collaboration with Ed Osborne, who also provided a helpful review of an early draft of this manuscript. Matt Surles assisted with design of the illustrations. Constructive reviews by Jim Hibbard and Adolph Yonkee are gratefully acknowledged. It is a privilege to contribute to this volume in honor of Ray Price, who has contributed so much to our knowledge of thrust belts.

REFERENCES CITED

Benvenuto, G.L., and Price, R.A., 1979, Structural evolution of the Hosmer thrust sheet, southeastern British Columbia: Bulletin of Canadian Petroleum Geology, v. 27, p. 360–394.

Boyer, S.E., and Elliott, D., 1982, Thrust systems: American Association of Petroleum Geologists Bulletin, v. 66, p. 1196–1230.

Dahlstrom, C.D.A., 1969, Balanced cross sections: Canadian Journal of Earth Sciences, v. 6, p. 743–757.

Groshong, R.H., Jr., and Epard, J.-L., 1994, The role of strain in area-constant detachment folding: Journal of Structural Geology, v. 16, p. 613–618, doi: 10.1016/0191-8141(94)90113-9.

Hatcher, R.D., Jr., Thomas, W.A., Geiser, P.A., Snoke, A.W., Mosher, S., and Wiltschko, D.V., 1989, Alleghanian orogen, in Hatcher, R.D., Jr., Thomas, W.A., and Viele, G.W., eds., The Appalachian-Ouachita Orogen in the United States: Geological Society of America, The Geology of North America, v. F-2, p. 233–318.

Hibbard, J.P., van Staal, C.R., Rankin, D.W., and Williams, H., 2006, Lithotectonic map of the Appalachian orogen, Canada—United States of America: Geological Survey of Canada, Map 2096A, scale 1:1,500,000.

Irvin, G.D., Osborne, W.E., and Ward, W.E., 2002, Geologic map of the Chelsea 7.5-minute quadrangle, Shelby County, Alabama: Alabama Geological Survey Quadrangle Series Map 22, plate 1, scale 1:24,000.

Kidd, J.T., and Neathery, T.L., 1976, Correlation between Cambrian rocks of the southern Appalachian geosyncline and the interior low plateaus: Geology, v. 4, p. 767–769, doi: 10.1130/0091-7613(1976)4<767:CBCROT>2.0.CO;2.

Mack, G.H., Thomas, W.A., and Horsey, C.A., 1983, Composition of Carboniferous sandstones and tectonic framework of southern Appalachian-Ouachita orogen: Journal of Sedimentary Petrology, v. 53, p. 931–946.

Mars, J.C., and Thomas, W.A., 1999, Sequential filling of a late Paleozoic foreland basin: Journal of Sedimentary Research, v. 69, p. 1191–1208.

Osborne, W.E., and Ward, W.E., II, 1996, Geologic map of the Helena 7.5-minute quadrangle, Jefferson and Shelby Counties, Alabama: Alabama Geological Survey Quadrangle Series Map 14, plate 1, scale 1:24,000.

Osborne, W.E., Szabo, M.W., Neathery, T.L., and Copeland, C.W., Jr., 1988, Geologic map of Alabama northeast sheet: Alabama Geological Survey Special Map 220, scale 1:250,000.

Osborne, W.E., Thomas, W.A., Astini, R.A., and Irvin, G.D., 2000, Stratigraphy of the Conasauga Formation and equivalent units, Appalachian thrust belt in Alabama, in Osborne, W.E., Thomas, W.A., and Astini, R.A., eds., The Conasauga Formation and Equivalent Units in the Appalachian Thrust Belt in Alabama: Alabama Geological Society Guidebook, 37th Annual Field Trip, p. 1–17.

Pashin, J.C., 1994, Flexurally influenced eustatic cycles in the Pottsville Formation (Lower Pennsylvanian), Black Warrior basin, Alabama, in Dennison, J.M., and Ettensohn, F.R., eds., Tectonic and Eustatic Controls on Sedimentary Cycles: Society of Economic Paleontologists and Mineralogists Concepts in Sedimentology and Paleontology, v. 4, p. 89–105.

Pashin, J.C., and Groshong, R.H., Jr., 1998, Structural control of coalbed methane production in Alabama: International Journal of Coal Geology, v. 38, p. 89–113, doi: 10.1016/S0166-5162(98)00034-2.

Price, R.A., 1981, The Cordilleran foreland thrust and fold belt in the southern Canadian Rocky Mountains, in McClay, K.R., and Price, N.J., eds., Thrust and Nappe Tectonics: Geological Society of London Special Publication 9, p. 427–448.

Raymond, D.E., 1993, The Knox Group of Alabama: An overview: Alabama Geological Survey Bulletin 152, 160 p.

Rich, J.L., 1934, Mechanics of low-angle overthrust faulting as illustrated by Cumberland thrust block, Virginia, Kentucky, and Tennessee: American Association of Petroleum Geologists Bulletin, v. 18, p. 1584–1596.

Rindsberg, A.K., Ward, W.E., Osborne, W.E., and Irvin, G.D., 2003, Geologic map of the Vandiver 7.5-minute quadrangle, Shelby and Jefferson Counties, Alabama: Alabama Geological Survey Quadrangle Series Map 24, plate 1, scale 1:24,000.

Suppe, J., 1983, Geometry and kinematics of fault-bend folding: American Journal of Science, v. 283, p. 684–721.

Suppe, J., and Medwedeff, D.A., 1990, Geometry and kinematics of fault propagation folding: Eclogae Geologicae Helvetiae, v. 83, p. 409–454.

Szabo, M.W., Osborne, W.E., and Copeland, C.W., Jr., 1988, Geologic map of Alabama northwest sheet: Alabama Geological Survey Special Map 220, scale 1:250,000.

Thomas, W.A., 1972, Mississippian stratigraphy of Alabama: Alabama Geological Survey Monograph 12, 121 p.

Thomas, W.A., 1988, The Black Warrior basin, in Sloss, L.L., ed., Sedimentary Cover—North American Craton: U.S.: Geological Society of America, The Geology of North America, v. D-2, p. 471–492, plate 8.

Thomas, W.A., 1990, Controls on locations of transverse zones in thrust belts: Eclogae Geologicae Helvetiae, v. 83, p. 727–744.

Thomas, W.A., 1991, The Appalachian-Ouachita rifted margin of southeastern North America: Geological Society of America Bulletin, v. 103, p. 415–431, doi: 10.1130/0016-7606(1991)103<0415:TAORMO>2.3.CO;2.

Thomas, W.A., 1995, Diachronous thrust loading and fault partitioning of the Black Warrior foreland basin within the Alabama recess of the late Paleozoic Appalachian-Ouachita thrust belt, in Dorobek, S.L., and Ross, G.M., eds., Stratigraphic Evolution of Foreland Basins: Society of Economic Paleontologists and Mineralogists Special Publication No. 52, p. 111–126.

Thomas, W.A., 2001, Mushwad: Ductile duplex in the Appalachian thrust belt in Alabama: American Association of Petroleum Geologists Bulletin, v. 85, p. 1847–1869.

Thomas, W.A., and Bayona, G., 2002, Palinspastic restoration of the Anniston transverse zone in the Appalachian thrust belt, Alabama: Journal of Structural Geology, v. 24, p. 797–826, doi: 10.1016/S0191-8141(01)00117-1.

Thomas, W.A., and Bayona, G., 2005, The Appalachian thrust belt in Alabama and Georgia: Thrust-belt structure, basement structure, and palinspastic reconstruction: Geological Survey of Alabama Monograph 16, 48 p., 2 plates.

Thomas, W.A., Astini, R.A., Osborne, W.E., and Bayona, G., 2000, Tectonic framework of deposition of the Conasauga Formation, in Osborne, W.E., Thomas, W.A., and Astini, R.A., eds., The Conasauga Formation and Equivalent Units in the Appalachian Thrust Belt in Alabama: Alabama Geological Society Guidebook, 37th Annual Field Trip, p. 19–40.

Ward, W.E., Bearce, D.N., Osborne, W.E., and Irvin, G.D., 2004, Geologic map of the Cooks Springs 7.5-minute quadrangle, St. Clair and Shelby Counties, Alabama: Alabama Geological Survey Quadrangle Series Map 32, plate 1, scale 1:24,000.

Whiting, B.M., and Thomas, W.A., 1994, Three-dimensional controls on subsidence of a foreland basin associated with a thrust-belt recess: Black Warrior basin, Alabama and Mississippi: Geology, v. 22, p. 727–730, doi: 10.1130/0091-7613(1994)022<0727:TDCOSO>2.3.CO;2.

Wiltschko, D., and Eastman, D., 1983, Role of basement warps and faults in localizing thrust fault ramps, *in* Hatcher, R.D., Jr., Williams, H., and Zietz, I., eds., Contributions to the Tectonics and Geophysics of Mountain Chains: Geological Society of America Memoir 158, p. 177–190.

Wiltschko, D.V., and Eastman, D.B., 1988, A photoelastic study of the effects of preexisting reverse faults in basement on the subsequent deformation of the cover, *in* Schmidt, C.J., and Perry, W.J., Jr., eds., Interaction of the Rocky Mountain Foreland and the Cordilleran Thrust Belt: Geological Society of America Memoir 171, p. 111–118.

Wood, G.H., Jr., and Bergin, M.J., 1970, Structural controls of the Anthracite region, Pennsylvania, *in* Fisher, G.W., Pettijohn, F.J., Reed, J.C., Jr., and Weaver, K.N., eds., Studies of Appalachian Geology: Central and Southern: New York, Interscience, p. 147–160.

MANUSCRIPT ACCEPTED BY THE SOCIETY 22 MARCH 2007

Links among Carolinia, Avalonia, and Ganderia in the Appalachian peri-Gondwanan realm

James P. Hibbard*
Department of Marine, Earth, and Atmospheric Sciences, North Carolina State University, Raleigh, North Carolina 27695, USA

Cees R. van Staal
Geological Survey of Canada, 101-605 Robson St., Vancouver, British Columbia V6B 5J3, Canada

Brent V. Miller
Department of Geology and Geophysics, Texas A&M University, College Station, Texas 77843-3115, USA

ABSTRACT

The eastern flank of the Appalachian orogen is composed of extensive Neoproterozoic–early Paleozoic crustal blocks that originated in a peri-Gondwanan setting. Three of these blocks record the evolution of Neoproterozoic magmatic-arc systems, including Carolinia in the southern Appalachians and Ganderia and Avalonia in the northern Appalachians. Relationships among these three crustal blocks are important for understanding both the accretionary history of the orogen and the evolution of the Iapetus and Rheic Oceans, first-order geographic features of the Paleozoic globe.

Traditionally, Carolinia and Avalonia have been considered to represent a single microcontinental magmatic arc that accreted to Laurentia in the middle to late Paleozoic. The early lithotectonic history (ca. 680–570 Ma) of the two blocks is obscure; however, their latest Neoproterozoic-Paleozoic histories are distinct. This disparity is manifest in the first-order features of (1) timing and style of magmatic-arc cessation and (2) the nature of their Paleozoic lithotectonic records. Magmatic arc activity ceased in Avalonia in the late Neoproterozoic (ca. 570 Ma), succeeded by extension-related magmatism and sedimentation that was transitional into a robust latest Neoproterozoic–Silurian platformal clastic sedimentary sequence. This platform was tectonically unperturbed until the Late Silurian–Early Devonian. In contrast, Carolinia records late Neoproterozoic tectonothermal events coeval with arc magmatism, which extended into the Cambrian; a relatively thin Middle Cambrian shallow-marine clastic sequence is preserved unconformably atop the Carolinia arc sequences. Subsequently, Carolinia experienced widespread Late Ordovician–Silurian deformation and metamorphism.

However, we note striking similarities between Carolinia and Ganderia; specifically, in Ganderia, like Carolinia, late Neoproterozoic tectonism was accompanied by arc magmatism that extended into the Cambrian. Ganderian arc rocks are capped

---
*jim_hibbard@ncsu.edu

unconformably by a Middle Cambrian to Early Ordovician clastic sequence, and they were tectonized in the Late Ordovician–Silurian, similar to relations in Carolinia. Independent studies indicate that the Late Ordovician–Silurian tectonism in both blocks was related to their accretion to Laurentia. Thus, Carolinia and Ganderia show parallel development of first-order lithotectonic characteristics for two endpoints in their global strain path, i.e., their Gondwanan source region and their accretion to Laurentia.

Consequently, we posit that Carolinia appears to be more closely affiliated with Ganderia than with Avalonia. The recognition of this linkage between Appalachian peri-Gondwanan realm crustal blocks in light of paleomagnetic and isotopic data leads to a unified model for the accretion of these blocks to the eastern margin of Laurentia.

Keywords: Appalachian peri-Gondwanan realm, Carolinia, Ganderia, Avalonia, accretion, paleogeography.

INTRODUCTION

The Appalachian peri-Gondwanan realm encompasses a group of crustal blocks along the eastern flank of the orogen that had a Gondwanan heritage prior to their accretion to eastern North America. Within the exposed orogen, the realm contains four major blocks, including Carolinia, Ganderia, Avalonia, and Meguma (Fig. 1). In the northern Appalachians, Avalonia and Meguma extend in the subsurface to the edge of the modern passive margin. In the southern Appalachians, another major peri-Gondwanan block, Suwanee, lies entirely in the subsurface immediately to the south of the exposed orogen; the limited nature of subsurface data beneath the Atlantic Coastal Plain allows for the potential that other distinct peri-Gondwanan blocks could lie outboard of Carolinia (e.g., Dennis et al., 2004). With the exception of Meguma, the exposed Appalachian peri-Gondwanan realm blocks record the history of Neoproterozoic Gondwanan magmatic-arc systems and their ensuing Paleozoic evolution; Meguma preserves a record of sustained Paleozoic quiescence along a passive margin until the latest Early Devonian (e.g., Schenk, 1997; Hicks et al., 1999). In this study, we focus on the exposed Appalachian peri-Gondwanan realm crustal blocks that contain Neoproterozoic magmatic-arc sequences, namely Carolinia, Avalonia, and Ganderia[1]. In this paper, the term "Appalachian peri-Gondwanan" realm will be used to collectively refer to these three blocks and will exclude Meguma.

The understanding of spatial, temporal, tectonic, and stratigraphic relationships among the Appalachian peri-Gondwanan realm blocks is requisite to (1) constraining the configuration and evolution of oceanic tracts that once bordered these crustal blocks, including the Iapetus Ocean, which once lay between Laurentia and the Appalachian peri-Gondwanan realm blocks, and the Rheic Ocean, which separated these blocks from Gondwana, (2) elucidating the accretionary history of the Appalachian orogen, and (3) providing control on the interpretation of geophysical surveys of Appalachian lithosphere, such as the imminent, National Science Foundation–sponsored EarthScope projects. Clearly, during the interpretation of such data, it is a decided asset to know which blocks might have shared a common lithosphere prior to their modification during Appalachian orogenesis.

Current ideas concerning the links among the Appalachian peri-Gondwanan realm crustal blocks are mainly grounded in correlations made more than 20 yr ago, when little was known of their detailed geological evolution. During the time since these interpretations were proffered, there have been substantial advances in our understanding of the stratigraphy, tectonics, geochronology, and isotopic character of the Appalachian peri-Gondwanan realm blocks. In this paper, we synthesize and compare the detailed geologic history and paleomagnetic and isotopic records in each of the three major exposed Appalachian peri-Gondwanan realm blocks in order to determine the relationships among them. We also present new results that serve to add to the growing bank of Nd isotopic data for Carolinia. The implications of our synthesis lead us to significantly different conclusions concerning relationships among the Appalachian peri-Gondwanan realm blocks than those currently popular, and we conclude the manuscript by exploring the implications of our findings. The time scale of McKerrow and van Staal (2000) is used herein.

DEFINITION AND DISTRIBUTION OF THE APPALACHIAN PERI-GONDWANAN REALM BLOCKS

Nomenclature of the Appalachian peri-Gondwanan realm blocks has been a long-standing source of ambiguity and confusion. This problem ultimately reflects differences in interpretations of relationships between Appalachian peri-Gondwanan realm blocks. For example, some workers include either or both Carolinia and portions of Ganderia as subdivisions of Avalonia, implicitly viewing it as a composite terrane (e.g., Williams and Hatcher, 1983; Keppie and Ramos, 1999; Nance et al., 2002; Murphy and Nance, 2002), or terming it Avalon sensu lato (e.g., O'Brien et al., 1996) or West Avalonia (e.g., Nance and Murphy, 1994). This linkage implies a common geological history for

---

[1]For simplicity, we herein drop terms commonly used to designate these extensive crustal blocks, such as "terrane," "superterrane," "composite terrane," "zone," or "belt" and define the terms "Carolinia," "Avalonia," and "Ganderia" in the text below.

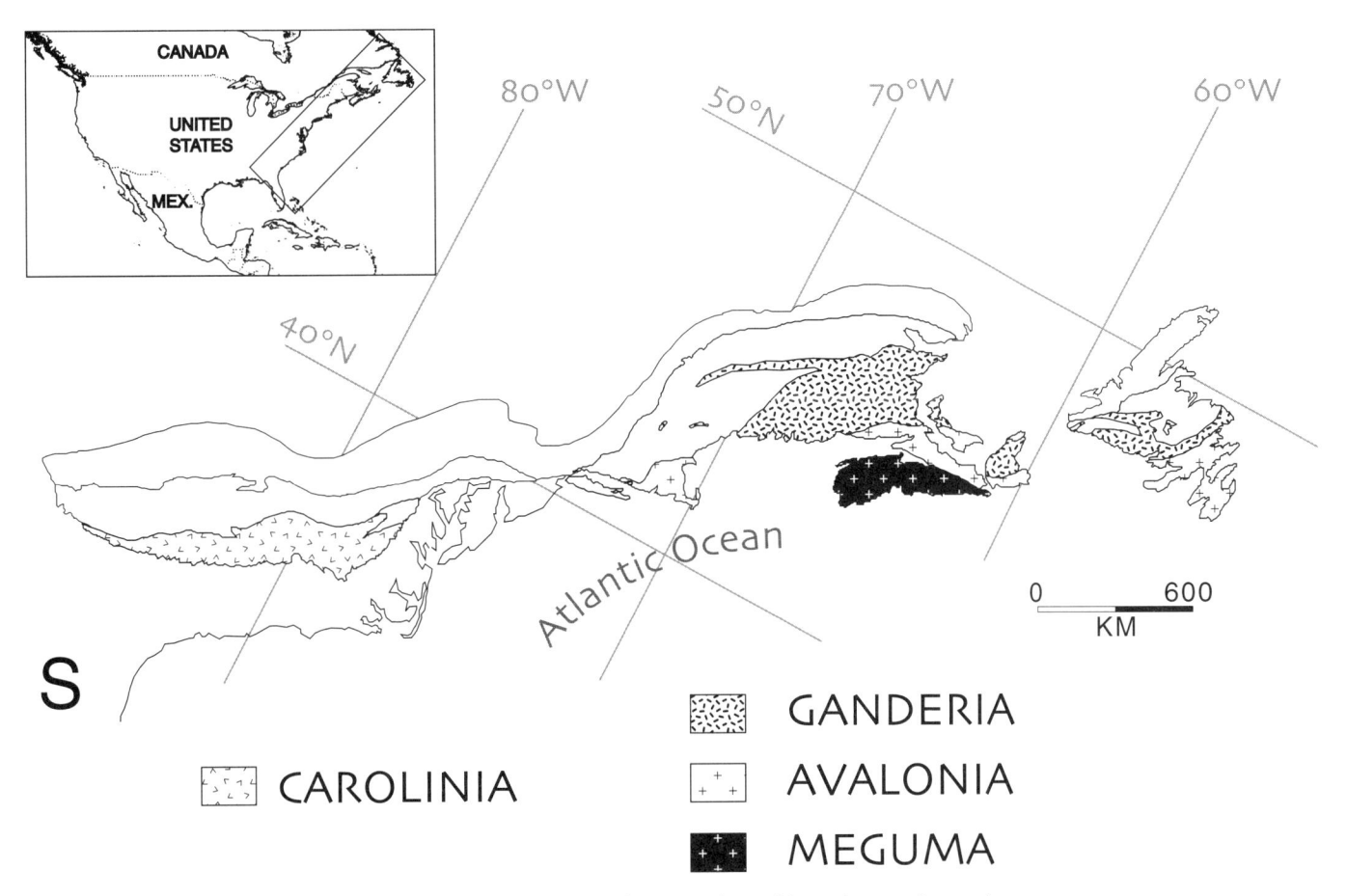

Figure 1. The Appalachian peri-Gondwanan realm and its major constituent elements.

these blocks, one which has not necessarily been explicitly demonstrated. Other workers have recognized additional blocks, such as Bras D'Oria (combined Brookville and Bras d'Or terranes) (Barr and White, 1996) and New River (Johnson and McLeod, 1996), which they consider to be independent blocks derived from a peri-Gondwanan source.

For clarity, in this section, we briefly outline the definitions that we use for the three Appalachian peri-Gondwanan realm blocks considered here, their distribution, and their relation to smaller blocks and terranes. Carolinia encompasses all of the exposed, proven peri-Gondwanan and closely affiliated blocks of the southern Appalachians; its two largest components are the Carolina and Charlotte terranes. It extends from the Atlantic Coastal Plain northwestward to an extensive late Paleozoic fault system termed the central Piedmont shear zone (Figs. 1 and 2) (Hibbard et al., 1998). This usage of "Carolinia" corresponds to the Carolina zone defined by Hibbard et al. (2002) and is similar to previous usage of the term "Carolinia" (e.g., Glover et al., 1997). Originally, "Carolinia" (Glover et al., 1997) included the Carolina zone (Hibbard et al., 2002) plus the Milton and Chopawamsic terranes (Fig. 2) (Horton et al., 1989). It has since been demonstrated that these latter terranes are temporally, structurally, and isotopically distinct from adjacent peri-Gondwanan rocks (Wortman et al., 1996; Hibbard et al., 1998; Coler et al., 2000), and therefore they are excluded here from Carolinia.

Avalonia, the easternmost of the northern Appalachian peri-Gondwanan realm blocks, extends northwestward to the Dover–Hermitage Bay faults in Newfoundland, the Georges River fault on Cape Breton Island, the Caledonia–Clover Hill fault in southern New Brunswick, and the Bloody Bluff–Lake Char–Honeyhill fault system in New England (Fig. 3). Rocks within these limits correspond to a crustal block commonly termed Avalon sensu stricto (e.g., O'Brien et al., 1996). This definition of Avalonia includes the type area of Avalon in Newfoundland, the Mira terrane (Barr and Raeside, 1989) of Cape Breton Island, the Antigonish and Cobequid Highlands of Nova Scotia, the Caledonia terrane (Barr and White, 1996) of southern New Brunswick, and the commonly termed "Boston Avalon" of southeastern New England.

We use the term "Ganderia" following its definition by van Staal et al. (1998), which includes classic early Paleozoic Gander passive-margin clastic sedimentary rocks that extend from eastern Newfoundland to the Miramichi Highlands of New

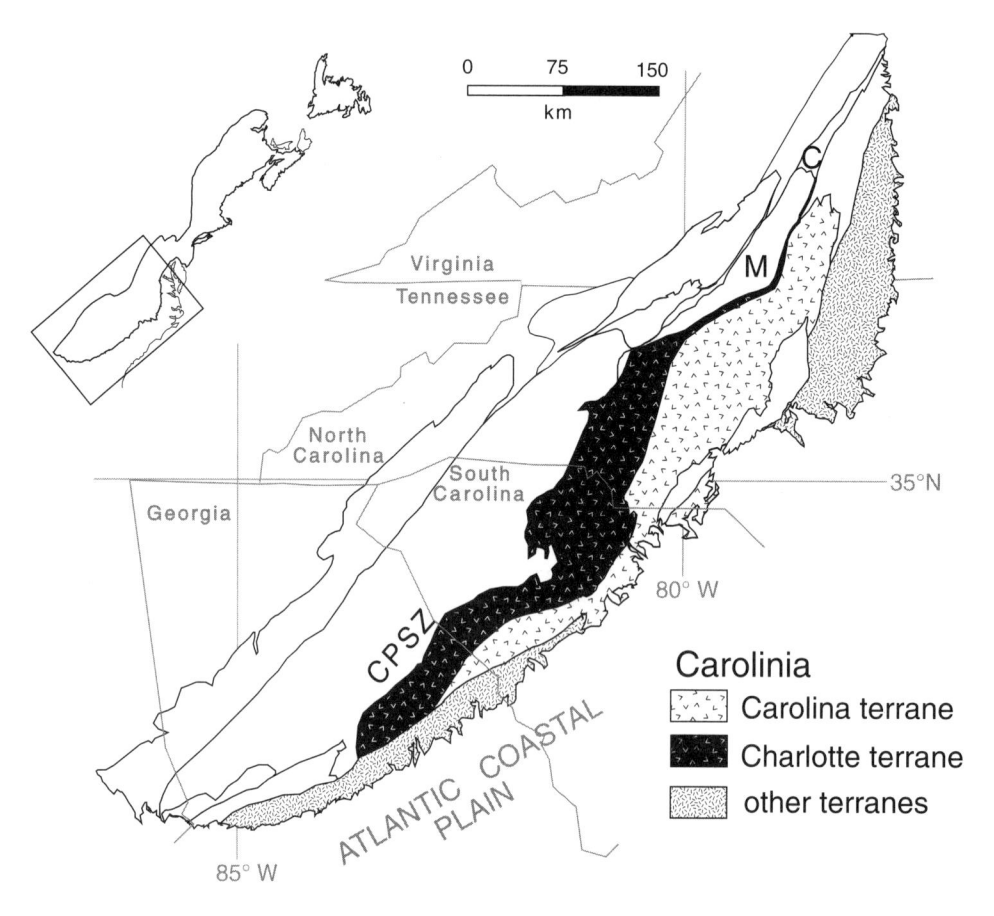

Figure 2. Lithotectonic elements of Carolinia and the surrounding region. C—Chopawamsic terrane, CPSZ—central Piedmont shear zone, M—Milton terrane.

Brunswick and into northern New England as well as blocks such as New River, Bras D'Oria, and Cinq Cerf–Grey River, which are interpreted to form the substrate to the classic Gander rocks (Dunning and O'Brien, 1989; van Staal et al., 1996; van der Velden et al., 2004). In addition, we include magmatic sequences that are closely associated with Ganderia, including the Victoria Lake Supergroup (Rogers et al., 2006) of central Newfoundland, the Balmoral Group in northern New Brunswick, the Meductic and Annidale Groups in southern New Brunswick, and the Ellsworth and Castine Formations of coastal Maine. These sequences represent the remnants of at least two arc–back-arc systems that were active at different times along the northern, leading edge of Ganderia between the Middle Cambrian and the Late Ordovician (e.g., van Staal et al., 1998). The northwest boundary of Ganderia approximately coincides with the Red Indian Line (Figs. 1 and 3), which is the Iapetan tectonic juncture between peri-Laurentian elements and peri-Gondwanan components. Easternmost Ganderian rocks form a series of inliers, such as the Pelham dome, Massebesic gneiss, the Lunksoos and Miramichi arches, and the Mt. Cormack dome. The contact between Neoproterozoic–Early Ordovician basement and younger cover ranges from locally faulted to unconformable.

PRESENT STATUS OF RELATIONSHIPS AMONG APPALACHIAN PERI-GONDWANAN REALM BLOCKS

Traditionally, the most common connection made between blocks of the Appalachian peri-Gondwanan realm has been the correlation of Carolinia with Avalonia (e.g., Williams and Hatcher, 1983). This correlation is rooted in broad lithologic similarities; i.e., both blocks contain substantial volumes of Neoproterozoic magmatic-arc related rocks overlain by a Cambrian clastic sedimentary sequence bearing an Acado-Baltic trilobite fauna, and their positions are similar along the eastern flank of the orogen (Williams and Hatcher, 1983). Notably, not all workers subscribe to this broadly accepted correlation; for example, Secor et al. (1983) described distinctions between the Paleozoic sequences of the two crustal blocks.

At the same time that Carolinia and Avalonia were being favorably compared, the prevailing thought was that Ganderia represented an early Paleozoic Andean margin, and it was thus considered to be distinct from the Avalonian Paleozoic clastic platform (Williams and Hatcher, 1983). Although not necessarily for the same reasons, some later workers (e.g., van Staal et al., 1996) have adhered to this separation of Avalonia and

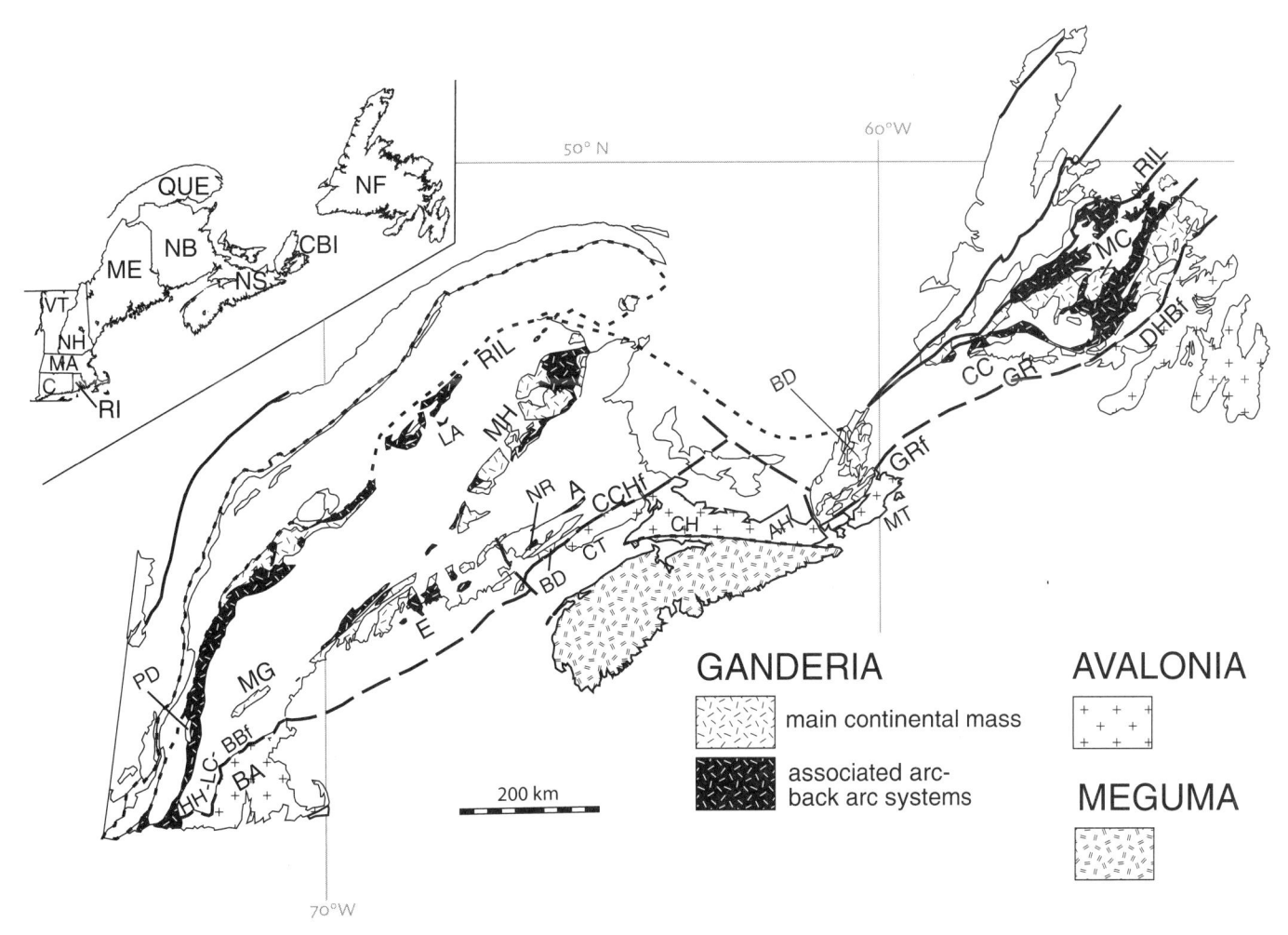

Figure 3. Lithotectonic elements of the northern Appalachian peri-Gondwanan realm. A—Annidale block, AH—Antigonish Highlands, BA—Boston Avalon, BD—Bras D'Oria, CBI—Cape Breton Island, CC—Cinq Cerf block, CCHf—Caledonia-Clover Hill fault, CH—Cobequid Highlands, CT—Caledonia terrane, DHBf—Dover–Hermitage Bay fault, E—Ellsworth terrane, GR—Grey River block, GRf—George's River fault, HH-LC-BB—Honey Hill–Lake Char–Bloody Bluff fault system, LA—Lunksoos arch, MC—Mount Cormack dome, MG—Massebesic gneiss, MH—Miramichi Highlands, MT—Mira terrane, NB—New Brunswick, NF—Newfoundland, NR—New River block, NS—Nova Scotia, PD—Pelham dome, RIL—Red Indian Line, QUE—Quebec.

Ganderia. In addition, Nd isotopic data have been used to support the distinction of Avalonia from Ganderia (e.g., Kerr et al., 1995; Whalen et al., 1996; van Staal et al., 1996; Samson et al., 2000). However, others have implicitly considered either all or portions of Ganderia to be included in Avalonia (e.g., Keppie and Ramos, 1999; Murphy and Nance, 2002; Nance et al., 2002).

The popular concept of correlating Carolinia and Avalonia, either as a single microcontinent, or as distinct, but closely linked crustal blocks, has been difficult to test because of the limited nature of data that have direct bearing on paleogeographic relationships. In recent years, however, there has been a surge in new stratigraphic, tectonic, geochronologic, and isotopic data from rocks of the Appalachian peri-Gondwanan realm; these new data permit a reevaluation of lithotectonic relationships of the Appalachian peri-Gondwanan realm blocks (Figs. 4 and 5). In the following three sections, we reexamine the widely held hypothesis that Carolinia is linked to Avalonia by synthesizing extant lithotectonic, paleomagnetic, and isotopic data that bear on this correlation. We will also compare Carolinia to Ganderia in order to explore the potential relationship between these two Appalachian peri-Gondwanan realm blocks.

LITHOTECTONIC ANALYSIS OF APPALACHIAN PERI-GONDWANAN REALM CRUSTAL BLOCKS

The goal of lithotectonic analysis is to use the geological record of displaced crustal blocks to attempt to track their tectonic travels from a source location to their present position. This type of analysis relies on the comparison of depositional, crystallization, and tectonothermal histories of crustal blocks; such lithotectonic analysis is the foundation to any comparison of crustal blocks because other data sets are rendered meaningless for com-

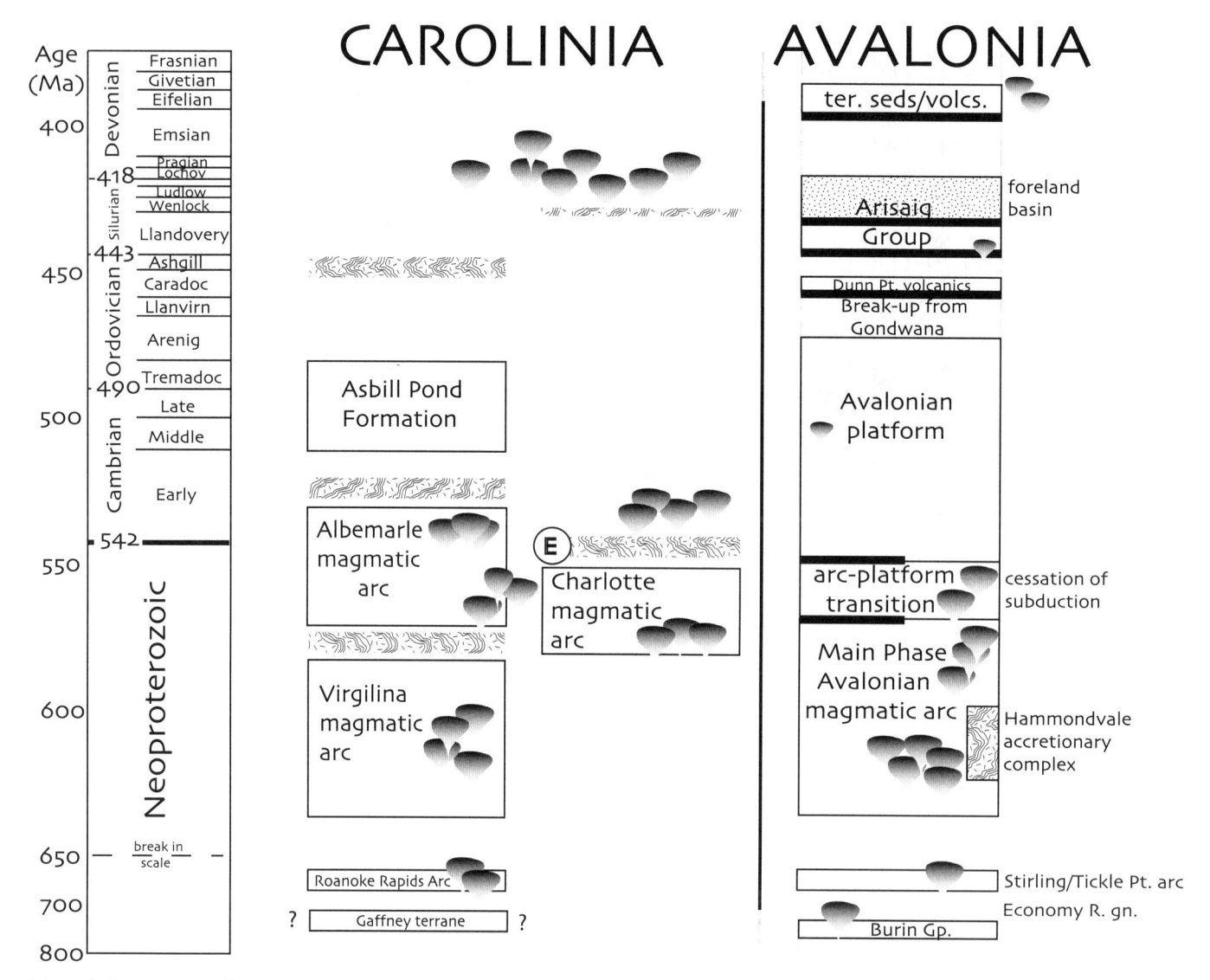

Figure 4. Comparison of lithotectonic elements (labeled) and magmatic (pluton symbol) and tectonothermal events (fold pattern) in Carolinia and Avalonia. Solid horizontal bar denotes an unconformity; E—eclogite occurrence. Compilation sources are given in text.

parative purposes if the lithotectonic histories of two blocks are not mutually compatible. In this section, we undertake a detailed lithotectonic analysis of Carolinia and Avalonia in order to test their mutual correlation, and we present the first comparison of Carolinia and Ganderia.

Lithotectonic Comparison of Carolinia and Avalonia

The traditional Carolinia-Avalonia correlation is based mainly on broad geologic similarities of the Neoproterozoic portions of each block. In detail (Fig. 4), these two blocks have early histories involving an older magmatic period ca. 685–670 and younger, main magmatic-arc activity in the span from ca. 635 to 570 Ma.

The oldest known magmatism in Carolinia is in the Roanoke Rapids complex, a low-grade volcanic-plutonic pile that is intruded by metagranodiorite dated at 672 ± 2 Ma (U-Pb zircon) (Coler and Samson, 2000). The complex is spatially associated with ophiolitic rocks that formed in a volcanic-arc setting (Kite and Stoddard, 1984), although relationships between the complex and the ophiolite are unexposed, and the age of the ophiolite is unknown. Magmatic-arc rocks of similar vintage are found in the Newfoundland and Cape Breton Island portions of Avalonia. In southeastern Newfoundland, volcanic rocks of the Tickle Point Formation have yielded a 683 ± 2 Ma U-Pb zircon age (Swinden and Hunt, 1991), and they are intruded by granite and gabbro of the Furby's Cove intrusive suite. U-Pb zircon analysis on granite from the suite indicates an age of 673 ± 3 Ma (O'Brien et al., 1996). Likewise, on Cape Breton Island, the Stirling belt contains volcanic-arc rocks that have been dated by U-Pb zircon methods at 681 +6/–2 (Barr, 1993; Bevier et al., 1993). In both blocks, this

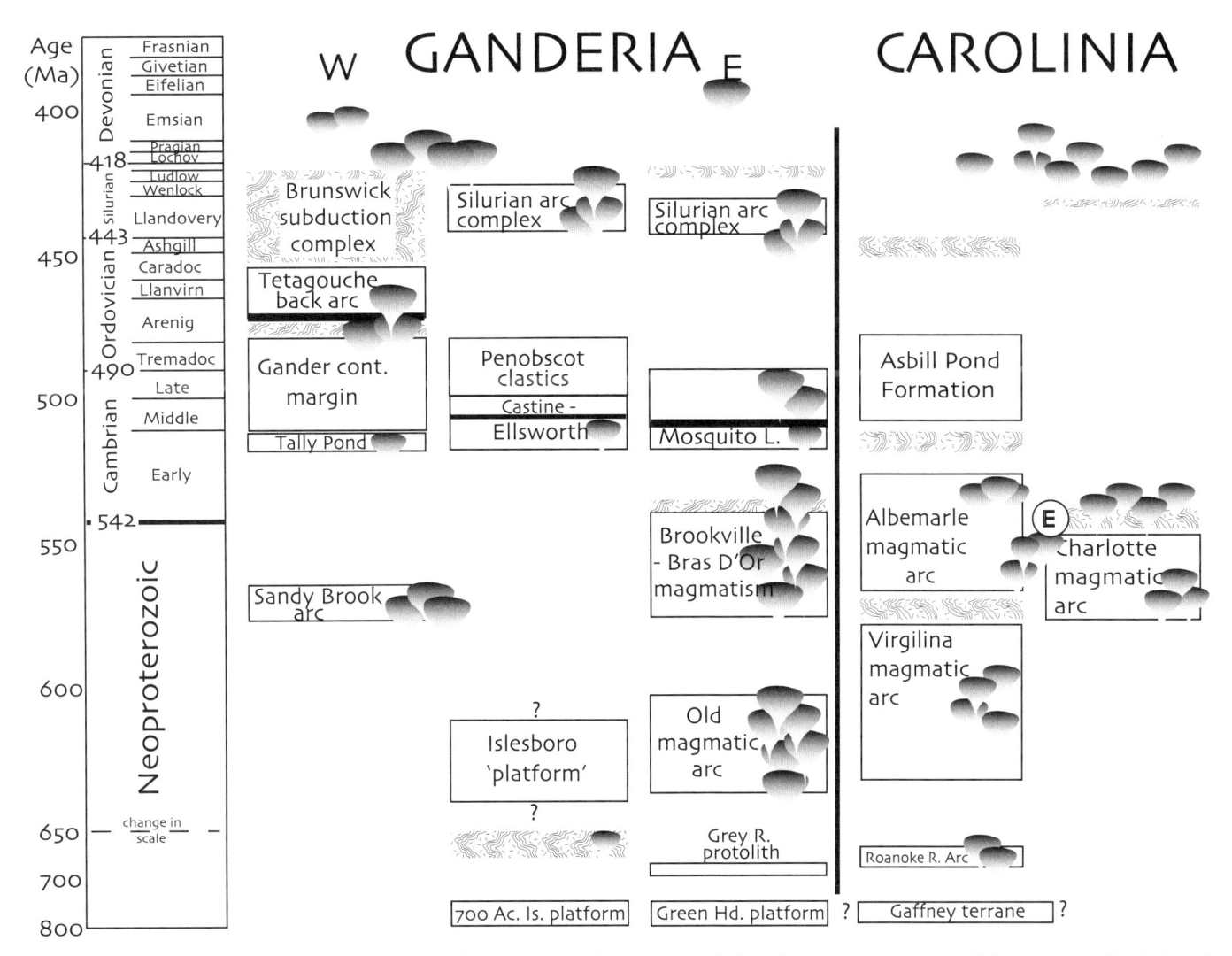

Figure 5. Comparison of lithotectonic elements (labeled) and magmatic (pluton symbol) and tectonothermal events (fold pattern) in Carolinia and Ganderia. Solid horizontal bar denotes an unconformity; E—eclogite occurrence. Compilation sources are given in text.

early magmatic activity was followed by an apparent hiatus of ~40 m.y., after which the main phase of arc magmatism began. In Carolinia, the Virgilina sequence records juvenile arc magmatism during a 20 m.y. span starting at ca. 633 Ma (Wortman et al., 2000), and it is overlain by associated sedimentary rocks at least as young as ca. 578 Ma, the age of the youngest recovered detrital zircon (Samson et al., 2001). Similar to peak magmatism in Carolinia, Avalonian main-phase arc magmatism ranged from ca. 635 to 570 Ma (e.g., Nance et al., 2002). The relationship between Carolinian and Avalonian magmatic arcs prior to 570 Ma is unclear due to the limited nature of the available relevant data.

After ca. 570 Ma, their geological evolution shows a clear-cut distinction between the two blocks. First-order disparities in their histories are manifest in (1) the timing and style of magmatic-arc cessation and (2) the nature of their Paleozoic records.

Main-phase Avalonian arc magmatism terminated diachronously from Boston to Newfoundland between ca. 605 and 570 Ma (e.g., Murphy et al., 1999; Nance et al., 2002). Avalonian arc cessation was marked by a transition into bimodal, within-plate magmatism (e.g., O'Brien et al., 1996; Barr and Kerr, 1997) accompanied by continental clastic sedimentation in extensional basins (e.g., Smith and Hiscott, 1984). Notably, there is a lack of evidence for collision between crustal blocks during the termination of main-phase arc magmatism; consequently, this transition has been interpreted as representing the migration of a trench-ridge-transform triple junction along the Avalonian margin, which terminated subduction and initiated a transform regime (Murphy et al., 1999; Nance et al., 2002).

The Avalonian transform regime was succeeded in the latest Neoproterozoic by a shallow-marine, shale-dominated, clastic platform sequence that extends into the Silurian (Lower Arisaig Group, Nova Scotia); on the Avalon Peninsula, Newfoundland, the Cambrian-Ordovician platform sequence attains a composite thickness on the order of 2.5–4.5 km (e.g., McCartney, 1967;

King et al., 1974; Ranger et al., 1984; Landing, 1996). Subsequently, Avalonia was not affected by orogenesis until the Late Silurian to Early Devonian (Dunning et al., 1993; Dallmeyer and Nance, 1994; Waldron et al., 1996).

In contrast to the relatively quiescent late Neoproterozoic arc cessation in Avalonia, late Neoproterozoic–early Paleozoic penetrative deformational events are recorded in Carolinia (e.g., Hibbard and Samson, 1995; Dennis and Wright, 1997; Barker et al., 1998). In North Carolina, foliated felsic volcanic rocks with a U-Pb zircon age of 612 +5/–2 Ma are cut by an undeformed granitoid with a U-Pb zircon age of 547 +3/–2 Ma (Hibbard and Samson, 1995; Wortman et al., 2000), thus constraining the timing of deformation between ca. 612 and 547 Ma. In South Carolina, penetratively foliated volcanic rocks with U-Pb zircon ages of 579 ± 4 Ma and 571 ± 16 Ma are intruded by an undeformed and unmetamorphosed diorite that has yielded a U-Pb zircon age of 535 ± 4 Ma, indicating deformation between ca. 570 and 535 Ma (Dennis and Wright, 1997). Similarly, Barker et al. (1998) showed that penetrative deformation elsewhere in South Carolina is constrained between 557 and 548 Ma. This late Neoproterozoic event, or series of events, appears to have been associated with an arc-arc collision involving the Charlotte terrane arc and the Carolina terrane arc (Barker et al., 1998; Hibbard et al., 2002). This collision has been considered to be responsible for the genesis of eclogite in the Charlotte terrane arc (Shervais et al., 2003).

Carolinian arc magmatism continued concomitantly with tectonism into at least the Early Cambrian before terminating. In particular, a zoned mafic-ultramafic plutonic complex, the Means Crossroads complex, has been interpreted as a manifestation of magmatic-arc rifting (Dennis and Shervais, 1991). Syn- and post-tectonic dioritic phases of this complex have yielded U-Pb zircon ages of 538 ± 5 Ma and 535 ± 4 Ma, respectively. In addition, felsic volcanic-arc rocks from the Flat Swamp Member of the Cid Formation have been dated by U-Pb zircon methods at 541 ± 1 Ma (Ingle et al., 2003). Carolinian arc-magmatic and associated sedimentary rocks are unconformably overlain by 1–2 km of fossiliferous, shallow-marine, Middle Cambrian mudstone-quartz graywacke of the Asbill Pond Formation (Dennis and Wright, 1997; Secor and Snoke, 2002).

The geologic evolution of Carolinia was also distinct from that of Avalonia during the Paleozoic, during which Carolinia was involved in regional deformation and low-grade metamorphism in the Late Ordovician–Silurian. The westerly Charlotte terrane arc was thrust with a sinistral component southeasterly over Carolina terrane volcanic-arc and volcaniclastic rocks of the Neoproterozoic–Early Cambrian Albemarle Group (Hibbard et al., 2003). Consequently, the Albemarle Group was overprinted by upright, subhorizontal regional folds that verge to the southeast and that have a penetrative steeply northwest-dipping cleavage accompanied by greenschist-facies metamorphism. The ^{40}Ar/^{39}Ar ages on biotite and muscovite have been interpreted as indicating peak metamorphism between ca. 455 and 443 Ma (Noel et al., 1988; Offield et al., 1995). This tectonothermal event appears to have also affected rocks in South Carolina, where a white mica ^{40}Ar/^{39}Ar date of 449 ± 2 Ma has been obtained (Ayuso et al., 1997). Amphibolite-facies rocks within the Charlotte terrane (Hibbard et al., 2002) yield younger ^{40}Ar/^{39}Ar amphibole ages of 425–430 Ma (Sutter et al., 1983); these ages likely reflect the same event recorded in the lower-grade rocks wherein the formerly deeper-seated Charlotte rocks cooled through the argon blocking temperature after higher-crustal-level rocks of the Albemarle Group. Late Ordovician to Silurian deformation is attributed to the sinistral oblique subduction of Carolinia beneath Laurentia (Hibbard, 2000; Hibbard et al., 2002, 2003).

In the Early Devonian, Carolinia was intruded by a suite of granitoids and gabbros that form a well-defined outcrop corridor less than 40 km wide; the suite trends NNE through North and South Carolina, skewed to the general northeasterly Carolinia structural trend. These plutons were initially thought to be Middle to Late Devonian in age on the basis of sparse Rb-Sr whole-rock ages (e.g., Fullagar, 1971; Butler and Fullagar, 1978). However, recent high-precision U-Pb zircon ages on some of these plutons indicate that they crystallized in a relatively short span between ca. 410 and 420 Ma (Samson and Secor, 2000; Miller et al., 2003). Their tectonic setting remains unknown (McSween and Harvey, 1997).

From these collective lithotectonic observations and interpretations of Carolinia and Avalonia, it is clear that their latest Neoproterozoic–Paleozoic developmental paths were distinct. Avalonian arc magmatism was terminated in the late Neoproterozoic without crustal collision, and it was succeeded by a transform regime that may have overlapped in time with deposition of an overlying platform sequence. In contrast, Carolinia arc cessation was heralded by late Neoproterozoic tectonism, perhaps related to arc-arc collision, after which arc magmatism continued into the Cambrian, on the order of 40–50 m.y. after the Avalonian magmatic-arc activity had ceased. Subsequently, Avalonia was unblemished by significant tectonothermal activity until the Late Silurian–Early Devonian (e.g., Dallmeyer and Nance, 1994; Waldron et al., 1996), whereas Carolinia was subjected to Late Ordovician–Silurian tectonism and Early Devonian magmatism.

In summary, Avalonia was initiated along an active Neoproterozoic margin, and it occupied a transform margin in the late Neoproterozoic and either a transform or passive margin in the early Paleozoic. In contrast, Carolinia was situated above an active margin into the Cambrian, was tectonized in the Late Ordovician–Silurian, and was magmatically active in the Early Devonian. Clearly, subsequent to 570 Ma, there are first-order differences in the geological development of Carolinia and Avalonia, and these differences are substantial enough to pose serious doubts that they constituted portions of a single ribbon continent.

Lithotectonic Comparison of Carolinia and Ganderia

The contrasts between Carolinia and Avalonia led us to compare Carolinia with the other major northern Appalachian peri-Gondwanan realm block—Ganderia. Traditionally, Ganderia has been viewed in terms of an Andean-type tectonic setting, distinct from that of Carolinia and Avalonia. However, the increased resolution of the geological history of the block during the past decade has made clear an earlier, more varied, Neoproterozoic magmatic history (Fig. 5).

The oldest rocks known in Ganderia are platformal clastic and carbonate units deposited on an unexposed basement in New Brunswick and Maine (Fig. 5). In New Brunswick, stromatolites and detrital zircon ages delimit the age range of the platformal sequence between 750 and 1230 Ma (White and Barr, 1996; Barr et al., 2003). The age of lithologically similar rocks in coastal Maine is constrained only by ca. 670–640 Ma ^{40}Ar/^{39}Ar metamorphic cooling ages and the observation that they are crosscut by a pegmatite dated at 647 ± 3 Ma (Stewart et al., 2001); thus they are pre–670 Ma. Although there are no known rocks of this vintage in Carolinia, similar rock types, including carbonates, metaclastics, and amphibolite outcrop in the Gaffney terrane along the western edge of Carolinia (Horton et al., 1989). Speculatively, these rocks could be correlative with the oldest sequences in Ganderia.

Ganderia, like Carolinia, records two main pulses of late Neoproterozoic magmatism (Fig. 5); one pulse spans from ca. 630 to 610 Ma, and the second pulse started at ca. 570–560 Ma and continued sporadically until ca. 525 Ma (e.g., Evans et al., 1990; Raeside and Barr, 1990; White and Barr, 1996; Johnson, 2001; White et al., 2002; Rogers et al., 2006). The older arc rocks are found in such units as the Lingley intrusive suite and possibly the Blacks Harbour granite and Brookville gneiss in New Brunswick (e.g., Bevier et al., 1990; Johnson, 2001; Barr et al., 2003). The younger phase of arc activity is represented by most of the younger bodies in the Golden Grove plutonic suite (White et al., 2002) in New Brunswick, plutons such as the Kathy Road dioritic suite in Cape Breton (e.g., Raeside and Barr, 1990), the Roti intrusive suite (Dunning and O'Brien, 1989; Dubé et al., 1995) in southern Newfoundland, and Crippleback and Valentine lake plutons and their volcanic envelope in central Newfoundland (Rogers et al., 2006). Some of the older, ca. 555–545 Ma magmatism during this second pulse records within-plate geochemical signatures, and these rocks have been interpreted as representing an extensional environment (e.g., Barr et al., 2003). In comparing Appalachian peri-Gondwanan realm crustal blocks, the most significant observation is that arc activity continued into the Paleozoic in both Carolinia and Ganderia (e.g., Johnson, 2001; Barr et al., 2003; Ingle et al., 2003; Rogers et al., 2006); that is, it continued into the time span when Avalonia was characterized by mainly clastic platformal deposition (Figs. 4 and 5).

Parallel to arc evolution in Carolinia, late Neoproterozoic–earliest Paleozoic tectonothermal events have been documented throughout Ganderia (Fig. 5). In New Brunswick, geochronologic studies of metamorphic tectonites indicate that deformation and metamorphism was active at ca. 564 Ma (Bevier et al., 1990), from ca. 550–540 Ma, and possibly as young as ca. 510 Ma (e.g., Dallmeyer and Nance, 1992; Nance and Dallmeyer, 1994; White and Barr, 1996). Likewise, on Cape Breton Island, ca. 550–540 Ma and ca. 490–500 Ma thermal events have been documented by geochronologic studies (Raeside and Barr, 1990; Dunning et al., 1990). It is unknown whether the events recorded in Ganderia represent a broad, continuous period of tectonothermal activity or short-lived, individual pulses of crustal mobility. Likewise, the cause of tectonism is ambiguous.

Similarities between Carolinia and Ganderia persist following the Cambrian cessation of subduction in each block. Analogous to deposition of the Middle Cambrian Asbill Pond Formation in Carolinia, Ganderian arc rocks are succeeded by the hallmark of Ganderia, a Middle Cambrian–Early Ordovician passive-margin sequence of quartzose and feldspathic clastic rocks (e.g., van Staal et al., 1996; van Staal, 2005). This metaclastic sequence is well preserved in such units as the Gander Group of Newfoundland, the Miramichi Group of New Brunswick, and the Penobscot Formation of Maine; however, the contact between these metaclastic rocks with the underlying arc-related rocks is only locally preserved in southern New Brunswick and coastal Maine (Johnson, 2001; Reusch et al., 2001). In New Brunswick, quartzose metaclastic rocks of the Matthews Lake Formation overlie arc-related volcanic rocks dated at ca. 515 Ma (Johnson and McLeod, 1996). These metaclastics bear strong resemblance to rocks in coastal Maine that underlie ca. 503 Ma felsic tuff of the Penobscot Formation, thus suggesting that the Matthews Lake Formation is Middle Cambrian (Johnson, 2001). In the Ellsworth terrane of coastal Maine, quartzose and feldspathic metaclastics are interlayered with intensely deformed volcaniclastic phyllites of the Ellsworth Formation; the deformation is such that the nature of the original contact between protoliths of the metaclastics and the phyllite cannot be determined. Felsic tuffs within the Ellsworth Formation have yielded U-Pb zircon ages of 508 ± 2 Ma and 509 ± 2 Ma (Stewart et al., 1995), suggesting that the metaclastic rocks are likely Middle Cambrian.

In Ganderia, as in Carolinia, the Neoproterozoic–early Paleozoic sequences were deformed and metamorphosed in the Late Ordovician–Silurian. Late Ordovician–Silurian deformation and metamorphism of Ganderia is well documented in New Brunswick. There, a structurally complex Late Ordovician–Early Silurian thrust belt with associated mélanges, blueschists, and foredeep deposits forms the northwestern margin of exposed Ganderian rocks; this plexus of tectonic elements, termed the Brunswick subduction complex, is responsible for telescoping an arc–back-arc system at the leading edge of Ganderia (van Staal, 1994). The ensialic Popelogan arc, built on extended Ganderian crust, was accreted to Laurentia in the Caradocian followed by destruction of the associated Exploits-Tetagouche back-arc system and ultimately the accretion of the bulk of Ganderia in the Silurian (van Staal et al., 1996, 1998, 2003). Thrusting and related deformation and metamorphism in the complex continued at least until 430 ± 4 Ma, which is the youngest ^{40}Ar/^{39}Ar age of low-temperature phengite related to the main phase of deformation (van Staal et al., 2003). Main-phase structures indicate that shortening was oblique, involving a sinistral component of displacement (van Staal, 1994). Thus, accretion of Ganderia was broadly coeval with that of Carolinia.

Following accretion to Laurentia, Ganderia recorded a plutonic event similar in nature and timing to, but volumetrically greater than, the Early Devonian plutonism in Carolinia. This suite in Ganderia is represented by the ca. 417–410 Ma North Bay intrusive suite, ca. 410 Middle Ridge granite, and the ca. 417 Ma Cape Freels pluton in Newfoundland (e.g., Kerr, 1997), and the bimodal Mt. Elizabeth complex and the North Pole Stream suite in New Brunswick (e.g., Whalen et al., 1996b). Both Kerr (1997) and Whalen et al. (1996b) considered these plutons to be Silurian on the basis of time scales now considered obsolete; clearly, utilizing more recent time scales (e.g., that of McKerrow and van Staal [2000] used here), they are Early Devonian and are synchronous with the Early Devonian magmatic pulse in Carolinia. Multiple investigators have hypothesized that this plutonism is related to postcollisional (accretion of Ganderia) lithospheric delamination (e.g., Whalen et al., 1994a; van Staal and deRoo, 1995; Kerr 1997; Schofield and D'Lemos, 2000).

In summary Carolinia and Ganderia share the following lithotectonic elements: (1) they record late Neoproterozoic suprasubduction-zone magmatism that continued into the Cambrian, and this magmatism was coincident with tectonothermal events that likely involved arc-arc collision in Carolinia, (2) a sequence of Cambrian clastic sedimentary rocks overlie magmatic-arc and associated rocks in each block, (3) they both record a Late Ordovician–Silurian tectonothermal overprint, and (4) following accretion to Laurentia, both blocks were host to Early Devonian, bimodal plutonism. We interpret these observations to indicate that both crustal blocks occupied the upper plate of an active, late Neoproterozoic convergent plate margin that shut down in the Early Cambrian. Both blocks record Middle Cambrian deposition of marine clastic sequences. These successions were subsequently subjected to Late Ordovician–Silurian tectonothermal remobilization related to the accretion of these blocks to Laurentia. Note that both blocks occupied the downgoing plate during the final phases of accretion to Laurentia.

The evolutionary tracks of Carolinia and Ganderia appear to have strayed from each other from the Middle Cambrian to the Middle Ordovician (Fig. 5). Following the ca. 570–525 Ma magmatic pulse in both blocks, there was a lull in arc magmatism; however, arc magmatism was renewed in Ganderia after a 10–15 m.y. hiatus, whereas arc magmatism of this vintage has yet to be documented in Carolinia. This young pulse of arc activity in Ganderia continued into the Early Ordovician and has been interpreted as recording the growth and evolution of an ensimatic arc–back-arc system termed the Penobscot arc (van Staal et al., 1998); this arc system was accreted to Ganderia in the Early Ordovician, just prior to development of the Popelogan-Tetagouche-Exploits arc–back-arc system. Middle Cambrian to Early Ordovician arc magmatism is manifest in such units as the ca. 513–509 Ma Tally Pond Group in Newfoundland (Dunning et al., 1991; Rogers et al., 2006) and the 515 +3/–2 Ma Mosquito Lake Road Formation in New Brunswick (Johnson, 2001), and the Tremadocian portions of the Wild Bight and Exploits Groups in Newfoundland (e.g., O'Brien et al., 1997; MacLachlan and Dunning, 1998). In addition, multiple volcanic units in Ganderia have geochemical compositions characteristic of an extensional setting. These units, including the ca. 494 Ma Coy Pond and Pipestone Pond ophiolites in Newfoundland (Jenner and Swinden, 1993), the 505 ± 3 Ma North Boisdale Hills volcanics in Cape Breton (White et al., 1994), the 493 ± 2 Ma and 497 ± 10 Ma Annidale volcanics in New Brunswick (McLeod et al., 1992), and the 508 ± 2 Ma and 509 ± 2 Ma Ellsworth Formation (Stewart et al., 1995), and the 503 ± 4 Ma Castine volcanics (Ruitenberg et al., 1993) in coastal Maine, are all compatible with the interpretation that they formed in an extensional arc or back-arc environment.

The apparent lack of this Middle Cambrian–Early Ordovician arc magmatism in Carolinia may be an artifact of either or both (1) the paucity of geochronological data or (2) exposure and preservation more than a reality of Paleozoic geology. For example, it is possible that Middle Cambrian, or younger, arc magmatism is represented by a suite of gabbro, the Stony Mountain gabbro (Hibbard et al., 2002), that intrudes the entire Carolina terrane sequence in North Carolina yet was also involved in Late Ordovician tectonism. The youngest unit in the terrane is the Yadkin Formation, which overlies earliest Cambrian volcaniclastic strata (Ingle et al., 2003); thus, the Stony Mountain gabbro is Early Cambrian or younger. Detailed geochemistry and geochronology of this suite are in progress in order to ascertain if it represents a Middle Cambrian pulse of arc magmatism in Carolinia.

In terms of exposure and preservation, the western limit of Carolinia, the central Piedmont shear zone, is a late Paleozoic thrust fault that is interpreted as originally surfacing within Carolinia and, thus, telescoping the bulk of Carolinia over the Late Ordovician suture zone of Carolinia to the orogen (Hibbard et al., 1998; Vines et al., 1998; Wortman et al., 1998). Thus, the leading edge of Carolinia that first became imbricated with Laurentia in a suture zone, and that might preserve any features such as associated arc–back-arc systems analogous to the Brunswick subduction complex, lies unexposed and inaccessible beneath the bulk of Carolinia.

Alternatively, the presence/absence of Middle Cambrian–Early Ordovician arc magmatism may be a real difference between Ganderia and Carolinia, suggesting that they may have been separated in the Middle Cambrian, perhaps by a transform boundary.

Later in the Paleozoic, Ganderia formed the substrate to a mainly Early Silurian magmatic arc–back-arc basin system from Maine to Newfoundland (O'Brien et al., 1991; Seaman et al., 1999, Keppie et al., 2000; Barr et al., 2002). Silurian rocks are conspicuously absent in Carolinia. This difference in the postaccretion history of Carolinia and Ganderia reflects heterogeneous tectonics along the middle Paleozoic Laurentian margin.

PALEOMAGNETIC ANALYSIS OF APPALACHIAN PERI-GONDWANAN REALM BLOCKS

There are scant available Neoproterozoic–mid-Paleozoic paleomagnetic data that have bearing on the relationships among Appalachian peri-Gondwanan realm blocks (Fig. 6). Data from

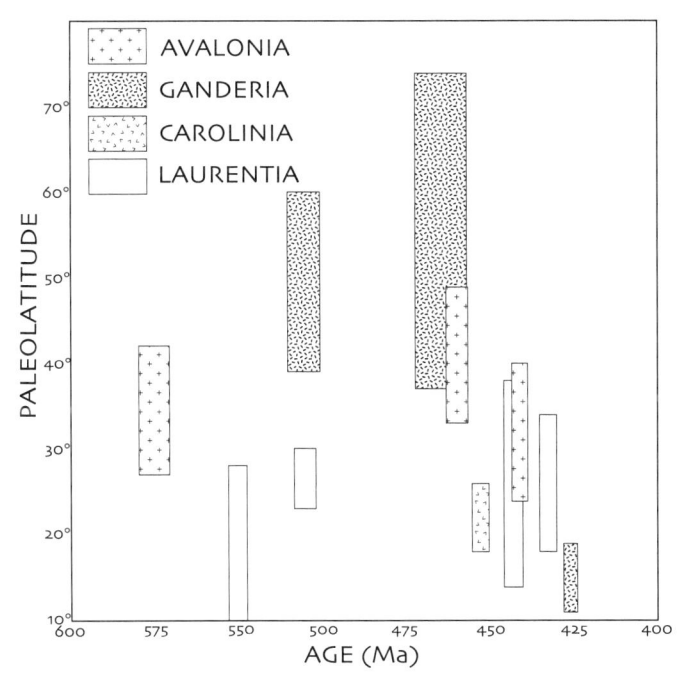

Figure 6. Available Neoproterozoic-Silurian paleolatitude data for the Appalachian peri-Gondwanan realm and eastern Laurentia (compiled from references given in the text).

two independent paleomagnetic studies undertaken in south-central North Carolina indicate that Carolinia was at 22°S ± 1° and 22°S ± 4° when the low-grade Albemarle Group acquired its Late Ordovician (ca. 455–450 Ma) thermal overprint (Noel et al., 1988; Vick et al., 1989). These results indicate that Carolinia was indistinguishable from the eastern Laurentian margin with respect to paleolatitude in the Late Ordovician (Noel et al., 1988; Vick et al., 1987).

Paleomagnetic data from the Marystown Group in southeastern Newfoundland indicate that Avalonia was situated at 34°S +8°/–7° paleolatitude at ca. 580–570 Ma (McNamara et al., 2001). It moved to high southerly latitudes (60°–54°) during the Late Cambrian (Trench et al., 1992; McCabe et al. 1992), but at the onset of the Late Ordovician (ca. 460 Ma), Avalonia had returned to intermediate latitudes (41° ± 8°) (Johnson and van der Voo, 1990; Hamilton and Murphy, 2004) and kept moving north such that it had achieved a latitude of 32°S ± 8° at the beginning of the Silurian (ca. 441 Ma) (Hodych and Buchan, 1998).

Fossiliferous sedimentary and volcanic rocks of the Bourinot Group on Cape Breton Island have preserved primary magnetizations that indicate that this portion of Ganderia was situated at 49°S ± 11° in the earliest Middle Cambrian (Johnson and van der Voo, 1985). Liss et al. (1993) determined an average paleolatitude of 52°S +21°/–16° for pillow lava samples from the Middle Ordovician Tetagouche Group, which overlie the Cambrian-Ordovician Ganderian passive-margin sequence in northern New Brunswick. Samples used in their averaged data set mainly ranged in age from 470 to 466 Ma.

The analytical uncertainties in many of these data are large and thus permit only the coarsest of resolution concerning the paleolatitudinal positions of, and thus linkages between, the Appalachian peri-Gondwanan realm during the late Neoproterozoic to Paleozoic (Fig. 6). The data are permissive of Avalonia and Ganderia having traveled from Gondwana to Laurentia in close latitudinal proximity, but they do not require this companionship. Avalonia and the main continental portion of Ganderia appear to have still been at intermediate latitudes at ca. 455 Ma, the time at which Carolinia was situated at low latitude, equivalent to that of the eastern Laurentian margin. Hypothetically, if Carolinia was contiguous with either Avalonia or Ganderia, these paleomagnetic data would imply that at the outset of the Late Ordovician, these Appalachian peri-Gondwanan realm blocks constituted a ribbon continent or archipelago that trended oblique to the Laurentian margin.

ISOTOPIC ANALYSIS OF APPALACHIAN PERI-GONDWANAN REALM BLOCKS

For more than a decade, isotopic characterization has informed our understanding of the relationships between elements of the Appalachian peri-Gondwanan realm (Ayuso and Bevier, 1991; Barr and Hegner, 1992; Fryer et al., 1992; Nance and Murphy, 1994, 1996; Kerr et al., 1995; Whalen et al., 1994, 1996; Samson, et al., 1995; Wortman et al., 2000; Samson et al., 2000). Isotopic signatures of Appalachian peri-Gondwanan realm crustal blocks have been used in attempts to determine their paleogeographic relationships to each other as well as their source continents (e.g., Nance and Murphy, 1994). The focus has been primarily on the use of Nd isotopic compositions of arc-related volcanic and plutonic rocks to distinguish segments of the Appalachian peri-Gondwanan realm with isotopically distinct components of recycled ancient (isotopically evolved) crust from those formed either from young continental crust or within an oceanic setting (isotopically juvenile). A few studies (e.g., Murphy and Nance, 2002; Samson et al., 2000) have also taken into account the Nd isotopic composition of early Paleozoic sedimentary rocks to constrain sediment source provenances and paleogeography. In this section, we add to the ever-growing Appalachian peri-Gondwanan realm Nd database, and we use the new and existing data to examine the potential for, and significance of, Nd isotopic discrimination among the three Appalachian peri-Gondwanan realm blocks.

Nd Isotopic Characteristics of Carolinia

Twenty new Nd isotopic analyses were conducted on rocks from Carolinia in North Carolina and South Carolina (Tables 1 and 2). These new data complement existing data (Fig. 7) in demonstrating the characteristically juvenile nature of the block, in accordance with the findings of previous workers (Samson et al., 1995; Mueller et al., 1996; Wortman et al., 1996; Coler et al., 1996; Fullagar et al., 1997; Ingle et al., 2003).

TABLE 1. NEW Nd ISOTOPIC DATA FROM CAROLINIA ZONE

Unit/Lithology (sample number; latitude, longitude)	Age (Ma)	Nd (ppm)	Sm (ppm)	$^{143}Nd/^{144}Nd$[†]	$^{147}Sm/^{144}Nd$	$\varepsilon_{Nd(i)}$[§]	T_{DM}[#] (Ga)
Carolina—Virgilina Plutons							
Abbotts Creek diorite (HRL02-06; 35.7483°N, 80.2414°W)	614	26.73	6.181	0.512667	0.1398	5.0	0.8
Pamona Quarry diorite (GHSZ01-05; 36.0527°N, 79.9272°W)	600	11.94	2.187	0.512600	0.1107	5.9	0.7
Carolina—Virgilina Volcanics							
Western sequence, felsic volcanic (HRL02-08; 35.4428°N, 80.4277°W)	613	10.54	1.940	0.512528	0.1113	4.6	0.8
Carolina—Uwharrie							
Uwharrie, Horse Trough Mtn. (UWH02-05; 35.3190°N, 80.0500°W)	580	33.17	6.502	0.512460	0.1185	2.3	1.0
Carolina—Albemarle Volcanics							
Tillery, Morrow Mtn. Rhyolite (UWH02-06; 35.3520°N, 80.0925°W)	569	26.92	6.115	0.512544	0.1373	2.5	1.0
Tillery, Shingle Trap Mtn. felsic volcanic (UWH02-03; 35.4027°N, 80.0367°W)	569	30.87	7.511	0.512566	0.1471	2.2	1.1
Tillery, Shingle Trap Mtn. felsic volcanic (UWH02-04; 35.4037°N, 80.0410°W)	569	32.53	7.576	0.512529	0.1408	1.9	1.1
Cid, felsic volcanic (HRL02-05; 35.7174°N, 80.2404°W)	542	15.73	2.742	0.512522	0.1054	4.1	0.8
Cid, Wolf Den Mtn. felsic volcanic (UWH02-01; 35.4124°N, 80.0707°W)	542	31.63	8.077	0.512596	0.1544	2.1	1.2
Cid, Wolf Den Mtn. felsic volcanic (UWH02-02; 35.4100°N, 80.0708°W)	542	31.39	7.849	0.512597	0.1512	2.4	1.1
Cid, basalt (SHS02-04; 35.7443°N, 80.1642°W)	542	8.11	2.045	0.512576	0.1526	1.8	1.2
Carolina—Albemarle							
Cid, Flat Swamp, fine crystal tuff (SHS02-05; 35.7044°N, 80.1218°W)	540	24.91	5.786	0.512547	0.1404	2.1	1.1
Carolina—Greensboro Igneous Suite							
Hicone quarry, massive diorite (GHSZ01-03a; 36.1602°N, 79.7002°W)	545	20.71	3.108	0.512439	0.0907	3.5	0.8
Hicone quarry, mafic phase (GHSZ01-04; 36.1602°N, 79.7002°W)	545	1.45	3.359	0.512582	0.1402	2.8	1.0
Gold Hill Pluton, granite (HRL02-07; 35.5493°N, 80.3501°W)	542	13.94	2.307	0.512497	0.1001	3.9	0.8
Mount Pleasant Pluton, granite (MP02-03; 35.4050°N, 80.4566°W)	542	17.95	4.472	0.512645	0.1506	3.3	1.0
Charlotte terrane plutons							
Edgemoor granite (GHSZ01-08; 34.8285°N, 80.9526°W)	508	13.92	2.785	0.512426	0.1210	0.8	1.0
Salisbury N, granite (Sals02-02; 35.6393°N, 80.4062°W)	416	11.72	8.252	0.513475	0.4257	4.2	0.3
Salisbury S, granite (Sals02-01; 35.5418°N, 80.4901°W)	416	17.24	9.191	0.513173	0.3223	3.8	0.1
Catawba granite (GHSZ01-06; 34.8571°N, 80.9181°W)	320	23.24	6.586	0.512543	0.1713	–0.8	1.7

[†]Measured ratio, normalized to $^{146}Nd/^{144}Nd = 0.7219$.
[§]Initial value was calculated at the known or estimated age using the following bulk earth parameters: $^{143}Nd/^{144}Nd = 0.512638$; $^{147}Sm/^{144}Nd = 0.1966$. Mean $^{143}Nd/^{144}Nd$ of standard JNdi-1 was 0.512110 ± 8 ($n = 25$). Analytical methods are described in Kylander-Clark et al. (2005).
[#]Depleted mantle model age of DePaolo (1981).

TABLE 2. SOURCES OF COMPILED Nd ISOTOPE DATA

Crustal block	References
Carolinia	1, 7, 12, 16, 23, 31, 32, 42
Avalonia	2, 3, 4, 10, 15, 19, 20, 24, 27, 29, 33, 34, 39
Ganderia	2, 3, 4, 9, 10, 17, 18, 19, 20, 30, 33, 34, 36, 37, 39, 40, 41
App. Grenville	2, 3, 5, 6, 8, 11, 12, 13, 14, 19, 21, 22, 25, 26, 28, 35, 38

References: (1) This work; (2) Barr and Hegner (1992); (3) Barr et al. (1998); (4) Barr et al. (2003); (5) Bream et al. (2004); (6) Carrigan et al. (2003); (7) Coler et al. (1997); (8) Dickin et al. (2004); (9) D'Lemos and Holdsworth (1995); (10) Dostal et al. (1996); (11) Fetter and Goldberg (1995); (12) Fullagar et al. (1997); (13) Goldberg and Dallmeyer (1997); (14) Hatcher et al. (2004); (15) Heatherington et al. (1996); (16) Ingle et al. (2003); (17) Jenner and Swinden (1993); (18) Keppie and Dostal (1998); (19) Keppie et al. (1997); (20) Kerr et al. (1995); (21) Miller and Barr (2000); (22) Miller and Barr (2004); (23) Mueller et al. (1996); (24) Murphy et al. (1996); (25) Owen et al. (1992); (26) Owens and Samson (2004); (27) Pe Piper and Piper (1998); (28) Pettingill et al. (1984); (29) Rabu et al. (1996); (30) Rogers et al. (2006); (31) Samson et al. (1995a); (32) Samson et al. (1995b); (33) Samson et al. (2000); (34) Samson (1995); (35) Su and Fullagar (1995); (36) Tomascak et al. (1996); (37) West et al. (2004); (38) Whalen et al. (1994a); (39) Whalen et al. (1994b); (40) Whalen et al. (1996a); (41) Whalen et al. (1996b); (42) Wortman et al. (1996).

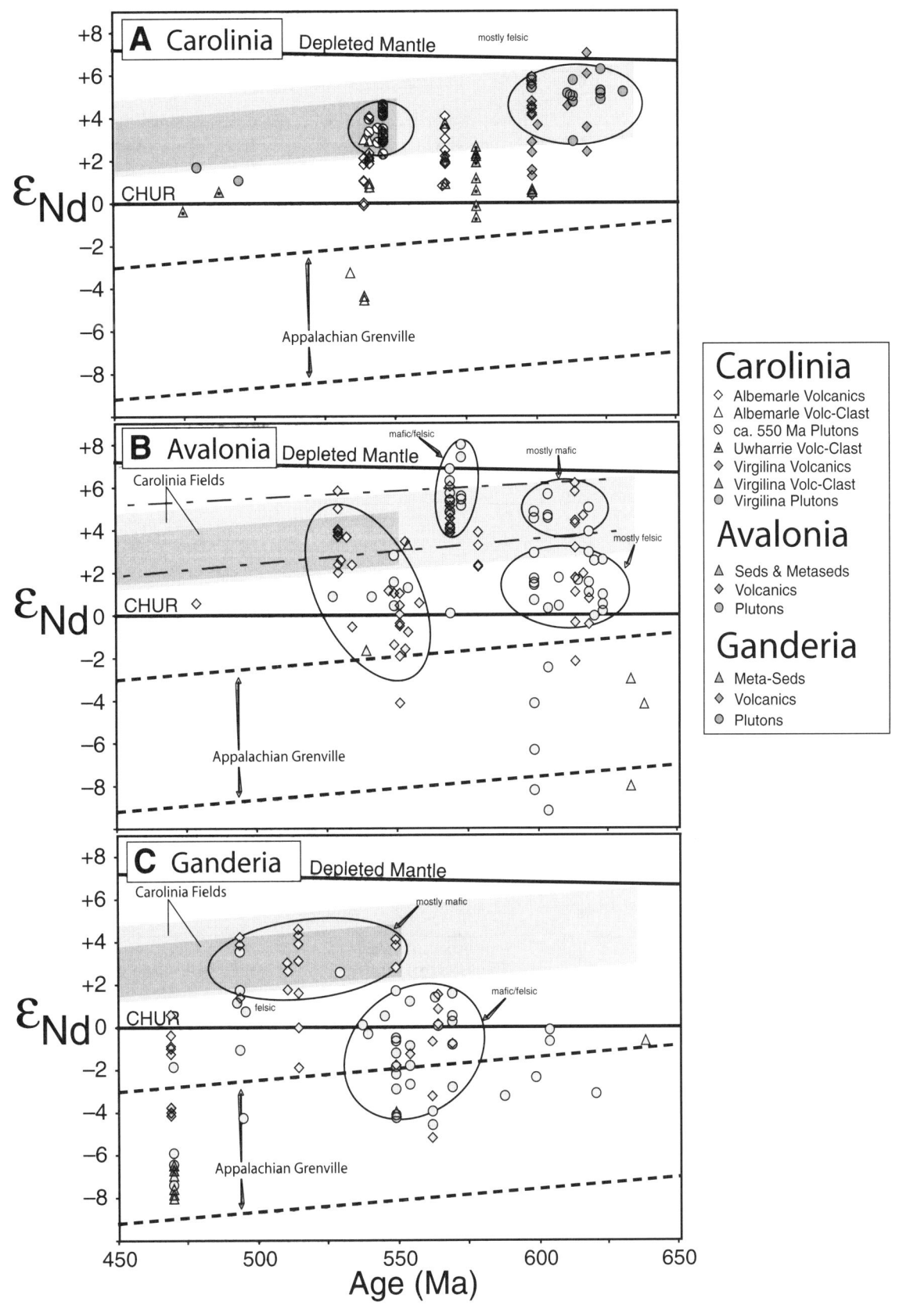

Figure 7. Compilation of Nd isotopic data for Neoproterozoic–early Paleozoic magmatic rocks of Carolinia, Avalonia, and Ganderia. Length and width of bars indicate range of analytical error for paleolatitude and age, respectively. CHUR—chondritic uniform reservoir.

The main, Virgilina, phase of arc-related plutonism produced juvenile intrusive rocks from ca. 635 to 600 Ma with ε_{Nd} values ranging mainly from +6.3 to +4.7, but one sample extends the range down to +2.9 (Fig. 7). Virgilina volcanic and volcaniclastic rocks have a wider spread of initial values, and they tend toward overall more-evolved compositions compared to the plutons (Fig. 7). The Virgilina rocks show a pattern of juvenile felsic plutonic rocks with contemporaneous volcanic and volcaniclastic rocks that are either more highly evolved or more contaminated with old crustal material. A later phase of Carolina terrane plutonism at ca. 550–540 Ma is exposed in central North Carolina (Fig. 7). The ε_{Nd} values from these rocks cluster in the range from +4.6 to +2.3. The data are consistent with derivation entirely from melting of the older Virgilina crust. Contemporaneous volcanic and volcaniclastic rocks show a wider range, to lower ε_{Nd} values, that is best explained by variable mixing of a sedimentary component, perhaps with a composition similar to that of the presumed Mesoproterozoic basement to Carolinia (e.g., Mueller et al., 1996). Three volcaniclastic rocks from this time frame fall within the field of Mesoproterozoic ("Grenvillian") basement rocks in the Appalachians (Fig. 7).

Plutons of both magmatic phases are consistently more juvenile than their age-equivalent volcanic and volcaniclastic rocks. This situation can be explained by different sources: a more mantle-derived source for the plutonic rocks, whereas the source of the volcanic rocks was more highly evolved crust not sampled by the plutons. Alternatively, both the plutonic and magmatic rocks could have been derived from similar juvenile basement sources, but the volcanic rocks could have entrained an evolved clastic cover during ascent and eruption. The evolved component could be unexposed Mesoproterozoic basement as suggested by Mueller et al. (1996). However, it may be significant that Mesoproterozoic inherited zircons commonly complicate U-Pb dating of volcanic and volcaniclastic rocks (Mueller et al., 1996; Ingle et al., 2003), whereas, to date, U-Pb dating has yet to reveal any obvious sign of Mesoproterozoic inheritance in any Neoproterozoic plutonic rock (e.g., Wright and Seiders, 1980; Samson et al., 1995; Samson, 1995; Dennis and Wright, 1997; Barker et al., 1998; Wortman et al., 2000; Dennis et al., 2004). This apparent "partitioning" of inherited zircons may be a result of sample bias in isotope-dilution–thermal ionization mass spectrometry (ID-TIMS) dating, in which zircon grain selection is purposely biased toward those least likely to contain inheritance. In other peri-Gondwanan terranes, however, inherited zircons in Neoproterozoic arc-related plutons goes hand-in-hand with more-evolved Nd isotopic compositions; both are commonly interpreted as a result of basement assimilation into the plutons (e.g., Samson, 1995; Miller et al., 1999; Samson and D'Lemos, 1998; Barr et al., 2003). Both the Nd isotopic character and the nature of zircon inheritance can be explained if Carolinia originated, not as an Andean-type continental-margin arc, but instead, as an oceanic arc adjacent to a denuded segment of a nonvolcanic rifted continental margin (e.g., West Iberian margin; Whitmarsh et al., 1996; Henning et al., 2004) with a sedimentary-volcaniclastic cover that was derived in part from adjacent Mesoproterozoic highlands.

Nd Isotopic Characteristics of Avalonia

The main arc phase of Avalonia plutonism (630–600 Ma) shows a distinct division between mainly mafic plutonic and volcanic rocks with ε_{Nd} values of +6 to +4 and mainly felsic plutonic and volcanic rocks in the range +3–0 (Fig. 7). A small group of plutonic rocks (and one volcanic rock) shows much more isotopically evolved compositions, with ε_{Nd} values falling in or below the field for Appalachian Grenville. Three analyses of sedimentary rocks from Avalonia also plot in this range. The plutons with lowest ε_{Nd} values are isotopically consistent with melts wholly derived from sediments of similar composition, and the group of mainly felsic rocks of this age between +3 and 0 could represent mixtures of melts derived from the more mafic sources that have been contaminated to varying degrees by these sedimentary rocks or continental basement of similar composition. A younger phase of plutonism and volcanism (ca. 580–570 Ma) shows mainly juvenile isotopic compositions in both mafic and felsic rocks, with ε_{Nd} values of +7 to +4. Between these two episodes, there is sparse evidence (four data points on Fig. 7) for an episode of volcanism with lower ε_{Nd} values than that of this later plutonic/volcanic phase. Other Neoproterozoic-Cambrian volcanic and plutonic rocks show a wide range in ε_{Nd} values, but mainly from +6 to −2, with the younger rocks in this group tending toward more juvenile compositions. Middle and late Paleozoic Avalonian plutons show a distinct lack of the more isotopically evolved compositions seen in the Neoproterozoic rocks.

The Nd isotopic composition of Avalonia is thus characterized by an early phase of mainly juvenile mafic plutonic and volcanic rocks with more-evolved contemporaneous felsic plutonic and volcanic rocks. A ca. 580–570 Ma phase of plutonism that has no known contemporaneous equivalent in Carolinia is highly juvenile and requires rejuvenation of mantle input during this time in Avalonia. The latest Neoproterozoic to Cambrian magmatism in Avalonia is isotopically consistent with melting of the earlier-formed crust.

Nd Isotopic Characteristics of Ganderia

The oldest Ganderia rocks (640–620 Ma) are poorly represented in the Nd database (Fig. 7), but they are isotopically evolved with ε_{Nd} values of ~0 to −3. In contrast to both Carolinia and Avalonia, there is no indication of the juvenile components in this early phase—although it must be reiterated that there are very few Ganderia data. A later Neoproterozoic phase (ca. 570–540 Ma) of plutonic and volcanic rocks, which includes both mafic and felsic components, has ε_{Nd} values in the range +1.7 to −4.6 (Fig. 7). Late Neoproterozoic–early Cambrian mainly mafic volcanic rocks have ε_{Nd} values, with a few exceptions, in the range +4.6 to +1.1. The Nd isotopic compositions of middle Paleozoic plutonic and volcanic rocks in Ganderia scatter widely (Fig. 7). The late Paleozoic plutonic rocks, however, are restricted to sources with isotopic compositions similar to those of the earliest mafic/felsic phase of plutonism and volcanism in Ganderia.

Summary of Nd Isotopic Character of the Appalachian Peri-Gondwanan Realm Blocks

The Appalachian peri-Gondwanan realm blocks are characterized by distinct Nd isotopic signatures: Carolinia is the most consistently juvenile and somewhat similar in character to the mafic juvenile rocks in Avalonia, but the felsic plutonic and volcanic rocks in Avalonia show consistently more-evolved compositions. The difference between the Nd isotopic signatures of the felsic rocks could reflect differences between the nature or thickness of the crust upon which the Carolinian and Avalonian segments of the arc were built. If our inference is correct that Carolinia was built along the Gondwanan continental margin in a rifted segment that had been denuded of continental crust before being blanketed by sediments derived from an adjacent Mesoproterozoic highlands, then Avalonia may have been located along the same arc, but possibly across a transform in the original rifted margin where the Mesoproterozoic basement had remained attached to Gondwana. Ganderia is consistently more evolved than Carolinia and Avalonia. Like Avalonia, the Nd characteristics of Ganderia imply that it was built on a basement of continental crust, but the thickness of that crust remains unclear.

Overall, these Nd isotopic characteristics demonstrate that, while the Appalachian peri-Gondwanan realm blocks share some aspects of petrogenetic and isotopic evolution, in detail there are important differences. Although the elements of the Appalachian peri-Gondwanan realm may have been built on different basements (e.g., Samson, 1995), these data do not necessarily preclude their formation in different segments of the same arc system along the Gondwanan continental margin (e.g., Nance and Murphy, 1994).

IMPLICATIONS FOR LINKAGES AMONG APPALACHIAN PERI-GONDWANAN REALM BLOCKS

Each one of the three data sets that we have compiled herein for the Appalachian peri-Gondwanan realm blocks, namely lithotectonic, paleomagnetic, and isotopic data, reflects a different aspect of the fundamental paleogeographic "grail" that is implied in our theme of linkages among these crustal blocks; additionally, these data sets represent an incomplete record that only provide glimpses of these paleogeographic relationships. In discussions of whether or not the components of the Appalachian peri-Gondwanan realm blocks are "the same" or "different," it should be realized that each of the data sets taken alone provides a plurality of permissible interpretations of original relationships of the Appalachian peri-Gondwanan realm elements and that all data sets must be integrated in order to reconstruct the history of these blocks. In this section, we summarize the relationships among Appalachian peri-Gondwanan realm blocks as elicited by all of the reviewed data sets, and we attempt to compose a model for the Appalachian peri-Gondwanan realm blocks that is consistent with these relationships.

Detailed comparisons of Carolinia, Avalonia, and Ganderia evoke the following first-order observations and interpretations that have bearing on the relationships among these blocks:

1. During the latest Neoproterozoic–Early Cambrian, both Carolinia and Ganderia were located above subducting plate margins; in contrast, Avalonia appears to have occupied either a transform or passive margin at this time, and it records the deposition of a platformal sedimentary sequence.

2. In both Carolinia and Ganderia, Early Cambrian magmatic-arc rocks are overlain by a marine clastic sequence. Deposition of the Avalonian platform sequence was more or less continuous through this time frame.

3. Ganderia records the evolution of a Middle Cambrian–Early Ordovician arc–back-arc system that is not preserved in exposed portions of Carolinia.

4. Carolinia and Ganderia were both overprinted by Late Ordovician to Silurian regional, foliation-forming tectonothermal events that record the docking of these Appalachian peri-Gondwanan realm blocks to eastern Laurentia. Accretion of Ganderia initiated with subduction of an associated Middle Ordovician arc–back-arc complex. Accretion of both blocks involved a sinistral component, with Laurentia occupying the upper plate. Avalonia does not record an Ordovician-Silurian tectonothermal event.

5. Carolinia and Ganderia record subsequent middle Paleozoic magmatism; Silurian magmatism in Ganderia was arc-related and absent in Carolinia. Early Devonian plutonism in both Carolinia and Ganderia is represented by compositionally bimodal bodies, and in Ganderia, it has been attributed to lithospheric delamination. Avalonia was unaffected by plutonism and tectonism until the Middle Devonian.

6. Construction of a Silurian magmatic arc on Ganderia after its accretion to Laurentia suggests that an oceanic tract lay between Ganderia and the more easterly, outboard, Avalonia in the Late Ordovician to Early Silurian.

7. Paleomagnetic studies indicate that during the Late Ordovician, Carolinia was at low latitudes, equivalent to Laurentia, whereas Avalonia and Ganderia appear to have occupied intermediate latitudinal positions.

8. Each Appalachian peri-Gondwanan realm block records a distinct Nd isotopic signature; Carolinia is the most juvenile, and Ganderia is the most evolved. These signatures likely reflect the nature of the basement beneath these blocks, where Carolinia probably represents an oceanic arc, and Ganderia was constructed on continental basement.

Collectively, these points demonstrate that Carolinia and Avalonia clearly have disparate lithotectonic, paleomagnetic, and Nd isotopic histories. Thus, the major implication of this analysis is that, contrary to popular modern orthodoxy, it is unlikely that Carolinia and Avalonia represent either a single microcontinent or even distinct, but correlative, crustal blocks. Our analysis also indicates that Carolinia and Ganderia record lithotectonic evolutionary paths that have much more in common, including some significant first-order overlaps. Specifically, Carolinia and Ganderia record parallel development at both their Neoproterozoic, Gondwanan source area and during their arrival at the Laurentian margin. During the late Neoproterozoic, they were both located above an actively subducting peri-Gondwanan margin, and subsequently, both blocks record Middle Cambrian marine deposition. During docking at the Laurentian end of their paths, both arrived at the Laurentian margin at approximately the same time and on a plate that was consumed beneath Laurentia.

Based on these first-order attributes, it is compelling to speculate that Carolinia and Ganderia occupied either the same, or closely affiliated, lithospheric plate margin(s) during the late Precambrian and Paleozoic; the available data may not allow for the resolution necessary to distinguish between these two hypotheses. Although Carolinia and Ganderia do exhibit second-order lithotectonic, as well as paleomagnetic, and isotopic differences, they are compatible with a hypothesis that the two Appalachian peri-Gondwanan realm blocks are closely related. Paleogeographic reconstructions for two key time frames (ca. 540 Ma and ca. 450 Ma) that accommodate a Carolinia-Ganderia linkage and that account for the differences between these blocks are depicted in Figure 8 and are discussed next.

Available lithotectonic, isotopic, and geochronologic data are compatible with Carolinia, Ganderia, and Avalonia having a peri-Gondwanan source from around the Amazon craton (e.g., Nance and Murphy, 1996; van Staal et al., 1996; Murphy et al., 1999; Ingle et al., 2003); our positioning of the Appalachian peri-Gondwanan realm blocks around Amazonia at ca. 540 (Fig. 8A) is modified from reconstructions presented in Hibbard et al. (2005) and Rogers et al. (2006). Differences in the Neoproterozoic-Cambrian Nd isotopic character of the Appalachian peri-Gondwanan realm blocks can be readily accommodated in our reconstruction involving a Carolinia-Ganderia linkage. The Nd isotopic character likely reflects the basement substrate of these blocks, where Carolinia represents a juvenile arc, and Ganderia was built upon extended continental crust. Thus, we represent Carolinia and Ganderia as forming an elongate archipelago that spanned Amazonian crust at its Ganderian end and oceanic crust or highly attenuated continental crust at its Carolinian end (Fig. 8A). Paleomagnetic data suggest that during their transit of Iapetus, Carolinia and Ganderia formed an elongate arc that trended athwart latitudes and, thus, oblique to the long axis of Iapetus, which trended roughly east-west. This interpretation is supported by the slightly different times of arrival of Carolinia and the bulk of Ganderia at the eastern Laurentian margin (Fig. 8B). Carolinia arrived first in the Late Ordovician, and collision was under way by ca. 450 Ma, whereas to the north (present coordinates), the Popelogan arc–Tetagouche-Exploits back-arc system arrived in the Late Ordovician, with the main continental mass of Ganderia following in the Early to Late Silurian (e.g., van Staal et al., 2003), when deeper portions of Carolinia (Charlotte terrane) were being uplifted and cooled.

Kinematic analyses along the length of the orogen indicate that accretion of these first Appalachian peri-Gondwanan realm blocks to arrive at the Laurentian margin involved sinistral

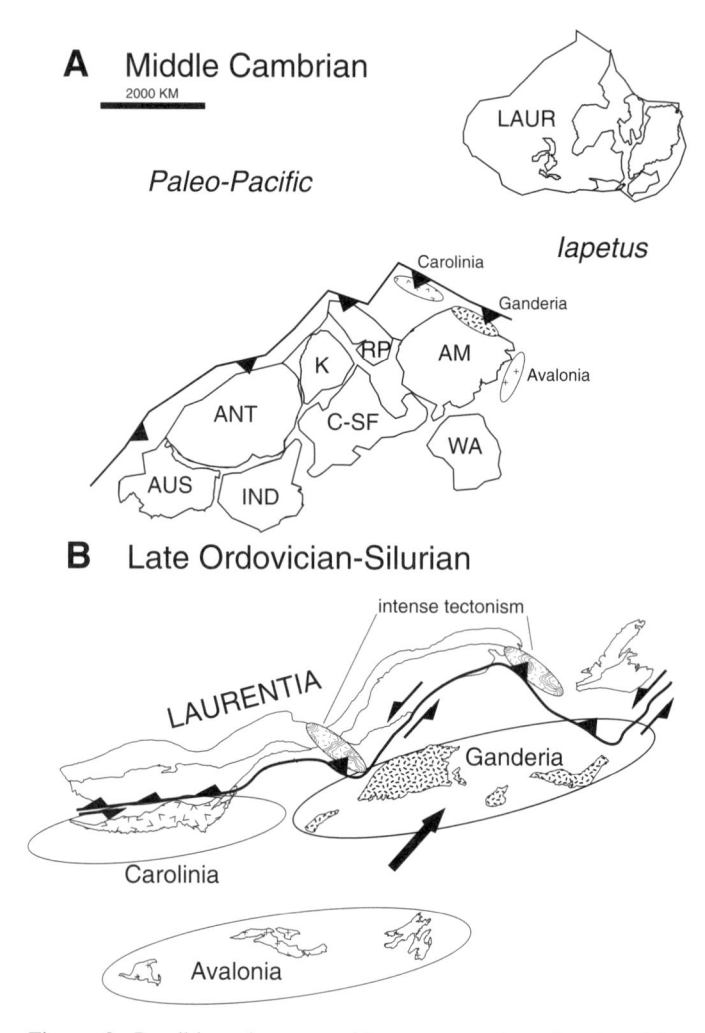

Figure 8. Possible paleogeographic reconstructions for Carolinia, Avalonia, and Ganderia during (A) the Middle Cambrian and (B) the Late Ordovician–Silurian. AM—Amazonia, ANT—Antarctica, AUS—Australia, C-SF—Congo-São Francisco, IND—India, K—Kalahari, LAUR—Laurentia, RP—Rio de la Plata, WA—West Africa.

transpression (e.g., Keppie, 1992; Hibbard, 1994; Dubé et al., 1994; Hibbard et al., 2002). This kinematic pattern is consistent with the distribution of intense Late Ordovician to Silurian tectonism in the orogen. The southern edge of the St. Lawrence promontory is the locus of Late Ordovician–Silurian blueschist metamorphism and ductile deformation in New Brunswick (van Staal et al., 1990). Recent U-Pb ages from xenotime, titanite, and monazite in southern New England (Sevigny and Hanson, 1995; Dietsch and Jercinovic, 2005; R. Tracy, 2004, personal commun.) indicate the presence of intense Silurian tectonism along the southern margin (present coordinates) of the New York promontory. We thus suggest that the orogenic promontories functioned as restraining bends that accumulated intense shortening during Late Ordovician–Silurian sinistral transpressional accretion of the Appalachian peri-Gondwanan realm blocks (Fig. 8B).

CONCLUSIONS

The traditional correlation of Carolinia with Avalonia (e.g., Williams and Hatcher, 1983) does not stand up to detailed analyses of available data sets. In particular, these two Appalachian peri-Gondwanan realm blocks differ with respect to the mode and timing of the termination of subduction in the Neoproterozoic-Cambrian, and their Paleozoic histories are different. Conceptually, Carolinia has "seceded" from Avalonia. In contrast, Carolinia shows a strong affinity toward Ganderia, and their first-order evolutionary paths overlap both at the Gondwanan margin and later at the Laurentian margin. Recognition of this linkage between Carolinia and Ganderia supports a unified model of the Late Ordovician–Silurian accretionary history along the entire length of the Appalachians. Clearly, there are still troublesome "loose strands" to be resolved, such as whether or not Carolinia was associated with an arc–back-arc system analogous to the Popelogan-Tetagouche-Exploits system. However, our new linkage of Carolinia and Ganderia poses new horizons to be approached in future studies. For example, did Carolinia and Ganderia occupy the same plate margin, or were they on separate lithospheric plates (i.e., how close is the linkage between them?)? Where in Gondwana is their source region? Following the accretion of Carolinia and Ganderia, what was the nature of the Laurentian margin? These and a myriad of other concerns require further data collection in the form of field data, focused detrital zircon studies, and geochronologic, paleomagnetic, seismic-reflection profiling, and paleontological investigations in order to comprehensively reconstruct the Appalachian peri-Gondwanan realm.

ACKNOWLEDGMENTS

Studies by Hibbard and Miller that led to this contribution were funded by U.S. National Science Foundation grants (EAR-9814273 and EAR-0228908 to Hibbard and EAR-0106112 to Hibbard and Miller). All of the authors thank Jeff Pollock, Neil Rogers, and Pablo Valverde-Vaquero for discussions and consultation that helped to focus our ideas. We are grateful to Damian Nance and an anonymous reviewer for reviews that sharpened our thinking as well as the presentation of our study. We also thank Tekla Harms, Carol Evenchick, and Jim Sears for providing the opportunity for us to participate in the Geological Society of America topical session and this volume honoring Ray Price.

REFERENCES CITED

Ayuso, R.A., and Bevier, M.L., 1991, Regional differences in lead isotopic compositions of feldspars from plutonic rocks of the northern Appalachian Mountains, USA and Canada: A geochemical method of terrane correlation: Tectonics, v. 10, p. 191–212.

Ayuso, R.A., Seal, R., II, Foley, N., Offield, T.W., and Kunk, M., 1997, Genesis of gold deposits in the Carolina slate belt, USA: Regional constraints from trace element, Pb-Nd isotopic variations and $^{40}Ar/^{39}Ar$ geochronology: Geological Society of America Abstracts with Programs, v. 29, no. 6, p. A-60.

Barker, C.A., Secor, D.T., Jr., Pray, J., and Wright, J., 1998, Age and deformation of the Long Town metagranite, South Carolina Piedmont: A possible constraint on the origin of the Carolina terrane: The Journal of Geology, v. 106, p. 713–725.

Barr, S., 1993, Geochemistry and tectonic setting of late Precambrian volcanic and plutonic rocks in southeastern Cape Breton Island, Nova Scotia: Canadian Journal of Earth Sciences, v. 30, p. 1147–1154.

Barr, S.M., and Hegner, E., 1992, Nd isotopic compositions of felsic igneous rocks in Cape Breton, Nova Scotia: Canadian Journal of Earth Sciences, v. 29, p. 650–657.

Barr, S.M., and Kerr, A., 1997, Late Precambrian plutons in the Avalon terrane of New Brunswick, Nova Scotia, and Newfoundland, in Sinha, A.K., Whalen, J.B., and Hogan, J.P., eds., The Nature of Magmatism in the Appalachian Orogen: Geological Society of America Memoir 191, p. 45–74.

Barr, S.M., and Raeside, R., 1989, Tectono-stratigraphic terranes in Cape Breton Island, Nova Scotia: Implications for the configuration off the northern Appalachian orogen: Geology, v. 17, p. 822–825, doi: 10.1130/0091-7613(1989)017<0822:TSTICB>2.3.CO;2.

Barr, S.M., and White, C.E., 1996, Contrasts in late Precambrian–early Paleozoic tectonothermal history between Avalon composite terrane sensu stricto and other possible peri-Gondwanan terranes in southern New Brunswick and Cape Breton Island, Canada, in Nance, R.D., and Thompson, M.D., eds., Avalonian and Related Peri-Gondwanan Terranes of the Circum–North Atlantic: Geological Society of America Special Paper 304, p. 95–108.

Barr, S.M., Raeside, R.P., and White, C.E., 1998, Geological correlations between Cape Breton Island and Newfoundland, northern Appalachian orogen: Canadian Journal of Earth Sciences, v. 35, p. 1252–1270, doi: 10.1139/cjes-35-11-1252.

Barr, S.M., White, C.A., and Miller, B., 2002, The Kingston terrane, southern New Brunswick, Canada: Evidence for a Silurian volcanic arc: Geological Society of America Bulletin, v. 114, p. 964–982, doi: 10.1130/0016-7606(2002)114<0964:TKTSNB>2.0.CO;2.

Barr, S.M., White, C.E., and Miller, B.V., 2003, Age and geochemistry of Late Neoproterozoic and Early Cambrian igneous rocks in southern New Brunswick: Similarities and contrasts: Atlantic Geology, v. 39, p. 55–73.

Bevier, M., White, C.A., and Barr, S., 1990, Late Precambrian U-Pb ages from the Brookville gneiss, southern New Brunswick: The Journal of Geology, v. 98, p. 955–965.

Bevier, M., Barr, S., White, C.A., and Macdonald, A., 1993, U-Pb geochronologic constraints on the volcanic evolution of the Mira (Avalon) terrane, southeastern Cape Breton Island, Nova Scotia: Canadian Journal of Earth Sciences, v. 30, p. 1–10.

Bream, B.R., Hatcher, R.D., Jr., Miller, C.F., and Fullagar, P.D., 2004, Detrital zircon ages and Nd isotopic data from the Southern Appalachian crystalline core, Georgia, South Carolina, North Carolina, and Tennessee: New provenance constraints for part of the Laurentian margin, in Tollo, R.P., Corriveau, L., McLelland, J., and Bartholomew, M.J., eds., Proterozoic Tectonic Evolution of the Grenville Orogen in North America: Geological Society of America Memoir 197, p. 459–475.

Butler, J.R., and Fullagar, P., 1978, Petrochemical and geochronological studies of plutonic rocks in the southern Appalachians: III. Leucocratic adamellites of the Charlotte belt near Salisbury, North Carolina: Geological Society of America Bulletin, v. 89, p. 460–466, doi: 10.1130/0016-7606(1978)89<460:PAGSOP>2.0.CO;2.

Carrigan, C.W., Miller, C.F., Fullagar, P.D., Bream, B.R., Hatcher, R.D., Jr., and Coath, C.D., 2003, Ion microprobe age and geochemistry of Southern Appalachian basement, with implications for Proterozoic and Paleozoic reconstructions: Precambrian Research, v. 120, p. 1–36, doi: 10.1016/S0301-9268(02)00113-4.

Coler, D., and Samson, S., 2000, Characterization of the Spring Hope and Roanoke Rapids terranes, southern Appalachians: A U-Pb geochronologic and Nd isotopic study: Geological Society of America Abstracts with Programs, v. 32, no. 1, p. A-11–A-12.

Coler, D., Samson, S., and Stoddard, E., 1996, Terrane correlation and characterization in the southern Appalachians: An Sm-Nd isotopic investigation: Eos (Transactions, American Geophysical Union), v. 77, p. 290.

Coler, D.G., Samson, S.D., and Speer, J.A., 1997, Nd and Sr isotopic constraints on the source of Alleghanian granites in the Raleigh metamorphic belt and Eastern slate belt, southern Appalachians, U.S.: Chemical Geology, v. 134, p. 257–275, doi: 10.1016/S0009-2541(96)00063-0.

Coler, D., Wortman, G., Samson, S., Hibbard, J., and Stern, R., 2000, U-Pb geochronologic, Nd isotopic, and geochemical evidence for the correlation of the Chopawamsic and Milton terranes, Piedmont zone, southern Appalachian orogen: The Journal of Geology, v. 108, p. 363–380, doi: 10.1086/314411.

Dallmeyer, R.D., and Nance, D., 1992, Tectonic implications of ^{40}Ar/^{39}Ar mineral ages from late Precambrian–Cambrian plutons, Avalon composite terrane, southern New Brunswick, Canada: Canadian Journal of Earth Sciences, v. 29, p. 2445–2462.

Dallmeyer, R.D., and Nance, D., 1994, ^{40}Ar/^{39}Ar whole-rock phyllite ages from late Precambrian rocks of the Avalon composite terrane, New Brunswick: Evidence of Silurian-Devonian thermal rejuvenation: Canadian Journal of Earth Sciences, v. 31, p. 818–824.

Dennis, A., and Shervais, J., 1991, Arc rifting of the Carolina terrane in northwestern South Carolina: Geology, v. 19, p. 226–229, doi: 10.1130/0091-7613(1991)019<0226:AROTCT>2.3.CO;2.

Dennis, A.J., and Wright, J.E., 1997, The Carolina terrane in northwestern South Carolina, USA: Age of deformation and metamorphism in an exotic arc: Tectonics, v. 16, p. 460–473, doi: 10.1029/97TC00449.

Dennis, A.J., Shervais, J.W., Mauldin, J., Maher, H.D., and Wright, J.E., 2004, Petrology and geochemistry of Neoproterozoic volcanic arc terranes beneath the Atlantic Coastal Plain, Savannah River site, South Carolina: Geological Society of America Bulletin, v. 116, p. 572–593, doi: 10.1130/B25240.1.

DePaolo, D.J., 1981, Neodymium isotopes in the Colorado Front Range and crust-mantle evolution in the Proterozoic: Nature, v. 291, p. 193–196, doi: 10.1038/291193a0.

Dickin A.P., Tollo, R.P., Corriveau, L., McLelland, J., and Bartholomew, M.J., 2004, Mesoproterozoic and Paleoproterozoic crustal growth in the eastern Grenville Province; Nd isotope evidence from the Long Range Inlier of the Appalachian Orogen, in Tollo, R.P., Corriveau, L., McLelland, J., and Bartholomew, M.J., eds., Proterozoic tectonic evolution of the Grenville Orogen in North America: Geological Society of America Memoir 197, p. 495–503.

Dietsch, C.A., and Jercinovic, M., 2005, Tectonic implications of electron microprobe monazite ages of 525–535, 500–510, and ~475 Ma from the Waterbury dome, southwestern New England: Geological Society of America Abstracts with Program, v. 37, no. 1, p. 65.

D'Lemos, R.S., and Holdsworth, R.E., 1995, Samarium-neodymium isotopic characteristics of the northeastern Gander zone, Newfoundland Appalachians, in Hibbard, J.P., van Staal, C.R., and Cawood, P.A., eds., Current Perspectives in the Appalachian-Caledonian Orogen: Geological Association of Canada Special Paper 41, p. 239–252.

Dostal, J., Keppie, J.D., Cousens, B.L., and Murphy, J.B., 1996, 550–580 Ma magmatism in Cape Breton Island (Nova Scotia, Canada): The product of NW-dipping subduction during the final stage of amalgamation of Gondwana: Precambrian Research, v. 76, no. 1–2, p. 93–113, doi: 10.1016/0301-9268(96)00040-X.

Dubé, B., Dunning, G., and Lauzier, K., 1995, Geology of the Hope Brook Mine, Newfoundland, Canada: A preserved Late Proterozoic high-sulfidation epithermal gold deposit and its implications for exploration: Economic Geology and the Bulletin of the Society of Economic Geologists, v. 93, p. 405–436.

Dunning, G., and O'Brien, S., 1989, Late Proterozoic–early Paleozoic crust in the Hermitage flexure, Newfoundland Appalachians: U-Pb ages and tectonic significance: Geology, v. 17, p. 548–551, doi: 10.1130/0091-7613(1989)017<0548:LPEPCI>2.3.CO;2.

Dunning, G.R., Barr, S.M., Raeside, R.P., and Jamieson, R.A., 1990, U-Pb zircon, titanite and monazite ages in the Bras d'Or and Aspy terranes of Cape Breton Island, Nova Scotia: Implications for magmatic and metamorphic history: Geological Society of America Bulletin, v. 102, p. 322–330, doi: 10.1130/0016-7606(1990)102<0322:UPZTAM>2.3.CO;2.

Dunning, G.R., Swinden, H.S., Kean, B.F., Evans, D.T., and Jenner, G.A., 1991, A Cambrian island arc in Iapetus; geochronology and geochemistry of the Lake Ambrose volcanic belt, Newfoundland Appalachians: Geological Magazine, v. 128, p. 1–17.

Dunning, G.R., O'Brien, S.J., Holdsworth, R.E., and Tucker, R.D., 1993, Chronology of Pan-African, Penobscot and Salinic shear zones on the Gondwanan margin, northern Appalachians: Geological Society of America Abstracts with Program, v. 25, no. 1, p. 421–422.

Evans, D.T., Kean, B.F., and Dunning, G.R., 1990, Geological Studies, Victoria Lake Group, Central Newfoundland: Current Research: Newfoundland Department of Mines and Energy, Geological Survey Branch Report 90-1, p. 131–144.

Fetter, A.H., and Goldberg, S.A., 1995, Age and geochemical characteristics of bimodal magmatism in the Neoproterozoic Grandfather Mountain rift basin: The Journal of Geology, v. 103, p. 313–326.

Fryer, B.J., Kerr, A., Jenner, G.A., and Longstaffe, F.J., 1992, Probing the crust with plutons: Regional isotopic geochemistry of granitoid plutonic suites across Newfoundland: Current Research: Newfoundland Department of Mines and Energy, Geological Survey Branch Report 92-1, p. 118–139.

Fullagar, P.D., 1971, Age and origin of plutonic intrusions in the Piedmont of the southeastern Appalachians: Geological Society of America Bulletin, v. 82, p. 2845–2862, doi: 10.1130/0016-7606(1971)82[2845:AAOOPI]2.0.CO;2.

Fullagar, P., Goldberg, S., and Butler, R., 1997, Nd and Sr isotopic characterization of crystalline rocks from the southern Appalachian Piedmont and Blue Ridge, North and South Carolina, in Sinha, K., Whalen, J., and Hogan, J., eds., The Nature of Magmatism in the Appalachian Orogen: Geological Society of America Memoir 191, p. 165–179.

Glover, L.G., Sheridan, R.E., Holbrook, W.S., Ewing, J., Talwani, M., Hawman, R.B., and Wang, P., 1997, Paleozoic collisions, Mesozoic rifting, and structure of the Middle Atlantic states continental margin: An EDGE project report, in Glover, L., and Gates, A., eds., Central and Southern Appalachian Sutures: Results of the EDGE Project and Related Studies: Geological Society of America Special Paper 314, p. 107–135.

Goldberg, S.A., and Dallmeyer, R.D., 1997, Chronology of Paleozoic metamorphism and deformation in the Blue Ridge thrust complex, North Carolina and Tennessee: American Journal of Science, v. 297, p. 488–526.

Hamilton, M., and Murphy, J.B., 2004, Tectonic significance of a Llanvirn age for the Dunn Point volcanic rocks, Avalon terrane, Nova Scotia, Canada: Implications for the evolution of the Iapetus and Rheic Oceans: Tectonophysics, v. 379, p. 199–209, doi: 10.1016/j.tecto.2003.11.006.

Hatcher, R.D., Jr., Bream, B.R., Miller, C.F., Eckert, J.O., Jr., Fullagar, P.D., and Carrigan, C.W., 2004, Paleozoic structure of internal basement massifs, southern Appalachian Blue Ridge, incorporating new geochronologic, Nd and Sr isotopic, and geochemical data, in Tollo, R.P., Corriveau, L., McLelland, J., and Bartholomew, M.J., eds., Proterozoic Tectonic Evolution of the Grenville Orogen in North America: Geological Society of America Memoir 197, p. 525–547.

Heatherington, A.L., Mueller, P.A., and Nutman, A.P., 1996, Neoproterozoic magmatism in the Suwannee terrane: Implications for terrane correlation, in Nance, R.D., and Thompson, M.D., eds., Avalonian and Related peri-Gondwanan Terranes of the Circum–North Atlantic: Geological Society of America Special Paper 304, p. 257–268.

Henning, A.T., Sawyer, D.S., and Templeton, D.C., 2004, Exhumed upper mantle within the ocean-continent transition on the northern West Iberia margin: Evidence from prestack depth migration and total tectonic subsidence analyses: Journal of Geophysical Research, v. 109, no. B5, B05103, 16 p.

Hibbard, J., 1994, Kinematics of Acadian deformation in the northern and Newfoundland Appalachians: The Journal of Geology, v. 102, p. 215–228.

Hibbard, J., 2000, Docking Carolina: Mid-Paleozoic accretion in the southern Appalachians: Geology, v. 28, p. 127–130, doi: 10.1130/0091-7613(2000)28<127:DCMAIT>2.0.CO;2.

Hibbard, J., and Samson, S., 1995, Orogenesis exotic to the Iapetan cycle in the southern Appalachians, in Hibbard, J., van Staal, C., and Cawood, P., eds., Current Perspectives in the Appalachian-Caledonian Orogen: Geological Association of Canada Special Paper 41, p. 191–205.

Hibbard, J., Shell, G., Bradley, P., Samson, S., and Wortman, G., 1998, The Hyco shear zone in North Carolina and southern Virginia: Implications for the Piedmont zone–Carolina zone boundary in the southern Appalachians: American Journal of Science, v. 298, p. 85–107.

Hibbard, J., Stoddard, E., Secor, D., and Dennis, A., 2002, The Carolina zone: Overview of Neoproterozoic to early Paleozoic peri-Gondwanan terranes along the eastern flank of the southern Appalachians: Earth-Science Reviews, v. 57, p. 299–339, doi: 10.1016/S0012-8252(01)00079-4.

Hibbard, J., Standard, I., Miller, B., Hames, W., and Lavallee, S., 2003, Regional significance of the Gold Hill fault zone, Carolina zone of North Carolina: Geological Society of America Abstracts with Programs, v. 35, no. 1, p. 24.

Hibbard, J., Miller, B., Tracy, R., and Carter, B., 2005, The Appalachian peri-Gondwanan realm: A paleogeographic perspective from the south, in Vaughan, A., and Leat, P., eds., Terrane Processes at the Pacific Margin

of Gondwana: Geological Society [London] Special Publication 246, p. 97–112.
Hicks, R., Jamieson, R., and Reynolds, P., 1999, Detrital and metamorphic ^{40}Ar/^{39}Ar ages from muscovite and whole-rock samples, Meguma Supergroup, southern Nova Scotia: Canadian Journal of Earth Sciences, v. 36, p. 23–32, doi: 10.1139/cjes-36-1-23.
Hodych, J., and Buchan, K., 1998, Palaeomagnetism of the ca. 440 Ma Cape St. Mary's sills of the Avalon Peninsula of Newfoundland: Implications for Iapetus Ocean closure: Geophysical Journal International, v. 135, p. 155–164, doi: 10.1046/j.1365-246X.1998.00263.x.
Horton, W., Drake, A., and Rankin, D., 1989, Tectonostratigraphic terranes and their Paleozoic boundaries in the central and southern Appalachians, in Dallmeyer, D., ed., Terranes in the Circum-Atlantic Paleozoic orogens: Geological Society of America Special Paper 230, p. 213–245.
Ingle, S., Mueller, P., Heatherington, A., and Kozuch, M., 2003, Isotopic evidence for the magmatic and tectonic histories of the Carolina terrane: Implications for stratigraphy and terrane affiliation: Tectonophysics, v. 371, p. 187–211, doi: 10.1016/S0040-1951(03)00228-2.
Jenner, G.A., and Swinden, H.S., 1993, The Pipestone Pond Complex, central Newfoundland: Complex magmatism in an eastern Dunnage zone ophiolite: Canadian Journal of Earth Sciences, v. 30, p. 434–448.
Johnson, R., and Van der Voo, R., 1985, Middle Cambrian paleomagnetism of the Avalon terrane in Cape Breton Island, Nova Scotia: Tectonics, v. 4, p. 629–651.
Johnson, R., and Van der Voo, R., 1990, Pre-folding magnetization reconfirmed for the Late Ordovician–Early Silurian Dunn Point volcanics, Nova Scotia: Tectonophysics, v. 178, p. 193–205, doi: 10.1016/0040-1951(90)90146-Y.
Johnson, S., 2001, Contrasting geology in the Pocologan River and Long Reach areas: Implications for the New River belt and correlations in southern New Brunswick and Maine: Atlantic Geology, v. 37, p. 61–79.
Johnson, S.C.A., and McLeod, M.J., 1996, The New River belt: A unique segment along the western margin of the Avalon composite terrane, southern New Brunswick, Canada, in Nance, R.D., and Thompson, M.D., eds., Avalonian and Related Peri-Gondwanan Terranes of the Circum–North Atlantic: Geological Society of America Special Paper 304, p. 149–164.
Keppie, J.D., 1992, Structure of the Canadian Appalachians: Nova Scotia Department of Natural Resources, Mines and Energy Branch Bulletin 7, 89 p.
Keppie, J.D., and Dostal, J., 1998, Birth of the Avalon arc in Nova Scotia, Canada; geochemical evidence for approximately 700–630 Ma back-arc rift volcanism off Gondwana: Geological Magazine, v. 135, p. 171–181, doi: 10.1017/S0016756898008322.
Keppie, J.D., and Ramos, V., 1999, Odyssey of terranes in the Iapetus and Rheic Oceans during the Paleozoic, in Ramos, V., and Keppie, D., eds., Laurentia-Gondwana Connections before Pangea: Geological Society of America Special Paper 336, p. 267–276.
Keppie, J.D., Dostal, J., Murphy, J.B., and Cousens, B.L., 1997, Palaeozoic within-plate volcanic rocks in Nova Scotia (Canada) reinterpreted; isotopic constraints on magmatic source and palaeocontinental reconstructions: Geological Magazine, v. 134, p. 425–447, doi: 10.1017/S001675689700719X.
Keppie, J.D., Dostal, J., Dallmeyer, R.D., and Doig, R., 2000, Superposed Neoproterozoic and Silurian magmatic arcs in central Cape Breton Island, Canada: Geochemical and geochronological constraints: Geological Magazine, v. 137, p. 137–153, doi: 10.1017/S0016756800003769.
Kerr, A., 1997, Space-time composition relationships among Appalachian cycle plutonic suites in Newfoundland, in Sinha, A.K., Whalen, J.B., and Hogan, J.P., eds., The Nature of Magmatism in the Appalachian Orogen: Geological Society of America Memoir 191, p. 193–220.
Kerr, A., Jenner, G.A., and Fryer, B.J., 1995, Sm-Nd isotopic geochemistry of Precambrian to Paleozoic granites and the deep-crustal structure of the southeast margin of the Newfoundland Appalachians: Canadian Journal of Earth Sciences, v. 32, p. 224–245.
King, A.F., Brueckner, W.D., Anderson, M.M., and Fletcher, T., 1974, Late Precambrian and Cambrian Sedimentary Sequences of Eastern Newfoundland: Geological Association of Canada Fieldtrip Manual B-6, 59 p.
Kite, L., and Stoddard, E., 1984, The Halifax County complex: Oceanic lithosphere in the eastern North Carolina Piedmont: Geological Society of America Bulletin, v. 95, p. 422–432, doi: 10.1130/0016-7606(1984)95<422:THCCOL>2.0.CO;2.
Kylander-Clark, A.R.C., Coleman, D.S., Glazner, A.F., and Bartley, J.M., 2005, Evidence for 65 km of dextral slip across Owens Valley, California, since 83 Ma: Geological Society of America Bulletin, v. 117, p. 962–968, doi: 10.1130/B25624.1.
Landing, E., 1996, Avalon: Insular continent by the latest Precambrian, in Nance, R.D., and Thompson, M.D., eds., Avalonian and Related Peri-Gondwanan Terranes of the Circum–North Atlantic: Geological Society of America Special Paper 304, p. 29–63.
Liss, M., van der Pluijm, B., and Van der Voo, R., 1993, Avalonian proximity of the Ordovician Miramichi terrane, northern New Brunswick, northern Appalachians: Paleomagnetic evidence for rifting and back-arc basin formation at the southern margin of Iapetus: Tectonophysics, v. 227, p. 17–30, doi: 10.1016/0040-1951(93)90084-W.
MacLachlan, K., and Dunning, G.R., 1998, U-Pb ages and tectonic setting of Middle Ordovician calc-alkaline and within plate volcanic rocks of the Wild Bight Group, northeastern Dunnage zone, Newfoundland Appalachians: Implications for the evolution of the Gondwanan margin: Canadian Journal of Earth Sciences, v. 35, p. 998–1017, doi: 10.1139/cjes-35-9-998.
McCabe, C.A., Channell, J., and Woodcock, N., 1992, Further paleomagnetic results from the Builth Wells Ordovician Inlier, Wales: Journal of Geophysical Research, v. 97, p. 9357–9370.
McCartney, W.D., 1967, Whitbourne Map-Area, Newfoundland: Geological Survey of Canada Memoir 341, 135 p.
McKerrow, W.S., and van Staal, C.A.R., 2000, The Palaeozoic time scale reviewed, in Franke, W., Haak, V., Oncken, O., and Tanner, D., eds., Orogenic Processes: Quantification and Modelling in the Variscan Belt: Geological Society [London] Special Publication 179, p. 5–8.
McLeod, M.J., Ruitenberg, A.A., and Krogh, T.E., 1992, Geology and U-Pb geochronology of the Annidale Group, southern New Brunswick: Lower Ordovician volcanic and sedimentary rocks formed near the southeastern margin of Iapetus Ocean: Atlantic Geology, v. 28, p. 181–192.
McNamara, A., MacNiocaill, C.A., van der Pluijm, B., and Van der Voo, R., 2001, West African proximity of the Avalon terrane in the latest Precambrian: Geological Society of America Bulletin, v. 113, p. 1161–1170, doi: 10.1130/0016-7606(2001)113<1161:WAPOTA>2.0.CO;2.
McSween, H., and Harvey, R., 1997, Concord plutonic suite: Pre-Acadian gabbro-syenite intrusions in the southern Appalachians, in Sinha, K., Whalen, J., and Hogan, J., eds., The Nature of Magmatism in the Appalachian Orogen: Geological Society of America Memoir 191, p. 221–234.
Miller, B.V., and Barr, S.M., 2000, Petrology and isotopic composition of a Grenvillian basement fragment in the northern Appalachian orogen; Blair River Inlier, Nova Scotia, Canada: Journal of Petrology, v. 41, p. 1777–1804, doi: 10.1093/petrology/41.12.1777.
Miller, B.V., and Barr, S.M., 2004, Metamorphosed gabbroic dikes related to opening of Iapetus Ocean at the St. Lawrence Promontory; Blair River Inlier, Nova Scotia, Canada: The Journal of Geology, v. 112, p. 277–288, doi: 10.1086/382759.
Miller, B.V., Samson, S.D., and D'Lemos, R.S., 1999, Time span of plutonism, fabric development, and cooling in a Neoproterozoic arc segment; U-Pb age constraints from syn-tectonic plutons, Sark, Channel Island, UK: Tectonophysics, v. 312, p. 79–95.
Miller, B.V., Hibbard, J.P., Standard, I.D., Hames, W.E., and Lavallee, S.B., 2003, U-Pb zircon and titanite ages from rocks associated with the Gold Hill fault zone, Carolina zone of North Carolina: Relationships with other peri-Gondwanan terranes: Geological Society of America Abstracts with Programs, v. 35, no. 1, p. 24.
Mueller, P., Kozuch, M., Heatherington, A., Wooden, J., Offield, T., Koeppen, R., Klein, T., and Nutman, A., 1996, Evidence for Mesoproterozoic basement in the Carolina terrane and speculation on its origins, in Nance, R.D., and Thompson, M.D., eds., Avalonian and Related Peri-Gondwanan Terranes of the Circum–North Atlantic: Geological Society of America Special Paper 304, p. 207–217.
Murphy, J.B., and Nance, D., 2002, Sm-Nd isotopic systematics as tectonic tracers: An example from West Avalonia in the Canadian Appalachians: Earth-Science Reviews, v. 59, p. 77–100, doi: 10.1016/S0012-8252(02)00070-3.
Murphy, J.B., Keppie, J.D., Dostal, J., and Cousens, B.L., 1996, Repeated late Neoproterozoic–Silurian lower crustal melting beneath the Antigonish Highlands, Nova Scotia: Nd isotopic evidence and tectonic interpretations, in Nance, R.D., and Thompson, M.D., eds., Avalonian and Related Peri-Gondwanan Terranes of the Circum–North Atlantic: Geological Society of America Special Paper 304, p. 109–120.
Murphy, J.B., Keppie, D., Dostal, J., and Nance, D., 1999, Neoproterozoic–early Paleozoic evolution of Avalonia, in Ramos, V., and Keppie, D., eds.,

Laurentia-Gondwana Connections before Pangea: Geological Society of America Special Paper 336, p. 253–266.

Nance, D., and Dallmeyer, R.D., 1994, Structural and ^{40}Ar/^{39}Ar mineral age constraints for the tectonothermal evolution of the Green Head Group and Brookville gneiss, southern New Brunswick, Canada: Implications for the configuration of the Avalon composite terrane: Geological Journal, v. 29, p. 293–322.

Nance, D., and Murphy, B., 1994, Contrasting basement isotopic signatures and the palinspastic restoration of peripheral orogens: Example from the Neoproterozoic Avalonian-Cadomian belt: Geology, v. 22, p. 617–620, doi: 10.1130/0091-7613(1994)022<0617:CBISAT>2.3.CO;2.

Nance, D., and Murphy, B., 1996, Basement isotopic signatures and Neoproterozoic paleogeography of Avalonian-Cadomian and related terranes in the circum–North Atlantic, in Nance, R.D., and Thompson, M.D., eds., Avalonian and Related Peri-Gondwanan Terranes of the Circum–North Atlantic: Geological Society of America Special Paper 304, p. 333–346.

Nance, D., Murphy, J.B., and Keppie, J.D., 2002, A Cordilleran model for the evolution of Avalonia: Tectonophysics, v. 352, p. 11–31, doi: 10.1016/S0040-1951(02)00187-7.

Noel, J., Spariosu, D., and Dallmeyer, D., 1988, Paleomagnetism and ^{40}Ar/^{39}Ar ages from the Carolina slate belt, Albemarle, North Carolina: Implications for terrane amalgamation with North America: Geology, v. 16, p. 64–68, doi: 10.1130/0091-7613(1988)016<0064:PAAAAF>2.3.CO;2.

O'Brien, B., O'Brien, S., and Dunning, G., 1991, Silurian cover, Late Precambrian–Early Ordovician basement, and the chronology of Silurian orogenesis in the Hermitage Flexure (Newfoundland Appalachians): American Journal of Science, v. 291, p. 760–799.

O'Brien, B.H., Swinden, H.S., Dunning, G.R., Williams, S.H., and O'Brien, F.H., 1997, A peri-Gondwanan arc-backarc complex in Iapetus: Early-Mid Ordovician evolution of the Exploits Group, Newfoundland: American Journal of Science, v. 297, p. 220–272.

O'Brien, S., O'Brien, B., Dunning, G., and Tucker, R., 1996, Late Neoproterozoic Avalonian and related peri-Gondwanan rocks of the Newfoundland Appalachians, in Nance, R.D., and Thompson, M.D., eds., Avalonian and Related Peri-Gondwanan Terranes of the Circum–North Atlantic: Geological Society of America Special Paper 304, p. 9–28.

Offield, T., Kunk, M., and Koeppen, R., 1995, Style and age of deformation, Carolina slate belt, central North Carolina: Southeastern Geology, v. 35, p. 59–77.

Owen, J.V., Greenough, J.D., Fryer, B.J., and Longstaffe, F.J., 1992, Petrogenesis of the Potato Hill Pluton, Newfoundland; transpression during the Grenvillian orogenic cycle?: Geological Society [London] Journal, v. 149, p. 923–935.

Owens, B.E., and Samson, S.D., 2004, Nd isotopic constraints on the magmatic history of the Goochland terrane, easternmost Grenvillian crust in the Southern Appalachians, in Tollo, R.P., Corriveau, L., McLelland, J., and Bartholomew, M.J., eds., Proterozoic Tectonic Evolution of the Grenville Orogen in North America: Geological Society of America Memoir 197, p. 601–608.

Pe Piper, G., and Piper David, J.W., 1998, Geochemical evolution of Devonian-Carboniferous igneous rocks of the Magdalen Basin, eastern Canada: Pb- and Nd-isotope evidence for mantle and lower crustal sources: Canadian Journal of Earth Sciences, v. 35, p. 201–221, doi: 10.1139/cjes-35-3-201.

Pettingill, H.S., Sinha, A.K., and Tatsumoto, M., 1984, Age and origin of anorthosites, charnockites, and granulites in the central Virginia Blue Ridge: Nd and Sr isotopic evidence: Contributions to Mineralogy and Petrology, v. 85, p. 279–291, doi: 10.1007/BF00378106.

Rabu, D., Thieblemont, D., Tegyey, M., Guerrot, C., Alsac, C., Chauvel, J.J., Murphy, J.B., and Keppie, J.D., 1996, Late Proterozoic to Paleozoic evolution of the St. Pierre and Miquelon islands: A new piece in the Avalonian puzzle of the Canadian Appalachians, in Nance, R.D., and Thompson, M.D., eds., Avalonian and Related Peri-Gondwanan Terranes of the Circum–North Atlantic: Geological Society of America Special Paper 304, p. 65–94.

Raeside, R., and Barr, S., 1990, Geology and tectonic development of the Bras d'Or suspect terrane, Cape Breton Island, Nova Scotia: Canadian Journal of Earth Sciences, v. 27, p. 1371–1381.

Ranger, M., Pickerill, R., and Fillon, D., 1984, Lithostratigraphy of the Cambrian?–Lower Ordovician Bell Island and Wabana Groups of Little Bell, and Kellys Islands, Conception Bay, eastern Newfoundland: Canadian Journal of Earth Sciences, v. 21, p. 1245–1261.

Reusch, D., van Staal, C.A., and Hibbard, J., 2001, The Surry Complex: Root zone of the Penobscot orogen?: Geological Society of America Abstracts with Programs, v. 33, no. 6, p. A-261.

Rogers, N., van Staal, C., McNicoll, V., Pollock, J., Zagorevski, A., Whalen, J., and Kean, B., 2006, Neoproterozoic and Cambrian arc magmatism along the eastern margin of the Victoria Lake Supergroup: A remnant of Ganderian basement in central Newfoundland?: Precambrian Research, v. 147, p. 320–341, doi: 10.1016/j.precamres.2006.01.025.

Ruitenberg, A., McLeod, M., and Krogh, T., 1993, Comparative metallogeny of Ordovician volcanic and sedimentary rocks in the Annidale-Shannon (New Brunswick) and Harborside-Blue Hill (Maine) areas: Implications of new U-Pb age dates: Exploration and Mining Geology, v. 2, p. 355–365.

Samson, S., 1995, Is the Carolina terrane part of Avalon?, in Hibbard, J., van Staal, C., and Cawood, P., eds., Current Perspectives in the Appalachian-Caledonian Orogen: Geological Association of Canada Special Paper 41, p. 253–264.

Samson, S.D., and D'Lemos, R.S., 1998, U-Pb geochronology and Sm-Nd isotopic composition of Proterozoic gneisses, Channel Islands, UK: Journal of the Geological Society, v. 155, p. 609–618.

Samson, S., and Secor, D.T., 2000, New U-Pb geochronological evidence for a Silurian magmatic event in central South Carolina: Geological Society of America Abstracts with Programs, v. 32, no. 2, p. A-71.

Samson, S.D., Coler, D.G., and Speer, J.A., 1995a, Geochemical and Nd-Sr-Pb isotopic composition of Alleghanian granites of the Southern Appalachians; origin, tectonic setting, and source characterization: Earth and Planetary Science Letters, v. 134, p. 359–376, doi: 10.1016/0012-821X(95)00124-U.

Samson, S., Hibbard, J., and Wortman, G., 1995b, Nd isotopic evidence for juvenile crust in the Carolina terrane, southern Appalachians: Contributions to Mineralogy and Petrology, v. 121, p. 171–184, doi: 10.1007/s004100050097.

Samson, S.D., Barr, S.M., and White, C.E., 2000, Nd isotopic characteristics of terranes within the Avalon zone, southern New Brunswick: Canadian Journal of Earth Sciences, v. 37, p. 1039–1052, doi: 10.1139/cjes-37-7-1039.

Samson, S.D., Secor, D.T., and Hamilton, M.A., 2001, Wandering Carolina: Tracking exotic terranes with detrital zircons: Geological Society of America Abstract with Programs, v. 33, no. 2, p. A-263.

Schenk, P., 1997, Sequence stratigraphy and provenance on Gondwana's margin: The Meguma zone (Cambrian to Devonian) of Nova Scotia, Canada: Geological Society of America Bulletin, v. 109, p. 395–409, doi: 10.1130/0016-7606(1997)109<0395:SSAPOG>2.3.CO;2.

Schofield, D., and D'Lemos, R., 2000, Granite petrogenesis in the Gander zone, NE Newfoundland: Mixing of melts from multiple sources and the role of lithospheric delamination: Canadian Journal of Earth Sciences, v. 37, p. 535–547, doi: 10.1139/cjes-37-4-535.

Seaman, S., Scherer, E., Wobus, R., Zimmer, J., and Sales, G., 1999, Late Silurian volcanism in coastal Maine: The Cranberry Island series: Geological Society of America Bulletin, v. 111, p. 686–708, doi: 10.1130/0016-7606(1999)111<0686:LSVICM>2.3.CO;2.

Secor, D., and Snoke, A., 2002, Geologic Map of the Batesburg and Emory Quadrangles, Lexington and Saluda Counties, South Carolina, with Explanatory Notes: Geological Society of America Maps and Charts Series MCH091, 32 p.

Secor, D.T., Samson, S., Snoke, A., and Palmer, A., 1983, Confirmation of the Carolina slate belt as an exotic terrane: Science, v. 221, p. 649–651, doi: 10.1126/science.221.4611.649.

Sevigny, J., and Hanson, G., 1995, Late Taconian and pre-Acadian history of the New England Appalachians of southwestern Connecticut: Geological Society of America Bulletin, v. 107, p. 487–498, doi: 10.1130/0016-7606(1995)107<0487:LTAPAH>2.3.CO;2.

Shervais, J., Dennis, A., McGee, J., and Secor, D., 2003, Deep in the heart of Dixie: Pre-Alleghanian eclogite and HP granulite metamorphism in the Carolina terrane, South Carolina, USA: Journal of Metamorphic Geology, v. 21, p. 65–80, doi: 10.1046/j.1525-1314.2003.00416.x.

Smith, S.A., and Hiscott, R.N., 1984, Latest Precambrian to Early Cambrian basin evolution, Fortune Bay, Newfoundland: Fault-bounded basin to platform: Canadian Journal of Earth Sciences, v. 21, p. 1379–1392.

Stewart, D., Tucker, R., and West, D., 1995, Genesis of Silurian composite terrane in northern Penobscot Bay, in Hussey, A., and Johnston, R., eds., Guidebook for Field Trips in Southern Maine and Adjacent New Hampshire: New England Intercollegiate Geological Conference, 87th Annual Meeting, October 6–8, 1995, Brunswick, Maine, p. A3-1–A3-21.

Stewart, D., Tucker, R., Ayuso, R., and Lux, D., 2001, Minimum age of the Neoproterozoic Seven Hundred Acre Island Formation and the tectonic setting of the Islesboro Formation, Islesboro block, Maine: Atlantic Geology, v. 37, p. 41–59.

Su, Q., and Fullagar, P.D., 1995, Rb-Sr and Sm-Nd isotopic systematics during greenschist facies metamorphism and deformation; examples from the southern Appalachian Blue Ridge: The Journal of Geology, v. 103, p. 423–436.

Sutter, J., Milton, D., and Kunk, M., 1983, ^{40}Ar/^{39}Ar age spectrum dating of gabbro plutons and surrounding rocks in the Charlotte belt of North Carolina: Geological Society of America Abstracts with Programs, v. 15, no. 2, p. 110.

Swinden, H.S., and Hunt, P.A., 1991, A U-Pb zircon age from the Connaigre Bay Group southwestern Avalon zone, Newfoundland: Implications for regional correlations and metallogenesis, in Radiogenic age and isotopic studies Report 4, Geological Survey of Canada Paper 90-2, p. 3–10.

Tomascak, P.B., Krogstad, E.J., and Walker, R.J., 1996, Nature of the crust in Maine, USA; evidence from the Sebago Batholith: Contributions to Mineralogy and Petrology, v. 125, p. 45–59, doi: 10.1007/s004100050205.

Trench, A., Torsvik, T.H., and McKerrow, W.S., 1992, The paleogeographic evolution of Southern Britain during early Palaeozoic times: A reconciliation of palaeomagnetic and biogeographic evidence: Tectonophysics, v. 201, p. 75–82.

van der Velden, A.J., van Staal, C.A.R., and Cook, F.A., 2004, Crustal structure, fossil subduction, and the tectonic evolution of the Newfoundland Appalachians: Evidence from a reprocessed seismic reflection survey: Geological Society of America Bulletin, v. 116, p. 1485–1498, doi: 10.1130/B25518.1.

van Staal, C.A., 1994, Brunswick subduction complex in the Canadian Appalachians: Record of the Late Ordovician to Late Silurian collision between Laurentia and the Gander margin of Avalon: Tectonics, v. 13, p. 946–962, doi: 10.1029/93TC03604.

van Staal, C.A., 2005, Northern Appalachians, in Selley, R.C., Cocks, R.M., and Plimer, I.R., eds., Encyclopedia of Geology: Oxford, Elsevier, v. 4, p. 81–91.

van Staal, C.A., and deRoo, J., 1995, Mid-Paleozoic tectonic evolution of the Appalachian central mobile belt in northern New Brunswick, Canada: Collision, extensional collapse and dextral transpression, in Hibbard, J., van Staal, C., and Cawood, P., eds., Current Perspectives in the Appalachian-Caledonian Orogen: Geological Association of Canada, Special Paper 41, p. 367–389.

van Staal, C., Ravenhurst, C., Winchester, J., Roddick, J., and Langton, J., 1990, Post-Taconic blueschist suture in the northern Appalachians of northern New Brunswick, Canada: Geology, v. 18, p. 1073–1077, doi: 10.1130/0091-7613(1990)018<1073:PTBSIT>2.3.CO;2.

van Staal, C.A., Sullivan, R., and Whalen, J., 1996, Provenance and tectonic history of the Gander zone in the Caledonian/Appalachian orogen: Implications for the origin and assembly of Avalon, in Nance, R.D., and Thompson, M.D., eds., Avalonian and Related Peri-Gondwanan Terranes of the Circum–North Atlantic: Geological Society of America Special Paper 304, p. 347–367.

van Staal, C., Dewey, J., MacNiocaill, C., and McKerrow, W., 1998, The Cambrian-Silurian tectonic evolution of the northern Appalachians and British Caledonides: History of a complex, west and southwest Pacific-type segment of Iapetus, in Blundell, D., and Scott, A., eds., Lyell: The Past is the Key to the Present: Geological Society [London] Special Publication 143, p. 199–242.

van Staal, C.A., Wilson, R., Rogers, N., Gyffe, L., Langton, J., McCutcheon, S., McNicoll, V., and Ravenhurst, C.A., 2003, Geology and tectonic history of the Bathurst Supergroup, Bathurst Mining Camp, and its relationships to coeval rocks in southwestern New Brunswick and adjacent Maine—A synthesis: Economic Geology Monograph, v. 11, p. 37–60.

Vick, H., Channell, J., and Opdyke, N., 1987, Ordovician docking of the Carolina slate belt: Paleomagnetic data: Tectonics, v. 6, p. 573–583.

Vines, J., Hibbard, J., and Shell, G., 1998, Structural geology of the High Rock granite: Implications for displacement along the Hyco shear zone, North Carolina: Southeastern Geology, v. 37, p. 163–176.

Waldron, J.W.F., Murphy, J.B., Melchin, M.J., and Davis, G., 1996, Silurian tectonics of western Avalonia: Strain-corrected subsidence history of the Arisaig Group, Nova Scotia: The Journal of Geology, v. 104, p. 677–694.

West, D.P., Jr., Coish, R.A., and Tomascak, P.B., 2004, Tectonic setting and regional correlation of Ordovician metavolcanic rocks of the Casco Bay Group, Maine; evidence from the trace element and isotope geochemistry: Geological Magazine, v. 141, p. 125–140, doi: 10.1017/S0016756803008562.

Whalen, J.B., Jenner, G.A., Hegner, E., Gariepy, C., and Longstaffe, F.J., 1994a, Geochemical and isotopic (Nd, O, and Pb) constraints on granite sources in the Humber and Dunnage zones, Gaspesie, Quebec, and New Brunswick; implications for tectonics and crustal structure: Canadian Journal of Earth Sciences, v. 31, p. 323–340.

Whalen, J.B., Jenner, G.A., Currie, K.L., Barr, S.M., Longstaffe, F.J., and Hegner, E., 1994b, Geochemical and isotopic of granitoids of the Avalon zone, southern New Brunswick: Possible evidence for repeated delamination events: The Journal of Geology, v. 102, p. 269–282.

Whalen, J.B., Fyffe, L.R., Longstaffe, F.J., and Jenner, G.A., 1996a, The position and nature of the Gander-Avalon boundary, southern New Brunswick, based on geochemical and isotopic data from granitoid rocks: Canadian Journal of Earth Sciences, v. 33, p. 129–139.

Whalen, J.B., Jenner, G.A., Longstaffe, F.J., and Hegner, E., 1996b, Nature and evolution of the eastern margin of Iapetus; geochemical and isotopic constraints from Siluro-Devonian granitoid plutons in the New Brunswick Appalachians: Canadian Journal of Earth Sciences, v. 33, p. 140–155.

White, C., and Barr, S., 1996, Geology of the Brookville terrane, southern New Brunswick, Canada, in Nance, R.D., and Thompson, M.D., eds., Avalonian and Related Terranes of the Circum–North Atlantic: Geological Society of America Special Paper 304, p. 133–147.

White, C., Barr, S., Bevier, M., and Kamo, S., 1994, A revised interpretation of Cambrian and Ordovician rocks in the Bourinot belt of central Cape Breton Island, Nova Scotia: Atlantic Geology, v. 30, p. 123–142.

White, C.E., Barr, S.M., Miller, B.V., and Hamilton, M.A., 2002, Granitoid plutons of the Brookville terrane, southern New Brunswick: Petrology, age, and tectonic setting: Atlantic Geology, v. 38, p. 53–74.

Whitmarsh, R.B., White, R.S., Horsefield, S.J., Sibuet, J.-C.A., Recq, M., and Louvel, V., 1996, The ocean-continent boundary off the western continental margin of Iberia: Crustal structure west of Galicia Bank: Journal of Geophysical Research, v. 101, no. B12, p. 28,291–28,314, doi: 10.1029/96JB02579.

Williams, H., and Hatcher, R.D., 1983, Appalachian suspect terranes, in Hatcher, R.D., Jr., Williams, H., and Zietz, I., eds., Contributions to the Tectonics and Geophysics of Mountain Chains: Geological Society of America Memoir 158, p. 33–54.

Wortman, G., Samson, S., and Hibbard, J., 1996, Discrimination of the Milton belt and the Carolina terrane in the southern Appalachians: The Journal of Geology, v. 104, p. 239–247.

Wortman, G., Samson, S., and Hibbard, J., 1998, Precise U-Pb timing constraints on the kinematic development of the Hyco shear zone, southern Appalachians: American Journal of Science, v. 298, p. 108–130.

Wortman, G., Samson, S., and Hibbard, J., 2000, Precise U-Pb zircon constraints on the earliest magmatic history of the Carolina terrane: The Journal of Geology, v. 108, p. 321–338, doi: 10.1086/314401.

Wright, J., and Seiders, V., 1980, Age of zircon from volcanic rocks of the central North Carolina Piedmont and tectonic implications for the Carolina volcanic slate belt: Geological Society of America Bulletin, v. 91, p. 287–294, doi: 10.1130/0016-7606(1980)91<287:AOZFVR>2.0.CO;2.

MANUSCRIPT ACCEPTED BY THE SOCIETY 22 MARCH 2007

Printed in the USA

Cat Square basin, Catskill clastic wedge: Silurian-Devonian orogenic events in the central Appalachians and the crystalline southern Appalachians

Allen J. Dennis*

Department of Biology and Geology, University of South Carolina, Aiken, South Carolina 29801-6309, USA

ABSTRACT

Recognition of the timing of peak metamorphism in the eastern Blue Ridge (ca. 460 Ma), Inner Piedmont (ca. 360 Ma), and Carolina terrane (ca. 540 Ma) has been critical in discerning the history of the collage of terranes in the hinterland of the southern Appalachian orogen. The Inner Piedmont consists of two terranes: the Tugaloo terrane, which is an Ordovician plutonic arc intruding thinned Laurentian crust and Iapetus, and the Cat Square paragneiss terrane, which is interpreted here as a Silurian basin that formed as the recently accreted (ca. 455 Ma) Carolina terrane rifted from Laurentia and was transferred to an oceanic plate. The recognition of an internal Salinic basin and associated magmatism in the southern Appalachian hinterland agrees with observations in the New England and Maritime Appalachians. Structural analysis in the Tugaloo terrane requires the Inner Piedmont to be restored to its pre-Carboniferous location, near the New York promontory. At this location, the Catskill and Pocono clastic wedges were deposited in the Devonian and Mississippian, respectively. Between the two wedges, an enigmatic formation (Spechty Kopf and its correlative equivalent Rockwell Formation) was deposited. Polymictic diamictites within this unit contain compositionally immature exotic clasts that may prove to have been derived from the Inner Piedmont. Following deposition of the Spechty Kopf and Rockwell Formations, the Laurentian margin became a right-lateral transform plate boundary. This continental-margin transform was subsequently modified and translated northwest above the Alleghanian Appalachian décollement. Thus, several critical recent observations presented here inspire a new model for the Silurian through Mississippian terrane dispersal and orogeny that defines southern Appalachian terrane geometry prior to emplacement of the Blue Ridge–Inner Piedmont–Carolina–other internal terranes as crystalline thrust sheets.

Keywords: Appalachian, Salinic, successor basin, Carolinia, Acadian, Inner Piedmont.

*dennis@sc.edu

Dennis, A.J., 2007, Cat Square basin, Catskill clastic wedge: Silurian-Devonian orogenic events in the central Appalachians and the crystalline southern Appalachians, *in* Sears, J.W., Harms, T.A., and Evenchick, C.A., eds., Whence the Mountains? Inquiries into the Evolution of Orogenic Systems: A Volume in Honor of Raymond A. Price: Geological Society of America Special Paper 433, p. 313–329, doi: 10.1130/2007.2433(15). For permission to copy, contact editing@geosociety.org. ©2007 The Geological Society of America. All rights reserved.

INTRODUCTION

Clastic wedges that disrupt passive-margin sedimentation are reliable indicators of tectonic activity on a margin (Thomas, 1977; Hoffman and Bowring, 1984), and they are commonly interpreted to indicate collisions of a continental margin with arc terranes, other continents, or continental fragments. These clastic wedges, which coarsen and thicken toward the hinterland and often are preceded by unconformities and black shales, are unambiguous indicators of events that may be diachronous along the length of an active margin (e.g., Ettensohn, 1987; Ferrill and Thomas, 1988). There are numerous clastic wedges in the Appalachians that range from Middle Ordovician to Upper Mississippian and Pennsylvanian in age (Thomas, 1977). They record the accretion or collapse of various Neoproterozoic, Cambrian-Ordovician, and younger arc terranes that accompanied the Paleozoic destruction of Iapetus, the closing of the Rheic Ocean, and the ultimate Gondwanan collision with Laurentia.

Additionally, successor basins (variously called overlap or overstep assemblages, or retroarc foreland basins) record accretion of one terrane to another either by: (1) a single stratigraphic sequence deposited upon different basement terranes, or (2) basin strata recording provenance from two different basement terranes, or both (Monger, 1977; Eisbacher, 1985; Ricketts et al., 1992; Graham et al., 1993; Williams, 1978, his 17c[1]). The middle Paleozoic history of the central and crystalline southern Appalachians may be clarified by the following interpretations: the Cat Square terrane paragneiss represents the fill of a late Silurian successor basin, and structural, metamorphic, and stratigraphic correlations between the Devonian clastic wedge of the central Appalachians and orogenic events preserved in the crystalline southern Appalachian Piedmont may indicate that these regions were once contiguous. This proposed history has been obscured by Devonian-age upper-amphibolite– to granulite-facies metamorphism of the Cat Square successor basin and significant Mississippian and younger strike-slip translation separating the Frasnian-Famennian Catskill clastic wedge from its orogenic source. Timing relationships recently described in the Appalachian hinterland may constrain a more complete model of middle Paleozoic orogeny in the central and southern Appalachians and extend our understanding of terrane dispersal on this long-lived orogenic belt.

[1]In his Tectonolithofacies Map of the Appalachian Orogen, Williams (1978) defines central and northern Appalachian successor basins (his unit 17) in this way: "MARINE TO CONTINENTAL ROCKS DEPOSITED IN TROUGHS AND BASINS ACROSS ENTIRE APPALACHIAN OROGEN: Middle to Late Ordovician, Silurian, and Devonian sedimentary and oceanic rocks... [and specifically] 17c) marine to terrestrial sedimentary rocks and mainly terrestrial bimodal volcanic rocks deposited across the deformed continental margins of Iapetus (Humber and Gander zones) ... and also across the Avalon and Meguma zones, unconformably overlying volcanic rocks (12) and ophiolite complexes (9) of the Dunnage zone except locally where deposited conformably as upward shoaling sequences upon marine Ordovician rocks..." (compare with Robinson et al., 1998, p. 121).

EVIDENCE THAT THE SOUTHERN APPALACHIAN INNER PIEDMONT WAS IN THE CENTRAL APPALACHIANS AT THE DEVONIAN-MISSISSIPPIAN BOUNDARY

Several observations suggest that the southern Appalachian Inner Piedmont may have originally collided with the Laurentian New York promontory to create the Catskill-Pocono wedge prior to dextral strike-slip translation and overthrust faulting to its present location:

1. Adjacent terranes have contrasting times of peak metamorphism. Peak metamorphism in the eastern Blue Ridge of North Carolina is reported at 457.6 ± 1.0 Ma (Figs. 1 and 2; granulite facies, Winding Stairs Gap; Moecher et al., 2004) and 459 +1.5/–0.6 Ma (eclogite facies, Lick Ridge; Busch et al., 2002). These ages correlate with the Middle Ordovician unconformity and (Blount) clastic wedge observed in central Tennessee (e.g., Rodgers, 1953; Thomas, 1977). This correlation permits the Blount wedge to represent a foreland basin linked to unroofing of an orogenic belt in the Blue Ridge. In the adjacent Inner Piedmont, however, peak metamorphic conditions occurred ca. 355–365 Ma, according to U-Pb thermal-ionization mass spectrometry (TIMS) analysis of monazite, and local resetting of these ages occurred ca. 320–330 Ma (Figs. 1 and 2; Dennis and Wright, 1997a). A middle Paleozoic age has been independently reported for the Cat Square charnockite (Kish, 1997), a metamorphic unit reported on Goldsmith et al.'s (1988) map of the Charlotte 2° sheet. Specifically, Kish (1997) reported $^{206}Pb/^{238}U$ ages ranging from 348 to 357 Ma and $^{207}Pb/^{206}Pb$ ages ranging from 360 to 380 Ma on air-abraded zircon fractions. While both the eastern Blue Ridge and the Inner Piedmont record Devonian plutonism, the character of this magmatism differs across the Brevard zone (e.g., Mapes, 2002; Table 1), and despite widespread Ordovician plutonism, **there is no evidence of an earlier Ordovician metamorphic event in the Inner Piedmont**. Furthermore, there is no evidence of a clastic wedge deposited at the Devonian-Mississippian boundary in the Tennessee-Georgia-Alabama Appalachians that could correspond to unroofing of the Inner Piedmont. These observations indicate that the Inner Piedmont was not in its present structural position in the Ordovician, because it lacks the Ordovician metamorphism recorded in the adjacent Blue Ridge, nor at the Devonian-Mississippian boundary, because no evidence of its erosion is seen in the southern foreland.

2. There is intense dextral strike-slip shearing in the Inner Piedmont. Structures adjacent to the Brevard zone preserve considerable evidence of dextral strike-slip motion in Inner Piedmont rocks, particularly within the western Inner Piedmont (Tugaloo terrane[2]). Hatcher and Bream (2002, and papers therein) stressed

[2]Hatcher (2001) defined the Tugaloo terrane to include the Smith River allochthon, Eastern Blue Ridge, and portions of the Inner Piedmont. Because there is no evidence for Middle Ordovician peak metamorphism in the Inner Piedmont (versus Middle Ordovician peak metamorphic conditions at Lick Ridge and Winding Stairs Gap), and notwithstanding the strong similarity of eastern Blue Ridge and western Inner Piedmont detrital zircon populations, I choose to restrict Tugaloo to the western Inner Piedmont segment only.

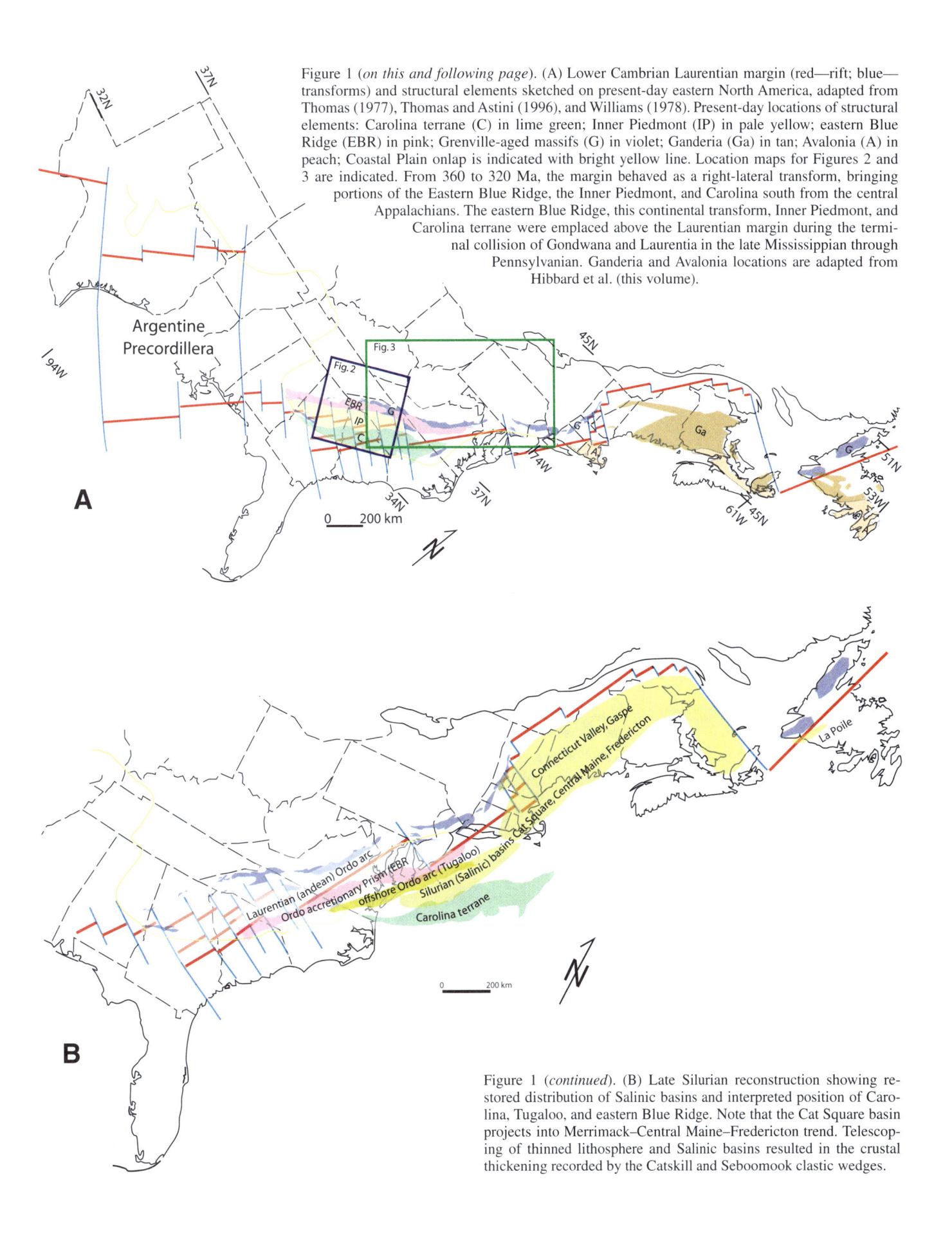

Figure 1 (*on this and following page*). (A) Lower Cambrian Laurentian margin (red—rift; blue—transforms) and structural elements sketched on present-day eastern North America, adapted from Thomas (1977), Thomas and Astini (1996), and Williams (1978). Present-day locations of structural elements: Carolina terrane (C) in lime green; Inner Piedmont (IP) in pale yellow; eastern Blue Ridge (EBR) in pink; Grenville-aged massifs (G) in violet; Ganderia (Ga) in tan; Avalonia (A) in peach; Coastal Plain onlap is indicated with bright yellow line. Location maps for Figures 2 and 3 are indicated. From 360 to 320 Ma, the margin behaved as a right-lateral transform, bringing portions of the Eastern Blue Ridge, the Inner Piedmont, and Carolina south from the central Appalachians. The eastern Blue Ridge, this continental transform, Inner Piedmont, and Carolina terrane were emplaced above the Laurentian margin during the terminal collision of Gondwana and Laurentia in the late Mississippian through Pennsylvanian. Ganderia and Avalonia locations are adapted from Hibbard et al. (this volume).

Figure 1 (*continued*). (B) Late Silurian reconstruction showing restored distribution of Salinic basins and interpreted position of Carolina, Tugaloo, and eastern Blue Ridge. Note that the Cat Square basin projects into Merrimack–Central Maine–Fredericton trend. Telescoping of thinned lithosphere and Salinic basins resulted in the crustal thickening recorded by the Catskill and Seboomook clastic wedges.

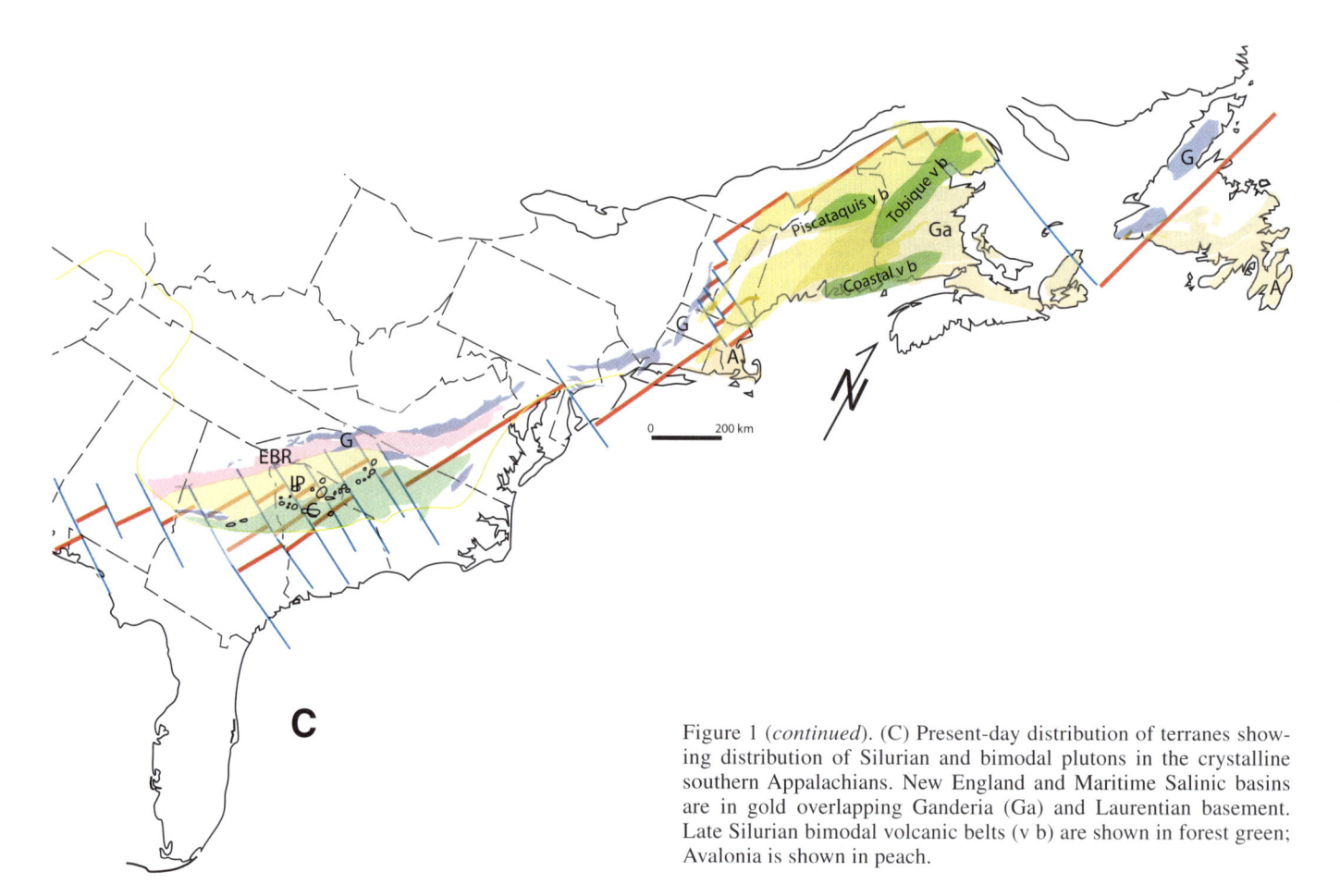

Figure 1 (*continued*). (C) Present-day distribution of terranes showing distribution of Silurian and bimodal plutons in the crystalline southern Appalachians. New England and Maritime Salinic basins are in gold overlapping Ganderia (Ga) and Laurentian basement. Late Silurian bimodal volcanic belts (v b) are shown in forest green; Avalonia is shown in peach.

the role of "Neoacadian" SW-directed sheath folds in the formation of the map pattern of rock units in the Chauga belt (e.g., Hatcher, 2002, p. 6, 10; see also Merschat et al., 2005). Abundant evidence for dextral strike-slip motion in these rocks includes composite planar fabric, asymmetric porphyroclasts, winged porphyroblasts, and crystallographic fabric in ribbon quartz lineations (e.g., Edelman et al., 1987; Bobyarchick et al., 1988). The mesoscopic sheath folds and the map-scale sheath fold map patterns represent high shear strains ($\gamma > 20$) over a >20-km-wide belt. This shear strain over this width is consistent with large southwestward displacement of the Inner Piedmont. Additionally, NW-directed structures (including sheath folds) southeast of the Chauga belt (westernmost, lower-grade Inner Piedmont) are transposed into parallelism with the SW tube axes and mineral lineations of the Chauga belt as they approach this zone. Hatcher and his students (articles in Hatcher and Bream, 2002) documented the transposition of structures into this zone from the core of the Inner Piedmont into the Chauga belt. Evidence for this dextral strike-slip shearing is not restricted to rocks southeast of the Brevard zone but in fact is also observed along the Middle Ordovician suture in the eastern Blue Ridge, the Burnsville fault (Adams et al., 1995; Trupe et al., 2003). A Devonian-Mississippian age for this shearing is indicated by the deformation of dated plutons.

3. Devonian-Mississippian clastic wedges occur in the central Appalachians. In the central Appalachians, deposition of the Devonian Catskill clastic wedge in New York State (Thomas, 1977; Woodrow and Sevon, 1985) and the latest Devonian through Mississippian Price-Pocono wedge from Pennsylvania through western Virginia records a major collisional event in the central Appalachians that correlates closely with the Inner Piedmont thermal peak (Figs. 3A and 3B).

4. The Devonian-Mississippian clastic wedge migrated progressively southward. The southward progression of black shale through time has been used to track the progress of the migrating foredeep of the Devonian-Mississippian collision from the St. Lawrence promontory (at ca. 410 Ma) to the Virginia promontory (at ca. 362 Ma; Fig. 3; Ettensohn, 1985, 1987). Ferrill and Thomas (1988) extended Ettensohn's model to a broader, dextral-transpressive model for the length of the Devonian Appalachians. This is consistent with displacement of the Inner Piedmont from a docking point near the Catskills to its present location during southward progression of the foreland basin.

5. The Devonian-Mississippian source terrane was removed before deposition of the Pennsylvanian clastic wedge of the central Appalachians. Gray and Zeitler (1997) conducted a detrital zircon provenance study of the Silurian Shawangunk Formation

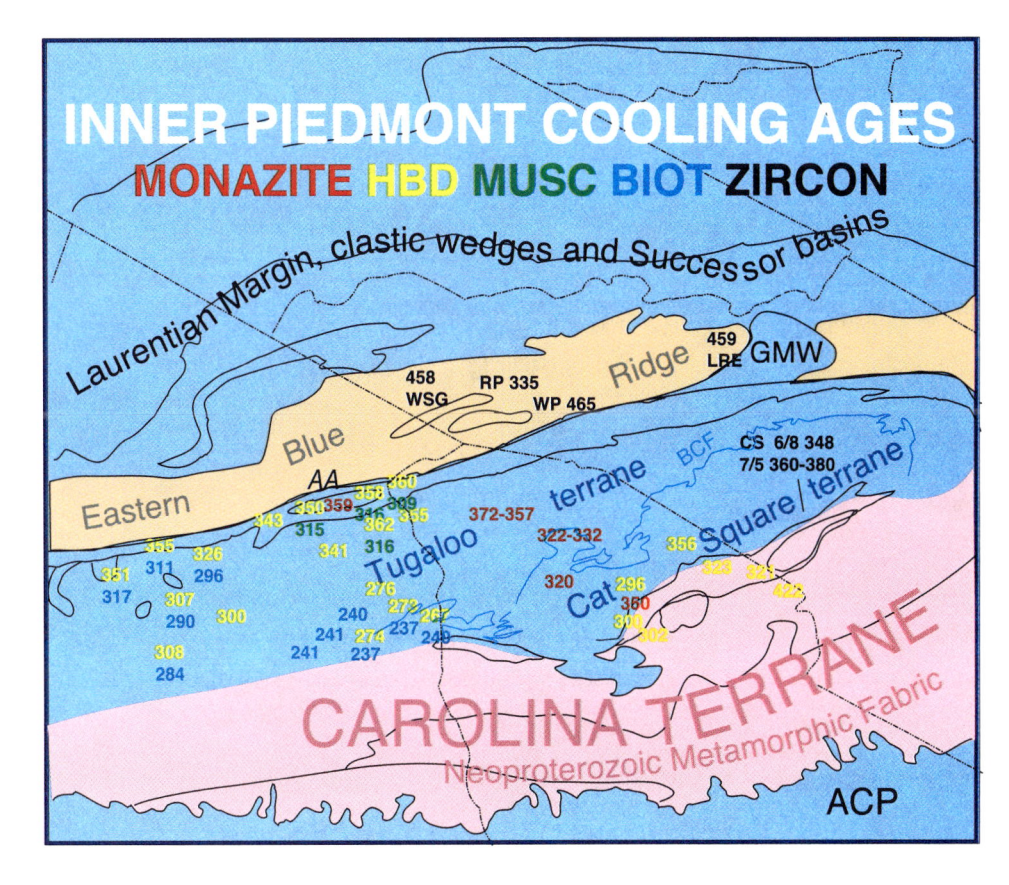

Figure 2. The primary purpose of this figure is to show that the thermal peak of the Inner Piedmont occurred ca. 355–365 Ma as revealed by monazite U-Pb thermal-ionization mass spectrometry (TIMS) and that this peak is different from that observed in adjacent terranes: eastern Blue Ridge (ca. 458 Ma granulite facies for Winding Stairs Gap; eclogite facies for Lick Ridge eclogite) and Carolina (Neoproterozoic; Dennis and Wright, 1997b). This figure was adapted from Dennis and Wright (1997a). Data are from Dennis and Wright (1997a), Dallmeyer et al. (1986), Dallmeyer (1988), Miller et al. (2000), Busch et al. (2002), Miller et al. (2006), Hatcher (2002), and Moecher et al. (2004). AA—Alto allochthon, ACP—Atlantic Coastal Plain, BCF—Brindle Creek fault, CS—Cat Square, GMW—Grandfather Mountain Window, LRE—Lick Ridge eclogite, RP—Rabun pluton, WP—Whiteside pluton, WSG—Winding Stairs Gap.

and the Pennsylvanian Pottsville Formation in eastern Pennsylvania. They conclude in their study of the Pottsville Formation that Devonian-Mississippian sources are underrepresented (p. 158), even as they cite detrital micas of this age recovered from the Pottsville; these micas may be reworked sediments. In light of these five observations, it may be the case that the source terrane from which Price-Pocono wedge material was derived had been tectonically removed by the time of Pottsville deposition. Postaccretionary terrane dispersal removed the highlands from which the Catskill clastic wedge had been derived.

Thus, a major clastic wedge in the central Appalachian foreland has no defined source, and a major composite terrane in the southern Appalachians has a distinct and different history from adjacent terranes. The Inner Piedmont experienced peak metamorphic conditions, however, coeval with the deposition of the Catskill clastic wedge. Abundant kinematic evidence indicates that the Inner Piedmont moved to its current position in the orogen during the time the Pocono clastic wedge was being deposited.

INNER PIEDMONT OF THE CAROLINAS AND ADJACENT GEORGIA

The Inner Piedmont is a composite crystalline terrane in the southern Appalachian hinterland (e.g., Hatcher, 2002). The lithology of the Inner Piedmont consists of orthogneisses and paragneisses and local amphibolite intruded by granites and granitic orthogneisses ranging in age from Ordovician to Devonian. Metamorphism generally achieved amphibolite facies, although in the west (Chauga belt), metamorphism is of upper greenschist facies (Butler, 1991). The structure of the Inner Piedmont is dominated by large recumbent folds and nappes (e.g., Alto allochthon, Six Mile thrust sheet; Hopson and Hatcher, 1988). As discussed earlier, approaching the Brevard zone from the east, earlier structures are transposed in a shear zone no less than 20 km in width with subhorizontal mineral lineations and sheath fold development.

Country Rock Protolith

The Brindle Creek fault subdivides the Inner Piedmont into the Cat Square and Tugaloo terranes. This feature was identified on the basis of truncation of units by Giorgis (1999), though this truncation is clearly visible on the maps of Goldsmith et al. (1988). Detrital zircon studies provide dramatic evidence of the significance of the Brindle Creek fault (Bream, 2002). In summary, the western Inner Piedmont (Tugaloo) terrane contains abundant detrital zircons of Grenville age. By contrast, the paragneiss of the Cat Square terrane, east of the Brindle Creek fault, contains a detrital zircon population that is unlike any other in the Blue Ridge or Inner Piedmont (see Table 1). Although it also contains Grenville detrital zircons, it may in part have been derived

TABLE 1. GENERAL CONTRASTING CHARACTERISTICS ACROSS THE EASTERN BLUE RIDGE (BR) AND INNER PIEDMONT (IP) IN THE CAROLINAS AND ADJACENT GEORGIA, AND CONTRASTS IN CHARACTERISTICS OF SILURIAN–DEVONIAN–MISSISSIPPIAN PLUTONS ACROSS THE BELTS

Eastern Blue Ridge		Tugaloo terrane (wIP)		Cat Square terrane (eIP)
Sparse Ordovician plutons, interspersed younger "Acadian" plutons	Brevard zone	Ordovician plutons into coherent metasedimentary and metavolcanic "stratigraphy"	Brindle Creek fault[†]	Devonian-Mississippian plutons into structurally complex migmatitic terrane
Depth of source of plutons interpreted to be deep (500–1000 km; by Sr/Yb proxy)		Depth of crustal source of plutons interpreted to be shallow (<100 km; Sr/Yb proxy)		Depth of crustal source of plutons interpreted to be 100–500 km (by Sr/Yb proxy)
Abundant "Grenville"-age detrital zircons		"Grenville"-age detrital zircons + additional comp. ca. 1300–1700 Ma		Unlike any other in the IP or BR: as young as 430 Ma; Ordovician zircons; ca. 590–600 Ma (Carolina terrane); "Grenville" population
		Detrital zircons consistent with Nd-derived T_{DM} (Fullagar et al., 1997, p. 172): "diverse Inner Piedmont basement."		
Basement massifs*		No basement massifs		

Characteristics of middle Paleozoic plutons	
Extreme inheritance, low zircon saturation temp. 680–720 °C	Minor/no inheritance, zircon saturation thermometry yields 760–880 °C
$\varepsilon_{Nd}(0)$ –3.4 to 4.9	$\varepsilon_{Nd}(0)$ –6.4 to –1.6
Initial $^{87}Sr/^{86}Sr$ 0.7040–0.7083	Initial $^{87}Sr/^{86}Sr$ 0.70540–0.7714
Rare earth element (REE) pattern generally flatter La 9–200× chondrite Lu 0.7–5× chondrite	REE pattern generally more enriched, and steeper; La 100–500× chondrite Lu 1–10× chondrite
Lower K_2O ≈ 0.6–4 wt%	3% < K_2O < 5%
Much lower concentration incompatible trace elements (Rb, Ba, light REE)	Higher incompatible trace-element concentrations
[Sr] 250–1150 ppm	Much lower [Sr] ≈ 50–250 ppm
$\delta^{18}O$ ≈ 7‰–9.5‰	$\delta^{18}O$ ≈ 10‰–10.5‰

Note: Compiled from Mapes (2002), Bream (2002), Hatcher (2002), and Fullagar et al. (1997); see also, Drummond et al. (1997) and Steltenpohl (2005).
*But see, for example, Livingston and McNiff (1967) and Rankin (1987).
[†]The feature separating Tugaloo and Cat Square terranes (Brindle Creek fault) is here interpreted to have formed as a Salinic unconformity, above Tugaloo and Carolina basement, covering the suture between these terranes. Later this Telychian or younger nonconformity may have been reactivated as a ductile fault.

---

from peri-Gondwanan terranes that have source rocks of similar ages. Notably the Cat Square terrane contains detrital zircons as young as 430 Ma (Bream, 2002). Thus, the Cat Square terrane may be more properly described as a Late Silurian sedimentary basin, and the Brindle Creek fault may therefore represent a Salinic unconformity (Boucot, 1962). The interpreted Brindle Creek unconformity may have been locally reactivated as a mylonitic fault; where present, this mylonitization is interpreted to be Alleghanian (Dennis and Wright, 1997a).

Magmatism

There are conspicuous differences in the magmatism recorded by the eastern Blue Ridge and Inner Piedmont. The eastern Blue Ridge contains sparse Ordovician plutons and interspersed "Acadian" plutons (Table 1; Fig. 4). The Tugaloo terrane (western Inner Piedmont) contains Ordovician plutons that intrude a coherent metavolcanic and metasedimentary "stratigraphy" and also contains a few Silurian plutons. The Cat Square terrane contains no plutons older than ca. 415 Ma, but it does contain abundant plutons dated between 380 and 360 Ma. A robust database of ion-microprobe ages, tracer isotopes, and major- and trace-element chemistry of the Devonian and younger plutons of the Inner Piedmont and eastern Blue Ridge prepared by Mapes (2002) shows this pattern.

Timing of Peak Metamorphism

Dennis and Wright (1997a, 1997b) used U-Pb zircon and monazite dating in areas of detailed mapping in the Carolina terrane and Inner Piedmont to test the assumption that the peak metamorphism in these terranes was Ordovician in age (Dennis, 1991). Their results demonstrate that peak metamorphic conditions occurred between 538 ± 5 Ma and 535 ± 4 Ma in the Carolina terrane in South Carolina but occurred at the Devonian-Mississippian boundary in the Inner Piedmont. Additionally, Dennis and Wright (1997a) recognized a local resetting of monazite ages in the Inner Piedmont ca. 335–325 Ma. Thus, the timing of peak metamorphism in the Inner Piedmont was contemporaneous with the age of the younger suite of plutons as reported

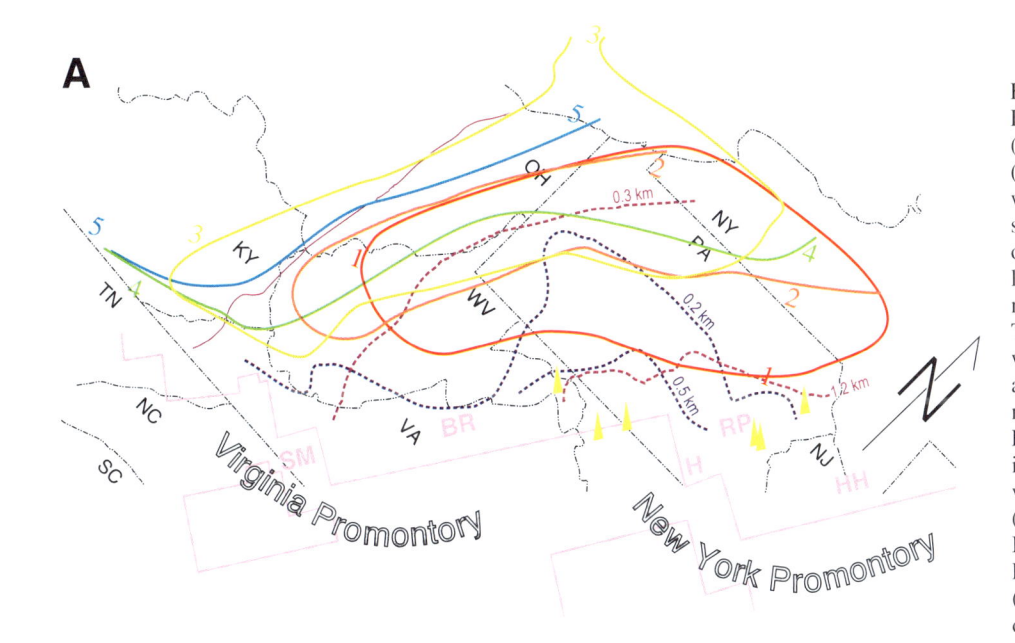

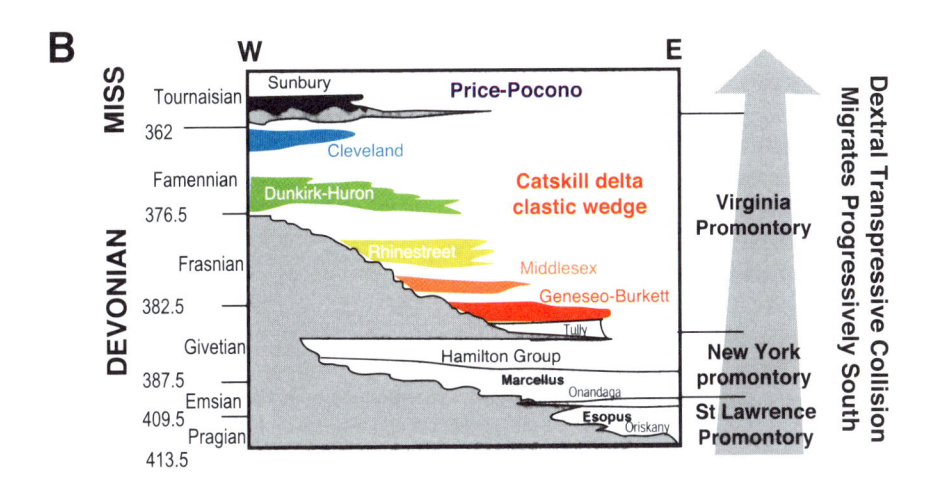

Figure 3. (A) Spatial distribution of Devonian black shale basins from oldest (1—red) Geneseo-Burkett to youngest (5—blue) Cleveland. Note southern and western movement over time. Basins stepped south as dextral transpressive orogeny progressively moved south. Pink lines indicate shape of rifted Laurentian margin and offshore Grenville blocks. The Cambrian-Ordovician rifted margin was reactivated in the middle Paleozoic as a transform, bringing the Inner Piedmont and other exotic terranes south. Purple dashed isopachs (1.2 km, 0.3 km) indicate thickness of Devonian Catskill wedge. Navy blue dashed isopachs (0.5 km, 0.2 km) indicate thickness of Mississippian Price-Pocono wedge. Figure was modified from Thomas (1977) and Ettensohn (1987). Yellow darts indicate location of several (but not all) critical exotic clast locations and the approximate eastern limit of Spechty Kopf–Rockwell Formation. The southern three darts indicate sites with abundant plutonic, gneissic, and volcanic boulders and cobbles. BR—Blue Ridge, H—Honeybrook Upland, HH—Hudson Highlands, RP—Reading Prong, SM—Sauratown Mountains. (B) Composite section showing temporal distribution of unconformity-bound Devonian black shales (colors as in A) interfingering with coarse clastic sediments (to right) and the progressively southern advance of collision. Devonian stage boundaries are from Tucker et al. (1998). Figure was modified from Ettensohn (1987).

by Mapes (2002). If the length of time required to heat Inner Piedmont rocks above the second sillimanite isograd is estimated to take between 30 and 10 m.y. (if enhanced by fluid flow), this collision (between the Inner Piedmont and Laurentia) may have begun between 370 Ma and 400 Ma.

Exhumation

U-Pb and Ar-Ar geochronology indicate that the high-grade metamorphic rocks of the Inner Piedmont did not reach shallow crustal levels until at least the late Mississippian (Fig. 2). For example, around the Alto allochthon in Figure 2 (AA), a monazite age of ca. 359 Ma was reported by Dennis and Wright (1997a). This location is surrounded by a ring characterized by hornblende $^{40}Ar/^{39}Ar$ cooling ages ranging from 362 to 341 Ma. Nominally, these plateaus record cooling through ~500 °C, depending on grain size, cooling rate, etc. Muscovite ages from the same area record cooling through ~400 °C at 316–309 Ma. Away from the westernmost edge of the Inner Piedmont and Brevard zone, hornblende, muscovite, and biotite (cooling through ~300 °C) $^{40}Ar/^{39}Ar$ ages in the Inner Piedmont tend to be much younger (younger than 300 Ma).

ALTERNATIVE EVIDENCE FROM THE SOUTHERN APPALACHIANS

Eastern Blue Ridge and Its Possible Relationship to the Inner Piedmont

There is little doubt that peak metamorphic conditions in the eastern Blue Ridge were achieved ca. 460 Ma as discussed already for the Winding Stairs Gap granulite and the

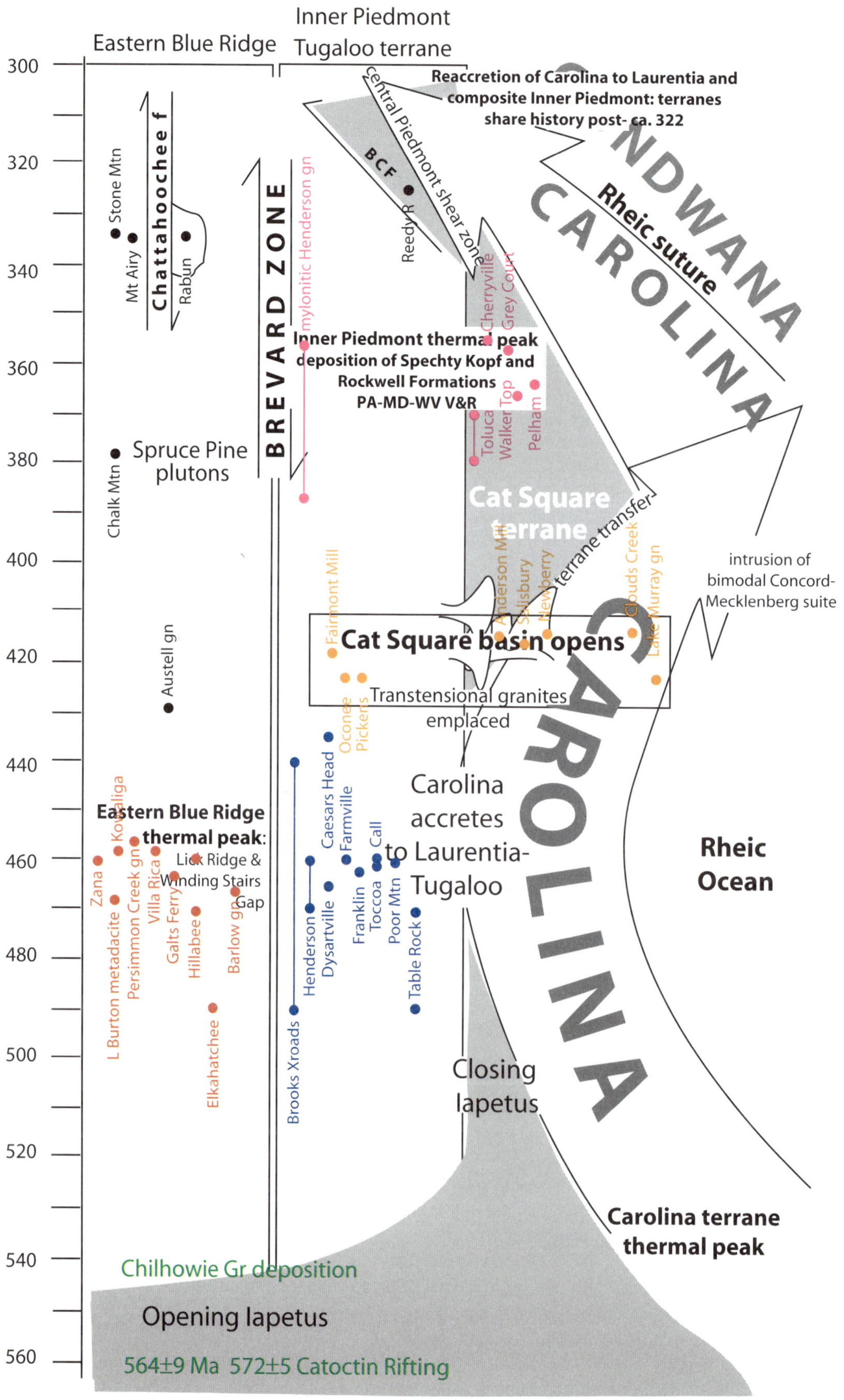

Eastern Blue Ridge
Inner Piedmont
Tugaloo terrane
300
320
340
360
380
400
420
440
460
480
500
520
540
560
Reaccretion of Carolina to Laurentia and composite Inner Piedmont: terranes share history post- ca. 322
central Piedmont shear zone
BCF
Reedy R
Stone Mtn
Mt Airy
Chattahoochee f
Rabun
BREVARD ZONE
mylonitic Henderson gn
Rheic suture
CAROLINA
Cherryville
Grey Court
Inner Piedmont thermal peak
deposition of Spechty Kopf and Rockwell Formations
PA-MD-WV V&R
Toluca
Walker Top
Pelham
Spruce Pine plutons
Chalk Mtn
Cat Square terrane
terrane transfer
Anderson Mill
Salisbury
Newberry
Clouds Creek
Lake Murray gn
intrusion of bimodal Concord-Mecklenberg suite
Fairmont Mill
Cat Square basin opens
Austell gn
Transtensional granites emplaced
Oconee
Pickens
Caesars Head
Farmville
Call
Carolina accretes to Laurentia-Tugaloo
Rheic Ocean
Eastern Blue Ridge thermal peak:
Lick Ridge & Winding Stairs Gap
Kowaliga
Zana
L Burton metadacite
Persimmon Creek gn
Villa Rica
Galts Ferry
Hillabee
Barlow gn
Elkahatchee
Brooks Xroads
Henderson
Dysartville
Franklin
Toccoa
Poor Mtn
Table Rock
Closing Iapetus
Carolina terrane thermal peak
Chilhowie Gr deposition
Opening Iapetus
564±9 Ma 572±5 Catoctin Rifting

Lick Ridge eclogite. There is, however, evidence for a Mississippian high-grade metamorphic overprint. Quinn and Wright (1993) recognized magmatic cores and metamorphic rims in a hornblende + biotite orthogneiss near Sylva, North Carolina. These authors reported TIMS ages of 334 ± 15 Ma (middle Mississippian or Visean) for residual metamorphic zircon rims following dissolution (in concentrated HF) of zoned (1147 ± 8 Ma) zircon cores. Thus, for an eastern Blue Ridge Grenville basement unit, Quinn and Wright (1993) indicated that the metamorphic peak was not "Taconic," but was instead consistent with an Alleghanian age.

Metamorphic rims have been recognized in cathodoluminescence (CL) and backscattered-electron (BSE) images of zircons from Grenville-aged basement in the eastern Blue Ridge (Toxaway gneiss), Ordovician plutons from the central Blue Ridge (Persimmon Creek gneiss, Whiteside pluton) and Inner Piedmont (Henderson gneiss, Dysartville pluton), and North Carolina and Georgia paragneisses (Tallulah Falls Formation) (Carrigan et al., 2001). Sensitive high-resolution ion microprobe (SHRIMP) analyses for individual rims range from 325 to 380 Ma. A probability plot combining data from the eastern Blue Ridge and Inner Piedmont is interpreted "with a well-defined maximum" at 352 Ma. Thus, these authors argued for a widespread late Acadian tectonothermal effect across both the eastern Blue Ridge and Inner Piedmont and the juxtaposition of these terranes prior to rim growth. The lack of a distinct rim age for the Rabun pluton may result from its relative youth (335.1 ± 2.8 Ma by ID-TIMS; Miller et al., 2006). The interpretation outlined here would not be falsified by a later metamorphic overprint in the eastern Blue Ridge. The eastern Blue Ridge and Inner Piedmont have distinct and different early tectonothermal histories.

Figure 4. Temporal relationships among metamorphism, plutonism, erosion, and deposition in the eastern Blue Ridge, Tugaloo terrane, Cat Square terrane (Cat Square + Tugaloo comprise the Inner Piedmont), and Carolina terrane. Catoctin rifting ages are from Aleinikoff et al. (1995). Additional pluton ages are from sources in text; additional data are from Odom and Fullagar (1973), Bond and Fullagar (1974), Sinha and Glover (1978), and Fullagar et al. (1997). Eastern Blue Ridge (red) and Tugaloo (blue) pluton ages: These terranes have different tectonothermal histories, and the zircon and trace-element chemistries of the plutons in the two terranes are quite different, validating their separation into two distinct terranes, despite the similarity of the plutons' ages in the two terranes. In this paper, I do not speculate on the nature of the coincidence of the eastern Blue Ridge thermal peak, eastern Blue Ridge and Tugaloo pluton ages, and the Middle to Late Ordovician accretion of Carolina to Laurentia. Between Telychian-Wenlock time (post–430 Ma), the Cat Square basin began to open, accepting detritus from Tugaloo and Carolina. Coeval with basin opening, between 424 and 414 Ma, plutons (orange) intruded Tugaloo and Carolina. These plutons have initial Sr ratios >0.706, inherited zircons 438–489 Ma, and $-2.5 < \varepsilon_{Nd} < 4.5$. The youngest plutons have the highest (+) ε_{Nd}. This event is interpreted to represent partial melting of Tugaloo-Carolina lower crust, transitional to an increasing mantle component. Late Silurian plutonism was followed by a bimodal alkalic-subalkalic gabbro-leucogranite-syenite suite with $+3 < \varepsilon_{Nd} < +4.5$ intrusive into the westernmost exposed Carolina terrane. Bimodal plutonic rocks are interpreted to represent significant lithospheric thinning accompanying rifting and postaccretionary terrane transfer and dispersal. Almost all of the 380–350 Ma plutons (pink) in the Inner Piedmont are found in the Cat Square terrane, coeval with upper-amphibolite– to granulite-facies metamorphism as revealed by thermal-ionization mass spectrometry (TIMS) and sensitive high-resolution ion microprobe–reverse geometry (SHRIMP-RG) dating of monazite, as well as zircon from the Cat Square charnockite. These plutons are restricted to the Cat Square terrane. These plutons are interpreted to have formed by anatexis of wet Cat Square basin rocks during telescoping of thinned Inner Piedmont lithosphere + Carolina during an event recorded in the foreland as the Catskill clastic wedge. The sole exception is the mylonitic, "Acadian" Henderson gneiss mapped along the western margin of Tugaloo. These ages are interpreted to record resetting during dextral transpression and shearing ($\gamma \geq 20$) along the western margin of Tugaloo and the eastern Blue Ridge from at least the Late Devonian through the Mississippian. At least 400 km post-Devonian dextral slip restores the Inner Piedmont to a location near the New York Promontory. Additional mid- to late Paleozoic dextral strike-slip deformation is recorded in the eastern Blue Ridge. These displacements require the Inner Piedmont to be restored even further (present-day) north. By 320 Ma, Carolina and the Inner Piedmont are interpreted to have been juxtaposed along the central Piedmont shear zone. Additional fault movement may have been localized along the Brindle Creek fault (BCF), which is interpreted to have formed as the Salinic unconformity beneath the Cat Square basin.

Unconformities and Successor Basins in the Southern Appalachian Blue Ridge and Valley and Ridge

Middle Paleozoic shear zones as wide as 20–30 km record shear strains high enough to create sheath folds in map pattern, with consistent evidence of dextral shear. These indicate that the Inner Piedmont and perhaps portions of the Blue Ridge should be restored a significant distance to the northeast. Thus, southern Appalachian Blue Ridge Paleozoic sedimentary basins deposited in angular unconformity above folded Precambrian-Cambrian strata (Tull and Groszos, 1990) probably do not record Inner Piedmont tectonism because the Inner Piedmont is interpreted to have been located well north of these basins when they were deposited. These olistostromal basins may record tectonism and subsequent dispersal of other, unknown (once), southern Appalachian exotic terranes. Ferrill and Thomas (1988) argued that the Silurian–Early Devonian Lay Dam Formation (Tull and Telle, 1989) represents a pull-apart basin formed between right-stepping faults on the dextral transpressional Laurentian plate boundary. Finally the sub-Chattanooga shale unconformity in Tennessee exposes rocks from Middle Ordovician (Caradoc-Trenton equivalent) to Middle Devonian age (Hermitage Formation, Pegram Formation; Wilson, 1949). This unconformity may be related to the same event responsible for deposition of the Foothills–Mineral Bluff–Lay Dam system of basins. The structural data require that, while it is coeval with Inner Piedmont orogeny, this system of unconformities and successor basins is not specifically related to accretion of Inner Piedmont rocks to Laurentia.

DISCUSSION

Timing of Accretion of the Carolina Terrane to Laurentia

Paleomagnetic data (Noel et al., 1988; Vick et al., 1987) strongly suggest that Carolinia shared the same latitude as the (southern) Laurentian margin in Middle to Late Ordovician time, (~22°S). The $^{40}Ar/^{39}Ar$ age of white micas in slates in the central North Carolina slate belt (Noel et al., 1988; Offield et al., 1995) correlates with the age of folding of an angular unconformity in South Carolina, which is post–Middle Cambrian, pre–414 +2.1/–1.7 Ma (intrusion of Clouds Creek meta-igneous complex into Delmar synclinorium; Dennis et al., 1993; Samson and Secor, 2000). Additional stratigraphic evidence from the Laurentian margin for the timing of accretion of Carolinia has been discussed by Hibbard (2000) and Dorsch et al. (1994). Recent reconstructions (e.g., Stampfli and Borel, 2002; Cocks and Torsvik, 2002) based on paleomagnetic data place Avalonia at 50°S latitude in the Middle Ordovician. Carolinia thus has a distinct and different history in tectonic reconstructions from Avalonia.

Postaccretionary Terrane Transfer and Dispersal and the Relation of the Cat Square Terrane to Carolinia (Fig. 1B)

Detrital zircon data have permitted the separation of the Inner Piedmont into the Cat Square terrane and the Tugaloo terrane (Bream, 2002). These terranes are separated along the Brindle Creek fault (Giorgis, 1999). Both the Inner Piedmont Tugaloo and Cat Square terranes contain abundant detrital Grenville zircons; however, the Cat Square also contains populations of ca. 590–600 Ma zircons (interpreted to be derived from Carolinia), as well as abundant Middle Ordovician zircons interpreted to have been eroded from voluminous arc plutons within both the Inner Piedmont and eastern Blue Ridge portions of the Tugaloo terrane (Bream, 2002). The youngest detrital zircon reported by Bream (2002) from the Cat Square terrane is 430 Ma, which sets a minimum age for deposition of the protolith of the Cat Square paragneiss. A further indication of the youth of the Cat Square terrane relative to the Tugaloo terrane is the absence of plutons older than Llandovery: the oldest Cat Square pluton is the Anderson Mill at 415 ± 3 Ma (Mapes, 2002). The (SHRIMP) age of the oldest metamorphic rim on zircon (400 Ma) limits the short depositional span of the Cat Square terrane (Bream, 2002). Bream (2002, p. 55) and Hatcher (2002, p. 13) interpreted the detrital zircon population of the Cat Square terrane to record the encroachment of Carolinia upon the Laurentian margin. The interpretation presented here is that the Cat Square terrane represents deposition in a basin that formed as the already accreted Carolinia rifted from Laurentia, in a nominal Gulf of California setting (e.g., Stock and Hodges, 1989; Oskin and Stock, 2003). Thus, I interpret the evidence to show that the Cat Square basin does not record Carolinia impinging on Laurentia, but instead Carolina's departure from Laurentia during postaccretionary terrane dispersal (Fig. 4). The Brindle Creek fault is interpreted to have formed as the sub-basin unconformity; there may be local fault reactivation of the unconformity separating the Cat Square basin from the peri-Laurentian portion of its basement.

Thus, Hibbard et al.'s (1998) reinterpretation of the central Piedmont suture as the central Piedmont shear zone is correct in that the suture between Carolinia and Inner Piedmont must have been between Carolinia and the Tugaloo terrane, because Carolinia accreted to the peri-Laurentian Inner Piedmont before rocks of the Cat Square basin had even been deposited; this suture is, however, nowhere preserved. This interpretation may help reconcile Middle-Late Ordovician accretion of Carolinia to Laurentia and the separate geologic histories of the Inner Piedmont and Carolinia in the interval between that accretion and telescoping of southern Appalachian crystalline terranes in late Mississippian–Pennsylvanian time with Gondwana-Laurentia collision.

Opening and Closing of the Cat Square Basin and Devonian-Carboniferous Laurentian Transform Plate Boundary

After 430 Ma, Carolinia is interpreted to have rifted away from Laurentia in the Late Silurian. Carolinia, the Cat Square basin, and the Tugaloo terrane were intruded by transtensional batholiths (in Carolinia: Newberry, Lake Murray, Clouds Creek, Salisbury; Samson and Secor, 2000), culminating with the intrusion of a bimodal suite of mantle-derived plutons in Carolinia: Southmont (leucogranites) and Concord-Mecklenburg (alkalic to subalkalic gabbros) suites (Dennis, 1991; McSween and Harvey, 1997). Over this same interval, the Cat Square basin accumulated detritus that is interpreted to have been derived from either side of the Iapetan suture: Carolinia on one side and the Ordovician arcs built on thinned Laurentian crust and the adjacent Iapetan Ocean floor (Tugaloo terrane) on the other.

The Acadian orogeny affected the Inner Piedmont. At this time, crustal thickening in the central Appalachians resulted in deposition of the earliest stages of the Catskill clastic wedge (Walton Formation, Frasnian; Rickard, 1975). Collapse of Salinic basins in the Maine Appalachians may have begun as early as Pridolian time based on the age of the Seboomook clastic wedge there (Hibbard and Hall, 1992; Pollock, 1987). The so-called Neoacadian orogeny at the Devonian-Mississippian boundary in the central and southern Appalachians is recognized by simultaneous (1) Inner Piedmont metamorphism to granulite facies (e.g., Dennis and Wright, 1997a, 1997b; Cat Square charnockite; Kish, 1997), (2) Inner Piedmont nappe and thrust sheet emplacement (e.g., Alto allochthon; Hopson and Hatcher, 1988; Dallmeyer, 1988), and (3) deposition of the Spechty Kopf and Rockwell Formations between the Catskill and Price-Pocono clastic wedges (as discussed next; Dennis, 2005a, 2005b, 2005c). Immediately following the crustal thickening that accompanied the collapse of the Cat Square basin, its thinned lithosphere, and Carolinia against the Laurentian margin, kinematics along the margin changed from oblique dextral convergence to dextral strike slip. This change in kinematics, its abruptness, and its simultaneity along strike suggest that the orogenic event at the close of the Devonian involved major plate reorganization. The change in kinematics on the Laurentian margin to a continental-margin transform setting is paralleled in the European Variscan belt by a transition to deformation in the autochthon (Dallmeyer et al., 1997; Martinez-Catalan and Arenas, 2005).

Comparison with Maritime Canadian and New England Results (Fig. 1C)

The model outlined here, developed from data compiled in the crystalline southern Appalachians, correlates with results of detailed geologic and geochronologic studies in Maritime Canada and the New England Appalachians. This could be anticipated based on the restoration of the southern Appalachian Piedmont to no less than 400 km north along strike at the Devonian-Mississippian boundary presented here. With this required structural restoration, the long axis of the Cat Square paragneiss projects into the Connecticut Valley synclinorium paragneisses–Gaspé belt and the Merrimack synclinorium–Central Maine synclinorium (Tremblay and Pinet, 2005; Dennis, 2006a, 2006b, 2006c). The Windsor Point (Dube et al., 1996) and the La Poile Groups (O'Brien et al., 1991) of southwestern Newfoundland, north of the St. Lawrence promontory, are part of the same tectonic system. These units are well-known variably deformed and metamorphosed late Silurian Salinic (Boucot, 1962) successor basins that formed after the accretion of Ganderia (e.g., van Staal et al., 1998; Hibbard et al., this volume) to Laurentia. The accretion of peri-Gondwanan Ganderia to Laurentia is well-documented as a Late Ordovician–Early Silurian event along the Brunswick subduction complex (van Staal, 1994). Thus, Carolinia and Ganderia were accreted to Laurentia in the Middle to Late Ordovician to Late Ordovician–Silurian, respectively (Hibbard et al., this volume). Postaccretionary terrane dispersal resulted in the formation of short-lived Wenlock-Ludlow-Pridoli sedimentary basins that accepted detritus from Laurentia and the recently accreted exotic terranes (Bourque et al., 2000, 2001; Tremblay and Castonguay, 2002; Castonguay and Tremblay, 2003; Wilson et al., 2004; LaVoie and Asselin, 2004; Tremblay and Pinet, 2005; Dennis, 2006a, 2006b, 2006c; Rankin et al., 2007). This is interpreted to be a sinistral transtensional event based on structural data from New England and the Maritime Appalachians (Goldstein, 1989; Nance and Dallmeyer, 1993; Hibbard, 1994; Holdsworth, 1994; Karabinos, 2002) and on the orogenic scale (e.g., Soper et al., 1992; Soper and Woodcock, 2003; Dewey and Strachan, 2003). There is little published evidence for middle Paleozoic sinistral deformation in the southern Appalachian Piedmont at this time (e.g., Lawrence and Foster, 2006; Hibbard et al., 2004).

As in the Inner Piedmont and Carolinia, formation of these Wenlock-Ludlow-Pridoli sedimentary basins was accompanied by the volcanic and intrusive activity that first melted the lower crust and then, within 10–15 m.y., became bimodal and mantle-derived (Dennis, 2006a, 2006b, 2006c). In the northern Appalachians, these igneous rocks include the Tobique (Dostal et al., 1989), the Coastal Maine volcanic belt (Seaman et al., 1999; van Wagoner et al., 2002; van Wagoner and Dadd, 2003), and the Piscataquis belts (Rankin, 1968) among others (e.g., Bédard, 1986). Igneous rocks related to the final stages of subduction and accretion overlapped temporally and spatially with those related to extension, as in the Basin and Range (Gans et al., 1989; see also Quesada, 2006). This magmatic activity extended into the Devonian.

Black shales related to the dextral transpressive orogeny recorded by the Catskill-Price-Pocono clastic wedges were deposited following a widespread Givetian unconformity (Fig. 3; Ettensohn, 1987). The oldest rocks of the Catskill wedge (Walton Formation) are Frasnian in age (Rickard, 1975). However, to the north in Maine and New Hampshire, the Pridoli-Pragian Seboomook Formation (Pollock, 1987; Hibbard and Hall, 1992; Bradley and Tucker, 2002) represents part of a clastic wedge formed in the same diachronous, southward-migrating crustal-thickening event. This crustal thickening is interpreted here to be the collapse of the internal Salinic basins and their thinned lithosphere, and the reaccretion of the Carolinia and Ganderia terrane blocks to Laurentia. Thus, in the interpretation presented here, Wenlock-Ludlow-(Pridoli?) sinistral transtension is followed by Pridoli–Early Devonian crustal thickening and dextral transpression; crustal thickening is interpreted to result from telescoping of Late Silurian thinned lithosphere between Laurentia and the exotic terranes accreted in the Late Ordovician–Silurian.

Spechty Kopf and Rockwell Formations

It may be possible to use an unusual formation deposited between the Catskill and Pocono Formations in the Valley and Ridge of Pennsylvania and adjacent Maryland as a piercing point to constrain the precise location of the crystalline southern Appalachian Piedmont at the Devonian-Mississippian boundary. The Spechty Kopf and Rockwell Formations were deposited at the Devonian-Mississippian boundary, separating rocks of the Catskill wedge below from those of the Pocono wedge above (Berg, 1999; Bjerstedt, 1986; Sevon et al., 1997, p. 49). These rocks and their correlatives, the Huntley Mountain Formation, the Cussewago Sandstone, the Price Formation of Virginia, and the Sunbury Shale of West Virginia, Kentucky, and Ohio, covered a significant area of Pennsylvania, West Virginia, and Maryland, as well as Ohio, Kentucky, and Virginia (Berg and Edmunds, 1979; Bjerstedt and Kammer, 1988; Kammer and Bjerstedt, 1986). An unusual diamictite is preserved in a narrow outcrop belt at the eastern edge of the exposed formation. The diamictite has been intensively studied since the 1960s to determine the facies or conditions under which it was deposited and the provenance of the abundant and sometimes quite large clasts (<50 cm) that it contains (e.g., Dennison, 1972; Sevon, 1969, 1979, 1985; Sevon et al., 1997; Bjerstedt, 1986; Bjerstedt and Kammer, 1988; Suter, 1991; Colstolnick, 1987; Cecil et al., 2002, 2004; Dennis, 2005a, 2005b, 2005c). The diamictite is unusual for a number of reasons, such as the diversity and unusual nature of the clasts found within it, including slates, schists, gneisses, granites, and metavolcanic rocks. One result of prior studies has been the recognition of at least eight discrete "sediment dispersal systems" into the underlying Catskill rocks, at the mouths of which the diamictite is found (e.g., Sevon, 1979). "Many of the clasts within the diamictite and pebbly mudstone are exotic lithologies not present in the underlying Catskill rocks and the overlying Mississippian and Pennsylvanian rocks in Pennsylvania. These clasts are interpreted to have been derived directly from the Acadian Mountains. As such they presumably provide an intimate real representation of the bedrock of the Acadian Mountains" (Sevon et al., 1997, p. 50).

Dennis (2005a, 2005b, 2005c) has suggested that these clasts were derived from higher structural levels of the presently exposed eastern Blue Ridge and Inner Piedmont (Table 1), before southwestward translation of these terranes, based on a preliminary collection and evaluation of clasts. Additionally, it may be possible that some Carolina terrane material is preserved as clasts within the Spechty Kopf. The collection sampled over a dozen sites, focusing particularly on the southern localities, LaVale (Maryland), Sideling Hill (Maryland), Town Hill (Maryland and Pennsylvania), and Crystal Springs (Pennsylvania). The clasts include granodiorite, gabbroic, and granite gneiss clasts as well as mafic and felsic volcanic and metavolcanic rocks (Fig. 5). No clasts collected to this point have been greater than 15 cm in the longest dimension. The greatest diversity of clasts has been found near the base of the formation in massive sandy beds. The thickness of these massive sandy beds is generally less than 5 m. In these beds, clasts are rare, <5% by volume. Earlier reconnaissance reports of larger clasts may simply reflect the tendency of these massive beds to weather spheroidally.

A consistent stratigraphy has been described for the base of the Spechty Kopf Formation, including (1) diamictite, (2) pebbly mudstone, (3) laminite and (4) sandstone (DMLS; Sevon et al., 1997). Above this distinctive lower stratigraphy, there is a thick fluvial sandstone-siltstone-shale and conglomerate system. In addition to the diamictite, exotic clasts are also found within the laminite (dropstones?; Sevon, 1969, 1973; Sevon et al., 1997; Cecil et al., 2004, p. 116; Dennis, 2005a, 2005b, 2005c). Thickness variations in the Spechty Kopf Formation are presumed to be a result of the incision of valleys of considerable relief in the underlying sandstones of the Catskill wedge as described previously. Thus, the thickness of the Spechty Kopf and its constituent DMLS and younger units varies considerably along strike, from ~20 m to >200 m thick (Sevon et al., 1997; Bjerstedt, 1986).

Examination of the cooling ages around the Alto allochthon (AA, Fig. 1) in the western Inner Piedmont introduces an important proviso to the hypothesis that it may be possible to fingerprint exotic clasts of the Spechty Kopf Formation. Hornblende $^{40}Ar/^{39}Ar$ ages are nearly identical to the TIMS and SHRIMP ages of monazite from the Alto allochthon, indicating rapid cooling from >700 °C through 500 °C at ca. 360 Ma (Dallmeyer, 1988; Dennis and Wright, 1997a, 1997b; Dennis, 2005a, 2005b, 2005c). However, these rocks did not cool through the blocking temperature for muscovite before the earliest Pennsylvanian, probably after they had already been emplaced in their current structural position. This means that if any of the boulders or cobbles of the Spechty Kopf–Rockwell were derived from the Inner Piedmont of the Carolinas and adjacent Georgia, they would have been eroded from much higher structural levels. These structural levels evidently are no longer preserved. It may be possible to find Devonian-Mississippian igneous and metamorphic cobbles, consanguineous with plutons and metaplutonic gneisses observed today in the Carolinas, crystallized at higher structural levels, or related volcanic rocks, or the same country rock, perhaps not as highly metamorphosed; or clasts eroded from Ordovician plutonic suites. This lattermost case is suggested by the results of McLennan et al. (2001) and the suite of 470–420 Ma detrital zircons they report from the Catskill delta. This is why the presence of rocks as clasts with a diverse set of characters is important to the solution of this problem versus single detrital (zircon) grains. A suite of characteristics from the Spechty Kopf–Rockwell clasts must be compared against the existing database of Inner Piedmont characteristics as represented in summary form in Table 1.

CONCLUSIONS

The eastern Blue Ridge, the Inner Piedmont, and the Carolina terrane of the southern Appalachian hinterland experienced peak metamorphic conditions at ca. 460 Ma, 360 Ma, and ca. 540 Ma, respectively; thus, terrane correlation between them is problematic. The Inner Piedmont of the Carolinas and adjacent Georgia was likely not in its current position along the strike of the orogen at the time it experienced peak metamorphic conditions at the Devonian-Mississippian boundary, ca. 360 Ma. The Inner Piedmont consists of the Tugaloo and Cat Square terranes. The Tugaloo terrane preserves a history of Ordovician arc plutonism. These plutons are intrusive into country rocks derived from the Laurentian margin, based on the detrital zircon populations collected from those country rocks (Bream, 2002). By contrast, the Cat Square terrane contains detrital zircons as young as 430 Ma and lacks plutons older than 415 Ma; additionally, the Cat Square terrane contains a population of ca. 590–600 Ma zircons, putatively derived from Carolinia. The paragneiss of the Cat Square terrane is interpreted here to record rifting of the recently accreted (ca. 455 Ma) Carolina terrane from Laurentia-Tugaloo, and deposition in a transtensional Wenlock-Ludlow basin, similar to the Gulf of California in some superficial respects. Thus, the Cat Square basin may signal the postaccretionary terrane dispersal of Carolinia. Such an interpretation squares existing paleomagnetic data for the Carolina terrane in the Ordovician with the lack of a shared history with the Inner Piedmont prior to ca. 320 Ma. The Cat Square paragneiss and coeval igneous rocks correlate with well-known internal Salinic basins and their associated volcanic rocks in New England and Canadian Maritime provinces. The structural arguments outlined here and elsewhere suggest that the Cat Square basin paragneiss may have been contiguous with rocks of the Connecticut Valley or Merrimack synclinoria.

Figure 5. Clasts from Devonian-Mississippian Spechty Kopf and Rockwell Formations, Pennsylvania and Maryland. More exotic clasts (A, B, C, D; granites, gneisses, greenstones, metavolcanic rocks) occur to the south (Sideling Hill, Maryland; LaVale, Maryland; Crystal Spring, Pennsylvania; Town Hill, Maryland-Pennsylvania), leading to the interpretation that the source of these clasts is the terrane responsible for the collision represented by the Catskill-Pocono clastic wedge. Diamictites are massive-bedded, often weathering spheroidally (E). This spheroidal weathering has resulted in past misidentification of large boulders in till. Bed is about 1.5 m thick. At Klingerstown, Pennsylvania (F, G, H), a very fine-grained laminite (varved clay?) may represent a glacial lake with dropstones. Other laminites are known from the Spechty Kopf–Rockwell section, but a glacial/dropstone origin may be best represented here. Some clasts within the diamictite are faceted and striated (I, J; ruler units in cm). These samples are from LaVale, Maryland. The surfaces of some LaVale clasts are covered with hematite. This hematite coats these striations. The striations are not pedogenic or the result of later (Alleghanian) tectonic deformation. In combination with the incised valleys, massive bedding, and unusual grain-size distribution, these characteristics may suggest a glacial origin for the diamictite and laminite portions of the Spechty Kopf and Rockwell Formations.

The recognition of extensive transposition and subhorizontal map-scale sheath folding accompanying dextral shear in the western Inner Piedmont has permitted the reconciliation of peak metamorphism at the Devonian-Mississippian boundary with the absence of a clastic wedge in the southern Appalachian foreland. These observations have the potential to tie orogeny in this southern Appalachian terrane to the Catskill-Pocono clastic wedges in the central Appalachians. The Spechty Kopf and Rockwell Formations of the Pennsylvania and Maryland Valley and Ridge were deposited at the Devonian-Mississippian boundary between the Catskill and Pocono clastic wedges. These distinctive formations contain exotic clasts, including granite, gabbro, gneiss, schist, and various volcanic and metavolcanic lithologies. Deposition of the Spechty Kopf–Rockwell Formation may signal the transition of the Laurentian margin to a Carboniferous transform plate boundary, and the translation of the Inner Piedmont and perhaps portions of the eastern Blue Ridge at least 400 km south along the Brevard zone. During the terminal closure of the Rheic Ocean and Gondwana collision, this continental transform margin was emplaced on the Laurentian margin as part of the Blue Ridge–Piedmont crystalline thrust sheet.

ACKNOWLEDGMENTS

This work was supported by the South Carolina Universities Research and Education Foundation, the Vice President for Research and Health Sciences at the University of South Carolina, the Executive Vice Chancellor for Academic Affairs at the University of South Carolina–Aiken, a 2005 sabbatical award from the University of South Carolina–Aiken, and the SCANA Chair in Physical Sciences at the University of South Carolina–Aiken. Evan Goldstein, Dwight Jones, and Mike Meredith assisted in the field and in the laboratory. I am very grateful to Art Boucot, John Dennison, Chris Hepburn, Jim Hibbard, Jim McLelland, Brendan Murphy, Doug Rankin, Scott Samson, Sheila Seaman, Bill Sevon, Vic Skema, Bill Thomas, Cees van Staal, Bob Wintsch, and Jim Wright for helpful discussions of these topics. Paul Karabinos, Don Wise, and Tekla Harms provided thorough and useful reviews. I remain responsible for all the views and interpretations presented here. I am grateful to the editors of this volume, for the opportunity to contribute to this volume, and Ray Price, who graciously encouraged me before I entered graduate school.

REFERENCES CITED

Adams, M.G., Stewart, K.G., Trupe, C.H., and Willard, R.A., 1995, Tectonic significance of high-pressure metamorphic rocks and dextral strike slip faulting in the southern Appalachians, in Hibbard, J.P., van Staal, C.R., and Cawood, P., eds., Current Perspectives in the Appalachian-Caledonian Orogen: Geological Association of Canada Special Paper 41, p. 21–42.

Bédard, J.H., 1986, Pre-Acadian magmatic suites of the southeastern Gaspé Peninsula: Geological Society of America Bulletin, v. 97, p. 1177–1191, doi: 10.1130/0016-7606(1986)97<1177:PMSOTS>2.0.CO;2.

Berg, T.M., 1999, Chapter 8. Devonian-Mississippian transition, in Shultz, C.H., ed., The Geology of Pennsylvania: Harrisburg, Pennsylvania Geological Survey and Pittsburgh Geological Society, p. 129–137.

Berg, T.M., and Edmunds, W.E., 1979, The Huntley Mountain Formation: Catskill-to-Burgoon transition in north central Pennsylvania: Pennsylvania Geologic Survey, 4th ser., Information Circular 83, 80 p.

Bjerstedt, T.J., 1986, Regional stratigraphy and sedimentology of the Lower Mississippian Rockwell Formation and Purslane Sandstone based on the New Sideling Hill Roadcut, Maryland: Southeastern Geology, v. 27, p. 69–94.

Bjerstedt, T.J., and Kammer, T.W., 1988, Genetic stratigraphy and depositional systems of the Upper Devonian–Lower Mississippian Price-Rockwell deltaic complex in the central Appalachians, U.S.A.: Sedimentary Geology, v. 54, p. 265–301, doi: 10.1016/0037-0738(88)90037-1.

Bobyarchick, A.R., Edelman, S.H., and Horton, J.W., Jr., 1988, The role of dextral strike slip in the displacement history of the Brevard zone, in Secor, D.T., ed, Southeastern Geological Excursions: Columbia, South Carolina, South Carolina Geological Survey, p. 53–154.

Bond, P.A., and Fullagar, P.D., 1974, Origin and age of the Henderson augen gneiss and associated cataclastic rocks in southwestern North Carolina: Geological Society of America Abstracts with Programs, v. 6, no. 4, p. 336.

Boucot, A., 1962, Appalachian Siluro-Devonian, in Coe, K., ed., Some Aspects of the Variscan Fold Belt: Manchester, UK, Manchester University Press, p. 155–163.

Bourque, P.-A., Malo, M., and Kirkwood, D., 2000, Paleogeography and tectono-sedimentary history at the margin of Laurentia during Silurian to earliest Devonian time: The Gaspe belt, Quebec: Geological Society of America Bulletin, v. 112, p. 4–20, doi: 10.1130/0016-7606(2000)112 <0004:PATSHA>2.3.CO;2.

Bourque, P.-A., Malo, M., and Kirkwood, D., 2001, Stratigraphy, tectonosedimentary evolution and paleogeography of the post-Taconian–pre-Carboniferous Gaspe belt: An overview: Bulletin of Canadian Petroleum Geology, v. 49, p. 186–201, doi: 10.2113/49.2.186.

Bradley, D., and Tucker, R., 2002, Emsian synorogenic paleogeography of the Maine Appalachians: The Journal of Geology, v. 110, p. 483–492, doi: 10.1086/340634.

Bream, B.R., 2002, The southern Appalachian Inner Piedmont: New perspectives based on recent detailed geologic mapping, Nd isotopic evidence and zircon geochronology, in Hatcher, R.D., and Bream, B.R., eds., Inner Piedmont Geology in the South Mountains–Blue Ridge Foothills and the Southwestern Brushy Mountains, Central Western North Carolina: Raleigh, North Carolina, North Carolina Geological Survey, Carolina Geological Society Annual Field Trip Guidebook, p. 45–63.

Busch, M.M., Miller, B.V., and Stewart, K.G., 2002, Disparate U-Pb and Sm-Nd ages from the Lick Ridge eclogite, eastern Blue Ridge Province, North Carolina: Geological Society of America Abstracts with Programs, v. 34, no. 6, p. 41.

Butler, J.R., 1991, Metamorphism, in Horton, J.W., and Zullo, V.A., eds., The Geology of the Carolinas: Knoxville, University of Tennessee Press, p. 127–141.

Carrigan, C.W., Bream, B.R., Miller, C.F., and Hatcher, R.D., 2001, Ion microprobe analyses of zircon rims from the eastern Blue Ridge and Inner Piedmont, NC-SC-GA: Implications for the timing of Paleozoic metamorphism in the southern Appalachians: Geological Society of America Abstracts with Programs, v. 33, no. 2, p. A-7.

Castonguay, S., and Tremblay, A., 2003, Tectonic evolution and significance of Silurian–Early Devonian hinterland-directed deformation in the internal Humber zone of the southern Quebec Appalachians: Canadian Journal of Earth Sciences, v. 40, p. 255–268, doi: 10.1139/e02-045.

Cecil, C.B., Skema, V., Stamm, R., and Dulong, F.T., 2002, Evidence for Late Devonian and Early Carboniferous global cooling in the Appalachian basin: Geological Society of America Abstracts with Programs, v. 34, no. 7, p. 500.

Cecil, C.B., Brezinski, D.K., and DuLong, F., 2004, The Paleozoic record of changes in global climate and sea level: Central Appalachian basin, in Southworth, S., and Burton, W., eds., Geology of the National Capital Region—Field Trip Guidebook: Joint Meeting of Northeast and Southeast Sections, Geological Society of America: U.S. Geological Survey Circular 1264, p. 77–135.

Cocks, L.R.M., and Torsvik, T.H., 2002, Earth geography from 500 to 400 million years ago: A faunal and paleomagnetic review: Geological Society [London] Journal, v. 159, p. 631–644.

Colstolnick, D.E., 1987, Sedimentology of the Devonian Mississippian Spechty Kopf Formation in Northeastern Pennsylvania [M.A. thesis]: Oneonta, State University of New York, 117 p.

Dallmeyer, R.D., 1988, Late Paleozoic tectonothermal evolution of the western Piedmont and eastern Blue Ridge: Controls on the chronology of terrane accretion and transport in the southern Appalachian orogen: Geological Society of America Bulletin, v. 100, p. 702–713, doi: 10.1130/0016-7606 (1988)100<0702:LPTEOT>2.3.CO;2.

Dallmeyer, R.D., Wright, J.E., Secor, D.T., and Snoke, A.W., 1986, Character of the Alleghanian orogeny in the southern Appalachians: Part II. Geochronological constraints on the tectonothermal evolution of the eastern Piedmont in South Carolina: Geological Society of America Bulletin, v. 97, p. 1329–1344.

Dallmeyer, R.D., Martinez-Catalan, J.R., Arenas, R., Gil-Ibarguchi, J.I., Gutierrez-Alonso, G., Farias, P., Bastida, F., and Aller, J., 1997, Diachronous Variscan tectonothermal activity in the NW Iberian Massif: Evidence from ^{40}Ar/^{39}Ar dating of regional fabrics: Tectonophysics, v. 277, p. 307–337, doi: 10.1016/S0040-1951(97)00035-8.

Dennis, A.J., 1991, Is the central Piedmont suture a low-angle normal fault?: Geology, v. 19, p. 1081–1084, doi: 10.1130/0091-7613(1991)019 <1081:ITCPSA>2.3.CO;2.

Dennis, A.J., 2005a, Provenance of polymictic conglomerates in the Catskill-Pocono clastic wedge: Implications for the Inner Piedmont of the Carolinas and Devonian Mississippian collisional tectonics in the central Appalachians: Geological Society of America Abstracts with Programs, v. 37, no. 1, p. 65.

Dennis, A.J., 2005b, Middle Paleozoic clastic wedge provenance, dextral collision and the southern Appalachian Inner Piedmont: The Spechty Kopf and Rockwell Formations of the central Appalachians as tectonic markers: Geological Society of America Abstracts with Programs, v. 37, no. 2, p. 5.

Dennis, A.J., 2005c, Mississippian-Devonian dextral collision in the central Appalachians: Evidence from the Catskill-Pocono clastic wedge and the southern Appalachian Inner Piedmont: Geological Society of America Abstracts with Programs, v. 37, no. 7, p. 304.

Dennis, A.J., 2006a, Cat Square terrane of the southern Appalachian Piedmont: Correlation with northern Appalachian Salinic basins and magmatism: Geological Society of America Abstracts with Programs, v. 38, no. 2, p. 73.

Dennis, A.J., 2006b, Cat Square terrane of the southern Appalachian Piedmont: A post-accretionary Salinic basin recording terrane dispersal: Geological Society of America Abstracts with Programs, v. 38, no. 3, p. 8.

Dennis, A.J., 2006c, Wenlock-Ludlow post-accretionary terrane dispersal in the central Appalachians: The Carolina and Cat Square terranes and the Salinic basins of the northern Appalachians, in Pereira, M.F., and Quesada, C., eds., International Geologic Correlation Project 497: The Rheic Ocean and Its Correlatives; Evora (Portugal) Meeting: Madrid, Instituto Geológico y Minero de España, p. 23–24.

Dennis, A.J., and Wright, J.E., 1997a, Middle and late Paleozoic monazite U-Pb ages, Inner Piedmont, South Carolina: Geological Society of America Abstracts with Programs, v. 29, no. 3, p. 12.

Dennis, A.J., and Wright, J.E., 1997b, The Carolina terrane in northwestern South Carolina, U.S.A.: Late Precambrian–Cambrian deformation and metamorphism in a peri-Gondwanan oceanic arc: Tectonics, v. 16, p. 460–473, doi: 10.1029/97TC00449.

Dennis, A.J., Wright, J.E., Barker, C.A., Pray, J.R., and Secor, D.T., 1993, Late Precambrian–Early Cambrian orogeny in the South Carolina Piedmont: Geological Society of America Abstracts with Programs, v. 25, no. 6, p. A-484.

Dennison, J.M., 1972, Stratigraphy, sedimentology, and structure of Silurian and Devonian rocks along the Allegheny Front in Bedford County, Pennsylvania, Allegheny County, Maryland, and Mineral and Grant Counties, West Virginia: Middletown, Pennsylvania, Pennsylvania Geological Survey, Pennsylvania Geologists 37th Annual Field Conference Guidebook, p. 61–66.

Dewey, J., and Strachan, R.A., 2003, Changing Silurian-Devonian relative plate motion in the Caledonides: Sinistral transpression to sinistral transtension: Geological Society [London] Journal, v. 160, p. 219–229.

Dorsch, J., Bambach, R.K., and Driese, S.G., 1994, Basin-rebound origin for the "Tuscarora Unconformity" in southwestern Virginia and its bearing on the nature of the Taconic orogeny: American Journal of Science, v. 294, p. 237–255.

Dostal, J., Wilson, R.A., and Keppie, J.D., 1989, Geochemistry of Siluro-Devonian Tobique volcanic belt in northern and central New Brunswick (Canada): Tectonic implications: Canadian Journal of Earth Sciences, v. 26, p. 1282–1296.

Drummond, M.S., Neilson, M.J., Allison, D.T., and Tull, J.F., 1997, Igneous petrogenesis and tectonic setting of granitic rocks form the eastern Blue Ridge and Inner Piedmont, Alabama Appalachians, in Sinha, A.K., Whalen, J.B., and Hogan, J.P., eds., The Nature of Magmatism in the Appalachian Orogen: Geological Society of America Memoir 191, p. 147–164.

Dube, B., Dunning, G.R., Lauziere, K., and Roddick, J.C., 1996, New insights into the Appalachian orogen from geology and geochronology along the Cape Ray fault zone, southwest Newfoundland: Geological Society of America Bulletin, v. 108, p. 101–116, doi: 10.1130/0016-7606(1996)108 <0101:NIITAO>2.3.CO;2.

Edelman, S.H., Liu, A., and Hatcher, R.D., Jr., 1987, The Brevard zone in South Carolina and adjacent areas: An Alleghanian orogen-scale dextral shear zone reactivated as a thrust fault: The Journal of Geology, v. 95, p. 793–806.

Eisbacher, G.H., 1985, Pericollisional strike-slip faults and synorogenic basins, Canadian Cordillera, in Biddle, K.T., and Christie-Blick, N.J., eds., Strike-Slip Deformation, Basin Formation and Sedimentation: Society of Economic Paleontologists and Mineralogists Special Publication 37, p. 265–282.

Ettensohn, F.R., 1985, The Catskill delta complex and the Acadian orogeny, in Woodrow, D.L., and Sevon, W.D., eds., The Catskill Delta: Geological Society of America Special Paper 201, p. 39–49.

Ettensohn, F.R., 1987, Rates of relative plate motion during the Acadian orogeny based on the spatial distribution of black shales: The Journal of Geology, v. 95, p. 572–582.

Ferrill, B.A., and Thomas, W.A., 1988, Acadian dextral transpression and synorogenic sedimentary successions in the Appalachians: Geology, v. 16, p. 604–608, doi: 10.1130/0091-7613(1988)016<0604:ADTASS> 2.3.CO;2.

Fullagar, P.D., Goldberg, S.A., and Butler, J.R., 1997, Nd and Sr isotopic characterization of crystalline rocks from the southern Appalachian Piedmont and Blue Ridge, North and South Carolina, in Sinha, A.K., Whalen, J.B., and Hogan, J.P., eds., The Nature of Magmatism in the Appalachian Orogen: Geological Society of America Memoir 191, p. 165–179.

Gans, P.B., Mahood, G.A., and Schermer, E., 1989, Synextensional Magmatism in the Basin and Range Province: A Case Study from the Eastern Great Basin: Geological Society of America Special Paper 233, 53 p.

Giorgis, S.D., 1999, Geology of the Northwestern South Mountains near Morganton, North Carolina [M.S. thesis]: Knoxville, University of Tennessee, 186 p.

Goldsmith, R.L., Milton, D.J., and Horton, J.W., Jr., 1988 Geologic map of the Charlotte 1° × 2° quadrangle, North Carolina and South Carolina: Reston, Virginia, U.S. Geological Survey Miscellaneous Investigations Series Map I-1251-E, scale 1:250,000.

Goldstein, A.G., 1989, Tectonic significance of multiple motions on terrane-bounding faults in the northern Appalachians: Geological Society of America Bulletin, v. 101, p. 927–938.

Graham, S.A., Hendrix, M.S., Wang, L.B., and Carroll, A.R., 1993, Collisional successor basins of western China: Impact of tectonic inheritance on sand composition: Geological Society of America Bulletin, v. 105. p. 323–344, doi: 10.1130/0016-7606(1993)105<0323:CSBOWC>2.3.CO;2.

Gray, M.B., and Zeitler, P.K., 1997, Comparison of clastic wedge provenance in the Appalachian foreland using U/Pb ages of detrital zircons: Tectonics, v. 16, p. 151–160, doi: 10.1029/96TC02911.

Hatcher, R.D., 2001, Terranes and terrane accretion in the Southern Appalachians; an evolved working hypothesis: Geological Society of America Abstracts with Programs, v. 33, no. 2, p. 65.

Hatcher, R.D., 2002, An Inner Piedmont primer, in Hatcher, R.D., and Bream, B.R., eds., Inner Piedmont Geology in the South Mountains–Blue Ridge Foothills and the Southwestern Brushy Mountains, Central Western North Carolina: Raleigh, North Carolina, North Carolina Geological Survey, Carolina Geological Society Annual Field Trip Guidebook, p. 1–18.

Hatcher, R.D., and Bream, B.R., eds., 2002, Inner Piedmont Geology in the South Mountains–Blue Ridge Foothills and the Southwestern Brushy Mountains, Central Western North Carolina: Raleigh, North Carolina, North Carolina Geological Survey, Carolina Geological Society Annual Field Trip Guidebook, 146 p. (guidebook available online as a pdf at www.carolinageologicalsociety.org/gb%202002.pdf).

Hibbard, J.P., 1994, Kinematics of Acadian deformation in the northern and Newfoundland Appalachians: The Journal of Geology, v. 102, p. 215–228.

Hibbard, J.P., 2000, Docking Carolina: Mid-Paleozoic accretion in the southern Appalachians: Geology, v. 28, p. 127–130, doi: 10.1130/0091-7613(2000) 28<127:DCMAIT>2.0.CO;2.

Hibbard, J.P., and Hall, S., 1992, Geology of the Saddle Pond–Grand Lake Seboeis Region, Maine: Maine Geological Survey Open-File Report 92-60, 26 p.

Hibbard, J.P., Shell, G.S., Bradley, P.J., Samson, S.D., and Wortman, G.L., 1998, The Hyco shear zone in North Carolina and southern Virginia: Implications for the Piedmont zone–Carolina zone boundary of the southern Appalachians: American Journal of Science, v. 298, p. 85–107.

Hibbard, J.P., Miller, B.V., Tracy, R.J., and Carter, B.T., 2004, The Appalachian peri-Gondwanan realm; a paleogeographic perspective from the south: Geological Society of America Abstracts with Programs, v. 36, no. 2, p. 104.

Hibbard, J.P., van Staal, C.R., and Miller, B.V., 2007, this volume, Links between Carolinia, Avalonia, and Ganderia in the Appalachian peri-Gondwanan realm, in Sears, J.W., Harms, T.A., and Evenchick, C.A., eds., Whence the Mountains? Inquiries into the Evolution of Orogenic Systems: A Volume in Honor of Raymond A. Price: Geological Society of America Special Paper 433, doi: 10.1130/2007.2433(14).

Hoffman, P.F., and Bowring, S.A., 1984, Short-lived 1.9 Ga continental margin and its destruction, Wopmay orogen, northwest Canada: Geology, v. 12, p. 68–72, doi: 10.1130/0091-7613(1984)12<68:SGCMAI>2.0.CO;2.

Holdsworth, R.E., 1994, Structural evolution of the Gander-Avalon terrane boundary: A reactivated transpression zone in the NE Newfoundland Appalachians: Geological Society [London] Journal, v. 151, p. 629–646.

Hopson, J.L., and Hatcher, R.D., 1988, Structural and stratigraphic setting of the Alto allochthon, northeast Georgia: Geological Society of America Bulletin, v. 100, p. 339–350, doi: 10.1130/0016-7606(1988)100 <0339:SASSOT>2.3.CO;2.

Kammer, T.W., and Bjerstedt, T.J., 1986, Stratigraphic framework of the Price Formation (Upper Devonian–Lower Mississippian) in West Virginia: Southeastern Geology, v. 27, p. 13–32.

Karabinos, P., 2002, Acadian extension around the Chester Dome, Vermont, in McLelland, J., and Karabinos, P., eds., Guidebook for Fieldtrips in New York and Vermont: Lake George, New York, New England Intercollegiate Geological Conference (94th Annual Meeting) and New York State Geological Association (74th Annual Meeting), p. C6-1–C6-21.

Kish, S.A., 1997, The Cat Square charnockite—A Paleozoic charnockite in the Inner Piedmont of North Carolina: Geological Society of America Abstracts with Programs, v. 29, no. 3, p. 28.

LaVoie, D., and Asselin, E., 2004, A new stratigraphic framework for the Gaspe Belt in southern Quebec: Implications for the pre-Acadian Appalachians of eastern Canada: Canadian Journal of Earth Sciences, v. 41, p. 507–525, doi: 10.1139/e03-099.

Lawrence, D.P., and Foster, D.M., 2006, Sinistral shear zones and terrane boundaries in the eastern Piedmont of South Carolina: Geological Society of America Abstracts with Programs, v. 38, no. 3, p. 21.

Livingston, J.L., and McNiff, J.M., 1967, Tallulah Falls dome, northeastern Georgia: Another window?, in Abstracts for 1967: Geological Society of America Special Paper 115, p. 485.

Mapes, R.W., 2002, Geochemistry and Geochronology of Mid-Paleozoic Granitic Plutonism in the Southern Appalachian Piedmont Terrane, North Carolina–South Carolina–Georgia [M.S. thesis]: Nashville, Vanderbilt University, 150 p.

Martinez-Catalan, J.R., and Arenas, R., 2005, The upper allochthon of NW Iberia: Structural constraints to a polyorogenic peri-Gondwanan terrane [abs.]: The International Geoscience Programme (IGCP) 497: The Rheic Ocean, Its Origin, Evolutions and Correlatives: Portsmouth, England, University of Portsmouth, p. 21.

McLennan, S.M., Bock, B., Compston, W., Hemming, S.R., and McDaniel, D.K., 2001, Detrital zircon geochronology of Taconian and Acadian foreland sedimentary rocks in New England: Journal of Sedimentary Research, v. 71, p. 305–317.

McSween, H.Y., and Harvey, R.P., 1997, Concord plutonic suite; pre-Acadian gabbro-syenite intrusions in the southern Appalachians, in Sinha, A.K., Whalen, J.B., and Hogan, J.P., eds., The Nature of Magmatism in the Appalachian Orogen: Geological Society of America Memoir 191, p. 221–234.

Merschat, A., Hatcher, R.D., and Davis, T.L., 2005, The northern Inner Piedmont, southern Appalachians, USA: Kinematics of transpression and SW-directed mid-crustal flow: Journal of Structural Geology, v. 27, p. 1252–1281, doi: 10.1016/j.jsg.2004.08.005.

Miller, B.V., Fetter, A.H., and Stewart, K.G., 2006, Plutonism in three orogenic pulses, eastern Blue Ridge Province, southern Appalachians: Geological Society of America Bulletin, v. 118, p. 171–184, doi: 10.1130/B25580.1.

Moecher, D.P., Samson, S.D., and Miller, C.F., 2004, Precise time and conditions of peak Taconian granulite facies metamorphism in the southern Appalachian orogen, U.S.A., with implications for zircon behavior during crustal melting events: The Journal of Geology, v. 112, p. 289–304, doi: 10.1086/382760.

Monger, J.W.H., 1977, Upper Paleozoic rocks of the western Canadian Cordillera and their bearing on Cordilleran evolution: Canadian Journal of Earth Sciences, v. 14, p. 1832–1859.

Nance, R.D., and Dallmeyer, R.D., 1993, ^{40}Ar/^{39}Ar amphibole ages from the Kingston Complex, New Brunswick: Evidence for Silurian-Devonian tectonothermal activity and implications for the accretion of the Avalon composite terrane: The Journal of Geology, v. 101, p. 375–388.

Noel, J.R., Spariosu, D.J., and Dallmeyer, R.D., 1988, Paleomagnetism and ^{40}Ar/^{39}Ar ages from the Carolina slate belt, Albemarle, North Carolina: Implications for terrane amalgamation in North America: Geology, v. 16, p. 64–68, doi: 10.1130/0091-7613(1988)016<0064:PAAAAF>2.3.CO;2.

O'Brien, B.H., O'Brien, S.J., and Dunning, G.R., 1991, Silurian cover, Late Precambrian–Early Ordovician basement, and the chronology of Silurian orogenesis in the Hermitage flexure (Newfoundland Appalachians): American Journal of Science, v. 291, p. 760–799.

Odom, A.L., and Fullagar, P.D., 1973, Geochronological and tectonic relationships between the Inner Piedmont, Brevard zone and Blue Ridge belts: American Journal of Science, v. 273, p. 133–149.

Offield, T.W., Kunk, M.J., and Koeppen, R., 1995, Style and age of deformation, Carolina slate belt, central North Carolina: Southeastern Geology, v. 35, p. 59–77.

Oskin, M., and Stock, J., 2003, Marine incursion synchronous with plate-boundary localization in the Gulf of California: Geology, v. 31, p. 23–26, doi: 10.1130/0091-7613(2003)031<0023:MISWPB>2.0.CO;2.

Pollock, S., 1987, The Lower Devonian slate problem of western and northern Maine revisited: Northeastern Geology, v. 9, p. 135–144.

Quesada, C., 2006, IV.1A. Introduction: The Ossa-Morena zone—From Neoproterozoic arc through early Palaeozoic rifting to late Palaeozoic orogeny, in Pereira, M.F., and Quesada, C., eds., Ediacaran to Visean Crustal Growth Processes in the Ossa-Moreno Zone (SW Iberia): The International Geoscience Programme (IGCP) 497: The Rheic Ocean: Its Origin, Evolution and Correlatives: Evora Meeting 2006: Madrid, Instituto Geológico y Minero de España, p. 27–48 and p. 103–110.

Quinn, M.J., and Wright, J.E., 1993, Extension of Middle Proterozoic (Grenville) basement into the eastern Blue Ridge of southwestern North Carolina: Results from U/Pb geochronology: Geological Society of America Abstracts with Programs, v. 25, no. 6, p. A-483–A-484.

Rankin, D.W., 1968, Volcanism related to tectonism in the Piscataquis volcanic belt and island arc of Early Devonian age in north central Maine, in Zen, E., et al., eds., Studies of Appalachian Geology, Northern and Maritime: New York, Wiley-Interscience, p. 355–369.

Rankin, D.W., 1987, The Jefferson terrane of the Blue Ridge tectonic province: An exotic accretionary prism: Geological Society of America Abstracts with Programs, v. 20, no. 4, p. 310.

Rankin, D.W., Coish, R.A., Tucker, R.D., Peng, Z.X., and Rouff, A.A., 2007, Silurian extension in the upper Connecticut Valley, United States, and the origin of middle Paleozoic basins in the Quebec embayment: American Journal of Science, v. 307, p. 216–264, doi: 10.2475/01.2007.07.

Rickard, L.V., 1975, Correlation of the Silurian and Devonian Rocks of New York State: New York State Museum Map and Chart Series 24, 16 p., 4 plates.

Ricketts, B.D., Evenchick, C.A., Anderson, R.G., and Murphy, D.C., 1992, Bowser basin, northern British Columbia: Constraints on the timing of initial subsidence and Stikinia–North America terrane interactions: Geology, v. 20, p. 1119–1122, doi: 10.1130/0091-7613(1992)020 <1119:BBNBCC>2.3.CO;2.

Robinson, P.R., Tucker, R.D., Bradley, D., Berry, H.N., IV, and Osberg, P.H., 1998, Paleozoic orogens in New England, USA: GFF, v. 120, p. 119–148.

Rodgers, J., 1953, Geologic Map of East Tennessee with Explanatory Text: Tennessee Division of Geology Bulletin 58, pt. 2, 168 p.

Samson, S.D., and Secor, D.T., 2000, New U-Pb geochronological evidence for a Silurian magmatic event in central South Carolina: Geological Society of America Abstracts with Programs, v. 32, no. 2, p. A-71.

Seaman, S.J., Scherer, E.E., Wobus, R.A., Zimmer, J.H., and Sales, J.G., 1999, Late Silurian volcanism in coastal Maine: The Cranberry Island series: Geological Society of America Bulletin, v. 111, p. 686–708, doi: 10.1130/0016-7606(1999)111<0686:LSVICM>2.3.CO;2.

Sevon, W.D., 1969, The Pocono Formation in northeastern Pennsylvania: Middletown, Pennsylvania, Pennsylvania Geological Survey, 44th Annual Field Conference of Pennsylvania Geologists Guidebook, 129 p.

Sevon, W.D., 1973, Glaciation and sedimentation in the Late Devonian and early Mississippian: Geological Society of America Abstracts with Programs, v. 6, no. 1, p. 218–219.

Sevon, W.D., 1979, Polymictic diamictites in the Spechty Kopf and Rockwell Formations, in Dennison, J.M., et al., eds., Devonian Shales in South Central Pennsylvania and Maryland: Middletown, Pennsylvania, Pennsylvania Geological Survey, 44th Annual Field Conference of Pennsylvania Geologists Guidebook, p. 61–66, 107–110.

Sevon, W.D., 1985, Nonmarine facies of the Middle and Late Devonian Catskill coastal alluvial plain, in Woodrow, D.L., and Sevon, W.D., eds., The Catskill Delta: Geological Society of America Special Paper 201, p. 79–90.

Sevon, W.D., Woodrow, D.L., Costolnick, D.E., Richardson, J.B., and Attrep, M., 1997, Convulsive geologic events and the origin of diamictite in the Spechty Kopf Formation in northeastern Pennsylvania, in Inners, J.D., ed., Geology of the Wyoming-Lackawanna Valley and its Mountain Rim, Northeastern Pennsylvania: Pennsylvania Geological Survey, 62nd Annual Field Conference of Pennsylvania Geologists Guidebook, p. 34–60, 120–126.

Sinha, A.K., and Glover, L., 1978, U-Pb systematics of zircons during dynamic metamorphism: Contributions to Mineralogy and Petrology, v. 66, p. 305–310, doi: 10.1007/BF00373414.

Soper, N.J., and Woodcock, N.H., 2003, The lost Lower Old Red Sandstone of England and Wales: A record of post-Iapetan flexure or Early Devonian transtension?: Geological Magazine, v. 140, p. 627–647, doi: 10.1017/S0016756803008380.

Soper, N.J., Strachan, R.A., Holdsworth, R.E., Gayer, R.A., and Greiling, R.O., 1992, Sinistral transpression and the Silurian closure of Iapetus: Geological Society [London] Journal, v. 149, p. 871–880.

Stampfli, G.M., and Borel, G.D., 2002, A plate tectonic model for the Paleozoic and Mesozoic constrained by dynamic plate boundaries and restored synthetic oceanic isochrones: Earth and Planetary Science Letters, v. 196, p. 17–33, doi: 10.1016/S0012-821X(01)00588-X.

Steltenpohl, M.G., ed., 2005, Southernmost Appalachian Terranes, Alabama and Georgia: Field Trip Guidebook for the Southeastern Section Geological Society of America 2005 Meeting: Tuscaloosa, Alabama Geological Society, 162 p.

Stock, J.M., and Hodges, K.V., 1989, Pre-Pliocene extension around the Gulf of California and the transfer of Baja California to the Pacific plate: Tectonics, v. 8, p. 99–115.

Suter, T.D., 1991, The Origin and Significance of Mississippian Polymictic Diamictites in the Central Appalachian Basin [Ph.D. thesis]: Morganton, West Virginia University, 370 p.

Thomas, W.A., 1977, Evolution of Appalachian salients and recesses from reentrants and promontories in the continental margin: American Journal of Science, v. 277, p. 1233–1278.

Thomas, W.A., and Astini, R.A., 1996, The Argentine Precordillera: A traveler from the Ouachita embayment of North America Laurentia: Science, v. 273, p. 752–757, doi: 10.1126/science.273.5276.752.

Tremblay, A., and Castonguay, S., 2002, Structural evolution of the Laurentian margin revisited (southern Quebec Appalachians): Implications for the Salinian orogeny and successor basins: Geology, v. 30, p. 79–82, doi: 10.1130/0091-7613(2002)030<0079:SEOTLM>2.0.CO;2.

Tremblay, A., and Pinet, N., 2005, Diachronous supracrustal extension in an intraplate setting and the origin of the Connecticut Valley—Gaspe and Merrimack troughs, northern Appalachians: Geological Magazine, v. 142, p. 7–22, doi: 10.1017/S001675680400038X.

Trupe, C.H., Stewart, K.G., Adams, M.G., Waters, C.L., Miller, B.V., and Hewitt, L.K., 2003, The Burnsville fault: Evidence for the timing and kinematics of southern Appalachian Acadian dextral transform tectonics: Geological Society of America Bulletin, v. 115, p. 1365–1376, doi: 10.1130/B25256.1.

Tucker, R.D., Bradley, D.C., Ver Straeten, C.A., Harris, A.G., Ebert, J.R., and McCutcheon, S.R., 1998, New U-Pb zircon ages and the duration and division of Devonian time: Earth and Planetary Science Letters, v. 158, p. 175–186, doi: 10.1016/S0012-821X(98)00050-8.

Tull, J.F., and Groszos, M.F., 1990, Nested Paleozoic "successor" basins in the southern Appalachian Blue Ridge: Geology, v. 18, p. 1046–1049, doi: 10.1130/0091-7613(1990)018<1046:NPSBIT>2.3.CO;2.

Tull, J.F., and Telle, W.R., 1989, Tectonic setting of olistostromal units and associated rocks in the Talladega slate belt, Alabama Appalachians, in Horton, J.W., and Rast, N., eds., Mélanges and Olistostromes of the U.S. Appalachians: Geological Society of America Special Paper 228, p. 247–269.

van Staal, C.R., 1994, Brunswick subduction complex in the Canadian Appalachians: Record of the Late Ordovician to Late Silurian collision between Laurentia and the Gander margin of Avalon: Tectonics, v. 13, p. 946–962, doi: 10.1029/93TC03604.

van Staal, C.R., Dewey, J.F., MacNiocaill, C., and McKerrow, W.S., 1998, The Cambrian-Silurian tectonic evolution of the northern Appalachians and British Caledonides: History of a complex, west and southwest Pacific-type segment of Iapetus, in Blundell, D.J., and Scott, A.C., eds., Lyell: The Past Is the Key to the Present: Geological Society [London] Special Publication 143, p. 199–242.

van Wagoner, N.A., and Dadd, K.A., 2003, A Silurian age for the Passamaquoddy Bay volcanic sequence in southwestern New Brunswick: Implications for regional corrections: Geological Society of America Abstracts with Programs, v. 35, no. 3, p. 79.

van Wagoner, N.A., Leybourne, M.I., Dadd, K.A., Baldwin, D.K., and MacNeil, W., 2002, Late Silurian bimodal volcanism of southwestern New Brunswick: Products of continental extension: Geological Society of America Bulletin, v. 114, p. 400–418, doi: 10.1130/0016-7606(2002)114<0400:LSBVOS>2.0.CO;2.

Vick, H.K., Channell, J.E.T., and Opdyke, N.D., 1987, Ordovician docking of the Carolina slate belt: Paleomagnetic data: Tectonics, v. 6, p. 573–583.

Williams, H., 1978, Tectonic Lithofacies Map of the Appalachian Orogen: Memorial University of Newfoundland Map 1, scale 1:1,000,000.

Wilson, C.W., 1949, Pre-Chattanooga Stratigraphy in Central Tennessee: Tennessee Division of Geology Bulletin 56, 407 p.

Wilson, R.A., Burden, E.T., Bertrand, R., Asselin, E., and McCracken, A.D., 2004, Stratigraphy and tectono-sedimentary evolution of the Late Ordovician to Middle Devonian Gaspe belt in northern New Brunswick: Evidence from the Restigouche area: Canadian Journal of Earth Sciences, v. 41, p. 527–551, doi: 10.1139/e04-011.

Woodrow, D.L., and Sevon, W.D., eds., 1985, The Catskill Delta: Geological Society of America Special Paper 201, 246 p.

MANUSCRIPT ACCEPTED BY THE SOCIETY 22 MARCH 2007

A stratigraphic unit converted to fault rocks in the Northland Allochthon of New Zealand: Response of a siliceous claystone to obduction

K. Bernhard Spörli*

School of Geography, Geology and Environmental Science (SGGES), The University of Auckland, Private Bag 92019, Auckland 1142, New Zealand

ABSTRACT

A spectacular, dense network of cataclastic faults characterizes the Late Cretaceous Ngatuturi Claystone, a massive and mechanically almost isotropic siliceous mudstone. It is part of a Cretaceous to late Oligocene shelf sequence deposited NE of New Zealand that was translated SW in the late Oligocene with the Northland Allochthon in an obduction event associated with southward propagation of a new convergent plate boundary. The allochthon was reactivated in the Miocene, forming the southward-moving substrate of the Waitemata piggyback basin. The cataclasites are submillimeter- to several centimeters–thick black seams that were formed without contemporaneous open tensile fractures, because any fault asperities were immediately ground away. Riedel shear patterns are prominent at all scales, due to multiple reactivation of preexisting fault surfaces. Some fault arrays are so closely spaced that they resemble a cleavage compatible with large-scale folds in the Ngatuturi Claystone. Movement on such faults has allowed formation of structures that appear mesoscopically ductile. More than twenty phases of cross-cutting structures (events E1–E22) are part of the following stages of tectonic development: (I) northeastward thrusting in an accretionary prism; (II) southward transport in the Northland Allochthon; (III) southwestward movement during the main phase of allochthon emplacement; (IV) renewed southward movement of the allochthon; (V) sliding during sedimentation of the Miocene Waitemata Group; and (VI) further intrabasinal thrusting to the south. During the pre-Miocene phases (I–IV), the cataclasites fault network allowed the Ngatuturi Claystone to deform in a macroscopically ductile manner, simultaneously acting as a dynamic aquiclude, thereby facilitating high fluid pressures in the surrounding rocks.

Keywords: siliceous mudstone, cataclasites, fault network, piggyback basin, accretionary complex, obduction, allochthon.

*kb.sporli@auckland.ac.nz

INTRODUCTION

Complex disruption of rocks associated with shortening (thrusting) occurs in three main environments of convergent tectonics: (1) deformation zones between colliding continents (Ellis et al., 2001; Hacker and Gans, 2005); (2) accretionary complexes at the sites of subduction of oceanic crust (Byrne, 1994; Pini, 1999); and (3) plate boundaries where obduction of oceanic crust is taking place (e.g., Boudier and Juteau, 2000; Cluzel et al., 2001). Because of historical and geographical vicissitudes, structures in collision zones have become very well understood, especially where deeper levels of the crust are involved. They have dominated the development of classical structural geology. However, in recent years, much data on shallow level deformation in accretionary complexes have also become available, leading to better insights into such features as mélanges, broken formation (Raymond, 1984), web structures (Lucas and Moore, 1986), scaly cleavage (Pini, 1999), and duplex underplating (Hashimoto and Kimura, 1999; Spaggiari et al., 2004). Conditions of deformation in obduction zones are somewhat similar to those in accretionary complexes, but there is an even larger component of deformation at high crustal levels, dominated by brittle rather than ductile behavior. This results in interesting modifications of the tenets of classical structural geology (e.g., Hancock, 1985; Groshong, 1988) and necessitates adaptations to the concepts of penetrativeness, continuity of structures, and scale invariance that allow the nature of macroscopic structures to be deduced directly from minor structures.

The rocks described in this paper spectacularly illustrate the reaction of a homogeneous sedimentary unit to obduction. They are exposed between Snells Beach and Algies Bay on Mahurangi Peninsula near Warkworth (Fig. 1), 50 km north of Auckland City. After first presenting their tectonic and stratigraphic setting, I will describe their intensively developed internal structures and subsequent deformations, then develop a likely tectonic scenario for their formation and discuss implications for deformation mechanisms, obduction tectonics, and techniques of structural analysis in rocks deformed at shallow levels in the crust.

TECTONIC FRAMEWORK

Northernmost New Zealand provides a unique example of the processes operating at the inception of a new subduction system. The New Zealand microcontinent rifted and drifted away from Gondwana between 80 Ma and 50 Ma (Spörli, 1989a). After this long period of extensional tectonics, a new convergent system propagated into New Zealand from the north, eventually forming the present Alpine Fault plate boundary (Fig. 1A) through New Zealand (King, 2000). The initial lithosphere break was accompanied by an obduction event marked by emplacement of the Northland Allochthon in the late Oligocene (Ballance and Spörli, 1979; Herzer and Mascle, 1996; Rait, 2000).

Northland Allochthon has a tectonic thickness ranging up to slightly more than 4 km and consists of (1) a Late Cretaceous to mid-Tertiary sedimentary sequence; (2) the Late Cretaceous–Early Tertiary Tangihua volcanics (=Tangihua Complex, Fig. 2); and (3) the Cretaceous–Early Tertiary rocks of the Mount Camel terrane (Toy et al., 2002) in northernmost Northland (Fig. 1B). The stratigraphic sequence of the Northland Allochthon (excluding Mount Camel terrane) is summarized in Figure 2. The older sedimentary sequences (=Mangakahia Complex) are dominated by terrigenous clastics in the lower part. Finer grained and more distal units follow through the Cretaceous-Tertiary boundary section, which includes the unit described in this paper. It was included in the regional Whangai Formation by Isaac et al. (1994); however, for the purpose of the present paper, the old, but still popular, name "Ngatuturi Claystone" is used. Eocene to Oligocene sediments (=Motatau Complex) consist of various mudstones, some of which are siliceous. Calcareous mudstones and micritic limestones occur in the Eocene and form the youngest, Oligocene, rocks (Mahurangi Limestone) in the sequence. The diagenetic history of the sequence is restricted to effects of their burial and was not modified by obduction (Aadil et al., 2001).

Overall, the thrust sheets comprising the allochthon consist of older rocks in the north while the younger rocks form the southern lower (and younger?) sheets. A thrust sheet dominated by the Ngatuturi Claystone underlies the Ngawha area (Fig. 1) of Northland (Bayrante and Spörli, 1989). Northland Allochthon sediments are usually assumed to have been deposited on the northeastern passive margin of New Zealand during its rifting away from Gondwana (Isaac et al., 1994); however, Toy et al. (2002) and Bradshaw (2004) have recently suggested that there was an accretionary prism.

The Tangihua volcanics structurally overlie the sediments and represent an ophiolite sequence that, except for one sliver at North Cape (Fig. 1), has been detached from its lower, ultramafic portions. They appear to have been formed in a suprasubduction environment (Whattam et al., 2005) and were considered to be Late Cretaceous–Paleocene in age until recently (Isaac et al., 1994); Whattam et al. (2005) consider a portion to be as young as Oligocene.

Still during or shortly after the obduction, the inception of the new subduction system led to formation of a calc-alkaline volcanic arc that is today preserved in the Waitakere Group (Hayward, 1993) and associated with an intra-arc turbidite basin in which the Miocene Waitemata Group (Fig. 2) was deposited. Continued movement of the allochthon led to establishment of a piggyback situation and to complex deformation in the Waitemata Group (Spörli, 1989b; Spörli and Rowland, 2007).

Shallow-water early Tertiary sedimentary sequences deposited on autochthonous basement of northern New Zealand (Figs. 1 and 2) contrast with the bathyal-pelagic nature of the youngest sedimentary rocks (Mahurangi limestone in Fig. 2) in the allochthon (Hayward, 2004). This indicates that just prior to obduction, the continental crust of New Zealand was pulled down to 1–2 km depth to allow low-angle overthrusting of the Northland Allochthon. Shallow-water basal Waitemata sediments record renewed uplift shortly after the emplacement. This was followed by a second episode of rapid subsidence followed by deposition of the

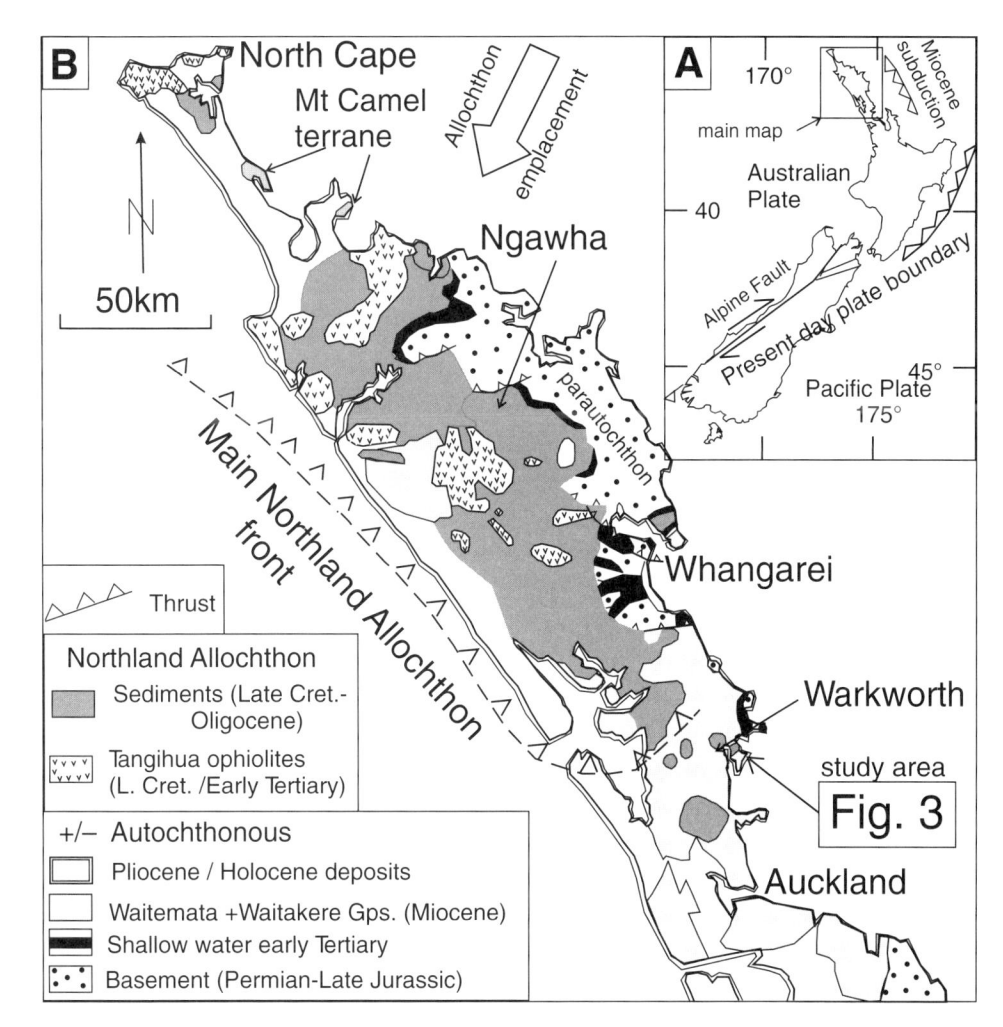

Figure 1. (A) Index map of New Zealand, showing location of Northland, the Miocene subduction zone, and the present plate boundary. (B) Geologic map of Northland showing the Northland Allochthon, as well as units predating, fringing, and postdating it (modified from Hayward et al., 1989). Unit patterns correspond to those in Figure 2, except for Mount Camel terrane.

main, bathyal part of the Waitemata Group (Hayward, 2004). These elevation oscillations may be due to transient down-dragging events during forced subduction (Gurnis et al., 2004; Stern, 2004) associated with such an initiation of a plate boundary.

Paleogeographic and tectonic considerations require a derivation of the allochthon from the E or NE, with emplacement mechanisms including both thrusting and gravity sliding. Rait (2000) proposed a uniform direction of movement from the NE, based on shear sense indicators in prominent shear zones. This, however, contrasts with evidence of NE-trending regional cleavage and fold trains implying other directions of movement (Spörli and Kadar, 1989; Bayrante and Spörli, 1989) and of reactivation of the allochthon during or just after deposition of the Waitemata Group (Spörli, 1989b; Isaac et al., 1994). The autochthon beneath and northeast of the allochthon ("parautochthon" in Fig. 1B) also shows evidence of reactivation (Hayward, 1986; Spörli and Harrison, 2004). Hayward (1986) interpreted the southernmost exposures of disturbed Cretaceous–Early Tertiary rocks (Fig. 1B) as blocks shed into the Waitemata basin off the front of the allochthon. There is no doubt that syn- or post-Waitemata movements were from NW to SE and even from W to E (Spörli, 1989b).

The question of how the smaller, outcrop-scale structures relate to overall transport direction in such an allochthon is one of the fascinating aspects of shallow-level deformation.

REGIONAL CONTEXT OF THE STUDY AREA

In the Warkworth area (Fig. 3), Northland Allochthon forms six inliers (Hayward, 1986) in the overlying Miocene Waitemata Group. Bedding strikes in the Waitemata Group are dominated by westerly trends. Dips are mostly to the north, but dip reversals represent prominent E-W–trending fold axes. However, the three northernmost inliers are dominated by NW strikes that may in part be an expression of a fault extending through the headwaters of Mahurangi Harbour. The two southern inliers (including the exposures dealt with in the present paper) are controlled by the E-W–trending structures, especially the southernmost one, which includes the important Pukapuka locality that demonstrates Miocene thrust-reactivation of the Northland Allochthon (Spörli, 1982; Hayward, 1986). The thickness of the Northland Allochthon in this region is unknown; however, basement emerges above sea level ~7 km to the east of the area (e.g., Tawharanui

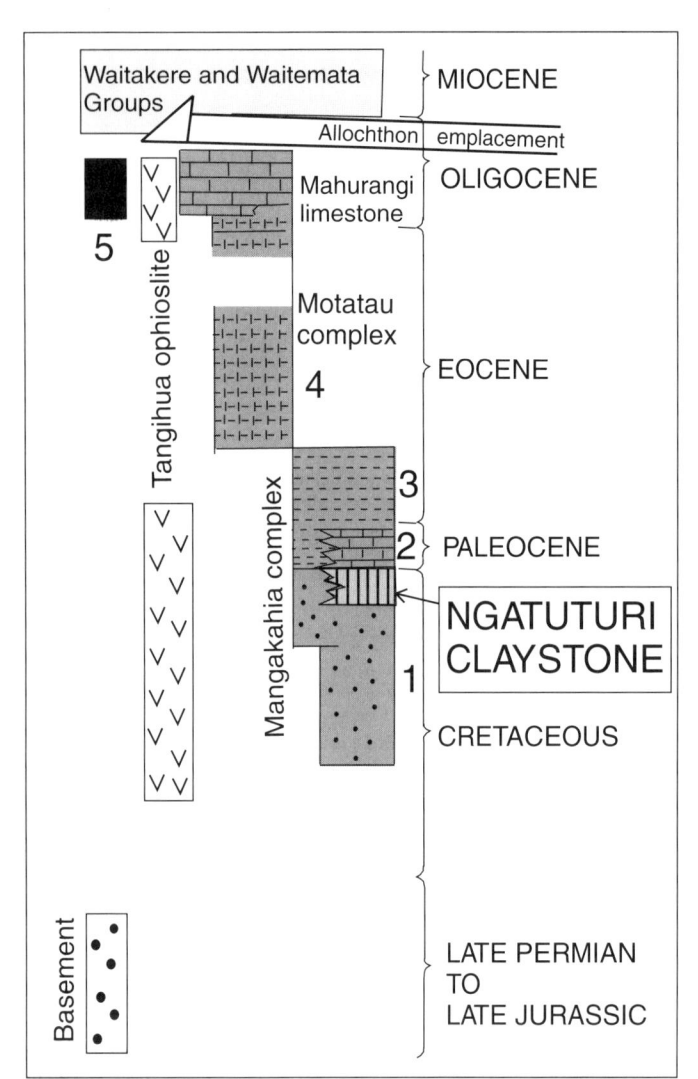

Figure 2. Simplified regional stratigraphy (after Hayward et al., 1989) and position of Ngatuturi Claystone. Lithologies in the Northland Allochthon: (1) terrigenous clastics, (2) muddy limestone, (3) non-calcareous mudstone, (4) calcareous mudstone. Autochthon: (5) shallow water limestone and siliciclastic rocks. Exposed basement is mainly the Permian to Late Jurassic Waipapa terrane (Spörli, 1989a).

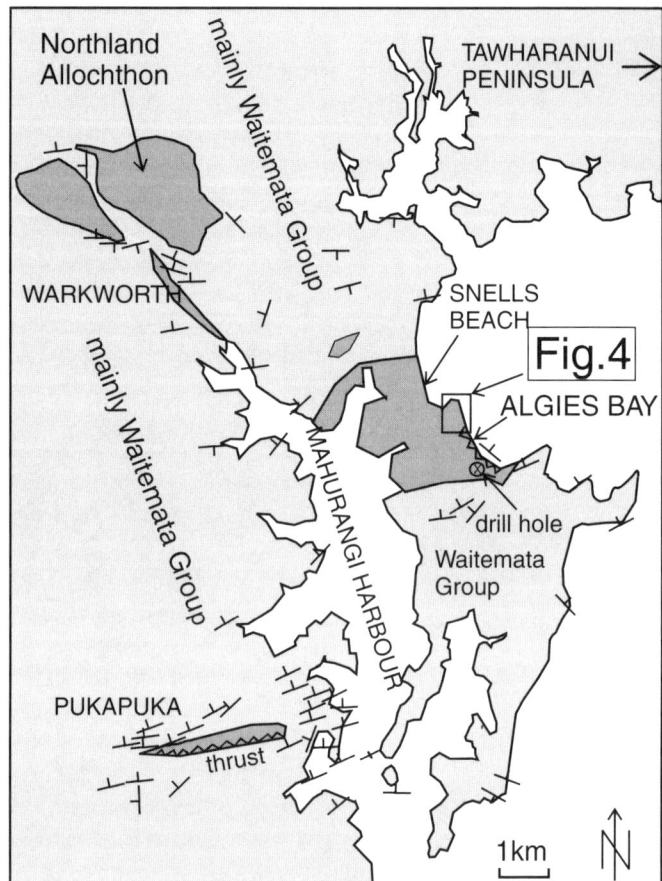

Figure 3. Simplified structural map of the Warkworth region (modified after Hayward, 1986) showing the location of the Algies Bay–Snells Beach area. Strike and dip symbols are all in the Waitemata Group. Note the predominance of ENE-striking beds.

Peninsula in Fig. 3), forming a NNW-striking basement high. Another basement high, striking ENE, closes off the area immediately north of Figure 3 and was also reactivated by thrusting in the Miocene (Spörli, 1982). While the other inliers contain a number of the units of the allochthon, especially the Oligocene Mahurangi Limestone (including its type locality), the Algies Bay–Snells Beach inlier is dominated by Ngatuturi Claystone.

LITHOLOGY OF THE NGATUTURI CLAYSTONE

The "Ngatuturi Claystone" is a very uniform unit of gray mudstones that weather to a characteristic very light gray to yellowish white color, accompanied by sulfur efflorescences and jarosite. The low-fertility, silica-rich, light-colored soils developed on this unit are a characteristic part of the Northland landscape.

In many localities, the rocks appear massive. However, an examination of cut surfaces invariably reveals millimeter-scale laminations outlined by subtle differences in grain-size, clay content, and blotchy bioturbation structures. In the northern part of the study area, this lamination becomes more pronounced and in favorable lighting can be seen macroscopically in the outcrop. There are also occasional layers that have been internally folded and disrupted by slumping.

These rocks have a clay-poor and a clay-rich variant, with the clay-poor type making up most of the cliffs and adjacent shore platform, and the clay-rich, more easily eroded type occupying lower areas (Fig. 4).

Orange weathering ferrocarbonate concretions are especially abundant in the middle part of the area (in the clay-poor unit). Their original shape was ellipsoidal to subspherical, with diameters ranging from a few tens of centimeters to ~1 m, or tabular, with their shape controlled by bedding. The tabular concretions originally were a few tens of centimeters thick and extended laterally up to several meters. Most of the concretions have been highly disrupted by faulting (see below).

In the clay-rich mudstones at the southern end of the study area is a horizon of white-weathering limestone in beds 5–10 cm thick. No fossils have yet been found to determine whether this horizon is actually part of the Ngatuturi Claystone or belongs to a younger unit.

OVERALL MAP PATTERN IN THE STUDY AREA

Between Snells Beach and Algies Bay, the Ngatuturi Claystone forms an elongate outcrop of shore platforms up to 140 m wide (Fig. 4) over a length of ~1 km and cliffs up to ~50 m high. The unit can be traced inland (to the west) and to the Mahurangi Harbour by its characteristic weathering color and soil development. The contact between it and the Waitemata Group is only well exposed at Algies Bay (Figs. 3 and 4). From there, the Waitemata Group continues farther south in good outcrop and has also been mapped inland.

PATTERN OF BEDDING IN THE NGATUTURI CLAYSTONE

Within the main rockmass, bedding is generally not detectable in outcrop, except in the northern portion of the area (see Lithology of the Ngatuturi Claystone above). Where bedding was not macroscopically visible, it was still possible to obtain data by collecting oriented samples that were slabbed and processed in the laboratory.

The dominant strike direction is NW with steep to moderately steep southwesterly dips. Generally, attitudes of bedding are quite constant over areas of several hundred square meters (Fig. 4) except in some major fault zones where changes at the centimeter scale are common. Considering this general strike and dip of bedding and the width of the outcrop, the thickness of Ngatuturi Claystone involved in the structures to be described is ~110 m. This is not a stratigraphic thickness, because there may be hidden repetitions and/or excisions within this stack of beds.

A major change in orientation to a dominant ENE strike and moderate to steep NNW dips occurs in the southern part of the area. Combined with the attitude in the northern part of the area, this suggests the presence of a major WNW-plunging fold (Fig. 5). A lesser pattern of gently NNW-plunging folds is suggested by the subordinate π-circle in Figure 5.

ALLOCHTHON–WAITEMATA GROUP CONTACT

At Algies Bay, Ngatuturi Claystone occupies the higher elevations behind the beach (Figs. 3 and 4), while Waitemata Group rocks occupy the lower elevations, indicating a low-angle, older-over-younger contact of units. This has been confirmed by drilling above the southern end of Algies Bay (Fig. 3).

In contrast to this, a subvertical, ENE-striking, south-facing stratigraphic contact is exposed at the northern end of the bay (Fig. 4). Proceeding southeast from a prominent sinistral strike-slip shear zone (fault BF, Figs. 4 and 6), the clay-rich Ngatuturi Claystone unit is replaced by an even more clay-rich mudstone with a characteristic splintery fabric and a few isolated sandstone 5–10-m thick-beds. This is overlain by an ~10-m-thick olistostrome mudstone with layers of allochthon debris conglomerate and rafts of deformed allochthon lithological units. The olistostrome is capped with a sharp contact by the first Waitemata turbidite bed, which is 80 cm thick and has a coarse basal horizon of allochthon pebbles. This bed is vertical to slightly overturned, strikes ESE, and is displaced by a sinistral shortening fault with a steeply plunging fault-bedding intersection line. The coarse sandstone bed is in turn overlain by a mudstone sequence (10–20 m thick) with occasional coarse sandstone layers. Attitude of bedding now changes to a very shallow northeastward dip along a gently E-plunging fold axis, and at least 5 m of thin-bedded alternating sandstones and mudstones overlie the mudstone-dominated sequence. This facies, which is common elsewhere in Waitemata Group, continues farther south along the shore platform, reaching a yet unknown thickness. It contains numerous dextral, NNE-plunging folds. The sinistral shortening fault mentioned above is a refolded equivalent of these structures, seen from the opposite end of their fold axis.

The Allochthon-Waitemata contact at Algies Bay therefore records the following events: (a) shedding of debris into a mudstone sequence of unknown age; (b) deposition of turbidites with additional allochthon debris; (c) asymmetric folding on a NNE trend, indicating thrusting toward the ESE; and (d) rotation to vertical of the Allochthon-Waitemata contact on a gently east-plunging fold axis, indicating thrusting toward the south.

PRINCIPAL STRUCTURAL FEATURES IN THE NGATUTURI CLAYSTONE

General Nature of the Intense Faulting

The Ngatuturi Claystone is cut by a dense network of black-seam faults with thicknesses ranging from <0.1 mm to several centimeters (Figs. 7 and 8), such that a 5 × 5 × 5 cm rock sample will contain at least one (but more generally 3–4) black seams. Such fault networks are so widespread and characteristic for the Ngatuturi Claystone throughout the Northland Allochthon (Spörli and Kadar, 1989; Bayrante and Spörli, 1989; Spörli and Harrison, 2004) that it is tempting to use them as a diagnostic for identifying this stratigraphic unit. However, the location described in this paper provides the best exposure in New Zealand, representing 140 m × 50 m × 1000 m = 7 million m³ of intensely faulted rock.

Usually, the rock between the black seams is intact, with no changes in bedding orientation, a characteristic of high-level, low-temperature deformation (Groshong, 1988). However, there are some occasional irregular patches where the rock breaks down into a breccia (Fig. 8). Rather than being random, the fault network is composed of distinct sets, with

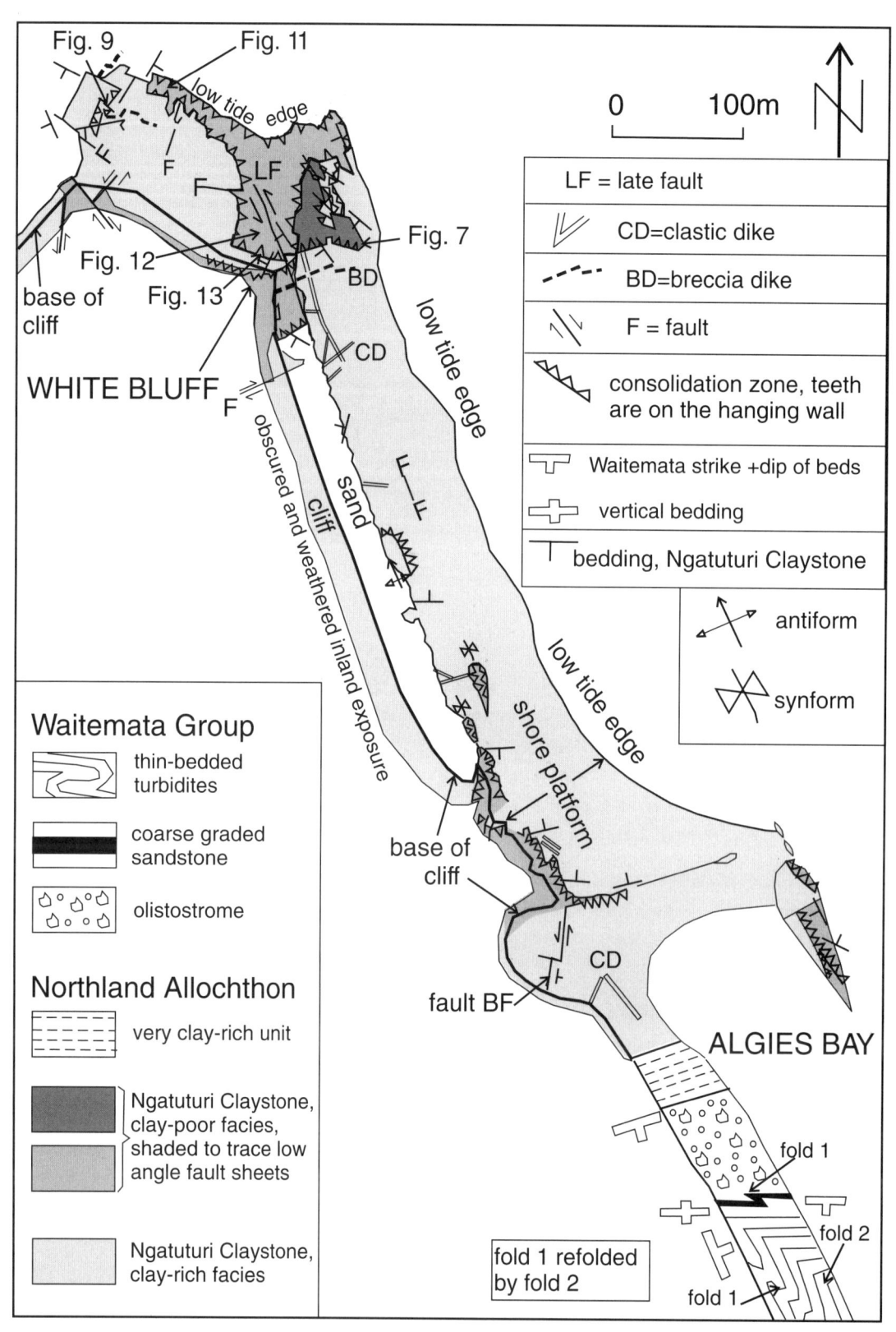

Figure 4. Geologic map of the White Bluff area north of Algies Bay (see Fig. 3 for location). Note the existence of low-angle fault sheets. Although a "tooth" symbol is used for the consolidation zones marking their boundaries, not all of these faults are thrusts. Some of them originated as or were reactivated as low-angle normal faults. The Waitemata Group geology is only shown very schematically to indicate general relationships and structures. The thickness of the coarse graded sandstone (shown in black) is exaggerated.

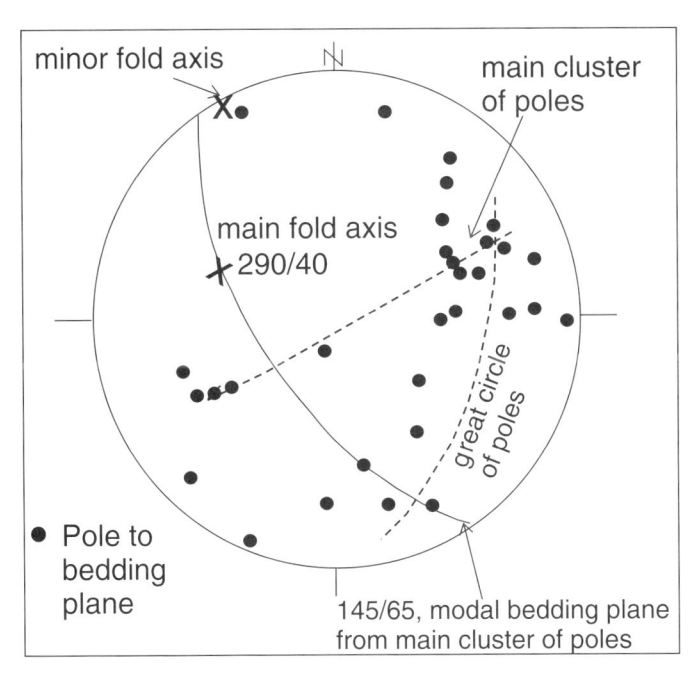

Figure 5. Bedding attitudes in the Ngatuturi Claystone. Lower hemisphere, equal area stereonet. Beds most commonly strike NW, with steep dips to the SW. Great circle π-patterns (shown by dashed lines) indicate the presence of a major WNW-plunging fold (290°/40°) and less pronounced folding on a gently NW-plunging axis.

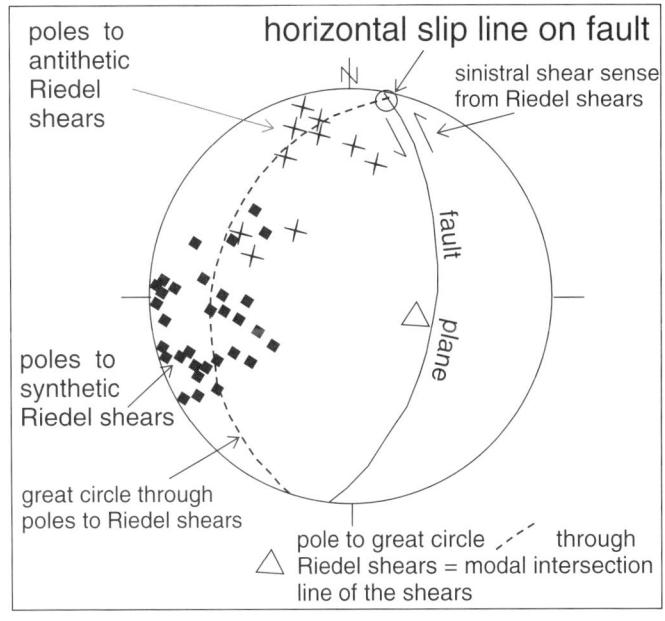

Figure 6. Determination of the slip vector and the sense of movement on a fault near the southern end of the Ngatuturi Claystone exposure (BF in Fig. 4). Lower hemisphere, equal area stereonet. The dashed line is the statistical π-circle for poles to synthetic and antithetic Riedel shear planes. The slip line is defined by the intersection of this great circle with the fault plane and indicates pure strike slip with NW-SE shortening for motion on this fault.

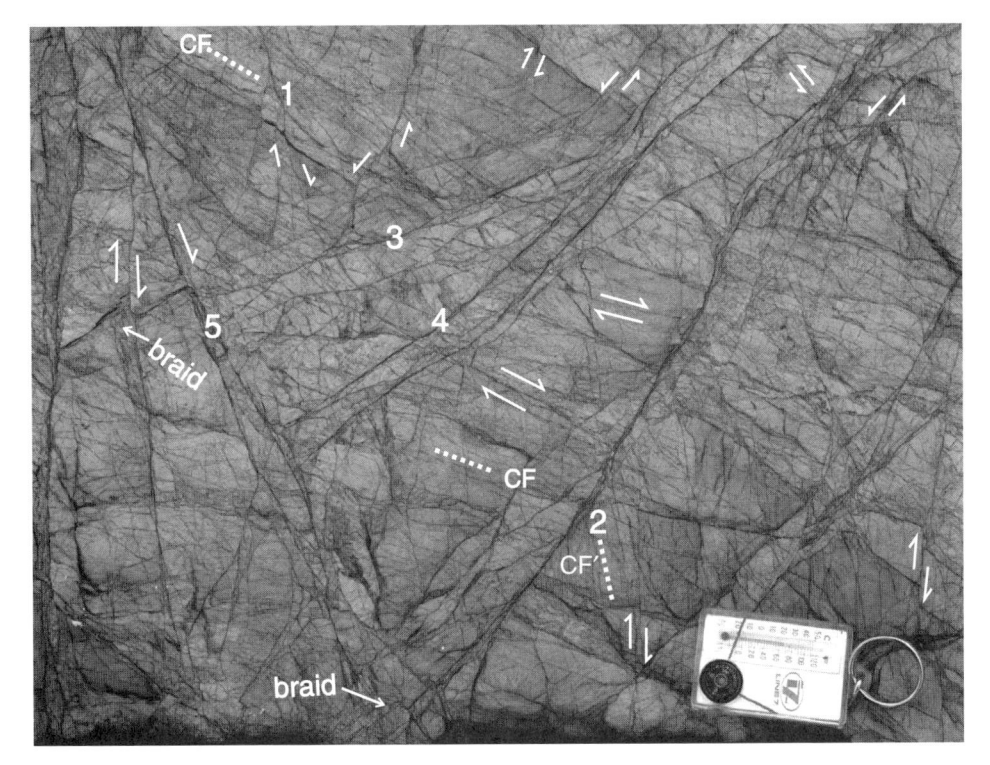

Figure 7. Typical example of the dense fault mesh developed in the Ngatuturi Claystone, exposed on a surface gently dipping toward the observer. All the black lines on this photo and in Figure 8 are faults with cataclasite seams; there is no shadowing. Arrow pairs indicate apparent movement in the exposure plane. Scale: the long edge of pocket compass is 53 mm long and is aligned N-S (north to the right). A NNE-trending, closely spaced fault fabric (CF) is cut by a sequence of steep faults (numbers 1–5 indicate local relative age of structures). Braiding indicates some interaction between the fault sets. CF′ = local development of a second phase of closely spaced faults.

338   Spörli

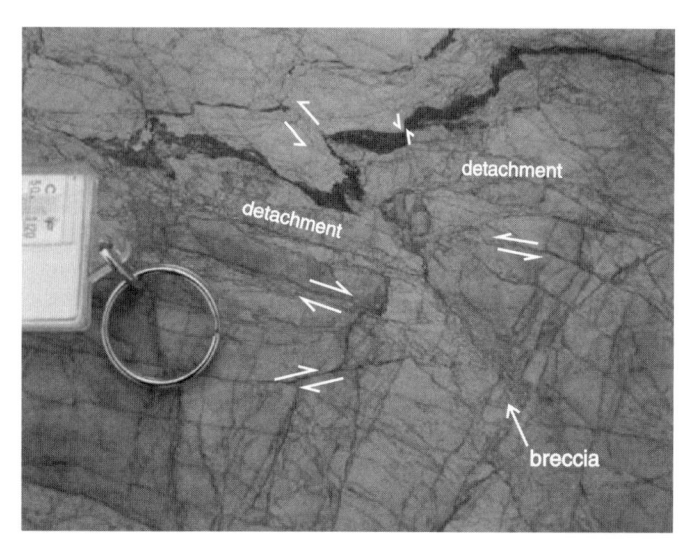

Figure 8. Fault mesh with movements causing sinistral folds and faults, exposed on a surface dipping moderately toward the observer. The short edge of the pocket compass is 33 mm long and is aligned E-W (west toward top). There is some development of breccia, allowing localized detachment of folds on a set of NE-trending, closely spaced faults. The thick black cataclasite seam experienced extensional faulting prior to folding.

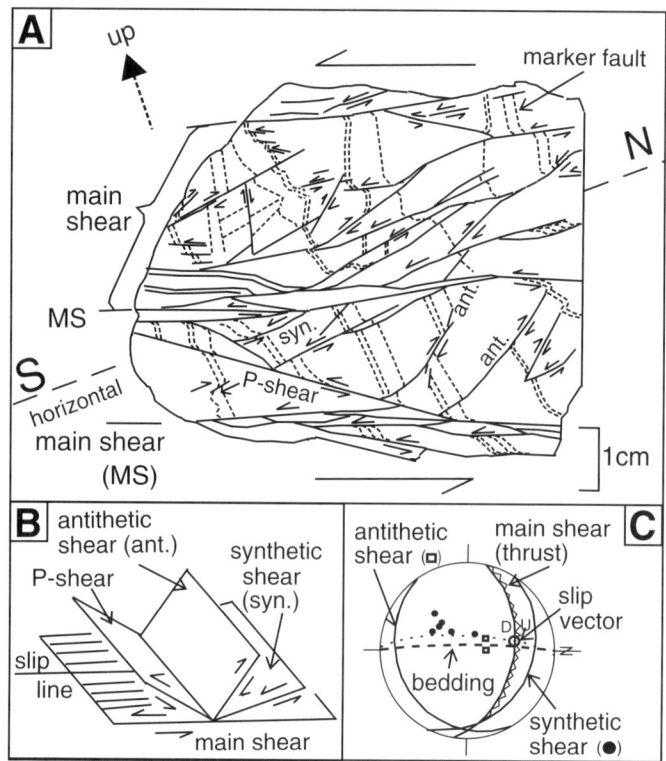

Figure 9. Fault pattern of a relatively simple consolidated zone: a southward thrust zone dominated by a Riedel shear pattern (see Fig. 4 for location). (A) Tracing of a slabbed, oriented specimen. Bedding is subparallel to the face shown. Note the incipient ductile drag on the faults. (B) Geometry of a Riedel shear and its relationship to the slip line. (C) Orientation of some of the faults in the specimen and construction of the slip vector (lower hemisphere equal area stereonet). Fault movement on slip vector: U = up; D = down.

age relationships that can be deduced from cross-cutting patterns, although numerous instances of fault reactivation are a complicating factor (Figs. 7 and 8).

Fault displacements can be analyzed (1) by using the ferrocarbonate concretions as markers, or (2) by using older faults as markers. Because of the lack of marker beds and the increasing complexity as one proceeds farther back in the structural sequence, it is very difficult to identify and analyze the initial phases of deformation. However the abundance of displaced preexisting faults allows numerous determinations of fault slip, from offsets of two or more differently oriented markers, in addition to the common conjugate fault couples and Riedel shear arrays (e.g., Fig. 9).

Nature of the Black Cataclastic Seams

The black seams consist of cataclasite and ultracataclasite in the sense of House and Gray (1982). When Ngatuturi Claystone is split with the hammer, conchoidal breaks cut across the host rock and the faults, indicating that the breaking strengths of both are now similar. If the contact between the cataclasite and the surrounding rock is uncovered, it often displays subtle groove striations that indicate the movement direction during faulting.

Isolated simple seams, deformation bands of Aydin and Johnson (1983), do not appear to have any damage zones in the sense of Caine et al. (1996). In the thicker seams (e.g., Fig. 8), lateral variations in seam thickness indicate that they were formed by bundling and coalescence of thinner individual strands, suggesting that there is a relationship between displacement and thickness. More complex fault bundles develop where faults are braiding (Fig. 7). Homogeneous fine-grained black seams can be considered to be ultracataclasites. Cataclasites contain milled fragments of Ngatuturi Claystone, indicating particulate flow (Cladouhos, 1999). Some cataclasites display an oblique foliation (Fig. 10) derived from synthetic Riedel shears (e.g., Dresen, 1991). This is a mechanism and shear sense indicator different from that of the S-C fabrics described by Lin (2001) but may be complementary to the P-shear foliation seen in some cataclasites (Cladouhos, 1999).

Fabric of Closely Spaced Faults

Individual fault sets usually have spacings of 0.5 m to several meters. However, there are arrays that are much more closely spaced at 3–10 cm (Fig. 7). The orientation of their fabric is reasonably constant, with strikes ranging from N to NNE and steep dips, giving them a cleavage-like aspect. Their orientation is distinctly oblique to the main NW orientation of bedding. In some areas, a second, less well-developed fabric with a different

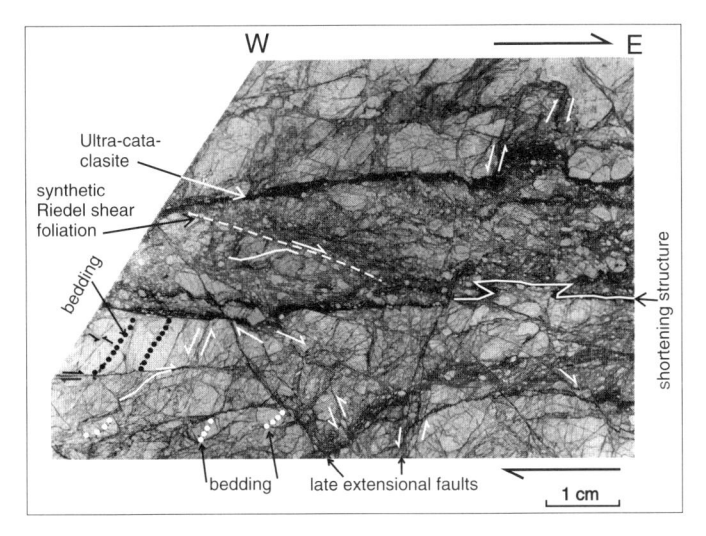

Figure 10. Example of the fabric in a complex consolidated fault zone with top-to-the-east movement. Slabbed, oriented specimen from the base of White Bluff (for location, see Fig. 4, ~10 m east of Fig. 13). All the black features are cataclasite and ultracataclasite. The gray material is Ngatuturi Claystone and shows brecciation and milling. The zone between the two horizontal black ultracataclasite seams in the center has a distinctive oblique fabric due to synthetic Riedel shears. A conjugate shortening structure affects the lower thick seam. Synthetic and antithetic Riedel shears are also present in the surrounding rocks (arrow pairs). Late faults are more widely spaced and dip more steeply. Bedding in the Ngatuturi Claystone is indicated by white or black dotted lines at the bottom left of photograph.

orientation occurs sporadically (Fig. 7). Very rarely, three closely spaced fault systems are superposed, switching from a NNE strike to an ESE strike and back to a NE strike.

Consolidated Fault Zones

Dark, tabular zones 5 cm to several tens of centimeters thick that are more resistant to erosion occasionally interrupt the general homogeneity of the Ngatuturi Claystone. Because of their persistence, often over several hundreds of meters, and their relatively constant thickness, it is possible to misinterpret them as bedding. However, analysis of oriented samples reveals that they are consistently oriented at high angles to bedding. Deformation of ferrocarbonate concretions along them clearly identify these zones as faults. The terms *consolidated fault zone* or *consolidation zone* will be used interchangeably for these features. Some of these low angle zones form the boundaries between clay-rich and clay-poor units of the Ngatuturi Claystone and bound relatively large but thin sheets of displaced rock (Fig. 4).

In the consolidation zones, the density of black seams is even higher than in the surrounding rocks (Fig. 10), rising to >5 seams per cubic centimeter. Some zones display a gradual decrease in seam density toward one side and an abrupt contact on the other (usually the lower) side of the zone, giving them a "graded" appearance.

A 10–15-cm-thick, relatively low intensity consolidated fault zone from a locality below White Bluff (Fig. 4) displays a simple Riedel shear pattern due to dip-slip thrusting to the south, parallel to the subvertical bedding (Fig. 9).

The specimen illustrated in Figure 10 is an example of the more usual (higher intensity) consolidated fault zones. Total thickness is 25–30 cm. Deflection of older black seam faults and of concretions indicates movement with a dextral (top to the NE) component. This is supported by analysis of micro-faults on the cut specimen: The zone is subdivided by a number of more prominent black seams up to several millimeters thick and subparallel to the fault zone boundaries. Some of the seams include angular to subrounded fragments of country rock of various sizes to produce a cataclastic texture (but without lamination). Locally, the fragments appear to be derived from irregular asperities in the country rock caused by geometric incompatibility created by movement on sets of microfaults adjacent to the thicker seams. Bedding traces in the intact matrix of the fault zone indicate rotation of individual fault blocks against each other, producing fold-like structures in the bedding. Groups of closely spaced micro faults have been deformed into sigmoidal shape, indicating shear sense. Locally, the country rocks assume a breccia texture.

Some consolidated shear zones represent one distinct shear zone with a consistent sense of displacement. Others follow several different structures. This is probably due to the movement on the youngest structure reactivating the older fault zones.

Folding of the Consolidated Fault Zones

While many consolidated fault zones are planar, there is an area in the north (Fig. 11) where a stack of these faults form a dome-and-basin pattern of folds. Less well-developed examples are visible in the south of the area (Fig. 4). These are open folds (interlimb angle range around 135°) with kink-like to rounded profiles. Axial planes are subvertical, and no clear asymmetry has been detected.

The dome-and-basin pattern in the north is mainly due to interference between folds with steep NNE-striking, NW-striking, NE-striking, and E-striking axial planes. In the south, folds with steep NW-striking axial planes interfere with folding on steep ENE-striking axial planes.

Formation of these folds has been facilitated by the dense network of preexisting micro- and meso-faults. The consolidated fault zones formed relatively late in the structural sequence (Table 1), and they are cut by several sets of relatively straight, extensional conjugate fault couples. Drag on these later faults is at least partially responsible for the fold structures.

Other Fold-like Structures

Northerly striking cataclasite seams are often folded into reasonably tight, sinistral folds up to several tens of centimeters in size (Fig. 8). These are mostly due to shearing of already closely faulted or brecciated regions. However, folds with various asymmetries also arise from transposition of earlier fault seams by subsequently formed, closely spaced, fault arrays.

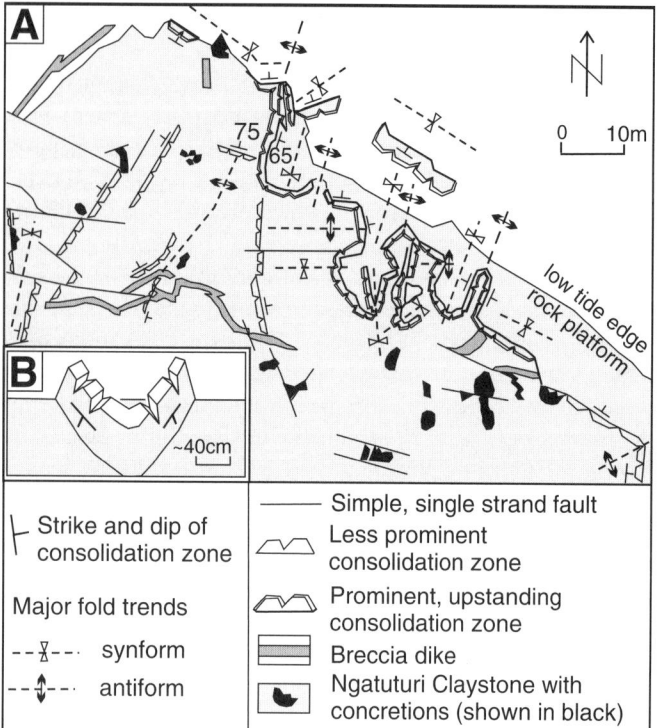

Figure 11. Area with the most distinct folding of consolidation zones, as exposed on a wave-cut rock platform (for location, see Fig. 4). (A) Pace and compass map showing relationship of folds to faults, breccia dikes, and ferrocarbonate concretions. Note the interference of NNE, NW, NE, and E-W fold trends. The concretions in the lower part of the map may all belong to a single low-angle sheet disrupted by faulting. (B) Schematic sketch of a typical exposure of an antiformally folded consolidation zone to explain map patterns used in A.

TABLE 1. SEQUENCE OF EVENTS

Event	Structures*
Pre–Waitemata Group Structures	
E1	~E-W–striking faults and other events
E2	Ductile dextral strike-slip extension, broken formation
E3	**Steeply plunging dextral folds** (Fig. 12)
E4	Low angle top-to-the-south extensional shear
E5	Low angle top-to-the-north extensional shear
E6	**Closely spaced faults, up to 3 subphases** (Figs. 7 and 8)
E7	Sinistral faulting and folding
E8	E-W–striking sinistral separation faults
E9	**Main NE-striking sinistral faults, folds** (Figs. 8, 12, and 13)
E10	E-W–striking, down-to-the-south normal faults (Fig. 13)
E11	NE-striking sinistral faults (Fig.13)
E12	Ductile (Fig. 13) and brittle (Fig. 12) top-to-the-east extensional shear
E13	N-S–striking conjugate faults with E-W extension (Fig. 13)
E14	E-W– and ENE-striking dextral strike-slip faults (Figs. 12 and 13)
E15	NW-SE–striking sinistral strike-slip faults (Fig. 12)
E16a	Eastward thrusting (Fig. 13)
E16b	Eastward thrusting (Fig. 13)
E16c	**Eastward thrusting, formation of consolidation zones** (Fig. 13)
E17	Southward thrusting, formation of consolidation zones (Fig. 9)
E18	Steep conjugate faults, at least three sets (CFS, CFE, CFN in Fig. 14)
Syn– and post–Waitemata Group Structures	
E19a	Intrusion of breccia dikes, allochthon slide into Waitemata basin
E19b	**NE- then NW-striking clastic dikes; olistostromes from allochthon** (Fig. 4)
E20	post-sedimentary thrusting to ESE (Fig. 14)
E21	**post-sedimentary thrusting to S** (Fig. 14)
E22	NW-striking sinistral strike-slip faults (Figs. 4 and 14)

*See Figure 14. Principal events emphasized by bold typeface.

Breccia Dikes and Sandstone Dikes

These dikes crosscut all the cataclastic seams in the Ngatuturi Claystone. Breccia dikes exclusively contain angular fragments up to 15 cm in diameter of Ngatuturi Claystone containing abundant cataclastic seams. The breccias are clast-supported without any fine-grained matrix. All the dikes dip steeply. The thickest dike (up to 2 m thick) cuts through White Bluff (Fig. 4) and strikes NE. Thinner breccia dikes NE of White Bluff (Figs. 4 and 11) strike NE and ESE.

Sandstone dikes consistently cross-cut or follow the breccia dikes. They contain mobilized Waitemata Group sand, sometimes with symmetrical zoning of different grain sizes. The dikes display most of the various shapes known from igneous dikes, such as segments, steps, buds, and horns (Pollard et al., 1975). The axes of segments and steps in the Waitemata clastic dikes plunge steeply, indicating subvertical intrusion, but, because of the nature of the exposures, it has not yet been possible to determine whether intrusion was from above or below. Clastic dikes dip steeply and strike NE, SE, NNE, and NNW. The NNW-striking dikes cut across all the other dikes.

Late Faults

Both the breccia dikes and the Waitemata Group–bearing clastic dikes are cut by NW-striking faults (Figs. 4 [LF] and 14) that display no development of black seams. Displacement of clastic dikes with different attitudes allows determination of the slip vector on fault LF in Figure 4, indicating that it is a sinistral strike-slip fault with a small normal component.

OVERALL STRUCTURAL GEOMETRY AND SEQUENCE OF EVENTS

From crosscutting of structures (such as black seams, consolidation zones, ferrocarbonate concretions, breccia and clastic dikes, and non-seamed faults), more than twenty different phases of deformational events have been recognized. Since these events cannot represent deformation phases in the classical sense, but more likely are substages of only a few phases, an *E1 ... En* scheme is used in the following descriptions (also see Table 1, where the major phases of deformation are shown in bold), instead of the commonly used D1 ... Dn indexing (e.g., Hobbs et al., 1976). The events recognized will later be set into a framework of tectonic stages of development (I–VI).

Observations on sequences of deformations were made at numerous localities (130+) and on oriented specimens (35+). To correlate between these, as a first step, structures that could be traced between exposures were identified in the field. Then a structural sequence diagram showing the orientation, the main characteristics of each structure, and the crosscutting relationships was prepared for each locality. In a third step, these were then arranged into an overall sequence of events using the structures traceable between outcrops as a starting point and continuing by correlating the others by their orientation, their characteristics, and their place in the local structural sequence.

While an overall sequence readily appeared, there were some structures that did not easily fit into the scheme. There are four possible causes for this: (1) the particular structure had not been adequately described; (2) in a pattern with only one cross-cutting relationship, too many events were missing between the structures present to allow correlation; (3) reactivation of an earlier formed structure was not recognized; and (4) any rare structures would appear as "background noise" rather than as well-defined events. Causes (3) and (4) are probably the most important. For the present paper, these problematic structures are neglected, and the emphasis is placed on the overall sequence that emerged.

To illustrate the type of analysis that was performed at individual localities, I present two examples of deformed ferrocarbonate concretions, because they show relationships between structures most clearly. One concretion is still parallel to bedding (Fig. 12) and therefore dips steeply, while the other (Fig. 13) has been rotated into low dips by shearing and has subsequently acted as a low-angle marker during further deformation. Other crosscutting patterns are also visible in Figures 7–10.

Deformation of a Bedding-Parallel Tabular Concretion (Fig. 12)

The first event that can be recognized in this exposure is E3 dextral folding of the concretion (southeastern part of Fig. 12A). This fold has an orientation similar to that of the main fold, deduced from bedding orientations over the whole area (Fig. 6). A sinistral shortening fault forms the acute southward terminations of the concretion ~50 cm to the NW and has been assigned to event E9 (main NE-striking faults and folds). It is possible that the dextral fault associated with the E3 fold is conjugate to the sinistral E9 fault. If so, this is an example of reactivation of earlier structures. The acute sinistral fault termination of the concretion appears twice on the map because it has been displaced by event E12 low angle, top-to-the-NE shearing (Fig. 12B), providing a slip vector for the E12 fault movement (Fig. 12A). A network of conjugate, dextral and sinistral, almost pure strike-slip faults (E14) are part of a system covering the entire shore platform, with offsets on the dextral faults consistently larger than those on the sinistral faults and shortening in an ENE-WSW direction. The latest structure is a NNW-striking, vertical strike-slip fault (E15) indicating a clockwise rotation of the shortening direction to a SE-NW orientation.

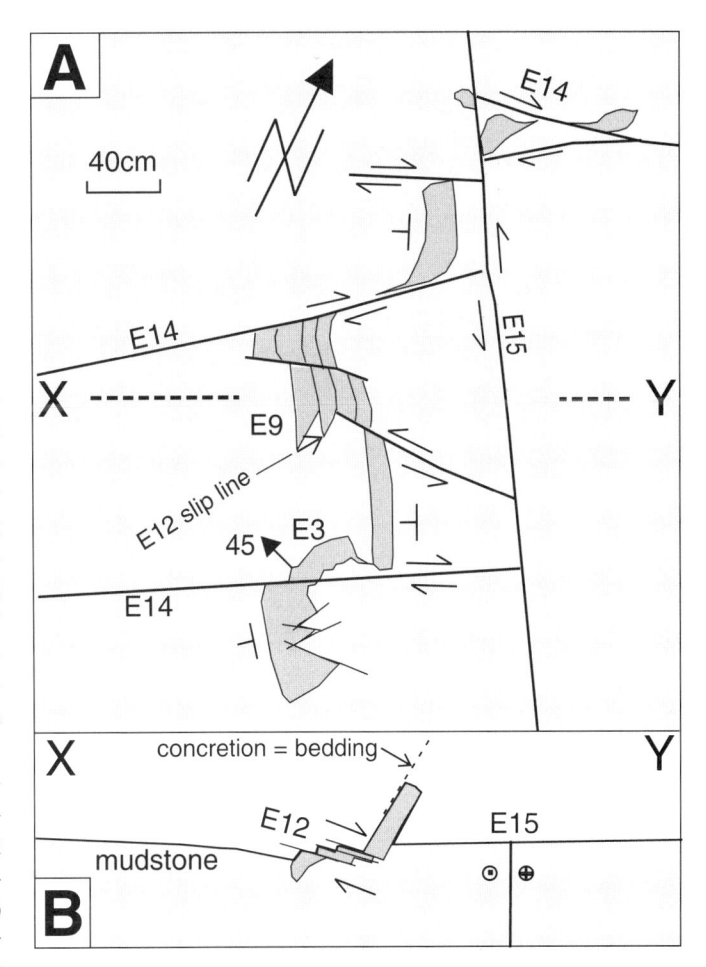

Figure 12. Deformation of a tabular, bedding-parallel concretion and labeling of deformation events. Exposure on a horizontal shore platform (for location, see Fig. 4). The concretion is shaded dark gray; Ngatuturi Claystone is light gray. (A) Plan view. The arrow with the 45° plunge indicates the axis of the earliest structure recognized here, an E3 dextral fold. The slip line of the E12 low-angle northeastward faulting was determined from the displacement of the intersection line of a sinistral E9 shortening fault with bedding. Strike-slip motion is dominant both in the NE-striking E14 and NW-striking E15 fault sets. (B) Cross section along line XY in A showing the northeastward displacement of the concretion during event E12.

The strike slip–dominated movement on the event E15 fault can be deduced from the identical spacing of the steeply dipping conjugate event E14 faults on either side of this later fault.

In Part Ductile Deformation of a Subhorizontal, Lensoid Concretion (Fig. 13)

This concretion lies immediately to the north of and has been influenced by movement on the major consolidated shear zone shown to the north of White Bluff in Figure 4 (this is also the southernmost fault shown in Fig. 13A). E9 sinistral shearing (also see Fig. 8) displaces a number of earlier fault sets, including E6 closely spaced faults (shown as markers for the E9 displace-

342   Spörli

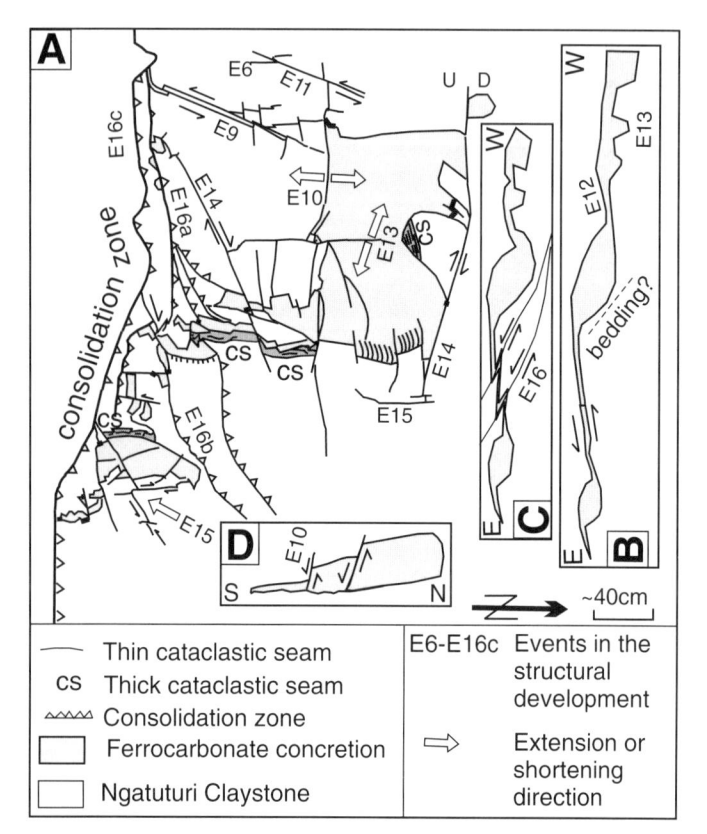

Figure 13. Deformation (in part ductile) of a subhorizontal, lensoid concretion and labeling of the sequence of events. See Figure 4 for location. (A) Map view of a horizontal surface. Only those faults of the dense network indicating clear relative ages (shown by E-indexes) or other structural information are shown. Asymmetric arrow couples indicate sense of movement on faults; D—down; U—up, indicating the vertical component of fault movement. (B) E-W cross section through concretions during the phase of ductile shearing (E12), which is indicated by the sigmoidal shear sense indicator shape that the concretion acquired. (C) The same E-W cross section after E16 thrusting. (D) N-S section through the concretion showing N-S extension and down-to-the-south faulting during event E10.

ment in the upper left hand corner of Figure 13A). Down-to-the-south normal faulting E10 (Fig. 13D) is followed by renewal of sinistral shearing (E11, top of Figure 13A). Ductile top-to-the-east shearing (in the present position) during event E12 reshaped the concretion into sigmoidal shear-sense indicators (Fig. 13B) as it was rotated from its original steep attitude into a lower dip. This was followed by normal conjugate faulting (E13) with an ESE-WNW extension direction, E14 conjugate strike-slip dominated faulting with an ~E-W shortening direction, and an E15 strike-slip–dominated conjugate system (lower left in Fig. 13A) with a NE-SW shortening direction. These three events indicate a change from extension to shortening in the E-W direction and back to extension again. Formation of the consolidated shear zone was the last stage. It can be divided into at least three subevents (E16A–C). It again involved top-to-the-east movement, this time associated with shortening (Fig. 13C), leading to tele-

scoping of the thin tails of the sigmoidal shear sense indicators in the map pattern (southeastern part of Fig. 13A). In their present orientation, the three consecutively formed consolidation zones E16a to E16c can be considered to be analogous to a leading imbricate fan, as known from thrust structures elsewhere (Boyer and Elliott, 1982).

Crosscutting Features in Other Illustrations

In Figure 7, two substages (local phases 1 and 2) in the development of the closely spaced fault set (E6) can be recognized. The crosscutting braided conjugate strike-slip faults are not precisely located in the sequence. Most likely they are minor structures on E14 (E- and ENE-striking dextral faults) major faults. In Figure 8, a thick black seam, part of the closely spaced fault set E9, is displaced by and folded by sinistral structures of the event E9 (main NE-striking sinistral faults and folds).

Figure 9 shows structures in an E17 southward thrust. Closely spaced earlier faults (E6) provide markers indicating the displacements. The consolidated fault zone in Figure 10 belongs to event E12 (brittle-ductile extensional shear) and was shortened during event E16 (eastward thrusting).

Overall Sequence of Deformation Phases

Figure 14 illustrates the orientation of the most important structures. Some of these are approximately in their correct position (e.g., deformation in the Waitemata Group); others are placed for convenience of illustrations (e.g., small block [E14–E15]) to the left of the main diagram).

The first event, E1, is not shown because it represents a group of structures. It has not yet been possible to positively identify the very first deformation to affect these rocks. E2 semi-ductile shearing (upper righthand corner of diagram) produced a lozenge fabric reminiscent of a broken formation in the sense of Raymond (1984). Its shear sense is shown as observed in the field. However, since it is not known whether the beds are overturned, the original sense cannot be reconstructed. The orientation of the slip vector during this deformation (theoretically perpendicular to the mean orientation of the lozenge axes) is also unknown; the axis orientation shown in Figure 14 is a gross simplification. E3 folds are usually dextral, but the vergence of the major structure controlling them is not known. Also, little is known about the fault sets E4 and E5. It is possible that E4 fault sets have slip vectors parallel to bedding and represent late shortening of the E3 folds.

The ~NE-SW–striking closely spaced faults (E6, middle of diagram) are the most penetrative structures of the whole sequence. It is tempting to interpret them as a spaced, fracture-type cleavage. By their orientation, they would be compatible with a large sinistral fold; however, they clearly pre-date the sinistral E9 folds (also see Fig. 8). One possibility is that both E3 and E9 folds are part of one and the same sinistral fold system, which had a protracted history of deformation, expressed in events E3 to E9. Event E10 (shown on E9 fold) represents down-to-the south

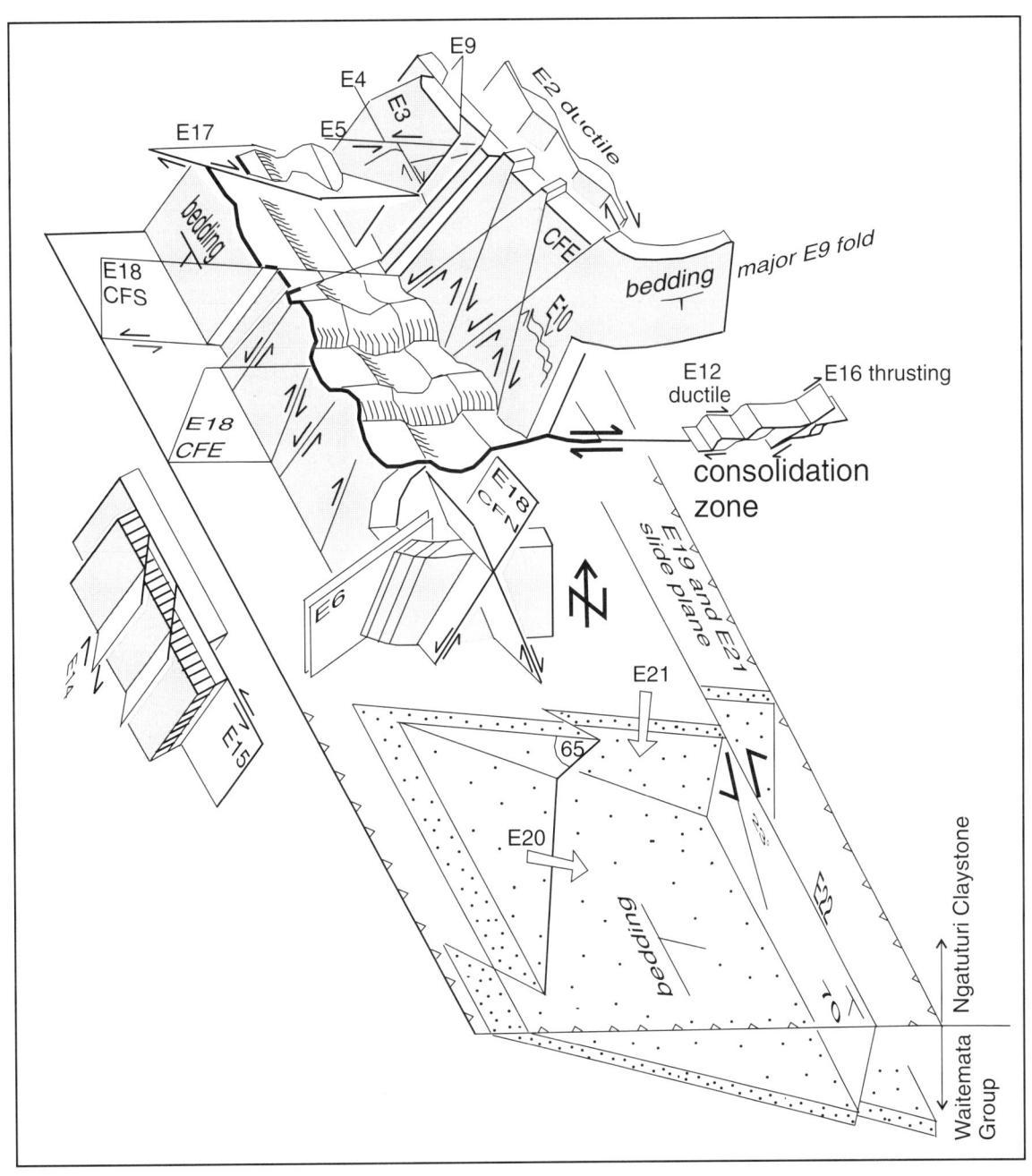

Figure 14. Geometry of the main structures in the study area, shown in a schematic block diagram (not to scale). Deformational events are identified by E-indexes. A low-angle fault (slide plane E19 and E21, identified by thrust symbols) separates the Ngatuturi Claystone block above from Waitemata Group (shaded) below. Structures shown in the Ngatuturi claystone mostly predate Waitemata Group deposition. The axis of the major sinistral E16 fold in the Ngatuturi Claystone has been simplified to a down-dip orientation for ease of graphic presentation but should be plunging NW (see Fig. 5). Portions of this fold have been cut away to facilitate illustration. The earliest structure shown is E2 extensional ductile dextral shearing (in the present orientation, top righthand corner of the diagram) causing a broken formation fabric. The following structures are also shown in previous figures: The E3 dextral fold and the E9 sinistral fold/fault in Figure 12; E6 closely spaced faults in Figures 7 and 8; E12 ductile eastward faulting and E16 thrusting in Figure 13, and E17 southward thrusting in Figure 9. CFS, CFE, and CFN are important late conjugate fault couples (E18). Some of these caused the folding shown in Figure 11. The number 65 in the Waitemata Group identifies the pitch of the refolded E20 fold axis in the E-W–trending sub-vertical Waitemata Group beds of the south-verging E21 fold (see Fig. 4 and corresponding text section); 23° is the slip vector pitch of the major sinistral E22 strike-slip dominated fault (LF in Fig. 4). Tilting of Ngatuturi Claystone to steep dips of bedding along a NW-SE–trending subhorizontal fold axis is assumed to have taken place between events E16 and E17. The geometric relationships of the pre-tilting structures can be visualized by looking at the figure from the righthand edge in "landscape" orientation.

faulting in the present orientation, or, with bedding returned to its horizontal attitude, a dextral rotation.

The consolidation zones are given prominence in the diagram (thick line) because they are also some of the most important structures in the field. They are the product of a number of events, in particular E12 ductile shearing and E16 thrusting (also see Fig. 13). The faults of events E14 and E15 (see small block diagram on the left side of Fig. 14) were active between these two events and display almost pure strike-slip movement in their present orientation (also see Fig. 12).

The folding of the consolidated fault zones (Figs. 11 and 14) is mostly due to the geometry of and displacement on three sets of conjugate faults which are, for the time being, all assigned to event E18: a strike-slip system (CFS in Fig. 14) with NE-SW shortening across the main bedding strike and two conjugate couples of normal faults, one (CFE) striking at high angle to and the other (CFN) striking subparallel to the main strike of the bedding. The relative age of these three sets is not known.

Southward thrusting during event E17 also produced consolidated fault zones, but these are less intensely developed (Fig. 9). For this event, eastward thrusting (E16), and the southward thrusting (E17), a few conjugate thrusts in the opposite direction have as well been found.

Structures affecting the Waitemata Group indicate initial overfolding and thrust movement of the whole block to the ESE during event E20, followed by overfolding and thrusting to the south in event E21. This caused the superposition of the Ngatuturi Claystone onto the Waitemata Group in the drill hole at Algies Bay (Fig. 3) on the slide plane shown in Figure 14. A progression from NNE (or NE) to E-W–trending fold-thrusts such as the one described above has been recognized elsewhere in the Waitemata Group (Spörli 1989b).

The latest structures are sinistral faults (E22), which cut both the breccia dikes and the clastic dikes filled with Waitemata Group sediment. These faults are marked by simple fractures without development of black cataclastic seams.

Events not especially mentioned above mostly represent adjustments of the rock mass (e.g., lateral extension or shortening) during or between more important events.

DISCUSSION: REGIONAL IMPLICATIONS AND TECTONIC REGIME

Tectonic Model (Fig. 14)

The following points need to be taken into consideration for any model: (1) the relatively late eastward thrusting during formation of the consolidated zones (E12–E16, Fig. 14; Table 1) is difficult to reconcile with the dominant southwesterly transport of the Northland Allochthon previously proposed (Spörli, 1982; Isaac et al., 1994; Rait, 2000); (2) the dominant northwesterly strikes and steep southwesterly dips of bedding in the Ngatuturi Claystone in the study area are anomalous for the Northland Allochthon (Spörli, 1982); (3) allowing tilting to steep dips to have occurred early in the scheme would assign all the presently sinistral shearing (E6 to E15) to a major strike-slip regime, for which there is no evidence elsewhere in the Northland Allochthon; and (4) as in the Mount Camel Terrane in northern Northland (Toy et al., 2002), it is easier to develop a tectonic model if initial eastward or northeastward thrusting, possibly in an accretionary regime (Bradshaw, 2004), is allowed.

A testable model that provides the most coherent picture results if the tilting is positioned after eastward shearing (E16) and before southward thrusting (E 17). Therefore, the main bedding attitude has to be regarded as subhorizontal until the end of E16 (Fig. 14). The stages of the model follow in subsequent sections.

Tectonic Stage I: Accretionary Prism Thrusting to NE at the Continental Edge of Northland (E2 and E3)

The structures of the E2 extensional shearing, resulting in blocks and lozenges, are reminiscent of a broken formation (Raymond, 1984; Toy et al., 2002) that then was folded coaxially by the presently dextral E3 folds. Preexisting E1 structures are probably also part of this regime and indicate that the accretionary prism had a complex initial history. The E5 and E6 structures may have been strike-slip faults in their original orientation and, although recorded in the field as a sequence, may in fact be a braided conjugate system with shortening approximately at a right angle to the E3 fold axes. They were either a late feature of stage I or an early feature of Stage II.

Tectonic Stage II: Allochthon Movement to SSW (Mainly E9 but Extending to E15)

Because of the number of events involved, the intensity of the E6 closely spaced faulting, the size range of the structures, and the extensive overlap with Stage III, this stage is considered to represent the main movement of the allochthon, dominated by SSW-verging folds. Although it could be geometrically explained as a late back-thrusting in the accretionary prism, this would apportion too much deformation to Stage I, which is rather difficult to detect in most other exposures of the Northland Allochthon.

Tectonic Stage III: Allochthon Movement to SW, Main Formation of Consolidated Fault Zones (Beginning at E12 but Extending to E16) and Subsequent Tilting to Steep Dips

Tilting of Ngatuturi Claystone to steep dips of bedding along a NW-SE–trending subhorizontal fold axis is assumed to have taken place between events E16 and E17. With bedding still subhorizontal, the E12/E16 faults initially acted as NW-SE–trending, NE-dipping normal faults formed as the Northland Allochthon sheet spread to the SW. The change from extensional to apparent shortening (Fig. 13) may represent a rotation of the major shortening axis form vertical toward the fault plane, as the motion acquired a thrust component. This thrusting culminated in folding on subhorizontal axes (possibly represented by the lesser π-circles of bedding poles in Fig. 5) and tilting of the entire sequence to steep dips. Sinistral shearing, a remnant of the movement to the SSW in Stage II, interfered with the southwestward movement and probably was accentuated by a displacement gradient toward the central axis of the Northland Allochthon. By now, so many defects had been formed in the Ngatuturi Claystone that the faulting could occasionally bundle itself into zones of preexisting weakness, leading to the formation of consolidated fault zones.

Tectonic Stage IV: Renewed Allochthon Movement to the South (E17)

These are the first structures that remain more or less in their original attitude. Consolidated fault zones continued to be formed (Fig. 9), but are less complex than those of Stage III, possibly because of some strain hardening. Event E18 strike-slip conjugate couple CFS (Fig. 14) with its NE-SW shortening may represent a remnant of Stage III. The two other conjugate couples (CFE and CFN) are extensional and represent further gravitational spreading of the allochthon.

Tectonic Stage V: Allochthon Movement Synchronous with Sedimentation of Miocene Waitemata Group (E19)

During this stage, the portion of the allochthon examined in the present study may have broached the sea floor for the first time, experiencing extremely low confining pressures and sliding over unconsolidated Waitemata sediments, leading to, first, intrusion of breccia dikes, then of clastic dikes filled with Waitemata Group. Both large blocks and finer debris were detached from the main mass, and some of the smaller fragments became rounded during transport to their site of deposition in the Waitemata basin.

Tectonic Stage VI: Allochthon Movement to the South, Late in the Sedimentation of the Waitemata Group, or Postdating It

The beach exposure at Algies Bay unequivocally demonstrates that subsequent thrusting (E20 and E21) affected allochthon pebble–bearing Waitemata Group beds that were already consolidated at the time of their deformation. The same structures affect a considerable thickness of Waitemata Group strata to the south of Algies Bay, indicating that these events took place under a considerable cover of Waitemata Group turbidites. It is unknown at present whether Waitemata sedimentation was still proceeding at the top of the sequence or had ceased. Sinistral NW-striking faults (E22) are kinematically compatible and may be directly associated with the E21 thrusting. Alternatively, they could be a separate later phase of yet unknown regional significance in the Waitemata basin.

DISCUSSION: GENERAL IMPLICATIONS

Deformation Mechanisms

These rocks have experienced an extraordinary amount of faulting. An initial deformation probably associated with early accretion (Stage I) provided the initial defects. While new faults were undoubtedly forming throughout the subsequent structural development, those already present continued to be reactivated. This may explain the common occurrence of Riedel shear patterns (e.g., Figs. 9 and 10).

In assessing the mechanical behavior of the Ngatuturi Claystone, the following points need to be taken in account: (1) absence of soft sediment deformation, (2) the mechanically isotropic nature of the Ngatuturi Claystone, (3) brittle versus ductile deformation during faulting in the Ngatuturi Claystone, and (4) absence of real cleavages.

(1) Absence of soft sediment deformation: The only soft sediment deformation seen in the Ngatuturi Claystone of the study area are very rare synsedimentary, bedding-parallel disruptions described earlier, in the sections on stratigraphy. In contrast to this, the deformations described in the present paper involve variously spaced fault arrays bounding blocks of undeformed claystone, indicating that the claystone was well lithified during these events. This is supported by the fact that the allochthon emplacement occurred in the late Oligocene–early Miocene, ca. 45 Ma after the deposition of these sediments in the Late Cretaceous. It can therefore be confidently stated that there was no soft-sediment deformation of the Ngatuturi Claystone during allochthon obduction.

(2) The mechanically isotropic nature of the Ngatuturi Claystone: At the mesoscopic scale, there is no refraction of faults or any sidestepping or other discontinuities across bedding. The Ngatuturi Claystone can therefore be considered to be mechanically isotropic at the outcrop scale. However, at the larger scale (see Fig. 14) and in the earlier deformations, bedding exercised some general control on fault orientation, but this disappeared in the later events because of the presence of numerous preexisting structural defects.

(3) Brittle versus ductile deformation in the Ngatuturi Claystone: The structural pattern involving intact blocks of Ngatuturi Claystone bounded by cataclastic faults mentioned under point (1) also demonstrates the brittle nature of the deformations (at the scale of the blocks). However, for event E12, ferrocarbonate concretions have been distorted into sigmoidal shear sense indicators (Fig. 13). The shape of these bodies indicates continuously distributed (i.e., ductile) deformation in a shear zone (e.g., Ramsay, 1967; Passchier and Trouw, 1996). There are two possible explanations for this:

(a) genuine ductile behavior even at the microscopic scale was caused by the pressure-temperature conditions in the rock mass; (b) local concentrations of meso- and micro-faults allows the material to mimic ductile geometry. Although an element of cause (b) may have influenced the structures and is definitely seen in other situations (e.g., Figs. 8 and 9), the development of long tails on the sigmoidal bodies (Fig. 13B) favors cause (a) where the presence of carbonate may have enhanced local ductility.

If the dense fault patterns allowed formation of ductile structures even at outcrop scale (Fig. 8), they must equally have enabled the entire slab of Ngatuturi Claystone to deform in a macroscopically ductile manner analogous to that described by Marques (2001) for clay samples. However, it is not possible at the moment to verify this because of the absence of markers traceable over long distances and the difficulty of mapping the shape of these blocks, given the lack of good inland exposure and the deep weathering in Northland.

(4) Absence of real cleavages: Although the E6 closely spaced faults have been likened to a cleavage, they do not have the microscopically penetrative fabric typical of cleavages found even in even low-grade metamorphic rocks (e.g., Ramsay, 1967; Hobbs et al., 1976). This suggests that in the E6 event the rocks may have behaved in a ductile manner at some scale above the spacing of the faults, but below that scale, brittle discontinuous deformation was the ruling pattern. On the other hand, the entire fault network in the Ngatuturi Claystone can be regarded as the equivalent, in an almost isotropic lithology, to the scaly foliation of even more clay-rich units (Pini, 1999) or the web structures of sandstones (Lucas and Moore, 1986) in mechanically anisotropic layered sequences.

In summary, Ngatuturi Claystone was in a lithified state during emplacement of the Northland Allochthon. Its deformation was brittle at the scale of the fault spacing, with the possible exceptions of the carbonate concretions. However, continuous, ductile-like deformation was facilitated at various scales and in a number of events by the presence of preexisting dense networks of faults. Further research is needed to decide whether the formation of these networks was due to seismic rupturing or to distributed aseismic creep (Blenkinsop and Sibson, 1992), although the overall geometry of the networks seems to favor the latter mechanism.

Fluid Regime

The conspicuous absence of veins in the Ngatuturi Claystone (except for some rare septarian veins within concretions and sparse calcite veins associated with the Waitemata Group clastic dike intrusions) indicates that fluids cannot have had a large influence on the deformation. Although fluids were no doubt present elsewhere in the stratigraphic column, no tensional cavities to create accessible permeability networks (Sibson, 1987) were sustained in the deforming claystone because any asperities on which these structures could form were immediately ground down due to the low strength of the claystone. It is likely that initial formation of the ferrocarbonate concretions was the last normal hydrologic event that these rocks saw before water entered them again in the Miocene through the Waitemata sandstone clastic dikes. Between these two events, any water extracted from the Ngatuturi claystone was efficiently removed from the formation by movement on the fault network. Such slabs of Ngatuturi claystone may have formed significant, dynamic, aquicludes in the structural architecture of the Northland Allochthon and may have played an important mechanical role in creating high pore fluid pressures around them during deformation.

Significance to Obduction Processes

The rocks described in the present paper have experienced the whole emplacement of the Northland Allochthon without acquiring a really pervasive cleavage. This indicates that they remained at very shallow levels throughout the obduction, a scenario that is supported by the observation of Aadil et al. (2001) that the rocks in the allochthon only show the effects of burial and were not affected by obduction. Although the fact that the youngest rocks in the allochthon, the Mahurangi limestones, display a well-developed, spaced, pressure solution cleavage (Spörli, 1982; Spörli and Kadar, 1989) may in part be due to the propensity of limestone for forming such cleavages, it may also indicate that the ophiolites were originally thrust onto or close to the Mahurangi limestone, and their juxtaposition onto the other allochthon rocks occurred later, under different, more shallow-level conditions.

The structural history of the Ngatuturi Claystone described herein, together with previous regional studies (e.g., Toy et al., 2002) suggest that southwestward obduction was preceded by an initial attempt at subduction in that direction, leading to formation of a short-lived accretionary prism in which the first structures were formed and causing the down-dragging of the Northland continental crust so that the rocks later emplaced by obduction did not have to surmount a significant step in elevation. Thus, the structural regime for the obduction was less collisional than would have been expected if the Northland crust had been at its present elevation.

Movements during emplacement were complex, involving multiple alternations between shortening and extension in and across the movement direction. Significant changes in this movement direction are suggested by the southwestward tilting of the main fold to its current attitude of a sinistral steeply plunging fold (Fig. 14), which indicates initial SSW movement followed by overfolding in a southwesterly direction. Further remobilization of the obducted mass occurred during and just after deposition of the Waitemata Group and involved initial movement to the ESE followed by movement to the south.

In response to obduction, Ngatuturi Claystone developed a unique style of deformation characterized by an almost evenly distributed network of cataclastic faults, allowing it to deform in a megascopically continuous, quasi-ductile manner. The nature of this faulting may have caused this formation to act as an aquiclude, facilitating high fluid pressures in adjacent units.

Significance to the Analysis of Structural Sequences in Rocks Deformed at Shallow Crustal Levels

It is hoped that the structures described in the present paper serve to demonstrate the difference between structural analysis in rocks at shallow levels in the crust and structural analysis in those buried at deeper levels. In the latter, the principles of structural analysis of metamorphic tectonites (e.g., Ramsay, 1967; Hobbs et al., 1976) can be applied, where (1) deformation is continuous and (2) homogenous over significant volumes of rocks, where, (3) explicitly stated or implied, a scale invariance of structures can be assumed (i.e., "minor structures represent the geometry of major structures"), and where (4) cross-cutting structures generally identify significantly distinct phases of deformation.

Most of these principles do not apply to structures in the Ngatuturi Claystone. (1) The faults clearly separate intact blocks of rocks, and therefore are discontinuities (although the fault mesh overall may mimic continuous deformation if looked at over larger dimensions). (2) At the scale of the fault spacings, the fabric is not homogeneous; however it may be homogenous seen over large volumes of rock. (3) The comments in points (1) and (2) clearly demonstrate that the properties of the deformation are not scale invariant. This leads to the case in which a structural fabric can be discontinuous and brittle at smaller dimensions but appears continuous and shows quasi-ductile deformation over larger dimensions. (4) The sheer number of deformational events recognized in the Ngatuturi Claystone suggests that many of these must be progressive steps within a few distinct phases of deformation; hence, the temporal transition between these events may be gradual rather than abrupt, causing difficulties for separating them rigorously. The common reactivation of preexisting structures also contributes to these problems.

For these reasons a "fuzzy" approach to the analysis of the time sequence was adopted by collecting a large number of good examples of crosscutting relationships of structures, then arranging them initially with the help of the most obviously correlatable structures and progressing further by using additional criteria such as nature and orientation of the fault surfaces. Invariably, a number of structures could not definitely be assigned to the time sequence that emerged. If they did not violate this sequence, they were neglected in the analysis and regarded as "noise" caused by the difficulties mentioned above.

An additional problem is that in rocks deformed at shallow levels, the main structures available are fault and joints. Unlike cleavage, which mostly occurs as one single set of parallel planar features, faults tend to form conjugate and even more complex (e.g., Riedel shear) arrays, so that a set belonging to one deformation phase can have two or more different orientations. This effect needs to be taken account when structures are correlated by their orientation. Furthermore, the faults in a conjugate couple not only must have the correct opposite displacement, they must also be proven to have moved simultaneously. This is best recognized by braiding of the two faults (e.g., Fig. 7). However, a further problem arises: within the braided structure, the two faults alternate in cross-cutting relationships; i.e., there is conflicting information on which is the younger fault. Therefore, in rocks deformed at shallow levels, beside the two classical relationships "older than…" and "younger than…," there is a third: "simultaneous with… (=conflicting crosscutting relationships)."

These comments illustrate that structural analysis of time sequences in shallow level rocks, while somewhat less straightforward than that in metamorphic tectonites, is well worth doing. However, development of the techniques is only in its infancy.

ACKNOWLEDGMENTS

Ray Price's work has been an inspiration to my students and to me by providing a sound foundation of structural principles for thrust belts. These helped us in our studies of the rather unorthodox geology of the New Zealand microcontinent. I salute the members of the 1993 graduate class in structural geology at Auckland University who fearlessly dealt with these complexities during some of the initial fieldwork at Snells Beach and Algies Bay. Thanks to Alison Sprott for diligent structural analyses of numerous oriented samples and helping to marshal the data during a summer scholarship held at the University of Auckland in 1994–1995. Svetlana Danilova assisted in attempts to look further into the fine structure of the cataclasites with the scanning electron microscope. Louise Cotterall helped with numerous versions of the figures. Andrea Pini provided useful comments in the field and the encouragement to write early drafts of this paper. Carol Evenchick, Michelle Markley, and an anonymous reviewer made numerous constructive suggestions that led to considerable improvement of the manuscript. Finally, I am grateful to Jim Sears for encouraging me to contribute this paper.

REFERENCES CITED

Aadil, N., Black, P.M., and Ballance, P.F., 2001, Interpretation of burial and diagenetic history of Northland Allochthon rocks, Northland Basin, New Zealand, in Hill, K.C., and Bernecker, T., eds., Eastern Australian basin symposium 2001; a refocused energy perspective for the future: Petroleum Exploration Society of Australia Special Publication, v. 1, p. 97–106.

Aydin, A., and Johnson, A.M., 1983, Analysis of faulting in porous sandstones: Journal of Structural Geology, v. 5, p. 19–31, doi: 10.1016/0191-8141(83)90004-4.

Ballance, P.F., and Spörli, K.B., 1979, Northland Allochthon: Journal of the Royal Society of New Zealand, v. 9, p. 259–275.

Bayrante, L.F., and Spörli, K.B., 1989, Structural observations in the autochthon and allochthon at Ngawha geothermal field, New Zealand, in Spörli, K.B., and Kear, D., eds., Geology of Northland—Accretion, allochthons and arcs at the edge of the New Zealand micro-continent: Royal Society of New Zealand Bulletin, v. 26, p. 183–194.

Blenkinsop, T.G., and Sibson, R.H., 1992, Aseismic fracturing and cataclasis involving reaction softening within core material from the Cajon Pass drill hole: Journal of Geophysical Research, v. 94, B4, p. 5135–5144.

Boudier, F., and Juteau, T., 2000, The ophiolite of Oman and United Arab Emirates: Marine Geophysical Researches, v. 21, p. 145–146, doi: 10.1023/A:1026754209262.

Boyer, S.E., and Elliott, D., 1982, Thrust systems: American Association of Petroleum Geologists Bulletin, v. 66, p. 1196–1230.

Bradshaw, J., 2004, Northland Allochthon: An alternative hypothesis of origin: New Zealand Journal of Geology and Geophysics, v. 47, p. 375–382.

Byrne, T., 1994, Sediment deformation, dewatering and diagenesis: Illustrations from selected mélange zones, in Maltman, A., ed., The geological deformation of sediments: London, Chapman & Hall, p. 239–260.

Caine, J.S., Evans, J.P., and Forster, C.B., 1996, Fault zone architecture and permeability structure: Geology, v. 24, no. 11, p. 1025–1028, doi: 10.1130/0091-7613(1996)024<1025:FZAAPS>2.3.CO;2.

Cladouhos, T.T., 1999, A kinematic model for deformation within brittle shear zones: Journal of Structural Geology, v. 21, p. 437–448, doi: 10.1016/S0191-8141(98)00124-2.

Cluzel, D., Aitchison, J., and Picard, C., 2001, Tectonic accretion and underplating of mafic terranes in the late Eocene intraoceanic fore-arc of New Caledonia (Southwest Pacific); geodynamic implications: Tectonophysics, v. 340, p. 23–59, doi: 10.1016/S0040-1951(01)00148-2.

Dresen, G., 1991, Stress distribution and the orientation of Riedel shears: Tectonophysics, v. 188, p. 239–247, doi: 10.1016/0040-1951(91)90458-5.

Ellis, S., Wissing, S., and Pfiffner, A., 2001, Strain localization as a key to reconciling experimentally derived flow-law data with dynamic models of continental collision, in Dresen, G., and Handy, M., eds., Deformation mechanism, rheology and microstructures: International Journal of Earth Sciences, v. 90, p. 168–180.

Groshong, R.H., 1988, Low-temperature deformation mechanisms and their interpretation: Geological Society of America Bulletin, v. 100, p. 1329–1360, doi: 10.1130/0016-7606(1988)100<1329:LTDMAT>2.3.CO;2.

Gurnis, M., Hall, C., and Lavier, L., 2004, Evolving force balance during incipient subduction: Geochemistry, Geophysics, Geosystems, v. 5, p. 7.

Hacker, B.R., and Gans, P.B., 2005, Continental collision and the creation of ultrahigh-pressure terranes; petrology and thermochronology of nappes in the central Scandinavian Caledonides: Geological Society of America Bulletin, v. 117, p. 117–135, doi: 10.1130/B25549.1.

Hancock, P.L., 1985, Brittle microtectonics; principles and practice: Journal of Structural Geology, v. 7, p. 437–457, doi: 10.1016/0191-8141(85)90048-3.

Hashimoto, Y., and Kimura, G., 1999, Underplating process from mélange formation to duplexing: Example from the Cretaceous Shimanto Belt, Kii Peninsula, southwest Japan: Tectonics, v. 18, p. 92–107, doi: 10.1029/1998TC900014.

Hayward, B.W., 1986, Onerahi Chaos Breccia at Mahurangi, Northland: New Zealand Geological Survey Record, v. 18, p. 41–47.

Hayward, B.W., 1993, The tempestuous 10 million year life of a double arc and intra-arc basin—New Zealand's Northland Basin in the early Miocene, in Ballance, P.F., ed., Sedimentary Basins of the World, Vol. 2: South Pacific Sedimentary Basins: Amsterdam, Elsevier, p. 113–142.

Hayward, B.W., 2004, Foraminifera-based estimates of paleobathymetry using Modern Analogue Technique, and the subsidence history of the early Miocene Waitemata Basin, New Zealand: New Zealand Journal of Geology and Geophysics, v. 47, p. 749–767.

Hayward, B.W., Brook, F.J., and Isaac, M.J., 1989, Cretaceous to middle Tertiary stratigraphy, paleogeography and tectonic history of Northland, New Zealand, in Spörli, K.B., and Kear, D., eds., Geology of Northland—Accretion, allochthons and arcs at the edge of the New Zealand micro-continent: Royal Society of New Zealand Bulletin, v. 26, p. 47–64.

Herzer, R.H., and Mascle, J., 1996, Anatomy of a continent-back arc transform—The Vening Meinesz Fracture Zone northwest of New Zealand: Marine Geophysical Researches, v. 18, p. 401–427, doi: 10.1007/BF00286087.

Hobbs, B.E., Means, W.D., and Williams, P.F., 1976, An outline of structural geology: New York, John Wiley, 571 p.

House, W.M., and Gray, D.R., 1982, Cataclasites along the Saltville thrust, U.S.A., and their implication for thrust sheet emplacement: Journal of Structural Geology, v. 4, p. 257–269, doi: 10.1016/0191-8141(82)90013-X.

Isaac, M.J., Herzer, R.H., Brook, F.J., and Hayward, B.W., 1994, Cretaceous and Cenozoic sedimentary basins of Northland, New Zealand: Institute of Geological and Nuclear Sciences Monograph, v. 8, 203 p.

King, P.B., 2000, Tectonic reconstructions of New Zealand: 40 Ma to the Present: New Zealand Journal of Geology and Geophysics, v. 43, p. 611–638.

Lin Aiming, 2001, S-C fabrics developed in cataclastic rocks from the Nojima fault zone, Japan and their implication for tectonic history: Journal of Structural Geology, v. 23, p. 1161–1178.

Lucas, S.E., and Moore, J.C., 1986, Cataclastic deformation in accretionary wedges: Deep Sea Drilling Project Leg 55, southern Mexico, and on-land examples form Barbados and Kodiak Islands: Geological Society of America Memoir 166, p. 89–103.

Marques, F.O., 2001, Flow and fracturing of clay; analogue experiments in bulk pure shear: Geological Society of America Memoir 193, p. 261–270.

Passchier, C.W., and Trouw, R.A.J., 1996, Micro-tectonics: Berlin, Springer-Verlag, 289 p.

Pini, G.A., 1999, Tectonosomes and olistostromes in the Argille Scagliose of the Northern Apennines, Italy: Geological Society of America Special Paper 335, 70 p.

Pollard, D.D., Muller, O.H., and Dockstader, D.R., 1975, The form and growth of fingered sheet intrusions: Geological Society of America Bulletin, v. 86, p. 351–363, doi: 10.1130/0016-7606(1975)86<351:TFAGOF>2.0.CO;2.

Rait, G.J., 2000, Thrust transport directions in the Northland Allochthon, New Zealand: New Zealand Journal of Geology and Geophysics, v. 43, p. 271–288.

Ramsay, J.G., 1967, Folding and fracturing of rocks: New York, McGraw-Hill, 567 p.

Raymond, L.A., 1984, Classification of mélanges, in Raymond, L.A., ed., Mélanges, their nature, origin and significance: Geological Society of America Special Paper 198, p. 7–20.

Sibson, R.H., 1987, Earthquake rupturing as a hydrothermal mineralizing agent: Geology, v. 15, p. 701–704, doi: 10.1130/0091-7613(1987)15<701:ERAAMA>2.0.CO;2.

Spaggiari, C.V., Gray, D.R., and Foster, D.A., 2004, Ophiolite accretion in the Lachlan Orogen, southeastern Australia: Journal of Structural Geology, v. 26, p. 87–112, doi: 10.1016/S0191-8141(03)00084-1.

Spörli, K.B., 1982, Review of paleostrain/stress directions in Northland, New Zealand and of the structure of the Northland Allochthon: Tectonophysics, v. 87, p. 25–36, doi: 10.1016/0040-1951(82)90219-0.

Spörli, K.B., 1989a, Tectonic framework of Northland, New Zealand, in Spörli, K.B., and Kear, D., eds., Geology of Northland—Accretion, allochthons and arcs at the edge of the New Zealand micro-continent: Royal Society of New Zealand Bulletin, v. 26, p. 3–14.

Spörli, K.B., 1989b, Exceptional structural complexity in turbidite deposits of the piggy-back Waitemata Basin, Miocene, Auckland/Northland, New Zealand, in Spörli, K.B., and Kear, D., eds., Geology of Northland—Accretion, allochthons and arcs at the edge of the New Zealand micro-continent: Royal Society of New Zealand Bulletin, v. 26, p. 183–194.

Spörli, K.B., and Harrison, R.E., 2004, Northland Allochthon infolded into basement, Whangarei area, northern New Zealand: New Zealand Journal of Geology and Geophysics, v. 47, p. 391–398.

Spörli, K.B., and Kadar, A., 1989, Structurally superposed sheets of Cretaceous–Tertiary sediments at Pahi, North Kaipara Harbour, New Zealand, in Spörli, K.B., and Kear, D., eds., Geology of Northland—Accretion, allochthons and arcs at the edge of the New Zealand micro-continent: Royal Society of New Zealand Bulletin, v. 26, p. 115–125.

Spörli, K.B., and Rowland, J.V., 2007, Superposed deformation in turbidites and syn-sedimentary slides of the tectonically active Miocene Waitemata Basin, northern New Zealand: Basin Research, v. 19, p. 199–216, doi: 10.1111/j.1365-2117.2007.00320.x.

Stern, R.J., 2004, Subduction initiation spontaneous and induced: Earth and Planetary Science Letters, v. 226, p. 275–292, doi: 10.1016/S0012-821X(04)00498-4.

Toy, V.G., Spörli, K.B., Black, P.M., and Nicholson, K.N., 2002, Mt. Camel terrane exotic source and new evidence for an accretionary precursor for the Northland Allochthon: Eos (Transactions, American Geophysical Union), Western Pacific Geophysics Meeting Supplement Abstract #SE41D-13, v. 83, p. 22.

Whattam, S.A., Malpas, J., Ali, J.R., Lo, C., and Smith, I.E.M., 2005, Formation and emplacement of the Northland ophiolite, northern New Zealand: SW Pacific tectonic implications: Journal of the Geological Society, v. 162, p. 225–241, doi: 10.1144/0016-764903-167.

MANUSCRIPT ACCEPTED BY THE SOCIETY 22 MARCH 2007

Printed in the USA

The Geological Society of America
Special Paper 433
2007

Strain rate in Paleozoic thrust sheets, the western Lachlan Orogen, Australia: Strain analysis and fabric geochronology

David A. Foster*
Department of Geological Sciences, University of Florida, Gainesville, Florida 32611, USA

David R. Gray[†]
School of Earth Sciences, University of Melbourne, Melbourne 3010, Victoria, Australia

ABSTRACT

Average orogenic strain rates may be calculated when it is possible to date mica cleavage or syndeformational veins and estimate finite strain. Deformation of accretionary-style thrust sheets in the western Lachlan Orogen occurred by chevron folding and faulting over an eastward propagating décollement. Based on ^{40}Ar/^{39}Ar dates of white micas, which grew below the closure temperature, this deformation started ca. 457 Ma in the west and ended ca. 378 Ma in the east, with apparent "pulses" of deformation ca. 440, 420, and 388 Ma. The ^{40}Ar/^{39}Ar data from thrust sheets in the Bendigo structural zone show that deformation progressed from early buckle folding, which started at 457–455 Ma, through to chevron fold lock-up and thrusting at 441–439 Ma. Based on retrodeformation, the total average strain for this thrust sheet is −0.67, such that the bulk shortening across the thrust sheet is 67%. This amount of strain accumulated over a duration of ~16 m.y. gives a minimum strain rate of 1.3×10^{-15} s^{-1} and a maximum strain rate of 5.0×10^{-15} s^{-1}, based on fan thickness considerations. The total shortening is between ~310 km and ~800 km, which gives a décollement displacement rate between ~19 mm yr^{-1} (minimum) and ~50 mm yr^{-1} (maximum). If deformation occurred in pulses ca. 457–455 and ca. 441–439 Ma, then the calculated strain rate would be on the order of 1×10^{-14} s^{-1}. These strain rates are similar to convergence rates in western Pacific backarc basins and shortening rates in accretionary prisms and turbidite-dominated thrust systems as in Taiwan.

Keywords: strain rate, Lachlan Orogen, ^{40}Ar/^{39}Ar thermochronology, strain analysis.

---

*dafoster@ufl.edu
[†]drgray@unimelb.edu.au

Foster, D.A., and Gray, D.R., 2007, Strain rate in Paleozoic thrust sheets, the western Lachlan Orogen, Australia: Strain analysis and fabric geochronology, *in* Sears, J.W., Harms, T.A., and Evenchick, C.A., eds., Whence the Mountains? Inquiries into the Evolution of Orogenic Systems: A Volume in Honor of Raymond A. Price: Geological Society of America Special Paper 433, p. 349–368, doi: 10.1130/2007.2433(17). For permission to copy, contact editing@geosociety.org. ©2007 The Geological Society of America. All rights reserved.

INTRODUCTION

In ancient orogens, quantifying strain rates and displacement rates has been more challenging than determining the timing of discrete geological events. Deformation within orogenic belts can be quantified as a strain rate ($e\cdot = e/t$, where e is the elongation, and t is the time in seconds) or as displacement rate (velocity) on structures or the deformation front. Short-term strain rates in active tectonic environments have been determined by geodic measurements using global positioning system (GPS) instrument arrays (e.g., Puntodewo et al., 1994; Bennett et al., 1999; Chang et al., 2003). In ancient orogens, however, determining rates of deformation has been problematical, because precise chronologies of fabric development are limited (Ramsay, 2000). Strain rates in inactive orogens have been estimated assuming that deformation pulses have an average duration of <5 m.y. (Pfiffner and Ramsay, 1982), and paleodisplacement rates on faults have been estimated by thermochronologic analyses (e.g., Foster and John, 1999; Wells et al., 2000; Brichau et al., 2006), but there are only a small number of direct estimates of internal deformation rates within the lower or middle crust of an orogen (e.g., Ramsay, 2000; Müller et al., 2000).

Accretionary orogens, or orogenic belts formed from the addition of material to continents from plate margin processes at convergent or transpressive plate boundaries, are one of the most important "factories" for producing and maturing continental crust (Şengör and Natal'in, 1996; Windley et al., 2001; Foster and Gray, 2000; Gray et al., 2007). Although the setting for accretionary orogens is relatively well understood, the rates of deformation within these orogens and in accretionary thrust systems is poorly defined. In this paper, we summarize data from the Lachlan Orogen of southeastern Australia, which is a Paleozoic, turbidite-dominated accretionary orogen (e.g., Gray and Foster, 2004). Our analysis of paleodeformation rates in the western Lachlan Orogen combines detailed geochronology of fabrics and faults with restoration of strain in metaturbidites. These data suggest that strain rates for the orogen were on the order of 10^{-14}–10^{-15} s^{-1} for regional-scale fold-thrust sheets.

GEOLOGIC SETTING

The eastern part of Australia formed along the Paleozoic margin of Gondwana due to accretion of crust of oceanic affinities, recycled crustal-derived turbidite, and oceanic volcanic arcs (Foster and Gray, 2000; Cawood, 2005). Eastern Australia is now >1000 km wide and consists of three distinct orogenic belts, the Delamerian, Lachlan/Thomson, and New England orogens, collectively referred to as the Tasmanides or the Tasman Orogenic Belt (Fig. 1) (Scheibner, 1978; Gray et al., 2006b). Accretion of these orogenic belts occurred in a stepwise fashion with an eastward decrease in the age of peak deformations from Cambrian through Triassic times.

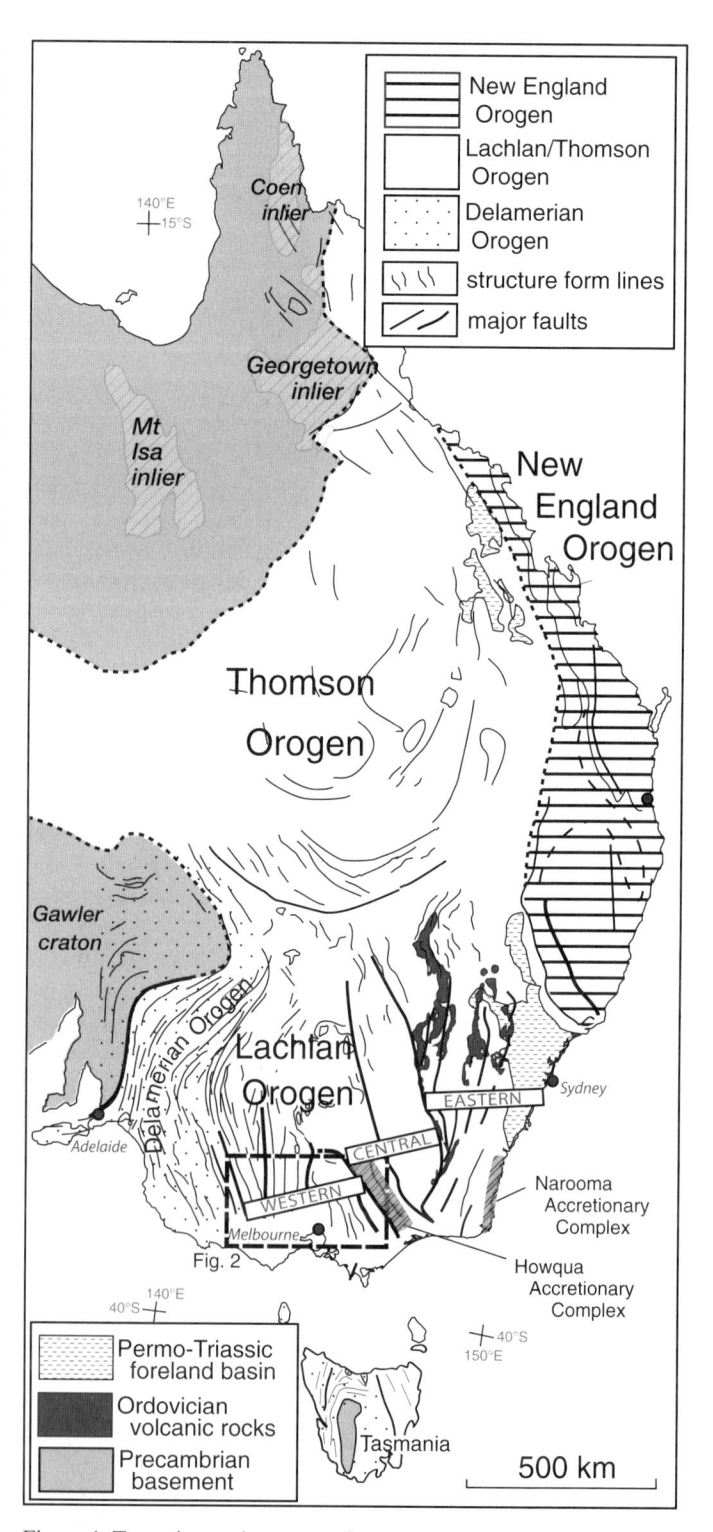

Figure 1. Tectonic province map of eastern Australia showing the orogenic belts of the Tasmanides. Western, central, and eastern refer to the three subprovinces of the Lachlan Orogen. The box outlined by the dashed line shows the area of the map in Figure 2.

Lachlan Orogen

The Lachlan Orogen is the central orogenic belt in the Tasmanides and consists of three subprovinces (Fig. 1) (e.g., Gray and Foster, 2004). The western and central Lachlan Orogen are dominated by a turbidite succession consisting of quartz-rich sandstones and black shales (Fig. 2) (VandenBerg et al., 2000). These are laterally extensive over 800 km present-width and have a current thickness upward of 10 km. The eastern Lachlan Orogen consists of andesitic volcanics, volcaniclastic rocks, and limestone, as well as quartz-rich turbidites and extensive black shale in the easternmost part (VandenBerg and Stewart 1992; Glen et al., 1998).

The turbidite fan was deposited within a Cambrian to early Ordovician oceanic backarc basin (Fig. 3) that was floored mainly by crust of oceanic affinity but may have also contained small continental ribbons and/or older rifted magmatic arcs (e.g., Foster et al., 2005). In the western province, Cambrian mafic volcanic rocks of oceanic affinities underlie the quartz-rich turbidite succession, whereas in the eastern province, the oldest rocks observed are Ordovician arc volcanic rocks and a late Cambrian–Early Ordovician chert-turbidite-mafic volcanic sequence. The three provinces of the Lachlan were juxtaposed along regional fault systems both within and bounding the central belt (Fergusson et al., 1986; Morand and Gray, 1991; VandenBerg and Stewart, 1992; Glen, 1992; Gray and Foster, 1998; Foster et al., 1999; Spaggiari et al., 2003, 2004a).

Closing of the Lachlan backarc basin caused Silurian to Devonian deformation of the turbidite fan and underlying mafic crust by underthrusting on both sides (Gray and Foster, 1997; Foster et al., 1999; Fergusson, 2003; Spaggiari et al., 2004b). This took place behind the major Pacific-Gondwana subduction zone, in a supra-subduction zone position (Foster et al., 1999; Collins, 2002). The thickest part of the sediment fan system, in the western Lachlan Orogen, was deformed into dominantly east-vergent structures over a west dipping décollement system associated with west-directed underthrusting and subduction (Fig. 3).

Major Structures in the Western Lachlan Orogen

The western Lachlan Orogen is composed of a deformed turbidite sequence cut by a series of major west-dipping, reverse faults that link to an inferred, gently west-dipping, mid-crustal décollement (e.g., Cox et al., 1991; Gray and Willman, 1991a, 1991b; Gray et al., 1991, 2006a, 2006b; Glen 1992, Fergusson and Coney 1992; Gray, 1995; Gray and Foster, 1998). Part of a composite, east-vergent, leading-imbricate fan fold- and thrust-belt, the reverse faults expose Cambrian mafic volcanics, cherts, and volcaniclastics in their immediate hanging walls (Gray and Willman, 1991b; Gray, 1995; Gray and Foster, 1998; Gray et al., 2006b). Thrust sheets largely consist of chevron-folded sandstone and mudstone layers that reflect up to 65% shortening above the mid-crustal décollement (Gray and Willman, 1991a, 1991b). Deep crustal seismic reflection profiles (Gray et al., 1991, 2006b; Korsch et al., 2002) and microseismicity studies (Gibson et al., 1981) indicate that the discontinuity is now at ~15–17 km depth (Fig. 2C).

The western Lachlan comprises three structural zones with variations in strike, style of deformation, and timing of deformation. These are, from west to east, the Stawell, Bendigo, and Melbourne zones (Fig. 2A). The western part of the Stawell zone is transitional to the Cambrian Delamerian Orogen through a series of faults between the Coongee and Moyston faults (Foster et al., 1999; Miller et al., 2005). Major strike-parallel faults in the structural zones have linear map traces from 25 to 100 km in length and generally dip at 60° to 70° west (Figs. 2A and 2B). Minor faults with throws of <100 m are abundant throughout the western Lachlan Orogen and show both dips to the east and to the west. NW-SE– and NE-SW–trending cross-faults, generally with strike displacements of meters to tens of meters, are also present within the zone (Gray and Foster, 1998).

Stawell Zone

The Stawell zone consists of Cambrian volcanic rocks (>4 km thickness) of tholeiite-boninite association and volcanogenic sediments overlain by unfossiliferous Cambrian-Ordovician, or most likely late Cambrian, quartz-rich turbidites with a distinct 500 Ma detrital zircon population (VandenBerg et al., 2000). West-dipping, high-strain zones separate regions of lower strain, characterized by symmetrical chevron folds and a single main cleavage (Wilson et al., 1992; Gray and Foster, 1998; VandenBerg et al., 2000; Gray et al., 2003).

Bendigo Zone

The Bendigo zone consists of an Ordovician turbidite package (~3–4 km original thickness), Upper Cambrian shales and cherts (0.9 km maximum original thickness), and the Mid- to Lower Cambrian volcanic and volcaniclastic rocks (2–2.5 km original thickness) (VandenBerg et al., 2000; Gray et al., 2003). The major faults dip west, juxtapose older rocks over younger rocks, and contain Lowermost Ordovician (Lancefieldian) strata in their immediate hanging walls. This suggests that fault propagation and detachment within the Lancefieldian strata was associated with an easterly transport of the folded and telescoped cover over the Cambrian metavolcanic succession. Fault-bounded slices of Cambrian rocks within the Heathcote fault zone (see Figure 3 *in* Spaggiari et al., 2004a), the leading fault, suggest duplexing and serial detachment in the Cambrian succession (Gray and Willman, 1991b). Inferred décollements developed at 4–6 km depth within the Lancefieldian strata (the Campbelltown, Muckleford, and Whitelaw faults are splays off this level) and at 7–10 km depth within the lower part of the Cambrian succession (Heathcote Fault Zone).

Melbourne Zone

The Melbourne zone consists of chevron-folded Silurian and Devonian interbedded mudstone and sandstone overlying Ordovician black shale and Cambrian andesitic lavas, agglomerates,

A

Stawell Zone | Bendigo Zone | Melbourne Zone | Tabberabbera Zone

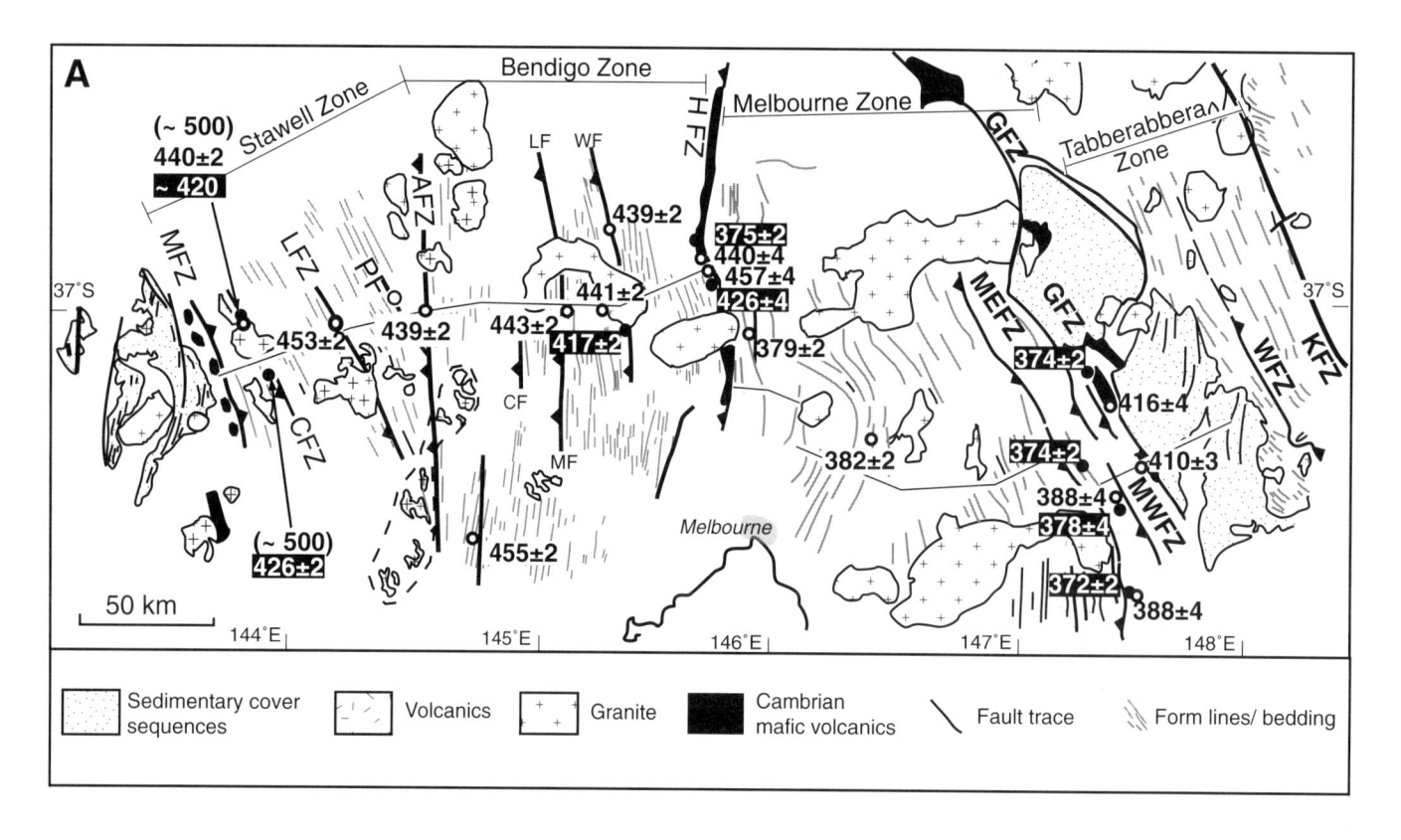

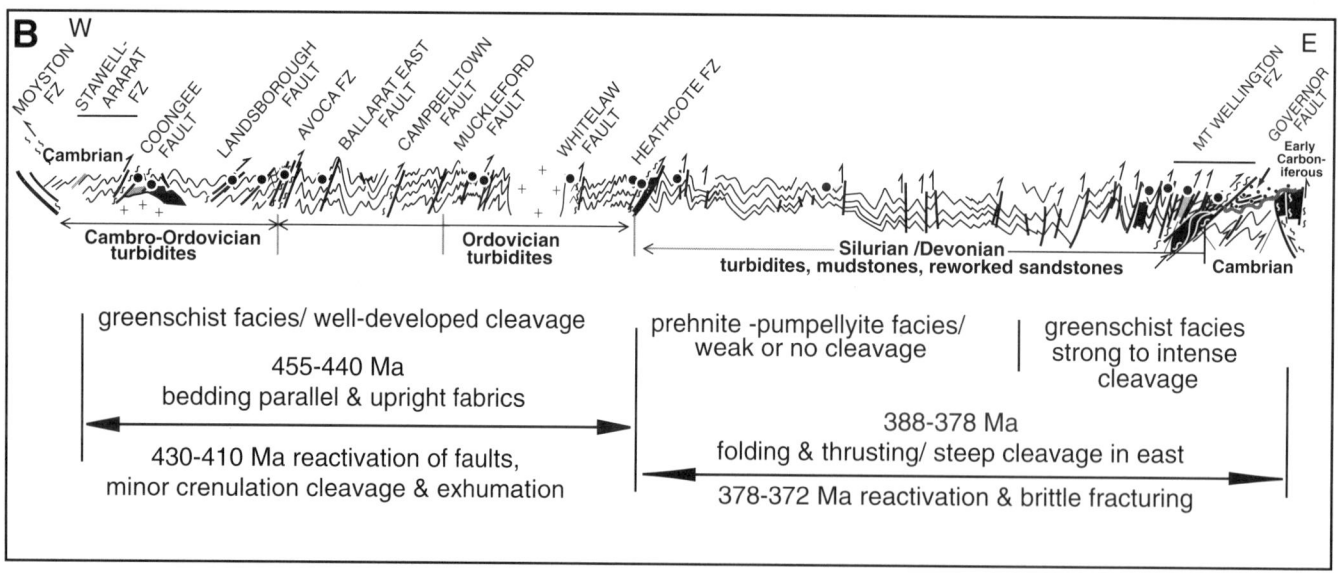

(~ 500) 440±2
~ 420
LF WF
439±2
375±2
440±4
457±4
426±4
453±2
439±2
441±2
443±2 417±2
379±2
374±2
416±4
382±2
374±2
410±3
388±4
455±2
378±4
372±2
388±4

MFZ, AFZ, LFZ, PF, CFZ, CF, MF, HFZ, GFZ, MEFZ, MWFZ, WFZ, KFZ

37°S

50 km

144°E | 145°E | 146°E | 147°E | 148°E

Melbourne

Legend:
- Sedimentary cover sequences
- Volcanics
- Granite
- Cambrian mafic volcanics
- Fault trace
- Form lines/ bedding

B

W — E

MOYSTON FZ | STAWELL-ARARAT FZ | COONGEE FAULT | LANDSBOROUGH FAULT | AVOCA FZ | BALLARAT EAST FAULT | CAMPBELLTOWN FAULT | MUCKLEFORD FAULT | WHITELAW FAULT | HEATHCOTE FZ | MT WELLINGTON FZ | GOVERNOR FAULT

Cambrian

Early Carboniferous

Cambro-Ordovician turbidites | Ordovician turbidites | Silurian/Devonian turbidites, mudstones, reworked sandstones | Cambrian

greenschist facies/ well-developed cleavage

455-440 Ma
bedding parallel & upright fabrics

430-410 Ma reactivation of faults, minor crenulation cleavage & exhumation

prehnite-pumpellyite facies/ weak or no cleavage

greenschist facies strong to intense cleavage

388-378 Ma
folding & thrusting/ steep cleavage in east

378-372 Ma reactivation & brittle fracturing

C

TWT (s)
0
4
8
12

MOHO

Cambrian andesite

?   ?

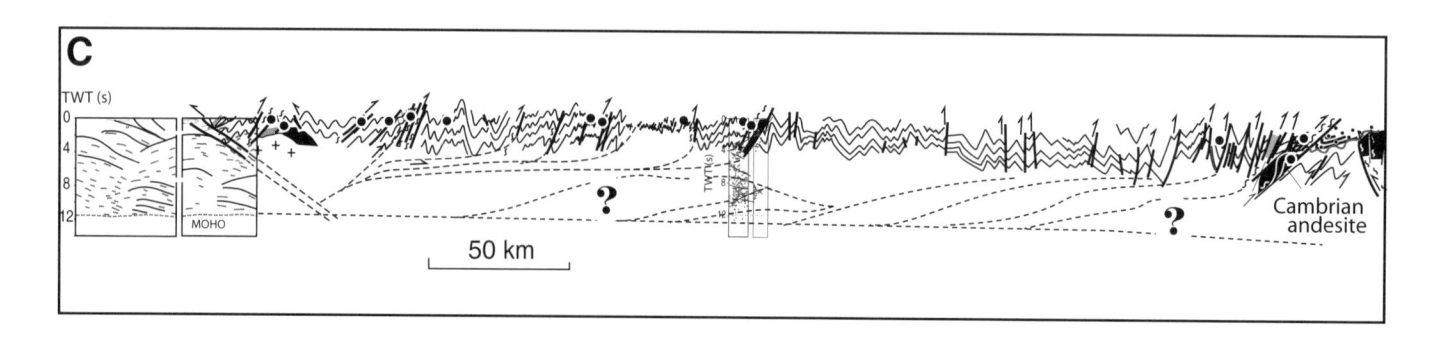

50 km

Figure 2. (A) Structural trend map and (B) structural profile, with degree of cleavage development and grade of metamorphism of the western the Lachlan Orogen incorporating the Stawell, Bendigo, and Melbourne zones (after Foster et al., 1999). Faults are shown as bold lines: MFZ—Moyston fault; CFZ—Congee fault; LFZ—Landsborough fault zone; PF—Percydale fault; AFZ—Avoca fault zone; BEF—Ballarat East fault; CF—Campbelltown fault; LF—Leichardt fault; MF—Muckleford fault; WF—Whitelaw fault; HFZ—Heathcote fault zone; MEFZ—Mount Easton fault zone; MWFZ—Mount Wellington fault zone; GFZ—Governor fault zone; WFZ—Wonnangatta fault zone. Circles show positions for $^{40}Ar/^{39}Ar$ samples with ages in million years and one-sigma errors: bold black type refers to primary mica cleavage or syntectonic vein mica age; white type on black shading refers to fault reactivation, or post-tectonic vein mica age. (C) Crustal cross section for the western Lachlan orogen based on surface exposures and geophysical data (from Gray et al., 2006b). TWT—two-way time.

boninites, pillowed and massive tholeiitic basalts, and gabbroic-doleritic intrusive rocks (VandenBerg et al., 2000; Gray et al., 2003). The Cambrian metavolcanic rocks occur as fault-bounded inliers within the structurally lowest part of the Mount Wellington fault zone and have a complex internally imbricated stratigraphy suggestive of the upper part of a duplex system (Gray and Foster, 1998; Spaggiari et al., 2003). Fault-bounded slices of Late Ordovician black mudstone and slate occur above the Cambrian inliers and are elongated subparallel to the regional strike with tapered ends (Gray, 1995; Gray and Foster, 1998).

Style of Folds, Faults, and Cleavage

Stawell Zone

Quartz-rich turbidites in the Stawell zone are tightly chevron-folded and cleaved and cut by major steeply (~60°) west-dipping NW-trending faults or high strain zones (Wilson et al., 1992; Cayley et al., 2002; Gray and Foster, 1998; Phillips et al., 2002; Miller et al., 2005). These include the Stawell-Ararat (incorporating the Mount Ararat, Cathcart, and Coongee faults), Landsborough, Percydale, St Arnaud, and Avoca fault zones, which are regularly spaced at ~15–20 km across the structural zone. The fault zones are characterized by broad (up to 2 km wide) zones of polydeformation with hanging-wall structures, including crenulation folds, crenulation cleavage, boudinage, and S-C fabrics indicating west-over-east displacement.

The variable geometry of the upright, north-trending chevron folds define large-scale synclinorial and anticlinorial closures (Cayley and MacDonald, 1995). Antiformal culminations are coincident with the hanging walls of the major faults where metamorphic grade is locally biotite zone. This succession has undergone 70%–85% shortening through regional folding and cleavage development, with an additional component due to thrust faulting within the sheet (Wilson et al., 1992; Cayley and MacDonald, 1995).

The consistent low-greenschist metamorphic grade indicates a flat enveloping surface for the tight folds, where the major faults have listric form, rooting into an underlying décollement. The lower (volcanic) parts of the stratigraphy are only brought to the surface on some of the major faults, including the Mount Ararat, Stawell-Ararat, and Avoca fault zones (see Figure 3 in Wilson et al., 1992).

Bendigo Zone

Regional anticlinoria and synclinoria in the Bendigo zone, as inferred from graptolite biostratigraphic zones, have wavelengths of 10–15 km and amplitudes on the order of 1–2 km (Cox et al., 1991; Gray and Willman, 1991a, 1991b; Gray et al., 2006a). Folding consists of extremely regular parallel trains of gently plunging, upright chevron folds with their axial surface traces spaced at ~8 per kilometer. They show several orders of folding, with lower order fold-wavelengths ranging from 100 to 300 m and fold-amplitudes from 50 to 100 m (Gray and Willman, 1991a, 1991b; Fowler and Winsor, 1996). Strike-lengths of the parallel chevron folds are on the order of tens of kilometers. The folds have typical chevron form with long, straight limbs and narrow hinge zones. Hinge zones are rounded in sandstone-dominant facies to a more angular form in mudstone-dominant sections (Fowler and Winsor, 1996). Limb-thrusts, bedded veins, and quartz saddle reefs are relatively common (Cox et al., 1991; Gray et al., 1991; Jessell et al., 1994; Fowler, 1996; Fowler and Winsor, 1996, 1997).

Cleavage varies from divergent-fanning slaty cleavage or primary crenulation cleavage (White and Johnston, 1981) in pelites to a convergent-fanning spaced cleavage in sandstones (Yang and Gray, 1994). It consists of thinly spaced zones of subparallel white mica and chlorite alternating with inter-cleavage zones of less-ordered quartz and white mica aggregates (Glasson and Keays, 1978; Stephens et. al., 1979; Gray and Willman, 1991a, 1991b; Yang and Gray, 1994). Abundant evidence of dissolution is shown by truncated quartz grains along cleavage, stripy differentiated layering, pressure shadows, and mica beards (cf. Glasson and Keays, 1978; Stephens et al., 1979; Waldron and Sandiford, 1988; Yang and Gray, 1994). Cleavage in pelitic layers at the structurally lowest levels of individual thrust sheets, or in the immediate hanging walls to major faults, is more intense with closer spacing of mica seams and low (<20°) bedding-cleavage angles such that slates have a distinct phyllitic sheen (Gray and Willman, 1991a, 1991b).

Major intra-zone faults of the Bendigo zone include the Campbelltown, Leichardt, Muckleford, Sebastian, and Whitelaw faults. The faults have N-S trends, spacing of ~20 km, steep (>60°) westerly dip, and throws on the order of 1–2 km (Gray and Willman, 1991a, 1991b; Gray and Foster, 1998). These faults generally place lowermost Ordovician (Lancefieldian) over either mid-Early Ordovician (Chewtonian-Castlemainian) or latest Early Ordovician (Darriwilian-Yapeenian) rocks. Chevron folds in the turbidite succession of their hanging walls show increased fold tightness (<30°) and markedly higher strains ($X/Z > 9:1$), as well as a change in axial surface dip direction within 3 km of the fault trace (Table 1) (see Figure 6 in Gray and Willman, 1991b).

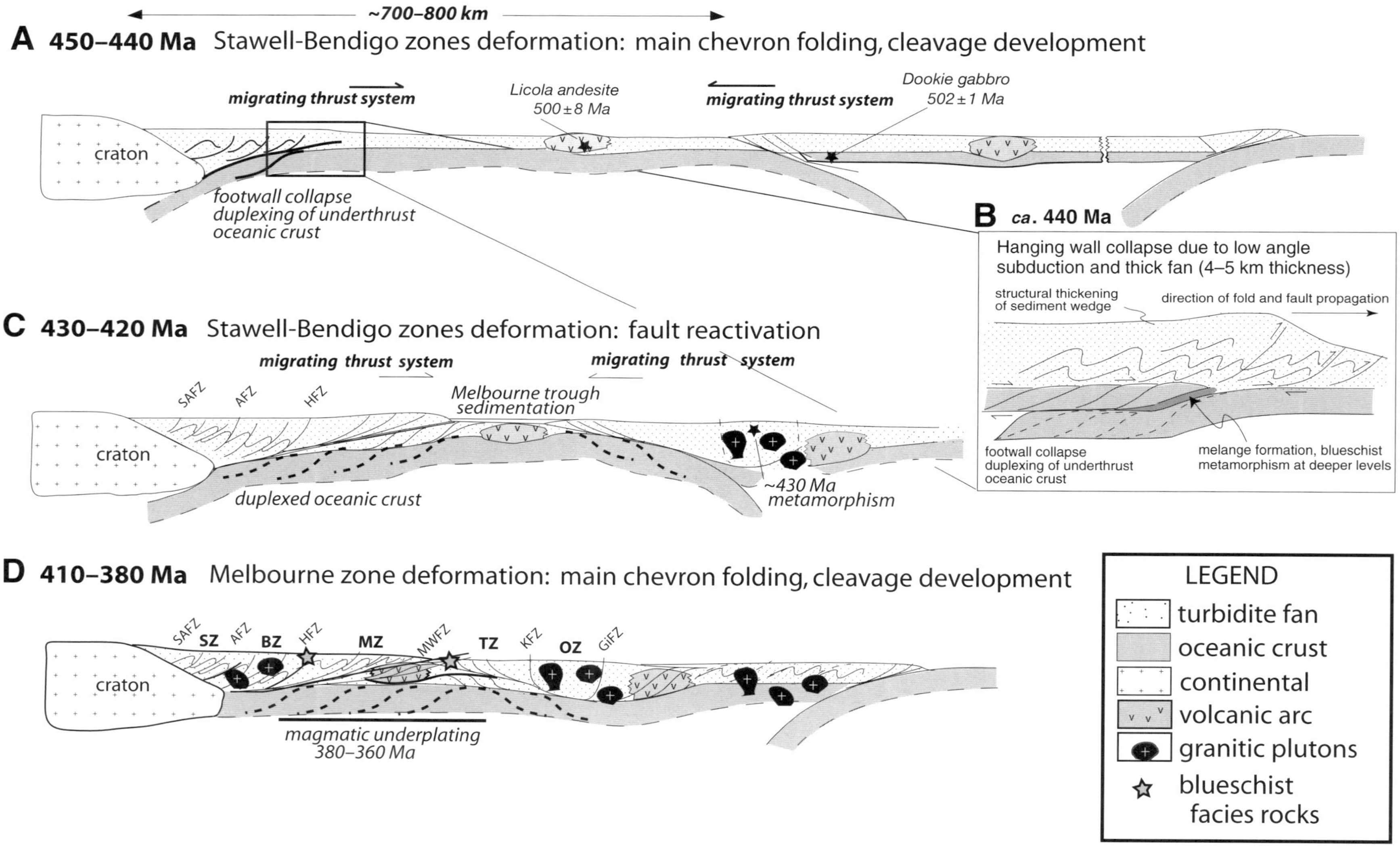

Figure 3. Schematic cartoons depicting the tectonic setting and development of the western Lachlan Orogen from Late Ordovician through Devonian times (modified from Foster and Gray, 2000; Spaggiari et al., 2004b). AFZ—Avoca fault zone; BZ—Bendigo zone; GFZ—Governor fault zone; HFZ—Heathcote fault zone; KFZ—Kiawa fault zone; MWFZ—Mount Wellington fault zone; MZ—Melbourne zone; OZ—Omeo zone; SAFZ—Stawell-Ararat fault zone; SZ—Stawell zone; TZ—Tabberabbera zone.

TABLE 1. THRUST WEDGE COMPONENTS, WESTERN LACHLAN OROGEN

	WESTERN BELT			EASTERN BELT	
W	Stawell Zone	Bendigo Zone	Segment B	Melbourne Zone	
	Segment A			Segment C	Segment D E
				Gooseneck	Mount Wellington
				synclinorium	Fault Zone
Age of sediments	Cambrian	Ordovician	Silurian, Devonian	Silurian, Devonian	Late Ordovician, Silurian
Deformation style	chevron folds, cleavage, reverse faults	chevron folds, cleavage, reverse faults	open chevron folds	close to tight chevron folds, cleavage	transposition layering, isoclinal folds
Fold interlimb angle					
Thrust sheet	30°–60°	30°–60°	60°–100°	50°–70°	
Basal fault zone	>30°	>30°			>30°
Background X/Z strain	(4:1)	4:1	zero	~1.4:1	~10:1
Maximum X/Z strain	undetermined	~34:1	—	1.4:1	~15:1
Percent shortening	(~66%)	~66%	~34 %	~40%	~69%
Deformed width	~62 km	~96 km	~85 km	~40 km	~16 km
Basal fault zone					
Lithology	metavolcanic rocks	Lower Ordovician turbidites	—	—	Ordovician-Silurian turbidites
X/Z strain	>15:1	34:1			15:1
Metamorphism	epizone	epizone	anchizone	epizone	epizone

Melbourne Zone

The Melbourne structural zone shows a transition from open, upright folds with weak or no cleavage to inclined and overturned, tight to isoclinal folds with strong axial surface cleavage, into a 10–15-km-wide zone of intense deformation as part of the Mount Wellington fault zone (Gray, 1995; Gray and Foster, 1998). Major anticlinoria and synclinoria are spaced at ~20 km intervals and have wavelengths of ~40 km and amplitudes of ~4 km. Regionally, the folds tend to be upright, open to tight chevron folds cut by steeply dipping reverse faults and have 70°–100° fold interlimb angles (Table 1) showing either a weak reticulate fissility or no cleavage fissility at all (Gray, 1995; Gray and Mortimer, 1996). The amount of internal strain recorded by regional folding, based on arc length determinations, and penetrative fabric development in the higher strained areas (see below), approximates ~32% shortening (Gray, 1995). Where cleavage is more strongly developed, the folds are tight, have more regular interlimb angles ranging between 60° and 70°, and reflect a shortening of ~53%. Where cleavage is intense and folds are cut by significant numbers of faults, the fold interlimb angles range between 18°–42°, and shortening approaches 64%. Within the Mount Wellington fault zone, narrow elongate fault-bounded slivers of Cambrian metavolcanic rocks and Ordovician black slate, up to 20 km in strike-length and 4 km outcrop width, occur associated with strongly foliated phyllonite (Gray, 1995; Gray and Foster, 1998).

P-T Conditions of Metamorphism and Deformation

The eastern boundary of the complex transition zone between the Delamerian and Lachlan orogenic belts (Korsch et al., 2002; Miller et al., 2005) is gradational into a wide zone of greenschist facies turbidites metamorphosed under medium-pressure conditions (Offler et al., 1998; Phillips et al., 2002). Exposed rocks in the eastern Stawell zone and the northwestern part of the Bendigo zone were metamorphosed to biotite zone conditions, with a biotite-muscovite-quartz-albite assemblage in pelitic rocks (Morand et al., 1995). Greenschist facies metabasalts along the Avoca fault zone are composed of actinolite-chlorite-epidote-albite-sphene or chlorite-sericite-albite ± carbonate assemblages. In most of the Bendigo zone, the metamorphic grade is chlorite zone with some subgreenschist facies (anchizone) rocks in the east (Offler et al., 1998).

Metamorphic pressure for the low-grade rocks (apart from contact zones of Devonian granitic intrusions) in the Stawell and Bendigo zones, based on lattice parameter (b_0) measurements of phengitic micas, was ~0.4 GPa, assuming a temperature (T) of ~360 °C, indicating an intermediate pressure (P) series (Offler et al., 1998). The Fe/Mg compositions of coexisting actinolite and chlorite from metabasalt in the Avoca fault zone indicate moderately high-P metamorphism (Offler et al., 1998). Most of the Heathcote fault zone metabasalt is prehnite-pumpellyite facies, with assemblages of albite-chlorite-prehnite-pumpellyite-titanite in the tholeiitic rocks (Spaggiari et al., 2002; Crawford et al., 2003). Metamorphic grade increases up to lower greenschist facies toward the base of the Heathcote fault zone (Spaggiari et al., 2002; Crawford et al., 2003). Blueschist facies metavolcanic rocks occur as blocks in serpentinite-matrix mélange in the central segment, where deeper structural levels are exposed (Spaggiari et al., 2002). They consist of winchite (Na-Ca blue amphibole) + albite + stilpnomelane + chlorite + Mg-Cr spinel, with accessories titanite + quartz ± apatite ± talc, which give estimated pressures and temperatures of 0.6–0.7 GPa and <450 °C and formed ca. 455–440 Ma (Spaggiari et al., 2002).

Rocks in the west of the Melbourne zone are unmetamorphosed (diagenetic zone) (Offler et al., 1998), and those in the east are greenschist grade. There is a transition from uncleaved or weakly cleaved in the west and north (Gray and Mortimer 1996) to strongly cleaved in the east in the Mount Wellington fault zone, where slate and phyllite are the dominant rocks (VandenBerg et al., 1995; Gray, 1995). Cambrian mafic to intermediate volcanic rocks in the Mount Wellington fault zone show pumpellyite-actinolite and lower greenschist facies assemblages (VandenBerg et al., 1995). Chlorite-actinolite assemblages defining foliated margins and shear zones within fault slices of metavolcanic rocks indicate temperatures of 350–400 °C and pressures of 0.2–0.4 GPa during deformation and internal fault duplexing, suggesting that the décollement initiated at a depth of 7–12 km.

Chronology of Deformation

The geochronology of metamorphism and deformation in the western Lachlan has been defined mainly by ^{40}Ar/^{39}Ar dating of white mica growth in the low- to intermediate-grade metamorphic rocks (Foster et al., 1996a, 1996b, 1998, 1999; Bierlein et al., 1999, 2001; Gray et al., 2003; Miller et al., 2005). The data come from several types of structures and lithologies within the thrust wedges, including mica concentrates and whole-rock analyses of mica-rich cleavage bands, mica separates, single grains from syndeformational quartz veins, and mica separates and single grains from fault fibers. Within the Bendigo and Stawell structural zones, data from the different structures are very consistent. The data from the Melbourne zone are more variable in quality due to the relatively lower grade of metamorphism and deformation outside of the frontal fault zone.

At low greenschist and prehnite-pumpellyite facies *P-T* conditions, metamorphic neocrystalline white mica grows at or below the argon closure temperature of ~330–400 °C. The ages obtained from white mica grown in these conditions, therefore, reflect the timing of deformation-induced crystallization or recrystallization (e.g., Dunlap et al., 1991, 1997; Kirschner et al., 1996; Foster et al., 1999; Mulch and Cosca, 2004). In the metaturbidites of the western Lachlan, there is a distinct mica-preferred orientation related to fabric development during thrust-sheet emplacement and to cleavage formation during folding (White and Johnston, 1981; Cox et al., 1991; Yang and Gray, 1994; Tan et al., 1995). The timing of deformation can be determined by isotopic dating of the metamorphic mica when samples contain only neocrystalline white mica. This requirement is met in some, but not all, of the low-grade rocks. Rocks with very high strains have the best chance of complete recrystallization and therefore most applications of this dating method for the western Lachlan orogen have concentrated on the fault zones. At higher metamorphic grades, and for muscovite growing in syntectonic quartz veins, individual grains of metamorphic white mica grow to larger sizes and are easily extracted and measured

Figure 4. Time-space diagram for the western Lachlan Orogen (modified from Gray and Foster, 2004). References for geochronology data (inset box): [1]Foster et al. (1998, 1999), Spaggiari et al. (2003), Miller et al. (2005); [2]Foster et al. (1996a, 1998, 1999), Bierlein et al. (1999, 2001); [3]Arne et al. (1998), Foster et al. (1996b), Bucher (1998), Spaggiari et al. (2003). bt—biotite; hb—hornblende; LFZ—Landsborough fault zone; MWFZ—Mount Wellington fault zone.

as pure phases, thereby removing potential mixed-ages caused by detrital mica. Complications arise when analyzing very fine grained, <10 micron-sized mica, which include recoil redistribution of ^{39}Ar during sample irradiation. Metamorphic mica grains for most of the greenschist facies rocks in the Bendigo and Stawell structural zone are generally large enough (>10–50 microns) that recoil has not been a problem. This is not the case with samples from parts of the Melbourne structural zone, where the finer-grain size resulted in many samples yielding discordant age spectra as a result of ^{39}Ar recoil.

The results summarized in this section are based upon data presented Foster et al. (1996a, 1996b, 1998, 1999), Bucher (1998), Bierlein et al. (1999, 2001), Spaggiari et al. (2003), and Miller et al. (2005). A summary of the results is given in Figure 4 and locations of selected samples and their ages are shown the map and cross section in Figure 2.

Stawell Zone

Ordovician ^{40}Ar/^{39}Ar plateau ages of ca. 453–439 Ma are recorded by cleavage mica from the Stawell-Ararat fault zone and Landsborough fault, and by sericite from syntectonic quartz veins in the Stawell gold mine (Foster et al., 1999; Bierlein et al., 2001; Miller et al., 2005). These late Ordovician to early Silurian white mica dates record the peak of metamorphism and tectonic shortening. Minimum ages for low-temperature release steps from the white micas in the Stawell zone range from ca. 426 to ca. 405 Ma and are related to fault reactivation, exhumation during Silurian to Early Devonian times, and heating by ca. 400 Ma post-tectonic plutons (Foster et al., 1996a; Arne et al., 1998; Bucher, 1998).

Bendigo Zone

The oldest ^{40}Ar/^{39}Ar plateau ages (ca. 457–455 Ma) for samples in the Bendigo zone are given by sericite from early quartz veins within thrust faults at Ballarat East (Foster et al., 1998; Bierlein et al., 1999) and cleavage mica from early fabrics in the Heathcote fault zone (Fig. 5) (Foster et al., 1999). Sericite samples from early laminated quartz veins in the Bendigo gold mine give ages ca. 445–442 Ma (Fig. 6) (Bierlein et al., 2001). Most samples from the Bendigo zone that show well-developed subvertical cleavage give ^{40}Ar/^{39}Ar plateau ages of ca. 440 Ma, including (1) phyllite and slate samples from the Avoca fault zone (ca. 439 Ma) (Fig. 7); (2) white mica from bedding-parallel quartz-mica fibers within the hanging wall of the Whitelaw thrust sheet (ca. 444 Ma) (Fig. 8); and (3) sericite from syntectonic quartz veins in the Central Deborah Mine and Wattle Gully Mine

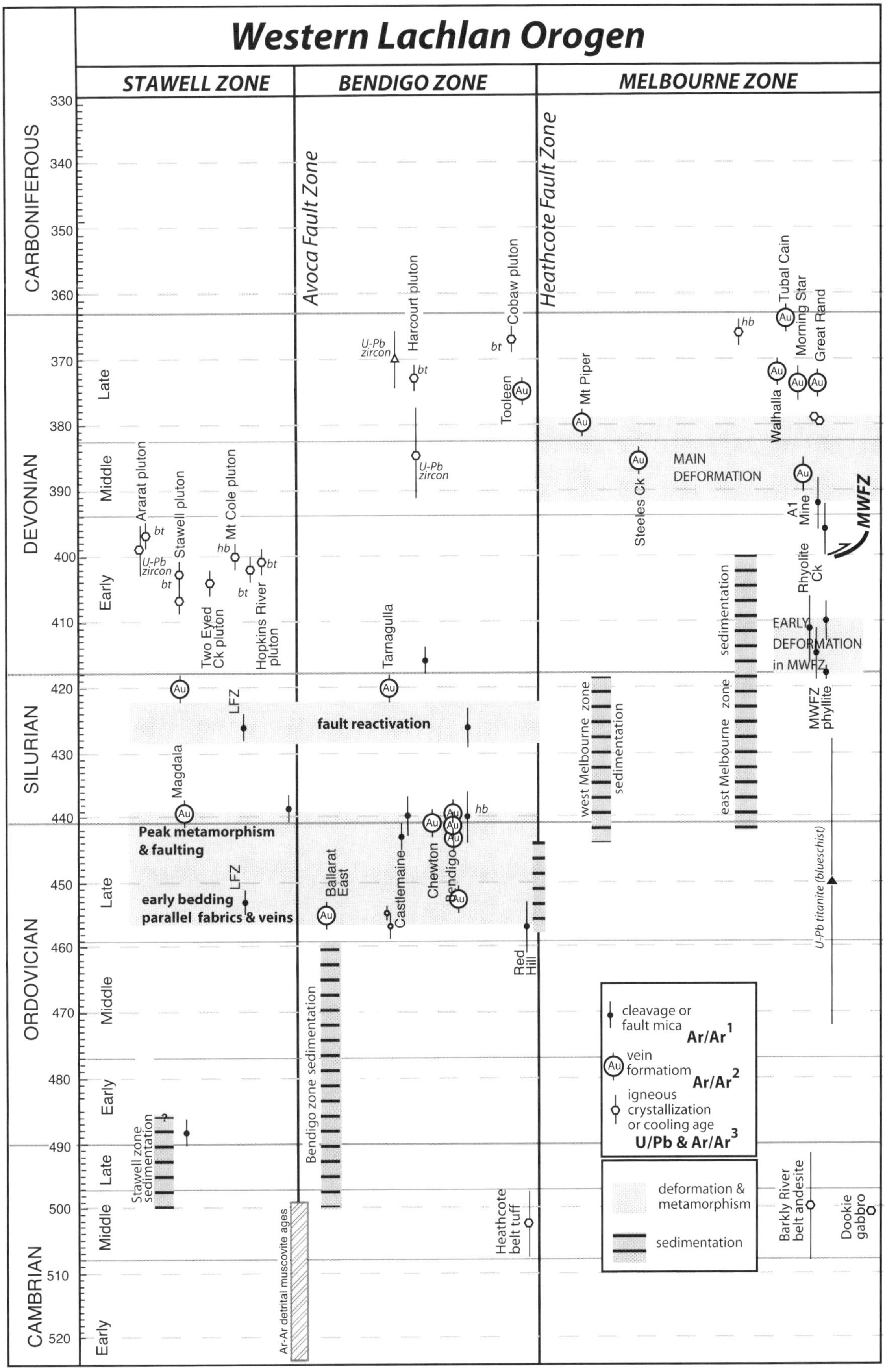
Western Lachlan Orogen
STAWELL ZONE
BENDIGO ZONE
MELBOURNE ZONE
CARBONIFEROUS
DEVONIAN
SILURIAN
ORDOVICIAN
CAMBRIAN
Late
Middle
Early
330
340
350
360
370
380
390
400
410
420
430
440
450
460
470
480
490
500
510
520
Avoca Fault Zone
Heathcote Fault Zone
Ararat pluton
Stawell pluton
Mt Cole pluton
Two Eyed Ck pluton
Hopkins River pluton
U-Pb zircon
bt
hb
LFZ
Magdala
Peak metamorphism & faulting
early bedding parallel fabrics & veins
fault reactivation
Stawell zone sedimentation
Harcourt pluton
Cobaw pluton
Tooleen
Tarnagulla
Ballarat East
Castlemaine
Chewton
Bendigo
Bendigo zone sedimentation
Ar-Ar detrital muscovite ages
Red Hill
Heathcote belt tuff
Mt Piper
Steeles Ck
MAIN DEFORMATION
Walhalla
Tubal Cain
Morning Star
Great Rand
A1 Mine
MWFZ
Rhyolite Ck
EARLY DEFORMATION in MWFZ
MWFZ phyllite
west Melbourne zone sedimentation
east Melbourne zone sedimentation
U-Pb titanite (blueschist)
Barkly River belt andesite
Dookie gabbro
cleavage or fault mica Ar/Ar[1]
vein formation Ar/Ar[2]
igneous crystallization or cooling age U/Pb & Ar/Ar[3]
deformation & metamorphism
sedimentation

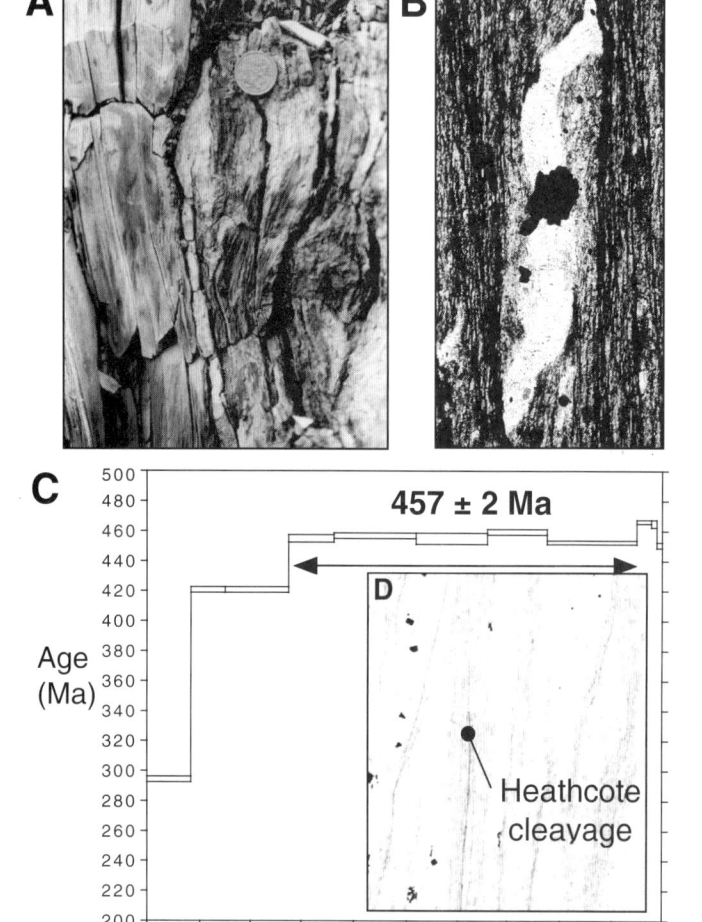

Figure 5. (A) Outcrop photo showing isoclinal folds in a quartz vein and strongly transposed fabric from the hanging wall of the Heathcote fault zone (Australian 20-cent coin for scale). (B) Photomicrograph (width is 800 microns) from the Heathcote fault zone showing curved pressure shadow on pyrite in slate. (C) ^{40}Ar/^{39}Ar age spectrum from a sample of cleavage mica from the Heathcote fault zone (Foster et al., 1999). Inset (D) photomicrograph (width is 10 mm) of the location of the cleavage micas analyzed to produce the age spectrum.

(ca. 440–439 Ma) (Foster et al., 1998, 1999). Cleavage mica from other fault zones in the Bendigo zone, including the Muckleford fault, also give plateau dates of ca. 440 Ma (Bucher, 1998). Minimum, low-temperature, ages for four of the age spectra from cleavage mica, the plateau age of one slate sample, and the minimum ages from many of the vein sericites from the Bendigo zone are between 417 and 426 Ma, suggesting a fault reactivation and exhumation history similar to the Stawell zone. Finally, a Devonian date of 375 ± 2 Ma (revised from the ca. 382 Ma date in Foster et al., 1999) is given by sericite alteration surrounding a late syntectonic dike intruding en-echelon fractures in the Heathcote fault zone. This is consistent with post-early Devonian thrusting along the eastern margin of the Heathcote fault zone, which is required to explain deformed strata of this age in the footwall where it overrides the Melbourne structural zone (Fig. 4) (VandenBerg, 1999).

In summary, the Bendigo zone ^{40}Ar/^{39}Ar data indicate that early (generally bedding-parallel) quartz veins and early cleavage, protected from reactivation in fold hinges, give ages of 457–455 Ma. Several samples from axial planar cleavage and syntectonic quartz veins give ages ca. 445 Ma. The most common age given by the highly transposed fabrics in the major fault zones, and by white mice separated from late-tectonic quartz veins that intrude brittle fractures in fold hinge zone, is 440–439 Ma (Fig. 9). We interpret this to indicate that folding of the turbidites began ca. 457–455 Ma and ended ca. 440–439 Ma.

Based on the style of deformation in the less deformed areas of the Melbourne zone, and the lack of evidence for fault-bend folding, deformation of the Bendigo zone began by sinusoidal buckle folding associated with bedding-parallel veins. Deformation then progressed to chevron folding and through to isoclinal folding and fabric transposition in the high strain zones, along with brittle thrusting in the hinge zones of chevron folds higher in the thrust sheets (Fig. 9). This progressive deformation occurred over a total of ~15–17 m.y., with most of the deformation in the thrust-sheet hanging walls occurring between 445 and 439 Ma, based on the ^{40}Ar/^{39}Ar geochronology.

Melbourne Zone

The lack of significant cleavage over most of the Melbourne structural zone precluded the growth of significant metamorphic mica and limits the ability to date deformation by the ^{40}Ar/^{39}Ar method. Stronger fabrics in the frontal thrusts within the Mount Wellington fault zone resulted in better data, but not of the quality obtained from the Stawell and Bendigo zones. Slate with strong transposition foliation from the basal parts of the Mount Wellington fault zone gives ages of ca. 410 Ma (Foster et al., 1998). The mica cleavage in this sample contains a small amount of detrital mica has some recoil problems and therefore, ca. 410 Ma is taken as a maximum age for the metamorphic mica growth. White mica from a folded metavolcanic rock in the basal zone gave an age of 416–411 Ma (Foster et al., 1999), and cleavage mica from phyllite just above the contact with the metavolcanic rocks gave a plateau age of 419 ± 1 Ma (Spaggiari et al., 2003), which probably records early deformation in the Mount Wellington fault zone. It is unclear how widespread the 419–411 Ma deformation was in the Melbourne zone, because an associated unconformity is only local in extent (VandenBerg, 1999). Foster et al. (1999) interpreted the 419–411 Ma deformation to indicate that the basal décollement, exposed in the Mount Wellington fault zone, was active before significant folding of the hanging wall.

Structurally higher samples of cleavage from the Mount Wellington fault zone give ages ca. 388 Ma (Foster et al., 1999); an age of 388 ± 2 Ma is also given by white mica in veins in a gold deposit in the eastern Melbourne zone (Foster et al., 1998). ^{40}Ar/^{39}Ar dates of hornblende from mafic dikes (Woods Point dike swarm) that intrude fractures aligned with the fold hinges give

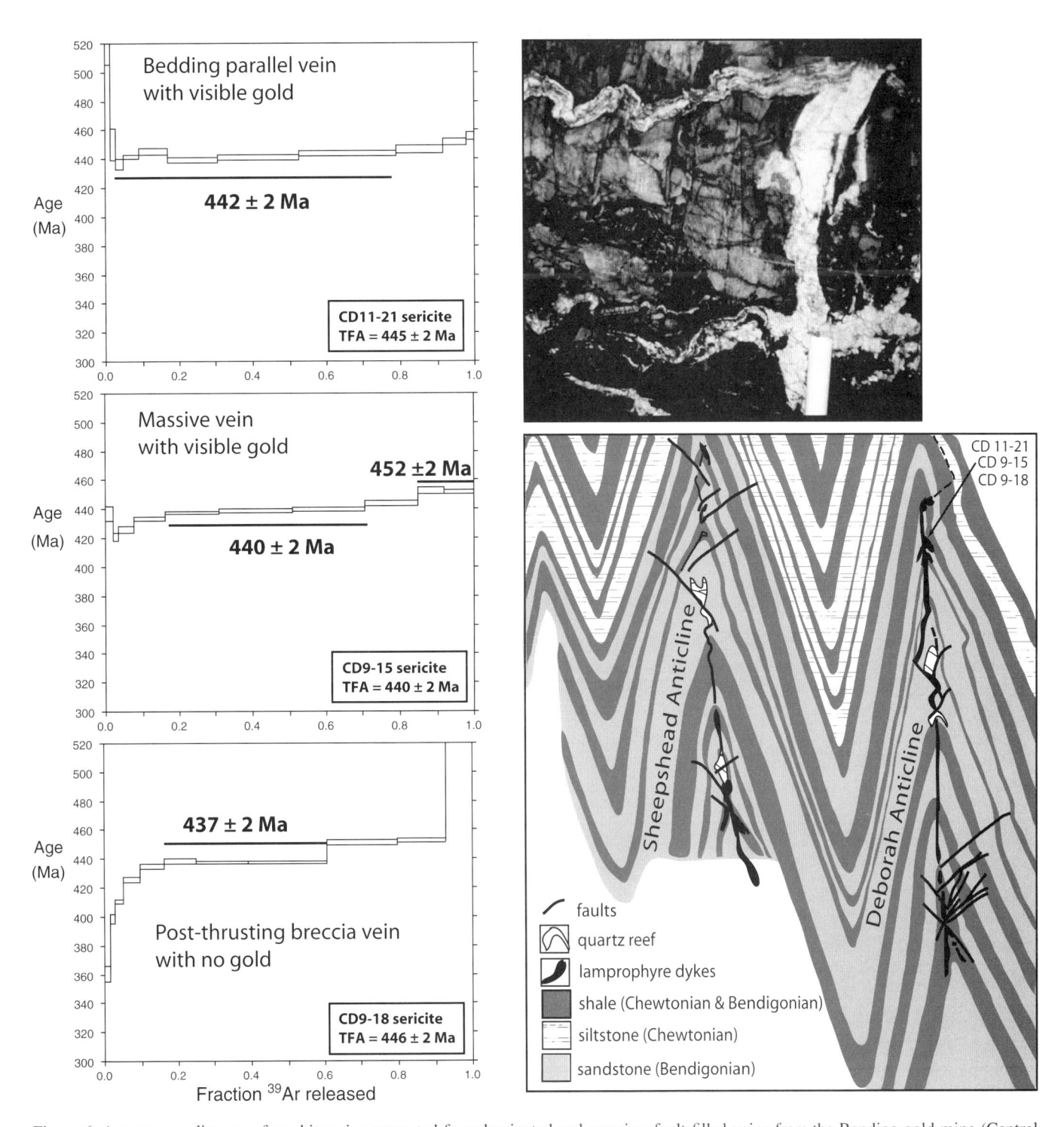

Figure 6. Age spectra diagrams for white mica separated from laminated and massive, fault-filled veins from the Bendigo gold mine (Central Deborah mine). Cross section shows the location of the samples from the hinge of the Deborah anticline (from Bierlein et al., 2001), and the photo shows multiple generations of laminated and massive synorogenic quartz veins from the mine (hammer handle for scale [4 cm wide]). Early bedding-parallel veins in this deposit give ages of ca. 445–442 Ma, massive veins in limb thrusts give ages ca. 440 Ma, and post-thrusting breccia veins give ages ca. 437 Ma. The older high-temperature release steps in sample CD9-18 are due to incorporation of some older mica during brecciation.

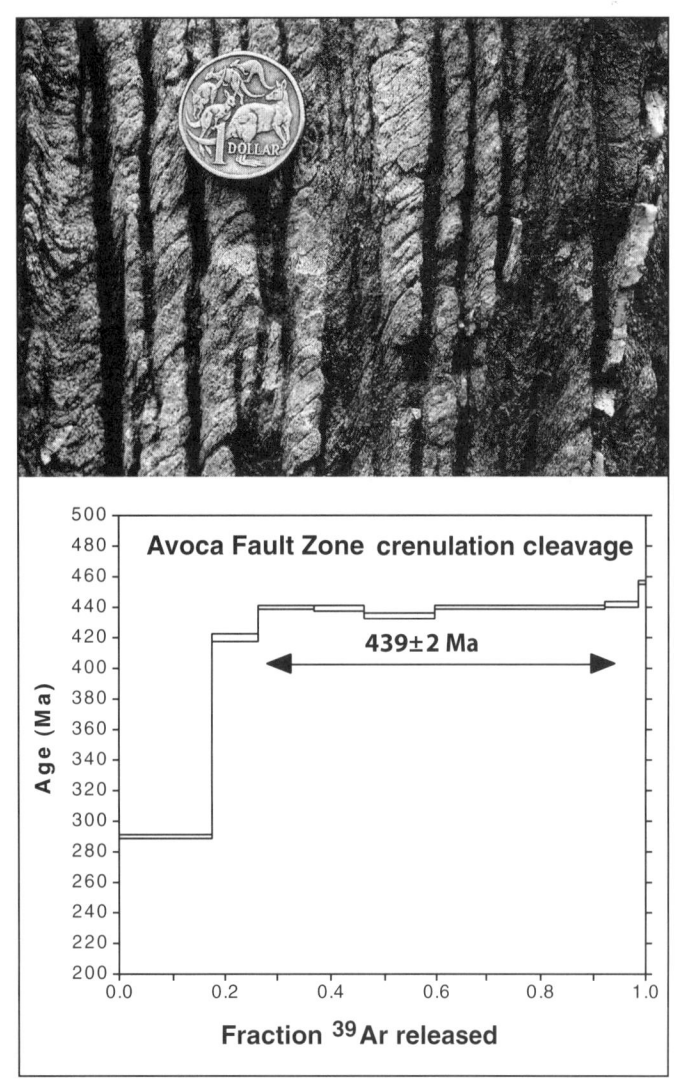

Figure 7. Field photo of strong crenulation cleavage from the Avoca fault zone. Diagram showing $^{40}Ar/^{39}Ar$ age spectrum from cleavage mica (data from Foster et al., 1998).

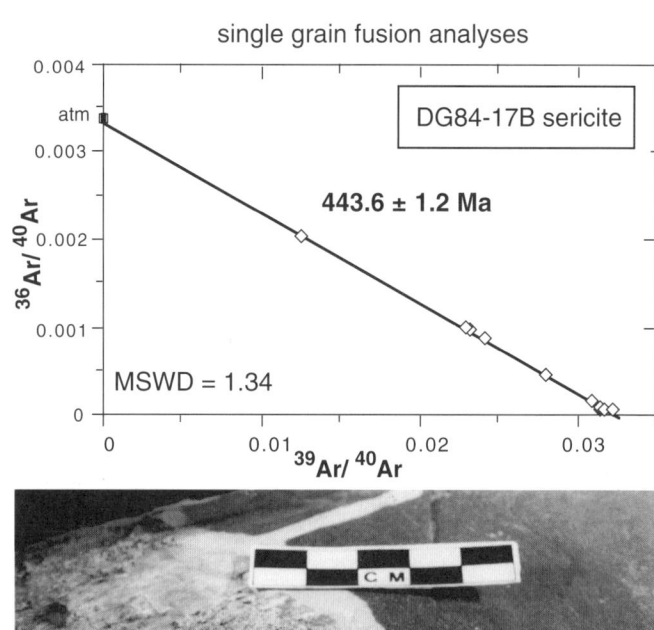

Figure 8. $^{40}Ar/^{39}Ar$ inverse isochron diagram for analyses of single grains of white mica extracted from a bedding plane fault (photo) from a quarry in Castlemaine, Victoria (Foster et al., 1999). MSWD—mean square of weighted deviations.

ages of 378–376 Ma (Bierlein et al., 2001), and white mica from mineralization associated with dikes of this swarm give ages of 378–374 Ma (Foster et al., 1996b, 1998, 1999). These data indicate that deformation had ended by ca. 378 Ma, which is consistent with the stratigraphy of the Melbourne zone (Fig. 4) (VandenBerg, 1999). The results from the Melbourne zone, therefore, suggest that deformation of the hanging wall of the thrust sheet took place between ca. 388 and 378 Ma, or over ~10 m.y. The error on this duration, however, could be as large as ± 5 m.y.

Strain States in the Western Lachlan Orogen

Strain markers within the western Lachlan Orogen include deformed pillows and amygdales in metavolcanics of the Stawell zone (Wilson et al., 1992), syntectonic quartz fibers in pressure shadows on pyrite and graptolites on bedding planes in the Bendigo zone (Gray and Willman, 1991b; Gray, 1997), and syntectonic quartz fibers in pressure shadows on pyrite and deformed cooling columns in metavolcanics of the Melbourne zone (Gray, 1995) (Fig. 10). Strain analysis techniques are after Durney and Ramsay (1973) and Ramsay and Huber (1983). For description of the methods see Wilson et al. (1992), Gray and Willman (1991b), and Gray (1995) for the Stawell, Bendigo, and Melbourne zones, respectively.

Stawell Zone Strain

Total XZ strain estimates are restricted to the Cambrian metavolcanic units (see Wilson et al., 1992, for data and location information). Deformed pillow shapes and quartz-calcite augens, formerly plagioclase phenocrysts, within porphyritic metavolcanics in the strongly deformed parts of steep, east-dipping fault slivers ("Waterloo" structures) at the Stawell gold mine give $X:Y:Z = 4.9:1:0.31$ with $X/Z = 16:1$ and a Flinn k value

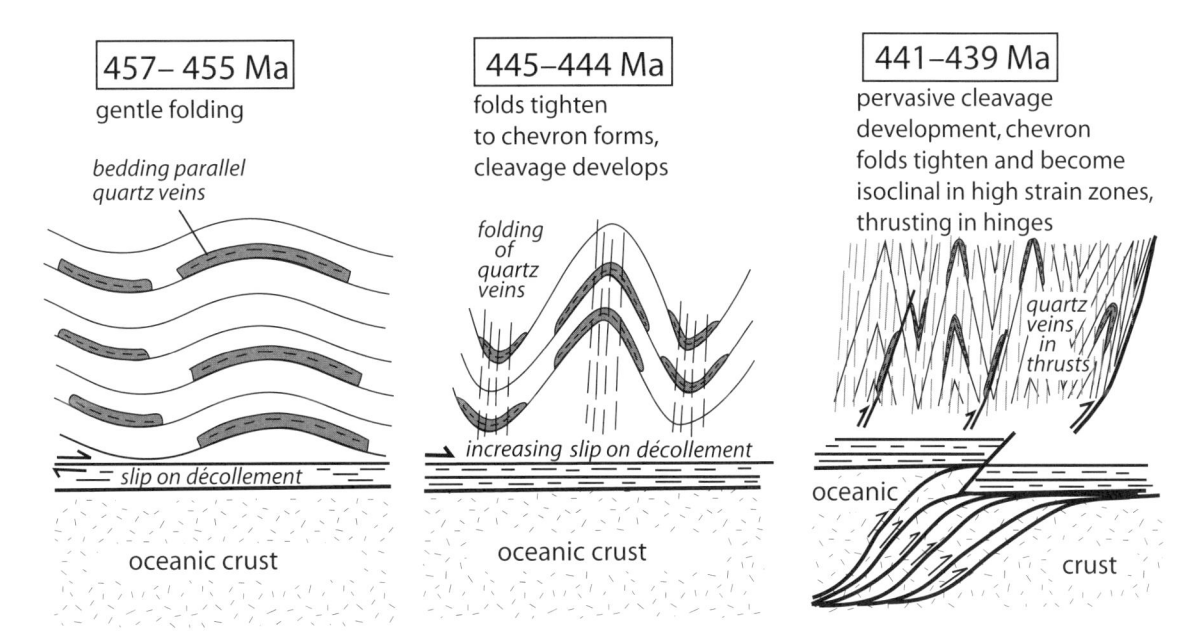

Figure 9. Timing and sequence of structural shortening within a typical Bendigo zone thrust sheet.

of ~0.25 (flattening field). Volume loss for the western Lachlan slates was <10% (Gray, 1997b) so that the strain remains in the flattening field close to plain strain, even assuming the maximum volume change. A background strain in the mine exposures of X/Z = ~1.5:1 is recorded within the less-deformed cores of the individual fault slices or shear lozenges (Wilson et al., 1992).

Amygdaloidal strain markers within the Mount Ararat metavolcanic rocks show a marked increase in strain toward the Ararat fault (Table 3 in Wilson et al., 1992), with X/Z strains increasing from 3:1, to 11:1, to 25:1 in the high strain zone of the fault. Flinn k values have k > 1 (constriction field), indicating significant stretching where the principal stretch X-axis is subparallel to the stretching lineation (Wilson et al., 1992).

Bendigo Zone Strain

Total XZ strain in slates (R), determined largely from syntectonic quartz fibers in pressure shadows on pyrite and graptolites on bedding planes (see Gray and Willman, 1991a, 1991b; Gray, 1997a, for methods and terminology), varies regionally across the thrust belt with XZ strain ranging from 2> R ≤ 40 (Gray and Willman, 1991b). The background X/Z strain values are relatively homogenous and range between 2 and 5 in slates away from fault zones. The highest strains (XZ > 40:1) are associated with the hanging walls of the major faults (Table 1) (Gray and Willman, 1991b). Strain magnitude also varies with lithology, because XZ strains in sandstones determined by the Fry method range between 1.3 and 1.7 and are significantly lower than those in mudstones (Gray and Willman, 1991a; Yang and Gray, 1994). Psammites show XZ strain magnitude variations dependent on position on folds; XZ strain estimates from fold hinges are generally higher (≈1.6–1.7) than limb estimates (≈1.3) from the same fold.

Melbourne Zone Strain

In the penetratively deformed, chevron-folded, cleaved turbidites of the Melbourne zone, total X/Z strain states range from ~6:1–15:1 (Gray, 1995). Strain magnitude increases from the upright fold zone (X/Z strains <6:1) into the inclined fold zone (X/Z strains between 10:1 and 15:1) and matches the increase in cleavage intensity into the Mount Wellington Fault Zone. The magnitude of total strain varies within the Cambrian mafic volcanic inliers, with less deformed parts giving X/Z strains <3.5:1, and the more strongly deformed, foliated parts show X/Z = 13:1. Quartz-fiber pressure shadows on pyrite in fault slices of Upper Ordovician black slate give XZ strains from 9:1–19:1.

Retrodeformation of the Western Lachlan Orogen

Shortening within western Lachlan Orogen thrusts sheets is partitioned between chevron folding and cleavage development in both mudstones and sandstones, as well as faulting (Cox et al., 1991; Gray and Willman, 1991a, 1991b; Gray et al., 2006a). Palinspastic restoration (unstraining) of the regional chevron-folded and faulted profile shown in Figure 10 includes a fault restoration followed by strain removal and then unfolding of the remaining buckle component (see Figure 10 in Gray and Willman, 1991b).

The fault restoration is problematical because of the uniformity of sedimentary facies and the lack of biostratigraphic markers over segments of the exposed thrust system. Where there is stratigraphic control on fault displacements, particularly in profiles between the intrazone faults, the calculated shortening components due to faulting are <5% (see Table 1 in Gray and Willman, 1991a). Displacement on the major zone-bounding faults that bring the underlying Cambrian metavolcanics to the present level

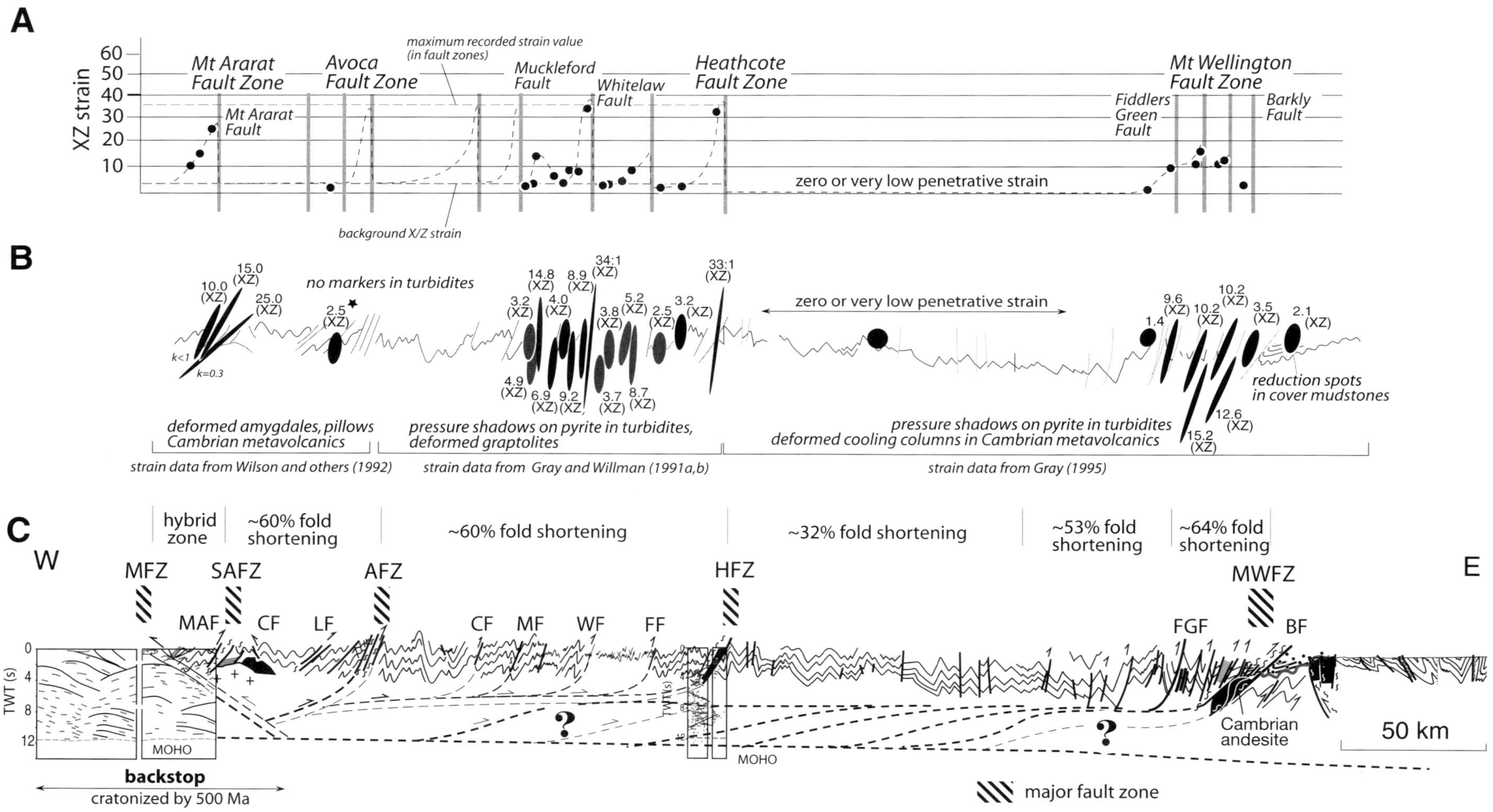

Figure 10. (A) West to east profiles through the western Lachlan Orogen showing strain data for the structural zones (see text for references). (B) Positions of major faults (Gray and Foster, 1998). (C) Crustal section (Gray et al., 2006b). MFZ—Moyston fault; MAF—Mount Ararat fault; SAFZ—Stawell-Ararat fault zone; CF—Campbelltown fault; LF—Leichardt fault; AFZ—Ararat fault zone; MF—Muckleford fault; WF—Whitelaw fault; FF—Fosterville fault; HFZ—Heathcote fault zone; FGF—Fiddlers Green fault; MWFZ—Mount Wellington fault zone; BF—Barkly fault; TWT—two-way time.

of exposure requires a minimum displacement (throw) of 17 km (i.e., the projection of the Cambrian metavolcanics above the inferred décollement depth for the Bendigo Zone), which equates to a heave of ~10 km for each of the Stawell-Ararat, Avoca, Heathcote, and Mount Wellington Fault Zones. This gives a minimum fault shortening across the profile of 11% (or $e = -0.11$, where e is the minimum elongation) by restoration of the Cambrian and thereby removing displacement on the major fault zones, given $L_1 = 350$ km and $L_0 = (350 + 40)$ km (where L is length).

Additional uncertainties from volume change and out-of-section motion are minor. Gray (1997b) showed that the maximum amount of volume change in the section is <10%. Out-of-section, strike-slip movement along the faults is very minor in the Lachlan Orogen and particularly in the western Lachlan Orogen, as demonstrated by Gray and Foster (1998).

Methods for Removing Strain

In strained rock sections, simple bed-length balancing techniques as part of section balancing (e.g., Dahlstrom, 1969) only provide a minimum estimate of the total rock shortening. The excess area method (Chamberlain, 1910; Gwinn, 1970) and the total area method (Dennison and Woodward, 1963; Hossack, 1979; Mitra and Namson, 1989) overcome this inadequacy (see Merle, 1986), but a more reliable method utilizes the input of strain data and involves the unstraining, segment by segment, of the structural profiles (e.g., Hossack, 1978; Cobbold, 1979; Woodward et al., 1986; Mitra and Namson, 1989; Gray and Willman, 1991b).

In this paper, we have applied the strain reversal technique to a 350-km-long regional profile across the western Lachlan Orogen (Fig. 11). The profile was split into four segments, A, B, C, and D, for unstraining (see Fig. 11, component line lengths), to accommodate regional differences in strain, particularly the marked gradient in XZ strain across the Melbourne Zone (Fig. 10). Segment A, or the western belt consisting of the Stawell and Bendigo Zones (see Fig. 10), has an assumed regional background X/Z strain of 4:1 and a deformed length of 200 km (maximum) or 150 km (minimum) (Table 1). The Melbourne Zone (see Fig. 10), or the eastern belt, was split into Segments B, C, and D, with the adopted values of zero X/Z strain (i.e., no penetrative strain), 1.4:1 strain, and 10:1 strain, respectively, over a combined total length of 150 km (Table 1).

Retrodeformation of the individual profile segments was undertaken in two steps (see Fig. 11). These were (1) removal of the penetrative strain components by applying the X-Y stretch tool (strain transformations) in the vector-based Adobe Illustrator drafting package to the line segments of the profile; and (2) removal of the buckle shortening components to the unstrained profile segment by the bed-length method involving comparison between the undeformed (layer arc length measurement determined by map measurer tool) and the deformed bed lengths.

Calculated Initial Lengths

Based on the above, the predeformation former basin width is ~740 km, given $e = (L_1-L_0)/L_0$, and $e = -0.58$, and $L_1 = 310$ km (see Fig. 11). Adding in the major fault displacements would give a former width of at least 780 km, and therefore on the order of 800 km.

Crustal scale area balancing of the western Lachlan Orogen, assuming strain compatibility between the upper and lower crust, however, requires a two-times fault duplication of the former turbidite fan thickness in the Western Belt (Stawell and Bendigo zones), which is ~6 km (see Gray et al., 2006a). This means that the Western Belt width should be doubled by fault restoration, and therefore the combined original width of the Stawell and Bendigo zones could approximate ~900–950 km.

Treating the western and eastern belts separately (Fig. 11) gives the following:

1. Western Belt (66% shortening) Present profile width is ~160 km and with $e = -0.66$ gives a ~470 km width of the former Stawell and Bendigo zones depositional region. This represents a minimum basin width, as fan thickness and area balance considerations suggest that the width may be as much as ~950 km (Gray et al., 2006a).

2. Eastern Belt (38% shortening) Present profile width is ~150 km and with $e = -0.38$ gives a ~242 km width of the former Melbourne zone depositional region.

Deformation Rate for the Western Lachlan Orogen

By utilizing the retrodeformation and geochronological data from the western Lachlan Orogen, we are able to quantify the timing and rate of deformation across the orogenic belt. Retrodeformation for the western part, the Stawell and Bendigo zones, gave a minimum shortening of 310 km and a maximum of 790 km (see above). Deformation of this zone took an average of 16 m.y. (455–439 Ma). This gives an average displacement rate for the basal décollement of ~19 mm yr^{-1} (minimum) and ~50 mm yr^{-1} (maximum). We have not assigned errors to these values due to the uncertainty in internal thrust shortening (see discussion in Gray et al., 2006a). The error in the duration of deformation amounts to ±1 mm yr^{-1}. The Melbourne zone shortened ~92 km over ~10 m.y., giving a displacement rate of ~9 mm yr^{-1}. The error in the duration of shortening is up to ±5 m.y., which corresponds with displacement rates ranging from 7 to 23 mm yr^{-1}.

The geochronology and strain data can also be used to calculate internal strain rates for the western Lachlan thrust wedges. With strain rate calculations, it is possible to calculate a conventional strain rate or a natural strain rate (Pfiffner and Ramsay, 1982). A conventional strain rate is elongation/time (e/t) where e is the elongation ($e = (L_1-L_0)/L_0$) and t is time (in seconds). Using this approach, strain estimates for the western belt (Stawell and Bendigo zones) and eastern belt (Melbourne zone) are as follows:

1. Western belt, where shortening of 66% occurred over 16 m.y. (0.51×10^{15} s), gives a strain rate equal to 1.3×10^{-15} s^{-1}.

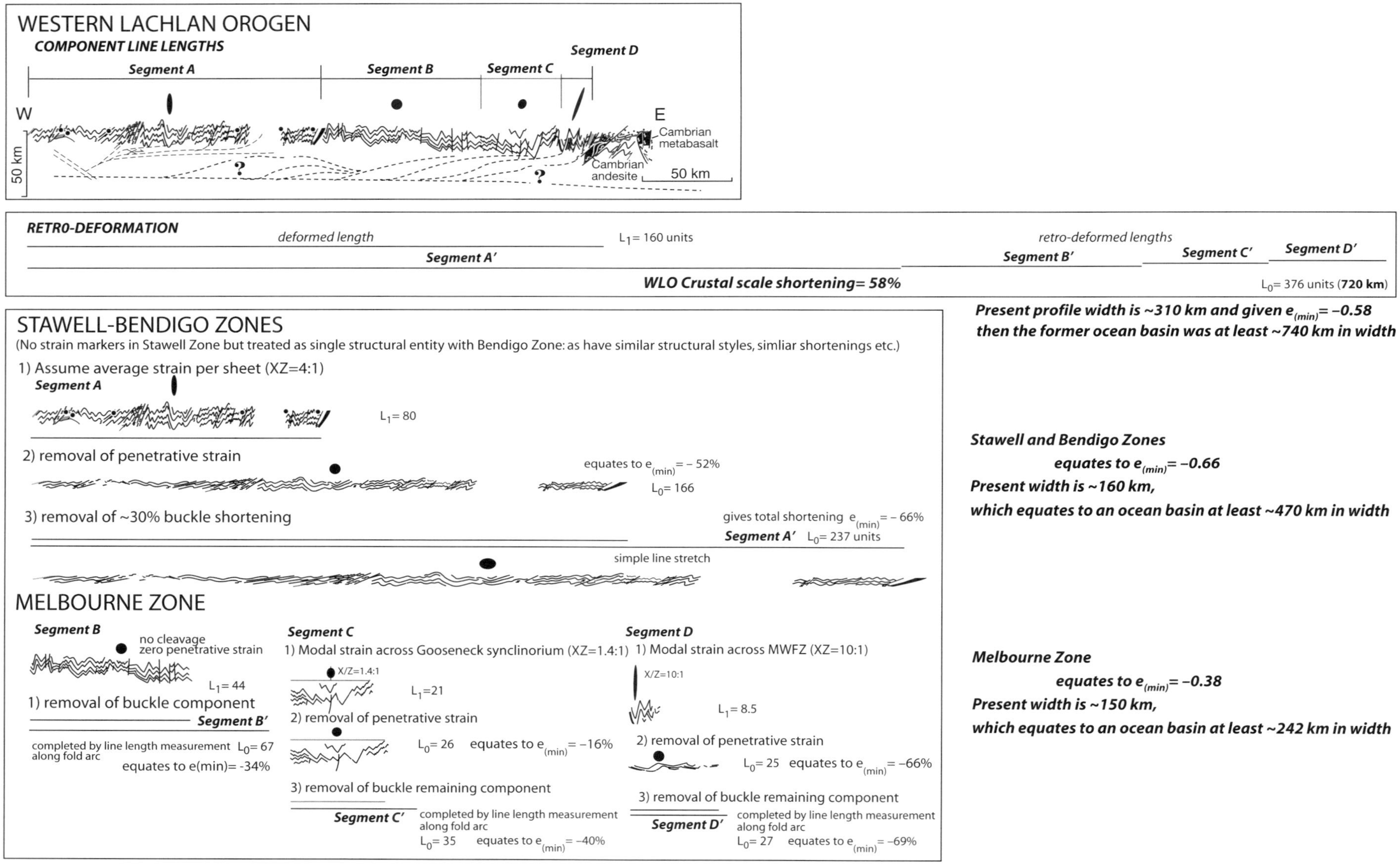

Figure 11. Deformed and retrodeformed crustal sections for the western Lachlan Orogen. Segment A includes the section in the Stawell and Bendigo zones. The Melbourne zone is divided into segments B, C, and D because of significant variations in strain. MWFZ—Mount Wellington fault zone; e—strain; L—length.

2. Eastern belt, where shortening of 43% occurred over 10 m.y. (0.32 × 10^{-15} s), gives a strain rate equal to 1.3 × 10^{-15} s^{-1}.

Another approach to derive strain rate involves displacement rate/distance (Twiss and Moores, 1992). Using this approach, the western belt, where the minimum displacement rate is 19 km m.y.$^{-1}$/310 km gives a minimum strain rate of 2.0 × 10^{-15} s^{-1}, and the maximum displacement rate is considered to be 50 km m.y.$^{-1}$/310 km gives a maximum strain rate of 5.0 × 10^{-15} s^{-1}. The eastern belt, where the average displacement rate is 9 km m.y.$^{-1}$/92 km, gives a strain rate of 3.1 × 10^{-15} s^{-1}.

With either calculation method, the data indicate that folding and thrusting of turbidite in this oceanic backarc basin setting occurred at average strain rates between 1 × 10^{-15} s^{-1} and 5 × 10^{-15} s^{-1}. The concentrations of $^{40}Ar/^{39}Ar$ vein and cleavage mica ages ca. 445–439 Ma in the Bendigo zone may indicate that most deformation in the thrust sheets occurred over a shorter interval of more rapid deformation. In fact, most of the intense cleavage development occurred ca. 441–440 Ma, and the limb thrust veins, which give ages of ca. 440–439 Ma, suggest that most deformation ended by 439 Ma. Given the errors in the analyses and ranges of ages, a pulse of deformation ~2 m.y. may have resulted in much of the 66% shortening. If this were the case, the strain rate would have been on the order of 1 × 10^{-14} s^{-1}.

DISCUSSION AND CONCLUSIONS

Modern deformation rates within orogenic belts range from ~5–15 mm yr^{-1} (Hyndman et al., 2005) based on GPS measurements and longer term geological studies (e.g., Hindle et al., 2002; McQuarrie and Wernicke, 2005). These rates are a fraction of plate tectonic rates, which typically range from 10 to 200 mm yr^{-1}, although individual faults and shear zones may take on a larger fraction of plate tectonic rates for short intervals of time (e.g., Carter et al., 2004). Corresponding strain rates for tectonic and orogenic processes, therefore, typically range from 10^{-12} to 10^{-15} s^{-1} (Pfiffner and Ramsay, 1982; van der Pluijm and Marshak, 2004).

Internal rates of deformation in the ductile middle and lower crust of orogens are more difficult to evaluate (Ramsay, 2000). Strain rate calculations for deformation within ancient orogenic belts are limited by the existing geochronology of the fabric-forming events, as well as knowledge of the incremental strains (Pfiffner and Ramsay, 1982; Ramsay, 2000; Müller et al., 2000). These difficulties have been addressed by a small number of studies that succeeded in combining structural and microstructural analysis with radioisotopic analyses on mineral phases grown during deformation. The challenge is finding areas where deformation and metamorphism occurred at temperatures below the closure temperature for the isotopic system in the phases of interest. This is a rapidly growing field of research with significant potential (Müller, 2003).

Our results from the western Lachlan Orogen give minimum average strain rates of 1 to 5 ×10^{-15} s^{-1}, which are at the low end of the range of deformation rates estimated for orogenic belts. The concentrations of $^{40}Ar/^{39}Ar$ vein and cleavage mica ages ca. 457–455 and 440–439 Ma in both the Stawell and Bendigo zones suggest that deformation occurred dominantly in two more rapid pulses of shortening. Most of the intense cleavage development occurred ca. 441–440 Ma, and the limb thrust veins, which give ages of ca. 440–439 Ma, suggest that deformation ended by 439 Ma. A pulse of deformation of ~2 m.y. duration may have resulted in a considerable fraction of the 66% shortening. If this were the case, the strain rate would have been ~1 × 10^{-14} s^{-1}.

The $^{40}Ar/^{39}Ar$ data from the thrust sheets in the Bendigo zone suggest that thrust sheet deformation occurred over a total of ~16 m.y. and progressed from early buckle folding at 457–455 Ma through to chevron fold lock-up and thrusting at 440–439 Ma. Deformation and fabric development in the thrust sheets were progressive during this time interval, and folding and ductile strain probably developed diachronously (cf. Pinan et al., 2004). It is also possible that the thrust sheet strain accumulated in a pulse-like manner due to a caterpillar-style movement on the basal décollement, linked to stop-start movements as the deforming sedimentary wedge achieved the necessary wedge taper requirements for slippage on the basal fault, but this requires further research.

Comparing the deformation rates with other estimates from Phanerozoic convergent orogens shows distinct similarities. For example, Kligfield et al. (1981, 1986) integrated measured finite strain and the total time of deformation, based on K/Ar and $^{40}Ar/^{39}Ar$ geochronology, to calculate average strain rates of 10^{-14} to 10^{-15} s^{-1} for the Northern Apennines. In the Apennine case, simple shear components are a major part of the incremental strain accumulation, whereas in the western Lachlan Orogen, the largest volume of the thrust sheets (up to 94% per volume) has undergone pure shear accumulation of incremental strains (see Figure 12 in Gray and Willman, 1991b).

Müller et al. (2000) used microscale Rb-Sr dating of fibrous strain fringes on pyrites in the Pyrenees, which gave strain rates up to 1.1 × 10^{-15} to 7.7 × 10^{-15} s^{-1}. Three separate studies of Rb-Sr and Nd-Sm dating of large metamorphic garnets have suggested shear strain rates ~2–3 × 10^{-14} s^{-1}. van der Pluijm et al. (2006) used $^{40}Ar/^{39}Ar$ dating of clay in fault gouge from the Canadian Rockies to suggest pulses of deformation with rates of ~10^{-14} s^{-1}.

The western Lachlan shortening rate estimates are similar to rates estimated for the Robertson Bay terrane of North Victoria Land, where $^{40}Ar/^{39}Ar$ ages of cleavage mica give convergence rates of ~4–10 mm yr^{-1} when fold shortening is considered (Dallmeyer and Wright, 1992). It is also similar to fold propagation and cleavage formation in flysch across the Rheinesche Schiefergebirge, where K-Ar ages of sericite indicate a rate ~5 mm yr^{-1} (Ahrendt et al., 1983). Moreover, the calculated displacement rates are within the range of modern plate tectonic velocities in similar backarc settings, like the western Pacific backarc basins and accretionary prisms (e.g., McCaffrey, 1996), as well as the rates of shortening in the turbidite-dominated thrust belt of Taiwan (Yu et al., 1997).

ACKNOWLEDGMENTS

Reviews by D. De Paor and C. Fergusson helped improve the clarity of the paper. Fieldwork for this research was supported by Australian Research Council Grant E8315675 and Monash University Special Research Funds (awarded to DRG). The $^{40}Ar/^{39}Ar$ geochronology was supported by the Australian Geodynamics Cooperative Research Centre (1995-97) and the National Science Foundation grant EAR0073638 to DAF. Support for the write-up for DRG was from an Australian Professorial Fellowship as part of Australian Research Council Discovery Grant DP0210178.

REFERENCES CITED

Ahrendt, H., Clauer, N., Hunziker, J.C., and Weber, K., 1983, Migration of folding and metamorphism in the Rheinisches Schiefergebirge deduced from K-Ar and Rb-Sr age determinations, in Martin, H., and Eder, F.W., eds., Intracontinental Fold Belts: Berlin, Springer-Verlag, p. 323–338.

Arne, D.C., Bierlein, F.P., McNaughton, N., Wilson, C.J.L., and Morand, V.J., 1998, Timing of gold mineralization in western and central Victoria, Australia: New constraints from SHRIMP II analysis of zircon grains from felsic intrusive rocks: Ore Geology Reviews, v. 13, p. 251–273, doi: 10.1016/S0169-1368(97)00021-8.

Bennett, R.A., Davis, J.L., and Wernicke, B.P., 1999, Large scale pattern of western U.S. Cordillera deformation: Geology, v. 27, p. 371–374, doi: 10.1130/0091-7613(1999)027<0371:PDPOCD>2.3.CO;2.

Bierlein, F.P., Foster, D.A., McKnight, S., and Arne, D.C., 1999, Timing of gold mineralization in the Ballarat Goldfield, central Victoria: Constraints from $^{40}Ar/^{39}Ar$ results: Australian Journal of Earth Sciences, v. 46, p. 301–309, doi: 10.1046/j.1440-0952.1999.00708.x.

Bierlein, F.P., Arne, D.C., Foster, D.A., and Reynolds, P., 2001, A geochronological framework for orogenic gold mineralization in central Victoria, Australia: Mineralium Deposita, v. 36, p. 741–767, doi: 10.1007/s001260100203.

Brichau, S., Ring, W., Ketcham, R.A., Carter, A., Stockli, D., and Brunel, M., 2006, Contrasting the long-term evolution of the slip rate for a major extensional fault system in the central Aegean, Greece, using thermochronology: Earth and Planetary Science Letters, v. 241, p. 293–306, doi: 10.1016/j.epsl.2005.09.065, doi: 10.1016/j.epsl.2005.09.065.

Bucher, M., 1998, Timing of deformation, plutonism and cooling in the western Lachlan fold belt, southeastern Australia [Ph.D. thesis]: Melbourne, Australia, La Trobe University, 294 p.

Carter, T.J., Kohn, B.P., Foster, D.A., and Gleadow, A.J.W., 2004, How the Harcuvar Mountains metamorphic core complex became cool: Evidence from apatite (U-Th)/He thermochronometry: Geology, v. 32, p. 985–988, doi: 10.1130/G20936.1, doi: 10.1130/G20936.1.

Cawood, P.A., 2005, Terra Australis Orogen: Rodinia breakup and development of the Pacific and Iapetus margins of Gondwana during the Neoproterozoic and Paleozoic: Earth-Science Reviews, v. 69, p. 249–279, doi: 10.1016/j.earscirev.2004.09.001.

Cayley, R.A., and MacDonald, P.A., 1995, Beaufort 1:100,000 scale map geological report: Geological Survey of Victoria Report 104, 145 p.

Cayley, R.A., Taylor, D.H., VandenBerg, A.H.M., and Moore, D.H., 2002, Proterozoic-Early Palaeozoic rocks and the Tyrrhenian Orogeny in central Victoria: The Selwyn Block and its tectonic implications: Australian Journal of Earth Sciences, v. 49, p. 225–254, doi: 10.1046/j.1440-0952.2002.00921.x.

Chang, C.-P., Chang, T.-Y., Angelier, J., Kao, H., Lee, J.-C., and Yu, S.-B., 2003, Strain and stress field in Taiwan oblique convergent system: constraints from GPS observation and tectonic data: Earth and Planetary Science Letters, v. 214, p. 115–127, doi: 10.1016/S0012-821X(03)00360-1.

Chamberlain, R.T., 1910, The Appalachian folds of central Pennsylvania: The Journal of Geology, v. 18, p. 228–251.

Cobbold, P.R., 1979, Removal of finite deformation using strain trajectories: Journal of Structural Geology, v. 9, p. 667–677.

Collins, W.J., 2002, Hot orogens, tectonic switching, and creation of continental crust: Geology, v. 30, p. 535–538, doi: 10.1130/0091-7613(2002)030<0535:HOTSAC>2.0.CO;2.

Cox, S.F., Wall, V.J., Etheridge, M.A., and Potter, T.F., 1991, Deformational and metamorphic processes in the formation of mesothermal vein-hosted gold deposits—Examples from the Lachlan Fold Belt in central Victoria, Australia: Ore Geology Reviews, v. 6, p. 391–423, doi: 10.1016/0169-1368(91)90038-9.

Crawford, A.J., Cayley, R.A., Taylor, D.H., Morand, V.J., Gray, C.M., Kemp, A.I.S., Wohlt, K.E., VandenBerg, A.H.M., Moore, D.H., Maher, S., Direen, N.G., Edwards, J., Donaghy, A.G., Anderson, J.A., and Black, L.P., 2003, Neoproterozoic and Cambrian, Chapter 3, in Birch, W.D., ed., Geology of Victoria: Geological Society of Australia Special Publication 23, p. 73–92.

Dahlstrom, C.D.A., 1969, Balanced Cross-Sections: Canadian Journal of Earth Sciences, v. 6, p. 743–757.

Dallmeyer, R.D., and Wright, T.O., 1992, Diachronous cleavage development in the Robertson Bay terrane, northern Victoria Land, Antarctica: Tectonic implications: Tectonics, v. 11, p. 437–448.

Dennison, J.M., and Woodward, H.P., 1963, Palinspastic maps of Central Alps: AAPG Bulletin, v. 47, p. 666–680.

Dunlap, W.J., Teyssier, C., McDougall, I., and Baldwin, S., 1991, Ages of deformation from K/Ar and $^{40}Ar/^{39}Ar$ dating of white micas: Geology, v. 19, p. 1213–1216, doi: 10.1130/0091-7613(1991)019<1213:AODFKA>2.3.CO;2.

Dunlap, W.J., Hirth, G., and Teyssier, C., 1997, Thermomechanical evolution of a ductile duplex: Tectonics, v. 16, p. 983–1000, doi: 10.1029/97TC00614.

Durney, D.W., and Ramsay, J.G., 1973, Incremental strains measured by syntectonic crystal growths, in De Jong, K.A., and Scholten, R., eds., Gravity and Tectonics: Wiley, New York, p. 67–96.

Fergusson, C.L., 2003, Ordovician-Silurian accretion tectonics of the Lachlan Fold Belt, southeastern Australia: Australian Journal of Earth Sciences, v. 50, p. 475–490, doi: 10.1046/j.1440-0952.2003.01013.x.

Fergusson, C.L., and Coney, P.J., 1992, Convergence and intraplate deformation in the Lachlan Fold Belt of southeastern Australia: Tectonophysics, v. 214, p. 417–439, doi: 10.1016/0040-1951(92)90208-N.

Fergusson, C.L., Gray, D.R., and Cas, R.A.F., 1986, Overthrust terranes in the Lachlan fold belt, southeastern Australia: Geology, v. 14, p. 519–522, doi: 10.1130/0091-7613(1986)14<519:OTITLF>2.0.CO;2.

Foster, D.A., and Gray, D.R., 2000, The structure and evolution of the Lachlan Fold Belt (Orogen) of eastern Australia: Annual Review of Earth and Planetary Sciences, v. 28, p. 47–80, doi: 10.1146/annurev.earth.28.1.47.

Foster, D.A., and John, B.E., 1999, Quantifying tectonic exhumation in an extensional orogen with thermochronology: Examples from the southern Basin and Range Province, in Ring, U., Brandon, M., Lister, G.S., and Willett, S.D., eds., Exhumation Processes: Normal Faulting, Ductile Flow, and Erosion: Geological Society [London] Special Publication 154, p. 356–378.

Foster, D.A., Gray, D.R., and Offler, R., 1996a, The Western Subprovince of the Lachlan Fold Belt: Structural style, geochronology, metamorphism, and tectonics: Sydney, Geological Society of Australia, Specialist Group in Geochemistry, Mineralogy, and Petrology Field Guide, no. 1, 89 p.

Foster, D.A., Kwak, T.A.P., and Gray, D.R., 1996b, Timing of gold mineralization and relationship to metamorphism, thrusting, and plutonism in Victoria: Australian Institute of Geologists Bulletin, v. 20, p. 49–53.

Foster, D.A., Gray, D.R., Kwak, T.A.P., and Bucher, M., 1998, Chronology and tectonic framework of turbidite-hosted gold deposits in the western Lachlan Fold Belt, Victoria: $^{40}Ar-^{39}Ar$ results: Ore Geology Reviews, v. 13, p. 229–250, doi: 10.1016/S0169-1368(97)00020-6.

Foster, D.A., Gray, D.R., and Bucher, M., 1999, Chronology of deformation within the turbidite-dominated Lachlan orogen: Implications for the tectonic evolution of eastern Australia and Gondwana: Tectonics, v. 18, p. 452–485, doi: 10.1029/1998TC900031.

Foster, D.A., Gray, D.R., and Spaggiari, C.V., 2005, Timing of subduction and exhumation along the Cambrian East Gondwana margin, and the formation of Paleozoic backarc basins: GSA Bulletin, v. 117, p. 105–116, doi: 10.1130/B25481.1.

Fowler, T.J., 1996, Flexural-slip generated bedding-parallel veins from central Victoria, Australia: Journal of Structural Geology, v. 18, p. 1399–1415, doi: 10.1016/S0191-8141(96)00066-1.

Fowler, T.J., and Winsor, C.N., 1996, Evolution of chevron folds by profile shape changes: Comparison between multilayer deformation experiments and folds of the Bendigo-Castlemaine goldfields, Australia: Tectonophysics, v. 88, p. 291–312.

Fowler, T.J., and Winsor, C.N., 1997, Characteristics and occurrence of bedding-parallel slip surfaces and laminated veins from the Bendigo-Castlemaine goldfields: Implications for flexural-slip folding: Journal of Structural Geology, v. 19, p. 799–815, doi: 10.1016/S0191-8141(97)00004-7.

Gibson, G., Wesson, V., and Cuthbertson, R., 1981, Seismicity of Victoria to 1980: Journal of Geological Society of Australia, v. 28, p. 341–356.

Glasson, M.J., and Keays, R.R., 1978, Gold mobilization during cleavage development in sedimentary rocks from the auriferous slate belt of central Victoria, Australia; some important boundary conditions: Economic Geology and the Bulletin of the Society of Economic Geologists, v. 73, p. 496–511.

Glen, R.A., 1992, Thrust, extensional and strike-slip tectonics in an evolving Palaeozoic Orogen of southeastern Australia: Tectonophysics, v. 214, p. 341–380, doi: 10.1016/0040-1951(92)90205-K.

Glen, R.A., Walshe, J.L., Barron, L.M., and Watkins, J.J., 1998, Ordovician convergent margin volcanism and tectonism in the Lachlan sector of east Gondwana: Geology, v. 26, p. 751–754, doi: 10.1130/0091-7613(1998)026<0751:OCMVAT>2.3.CO;2.

Gray, D.R., 1995, Thrust kinematics and transposition fabrics from a basal detachment zone, eastern Australia: Journal of Structural Geology, v. 17, p. 1637–1654, doi: 10.1016/0191-8141(95)00068-O.

Gray, D.R., 1997a, Tectonics of the southeastern Australian Lachlan Fold Belt: Structural and thermal aspects, in Burg, J.-P., and Ford, M., eds., Orogeny through Time: Geological Society [London] Special Publication 121, p. 149–177.

Gray, D.R., 1997b, Volume loss and slaty cleavage development, in Sengupts, S., ed., Evolution of Geological Structures in Micro- to Macro-Scales: London, Chapman and Hall, p. 273–288.

Gray, D.R., and Foster, D.A., 1997, Orogenic concepts—application and definition: Lachlan Fold Belt, eastern Australia: American Journal of Science, v. 297, p. 859–891.

Gray, D.R., and Foster, D.A., 1998, Character and kinematics of faults within the turbidite-dominated Lachlan Orogen: Implications for tectonic evolution of eastern Australia: Journal of Structural Geology, v. 20, p. 1691–1720, doi: 10.1016/S0191-8141(98)00089-3.

Gray, D.R., and Foster, D.A., 2004, Tectonic evolution of the Lachlan Orogen, southeast Australia: Historical review, data synthesis and modern perspectives: Australian Journal of Earth Sciences, v. 51, p. 773–817, doi: 10.1111/j.1400-0952.2004.01092.x.

Gray, D.R., and Mortimer, L., 1996, Implications of overprinting deformations and fold interference patterns in the Melbourne Zone, Lachlan Fold Belt: Australian Journal of Earth Sciences, v. 43, p. 103–114.

Gray, D.R., and Willman, C.E., 1991a, Thrust-related strain gradients and thrusting mechanisms in a chevron-folded sequence, south-eastern Australia: Journal of Structural Geology, v. 13, p. 691–710, doi: 10.1016/0191-8141(91)90031-D.

Gray, D.R., and Willman, C.E., 1991b, Deformation in the Ballarat Slate Belt, Central Victoria and implications for the crustal structure across SE Australia: Australian Journal of Earth Sciences, v. 38, p. 171–201.

Gray, D.R., Wilson, C.J.L., and Barton, T.J., 1991, Intracrustal detachments and implications for crustal evolution within the Lachlan Fold Belt, southeastern Australia: Geology, v. 19, p. 574–577, doi: 10.1130/0091-7613(1991)019<0574:IDAIFC>2.3.CO;2.

Gray, D.R., Foster, D.A., Morand, V.J., Willman, C.E., Cayley, R.A., Spaggiari, C.V., Taylor, D.H., Gray, C.M., VandenBerg, A.H.M., Hendrickx, M.A., and Wilson, C.J.L., 2003, Structure, metamorphism, geochronology and tectonics of Palaeozoic Rocks—interpreting a complex, long-lived orogenic system, Chapter 2, in Birch, W.M., ed., Geology of Victoria: Geological Society of Australia Special Publication 23, p. 15–71.

Gray, D.R., Willman, C.E., and Foster, D.A., 2006a, Crust restoration for the western Lachlan Orogen using the strain-reversal, area-balancing technique: Implications for crustal components and original thicknesses: Australian Journal of Earth Sciences, v. 53, p. 329–341, doi: 10.1080/08120090500499305.

Gray, D.R., Foster, D.A., Korsch, R.J., and Spaggiari, C.V., 2006b, Structural style and crustal architecture of the Tasmanides of eastern Australia, example of a composite accretionary orogen, in Mazzoli, S., and Butler, B., eds., Styles of continental compression: Geological Society of America Special Paper 414, p. 119–232.

Gray, D.R., Foster, D.A., Maas, R., Spaggiari, C.V., Gregory, R.T., Goscombe, B., and Hoffmann, K.II., 2007, Continental growth and recycling by accretion of deformed turbidite fans and remnant ocean basins: Examples from Neoproterozoic and Phanerozoic orogens, in Hatcher, R.D., Jr., Carlson, M.P., McBride, J.H., and Martínez Catalán, J.R., eds., 4-D Framework of Continental Crust: Geological Society of America Memoir 200, doi: 10.1130/2007.1200(05).

Gwinn, V.E., 1970, The Valley and Ridge and Appalachian Plateau; structure and tectonics; kinematic patterns and estimates of lateral shortening, Valley and Ridge and Great Valley provinces, central Appalachians, south-central Pennsylvania, in Fisher, G.W., Pettijohn, F.J., and Reed, J.C., Jr., eds., The Valley and Ridge and Appalachian Plateau; Structure and Tectonics; Kinematic Patterns and Estimates of Lateral Shortening, Valley and Ridge and Great Valley Provinces, Central Appalachians, South-Central Pennsylvania: New York, Interscience Publishers, p. 127–146.

Hindle, D., Kley, J., Klosko, E., Stein, S., Dixon, T., and Norabuena, E., 2002, Consistency of geologic and geodetic displacements during Andean orogenesis: Geophysical Research Letters, v. 29, p. 1188, doi: 10.1029/2001GL013757.

Hossack, J.R., 1978, The correction of stratigraphic sections for tectonic finite strains in the Bygdin area, Norway: Geological Society [London] Journal, v. 135, p. 229–241.

Hossack, J.R., 1979, The use of balanced cross sections in the calculation of orogenic translation: A review: Geological Society [London] Journal, v. 136, p. 705–711.

Hyndman, R.A., Flück, P., Mazzotti, S., Lewis, T.J., Ristau, J., and Leonard, L., 2005, Current tectonics of the northern Canadian Cordillera: Canadian Journal of Earth Sciences, v. 42, p. 1117–1136, doi: 10.1139/e05-023.

Jessell, M.W., Willman, C.E., and Gray, D.R., 1994, Bedding parallel veins and their relationship to folding: Journal of Structural Geology, v. 16, p. 753–767, doi: 10.1016/0191-8141(94)90143-0.

Kirschner, D.L., Cosca, M.A., Masson, H., and Hunziker, J.C., 1996, Staircase ^{40}Ar/^{39}Ar spectra of fine-grained-white mica: Timing and duration of deformation and empirical constraints on argon diffusion: Geology, v. 24, p. 747–750, doi: 10.1130/0091-7613(1996)024<0747:SAASOF>2.3.CO;2.

Kligfield, R., Carmignani, L., and Owens, W.H., 1981, Strain analysis of a Northern Apennine shear zone using deformed marble breccias: Journal of Structural Geology, v. 3, p. 421–436, doi: 10.1016/0191-8141(81)90042-0.

Kligfield, R., Hunziker, J., Dallmeyer, R.D., and Schamel, S., 1986, Dating of deformation phases using K-Ar and ^{40}Ar/^{39}Ar techniques: Results from the Northern Apennines: Journal of Structural Geology, v. 8, p. 781–798, doi: 10.1016/0191-8141(86)90025-8.

Korsch, R.J., Barton, T.J., Gray, D.R., Owen, A.J., and Foster, D.A., 2002, Geological interpretation of a deep seismic-reflection transect across the boundary between the Delamerian and Lachlan Orogens, in the vicinity of the Grampians, western Victoria: Australian Journal of Earth Sciences, v. 49, p. 1057–1075, doi: 10.1046/j.1440-0952.2002.00963.x.

McCaffrey, R., 1996, Slip partitioning at convergent plate boundaries of SW Asia, in Hall, R., and Blundell, D., eds., Tectonic Evolution of Southeast Asia: Geological Society [London] Special Publication 106, p. 3–18.

McQuarrie, N., and Wernicke, B.P., 2005, An animated tectonic reconstruction of southwestern North America since 36 Ma: Geosphere, v. 1, p. 147–172, doi: 10.1130/GES00016.1.

Merle, O., 1986, Patterns of stretch trajectories and strain rates within spreading-gliding nappes: Tectonophysics, v. 124, p. 211–222, doi:10.1016/0040-1951(86)90201-5.

Miller, J.McL., Phillips, D., Wilson, C.J.L., and Dugdale, L.J., 2005, Evolution of a reworked orogenic zone: The boundary between the Delamerian and Lachlan Fold Belts, southeastern Australia: Australian Journal of Earth Sciences, v. 52, p. 921–940, doi: 10.1080/08120090500304265.

Mitra, S., and Namson, J.S., 1989, Equal-area balancing: American Journal of Science, v. 289, p. 563–599.

Morand, V.J., and Gray, D.R., 1991, Major fault zones related to the Omeo Metamorphic Complex, northeastern Victoria: Australian Journal of Earth Sciences, v. 38, p. 203–221.

Morand, V.J., Ramsay, W.R.H., Hughes, M., and Stanley, J.M., 1995, The southern Avoca Fault Zone: Site of a newly identified 'greenstone' belt in western Victoria: Australian Journal of Earth Sciences, v. 42, p. 133–143.

Mulch, A., and Cosca, M.A., 2004, Recrystallization or cooling ages?—In situ UV-laser ^{40}Ar/^{39}Ar geochronology of muscovite in mylonitic rocks: Journal of the Geological Society, v. 161, p. 573–582.

Müller, W., 2003, Strengthening the link between geochronology, textures and petrology: Earth and Planetary Science Letters, v. 206, p. 237–251, doi: 10.1016/S0012-821X(02)01007-5.

Müller, W., Aerden, D., and Halliday, A.N., 2000, Isotopic dating of strain fringe increments: Duration and rates of deformation in shear zones: Science, v. 288, p. 2195–2198, doi: 10.1126/science.288.5474.2195.

Offler, R., McKnight, S., and Morand, V., 1998, Tectonothermal history of the western Lachlan Fold Belt, Australia—insights from white mica studies: Journal of Metamorphic Geology, v. 16, p. 531–540, doi: 10.1111/j.1525-1314.1998.00153.x.

Pfiffner, O.A., and Ramsay, J.G., 1982, Constraints on geological strain rates; arguments from finite strain states of naturally deformed rocks: Journal of Geophysical Research, v. 87, p. 311–321.

Phillips, G., Miller, J.McL., and Wilson, C.J.L., 2002, Structural and metamorphic evolution of the Moornambool metamorphic complex, western Lachlan Fold Belt, southeastern Australia: Australian Journal of Earth Sciences, v. 49, p. 891–913, doi: 10.1046/j.1440-0952.2002.00958.x.

Pinan, A., DePaor, D., and Simpson, C., 2004, 4-dimensional model of diachronous chevron folding in an accretionary prism setting: Geological Society of America Abstracts with Programs, v. 36, no. 5, p. 253.

Puntodewo, S.S.O., McCaffrey, R., Calais, E., Bock, Y., Rais, J., Subarya, C., Poewariardi, R., Stevens, C., Genrich, J., Fauzi, C., Zwick, P., and Wdowinski, S., 1994, GPS measurements of crustal deformation within the Pacific-Australia plate boundary zone in Irian Jaya, Indonesia: Tectonophysics, v. 237, p. 141–153.

Ramsay, J.G., 2000, A strained Earth, past and present: Science, v. 288, p. 2139–2141, doi: 10.1126/science.288.5474.2139.

Ramsay, J.G., and Huber, M.I., 1983, The Techniques of Modern Structural Geology. Volume 1: Strain Analysis: New York, Academic Press, 307 p.

Scheibner, E., 1978, Tasman fold belt system or orogenic system: Tectonophysics, v. 48, p. 153–157, doi: 10.1016/0040-1951(78)90117-8.

Şengör, C.A.M., and Natal'in, B.A., 1996, Turkic-type orogeny and its role in the making of the continental crust: Annual Review of Earth and Planetary Sciences, v. 24, p. 263–337, doi: 10.1146/annurev.earth.24.1.263.

Spaggiari, C.V., Gray, D.R., and Foster, D.A., 2002, Blueschist metamorphism during accretion in the Lachlan Orogen, southeastern Australia: Journal of Metamorphic Geology, v. 20, p. 711–726, doi: 10.1046/j.1525-1314.2002.00405.x.

Spaggiari, C.V., Gray, D.R., Foster, D.A., and McKnight, S., 2003, Evolution of the boundary between the western and central Lachlan Orogen: Implications for Tasmanide tectonics: Australian Journal of Earth Sciences, v. 50, p. 725–749, doi: 10.1111/j.1440-0952.2003.01022.x.

Spaggiari, C.V., Gray, D.R., and Foster, D.A., 2004a, Ophiolite accretion in the Lachlan Orogen, southeastern Australia: Journal of Structural Geology, v. 26, p. 87–112, doi: 10.1016/S0191-8141(03)00084-1.

Spaggiari, C.V., Gray, D.R., and Foster, D.A., 2004b, Lachlan Orogen subduction-accretion systematics revisited: Australian Journal of Earth Sciences, v. 51, p. 549–553, doi: 10.1111/j.1400-0952.2004.01073.x.

Stephens, M.B., Glassons, M.J., and Keays, R.R., 1979, Structural and chemical aspects of metamorphic layering development in metasediments from Clunes, Australia: American Journal of Science, v. 279, p. 129–160.

Tan, B.K., Gray, D.R., and Stewart, I., 1995, Volume change accompanying cleavage development in graptolitic shales from Gisborne, Victoria, Australia: Journal of Structural Geology, v. 17, p. 1387–1394, doi: 10.1016/0191-8141(94)00036-Y.

Twiss, R.J., and Moores, E.M., 1992, Structural Geology: New York, Freeman and Company, 532 p.

VandenBerg, A.H.M., 1999, Timing of orogenic events in the Lachlan Orogen: Australian Journal of Earth Sciences, v. 46, p. 691–701, doi: 10.1046/j.1440-0952.1999.00738.x.

VandenBerg, A.H.M., and Stewart, I.R., 1992, Ordovician terranes of the southeastern Lachlan fold belt: Stratigraphy, structure and palaeogeographic reconstruction: Tectonophysics, v. 214, p. 159–176, doi: 10.1016/0040-1951(92)90195-C.

VandenBerg, A.H.M., Willman, C.E., Hendrickx, M.A., Bush, M.D., and Sands, B.C., 1995, The geology and prospectivity of the 1993 Mount Wellington Airborne Survey area: Geological Survey of Victoria VIMP Report 2, 165 p.

VandenBerg, A.H.M., Willman, C.E., Maher, S., Simons, B.A., Cayley, R.A., Taylor, D.H., Morand, V.J., Moore, D.H., and Radojkovic, A., 2000, The Tasman Fold Belt System in Victoria: Geological Survey of Victoria Special Publication, 462 p.

van der Pluijm, B.A., and Marshak, S., 2004, Earth Structure: An Introduction to Structural Geology and Tectonics, Second Edition: New York, W.W. Norton and Company, 656 p.

van der Pluijm, B.A., Vrolijk, P.J., Pevear, D.R., Hall, C.H., and Solum, 2006, Fault dating in the Canadian Rocky Mountains: Evidence for late Cretaceous and early Eocene orogenic pulses: Geology, v. 34, p. 837–840, doi: 10.1130/G22610.1.

Waldron, H.M., and Sandiford, M., 1988, Deformation volume and cleavage development in metasedimentary rocks from the Ballarat Slate Belt: Journal of Structural Geology, v. 10, p. 53–62, doi: 10.1016/0191-8141(88)90127-7.

Wells, M.L., Snee, L.W., and Blythe, A.E., 2000, Dating of major normal fault systems using thermochronology; an example from the Raft River detachment, Basin and Range, Western United States: Journal of Geophysical Research, v. 105, p. 16,303–16,327, doi: 10.1029/2000JB900094.

White, S.H., and Johnston, D.C., 1981, A microstructural study and microchemical study of cleavage lamellae in a slate: Journal of Structural Geology, v. 3, p. 279–290, doi: 10.1016/0191-8141(81)90023-7.

Wilson, C.J.L., Will, T.M., Cayley, R.A., and Chen, S., 1992, Geologic framework and tectonic evolution in western Victoria, Australia: Tectonophysics, v. 214, p. 93–127, doi: 10.1016/0040-1951(92)90192-9.

Windley, B.F., Badarch, G., Cunningham, W.D., Kroener, A., Buchan, A.C., Tomurtogoo, O., and Salnikova, E.B., 2001, Subduction-accretion history of the central Asian orogenic belt: Constraints from Mongolia: Gondwana Research, v. 4, p. 825–826, doi: 10.1016/S1342-937X(05)70610-9.

Woodward, N.B., Gray, D.R., and Spears, D.B., 1986, Including strain data in balanced cross-sections: Journal of Structural Geology, v. 8, p. 313–324.

Yang, X., and Gray, D.R., 1994, Strain, cleavage and microstructure variations in sandstone: Implications for stiff layer behavior in chevron folding: Journal of Structural Geology, v. 16, p. 1353–1365, doi: 10.1016/0191-8141(94)90002-7.

Yu, S.A., Chen, H.Y., and Kue, L.C., 1997, Velocity field of GPS stations in the Taiwan area: Tectonophysics, v. 274, p. 41–57, doi: 10.1016/S0040-1951(96)00297-1.

MANUSCRIPT ACCEPTED BY THE SOCIETY 22 MARCH 2007

The Geological Society of America
Special Paper 433
2007

Cenozoic tectonic evolution of Qaidam basin and its surrounding regions (part 2): Wedge tectonics in southern Qaidam basin and the Eastern Kunlun Range

An Yin*
Department of Earth and Space Sciences and Institute of Geophysics and Planetary Physics,
University of California, Los Angeles, California 90095-1567, USA, and
School of Earth Sciences and Resources, China University of Geosciences, Beijing 100083, China

Yuqi Dang
Min Zhang
Qinghai Oilfield Company, PetroChina, Dunhuang, Gansu Province, China

Michael W. McRivette
W. Paul Burgess
Department of Earth and Space Sciences and Institute of Geophysics and Planetary Physics,
University of California, Los Angeles, California 90095-1567, USA

Xuanhua Chen
Institute of Geomechanics, Chinese Academy of Geological Sciences, Beijing 100081, China

ABSTRACT

Our inability to determine the growth history and growth mechanism of the Tibetan plateau may be attributed in part to the lack of detailed geologic information over remote and often inaccessible central Tibet, where the Eastern Kunlun Range and Qaidam basin stand out as the dominant tectonic features. The contact between the two exhibits the largest and most extensive topographic relief exceeding 2 km across their boundary inside the plateau. Thus, determining their structural relationship has important implications for unraveling the formation mechanism of the whole Tibetan plateau. To address this issue, we conducted field mapping and analysis of subsurface and satellite data across the Qimen Tagh Mountains, part of the western segment of the Eastern Kunlun Range, and the Yousha Shan uplift in southwestern Qaidam basin. Our work suggests that the western Eastern Kunlun Range is dominated by south-directed thrusts that carry the low-elevation Qaidam basin over the high-elevation Eastern Kunlun Range. Cenozoic contraction in the region initiated in the late Oligocene and early Miocene (28–24 Ma) and has accommodated at least

*yin@ess.ucla.edu

Yin, A., Dang, Y., Zhang, M., McRivette, M.W., Burgess, W.P., and Chen, X., 2007, Cenozoic tectonic evolution of Qaidam basin and its surrounding regions (part 2): Wedge tectonics in southern Qaidam basin and the Eastern Kunlun Range, *in* Sears, J.W., Harms, T.A., and Evenchick, C.A., eds., Whence the Mountains? Inquiries into the Evolution of Orogenic Systems: A Volume in Honor of Raymond A. Price: Geological Society of America Special Paper 433, p. 369–390, doi: 10.1130/2007.2433(18). For permission to copy, contact editing@geosociety.org. ©2007 The Geological Society of America. All rights reserved.

48% upper-crustal shortening (i.e., ~150 km shortening) since that time. In order to explain both the high elevation of the Eastern Kunlun Range and the dominant south-directed thrusts across the range, we propose that the Cenozoic uplift of the range has been accommodated by large-scale wedge tectonics that simultaneously absorb southward subduction of Qaidam lower crust below and southward obduction of Qaidam upper crust above the Eastern Kunlun crust. In the context of this model, the amount of Qaidam lower-crust subduction should be equal to or larger than the amount of shortening across Qaidam upper crust in the north-tapering thrust wedge system, thus implying at least 150 km of Qaidam lower-crust subduction since the late Oligocene and early Miocene. The late Oligocene initiation of contraction along the southern margin of Qaidam basin is significantly younger than that for the northern basin margin in the Paleocene to early Eocene between 65 and 50 Ma. This temporal pattern of deformation indicates that the construction of the Tibetan plateau is not a simple process of northward migration of its northern deformation fronts. Instead, significant shortening has occurred in the plateau interior after the plateau margins were firmly established at or close to their current positions. If Cenozoic crustal deformation across the Eastern Kunlun and Qaidam regions was accommodated by pure-shear deformation, our observed >48% upper-crustal shortening strain is sufficient to explain the current elevation and crustal thickness of the region. However, if the deformation between the upper and lower crust was decoupled during the Cenozoic Indo-Asian collision, lower-crustal flow or thermal events in the mantle could be additional causes of plateau uplift across the Eastern Kunlun Range and Qaidam basin.

Keywords: Qaidam basin, Eastern Kunlun Range, Tibetan plateau, wedge tectonics, Qilian Shan-Nan Shan thrust belt.

INTRODUCTION

Despite its importance in determining dynamics of continental deformation, the growth mechanism and constructional history of the Tibetan plateau during the Cenozoic Indo-Asian collision remain poorly understood. This incomplete knowledge may be attributed in part to the lack of detailed geologic information about remote and inaccessible central Tibet between the Bangong-Nujiang suture in the south and the Qilian Shan in the north (Fig. 1). In this area, the Eastern Kunlun Range and Qaidam basin stand out as the most dominant Cenozoic tectono-geomorphological features; the former has an average elevation of ~5000 m with peaks exceeding 7000 m, whereas the latter has an average elevation of ~2800 m with little internal relief. The contact between the two tectonic features forms the most extensive topographic front (>1000 km long) and the largest relief (>2.5 km on average) inside the plateau (Fig. 2). Understanding structural relationships between the development of the Eastern Kunlun Range and Qaidam basin has important implications for unraveling the formation mechanism and growth history of the whole Tibetan plateau.

In the past two decades, several tectonic studies have addressed this question. Burchfiel et al. (1989a) propose that the uplift of the Eastern Kunlun Range was induced by motion on a major south-dipping thrust; the fault marks the southern boundary of a north-directed thrust belt across Qaidam basin and the Qilian Shan to the north. The concept of the Eastern Kunlun Range being thrust over Qaidam basin was later expanded by Mock et al. (1999), who speculated that northward thrusting across the Eastern Kunlun Range was associated with vertical-wedge extrusion starting ca. 30–20 Ma.

In contrast to the above models emphasizing the role of dip-slip faulting across the Eastern Kunlun Range, Meyer et al. (1998), Tapponnier et al. (2001), Jolivet et al. (2003), and Wang et al. (2006) suggest that Cenozoic left-slip deformation has dominated the region. Specifically, they suggest that the Eastern Kunlun Range consists of a large transpressional system with the left-slip Kunlun fault in the south and a north-directed thrust along the northern flank of the Eastern Kunlun Range.

Deep crustal structures below the Eastern Kunlun Range have been explored by seismological studies. Based on interpretation of fault-plane solutions, focal-depth distributions, and active fault distributions, Tapponnier et al. (1990) and Chen et al. (1999) suggest that the Qaidam crust has been subducted southward below the Eastern Kunlun Range along a low-angle south-dipping thrust to a mantle depth. Examination of teleseismic P-wave arrivals by Zhu and Helmberger (1998) reveals that the Moho depth decreases abruptly across the topographic front between the Eastern Kunlun Range and Qaidam basin near Golmud, from ~50 km below the Eastern Kunlun Range to ~35 km below Qaidam basin. They attribute this staircase-like sharp decrease in Moho depth to the presence of a ductile-flow channel in the lower crust of the Eastern Kunlun Range that terminates abruptly in the north below the range front. This

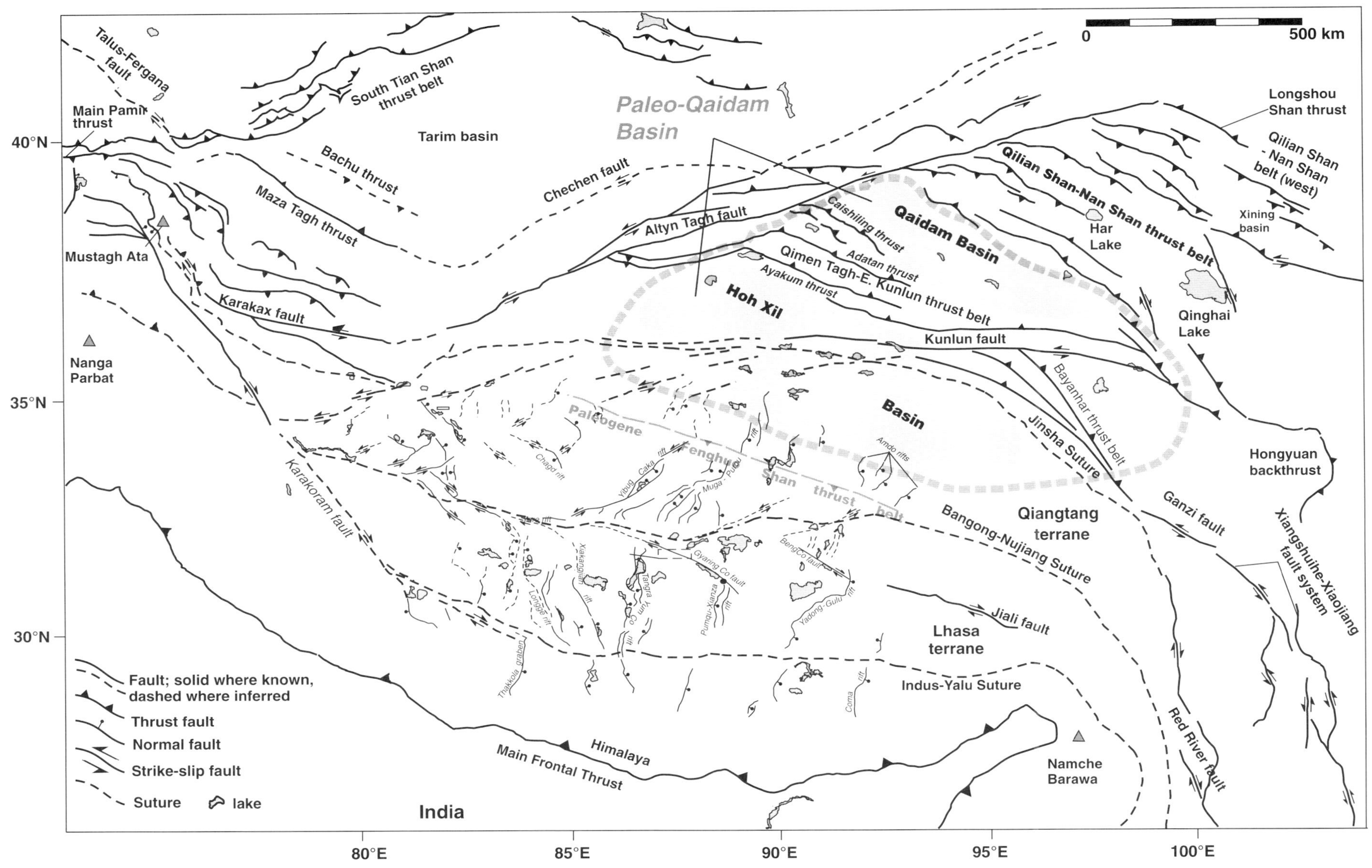

Figure 1. Tectonic map of major Cenozoic faults and sutures across the Tibetan plateau, modified after Taylor et al. (2003).

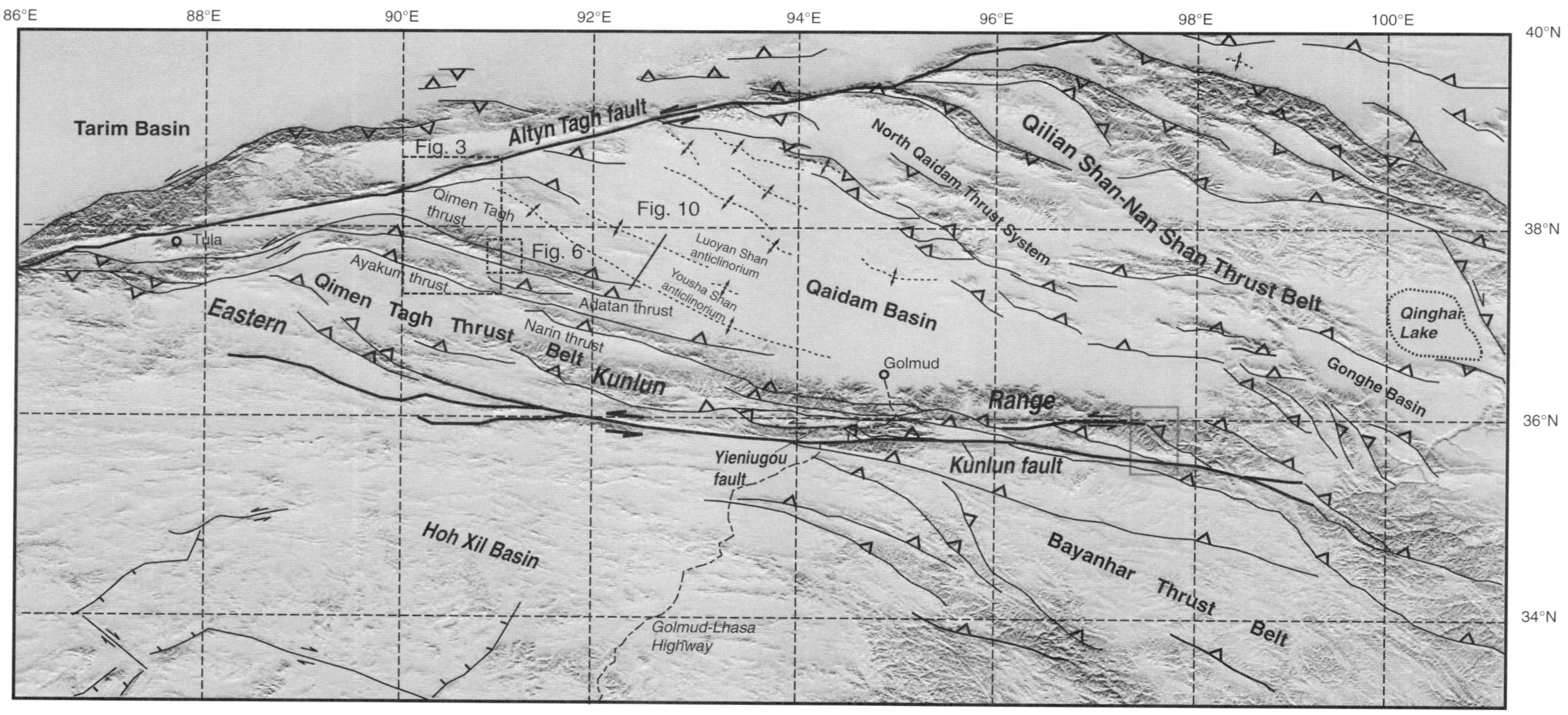

Figure 2. Digital topographic map of northern and central Tibet, showing major Cenozoic faults and locations of Figures 3, 6, and 10.

proposal is consistent with the predictions of the channel-flow model of Royden et al. (1997) and Clark and Royden (2000) for the development of the Tibetan plateau.

Except the model of Zhu and Helmberger (1998), all aforementioned structural models require the presence of a major south-dipping and north-directed thrust system that carries the Eastern Kunlun Range over Qaidam basin. This inference appears to be plausible, as it provides a simple explanation for the large elevation difference between the Eastern Kunlun Range and Qaidam basin. However, the existence of the *inferred* structure has never been established in the field. For example, the tectonic maps of Meyer et al. (1998) and Jolivet et al. (2003) place a south-dipping thrust along the northern front of the Eastern Kunlun Range, but these authors did not provide any field documentation of the structure. To the contrary, the existing regional geologic map (e.g., Liu, 1988) and fieldwork performed by Dewey et al. (1988) during the Sino-Anglo Geotraverse have all failed to document such a structure along the northern margin of the Eastern Kunlun Range.

Although focal mechanisms and focal-depth distribution have been used to support the presence of a south-dipping thrust below the Eastern Kunlun Range (Chen et al., 1999), it should be kept in mind that the data projected on the single cross section were collected along the whole Eastern Kunlun Range, which is more than 1000 km long. In addition, the fault plane solutions could not differentiate which of the two possible fault planes actually created the earthquakes.

The contradiction between having a major south-dipping range-bounding thrust along the northern margin of the Eastern Kunlun Range and the expected thickness distribution of Cenozoic strata across Qaidam basin has long been noted; Cenozoic strata and their individual units consistently thicken toward the basin center and thin toward the basin margins (Bally et al., 1986; Huang et al., 1996; Sobel et al., 2003; Wang et al., 2006; Yin et al., 2008a, 2008b). This isopach pattern contradicts the prediction of a classic foreland-basin model that requires the maximum thickness of foreland sedimentation to be restricted to the bounding thrust (e.g., Jordan, 1981). For this reason, Bally et al. (1986) speculated that the Cenozoic Qaidam basin formed during the development of a large synclinorium.

In order to determine the geologic relationship between the Cenozoic uplift of the Eastern Kunlun Range and the evolution of Qaidam basin, we conducted geologic research in the region in the past decade by integrating surface mapping with interpretations of subsurface seismic, drill-hole, and sedimentological data. Due to the large geographic extent of the area (the Eastern Kunlun Range alone is larger than California!), we report our results in three separate papers. The first paper by Yin et al. (2008a) deals with the Cenozoic structural evolution of the North Qaidam thrust system and the southern Qilian Shan-Nan Shan thrust belt. The second paper, presented here, documents the geology of the western segment of the Eastern Kunlun Shan and southwestern Qaidam basin and discusses its implications on the mode and timing of plateau construction across central Tibet. Our third paper by Yin et al. (2008b) examines Cenozoic sedimentation and structural evolution of the whole Qaidam basin and its relationship to slip distribution along the left-slip Altyn Tagh fault.

Our work in the western segment of the Eastern Kunlun Range and southern Qaidam basin is based on detailed field mapping, interpretation of subsurface data from petroleum drill holes and seismic reflection profiles, and systematic analysis of satellite images. Our work, in conjunction with the existing seismic reflection profiles across the region, suggests that (1) the contact between the western Eastern Kunlun Range and southwestern Qaidam basin is a north-dipping unconformity, (2) the major structures across the Eastern Kunlun Range are dominated by *south-directed* Cenozoic thrusts carrying the low-altitude Qaidam basin over the high-altitude Eastern Kunlun Range, and (3) Cenozoic contraction in the region initiated in the late Oligocene to early Miocene (28–24 Ma), much later than the Paleocene–early Eocene (65–50 Ma) initiation age of Cenozoic deformation along the northern margin of Qaidam basin. In the following sections, we outline the regional geologic framework of Qaidam basin and the Eastern Kunlun Range. This is followed by a detailed documentation of our field observations and subsurface data from our study area.

GEOLOGY OF QAIDAM BASIN AND THE EASTERN KUNLUN RANGE

The morphologically defined Eastern Kunlun Range is bounded by Qaidam basin to the north and the Hoh Xil basin to the south (Fig. 1). The boundary between the Hoh Xil basin and the Eastern Range is marked by the left-slip Kunlun fault. Qaidam basin is the largest topographic depression inside Tibet and contains an Eocene to Quaternary continental sedimentary sequence (Bally et al., 1986; Wang and Coward, 1990; Song and Wang, 1993; Huang et al., 1996; Zhang, 1997; Métivier et al., 1998; Xia et al., 2001; Yin et al., 2002; Dang et al., 2003; Sobel et al., 2003; Sun et al., 2005; Rieser et al., 2005, 2006a, 2006b; Zhou et al., 2006). The thickest Cenozoic strata are located in the basin center with a depth >15 km; the strata thin toward the basin margins to ~<3 km (Huang et al., 1996). Sedimentological development of the basin has been established by analyzing thickness distribution (Huang et al., 1996), paleocurrent directions (Hanson, 1998), lithofacies patterns (e.g., Zhang, 1997), sandstone petrology (Rieser et al., 2005), $^{40}Ar/^{39}Ar$ ages of detrital micas (Rieser et al., 2006a, 2006b), and fission-track ages of detrital apatites (Qiu, 2002). These studies suggest that the Cenozoic Qaidam basin has expanded eastward since the Eocene, with its depocenters consistently located along the axis of the basin.

Cenozoic stratigraphic division and age assignments across Qaidam basin are based on the correlation of outcrop geology with subsurface data (seismic profiles and drill cores), terrestrial fossils (i.e., spores, ostracods, and pollen), basin-wide stratigraphic correlation via a dense network of seismic reflection profiles, magnetostratigraphic studies, and fission-track and $^{40}Ar/^{39}Ar$ dating of detrital grains (Huo, 1990; Qinghai, B.G.M.R., 1991; Yang et al., 1992; Song and Wang, 1993; Huang et al.,

1996; Xia et al., 2001; Qiu, 2002; Sun et al., 2005; Rieser et al., 2006a, 2006b). Major Cenozoic stratigraphic units in Qaidam basin include (Table 1): Paleocene (?) and early Eocene Lulehe Formation (E1+2; >54–49 Ma) (Yang, 1988; Huo, 1990; Yang et al., 1992; Rieser et al., 2006a, 2006b), middle and late Eocene Lower Xiaganchaigou Formation (E3-1; 49–34 Ma) (Yang et al., 1992; Sun et al., 2005), early Oligocene Upper Xiaganchaigou Formation (E3-2; 34–28.5 Ma) (Sun et al., 1999), late Oligocene Shangganchaigou Formation (N1; 28.5–24 Ma) (Sun et al., 1999), early to middle Miocene Xiayoushashan Formation (N2-1; 24–11 Ma) (Sun et al., 1999), late Miocene Shangyoushashan Formation (N2-2; 11–5.3 Ma), Pliocene Shizigou Formation (N2-3; 5.3–1.8 Ma) (Sun et al., 1999), and Quaternary Qigequan and Dabuxun Yanqiao Formations (Q2) (Yang et al., 1997; Sun et al., 1999). Our age division closely follows that of Rieser et al. (2006a, 2006b) except the age assignment of the Xiaganchaigou Formation, for which we rely on the result of a more recent magnetostratigraphic study by Sun et al. (2005).

The Kunlun fault was initiated at 15–7 Ma, with a Quaternary slip rate of 11–16 mm/yr and a total displacement of ~75 km (Kidd and Molnar, 1988; van Der Woerd et al., 2002; Jolivet et al., 2003; Lin et al., 2006). The bedrock of the Eastern Kunlun Range is dominated by two phases of plutonism at the Ordovician-Silurian and Permian-Triassic (Liu, 1988). Due to the extensive exposure of plutons, the area is often referred to as the Kunlun Batholith Belt (e.g., Chang and Zheng, 1973; Allègre et al., 1984; Harris et al., 1988; Dewey et al., 1988). The Paleozoic and early Mesozoic igneous rocks were intruded into Precambrian gneiss, Neoproterozoic metasediments, and Devonian-Carboniferous marine strata; they were in turn overlain by Jurassic to Cenozoic continental deposits (Liu, 1988; Coward et al., 1988; Harris et al., 1988; Mock et al., 1999; Cowgill et al., 2003; Robinson et al., 2003; Roger et al., 2003). The Sino-Anglo Geotraverse along the Golmud-Lhasa Highway (Dewey et al., 1988; Coward et al., 1988; Harris et al., 1988) and the reconnaissance work of others at the western end of the Eastern Kunlun Range (Molnar et al., 1987; Burchfiel et al., 1989b; McKenna and Walker, 1990) have revealed the general style of deformation, extent of major Mesozoic sutures, and existence of Cenozoic igneous activities across the region. ^{40}Ar/^{39}Ar thermochronological studies reveal widespread Mesozoic cooling events across the range, which were overprinted locally by a prominent Cenozoic cooling event at 30–20 Ma (Mock et al., 1999; Liu et al., 2005). Apatite fission-track studies suggest that the Eastern Kunlun region experienced rapid and widespread cooling ca. 20–10 Ma, possibly related to the initiation of Cenozoic uplift (Jolivet et al., 2001; Wang et al., 2004; Liu et al., 2005; Yuan et al., 2006). Due to the reconnaissance nature of the above investigations and their strong emphasis on thermochronology, the structural relationships and the evolution of the geologic contact between the Eastern Kunlun Range and Qaidam basin are essentially unconstrained. Below we address this issue by presenting new field observations and interpretation of newly available subsurface data from the Qimen Tagh Mountains of the western Eastern Kunlun Range and southwestern Qaidam basin (Fig. 2).

TABLE 1. CENOZOIC STRATIGRAPHY OF QAIDAM BASIN

Unit name	Symbol	Geologic time	Age	Reflectors
Dabuxun Yanqiao Formation	Q2	Holocene	0.01–present	
Qigequan Formation	Q1	Pleistocene	1.8–0.01 Ma	T0
Shizigou Formation	N2-3	Pliocene	5.3–1.8 Ma	T1
Shangyoushashan Formation	N2-2	Late Miocene	11.2–5.3 Ma	T2'
Xiayoushashan Formation	N2-1	Early and middle Miocene	23.8–11.2 Ma	T2
Shangganchaigou Formation	N1	Late Oligocene	28.5–23.8 Ma	T3
Upper Xiaganchaigou Formation	E3-2	Early Oligocene	37–28.5 Ma	T4
Lower Xiaganchaigou Formation	E3-1	Middle Eocene to late Eocene	49–37 Ma	T5
Lulehe Formation	E1+2	Paleocene to early Eocene	>54.8–49 Ma	TR
(Dominantly Jurassic strata locally with Cretaceous beds on top)	Jr	Jurassic-Cretaceous	206–65 Ma	T6

THE QIMEN TAGH AND SOUTHWEST QAIDAM BASIN

Structural Geology from Surface Mapping

The Qimen Tagh Mountains and southwestern Qaidam basin expose from north to south the following major Cenozoic structures: (1) the active left-slip Altyn Tagh fault, (2) the north-dipping Yiematan thrust, (3) the north-dipping Caishiling thrust, (4) the south-verging and eastward-plunging Yousha Shan anticlinorium, (5) the Gas Hure syncline, (6) the north-dipping Qimen Tagh imbricate thrust zone, (7) the north-dipping Adatan thrust, and (8) the north-dipping Ayakum thrust (Figs. 3 and 4).

The left-slip Altyn Tagh fault makes a right step in our study area and forms a prominent restraining bend. Several south-directed thrusts and folds are developed in this transpressional zone, including the Yiematan and Caishiling thrusts and the Yousha Shan anticlinorium. The Yiematan thrust juxtaposes Proterozoic gneisses over Neogene strata, whereas the Caishiling thrust places Ordovician metavolcanics and metagraywacke over Jurassic fluvial and lacustrine deposits and Cenozoic red beds (Fig. 3).

The Yousha Shan anticlinorium is a south-verging fold, with a steep to overturned southern limb (Figs. 4 and 5). Contractional structures in this fold complex have an eastward decreasing magnitude of shortening, which is expressed by a progressive decrease in stratigraphic throw across thrusts and eastward plunging fold axes (Figs. 3 and 5). The fold is active and has been propagating eastward in the Quaternary, which is expressed by sequential eastward uplift and tilting of Quaternary terrace surfaces (Fig. 5). In addition, the eastward growth of the anticline has caused a systematic deflection of drainages that flow around the fold nose (Fig. 5).

The Gas Hure syncline is the southern continuation of the active Yousha Shan anticlinorium and controls an active depocenter zone expressed by the development of Gas Hure Lake (Figs. 3 and 5). The Cenozoic development of the syncline has produced a northward-thickening sequence across its southern limb, which is evident from well data indicating a systematic northward increase in depth to crystalline basement (Fig. 4). The north-directed Arlar thrust is a minor blind structure well-imaged seismically below the Gas Hure syncline (Fig. 4). Motion on the thrust has produced a fault-propagation fold in its hanging wall and an early Miocene to Quaternary growth-strata sequence, indicating the thrust initiated in the early Miocene and Pliocene (Fig. 4).

The northern Qimen Tagh Mountains are dominated by south-directed imbricate thrusts, which juxtapose Proterozoic gneiss and early Paleozoic granites over Ordovician and Carboniferous strata, and locally place Precambrian and Paleozoic rocks directly over Eocene strata (Figs. 3 and 4). Due to the lack of matching units across these faults, the exact slip on these faults is unknown. In contrast to the thrusts to the north that are mainly developed in the bedrock of the range, the southern Qimen Tagh Mountains are dominated by the Adatan and Ayakum thrusts that place Proterozoic, Paleozoic, and Triassic rocks directly over Quaternary sediments. The morphological expression of the active Adatan fault is particularly striking in Advanced Spaceborne Thermal Emission and Reflection Radiometer (ASTER) images and in the field; it cuts alluvial fan surfaces, forming prominent fault scarps (Figs. 6, 7, and 8A). Ordovician graywacke in the Adatan hanging wall experienced two phases of folding. The first was isoclinal folding that transposed bedding and was associated with well-developed slaty cleavage. The second phase of folding is tight and commonly associated with brittle thrusting forming fault-bend or fault-propagation folds (Fig. 8D). These folds are more widely spaced and are defined by warping of the slaty cleavage from the first-phase folding (Fig. 7). The style of the deformation for the second-phase folding is compatible with that in the Cenozoic strata of southern Qaidam basin. For this reason, we suggest that the later folding event was caused by Cenozoic tectonics.

The Ayakum thrust zone is an active structure and has a strong topographic expression (Fig. 2). It consists of two fault branches in the study area (Figs. 3 and 4), with a northern thrust placing Paleozoic granite over Triassic strata and a southern thrust placing Triassic rocks over Neogene-Quaternary strata. The Ayakum fault strikes east and makes a sharp turn to the southwest at its western end; the southwest-trending fault segment in turn is linked with a north-directed Cenozoic thrust system to the south near Tula (Robinson et al., 2003; Dupont-Nivet et al., 2004) (Figs. 2 and 3).

To determine the contact relationship between the western Eastern Kunlun Range and Qaidam basin, we systematically examined the geology across the topographic boundary between the two tectonic features. Our field observations reveal no disrupted Quaternary fans along the range front, indicating no Quaternary active faults that bound the Eastern Kunlun Range, as speculated by many early workers in the region. This observation applies to the whole northern front of the Eastern Kunlun Range.

The contact between the range and Qaidam basin is concealed in most places by Quaternary fan deposits that overlap bedrock of the Eastern Kunlun Range. In a few places, the contact between the Paleozoic rocks of the Eastern Kunlun Range and Cenozoic strata of southern Qaidam basin is exposed in deeply cut canyons, where we observed that north-dipping Cenozoic strata rest unconformably on top of Paleozoic rocks (Fig. 8B). The Cenozoic strata above the unconformity are folded and cut by south-directed thrusts (Fig. 9). Locally, the contact between the Eastern Kunlun Range and Qaidam basin is defined by gentle north-dipping Pliocene strata overlapping north-dipping Carboniferous beds (Fig. 8C). Similar relationships are also seen from seismic reflection profiles across the southern margin of Qaidam basin (Fig. 10).

Structural Geology from Subsurface Data

We have examined a series of seismic-reflection profiles and drill-hole data from southwestern Qaidam basin provided to us by the Qinghai Oilfield Company. These data help constrain the geometry of major structures in the basin and their relationship to the Eastern Kunlun uplift. Figure 10 shows an example

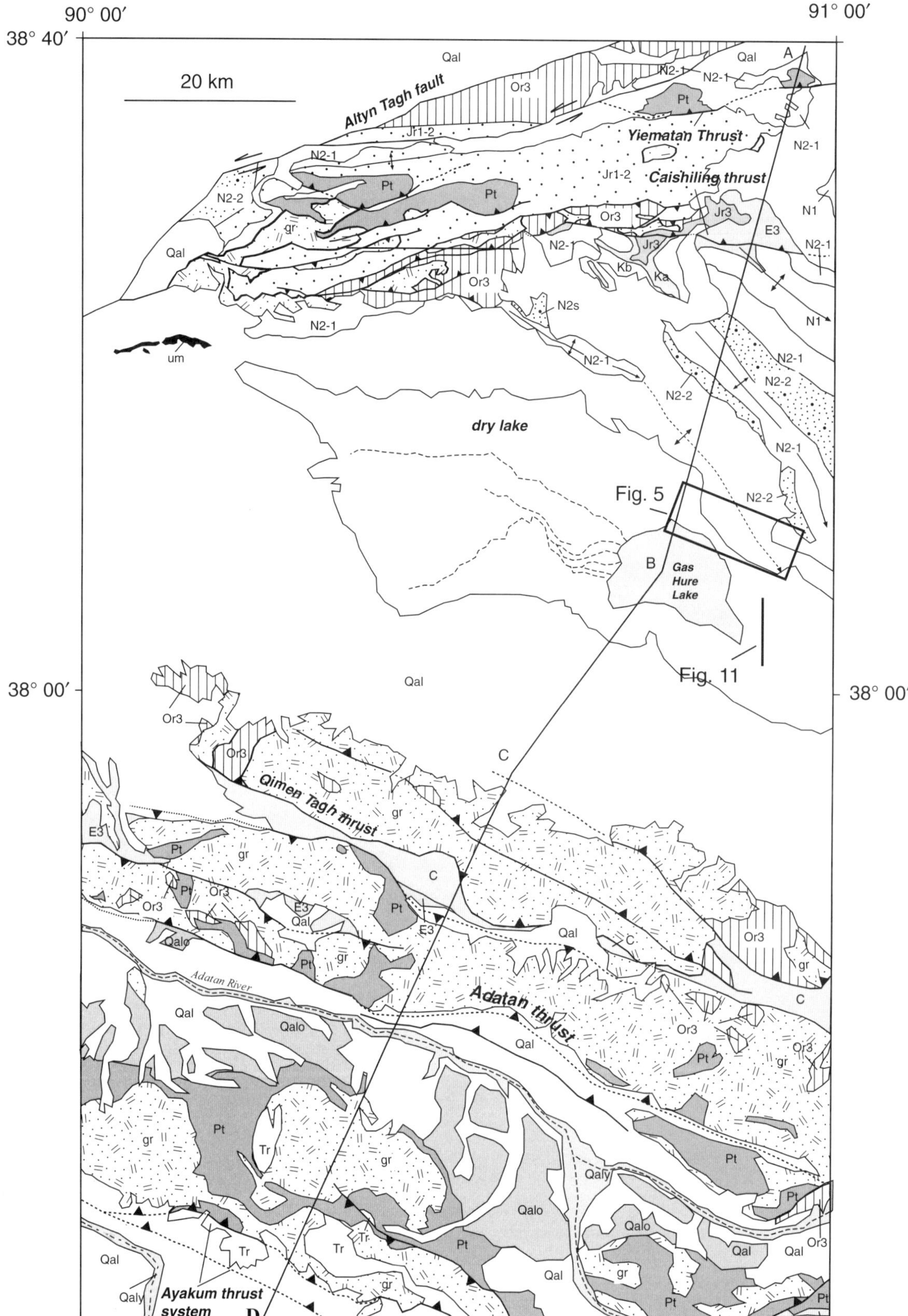
90° 00′
91° 00′
38° 40′
38° 00′
37° 20′
20 km
Altyn Tagh fault
Yiematan Thrust
Caishiling thrust
Qal
Or3
Pt
Jr1-2
N2-1
N2-2
N2s
N1
Jr3
E3
Kb
Ka
gr
um
A
B
C
D
dry lake
Fig. 5
Fig. 11
Gas Hure Lake
Qimen Tagh thrust
Adatan thrust
Adatan River
Qalo
Qaly
Tr
Ayakum thrust system
Yiqiandaba River

Figure 3. Geologic map of the Qimen Tagh and Yousha Shan area; see Figure 2 for location. Major lithologic units: Qaly—active fluvial channel deposits; Qal—Quaternary alluvial deposits; Qalo—older Quaternary alluvial deposits; N1—late Oligocene strata; N2-1 and N2-2—older and younger Pliocene sedimentary units; E3—Eocene-Oligocene strata; Ka—older Cretaceous unit; Kb—younger Cretaceous unit that also includes Paleocene–early Eocene strata; Jr1-2 and Jr3—Jurassic sedimentary units; Tr—Triassic sedimentary and volcanic rocks; C—Carboniferous sedimentary unit; Or3—Upper Ordovician metavolcanic and metagraywacke sequences; Pt—Proterozoic gneisses; gr—early Paleozoic granites; um—ultramafic rocks. The map also shows locations of Figures 5 and 11.

of such seismic profiles, this one lying across the eastern tip of the north-dipping Qimen Tagh thrust zone that we mapped in the field to the west (Fig. 2). The thrust fault in the seismic section offsets the base of the Cenozoic strata for ~5 km. Displacement on this fault induced a south-verging fault-bend fold in the hanging wall. The north-dipping Cenozoic strata shown in the seismic line along the southern basin margin are consistent with our field observations discussed earlier, and so is the folding of the Cenozoic strata (cf. Figs. 8B and 9). More complete data sets on seismic-reflection profiles across the whole Qaidam basin are reported in Yin et al. (2008b).

In addition to the seismic profiles we obtained from Qinghai Oilfield Company, we also reinterpret a published seismic profile by Song and Wang (1993) (Fig. 11). In the original interpretation, Song and Wang (1993) failed to recognize the presence of a fault-bend fold above the Arlar thrust. This thrust is a backthrust in the forelimb of the south-verging Yousha Shan anticlinorium. A similar backthrust is also shown in Figure 10 across the eastern end of the Yousha Shan anticlinorium.

Across the eastern segment of the Yousha Shan anticlinorium, the backlimbs of the south-directed Yousha Shan and north-directed Luoyan Shan anticlinoriums are flat. Based on the classic fault-bend fold model of Suppe (1983), this suggests the presence of a flat décollement in the middle crust when the detailed geometric relationships between the two high-angle thrusts and their related fold complexes in the respective hanging wall are considered (Fig. 10). Across the western segment of the Yousha Shan anticlinorium, surface geology and its downward projection requires the presence of a major thrust ramp below the anticlinorium when cross section balance is considered. This thrust ramp is also exposed in the eastern section in Figure 10, but it lies at a much shallower depth.

Our inferred décollement at a middle crustal level below Qaidam basin is consistent with the early proposal of Burchfiel et al. (1989a) based on two lines of argument: (1) there are no lower crustal rocks exposed across Qaidam basin and the Qilian Shan–Nan Shan thrust belt, and (2) continental middle crust tends to localize ductile shear zones, as exemplified by many extensional detachment fault zones around the world. We admit that our location of the décollement is highly tentative. For example, the fault could be located at greater depth than we assume

here, such as along the base of the lower crust immediately above the Moho, where the crustal strength is presumably weakest, as inferred from depth-dependent crustal rheology (e.g., Chen and Molnar, 1983). However, determining the existence and depth of the proposed middle or lower crustal décollement is beyond the scope of this study; we hope our proposal will inspire more geophysical research across the area to resolve this issue.

Magnitude of Cenozoic Shortening

Due to the lack of matching units across many Cenozoic thrusts within the Qimen Tagh Mountains, it is difficult to estimate their exact offsets and thus the Cenozoic shortening strain across the range. However, a relatively accurate estimate of crustal shortening can be obtained across southwestern Qaidam basin, where surface geology is well mapped and seismic-reflection profiles provide tight constraints on the geometry of the deeper structures. The cross section of southwest Qaidam basin, from point A to point B on Figure 4, was constructed using the kink-bend method of Suppe (1983) and reinterpretation of a seismic profile published by Song and Wang (1993). Line balancing and summation of minimum fault slips across this section suggest shortening of >58 km, which yields a shortening strain of >48%. Given the significantly higher elevation of the Qimen Tagh Mountains (~5000 m) relative to the Yousha Shan anticlinorium (~3200 m) and field evidence for a denser distribution of Cenozoic thrusts and folds across the range, the shortening strain from southwest Qaidam basin may only be a lower bound for the shortening strain across the Qimen Tagh Mountains to the south.

Timing of Cenozoic Deformation

Three lines of evidence suggest that Cenozoic deformation across the Qimen Tagh Mountains and southwestern Qaidam basin was initiated in or after the latest Oligocene–early Miocene. First, Oligocene lacustrine deposits are scattered in the Qimen Tagh Mountains and have been correlated with coeval deposition of lacustrine strata in southwestern Qaidam (Zhong et al., 2004; Guo et al., 2006) (Fig. 3). This observation implies the presence of a unified but much larger basin extending from the present southwestern Qaidam basin into the Qimen Tagh Mountains. The basin did not become fragmented until after Oligocene lacustrine deposition across the region, which implies that the Qimen Tagh was not uplifted until after the end of the Oligocene.

The second line of evidence for a Neogene initiation of contraction in the region comes from growth strata relationships imaged in seismic-reflection profiles (Figs. 10 and 11). In Figure 10, stratigraphic thickening occurs in unit N2-1 (early and middle Miocene; see Table 1) and the younger strata above it in a synclinal position, whereas the thickness of unit N1 (late Oligocene; see Table 1) and older strata are not affected by folding. This suggests that the eastern segment of the Yousha Shan anticlinorium did not start to develop until after the early Miocene. In contrast to the eastern section, the seismic section across the western Yousha

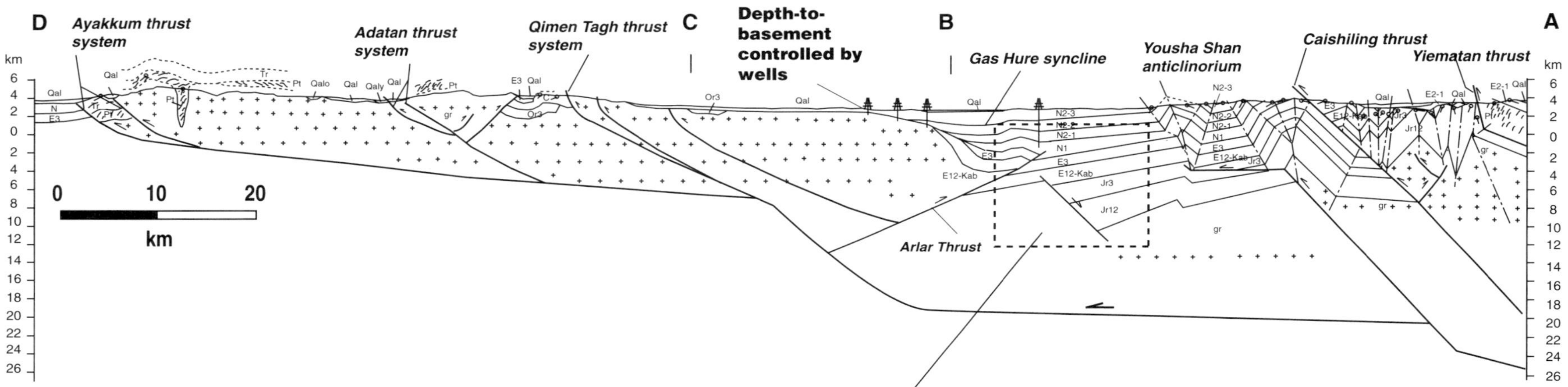

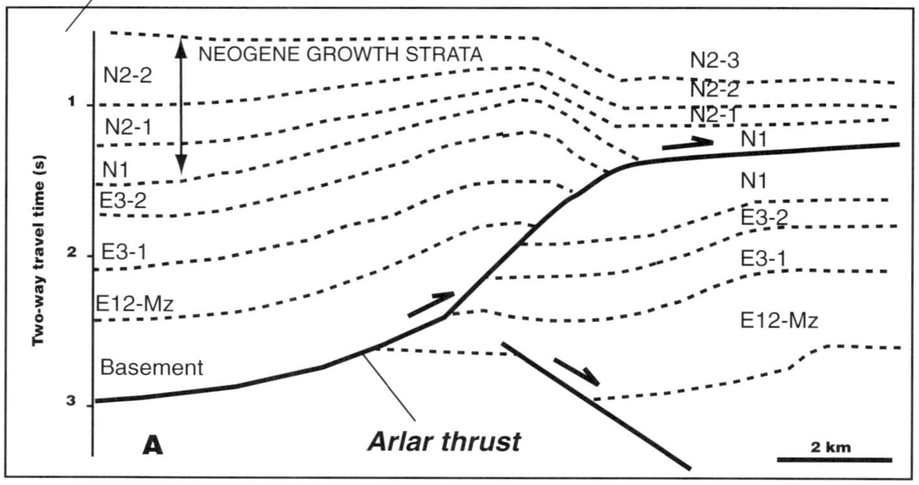

Figure 4. Geologic cross section across the Qimen Tagh–Yousha Shan thrust belt. Structures in the Yousha Shan region are constructed using surface geology and the kink-bend method of Suppe (1983). Geology across the Gas Hure syncline and the northern flank of the Qimen Tagh Mountains is constrained by drill-hole and seismic data. See text for details. Lithologic units: Qaly—active fluvial channel deposits; Qal—Quaternary alluvial deposits; Qalo—older Quaternary alluvial deposits; N2-1 and N2-2—older and younger Miocene sedimentary units; N1—late Oligocene strata; E3—Eocene-Oligocene strata; E12-Kab—Cretaceous and lower Eocene strata; Jr1-2 and Jr3—Jurassic sedimentary units; Tr—Triassic sedimentary and volcanic rocks; C—Carboniferous sedimentary unit; Or3—Upper Ordovician metavolcanic and metagraywacke sequences; Pt—Proterozoic gneisses; gr—early Paleozoic granites. Inset cross section: Detailed structural relationships interpreted from seismic line in Figure 11 across Gas Hure syncline.

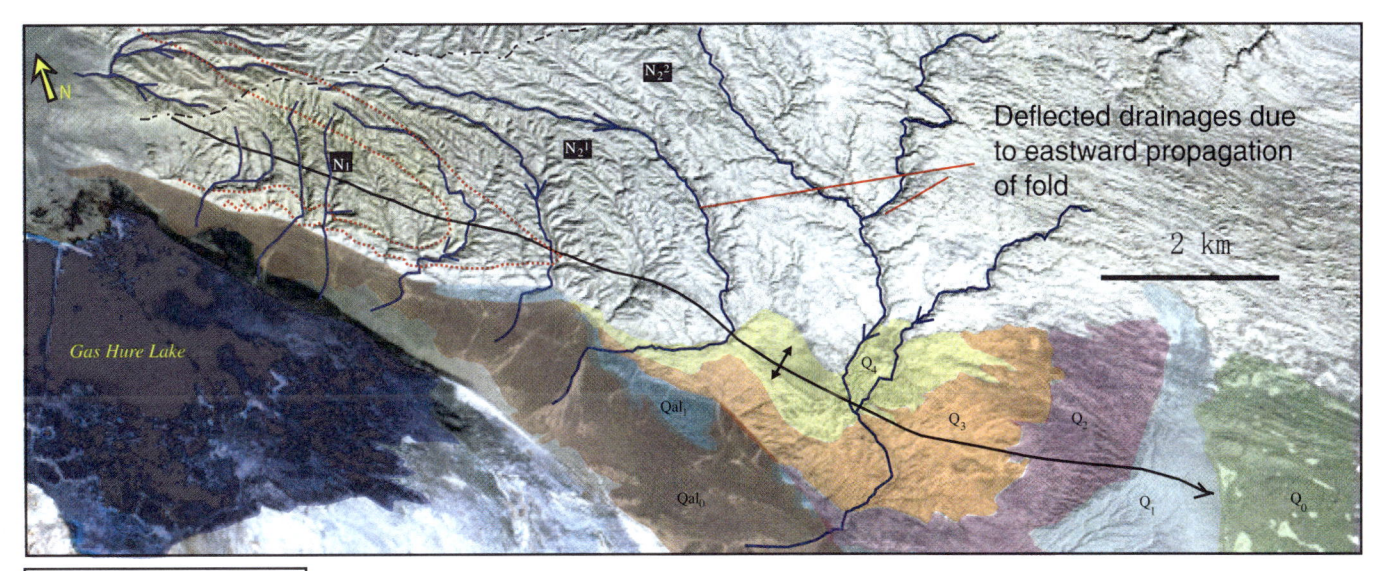

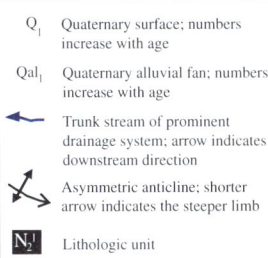

Figure 5. Advanced Spaceborne Thermal Emission and Reflection Radiometer image of the Yousha Shan anticlinorium. Note that south-flowing drainages are systematically deflected and make U-shaped turns around the anticlinal nose. The tightness of the turns decreases eastward, which is associated with eastward younging of tilted Quaternary fan surfaces as indicated by an eastward decrease in drainage density and incision. We interpret the above observations as results of eastward propagation of the Yousha Shan anticlinorium. Also note that the drainage divide does not coincide with the anticline axis. This may be caused by a lower base level on the south side of the anticline than that on the north side, forcing the drainage divide to have propagated northward as the anticline grew to the east.

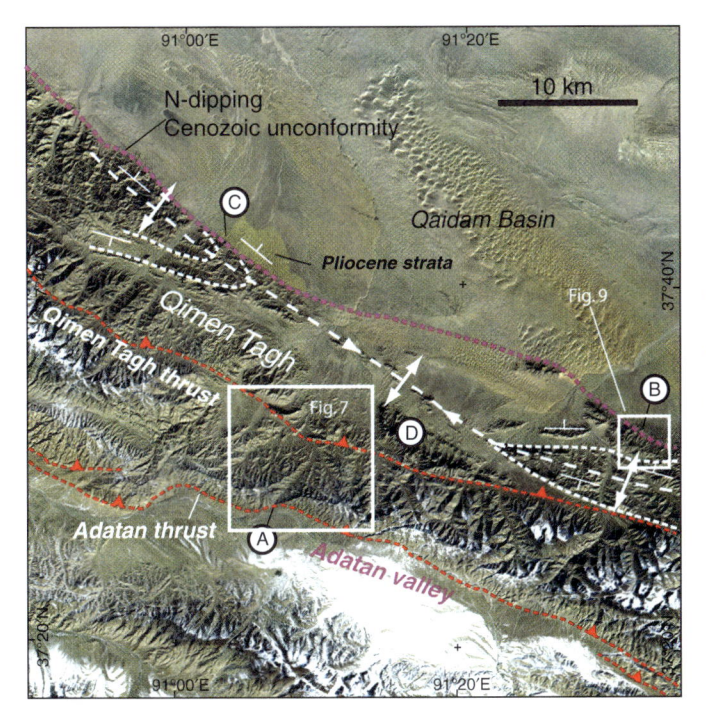

Figure 6. Advanced Spaceborne Thermal Emission and Reflection Radiometer image of the central Qimen Tagh area. The sharp contact between Qaidam basin and bedrock in the range to the south has been interpreted as evidence for active north-directed thrusting (e.g., Meyer et al., 1998; Jolivet et al., 2003). However, field examination indicates that this is a steeply north-dipping unconformity between a resistant Carboniferous marble unit in the south- and north-dipping Cenozoic strata in the north. The image also shows the trace of the active Adatan thrust. Locations of field photos in Figure 8 are indicated on this image as A, B, C, and D. Locations of Figures 7 and 9 are also indicated.

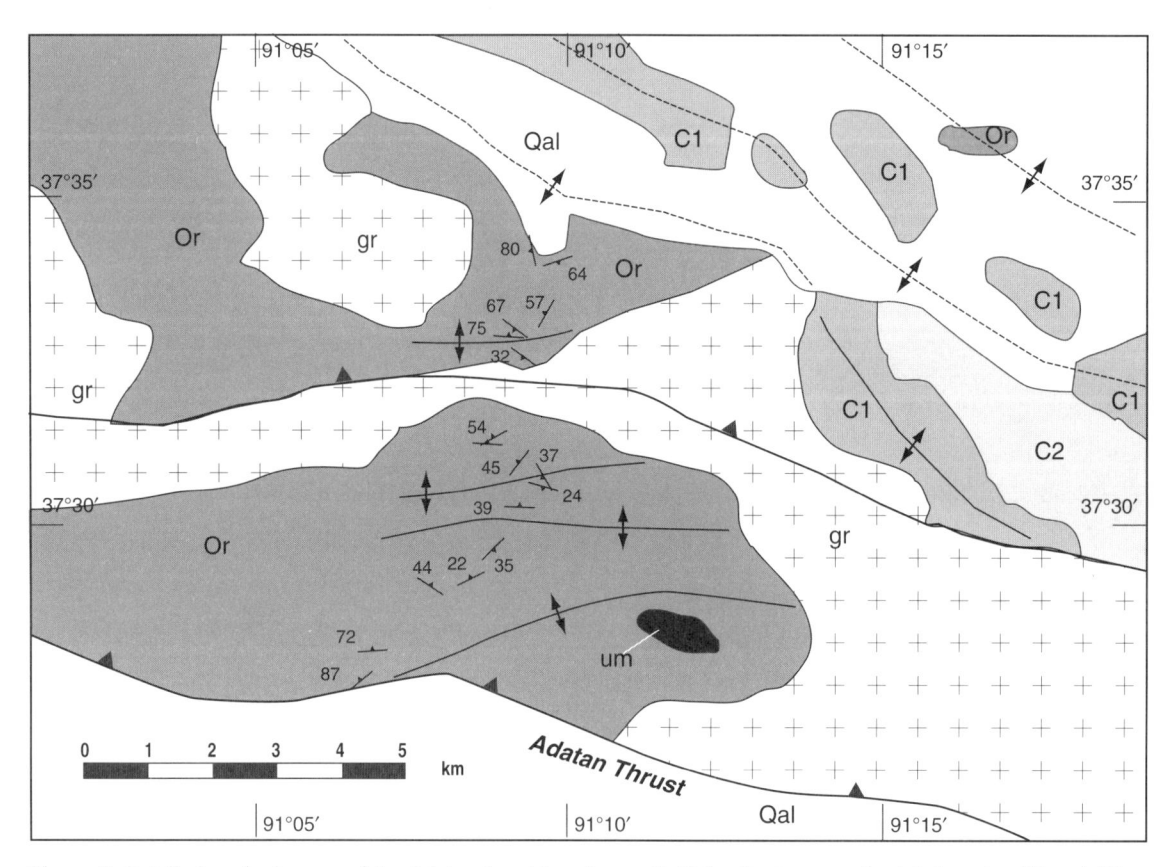

Figure 7. Detailed geologic map of the Adatan thrust hanging wall. Qal—Quaternary alluvial deposits; C1 and C2—Carboniferous units; gr—early Paleozoic granites; Or—Ordovician unit; um—ultramafic rocks. The major folds shown in the map are defined by cleavage that was developed during an early phase of isoclinal folding that transposed most of the original bedding.

Shan anticlinorium indicates an earlier initiation age of deformation. This is shown by synfolding and synthrusting deposition of unit N1; its thickness changes across the fault-bend fold above the Arlar thrust. Because the Arlar thrust is part of the forelimb structure of the Yousha Shan anticlinorium, we suggest that Cenozoic contraction across the western segment of the anticlinorium did not begin until the late Oligocene. The combination of these age interpretations brings us to the conclusion that the Yousha Shan anticlinorium has grown from west to east. This interpretation is consistent with the conclusion reached by our independent geomorphologic arguments (as discussed earlier; see Fig. 5).

The third line of evidence comes from ^{40}Ar/^{39}Ar and apatite fission-track thermochronology across the Eastern Kunlun Range. Existing ^{40}Ar/^{39}Ar mica and biotite ages indicate that the range experienced a profound Mesozoic cooling, but no Cenozoic signals were detected (Wang et al., 2004; Liu et al., 2005). By conducting ^{40}Ar/^{39}Ar multidomain analysis of K-feldspar samples collected across the central Eastern Kunlun Range, Mock et al. (1999) revealed a highly localized rapid cooling event at 30–20 Ma (late Oligocene and early Miocene). In contrast to the general lack of Cenozoic cooling signals from high-temperature thermochronology, apatite fission-track studies across the Eastern Kunlun region consistently point to a Neogene initiation of range uplift. Jolivet et al. (2001) show that rapid cooling in the western segment of the Eastern Kunlun Range did not begin until after 15 Ma. Similarly, Yuan et al. (2006) show initial rapid cooling to have occurred in the eastern segment of the Eastern Kunlun Range between 20 and 10 Ma. If Cenozoic cooling detected by apatite fission-track thermochronology reflects crustal thickening and its induced exhumation, then these thermochronological data suggest that the Eastern Kunlun Range has been uplifted since 20–10 Ma, with local areas uplifted slightly earlier, ca. 30–20 Ma. This age range is significantly younger than the initiation age of contractional deformation along the northern margin of Qaidam basin (Jolivet et al., 2001; Yin et al., 2002).

DISCUSSION

Our field studies, together with the existing observations, lead to the following findings: (1) the western segment of the Eastern Kunlun Range is dominated by south-directed thrusts carrying the low-elevation Qaidam basin over the high-elevation Eastern Kunlun Range; (2) shortening strain across southwestern Qaidam basin exceeds 48%; and (3) Cenozoic contraction in

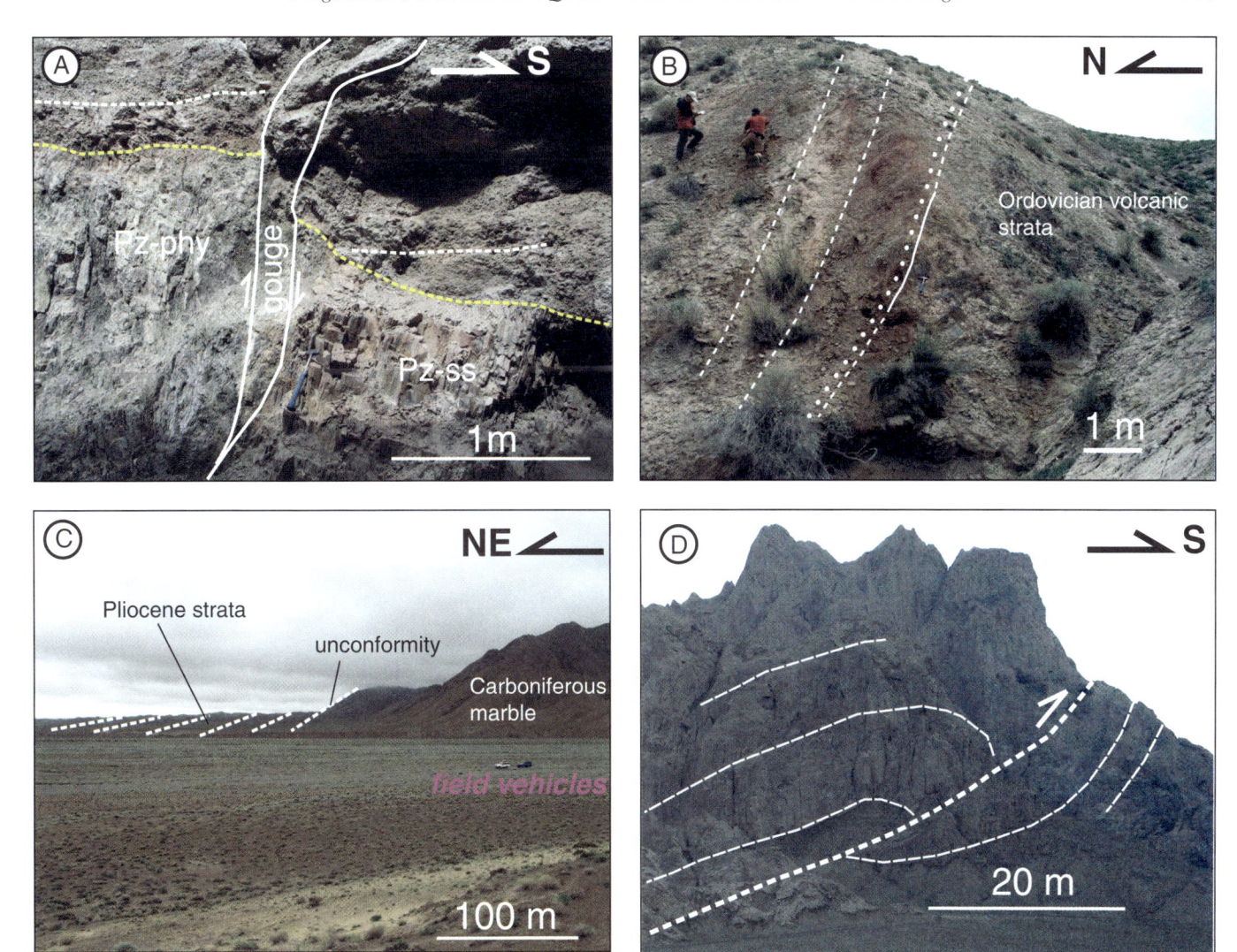

Figure 8. Field photos from the Qimen Tagh area. (A) Active Adatan thrust zone, which cuts Quaternary alluvial deposits. Pz-phy—Paleozoic phyllite; Pz-ss—Paleozoic sandstone. (B) Unconformity between north-dipping Cenozoic strata of Qaidam basin and Ordovician metavolcanic rocks of the Eastern Kunlun Range. (C) Pliocene-Quaternary sedimentary strata unconformably rest on top of a Carboniferous marble unit. Due to the resistant nature of the marble unit below, this contact forms a prominent linear topographic feature, as imaged in Figure 6. (D) Fault-bend fold in Carboniferous limestone in the Qimen Tagh Mountains.

the Qimen Tagh Mountains and southwestern Qaidam basin was initiated in the late Oligocene and early Miocene. We address the implications of these findings with respect to the history and mechanism of Cenozoic Tibetan uplift in the following sections.

Wedge Tectonics across the Eastern Kunlun Range

The discovery of a north-dipping unconformity along the northern edge of the Eastern Kunlun Range and dominant south-directed thrusts across the Qimen Tagh–Yousha Shan thrust belt is counterintuitive and contrary to the existing structural models reviewed in this paper. However, this type of structure is not uncommon in fold-thrust belts around the world; it is typically associated with the development of large triangle zones as a result of flake or wedge tectonics either during continental collision or intracontinental deformation (Oxburgh, 1972; Price, 1981, 1986). We propose that the development of the south-directed thrusts across the Qimen Tagh and southern Qaidam regions were induced by a similar tectonic process. Specifically, we envision that the lower crust and mantle lithosphere of Qaidam basin has been subducting southward below the Eastern Kunlun Range since the late Oligocene and early Miocene, whereas Qaidam upper crust has been carried over the range by southward thrusting (Fig. 12). This model is not only consistent with the observed structural geometry across the western Eastern Kunlun Range but also predicts simultaneous uplift of the range in both the footwall and hanging wall of the Qimen Tagh–Yousha Shan thrust belt. An important geometric property of this model is that crustal shortening across

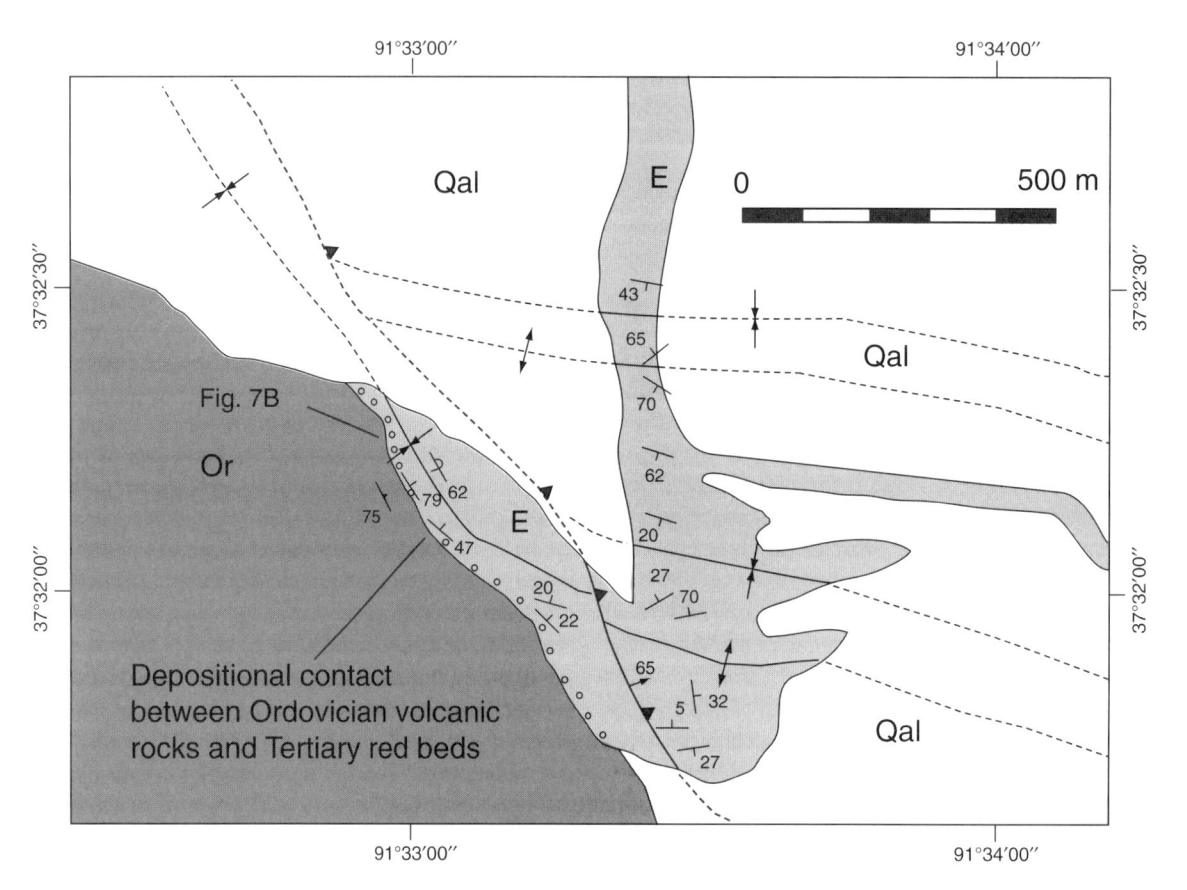

Figure 9. Detailed geologic map showing the unconformity between Paleogene strata (E) and Ordovician metavolcanics (Or). The area is mostly covered by Quaternary alluvial and loess deposits (Qal), except the deep-cut gullies. Note that the Cenozoic strata themselves are folded and cut by south-directed thrusts.

the Qimen Tagh–Yousha Shan thrust belt provides a minimum estimate for the amount of southward subduction of the Qaidam basement. That is, if the Eastern Kunlun tectonic wedge terminates below the range front where the basal thrust and backthrust meet, then the amount of fault slip on the two faults above and below the wedge must be the same. However, if a frontal décollement is present and extends beyond the tip point of the wedge, as shown in Figure 12, then slip on the basal thrust could be much greater than that on the backthrust above. If we assume that the calculated 48% Cenozoic shortening strain from southwestern Qaidam basin has been applied uniformly across the 150-km-wide Qimen Tagh–Yousha Shan thrust belt, then the minimum amount of southward overthrusting of the Qaidam basement is ~150 km. Because the total amount of shortening across the western end of the Eastern Kunlun Range is 250–300 km, as deduced from Yin and Harrison (2000) based on differential Cenozoic slip along the Altyn Tagh fault, the difference suggests that the frontal décollement below Qaidam basin has a total slip of ~100–150 km (i.e., $S2 = S - S1$, where $S = 250–300$ km and $S1 = 150$ km; see inset in Figure 12 for this geometric relationship).

There are at least two possible kinematic histories that could lead to the currently observed structural configuration across the Qimen Tagh Mountains and southwestern Qaidam basin. The first scenario is that the initial Cenozoic uplift of the Eastern Kunlun Range was produced by north-directed thrusting and associated with foreland-basin development (Fig. 13). Later development of a south-directed thrust in the foreland caused partitioning and progressive closure of the early foreland basin. Continuous motion on the younger south-directed thrust allowed its hanging wall to override the older north-directed thrusts and to create a new foreland basin at a high altitude on top of the older north-directed

Figure 10. Interpreted seismic line across the eastern tip of the north-dipping Qimen Tagh thrust zone and the eastern segments of the Yousha Shan anticlinorium. The main thrust offsets the base of Cenozoic strata for ~5 km; its motion induced the development of the south-verging fault-bend fold in the hanging wall, expressed as the Yousha Shan anticlinorium on the surface. The north-dipping geometry of Cenozoic strata shown in the seismic line along the southern margin of Qaidam basin is consistent with our field observations. E1+2—Paleocene–early Eocene strata; E3-1—middle to late Eocene strata; E3-2—early Oligocene strata; Jr—Jurassic strata; pre-Jr—pre-Jurassic rocks; N1—late Oligocene strata; N2-1 and N2-2—Miocene strata sedimentary units; TWTT—two-way travel time.

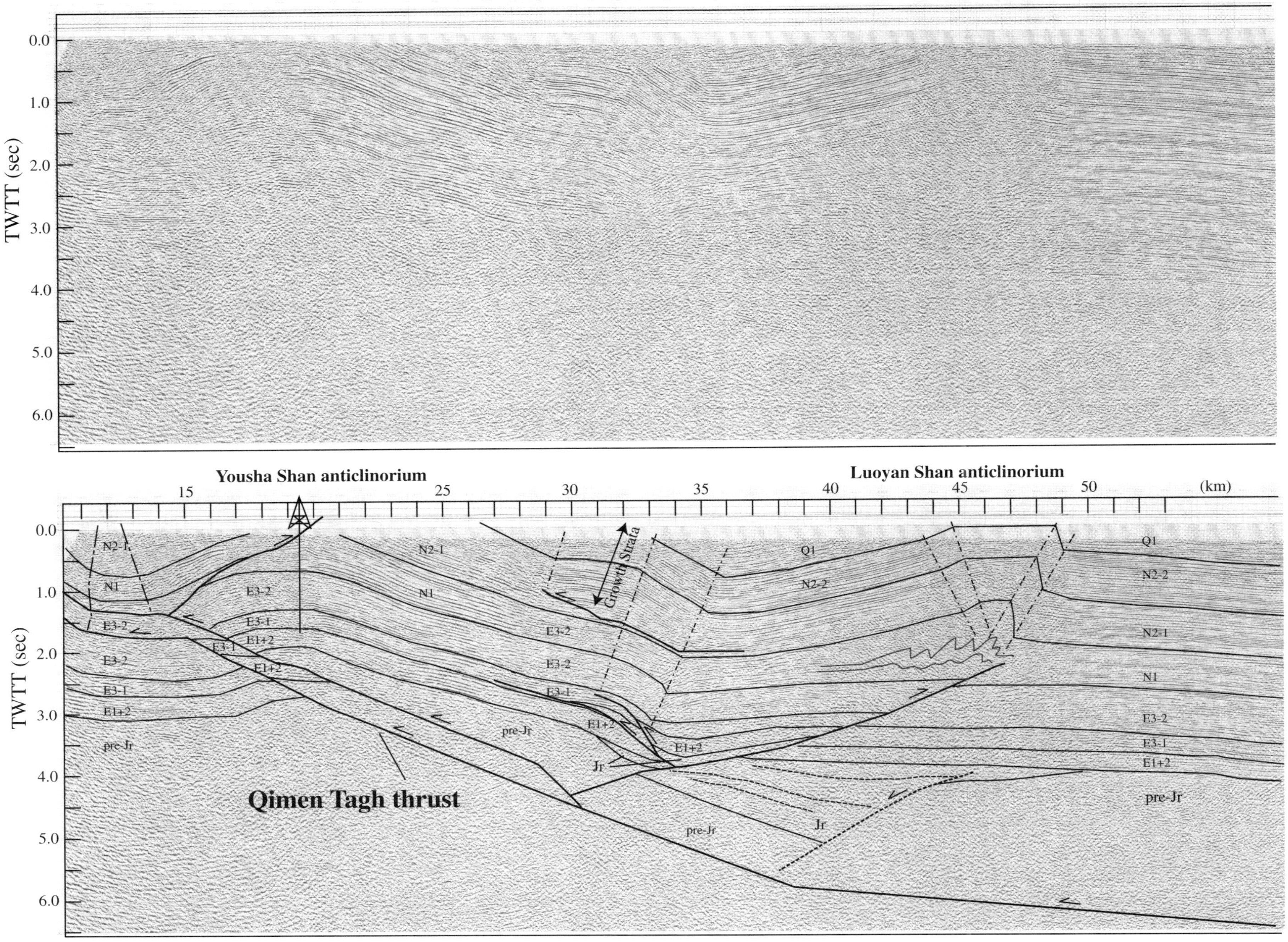
TWTT (sec)
0.0
1.0
2.0
3.0
4.0
5.0
6.0
Yousha Shan anticlinorium
Luoyan Shan anticlinorium
15
25
30
35
40
45
50
(km)
Growth Strata
Qimen Tagh thrust
Q1
N2-2
N2-1
N1
E3-2
E3-1
E1+2
Jr
pre-Jr

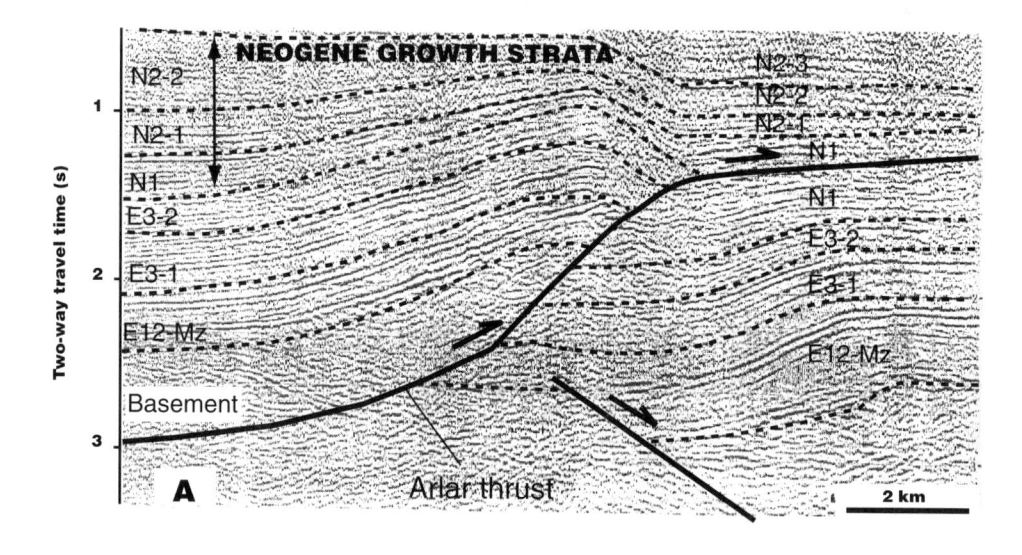

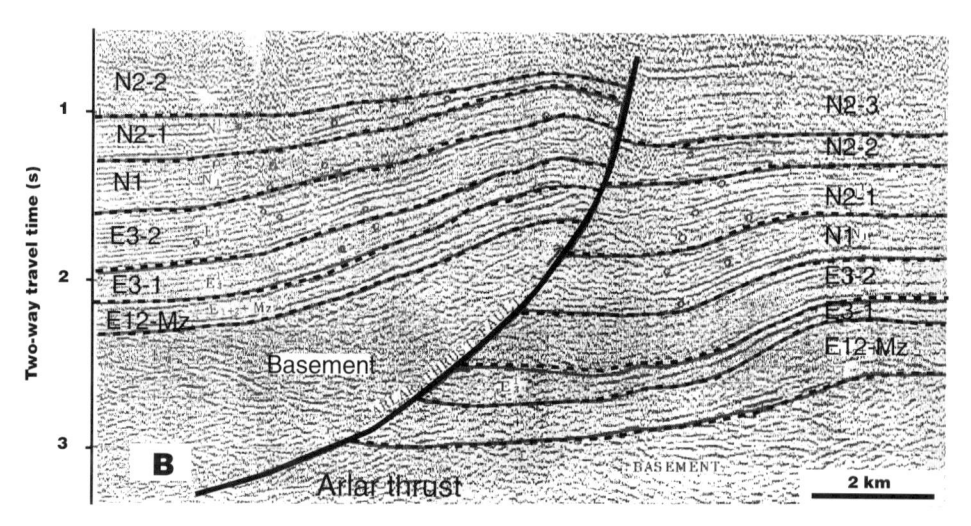

Figure 11. Seismic profile of Song and Wang (1993) along the eastern edge of Gas Hure Lake; see Figure 3 for location. (A) Reinterpreted seismic section of Song and Wang (1993). Note that growth strata started to develop during initial deposition of late Oligocene unit N1, which suggests that the Arlar thrust initiated during the deposition of this unit. The abrupt thickness change in Mesozoic strata is induced by a north-dipping normal fault. (B) Originally interpreted seismic section by Song and Wang (1993). This interpretation forces the Arlar thrust to cut through a series of continuous reflectors in the upper stratigraphic units and implies that the thrust is exposed at the surface. The latter is inconsistent with our field observations. N1—late Oligocene strata; N2-1—early and middle Miocene strata; N2-2—late Miocene sedimentary units; N2-3—Pliocene strata; E3-1—middle to late Eocene strata; E3-2—early Oligocene strata; E12-Mz—Middle to late Eocene strata.

thrust. This may explain the development of active foreland basins in the footwalls of the Ayakum and Adatan thrusts at elevations >4500 m, some 1500 m higher than Qaidam basin. We refer to this kinematic history as the *push-together process*. The model implies that large-magnitude upper-crustal shortening could be completely concealed due to long-distance travel of large thrusts.

Alternatively, the uplift of the Eastern Kunlun Range may have been accomplished by accretion of the Qaidam crust in a piecemeal fashion. We refer to this kinematic history as the *piecemeal-accretion process* (Fig. 14). That is, the uplift of the Eastern Kunlun Range was accomplished by northward motion of a tectonic wedge below, with the wedge tip line propagating northward. The tip-line propagation resulted in sequential northward creation of south-directed thrusts in southern Qaidam basin. These thrusts accreted Qaidam crust incrementally onto the Eastern Kunlun Range, causing the range to be uplifted and enlarged progressively northward with time.

The Youshan Shan anticlinorium and its underlying south-directed blind thrust system could be regarded as an incipient south-directed thrust in our models; that is, its development could

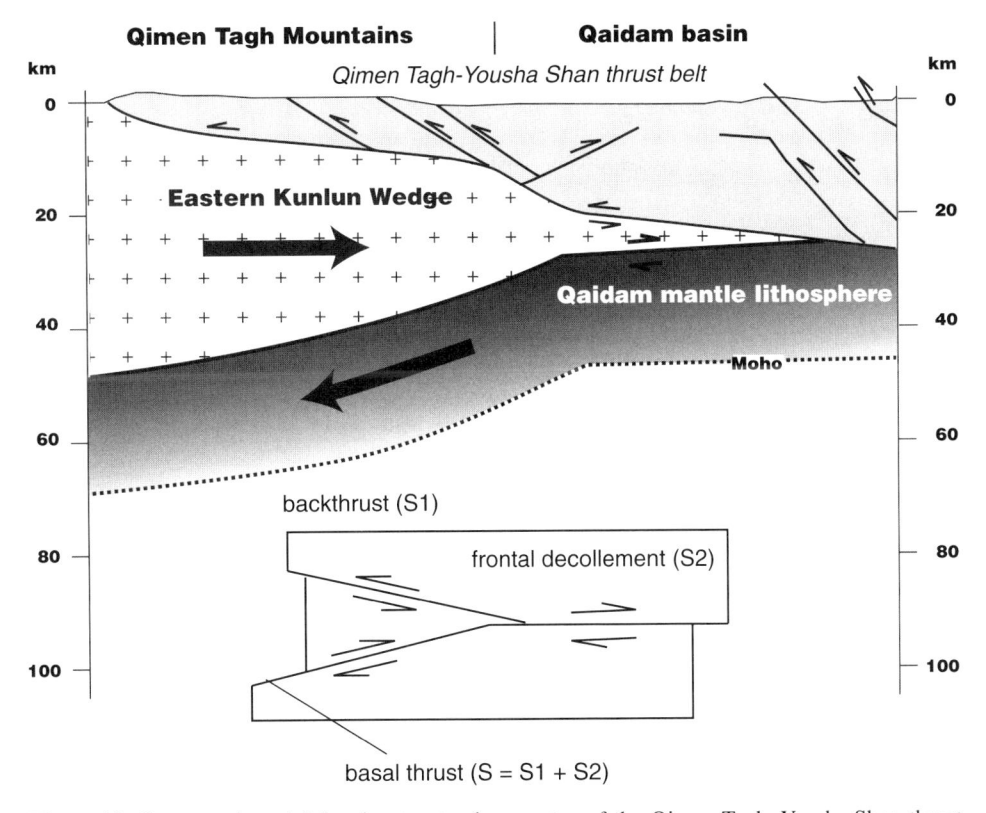

Figure 12. Conceptual model for the structural geometry of the Qimen Tagh–Yousha Shan thrust belt. We infer the presence of a large tectonic wedge below the Eastern Kunlun Range. Its northward extrusion is accommodated by south-directed thrusting in the upper crust and north-directed thrusting in lower crust. This model predicts synchronous uplift of the footwall and hanging wall of the south-directed Qimen Tagh–Yousha Shan thrust belt. Inset shows the relationship between total slip on the basal thrust ($S = S1 + S2$) and slip on the backthrust ($S1$) and frontal décollement ($S2$).

follow either of the kinematic paths shown in Figures 13 and 14. The ultimate test of the two models is elucidation of the deep structures below the Eastern Kunlun Range by either projecting map-view relationships onto deep crustal sections, or more directly, by conducting deep seismic studies to examine if low-density and low-velocity sedimentary basins are buried below major thrust sheets.

It has been long noted that Cenozoic depocenters of Qaidam basin have been persistently located near the axis of the basin rather than along its edges (Bally et al., 1986; Huang et al., 1996; Wang et al., 2006). The modern expression of this situation is the presence of a series of large basins along the basin axis, where all large rivers flowing out of the Eastern Kunlun Range and Qilian Shan terminate. Our observation that Cenozoic strata thicken toward basin centers is consistent with this early conclusion but is inconsistent with the classic foreland-basin model that predicts flexural bending of the foreland basement toward basin-bounding thrusts and thickening of foreland sediments toward the basin margins. The presence of the south-directed thrusts across the western Eastern Kunlun Range and a triangle zone along the northern edge of Qaidam basin (Yin et al., 2008a) indicate that the development of the synclinal trough across Qaidam basin is caused by the uplift of the two structural systems on the basin margins. This deformational process produces an apparent synclinorium across Qaidam basin, as first recognized by Bally et al. (1986).

Out-of-Sequence Thrusting and Growth History of the Tibetan Plateau

Studies along the northern margin of Qaidam basin indicate that its bounding contractional structures started to develop in the early to middle Eocene, ca. 55–45 Ma (Jolivet et al., 2001; Yin et al., 2002, 2008a). This timing of deformation is significantly older than the late Oligocene–early Miocene initiation of deformation along the southern margin of Qaidam basin. Neogene initiation of contraction at 20–10 Ma appears to be a widespread phenomenon across the whole Eastern Kunlun Range as well, which is revealed by low-temperature fission-track thermochronometry (Jolivet et al., 2003; Yuan et al., 2006). Even the locally determined older initiation age of 30–20 Ma (Mock et al., 1999) is significantly younger than the age of initial Cenozoic deforma-

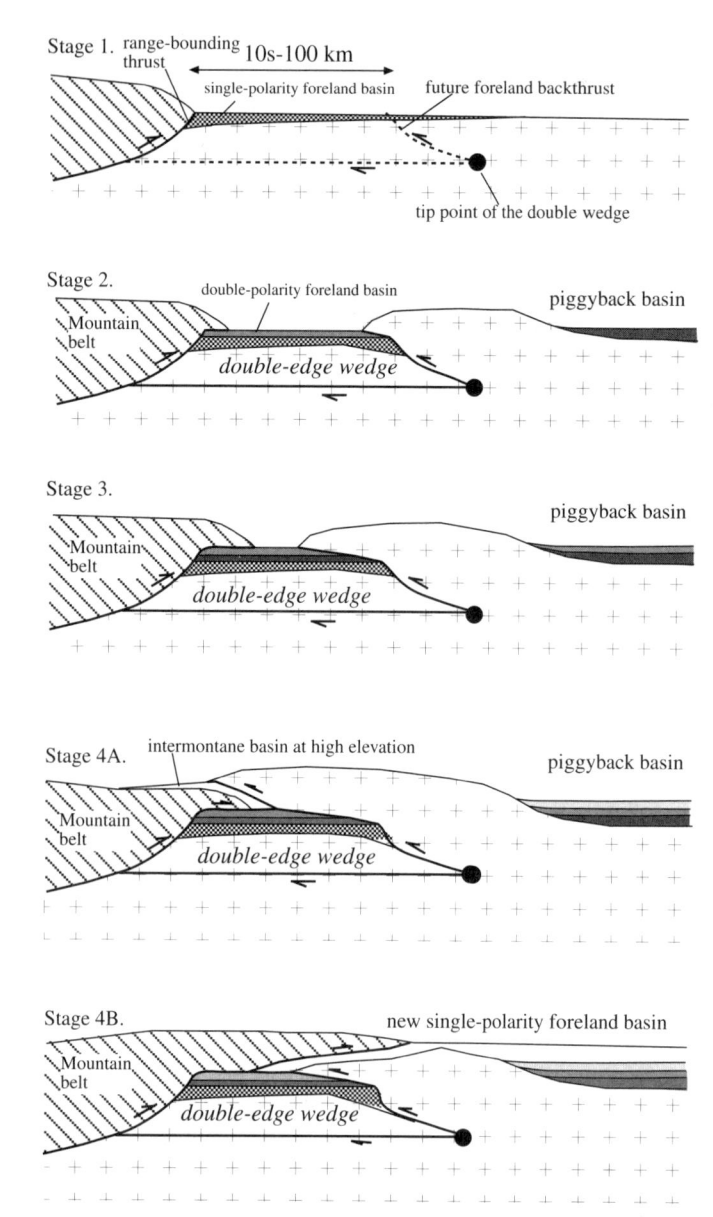

Figure 13. A push-together model for the development of the south-directed Qimen Tagh–Yousha Shan thrust belt. This model predicts that early north-directed thrusts are overridden by later south-directed thrusts across the Eastern Kunlun Range. Stage 1: A north-directed thrust is developed, generating a southward-thickening foreland basin. Stage 2: A south-directed thrust is developed above the tip line of a north-extruding wedge. As a result, the original single-polarity foreland basin is modified to become a double-polarity foreland basin, with its strata thickening to the basin edges toward the two bounding thrusts. Corresponding to the double-polarity basin is the double-edge wedge block below, which is extruding northward with northward migration of its tip line. Stage 3: Continuous motion on the south-directed thrust progressively closes the double-polarity foreland basin. Stage 4A: The south-directed thrust may eventually override the north-directed thrust, placing the older foreland basin entirely below its footwall. Stage 4B: Alternatively, if motion on the north-directed thrust has been active coevally with the south-directed backthrusting, the north-directed thrust could override the south-directed thrust. The most important implication of this model is that a large amount of crustal shortening may not be detectable by surface mapping only.

tion along the northern margin of Qaidam basin. This leads us to conclude that the northern and southern bounding structures of Qaidam basin were initiated diachronously, with the north started first followed by the south. This inference raises the question of where the southern margin of the Eocene Qaidam basin was located. Eocene lacustrine sediments are stranded and scattered across the Eastern Kunlun Range (Fig. 3), so it is possible that the Eocene to early Oligocene southern boundary of Qaidam basin was located south of the Eastern Kunlun Range. If this is the case, then it is highly likely that the Eocene Qaidam basin and the Eocene Hoh Xil basin of Liu et al. (2001, 2003) were parts of a once contiguous basin that was later partitioned by the Neogene uplift of the Eastern Kunlun Range. This size of this basin is comparable to that of Tarim basin between the Tian Shan and Tibetan plateau, and we refer to this inferred large Eocene basin across central Tibet as the *Paleo-Qaidam basin* (Fig. 1).

Our inferred late Oligocene and early Miocene age of initial deformation for the south-directed thrust belt across the western Eastern Kunlun Range is older than the initiation age of the Kunlun fault at 15–7 Ma (Kidd and Molnar, 1988; Jolivet et al., 2003). If southward thrusting was related to left-slip deformation along the Kunlun fault, then deformation must have been propagated from the west to the east. This implies a more complicated deformation history for the development of the Eastern Kunlun transpressional system. That is, the Qimen Tagh thrust belt propagated eastward first and was later linked with the nearly nucleated Kunlun fault that serves as a transfer zone to pass on the contraction to the thrust belt in the Bayanhar Mountains south of the Kunlun fault (Fig. 2).

The younger age of initiation of deformation along the southern margin of Qaidam basin relative to the northern margin suggests that the development of contractional structures across the basin is *out of sequence*, following the terminology of thrust tectonics (e.g., Boyer and Elliott, 1982). This temporal pattern of deformation is inconsistent with the prediction of the thin-viscous-sheet model, at least in its original form, assuming uniform material properties across the whole Tibetan plateau (England and Houseman, 1986). A revised thin-viscous-sheet model, allowing heterogeneous distribution of mechanical strength and the presence of preexisting weakness, may explain the irregular deformation across northern Tibet as we observed across Qaidam basin (e.g., Neil and Houseman, 1997; Kong et al., 1997).

Mechanisms of Tibetan-Plateau Uplift

Determining the magnitude of Cenozoic shortening is a key to evaluating whether lower-crustal flow or thermal events in the upper mantle were responsible for the uplift of the Tibetan plateau (Royden et al., 1997; Clark and Royden, 2000; Molnar et al., 1993). Our estimated crustal shortening in southwestern Qaidam basin is >48%. As discussed above, a similar or even larger magnitude of shortening is expected across the Eastern Kunlun Range. The crustal thickness of Qaidam basin is ~45 km, which is 15–20 km thinner than that for the Eastern Kunlun Range (Zhu

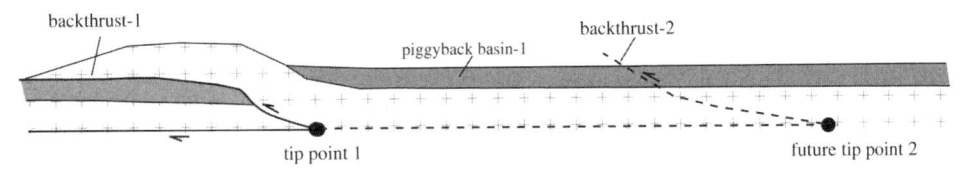

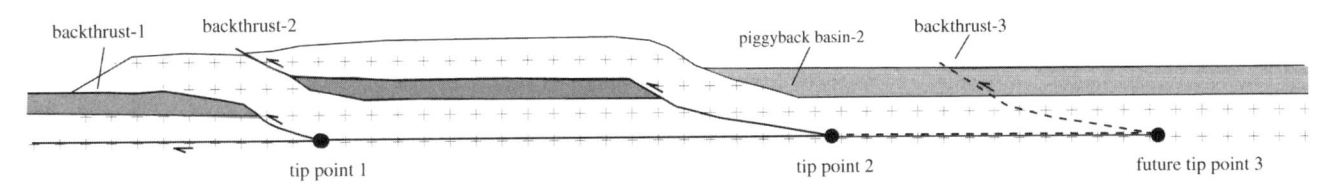

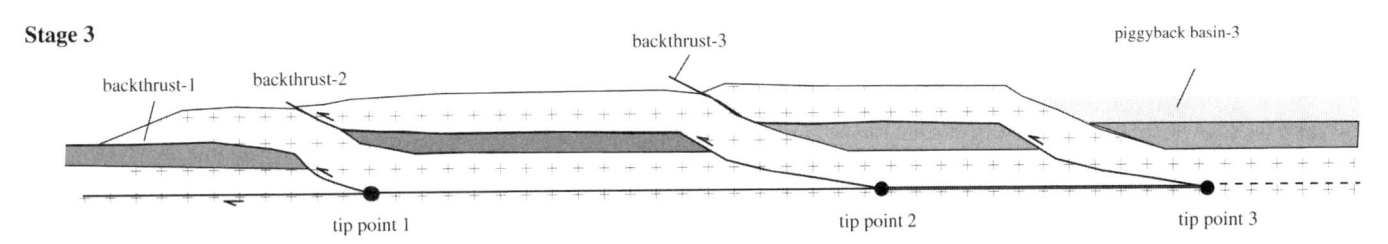

Figure 14. A piece-meal accretion model for the uplift history of the Qimen Tagh Mountains and the development of the south-directed Qimen Tagh–Yousha Shan thrust belt. In this model, a large piggyback basin was first developed on the backlimb of a large backthrust that had climbed up the range front. The piggyback basin was subsequently fragmented by younger backthrusts developed to the north in the foreland as a result of northward propagation of a wedge tip.

and Helmberger, 1998; Zhao et al., 2006; S.L. Li et al., 2006; Y.H. Li et al., 2006). Thickening crust from an original thickness of 35 km to the present 45 km requires shortening strain of 29% (i.e., assuming Qaidam basin was at sea level prior to Cenozoic deformation). If upper crustal shortening, as we observed in the field and inferred from the subsurface data, is accommodated by pure-shear contraction in the Qaidam lower crust, then there is no need for lower crustal channel flow or thermal events in the mantle to raise the Eastern Kunlun Range and southern Qaidam basin. However, if there is a detachment surface separating deformation between the upper and lower crust across the Qaidam and Eastern Kunlun regions, other complicated processes not detectable by surface geologic investigations, such as channel flow or mantle thermal events, need to be considered.

CONCLUSIONS

Our field mapping and analysis of satellite and subsurface data suggest that the western segment of the Eastern Kunlun Range is dominated by south-directed thrusts, carrying the low-elevation Qaidam basin over the high-elevation Eastern Kunlun Range. Contractional deformation initiated in the late Oligocene to early Miocene (28–24 Ma) and has accommodated at least 48% upper-crustal shortening. We suggest that south-directed thrusting is an expression of a large-scale wedge tectonic system across the Eastern Kunlun region; the operation of this system has simultaneously accommodated southward subduction of the Qaidam lower crust and mantle lithosphere and southward thrusting of Qaidam basin over the Eastern Kunlun Range. The early Neogene initiation of deformation along the southern margin of Qaidam basin is significantly younger than that for the northern basin margin. This temporal pattern of deformation suggests that the construction of the Tibetan plateau was not by focusing deformation along its northward expanding margin. Instead, high strain zones were developed inside the Tibetan plateau and remained active during late stages of the Indo-Asian collision after the plateau margins had already been established. If Cenozoic deformation across the Eastern Kunlun and Qaidam regions has been accommodated by pure shear deformation over the whole crust, then our observed > 48% upper-crustal shortening strain is more than sufficient to explain the current elevation and crustal thickness of the region. However, if contraction between the upper and lower crust has been decoupled, then lower crustal flow and thermal events in the upper mantle could be additional causes of plateau uplift across the Eastern Kunlun Range and Qaidam basin.

ACKNOWLEDGMENTS

Reviews by Peter Molnar and Becky Bendick are greatly appreciated. We are also grateful for the constructive suggestions made by the editors of this volume, James Sears, Carol Evenchick, and Tekla Harms, who helped improve the clarity and the quality of this paper. This research is supported by a U.S. National Science Foundation grant, the Qinghai Petroleum Research Institute of PetroChina, and the China University of Geosciences (Beijing).

REFERENCES CITED

Allègre, C.J., and 34 others, 1984, Structure and evolution of the Himalayan-Tibet orogenic belt: Nature, v. 307, p. 17–22, doi: 10.1038/307017a0.

Bally, A.W., Chou, I.-M., Clayton, R., Eugster, H.P., Kidwell, S., Meckel, L.D., Ryder, R.T., Watts, A.B., and Wilson, A.A., 1986, Notes on sedimentary basins in China—Report of the American Sedimentary Basins delegation to the People's Republic of China: U.S. Geological Survey Open-File Report 86-327, 108 p.

Boyer, S.E., and Elliott, D., 1982, Thrust systems: AAPG Bulletin, v. 66, p. 1196–1230.

Burchfiel, B.C., Deng, Q., Molnar, P., Royden, L.H., Wang, Y., Zhang, P., and Zhang, W., 1989a, Intracrustal detachment with zones of continental deformation: Geology, v. 17, p. 748–752, doi: 10.1130/0091-7613(1989)017<0448:IDWZOC>2.3.CO;2.

Burchfiel, B.C., Molnar, P., Zhao, Z., Liang, K., Wang, S., Huang, M., and Sutter, J., 1989b, Geology of the Ulugh Muztagh area, northern Tibet: Earth and Planetary Science Letters, v. 94, p. 57–70, doi: 10.1016/0012-821X(89)90083-6.

Chang, C.-F., and Zheng, S.-L., 1973, Tectonic features of the Mount Jolmo Lungma region in southern Tibet, China: Scientia Geologica Sinica, v. 16, p. 1–12.

Chen, W.P., and Molnar, P., 1983, Focal depths of intracontinental and intraplate earthquakes and their implications for the thermal and mechanical properties of the lithosphere: Journal of Geophysical Research, v. 88, p. 4183–4214.

Chen, W.P., Chen, C.Y., and Nabelek, J.L., 1999, Present-day deformation of the Qaidam basin with implications for intra-continental tectonics: Tectonophysics, v. 305, p. 165–181, doi: 10.1016/S0040-1951(99)00006-2.

Clark, M.K., and Royden, L.H., 2000, Topographic ooze: Building the eastern margin of Tibet by lower crustal flow: Geology, v. 28, p. 703–706, doi: 10.1130/0091-7613(2000)28<703:TOBTEM>2.0.CO;2.

Coward, M.P., Kidd, W.S.F., Yun, P., Shackleton, R.M., and Hu, Z., 1988, Folding and imbrication of the Indian crust during Himalayan collision: Philosophical Transactions of the Royal Society of London Series A, v. 327, p. 89–116.

Cowgill, E., Yin, A., Harrison, T.M., and Wang, X.-F., 2003, Reconstruction of the Altyn Tagh fault based on U-Pb geochronology: Role of back thrusts, mantle sutures, and heterogeneous crustal strength in forming the Tibetan plateau: Journal of Geophysical Research, v. 108, p. 2346, doi: 10.1029/2002JB002080

Dang, Y., Hu, Y., Yu, H., Song, Y., and Yang, F., 2003, Petroleum geology of northern Qaidam basin: Beijing, Geological Publishing House, 187 p.

Dewey, J.F., Shackleton, R.M., Chang, C., and Sun, Y., 1988, The tectonic evolution of the Tibetan Plateau: Philosophical Transactions of the Royal Society of London Series A, v. 327, p. 379–413.

Dupont-Nivet, G., Robinson, D., Butler, R.F., Yin, A., and Melosh, H.J., 2004, Concentration of crustal displacement along a weak Altyn Tagh fault: Evidence from paleomagnetism of the northern Tibetan Plateau: Tectonics, v. 23, TC1020, doi: 10.1029/2002TC001397.

England, P., and Houseman, G., 1986, Finite strain calculations of continental deformation 2; Comparison with the India-Asia collision zone: Journal of Geophysical Research, v. 91, p. 3664–3676.

Guo, X.P., Wang, N.W., Ding, X.Z., Zhao, M., and Wang, D.N., 2006, Discovery of Paleogene palynological assemblages from the Wanbaogou Group-complex in western part of the Eastern Kunlun orogenic belt and its geological significance: Science in China Series D—Earth Sciences, v. 49, p. 358–367, doi: 10.1007/s11430-006-0358-9.

Hanson, A.D., 1998, Organic geochemistry and petroleum geology, tectonics and basin analysis of southern Tarim and northern Qaidam basins, northwest China [Ph.D. thesis]: Palo Alto, California, Stanford University, 316 p.

Harris, N.B.W., Xu, R., Lewis, C.L., Hawkeworth, C.J., and Zhang, Y., 1988, Isotope geochemistry of the 1985 Tibet Geotraverse, Lhasa to Golmud: Philosophical Transactions of the Royal Society of London Series A, v. 327, p. 263–285.

Huang, H., Huang, Q., and Ma, Y., 1996, Geology of Qaidam Basin and its petroleum prediction: Beijing, Geological Publishing House, 257 p.

Huo, G.M., ed., 1990, Petroleum geology of China: Oil fields in Qianghai and Xizang: Chinese Petroleum Industry Press, v. 14, 483 p.

Jolivet, M., Brunel, M., Seward, D., Xu, Z., Yang, J., Roger, F., Tapponnier, P., Malavieille, J., Arnaud, N., and Wu, C., 2001, Mesozoic and Cenozoic tectonics of the northern edge of the Tibetan plateau: Fission-track constraints: Tectonophysics, v. 343, p. 111–134, doi: 10.1016/S0040-1951(01)00196-2.

Jolivet, M., Brunel, M., Seward, D., Xu, Z., Yang, J., Malavieille, J., Roger, F., Leyreloup, A., Arnaud, N., and Wu, C., 2003, Neogene extension and volcanism in the Kunlun Fault Zone, northern Tibet: New constraints on the age of the Kunlun Fault: Tectonics, v. 22, p. 1052, doi: 10.1029/2002TC001428

Jordan, T.E., 1981, Thrust loads and foreland basin evolution, Cretaceous, western United States: AAPG Bulletin, v. 65, p. 2506–2520.

Kidd, W.S.F., and Molnar, P., 1988, Quaternary and active faulting observed on the 1985 Academia Sinica-Royal Society Geotraverse of Tibet: Philosophical Transactions of the Royal Society of London Series A, v. 327, p. 337–363.

Kong, X., Yin, A., and Harrison, T.M., 1997, Evaluating the role of pre-existing weakness and topographic distributions in the Indo-Asian collision by use of a thin-shell numerical model: Geology, v. 25, p. 527–530, doi: 10.1130/0091-7613(1997)025<0527:ETROPW>2.3.CO;2.

Li, S.L., Mooney, W.D., and Fan, J.C., 2006, Crustal structure of mainland China from deep seismic sounding data: Tectonophysics, v. 420, p. 239–252, doi: 10.1016/j.tecto.2006.01.026.

Li, Y.H., Wu, Q.J., An, Z.H., Tian, X.B., Zeng, R.S., Zhang, R.Q., and Li, H.G., 2006, The Poisson ratio and crustal structure across the NE Tibetan Plateau determined from receiver functions: Chinese Journal of Geophysics—Chinese Edition, v. 49, p. 1359–1368.

Lin, A.M., Guo, J.M., Kano, K., and Awata, Y., 2006, Average slip rate and recurrence interval of large-magnitude earthquakes on the western segment of the strike-slip Kunlun fault, northern Tibet: Bulletin of the Seismological Society of America, v. 96, p. 1597–1611, doi: 10.1785/0120050051.

Liu, Y.J., Genser, J., Neubauer, F., Jin, W., Ge, X.H., Handler, R., and Takasu, A., 2005, $^{40}Ar/^{39}Ar$ mineral ages from basement rocks in the Eastern Kunlun Mountains, NW China, and their tectonic implications: Tectonophysics, v. 398, p. 199–224, doi: 10.1016/j.tecto.2005.02.007.

Liu, Z.F., Wang, C.S., and Yi, H.S., 2001, Evolution and mass accumulation of the Cenozoic Hoh Xil basin, northern Tibet: Journal of Sedimentary Research, v. 71, p. 971–984.

Liu, Z.F., Zhao, X.X., Wang, C.S., Liu, S., and Yi, H.S., 2003, Magnetostratigraphy of Tertiary sediments from the Hoh Xil Basin: Implications for the Cenozoic tectonic history of the Tibetan Plateau: Geophysical Journal International, v. 154, p. 233–252, doi: 10.1046/j.1365-246X.2003.01986.x.

Liu, Z.Q., 1988, Geologic Map of the Qinghai-Xizhang Plateau and its Neighboring Regions: Beijing, Chengdu Institute of Geology and Mineral Resources, Geologic Publishing House, scale 1:1,500,000.

McKenna, L.W., and Walker, J.D., 1990, Geochemistry of crustally derived leucogranitic igneous rocks from the Ulugh Muztagh area, northern Tibet and their implications for the formation of the Tibetan Plateau: Journal of Geophysical Research, v. 95, p. 21,483–21,502.

Métivier, F., Gaudemer, Y., Tapponnier, P., and Meyer, B., 1998, Northeastward growth of the Tibet plateau deduced from balanced reconstruction of two depositional areas: The Qaidam and Hexi Corridor basins, China: Tectonics, v. 17, p. 823–842, doi: 10.1029/98TC02764.

Meyer, B., Tapponnier, P., Bourjot, L., Métivier, F., Gaudemer, Y., Peltzer, G., Shunmin, G., and Zhitai, C., 1998, Crustal thickening in Gansu-Qinghai, lithospheric mantle subduction, and oblique, strike-slip controlled growth of the Tibet plateau: Geophysical Journal International, v. 135, p. 1–47, doi: 10.1046/j.1365-246X.1998.00567.x.

Mock, C., Arnaud, N.O., and Cantagrel, J.M., 1999, An early unroofing in northeastern Tibet? Constraints from $^{40}Ar/^{39}Ar$ thermochronology on granitoids from the eastern Kunlun range (Qianghai, NW China): Earth and Planetary Science Letters, v. 171, p. 107–122, doi: 10.1016/S0012-821X(99)00133-8.

Molnar, P., Burchfiel, B.C., Zhao, Z., Lian, K., Wang, S., and Huang, M., 1987, Geologic Evolution of Northern Tibet: Results of an expedition to Ulugh Muztagh: Science, v. 235, p. 299–305, doi: 10.1126/science.235.4786.299.

Molnar, P., England, P., and Martinod, J., 1993, Mantle dynamics, the uplift of the Tibetan Plateau, and the Indian monsoon: Review of Geophysics, v. 31, p. 357–396, doi: 10.1029/93RG02030.

Neil, E.A., and Houseman, G.A., 1997, Geodynamics of the Tarim Basin and the Tian Shan in central Asia: Tectonics, v. 16, p. 571–584, doi: 10.1029/97TC01413.

Oxburgh, E.R., 1972, Flake Tectonics and continental collision: Nature, v. 239, p. 202–205, doi: 10.1038/239202a0.

Price, R.A., 1981, The Cordilleran foreland thrust and fold belt in the southern Canadian Rocky Mountains, *in* Coward, M.P., and McClay, K.R., eds., Thrust and Nappe Tectonics: London, Geological Society Special Publication 9, p. 427–448.

Price, R.A., 1986, The southeastern Canadian Cordillera: Thrust faulting, tectonic edging, and delamination of the lithosphere: Journal of Structural Geology, v. 8, p. 239–254, doi: 10.1016/0191-8141(86)90046-5.

Qinghai, B.G.M.R., 1991, Regional Geology of Qinghai Province: Beijing, Qinghai Bureau of Geology and Mineral Resources, Geological Publishing House, 662 p.

Qiu, N., 2002, Tectono-thermal evolution of the Qaidam Basin, China: Evidence from Ro and apatite fission track data: Petroleum Geoscience, v. 8, p. 279–285.

Rieser, A.B., Neubauer, F., Liu, Y.J., and Ge, X.H., 2005, Sandstone provenance of north-western sectors of the intracontinental Cenozoic Qaidam basin, western China: Tectonic vs. climatic control: Sedimentary Geology, v. 177, p. 1–18, doi: 10.1016/j.sedgeo.2005.01.012.

Rieser, A.B., Liu, Y.J., Genser, J., Neubauer, F., Handler, R., Friedl, G., and Ge, X.H., 2006a, $^{40}Ar/^{39}Ar$ ages of detrital white mica constrain the Cenozoic development of the intracontinental Qaidam Basin: Geological Society of America Bulletin, v. 118, p. 1522–1534, doi: 10.1130/B25962.1.

Rieser, A.B., Liu, Y.J., Genser, J., Neubauer, F., Handler, R., and Ge, X.H., 2006b, Uniform Permian $^{40}Ar/^{39}Ar$ detrital mica ages in the eastern Qaidam Basin (NW China): Where is the source?: Terra Nova, v. 18, p. 79–87, doi: 10.1111/j.1365-3121.2005.00666.x.

Robinson, D.M., Dupont-Nivet, G., Gehrels, G.E., and Zhang, Y.Q., 2003, The Tula uplift, northwestern China: Evidence for regional tectonism of the northern Tibetan Plateau during late Mesozoic-early Cenozoic time: Geological Society of America Bulletin, v. 115, p. 35–47, doi: 10.1130/0016-7606(2003)115<0035:TTUNCE>2.0.CO;2.

Roger, F., Arnaud, N., Gilder, S., Tapponnier, P., Jolivet, M., Brunel, M., Malavieille, J., Xu, Z.Q., and Yang, J.S., 2003, Geochronological and geochemical constraints on Mesozoic suturing in east central Tibet: Tectonics, v. 22, 1037, doi:10.1029/ 2002TC001466.

Royden, L.H., Burchfiel, B.C., King, R.W., Wang, E., Chen, Z., Shen, F., and Liu, Y., 1997, Surface deformation and lower crustal flow in eastern Tibet: Science, v. 276, p. 788–790, doi: 10.1126/science.276.5313.788.

Sobel, E.R., Hilley, G.E., and Strecker, M.R., 2003, Formation of internally drained contractional basins by aridity-limited bedrock incision: Journal of Geophysical Research, v. 108, p. 2344, doi: 10.1029/2002JB001883.

Song, T., and Wang, X., 1993, Structural styles and stratigraphic patterns of syndepositional faults in a contractional setting: Examples from Qaidam basin, northwestern China: AAPG Bulletin, v. 77, p. 102–117.

Sun, Z., Feng, X., Li, D., Yang, F., Qu, Y., and Wang, H., 1999, Cenozoic Ostracoda and palaeoenvironments of the northeastern Tarim Basin, western China: Palaeogeography, Palaeoclimatology, Palaeoecology, v. 148, p. 37–50, doi: 10.1016/S0031-0182(98)00174-6.

Sun, Z.M., Yang, Z.Y., Pei, J.L., Ge, X.H., Wang, X.S., Yang, T.S., Li, W.M., and Yuan, S.H., 2005, Magnetostratigraphy of Paleogene sediments from northern Qaidam basin, China: Implications for tectonic uplift and block rotation in northern Tibetan plateau: Earth and Planetary Science Letters, v. 237, p. 635–646, doi: 10.1016/j.epsl.2005.07.007.

Suppe, J., 1983, Geometry and kinematics of fault-bend folding: American Journal of Science, v. 283, p. 684–721.

Tapponnier, P., Meyer, B., Avouac, J.P., Peltzer, G., Gaudemer, Y., Guo, S., Xiang, H., Yin, K., Chen, Z., Cai, S., and Dai, H., 1990, Active thrusting and folding in the Qilian Shan, and decoupling between upper crust and mantle in northeastern Tibet: Earth and Planetary Science Letters, v. 97, p. 382–403, doi: 10.1016/0012-821X(90)90053-Z.

Tapponnier, P., Xu, Z.Q., Roger, F., Meyer, B., Arnaud, N., Wittlinger, G., and Yang, J.S., 2001, Oblique stepwise rise and growth of the Tibet plateau: Science, v. 294, p. 1671–1677, doi: 10.1126/science.105978.

Taylor, M., Yin, A., Ryerson, F.J., Kapp, P., and Ding, L., 2003, Conjugate strike-slip faulting along the Bangong-Nujiang suture zone accommodates coeval east-west extension and north-south shortening in the interior of the Tibetan Plateau: Tectonics, v. 22, p. 1044, doi: 10.1029/2002TC001361.

van Der Woerd, J., Tapponnier, P., Ryerson, F.J., Meriaux, A.S., Meyer, B., Gaudemer, Y., Finkel, R.C., Caffee, M.W., Zhao, G.G., and Xu, Z.Q., 2002, Uniform postglacial slip-rate along the central 600 km of the Kunlun Fault (Tibet), from ^{26}Al, ^{10}Be, and ^{14}C dating of riser offsets, and climatic origin of the regional morphology: Geophysical Journal International, v. 148, p. 356–388, doi: 10.1046/j.1365-246x.2002.01556.x.

Wang, E., Xu, F.Y., Zhou, J.X., Wan, J.L., and Burchfiel, B.C., 2006, Eastward migration of the Qaidam basin and its implications for Cenozoic evolution of the Altyn Tagh fault and associated river systems: Geological Society of America Bulletin, v. 118, p. 349–365, doi: 10.1130/B25778.1.

Wang, F., Lo, C.H., Li, Q., Yeh, M.W., Wan, J.L., Zheng, D.W., and Wang, E.Q., 2004, Onset timing of significant unroofing around Qaidam basin, northern Tibet, China: constraints from $^{40}Ar/^{39}Ar$ and FT thermochronology on granitoids: Journal of Asian Earth Sciences, v. 24, p. 59–69, doi: 10.1016/j.jseaes.2003.07.004.

Wang, Q., and Coward, M.P., 1990, The Chaidam Basin (NW China): Formation and hydrocarbon potential: Journal of Petroleum Geology, v. 13, p. 93–112.

Xia, W., Zhang, N., Yuan, X., Fan, L., and Zhang, B., 2001, Cenozoic Qaidam basin, China: A stronger tectonic inversed, extensional rifted basin: AAPG Bulletin, v. 85, p. 715–736.

Yang, F., 1988, Distribution of the brackish-salt water ostracod in northwestern Qinghai Plateau and its geological significance, *in* Hanai, T., Ikeya, T., and Ishizaki, K., eds., Evolutionary biology of ostracoda: Shizuoka, Japan, Proceedings of the 9th International Symposium on Ostracoda, p. 519–530.

Yang, F., Ma, Z., Xu, T., and Ye, S., 1992, A Tertiary paleomagnetic stratigraphic profile in Qaidam basin: Acta Petrologica Sinica, v. 13, p. 97–101.

Yang, F., Sun, Z., Cao, C., Ma, Z., and Zhang, Y., 1997, Quaternary fossil ostracod zones and magnetostratigraphic profile in the Qaidam basin: Acta Micropalaeontologica Sinica, v. 14, p. 378–390.

Yin, A., and Harrison, T.M., 2000, Geologic evolution of the Himalayan-Tibetan orogen: Annual Review of Earth and Planetary Sciences, v. 28, p. 211–280, doi: 10.1146/annurev.earth.28.1.211.

Yin, A., Rumelhart, P., Cowgill, E., Butler, R., Harrison, T.M., Ingersoll, R.V., Cooper, K., Zhang, Q., and Wang, X.-F., 2002, Tectonic history of the Altyn Tagh fault in northern Tibet inferred from Cenozoic sedimentation: Geological Society of America Bulletin, v. 114, p. 1257–1295, doi: 10.1130/0016-7606(2002)114<1257:THOTAT>2.0.CO;2.

Yin, A., Chen, X., McRivette, M.W., Wang, L., and Jiang, W., 2008a, Cenozoic tectonic evolution of Qaidam Basin and its surrounding regions (part 1): The Southern Qilian Shan-Nan Shan Thrust Belt and Northern Qaidam Basin: Geological Society of America Bulletin, (in press).

Yin, A., Chen, X., McRivette, M.W., Wang, L., and Jiang, W., 2008b, Cenozoic tectonic evolution of Qaidam Basin and its surrounding regions (part 3): Structural geology, sedimentation, and tectonic reconstruction (in press).

Yuan, W.M., Dong, J.Q., Wang, S.C., and Carter, A., 2006, Apatite fission track evidence for Neogene uplift in the eastern Kunlun Mountains, northern Qinghai-Tibet Plateau: Journal of Asian Earth Sciences, v. 27, p. 847–856, doi: 10.1016/j.jseaes.2005.09.002.

Zhang, Y.C., 1997, Prototype of analysis of Petroliferous basins in China: Nanjing, Nanjing University Press, 434 p.

Zhao, J.M., Mooney, W.D., Zhang, X.K., Li, Z.C., Jin, Z.J., and Okaya, N., 2006, Crustal structure across the Altyn Tagh Range at the northern margin of the Tibetan plateau and tectonic implications: Earth and Planetary Science Letters, v. 241, p. 804–814, doi: 10.1016/j.epsl.2005.11.003.

Zhong, J.H., Wen, Z.F., Guo, Z.Q., Wang, H.Q., and Gao, J.B., 2004, Paleogene and early Neogene lacustrine reefs in the western Qaidam Basin, China: Acta Geologica Sinica—English Edition, v. 78, p. 736–743.

Zhou, J.X., Xu, F.Y., Wang, T.C., Cao, A.F., and Yin, C.M., 2006, Cenozoic deformation history of the Qaidam Basin, NW China: Results from cross section restoration and implications for Qinghai-Tibet Plateau tectonics: Earth and Planetary Science Letters, v. 243, p. 195–210, doi: 10.1016/j.epsl.2005.11.033.

Zhu, L.P., and Helmberger, D.V., 1998, Moho offset across the northern margin of the Tibetan Plateau: Science, v. 281, p. 1170–1172, doi: 10.1126/science.281.5380.1170.

MANUSCRIPT ACCEPTED BY THE SOCIETY 22 MARCH 2007

Printed in the USA

Defining the eastern boundary of the North Asian craton from structural and subsidence history studies of the Verkhoyansk fold-and-thrust belt

Andrei K. Khudoley*
St. Petersburg State University, Geological Department, University nab. 7/9, St. Petersburg 199034, Russia

Andrei V. Prokopiev
Diamond and Precious Metal Geology Institute SB RAS, Lenin Avenue 39, Yakutsk 677980, Republic Sakha (Yakutia), Russia

ABSTRACT

The Verkhoyansk fold-and-thrust belt developed along the east margin of the North Asian craton during Cretaceous time. Imbricate thrust fans characterize the structural style of the outer (west) zone of the Verkhoyansk fold-and-thrust belt. In the inner (east) zone of the Verkhoyansk fold-and-thrust belt, thrust tectonics is less pronounced, and parallel, open folds predominate. The sedimentary succession of the outer zone contains rock units from Mesoproterozoic to Mesozoic in age, whereas in the inner zone only Permian and younger rocks are exposed. According to the interpretation of limited seismic data, along with gravity and magnetic data, depth to the basement in the outer zone varies from 18 km to 14 km, whereas the inner zone contains a wide basement uplift with depth to the basement ranging from 10 km to 8 km.

Total and tectonic subsidence estimations show a similarity between the Early Carboniferous to Middle Jurassic sedimentary history of the outer Verkhoyansk fold-and-thrust belt and the subsidence history of the Mesozoic Atlantic margin of North America. Application of the thrust wedge model to the outer zone of the Verkhoyansk fold-and-thrust belt shows reasonable agreement between observations and theory. However, application of this model to the inner zone results in problems related to the very small wedge angle. Widespread distribution of Devonian evaporites is inferred to be the main factor that facilitates displacement of a narrow thrust wedge developed in the inner zone.

Both structural and subsidence studies define a boundary between the outer and the inner zones of the Verkhoyansk fold-and-thrust belt that appears to separate crustal blocks that have experienced different structural, sedimentary, and

*khudoley@ah3549.spb.edu

Khudoley, A.K., and Prokopiev, A.V., 2007, Defining the eastern boundary of the North Asian craton from structural and subsidence history studies of the Verkhoyansk fold-and-thrust belt, in Sears, J.W., Harms, T.A., and Evenchick, C.A., eds., Whence the Mountains? Inquiries into the Evolution of Orogenic Systems: A Volume in Honor of Raymond A. Price: Geological Society of America Special Paper 433, p. 391–410, doi: 10.1130/2007.2433(19). For permission to copy, contact editing@geosociety.org. ©2007 The Geological Society of America. All rights reserved.

tectonic evolution. We interpret this fundamental boundary as the eastern limit of North Asian cratonic basement. The inner zone of the Verkhoyansk fold-and-thrust belt is underlain by blocks with transitional continental crust separated by rifts and rift-related basins.

Keywords: Verkhoyansk, stratigraphy, structural styles, tectonic evolution.

INTRODUCTION

The Verkhoyansk fold-and-thrust belt borders the eastern margin of the Siberian platform for 2000 km, from the Arctic Ocean to the Sea of Okhotsk, and is up to 500 km in cross-sectional width (Fig. 1). In this contribution, we follow tectonic nomenclature developed by Parfenov (1995, 2001), which defines the Siberian platform as a tectonic domain underlain by an ancient crystalline basement and its undeformed cover sequence. However, a larger area that occupies both the Siberian platform and the surrounding fold-and-thrust belt is underlain by the same basement as the Siberian platform, and is defined as the North Asian craton (Fig. 1). According to the suggested terminology, the Verkhoyansk fold-and-thrust belt represents a passive margin of the North Asian craton that developed during several stages of rifting and that was deformed and transformed into a fold-and-thrust belt in the Cretaceous. Deformation is believed to be related to collision between the North Asian craton and the Kolyma-Omolon superterrane in the east and to accretionary processes along the Okhotsk active continental margin in the south. The Verkhoyansk fold-and-thrust belt is predominantly composed of sedimentary rocks ranging in age from Mesoproterozoic to Cenozoic, although mafic volcanic rocks and sills are found throughout the succession (Parfenov, 1984; Kovalskiy, 1985; Til'man and Bogdanov, 1992; Nokleberg et al., 1994; Prokopiev et al., 2001a, 2001b; Khudoley and Guriev, 2003). The dominant structural style of the Verkhoyansk fold-and-thrust belt is one of thin-skinned tectonics (e.g., Parfenov et al., 1995; Prokopiev and Deikunenko, 2001).

During the past three decades, several thermal and mechanical models have been developed to explain the origin and evolution of passive margins and their subsequent deformation after transformation into active margins. Post-rifting evolution of sedimentary basins may be quantified using a subsidence model developed by Bond and Komniz (1984) and Steckler et al. (1988). According to Chapple (1978), Davis et al. (1983), and Dahlen et al. (1984), a critical Coulomb-wedge theory successfully explains the geometry of accretionary wedges and foreland fold-and-thrust belts. The main task of this study is to discuss basic structural and stratigraphic features of the Verkhoyansk fold-and-thrust belt and to apply subsidence and thrust-wedge models to explain its post-Devonian sedimentary evolution and its modern structure.

GEOLOGICAL FRAMEWORK

The stratigraphy and structure of the Verkhoyansk fold-and-thrust belt has been discussed in many papers (Gusev, 1979; Kovalskiy, 1985; Nokleberg et al., 1994, 1997; Parfenov et al., 1995; Prokopiev et al., 2001a, 2001b; Prokopiev and Deikunenko, 2001; Khudoley and Guriev, 1994, 2003). Two major zones with different stratigraphy and structural style are identified in the Verkhoyansk fold-and-thrust belt—the outer (western) zone and the inner (eastern) zone (Fig. 2). The outer zone of the Verkhoyansk fold-and-thrust belt is divided into the West Verkhoyansk and South Verkhoyansk sectors with sev-

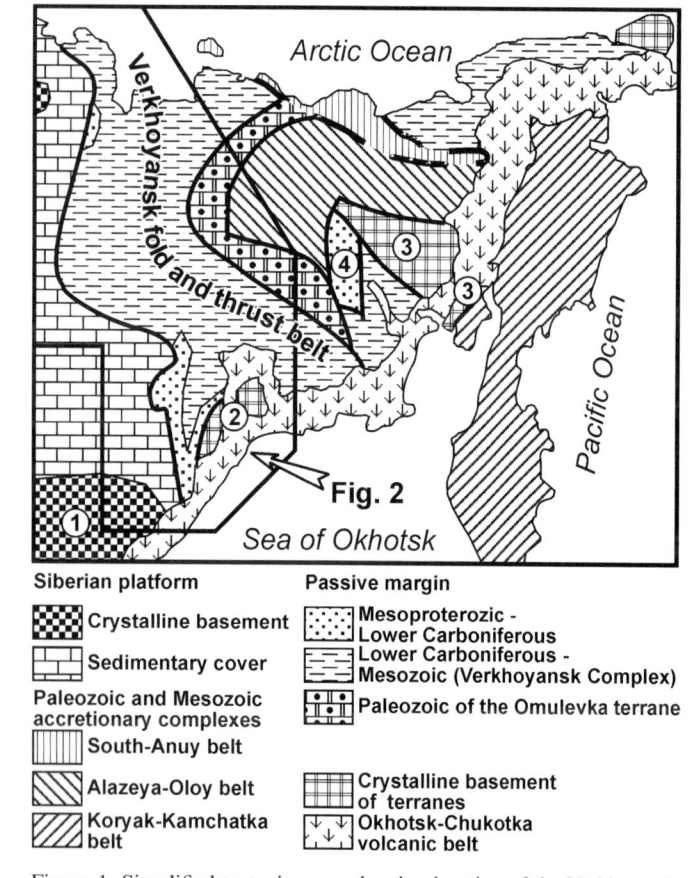

Figure 1. Simplified tectonic map, showing location of the Verkhoyansk fold-and-thrust belt. Circled numbers: 1—Aldan shield, 2—Okhotsk terrane, 3—Omolon terrane, 4—Prikolyma terrane. According to Parfenov (1995, 2001), Omulevka terrane, Alazeya-Oloy belt, Prikolyma terrane, and Omolon terrane form the Kolyma-Omolon superterrane, whereas Siberian platform, Okhotsk terrane, and Verkhoyansk fold-and-thrust belt form the North Asian craton.

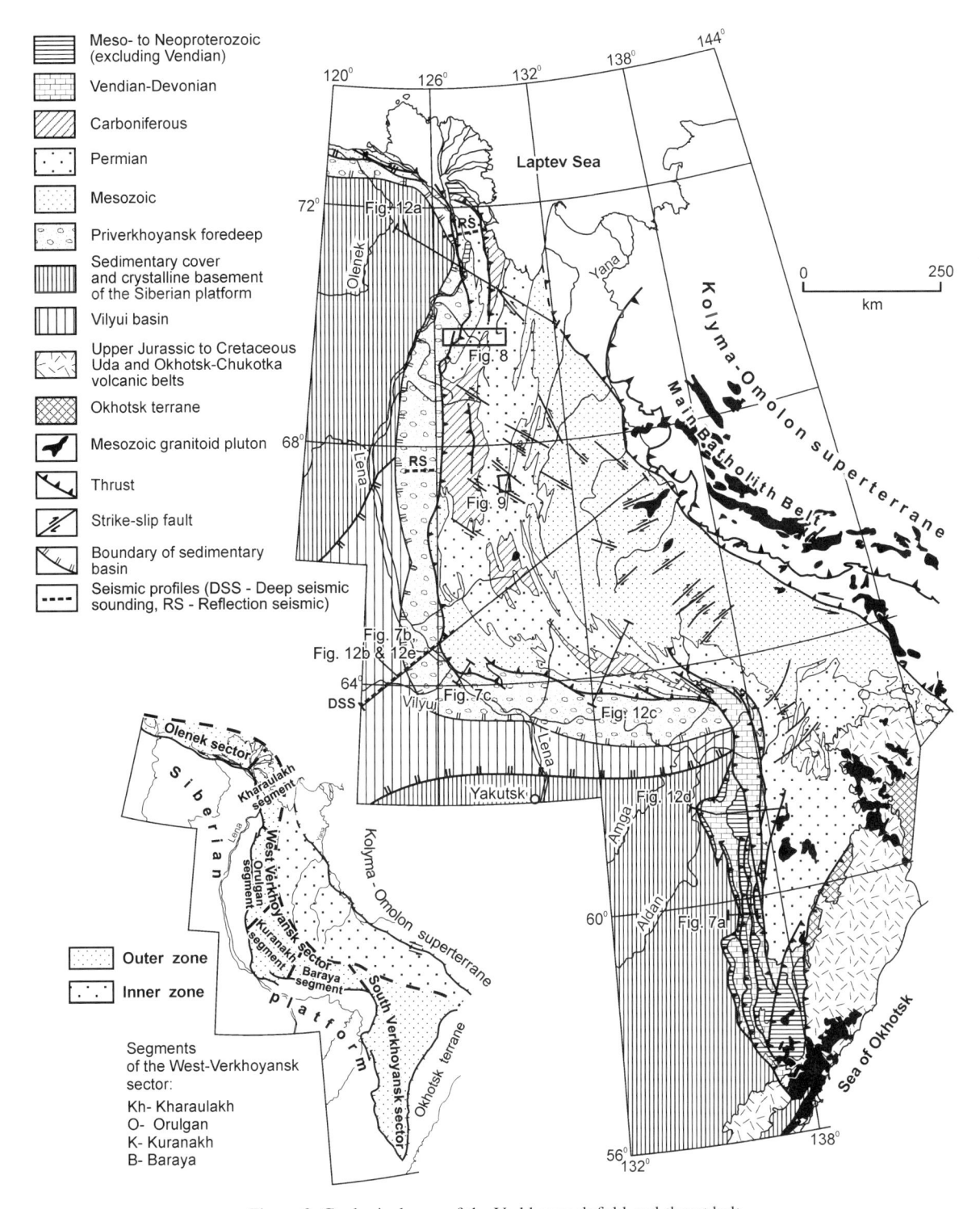

Figure 2. Geological map of the Verkhoyansk fold-and-thrust belt.

eral segments characterized by specific frontal thrust structures (Parfenov et al., 1995; Prokopiev and Deikunenko, 2001). In the north, the West Verkhoyansk sector is bounded by the Olenek sector, which is not discussed in this paper. In the west, the West Verkhoyansk sector is bounded by a foredeep basin, which is filled with Upper Jurassic to Cretaceous terrigenous rocks as thick as 7 km. In the South Verkhoyansk sector, the Mesozoic foredeep diminishes in thickness and pinches out, and thrust sheets are in direct contact with the Siberian platform with no intervening foredeep sediments.

Basement of the Verkhoyansk Fold-and-Thrust Belt

Interpretation of limited reflection and deep seismic sounding (DSS) profiles (Fig. 2) together with gravity data show that within the Priverkhoyansk foredeep the crystalline basement gradually deepens eastward from 2 km to 7 km, and locally, in the eastern part of the Vilyui basin, up to 14 km. According to interpretation of the gravity data, within the Verkhoyansk fold-and-thrust belt the basement structure is assumed to be more complicated (Tretyakov, 2004). In the outer zone of the Verkhoyansk fold-and-thrust belt, depth to basement increases from the frontal thrusts eastward, typically reaching 14–16 km and locally is as much as 18 km. In the inner part, depth to basement varies greatly. In the central and southern parts of the inner Verkhoyansk fold-and-thrust belt, depth to basement is typically 16–18 km, whereas in the northern part of the inner Verkhoyansk fold-and-thrust belt, there is a 200-km-wide basement high where basement is only 8–10 km deep. Increase of the depth to basement typically is not gradual, and some basement ramps have been identified. The nature of the ramps is not clear. According to Tretyakov (2004), the ramps are related to basement faults that formed during Devonian rifting. Parfenov et al. (1995), Prokopiev (2000), and Prokopiev and Deikunenko (2001), however, interpreted the structure of the Verkhoyansk fold-and-thrust belt in terms of thin-skinned tectonics without significant involvement of basement rock units. The contacts of the outer and inner zones are typically marked by wide depressions (Tretyakov, 2004).

Important information on the nature of the basement below the Verkhoyansk fold-and-thrust belt comes from $^{87}Sr/^{86}Sr$ isotopic study of the Mesozoic granitoid plutons (Fig. 3). Typically, the Mesozoic granite plutons exhibit very high (0.706–0.710) initial $^{87}Sr/^{86}Sr$ ratios, indicating involvement of continental crust. However, a few plutons from the Main Batholith Belt as well as some granite plutons located in the inner zone of the Verkhoyansk fold-and-thrust belt (Fig. 3) have initial $^{87}Sr/^{86}Sr$ ratios as low as 0.703–0.705 (Nenashev and Zaitsev, 1985; Gamyanin et al., 2003). These low values are usually interpreted as a result of contamination of the magmas by Upper Triassic sediments that had very low $^{87}Sr/^{86}Sr$ ratios at the time of granite intrusion, but the low ratios might also represent existence of a mantle component.

Stratigraphy

Verkhoyansk Fold-and-Thrust Belt

The sedimentary succession of the Verkhoyansk fold-and-thrust belt is divided into three major parts corresponding to the main stages in tectonic evolution of the eastern margin of the North Asian craton (Fig. 4). The first subdivision of the succession is Meso- to Neoproterozoic in age; the second is latest Neoproterozoic (Vendian) to Early Devonian in age; the third is Middle Devonian to Jurassic in age. Strata of the first and second subdivisions as well as pre-Middle Carboniferous strata of the third subdivision are mapped only in the outer zone of the Verkhoyansk fold-and-thrust belt and underlie ~15% of the total area of the Verkhoyansk fold-and-thrust belt, whereas the Middle Carboniferous to Jurassic terrigenous rocks underlie ~85% of its total area and occur in both outer and inner zones of the Verkhoyansk fold-and-thrust belt.

The first subdivision (Meso- to Neoproterozoic) is exposed in the South Verkhoyansk sector and Kharaulakh segment of the West Verkhoyansk sector (Fig. 2). It consists of alternating carbonate and terrigenous units with a total composite thickness of ~12–14 km. Several unconformity-bounded terrigenous-carbonate successions are identified. Mafic sills and volcanics are widespread in the upper part of the succession. Near-shore to shallow marine environments predominated, with a few thick terrigenous units deposited in fluvial and basinal environments (Khudoley et al., 2001, and references therein). Deposition occurred in an intracratonic sedimentary basin but was affected by rifting and local compression in late Mesoproterozoic and Neoproterozoic time (Khudoley and Guriev, 2003). The second subdivision of the succession includes uppermost Neoproterozoic (Vendian) to Lower Devonian strata and is exposed in the South Verkhoyansk sector and Kharaulakh segment of the West Verkhoyansk sector, as well as in a few tectonic slices at the basal detachment that separates the Verkhoyansk fold-and-thrust belt from the foredeep in the Orulgan segment (Figs. 2 and 4). The uppermost Neoproterozoic to Lower Devonian succession is ~11 km in total composite thickness and consists mainly of carbonate with some shale units. Shallow-marine deposition was most typical, but for Cambrian and Ordovician strata, the depositional environments varied eastward from the Siberian platform from shallow-marine to basinal (Khudoley et al., 1991; Prokopiev et al., 2001b). Several stages of mafic magmatic activity have been documented. After the early Cambrian rifting, deposition occurred on a wide, passive margin (Khudoley and Guriev, 2003; Prokopiev et al., 2001b).

The third subdivision (Middle Devonian to Jurassic) consists of predominantly terrigenous rocks with some carbonate and evaporite units in the lower part of the succession (Fig. 4). The beginning of deposition of the third subdivision was marked by Middle to Late Devonian rifting, which was a very widespread tectonic event recognized in different parts of the Siberian platform, the Verkhoyansk fold-and-thrust belt, and the Okhotsk and Omulevka terranes (Fig. 1). In the South Verkhoyansk sector, the rifting event locally began in latest Early Devonian (Alkhovik and Baranov, 2001). On the eastern margin of the North Asian craton, including the outer part of the Verkhoyansk fold-and-thrust belt, Upper Devonian rocks typically contain tholeiitic and alkali basalt flows that are intercalated with locally derived clastics, evaporites, and carbonate units that define coarsening-upward cycles. These exhibit sharp facies changes, contain numerous unconformities, and were deposited in continental, lagoonal, and shallow-marine environments that characterize rift successions (Parfenov, 1984, 1995). The Middle Devonian strata have local distribution, whereas the Upper Devonian rift-related

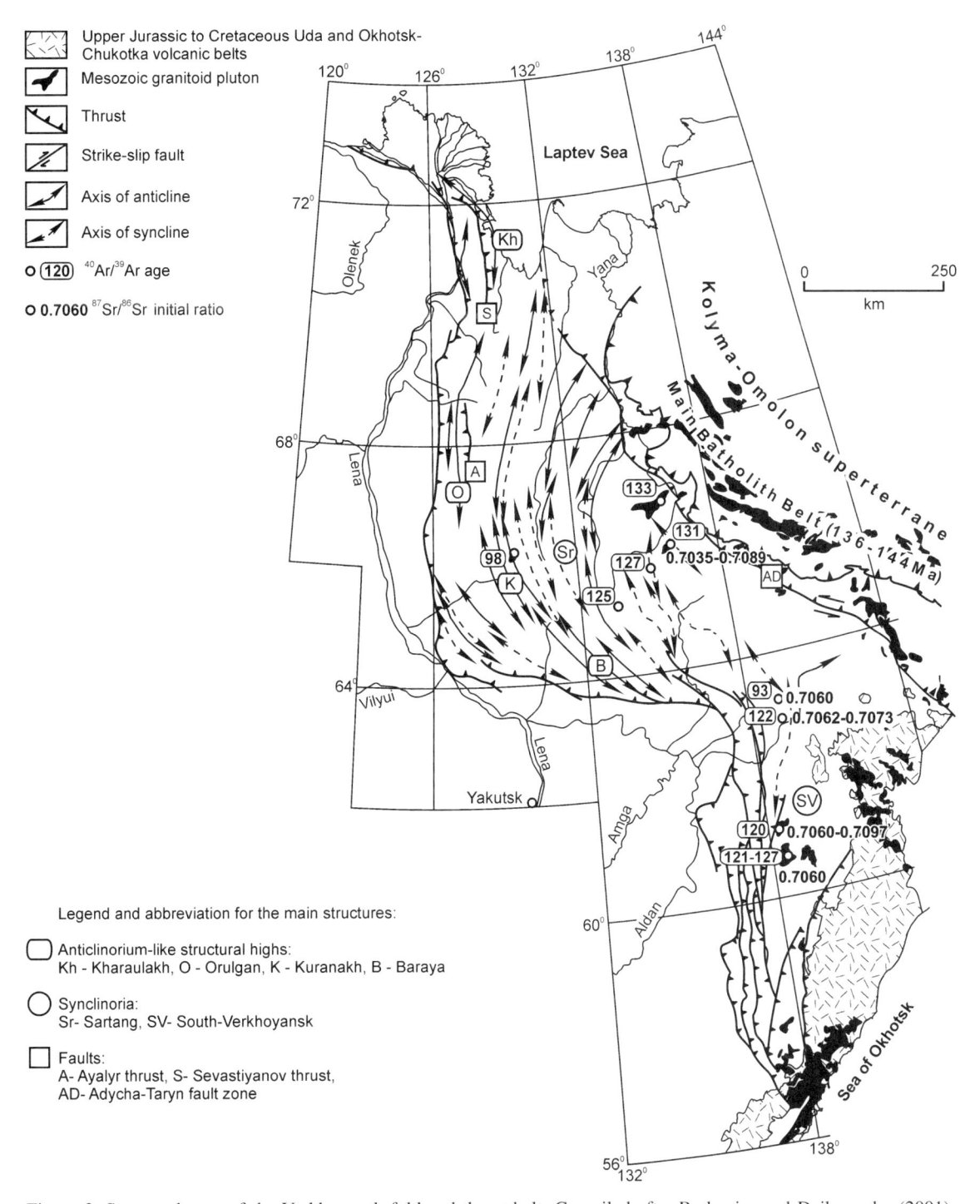

Figure 3. Structural map of the Verkhoyansk fold-and-thrust belt. Compiled after Prokopiev and Deikunenko (2001), simplified and modified. Data on $^{40}Ar/^{39}Ar$ ages from Layer et al. (2001) and Prokopiev et al. (2003), data on $^{87}Sr/^{86}Sr$ initial ratios from Nenashev and Zaitsev (1985) and Gamyanin et al. (2003).

strata are recognized throughout the east margin of the North Asian craton (Parfenov, 1984; Kovalskiy, 1985; Khudoley and Guriev, 1994, 2003; Prokopiev et al., 2001b).

Lower Carboniferous (Tournaisian and Visean) rocks unconformably overlie Devonian and older rocks, indicating regional subsidence and the development of a widespread sedimentary basin. The lowermost units consist mainly of limestones that were deposited in an open, shallow-marine carbonate-platform setting. However, overlying rock units contain both shallow-marine limestones and slope-to-basinal

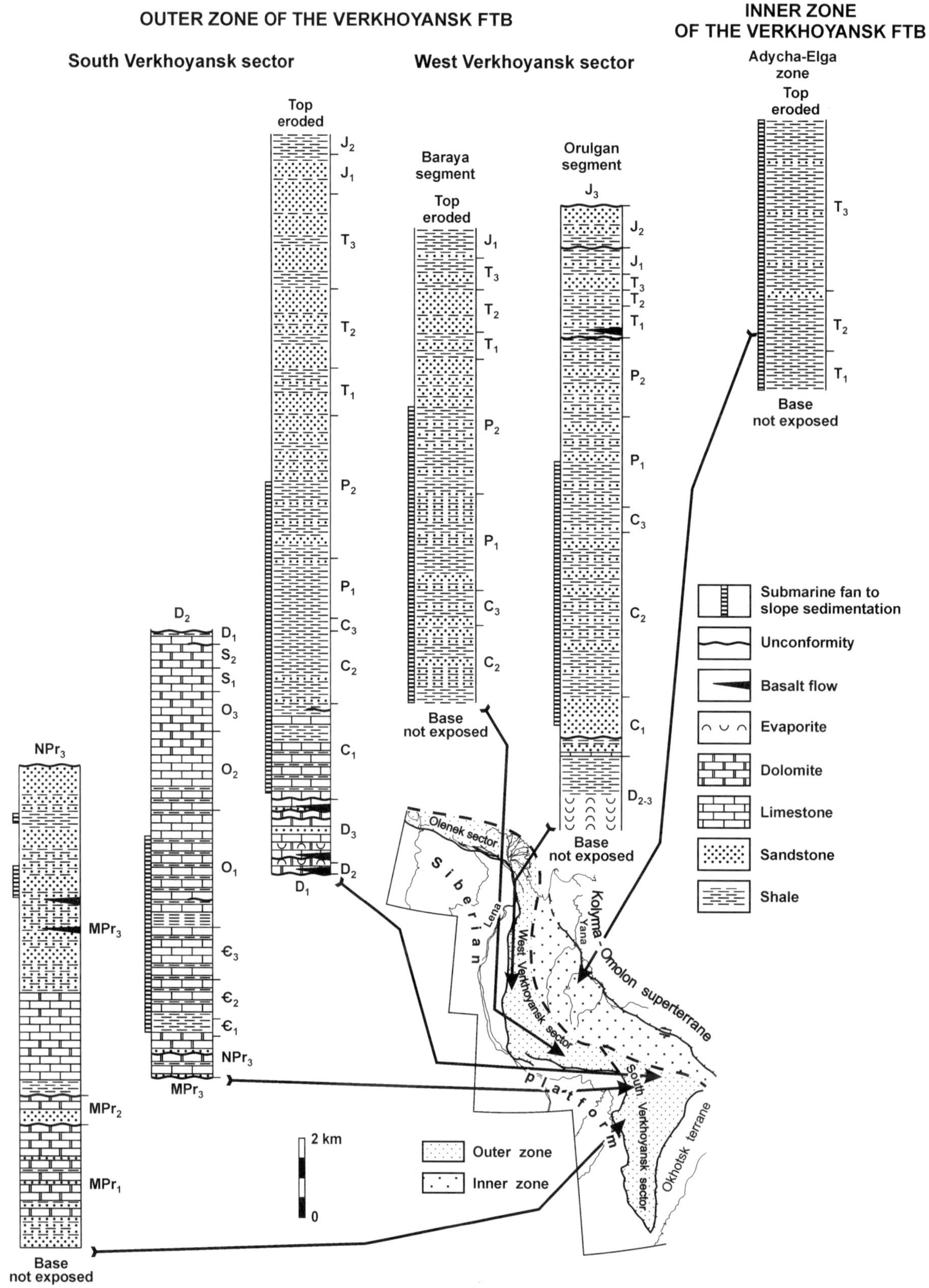

Figure 4. Stratigraphy of the Verkhoyansk fold-and-thrust belt (FTB) (after Prokopiev et al., 1999, 2001b; Khudoley and Guriev, 1994, 2003, and references therein). MPr—Mesoproterozoic; NPr—Neoproterozoic; Є—Cambrian; O—Ordovician; S—Silurian; D—Devonian; C—Carboniferous; P—Permian; T—Triassic; J—Jurassic.

calcareous turbidites, shales, and tuffs. Carbonate-free shales with sponge spicules and radiolaria are also present, and, in the South Verkhoyansk sector, they commonly contain large olistoliths of Devonian carbonates. These characteristics are typical of carbonate platform margins and adjacent basins (Khudoley and Guriev, 1994, and references therein). Evidence for widespread normal faulting in the Late Devonian and the Early Carboniferous is documented on the eastern margin of the North Asian craton including the outer part of the Verkhoyansk fold-and-thrust belt (Khudoley et al., 1991, Prokopiev et al., 2001b, Khudoley and Guriev, 2003).

At the end of the Early Carboniferous (late Visean to Serpukhovian), deposition of siliciclastic strata marked the beginning of significant changes in depositional environments. Lower Carboniferous to Jurassic siliciclastic strata, collectively known as the Verkhoyansk Complex, form a succession with a total composite thickness of ~14–16 km. The lower part of the succession consists of mainly fine-grained terrigenous rocks with sedimentary structures typical of fine-grained turbidites and, rarely, contourites. Upward in the succession, more coarse-grained terrigenous rocks that were deposited in shallow-marine to deltaic environments become abundant. Deposition continued into the Late Jurassic without significant unconformities, although some erosion is locally documented in the Mesozoic part of the succession (Fig. 4). Deposition of the Verkhoyansk Complex was controlled by several giant Mississippi-size submarine fan–delta systems (Fig. 5), which supplied sedimentary basins on the North Asian craton margins with clastic material (Parfenov, 1984; Egorov, 1993; Khudoley and Guriev, 1994).

In general, exposed rocks become younger eastward. In the inner zone of the Verkhoyansk fold-and-thrust belt, only Upper Permian and Mesozoic rocks of the Verkhoyansk Complex are exposed. Comparison of synchronous rock units from different parts of the Verkhoyansk fold-and-thrust belt shows that the volume of fine-grained terrigenous rocks deposited mainly in the lower parts of submarine fans increases from the outer to the inner part of the belt (Fig. 5). Thickness variation of synchronous units in the outer and inner zones of the Verkhoyansk fold-and-thrust belt reflects deposition in different parts of submarine fans and basin plains (Egorov, 1993; Khudoley and Guriev, 1994).

The Middle Devonian to Middle Jurassic evolution of the Verkhoyansk fold-and-thrust belt area is commonly interpreted in terms of rifting and passive margin formation, although details of the story are controversial (e.g., Parfenov, 1984; Khudoley and Guriev, 1994; Prokopiev, 2000). It is generally assumed that rifting occurred and initiated a transformation of rift basins into a wide sedimentary basin from the Devonian to the Visean (Early Carboniferous) (Khudoley and Guriev, 1994). Deposition of the Verkhoyansk Complex corresponds to post-rift subsidence of a large area. The huge composite thickness of the Verkhoyansk Complex (Figs. 4 and 5) points to significant subsidence of the basement that can be better quantified by construction of subsidence curves.

According to interpretation of gravity data, the crystalline basement surface reaches a depth as great as ~18 km (Parfenov et al., 1995; Tretyakov, 2004). However, the total composite thickness of the Mesoproterozoic to Mesozoic sedimentary succession is ~35 km. That is twice as great as the largest estimated depth to crystalline basement. The most likely explanation of the contradiction between depth to basement and total thickness of sedimentary succession is that the Meso- to Neoproterozoic, latest Neoproterozoic to Early Devonian, and Middle Devonian to Jurassic subdivisions of the succession were deposited in successive sedimentary basins. In this interpretation, each succession had a lens shape in cross-sectional view. The deep parts of successive sedimentary basins overlapped the thin margins of previous basins. The actual thickness is thus much less than the composite thickness. The eastward-facing successions (Fig. 2) indicate eastward migration of the depocenters from Mesoproterozoic to Mesozoic time, as reported by Khudoley and Guriev (2003) for the South Verkhoyansk sector of the belt and by Parfenov (1984) and Yapaskurt (1992) for the West Verkhoyansk sector.

Mesozoic Foredeep

The beginning of the Mesozoic foredeep deposition on the eastern margin of the Siberian platform was marked by Tithonian terrigenous rocks that have erosional contacts at the base. In contrast to underlying terrigenous marine units, they contain significant volumes of rocks that were deposited in predominantly continental environments. They exhibit increasing thickness eastward toward the Verkhoyansk fold-and-thrust belt (Prokopiev

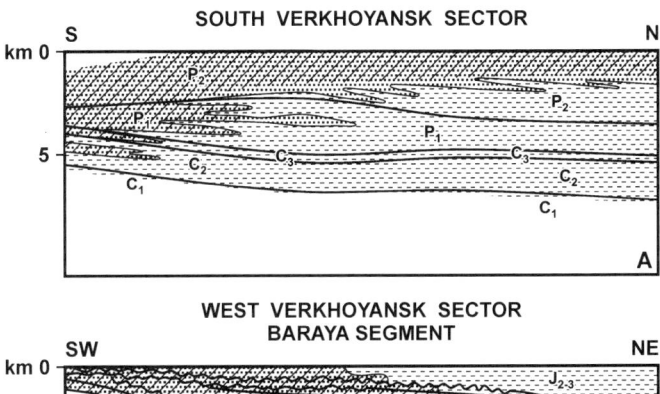

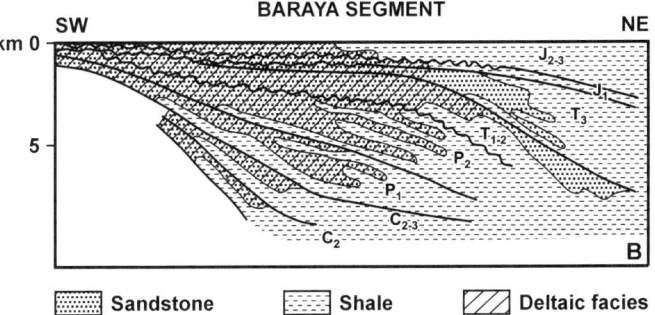

Figure 5. Submarine fan–delta systems in the Verkhoyansk Complex. (A) South Verkhoyansk sector. (B) Baraya segment (after Parfenov, 1984, modified). Note finger-like relation between sandstone and fine-grained clastics and thickness variation of synchronous stratigraphic units. C—Carboniferous; P—Permian; T—Triassic; J—Jurassic.

et al., 2001b, and references therein). Terrigenous rocks in the foredeep form a coarsening-upward succession in which the Upper Jurassic and most of the Lower Cretaceous rocks consist of shales and sandstones, whereas the Upper Cretaceous rocks are predominantly sandstones. Coal beds and thin lenticular conglomerate units are reported throughout the succession. Deposition of the Upper Jurassic and most of the Lower Cretaceous rocks occurred in fluvial, lacustrine, and shallow-marine environments, whereas in Albian and Late Cretaceous time, fluvial and alluvial-fan environments were predominant. The westward migration of the sedimentary basin depocenter as well as the distribution of coarse-grained terrigenous rocks reflects westward progradation of the Verkhoyansk fold-and-thrust belt in Cretaceous time (Fig. 6) (Prokopiev et al., 1999, 2001b). Oligocene rocks have local distribution and consist of sandstone, shale, and conglomerate units as thick as ~800 m (Parfenov et al., 2001).

No angular unconformities are known in the foredeep succession, although erosional surfaces have been documented at different stratigraphic levels. Available paleontological data show that deposition during Late Jurassic and most of the Early Cretaceous was almost continuous. However, the Albian and the Upper Cretaceous succession only contains fossil flora that do not permit high-resolution local stratigraphy or estimation of the duration of a hiatus. Fluvial cross-bedding measurements together with sandstone and pebble composition analyses show that in Late Jurassic–Hauterivian time the main sediment source area was the Siberian platform. The Verkhoyansk fold-and-thrust belt did not become the predominant source for the foredeep clastic material until late Hauterivian–Barremian time (Fig. 6) (Galabala, 1971; Yapaskurt, 1992; Parfenov et al., 1995).

Structural Styles of the Verkhoyansk Fold-and-Thrust Belt

Outer Zone of the Verkhoyansk Fold-and-Thrust Belt

The structural style along the thrust front of the Verkhoyansk fold-and-thrust belt varies greatly in different segments of the belt and, following terminology developed by Morley (1986), includes both emergent and buried styles (Prokopiev and Deikunenko, 2001; Prokopiev et al., 2001a). The frontal thrust faults, where exposed, change stratigraphic levels in different segments. The frontal thrust lies within the Precambrian succession in the South Verkhoyansk and the central part of the Kharaulakh segments. It is in the Devonian to Carboniferous strata in the Orulgan segment of the belt and in the Permian to Jurassic beds along the remainder of the frontal ranges (Fig. 2). Kilometer-scale changes in the stratigraphic location of the basal detachment are clearest in the South Verkhoyansk sector, where transverse footwall ramps control the structural style of the thrust front (Parfenov et al., 1995; Prokopiev and Deikunenko, 2001; Prokopiev et al., 2001a; Khudoley and Guriev, 2003). Continuation of these lateral ramps onto the Siberian platform basement and sedimentary cover illustrates that they correspond to preexisting basement faults that probably were active during Devonian rifting. Several levels of detachment are inferred to occur within the sedimen-

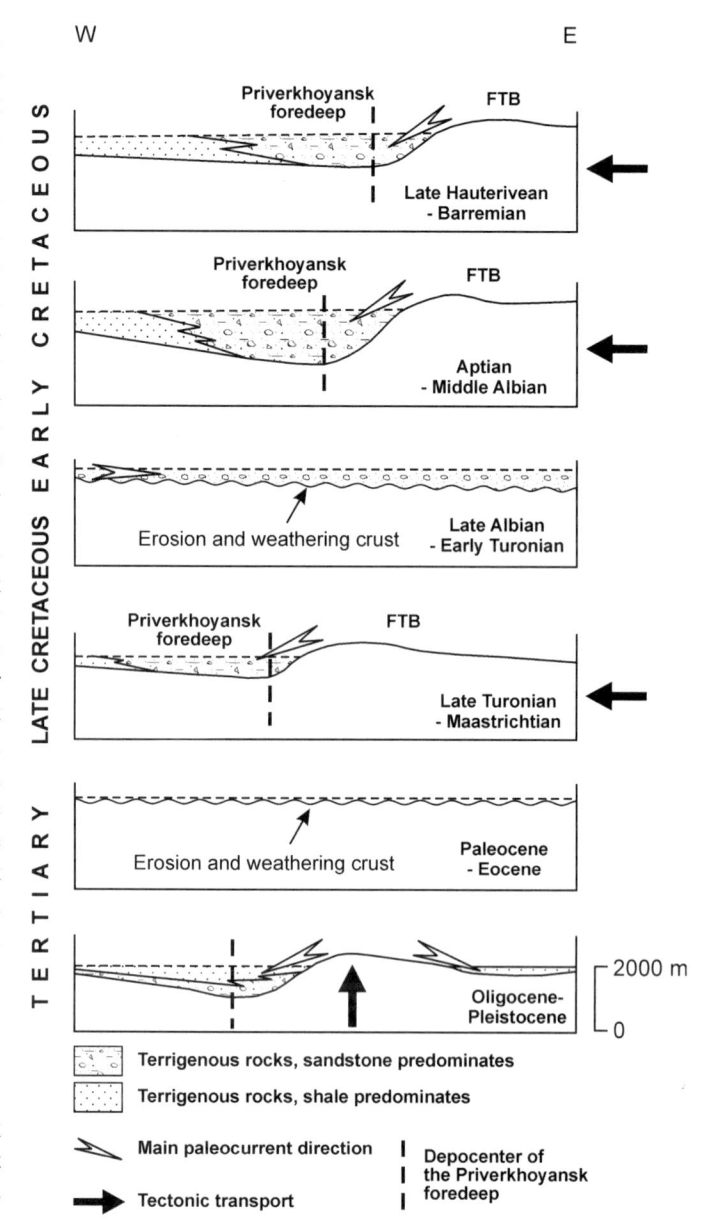

Figure 6. Cretaceous to Pleistocene evolution of the Priverkhoyansk foredeep. FTB—fold-and-thrust belt.

tary succession. The most significant one separated fine-grained terrigenous rocks of the Verkhoyansk Complex from underlying massive carbonate units.

Cross sections showing typical structural styles of the outer zone of the belt are presented in Figure 7. Imbricate thrust fans are typical of the frontal ranges of the Verkhoyansk fold-and-thrust belt, and these are best expressed in the frontal ranges of the South Verkhoyansk sector of the belt (Fig. 2), where thick, massive carbonate and sandstone units of Meso- to Neoproterozoic age dominate (Fig. 7A). In the West Verkhoyansk sector (Fig. 2) at a distance of 50 km to 180 km from the thrust front, there is a set of anticlinoria-like structural highs composed of Carboniferous rock

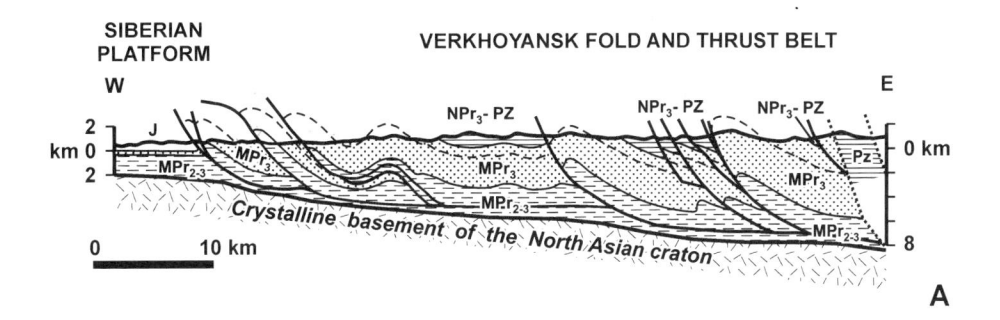

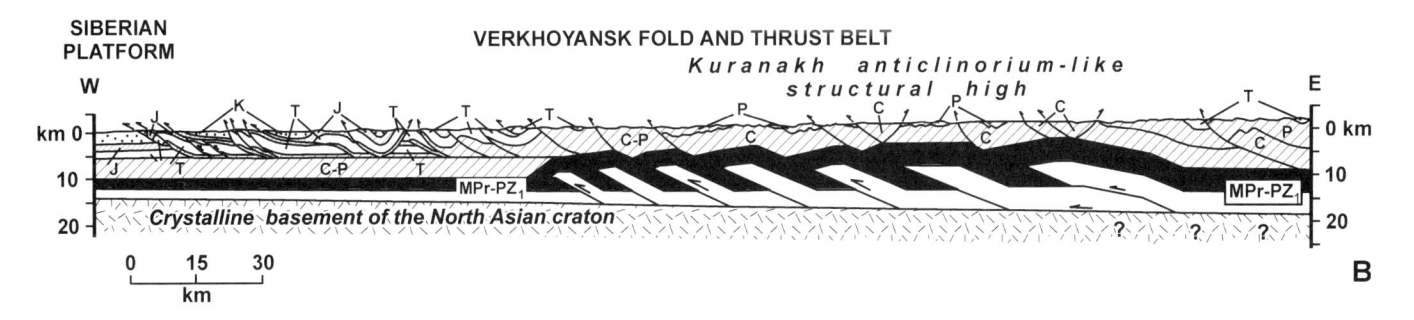

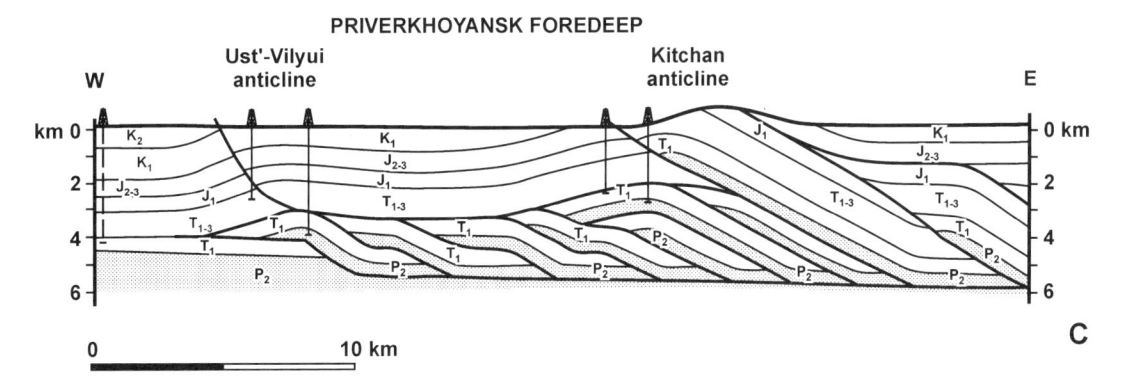

Figure 7. Cross sections through the outer Verkhoyansk fold-and-thrust belt and the Priverkhoyansk foredeep showing typical structural styles (after Prokopiev and Deikunenko, 2001; Khudoley and Guriev, 2003; simplified and modified). (A) South Verkhoyansk sector; (B) West Verkhoyansk sector, Kuranakh segment; (C) Priverkhoyansk foredeep. See location in Figure 2. MPr—Mesoproterozoic; NPr—Neoproterozoic; PZ—Paleozoic; C—Carboniferous; P—Permian; T—Triassic; J—Jurassic; K—Cretaceous.

units (Figs. 2 and 7B). Concentric linear folds up to 150–200 km in length and 20–30 km in width are typical for the outer zone of the belt as well. Within the Priverkhoyansk foredeep, thrusts dip both to the east and west, geometrically resembling triangle zones. The existence of triangle zones is also supported by interpretation of borehole and gravity data (Fig. 7C).

Penetrative cleavage and high-strain shear zones are locally distributed in the West Verkhoyansk sector and are mainly related to weak units close to thrust faults. Thick, massive sandstone units typically do not show evidence for penetrative strain and form parallel folds where bedding length is maintained. These properties allow line-length balancing in cross section restoration. In contrast, the South Verkhoyansk sector of the belt contains a strike-parallel zone characterized by penetrative cleavage, stretched pebbles, and strained fauna. The width of the cleavage zone varies greatly but locally is as great as 40 km.

Within this zone, the measured extension axes are typically subparallel to the strike of the belt, implying only a small vertical component of extension.

Estimates of shortening from balanced sections across the outer zone of the belt range from 15% to 50% but are most typically close to 30%. In the Kharaulakh segment (Fig. 2), shortening of Carboniferous and younger rocks (30%) is estimated to be much higher than shortening of older rock units (15%–20%), indicating detachment at the base of the Verkhoyansk Complex (Parfenov et al., 1995; Prokopiev and Deikunenko, 2001; Khudoley and Guriev, 2003).

The Inner Zone of the Verkhoyansk Fold-and-Thrust Belt

The inner zone of the Verkhoyansk fold-and-thrust belt lies east of a line of structural anticlinoria. The zone is underlain mainly by Triassic rocks, and its structural style is largely controlled by the predominance of shales in the succession (Fig. 4). Various outcrop-scale deformations include thrusts, shear zones, overturned and recumbent folds, and mélange-like units (Gusev, 1979; Parfenov et al., 1995; Prokopiev and Deikunenko, 2001). However, at the regional scale, the dominant structures are linear, wide, symmetrical, open folds with limb-dip angles of only 20°–30°, locally complicated by shear zones with minor, tight folds. Thrust faults with significant displacement have been reported (Parfenov et al., 1995; Prokopiev and Deikunenko, 2001); however, the evidence for their large displacement is based on the juxtaposition contrasting sedimentary facies in hanging wall and footwall blocks. The largest structure is the Sartang synclinorium, ~450 km long and 100 km wide, located in the northern part of the inner zone of the belt (Fig. 3). The contact between the outer and the inner zones of the belt is typically gradational (Fig. 8), but locally it is marked by thrusts (e.g., the eastward-dipping Sevastyanov thrust and westward-dipping Ayalyr thrust; Fig. 3).

A peculiar feature of the inner zone of the Verkhoyansk fold-and-thrust belt is the wide distribution of northwest-trending sinistral strike-slip faults and northeast-trending dextral strike-slip faults that form a conjugate fault system (Fig. 9) (Gusev, 1979; Parfenov et al., 1995). The strike-slip faults cut folds and thrusts and were likely formed during the latest stages of Cretaceous deformation. Despite local variations in their trends, the compressional axis inferred from the conjugate strike-slip faults was normal to fold axes and thrusts. In the northern part of the inner zone, sinistral strike-slip faults are more abundant, whereas in the southern part of the inner zone, dextral strike-slip faults are more abundant. Displacement along strike-slip faults typically varies from several hundred meters to a few kilometers but locally reaches 15–20 km. Dextral strike-slip faults with displacements of ~15–20 km form zones that are traced by granite intrusions across the inner zone of the Verkhoyansk fold-and-thrust belt and likely affected basement as well as the sedimentary succession (Prokopiev and Deikunenko, 2001).

Timing of Deformation

The age of deformation in the West Verkhoyansk sector of the belt becomes younger westward from the inner zone of the belt toward the frontal thrust. In the easternmost part of the Verkhoyansk fold-and-thrust belt, folds and faults are cut by 160 Ma to 135 Ma (^{40}Ar/^{39}Ar) granite plutons of the Main Batholith Belt, although most plutons range in age from 144 Ma to 136 Ma (Layer et al., 2001). Granite plutons become younger westward, and the granite pluton closest to the frontal thrust yielded an ^{40}Ar/^{39}Ar age of 98 Ma (Fig. 3). On the western margin of the Verkhoyansk fold-and-thrust belt, rock units as young as Cenomanian and Turonian (ca. 100–89 Ma) were involved in thrusting. Granite plutons with ^{40}Ar/^{39}Ar ages of 127–120 Ma cut sinistral strike-slip faults that, in turn, cut regional-scale folds (Fig. 9). The magmatic history of the belt corresponds approximately to the evolution of the Priverkhoyansk foredeep, which began to form in Tithonian (ca. 151–146 Ma). However, the first synorogenic clastic material arrived in the foredeep from the eroded Verkhoyansk fold-and-thrust belt in late Hauterivian–Barremian time (ca. 132–125 Ma) and continued, probably with some interruptions, into Maastrichtian time (ca. 71–66 Ma) (Fig. 6 and related discussion). Deformation in the South Verkhoyansk sector occurred in two stages: at 151 Ma and 120 Ma based on ^{40}Ar/^{39}Ar age of synmetamorphic mica (Prokopiev et al., 2003; Toro et al., 2004). Folds and faults in the easternmost part of the South Verkhoyansk sector of the belt

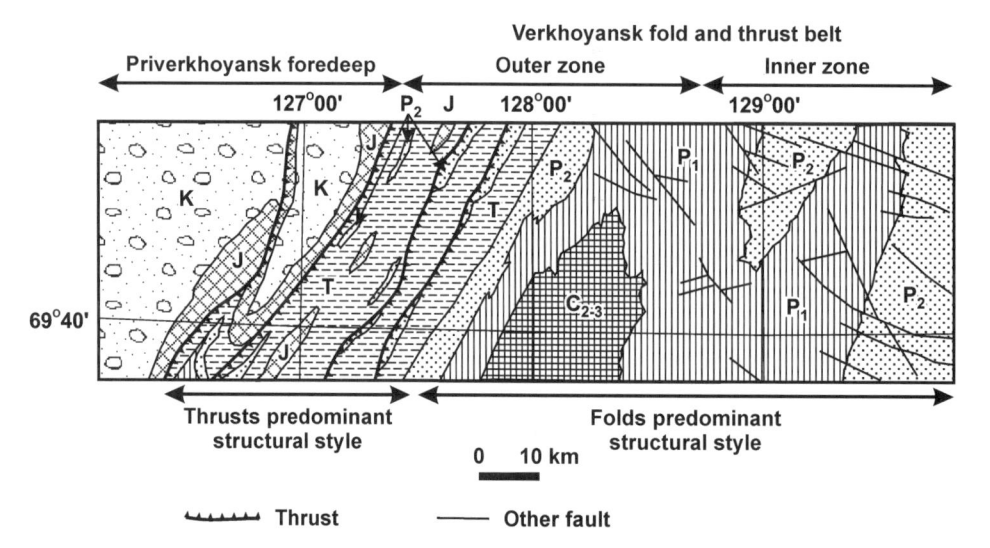

Figure 8. Contact between the outer and the inner zones of the Verkhoyansk fold-and-thrust belt with correspondent transition from thrust-predominant to fold-predominant structural style. After Prokopiev et al. (1999), simplified. See location in Figure 2. C—Carboniferous; P—Permian; T—Triassic; J—Jurassic; K—Cretaceous.

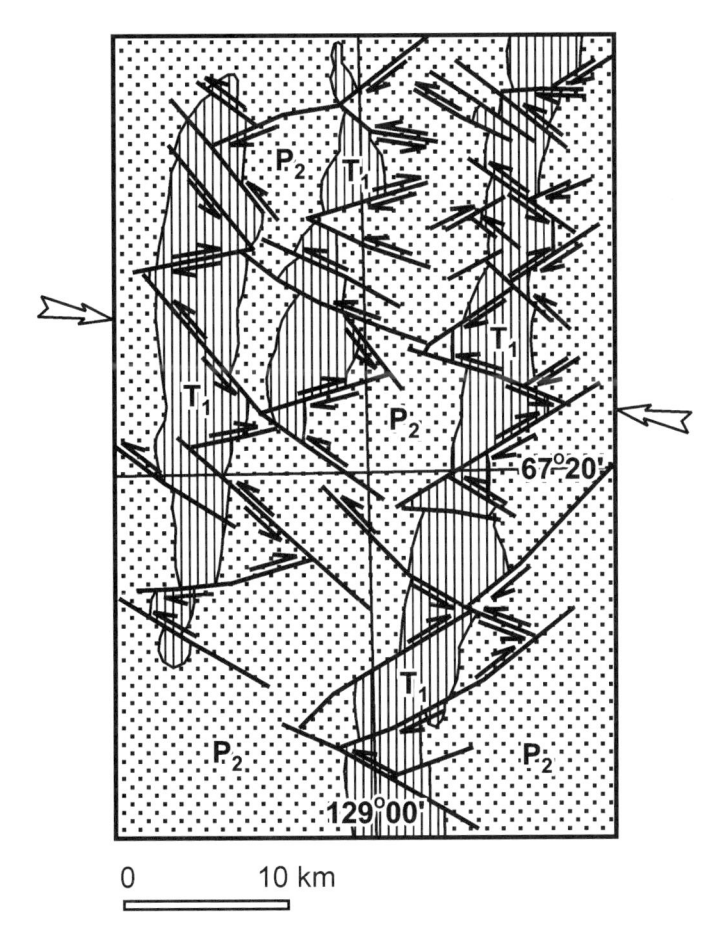

Figure 9. Conjugate strike-slip faults and their relationship with fold structures in the inner zone of the Verkhoyansk fold-and-thrust belt. Arrows show approximate direction of compressional axis, inferred from the conjugate strike-slip faults. After Prokopiev et al. (1999), simplified and modified. See location in Figure 2. P—Permian; T—Triassic.

were cut by granite plutons that typically yielded ca. 127–92 Ma ^{40}Ar/^{39}Ar ages (Layer et al., 2001). Modern mountain building and the latest reactivation of thrusts occurred in Oligocene and Pleistocene time and resulted, specifically, in deposition of a clastic unit up to 800 m thick (Fig. 6) (Parfenov et al., 2001).

SEDIMENTARY BASIN SUBSIDENCE STUDY

Subsidence Curve Construction

A back-stripping technique was developed to estimate the tectonic component of subsidence with respect to the total subsidence of a basin (e.g., Bond and Kominz, 1984; Bond et al., 1995, and references therein). The procedure is based on restoration of the precompaction thickness of each sedimentary unit and placing its top at a depth below sea level, corresponding to the average depth of water in which the unit was deposited. This procedure is repeated for all units in the section. The final graphical relationship between the depth to basement and the stratigraphic age of each unit represents the total subsidence curve. Subtraction of isostatic subsidence caused by sediment loading permits estimation of subsidence caused by tectonic forces associated with formation of the basin.

Accurate determination of the total and tectonic subsidence curve for the stratigraphic succession depends on exact knowledge of lithologies deposited throughout the interval and their age spans, porosity-depth relations, water depth of deposition of each unit, and the magnitude of eustatic sea-level changes. Construction and interpretation of tectonic subsidence curves also depend on the relationship between isostatic and flexural subsidence of the crust, the model used for stretching the crust, and its thermal history (Steckler et al., 1988; Keen and Beaumont, 1990; Bond et al., 1995). Most of these parameters cannot be well constrained for the Verkhoyansk fold-and-thrust belt, and our results are merely a first approximation to quantitatively model the evolution of the sedimentary basin.

Evolution of the Sedimentary Basin of the Outer Verkhoyansk Fold-and-Thrust Belt

The main task of the subsidence calculations is to quantify the post–Middle Devonian rifting evolution of the sedimentary basin that is located within the outer Verkhoyansk fold-and-thrust belt. Upper Devonian to Visean rock units, however, show contrasting amounts of subsidence in adjacent areas, and the fault-related deposition of most of this strata occurred in local half-graben sedimentary basins (Khudoley and Guriev, 2003); therefore, we do not include them in our subsidence analysis. The Devonian to Visean succession also contains a number of unconformities that represent an unknown amount of time and erosion. The Lower Carboniferous to Middle Jurassic Verkhoyansk Complex seems to be the best candidate for the construction of subsidence curves, and we selected three sections from the outer part of the Verkhoyansk fold-and-thrust belt located in the Orulgan and Baraya segments of the West Verkhoyansk sector and in the northern South Verkhoyansk sector (Fig. 2 and 4). The composite sections in Figure 4 were compiled from locally measured sections within structurally coherent domains not cut by significant faults. All local sections are situated close to each other to avoid errors due to the thickness changes that are typical of sedimentary units in submarine fan systems or to uncertain correlations of units across faults. Terrigenous rock units contain enough fauna to be correlated with an international stratigraphic scale that has good radiometric age control (Prokopiev et al., 1999, 2001b, and references therein). In the Orulgan segment and the South Verkhoyansk sector, the succession of the Verkhoyansk Complex is complete, whereas in the Baraya segment, the contact between the Verkhoyansk Complex and underlying units is hidden and the lowermost exposed units are of Middle Carboniferous age (Fig. 4).

A passive margin interpretation of the Verkhoyansk Complex sedimentary succession permits the use of an isostatic model for the latest Early Carboniferous to Middle Jurassic subsidence history. Relative amounts of sandstone, siltstone, and shale in each unit that was involved in the subsidence calculation was estimated using local and composite sections discussed by Khudoley and Guriev (1994), Khudoley et al. (1995), Prokopiev et al. (1999). No sea-level correction has been made, but sea-level changes seem to be negligible in comparison to the composite thickness of the Verkhoyansk Complex (Fig. 4) and related tectonic subsidence. Water depth of deposition is poorly constrained and we do not incorporate this correction in the subsidence curves. Study of fauna and sedimentary structures show that Upper Permian and Lower Triassic rock units in the outer zone of the belt are typically deltaic or shelfal in origin with water depth that ranged from several tens of meters to ~200 m. However, fine-grained turbidites and pelagic shales are widespread in Carboniferous rock units, indicating basinal environments (Parfenov, 1984; Khudoley and Guriev, 1994). This factor significantly underestimates the amounts of subsidence in the Carboniferous, but for the Late Permian and Early Triassic, subsidence should not be underestimated by more than 200 m; that is significantly less than the cumulative thickness of the Verkhoyansk Complex and its related tectonic subsidence.

Total subsidence curves and tectonic subsidence curves for three Verkhoyansk Complex successions are shown in Figure 10. Here we present subsidence analysis for the Paleozoic and Lower Triassic part of the Verkhoyansk Complex that corresponds to the post-rift stage of evolution of the east margin of the North Asian craton. This part of the section was deposited during an ~80 m.y. period from ca. 327 Ma to 245 Ma. Total subsidence curves and tectonic subsidence curves of wells located on the North American Atlantic shelf (Mississauga H-54) and near the transitional continental crust–oceanic crust boundary (COST-B2) are shown for comparison (Keen et al., 1990; Bond et al., 1995). The subsidence history of successions from these wells corresponds to the post-rift stage for the Mesozoic to modern North American Atlantic margin. To facilitate comparison of the evolution of the Paleozoic sedimentary basin on the east margin of the North Asian craton and the Mesozoic sedimentary basin on the Atlantic margin of North America, we present results of the subsidence calculations in terms of time since the beginning of the post-rift stage (Fig. 10).

Total subsidence for the Verkhoyansk Complex successions after ~80 m.y. of deposition was significant and, by the end of the Permian, varied from 11 km to 13 km. Theoretical curves shown on the tectonic subsidence plot were calculated for the uniform-stretching model and correspond to post-rift thermal subsidence with different stretching coefficients (Bond et al., 1995, and references therein). Both total subsidence curves and tectonic subsidence curves for the Verkhoyansk Complex successions are located on the plots in Figure 10 between curves for Mississauga and COST-B2 wells. During the first 60 m.y. after the beginning of post-rift stage, deposition of the Verkhoyansk Complex (Middle Carboniferous to Early Permian) occurred in basinal environments, whereas deposition of the COST-B2 succession occurred in shallow-marine environments (Steckler et al., 1988; Bond et al., 1995). Absence of the water depth correction for the Verkhoyansk Complex successions implies that their actual tectonic subsidence was significantly higher than is shown in Figure 10 and that the tectonic subsidence history of the Verkhoyansk Complex successions was closer to that of COST-B2 rather than that of Mississauga well. During the next 40 m.y., from 60 m.y.

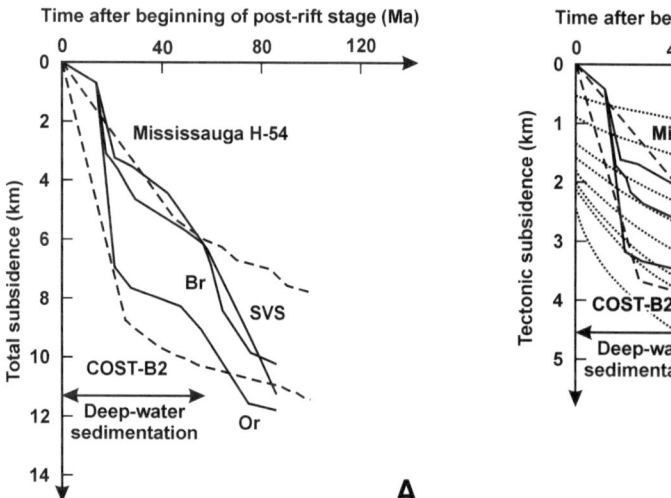

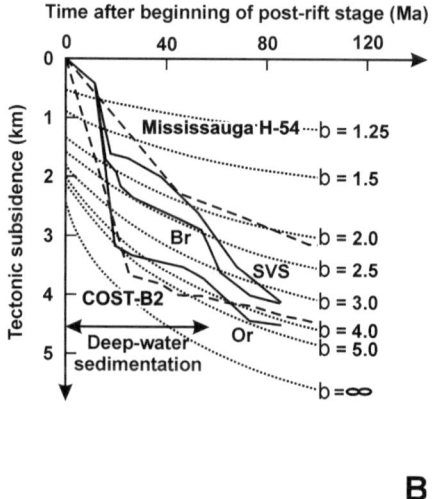

Figure 10. Total (A) and tectonic (B) subsidence curves for the Verkhoyansk Complex (solid line) in comparison with those from the modern Atlantic margin (dashed line); dotted line is tectonic subsidence curve calculated from the uniform stretching model (see Keen and Beaumont, 1990; and Bond et al., 1995, for discussion); b is stretching coefficient. COST-B2 and Mississauga H-54 wells data are after Steckler et al. (1988), Keen et al. (1990), and Bond et al. (1995). COST-B2 well is located near the transitional continental crust-oceanic crust boundary; Mississauga H-54 well is on the Nova Scotia shelf. Abbreviations for the studied sections from the Verkhoyansk fold-and-thrust belt: Or—Orulgan segment; Br—Baraya segment; SVS—South Verkhoyansk sector.

to 100 m.y. after the beginning of the post-rift stage, in all successions from the east margin of the North Asian craton and the Atlantic margin of the North America, deposition occurred in deltaic to shelf environments. For this time interval, total and tectonic subsidence curves calculated for the Verkhoyansk Complex successions were close to those calculated for COST-B2. We interpret the similarity in the tectonic subsidence history of the Verkhoyansk Complex successions and COST-B2 to indicate similarity in their tectonic setting. Since COST-B2 is located near the boundary between transitional continental crust and oceanic crust, it is likely that the Carboniferous to Early Triassic Verkhoyansk Complex successions from the outer zone of the Verkhoyansk fold-and-thrust belt were located on the transitional continental crust of the east margin of the North Asian craton, close to its boundary with oceanic crust.

STRUCTURAL STUDY

Critical Wedge Theory

The general theory of tapered orogenic wedges was developed by Chapple (1978) and successfully applied to accretionary complexes and foreland fold-and-thrust belts by Davis et al. (1983) and Dahlen et al. (1984). According to this theoretical model, the mechanical development of a fold-and-thrust belt can be described in terms of Mohr-Coulomb failure theory that requires a wedge shape of the fold-and-thrust belt. A thrust belt can advance under horizontal compression when the wedge angle (τ) (Fig. 11) is of sufficient magnitude ($\tau = \tau_c$). The wedge angle (τ) is sum of the mean topographic slope (α) and the basal detachment slope (β) (Fig. 11). The relation between α and β predicted by the theory is $\alpha + R\beta = F$, where R and F are functions defined by Davis et al. (1983) and depend on the mechanical properties of rocks in thrust sheets, as well as fluid pressure, which decreases frictional resistance to sliding. For a dry, cohesionless sand wedge, $R = 0.66$ and $F = 5.9°$. For most thrust belts $R = 0.55$ and $F = 6.0°$ give a better fit to the observed wedge geometry, although thrust belts in supercritical and subcritical states are reported (Davis et al., 1983;

Boyer, 1995). If $\tau < \tau_c$, the wedge must thicken to increase τ to a critical value to allow the wedge to move forward. If $\tau > \tau_c$, the wedge will tend to thin to decrease τ to a critical value. The mechanisms for thrust-wedge building and thickening include displacement along thrusts faults and ductile strain within thrust sheets, whereas wedge thinning can result from surface erosion, extensional faulting, or accretion of additional imbricate sheets to the tip of thrust wedge. Other parameters, like the strength of the wedge, flexural subsidence under the weight of advanced thrust sheets, fluid pressure, and weakness of the basal layer also affect shape of the thrust wedge (Davis et al., 1983; Woodward, 1987; Boyer, 1995; DeCelles and Mitra, 1995; Lohrmann et al., 2003; Ellis et al., 2004).

Application of the critical wedge model to foreland fold-and-thrust belts, although generally successful, has met some problems that required modifications of theory (e.g., Price, 1988; Bombolakis, 1994). Woodward (1987) found that actual shortening in the Cordilleran and Appalachian foreland fold-and-thrust belts was commonly significantly lower than that predicted by theory. Davis and Engelder (1985) discussed very narrow thrust wedges ($\tau \sim 1°$) with the basal detachment of the wedge within a salt layer. They pointed out that thrust wedges with a basal salt layer typically contain very long, symmetric anticlines and broad, flat-bottomed synclines. When the salt layer is thicker than 100 m and deeper than 3 km, salt-cored anticlines are common. Observations of the structural style of natural fold-and-thrust belts underlain by weak detachments are in reasonable agreement with analogue modeling (Costa and Vendeville, 2002).

Application of the Critical Wedge Model to the Verkhoyansk Fold-and-Thrust Belt

We use published balanced and restored cross sections to estimate the main characteristics of the Verkhoyansk thrust wedge (Parfenov et al., 1995; Prokopiev and Deikunenko, 2001; Khudoley and Guriev, 2003). These balanced cross sections utilized available seismic and gravity data as well as information on the structural style and thickness of sedimentary units. Figure 12 presents an outline of balanced cross sections of the foreland of

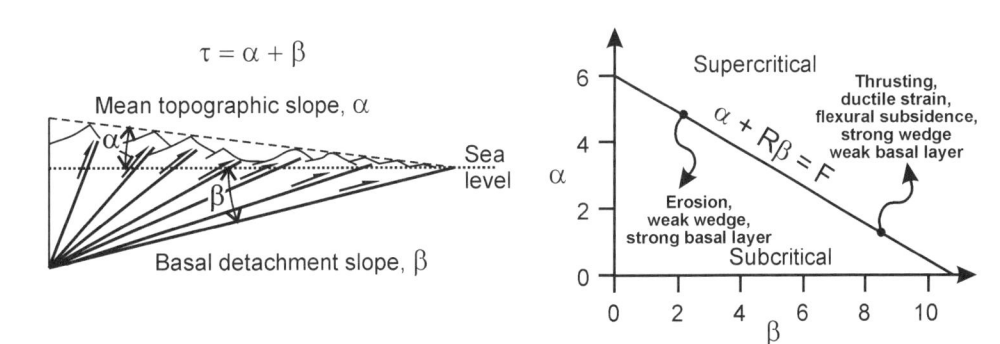

Figure 11. Critical Coulomb wedge model showing the main parameters (α, β) and the main factors that affect the shape of the thrust wedge. See text for discussion and definition of R and F.

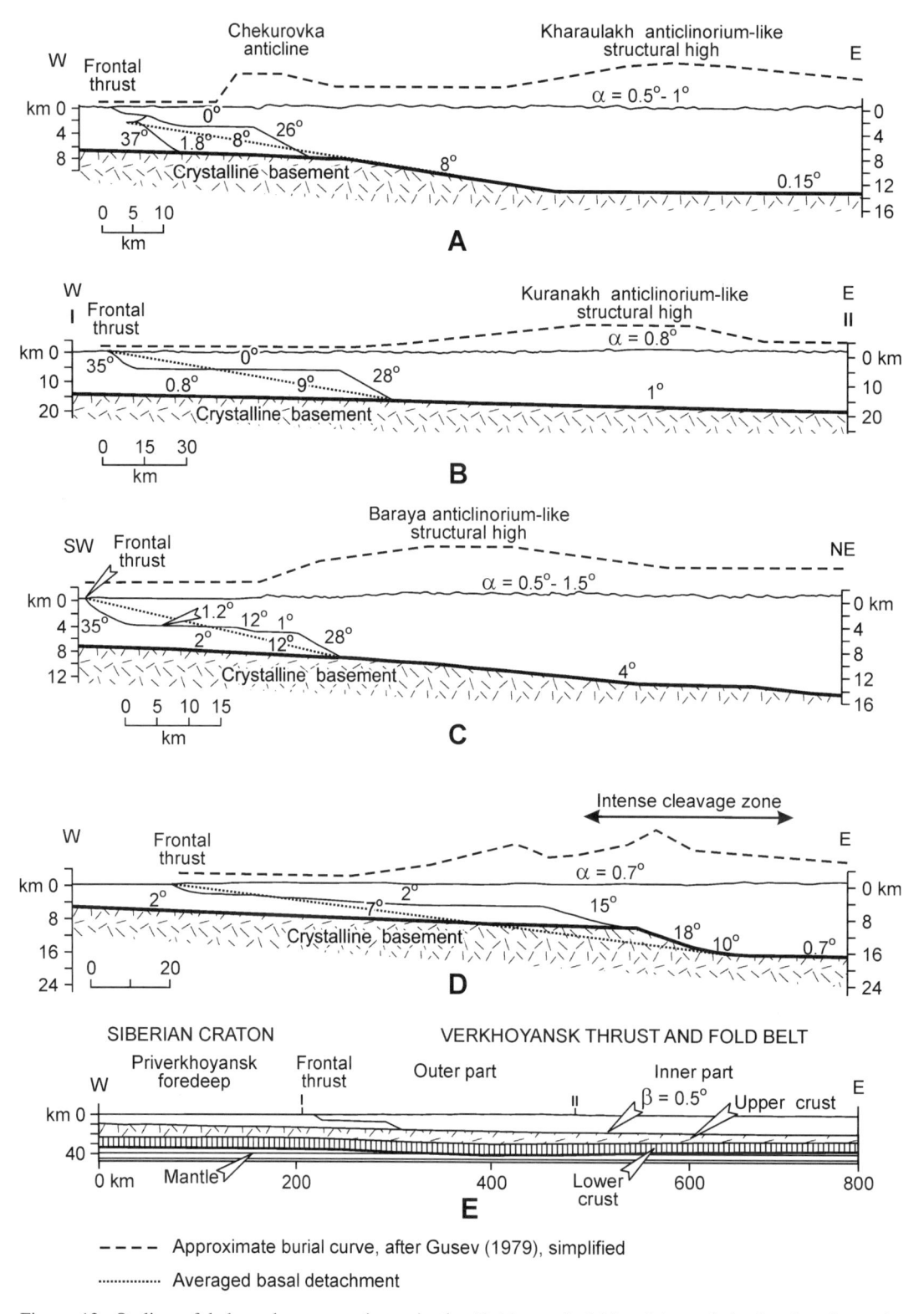

Figure 12. Outline of balanced cross sections via the Verkhoyansk fold-and-thrust belt showing the main parameters of thrust wedges. Data source for detachment and crystalline basement geometry are from Parfenov et al. (1995) and Prokopiev and Deikunenko (2001). See location in Figure 2. Horizontal and vertical scales are the same in all cross sections. (A–D) Outer zone of the Verkhoyansk fold-and-thrust belt; (A–C) West Verkhoyansk sector. (A) Kharaulakh segment, northern part. (B) Kuranakh segment, southern part. (C) Baraya segment. (D) South Verkhoyansk sector. (E) Cross section via the outer and inner parts of the Verkhoyansk fold-and-thrust belt based on interpretation of the gravity field. I–II corresponds to cross section B.

the Verkhoyansk fold-and-thrust belt and a regional-scale gravity field-based cross section of the entire Verkhoyansk fold-and-thrust belt showing the main characteristics of these thrust wedges.

Outer Zone of the Verkhoyansk Fold-and-Thrust Belt

In the outer zone of the Verkhoyansk fold-and-thrust belt, the topographic slope α is difficult to estimate from available data, but it may be constrained by the study of Mesozoic and Cenozoic sediments in the Priverkhoyansk foredeep (see Fig. 6 and related discussion). Although the foredeep began to form in Late Jurassic time, the Verkhoyansk fold-and-thrust belt became the predominant source region for foredeep clastics only since late Hauterivian–Barremian time. No significant erosion is recorded in the Barremian and Albian succession, showing that the early stages of the Verkhoyansk fold-and-thrust belt evolution are well documented by synchronous sediments in the foredeep. Cretaceous terrigenous rocks are mainly sandstones with shale and conglomerate interbeds. The lack of thick sequences of coarse-grained clastic rocks in the Lower Cretaceous succession is interpreted as evidence that at the early stages of the thrust belt formation, the orogeny was not accompanied by significant mountain building and high relief. The modern topography of the Verkhoyansk fold-and-thrust belt was apparently formed only in the Pleistocene and thus does not reflect Cretaceous folding and thrusting. This is interpreted as indication that high mountains did not exist within the Verkhoyansk fold-and-thrust belt in Cretaceous and that synorogenic relief was lower than modern relief (Parfenov et al., 1995). The modern Verkhoyansk Mountains are characterized by elevations up to 2 km, corresponding to α varying from 0.5° to 1° and, locally, in the South Verkhoyansk sector are as much as 2°. These values mark the upper limit for Cretaceous relief and topographic slope in the outer zone of the belt of 1° or lower.

The basal detachment of the Verkhoyansk fold-and-thrust belt typically coincides with the sedimentary-crystalline basement interface and only in the frontal thrusts does it cut upward into the sedimentary succession (Figs. 7 and 12). The modern dip of the sediment-basement interface measured directly from cross sections can be higher than its initial angle due to flexural subsidence of the basement under the weight of the advancing thrust wedge. However, erosion decreases the total weight of the thrust sheets, which, in turn, decreases the basement dip angle. The most detailed study of the total amount of erosion in the Verkhoyansk fold-and-thrust belt was carried out by Gusev (1979) and was based mainly on data on the rock unit thicknesses, stratigraphic correlations, and their pre-Cretaceous distributions. A recent study of conodont alteration indices in the western part of the South Verkhoyansk sector of the belt and vitrinite reflectance from the Priverkhoyansk foredeep are in reasonable agreement with Gusev's (1979) data (Prokopiev and Deikunenko, 2001; Toro et al., 2004). Gusev's (1979) estimate is presented in Figures 12A–12D and shows that in the outer part of the Verkhoyansk fold-and-thrust belt, a significant part of the thrust wedge was eroded. Conglomerate beds in the upper part of the Lower Cretaceous succession contain pebbles derived from Permian rocks of the Verkhoyansk fold-and-thrust belt, showing that significant erosion of the belt occurred in Cretaceous time (Galabala, 1971). Since high mountains did not exist in the outer Verkhoyansk fold-and-thrust belt in Cretaceous time, we infer that despite significant erosion, the total thickness of the thrust wedge in the Cretaceous was similar to that observed in its modern structure. Although there are no reliable data to quantify changes in the basement dip angle related to the periods of mountain building and subsequent erosion, we use modern dip angles as a first-order approximation of the initial ones.

The shape of the thrust wedge estimated from the cross sections in Figures 12A–12D is in reasonable agreement with theory. Averaged basal detachment slopes β, in the front of the Verkhoyansk fold-and-thrust belt, ranged from 7° to 12°, whereas topographic slopes α were <1°. According to the relation between β and α, the thrust wedge in the outer zone of the belt was at a critical state (Fig. 13) and could even move with very low topographic slopes. This helps to explain the sedimentological data from the Priverkhoyansk foredeep that suggests that the advancing thrust sheets during the Mesozoic did not build a significant mountain range with high relief. Displacement of high-angle wedges does not require significant internal deformation of thrust sheets, and this corresponds to the observed absence of ductile strain, low amount of shortening (30%), and relatively simple structures of the frontal thrusts of the Verkhoyansk fold-and-thrust belt compared with those in corresponding tectonic units or zones of Cordillera or Appalachians where the slope of the basal detachment, β, is much less than in the frontal thrust zone of the Verkhoyansk fold-and-thrust belt (e.g., Price, 1981; Evans, 1989; Price and Sears, 2000).

Inner Zone of the Verkhoyansk Fold-and-Thrust Belt

The estimated total amount of erosion in the inner zone of the Verkhoyansk fold-and-thrust belt typically varies from 2 km to 4 km, corresponding with the absence of significant structural depressions and highs in the modern structure of the inner zone

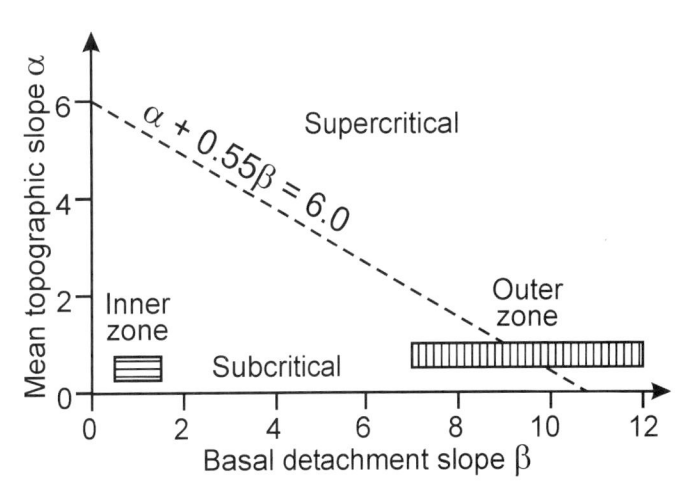

Figure 13. Relationship between shape of the thrust wedges in the outer and the inner zones of the Verkhoyansk fold-and-thrust belt.

of the belt (Gusev, 1979). There is no data to constrain how much erosion occurred in the Cretaceous, but very gradual changes in the amount of erosion suggest that the topographic slope was always very low and was likely lower than in the outer zone where changes in amount of erosion were much more abrupt (Fig. 12). The upper limit for Cretaceous relief and topographic slope in the outer zone of the belt has been estimated to be 1°, and this implies that in the inner zone, topographic slope was <1°. According to interpretations of gravity data, the mean slope of the basement in the inner zone is very low and typically does not exceed 1° (Fig. 12E) (Parfenov et al., 1995; Prokopiev and Deikunenko, 2001; Tretyakov, 2004). Data on topographic and basement slope combined together show that the thrust wedge in the inner zone of the belt is much thinner than that in the outer zone of the belt, and it is unlikely that they form a single thrust-related Coulomb wedge (Fig. 13).

Variation in structural style between the outer and inner parts of the Verkhoyansk fold-and-thrust belt is inconsistent with their origins as a single thrust-related Coulomb wedge. It is, however, very likely that formation of the outer and inner parts of the Verkhoyansk fold-and-thrust belt resulted from the evolution of two separate thrust wedges. The boundary between the two thrust wedges or thrust systems is probably represented by the westward-vergent Sevastiyanov thrust that displaces deep-water Carboniferous shales over shallow-water Permian clastics on the eastern limb of the Kharaulakh segment of the belt (Fig. 3). On the eastern limb of the Orulgan segment, eastward-vergent thrusts (e.g., Ayalyr thrust) may represent back-thrusts in a triangle-zone at the front of the inner zone thrust wedge (Fig. 3). To the south of the Orulgan segment, significant thrusts have not been mapped within the Verkhoyansk fold-and-thrust belt, and the boundary between the inner and outer thrust wedges is only approximated.

The occurrence of two distinct thrust wedges within a single fold-and-thrust belt is not a very widespread phenomenon but has been documented, for example, in Spitsbergen and, locally, in the Sevier fold-and-thrust belt of North America (Mitra, 1997; Braathen et al., 1999). However, in the Spitsbergen and Sevier belts, the inner thrust wedge consists of stronger rocks than the outer wedge. In the Verkhoyansk fold-and-thrust belt, both outer and inner wedges are composed of sedimentary rocks with similar lithology (Fig. 4).

Evolution of a thrust wedge with a mean detachment slope angle of 1° requires ~60%–65% of internal shortening (Boyer, 1995). This is much higher than the shortening inferred from structural style of the internal Verkhoyansk fold-and-thrust belt with a predominance of open folds with limb dip angles of 20°–30°. Moving the thrust wedge on a detachment that dips so shallowly requires specific conditions along the basal detachment; namely, it must be weak enough for thrusting to occur. Depth to basement in the south of the inner Verkhoyansk fold-and-thrust belt is typically >12 km, and the sedimentary cover–basement contact is likely located close to or within the ductile-brittle transition zone, which might act as a weak basal layer, facilitating detachment of the sedimentary cover from crystalline basement. However, in the northern part of the inner Verkhoyansk fold-and-thrust belt, there is an ~200-km-wide zone where depth to basement is <10 km (Tretyakov, 2004). No evidence for a high geothermal gradient is reported in the inner Verkhoyansk fold-and-thrust belt, making it unlikely that temperature-induced cataclastic flow could take place at such shallow depth.

A more reasonable explanation for the occurrence of a wide, low-deformed, narrow thrust wedge in the inner Verkhoyansk fold-and-thrust belt may be related to salt and evaporite deposits at the bottom of the sedimentary succession. The overall shape of the inner Verkhoyansk fold-and-thrust belt wedge and its structural style are in general agreement with those predicted for thin-skinned fold-and-thrust belts underlain by salt, although in this case, the evaporite unit is probably not thick enough to be exposed in cores of anticlines (e.g., Davis and Engelder, 1985; Costa and Vendeville, 2002). Evaporite units in the Middle and Upper Devonian succession have been found in many places within and around the Verkhoyansk fold-and-thrust belt and seem to be the only candidates that might localize a basal detachment in the inner Verkhoyansk fold-and-thrust belt (Gusev, 1979; Kovalskiy, 1985; Khudoley and Guriev 1994). Moreover, Devonian evaporites form the basal detachment surface of the Verkhoyansk fold-and-thrust belt in the northern Orulgan segment of the belt (Fig. 2), where the frontal thrust of Verkhoyansk fold-and-thrust belt has been estimated to have a displacement of ~20 km (Naumov, 1962). In the southern Orulgan segment, gypsum-anhydrite diapirs with fragments of basalts and fossil-bearing Devonian limestone are known as well (Sborshchikov and Natapov, 1969). However, Devonian evaporite units do not form a continuous stratigraphic horizon, and in the outer Verkhoyansk fold-and-thrust belt, the basal detachment layer is often located within other stratigraphic levels. The wide distribution of Devonian evaporites and their locations at the base of the sedimentary succession in the inner Verkhoyansk fold-and-thrust belt, inferred from the thrust wedge shape, imply the absence of or a significantly reduced thickness of older rocks. This accentuates yet another significant difference between the inner and the outer Verkhoyansk fold-and-thrust belt, where thick successions of Mesoproterozoic to lower Paleozoic rocks are documented (e.g., Figure 4).

Discussion

Results of studies presented in this paper can be summarized by the following main points:

- The Verkhoyansk fold-and-thrust belt consists of two structurally different domains, the outer and inner zones. The boundary between the two zones is located ~100 km to 180 km to the east of the modern frontal thrust separating the Siberian platform and the Verkhoyansk fold-and-thrust belt. Before ~30% shortening in the Mesozoic, this boundary was located 140–250 km east of the modern frontal thrust of the Verkhoyansk fold-and-thrust belt.

- The total thickness of the Mesoproterozoic to Mesozoic sedimentary succession is ~35 km, twice the maximum estimated depth to crystalline basement. The most likely explanation of this contradiction is that various parts of the succession were deposited in different sedimentary basins with an eastward migration of the depocenter from Mesoproterozoic to Mesozoic time.
- A major rifting event occurred in the Middle to Late Devonian on the eastern margin of the North Asian craton and within the Verkhoyansk fold-and-thrust belt. The Middle Devonian to Jurassic succession is typical of those deposited on passive margins (Parfenov, 1984; Khudoley and Guriev, 1994). The post-rift thermal subsidence stage corresponds to deposition of the Lower Carboniferous to Middle Jurassic Verkhoyansk Complex.
- The amount of total and tectonic subsidence calculated from sections located in the central and eastern parts of the outer zone of the Verkhoyansk fold-and-thrust belt is much higher than estimated for the Atlantic shelf of North America and is similar to that estimated near the boundary between transitional continental crust and oceanic crust on the Atlantic margin of North America.
- Variations in structural style and basement dip-angle make it difficult to interpret the outer and inner zones of the Verkhoyansk fold-and-thrust belt as a single thrust wedge. Here we recognized two thrust wedges, one in the outer and another in the inner zone of the belt.
- Displacement of a low-deformed thin thrust wedge corresponding to the inner Verkhoyansk fold-and-thrust belt requires a weak horizon at the base of the thrust wedge. The widespread distribution of a hypothetical salt or evaporite unit at the base of the sedimentary succession is inferred as the most viable option to explain the geometry of this part of the belt. The structural style and geometry of the inner zone of the Verkhoyansk fold-and-thrust belt corresponds well with those of fold-and-thrust belts around the world underlain by salt or evaporite units.
- Middle to Upper Devonian strata are the only parts of the sedimentary succession that are known to contain evaporite units. We assume that in the inner thrust wedge they host a basal detachment that separates basement from structurally-detached thrust sheets. Depth to basement, inferred from limited seismic and gravity data, corresponds approximately to the known thickness of the Middle Devonian to Middle Jurassic succession, ~14–18 km. These observations imply that within the inner zone pre–Middle Devonian rocks are thin or absent.

The main conclusion that comes from these data is that the outer and the inner zones of the Verkhoyansk fold-and-thrust belt are different not only in structure but also in stratigraphy and tectonic history. In the outer zone, very thick successions of Mesoproterozoic to Lower Paleozoic rocks are exposed, whereas in the inner zone it seems that they are thin or absent, with Middle to Upper Devonian evaporites comprising the lower unit of the sedimentary succession. In the outer zone, Devonian evaporites have discontinuous distribution, forming hundred-meter-scale units in some areas and absent in others (Prokopiev et al., 2001b, and references therein). In the inner zone, evaporites are inferred to be much more widely distributed in order to facilitate the displacement of very thin thrust wedges.

Interpretation of both structural and subsidence history data shows that the outer and inner zones of the Verkhoyansk fold-and-thrust belt also have a different basement. The outer zone mainly overlies the slope of the North Asian cratonic basement. Traditionally, the Verkhoyansk fold-and-thrust belt is interpreted as underlain by a continental basement (e.g., Parfenov, 1984, 1995; Prokopiev, 2000; Prokopiev and Deikunenko, 2001). The similarity of the Carboniferous-to-Mesozoic subsidence history of the Verkhoyansk fold-and-thrust belt outer zone and the Mesozoic-to-Cenozoic subsidence history of the North American Atlantic margin based on data from the COST-B2 well, which is located very close to the continent-ocean crust boundary (Steckler et al., 1988; Bond et al., 1995), supports a passive-margin setting for the Verkhoyansk Complex sedimentary succession (Parfenov, 1984; Khudoley and Guriev, 1994). In this interpretation, the outer zone corresponds to a zone of transitional continental crust ~140–250 km across, which is also comparable with the Atlantic margin of North America (e.g., Klitgord et al., 1988). The width of the Verkhoyansk fold-and-thrust belt, however, is at least 500 km, which is too broad to suppose that its entire basement is a uniform transitional continental crust extending this distance oceanward from the craton. The Verkhoyansk Complex is composed of a thick succession of terrigenous rocks, deposited in large Mississippi River–sized submarine fan–delta systems where significant amounts of sediment were deposited in deep-water environments (Fig. 5) (Khudoley and Guriev, 1994). On modern passive margins, a significant part of such a large submarine fan rests on oceanic basement. In view of these data, the inner zone of the Verkhoyansk fold-and-thrust belt may be interpreted as a fragment of an oceanic basin. In previous interpretations, an oceanic basin was recognized only to the east of the Verkhoyansk fold-and-thrust belt (Parfenov, 1995).

The conclusion that a large region of the Verkhoyansk fold-and-thrust belt is underlain by oceanic crust is not supported by magnetic data, as the region lacks strong positive magnetic anomalies. The structural style of the inner Verkhoyansk fold-and-thrust belt, dominated by open folds separated by rare thrusts, is very different from that of accretionary wedges developed in deeper-water sequences of sandstones, shales, and cherts that result in thin tectonic slices with complicated internal structures as well as disrupted mélange units. No magmatic rocks that might have resulted from subduction of the oceanic crust are known within the Verkhoyansk fold-and-thrust belt.

We believe that all of these observations could be put together into a common model (Fig. 14). Here we discuss it for the West Verkhoyansk sector and the inner zone, but for the South Verkhoyansk sector, a similar interpretation has already

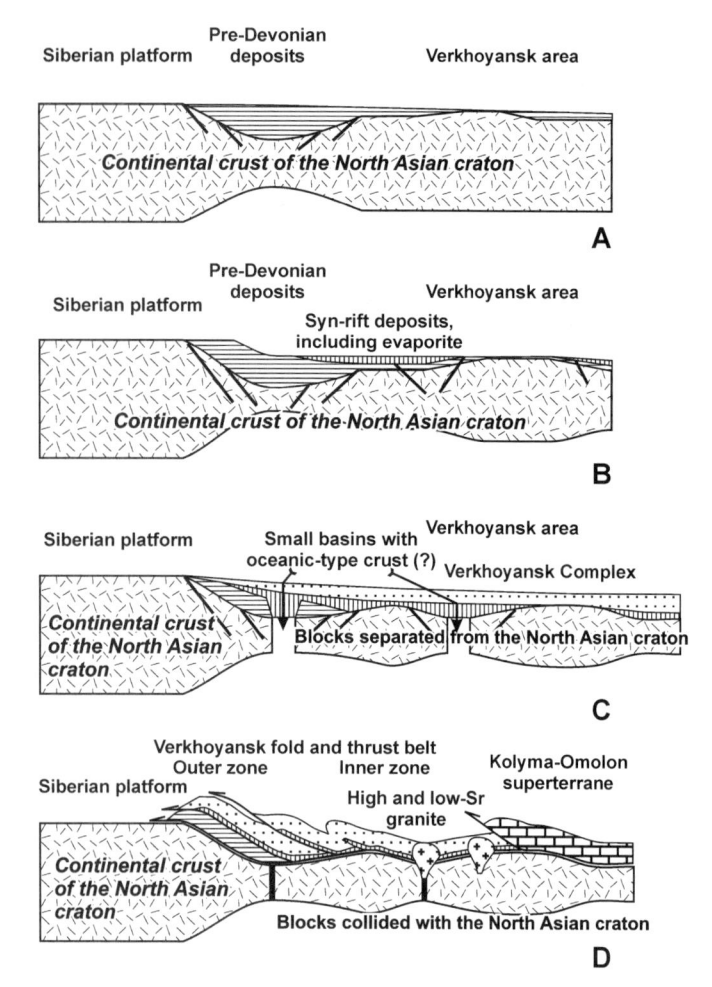

Figure 14. A model of evolution of the West Verkhoyansk sector and the inner zone of the Verkhoyansk fold-and-thrust belt. (A) Pre-Devonian time. (B) Early stage of the Middle to Late Devonian rifting. (C) Carboniferous to Jurassic passive margin stage. (D) Cretaceous orogeny and formation of the fold-and-thrust belt. See text for discussion.

been presented by Khudoley and Guriev (2003). Before Middle to Late Devonian rifting, both the outer and the inner zones were underlain by the North Asian craton crust, although it was probably attenuated below the inner zone due to older rifting events. Wide depressions in the basement along the contact between the outer and inner Verkhoyansk fold-and-thrust belt recognized by Tretyakov (2004) also represent old pre-Devonian structures. The depression filled with sediments that did not spread far to the east (Fig. 14A). Middle to Late Devonian rifting renewed some older rifts and formed new ones. Synrift deposits included evaporites (Fig. 14B). Middle to Late Devonian rifting resulted in significant attenuation and local breakup of the continental crust in the inner zone of the Verkhoyansk fold-and-thrust belt, where some rift-related basins probably contained oceanic-type crust (Fig. 14C). A new margin of the North Asian craton approximately corresponds to the boundary between the outer and the inner zones of the Verkhoyansk fold-and-thrust belt. However, details of Middle to Late Devonian evolution of the inner Verkhoyansk fold-and-thrust belt remain unknown because related structures are hidden below the Verkhoyansk Complex succession. Existing rifts that separated blocks with highly attenuated continental crust led to a subsidence history of the Verkhoyansk Complex that was similar to the continental margin and slope, but not like shelf areas, even in the outer zone of the Verkhoyansk fold-and-thrust belt. During Mesozoic shortening, the Kolyma-Omolon superterrane collided with the North Asian craton, forming the Verkhoyansk fold-and-thrust belt (Fig. 14D). Small rift-related basins were closed and small crustal blocks converged and resulted in relatively small shortening (averaged at 30%) reported from the outer Verkhoyansk fold-and-thrust belt. In the inner Verkhoyansk fold-and-thrust belt, shortening was less than that in the outer zone, but wide distribution of rift-related evaporites facilitated formation of a very thin thrust wedge. Collision of the Kolyma-Omolon superterrane also resulted in granite intrusion, especially to the east of the Verkhoyansk fold-and-thrust belt. Typically, they intruded transitional continental crust and had high $^{87}Sr/^{86}Sr$ ratios. However, where the granite bodies cut relics of basins with oceanic-type crust, they displayed low $^{87}Sr/^{86}Sr$ ratios (Fig. 3).

CONCLUSION

The structural and subsidence histories of the Verkhoyansk fold-and-thrust belt outline significant and unusual differences between the outer and inner zones of this thrust belt. Imbricate thrust fans are the dominant structures in the outer zone of this thrust belt, whereas the inner zone is characterized by concentric open folds. The outer zone is located above the basement slope of the North Asian craton, which dips eastward beneath the thrust belt and controls the slope of the basal detachment that separates allochthonous thrust sheets from the autochthonous basement and its sedimentary cover. The thrust wedge angle τ, the sum of basal detachment slope β, and the inferred topographic slope α are in reasonable agreement with that predicted by the theory of tapered orogenic wedges (e.g., Davis et al., 1983; Boyer, 1995). However, below the inner Verkhoyansk fold-and-thrust belt, the basement surface is subhorizontal, and the related thrust wedge is very thin. Its subcritical shape requires a weak basal detachment in order to explain the displacement of thrust sheets. Because of its relatively shallow depth (~10 km), cataclastic flow and other temperature-induced ductile deformation mechanisms are unlikely, and a wide distribution of salt or evaporite units at the base of sedimentary succession is considered as the most probable option. The Middle to Upper Devonian succession is the only one that contains evaporites; we infer that it hosts the basal detachment in the inner Verkhoyansk fold-and-thrust belt. This interpretation is supported by observations from the Orulgan segment of the Verkhoyansk fold-and-thrust belt, where the deepest sedimentary units of the frontal thrust sheets contain evaporite units. Depth to basement (18–10 km), inferred from gravity data, approximates the thickness of the Middle Devonian to Middle Jurassic succession. This implies that within the inner zone, pre-Middle Devonian rocks are thin or absent, indicating significant differences in its tectonic history compared to the outer zone, where Mesoproterozoic to Devonian rocks form a very thick succession.

Middle to Late Devonian rifting is a well-documented event that occurred along the eastern margin of the North Asian craton and within the Verkhoyansk fold-and-thrust belt. Post-rift thermal subsidence corresponds to the time span of deposition of the Lower Carboniferous to Middle Jurassic Verkhoyansk Complex, and the amount of total (13–11 km) and tectonic (4.5–4 km) subsidence calculated from sections located in the outer zone of the Verkhoyansk fold-and-thrust belt is comparable to that estimated from the transitional continental crust–oceanic crust transition zone on the Atlantic margin of the North American continent.

Both interpretations of the inner zone basement as a continuation of North Asian craton basement (Parfenov, 1984; Prokopiev and Deikunenko, 2001) or as oceanic crust (Khudoley and Guriev, 1994) present problems. We assume that the Middle to Late Devonian rifting led to high attenuation and partial breakup of the continental crust below the inner zone of the belt, marking the boundary between the outer and the inner zones of the belt as a new boundary of the North Asian craton. The Mesozoic collision resulted mainly in formation of the orogenic belts to the east of the Verkhoyansk fold-and-thrust belt; however, within the Verkhoyansk fold-and-thrust belt, some shortening of rift depressions and closure of small remnant basins led to formation of a moderately deformed fold-and-thrust belt and intrusion of rare granite plutons with variable $^{87}Sr/^{86}Sr$ ratio.

ACKNOWLEDGMENTS

Thorough reviews by Elizabeth Miller and Mike Cecile greatly refined this manuscript. Numerous discussions with Jim Sears were very stimulating. Subsidence curves were calculated using software developed in the Geological Department at Moscow State University (A.M. Nikishin's research team). The study was partly supported by the Russian Foundation for Basic Research grants 04-05-64711, 05-05-65327, 06-05-96070, 06-05-64369, 07-05-00743, and integration projects of RAS ONZ 10.2. We acknowledge our late friend, Georgiy Guriev, one of the best experts of Devonian to Lower Carboniferous successions of NE Russia, for joint field trips and numerous discussions on geology of the Verkhoyansk Range.

REFERENCES CITED

Alkhovik, T.S., and Baranov, V.V., 2001, Stratigraphy of the Lower Devonian of eastern Yakutia (north-east of Russia): Yakutsk, Siberian Branch of the Russian Academy of Sciences Publishing House, 149 p. (in Russian).
Bombolakis, E.G., 1994, Applicability of critical-wedge theories to foreland belts: Geology, v. 22, p. 535–538, doi: 10.1130/0091-7613(1994)022 <0535:AOCWTT>2.3.CO;2.
Bond, G.C., and Kominz, M.A., 1984, Construction of tectonic subsidence curves for the early Paleozoic miogeocline, southern Canadian Rocky Mountains: Implications for subsidence mechanisms, age of breakup, and crustal thinning: Geological Society of America Bulletin, v. 95, p. 155–173, doi: 10.1130/0016-7606(1984)95<155:COTSCF>2.0.CO;2.
Bond, G.C., Kominz, M.A., and Sheridan, R.E., 1995, Continental terraces and rises, in Busby, C.J., and Ingersoll, R.V., eds., Tectonics of sedimentary basins: Cambridge, Massachusetts, Blackwell Scientific Publications, p. 149–178.
Boyer, S.E., 1995, Sedimentary basin taper as a factor controlling the geometry and advance of thrust belts: American Journal of Science, v. 295, p. 1220–1254.
Braathen, A., Bergh, S.G., and Maher, H.D., Jr., 1999, Application of a critical wedge taper model to the Tertiary transpressional fold-thrust belt on Spitsbergen, Svalbord: Geological Society of America Bulletin, v. 111, p. 1468–1485, doi: 10.1130/0016-7606(1999)111<1468:AOACWT> 2.3.CO;2.
Chapple, W.M., 1978, Mechanics of thin-skinned fold-and-thrust belts: Geological Society of America Bulletin, v. 89, p. 1189–1198, doi: 10.1130/ 0016-7606(1978)89<1189:MOTFB>2.0.CO;2.
Costa, E., and Vendeville, B.C., 2002, Experimental insights on the geometry and kinematics of fold-and-thrust belts above weak, viscous evaporitic décollement: Journal of Structural Geology, v. 24, p. 1729–1739, doi: 10.1016/S0191-8141(01)00169-9.
Dahlen, F.A., Suppe, J., and Davis, D., 1984, Mechanics of fold-and-thrust belts and accretionary wedges: Cohesive Coulomb theory: Journal of Geophysical Research, v. 89, B12, p. 10,087–10,101.
Davis, D., and Engelder, T., 1985, The role of salt in fold-and-thrust belts: Tectonophysics, v. 119, p. 67–82, doi: 10.1016/0040-1951(85)90033-2.
Davis, D., Suppe, J., and Dahlen, F.A., 1983, Mechanics of fold-and-thrust belts and accretionary wedges: Journal of Geophysical Research, v. 88, B2, p. 1153–1172.
DeCelles, P.G., and Mitra, G., 1995, History of the Sevier orogenic wedge in terms of critical taper models, northeast Utah and southwest Wyoming: Geological Society of America Bulletin, v. 107, p. 454–462, doi: 10.1130/ 0016-7606(1995)107<0454:HOTSOW>2.3.CO;2.
Egorov, A.Yu., 1993, Avalanche sedimentation—the main process in the Verkhoyansk Complex formation: Transactions (Doklady) of Russian Academy of Sciences, v. 332, no. 3, p. 346–351 (in Russian).
Ellis, S., Schreurs, G., and Panien, M., 2004, Comparisons between analogue and numerical models of thrust wedge development: Journal of Structural Geology, v. 26, p. 1659–1675, doi: 10.1016/j.jsg.2004.02.012.
Evans, M.A., 1989, The structural geometry and evolution of the thrust system, northern Virginia: Geological Society of America Bulletin, v. 101, p. 339–354, doi: 10.1130/0016-7606(1989)101<0339:TSGAEO> 2.3.CO;2.
Galabala, P.O., 1971, On the orogenesis in the western Verkhoyansk region, in Shilo N.A., ed., Mesozoic tectonogenesis: Magadan, Russian Academy of Sciences, North-East Interdisciplinary Scientific Research Institute Press, p. 61–68 (in Russian).
Gamyanin, G.N., Goryachev, N.A., Bakharev, A.G., Kolesnichenko, P.P., Zaitsev, A.I., Diman, E.N., and Berdnikov, N.V., 2003, Conditions of origin and evolution of granitoid gold-ore-magmatic systems in Mesosoides of North-East Asia: Magadan, Russian Academy of Sciences, North-East Interdisciplinary Scientific Research Institute Press, 196 p. (in Russian).
Gusev, G.S., 1979, Fold structures and faults of the Verkhoyansk-Kolyma Mesozoides: Moscow, Nauka, 207 p. (in Russian).
Keen, C.E., and Beaumont, C., 1990, Geodynamics of rifted continental margins, in Keen, M.J., and Williams, G.L., eds., Geology of the continental margin of eastern Canada: Ottawa, Geological Survey of Canada, Geology of Canada no. 2, p. 391–472.
Keen, C.E., Loncarevic, B.D., Reid, I., Woodside, J., Haworth, R.T., and Williams, H., 1990, Tectonic and geophysical overview, in Keen, M.J., and Williams, G.L., eds., Geology of the continental margin of eastern Canada: Geological Survey of Canada, Geology of Canada no. 2, p. 31–85.
Khudoley, A.K., and Guriev, G.A., 1994, The formation and development of late Paleozoic basin on the passive margin of the Siberian paleocontinent, in Beauchamp, B., Embry, A.F., and Glass, D., eds., Pangea: Global environments and resources: Calgary, Alberta, Canadian Society of Petroleum Geologists Memoir 17, p. 131–143.
Khudoley, A.K., and Guriev, G.A., 2003, Influence of synsedimentary faults on orogenic structure: Examples from the Neoproterozoic-Mesozoic east Siberian passive margin: Tectonophysics, v. 365, p. 23–43, doi: 10.1016/ S0040-1951(03)00016-7.
Khudoley, A.K., Guriev, G.A., and Zubareva, E.A., 1991, Density flow sediments in the carbonate complex of the Sette-Daban (South Verkhoyansk): Lithology and Mineral Deposits, no. 5, p. 106–116 (in Russian).

Khudoley, A.K., Guriev, G.A., and Ganelin, V.G., 1995, The Southern Verkhoyansk: Composition and evolution of the late Paleozoic basin with terrigenous sedimentation: Lithology and Mineral Deposits, no. 4, p. 421–432 (in Russian).

Khudoley, A.K., Rainbird, R.H., Stern, R.A., Kropachev, A.P., Heaman, L.M., Zanin, A.M., Podkovyrov, V.N., Belova, V.N., and Sukhorukov, V.I., 2001, Sedimentary evolution of the Riphean-Vendian basin of southeastern Siberia: Precambrian Research, v. 111, p. 129–163, doi: 10.1016/S0301-9268(01)00159-0.

Klitgord, K.D., Hutchinson, D.R., and Schouten, H., 1988, U.S. Atlantic continental margin; Structural and tectonic framework, in Sheridan, R.E., and Grow, J.A., eds., The Atlantic continental margin: U.S.: Boulder, Colorado, Geological Society of America, Geology of North America, v. I-2, p. 19–55.

Kovalskiy, V.V., ed., 1985, Structure and evolution of the earth crust of Yakutia: Moscow, Nauka, 247 p. (in Russian).

Layer, P.W., Newberry, R., Fujita, K., Parfenov, L.M., Trunilina, V.A., and Bakharev, A.G., 2001, Tectonic setting of the plutonic belts of Yakutia, northeast Russia, based on ^{40}Ar/^{39}Ar geochronology and trace element geochemistry: Geology, v. 29, p. 167–170, doi: 10.1130/0091-7613(2001) 029<0167:TSOTPB>2.0.CO;2.

Lohrmann, J., Kukowski, N., Adam, J., and Oncken, O., 2003, The impact of analogue material properties on the geometry, kinematics, and dynamics of convergent sand wedges: Journal of Structural Geology, v. 25, p. 1691–1711, doi: 10.1016/S0191-8141(03)00005-1.

Mitra, G., 1997, Evolution of salients in a fold-and-thrust belt: The effects of sedimentary basin geometry, strain distribution and critical taper, in Sengupta, S., ed., Evolution of geological structures in micro- to macroscales: London, Chapman and Hall, p. 59–90.

Morley, C.K., 1986, A classification of thrust fronts: American Association of Petroleum Geologists Bulletin, v. 70, p. 12–25.

Naumov, A.N., 1962, On the nature of junction of the Verkhoyansk fold system and the Priverkhoyansk foreland basin in the area of the western slope of the Orulgan Range: Leningrad, Transactions of the Research Institute of Geology of the Arctic (NIIGA), v. 130, no. 19, p. 91–98 (in Russian).

Nenashev, I.N., and Zaitsev, A.I., 1985, Evolution of granitoid magmatism in the Yana-Kolyma fold belt: Institute of Geology, Russian Academy of Sciences, Yakutsk, 176 p. (in Russian).

Nokleberg, W.J., Parfenov, L.M., Monger, J.W.H., Baranov, B.V., Byalobzhesky, S.G., Bundtzen, T.K., Feeney, T.D., Fujita, K., Gordey, S.P., Grantz, A., Khanchuk, A.I., Natal'in, B.A., Natapov, L.M., Norton, I.O., Patton, W.W., Jr., Plafker, G., Scholl, D.W., Sokolov, S.D., Sosunov, G.M., Stone, D.B., Tabor, R.W., Tsukanov, N.V., Vallier, T.L., and Wakita, K., 1994, Circum-North Pacific tectonostratigraphic terrane map, scale 1:5,000,000: U.S. Geological Survey Open-File Report 94-714, 433 p.

Nokleberg, W.J., Parfenov, L.M., Monger, J.W.H., Baranov, B.V., Byalobzhesky, S.G., Bundtzen, T.K., Feeney, T.D., Fujita, K., Gordey, S.P., Grantz, A., Khanchuk, A.I., Natal'in, B.A., Natapov, L.M., Norton, I.O., Patton, W.W., Jr., Plafker, G., Scholl, D.W., Sokolov, S.D., Sosunov, G.M., Stone, D.B., Tabor, R.W., Tsukanov, N.V., and Vallier, T.L., 1997, Circum-North Pacific tectonostratigraphic terrane map, scale 1:10,000,000: U.S. Geological Survey Open-File Report 96-727, 1 sheet.

Parfenov, L.M., 1984, Continental margins and island arcs of Mesozoides of north-East Asia: Nauka, Novosibirsk, 192 p. (in Russian).

Parfenov, L.M., 1995, Terranes and history of formation of Mesozoic orogenic belts of the eastern Yakutia: Pacific Geology, v. 14, no. 6, p. 3–10.

Parfenov, L.M., 2001, Tectonic analysis, in Parfenov, L.M., and Kuzmin, M.I., eds., Tectonics, geodynamics and metallogeny of the Sakha Republic (Yakutia): Moscow, International Academic Publishing Company "Nauka/Interperiodica," p. 69–80 (in Russian).

Parfenov, L.M., Prokopiev, A.V., and Gaiduk, V.V., 1995, Cretaceous frontal thrusts of the Verkhoyansk fold belt, eastern Siberia: Tectonics, v. 14, p. 342–358, doi: 10.1029/94TC03088.

Parfenov, L.M., Prokopiev, A.V., and Spector, V.B., 2001, Geodynamics of the Eastern Yakutia mountainous province and opening of the Eurasian basin: Russian Geology and Geophysics, v. 42, no. 4, p. 670–686.

Price, R.A., 1981, The Cordilleran foreland thrust and fold belt in the southern Canadian Rocky Mountains, in McClay, K.R., and Price, N.J., eds., Thrust and nappe tectonics: Geological Society [London] Special Publication 9, p. 427–448.

Price, R.A., 1988, The mechanical paradox of large overthrusts: Geological Society of America Bulletin, v. 100, p. 1898–1908, doi: 10.1130/0016-7606(1988)100<1898:TMPOLO>2.3.CO;2.

Price, R.A., and Sears, J.W., 2000, A preliminary palinspastic map of the Mesoproterozoic Belt–Purcell Supergroup, Canada and USA: Implications for the tectonic setting and structural evolution of the Purcell Anticlinorium and the Sullivan Deposit, in Lydon, J.W., Höy, T., Slack, J.E., and Knapp, M.E., eds., The geological environment of the Sullivan Deposit, British Columbia: Geological Association of Canada, Mineral Deposit Division, Special Volume no. 1, p. 61–81.

Prokopiev, A.V., 2000, Verkhoyansk-Chersky collisional orogen: Geology of the Pacific Ocean, v. 15, p. 891–904.

Prokopiev, A.V., and Deikunenko, A.V., 2001, Deformational structures of fold-and-thrust belts, in Parfenov, L.M., and Kuzmin, M.I., eds., Tectonics, geodynamics and metallogeny of the Sakha Republic (Yakutia): Moscow, International Academic Publishing Company "Nauka/Interperiodica," p. 156–198 (in Russian).

Prokopiev, A.V., Fridovsky, V.Yu., and Deikunenko, A.V., 2001a, Some aspects of the tectonics of the Verkhoyansk fold-and-thrust belt (northeast Asia) and structural setting of the Dyandi gold ore cluster: Polarforschung, v. 69, p. 169–176.

Prokopiev, A.V., Parfenov, L.M., Tomshin, M.D., and Kolodeznikov, I.I., 2001b, Sedimentary cover of the Siberian platform and adjacent fold-and-thrust belts, in Parfenov, L.M., and Kuzmin, M.I., eds., Tectonics, geodynamics and metallogeny of the Sakha Republic (Yakutia): Moscow, International Academic Publishing Company "Nauka/Interperiodica," p. 113–155 (in Russian).

Prokopiev, A.V., Bakharev, A.G., Toro, J., Miller, E.L., Hourigan, J.K., and Dumitru, T.A., 2003, Middle Paleozoic continental margin magmatism and Mesozoic metamorphic events in the zone of junction of the North Asian craton and the Okhotsk terrane: New geochemical and geochronological data and their geodynamic interpretation: Otechestvennaya Geologiya, no. 6, p. 57–63 (in Russian).

Prokopiev, V.S., Urzov, A.S., Budeleeva, S.S., Slastenov, Y.L., and Yuganova, L.A., 1999, Geological map of Yakutia at scale 1:500,000: St. Petersburg, All Russian Geological Research Institute (VSEGEI) Press, The West Verkhoyansk set, 19 sheets (in Russian).

Sborshchikov, I.M., and Natapov, L.M., 1969, Deformations related to the gypsum-anghydrite suite in the West Verkhoyansk region: Transactions (Doklady) of USSR Academy of Sciences, v. 186, no. 5, p. 1150–1153.

Steckler, M.S., Watts, A.B., and Thorne, J.A., 1988, Subsidence and basin modeling at the U.S. Atlantic passive margin, in Sheridan, R.E., and Grow, J.A., eds., The Atlantic Continental Margin: U.S.: Boulder, Colorado, Geological Society of America, Geology of North America, v. I-2, p. 399–416.

Til'man, S.M., and Bogdanov, N.A., 1992, Tectonic map of northeastern Asia, scale 1:5,000,000: Institute of Lithosphere, Russian Academy of Sciences, and Circum-Pacific Council for Energy and Mineral Resources, 1 sheet.

Toro, J., Prokopiev, A.V., Colgan, J., Dumitru, T., Hourigan, J., and Miller, E.L., 2004, Apatite fission-track thermochronology of the Southern Verkhoyansk fold-and-thrust belt, Russia [abs.]: Eos (Transactions, American Geophysical Union), v. 85, no. 47.

Tretyakov, F.F., 2004, Middle Paleozoic rift structures in the basement of the Verkhoyansk fold belt: Otechestvennaya Geologiya, no. 4, p. 57–60 (in Russian).

Woodward, N.B., 1987, Geological applicability of critical-wedge thrust-belt models: Geological Society of America Bulletin, v. 99, p. 827–832, doi: 10.1130/0016-7606(1987)99<827:GAOCTM>2.0.CO;2.

Yapaskurt, O.V., 1992, Lithogenesis and mineral deposits of miogeosynclines: Moscow, Nedra, 224 p.

MANUSCRIPT ACCEPTED BY THE SOCIETY 22 MARCH 2007

Index

A

Abemarle Group, 298
Absaroka thrust, 190–191, 194
ACCRETE transect, 99, 100, 109
accretion, 100–101, 121, 384
accretionary wedges, 5–7, 344
Adatan thrust zone, 375, 380, 381, 384
ADCOH. *See* Appalachian Ultradeep Core Hole Project
Adel Mountain volcanic, 157
Aegean-Turkey plate, 10
Ailao Shan, 52, 54
Akamina syncline, 172
Alabama basement fault, 256. *See also* Tusquittee fault
Alaskan Cordillera
    continental crustal-scale décollements of, 102–104
    crustal thickness of, 110
    geological map of, 101
    geotectonics of, 101–102
    Moho beneath, 110–111
    orogen-parallel deformation in, 102
    orogen-parallel translation, transpression, crustal-penetrating faults and, 107–110
    overview of, 99–100, 112
    seismic surveys of, 102–104
    subduction megathrust décollements of, 102–104
    summary of controlled-source seismic experiments in, 103
    temperature, lithosphere thickness of, 111–112
Albermarle Group, 301, 302
Alberta Foreland basin, 134, 136–137, 205–206, 208
Alberta Group, 176, 179
Alberta syncline, 81, 91
Aldridge Formation, 151
Aleutian-Wrangell subduction zone, 106
Alexander terrane, 101, 102, 132
Algies Bay, 332, 333–334, 336
Alpine Fault plate boundary, 332
alpine orogenic systems, 28
Alto allochthon, 319, 324
Altyn Tagh fault, 373, 375
Anaconda Complex, 212
Ancona thrust, 81, 91
aneurysms, tectonic, 43
Annidale Group, 294
Anniston transverse zone, 283, 285
Antarctica-fixed framework, 1, 11, 18, 20–21
apatite fission-track annealing, 168, 171
Appalachian fold-thrust belt
    boundaries of deformation in, 256–264
    Coosa deformed belt and, 281, 283–285
    Eden fault and, 285–286
    Helena thrust sheet and, 281, 283
    internal deformation of, 266–269
    Jones Valley thrust sheet and, 282–283
    measurement of tectonic shortening in, 286–288
    overview of, 243–244, 269–273, 277, 278–279, 288
    Palmerdale mushwad and, 281, 282–283, 288
    primary and secondary basement controls of, 265–266
    regional stratigraphic framework of, 252–253
    regional structural framework of, 253–256
    stratigraphy of, 279–282
    tectonic history, current structural fronts of, 250–252
    timing of basement fault formation in, 265
    Vandiver mushwad and, 281, 284, 285–286, 288
Appalachian gravity gradient, 263
Appalachian Ultradeep Core Hole Project (ADCOH), 258, 263
Appekunny Formation, 168, 173, 174, 176–179
Aquarius, 221
aquicludes, 331
arc magmatism, 297–298, 299, 300
argon dating, 232–236
Arisaig Group, 297
Arlar thrust, 375
Asbill Pond Formation, 298, 299
assimilation, 16
Atlantic Ocean, 1, 2–3, 17
Avalonia
    comparison of to Carolinia and Ganderia, 294–295
    lithotectonic comparison of to Carolinia, 296–298
    Nd isotopic characteristics of, 303, 304–305
    overview of, 291–293, 305–307
    paleomagnetic data for, 300–301
Avoca fault zone, 353, 356, 363
Ayakum thrust zone, 375, 384

B

back-arc basins, 140
Bajocian fractures, 198, 199–203
Ballarat East, 356
Balmoral Group, 294
"banana-doughnut" tomography, 19
Banda arc, 4
Baoshan unit, 60
barriers, whole-mantle circulation and, 16
Bayanhar Mountains, 386
Bayonne Suite, 125
Bearpaw Formation, 171
Bear thrust, 186, 190–191, 194
Beartooth fault, 152, 153, 156
Beaver Creek fault, 109, 214, 237
Beaver Valley fault, 267
Belly River strata, 179
Belt-Purcell Basin
    central horst of, 147, 155
    continental breakup and, 159
    Cordilleran orogenesis, thrust rotations and, 159–161
    northern half-graben of, 147, 149–155
    overview of, 147–149, 154, 161
    palinspastic restoration of, 152
    Siberian craton and, 147, 157–159
    southern graben of, 147, 156–157
    tectonic map of, 148
    tectonic setting of, 149
Bendigo gold mine, 356, 359
Bendigo zone of Lachlan orogen, 351, 353, 356–358, 361
Benton Creek fault, 214
Bessemer transverse zone, 282, 283
Bighorn syncline, 81, 91
biotite, 222, 224
Birmingham anticlinorium, 253
Bitterroot batholith, 156, 157
Bitterroot complex, 212, 213, 219, 236
black cataclastic seams of Ngatuturi Claystone, 338
Blackdome Peak, 229
Black Mountain syncline, 81, 91
Black River Group. *See* Stones River Group
Blacks Harbour granite, 299
Black Warrior basin, 280–281
Blairmore Group, 172, 176, 179
Bloody Bluff fault, 293
Blount Mountain syncline, 282
Blount wedge, 314
Blue Ridge fault, 252
Blue Ridge–Piedmont sheet, 243, 245, 250, 263–265
Boehls Butte–Goat Mountain. *See* Clearwater complex
Bohemian Massif, 39, 41
Bonin arc, 21
Boulder batholith, 156
Bourinot Group, 301
Bowser basin, 118–120, 125–129, 133–139
Bowser Lake Group, 129, 132
Bras D'Oria, 293, 294
Brazeau syncline, 81, 91
breccia dikes, 340
Brevard fault zone, 255, 264–265, 266, 314
Brindle Creek fault, 317–318, 322
British Columbia, 41
brittle deformation, 65, 250
Brooks Range, 102, 103, 107, 110–111
Brookville gneiss, 299
Brothers Peak Formation, 132
Bungalow pluton, 214, 237
buoyancy, 16, 30–31, 34, 35
Burnsville fault, 316
Burnt Timber syncline, 81, 91
Butte porphyry copper deposit, 156

C

Cabin Creek, 174
Cache Creek terrane, 101, 118, 120, 125, 127, 133–134
Cahaba synclinorium, 282
CAI. *See* conodont alteration index
Caishiling thrust, 375

Caledonia terrane, 293
Campbelltown fault, 351, 353
Canadian Cordillera
　continental crustal-scale décollements of, 104–106
　cross sections of evolution of, 120, 126, 130–131
　crustal thickness of, 110
　geological map of, 101
　geologic description of, 121
　geotectonics of, 100–101
　Mesozoic tectonic evolution of, 133–139
　Moho beneath, 110–111
　orogen-parallel deformation in, 102
　orogen-parallel translation, transpression, crustal-penetrating faults and, 107–110
　overview of, 99–100, 112, 118–121, 139–141
　seismic surveys of, 102–104
　subduction megathrust décollements of, 106–107
　summary of controlled-source seismic experiments in, 103
　temperature, lithosphere thickness of, 111–112
　See also Coast belt; Foreland belt; Insular belt; Lewis thrust; southern Omineca belt
Canadian Rocky Mountains, 64–65, 81, 91, 96
Canyon fault, 214, 221
Cardium Formation, 177
Cariboo Mountains, 135
Carolina terrane, 293–294, 322
Carolinia
　comparison of to Avalonia, 294–295, 296–298
　comparison of to Ganderia, 298–300
　Nd isotopic characteristics of, 301–304, 305
　overview of, 291–293, 305–307
　paleomagnetic data for, 300–301
Cartersville fault, 252
Cascadia subduction zone, 106
Cassiar batholith, 133, 137
Cassiar-Kutcho-Thibert system, 133
Castine Formation, 294
Cate Creek window, 171, 172
Cathcart fault, 353
Catskill-Pocono wedge, 314
Cat Square basin
　accretion of Carolina Terrane to Laurentia and, 322
　Carolinia and, 322
　collision of Appalachian Inner Piedmont with New York promontory and, 314–317
　Country Rock Protolith and, 317–318
　eastern Blue Ridge, Inner Piedmont and, 319–321
　exhumation in, 318
　magmatism in, 319–320
　maritime Canadian and New England results versus, 313, 323
　opening and closing of, 322–323
　overview of, 313–314, 324–326
　Spechty Kopf, Rockwell Formations and, 323–324
　unconformities, successor basins and, 321–322
Cedar Creek, 216, 232
Central Deborah mine, 356, 359

central Intermontane belt
　geologic description of, 122–124, 125–132
　Mesozoic tectonic evolution of, 133–139
　tectonic evolution of, 126
　See also Bowser basin; Skeena fold belt; Sustut basin
Central Montana trough, 156
central Piedmont shear zone, 293
Chancellor Basin, 153, 159
Changbai volcano, 13
Charlotte Terrane, 293–294, 298, 302
Chattanooga Shale, 252–253
Chauga belt, 316, 317
Chenghai fault, 54, 56
Chengling-Mengliang unit, 60
Chickamauga Group, 252
Chief Joseph batholith, 149, 156, 157
Chilhowee Group, 250, 256
Chugwater Formation, 186, 193–194
Cid Formation, 298
Cima de Gagnone body, 39
Cincinnati arch, 265
Cinq Cerf–Grey River, 294
circulation, whole-mantle. See dual circulation
clastic wedges, 314
claystone. See Ngatuturi Claystone
Clearwater complex
　argon dating of, 232–236
　external zone of, 215, 229, 231–232
　formation of, 236–238
　geologic setting of, 212–213
　internal zone of, 214–215, 228–231, 235
　Jug Rock shear zone of, 215–220
　metamorphic thermobarometry of, 222–228
　overview of, 211–212, 214, 238–239
　Smith Ridge Syncline of, 221–222
　southern culmination of, 221
　U-Pb SHRIMP dating of, 228–230
Clearwater corner, 156
Clearwater River, 221
Cliff Creek thrust. See Prospect thrust
Clinchport fault, 267
Clingman aeromagnetic lineament, 265
clinoenstatite, 37–38
clinopyroxene, 28–29
Clugs Jumpoff Fault, 214, 221–222, 237
Coast belt, 101–102, 107–109, 119, 126, 132–139
Coeur d'Alene district, 155
Collins Creek fault, 214, 215, 219, 237
Conasauga Group, 252, 266, 280, 282
conodont alteration index (CAI), 263, 264
consolidated fault zones, 339
Contact fault, 109
continental crust, 51–61
contractional structures, 31, 92
convection, 1–2, 140
convergence, 99
convergent margins, 4, 28
cooling, 33–37, 43
cooling (top-down). See top-down subduction
Coongee fault, 351, 353
Coosa deformed belt, 281, 283–285
Cordillera. See Alaskan Cordillera; Canadian Cordillera
Coryell Suite, 125

Coulomb-wedge theory, 392
Country Rock Protolith, 317–318
Crawford thrust, 194
Crazy Mountain basin, 156
creeping-flow deformation, 65
Crescendo Peak, 215
critical wedge models, 64, 92–94, 185, 403–406
cross-strike links, 277
Crowsnest deflection, 153
crystallization, 16, 42
Cumberland-Allegheny Plateau, 250, 252–253
Cumberland Plateau fault, 265, 267
Cussewago sandstone, 323

D

Dabieshan region, 39
Dabie-Sulu belt, 32, 34, 43
Dabuxun Yanquiao Formation, 374
Dali fault system, 51, 53–54, 56, 60
Darby thrust, 190–191, 194
decapitation, 30
décollement thrusting, 92, 99, 104–107, 349
décollement zones, defined, 65
decompression, 33
deep-mantle subduction, 11–16
deformation
　boundaries of in Appalachian fold-thrust belt, 256–264
　development of Canadian Cordillera and, 100–101
　in fold-thrust belts, 244–245, 250
　of forearc basins, 9
　internal in Appalachian fold-thrust belt, 266–269
　in Lachlan orogen, 349, 363–365
　in Lanping-Simao belt, 51–52
　in Ngatuturi Claystone, 345–346, 347
　in obduction zones, 332
　in orogenic belts, 350
　orogen-parallel in North American Cordillera, 102
　in Qaidam basin, 377–380
　in southern Yangtze platform and Xianshuihe-Xiaojiang fault system, 57
　in Verkhoyansk fold-and-thrust belt, 392, 400–401, 407
　without obvious through-going structures, 52
Delamerian orogen, 350, 351
delamination, syncollisional, 105
Denali fault system, 102, 107, 109
depleted mantle, 16
detachment zones, 65, 277
diamictite, 323–324, 325
Dien Bien Phu fault, 52, 57
dikes, 340
discontinuity, 660 km, 1, 12–15
Dogtooth structure, 135
Dominican Republic, 41
Dora Maira Massif, 32, 35
Dover fault, 293
dual circulation, 16
Dumplin Valley, 267, 268
Dworshak Reservoir, 221
Dysartville pluton, 321

E

Eager trough, 153, 159
EarthScope, 292
Eastern Kunlun Range
    geology of, 373–374
    overview of, 369–373, 387
    thrusting and growth history of, 385–386
    wedge tectonics across, 381–385
    See also Qaidam basin
Eastern Mojave shear zone, 52
eclogites, 1, 28–29
Eden fault, 285–286
elastic flexural model, 205
Eldorado thrust, 151
Elkhorn Mountains, 157
Ellis Group, 208
Ellsworth Formation, 294, 299, 300
Emory River fault, 267
erosion, 69
erosional decapitation, 30
Euler poles, 20
Eureka fault, 134
exhumation, 32, 37, 43
Exploits Group, 300
Exploits-Tetagouche back-arc system, 299, 300, 307
Explorer–Juan de Fuca–Gorda plate. *See* Farallon plate
extensional arc systems, overview of, 9–11

F

Fall Creek Canyon, 186, 190
Fang pluton, 136
Farallon anomaly, 14–15
Farallon plate, 101, 132, 183
faulting, 335–339
faults, 94–95. *See also* through-going faults
Fernie area, 171, 202
Flathead fault, 151, 170, 171, 172, 174
Flat Swamp Member, 298
Floodwood Creek, 219–221
fold-and-thrust belts, 244–250. *See also* Appalachian fold-thrust belt; thin-skinned fold-and-thrust belts; Verkhoyansk fold-and-thrust belt
folded mountain belts, fundamental characteristics of, 64–65
Foothills region, 198
forbidden zone, 38, 42
forearc systems, 4–9
forebulge model, 198, 202, 204–206, 207
Foreland belt
    evolution of, 118, 122–124
    folding and thrusting in, 171
    geologic description of, 121
    location of, 119
    Mesozoic tectonic evolution of, 133–139
    overview of, 100, 102
    tectonic evolution of, 126
Fourmile Creek, 285
fractional melting, 16
Franciscan complex, 34
Fraser–Straight Creek fault, 109
frontal ramps, 277

Front Ranges, 198
Fungo Hollow deformed zone, 285
Furby's Cove, 296

G

Gadsden mushwad, 282
GAHB. *See* garnet-hornblende thermometer
Gander Group, 294, 299
Ganderia
    comparison of to Carolinia, 298–300
    distinction of from Avalonia, 294–295
    Nd isotopic characteristics of, 303, 304–305
    overview of, 291–294, 305–307
    paleomagnetic data for, 300–301
GARB. *See* garnet-biotite exchange thermometer
Garlock fault, 52
garnet-biotite (GARB) exchange thermometer, 222
garnet-hornblende barometer (GHPQ), 222
garnet-hornblende thermometer (GAHB), 222
garnet mixing model, 222
garnet of Clearwater complex, 223
garnet peridotites, 37–42
Garrison depression, 156
Gas Hure syncline, 375
Gaspé belt, 323
geochronology, 32
Georges River fault, 293
GHPQ. *See* garnet-hornblende barometer
Gizzard Group, 253
Glacier National Park, 151
global heat flow, 15
Goat Mountain, 215, 217, 221, 235. *See also* Clearwater complex
Gog Group, 65, 159
Grainger Formation, 253
Grandfather Mountain window, 255, 263, 265
Granite Creek, 186
Granite-Rhyolite province, 153
gravitational instability, 31
Grease Creek syncline, 81, 91
Great Smoky fault, 252, 256, 266, 271
Greensboro Igneous Suite, 302
grikes, 197, 203–204, 206–209
Grinnel Formation, 168, 173, 174, 176–179
Gunbarrel igneous province, 159
Gwillim Creek shear zone, 138

H

Haig Brook window, 171, 172
Hall Lake fault, 151, 153
Hamill-Gog Group, 159
Harpersville transverse zone, 282, 283–284, 285–286
Hawaii, 19
Heathcote fault zone, 351, 355, 358, 363
heat transfer, 1, 15
Hefty thrust, 151
Heihe fault, 57
Helena fault, 281, 283, 284
Helena Formation, 153, 156, 283
Hellenic Trench, 10
Helvetic Alps, 249, 267
Henderson gneiss, 321
Hermitage Bay fault, 293

Hermitage Formation, 321
Himalayan Mountains, 34
Himalayan syntaxis. *See* western Himalayan syntaxis
hinge rollback
    accretionary wedges and, 5–7
    backarc basins and, 9–11
    complex flow and, 9–11
    extensional arc systems and, 9–11
    forearc basin deformation and, 9
    forearc basins and, 7–9
    forearcs and, 4–5
    interseismic locking and, 11
    plate tectonics and, 17
    subduction and, 3–4
Hoadley thrust, 151
Hoback River Canyon. *See* Snake River–Hoback River Canyon
Hobson Lake pluton, 136
Hoh Xil basin, 373, 386
Holston–Iron Mountain fault, 246, 252, 253, 254
Honeyhill fault, 293
Hong'an-Dabie terrane, 34, 36
Hope fault, 155, 161
horizontal shear, 51–52
hornblende, 225–226
hotspot reference frame, 19
hotspots, 15, 18–19
Howell Creek, 172
Hunter Valley fault, 267
Huntley Mountain Formation, 323
hydrocarbons, 206–209
hypersolidus temperatures, 29

I

Iapetus Ocean, 291, 292
Iceland thermal anomaly, 19
Idaho Batholith, 212, 213
illite neomineralization, 183–186, 193, 194
Indian Ocean, 3
Inner Piedmont, 313, 314–319
Insular belt, 119, 126, 132–139
Insular superterrane, 101, 102, 109, 110
Interior Platform, 171
Intermontane belt, 119. *See also* central Intermontane belt

J

Jacksboro fault, 263, 267
Jianchuan fault, 54
Jocko line, 155, 156
Jones Valley fault, 282–283, 286–288
Jug Rock shear zone, 215–218, 235–238
Jura Mountains, 249, 267

K

Kaghan Valley, 33, 34
Kaltag fault, 102
Kathy Road dioritic suite, 299
Kazakhstan, 40
Kechika-Sifton fault, 133
Kelly Forks fault, 214, 221, 237, 238
Kettle Dome, 212

Khastakh trough, 159
Kimberley fault, 153
Kimmeridgian strata, 121
King Salmon fault, 125
Kishenehn Formation, 171–172, 179
Knox Group, 252, 253, 266, 268, 280
Kokchetav Belt, 32, 34
Kolyma-Omolon superterrane, 408
Kootenay arc, 135, 151, 159, 174, 206
Kootenay Group, 168, 173
Kootenay terrane, 125, 127
Kula plate, 102, 132, 183
Kula-Pacific plate, 102
Kunlun fault, 373, 374
Kuskanax Suite, 125, 134
kyanite-phengite eclogites, 29

L

Lachlan orogen
  Bendigo zone of, 351, 353, 356–358, 361
  chronology of deformation in, 356–360
  deformation rate for, 363–365
  folds, faults, and cleavage of, 353–355
  geologic setting of, 350–351
  Melbourne zone of, 351–353, 355, 358–360, 361
  overview of, 349–350, 365
  P-T conditions of metamorphism and deformation in, 355–356
  retrodeformation of, 361–363
  Stawell zone of, 351, 353, 356, 360–361
  strain rates in, 360–361
Lahood Formation, 156
Lake Char fault, 293
Landsborough fault zone, 353
Lanping-Simao fold belt, 51–58, 60
La Poile Group, 323
Lardeau Group, 159
"large-hot" orogen, 118, 140
Lau basin, 10
Laurentia, 300–301
Lay Dam Formation, 321
Leichardt fault, 353
Lenoir Limestone, 252
Lepontine Alps, 39
Lewis and Clark Line, 151, 161, 211–212, 214
Lewis thrust
  Belt-Purcell Basin and, 151, 153–155
  constraints on maximum temperatures and, 177–178
  geologic setting of, 168–172
  overview of, 167–168, 179
  paleogeothermal gradient, burial depth at maximum temperature and, 178–179
  time of maximum paleotemperatures and, 174–177
  vitrinite reflectance analysis of, 172–173, 175
  zircon fission-track analysis of, 173–174, 175
Lianchang unit, 60
Libby thrust, 151
Lijiang fault, 54
Lithoprobe Southern Cordillera transect, 99, 100
lithosphere, causes of subduction of, 11, 12
Little Goat Mountains, 215

Logan Pass, 159
Lombard thrust plate, 156
Longmen Shan, 51–52, 55, 58
Lookout Mountain conglomerate, 190
Lower Belt-Purcell Group, 149
Lulehe Formation, 374
Lunksoos arch, 294
Luoyan Shan anticlinorium, 377
Luttrell basin, 252

M

Mackenzie Mountains, 110, 140
Madison Group, 197, 198
magma temperature, plate tectonics and, 15
magmatic arcs, subduction and, 4–5
Magurangi Limestone, 332
Mahurangi limestone, 332
majoritic garnet, 37–38
Malton Complex, 125, 137
Malton Gneiss basement slice, 137
Mangakahia Complex, 332
mantle circulation
  arguments against whole-mantle circulation and, 16
  dual circulation and, 16
  geochemistry and, 16
  heat transfer and, 15
  lack of subduction into lower mantle and, 13–15
  overview of, 11
  slabs in upper mantle and, 11–13
Mariana Trench, 10, 21
Marystown Group, 301
Matthews Lake Formation, 299
Meade thrust, 194
Means Crossroads complex, 298
Meductic Group, 294
Meguma, 292–293
Melbourne zone of Lachlan orogen, 351–353, 355, 358–360, 361
Mengliang fault, 57
Mengxing fault, 57, 60
Merrimack synclinorium, 323
mid-ocean ridges, 3
Miller Cove fault, 246, 252, 256
Miramachi Formation, 294, 299
Missoula Group, 149, 151, 153, 155, 156
Mist Mountain Formation, 168, 173, 174, 177–179
mixture modeling, 174
Moccasin Formation, 252
Moho depths, 99, 110–111
Monashee Complex, 125, 138
Monashee Mountains, 135
"Montania," 156
Monte Duria body, 39
Monumental Buttes, 215, 216, 217, 232
MORB (mid-ocean-ridge basalt), 16
Moses Butte, 228
Mosquito Road Formation, 300
Motatau Complex, 332
Mountain City window, 253, 255
Mount Ararat fault, 353, 361
Mount Camel terrane, 332, 333, 344
Mount Cormack dome, 294
Mount Forster thrust, 151

Mount Wellington fault zone, 353, 355–356, 358–359, 361
Moyie fault, 153, 155, 157, 158, 159
Moyston fault, 351
Muckleford fault, 351, 353, 358
muscovite, 224
mushwad, 280–288
mylonite, 218, 220

N

Nanga Parbat, 38
Nankai Trench, 21
Nantinghe fault, 57
Nd isotopic analysis, 301–305
Neihart Quartzite blanket, 153
Nelson Suite, 125, 134, 135
New England orogen, 350
New Hebrides arc, 10, 20
New River, 293, 294
New Zealand. See Northland Allochthon
Ngatuturi Claystone
  deformation at shallow crustal levels and, 347
  deformation mechanisms in, 345–346
  fluid regime in, 346
  obduction processes and, 346–347
  overview of, 334–335
  pattern bedding in, 335
  structural features of, 335–340
  structural geometry and sequence of events in, 340–344
  tectonic regime for, 344–345
  Waitemata Group and, 335
Ngawha area, 332, 333
Nicol Creek, 153, 159
Nile abyssal fan, 10
no-net-rotation framework, 19–20
North American Cordillera, 147, 149, 159–161. See also Alaskan Cordillera; Canadian Cordillera
North Asian Craton. See Verkhoyansk fold-and-thrust belt
Northern Rocky Mountain Trench fault, 133, 140
Northland Allochthon
  deformation at shallow crustal levels and, 347
  deformation mechanisms in, 345–346
  fluid regime in, 346
  obduction processes and, 346–347
  overview of, 331–332
  regional setting of, 333–334
  structural geometry and sequence of events in, 340–344
  tectonic framework of, 332–333
  tectonic regime for, 344–345
  See also Ngatuturi Claystone
North Slope Block, 102, 104

O

obduction zones, 332, 346–347
Ocoee Supergroup, 256
Okhotsk terrane, 394
olivine, 37–38
Omineca belt, 100, 119, 122–124, 171. See also central Intermontane belt; southern Omineca belt

Omulevka terrane, 394
Opossum Valley thrust fault, 282, 286–288
orogenic belts, deformation in, 350
Orphan Point, 219
Ouachita orogenesis, 280–281
out-of-sequence deformation, 385–386
Ovando graben, 155, 157
overlap assemblages, 314. *See also* successor basins
overstep assemblages, 314. *See also* successor basins
overthrusts, 94

P
Pacific Ocean, 1–4, 17
Pacific plate, 10
Palau Trench, 21
Paleo-Qaidam basin, 386
Palmerdale mushwad, 281, 282–283, 288
Paris thrust, 194
Pegram Formation, 321
Pelham dome, 294
Pell City fault, 285, 288
Peninsular terrane, 102
Penobscot Formation, 299, 300
Percydale fault zone, 353
peridotes, 37–42
Perma culmination, 151
Perry Line, 149, 156, 161
Persimmon Creek gneiss, 321
Peru Trench, 9
Philippine Sea plate, 21
piecemeal-accretion process, 384
Piegan Group, 149, 153, 155, 156
piggy-back basins, 81, 94, 95–96
Pinchi fault, 127
Pine Mountain fault, 253, 257, 263, 265, 267
Pinkham thrust, 151
plagioclase, 225
plastic yield stress, 65. *See also* brittle deformation
plate tectonics
 diagram of popular misconceptions about, 2
 drivers of, 1
 framework of, 18–21
 global spreading patterns and, 2–3
 hinge rollback and, 3–11
 mantle circulation and, 11–16
 mechanism of, 17
 overview of driving mechanism of, 1–2, 21
 subduction causes and, 11
plumes, plate tectonics and, 18–19
Po Plain, 81, 91, 92
Porcupine Creek, 151, 153, 161
positive feedback, 80
Pottsville Formation, 317
Price Formation, 323
Price-Pocono wedge, 316–317
Prichard Formation, 151, 215
Priest River complex, 172, 212, 213, 236
primitive mantle, 16
Priverkhoyansk foredeep, 399, 400
Prospect thrust, 190–192, 194

Pulaski fault, 251, 266
Pundata fault, 134
Purcell Basin. *See* Belt-Purcell Basin
Purcell Mountains, 134, 138
push-together process, 384, 386

Q
Qaidam basin
 Cenozoic deformation in, 377–380
 Cenozoic shortening in, 377
 geology of, 373–374
 mechanisms of uplift in, 386–387
 overview of, 369–373, 387
 structural geology of, 375–377
 thrusting and growth history of, 385–386
 See also Eastern Kunlun Range
Qigequan Formation, 374
Qilan Shan, 370, 373
Qimen Tagh Mountains, 369, 375–380
Qinghai Oilfield Company, 377
Qinghai province, 58
Queen Charlotte–Fairweather fault system, 102, 107, 109, 111
Quesnellia terrane, 101, 125, 127, 133, 135

R
Ravalli Group, 149, 151, 153, 155
recrystallization, 29, 32, 35
recycling, 16, 38
Red Creek, 190
Red Indian Line, 294
Red River fault, 54
Red River valley, 52
regional décollements, 277–278, 279–280, 288
Resurrection plates, 132
retroarc foreland basins, 314. *See also* successor basins
retrodeformation, 361–363
Revett Formation, 153
Rheic Ocean, 291, 292
Rheinesche Schiefergebirge, 365
Ridge push, 11
ridges, 3
Riedel shear, 331, 338, 339
Rierdon Formation, 206
Roanoke Rapids complex, 296
Robertson Bay terrane, 365
Rockwell Formation, 313, 322, 323–326
Rocky Mountain fold-and-thrust belt, 147, 168, 197–198. *See also* Swift Reservoir
Rocky Mountain Formation, 173, 174
Rocky Mountain Trench, 109, 149, 151
rollback. *See* hinge rollback
Rome Formation, 252, 256–258, 266, 280, 283
Rosman fault, 264–265, 266
rotation
 horizontal shear and, 52
 in North American Cordillera, 99, 101–102, 147, 149, 159–161
Roti intrusive suite, 299
Roundtop pluton, 214, 232
Russell Fork fault, 267

S
Sage Creek, 174
Saint Arnaud fault zone, 353
Saint Joe fault, 214, 237
Saint Mary fault, 153, 157, 158, 160
Saltville thrust, 249, 267
sampling error, 15
Sanbagawa belt, 34
sandstone dikes, 340
Sandsuck Formation, 250, 256
Sapphire thrust plate, 156
Sauratown Mountain window, 263
Sawtooth Formation, 197, 199–201, 204
Scrip nappe, 135
Sebastian fault, 353
Seboomook Formation, 322, 323
sedimentary basin subsidence, 401–403
sedimentation, 69
Seismic Hazard in Puget Sound (SHIPS) data set, 107
seismic locking, 20
Selkirk fan, 135
Selkirk Mountains, 134, 138
sensitive high-resolution ion microprobe. *See* SHRIMP analysis
Sequatchie Valley thrust sheet, 288
Sette-Daban trough, 157
Sevastiyanov thrust, 406
Sevier Shale, 252
Seview orogeny, 183–185
Shady Dolomite, 256
Shady Valley thrust sheet, 246, 252, 254, 256
Shangyoushashan Formation, 374
Shawangunk Formation, 316–317
Sheep Creek syncline, 81, 91
SHIPS data set. *See* Seismic Hazard in Puget Sound data set
Shizigou Formation, 374
Shooting Creek window, 263
shortening, 11, 286–288, 349, 377
SHRIMP analysis, 228–230
Shuswap complex, 236
Siberian craton, 157–159
Sichuan province, 58
Six Mile thrust sheet, 317
Skeena fold belt, 119, 120, 132
Skeena Group, 127
slab pull, 11
slab rollback, 4. *See also* hinge rollback
slabs, sinking of, 1
Slave-Northern Cordillera Lithosphere Evolution (SNORCLE) transect, 99, 100, 108
Slide Mountain terrane, 125, 127, 136
"small-cold" orogen, 118, 140
smectite neomineralization, 185–186
Smith Ridge Syncline, 221–222
Snake River–Hoback River Canyon
 Absaroka thrust and, 190–191
 Bear thrust and, 190–191
 clay characterizations for, 188–190
 cross section of, 184
 Darby thrust and, 190–191
 direct dating of fault rocks of, 185–186

illite, illite-smectite neomineralization character and, 185
    overview of, 183–185, 190–195
    Prospect thrust and, 190–191
    sample analysis for, 186–188
Snake River Plain, 19
Snells Beach, 332, 333–334
SNORCLE transect. *See* Slave-Northern Cordillera Lithosphere Evolution transect
Snowshoe fault, 149, 151, 155, 161
Songpan Ganze basin, 58
southern Omineca belt, 122–126, 133–139
Spechty Kopf Formation, 313, 322, 323–326
Stawell gold mine, 360–361
Stawell zone of Lachlan orogen, 351, 353, 356, 360–361
Stikinia terrane, 100, 120, 125
Stirling belt, 296
Stone Mountain fault, 253
Stones River Group, 252
Stony Mountain, 300
structural-lithic units (SLU), 245
Stubbs fault, 134
subduction
    causes of in oceanic lithosphere, 11
    convergent lithospheric plate-boundary diagram for, 37
    density inversions and, 1
    drivers of plate tectonics and, 17
    hinge retreat and, 4
    hinge rollback and, 3–4
    lack of evidence for deep-mantle, 11–16
    shallow earthquakes of, 5
subduction zones, 20, 31, 106
subsidence, 401–403, 407
successor basins, 314, 321
Sulu, 40
Sumatra-Java-Banda arc system, 4–5
Sun River, 155, 197, 199–201, 203–204
surface processes, 69
Sustut basin, 118, 119, 129–132, 137
Sustut Group, 137
Suwanee, 292
Sweetgrass arch, 202, 208, 209
Swift Formation, 206
Swift Reservoir
    Bajocian grikes of, 197, 203–204
    Bajocian unconformity and, 198, 199–203
    forebulge model and, 204–206
    hydrocarbon implications and, 206–209
    overview of, 197–198, 209
    structural setting of, 198
syncollisional delamination, 105
syndeformational deposition, 80, 81, 92, 94–95
Syringa terrane, 156

T

TACT. *See* Trans-Alaska Crustal Transect program
Talladega fault, 252
Tallulah Falls, 263, 268, 321
Tally Pond Group, 300
Tangihua Complex, 332
Tango Creek Formation, 129, 132, 138
Tasman orogen, 350

tectonic aneurysms, significance of, 43
tectonic wedge model, 125
Tennessee embayment, 256
Tensleep sandstone, 186
Tetagouche Group, 301
Teton Canyon, 204
Thibert fault, 127
thin-skinned fold-and-thrust belts
    fault strength and, 94–95
    model design for, 65–69
    model limitations and, 95
    model properties, parameters, and options for, 68
    natural examples of, 64–65
    one-detachment model for, 66
    overview of, 63–64, 95–96
    specific models for, 70–81
    summary of models for, 69
    surface processes, natural structures and, 81, 91
    three-detachment model for, 66
    two-detachment model for, 66
    whole-wedge solutions versus, 92–94
    See also Appalachian fold-thrust belt
through-going faults, 51–52, 57, 59–61
through-the-mantle plumes, 19
Tibet, 58–59
Tibetan Plateau, 53, 385–387. *See also* Eastern Kunlun Range; Qaidam basin
Tickle Point formation, 296
Tintina fault, 102, 107–110, 133
Tobique rocks, 323
tomography, 12–13, 14–15, 19
Tonga arc, 10, 20
top-down subduction, 1, 17, 18, 21
Toxaway gneiss, 321
Trans-Alaska Crustal Transect (TACT) program, 99, 100, 105
transition zone, 13
transpressional rotation, 99, 107
trench retreat. *See* hinge rollback
Tuckaleechee Cove, 246
Tugaloo terrane, 313, 317–318, 320, 322, 324
turbidite fans, 351
Turner Valley field, 198
Tusquittee fault, 243, 256–257, 263, 265–266, 268
Two Medicine volcanic, 157
Tyrrhenian-Apennine arc system, 10

U

Udzga-Khastakh Basin, 157
ultrahigh-pressure conditions, defined, 28
ultra-high pressure (UHP) metamorphism
    conductive cooling of continental complexes of, 33–37
    continental complexes ascent rates and, 32
    garnet peridotites of, 37–42
    generation and exhumation of, 29–32
    overview of, 27–28
    Phaneroic contractional orogens and, 42–43
    pressure-temperature conditions for, 28–29
    sketch map of continental-crustal belts of, 42
    specific complexes of, 32
UNAVCO Web site, 18
underthrusting, 31
Unicoi Formation, 253

U-Pb SHRIMP dating, 228–230
upwelling, 3
Uwharrie, 302

V

Valhalla complex, 138
Vandiver mushwad, 281, 284, 285–286, 288
Verkhoyansk Complex, 397, 398, 401–403, 407–408
Verkhoyansk fold-and-thrust belt
    basement of, 394
    critical wedge model and, 401–403
    evolution of sedimentary basin of, 401–403
    inner zone of, 399–400, 405–406
    outer zone of, 398–399, 405
    overview of, 391–393, 406–409
    stratigraphy of, 394–398
    timing of deformation in, 400–401
Victoria Creek fault, 109
Victoria Lake Supergroup, 294
Virgilina sequence, 297, 302, 304
vitrinite reflectance analysis, 167, 172–173, 175
Vulcan suture zone, 153

W

Waitakere Group, 332
Waitemata Group, 332–333, 335–336, 340–345
Waitemata piggyback basin, 331
Wallace Formation, 153
Wallen Valley fault, 267, 268
Walton Formation, 322
Wanding fault, 57, 60
Wanling Shan, 57
Wartburg basin, 253, 265, 267
Waterton field, 198
Waterton National Park, 151
Wattle Gully Mine, 356–358
Wenchuan-Maowen fault zone, 59
Western Alps, 35, 41
Western Gneiss Region
    coesite-bearing thrust sheets of, 34
    data for, 32
    garnet peridotites of, 39, 42
    HP-UHP domains in, 36
    UHP processes in, 43
western Himalayan syntaxis
    active faults of, 54
    conductive cooling of, 33–34
    data for, 32
    GPS velocities around, 55
    rotation of fault systems around, 51, 52–53
Whangai Formation, 332
White Bluff area, 336, 341
White Rock fault, 214, 215, 217, 219, 236
Whitelaw fault, 351, 353, 356
Whiteoak Mountain fault, 267, 268
Whiteside pluton, 321
whole-mantle circulation, 13–14
whole-wedge solutions. *See* critical wedge models
Widow Mountain shear zone, 217, 219, 236
Wigwam thrust, 151
Wild Blight Group, 300
WILDFIRE program, 188–189

Williams Creek syncline, 81, 91
Windermere Supergroup, 159–160
Winding Stair Gap, 319–321
Windsor Point Group, 323
Woods Point dike swarm, 358–360
Wrangellia terrane, 101, 102, 132
Wrangell subduction zone, 106
Wyoming thrust belt, 183–185, 187. *See also* Snake River–Hoback River Canyon

X

Xiaganchaigou Formation, 374
Xianshuihe-Xiaojiang fault system, 51, 52–54, 57
Xiayoushashan Formation, 374

Y

Yadkin Formation, 300
Yakutat terrane, 101, 102, 110
Yangtze platform, 52, 56, 60
Yap Trench, 21
Yellowleaf fault, 283–284
Yellowstone, 19
Yiematan thrust, 375
Yousha Shan fold, 369, 375, 377, 379, 381–382, 384–385
Yukon-Tanana terrane, 110, 111
Yunnan, China, 52–58

Z

Zhe Da fault, 59
Zhongdian fault, 54
zircon fission-track (ZFT) data, 167, 168, 173–174, 177, 178
zircons, 229